Inner Product Structures

Vasile Ion Istrăţescu

Inner Product Structures

Theory and Applications

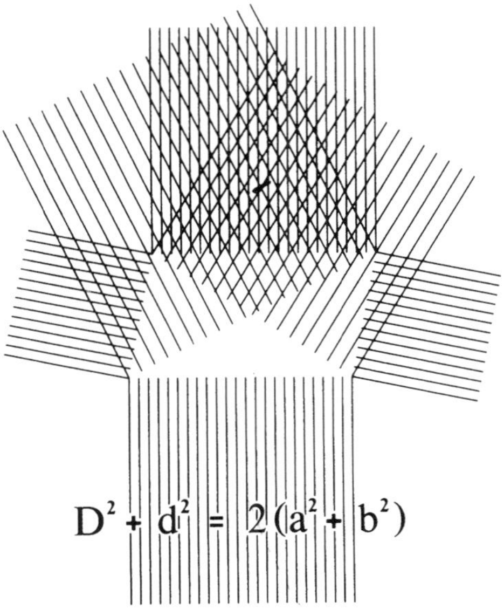

D. Reidel Publishing Company

A MEMBER OF THE KLUWER ACADEMIC PUBLISHERS GROUP

Dordrecht / Boston / Lancaster / Tokyo

Library of Congress Cataloging in Publication Data CIP

Istrăţescu, Vasile I.
Inner product structures.

(Mathematics and its applications)
Bibliography: p.
Includes indexes.
1. Inner product spaces. I. Title. II. Series:
Mathematics and its applications (D. Reidel Publishing Company)
QA322.4.I84 1987 515.7'33 87-9456

ISBN-13: 978-94-010-8162-7 e-ISBN-13: 978-94-009-3713-0
DOI: 10.1007/978-94-009-3713-0

Published by D. Reidel Publishing Company,
P.O. Box 17, 3300 AA Dordrecht, Holland

Sold and distributed in the U.S.A. and Canada
by Kluwer Academic Publishers,
101 Philip Drive, Norwell, MA 02061, U.S.A.

In all other countries, sold and distributed
by Kluwer Academic Publishers Group,
P.O. Box 322, 3300 AH Dordrecht, Holland

To my parents
Paraschiva (Gica) Istrăţescu
and
Ion (Ilie) Istrăţescu

SERIES EDITOR'S PREFACE

Approach your problems from the right end and begin with the answers. Then one day, perhaps you will find the final question.

'The Hermit Clad in Crane Feathers' in R. van Gulik's *The Chinese Maze Murders*.

It isn't that they can't see the solution. It is that they can't see the problem.

G.K. Chesterton. *The Scandal of Father Brown* 'The point of a Pin'.

Growing specialization and diversification have brought a host of monographs and textbooks on increasingly specialized topics. However, the "tree" of knowledge of mathematics and related fields does not grow only by putting forth new branches. It also happens, quite often in fact, that branches which were thought to be completely disparate are suddenly seen to be related.

Further, the kind and level of sophistication of mathematics applied in various sciences has changed drastically in recent years: measure theory is used (non-trivially) in regional and theoretical economics; algebraic geometry interacts with physics; the Minkowsky lemma, coding theory and the structure of water meet one another in packing and covering theory; quantum fields, crystal defects and mathematical programming profit from homotopy theory; Lie algebras are relevant to filtering; and prediction and electrical engineering can use Stein spaces. And in addition to this there are such new emerging subdisciplines as "experimental mathematics", "CFD", "completely integrable systems", "chaos, synergetics and large-scale order", which are almost impossible to fit into the existing classification schemes. They draw upon widely different sections of mathematics. This programme, Mathematics and Its Applications, is devoted to new emerging (sub)disciplines and to such (new) interrelations as exempla gratia:

- a central concept which plays an important role in several different mathematical and/or scientific specialized areas;
- new applications of the results and ideas from one area of scientific endeavour into another;
- influences which the results, problems and concepts of one field of enquiry have and have had on the development of another.

The Mathematics and Its Applications programme tries to make available a careful selection of books which fit the philosophy outlined above. With such books, which are stimulating rather than definitive, intriguing rather than encyclopaedic, we hope to contribute something towards better communication among the practitioners in diversified fields.

Inner product spaces or pre-Hilbert spaces are, inevitably, vector spaces equipped with an inner product and hence a norm. They are not necessarily complete. Such spaces occur in almost all parts of mathematics and completing them may not always be natural and/or suitable. Thus a systematic authoritative study of what is known about inner product spaces such as this fits almost ideally in the MIA program.

The unreasonable effectiveness of mathematics in science ...

Eugene Wigner

Well, if you know of a better 'ole, go to it.

Bruce Bairnsfather

What is now proved was once only imagined.

William Blake

As long as algebra and geometry proceeded along separate paths, their advance was slow and their applications limited.

But when these sciences joined company they drew from each other fresh vitality and thenceforward marched on at a rapid pace towards perfection.

Joseph Louis Lagrange.

Bussum, May 1987

Michiel Hazewinkel

Contents

Preface

The purpose of this book is to give a comprehensive introduction to the theory of inner product structures together with some applications to diverse fields, emphasizing the significance of the relationships between the general theory and its applications.

The book is written for people interested in research as well as for those who are interested in applications. Many references are scattered throughout the book and are included primarily for the reader's convenience. No attempt is made to give a specific reference for every result mentioned nor to give priority credits in every case. Then again, most of the sections contain remarks directed to the literature involving related results and applications or further developments of the material covered in the sections.

There is no claim made for completeness and it is certain that a number of readers will find their favourite area missing. The omissions in some instances are deliberate and in others are, no doubt, oversights. However, it is hoped that enough information has been included which may be of interest to the reader. A detailed analysis of many of the most important applications involves a great deal of material in the particular area from which they are taken. Therefore, because of space limitations, we must necessarily omit most of the details and an attempt has been made to include enough references to the literature to enable the interested reader to find as much of the omitted material as he or she wishes.

It is a pleasure to express my sincere thanks to many people for reprints and preprints as well as lecture notes which were of great help to me. I am also indebted to Professor M. Hazewinkel for his interest in this book as well as to the editors of D. Reidel Publishing Company, especially to Dr. D. J. Larner, for their attentiveness.

Vasile Ion Istrăţescu

CHAPTER 1

General Topology Topological Spaces

1.0. Introduction

The purpose of this chapter is to present very rapidly some of the basic, concise results in general topology – mostly without proofs. The reader interested in this topic may consult the works of N. Bourbaki, J. L. Kelley and K. Kuratowski for a detailed account.

We first state some facts about sets and partial orderings in sets, and give that useful result of Max Zorn – the so-called Zorn lemma. Further, in a discussion of the basic notions of topological spaces, we present results about topologies (including the so-called product topologies (or Tichonov topologies)) together with elements of convergence theory, first using nets and second using the filters. Some classes of sets and topological spaces are also presented. We then consider the class of compact spaces and some properties are given related to the finite intersection property. The class of metric spaces is presented in detail since it occupies a special position in analysis. Complete metric spaces are presented with a characterization (of Cantor). We finally present the Kuratowski measure of noncompactness which is, historically, the first measure of noncompactness that was defined by Kuratowski in 1929 and which plays (with other measures of noncompactness) an important role in modern researches of linear and nonlinear functional analysis.

1.1. Sets. Functions

We adopt the intuitive approach in set theory; and the purpose of our introductory sections is to enumerate facts and establish the notation and terminology that we shall use; only a few results are proved and the reader interested in the axiomatic approach may consult books dedicated to this important subject.

If A is a set then $x \in A$ means that an object x is an element of A; the symbol

$$\{x,\ x \text{ has property } P\}$$

means the set of all objects x having property P. We consider a set A to be given if we have a list of all its elements,

$$A = \{a, b, c, \ldots\},$$

or we have a property P such that

$$A = \{x, x \text{ has property } P\}.$$

If x is a single object then $\{x\}$ denotes the set that has only one element, x. A set with no members is called an empty set (void set) and is denoted by $\emptyset$. We use the symbol $\subseteq$ to denote inclusion; if A and B are two sets then $A \subseteq B$ means that every element of A is an element of B. In this case we say that A is a subset of B or that B contains A. We consider an empty set to be a subset for any set. Two sets are equal if they have the same elements, i.e. $A = B$ if and only if $A \subseteq B$ and $B \subseteq A$. The subset A of a set B is said to be proper if there is at least one element of B not in A. If A is a set and x is an object not in A we indicate this as $x \notin A$. If A and B are two sets then the complement of A relative to B is the set $\{x, x \in B, x \notin A\}$. This set is denoted by $B \setminus A$.

Suppose now that I is a set which we call an index set and that for each $i \in I$ we have a set M_i. Then the set of all elements x such that $x \in M_i$ for some i in I is called the union of the family $\{M_i\}_{i \in I}$, and for this set we use the notation $\cup_{i \in I} M_i$. The subset of the union of the family of the set $\{M_i\}_{i \in I}$ of all x such that $x \in M_i$ for all $i \in I$ is said to be the intersection of the family and is denoted by $\cap_{i \in I} M_i$.

If I is a finite set then we sometimes use the notation

$$A_1 \cup A_2 \cup A_3 \cup \cdots \cup A_n$$

for the union and

$$A_1 \cap A_2 \cap A_3 \cap \cdots \cap A_n$$

for the intersection.

Two sets A and B are disjoint if $A \cap B = \emptyset$; a family of sets is a pairwise disjoint family if for any two indices i and j in I, $i \neq j$, $M_i \cap M_j = \emptyset$ if the family is $\{M_i\}_{i \in I}$.

Suppose now that we have a family of sets $\{M_i\}_{i \in I}$ such that each M_i, $i \in I$, is a subset of the set S. If M is an arbitrary subset of S then the set $C_M = S \setminus M$ is called the *complement* set. Then it is not difficult to see that the following formulas hold:

$$(1)\quad C_{\bigcup_{i \in I} M_i} = \bigcap_{i \in I} C_{M_i},$$

$$(2)\quad C_{\bigcap_{i \in I} M_i} = \bigcup_{i \in I} C_{M_i}.$$

If A and B are two sets then a function (mapping, correspondence, transformation) $f: A \to B$ has a domain A and range B if for each x in A there is only one element of B to which it corresponds; that element is denoted by $f(x)$. If $g: B \to C$ is another function then we can define a new function $h: A \to C$ by the relation

$$h(x) = g(f(x)).$$

If $f: A \to B$ is a function then the subset

$$\{b, b \in B, \text{ there exists } a \text{ in } A \text{ such that } f(a) = b\}$$

is denoted by $f(A)$. The subset

$$\{a, a \in A, \text{ there exists } b \text{ in } B \text{ such that } f(a) = b\}$$

is denoted by $f^{-1}(B)$.

Suppose that $f: A \to B$ is a function. Then we say that:

(1) f is onto (or surjective, or a surjection) if $f(A) = B$;
(2) f is injective if u, v are arbitrary elements of A, u is not equal to v ($u \neq v$) then $f(u) \neq f(v)$;
(3) if $B = A$ and A_1 is a subset of A such that $f(A_1) \subseteq A_1$ then we say that A_1 is invariant under f;
(4) if f is onto and injective then f is said to be one-to-one or to have an inverse.

The inverse function has the domain B and range A and is defined as follows: if b is an element of B then there exists only one element a of A such that $f(a) = b$. Then we set $b \to a$ and the function is clearly well defined. For this we use the notation $f^{-1}: B \to A$.

Suppose, again, that A and B are two sets and that $f: A \to B$ is a function. Suppose further that A_1 is a subset of A and $f_1: A_1 \to B$ is a function.

If $f(x) = f_1(x)$ for all x in A_1 then we say that f_1 is a restriction of f to the set A_1 or that f is an extension of f.

Let us now consider the important and useful notion of a Cartesian product. Suppose that we have a family of sets $\{M_i\}_{i\in I}$. Then the Cartesian product of this family of sets is the set of all functions

$$f: I \to \bigcup_{i\in I} M_i.$$

This is denoted by

$$\prod_{i\in I} M_i \quad \text{or} \quad \prod_{i=1}^{\infty} M_i \quad \text{if } I = (1, 2, 3, \ldots).$$

If I is a finite set, $I = (n_1, \ldots, n_m)$, then we also use the notation

$$A_{n_1} \times A_{n_2} \times \cdots \times A_{n_m}.$$

For each $f \in \Pi_{i\in I} M_i$, $f(i)$ is said to be the ith coordinate of the point f; the element of the product of which the point $x_i \in M_i$ is the ith coordinate is denoted by $\{x_i\}_{i\in I}$ or, briefly, $\{x_i\}$. If the index set I is $(1, 2, 3, \ldots)$ we use the notation $\{x_n\}$.

A relation R on a set S, i.e. a subset of the Cartesian product $S_\times S$, satisfying the following properties:

(1) $x \, R \, x$ (reflexivity) for all x in S,
(2) $x \, R \, y$ and $y \, R \, z$ imply $x \, R \, z$ (transitivity),
(3) $x \, R \, y$ implies $y \, R \, x$

is said to be an equivalence relation.

If R is a relation which is an equivalence relation on S then R determines a decomposition of S into disjoint subsets; for each s in S we set

$$M_s = \{t, t \in S, s \, R \, t\}.$$

Then, clearly, $M_s \cap M_t = \emptyset$ if $t \neq s$ and obviously

$$S = \bigcup_{s\in S} M_s.$$

Now if S is a set and $\{M_i\}_{i\in I}$ is a family of subsets of S whose union is S and

$M_i \cap M_j = \emptyset$ for $i \neq j$ determines on S an equivalence relation, it suffices to set $s R t$ if and only if s and t are in the same subset M_i for some i in I.

If A and B are two sets then we say that $A R B$ if and only if there exists a mapping $f: A \to B$ which is one-to-one and surjective. It is easy to see that this is an equivalence relation in the family of sets. The equivalence class of the set A is denoted by $\hat{A}$ and means the family of all sets which are equivalent to A. We say that the elements in the above equivalence class have the same power. For every set A we associate a cardinal number (denoted by $\bar{\bar{A}}$) which is called the power of the set. We say that $\bar{\bar{A}} = \bar{\bar{B}}$ if and only if A and B are equivalent sets.

If the set A is finite then the power of the set is exactly the number of its elements; if the set A is $\{1,2,3,4,\dots\}$ then its cardinal number is denoted by $\aleph_0$ and any set that is finite or whose power is $\aleph_0$ is called *countable*.

If m is a cardinal number then we define 2^m as the cardinal of the set of all subsets of a set whose power is m.

Of fundamental importance in the study of spaces are the so-called 'order relations'. If S is a set we say that on S we have an ordering if on S there is a relation '$<$' defined such that the following assertions are true:

(1) If $x < y$ then $y < x$ does not hold.
(2) $x < y$ and $y < z$ imply $x < z$.
(3) If x is not equal to y $(x \neq y)$ then $x < y$ or $y < x$.

We also use the notation $y > x$ for $x < y$. If A is a set with an order relation then we can define the first and last elements of the set as follows: the element m is the first element of A if $m < a$ for all a in A not equal to m, and the last element M has the property $M > a$ for all a in A not equal to M. Using such elements we now define an important class of sets. We say that a set M with an order relation is well-ordered if the following property holds: *If A_1 is a nonempty subset of A then A_1 has a first element.*

We now define a notion which is very useful in many considerations. We say that a relation on a set A, denoted by '$\leqslant$', is a partial ordering if the following assertions hold:

(1) If $x \leqslant y$ and $y \leqslant z$ then $x \leqslant z$.
(2) $x \leqslant x$.
(3) If $x \leqslant y$ and $y \leqslant x$ then $x = y$.

It is to be noted that in this definition we do not suppose that our relation is defined for all pairs (x, y) of elements of a Cartesian product. If S is a set with a partial ordering '$\leqslant$' then we say that a subset S_1 of S is linearly ordered if whenever x, y are in S_1 then $x \leqslant y$ or $y \leqslant x$. If S is a set with an ordering then we can define a partial ordering as $s \leqslant t$ if $s < t$ or $s = t$.

If S is a set with a partial ordering we can distinguish several classes of elements which are very important in applications. If S is a set with a partial ordering '$\leqslant$' then we say that the element s_0 is the least upper bound for the subset S_1 of S if the following conditions are satisfied:

(1) $s \leqslant s_0$ for all s in S_1.

(2) If u is an element of S such that $s \leqslant u$ then $s_0 \leqslant u$.

We leave the reader to define the notion of the greatest lower bound of a subset of S.

If S_1 is a subset of S which has partial ordering then we use the notation glb S_1 for the greatest lower bound of S_1 and lub S_1 for the least upper bound of S_1.

Another formulation for glb S_1 and lub S_1 may be given using the notion of upper bound. If S_1 is a subset in S_1 which is a set with a partial ordering then an upper bound for S_1 is an element of S, say u, with the property that for all $s_1 \in S_1$ we have $s_1 \leqslant u$. Then clearly lub S_1 is the element m of S with the following properties:

(1) m is an upper bound for S_1.
(2) Every upper bound b of S_1 satisfies the relation $m \leqslant b$.

Similarly we can proceed for glb S_1.

Another important class of elements in a set with partial ordering is the class of maximal elements (and, of course, the class of minimal elements). If S is a set with a partial ordering '$\leqslant$' then an element M of S is a maximal element if $M \leqslant x$ then $x \leqslant M$. Similarly, we define the notion of minimal element for the set S.

In applications the following result of Max Zorn plays a fundamental rôle. It is equivalent to a result of K. Kuratowski obtained earlier.

ZORN'S LEMMA. *If S is a set with a partial ordering '$\leqslant$' and has the property that every subset S_1 which is totally ordered has an upper bound, then S has a maximal element.*

REMARK. Zorn's lemma admits a formulation using the family of minimal elements.

We note that the above lemma is applied to prove the existence of some objects with certain properties. We shall have occasion to apply this lemma several times in this book.

1.2. Topology and Topological Spaces

We consider a set S, and $P(S)$ denotes the family of all subsets of S including S and the empty set (i.e. the set without elements, which is denoted by $\emptyset$).

DEFINITION 1.2.1. A topology on the set S is a family τ of subsets in $P(S)$ such that the following assertions hold:

(1) S and $\emptyset$ are in τ.
(2) If G_1 and G_2 are in τ then $G_1 \cap G_2$ is in τ.
(3) If I is an index set and $\{G_i\}_{i \in I}$ is a family of elements of τ then $\cup_{i \in I} G_i$ is in τ.

DEFINITION 1.2.2. The pair $\{S, \tau\}$ is called a topological space. We say, briefly, that S is a topological space. The elements of S are called points and the elements of τ the open sets of the topological space S. The definition of a topology τ on S may be formulated in the following manner: the family τ of subsets of S is a topology if the following conditions are satisfied:

(a) The empty set and S are in τ.
(b) The intersection of any finite family of elements of τ is in τ.
(c) The union of an arbitrary number of elements of τ is an element of τ.

We now define an important notion used in convergence theory and continuity.

DEFINITION 1.2.3. A neighborhood of the point s of the topological space S is any set V_s which contains an open set G with the property $s \in G \subseteq V_s$.

We have the following result.

PROPOSITION 1.2.4. *The set M is open iff for each point s of M there exists a neighborhood V_s such that $V_s \subseteq M$.*

In the construction of topologies we frequently use bases and subbases.

DEFINITION 1.2.5. Let S be a topological space with a topology τ. Then the family $\mathscr{B}$ of subsets in τ is called a base for the topology τ of S if every nonempty set G of τ is the union of a number of sets in $\mathscr{B}$.

DEFINITION 1.2.6. The family $\mathscr{B}$ of subsets on τ is said to be a subbase of the topology of S if the family of all finite intersections $G_{i_1} \cap G_{i_2} \cap \cdots \cap G_{i_m} \cap G_{i_j}$ in $\mathscr{B}$ is a base for the toplogy τ of S.

Suppose now that $\{S, \tau\}$ is a topological space and consider an arbitrary point s in S. We now consider the family $\mathscr{B}(s)$ of the neighborhoods of s with the following property: for each open set G containing s there exists V_s in $\mathscr{B}(s)$ such that $s \in V_s \subseteq G$. We call such a family the base of S at s. Then we have the following result.

PROPOSITION 1.2.7. *The collection $\{\mathscr{B}(s)_{s \in S}\}$ of bases of the space S at the points of S has the following properties:*

(1) *$\mathscr{B}(s)$ is not empty and for each $s \in G$, G open, $G \in \mathscr{B}(s)$.*
(2) *If s is in $G \in \mathscr{B}(t)$ then there exists $H \in \mathscr{B}(s)$ such that $H \subseteq G$.*
(3) *If $G_1 \in \mathscr{B}(s)$, $G_2 \in \mathscr{B}(s)$ then there exists G in $\mathscr{B}(s)$ such that $G \subseteq G_1 \cap G_2$.*

DEFINITION 1.2.8. The collection $\{\mathscr{B}(s)_{s \in S}\}$ is called the neighborhood system of the topological space S.

If S is a topological space then we can consider the following very important class of subsets of S.

DEFINITION 1.2.9. The set F in the topological space S is said to be closed if the set $C_F = S \setminus F$ is open.

From the de Morgan formulas we clearly have the following assertion:

PROPOSITION 1.2.10. *The family of closed subsets of the topological space* S *has the following properties:*

(1) *The empty set and the whole space are closed sets.*
(2) *The union of a finite number of closed sets is closed.*
(3) *The intersection of arbitrary many closed sets is a closed set.*

It is clear that we can formulate the definition of a topology using closed sets as well as other notions. Suppose now that $\{S, \tau\}$ is a topological space and that M is an arbitrary subset in S. Associated with M we can consider the family F_M of all closed subsets of S which contains M. Obviously this is not empty (S is in F_M). On F_M we consider a partial ordering (by inclusion) and then we see easily that we can apply Zorn's lemma to obtain a minimal element of F_M. This closed set, which contains M, is denoted by $\bar{M}$.

DEFINITION 1.2.11. For any set M in S the set $\bar{M}$ is called the closure of M.

NOTE 1.2.12. Sometimes we also use the notation $\bar{M} = \mathrm{Cl}\, M$. The closure operator defined on a topological space is a mapping defined on $P(S)$ which associates to each M in $P(S)$ the element of $P(S)$ defined by $\bar{M}$. It is not difficult to remark that the following properties hold:

(1) $\bar{\emptyset} = \emptyset$.
(2) $M \subseteq \bar{M}$.
(3) If $M \subseteq N$ then $\bar{M} \subseteq \bar{N}$.
(4) For any sets A and B, $\overline{(A \cup B)} = \bar{A} \cup \bar{B}$.
(5) $\{\bar{\bar{M}}\} = \bar{M}$.

An important and useful notion related to the closure is that of the cluster point of a set.

DEFINITION 1.2.13. The point s is called a cluster point for the set M of S if every neighborhood of s contains a point of M.

Then we have clearly the following result.

PROPOSITION 1.2.14. *For any set M of S, $\bar{M}$ is exactly the set of cluster points of M.*

A dual notion of that of a cluster point is the notion of an interior point.

DEFINITION 1.2.15. Let M be a set in S. The point s of M is said to be an interior point if there exists an open set $\mathscr{G}$ containing s and G is in M. The set of all interior points of M is denoted by $\mathring{M}$ and is called the interior of M. For $\mathring{M}$ we sometimes also use the notation Int M.

From the definition of interior points we obtain the following result.

PROPOSITION 1.2.16. *A set M in S is open iff $\mathring{M} = M$.*

The relation between the interior and closure operators is given in the following proposition.

PROPOSITION 1.2.17. *For any set M of S the following relation holds:*

$$\mathrm{Int}\, M = C_{\bar{C}_M}.$$

Proof. From property (2) of the closure operator we have $C_M \subseteq \bar{C}_M$ and thus $C_{\bar{C}_M} \subseteq C_{C_M} = M$. But $C_{\bar{C}_M}$ is open and thus $C_{\bar{C}_M} \subseteq \mathrm{Int}\, M$. Now if G is open and $G \subseteq M$ then we get $C_M \subseteq C_G = \bar{C}_G$ which implies that $\bar{C}_M \subseteq C_G$ and $G \subseteq C_{\bar{C}_M}$ from which we obtain that $\text{Int}\, M \subseteq C_{\bar{C}_M}$ and thus $\text{Int}\, M = C_{\bar{C}_M}$. □

From the properties of the closure operator and the de Morgan formulas we have the following result.

PROPOSITION 1.2.18. *The 'interior' operator defined on the nonempty subsets of S which associates to each nonempty subset M the set* $\text{Int}\, M$, *has the following properties:*

(1) $\text{Int}\, S = S$.
(2) *For any* M, $\text{Int}\, M \subseteq M$.
(3) *For any* M, N, $\text{Int}(M \cap N) = \text{Int}\, M \cap \text{Int}\, N$.
(4) $\text{Int}(\text{Int}\, M) = \text{Int}\, M$.

We now give some very simple examples of topological spaces. Later we shall present more complicated examples.

EXAMPLE 1.2.19. Let S be an arbitrary nonempty set and let $\tau = P(S)$. Then clearly this is a topology on S which is called a discrete topology, and S with this topology is called a discrete topological space S.

EXAMPLE 1.2.20. Again S is an arbitrary nonempty set and τ is the family $(\emptyset, S)$ which obviously defines a topology on S. This topology is called an indiscrete topology and S with this topology is an indiscrete topological space S.

EXAMPLE 1.2.21. Let $S = R$ be the set of all real numbers and we say that a subset G of R is open if for any point x in G there exists $r > 0$ such that the segment $(x - r, x + r)$ is contained in G. Then $\tau = (G, G \text{ open in } R)$ is a topology on R.

EXAMPLE 1.2.22. Let us consider $S = R_{\times}R$ and we consider a subset G in S to be open if for any point (x, y) of G there exists $r > 0$ such that $(x - r, x + r)_{\times} (y - r, y + r)$ is contained in G. Then $\tau = (G, G \text{ open in } R_{\times}R)$ defines a topology on $R_{\times}R$. If S is a set then it is possible to have many topologies on S. We now introduce a partial ordering on the family of all topologies on a set S.

DEFINITION 1.2.23. Suppose that on the set S we have two topologies τ_1 and τ_2. We say that τ_1 is stronger than τ_2 (or that τ_2 is weaker than τ_1) if $\mathscr{D}_1$ contains $\mathscr{D}_2$, where $\tau_1 = \mathscr{D}_1$, $\tau_2 = \mathscr{D}_2$.

From this definition we see that the strongest topology is a discrete topology and the weakest topology is an indiscrete topology. Two topologies are called noncomparable if one is not stronger than the other.

If S is a topological space then for each subset M of S we can define the following useful subset, called the boundary of the subset M (and denoted by $\text{Fr}\, M$ of ∂M).

DEFINITION 1.2.24. For a subset M in the topological space S the boundary $\text{Fr}\, M$ (or ∂M) is the subset of M defined by the relation $\text{Fr}\, M = \bar{M} \cap \bar{C}_M$. Then we have the following assertions.

PROPOSITION 1.2.25. *The mapping defined on* $P(S)$ *which associates to each subset* M *the subset* $\text{Fr}\, M$ *has the following properties:*

(1) $\operatorname{Int} M = M \setminus \operatorname{Fr} M$.
(2) $\bar{M} = M \cup \operatorname{Fr} M$.
(3) $\operatorname{Fr}(M \cup N) \subseteq \operatorname{Fr} M \cup \operatorname{Fr} N$.
(4) $\operatorname{Fr}(M \cap N) \subseteq \operatorname{Fr} M \cap \operatorname{Fr} N$.
(5) $\operatorname{Fr} C_M = \operatorname{Fr} M$.
(6) *If M is an open subset of S then* $\operatorname{Fr} M = \bar{M} \setminus M$.
(7) $\operatorname{Fr} M = \emptyset$ *iff M is open and closed.*

We now consider several classes of points associated with a subset M of a topological space S.

DEFINITION 1.2.26. Let M be a subset in the tópological space S and let s be a point in S. We say that s is an accumulation point for the subset M if $s \in \{\overline{M \setminus \{s\}}\}$.

DEFINITION 1.2.27. The set of all accumulation points of M is called the derived set of M, and for this set we use the notation M^d.

REMARK 1.2.28. Sometimes we denote the subset M^d by M'.

DEFINITION 1.2.29. For any set M in S the points of the set $M \setminus M^d$ are called the isolated points of M.

From the definition of accumulation points of a subset M of S it is not difficult to see that a point s is an accumulation point for the subset M iff every neighborhood V_s of s contains a point of M (which is different from s); the point s of M is an isolated point of M iff there exists a neighborhood V_s of s such that $V_s \setminus \{x\} \cap M = \emptyset$.

The properties of derived sets are given in the following proposition.

PROPOSITION 1.2.30. *The derived operator which associates to any set M of S the derived subset M^d has the following properties:*

(1) $\bar{M} = M \cup M^d$.
(2) *If* $M \subseteq N$ *then* $M^d \subseteq N^d$.
(3) $(M \cup N)^d = M^d \cup N^d$.
(4) $\{\cup_{i \in I} M_i\}^d = \cup_{i \in I} M_i^d$.

Using the operators closure, interior and boundary we can introduce several important classes of sets in a topological space.

DEFINITION 1.2.31. Let S be a topological space. The subset M of S is said to be:

(1) dense in N, $N \subseteq S$ if $\bar{M} \subseteq N$, dense in itself if $M \subseteq M^d$;
(2) a boundary set in S if $\operatorname{Int} M = \emptyset$;
(3) nowhere dense in S if $\bar{M}$ is a boundary set.

An important class of topological spaces can be considered using the notion of a dense set.

DEFINITION 1.2.32. Let m be a cardinal. We say that a topological space S is m-separable if in S there exists a subset M, card $M = m$, with the property that $\bar{M} = S$. If $m = \aleph_0$ (aleph zero) we say that S is separable.

Suppose now that we have two topological spaces (S, τ) and (T, ν) and a mapping $f: S \to T$. Among the mappings defined on S and with values in T are some having certain topological properties. We now consider one of the most important classes of mappings.

DEFINITION 1.2.33. We say that the mapping $f: S \to T$ is continuous at the point s_0 of S if whenever we have a neighborhood $V_{f(s_0)}$ of the point $f(s_0)$ then there exists a neighborhood V_{s_0} of the point s_0 with the property that $f(s) \in V_{f(s_0)}$ for all s in V_{s_0}. The mapping f is called continuous on S if it is continuous at each point of S.

Another important class of mappings is considered in the following definition.

DEFINITION 1.2.34. The mapping $g: S \to T$ is said to be an open mapping if for each open subset G of S the subset $f(G) = \{f(s): s \in G\}$ is an open set in T. Similarly, we define the class of closed mappings.

An equivalence relation can be defined on the family of all topological spaces using the so-called homeomorphism mappings considered in the following defintion.

DEFINITION 1.2.35. The mapping $h: S \to T$ is called a homeomorphism if the following conditions are satisfied:

(1) h is continuous.
(2) h is surjective and injective.
(3) h^{-1} is continuous.

Two topological spaces are called homeomorphic if between them there exists a homeomorphism. Obviously $S \xrightarrow{h} T$, where h is a homeomorphism, defines an equivalence relation on the family of all topological spaces.

We now present some useful remarks concerning topologies on sets. Suppose that we have a set S and (T, τ_T) is a topological space. Further, suppose that we have a mapping defined on S and whose values are in T. We can define a topology τ_S on the set S such that the following conditions are satisfied:

(1) (S, τ_S) is a topological space.
(2) $f: S \to T$ is continuous.
(3) τ_S is the weakest topology on S with properties (1) and (2).

In fact we can give a slightly more general result which is very useful when we consider the Cartesian products.

PROPOSITION 1.2.36. *Suppose that S is a set and $(S_i, \tau_i)_{i \in I}$ is a family of topological spaces and that for each i in I we have a mapping $f_i: S \to S_i$. Then there exists on S a topology τ with the following properties*:

(1) *(S, τ) is a topological space.*
(2) *Each f_i is continuous.*
(3) *τ is the weakest topology on S with properties* (1) *and* (2).

Proof. We define the topology τ as follows. We take as a base for this topology the following family of subsets of S:

$$\bigcap_{k=1}^{n} f_{i_k}^{-1}(G_{i_k}),$$

where $i_k \in I$ and G_{i_k} are open subsets in S_{i_k}, $n < \infty$. Then it is easy to see that this defines a topology on S which satisfies the requirements.

REMARK 1.2.37. The method presented in Proposition 1.2.36, of constructing topologies on a set, is one of the methods that are most frequently used.

We now present an application of the above result. We consider the Cartesian product of the family of topological spaces $\{S_i\}_{i \in I}$ with the set S as the Cartesian product of the family. Then for each i in I we have the mapping which associates to each point s in the Cartesian product the ith coordinate.

$$p_i(\{x_j\}_{j \in I}) = x_i.$$

Then the product topology (or the Tichonov (=Tychonoff) topology) on the Cartesian product is exactly the topology which results when we apply Proposition 1.2.36 to S, S_i and the functions p_i (called the projections).

The notion of a topological space is very general; however, in applications it is important to have some special classes of topological spaces which are more appropriate to the problem being considered. In what follows we consider some of these classes.

DEFINITION 1.2.38. A topological space S is said to be a Hausdorff space (or a T_2-space) if for each pair of distinct points s_1 and s_2 of S there exist open subsets G_1 and G_2 such that $G_1 \cap G_2 = \emptyset$, $s_1 \in G_1$, $s_2 \in G_2$.

DEFINITION 1.2.39. A topological space S is said to be a T_1-space iff for each s in S the set $\{s\}$ is a closed subset of S (or, equivalently, for each pair of distinct points s_1 and s_2 of S there exists an open subset G with the properties $s_1 \in G$ and $s_2 \notin G$).

We now consider a class of topological spaces where the separation condition is formulated using continuous real-valued functions.

DEFINITION 1.2.40. A topological space S is called completely regular (or a Tichonov space) if the following conditions are satisfied:

(1) S is a T_1 topological space.
(2) If s is an arbitrary point of S and F is a closed subset of S with $s \notin F$ then there exists a real-valued continuous function with the following properties:

   (a) $0 \leqslant f(u) \leqslant 1$ for all u in S,
   (b) $f(s) = 1$ and $f(t) = 0$ for all $t \in F$.

Another important class of spaces is considered in the following definition.

DEFINITION 1.2.41. A topological space S is said to be a normal space if for each pair of disjoint closed subsets F_1 and F_2 of S there exist open subsets G_1 and G_2 in S

such that:

(1) $F_i \subseteq G_i, \quad i = 1, 2;$
(2) $G_1 \cap G_2 = \emptyset$.

The following lemma is a basic result for the study of normal spaces

LEMMA 1.2.42 (P. S. Uryson). *Let S be a normal space and let F_1, F_2 be two disjoint closed subsets of S. Then there exists a continuous real-valued function $f: S \to I = [0, 1]$ with the property that*

$$f(s) = 0, \quad s \in F_1, \qquad f(t) = 1, \quad t \in F_2.$$

A basic tool in the study of classical spaces is the use of convergence notions. In order to treat adequately the convergence problems in topological spaces (using to some extent, as a model, the convergence theory in classical spaces) we consider now the notion of 'net'.

DEFINITION 1.2.43. Let S be a topological space. A net in S is an arbitrary function defined on a directed set I with values in S. The value of the function at i of I is denoted by x_i and we say that $\{x_i\}_{i\in I}$ is a net.

If M is a subset of S then we say that the net $\{x_i\}_{i\in I}$ is eventually in M if there exists i_0 in I such that x_i is in M for all $i \geqslant i_0$. Further, we say that the net $\{x_i\}_{i\in I}$ is frequently in M if for each j in I there exists j_1 in I with the property that $j_1 \geqslant j$ and $x_{j_1} \in M$.

Using these notions we can distinguish certain classes of points related to nets.

DEFINITION 1.2.44. Let $\{x_i\}_{i\in I}$ be a net in a topological space S. Then a point u of S is said to be a cluster point of the net $\{x_i\}_{i\in I}$ if $\{x_i\}_{i\in I}$ is frequently in every neighborhood of u.

DEFINITION 1.2.45. The point u is said to be a limit point for the net $\{x_i\}_{i\in I}$ if $\{x_i\}_{i\in I}$ is eventually in every neighborhood of u; in this case we say that the net $\{x_i\}_{i\in I}$ converges to u. If u is the unique limit point of the net $\{x_i\}_{i\in I}$, we write this as $u = \lim_{i\in I} x_i$.

Using nets we have the following characterization of Hausdorff spaces.

PROPOSITION 1.2.46. *A topological space S is Hausdorff iff (if and only if) every net in S has at most one limit.*

DEFINITION 1.2.47. Suppose that $\{x_i\}_{i\in I}$ is a net in a topological space S. Suppose further that J is a cofinal subset of I. Then $\{x_j, j \in J\}$ is called a cofinal subnet of the net $\{x_i\}_{i\in I}$. We say that a subset J of the set I with a partial ordering '$\leqslant$' is cofinal iff for any i in I there exists j in J with the property that $j \geqslant i$.

We now consider a class of points for a net using the class of cofinal subnets.

DEFINITION 1.2.48. Suppose that $\{x_i\}_{i\in I}$ is a net. Then the point s of S is an adherent point for the net $\{x_i\}_{i\in I}$ if every neighborhood of s contains a cofinal subnet of the net $\{x_i\}_{i\in I}$.

Using the notion of net, the continuity of mappings may be formulated as follows.

PROPOSITION 1.2.49. *Suppose that S and T are two topological spaces and $f: S \to T$ is a mapping. Then f is continuous at s iff for each net $\{s_i\}_{i \in I}$ with $\lim_{s \in I} s_i = s$ we have*

$$\lim_{i \in I} f(s_i) = f(s).$$

In this respect it is useful to remark that we can use nets formed using neighborhoods of a base of neighborhoods of s (since these form a directed set for the partial ordering defined using the (set) inclusion).

Another interesting approach to convergence theory was proposed by H. Cartan in 1937 using the notion of filter.

DEFINITION 1.2.50 (Cartan, 1937). Let S be a nonempty set. A nonempty family of subsets of S, $F = \{F_i\}_{i \in I}$, I is an index set, is said to be a filter on S if the following conditions are satisfied:

(1) Every subset of S containing an F_i is in F.
(2) The intersection of finitely many F_i is in F.
(3) The empty set does not belong to F.

DEFINITION 1.2.51. A filter base on S is a nonempty family B of subsets of S satisfying the following conditions:

(1) If B_1 and B_2 are in B then there exists B_3 in B such that $B_3 \subseteq B_1 \cap B_2$.
(2) The empty set is not in B.

Using filter bases we can construct filters on S as follows.

PROPOSITION 1.2.52. *Suppose that on the set S we have a filter base B. Then the family F, defined as,*

$$F = \{M, M \in P(S), \text{ there exists } b \text{ in } B \text{ such that } b \subseteq M\} \text{ is a filter on } S.$$

The notion of convergence using filters is defined as follows.

DEFINITION 1.2.53. Let s be a point of S and let F be a filter on S. We say that the point s is a limit point of the filter F if

$$\{V_s, V_s \text{ is a neighborhood of } s\}$$

is in F.

DEFINITION 1.2.54. We say that the filter F converges to s if s is a limit point of F, and we write this as $s = \lim F$.

REMARK 1.2.55. We have considered only Hausdorff spaces in Definitions 1.2.51–53.

On the family of filters on the set S we can define a partial ordering in the usual way, by inclusion: the filter F_1 is finer than the filter F_2 if $F_1 \subseteq F_2$ (as families of subsets).

The connection between filters and nets is as follows. Suppose that F is a filter on

S. Then we can consider the family of subsets:

$$U = \{(s, M), s \in M \in F\}$$

and a partial ordering '$\leqslant$' defined as follows on U,

$$(s, M) \leqslant (t, N) \quad \text{if } N \subseteq M.$$

Then we have the following result.

PROPOSITION 1.2.56. *U is a directed set (with respect to the partial ordering defined above).*

Using the set U we can associate to the filter F the following net. For each (s, M) in U we set $x_M = s$. Then we have

PROPOSITION 1.2.57. *The net $\{x_m\}_{M \in F}$ converges to a point s iff F converges to s.*

REMARK 1.2.58. There is another net which we can associate to a filter F, which in some cases seems to be more natural, namely: for each M in F we choose x_M in M and then $\{x_M\}_{M \in F}$ is a net. The reader may note that this net has the property that it may converge, but F does not.

An important class of filters with many applications is the class of so-called 'ultrafilters'.

DEFINITION 1.2.59. An ultrafilter on S is a maximal filter. Of course, we consider the order relation introduced above (one filter is finer than another).

From Zorn's lemma we obtain the following result:

PROPOSITION 1.2.60. *Every filter F on S is contained in an ultrafilter on S.*

EXAMPLE 1.2.61. Suppose that we have a set S, and let us consider an arbitrary point s of S. Set

$$F_s = \{M, M \in P(S), s \in M\}.$$

Then, clearly, F_s is an ultrafilter.

For more information on filters and ultrafilters, and for more information on the classes of topological spaces, we refer the reader to books on general topology.

1.3. Compactness in Topological Spaces

We now present the notion of compactness which permits us to consider certain classes of topological spaces that have many applications. First we note that the notion of compactness is an abstraction of a property discovered by E. Borel and H. Lebesgue for the segments of the real line. In order to define precisely the notion of compactness we require some facts on 'coverings'.

DEFINITION 1.3.1. Suppose that S is a nonempty set and $\{M_i\}_{i \in I}$ is a family of subsets of S. We say that this family is a covering of S as the union of this family coincides with S.

DEFINITION 1.3.2. Suppose that we have on the set S two families of subsets, say $\{M_i\}_{i\in I}$ and $\{N_j\}_{j\in J}$ respectively, where I and J are index sets. Suppose that both are coverings of S. Then we say that $\{N_j\}_{j\in J}$ is a subcovering of $\{M_i\}_{i\in I}$ if for each j of J there exists i_j in I with the property that $M_{i_j} = N_j$.

DEFINITION 1.3.3. A covering $\{M_i\}_{i\in I}$ of a topological space S is said to be an open covering if all subsets M_i are open.

We now consider one of the most important classes of topological spaces, namely the class of compact spaces.

DEFINITION 1.3.4. A topological space S is said to be compact if the following conditions are satisfied:

(1) S is a Hausdorff topological space.
(2) If $\{G_i\}_{i\in I}$ is an open covering of S then there exists a finite subcovering, i.e. there exists a finite number of subsets $G_{i_1},\ldots,G_{i_m}$, $m<\infty$, such that

$$S = \bigcup_{j=1}^{m} G_{i_j}.$$

A useful property of compact spaces can be formulated using the so-called intersection properties.

DEFINITION 1.3.5. The family of subsets of S, $\{M_i\}_{i\in I}$, is said to have a finite intersection property if the following assertions hold:

(1) $\{M_i\}_{i\in I}$ is nonempty.
(2) If for every finite system $i_1,\ldots,i_m$ of elements of I $M_{i_1}\cap M_{i_2}\cap\cdots\cap M_{i_m}\neq\emptyset$ then

$$\bigcap_{i\in I} M_i \neq \emptyset.$$

Then we have the following important characterization of compact spaces.

PROPOSITION 1.3.6. *A topological space S is compact iff any family $\{F_i\}_{i\in I}$ of closed subsets has a finite intersection property.*

We now present the characterization of compact spaces using nets and filters.

PROPOSITION 1.3.7. *A topological space S is compact iff every net of S has a cluster point or every filter on S has a cluster point.*

Another useful result in the theory of compact spaces is the famous theorem of A. N. Tichonov which has the following formulation.

PROPOSITION 1.3.8. *The Cartesian product of a family of topological spaces $\{S_i\}_{i\in I}$ is compact (with respect to Tichonov toplogy) iff each S_i, $i\in I$, is a compact space.*

We now consider a useful class of topological spaces related to compact spaces.

DEFINITION 1.3.9 (Alexandrov and Uryson, 1929). A topological space S is said to be a locally compact space if for every s is S there exists a neighborhood V_s of s such that $\bar{V}_s$ is a compact subset of S.

A subset of a topological space is said to be compact iff the subset with the topology induced by the topology of the space is a compact space; if S is a topological space and S_1 is an arbitrary subset of S, then we can define a topology on S_1 as:

$$\tau_1 = (G \cap S_1, G \text{ open in } S).$$

1.4. Metric Spaces. Examples and Some Properties

In this section we present a class of topological spaces which is in many cases the most natural setting for generalizations and extensions of classical results of analysis, namely the class of metric spaces first considered by Maurice Fréchet in 1906. Further important contributions to the theory of metric spaces we owe to F. Hausdorff, K. Kuratowski, P. S. Uryson, P. S. Alexandrov, A. N. Tichonov, A. Stone, Yu. M. Smirnov and many others.

DEFINITION 1.4.1. A nonempty set S is said to be a metric space if on $S_\times S$ there exists a function d with values in R (the set of all real numbers) satisfying the following conditions:

(1) $d(x, y) = 0$ iff $x = y$,
(2) $d(x, y) = d(y, x)$ for all x, y in S,
(3) $d(x, z) \leqslant d(x, y) + d(y, z)$ for all x, y, z in S.

A function d with the above properties is called the metric of the space, and for any two points x, y in S the number $d(x, y)$ is said to be the distance between x and y. Also, condition (3) is called the 'triangle inequality'. We remark that the values of d are in fact in $[0, \infty]$. Indeed, if we take in (3) the point $x = z$ then we have

$$0 \leqslant d(x, y) + d(y, x) = 2d(x, y)$$

and the assertion follows.

If S is a metric space with the metric d and S_1 is an arbitrary subset of S, then S_1 can be organized as a metric space as follows: a metric d_1 on S_1 is defined for any pair of points of S_1 by the formula

$$d_1(x, y) = d(x, y).$$

The metric d_1 is the so-called induced metric on S_1.

An interesting subclass of metric spaces is the class of so-called non-Archimedean metric spaces considered in the following definition.

DEFINITION 1.4.2. A set S is said to be a non-Archimedean metric space if on $S_\times S$ there exists a function d with values in R satisfying the following conditions:

(1) $d(x, y) = 0$ iff $x = y$.
(2) $d(x, y) = d(y, x)$ for all x, y in S.
(3) $d(x, z) \leqslant \max\{d(x, y), d(y, z)\}$ for all x, y, z in S.

It is easy to see that every non-Archimedean metric space is a metric space.

We now present some examples of metric spaces.

EXAMPLE 1.4.3. (1) The set Q of all rational numbers with the metric d defined by the formula

$$d(p, q) = |p - q|$$

for all rationals p, q.

(2) The set R of all real numbers with the metric defined by the formula

$$d(s, t) = |s - t|$$

for all real numbers s, t.

(3) Let n be a fixed finite integer and consider a space R^n of all elements $x = (x_1, \ldots, x_n)$, where $x_i \in R$, $i = 1, 2, \ldots, n$. If $y = (y_1, \ldots, y_n)$ is another point in R^n then we set

$$d(x, y) = \left(\sum_{i=1}^{n} |x_i - y_i|^p \right)^{1/p}$$

where p is a fixed real number in $[1, \infty)$. Using the Minkowski inequality we get that d defined by the above formula is a metric on R^n.

(4) Let n be as above and consider the space C^n of all elements of the form $x = (x_1, \ldots, x_n)$, $x_i \in C$, C is the set of all complex numbers, and if $y = (y_1, \ldots, y_n)$ is another element of C^n we set

$$d(x, y) = \left(\sum_{i=1}^{n} |x_i - y_i|^p \right)^{1/p},$$

where p is as above. In this case we get that C^n is a metric space.

(5) Let p be as above and consider the set l^p of all elements $x = (x_i)_1^\infty$, $x_i \in R$, with the property that

$$\sum_{i=1}^{\infty} |x_i|^p < \infty.$$

If $y = (y_i)_1^\infty$ is another element of l^p then we set

$$d(x, y) = \left(\sum_{i=1}^{\infty} |x_i - y_i|^p \right)^{1/p}.$$

Using the Minkowski inequality we get that d is a well-defined function and that it is in fact a metric on l^p.

(6) Similarly as above, but we consider $x_i \in C$.

(7) We consider the space m_I of all sequences $x = (x_i)_{i \in I}$ such that $\sup |x_i| < \infty$, where I is an arbitrary index set. If $y = (y_i)_{i \in I}$ is another point in m_I then we set

$$d(x, y) = \sup_{i \in I} |x_i - y_i|$$

and it is easy to see that this is a metric on m_I. If card $I = \aleph_0$ then we write simply m (instead of m_I).

(8) Let us consider a segment $I = [0, 1]$ and let C_I be the set of all continuous real-valued functions on I. If f, g are two elements of C_I then we set

$$d(f, g) = \sup_{t \in I} |f(t) - g(t)|$$

and it is not difficult to prove that this defines a metric on C_I.

(9) Let C_I be as above, and for any two points f, g in C_I we set

$$d(f, g) = \left(\int_0^1 |f(s) - g(s)|^p \, \mathrm{d}s \right)^{1/p}$$

for p fixed in $[1, \infty)$. Then this is a metric on C_I.

(10) Let us consider a space $R_\times R$ where R is the set of all real numbers. If (x, y) and (u, v) are two points in this space then we set

$$d((x, y), (u, v)) = \begin{cases} |y - v| & \text{if } x = u \\ |y| + |v| + |x - u| & \text{if } x \neq u \end{cases}$$

and we can verify that this is a metric on $R_\times R$. This is the so-called 'river metric' on $R_\times R$.

In order to consider topological properties of metric spaces we now define several important classes of subsets of a metric space.

DEFINITION 1.4.4. Let S be a metric space with the metric d and let x be an arbitrary point of S. A subset $B(x, r)$, $r > 0$, is defined as

$$\{y, y \in S, d(x, y) < r\}$$

and is called an (open) ball with center at x and radius r. Using this subset we now show how we can introduce a topology on a metric space.

DEFINITION 1.4.5. We say that a subset G of a metric space S is an open subset of S if for each x in G there exists $r > 0$ such that $B(x, r) \subseteq G$.

Then we have

PROPOSITION 1.4.6. *For a metric space S the family τ is defined by*

$$\tau = \{G, G \text{ open in } S\}.$$

The topology defined above is called the metric topology of S and is the smallest topology on S which contains the family of subsets $\{B(x, r) : x \in S, r > 0\}$.

DEFINITION 1.4.7. A topological space S is said to be metrizable if on S there exists a metric such that the metric topology associated with that metric (as in 1.4.6) coincides with the topology of S.

DEFINITION 1.4.8. Let S be an arbitrary set and suppose that on S we have two metrics. Then we say that these metrics are equivalent if the metric topologies they define coincide. It is obvious that this is an equivalence relation in the family of all metrics on a space S.

We now consider the convergence notion in a metric space.

DEFINITION 1.4.9. Let S be a metric space with the metric d. The sequence $(x_n) \subseteq S$ is said to converge to the element x of S if $\lim d(x_n, x) = 0$. The point x is called the limit of the sequence $\{x_n\}$ and we write this as $x = \lim x_n$.

An important class of sequences in a metric space is considered in the following definition.

DEFINITION 1.4.10. A sequence $\{y_n\}$ in a metric space S is said to be a Cauchy sequence if for each $\varepsilon > 0$ there exists an integer N_ε such that for all $n, m \geqslant N_\varepsilon$ the following inequality holds:

$$d(x_n, x_m) \leqslant \varepsilon.$$

It is easy to see that every convergent sequence is a Cauchy sequence, but there exist metric spaces in which we have Cauchy sequences which do not converge. This motivates (to some extent) the introduction of the following important class of metric spaces.

DEFINITION 1.4.11. A metric space S is said to be a complete metric space if every Cauchy sequence of S converges to a point of S.

The spaces in (2)–(8) and (10) of Example 1.4.3 are complete metric spaces; (1) and (9) are not complete metric spaces. The following proposition gives a connection between complete metric spaces and incomplete metric spaces. First we give a definition.

DEFINITION 1.4.12. Let (S, d) and (S_1, d_1) be two metric spaces and let $f: S_1$ be a mapping. We say that f is an isometry if for all x, y in S the following relation holds:

$$d_1(f(x), f(y)) = d(x, y).$$

Then the above-mentioned connection is as follows.

PROPOSITION 1.4.13. *Let (S,d) be an incomplete metric space. Then there exists a complete metric space (S_1, d_1) and a mapping $f: S \to S_1$ with the following properties:*

(1) *f is an isometry.*
(2) *$f(S)$ is dense in S_1.*

The space (S_1, d_1) given in this proposition is called the completion of the metric space (S,d). We note that the construction of (S_1, d_1) is an abstraction of the construction of real numbers from rational numbers.

We now present some interesting characterizations of complete metric spaces. First we give

DEFINITION 1.4.14. Let M be a subset of the metric space (S, d). The number

$$\sup\{d(x, y): x, y \in M\} = d(M)$$

is called the diameter of the subset M. The set M is said to be bounded if

$$d(M) < \infty.$$

Then we have the following result.

THEOREM 1.4.15 (G. Cantor). *A metric space (S, d) is a complete metric space iff the following property holds: for every sequence (F_n) of closed subsets of S with*

(1) $F_1 \supseteq F_2 \supseteq \cdots \supseteq F_n \supseteq F_{n+1} \supseteq \cdots;$
(2) $\lim d(F_n) = 0$

then

$$\bigcap_{i=1}^{\infty} F_n \neq \emptyset.$$

Later we shall present a very interesting and important extension of this characterization using measures of noncompactness. The notion of accumulation points and cluster points may be formulated as follows.

DEFINITION 1.4.16. Let S be a metric space. Then

(1) The point s of S is a limit point or a point of accumulation for the subset M of S if every neighborhood of s contains a point t of $M, t \neq s$.
(2) If $\{x_n\}$ is a sequence in S then the point s is a cluster point of (x_n) if for every neighborhood V_s of s there exists an integer N such that for all $n \geqslant N. x_n \in V_s$.

We note that in the case of metric spaces the continuity of mappings can be characterized using only sequences.

PROPOSITION 1.4.17. *Let (S, d) and (S_1, d_1) be two metric spaces and let $f: S \to S_1$ be a mapping. Then f is continuous at $s_0 \in S$ iff for any sequence (s_n) of S with* lim $s_n = s_0$ *we have* $\lim f(s_n) = f(s_0)$.

Also, the compact subsets of a metric space can be characterized using only sequences of points. First we give a definition.

DEFINITION 1.4.18. If S is a topological space then we say that S is sequentially compact if for any sequence $(s_n)_1^\infty \subset S$, there exists a subsequence which converges to some point of S.

The characterization of metric spaces which are compact (with respect to the metric topology) is as follows.

PROPOSITION 1.4.19. *Let S be a metric space and let M be a subset of S. Then M is a compact subset of S iff the following assertions hold:*

(1) *M is closed.*
(2) *M is sequentially compact.*

An interesting class of subsets of a metric space is considered in the following definition.

DEFINITION 1.4.20. Let S be a metric space with the metric d. The subset M of S

is said to be totally bounded if the following condition is satisfied: for every $\varepsilon > 0$ there exists a finite subset E_ε in M with the property that for each x of M there exists x_ε in F_ε such that $d(x, x_\varepsilon) < \varepsilon$, i.e. x is in $B(x_\varepsilon, \varepsilon)$.

This class of subsets can be characterized in the following way.

PROPOSITION 1.4.21. *If S is a metric space and M is a subset of S then M is a totally bounded subset iff one of the following equivalent assertions hold*:

(1) *M is sequentially compact.*
(2) *$\bar{M}$ is compact.*

Using the notion of a nowhere-dense set we now consider a very important class of subsets in topological spaces.

DEFINITION 1.4.22. A set M in a topological space S is said to be of first category if M is the union of a countable family of nowhere-dense subsets of S. A set N is said to be of second category if it is not of first category.

One of the basic results which has many and deep applications is the Baire theorem on sets of second category.

THEOREM 1.4.23 (Baire, 1904). *A complete metric space is a set of second category.*

The proof of this result follows easily from the following lemma.

LEMMA 1.4.24. *Suppose that S is a complete metric space and $\{G_n\}_1^\infty$ is a countable family of open dense sets in S (i.e. for each n, G_n is open and dense in S). Then*

$$\bigcap_{n=1}^{\infty} G_n \neq \emptyset.$$

Proof. To prove the assertion of the lemma it suffices to show that if V is an arbitrary nonempty open subset then $V \cap (\cap G_n)$ is also nonempty. □

Now G_1 is dense in S and thus there exists x_1 and $\varepsilon_1 > 0$ such that $B(x_1, \varepsilon_1)$ is in $V \cap G_1$. We may suppose, without loss of generality, that ε_1 is in $(0, 1)$. Now we define a sequence $(B(x_n, \varepsilon_n)\}$ as follows: suppose that $(B(x_1, \varepsilon_1), \ldots\ B(x_{n-1}, \varepsilon_{n-1}))$ are defined, and since G_n is dense in S,

$$G_n \cap B(x_{n-1}, \varepsilon_{n-1})$$

is a nonempty open subset. Then we find x_n and ε_n in $(0, 1)$ such that

$$B(x_n, \varepsilon_n) \subseteq G_n \cap B(x_{n-1}, \varepsilon_{n-1}).$$

We may suppose, without loss of generality, that ε_n is with the property that $\varepsilon_n < 1/n$. Thus we get a sequence $\{B(x_n, \varepsilon_n)\}$ and we now consider the sequence of elements $(x_n)_1^\infty$. We show that $(x_n)_1^\infty$ is a Cauchy sequence. Indeed, let x_n and x_m be two points of this sequence and suppose that $n > m$. Suppose now that $\varepsilon > 0$ is given and m has the property that $2/m < \varepsilon$. Since x_n is in $B(x_m, \varepsilon_m)$ we have

$$d(x_n, x_m) \leqslant 2/_n m = 2/m \leqslant \varepsilon.$$

Thus $(x_n)_1^\infty$ is a Cauchy sequence, and since S is supposed to be complete we can consider $\lim x_n = u$. Since, as noted above, $x_n \in B(x_m, \varepsilon_m)$ for $n \to \infty$, we obtain that u is in $B(x_m, \varepsilon_m)$ and thus is in V. Thus, it is also in $\cap_1^\infty G_n$ and the lemma is proved.

Now the proof of Theorem 1.4.23, using Lemma 1.4.24, is as follows. Suppose that the assertion of the theorem is false and thus S is of the form $S = \cup G_n$, where G_n is a sequence of nowhere-dense subsets. This further gives that $(C_{\bar{G}_n})_1^\infty$ is a sequence of open subsets with the property that $\bar{C}_{\bar{G}_n} = S$ for all n. Then, from the above lemma we get that $\cap C_{\bar{G}_n} \neq \emptyset$. But this clearly implies that $\cup_{n=1}^\infty G_n$ is not equal to S. This contradiction proves the theorem.

REMARK 1.4.25. It is clear from the proof that we may prove the following assertion: in a complete metric space S every set of the form $(y, d(x, y) \leqslant r)$ is a set of the second category for all $r > 0$.

NOTE 1.4.26. We shall give later an interesting application of this theorem of R. Baire to obtain some of the basic results of linear functional analysis.

1.5. Measures of Noncompactness in Metric Spaces

Suppose that (S, d) is a metric space and M is a bounded set in S. In 1929, K. Kuratowski defined a number associated with M which, in some sense, measured how far the set M was to have the property that $\bar{M}$ was a compact set. This number is now called the Kuratowski measure of noncompactness and is defined as follows.

DEFINITION 1.5.1. For any bounded set M is a metric space $S, \alpha(M)$ is the infimum of all $\varepsilon > 0$ such that there exists a finite system of subsets of $S, A_1, \ldots, A_m$, with the following properties:

(1) $(A_1, \ldots, A_m)$ is a covering of M.
(2) For each $i = 1, 2, 3, \ldots, m, d(A_i) < \varepsilon$.

It is not difficult to see that the following properties hold:

(1) $\alpha(M) \leqslant d(M)$.
(2) $\alpha(M \cup N) = \max\{\alpha(M), \alpha(\mathrm{N})\}$.
(3) $\alpha(M) = 0$ iff $\bar{M}$ is compact.
(4) If $M \subseteq N$ then $\alpha(M) \leqslant \alpha(N)$.

Now we have the following characterization of completeness of a metric space using the Kuratowski measure of noncompactness.

PROPOSITION 1.5.2 (Kuratowski, 1929). *A metric space (S, d) is complete iff the following assertion holds: if $(F_n)_1^\infty$ is a sequence of closed subsets of S with the properties:*

(a) $F_1 \supseteq F_2 \supseteq \cdots \supseteq F_n \supseteq F_{n+1} \supseteq \cdots$,
(b) $\lim \alpha(F_n) = 0$

then $\cap_{n=1}^\infty F_n$ is nonempty (and compact).

For an interesting proof of this result, see Pasicki (1979). For other measures of noncompactness, as well as for their use in fixed point theory, the reader may consult the author's book on fixed point theory, Istrăţescu (1981).

1.6. Some Historical Remarks

The concept of abstract spaces based on the notion of convergence and a derived set were first considered in 1906 by the French mathematician M. Fréchet, who also has introduced the notion of a metric space (1906). Other interesting contributions to this subject were made by F. Riesz in 1907 and 1909. An important step in the development of the general topology was marked by the appearance in 1914 of the book by F. Hausdorff, in which the concept of a topological space was defined in sufficient generality. The theory of limits was essentially constructed by the Italian mathematician M. Picone in 1923 (see the excellent lectures of Fichera (1954) and by the American mathematicians Moore and Smith (1922)). Important contributions to the theory of limits were made by H. Cartan in 1937, who introduced the notion of 'filter', and by J. L. Kelley on the theory of nets (to whom the name 'net' is attributed (1950)).

For more details on this subject we refer the reader to the excellent paper by McShane (1962) and to books on general topology.

CHAPTER 2

Banach Spaces and Complete Inner Product Spaces

2.0. Introduction

The fundamental work of the Italian mathematician Vito Volterra (from 1887 onwards when he published his earliest paper 'Functions Depending on Other Functions') has shown a definite advantage of the operational techniques in the study of families of functions. Later work in this direction by Ivar Fredholm, Frigyes Riesz, E. Hellinger and O. Toeplitz (as well as many others) emphasized the direction inaugurated by Volterra. This work led to the consideration of some classes of spaces which are of fundamental importance. First was considered by Stefan Banach, E. Helly, H. Hahn and Norbert Wiener (independently of each other) the class of normed linear spaces (1922). It is worth mentioning that the axioms for linear spaces as well as some notions about operators on such spaces, are given in the book by Salvatore Pincherle (1901; in collaboration with Ugo Amaldi) in a form almost identical with that now in use.*

In the theory of normed linear spaces the contribution of S. Banach is of fundamental importance; other mathematicians have contributed with profound results, among them we mention Frigyes Riesz, O. Toeplitz, E. Hellinger, J. Schauder, N. Dunford, W. Orlicz, J. Dieudonné, L. Schwartz, S. Mazur, A. E. Taylor, I. M. Gelfand, N. Dunford, and A. Grothendieck. The reader interested in further information about normed linear spaces and topological linear spaces may consult the excellent treatises of N. Bourbaki and of N. Dunford and J. Schwartz. The reader may take some care in reading the Notes and Remarks in the monumental work of N. Dunford and J. Schwartz since these are carelessly compiled on some points. For example, in giving an account of a conjecture of I. Kaplansky about the commutators of two elements in a Banach algebra, the compiler indicates some partial results, but at the time of writing the Notes and Remarks the conjecture was solved; in discussing the characterization of inner product spaces in the class of metric spaces the compiler does not mention the fundamental work of M. Fréchet which is at the origin of such characterizations and, as P. Jordan and J. von Neumann mention in their work, their result is only an improvement on Fréchet's result.

In what follows we present an account of the theory of normed linear spaces for future use in this book.

*This interesting information was given to me by the late Professor Beniamino Segre.

2.1. Linear Spaces. Sets in Linear Spaces

Let S be a set and let K be a field (R or C, the set of all real numbers or the set of all complex numbers, respectively). The basic notion of a linear space (or of a vector space) is considered in the following definition.

DEFINITION 2.1.1. The set S is said to be a linear space over K if there exist two mappings defined on $S_\times S$ and on $K_\times S$ with values in S, and denoted by '+' (the addition mapping) and '·' (the scalar multiplication), respectively, such that the conditions

(1) $(x + y) + z = x + (y + z)$,
(2) $x + y = y + x$,
(3) there exists an element, denoted by 0, of S such that $x + 0 = 0$,
(4) $(k_1 + k_2)\cdot x = k_1 \cdot x + k_2 \cdot x$,
(5) $k\cdot(x + y) = k\cdot x + k\cdot y$,
(6) $(k_1 k_2)\cdot x = k_1\cdot(k_2\cdot x)$,
(7) $1\cdot x = x$,
(8) $0\cdot x = 0$

are satisfied for all x, y, z in S and k, k_1, k_2 in K. The linear space S over K is said to be a real linear space if $K = R$; if $K = C$ then we say that S is a complex linear space.

We now present some examples of linear spaces.

EXAMPLE 2.1.2. Let us consider the set $S = R_\times R$ with mappings '+' and '·': if (x, y) and (x_1, y_1) are two elements in S and k is a real number then we set,

$$(x, y) + (x_1, y_1) = (x + x_1, y + y_1), \qquad k\cdot(x, y) = (kx, ky).$$

Then it is not difficult to see that this is a real linear space if we set $0 = (0, 0)$.

EXAMPLE 2.1.3. Let us consider the set of all sequences of complex numbers, denoted by s. If $x = (x_n)_1^\infty$ and $y = (y_n)_1^\infty$ are two elements of s and c is a complex number then we define the mappings '+' and '·' by the relations

$$x + y = (x_n + y_n)_1^\infty, \qquad c\cdot x = (cx_n)_1^\infty .$$

If we set $0 = (u_n)_1^\infty$, $u_n = 0$, $n = 1, 2, 3, \ldots$, then we can verify immediately that this is a complex linear space.

EXAMPLE 2.1.4. Suppose that M is a set and consider the space S of all functions defined on S and whose values are in R (or C). If f, g are two elements of S then we set

$$(f + g)(m) = f(m) + g(m), \qquad (k\cdot f)(m) = kf(m)$$

for all $m \in M$ and k in R (or C). Then these define the '+' and '·' mappings such that S is clearly a real (complex) linear space.

EXAMPLE 2.1.5. Let S be the linear space considered above (2.1.4) and consider m_0 to be an arbitrary but fixed point in M. We consider the space of all functions in S

with the property that the value at m_0 is the number 0. Then with the addition and multiplication defined exactly as in 2.1.2, this is a real (complex) linear space.

We shall now present some important classes of sets in linear spaces.

DEFINITION 2.1.5. Let S be a linear space over K and let C be a nonempty subset in S. The set C is called convex if whenever x, y are in C then for all t in $(0, 1)$ the element

$$z_t = t \cdot x + (1 - t) \cdot y$$

is in C.

NOTE 2.1.6. Concerning the multiplication '$\cdot$' we shall simply use the notation $k \cdot x = kx$ when no confusion can arise. Also, for the element $(-1) \cdot x$ we shall use the notation $-x$.

DEFINITION 2.1.7. The set C in a linear space S is said to be symmetric if

$$C = -C = (-x, x \in C).$$

DEFINITION 2.1.8. The subset C of a linear space S is said to be:

(1) absorbing, if for each x in S there exists $t_x > 0$ such that $tx \in C$ whenever $|t| \leqslant t_x$,
(2) balanced, if $tC = (tx, x \in C) \subseteq C$ for $|t| \leqslant 1$,
(3) affine, if $tC + (1 - t)C = \{tx + (1 - t)y, x, y \in C\} \subseteq C$ for all t in R,
(4) a line through x and y if $C = \{tx + (1 - t)y; t \in R\}$,
(5) a segment with endpoints x and y if $C = \{tx + (1 - t)y : t \in [0, 1]\}$.

Another important class of sets is defined below.

DEFINITION 2.1.9. Let S be a linear space over K and let $(s_1, s_2, \ldots, s_n)$ be elements of S with $n < \infty$. We say that the elements $s_1, \ldots, s_n$ are linearly independent if

$$\sum_1^{\infty} k_i s_i = 0$$

for some $k_i \in K$, then necessarily $k_i = 0$, $i = 1, 2, 3, \ldots, n$.

DEFINITION 2.1.10. Let S be a linear space over K and let C be a subset of S. We say that C is a linearly independent set in S if any finite part of it is a linearly independent set.

We now present some examples of subsets in linear spaces.

EXAMPLE 2.1.11. If S is the space in Example 2.1.2, then the set

$$C = \{(x, y), x^2 + y^2 \leqslant 1\}$$

is clearly convex, symmetric, absorbing and balanced. The subset

$$C = \{(x, y), x^2 + y^2 = 1\}$$

is obviously symmetric. It is clearly not convex and not absorbing. The subset

$$C = \{(x, 0), 0 \leqslant x \leqslant 1\}$$

is a segment in S; the subset

$$C = \{(x, 0), x \in R\}$$

is clearly a line in S.

We leave the interested reader to construct, in other spaces, similar examples of subsets.

EXAMPLE 2.1.12. We consider the space S of all continuous real-valued functions on a segment $[0, 1]$. If f, g are two elements of S then we set

$$(f + g)(t) = f(t) + g(t), \qquad (rf)(t) = rf(t)$$

for all $r \in R$. Then with the addition and multiplication defined as above, S is a linear space. We consider the subset

$$C = \{1, t, t^2, t^3, \ldots\},$$

where 1 means the function f defined on $[0, 1]$ by the relation

$$f(t) = 1, t \in I = [0, 1].$$

Then from the known properties of polynomials we deduce that C is a linearly independent subset of S.

In the following definition we consider an interesting and important class of subsets of linear spaces.

DEFINITION 2.1.13. Let S be a linear space over K and let S_1 be a subset of S. We say that S_1 is a linear subspace of S if the following conditions are satisfied:

(1) If x, y are in S_1 then $x + y \in S_1$.
(2) If x is in S_1 then for all k in K, $kx \in S_1$.

If S is a linear space over K and M is a subset in S then we can consider the family $\mathscr{F}_1$ of all subspaces of S containing M. This family is clearly nonempty (it contains S). With the order relation defined by set inclusion, we see immediately that we can apply Zorn's lemma to $\mathscr{F}_1$. The minimal element we get is called the subspace generated by the subset M.

We now consider an important and interesting class of subsets in a linear space.

DEFINITION 2.1.14. Let S be a linear space over K and let B be a subset of S. We say that B is a linear base for S if the following conditions are satisfied:

(1) B is a linearly independent subset of S.
(2) The subspace generated by B is exactly S.

From Zorn's lemma we obtain the following result.

PROPOSITION 2.1.15. *Every linear space over K has a linear basis. Let S be a linear space over K and let B be a basis of S. The cardinal number* card B *is called the* *algebraic dimension of S. This definition is justified by the fact (easy to prove) that given two linear bases of S, B_1 and B_2, then* card B_1 = card B_2.

We now consider an important class of linear spaces.

DEFINITION 2.1.16. Let S be a linear space over K. We say that S is a finite dimensional space over K iff, for some linear basis B of S, card $B < \infty$.

2.2. Normed Linear Spaces and Banach Spaces

Let S be a linear space over K.

DEFINITION 2.2.1. The function $p: S \to R$ is said to be a seminorm if the properties

(1) $p(x + y) \leqslant p(x) + p(y)$,
(2) $p(kx) = |k| p(x)$

hold for all x, y in S and k in K. The seminorm p is said to be a norm if $p(x) = 0$ then $x = 0$. We remark that the values of a seminorm are in $R^+ = [0, \infty)$. Indeed, we have clearly that $p(0) = 0$, and thus for any x in S,

$$p(0) = p(x + (-x)) \leqslant p(x) + p(-x) = 2p(x)$$

which implies our assertion.

EXAMPLE 2.2.2. Let S be the space of all k-valued continuous functions on $I = [0, 1]$. If, for f in S (denoted as usual by C_I), we set

$$p(f) = \sup(|f(t)|, t \in I)$$

then this is obviously a norm on C_I. Suppose t_0 is an arbitrary but fixed point in I, then if we set

$$p(f) = |f(t_0)|$$

this is a seminorm and the reader may verify that it is not a norm.

NOTE 2.2.3. If S is a linear space over K and p is a norm then we use the (now standard) notation $p(x) = \|x\|$. Different norms on the same (or different spaces) will be denoted by such symbols as $\|\,,\|_1$, $\|\,,\|_+$, etc.

EXAMPLE 2.2.4. Let S be the space of all sequences $x = (x_n)$, $x_n \in K$ for all n. Let us consider p to be a fixed number in $[1, \infty)$ and for each j we set

$$p_j((x_n)) = \left(\sum_{n=1}^{j} |x_n|^p\right)^{1/p}$$

Then this is a seminorm.

DEFINITION 2.2.5. Let S be a linear space over S. If $\|\,,\|$ is a norm on S then the pair $(S, \|\,,\|)$ is called a normed linear space. We say, briefly, that S is a normed linear space. If S is a normed linear space then we can define a metric on S by the formula

$$d(x, y) = \|x - y\|.$$

(We leave the reader to verify that this is indeed a metric on S.) This is called the metric induced by the norm of S.

The most important class of normed linear spaces is considered in the following definition.

DEFINITION 2.2.6. A normed linear space S is said to be a Banach space if the metric induced by the norm is a complete metric space.

EXAMPLE 2.2.7. The space C_I considered above with the norm

$$\|f\| = \sup(|f(t)|, t \in I)$$

is a Banach space.

EXAMPLE 2.2.8. Let us again consider C_I. The norm on C_I is defined by the formula

$$f = \int_0^1 |f(s)|\, ds.$$

Then this is clearly a normed linear space but it is not a Banach space. (We leave the reader to construct a Cauchy sequence in C_I (with respect to this norm) which does not converge to an element of C_I.)

EXAMPLE 2.2.9. Let p be a fixed real number in $[1, \infty)$ and consider a probability space (Ω, B, P). We consider the space $L^p(\Omega, B, P)$ of a function (classes of functions) f with the property that

$$\|f\| = \left(\int_\Omega |f(s)|^p\, dP(s)\right)^{1/p} < \infty. \qquad (+)$$

With the well-known addition and multiplication, this is a Banach space with respect to the norm defined by $(+)$.

For details of these spaces (and for the case of measures) we refer to books on measure theory.

Using the theorem of the completion of a metric space and the fact that the completion of a normed linear space can be organized as a linear space, we get the following result.

PROPOSITION 2.2.10. *Let* $(S, \|\,,\|)$ *be a normed linear space over* K *which we suppose to be not a Banach space. Then there exists a Banach space* $(\bar{S}, \|\,,\|)$ *and an isometry* $f: S \to \bar{S}$ *with the following properties:*

(1) f *is linear (i.e.* $f(x + y) = f(x) + f(y), f(kx) = kf(x)$).
(2) $f(S)$ *is dense in* $\bar{S}$.

The space $\bar{S}$ is called the completion of the normed space S. Among the subsets of a normed linear space (or a Banach space) the subsets which are convex or linear subspaces play a crucial role in some problems.

We now present some facts about certain classes of subspaces. Let S be a Banach space over K and S_1, S_2 be two closed linear subspaces in S.

DEFINITION 2.2.11. The subspaces S_1 and S_2 are called complementary if the following assertions hold:

(1) $S_1 \cap S_2 = \{0\}$.
(2) For any x in S there exists two elements $s_i \in S_i$, $i = 1, 2$, with the property that $x = s_1 + s_2$.

We say that the subspace $S_1(S_2)$ is complemented in S if $S_{i=1,2}$ are closed subspaces.

If S is a Banach space and S_1 is a closed subspace of S then it is not necessarily a complemented subspace. For a famous example of an uncomplemented subspace we refer to the paper of Newman (1961). See also Rudin (1962).

We note that if S is a linear space over K and S_1 is a linear subspace then there exists a linear subspace S_2 such that

(1) $S_1 \cap S_2 = \{0\}$,
(2) every s in S is of the form $s = s_1 + s_2$ with $s_i \in S_i$, $i = 1, 2$.

This may be realized using Zorn's lemma as follows. We consider a linear basis of S_1 and we consider further all families of elements of S which contain B_1 and are linearly independent. Then with the partial ordering given by set inclusion, we show easily that we can apply Zorn's lemma. Then a maximal element is a linear basis of S. If B_2 is the complement of B_1 in the linear basis of S, then the subspace it generates satisfies the requirements. The only difference between the two decompositions of S into such subspaces is that in the first case the mapping $s \to s_1$ is continuous.

2.3. The Extension Theorems

Let S be a Banach space over K and suppose that S_1 is a linear subspace of S (not necessarily closed). Further, suppose that on S_1 we have a function $f: S_1 \to K$ with the following properties:

(1) $f(ax + by) = af(x) + bf(y)$, a, b in K and x, y in S_1.
(2) $\sup(|f(x)|, x \in S_1, \|x\| \leqslant 1) = \|f\|_{S_1} < \infty$.

We are interested in the following problem:

PROBLEM 2.3.1. There exists a function $F: S \to K$ with the following properties:

(1) $F(ax + by) = aF(x) + bF(y)$.
(2) $\sup(|F(x)|, x \in S, \|x\| \leqslant 1) = \|F\|_S = \|f\|_{S_1}$.
(3) $f(x) = F(x)$, $x \in S_1$.

We call F (if it exists) an extension of f and we say that f is the restriction of F to S_1.

The existence of extensions was proved for real spaces by S. Banach and H. Hahn and for the case of complex spaces by H. F. Bohnenblust and A. Sobczyc and G. A. Sukhomlinov (=G. A. Soukhomlinov). In the paper of Sukhomlinov the case of quaternionic spaces is also considered.

We present below the extension theorems for the case of real and complex spaces. We begin with the case of real spaces.

THEOREM 2.3.2. *Let S be a real linear space and let S_1 be a linear subspace of S (not necessarily closed). Suppose that $f: S_1 \to R$ is a function on S_1 satisfying the following conditions:*

(1) $f(ax + by) = af(x) + bf(y)$, *$a$, b in R, x, y in S_1.*
(2) $f(x) \leqslant p(x)$, *$x \in S_1$, p is a seminorm on S.*

Then there exists a function $F: S \to R$ with the properties:

(1) $F(au + bv) = aF(u) + bF(v)$, *$a$, b in R, u, v in S.*
(2) $F(x) \leqslant p(x)$, $\quad x \in S$.
(3) $f(x) = F(x)$, $\quad x \in S_1$.

If S is a linear space over K then any function $f: S \to K$ with the property

$$f(ax + by) = af(x) + bf(y) \quad \text{for } a, b \text{ in } K \text{ and } x, y \text{ in } S$$

is called a linear functional; if V is another linear space over K then any function $T: S \to V$ with the property

$$T(ax + by) = aT(x) + bT(y) \quad \text{for } a, b \text{ in } K \text{ and } x, y \text{ in } S$$

is called a linear operator. When no confusion can arise we write $T(x) = Tx$ for simplicity. It is clear that the linear functionals are linear operators.

Proof. We can suppose without loss of generality that S_1 is not equal to S and thus there exists x_0 in S which is not in S_1. We consider now the linear subspace generated by x_0 and S_1, say S_1^-. It is easy to see that any element z of S_1^- is of the form

$$z = ax_0 + v_1, \qquad a \in R, \quad v_1 \in S_1.$$

Since x_0 is not in S_1 the above decomposition of z is unique. Indeed, on the contrary, suppose that $z = a_1 x_0 + v_1'$, $a_1 \in R$, $v_1' \in S_1$. Then clearly we have $a \neq a_1$ and this implies that x_0 is in S_1. Now, since the above decomposition is unique we define a linear functional on S_1^- by

$$f^-(z) = f(v) + ac,$$

where c is a real number which will be defined more precisely later, but which has the property that

$$f^-(z) \leqslant p(z).$$

We shall now show that such a number c exists. First, let us suppose that c does exist and thus we have

$$f(v/a) + c \leqslant p(v/a + x_0)$$

for all $v \in S_1$ and $a \geqslant 0$. Further, we get

$$f(v/a) + c = -p(-v/a - x_0)$$

for all $v \in S_1$ and $a \leqslant 0$. From these we find that c satisfies the inequalities

$$f(v) - p(v - x_0) \leqslant c \leqslant -f(u) + p(u + x_0)$$

for all u, v in S_1. Since we have for any u, v in S_1,

$$f(u) + f(v) = f(u + v) \leqslant p(u + v) \leqslant p(v - x_0 + x_0 + u) \leqslant p(v - x_0) + p(u + x_0)$$

and if we set

$$m = \sup(f(v) - p(v - x_0): v \in S_1)$$
$$M = \inf(-f(v) + p(v + x_0): v \in S_1)$$

then the following inequality holds:

$$m \leqslant M.$$

Now it is clear that we can take as c any number in the interval $[m, M]$. Thus f^- satisfies the inequality $f(z) \leqslant p(z)$ for all z in S_1^-. Now we shall show the existence of an extension of f using Zorn's lemma. For this we consider the family of all extensions of f of the form (f_V, V) where f_V is an extension of f to the subspace V containing S_1. We define a partial order relation in this family as follows: we say that $(f_V, V) \leqslant (f_U, U)$ if V is a subspace of U and f_V is the restriction of f_U to V. From the above result with x_0 we see that this family is nonempty. Now it is easy to see that we can apply Zorn's lemma and thus we can consider a maximal pair, say (f_L, L). We show now that $L = S$ and thus the extension theorem is proved. Indeed, if L is not S then there exists an element, say y_0 in S, which is not in L. We can repeat the argument as with x_0 and we get an extension of (f_L, L) which is obviously an extension of f. This contradicts the maximality of the pair (f_L, L) and thus we must have $L = S$. □

We now note some important and useful corollaries of this theorem.

COROLLARY 2.3.3. *Let S be a real Banach space and let S_1 be a subspace of S. If $f: S_1 \to R$ is a continuous linear functional then there exists $F: S \to R$ such that the following relations hold:*

(1) $\|F\| = \sup\{|F(x)|, \|x\| \leqslant 1\} = \|f\|_{S_1}$,
(2) $f(x) = F(x)$ *for all* $x \in S_1$.

Proof. We consider the following seminorm on S:

$$p(x) = \|f\|_{S_1}\|x\|$$

and it is clear that $f(x) \leqslant p(x)$ holds for all $x \in S_1$. Then by the above theorem there exists a functional $F: S \to R$ such that $f(x) = F(x)$ for all $x \in S_1$ and $F(x) \leqslant p(x)$. This clearly implies that

$$\sup\{|F(x)|, \|x\| \leqslant 1\} \leqslant \|f\|_{S_1}.$$

Since

$$\|F\| = \sup\{|F(x)|, \|x\| \leqslant 1\} \geqslant \sup\{|f(x)|, x \in S_1, \|x\| \leqslant 1\} \geqslant \|f\|_{S_1}$$

the equality follows, and the corollary is proved. □

REMARK 2.3.4. We note that the extension theorem asserts the existence of at least one extension and nothing about the uniqueness.

COROLLARY 2.3.5. *Let S be a real Banach space and $x \in S$, $x \neq 0$. Then there exists a linear functional $f: S \to R$ such that:*

(1) $\|f\| = \sup\{(f(x)|, \|x\| \leqslant 1\} = 1$,
(2) $f(x) = \|x\|$.

Proof. We take $S_1 = \{ax\}$ and define $f_1(x) = a\|x\|$ which obviously satisfies the conditions of the extension theorem. The extension f of f_1 to the space S satisfies the corollary. □

COROLLARY 2.3.6. *A necessary and sufficient condition for a subspace S_1 of a Banach space S to be dense in S is that for any continuous linear functional f which has the property that $f(x) = 0$ for all $x \in S_1$ then necessarily $f \equiv 0$.*

Proof. We consider the subspace $\mathrm{Cl}\, S_1$ and suppose first that this is not S and we show that there exists a nonzero continuous functional f which has the property that $f(x) = 0$ for all $x \in \mathrm{Cl}\, S_1$. Indeed, since $\mathrm{Cl}\, S_1$ is not equal to S there exists x_0 such that

$$d = \inf\{\|x - x_0\|, x \in S_1\} > 0$$

and we consider on the space generated by $\mathrm{Cl}\, S_1$ and x_0 the following functional: if z is in this space then z is of the form

$$z = ax_0 + v$$

and this decomposition is unique; this makes it possible to define f by the formula $f(z) = a$.

Obviously $f(v) = 0$ for all $v \in \mathrm{Cl}\, S_1$ and

$$|a|d = |a|\,\|v/a + x_0\| = \|v + ax_0\|,$$

which implies that f is continuous on $\mathrm{Cl}\, S_1$. The extension of f to S clearly satisfies the condition of the corollary, since in the other direction the assertion is obvious. □

COROLLARY 2.3.7. *Let S be a real Banach space and let $x \in S$. Then*

$$\|x\| = \sup_{\|f\| \leqslant 1} |f(x)|$$

where f is a linear continuous function and $\|f\| = \sup\{|f(x)|, \|x\| \leqslant 1\}$. If X is a Banach space then we denote by X^* the set of all continuous linear functionals of X. If we set

$$f \to \|f\| = \sup\{|f(x)|, \|x\| \leqslant 1\}$$

then this is a norm on X^* and $(X^*, \|\,,\|)$ is a Banach space. More generally, if X and Y are two Banach spaces over K, we can consider all functions $T: X \to Y$ with the following properties:

(1) $T(ax + by) = aTx + bTy$,
(2) *T is continuous.*

We denote by $L(X, Y)$ all functions T satisfying the above properties.

We have the following two results concerning elements of $L(X, Y)$ which are called linear continuous operators.

LEMMA 2.3.8. *A linear operator $R: X \to Y$ (i.e. R satisfies only condition (1) above) is continuous if and only if it is continuous at some point $x_0 \in X$.*

Proof. Since the proof is simple it is omitted. □

LEMMA 2.3.9. *A linear operator $R: X \to Y$ is continuous if and only if*

$$\|R\| = \sup\{\|Rx\|, \|x\| \leqslant 1\} < \infty$$

(i.e. R is bounded and the number $\|R\|$ is called the norm of R).

Proof. It is clear that any bounded operator is continuous. Conversely, if R is continuous we show that it is bounded. Suppose that this is not true and thus for each n there exists $x_n \in X$, $\|x_n\| = 1$ such that $\|Tx_n\| \geqslant n$. We now consider the sequence (y_n), $y_n = x_n/n$ which clearly converges to zero. But $\|Ty_n\| \geqslant 1$ and thus we get a contradiction and the lemma is proved. □

We leave the reader to prove the following result.

LEMMA 2.3.10. *$(L(X, Y), \|\,,\|)$ is a Banach space.*

Now if S is a real Banach space we can consider S^*, which is called the dual of S, and further $(S^{*n}) = (S^{*(n-1)})^*$.

For any x is S fixed we consider the following element of S^{**},

$$F_x(x^*) = x^*(x)$$

and then we have a mapping $J: S \to S^{**}$. This is clearly linear and from the above results it is an isometry, i.e. $\|x\| = \|F_x\|$ for all $x \in S$. Using this mapping we can define an important class of Banach spaces.

DEFINITION 2.3.11. A real Banach space X is called reflexive if $J(X) = X^{**}$.

We now consider the case of complex spaces.

THEOREM 2.3.12 (Bohnenblust–Sobczyc–Sukhomlinov). *Let S be a complex Banach space and let S_1 be a linear subspace of S and suppose that there exists a linear functional $f: S_1 \to C$ with the properties:*

(1) $f(ax + by) = af(x) + bf(y)$, $\quad x, y \in S_1$; $\quad a, b \in C$.
(2) $|f(x)| \leqslant \|x\|$, $\quad x \in S_1$.

Then there exists a function $F: S \to C$ with the following properties:

(1) $F(ax + by) = aF(x) + bF(y)$, $\quad x, y \in S$; $\quad a, b \in C$,

(2) $F(x) = f(x), \quad x \in S_1$,
(3) $\|f\|_{S_1} = \|F\|$.

Proof. We consider the functional f on the space S_1 as a real Banach space and, of course, we consider S on the same field of scalars. We consider the function

$$u(x) = \operatorname{Re} f(x)$$

and we remark that it is a linear functional on S_1 (with real scalars). Of course, it is easy to see that

$$f(x) = u(x) - iu(ix)$$

and now S_1 being a subspace of the real space S there exists an extension U of u to the space S. But we have $u(x) = f(x)$, which implies that

$$\|u\|_{S_1} = \|f\|_{S_1}.$$

Since for each x there exists a complex number a_x such that $|f(x)| = f(a_x x)$, we get

$$|f(x)| = a_x f(x) = f(a_x x) \leqslant u(a_x x) \leqslant \|u\| \|a_x x\| = \|u\| \|x\|$$

i.e. $\|f\| = \|u\|$. Now it is obvious that F, defined by the formula

$$F(x) = U(x) - iU(ix),$$

satisfies the theorem. □

REMARK 2.3.13. All corollaries of the extension theorem for real spaces hold for the case of complex spaces.

2.4. Linear Operators on Banach Spaces. Classes of Linear Operators

Let X and Y be two Banach spaces over K. As mentioned above, a linear bounded operator $T: X \to Y$ is any function satisfying the conditions

(1) $T(ax + by) = aTx + bTy$,
(2) $\|T\| = \sup(\|Tx\|, \|x\| \leqslant 1) < \infty$.

The set of all bounded linear operators defined on X and with values in Y is denoted by $L(X, Y)$ and is a Banach space if we take as the norm of $T \in L(X, Y)$ the number $\|T\|$.

Now if $X = Y$ we write $L(X, Y) = L(X)$ and, in this case, on $L(X)$ there exists a product defined by

$$(TS)(x) = T(Sx)$$

and we have the inequality

$$\|TS\| \leqslant \|T\| \|S\|.$$

Suppose now that $T \in L(X, Y)$ and take $y^* \in Y^*$; then we can define the following functional:

$$x \to y^*(Tx).$$

Since

$$|y^*(Tx)| \leqslant \|y^*\| \, \|T\| \, \|x\|$$

we see that this is a continuous linear functional on X and thus it is an element of X^*, say u^*. Thus, using the operator T we can define a mapping $T^*: Y^* \to X^*$ by the formula $Ty^* = u^*$. It is easy to see that this is a linear operator, and from Corollary 2.3.5 we find that it is continuous and $\|T^*\| = \|T\|$.

PROPOSITION 2.4.1. *If* $T \in L(X, Y)$, $S \in L(Y, Z)$ *then* $(ST)^* = T^*S^*$.

Proof. Since the proof is simple it is omitted. □

We now consider certain important classes of operators.

DEFINITION 2.4.2. An operator $T \in L(X, Y)$, where X, Y are two Banach spaces, is said to be finite dimensional if there exist $y_1, \ldots, y_n$ in y and $x_1^*, \ldots, x_n^*$ in X^* such that for all $x \in X$,

$$Tx = \sum_{i=1}^{n} x_i^*(x) y_i.$$

It is not difficult to see that if $T \in L(X, Y)$ is finite dimensional then $T^* \in L(Y^*, X^*)$ is also finite dimensional.

DEFINITION 2.4.3. An operator $T \in L(X, Y)$ is said to be completely continuous if there exists a sequence of finite dimensional operators (T_n) such that $\lim \|T_n - T\| = 0$.

DEFINITION 2.4.4. An operator $T \in L(X, Y)$ is called compact if for any bounded set M in X, $\overline{TM} = \overline{\{Tx, x \in M\}}$ is a compact set in Y.

It is not difficult to see that any completely continuous operator is compact and that for any bounded operator R and any compact operator S, RS and SR are compact operators.

THEOREM 2.4.5. *If* $T \in L(X, Y)$ *is compact then* $T^* \in L(Y^*, X^*)$ *is compact.*

Proof. We remark that it suffices to prove that $T^* B_1^*(0)$ is relatively compact, i.e. its closure is compact, where for any Banach space $B_1(0)$ is the open closed unit ball, i.e. the set $\{x, \|x\| < 1\}$. Let $\varepsilon > 0$ and since T is compact there exists a finite number of sets of the form $(y, \|y - y_i\| < \varepsilon/2)$ which cover $\overline{TB_1(0)}$. We now consider $\bar{B}_1(0^*)$ and the mapping

$$U^*(y^*) = (y^*(y_1), \ldots, y^*(y_n))$$

where y_i are those considered above. □

Since this mapping U^* is clearly compact, the closure of the image of $\bar{B}_1(0^*)$ through U^* is compact and thus may be covered by a finite number of sets like those for $\overline{T\bar{B}_1(0)}$ but with centers at some $U^*(y_i^*)$, $i = 1, 2, \ldots, m$. This implies that for all $y^* \in \bar{B}_1(0^*)$ there exists y_l, $l = 1, 2, 3, \ldots, n$ such that

$$|y^*(y_l) - y_i^*(y_l)| < \varepsilon/4$$

for $l = 1, 2, 3, \ldots, n$.

Now since for each $x \in B_1(0)$ we find y_i such that $\|Tx - y_i\| < \varepsilon/4$ we get

$$|y^*(Tx) - y_i^*(Tx)| \leqslant |\, y^*(Tx) - y^*(y_i) + y^*(y_i) - y_i^*(y_i) + \\ + y_i^*(y_i) - y_i^*(Tx)| \leqslant \tfrac{3}{4}\varepsilon$$

and thus

$$\|T^*y^* - T^*y_i\| = \sup_{\|x\| \leqslant 1} |y^*(Tx) - y_i^*(Tx)| < \varepsilon,$$

which implies that

$$T^*\bar{B}_1(0^*) \subset \bigcup_{i=1}^{n} (x^*, \|x^* - T^*y_i^*\| < \varepsilon).$$

Thus T is compact.

We give now some results about the class of compact operators.

THEOREM 2.4.6. *Let $T \in L(X)$ be a compact operator and suppose that the operator $T - zI = T - z$ (I is the identity operator $Ix = x$ for all $x \in X$) is injective. Then for all $z \neq 0$ the range of $T - z$, $(T - z)X$ is a closed subspace.*

Proof. It is clear that $(T - z)X$ is a linear subspace of X, and to show the closedness we consider a sequence $(y_n) \subset (T - z)X$ which is convergent to an element, say $y \in X$. We must show that $y \in (T - z)X$. Since $y_n \in (T - z)X$ there exists $x_n \in X$ such that $y_n = Tx_n - zx_n$ for each n. Since T is compact we have to consider two cases: □

CASE 1. (x_n) contains a bounded subsequence and thus we can suppose without loss of generality that this is just (x_n), and from the compactness of T we can choose a convergent subsequence, say (Tx_{n_k}). This implies obviously that (x_{n_k}) is a convergent sequence and thus if $x = \lim x_{n_k}$ then $y = (T - z)x$.

CASE 2. (x_n) has no bounded subsequences and thus we can consider the sequence $x_n' = x_n/\|x_n\|$ which has the property that $\|x_n'\| = 1$ and $(T - z)x_n' = y_n/\|x_n\|$ gives that $((T - z)x_n')$ converges to zero, and T being compact and (x_n') bounded we get immediately that (x_n') is convergent to a point of X, say x'. Then clearly we have $Tx' - zx' = 0$, which contradicts the fact that $T - z$ is injective. □

For the result we give about compact operators we need a result, known as 'the Riesz lemma', which has many interesting applications.

THEOREM 2.4.7 (the Riesz lemma). *Let X be a Banach space and X_1 be a closed linear subspace of X. If $\varepsilon > 0$ then there exists $y \in X$, $y \notin X_1$ such that*

(1) $\|y - x\| > 1 - \varepsilon$ *for all* $x \in X_1$;
(2) $\|y\| = 1$.

Proof. We consider b in X and not in X_1 (of course, we suppose that $X_1 \neq X$) and since

$$\inf(\|b - x\|, x \in X_1) > 0$$

we set $d = \inf(\|b - x\|, x \in X_1)$ and take $a \in X_1$ such that

$$\|b - a\| < d(1 + \varepsilon).$$

Set $b_1 = b - a$ and thus

$$(1+\varepsilon)\inf(\|b_1 - x\|) = (1+\varepsilon)d > \|b_1\|.$$

We define $y = b_1/\|b_1\|$ and since

$$\begin{aligned}\|y - a\| &= \|b_1/\|b_1\| - a\| \\ &= 1/\|b_1\|\,\|b_1 - a'\| > 1/(d(1+\varepsilon))\|b_1 - a'\| > 1/(1+\varepsilon) > 1 - \varepsilon\end{aligned}$$

with the obvious relation $\|y\| = 1$, the assertion of the Riesz lemma is proved. □

From this result we obtain obviously the following characterization of finite dimensional Banach spaces.

COROLLARY 2.4.8. *A Banach space X is finite dimensional if and only if $\bar{B}_1(0)$ is compact.*

We now give another result concerning compact operators.

COROLLARY 2.4.9. *Let $T \in L(X)$ be a compact operator and suppose that there exists a sequence of complex numbers (z_n) $(z_n \neq z_m$ for $n \neq m)$ and a sequence of elements $(x_n) \subset X$, $\|x_n\| = 1$ such that $Tx_n - z_n x_n = 0$. Then (z_n) converges to zero.*

Proof. Suppose that the assertion is false and thus there exists $\varepsilon > 0$ such that for some sequence of integers (n_k), $|z_{n_k}| \geqslant \varepsilon$. We set $x_{n_k} = u_k$, $z_{n_k} = a_k$, $k = 1, 2, 3, \ldots$, and for each n we consider the subspace generated by the elements $u_1, \ldots, u_n$, which is denoted by E_n. We now show that we have the following sequence of strict inclusions:

$$E_1 \subset E_2 \subset E_3 \subset \cdots \subset E_n \subset E_{n+1} \subset \cdots.$$

It is clear that $E_n \subset E_{n+1}$ holds for all n. If at n we have $E_{n+1} = E_n$ then u_{n+1} is of the form,

$$u_{n+1} = \sum_{i=1}^{n} \alpha_i u_i$$

and applying T we get

$$0 = Tu_{n+1} - a_{n+1}u_{n+1} = \sum_{i=1}^{n} \alpha_i Tu_i - a_{n+1}\sum_{i=1}^{n} \alpha_i u_i = \sum_{i=1}^{n} (\alpha_i a_i - a_{n+1}\alpha_i)u_i$$

and since (u_i) is a linearly independent set we get that

$$\alpha_1 a_1 - a_{n+1}\alpha_1 = \cdots = \alpha_n a_n - a_{n+1}\alpha_n$$

and since $a_i \neq a_j$ for $i \neq j$ we must have $\alpha_i = 0$, $i = 1, 2, 3, \ldots, n$. This proves that all the inclusions are strict. □

We can now apply the Riesz lemma and get for each n an element z_{n+1} in E_{n+1} with the following properties:

(1) $\|z_{n+1} - x\| \geqslant \frac{1}{2}$ for all $x \in E_n$.

(2) $\|z_{n+1}\| = 1$.

But $z_{n+1} = \Sigma_{i=1}^{n+1} \alpha_i u_i$ and thus $(T - a_{n+1})z_{n+1} \in E_n$. Indeed, we have,

$$(T - a_{n+1})z_{n+1} = \sum_{i=1}^{n+1} \alpha_i Tu_i - a_{n+1} \sum_{i=1}^{n+1} \alpha_i u_i = \sum_{i=1}^{n+1} \alpha_i a_i u_i - \sum_{i=1}^{n} a_{n+1} \alpha_i u_i$$

$$= \sum_{i=1}^{n+1} (\alpha_i a_i - \alpha_i a_{n+1}) u_i = \sum_{i=1}^{n} (\alpha_i a_i - \alpha_i a_{n+1}) u_i$$

and the assertion is proved. □

Let us consider $n > m$ and the elements $z_{n,m}$ defined as

$$z_{n,m} = (z_{n+1} - a_{n+1}^{-1} Tz_{n+1}) + a_m^{-1} Tz_m.$$

Then $z_{n,m}$ are in E_n for each n. By the Riesz lemma we have

$$\| T(a_{n+1}^{-1} z_{n+1}) - T(a_m^{-1} z_m) \| = \| z_{n+1} - z_{n,m} \| \geqslant \tfrac{1}{2}$$

which implies that the sequence $(T(a_m^{-1} z_m))$ has no convergent subsequences. But $(a_m^{-1} z_m)$ is bounded, and this contradicts the compactness of T. (We note that boundedness of the sequence $(a_m^{-1} z_m)$ follows simply from the facts that $\|z_n\| = 1$ and $|a_m| \geqslant \varepsilon$, which imply that $\|a_m^{-1} z_n\| \leqslant 1/\varepsilon$.)

An interesting class of operators containing the class of compact operators may be defined using the measures of noncompactness. Below, we use the Kuratowski measure of noncompactness, defined as follows.

DEFINITION 2.4.10. (Kuratowski, 1929). Let M be a bounded set in the metric space (X, d). The Kuratowski measure of noncompactness of M, $\alpha(M)$, is the infimum of all $\varepsilon > 0$ such that M can be covered with a finite number of subsets with a diameter less than ε.

We recall that if S is a subset of the metric space (X, d) then the diameter $d(S)$ of the subset S in the number

$$\sup (d(x, y): x, y \in S).$$

The diameter $d(S)$ of the set S may be infinite; in that case we say that S is unbounded. In the contrary case the subset S is called bounded.

We now list the basic properties of the Kuratowski measure of noncompactness.

PROPOSITION 2.4.11. *Let (X, d) be a complete metric space and let $\alpha(M)$ be the Kuratowski measure of noncompactness of the bounded subset M. Then the following assertions hold:*

(1) $0 \leqslant \alpha(M) \leqslant d(M)$.
(2) $\alpha(M) = 0$ *iff* $\bar{M}$ *is a compact subset.*
(3) $\alpha(M \cup N) = \max(\alpha(M), \alpha(N))$.
(4) $\alpha(M \cup (x)) = \alpha(M)$.
(5) $M \subseteq N \rightarrow \alpha(M) \leqslant \alpha(N)$.
(6) $\alpha(M) = \alpha(\bar{M})$.
(7) *If X is a Banach space then for any bounded subset M*, $\alpha(M) = \alpha(\text{conv } M)$.

For a proof of these assertions, as well as for other information about measures of noncompactness, we refer to the author's book (1981).

The class of operators which contains the class of compact operators is considered in the following definition.

DEFINITION 2.4.12. An element $T \in L(X)$, where X is a Banach space, is said to be a k-set contraction if for any bounded noncompact subset M of X

$$\alpha(TM) \leqslant k\alpha(M)$$

for some fixed k in $(0, 1)$.

DEFINITION 2.4.13. If $T \in L(X)$ then the modulus of k-set contraction for T is the infimum of all k with the property that for any noncompact bounded subset M of X the following inequality holds:

$$\alpha(TM) \leqslant k\alpha(M).$$

This modulus is denoted by α_T. Of course, T is a k-set contraction if this modulus is in $(0, 1)$.

The most interesting property of this modulus is given in the following proposition.

PROPOSITION 2.4.14. *For any* $T \in L(X)$, $\alpha_T = \alpha_{T^*}$.

This result contains the well-known theorem of J. Schauder that an operator T in $\mathscr{L}(X)$ is compact iff T^* is compact.

EXAMPLE 2.4.15. Let X be a Banach space, let T be a compact operator in $L(X)$ and let $S \in L(X)$ be a strict contraction (i.e. $\|S\| < 1$). Then the operator $T + S$ is noncompact and is a k-set contraction. In fact, it is easy to see that

$$\alpha_{T+S} \leqslant \|S\|.$$

For other classes of operators related to k-set contractions and defined using measures of noncompactness, we refer to the author's book (1981).

In order to introduce another class of operators related to the class of compact operators we need the following notion.

DEFINITION 2.4.16. Let X be a Banach space and let M, N be two closed linear subspaces of X. We say that the pair (M, N) reduce $T \in L(X)$ if the following properties hold:

(1) $M \oplus N$.
(2) M, N are invariant subspaces for T (i.e. $TM \subseteq M$, $TN \subseteq N$).

We also need the notion of a Riesz point of an operator T in $L(X)$.

DEFINITION 2.4.17. A complex number z is said to be a Riesz point for the operator T in $L(X)$ if the following properties hold:

(1) $N(T - z) = \{x, (T - z)x\}$ is finite dimensional.
(2) $(T - z)/N(T - z)$ is quasinilpotent (i.e. the spectrum of $(T - z)/N(T - z)$ reduces to (0)).

(3) $(T - z)/R(T - z)$, where $R(,)$ denotes the range of $(,)$, is a homeomorphism.

Now the class is defined as follows.

DEFINITION 2.4.18. An operator T in $L(X)$ is said to be a Riesz operator if each point $z \neq 0$ in $\sigma(T)$ is a Riesz point for T.

One of the most interesting and important classes of operators in $L(X)$ is the so-called class of Fredholm operators. This class is considered in the following definition.

DEFINITION 2.4.19. An operator T in $L(X)$ is said to be a Fredholm operator if the following properties hold:

(1) $N(T)$ is finite dimensional.
(2) $R(T)$ is closed and its codimension (i.e. the dimension of $X/R(T)$) is finite.

If T is a Fredholm operator then we define the following number, which is called the index of T and is given by the formula

$$\text{ind } T = \dim N(T) - \dim(X/R(T)).$$

For each integer p, $F_p(X, y)$ is the set of all Fredholm operators with the index equal to p. We remark that the above definition of a Fredholm operator can be given for operators in $L(X, Y)$. Then we have the following result.

PROPOSITION 2.4.20. *If T in $L(X)$ is a compact operator then $I + T$ is a Fredholm operator in $L(X)$.*

Proof. The assertion follows from the properties of compact operators given above. □

If X, Y are two Banach spaces then $F(X, Y)$ denotes the set of all Fredholm. Given two Banach spaces X, Y then in $L(X, Y)$ we can distinguish two important classes of operators namely $K_f(X, Y)$ which is the closure in $L(X, Y)$ of elements in $L(X, Y)$ which are of finite rank and the class of compact operators in $L(X, Y)$ denoted by $K(X, Y)$. From Enflo's example we know that in some spaces we have that $K_f(X, Y) \neq K(X, Y)$.

In order to formulate a characterization of Fredholm operators using the classes K_f or K we use the notation I_Z to denote the operator on the Banach space Z by the formula $I_Z z = z$ for all z in Z.

DEFINITION 2.4.21. We say that an operator T in $L(X, Y)$ is invertible modulo the operators of finite rank (or the compact operators, respectively) if there exists S in $L(Y, X)$ such that

$$ST = I_X + C_X, \quad TS = I_Y + C_Y$$

with C_X, C_Y operators of finite rank (compact operators, respectively). Then we have the following characterization of Fredholm operators.

THEOREM 2.4.22. *An operator T in $L(X, Y)$ is a Fredholm operator iff it is invertible modulo the operators of finite rank or compact operators.*

Proof. Suppose that T is a Fredholm operator. Then $N(T)$ and $R(T)$ are closed and $\dim N(T)$ and $\operatorname{codim} R(T)$ are finite. Then we may write

$$X = N(T) \oplus X_1, \quad Y = R(T) \oplus Y_1.$$

The operator $u: X_1 \to R(T)$ induced by T is an isomorphism and then we define the operator S in $L(Y, Y)$ as follows:

$$S/R(T) = u^{-1}, \quad S/Y_1 = 0.$$

Then $I_X - ST$ is the projection of X onto $N(T)$ and $I_Y - TS$ is the projection of Y onto Y_1. This clearly indicates that the condition of the theorem is necessary. □

Suppose now that the condition is satisfied. Then for some S_1, S_2 in $K(X, Y)$ we have the relations:

$$I_X - S_1 T \in K(X), \qquad I_Y - TS_2 \in K(Y).$$

From these we obtain that $N(T)$ is of finite dimension and that $R(T)$ is of finite codimension. It is easy, using the above relations, to show that $R(T)$ is closed. The theorem is proved. □

The following result is not difficult to prove.

PROPOSITION 2.4.23. *Suppose that T is in $L(X, Y)$, S is in $L(Y, Z)$. Consider $R = ST$. Then, if one of the pairs (T, S), (T, R) or (R, S) is a pair of Fredholm operators, then the operator which is not in the pair is a Fredholm operator and the following relation holds:*

$$\operatorname{ind} R = \operatorname{ind} T + \operatorname{ind} S.$$

The following result may be used to obtain information about the function $T \to \operatorname{ind} T$ defined on $F(X, Y)$.

PROPOSITION 2.4.24. *The set $F_p(X, Y)$ is open in $L(X, Y)$.*

Proof. Let T be in $F_p(X, Y)$. Then $X = N(T) \oplus X_1$ and $Y = R(T) \oplus Y_1$. We set

$$u: X_1 \oplus Y_1 \to X \quad \text{with} \quad u(x, y) = Sx + y.$$

Here S is in $L(X, Y)$ with $T - S$ sufficiently small. Since for $S = T$ the above application is an isomorphism, it remains an isomorphism if the perturbation is (in norm) sufficiently small. Then $u X_1$ is closed and $\dim Y/v(X_1)$ is finite and in fact equal to $\dim Y/R(T)$. Then $N(S) \subseteq N(T)$ and if N_1 is a complement of $N(S)$ in $N(T)$ we have

$$S(X) = S(N_1 \oplus X_1)\, S(X_1)$$

which gives that the codimension of $S(X)$ is finite and the fact that $S(X)$ is a closed subspace. Then S is a Fredholm operator and

$$\operatorname{codim} S(X) = \operatorname{codim} S(X_1) - \dim N_1.$$

But as $\dim N(S) = \dim N(T) - \dim N_1$ and as S is an isomorphism we have

$$\begin{aligned} \operatorname{ind} S &= (\dim N - \dim N_1) - (\dim X_1 - \dim N_1) \\ &= \dim N(T) - \dim X_1 = \operatorname{ind} T. \end{aligned}$$

Thus S is in F_p and the proposition is proved. □

This clearly implies the following result:

PROPOSITION 2.4.25. *The mapping* $T \to \text{ind } T$ *is locally constant.*

For very interesting results concerning the set $GL(X)$, X is a Banach space, we refer to Douady (1965), Kuiper (1965), and to the excellent survey article of Mitiagin (1970).

2.5. Three Basic Theorems of Linear Functional Analysis

We now present three basic theorems of linear functional analysis which, together with the extension theorems, are the most important results of linear functional analysis.

The proof that we give for these theorems is that proposed by P. Zabreiko and has the advantage of deriving these theorems from an interesting lemma obtained from Baire's theorem. Also we note another famous consequence, the so-called Gelfand lemma. First we give the following:

DEFINITION 2.5.1. Let E be a Banach space and let $f: E \to R$. We say that f is subadditive if for all x, y in E,

$$f(x) + f(y) \geqslant f(x + y).$$

The following result about subadditive functions will be used in the proof of Zabreiko's lemma.

PROPOSITION 2.5.2. *Let* $F : E \to R$ *be a subadditive function and suppose, for some* $M > 0$, *that if* $\|x\| \geqslant M$ *then* $f(x) \geqslant 0$; *if* f *is subadditive, positive and continuous at the origin then* f *is continuous on* E.

Proof. Since

$$f(0) \leqslant 2f(0)$$

we get that $f(0) \geqslant 0$. Let m be the smallest integer greater than M. Then we have, for $z = mx$,

$$0 \leqslant f(0) \leqslant mf(x)$$

which implies that f is positive. □

Let us consider x_0 arbitrary in E and (x_n) converging to x_0. Since $(x_n - x_0)$ converges to zero, from the continuity at the origin, and using the inequalities

$$f(x_0) - f(x_n) \leqslant f(x_0 - x_n)$$
$$f(x_n) - f(x_0) \leqslant f(x_0 - x_n),$$

we get that

$$\lim(|f(x_0) - f(x_n)|) \leqslant \lim f(x_0 - x_n) = 0.$$ □

Thus the proposition is proved.

LEMMA 2.5.3. (Zabreiko, 1969). *Let X be a Banach space and let $F: X \to R$ be a function with the following properties:*

(1) *For each $x \in X$, $\lim_{t\to 0} F(tx) = 0$.*
(2) *If $\Sigma_{n=1}^{\infty} x_n$ converges then $F(\Sigma_{n=1}^{\infty} x_n) \leqslant \Sigma_{n=1}^{\infty} F(x_n)$.*

Then F is continuous.

Proof. Obviously F is subadditive and thus to prove the lemma it suffices to prove the continuity at the origin of X. □

Let $\varepsilon > 0$ and consider the sets

$$G_\varepsilon = (x, x \in X, F(x) + F(-x) \leqslant \varepsilon/2)$$

which are clearly symmetric subsets in X. From property (1) of the function F we get that there exists n_x such that $x \in n_x G_\varepsilon$. Indeed, since we have

$$F(x) = F(n \cdot 1/n \cdot x) \leqslant n \cdot F(1/n \cdot x)$$

our assertion follows. Thus we get that $X = \cup\, nG_\varepsilon$. But **X** is supposed to be a Banach space and since $X = \cup \operatorname{Cl}(nG_\varepsilon) = \cup\, n \operatorname{Cl} G_\varepsilon$, we obtain that there exists an n_0 such that $n_0 \operatorname{Cl} G_\varepsilon$ contains a set of the form $B(r, x_0)$. Since G_ε are symmetric sets we get that it also contains $-B(r, x_0)$. From these we get that $n_0 \operatorname{Cl} G_\varepsilon$ contains a set of the form $B(r_\varepsilon, 0)$. We can suppose without loss of generality that $r_\varepsilon < \varepsilon$. Thus if $x \in B(r_\varepsilon, 0)$ we have $F(x) \leqslant 2\varepsilon$. Indeed, let $G'_\varepsilon = G_\varepsilon \cap B(r, 0)$ which is dense in $B(r_\varepsilon, 0)$. Thus we can take x_1 in G'_ε such that $x - x_1 \in B(r_\varepsilon, 0)$. Suppose now that $x_1, \ldots, x_n$ are chosen and we choose $x_{n+1} \in G_{\varepsilon/2^n}$ such that $x - x_1 - \cdots - x_{n+1} \in B(r_{\varepsilon/2^{n+1}}, 0)$. Then clearly $x = \Sigma_{i=1}^{\infty} x_i$ is convergent and by the second property of F,

$$F(x) = F\left(\sum_{i=1}^{n} x_i\right) \leqslant \sum_{i=1}^{\infty} F(x_i) \leqslant 2\varepsilon.$$

This clearly implies that F is continuous at the origin and thus, by the above result about subadditive functions, it is continuous on X.

REMARK 2.5.4. Condition (2) for F is satisfied, for example, if F is subadditive and lower semicontinuous. Indeed, if F is subadditive and lower semicontinuous, we have for all $x = \Sigma_{i=1}^{\infty} x_i$,

$$F(x) = \liminf F\left(\sum_{i=1}^{n} x_i\right) = \lim_{n\to\infty} \sum_{i=1}^{n} F(x_i) \leqslant \sum_{i=1}^{\infty} F(x_i)$$

and the assertion is proved.

COROLLARY 2.5.5. *Let I be an index set and for each $\alpha \in I$, $F_\alpha: X \to R$ such that:*

(1) *F_α is continuous, positive and subadditive;*
(2) *$\lim_{t\to 0} \sup_{\alpha\in I} F_\alpha(tx) = 0$ for each $x \in X$.*

Then

$$\lim_{\|x\|\to 0} \sup F_\alpha(x) = 0.$$

Proof. For each $x \in E$ we set

$$F(x) = \sup_{\alpha} F_{\alpha}(x)$$

which is subadditive and lower semicontinuous. Thus Zabreiko's result applies and this proves the corollary. □

COROLLARY 2.5.6 (Gelfand's lemma). *Let $F: E \to R$ be a lower semicontinuous subadditive function. Suppose that $F(ax) = |a| F(x)$ for all $x \in E$. Then there exists a positive constant K such that $F(x) \leqslant K\|x\|$.*

Proof. This follows immediately from Zabreiko's result. □

We now give the proof of three basic results of linear functional analysis: the principle of uniform boundedness, the open mapping theorem, and the closed graph theorem. □

THEOREM 2.5.7 (the principle of uniform boundedness). *Let I be an index set and for each $\alpha \in I$ a bounded linear operator $T_{\alpha}: X \to Y$, where X, Y are Banach spaces. Then either:*

(1) *there exists $\infty > M > 0$ such that $\sup_{\alpha} \|T_{\alpha}\| = M < \infty$, or*
(2) *$\sup_{\alpha} \|Tx\| = \infty$ when $x \in G$, G is a G_{δ}-subset of X.*

Proof. First we note that a set G in a metric space is a G_{δ}-set if it is the intersection of a countable family of open subsets. We define the function

$$x \to F(x) = \sup_{\alpha} \|T_{\alpha} x\|$$

and we remark that it is subadditive and lower semicontinuous. Now if the first assertion of the theorem does not hold then for each n,

$$G_n = \left(x, \sup_{\alpha} \|T_{\alpha} x\| > n \right)$$

is a dense subset of X and clearly is open. Then it is obvious that on $G = \cap_n G_n$ we have (2). Thus the theorem is proved. □

THEOREM 2.5.8 (the open mapping theorem). *Let $T: X \to Y$ be a linear mapping in which X and Y are supposed to be Banach spaces. Now if T is a continuous mapping with the property that T is surjective, i.e. for each $y \in Y$ there exists at least one $x \in X$ such that $Tx = y$, then there exists $r > 0$ such that*

$$TB^X(1, 0) \supset rB^Y(1, 0).$$

($B^Z(r, \ )$ means that $B(r, \ )$ is considered in the space Z).

Proof. We define the following function on X,

$$x \to F(x) = \inf(\|y\|, Tx = y)$$

which is well defined and satisfies properties (1) and (2) in Zabreiko's lemma. Thus F is continuous and the continuity clearly implies the assertion of the theorem. □

For the theorem which follows we need the important notion of a 'closed operator' and this is considered in the following definition.

DEFINITION 2.5.9. Let X, Y be two Banach spaces and let D_T be a linear subspace of X. Let us suppose, for simplicity, that D_T is dense in X. Suppose also that we have a linear mapping $T: D_T \to Y$. Then we say that T is closed if the following property holds: if $(x_n) \subset D_T$ converges to x and (Tx_n) converges to y then necessarily $y = Tx$.

THEOREM 2.5.10 (the closed graph theorem). *If $T: X \to Y$ is a linear and closed operator then T is continuous (i.e. T is bounded).*

Proof. We define the following function F on X,

$$x \to F(x) = \|Tx\|$$

and we remark that property (1) in Zabreiko's lemma is obvious. For the second property, suppose that (x_n) has the property that $\Sigma F(x_n)$ converges (in the opposite case there is nothing to prove). Let

$$y = \sum Tx_n = \lim_{n\to\infty} \sum_{i=1}^{n} \mathrm{T}x_i$$

and since T is a closed operator we have

$$T\left(\sum_1^\infty x_n\right) = \sum_{i=1}^\infty Tx_n$$

and thus

$$\left\| T\left(\sum_{n=1}^\infty x_n\right) \right\| = \left\| \sum_{n=1}^\infty Tx_n \right\| \leqslant \sum \|Tx_n\|$$

which gives the second property of F. Thus F is continuous and this is clearly equivalent with the continuity of T. □

We now give briefly some applications of these theorems. For more details, and for other applications, see Dunford and Schwartz (1958).

COROLLARY 2.5.11. *Let $(T_n)_1^\infty$ be a sequence of linear and bounded operators defined on X with values in Y (X, Y are Banach spaces). Suppose that for each $x \in X$,*

$$\sup \|T_n x\| < \infty.$$

Then

$$\lim_n \sup_{\|x\|=1} \|T_n x\| < \infty.$$

Proof. We consider the following function F defined on X,

$$x \to F(x) = \sup_n \|T_n x\|$$

and this clearly satisfies Zabreiko's lemma. The continuity of F is clearly equivalent with the assertion. □

COROLLARY 2.5.12. *Let* (T_n) *be a sequence of bounded linear operators* $T_n\colon X \to Y$ *(X, Y are Banach spaces). Suppose that for each* $x \in X$,

$$\lim T_n x$$

exists. Then $Tx = \lim T_n x$ *is a bounded linear operator.*

Proof. This follows from 2.5.11. □

COROLLARY 2.5.13 (Toeplitz's theorem). *Let* $a = (a_1, a_2, \ldots, a_n, \ldots)$ *be a sequence of complex numbers such that for all* $b = (b_1, b_2, \ldots, b_n, \ldots)$ *in* $l^q (q > 1)$ *the series* $\Sigma_{n=1}^{\infty} a_n b_n$ *converges. Then* $a \in l^p (1/p + 1/q = 1)$.

Proof. We consider the following functional defined on l^p.

$$F(b) = \sum_{i=1}^{\infty} a_i b_i$$

and we remark that it is the pointwise limit of the sequence of functionals (F_n), defined by

$$F_n(b) = \sum_{i=1}^{n} a_i b_i.$$

Now it is easy to see that the norm of F_n is $(\Sigma_{i=1}^{n} |a_i|)^{1/p}$. From 2.5.12 we get that $\Sigma_{i=1}^{n} |a_i|^p$ converges and thus a is in l^p. □

COROLLARY 2.5.14 (Banach generalized limits). *Let m be the space of all sequences of real numbers which are bounded. There exists a function on m denoted by 'glim' which satisfies the following properties:*

(1) *If* $a = (a_i)$, $a_i \geqslant 0$ *then* $\text{glim}(a_i) \geqslant 0$,
(2) $\text{glim}(a_{i+1}) = \text{glim}(a_i)$, *i.e. the translation invariance property.*
(3) *If* $a = (a_i)$ *is convergent then* $\text{glim}(a_i) = \lim a_i$.
(4) $\underline{\lim}\, a_i \leqslant \text{glim}\,(a_i) \leqslant \overline{\lim}\, a_i$.

Proof. We consider on m the function

$$p(x, n_1, \ldots, n_p) = \overline{\lim_{n\to\infty}} \frac{1}{p}(x_{n+n_1} + x_{n+n_2} + \cdots + x_{n+n_p})$$

and set

$$p(x) = \inf p(x, n_1, \ldots, n_p),$$

where the infimum is taken over all systems $(n_1, \ldots, n_p)$. Clearly the following properties are satisfied:

(1) $p(x + y) \leqslant p(x) + p(y)$.
(2) $p(ax) = ap(x)$, $a \geqslant 0$. □

Since the first property is obvious we prove only the second. First we remark that if $\varepsilon > 0$ and $n_1, \ldots, n_p$ and $n_1, \ldots, n_q$ are two systems of integers such that

$$p(x, n_1, \ldots, n_p) < p(x) + \varepsilon/2 \quad \text{and} \quad p(y, n_1, \ldots, n_q) < p(y) + \varepsilon/2,$$

then if we consider the system $(n_i + n_j;\ i = 1, 2, 3, \ldots, p{:}\ j = 1, 2, 3, \ldots, q)$ we get

$$p(x + y, (n_i + n_j)) = 1/(pq)\, \overline{\lim}\Big(\sum_{i,j} (x_{n+n_i+n_j} + y_{n+n_i+n_j})\Big)$$

$$\leqslant 1/(pq)\, \overline{\lim}\Big(\sum_{i,j}{}' x_{n+n_i+n_j}\Big) + 1/(pq)\, \overline{\lim}\Big(\sum_{i,j} y_{n+n_i+n_j}\Big)$$

$$\leqslant 1/p\, \overline{\lim}\Big(\sup 1/q\Big(\sum_i x_{n+n_i+n_j}\Big)\Big) +$$

$$+ 1/q\, \overline{\lim}\Big(\sup 1/p\Big(\sum_j y_{n+n_i+n_j}\Big)\Big)$$

$$\leqslant p(x, n_1, \ldots, n_p) + p(y, n_1, \ldots, n_q) = p(x) + p(y) + \varepsilon$$

and since $\varepsilon > 0$ is arbitrary we obtain that

$$p(x + y) \leqslant p(x) + p(y).$$

Now we remark that in the proof of the extension theorems the only fact used for $p(ax) = |a|p(x)$ was for the case of $a \geqslant 0$ and thus the extension theorem is valid for functions p with this property.

Also, in m there exists a closed subspace c of all convergent sequences. A continuous functional on c is given by

$$f(x) = \lim x_i, \quad x = (x_i).$$

Clearly we have $f(x) \leqslant p(x)$ and thus by the extension theorem there exists an extension of f to m. We denote this extension by 'glim'. Since

$$-p(-x) \leqslant \operatorname{glim} x \leqslant p(x)$$

we get that glim satisfies properties (1), (3) and (4). For (2) we note that if $x = (x_n)$ is in m then the sequence $y_n = (x_{n+1} - x_n)$ has the property that $p(y) = 0 = p(-y)$, which implies (2).

REMARK 2.5.15. The notion of 'glim' has numerous applications; in this book we give only one application to operator theory.

An interesting application of the extension theorem was given by G. Lumer and refers to the existence of the so-called 'semi-inner products' on Banach spaces.

DEFINITION 2.5.16. Let X be a Banach space over K. Then there exists a mapping $[\,,\,]$ defined on $X \times X$ with values in K such that the properties

(1) $[ax + y, z] = a[x, z] + [y, z]$,
(2) $[x, x] > 0, \quad x \neq 0$,
(3) $|[x, y]|^2 \leqslant [x, x]\cdot[y, y]$,
(4) $[x, by] = b^*[x, y]$ (b^* denotes the conjugate of b)

hold for all x, y, z in X and a, b in K. This mapping $[\,,]$ is called a semi-inner product.

THEOREM 2.5.17 (Lumer, 1961). *On every Banach space there exist semi-inner products.*

Proof. For each $x \in X$ we consider $x^* \in X^*$ with the following properties:

(1) $\|x^*\| = \|x\|$.
(2) $x^*(x) = \|x\|^2$.

From the extension theorems such elements in X^* exist. Then we set

$$[x, y] = y^*(x)$$

and for zx with $z \in K$ we set z^*x^*. Then it is easy to see that $[\,,]$ satisfies properties (1)–(4). □

REMARK 2.5.18. Property (4) was first considered by Giles (1967). Later we shall give a characterization of complete inner product spaces using the semi-inner products of G. Lumer.

The proof of the following proposition is left to the reader.

PROPOSITION 2.5.19. *If X is a Banach space and $[\,,]$ is a semi-inner product on X then the function $x \to [x, x]^{1/2}$ is a norm on X.*

We note that a semi-inner product is said to be compatible with the norm if

$$x \to \|x\| = [x, x]^{1/2}.$$

2.6. Inner Product Spaces. Definitions and Some Examples

Let X be a linear space over K. The basic notion we consider is that of an inner product. This is defined as follows.

DEFINITION 2.6.1. An inner product on a linear space X is a function $\langle\,,\rangle$ defined on $X \times X$ with values in K and satisfying the properties

(1) $\langle ax + y, z\rangle = a\langle x, z\rangle + \langle y, z\rangle$,
(2) $\langle x, y\rangle = \overline{\langle y, x\rangle}$,
(3) $\langle x, x\rangle 0$ if $x \neq 0$

for all x, y, z in X and $a \in K$.

From the definition we obtain the following.

PROPOSITION 2.6.2. *If $\langle\,,\rangle$ is an inner product on the linear space over K then:*

(1) $\langle 0, x\rangle = 0, \quad x \in X$.
(2) $\langle x, by\rangle = \bar{b}\langle x, y\rangle$, *$x$, y in X and $b \in K$* (the bar denotes the conjugate).
(3) $\langle x, su + tv\rangle = \bar{s}\langle x, u\rangle + \bar{t}\langle x, v\rangle$, *$x$, u, v in X, t, s in K.*

DEFINITION 2.6.3. If X is a linear space over K with an inner product $\langle\,,\rangle$ then we define on X the function

$$x \to \|x\| = \langle x, x \rangle^{1/2}.$$

We used in this definition for the function, a notation which we used for norms. We prove now that, indeed, the function defined above is a norm on X. First we give the following inequality.

THEOREM 2.6.4 (the Cauchy–Buniakowsky inequality). *If $\langle\ ,\ \rangle$ is an inner product on the linear space X over K then for any x, y in X the following inequality holds:*

$$|\langle x, y \rangle| \leqslant \|x\| \, \|y\|$$

(*with equality iff $x = ay$ for some a in K*).

Proof. Let us consider a real number t with the property that

$$e^{it}\langle x, y \rangle = |\langle x, y \rangle|.$$

Then we have,

$$0 \leqslant \langle e^{it}x - y, e^{it}x - y \rangle = \langle x, x \rangle - e^{it}\langle x, y \rangle - \langle y, e^{it} x \rangle + \langle y, y \rangle$$
$$= \|x\|^2 + \|y\|^2 - 2|\langle x, y \rangle|.$$

This obviously implies the inequality: for all x, y in X,

$$|\langle x, y \rangle| \leqslant \tfrac{1}{2}(\|x\|^2 + \|y\|^2).$$

Now from this inequality the Cauchy–Buniakowsky inequality follows immediately. Indeed, we may suppose without loss of generality that both x and y are nonzero. In this case, if we consider

$$u = x/\|x\|, \; v = y/\|y\|$$

then $|\langle u, v|| \leqslant 1$, and this is equivalent with the Cauchy–Buniakowsky inequality. The second assertion follows from the fact that $\langle x, x \rangle = 0$ iff x is 0.

Using the Cauchy–Buniakowsky inequality we can easily prove that $x \to \|x\|$ is indeed a norm on X.

THEOREM 2.6.5. *Let X be a linear space over K with an inner product $\langle\ ,\ \rangle$. Then*

$$x \to \|x\| = \langle x, x \rangle^{1/2}$$

is a norm on X.

Proof. We have to prove the fact that $x \to \|x\|$ has the following properties:

(1) $\|x\| \geqslant 0$, $\|x\| = 0$ iff $x = 0$.
(2) $\|ax\| = |a| \, \|x\|$, $x \in X$, $z \in K$.
(3) $\|x + y\| \leqslant \|x\| + \|y\|$.

The first two properties are obvious. We prove only the third property. Let x, y be arbitrary elements of X and thus,

$$\|x + y\|^2 = \langle x + y, x + y \rangle = \|x\|^2 + \|y\|^2 + 2\,\mathrm{Re}\langle x, y \rangle \leqslant \|x\|^2 + \|y\|^2 +$$
$$+ 2\|x\| \, \|y\| = (\|x\| + \|y\|)^2,$$

which is equivalent with the required inequality. □

We consider now an important class of inner product spaces which corresponds to the class of Banach spaces.

DEFINITION 2.6.6. A linear space over K is said to be a complete inner product space if the following conditions are satisfied:

(1) X is a linear space with an inner product $\langle\, ,\rangle$.
(2) $(X, \|\, , \|)$ is a Banach space.

Using the fact that normed linear spaces which are not complete can be embedded in a complete space we have the following result.

PROPOSITION 2.6.7. *Let X be a linear space over K with an inner product $\langle\, ,\rangle$. Then there exists a complete inner product space $(X_1, \langle\, ,\rangle_1)$ with the following properties:*

(1) *$(X_1, \|, \|_1)$, $\|, \|_1$ being the norm induced by the inner product $\langle\, ,\rangle_1$, is the completion of the normed linear space $(X, \|\, , \|)$.*
(2) *X_1 is an inner product space.*

Since the proof of this result is simple it is omitted. We now present some examples of linear spaces with inner products.

EXAMPLE 2.6.8. Let us consider the space l^2 of all sequences $x = (x_n)_1^\infty$ with the property that

$$\sum_1^\infty |x_n|^2 < \infty.$$

If x, y are two elements in l^2 then we set

$$\langle x, y\rangle = \sum_{n=1}^\infty x_n \bar{y}_n.$$

From the Hölder inequality we see that $\langle\, ,\rangle$ is a well-defined function on $l^2 \times l^2$ and properties (1)–(3) of the inner product are clearly satisfied. Also, this space is a complete inner product space.

EXAMPLE 2.6.9. Let us consider a probability space $(\Omega, \mathscr{B}, P)$ and let $L^2(\Omega, \mathscr{B}, P)$ be the family of all (classes of) functions on Ω such that

$$\int_\Omega |f(s)|^2 \,\mathrm{d}P(s) < \infty.$$

Then if u, v are two (classes of) functions in $L^2 = L^2(\Omega, \mathscr{B}, P)$ then

$$\langle u, v\rangle = \int_\Omega u\bar{v} \,\mathrm{d}P(s)$$

defines an inner product on L^2. Again this is a complete inner product space.

EXAMPLE 2.6.10 (A. Brown). Let us consider the segment $[0, 1]$ and let $\mathrm{d}x$ be the

Lebesgue measure on $I = [0, 1]$. We consider the space of all functions defined on $[0, 1]$ with the property that they are absolutely continuous functions and for each f in the space f' is in $L^2(I, \mathscr{B}^1, dx) = L_I^2$.

If f, g are two elements of the space then we set

$$\langle f, g \rangle = f(0)\overline{g(0)} + \int_0^1 f'(s)\bar{g}'(s)\, ds.$$

Then it is easy to see that this defines an inner product on the above space which is a linear space. Since L_I^2 is a complete inner product space we get immediately that our space is a complete inner product space.

REMARK 2.6.11. This is an example of a complete inner product space with the property that if f, g are in the space then fg is an element of the space.

EXAMPLE 2.6.12. Let us consider the space C^n and R^n. If x, y are two arbitrary elements of $C^n(R^n)$ then we set

$$\langle x, y \rangle = x_1 \bar{y}_1 + x_2 \bar{y}_2 + \cdots + x_n \bar{y}_n,$$

where

$$x = (x_1, x_2, \ldots, x_n),\ y = (y_1, y_2, \ldots, y_n).$$

Then it is easy to see that this defines an inner product and the Cauchy–Buniakowsky inequality reduces to the well-known inequality of Cauchy. Later we shall present more examples of spaces with an inner product.

We defined on a linear space over K an inner product $\langle\, , \rangle$ as a function on the Cartesian product of the space with values in K and satisfying certain axioms. In an interesting paper, Vidav (1959) remarked that if on a set S we have two mappings, say '+' and '·' defined on $S_\times S$ and $K_\times S$ respectively, with values in S and where K is a ring with identity (denoted 1) then, under some conditions, S is a K-module. To formulate the conditions we first define the so-called K-forms on S.

DEFINITION 2.6.13. A mapping $f: S \to K$ is said to be a K-form if the properties

(a) $f(k \cdot x + y) = k \cdot f(x) + f(y)$
(b) $f(1 \cdot x) = f(x)$

are satisfied for all x, y in S and k in K.

Then we have the following result.

THEOREM 2.6.14 (Vidav, 1959). *Suppose that the set S has the above two mappings and also the following assertions hold:*

(1) $(-1)\cdot x + x = (-1)\cdot x + y$ *implies* $x = y$.
(2) *Except possibly for one element u of S there exists for each s of S a K-form such that* $f(s) \neq 0$.

Then S in a K-module.

Proof. Let x be an arbitrary element of S and, using (1), we get

$$f((-1)\cdot x + x) = -f(x) + f(x) = 0$$

for all K-forms. Then $(-1)\cdot x + x = u$ and $u = (-1)\cdot x + 1\cdot x$ implies that

$$(-1)\cdot x + x = (-1)\cdot x + 1\cdot x$$

and using (1) again we obtain that

$$1\cdot x = x.$$

Similarly, we obtain that $x + u = x$ for all x in S. This implies that for any K-form f we have

$$f(a\cdot x) = f(a\cdot x + u) = af(x) + f(u) = af(x)$$
$$f(x + y) = f(1\cdot x + y) = f(x) + f(y)$$

i.e. any K-form is additive and homogeneous. □

In the same way we obtain that

$$u + x = x$$

and that for all K-forms f,

$$f(0\cdot x) = 0\cdot f(x) = 0.$$

Also

$$0\cdot x = 0.$$

Thus the exceptional element u of the set S has the usual properties of the zero element. Thus we can use the notation 0 for u. Further, we remark that if for all K-forms f,

$$f(x) = f(y)$$

then necessarily $x = y$. Thus we have that the mappings '+' and '·' satisfy the following rules:

(1) $x + y = y + x$.
(2) $(x + y) + z = x + (y + z)$.
(3) $k\cdot(x + y) = k\cdot x + k\cdot y$.
(4) $(k_1 + k_2)\cdot x = k_1\cdot x + k_2\cdot x$.
(5) $k_1\cdot(k_2\cdot x) = (k_1 k_2)\cdot x$.
(6) $1\cdot x = x$.

This proves the assertion of the theorem. From this we obtain the following result. □

COROLLARY 2.6.15 (Vidav, 1959). *Suppose now that K is the field of real or complex numbers and S is a set on which we have the mappings '$-$' and '$\cdot$' satisfying condition* (1) *of Theorem* 2.6.14 *and there exists a K-valued mapping* $[\,,]$ *defined on* $S \times S$

such that:

(1) $[a \cdot x + y, z] = a[x, z] + [y, z]$,
(2) $[1 \cdot x, y] = [x, y]$,
(3) $[x, y] = \overline{[y, x]}$
(4) *there exists at most one element u in S such that* $[u, u] = 0$. *Then S is a linear space over K and* $\varepsilon[\,,\,]$ *is an inner product on S,* $\varepsilon = \pm 1$.

Proof. For each a in S we consider the mapping $f_a(x) = [x, a]$ which is clearly a K-form. Then from Theorem 2.6.14 we obtain that S is a K-module over K, i.e. S is a linear space.

Since $[x, y] = \overline{[y, x]}$, we get that $[x, x]$ is a real number. We now show that $[x, x] \geqslant 0$ or $[x, x] < 0$ for all x in S.

Indeed, in the opposite case we have two elements, say x and y with the property that $[x, x] > 0$ and $[y, y] < 0$.

For any t in R we consider the number

$$f(t) = [t \cdot x - y, tx - y] = t^2[x, x] - t([x, y] + [y, x]) + [y, y].$$

Since $[x, x] \cdot [y, y] < 0$ the equation $f(t) = 0$ has two distinct (real) roots, say t_1 and t_2. Then we have clearly the relations

$$t_1 \cdot x - y = 0, \qquad t_2 \cdot x - y = 0$$

which imply that $[x, x] = 0$. Thus, we must have $[x, x] \leqslant 0$ or $[x, x] \leqslant 0$, which obviously implies our assertion.

Suppose now that X is a linear space over K and we are interested in the class of functions defined on $X_\times X$ with the property that they may serve as inner products. This interesting problem was considered in the work of Heuvers (1973) which is related to some earlier results obtained by Riesz (1958). Following K. J. Heuvres we give first the following definition.

DEFINITION 2.6.16. A function $\langle\langle\,,\,\rangle\rangle$ on $X_\times X$ is called a generalized h-inner product if the following conditions are satisfied:

(1) For any $y \neq 0$ of X and z in K there exists x in X with the property that $\langle\langle x, y\rangle\rangle = z$.
(2) $\langle\langle x_1 + x_2, y_1 + y_2\rangle\rangle = \langle\langle x_1, y_1\rangle\rangle + \langle\langle x_1, y_2\rangle\rangle + \langle\langle x_2, y_1\rangle\rangle + \langle\langle x_2, y_2\rangle\rangle$
(3) $\langle\langle y, x\rangle\rangle = f(\langle\langle x, y\rangle\rangle)$, f is a continuous function.
(4) $\langle\langle x, zy\rangle\rangle = g(z, \langle\langle x, y\rangle\rangle)$, g is a function.

Thus the problem of the existence of such generalized h-inner product reduces to one of determining the functions f and g. After some computations (and supposing $K = C$) K. J. Heuvres shows that f and g must satisfy the following functional equations:

$$1 \cdot f(u + v) = f(u) + f(v)$$

for all complex numbers u, v (this is the so-called Cauchy functional equation), and

$$g(u + v, w) = g(u, w) + g(v, w),$$
$$g(u, v + w) = g(u, v) + g(v, w)$$

for all complex numbers u, v and w.

K. J. Heuvres discusses the solutions of these functional equations and the interested reader may want to consult the original paper for more detail.

We conclude this section with some inequalities related to the Cauchy–Buniakowsky inequality. Moore (1973) has presented an interesting inequality related to the Cauchy–Buniakowsky inequality which says something about the angles between the elements. The result is as follows.

THEOREM 2.6.17 (Moore, 1973). *Let X be a real linear space with an inner product $\langle , \rangle$. Suppose that $\varepsilon > 0$ (small) and x, y, z are elements in X such that*

$$|\langle x, y\rangle| \geqslant (1 - \varepsilon)\|x\| \|y\|$$
$$|\langle x, z\rangle| \geqslant (1 - \varepsilon)\|x\| \|z\|$$

then

$$|\langle y, z\rangle| \geqslant r(\varepsilon)\|y\| \|z\|$$

where

$$r(\varepsilon) = 1 - \varepsilon - (2\varepsilon)^{1/2}.$$

Proof. Since the inequalities are homogeneous we may suppose without loss of generality that $\|x\| = \|y\| = \|z\| = 1$. Also we may suppose that $\langle x, y\rangle \geqslant 0$ is satisfied. In this case we have

$$|\langle y, z\rangle| = |\langle y - x + x, z\rangle| = |\langle y - x, z\rangle + \langle x, z\rangle| \geqslant |\langle x, z\rangle - |\langle y - x, z\rangle| \geqslant 1 - \varepsilon - |\langle y - x, z\rangle|.$$

Now, by the Cauchy–Buniakowsky inequality,

$$|\langle y - x, z\rangle| \leqslant \|y - x\| \|z\| = \|y - x\|$$

and

$$\|y - x\|^2 = \langle y - x, y - x\rangle = 2 - 2\langle x, y\rangle \geqslant 2 - 2(1 - \varepsilon) = 2\varepsilon.$$

Thus

$$|\langle y, z\rangle| = 1 - \varepsilon - (2\varepsilon)^{1/2}$$

and the inequality is proved.

In the same paper M. H. Moore rephrased this inequality to obtain another inequality. This result is as follows.

THEOREM 2.6.18 (Moore, 1973). *Let X be a real linear space with an inner product $\langle , \rangle$. Suppose that y, z are elements in $X \setminus 0$. Then for any x in $X \setminus 0$ the following inequality holds:*

$$|\langle y, z\rangle| \geqslant r_x \|y\| \|z\|,$$

where

$$r_x = \max\{4\min(|\langle x,y\rangle|/(\|x\|\,\|y\|), |\langle x,z\rangle|/(\|x\|\,\|z\|)) - 3, 0\}.$$

For a proof of this inequality we refer to Moore's paper.

REMARK 2.6.19. Is it possible to prove an inequality of this type for Banach spaces?

Another interesting inequality related to the Cauchy–Buniakowsky inequality was proved by M. L. Buzano. The result is as follows.

THEOREM 2.6.20 (Buzano, 1971–73, 1974). *Let X be a complete inner product space and let x, a, b be elements of $X, x \neq 0$. Then the following inequality holds:*

$$\{|\langle x,a\rangle\langle x,b\rangle|\}/\|x\|^2 \leqslant \tfrac{1}{2}(\|a\|\,\|b\| - |\langle a,b\rangle|).$$

If X is a real linear space then the following inequality holds:

$$-\tfrac{1}{2}(\|a\|\,\|b\| - \langle a,b\rangle) \leqslant (\langle x,a\rangle\langle x,b\rangle)/\|x\|^2 \leqslant \tfrac{1}{2}(\|a\|\,\|b\| - \langle a,b\rangle).$$

Since the proof of this result requires some facts about the orthogonal decomposition of a complete inner product space, in Chapter 3 we present the proof of the above inequalities.

We now present an inequality which is very useful in number theory; the reader interested in such applications may want to consult the excellent exposition of Bombieri (1974).

2.6.21. THE SELBERG INEQUALITY. *Let X be a linear space with an inner product $\langle\,,\rangle$. Suppose that $x, x_1, \ldots, x_n (n < \infty)$ are elements of X. Then the following inequality holds:*

$$\sum_{i=1}^{n}\left[|\langle x, x_i\rangle|\Big/\left(\sum_{j=1}^{n}|\langle x_i, x_j\rangle\right)\right] \leqslant \|x\|^2.$$

Proof. Let $c_1, \ldots, c_n$ be in K and then we have that

$$\|x - \sum_{i=1}^{n} c_i x_i\| \geqslant 0$$

which gives that

$$x^2 - 2\,\mathrm{Re}\left(\sum_i \bar{c}_i\langle x, x_i\rangle\right) + \sum_{i,j} c_i\bar{c}_j\langle x_i, x_j\rangle \geqslant 0.$$

But if u, v are complex numbers then

$$|uv| \leqslant \tfrac{1}{2}(|u|^2 + |v|^2)$$

and thus

$$\sum_{i,j} c_i\bar{c}_j\langle x_i, x_j\rangle = \tfrac{1}{2}\cdot\sum_{i,j}(|c_i|^2 + |c_j|^2)|\langle x_i, x_j\rangle|$$
$$\leqslant \sum_{i,j}|c_i|^2|\langle x_i, x_j\rangle.$$

Thus

$$x^2 - 2\,\mathrm{Re}\sum_i \bar{c}_i\langle x, x_i\rangle + \sum_i c_i\left(\sum_j |\langle x_i, x_j\rangle|\right) \geqslant 0$$

which for

$$\frac{c_i = \langle x, x_i\rangle}{\Sigma_{j=1}^n |\langle x_i, x_j\rangle|}$$

gives the Selberg inequality.

An interesting generalization of the Cauchy–Buniakowsky inequality was obtained by Čjan Shi Siun (1957) and has applications in the theory of integral equations.

THEOREM 2.6.22 (Čjan Shi Siun, 1957). *If* $\{f_i\}_1^n$ *and* $\{g_i\}_1^n$ *are arbitrary elements in an inner product space* X *with an inner product* $\langle\,,\rangle$ *then*

$$|\det(\langle f_i, f_j\rangle)|^2 \leqslant \det\{\langle f_i, f_j\rangle\}\ \det\{\langle g_i, g_j\rangle\}$$

and if $K \in \mathrm{h}(X)$ *and is bounded and positive then*

$$|\det(\langle Kf_i, g_j\rangle)|^2 \leqslant \det\{\langle Kf_i, f_j\rangle\}\det\{\langle Kg_i, g_j\rangle\}.$$

For a proof as well as for the applications we refer to Čjan Shi Siun's paper (1957).

The reader interested in more information about the Cauchy–Buniakowsky inequality (and related matters) should find the following references to be of interest: Cauchy (1921), Buniakowsky (1859), Beurling (1932), Calleabaut (1965), Eliezer and Daykin (1967, 1969), Kurepa (1966), Lorch (1942), McLaughlin (1967), Precupanu (1976), Rao (1976), Steiger (1969), Textorius (1974), and Zarantonello (1976). We also present in Chapter 14, Section 14.1, certain interesting applications of the Cauchy–Buniakowsky inequality to some extremal problems.

NOTE 2.6.23. The class of complete inner product spaces was first defined by John von Neumann (only separable spaces were considered) in his 1929 paper on the general properties of hermitian operators (pp. 64–66, 1932); nonseparable complete inner product spaces were first considered by H. Löwig in his important paper in *Acta Math. Sci.* (*Szeged*), 1934.

Due to the fundamental contributions of John von Neumann to the theory of inner product spaces, as well as to the theory of complete inner product spaces and to the theory of linear operators on such spaces, we take the liberty of suggesting that the class of complete inner product spaces should also be called 'von Neumann inner product spaces' (or, indeed, should *only* be called 'von Neumann spaces').

2.7. Von Neumann Generalized Direct Sums

We now present a method of constructing complete inner product spaces; some spaces mentioned in Section 2.6 are then particular cases. We consider a locally compact space S and a positive Radon measure μ such that $\mu(S) < \infty$.

DEFINITION 2.7.1. A field of vectors on S is a mapping f defined on S with the property that for each $s \in S, f(s) \in E_s$, where E_s is a complete inner product space. Let $\mathscr{C}$ be the set of all field of vectors on S. Then we have the following assertion.

PROPOSITION 2.7.2. *$\mathscr{C}$ is a linear space with the operations defined as*

(1) $(f+g)(s) = f(s) + g(s)$,
(2) $(af)(s) = af(s)$

for all f, g and $a \in C$.

Proof. Obvious. □

PROPOSITION 2.7.3. *$\mathscr{C}$ is a module over $C(S)$, the algebra of all complex-valued continuous functions on S.*

Proof. Obvious. □

DEFINITION 2.7.4. An integral is a quadruple $(I, H, f \to f')$ where

(1) I is a family of fields of vectors such that $f \in I$ if and only if the following conditions are satisfied:
  (a) for any $g \in I$, $s \to \langle f(s), g(s) \rangle$ is μ-measurable,
  (b) $s \to \|f(s)\|^2$ is μ-integrable when it is measurable.
(2) H is a complete inner product space and $f \to f'$ defined on I with values in H has the property that

$$\langle f', g' \rangle = \int_S \langle f(s), g(s) \rangle \, d\mu(s)$$

where $f \to f'$ and $g \to g'$.

DEFINITION 2.7.5. If $(I, H, f \to f')$ is an integral then we say that the field of vectors in I are μ-integrable and f' is denoted by

$$f' = \int_S f(s) \, d\sqrt{\mu(s)}.$$

DEFINITION 2.7.6. We say that a complete inner product space K is a direct sum of the spaces $K(s)$, s in S, S is a locally compact Hausdorff space if there is an integral $(I, H, f \to f')$ such that

$$K = (f', f \in I)$$

and in this case we write this as

$$K = \int_S K(s) \, d\sqrt{\mu(s)}.$$

PROPOSITION 2.7.7. *Let $(I, H, f \to f')$ be an integral. Then the following assertions hold:*

(1) *I is a linear space.*
(2) *$f \to f'$ is a linear mapping.*

Proof. The fact that I is a linear subspace of $\mathscr{C}$ follows immediately from properties (a) and (b) in Definition 2.7.4. Now the linearity of $f \to f'$ may be proved as follows:

Let $f \to f'$, $g \to g'$ and we consider $h \to h'$ to be an arbitrary element. Then we have, by condition (2) in Definition 2.7.4,

$$\left\langle \int_S f(s) \right\rangle \mathrm{d}\sqrt{\mu(s)} + \int_S g(s)\,\mathrm{d}\sqrt{\mu(s)},\ \int_S h(s)\,\mathrm{d}\sqrt{\mu(s)} -$$

$$-\left\langle \int_S (f(s) + g(s))\,\mathrm{d}\sqrt{\mu(s)},\ \int_S h(s)\,\mathrm{d}\sqrt{\mu(s)} \right\rangle$$

$$= \int_S \langle f(s), g(s)\rangle\,\mathrm{d}\mu(s) + \int_S \langle g(s), h(s)\rangle\,\mathrm{d}\mu(s) -$$

$$- \int_S \langle (f(s) + g(s)), h(s)\rangle\,\mathrm{d}\mu(s) = 0$$

and

$$\left\langle a\int_S f(s)\,\mathrm{d}\sqrt{\mu(s)} - \int_S h(s)\,\mathrm{d}\sqrt{\mu(s)} - \left\langle \int_S (af(s)\,\mathrm{d}\sqrt{\mu(s)},\ \int_S h(s)\,\mathrm{d}\sqrt{\mu(s)} \right\rangle\right\rangle$$

$$= a\int_S \langle f(s), h(s)\rangle\,\mathrm{d}\mu(s) - \int_S \langle af(s), h(s)\rangle\,\mathrm{d}\mu(s) = 0$$

and thus

$$(f + g) \to f' + g', \qquad (af)' \to af'$$

and the assertion is proved.

The following result refers to completeness. □

PROPOSITION 2.7.8. *Let (f_n) be a sequence of fields of vectors such that*

$$\lim_{n,m\to\infty} \|f_n - f_m\|_2 = 0.$$

Then there exists an integrable field of vectors f such that

$$\lim_{n\to\infty} \|f - f_n\|_2 = 0.$$

Proof. The condition

$$\lim_{n,m\to\infty} \|f_n - f_m\|_2 = 0$$

means that

$$\lim_{n,m\to\infty} \int_S |f_n(s) - f_m(s)|^2\,\mathrm{d}\mu(s) = 0$$

and thus we can take a sequence of integers (n_k) such that

$$\|f_{n_{k+1}} - f_{n_k}\| \leqslant \frac{1}{2^k}.$$

In this case we have that

$$\sum_{k=1}^{\infty} \|f_{n_{k+1}} - f_{n_k}\|_2 < \infty$$

and thus we define

$$f = \sum_{k=1}^{\infty} (f_{n_{k+1}} - f_{n_k}).$$

Since

$$\left(\int_S \|f\|^2 \mathrm{d}\mu(s)\right)^{1/2} < \infty$$

is majorized by $\Sigma_{k=1}^{\infty} \|f_{n_{k+1}} - f_{n_k}\|_2$, we get that f is finite (a.e. μ). Then we set

$$f^{-}(x) = \sum (f_{n_{k+1}}(x) - f_{n_k}(x))$$

which converges absolutely (a.e. μ). Then we define further

$$f_1^{-}(x) = \begin{cases} f^{-}(x) \text{ if } \Sigma_{k=1}^{\infty} (f_{n_{k+1}}(x) - f_{n_k}(x)) \text{ converges} \\ 0 \quad \text{in the contrary case.} \end{cases}$$

In this case, for any vector field k

$$\langle f_1^{-}(x), k(x)\rangle = \sum_{k=1}^{\infty} \langle f_{n_{k+1}}(x), k(x)\rangle$$

which implies that

$$f_1^{-} = \lim f_{n_k} \quad \text{(a.e. } \mu).$$

Let $\varepsilon > 0$ and N_ε such that for all $n, m \geqslant N_\varepsilon$

$$\left(\int_S |f_n - f_m|^2 \,\mathrm{d}\mu\right)^{1/2} < \varepsilon$$

which gives that $x \to \langle f_1^{-}(x), k(x)\rangle$ is measurable. But

$$f_{n_i} + \sum_{i=1}^{k} (f_{n_{i+1}} - f_{n_i}) = f_{n_i},$$

which implies that

$$f_1^{-}(x) = \lim f_{n_i}(x) \quad \text{(a.e. } \mu).$$

For $m > N$ we get

$$\int_S (\|f_1^{-}(x) - f_m(x)\|^2 \,\mathrm{d}\mu(x))^{1/2} = \lim\inf \left(\int_S \|f_{n_i} - f_m\|^2 \,\mathrm{d}\mu\right)^{1/2}$$

and thus

$$\int_S \|f_1^{-}(x)\|^2 \,\mathrm{d}\mu(x) < \infty.$$

For $m \to \infty$ we obtain that

$$\lim \int_S \|f_1^-(x) - f_m(x)\|^2 \, d\mu(x) = 0$$

and this proves the assertion of the proposition.

REMARK 2.7.9. We can consider more general fields of vectors, namely with the property that

$$\int_S \|f(x)\|^p \, d\mu(x) < \infty$$

for some fixed p in $(1, \infty)$.

For some very interesting and important applications we refer to the elegant paper of Garding (1954).

2.8. Tensor Products of Banach Spaces and of Complete Inner Product Spaces

One way to obtain new Banach spaces or complete inner product spaces from already known Banach spaces or complete inner product spaces is to consider the tensor products. The tensor products of Banach spaces play a fundamental role in much of the work in the structure theory of Banach spaces as well as in the theory of operators on Banach spaces, and the usefulness of the tensor products was demonstrated by the beautiful results of R. Schatten, John von Neumann and especially by the monumental work of A. Grothendieck. For an excellent presentation of some of the results of Grothendieck we refer to Lindenstrauss and Petczyński (1969).

Let X, Y be two Banach space over K (R or C) and let $B(X, Y)$ be the set of all bilinear forms on $X_{\times}Y$ (i.e. functions defined on $X_{\times}Y$ with values in K and which are linear in both arguments).

For (x, y) in $X_{\times}Y$ we define a mapping on $B(X, Y)$ by the formula

$$(x, y) \to f((x, y))$$

which is clearly a linear form on $B(X, Y)$, which we denote by $b_{x,y}$. The mapping

$$(x, y) \to b_{x,y} = \aleph((x, y))$$

is then a bilinear mapping with values in the algebraic dual of $B(X, Y)$.

DEFINITION 2.8.1. The linear subspace generated by $(\aleph(x, y), (x, y) \in X_{\times}Y)$ is called the tensor product of X and Y and is denoted by $X \otimes Y$. The mapping $\aleph$ defined in (*) is called the canonical map.

Also we use the standard notation for the element $b_{x,y}, x \otimes y$.

With the usual convention (i.e. the sum over the empty set is zero) every element of $X \otimes Y$ is of the form

$$\sum_{i=1}^{n} k_i \cdot x_i \otimes y_i = \sum_{i=1}^{n} k_i x_i \otimes y_i,$$

where $k_i \in K$.

From just the definitions we see that the following rules hold:

(1) $(kx) \otimes y = x \otimes (ky) = k \cdot x \otimes y$,
(2) $(x + y) \otimes z = x \otimes z + y \otimes z$,
(3) $x \otimes (y + z) = x \otimes y + x \otimes z$.

We remark that if

$$b = \sum_{i=1}^{n} a_i x_i \otimes y_i$$

represents an element of $X \otimes Y$, this representation is in general not unique; we may suppose without loss of generality that the subsets of elements (x_i) and (y_i) are linearly independent in X and Y respectively. The numbers of elements of (x_i) (which is the same as that of (y_i)) is called the rank of b.

DEFINITION 2.8.2. A norm on the tensor product of two Banach spaces X and Y, α, is said to be reasonable if

$$\alpha(x \otimes y) = \|x\| \|y\|$$

and

$$\alpha'(x^* \otimes y^*) = \|x^*\| \|y^*\|$$

where α' is the dual norm of α on $X^* \otimes Y^*$ (the norm induced by α on the dual of $X \otimes Y$).

DEFINITION 2.8.3. For any reasonable norm $\alpha, X \overset{\alpha}{\otimes} Y$ is the Banach space obtained by the completion of $X \otimes Y$ with respect to the norm α.

DEFINITION 2.8.4. The greatest reasonable norm '$\wedge$' and the smallest reasonable norm '$\vee$' on $X \otimes Y$ are the norms defined as follows:

$$\|u\|_\vee = \sup\{\langle u, x^* \otimes y^* \rangle|, \|x^*\| \leqslant 1, \|y^*\| \leqslant 1)$$
$$\|u\|_\wedge = \sup|\langle u, v \rangle|, \quad v \in B_c(X, y), \|v\| \leqslant 1,$$

where B_c is the space of all continuous bilinear forms on $X_\times Y$ with the bilinear norm.

REMARK 2.8.5. From the definition of the greatest reasonable norm '$\wedge$' it follows that

$$(X \overset{\wedge}{\otimes} Y)^* = B_c(X, Y).$$

We now consider the family F of all Banach spaces which are isomorphic to R^n or C^n, $n < \infty$.

DEFINITION 2.8.6. A norm α defined on all ordered pairs (X, Y) of elements of F is said to be a tensor norm (in notation $\otimes$-norm) if the following properties hold:

(1) α is a reasonable norm, on $X \otimes Y, X \in F, Y \in F$,
(2) If $u_i: x_i \to Y_i$, $i = 1, 2, x_i, y_i \in F$, then $u_1 \otimes u_2$ is a mapping of $x_1 \otimes x_2$ into $y_1 \otimes y_2$ which satisfies the following inequality:

$$\|u_1 \overset{\alpha}{\otimes} u_2\| \leqslant \|u_1\| \, \|u_2\|.$$

DEFINITION 2.8.7. The tensor product of X and Y is $X \overset{\alpha}{\otimes} Y$ with α a tensor norm. If α is a tensor norm (we write this as 'α is a $\otimes$-norm') then we can associate the norm α^t by the relation

$$X \overset{\alpha^t}{\otimes} Y = Y \overset{\alpha}{\otimes} X.$$

This norm is called the $\otimes$-transposed norm of α. If $\alpha^t = \alpha$ then we say that α is a symmetric norm.

DEFINITION 2.8.8. The dual norm of a $\otimes$-norm α is defined as follows. We consider $X \otimes Y$ as the dual space of $X^* \otimes Y^* (X, Y \text{ in } F)$ and then the dual norm α' is the norm on $X \otimes Y$ induced by $X^* \overset{\alpha}{\otimes} Y^*$.

From the definition it is clear that the following properties hold:

(1) ${}^t({}^t\alpha) = \alpha$,
(2) $(\alpha')' = \alpha$,
(3) ${}^t(\alpha') = ({}^t\alpha)'$.

DEFINITION 2.8.9. If α and β are two tensor norms, then we say that $\alpha \leqslant \beta$ if for all u in $X \otimes Y$, $\|u\|_\alpha \leqslant \|u\|_\beta$.

REMARK 2.8.10. The norms '$\wedge$' and '$\vee$' are $\otimes$-norms.

We now define these notions for arbitrary Banach spaces.

Let α be a $\otimes$-norm and let X, Y be arbitrary Banach spaces. We consider two families of subspaces of X and Y which are finite dimensional; we denote these families by F_X and F_Y respectively. Then we consider the family $(m_X \otimes m_Y, m_X \in F_X, m_Y \in F_Y)$. This may be ordered by the inclusion and we remark that we have $X \otimes Y = \cup\{m_X \otimes m_Y : m_X \in F_X, m_Y \in F_Y\}$.

If u is in $X \otimes Y \cap (m_X \otimes m_Y)$ then we can consider $\|u\|_{m_X \overset{\alpha}{\otimes} m_Y}$ and since we suppose that α is a $\otimes$-norm, applying this to the embedding mapping we have, for

$$(m_X, m_Y) \leqslant (m_X^1, m_Y^1)$$

that the following inequality holds:

$$\|u\|_{m_X \overset{\alpha}{\otimes} m_Y} = \|u\|_{m_X^1 \overset{\alpha}{\otimes} m_Y^1}.$$

Then we can define

$$u \to \|u\|_\alpha = \inf_{(m_X \otimes m_Y)} \{\|u\|_{m_X \overset{\alpha}{\otimes} m_Y}\}$$

which is clearly a reasonable norm on $X \otimes Y$.

DEFINITION 2.8.11. The norm $u \to \|u\|_\alpha$ defined above is called the $\otimes$-norm of the Banach spaces X and Y.

Suppose now that X_i, Y_i are Banach spaces, $i = 1,2$, and let $u_i: X_i \to Y_i$ be continuous linear operators. Then

$$u_1 \otimes u_2: X_1 \otimes X_2 \to Y_1 \otimes Y_2$$

induces a continuous linear operator, denoted by $u_1 \otimes u_2$ on $X_1 \otimes X_2$ with values in $Y_1 \otimes Y_2$ and the following inequality holds:

$$\|u_1 \overset{\alpha}{\otimes} u_2\| \leqslant \|u_1\| \|u_2\|.$$

Let us consider X, Y be two Banach spaces and α be a $\otimes$-norm. We can then consider $X_t \otimes Y$ and $B^\alpha(X^*, Y^*)$ (for two Banach spaces E and $F, B^\alpha(E, F)$ is the dual space of $E \overset{t_\alpha}{\otimes} F$) which are spaces containing every element u of $X \otimes Y$. We set $|u|_\alpha$ as the norm of u as an element of $X \overset{\alpha}{\otimes} Y$ and $\|u\|_\alpha$ as the norm of u as an element of $B^\alpha(X^*, Y^*)$.

DEFINITION 2.8.12. A tensor norm α is accessible if $\|u\|_\alpha = |u|_\alpha$ when either X or Y is a finite dimensional space.

Let X, Y be two Banach spaces. We prove now that $X \otimes Y$ can be identified with a subspace of $L(X^*, Y)$.

Indeed, if $u = \Sigma_{i=1}^n x_i \otimes y_i$, then we set $T_x: X^* \to Y$ defined by the formula:

$$(T_x)x^* = \sum_{i=1}^n x^*(x_i)\, y_i.$$

As we remarked above, we can suppose that (x_i) (and (y_i)) is a linear independent subset whose cardinal is n. If $x \neq 0$ we see that T_x is not 0 and thus $x \to T_x$ is an injective linear mapping. The adjoint of $T_x, (T_x)^*$, is given by the formula

$$(T_x)^* y^* = \sum_{i=1}^n y^*(y_i)x_i$$

and the assertion is proved.

We now consider the special case when the Banach spaces X, Y have norms derived from an inner product.

For any u, v of $X \otimes Y$,

$$u = \sum_{i=1}^n x_i \otimes y_i, \qquad v = \sum_{j=1}^m \bar{x}_j \otimes \bar{y}_j$$

we set

$$\langle u, v\rangle = \sum_{i=1}^{n}\sum_{j=1}^{m} \langle x_i, \bar{x}_j\rangle\langle y_i, \bar{y}_j\rangle = f(u, v)$$

and then we have the following result whose proof is omitted.

LEMMA 2.8.13. *The mapping*

$$\langle u, v\rangle = f(u, v)$$

has all the properties of an inner product except the fact that $X \otimes Y$ with the norm induced by this may be not complete.

DEFINITION 2.8.14. The tensor product of two complete inner product spaces X and Y is the completion of the algebraic product of X and Y with respect to the norm induced by the seminorm defined in Lemma 2.8.13.

If Y is a complete inner product space then we can consider an orthonormal basis of Y, say $(e_i^Y)_{i\in I}$, I an index set. We can consider the complete inner product space $l_I^2(Y)$ (i.e. the space of all elements of the form $(y_i)_{i\in I}$, $y_i \in Y$ and $\Sigma_{i\in I}\|y_i\|^2 < \infty$). Then we can prove the following result.

LEMMA 2.8.15. *$l_I^2(X)$ is isometrically isomorphic to a subspace of $X \otimes Y$.*

Proof. Let $(x_i)_{i\in I}$ be an element of $l_I^2(X)$. For any finite part J of I we have

$$\sum_{i\in J}\|x_i \otimes e_i^Y\|^2 = \sum_{i,j\in I}\langle x_i, x_j\rangle\langle e_i^Y, e_j^Y\rangle = \sum_{i\in I}\|x_i\|^2$$

which implies that

$$u\colon l_I^2(X) \to X \otimes Y$$

defined by the formula

$$u((x_i)_{i\in I}) = \sum x_i \otimes e_i^Y$$

can be extended to an isometry of $l_I^2(X)$ on a subspace of $X \otimes Y$. □

Let us consider T in $L(X)$ and S in $L(Y)$ then we define the following operator on $X \otimes Y$ by the formula

$$(T \otimes S)(x \otimes y) = Tx \otimes Sy.$$

If $S = I$ then we have that $T \otimes I$ has the following property:

$$\left\|(T\otimes I)\Big(\sum_{i\in I} x_i \otimes e_i^Y\Big)\right\|^2 = \sum_{i\in I}\|Tx_i \otimes e_i^Y\|^2 \leqslant \sum_{i\in I}\|Tx_i\|^2$$

$$\leqslant \|T\|^2\sum_{i\in I}\|x_i\|^2 \leqslant \|T\|^2\sum_{i\in I}\|x_i \otimes e_i^Y\|^2$$

which implies that $T \otimes I$ is a bounded operator. Similarly, we show that $I \otimes S$ is a bounded operator. But

$$T \otimes S = (T \otimes I)(I \otimes S)$$

and thus $T \otimes S$ is a bounded linear operator.

The proof of the following assertions is left to the reader.

(1) $(aT + bS) \otimes R = a(T \otimes R) + b(T \otimes R)$.
(2) $T \otimes (S_1 + S_2) = T \otimes S_1 + T \otimes S_2$.
(3) $(T \otimes S)(T_1 \otimes S_1) = (TT_1 \otimes SS_1)$.
(4) $\|T \otimes S\| = \|T\| \cdot \|S\|$.

CHAPTER 3

Orthogonality and Bases

3.0. Introduction

Let X be a linear space over K and suppose that on X we have an inner product $\langle , \rangle$. The basic notion defined on X by the inner product $\langle , \rangle$ is the notion of orthogonality. Using this notion we consider certain families of orthogonal elements which are on the surface of the unit ball of X (considered as a normed linear space $(X, \|,\|)$). Further, we present the notion of c-completeness and the notion of base in complete inner product spaces. Also certain results about subspaces in X and the orthogonal decomposition of a complete inner product space are given and the dual space is then examined. We also present some examples of nonseparable complete inner product spaces.

3.1. Orthogonality in Linear Spaces with an Inner Product

Let X be a linear space over K and suppose that X has an inner product $\langle , \rangle$.

DEFINITION 3.1.1. Let M, N be two subsets of X. We say that M is orthogonal on the subset N if $\langle m, n \rangle = 0$ holds for all $m \in M$ and all $n \in N$.

We write this briefly as $M \perp N$, and if M reduces to a single element m and N reduces to a single element n then we say that the elements m and n are orthogonal, for which we use the notation $m \perp n$.

Using the notion of orthogonality we can distinguish several classes of interesting subsets of X. Some of them will be presented below.

DEFINITION 3.1.2. A family $F = \{x_i\}_{i \in I}$, I an index set, is said to be an orthonormal family if

$$\langle x_i, x_j \rangle = \delta_{ij} = \begin{cases} 1 & i = j \\ 0 & i \neq j \end{cases}.$$

On the set of all orthonormal families of X we consider a partial ordering which is defined by set inclusion.

Some classes of orthonormal families are of fundamental importance in the study of linear spaces with an inner product. Some of these are considered in the following definition.

DEFINITION 3.1.3. An orthonormal family $F = \{x_i\}_{i \in I}$ in a linear space X with an inner product $\langle , \rangle$ is said to be:

(1) maximal, if it is a maximal element in the set of orthonormal families (with the partial ordering considered above);
(2) total, if the smallest closed subspace that includes F is X;
(3) a base, if any $x \in X$ has the form

$$x = \sum_{i \in I} c_i x_i$$

(and where the series on the right-hand side is summable with respect to the norm convergence in X).

Suppose that $F = \}x_i\}_{i\in I}$ is an orthonormal family of elements of X. For each x in X we can define the numbers

$$c_i(x) = \langle x, x_i \rangle, \quad i \in I.$$

DEFINITION 3.1.4. The numbers $\{c_i(x)\}_{i\in I}$ are called the Fourier coefficients of x with respect to the orthonormal family F.

The following result is very useful in obtaining some results about orthonormal families in X.

THEOREM 3.1.5. *Let X be a linear space over X with an inner product and let F be a finite orthonormal family in X, i.e. F is*

$$F = \{e_1, e_2, \ldots, e_n\}$$

for some $n < \infty$. Then for each x in X,

$$\inf_{c_i \in K} \|x - c_1 e_1 - c_2 e_2 - \cdots - c_n e_n\| = \|x - \sum_{i=1}^{n} \langle x, e_i \rangle e_i\|.$$

Proof. First we remark that the following identity holds: for all numbers $c_1, c_2, \ldots, c_n$ in K,

$$0 \leqslant \|x - \sum_{i=1}^{n} c_i e_i\|^2 = \left\langle x - \sum_{i=1}^{n} c_i e_i, x - \sum_{j=1}^{n} c_j e_j \right\rangle = \|x\|^2 - \sum_{i=1}^{n} \langle x, c_i \rangle e_i -$$

$$- \sum_{i=1}^{n} c_i \langle e_i, x \rangle + \sum_{i,j=1}^{n} c_i \bar{c}_j \langle e_i, e_j \rangle$$

$$= x^2 - 2\,\mathrm{Re}\left(\sum_{i=1}^{n} c_i \overline{\langle x, e_i \rangle} \right) + \sum_{i=1}^{n} |c_i|^2$$

$$= \sum_{i=1}^{n} (|c_i|^2 - 2\,\mathrm{Re}(c_i \overline{x, e_i}) + |\langle x, e_i \rangle|^2) + \|x\|^2 - \sum_{i=1}^{n} |\langle x, e_i \rangle|^2.$$

This obviously implies that

$$\left\| x - \sum_{i=1}^{n} c_i e_i \right\|$$

has a minimum for $c_i = \langle x, e_i \rangle$, $i = 1, 2, 3, \ldots, n$. This clearly implies the assertion of the theorem. □

COROLLARY 3.1.6. *For any orthonormal family* $F = \{e_i\}_{i \in I}$ *and any* $x \in X$ *the following inequality holds:*

$$\sum_{i \in I} |c_i(x)|^2 \leqslant \|x\|^2.$$

NOTE 3.1.7. The above inequality is sometimes called the 'Bessel inequality'.

Proof of Corollary 3.1.6. Let J be an arbitrary finite part of I. Then, by Theorem 3.1.5 we have the inequality

$$\sum_{j \in J} |c_j(x)|^2 \leqslant \|x\|^2.$$

Obviously this implies the assertion of the corollary.

COROLLARY 3.1.8. *For any* $x \in X$, $c_i(x) = 0$ *except for countably many* $i \in I$.

We now give some properties of the classes of orthonormal families considered above.

THEOREM 3.1.9. *Let* X *be a linear space over* K *and let* $\langle , \rangle$ *be an inner product on* X. *Suppose that* $F = \{e_i\}_{i \in I}$ *and* $G = \{f_j\}_{j \in J}$ *are two orthonormal families which are maximal. Then*

$$\text{card } I = \text{card } J.$$

Proof. First we remark that if card $I < \infty$ then X is a finite dimensional space over K and then clearly we have

$$\text{card } I = \text{card } J.$$

Thus we may suppose that both are infinite. For i fixed in I we consider the set

$$B_i = (f_j, f_j \in G, \langle f_j, e_i \rangle \neq 0)$$

which is, by Corollary 3.1.8, countable. Since $G \subseteq \cup_{i \in I} B_i$ we have that

$$\text{card } J \leqslant \text{card } I \cdot \aleph_0 = \text{card } I.$$

Since the roles of F and G are symmetric we have the inequality

$$\text{card } I \leqslant \text{card } J$$

and thus

$$\text{card } I = \text{card } J.$$

THEOREM 3.1.10. *Let* X *be a linear space over* K *with an inner product. Then for an orthonormal family* $F = \{e_i\}_{i \in I}$ *of elements of* X *the following assertions are equivalent:*

(1) *F is a base (of X).*
(2) *F is total (in X).*

(3) *For any* $x \in X$,

$$\sum_{i \in I} |c_i(x)|^2 = \|x\|^2.$$

Proof. It is easy to see that (1) → (2) → (3). Suppose now that (3) holds and then we consider the element

$$x = \sum_{i \in I} c_i(x) e_i = \sum_{k=1}^{\infty} c_{i_k}(x) e_{i_k}.$$

Then we have,

$$\left\| x - \sum_{k=1}^{n} c_{i_k}(x) e_{i_k} \right\|^2 = \sum_{k=n+1}^{\infty} |c_{i_k}(x)|^2$$

and since

$$\lim_{n \to \infty} \sum_{k=n+1}^{\infty} |c_{i_k}(x)|^2 = 0$$

we obtain that F is a base for X. □

EXAMPLE 3.1.11 (Gudder, 1974). There exists an orthonormal family F which is maximal and not a base.

We consider the space l^2 of all sequence of numbers in K with the property that

$$\|x\|^2 = \sum_{i=1}^{\infty} |x_i|^2 < \infty,$$

where $x = (x_1, x_2, \ldots)$. Then this is a complete inner product space. We consider in l^2 the subspace X generated by the family of elements

$$\left(\sum_{n=1}^{\infty} n^{-1} e_n, e_2, e_3, e_4, \ldots \right)$$

with the inner product induced by the inner product of l^2. In this space X the family $F = \{e_2, e_3, \ldots\}$ is orthonormal and maximal. It is not a base for S since the element $\Sigma_{n=1}^{\infty} n^{-1} e_n$ is not representable under form $c_2 e_2 + c_3 e_3 + \cdots$.

We now consider another class of orthonormal subsets in a linear space with an inner product, the so-called complete systems.

DEFINITION 3.1.12. Let X be a linear space over K and suppose that X has an inner product. The orthonormal family $F = \{x_i\}_{i \in I}$ of elements of I is said to be a complete system if whenever x is in X with the property that $c_i(x) = 0$ for all $i \in I$, then $x = 0$.

We recall that a subset M in a linear topological space is said to be linearly dense if the closure of the space generated by M is the whole space. Then we have the following result.

PROPOSITION 3.1.13. *If* $F = \{e_i\}_{i \in I}$ *is an orthonormal family in the linear space over* K *with an inner product* $\langle , \rangle$, F *is complete whenever* F *is linearly dense in* X.

Proof. Suppose that $x \in X$ and has the property that $c_i(x) = 0$ for all $i \in I$. Let $\varepsilon > 0$ be given. Since the family F is linearly dense in X we find an element x_ε such that:

(1) $x_\varepsilon = c_1 e_{i_1} + \cdots + c_n e_{i_n}$
(2) $\|x - x_\varepsilon\| \leqslant \varepsilon$.

Then using the properties of the inner product which defines the norm we easily find that

$$\|x\| \leqslant \varepsilon.$$

Since ε is arbitrary we must have $x = 0$. Thus F is a complete system.

We may ask whenever the converse assertion is true, i.e. if F is a complete system, is it linearly dense in X? We shall present later an example by M. Golomb which shows that, in general, this is not the case. Motivated in part by this fact M. Golomb has introduced the following interesting notion.

DEFINITION 3.1.14 (Golomb, 1972). Let X be a linear space over K and let $\langle\,,\rangle$ be an inner product on X. Let $F = \{e_i\}_{i\in I}$ be an orthonormal family in X. We say that F is a c-complete system if for every Cauchy net $(u_j)_{j\in J}$ in X for which

$$\lim_{j\in J} \langle u_j, e_i\rangle = 0, \quad i \in I$$

then

$$\lim_{j\in J} \|u_j\| = 0.$$

REMARK 3.1.15. In his paper M. Golomb considered only sequence, but the extension to nets is standard.

THEOREM 3.1.16 (Golomb, 1872). *The orthonormal family $F = \{e_i\}_{i\in I}$ is c-complete iff F is linearly dense in X.*

Proof. Suppose first that F is a c-complete system in X and let x be an arbitrary element of X. For any finite part J of I we set

$$x_J = x - \sum_{i\in J} \langle x, e_i\rangle e_i$$

and then clearly

$$\langle x_J, e_i\rangle = 0 \quad \text{for all } i \in J.$$

We now consider the net (x_J) which is in fact a Cauchy net. Indeed, this follows immediately from the inequality in 3.1.6 since this implies that $\{c_i(x)\}_{i\in I}$ is with the property that $\{c_i(x)^2\}_{i\in I}$ is summable. Since F is c-complete we must have $\lim x_J = 0$, which is what we need. For the converse we need a famous result which is known as the Parseval identity.

THEOREM 3.1.17 (the Parseval identity). *Let X be a linear space over K with an inner product and F be a base for X. If $F = \{e_i\}_{i\in I}$ then for any x, y in X the following* *identity holds:*

$$\langle x, y\rangle = \sum_{i\in I} c_i(x)c_i(y).$$

Proof. Let $\varepsilon > 0$ and since x and y are of the form

$$x = \sum_{i\in I} c_i(x)e_i$$

$$y = \sum_{i\in I} c_i(y)e_i$$

we find a finite subset $H \subseteq I$ with the property that

$$\left\| x - \sum_{i\in H} c_i(x)e_i \right\| \leqslant \varepsilon.$$

In this case

$$\langle x, y\rangle = \left\langle x - \sum_{i\in H} c_i(x)e_i + \sum_{i\in I\setminus H} c_i(x)e_i, y\right\rangle =$$

or that

$$\left| \langle x, y\rangle - \sum_{i\in H} c_i(x)c_i(y) \right| \leqslant M\cdot\varepsilon.$$

This clearly implies our assertion. □

REMARK 3.1.18. The Parseval identity was first proved by Ch. de la Vallée Poussin in 1893 for integrable Riemann functions (on the interval $[0, 2\pi]$). For functions in $L^2_{[0,2\pi]}$ it was established by P. Fatou in 1906 and for Sturm–Liouville eigenfunctions by V. A. Steklov in 1901.

Now we have the following result.

PROPOSITION 3.1.19. *An orthonormal family F is linearly dense (in X) iff the Parseval identity holds.*

Now the proof of Theorem 3.1.16 may be completed as follows. Suppose that the orthonormal family F is linearly dense in X. Then the identity of Parseval holds. Suppose now that we have a net $(u_j)_{j\in J}$ which is Cauchy and $\lim\langle u_j, e_i\rangle = 0$ for all $i\in I$. We must show that $\lim\|u_j\| = 0$. Indeed, let $\varepsilon > 0$ and since (u_j) is a Cauchy net we find $j_0\in J$ such that whenever j_1, j_2 in I have the property that $j_i \geqslant j_0$ then $\|u_{j_1} - u_{j_2}\| \leqslant \varepsilon$. Since F is linearly dense we find $(e_{i_1}, \ldots, e_{i_n})$ with the property that

$$\left\| u_j - \sum_{i\in H} c_i(u_j)e_i \right\| = \left(\sum_{i\in I\setminus H} |\langle u_j, e_i\rangle|^2 \right)^{1/2}$$

$$\leqslant \|u_j - u_{j_0}\| + \left(\sum_{i\in I\setminus H} |\langle u_{j_0}, e_i\rangle|^2 \right)^{1/2}.$$

If H is a finite subset with the property that the last term in the above sum is less than ε then

$$\left\| u_j - \sum_{i \in H} \langle u_j, e_i \rangle e_i \right\| \leqslant 2\varepsilon.$$

Now the net (u_j) has the property that

$$\lim \langle u_j, e_i \rangle = 0, i \in I$$

and since H is a finite set we obtain that if $j \geqslant j_0$ the following estimate holds:

$$\|u_j\| \leqslant 3\varepsilon.$$

This clearly implies that $\lim u_j = 0$.

COROLLARY 3.1.20. *The orthonormal family $F = \{e_i\}_{i \in I}$ is c-complete iff for each x in X the following relation holds:*

$$\|x\|^2 = \sum_{i \in I} |c_i(x)|^2.$$

Later in this chapter, in some concrete complete inner product spaces, we shall present some examples of families of orthonormal subsets of the types considered above.

3.2. Bases in Complete Inner Product Spaces

As we remarked above, if X is a linear space with an inner product then there exists in X a maximal orthonormal family F which is not a base (for some specific X, which is not complete).

We now present some results about bases and completeness. First we have the following result.

THEOREM 3.2.1. *Let X be a complete inner product space (over K). Then the following assertions are equivalent for an orthonormal family $F = \{e_i\}_{i \in I}$ of elements in X:*

(1) *F is maximal.*
(2) *F is a base.*

Proof. It is obvious that if F is a base then it is a maximal orthonormal family in X. Suppose now that F is a maximal family in X. We must show that F is a base for X. Let x be an arbitrary element of X and consider the family of numbers $(c_i(x))_{i \in I}$. From the inequality in 3.1.6 we obtain that only countably many $c_i(x)$ are nonzero. Thus we may write this family as $(c_n(x))_1^{\infty}$. The inequality in 3.1.6 gives also that

$$\sum_{n=1}^{\infty} |c_n(x)|^2 < \infty.$$

We now consider the sequence $(v_n)_1^{\infty}$, where $v_n = x - u_n$ with

$$u_n = c_1(x)e_1 + \cdots + c_n(x)e_n$$

and then we clearly have the relation

$$\|v_n - v_m\|^2 = \|u_m - u_n\|^2 = \sum_{n+1}^{m} |c_i(x)|^2$$

(if $m > n$, which we may suppose without loss of generality). Since the above series converges we get that (v_n) is a Cauchy sequence in X. But X is supposed to be complete and thus we may consider $v = \lim v_n$. Now we show that $v = 0$. Indeed the element v is orthogonal to F and thus it must be 0. Thus

$$x = \sum_{i \in I} c_i(x) e_i$$

which is the fact that F is a base for X.

Using the result in Theorems 3.1.10 and 3.1.16 we conclude that in complete inner product spaces the following assertion holds.

THEOREM 3.2.2. *If X is a complete inner product space then for an orthonormal family $F = (e_i)_{i \in I}$ the following properties are equivalent:*

(1) *F is a base.*
(2) *F is total.*
(3) *F is maximal.*
(4) *F is a c-complete system.*
(5) *For any x, y in X, $\langle x, y \rangle = \Sigma_{i \in I} c_i(x) c_i(y)$.*

Now we prove that bases exist in complete inner product spaces.

THEOREM 3.2.3. *Let X be a complete inner product space over K. Then in X there exists a base.*

Proof. We suppose, of course, that X is not equal to (0). We consider the set of all families of orthonormal elements in X and on this set we consider a partial ordering defined by the set inclusion. In this case we can apply Zorn's lemma. Indeed, if $\{F_k\}_{k \in K}$ is a linearly ordered set of families of orthonormal elements in X then the family

$$F = \{e, e \text{ is in } F_k \text{ for some } k\} = \bigcup_{k \in K} F_k$$

is clearly an upper bound. □

Thus we get that our set has a maximal element. By the above result this maximal element is a base for X.

Using the notion of base we can prove the following isomorphism theorem which gives the structure of a complete inner product space. First we give a definition.

DEFINITION 3.2.4. Let X be a linear space over K with an inner product. The dimension of X is the cardinal of any maximal orthonormal family of elements of X.

From Theorem 3.1.9 we see that the dimension is a well-defined cardinal number. We show that there exists linear spaces over K with an inner product of arbitrarily given dimension.

EXAMPLE 3.2.5. Let I be an arbitrary set and consider all functions x on I with values in K which vanish except a countable subset of I and such that

$$\sum_{i\in I} |x(i)|^2 < \infty.$$

If we define addition '+' and multiplication '·' pointwise then this is clearly a linear space over K. If y is another element in this space then we can set

$$\langle x, y\rangle = \sum_{i\in I} x(i)\bar{y}(i)$$

which is obviously well defined and is an inner product. We consider the following family of elements in this space $F = \{x_i\}_{i\in I}$:

$$x_i(j) = \begin{cases} 1 & i = j \\ 0 & i \neq j. \end{cases}$$

Then this is an orthonormal family and is maximal. Obviously card F = card I. This gives our assertion.

REMARK 3.2.6. If we take only those functions defined on I which vanish on I except for finitely many i in I then we get a linear space over K which is with an inner product (defined as in Example 3.2.5) and this is not complete. Thus there exists linear spaces over K with an inner product of arbitrary dimension.

If the dimension of the complete inner product space in $\aleph_0$ then the space is separable.

Associated with the complete inner product space X we have the following cardinal number: we consider the following set of cardinal numbers c for which there exists a subset in X whose cardinal is c and which is dense in X. The number we consider is

$$c(X) = \inf(c).$$

Then we have the following result.

THEOREM 3.2.7. *For any complete inner product space $c(x)$ is exactly the dimension of X.*

Proof. Let $(e_i)_{i\in I}$ be a base for X. Then the dimension of X is card I. We may suppose that the dimension of X is not finite since in this case the assertion is clear.

We consider the set of all combinations

$$a_1 e_{i_1} + \cdots + a_n e_{i_n}$$

where a_i are rational numbers (if K is C then we call a complex number rational if it is of the form $r + is$, $i^2 = -1$, r, s rational (real) numbers). Then clearly the cardinal of this set is equal to card I. Thus

$$c(X) \geqslant \text{card } I.$$

Now if $(e_i)_{i\in I}$ is the above basis then for each i in I we consider an element u_i of X

with the property that

$$\|e_i - u_i\| \leqslant \tfrac{1}{2}.$$

We now define a mapping on I with values in $(u_i)_{i\in I}$ by the relation $i \to u_i$.

Since $(u_i)_{i\in I}$ we can consider this mapping as defined on I and with values in X. This is an injective mapping. Indeed, suppose that this is not true and thus we have two distinct elements i and j of I with the property that $u_i = u_j$. But then we have

$$\|e_i - e_j\| = \|e_i - u_i + u_i - e_j\| \leqslant 1$$

and

$$\|e_i - e_j\| = (2)^{1/2}.$$

This contradiction proves our assertion. This obviously implies that

$$\text{card } I \leqslant c(X).$$

But using the above inequality we get that $c(X) = \text{card } I$ and the theorem is proved. □

We now present a method for constructing orthonormal bases in separable complete inner product spaces.

Suppose that X is a separable complete inner product space (according to the above result we may consider that this means that in X there exists a subset D which is countable and dense in X) and $(x_n)_1^\infty$ be a sequence of elements of X. Of course we can consider the smallest closed subspace of X which contains this sequence. We show that we can construct an orthonormal family from the sequence (more precisely the resulting family will be clearly a sequence) which generates the same subspace.

We define recurrently the following sequences:

$$\begin{aligned}
&y_1 = x_1 && z_1 = y_1/\|y_1\|,\\
&y_2 = x_2 - \langle x_2, x_1\rangle && z_2 = y_2/\|y_2\|,\\
&y_3 = x_3 - \langle x_3, z_2\rangle - \langle x_3, z_1\rangle && z_3 = y_3/\|y_3\|,\\
&\vdots && \\
&y_{n+1} = x_{n+1} - \langle x_{n+1}, z_n\rangle - \cdots - \langle x_{n+1}, z_1\rangle && z_{n+1} = y_{n+1}/\|y_{n+1}\|.\\
&\vdots && \vdots
\end{aligned}$$

The sequence $(z_n)_1^\infty$ has clearly the property that the space it generates is equal to the space generated by $(x_n)_1^\infty$. Then the closed subspaces they generate are the same and (z_n) is a base of this subspace.

REMARK 3.2.8. If we have a complete inner product space which is separable and (x_n) is a linearly independent set which is linearly dense in X, then by the above procedure we obtain an orthonormal basis of X.

We now present the theorem which gives the structure of complete inner product spaces. First we give a definition.

DEFINITION 3.2.9. Let $(X, \|\,,\|)$ and $(Y, \|\,,\|)$ be two normed linear space over the field K. We say that X and Y are isomorphically isometric if there exists a mapping $T: X \to Y$ with the following properties:

(1) T is surjective.
(2) For any $x \in X$, $\|Tx\| = \|x\|$.
(3) For any a, b in K and x, y in X, $T(ax + by) = aT(x) + bT(y)$.

Then our result is as follows.

THEOREM 3.2.10. *Let X be a complete inner product space and let $(e_i)_{i \in I}$ be an orthonormal base of X. Let us consider the complete inner product space as in Example 3.2.5 for the set I. We denote this space by l_I^2. Then $(X, \|, \|)$ and $(l_I^2, \|, \|)$ are isomorphically isometric normed linear spaces.*

Proof. We consider for each element x in X the element of l_I^2 defined by

$$\hat{x}(i) = c_i(x) \quad \text{for all } i \in I.$$

Since $(|c_i(x)|^2)$ is a summable family we obtain that the mapping $x \to \hat{x}$, where $\hat{x}$ denotes the above function, is well defined. It is obviously with the property that

$$\|x\|^2 = \|\hat{x}\|^2$$

and thus, since

$$ax + by \to a\hat{x} + b\hat{y},$$

we obtain that the above mapping satisfies requirements (2) and (3) of the theorem. The surjectivity is obvious since if $\hat{x}$ is a function in l_I^2 then the element in X, $\Sigma_{i \in I}\, \hat{x}(i)e_i$, has the property that its image through the above mapping is exactly $\hat{x}$. □

REMARK 3.2.11. If X is a separable complete inner product space over K then X is isomorphically isometric with $l^2(K)$, i.e. l^2 constructed with sequences $x = (x_n)$, $x_n \in K$.

3.3. Subspaces in Spaces with an Inner Product. The Orthogonal Decomposition

Let X be a linear space over K and suppose that on X we have an inner product $\langle\,,\rangle$.

DEFINITION 3.3.1. Let M be a subset in X. Then the orthogonal complement of M is the set denoted by $M^\perp$ and defined as

$$M^\perp = \{x \in X, \langle x, m\rangle = 0 \text{ for all } m \in M\}.$$

From the definition we have the following results.

PROPOSITION 3.3.2. *For any subset M of X, $M^\perp$ is a linear space.*

PROPOSITION 3.3.3. *If $M \subseteq N$ then $N^\perp \subseteq M^\perp$.*

One of the most basic and useful results in the theory of complete inner product spaces is the so-called 'projection theorem' which is given in the following result.

THEOREM 3.3.4 (B. Levi's theorem; Riesz, 1934). *Let X be a complete inner product space and let M be a closed linear subspace of X. Then*

$$M \oplus M^{\perp} = X$$

in the sense that every x in X has a unique decomposition

$$x = x_1 + x_2$$

with $x_1 \in M$ and $x_2 \in M^{\perp}$ (and this decomposition is unique).

There are at least two methods to prove this important result. The first one, which we use now, is based on the existence of bases; the second is based on certain properties of closed convex sets in complete inner product spaces. We shall present this later.

The subspace M being closed is clearly a complete inner product space and thus it has an orthonormal basis, say $\{e_i\}_{i\in I}$.

We now consider all orthonormal families in X with the property that they contain $\{e_i\}_{i\in I}$. We consider on this set a partial ordering by set inclusion and then we can verify that for this set we can apply Zorn's lemma. Then we get an orthonormal family which is maximal. Let us denote this family by $\{f_s\}_{s\in S}$. Let us consider the family $\{f_s\}_{s\in S}\setminus\{e_i\}_{i\in I} = \{h_k\}_{k\in K}$. Then this family $\{h_k\}_{k\in K}$ is a base of a subspace of X which is clearly closed. We show that this is exactly the required space. Indeed, it is not difficult to prove that $\{f_s\}_{s\in S}$ is a base for the space X. Then every element x in X is of the form

$$x = \sum_{i\in I} c_i(x)e_i + \sum_{k\in K} c_k(x)h_k = x_1 + x_2.$$

It is obvious that x_i satisfies the assertion of the theorem.

In some of the above theorems the fact that the space is a complete (metric) space was essential for the validity of the results. We now present some properties which are equivalent to the completeness.

THEOREM 3.3.5. *Let X be a linear space over K and suppose that on X we have an inner product $\langle\, ,\rangle$. Then the following assertions are equivalent:*

(1) *X is a complete inner product space.*
(2) *If M is a closed subspace of X then $M \oplus M^{\perp} = X$.*
(3) *If M is a closed subspace of X then $M = (M^{\perp})^{\perp} = M^{\perp\perp}$.*
(4) *If M is a proper closed subspace of X then $M^{\perp}$ is proper.*
(5) *If f is a continuous linear functional on X then there exists x_0 in X with the property that*

$$f(x) = \langle x, x_0\rangle.$$

Proof. The assertion (1) → (2) is exactly Theorem 3.3.4. Suppose now that (2) holds. First we remark that $M \subseteq M^{\perp\perp}$. Now, by Theorem 3.3.4 we have that for any

x in $M^{\perp\perp}$,

$$x = x_1 + x_2$$

with $x_1 \in M$, $x_2 \in M^{\perp}$. But then $x - x_1 \in M^{\perp\perp}$, which gives that $x_2 = 0$ and thus $x = x_1$. Thus we have (2) → (3). □

Suppose now that (3) holds. Let M be a proper closed subspace of X (i.e. a closed subspace which is not X or $\{0\}$) and suppose that $M^{\perp} = \{0\}$. Then we have

$$X = M \oplus M^{\perp} = M$$

which is a contradiction. Thus (3) → (4).

Suppose now that (4) holds. Suppose further that f is a linear continuous functional on X. If $f = 0$ then we can take $x_0 = 0$. If this is not the case we consider the subspace M defined by

$$M = \{x, f(x) = 0\}.$$

It is obviously that M is a closed linear subspace of X and from our assumption about f is a proper subspace. Then we consider the subspace $M^{\perp}$. We consider there an element u_0 of norm 1. Then if x is an arbitrary element of X, the element

$$x - f(u_0)^{-1} f(x) u_0$$

is obviously in M. Then we have

$$f(x) = \langle x, x_0 \rangle$$

where

$$x_0 = \overline{f(u_0)}^{-1} u_0.$$

Thus (4) → (5).

Suppose now that (5) holds. We consider the completion of X which we denote by $\hat{X}$. Let y be an element of $\hat{X}$. The function $x \to \langle x, y \rangle$ is a continuous linear functional on X. Since we suppose that (5) holds we find an element u in X with the property that

$$\langle x, y \rangle = \langle x, u \rangle.$$

Since this is true for all x in X we must have $u = y$ and thus X is complete. □

REMARK 3.3.6. The fact proved above that on a complete inner product space X every continuous linear functional f is of the form

$$f(x) = \langle x, x_0 \rangle$$

for some x_0 (depending upon f) is a result proved first by Maurice Fréchet for l^2 and generally by Frigyes Riesz. It is known as the Fréchet–Riesz representation theorem of a continuous linear functional on a complete inner product space.

Using the above result we now prove that another property is equivalent to completeness.

THEOREM 3.3.7. *Let X be a linear space over K and suppose that on X we have an* *inner product. Then X is a complete inner product space iff every maximal orthonormal family of elements of X is a base (of X).*

Proof. If X is a complete inner product space then every maximal orthonormal family of elements of X is a base. Suppose now that X has the property that every maximal orthonormal family of elements of X is a base. Then we must show that X is complete. It suffices for this to use Theorem 3.3.5(5). Let f be a continuous linear functional on X. We may suppose without loss of generality that f is not identically zero. Thus we may consider the subspace M defined by

$$M = \{x \in X, f(x) = 0\}$$

which is closed (by the continuity of f). We consider an orthonormal family of elements of M, say $\{e_i\}_{i \in I}$ which is maximal. Then as in the proof of Theorem 3.3.4 we extend this to an orthonormal family of X which has the property that it is maximal. If $\{e_i\}_{i \in I} \cup \{f_j\}_{j \in J}$ is this orthonormal maximal family then J is not the empty set. It is also clear that each f_j is not in M. Let u, v be two elements in the subspace generated by $\{f_j\}$. It is clear that

$$u - f(u)/f(v) \cdot v$$

is in M and also in the space generated by f_j. This implies that we must have

$$u - f(u)/f(v) \cdot v = 0.$$

Thus the space generated by $\{f_j\}_{j \in J}$ is one-dimensional. Let us consider f_0 be an element of norm 1 in this space and thus every x in X is of the form (since our set is a base)

$$x = \sum_{i \in I} c_i(x) e_i + c_0(x) f_0$$

which implies that

$$f(x) = \langle x, f_0 \rangle.$$

Then by Theorem 3.3.4, X is a complete space. □

REMARK 3.3.8. The above result is proved in Gudder and Holland (1975).

Let us consider a complete linear space X (over K) and let M be a closed linear subspace of X. We now consider the following operator associated with the subspace M.

DEFINITION 3.3.9. The projection associated with the closed linear subspace M of X is the linear operator defined on X by the formula: if $x \in X$ then

$$x = x_1 + x_2$$

where $x_1 \in M$, $x_2 \in M^\perp$; the projection P_M has the values at x is x_1. Thus

$$P_M x = x_1.$$

From the properties of the decomposition $X = M \oplus M^\perp$ we obtain P_M as follows:

PROPOSITION 3.3.10. *The function P_M defined above has the following properties:*

(1) $P_M(x + y) = P_M x + P_M y$.
(2) $P_M(kx) = kP_M x$ *for all* $k \in K$, $x \in X$.
(3) *For each* $x \in X$, $\|P_M x\| \leqslant \|x\|$.
(4) *For any* x, y *in* X *the* $P_M x, y = x, P_M y$.
(5) $P_M + P_{M^\perp} = I$.
(6) $P_M \circ P_{M^\perp} = 0$.

For closed subspaces of a complete inner product space the following result holds.

THEOREM 3.3.11. *Let* M, N *be two closed orthogonal linear subspaces of the complete inner product space* X. *Then*

$$M + N = (m + n : m \in M, n \in N)$$

is a closed subspace of X.

Since the proof is easy it is omitted.

REMARK 3.3.12. There are examples of closed linear subspaces of a complete inner product spaces whose sum is not closed (and thus the necessity of the orthogonality in Theorem 3.3.11).

For a family of subspaces of a subspace X then we are interested in the connection between orthogonal subspaces of the family. Then we have the following result.

PROPOSITION 3.3.13. *If* $F = (M_i)_{i\in I}$ *is a family of closed subspaces of the complete inner product space* X *then*

$$\bigcap_{i\in I} M_i = \left(\sum_{i\in I} \oplus M_i^\perp\right)$$

where the symbol $\Sigma_{i\in I} \oplus X_i$ *means the closed subspace generated by* $\cup_{i\in I} M_i$ *(i.e. the smallest closed subspace of* X *which contains the above union).*

Proof. Since the proof is not difficult it is omitted. □

COROLLARY 3.3.14. *Suppose that* $F = (M_i)_{i\in I}$ *is a family of closed linear subspaces of the complete inner product space* X *and for* i *and* j *in* I, $i \neq j$, $M_i \perp M_j$ *then*

$$\sum_{i\in I} \oplus M_i = \left(x, x = \sum_{i\in I} x_i, x_i \in I\right)$$

and

$$\|x\|^2 = \sum_{i\in I} \|x_i\|^2$$

An interesting characterization of the completeness of a linear space with an inner product (i.e. to be a complete inner product space) was obtained by Amemya and Araki (1966). We consider the set of all closed linear subspaces of X, which we denote by $L_c(X)$. Then the result is as follows.

THEOREM 3.3.15. (Amemya and Araki, 1966). *X is a complete linear space iff $L_c(X)$ is an orthomodular lattice.*

For a notion of lattice see Ch. 4, §4.1 of this book or the book of Birkhoff or Maeda and Maeda (1970). Another characterization of the completeness of an inner product space was obtained by Holland (1969). For a detailed discussion we refer to the paper of S. S. Holland or F. Maeda and E. Maeda.

3.4. Some Applications of the Fréchet–Riesz Representation Theorem

An interesting application of the Fréchet–Riesz representation theorem was found by John von Neumann, namely a proof of the famous result in measure theory, the Lebesgue–Nikodym theorem. For the reader's convenience we repeat some definitions. Let S be a nonempty set and let B be a σ-algebra of subsets of S. If μ and ν are two positive measures on B we say that ν is absolutely continuous with respect to μ whenever $\mu(E) = 0$, then $\nu(E) = 0$.

The Lebesgue–Nikodym theorem gives the following result.

THEOREM 3.4.1. *If μ and ν are two positive measures on B and ν is absolutely continuous with respect to μ then there exists f in $L^1(S, B, \mu)$ such that for all E in B,*

$$\nu(E) = \int_E f(s)\, d\mu(s).$$

Proof. We consider on B the following measure $\xi = \mu + \nu$ and without loss of generality we may suppose that $\mu(S) < \infty$ and $\nu(S) < \infty$. We consider now the space $L^2(S, B, \xi)$ which is a complete inner product space. We consider the following function of $L^2 = L^2(S, B, \xi)$,

$$f \to \int_S f\, d\nu(s)$$

which is, from the form of the measure ν, a continuous linear functional. Then by the Fréchet–Riesz theorem there exists a (unique) element g in L^2 with the property that

$$\int_S f(s)\, d\nu(s) = \int_S f(s)g(s)\, d\xi.$$

We consider now a particular f, namely the function $\aleph_E$ where E is in B (and we suppose without loss of generality that $\xi(E) > 0$). Then

$$\xi(E)^{-1} \int_E g(s)\, d\xi(s) = 1$$

which implies that $g(s) \in [0, 1]$ (a.e. ξ) on E. Then we can suppose without loss of generality that $g(s) \in [0, 1]$ (a.e. ξ) on S since E was an arbitrary subset of S in B. We consider now two measurable subsets S_0 and S_1 defined as follows:

$$S_0 = (s, g(s) \subset [0, 1]),$$
$$S_1 = (s, g(s) = 1)$$

(since we may suppose that $g(s) \in ([0, 1], s \in S)$.

Now we remark that we have the relations:

$$\int_S (1-g)\aleph_{S_1}\,\mathrm{d}v = \int_S \aleph_{S_1}\,\mathrm{d}\xi$$

which implies that $\xi(S_1) = 0$. Let E be an arbitrary element of B and define

$$h_n = (1 + g + g^2 + \cdots + g^n)_E$$

and then

$$\int_E (1 - g^{n+1})\,\mathrm{d}v = \int_E h_n\,\mathrm{d}\xi.$$

But $g^m \to 0$ monotonically on S_0 and $g^m = 1$ on S_1 which implies that (gh_n) converges to a function u which is positive and

$$\lim \int_E (1 - g^{n+1})\,\mathrm{d}v = \int_E u\,\mathrm{d}\mu.$$

But

$$\lim \int_E (1 - g^{n+1})\,\mathrm{d}v = \mu(E)$$

and this gives that

$$\mu(E) = \int_E u(s)\,\mathrm{d}\mu(s).$$

Thus the Lebesgue–Nikodym theorem is proved.

We now present an application of the Fréchet–Riesz representation theorem to the theory of complete inner product spaces itself. The main result may be formulated as follows. □

THEOREM 3.4.2. *The dual space of a complete inner product space X may be organized as a complete inner product space.*

Proof. We consider x^* an arbitrary element of X^*. Then by the Fréchet–Riesz representation theorem there exists a unique element x_0 in X with the property that

$$x^*(x) = \langle x, x_0 \rangle.$$

It is obvious that

$$\|x_0\| = \sup\{|x(x)|, \|x\| \leqslant 1\} = \|x^*\|.$$

We now define a mapping on X^* with values in X by the relation

$$\varphi(x^*) = x_0.$$

Then we see immediately that this mapping has the following properties:

(1) φ is an isometry (i.e. $\|\varphi(x^*)\| = \|x^*\|$).
(2) φ is injective.

(3) For any a, b in K and x^*, y^* in X^*,

$$\varphi(ax^* + by^*) = \bar{a}\varphi(x^*) + \bar{b}\varphi(y^*).$$

Using the above mapping φ we can define an inner product $\langle\, , \rangle$ on X^* as follows: if y^* is another element in X^* with $y^*(x) = \langle x, y_0 \rangle$ then we set

$$\langle x^*, y^* \rangle = \overline{\langle \varphi(x^*), \varphi(y^*) \rangle}.$$

First we prove that the above formula defines an inner product on X. Indeed, if a, b are in X and x, y in X then,

$$\langle ax^* + by^*, z^* \rangle = \overline{\langle \varphi(ax^* + by^*), \varphi(z^*) \rangle} = \overline{\langle \bar{a}\varphi(x^*) + \bar{b}\varphi(y^*), \varphi(z^*) \rangle}$$
$$= a \overline{\langle \varphi(x^*), \varphi(z^*)} + b\overline{\langle \varphi(y^*), \varphi(z^*) \rangle} = a\langle x^*, z^* \rangle + b\langle y^*, z^* \rangle.$$

Further, we have,

$$\langle x^*, y^* \rangle = \overline{\langle \varphi(x^*), \varphi(y^*) \rangle} = \langle \varphi(y^*), \varphi(x^*) \rangle = \overline{\langle y^*, x^* \rangle}$$

and also

$$\langle x^*, x^* \rangle = \|\varphi(x^*)\|^2 = \|x^*\|^2.$$

This proves the theorem. □

From this result we get the following important property of complete inner product spaces.

COROLLARY 3.4.3. *Every complete inner product space is a reflexive Banach space.*

Proof. Let x^{**} be an arbitrary element of X^{**} and since X^* is a complete inner product space we find a (unique) element x_0^* of X^* such that for all x^* in X^* the following relation holds:

$$x^{**}(x^*) = \langle x^*, x_0^* \rangle.$$

Further, we get that

$$x^{**}(x^*) = \langle x^*, x_0^* \rangle = \overline{\langle \varphi(x^*), \varphi(x_0^*) \rangle} = \langle \varphi(x_0^*), \varphi(x^*) \rangle$$

and if

$$x_{x_0}^{**}(x^*) = x^*(x_0) = \langle x_0, \varphi(x^*) \rangle$$

we have that

$$x^{**} = x_{x_0}^{**}$$

since the values are the same at each point. Thus X^{**} is exactly the set of all functional of the form

$$x_{x_0}^{**}$$

for some x_0 in X. But this means exactly the reflexivity of X. □

REMARK 3.4.4. The reflexivity of complete inner product spaces may be obtained using a general theorem on uniformly convex spaces, namely the theorem of

Milman–Pettis which asserts that every Banach space which is uniformly convex is reflexive and the following result given in.

PROPOSITION 3.4.5. *Every complete inner product space is uniformly convex.*

Proof. Let $\varepsilon > 0$ and if x, y are two elements of the complete inner product space X with

(1) $\|x\| = \|y\| = 1$,
(2) $\|x - y\| \geqslant \varepsilon$

then from the identity

$$\|x + y\|^2 + \|x - y\|^2 = 2(\|x\|^2 + \|y\|^2) = 4$$

we get the inequality

$$\|x + y\|^2 \leqslant 4 - \varepsilon^2.$$

Thus if we set

$$\delta(\varepsilon) = 1 - (1 - (\varepsilon/2)^2)^{1/2}$$

we have clearly that

$$\|(x + y)/2\| \leqslant 1 - \delta(\varepsilon)$$

and the assertion of the proposition is proved. □

For the Milman–Pettis theorem and other results about uniformly convex spaces the reader may want to consult the author's book (1984).

3.5. Some Examples of Bases in Concrete Complete Inner Product Spaces

In what follows we present some examples of bases in certain classical complete inner product spaces.

EXAMPLE 3.5.1. Let us consider a space K^n of all elements x of the form $x = (x_1, \ldots, x_n)$, $x_i \in K$. This is a complete inner product space where the inner product is defined by

$$\langle x, y \rangle = x_1 \bar{y}_1 + \cdots + x_n \bar{y}_n$$

where

$$y = (y_1, \ldots, y_n).$$

Let us consider the following family of elements of K^n, $F = (e_1, \ldots, e_n)$ with

$$e_i = (0, 0, 0, \ldots, 0, 1, 0, 0, \ldots, 0)$$

and then F is a base for K^n.

EXAMPLE 3.5.2. Let us consider an index set I and a space $l_I^2(K)$ of all functions on I with the property that $\Sigma_{i\in I}|x(i)|^2 < \infty$. Then we consider the family $F = \{e_i\}_{i\in I}$

where

$$e_i(j) = \begin{cases} 1 & i = j \\ 0 & i \neq j. \end{cases}$$

Then F is an orthonormal base for $l_I^2(K)$.

EXAMPLE 3.5.3. Let us consider a space $L^2([-\pi, \pi]\, \mathrm{d}x/2\pi)$, $\mathrm{d}x$ being the Lebesgue measure.

We now consider all continuous functions on $[-\pi, \pi]$ with the property that $f(-\pi) = f(\pi)$. Then clearly we can consider f as defined on R and satisfying the relation $f(x + 2\pi) = f(x)$.This space is clearly a Banach space with the norm

$$\|f\| = \sup\{|f(t)|, t \in [-\pi, \pi]\}.$$

We now consider the following family of functions $F = \{e^{int}\}_{-\infty}^{\infty}$. It is not difficult to see that F is an orthonormal family in $L^2([-\pi, \pi]\, \mathrm{d}x/2\pi)$. We associate the Fourier coefficients $c_n(f)$ by the relations

$$c_n(f) = \frac{1}{2\pi}\int_{-\pi}^{\pi} f(s)\mathrm{e}^{-ins}\,\mathrm{d}s, \qquad n = 0, \pm 1, \pm 2, \ldots,$$

and we consider Cesàro's sums

$$\sigma_n(t) = 1/(n+1)(s_0(t) + s_1(t) + \cdots + s_n(t)),$$

where

$$s_n(t) = \sum_{k=-n}^{n} c_n(f)\,\mathrm{e}^{itk}.$$

Then we have

$$s_n(t) = \sum_{k=-n}^{n} c_n(f)\,\mathrm{e}^{ikt} = \frac{1}{2\pi}\int_{-\pi}^{\pi}\left(\sum_{k=-n}^{n} \mathrm{e}^{-iks}\mathrm{e}^{ikt}\right) f(s)\,\mathrm{d}s$$

$$= \frac{1}{2\pi}\int_{-\pi}^{\pi} f(s)\left(\sum_{k=-n}^{n} \mathrm{e}^{i(t-s)k}\right)\mathrm{d}s = \frac{1}{2\pi}\int_{-\pi}^{\pi} f(s)D_n(t-s)\,\mathrm{d}s$$

and thus

$$\sigma_n(t) = \frac{1}{2\pi}\int_{-\pi}^{\pi} f(s)((D_0(t-s) + \cdots + D_n(t-s))/n)\,\mathrm{d}s$$

$$= \frac{1}{2\pi}\int_{-\pi}^{\pi} f(s)K_n(t-s)\mathrm{d}s$$

where K_n is the function

$$\begin{aligned} 1/n\left(1 + \sum_{-1}^{1}\mathrm{e}^{ikt} + \cdots + \sum_{-n}^{n}\mathrm{e}^{ikt}\right) &= 1/n(1 + (\mathrm{e}^{it} + \mathrm{e}^{-it}) + \cdots + \\ &\quad + (\mathrm{e}^{it} + \mathrm{e}^{-it} + \cdots + \mathrm{e}^{int} + \mathrm{e}^{-int}) \\ &= 1/n(1 + 2\cos t + \cdots + 2\cos nt) \\ &= \frac{1}{n}\left[\frac{\sin\frac{n}{2}t}{\sin\frac{1}{2}t}\right]^2. \end{aligned}$$

Taking $f = 1$ then clearly $s_n(t) = 1$ and thus

$$\frac{1}{2\pi}\int_{-\pi}^{\pi} K_n(t)\mathrm{d}t = 1$$

and then since

$$U_n(-t) = U_n(t)$$

we obtain that

$$\frac{1}{\pi}\int_{0}^{\pi} U_n(t)\mathrm{d}t = 1.$$

The family of functions $\{K_n(t)\}_0^\infty$ is called the Fejér kernel. The most important properties of the Fejér kernel are given in the following proposition.

PROPOSITION 3.5.4. *For the Fejér kernel* $\{K_n\}_0^\infty$ *the following assertions hold:*

(1) $K_n(t) \geqslant 0.$

(2) $1/\pi \int_0^\pi K_n(t)\,\mathrm{d}t = \frac{1}{2}.$

(3) *For each* $0 < d \leqslant \pi$, *if* $C_n(d) = \max(K_n(t)\colon d \leqslant t \leqslant \pi)$ *then*

$$\lim_{n\to\infty} C_n(d) = 0.$$

Proof. First we remark that if $t \in [d, \pi]$ then the following inequality holds:

$$(\sin t/2)^{-2} \leqslant (\sin d/2)^{-2}.$$

Then we have

$$K_n(t) \leqslant (2(n+1))^{-1}(\sin d/2)^{-2} \quad \text{for } t \text{ in } [d, \pi]$$

which obviously implies property (3) of the proposition. The other assertions are clear from the above considerations. □

Using the Fejér's kernels we prove the following famous approximation theorem of L. Fejér.

THEOREM 3.5.5. *If f is a continuous function on* $[-\pi, \pi]$ *with* $f(-\pi) = f(\pi)$ *then*

$$\lim \sigma_n(t) = f(t)$$

uniformly.

Proof. We remark that we have the following identity:

$$\sigma_n(t) - f(t) = \frac{1}{2\pi}\int_{-\pi}^{\pi} (f(s+t) - f(t))K_n(s)\,\mathrm{d}s.$$

Let $\varepsilon > 0$ be given. Since the function f is supposed to be continuous it is in fact uniformly continuous on any compact interval of R (we consider f extended to R by the relation $f(t + 2\pi) = f(t)$). Thus we find $d_\varepsilon > 0$ such that whenever u, v satisfy the relation $|u - v| \leqslant d_\varepsilon$ then

$$|f(u) - f(v)| \leqslant \varepsilon.$$

In this case we get

$$|\sigma_n(t) - f(t)| \leqslant \frac{1}{2\pi}\int_{|s| \leqslant d_\varepsilon} |f(s + t) - f(t))|K_n(s)\,ds +$$

$$+ \frac{1}{2\pi}\int_{|s| \leqslant d_\varepsilon} (|f(t + s) - f(t)|)K_n(s)\,ds \leqslant \varepsilon + 2\|f\|C_n(d_\varepsilon).$$

Now using property (3) of Proposition 3.5.4 we find an integer N_ε such that for all $n \geqslant N_\varepsilon$, $C_n(d_\varepsilon) \leqslant \varepsilon/(2\|f\|)$. Then for $n \geqslant N_\varepsilon$ we obtain that the following estimate holds:

$$|\sigma_n(t) - f(t)| \leqslant \varepsilon + \varepsilon = 2\varepsilon.$$

This implies that the convergence of the sequence $(\sigma_n(t))$ to f is uniformly on $[-\pi, \pi]$ (as well as on R since $f(t + 2\pi) = f(t)$). □

From Fejér's approximation theorem we obtain the following important consequence.

COROLLARY 3.5.6. *If f is a continuous function on $[-\pi, \pi]$ with the following properties:*

(1) $f(-\pi) = f(\pi)$.

(2) $c_n(f) = 0, \quad n = 0, \pm 1, \pm 2, \ldots$

then $f \equiv 0$.

Proof. We remark that in this case $\sigma_n(t) = 0$ and the assertion follows from the uniform convergence given by Fejér's theorem. Now we prove, using the above results, that the family of functions

$$F = \{e^{int}\}_{-\infty}^{\infty}$$

is a base for the space $L^2([-\pi, \pi], \frac{1}{2\pi}\cdot dt) = L^2$. □

THEOREM 3.5.7. *The family F is a base for L^2.*

Proof. For any f in L^2 we associate the Fourier coefficients $c_n(f)$ defined by

$$c_n(f) = \frac{1}{2\pi}\int_{-\pi}^{\pi} f(s)e^{ins}\,ds.$$

We now consider the function on $[-\pi, \pi]$ defined by the relation

$$g(t) = a + \int_{-\pi}^{t} f(s)\,ds$$

which is continuous.

The constant a is defined by the relation

$$g(-\pi) = g(\pi)$$

and thus the value is

$$\frac{-1}{2\pi}\int_{-\pi}^{\pi}\left(\int_{-\pi}^{t} f(s)\,\mathrm{d}s\right)\mathrm{d}t.$$

From the definition we see that

$$\int_{-\pi}^{\pi} g(t)\,\mathrm{d}t = 0.$$

Now if f is an element in L^2 with $c_n(f) = 0$ then we can easily verify that the Fourier coefficients of the associated function g, defined as above, satisfy the relations

$$c_n(g) = 0 \quad \text{for } n = 0, \pm 1, \pm 2, \ldots$$

which implies that $g = 0$. Now since

$$g'(t) = f(t) \quad \text{a.e. } (\mathrm{d}t)$$

we obtain that $f = 0$ a.e. $(\mathrm{d}t)$. This clearly shows that F is a base for L^2. □

COROLLARY 3.5.8. *The space H^2 of all functions f in L^2 with $c_n(f) = 0$ for $n = -1, -2, -3, \ldots$ has as a base the family*

$$F_1 = \{e^{int}\}_{n=0}^{\infty}.$$

From more information about H^2 (the so-called Hardy space) we refer to the book of Hofman (1962) and to the book of Duren (1970).

EXAMPLE 3.5.9. Let X be the space $L^2([-1, 1], \mathrm{d}t)$. We consider in X the family of the Legendre polynomials $\{P_n\}_0^{\infty}$, where P_n is defined as

$$P_n(t) = c_n \cdot \mathrm{d}^n/\mathrm{d}t^n(1 - t^2)^n$$

with

$$c_n = ((2n + 1)/2)(n!2^n)^{-1}.$$

First we remark that this is an orthonormal family in X. Indeed,

$$\langle P_n, P_m\rangle = \int_{-1}^{1} P_n(s)\bar{P}_m(s)\,\mathrm{d}s = c_n\bar{c}_m\int_{-1}^{1} (\mathrm{d}^n/\mathrm{d}s^n(1 - s^2)^n.$$

$$\mathrm{d}^m/\mathrm{d}s^m(1 - s^2)^m\,\mathrm{d}s = c_n\bar{c}_m\int_{-1}^{1} \mathrm{d}^{n+m}/\mathrm{d}s^{n+m}(1 - s^2)^m\,\mathrm{d}s = 0$$

for $n \neq m$ and is equal to 1 for $n = m$.

We show now that this is a base for L^2. First we remark that $P_n(t)$ is a polynomial of degree n. For $n = 0$ this is trivial and the assertion follows easily using an induction argument (which we leave to the reader). Again an induction argument shows that any polynomial of degree n is a linear combination of the polynomials $P_0(t), \ldots, P_n(t)$. Then from the approximation theorem we get immediately that $\{P_n\}_0^{\infty}$ is a base for L^2.

For more information about orthogonal polynomials we refer to the excellent books of Szegö (1967), Bari (1961) and Yeronimus (1962).

3.6. Perturbation of Bases in Complete Inner Product Spaces

Let X be a Banach space and suppose that X has a basis, i.e. there exists a sequence of elements of X, $(e_n)_1^\infty$ such that every element x of X has a unique representation

$$x = \sum_{n=1}^{\infty} a_n(x)e_n.$$

Suppose now that we have another sequence $(f_n)_1^\infty$ of elements of X which is 'near' to the basis $(e_n)_1^\infty$ in a certain sense. The main problem is about the properties of (f_n) which are transmitted from (e_n). More generally, if we consider two sequences of elements of X, say (x_n) and (y_n) which are near, we may ask about the properties of the sequence (y_n) which are transmitted from the sequence (x_n). The first result in this direction was a famous theorem of Paley and Wiener (1934) which obtained a result about the perturbation of orthonormal basis $(e^{int})_{-\infty}^{\infty}$ of the space $L^2 = L^2_{[-\pi,\pi]}\,dx$, dx being the Lebesgue measure on $[-\pi, \pi]$.

The proof given by Paley and Wiener is essentially true in any Banach space, as was noted by R. P. Boas in 1940. An improvement of the Paley–Wiener result was obtained by R. J. Duffin and J. J. Eachus in 1942. The result of Paley and Wiener has attracted the attention of many mathematicians, who have obtained interesting and important extensions. Among the contributors we mention S. Hilding, M. Schaeffer, M. I. Kadets, V. D. Golovin, J. R. Retherford, J. R. Holub, H. Pollard, N. Levinson, etc.

In what follows we present some developments related to the above-mentioned result of R. E. A. C. Paley and N. Wiener.

Let X be a Banach space and let $(x_n)_1^\infty$ and $(y_n)_1^\infty$ be two sequences of elements of X. In order to present results extending or related to the Paley–Wiener theorem we introduce several conditions in the following definition.

DEFINITION 3.6.1. (Retherford and Holub, 1971). We say that the pair of the sequence (x_n) and (y_n) has the following properties:

(1) (P–W for Paley–Wiener) For some k in $(0, 1)$ and all scalars $a_1, \ldots, a_n$, $n < \infty$, the following inequality holds:

$$\left\| \sum_{i=1}^{n} a_i(x_i - y_i) \right\| \leqslant k \left\| \sum_{i=1}^{n} a_i x_i \right\|,$$

(2) (P–H for Pollard–Hilding) For each positive real number k there exist k_1 and k_2 in $[0, 2^{1-1/k}]$ such that for all scalars $a_1, \ldots, a_n$, $n < \infty$, the following inequality holds:

$$\left\| \sum_{i=1}^{n} a_i(x_i - y_i) \right\| = \left(k_1 \left\| \sum_{i=1}^{n} a_i x_i \right\|^k + k_2 \left\| \sum_{i=1}^{n} a_i y_i \right\|^k \right)^{k^{-1}},$$

(3) For some r, s, t, $r \in [0, 1)$, $s \in [0, 1)$, $t^2 \in [0, (1-r)(1-s)]$ and for all scalars $a_1, \ldots, a_n$ the following inequality holds:

$$\left\| \sum_{i=1}^{n} a_i(x_i - y_i) \right\|^2 \leqslant r \left\| \sum_{i=1}^{n} a_i x_i \right\|^2 + s \left\| \sum_{i=1}^{n} a_i y_i \right\|^2 + $$

$$+ t \left\| \sum_{i=1}^{n} a_i x_i \right\| \cdot \left\| \sum_{i=1}^{n} a_i y_i \right\|,$$

(4) (R_1 for Retherford) For some k in $(0, 1)$ and for any finite set of indices F and arbitrary scalars a_i, $i \in F$, there exist scalars b_i, $i \in F$, such that for any subset F_1 of F the following inequality holds:

$$\left\| \sum_{i \in F_1} a_i y_i - b_i x_i \right\| \leqslant k \left(\left\| \sum_{i \in F_1} a_i y_i \right\| + \left\| \sum_{i \in F_1} b_i x_i \right\| \right),$$

(5) (R_2 for Retherford) For some k in $(0, 1)$ and for any scalars $a_1, \ldots, a_n$, $n < \infty$, there exist the scalars $b_1, \ldots, b_n$ such that for $m \leqslant n$ the following inequality holds:

$$\left\| \sum_{i=1}^{m} (a_i y_i - b_i x_i) \right\| \leqslant k \left(\left\| \sum_{i=1}^{m} a_i y_i \right\| + \left\| \sum_{i=1}^{m} b_i x_i \right\| \right).$$

Concerning sequences of elements of a Banach space we consider several objects related to them.

DEFINITION 3.6.2. A sequence (x_n) of elements of a Banach space X is said to be basic if (x_n) is a basis for the space generated by these elements (i.e. the smallest closed linear subspace of X which contains (x_n); it is denoted by $[x_n]$).

DEFINITION 3.6.3 (Fage, 1950). The sequence of closed linear subspaces (M_n) of X is said to be a basis of subspaces of X if each x in X has a unique representation

$$x = \sum_{n=1}^{\infty} x_n \tag{+}$$

with $x_n \in M_n$.

The basis of subspaces is said to be unconditional if the convergence in (+) is unconditional.

If (M_n) is a basis of subspaces then for each n we define the mapping $P_n: X \to X$ by the relation

$$P_n(x) = x_n.$$

It is obvious that P_n is linear and that

$$P_n P_m = \delta_{nm} P_n, \quad \delta_{nm} = \begin{cases} 0, & n \neq m \\ 1, & n = m \end{cases}$$

is the Kronecker symbol.

We say that the basis of subspaces is a Schauder basis of subspaces if for each n, P_n is a continuous mapping.

DEFINITION 3.6.4 (Retherford, 1964). A sequence (M_n) of subspaces of X is said to be a (unconditional) basic sequence of subspaces if (M_n) is a (unconditional) basis of subspaces of $[M_n]$, the closed linear subspace of X generated by $\cup_n M_n$. Then we have the following result.

LEMMA 3.6.5 (Retherford, 1964). *Let (M_n), (N_n) be two sequences of subspaces of a Banach space X. Suppose that for each n there exists a linear mapping*

$$T_n: M_n \to N_n, \quad TM_n = N_n$$

such that for some positive constants m, M,

$$m \left\| \sum_{i=1}^{p} x_i \right\| \leqslant \left\| \sum_{i=1}^{p} Tx_i \right\| \leqslant M \left\| \sum_{i=1}^{p} x_i \right\|$$

for arbitrary x_i in M_n, $i = 1, 2, 3, \ldots, p$. Then there exists a linear homeomorphism $T: [M_n] \to [N_n]$ such that:

(1) $T/M_n = N_n$;
(2) $m\|x\| \leqslant \|Tx\| \leqslant M\|x\|$, $x \in [M_n]$;
(3) *(M_n) is a (unconditional) basic sequence of subspaces of $[M_n]$ iff (N_n) is a (unconditional) basic sequence of subspaces for $[N_n]$.*

Proof. We consider a subspace M_0 of X which consists of all finite linear combinations of elements in M_n, i.e. the set of all elements of the form

$$u = \sum_{i=1}^{k} a_{n_i} x_{n_i}, \quad x_{n_i} \in M_{n_i}.$$

We define on M_0, the mapping T by the following relation

$$T: M_0 \to [N_n], \quad Tu = \sum_{i=1}^{k} a_{n_i} T_{n_i} x_{n_i}.$$

Since

$$m\|u\| \leqslant \|Tu\| \leqslant M\|u\|$$

the mapping T has a unique extension to M_n. The extension is denoted by the same letter. Clearly T^{-1} exists and is continuous. It remains to show that $T[M_n] = [N_n]$. Let us consider an arbitrary element y in $[N_n]$. Then $y = \lim y_n$, where $y_n = \Sigma_{i=1}^{p} y_i^n$, $y_i^n \in M_i$. But then $y_i^n = T_i x_i^n$, and $x_n = \Sigma_{i=1}^{p} x_i^n$ has the property that $x = \lim x_n$ exists since (x_n) is a Cauchy sequence. Obviously $Tx = y$ and thus property (1) is proved.

Suppose now that (M_n) is a basic sequence of subspaces. Take y in N_n and thus $y = Tx$, x in M_n. Since $x = \Sigma x_i$ we have $Tx = y = \Sigma Tx_i$, which clearly implies that N_n is basic. If (M_n) is unconditional then (N_n) is unconditional and the lemma is proved. □

We have now the following Paley–Wiener type of result.

THEOREM 3.6.6 (Retherford, 1964). *Let X be a Banach space and let (M_n), (N_n) be two sequences of nontrivial subspaces of X with the following property: for each n there exists $T_n: M_n \to N_n$ is an injective linear mapping such that for some k in $(0, 1)$ and all*

$x_i \in M_i$, $i = 1, 2, 3, \ldots, p$ the following inequality holds:

$$\left\| \sum_{i=1}^{p} (x_i - T_i x_i) \right\| \leqslant k\left(\left\| \sum_{i=1}^{p} x_i \right\| + \left\| \sum_{i=1}^{p} T_i x_i \right\| \right).$$

Then there exists an isomorphism $T: M_n \to N_n$ such that:

(1) $T/M_n = T_n$;
(2) $(1-k)(1+k)^{-1}\|x\| \leqslant \|Tx\| \leqslant (1+k)(1-k)^{-1}\|x\|, x \in M_n$;
(3) *(M_n) is a (unconditional) basic sequence of subspaces iff (N_n) is a (unconditional) basic sequence of subspaces.*

Proof. Let $(x_i)_1^p$, $x_i \in M_i$, and thus

$$\left\| \sum_{i=1}^{p} T_i x_i \right\| = \left\| \sum_{i=1}^{p} (T_i x_i - x_i + x_i) \right\| \leqslant \left\| \sum_{i=1}^{p} (T_i x_i - x_i) \right\| + \left\| \sum_{i=1}^{p} x_i \right\|$$

$$\leqslant (1+k)\left\| \sum_{i=1}^{p} x_i \right\| + k\left\| \sum_{i=1}^{p} T_i x_i \right\|.$$

Thus

$$\left\| \sum_{i=1}^{p} T_i x_i \right\| \leqslant (1+k)(1-k)^{-1}\left\| \sum_{i=1}^{p} x_i \right\|$$

and, similarly,

$$\left\| \sum_{i=1}^{p} x_i \right\| \leqslant (1+k)(1-k)^{-1}\left\| \sum_{i=1}^{p} T_i x_i \right\|$$

which gives that

$$(1-k)(1+k)^{-1}\left\| \sum_{i=1}^{p} x_i \right\| \leqslant \left\| \sum_{i=1}^{p} T_i x_i \right\| \leqslant (1+k)(1-k)^{-1}\left\| \sum_{i=1}^{p} x_i \right\|.$$

Thus we may apply Lemma 3.6.5 and the assertion of the theorem follows.

In order to apply this theorem for sequences of subspaces satisfying certain conditions, as in Definition 3.6.1, we need the following result. □

LEMMA 3.6.7. *If (x_n), (y_n) are two sequences of elements in the Banach space X satisfying either of properties* (P–W), (P–H) *or* (N), *then for some k in* $(0, 1)$ *and all finite sets of scalars $a_1, \ldots, a_p$, the following inequality holds:*

$$\left\| \sum_{i=1}^{p} a_i(x_i - y_i) \right\| \leqslant k\left(\left\| \sum_{i=1}^{p} x_i \right\| + \left\| \sum_{i=1}^{p} y_i \right\| \right).$$

Proof. If condition (P–W) is satisfied then the assertion is clear. If (P–H) holds then we take

$$k = \max(k_1 k_2)^{1/k}$$

and if (N) is satisfied we take

$$k^2 = \max(r, s, t).$$ □

COROLLARY 3.6.8 (Retherford, 1964). *Suppose that* (M_n), (N_n) *are two sequences of nontrivial subspaces of a Banach space X. Suppose further that for each n we have an injective linear mapping* $T_n: M_n - N_n$ *and that the sequences* (x_n), $(T_n x_n)$ *have either the property* (P–W), (P–H) *or* (N) *for arbitrary* (x_n), $x_n \in M_n$. *Then there exists a linear isomorphism* $T: [M_n] \to [N_n]$ *such that the following assertions hold:*

(1) $T/M_n = T_n$.
(2) $m\|x\| \leqslant \|Tx\| \leqslant M\|x\|, \quad x \in M_n$.
(3) *If* (M_n) *is a (unconditional) basic sequence of subspaces then* (N_n) *is a (unconditional) basic sequence of subspaces.*

Proof. The assertion of the corollary follows from Theorem 3.6.6 and Lemma 3.6.7. □

COROLLARY 3.6.9. *Suppose that* (x_n), (y_n) *are nontrivial sequences of elements of a Banach space X, and for some k in* (0, 1) *and each p for any scalars* $a_1, \ldots, a_p$ *the following inequality holds:*

$$\left\| \sum_{i=1}^{p} a_i(x_i - y_i) \right\| \leqslant k\left(\left\| \sum_{i=1}^{p} a_i x_i \right\| + \left\| \sum_{i=1}^{p} a_i y_i \right\| \right).$$

Then the following assertions hold:

(1) *There exists a linear isomorphism* $T: [x_n] \to [y_n]$ *such that* $Tx_n = y_n$, $n = 1, 2, 3, \ldots$
(2) (x_n) *is a (unconditional) basic sequence iff* (y_n) *is a (unconditional) basic sequence.*

Proof. In the above result we take $M_n = [x_n]$, $N_n = [y_n]$ and then we set $T_n x_n = y_n$, and the assertions follow. □

COROLLARY 3.6.10. *Suppose that the sequences* (x_n), (y_n) *of a Banach space X satisfy either the property* (P–W), (P–H) *or* (N). *Then the conclusions of Corollary 3.6.9 hold in this case too.*

REMARK 3.6.11. The above results are related and, to some extent, are extensions of results proved by H. Pollard and S. Hilding. Using arguments similar to those of Grinblyum (1949; see also Wilansky, 1964, where the arguments of Grinblyum are given), we have the following results, which were noted by C. W. McArthur.

LEMMA 3.6.12. *Let X be a Banach space and let* (M_n) *be a sequence of nontrivial subspaces. Then* (M_n) *is a Schauder basis of subspaces for* M_n *iff for some* $m \geqslant 1$ *and* app *p, q in N,* $p \leqslant q$, *the following inequality holds: if* $x_i \in M_i$,

$$\left\| \sum_{i=1}^{p} x_i \right\| \leqslant m \left\| \sum_{i=1}^{q} x_i \right\|.$$

LEMMA 3.6.13. *Let X be a Banach space and let* (M_n) *be a sequence of nontrivial subspaces. Then* (M_n) *is an unconditional Schauder basis of subspaces for* M_n *iff for some* $m \geqslant 1$ *and for arbitrary finite sets of positive integers* J_1 *and* J_2, $J_1 \subseteq J_2$, *for*

$x_i \in M_i$ the following inequality holds:

$$\left\| \sum_{i \in J_1} x_i \right\| \leqslant m \left\| \sum_{i \in J_2} x_i \right\|.$$

Then we have,

THEOREM 3.6.14 (Retherford, 1964). *Suppose that the sequences (x_n) and (y_n) of elements of X have the property that for some k in $(0,1)$ and for any integer m and for arbitrary $y_i \in N_i$, $i = 1, 2, 3, \ldots, m$, there exist $x_1, \ldots, x_m$, $x_i \in M_i$, and the following inequality holds:*

$$\left\| \sum_{i=1}^{n} (x_i - y_i) \right\| \leqslant k\left(\left\| \sum_{i=1}^{n} x_i \right\| + \left\| \sum_{i=1}^{n} y_i \right\| \right)$$

for all $n \leqslant m$. Then (N_n) is a (Schauder) basic sequence of subspaces iff (M_n) is a (Schauder) basic sequence of subspaces.

If the sequences (x_n) and (y_n) of elements of X have the property that for some k in $(0,1)$ and for any finite set of integers J_2, and arbitrary $y_i \in M_i$, $i \in J_2$, there exist $x_i \in M_i$, $i \in J_2$, such that for all finite subsets J_1 of J_2 the following inequality holds:

$$\left\| \sum_{i \in J_1} (x_i - y_i) \right\| \leqslant k\left(\left\| \sum_{i \in J_1} x_i \right\| + \left\| \sum_{i \in J_2} y_i \right\| \right).$$

Then (N_n) is an unconditional (Schauder) basic sequence of subspaces iff (M_n) is an unconditional (Schauder) basic sequence of subspaces.

Proof. We give the proof of the first assertion; the second assertion may be proved similarly using Lemma 3.6.13.

Suppose that (M_n) is a basis of subspaces for M_n. Then we find $M \geqslant 1$ such that for any m, n, $m \leqslant n$ and arbitrary x_i in M_i,

$$\left\| \sum_{i=1}^{m} x_i \right\| \leqslant M \left\| \sum_{i=1}^{n} x_i \right\|.$$

But

$$\left\| \sum_{i=1}^{m} y_i \right\| = \left\| \sum_{i=1}^{m} (y_i - x_i) + \sum_{i=1}^{m} x_i \right\| \leqslant \left\| \sum_{i=1}^{m} (y_i - x_i) \right\| + \left\| \sum_{i=1}^{m} x_i \right\|$$

and thus from the condition of the theorem we have,

$$\left\| \sum_{i=1}^{m} y_i \right\| \leqslant (1+k)(1-k)^{-1} \left\| \sum_{i=1}^{m} x_i \right\|$$

which implies that

$$\left\| \sum_{i=1}^{m} y_i \right\| \leqslant ((1+k)/(1-k))^2\, m \cdot \left\| \sum_{i=1}^{n} y_i \right\|$$

and Lemma 3.6.9 gives that (N_n) is a basis of subspaces for N_n. □

COROLLARY 3.6.15. *Suppose that the sequences (x_n) and (y_n) are nontrivial and if they are property (R_1) then (y_n) is an unconditional (Schauder) basic sequence iff (x_i)*

is an unconditional (Schauder) basic sequence; if they have property (R_2) *then* (y_n) *is a (Schauder) basic sequence iff* (x_n) *is a (Schauder) basic sequence.*

Another interesting and related point of view concerning the perturbation of bases in complete inner product spaces was initiated by Bari (=N. K. Bary) (1944), and this has attracted the attention of many mathematicians, among whom we mention I. M. Gelfand, M. G. Krein, A. Kostjučenko, A. Skorohod, V. D. Milman and B. E. Veǐc.

In order to present results obtained in the above-mentioned direction we need some notions which are presented below. We mention that the terminology is not uniform. In what follows X denotes a complete inner product space.

DEFINITION 3.6.16. A basis $(e_n)_1^\infty$ of X is said to be a Riesz basis if for some constants $0 < m \leqslant M < \infty$ and any $x = \Sigma a_i e_i$ the following inequality holds:

$$m\left(\sum_{i=1}^{\infty} |a_i|^2\right)^{1/2} \leqslant \|x\| \leqslant M\left(\sum_{i=1}^{\infty} |a_i|^2\right)^{1/2}.$$

DEFINITION 3.6.17. The system of elements (u_n) in X is said to be w-linear independent if whenever

$$\sum_{n=1}^{\infty} c_n u_n = 0$$

then $c_n = 0, n = 1, 2, 3, \ldots$

DEFINITION 3.6.18. Two systems of elements $(x_n), (y_n)$ in X are called p-near if

$$\sum_{n=1}^{\infty} \|x_n - y_n\|^p < \infty.$$

If $p = 2$ we also use the term 'quadratically near'.

DEFINITION 3.6.19. A Bari basis for X is any basis of X which is quadratically near to an orthonormal basis of X.

Concerning Bari bases we have obviously the following result.

PROPOSITION 3.6.20. *If* (u_n) *is a Bari basis for* X *then* $(u_n/\|u_n\|)$ *is a Bari basis of* X *too.*

Proof. Suppose that (e_n) is the orthonormal basis of X to which (u_n) is quadratically near. Then

$$\sum \|u_n - e_n\|^2 < \infty.$$

We choose $t_i, |t_i| = 1$, such that

$$\|u_i/\|u_i\| - t_i e_i\|^2 = 2(1 - |\langle e_i, u_i/\|u_i\|\rangle|).$$

Since

$$\|u_i/\|u_i\| - t_i e_i\|^2 = 2 \min_s \|e_i - s u_i\| \leqslant 2\|e_i - u_i\|^2$$

the assertion of the proposition follows.

An interesting result concerning Bari bases was obtained by M. G. Krein and refers to ways to generate Bari bases from known ones. □

THEOREM 3.6.21 (Krein, 1957). *Let (u_n) be a Bari basis for X and let (f_n) have the property that*

$$\langle f_i, u_j \rangle = \delta_{ij}.$$

Then (f_n) is a Bari basis for X.

Proof. In order to prove this result we recall some notions from operator theory. An operator T in $L(X)$ is said to be in the class c_2 if for some (and thus for all) orthonormal bases the following inequality holds:

$$\sum_{n=1}^{\infty} \|Te_n\|^2 < \infty,$$

where (e_n) is an orthonormal basis for X.

It is not difficult to see that the set c_2 is an ideal in $L(X)$. An important property of this ideal (as well as of all c_p (the Schatten–von Neumann ideals)) is as follows: if T is in c_2 and $I + T$ is invertible then $(I + T)^{-1} = I + T_1$, where T_1 is in c_2. Since the proof of these facts is not difficult it is omitted; for an excellent account of the Schatten–von Neumann ideals we refer to McCarthy (1967).

Now the proof of Theorem 3.6.21 follows. We define an operator $T: X \to X$ by

$$T\left(\sum_{i=1}^{\infty} a_i e_i\right) = \sum_{i=1}^{\infty} a_i\left(u_i - e_i\right).$$

Here (e_i) is an orthonormal basis for X. From the definition it is clear that T is in c_2 and it is easy to see that $I + T$ is an invertible operator. Then, since $I + T$ is invertible and T^* is in c_2 we have that

$$(I + T^*)^{-1} = I + T_1^*.$$

Suppose now that (u_n) is a Bari basis and (f_n) satisfies the conditions

$$\langle f_i, u_j \rangle = \delta_{ij}$$

and thus

$$\langle e_i, (I + T^*)f_j \rangle = \langle (I + T)e_i, f_j \rangle = \langle u_i, f_j \rangle = \delta_{ij}$$

we must have

$$(I + T^*)f_j = e_j.$$

Since $I + T$ is invertible and (e_i) is an orthonormal basis of X, (f_n) is a Bari basis, as an estimate easily shows. □

DEFINITION 3.6.22 (Krein, 1957). A matrix (a_{ij}) is said to be square-summable if

$$\sum_{i,j} |a_{ij}|^2 < \infty.$$

Then we have the following characterization of Bari bases.

THEOREM 3.6.23 (Krein, 1957). *The complete system* (x_n) *in* X *is a Bari basis of* X *iff*:

(1) (x_n) *is w-linearly independent*,
(2) *the matrix* $(\langle x_i, x_j\rangle - \delta_{ij})$ *is square-summable*.

For a proof we refer to Krein's paper where other interesting facts are given.

THEOREM 3.6.24 (Bari, 1944). *Suppose that* (e_i) *and* (f_i) *are two orthonormal systems of elements of* X *and are quadratically near*,

$$\sum_{i=1}^{\infty} \|e_i - f_i\|^2 < \infty.$$

Then (e_i) *is complete iff* (f_i) *is complete*.

For an excellent proof of this result we refer to the paper by Kostjučenko and Skorohod (1953) which contains an interesting application to the eigenvalues of Sturm–Liouville operators. Here, we present another proof of Theorem 3.6.24.

Suppose that the systems (e_i) and (f_i) are as above but the condition (*) is a strong one, namely

$$\sum_{i=1}^{\infty} \|e_i - f_i\|^2 < 1. \tag{*}$$

Then (f_i) is a basis for H.

Proof. Indeed, in the opposite case, there exists in H an element $h \neq 0$ which is orthogonal to all elements f_i. Now,

$$\langle h, e_i\rangle = \langle h, (e_i - f_i)\rangle + \langle h, f_i\rangle = \langle h, (e_i - f_i)\rangle$$

and thus applying the Cauchy–Buniakowsky inequality we obtain that

$$|\langle h, e_i\rangle| \leqslant \|h\|\,\|e_i - f_i\|$$

and thus, since (e_i) is a basis for H, we have

$$\|h\|^2 = \sum_{i=1}^{\infty} |\langle h, e_i\rangle|^2 \leqslant \|h\|^2 \sum_{1}^{\infty} \|e_i - f_i\|^2 < \|h\|^2.$$

But this clearly represents a contradiction and the assertion is proved. □

Suppose that the system (f_i) satisfies the condition: for some N we have the inequality

$$\sum_{i=N}^{\infty} \|e_i - f_i\|^2 < 1. \tag{**}$$

Then if h is any element of H which is orthonormal to the system $(f_i)_{N+1}^{\infty}$ and to the elements $e_1, \ldots, e_N$, it is necessarily 0.

From the above inequalities we have that

$$\|h\|^2 = \sum_{i=1}^{N} |\langle h, e_i\rangle|^2 + \sum_{i=N+1}^{\infty} |\langle h, e_i\rangle|^2$$

$$= \sum_{i=N+1}^{\infty} |\langle h, e_i\rangle|^2 \leqslant \|h\|^2 \sum_{i=N+1}^{\infty} \|e_i - f_i\|^2 < \|h\|^2$$

which is a contradiction if $h \neq 0$.

Let (e_i), (f_i) *as in* (**). *Then any element h of H which is orthogonal to* $(f_i)_{N+1}^{\infty}$ *and to the elements*

$$g_i = e_i - \sum_{i=N+1}^{\infty} \langle e_i, f_k\rangle f_k, \quad i = 1,2,3,\ldots, N$$

is necessarily 0.

Proof. The element satisfies the conditions of the above remark and thus it is necessarily 0. □

Now the proof of the Theorem 3.6.24 is as follows. Since condition (*) is satisfied there exists N with the property that (**) holds. We consider now the subspace H_1 of all elements of H which are orthogonal to $f_i, i = N+1, N+2, \ldots$. This space H_1 clearly contains $f_1, \ldots, f_N$. It is clear that this is a linear subspace which contains the elements $g_i, i = 1,2,3,\ldots, N$. Since H_1 contains only these elements its dimension is not greater than N. But obviously $f_1, \ldots, f_N$ are linearly independent elements. This gives that $f_1, \ldots, f_N$ is a basis for H_1. Now, this implies that g_i are linear combinations of $f_i, i = 1,2,\ldots, N$, which further implies that any element which is orthogonal to $(f_i)_1^{\infty}$ is necessarily 0. This clearly gives that (f_i) is a basis for H. □

REMARK 3.6.25. The above elegant proof of Theorem 3.6.24 is given by Birkhoff and Rota (1960).

Suppose that we have the complete inner product space $L^2 = L^2_{[-\pi,\pi]}\,dx$ where dx is the Lebesgue measure on $[-\pi, \pi]$, and there is an orthonormal basis $(e^{int})_{-\infty}^{\infty}$. Suppose that we have the following system of functions

$$\{e^{i\lambda_k t}\}_{-\infty}^{\infty},$$

where $\lambda_k = k + \delta_k$, δ_k real numbers. Define $d = \sup|\delta_k|$. Then we have the following result which is at the origin of the results presented in Section 3.6.

THEOREM 3.6.26 (Paley and Wiener, 1934). *Suppose that the number d satisfies the inequality* $d \leqslant \pi^{-2}$. *Then the system* $(e^{i\lambda_k t})$ *is a basis for* L^2.

Proof. We remark that we have the relation

$$e^{i\lambda_n t} - e^{int} = \sum_{k=1}^{\infty} (i(\lambda_n - n))^k/k! \quad t^k e^{int}$$

and

$$\|t^k g(t)\| \leqslant \pi^k \|g\|$$

which implies that for any scalars $a_1, \ldots, a_n$,

$$\left\| \sum_n a_n(e^{int} - e^{i\lambda_n t}) \cdot (2\pi)^{-1/2} \right\| = \left\| \sum_n a_n \sum_k (i(\lambda_n - n)^k / k! \cdot t^k \cdot (2\pi)^{-1/2} \cdot e^{int} \right\|$$

$$\leqslant \sum_{k=1}^{\infty} \pi^k / k! \left(\sum_n |a_n|^2 (|\lambda_n - n|)^{2k} \right)^{1/2}$$

$$\leqslant \sum_{k=1}^{\infty} \pi^k / k! d^k \left(\sum_n |a_n|^2 \right)^{1/2}$$

$$= (e^{d\pi} - 1) \left(\sum_n |a_n|^2 \right)^{1/2}.$$

Now from the fact that $d \leqslant \pi^{-2}$ we see that $e^{d\pi} - 1$ is in (0.1) and then the sequences $(2\pi)^{-1/2} e^{int}$ and $(2\pi)^{-1/2} e^{i\lambda_n t}$ satisfy the (P–W) property, and thus the assertion of the theorem follows from the above results.

REMARK 3.6.27. In fact the above proof shows that if $d < \ln 2/\pi$ then the conclusion of Theorem 3.6.26 follows. This is a result obtained by Duffin and Eachus (1942). Levinson (1936) has given an example to show that if $d > 0.25$ then the conclusion of Theorem 3.6.26 is not true. See also Ingham (1936) and Young (1974). Further, Golovin (1963) has improved the above results to show that the conclusion of Theorem 3.6.26 remains valid if $d < 0.24$.

In an interesting paper Kadets (= M. I. Kadec) (1964) has proved the fact that the best constant for d is 0.25. The assertion is a consequence of the following result.

THEOREM 3.6.28 (Kadets, 1964). *If* $d < 0.25$ *then* $(e^{i\lambda_n t})$ *is a Riesz basis for* L^2.

Proof. We must estimate the following number

$$r = \left\| \sum_n a_n (1 - e^{i\lambda_n t}) e^{int} \right\| = \left((2\pi)^{-1} \int_{-\pi}^{\pi} \left| \sum a_n (1 - e^{i\lambda_n t}) e^{int} \right|^2 dt \right)^{1/2}$$

for scalars $a_1, \ldots, a_p$, $p < \infty$ and

$$\sum |a_n|^2 \leqslant 1.$$

Now since the Fourier series of $1 - e^{i\lambda_n t}$ with respect to the orthogonal system

$$(1, \cos nt, \sin (n - \tfrac{1}{2})t : n = 1, 2, 3, \ldots) \text{ is}$$

$$1 - e^{ist} = (1 - (\sin \pi s)/\pi s) + \sum_{k=1}^{\infty} (-1)^k 2s \sin \pi s/(\pi k^2 - s^2) \cos nt +$$

$$+ i \sum_{k=1}^{\infty} (-1)^k 2s \cos \pi s/(\pi(k - \tfrac{1}{2})^2 - s^2) \cdot \sin (k - \tfrac{1}{2})t.$$

In this case we get the following estimates,

$$r \leqslant \Bigg\| \sum_n (1 - \sin \pi\delta_n)/(\pi\delta_n) a_n e^{int} +$$

$$+ \sum_{n=1}^{\infty} \cos nt \sum_{k=1} (-1)^n \cdot 2\delta_k \sin \pi\delta_k \cdot (\pi(n^2 - \delta_k^2)^{-1}) a_k e^{ikt} +$$

$$+ i \sum_{n=1} \sin(n - \tfrac{1}{2})t \sum_k (-1)^n \cdot 2\delta_k \cos \pi\delta_k \times$$

$$\times (\pi(n - \tfrac{1}{2})^2 - \delta_k^2) a_k e^{ikt} \Bigg\|.$$

Since

$$\Bigg\| \sum_{k=1} 1 - \sin \pi\delta_k/(\pi\delta_k) a_k e^{ikt} \Bigg\| \leqslant (1 - \sin \pi d)/(d\pi) \cdot \Bigg\| \sum_k a_k e^{ikt} \Bigg\|$$

and

$$\Bigg\| \cos nt \sum_k (-1)^n \cdot 2\delta_k \cdot \sin \pi\delta_k / (\pi(n^2 - \delta_k^2) \cdot a_k e^{ikt} \Bigg\|$$

$$\leqslant 2d \cdot \sin \pi d / (\pi(n^2 - d^2)) \Bigg\| \sum_k a_k e^{ikt} \Bigg\|$$

as well as

$$\Bigg\| \sin(n - \tfrac{1}{2})t \sum_k (-1)^n \cdot 2\delta_k \sin \pi\delta_k \cdot (\pi(n - \tfrac{1}{2})^2 - \delta_k^2) a_k e^{ikt} \Bigg\|$$

$$\leqslant (2d \cos \pi d) \cdot (\pi(n - \tfrac{1}{2})^2 - d^2)^{-1} \cdot \Bigg\| \sum_k a_k e^{ikt} \Bigg\|$$

we get that

$$r \leqslant (1 - \sin \pi d)/(d\pi) + \sin \pi d \sum_{n=1}^{\infty} 2d \cdot (\pi(n^2 - d^2))^{-1} +$$

$$+ \cos \pi d \cdot \sum_{k=1}^{\infty} 2d \cdot (\pi(n - \tfrac{1}{2})^2 - d^2)^{-1} \Bigg\| \sum_k a_k e^{ikt} \Bigg\|$$

$$= (1 - \sin \pi d/(\pi d) + \sin \pi d(1/(\pi d) - \operatorname{cotan} \pi d) + \cos \pi d \cdot \tan \pi d) \cdot \left(\sum_k |a_k|^2 \right)^{1/2}$$

$$= (1 - \cos \pi d + \sin \pi d) \cdot \left(\sum_k |a_k|^2 \right)^{1/2}.$$

This implies that if $d < 0.25$, $\sup(r, \Sigma |a_n|^2 \leqslant 1)$ is in $(0,1)$. This clearly gives the assertion of the theorem.

Another interesting approach to the study of the perturbation of bases was given by J. R. Retherford and J. R. Holub and consists essentially in exploiting some properties of contraction and compact operators, such as:

(1) If $T \in L(X)$ and $\|T\| < 1$ then $\|T^*\| < 1$.
(2) $T \in L(X)$ is compact iff T^* is compact.

(3) If $\|T\| < 1$ or $T \in L(X)$ is compact and -1 is not an eigenvalue of T then $I - T$ is an isomorphism (of X onto X).

The proofs of these results are not very difficult; they follow easily from a result we give later. We mention that (2) is the so-called Schauder theorem about the adjoint of an operator.

The main difficulty with these properties is to define, given a pair of sequences in a Banach space or in a complete inner product space, an operator T which has one of the above properties. In what follows we illustrate the construction of several operators which have one of these properties.

Suppose that X is a (separable) Banach space and (x_n) is a Schauder basis for X. Then we can associate in a canonical way the elements (f_n) in X defined by the formula:

$$f_n(x_m) = \delta_{nm}, \quad \text{where } \delta_{nm} \text{ is the Kronecker symbol.}$$

Now if (y_n) is a sequence of elements in X then we can define the expressions:

(a) $\sum_{i=1}^{\infty} x^*(x_i - y_i)f_i, \quad x^* \in X^*,$

(b) $\sum_{i=1}^{\infty} f_i(x)(x_i - y_i), \quad x \in X.$

Then we have the following result.

LEMMA 3.6.29 (Retherford and Holub, 1971). *Suppose that for each* x (a) *converges. Then*

$$x^* \to Tx^* = \sum_{i=1}^{\infty} x^*(x_i - y_i)f_i$$

is an element of $L(X^*)$. *If*

$$\sum_{i=1}^{\infty} x^*(x_i - y_i)f_i$$

converges uniformly for $\|x^*\| \leqslant 1$ *then*

$$T^*\eta x = \sum_{i=1}^{\infty} f_i(x)(x_i - y_i)$$

where η *is the mapping* $\eta: X \to X^{**}$ *defined by* $\eta(x)(x^*) = x^*(x)$.

Proof. The principle of uniform boundedness clearly gives the first assertion. For the second we remark that

$$\sup_{\|x^*\| \leqslant 1} \left\| \sum_{n}^{m} x^*(x_i - y_i)f_i \right\| \to 0, \qquad \text{as } n, m \to \infty$$

which gives that

$$\sum_{i=1}^{\infty} f_i(x)x(x_i - y_i)$$

converges for each x. Then again the principle of uniform boundedness implies that

$$x \to \sum_{i=1}^{\infty} f_i(x)(x_i - y_i) = Tx$$

is a bounded linear operator. Since for each x we have

$$T^*\eta x(x) = T^*x^*(x) = \sum_{i=1}^{\infty} x^*(x_i - y_i) f_i(x)$$

we have that

$$T^*\eta x = \sum_{i=1}^{\infty} f_i(x)(x_i - y_i).$$

We consider now some properties of a pair of sequences $(x_n), (y_n)$ in a Banach space X, where (x_n) is supposed to be a Schauder basis. We recall that the series u_n of elements in a Banach space X is said to be weakly unconditionally Cauchy (w.u.c.) if for each x in X,

$$\sum_{i=1}^{\infty} |x^*(u_i)| < \infty, \quad x^* \in X^*$$

DEFINITION 3.6.30. We say that the pair $(x_n), (y_n)$ has the property:

(A) $\|x_n\| \to \infty$, as $n \to \infty$, (y_n) is w-linearly independent and $\Sigma(x_n - y_n)$ is w.u.c.

(B) f_i is w.u.c. and $\lim_{n\to\infty} \|x_n - y_n\| = 0$ (here f_n is defined as above),

(C) $$\sup_{\|x^*\| \leqslant 1} \left(\left\| \sum_m^n x^*(x_i - y_i) f_i \right\| \right) \to 0, \quad \text{as } n, m \to \infty,$$

(D) $$\sup_{\|x^*\| \leqslant 1} \left(\left\| \sum_n^m x^*(x_i - y_i) f_i \right\| \right) < 1,$$

(K–L for Krein and Lusternik)
(y_n) is w-linearly independent and

$$\sum_{i=1}^{\infty} \|f_i\| \, \|x_i - y_i\| < \infty,$$

(K–M–R for Krein, Milman and Rutman)

(M for Milman)

$$\|x_n - y_n\| \leqslant a_n \quad \text{and} \quad \sup_n \sup_{\varepsilon_i = \pm 1} \left(\left\| \sum_{i=1}^n a_i f_i \varepsilon_i \right\| \right) < 1,$$

(M–R for McArthur and Retherford),
$\sup \|f_n\| = K < \infty$ and

$$\sup_\sigma \left\| \sum_{i\in\sigma} x_i - y_i \right\| < \tfrac{1}{4}K, \quad \sigma \text{ is a finite set of indices},$$

(V for Veits (Veic))

$$0 < \inf \|x_n\| \leqslant \sup \|y_n\| < \infty, (y_n) \text{ is w-linearly independent and}$$

$$\sum (x_n - y_n) \text{ is w.u.c.}$$

Then we have the following result which has many applications in obtaining sufficient conditions of (y_n) to be a basis.

THEOREM 3.6.31 (Retherford and Holub, 1971). *Suppose that* $(x_n), (y_n)$ *are two sequences of elements in* X *and* (x_n) *is a Schauder basis for* X. *Further we suppose that* (f_n) *is the associated sequence to the basis* (x_n) *(i.e.* f_n *in* X *is defined by* $f_n(x_m) = \delta_{nm}$, δ_{nm} *being the Kronecker symbol). If the pairs* (x_n), (y_n) *satisfy any of* (A), (B), (C), (D), (K–L), (K–M–R), (M), (M–R) *or* (V) *then* (y_n) *is a Schauder basis of* X *and there exists an isomorphism* $T: X \to X$ *such that* $Tx_n = y_n, n = 1, 2, 3, \ldots$

Proof. If (A) holds then we set

$$Tx = \sum_{i=1}^{\infty} \|f_i\| \cdot f_i(x) / \|f_i\| \cdot (x_i - y_i) = \sum_{i=1}^{\infty} f_i(x)(x_i - y_i)$$

which is a compact linear operator. Since (y_n) is supposed to be w-linearly independent, -1 is not an eigenvalue for this operator. Then $I - T$ is an isomorphism and this implies the assertion. □

Now, if (B) holds then we set

$$Tx^* = \sum_{i=1}^{\infty} \|x_i - y_i\| \eta(x_i - y_i) / \|x_i - y_i\| (x^*) f_i = \sum_{i=1}^{\infty} x^*(x_i - y_i) f_i$$

which is a bounded linear operator which is also compact. Then T^* is also compact and -1 is not an eigenvalue for $T^*\eta$. Then $I - T^*\eta$ is an isomorphism of X and the assertion follows.

The assertions in the cases when (C) or (D) hold are obvious from the cases considered before.

Suppose now that (K–L) holds. In this case we set

$$x \to Tx = \sum_{i=1}^{\infty} f_i(x)(x_i - y_i)$$

and we get a compact linear operator. From the w-linear independence of (y_n) we obtain that -1 is not an eigenvalue for T. Then $I - T$ is an isomorphism and the assertion follows.

If (K–M–R) holds then we set

$$x \to Tx = \sum_{i=1}^{\infty} f_i(x)(x_i - y_i)$$

$$x - Tx = \sum_{i=1}^{\infty} f_i(x)(x_i - y_i)$$

which is a contraction. Then $I - T$ is an isomorphism and the assertion follows.

If (M) holds then if we set

$$a = \sup_n \sup_{\varepsilon_i = \pm 1} \left\| \sum_{i=1}^{n} a_i f_i \varepsilon_i \right\|$$

we have

$$a = \sup_n \sup \left(\sum_{i=1}^{\infty} a_n |x^{**}(f_i)|; x^{**} \in X^{**} \right)$$

which gives that

$$\sup \left(\sum_{i=1}^{\infty} |x^{**}(f_i)| \|x_i - y_i\|; \|x^{**}\| \leqslant 1 \right) < 1.$$

Then the operator

$$x \to Tx = \sum_{i=1}^{\infty} f_i(x)(x_i - y_i)$$

is bounded and

$$\|T\| \leqslant a.$$

Thus T is a contraction operator and $I - T$ is an isomorphism. Then clearly the assertion follows.

Suppose now that (M–R) holds. Then

$$\sup_\sigma \left\| \sum_{i \in \sigma} x_i - y_i \right\| < \infty$$

and thus $\Sigma(x_i - y_i)$ is w.u.c. Then

$$\sup_{\|x^*\| \leqslant 1} \left\| \sum_{i=1}^{\infty} x^*(x_i - y_i) f_i \right\| \leqslant 4K \sup \left\| \sum_{i \in \sigma} x_i - y_i \right\| < 1$$

which implies that

$$x^* \to Tx^* = \sum x^*(x_i - y_i) f_i$$

defines an operator which is a contraction. The operator T is clearly a contraction operator and thus $I - T$ is an isomorphism which then gives the assertion.

We note that in the above proof we used certain inequalities from McArthur and Retherford (1969).

Finally, suppose that (V) holds. Then $\Sigma(x_i - y_i)$ being unconditionally convergent and $\sup \|f_i\| < \infty$, the mapping

$$x \to Tx = \sum_{i=1}^{\infty} f_i(x)(x_i - y_i)$$

is compact.

The w-linear independence of (y_n) gives that -1 is not an eigenvalue for T and thus $I - T$ is an isomorphism. This implies the assertion.

For the relationship between conditions (A) to (V) considered above we refer to Retherford–Holub's paper (1971).

We now present some results for the case when the space is a complete inner product space. As a corollary of the above results we have K. N. Bari's classical theorem:

THEOREM 3.6.32 (Bari, 1944). *Suppose that (e_i) is an orthonormal basis of the complete inner product space X and let (y_n) be a w-linearly independent sequence which is quadratically near to (e_i) such that*

$$\sum_{i=1}^{\infty} \|e_i - y_i\|^2 < 1.$$

Then (y_n) is a Bari basis for X.

Proof. We set

$$x \to Tx = \sum \langle x,(e_i - y_i)\rangle e_i$$

and then clearly $\|T\| < 1$.

The assertion follows since -1 is not an eigenvalue for T^*. Then $I - T^*$ is an isomorphism and this gives that (y_n) is a Bari basis for X.

THEOREM 3.6.33 (Retherford and Holub, 1971). *Suppose that (e_i) is an orthonormal basis of X and (y_n) is a sequence of elements of X satisfying the following property: if (h_n) is a sequence in X converging weakly to zero then*

$$\sum_{i=1}^{\infty} |\langle h_n, e_i - y_i\rangle|^2 \to 0 \quad \text{as } n \to \infty.$$

Then (y_n) is a basis for X.

Proof. We set

$$x \to Tx = \sum_i \langle x, e_i - y_i\rangle e_i$$

which is clearly a compact operator. Since T^* is given by the formula

$$T^*x = \sum_{i=1}^{\infty} \langle x, e_i\rangle(e_i - y_i)$$

we obtain that -1 is not an eigenvalue for T^*. Thus $I - T^*$ is an isomorphism and the assertion of the theorem follows.

COROLLARY 3.6.34 (Bari, 1944). *Suppose that (e_i) is an orthonormal basis for X and (y_n) is a sequence which is w-linearly independent and quadratically near to (e_i). Then (y_n) is a Bari basis for X.*

Proof. Let $\varepsilon > 0$ and since the sequence (h_n) has the stated property we find an integer N with the property that

$$\sum_{i=N+1}^{\infty} \|(e_i - y_i)\|^2 < \varepsilon^2/4$$

and, for $n \geqslant M$,

$$|\langle f_n, e_i - y_i \rangle| < (\varepsilon/2N)^{1/2}, \quad i = 1,2,3,\ldots,N.$$

Then for any $n \geqslant N$ we have

$$\sum_{i=1}^{\infty} |\langle h_n, e_i - y_i \rangle|^2 < \varepsilon$$

and thus we may apply Theorem 3.6.33. This gives then the assertion of the corollary. □

We refer to Retherford–Holub's paper (1971) for the relationship between the conditions considered above as well as for other results concerning the perturbation of bases in complete inner product spaces.

Concerning unconditional bases we mention the very interesting results of Bari (1951) and Gelfand (1951) which assert that in L^2 each orthogonal basis is equivalent to an orthogonal basis. Thus if (b_n) is an unconditional basis of L^2 with the property

$$0 < \inf_n \|b_n\| \leqslant \sup_n \|b_n\| < \infty \tag{*}$$

then it is equivalent with the unit vector basis in l^2.

Pelczyński (1960) has proved the following important characterization of L^2 (respectively l^2) among the spaces $L^p(p > 1)$ $(l^p, p > 1)$ as a space with all unconditional bases satisfying (*) equivalent. We refer to A. Pelczyński's paper for a proof as well as for other related information.

3.7. Some Classes of Bases (Hardy Bases) and the Theory of Communication

The functions of the type

$$x \rightarrow \sum_{r=-\infty}^{\infty} f(a + rw) \sin((x - a - rw)/w)/((x - a - rw)/w)$$

were named by Whittaker (1915) 'cardinal functions'. This class of functions was considered in many papers from the point of view of interpolation theory. An interesting and basic observation was made by Hardy (1941); namely, that if we set

$$v_r(x) = \begin{cases} (\sin(x - r))/((x - r)) & x \neq r \\ 1 & x = r \end{cases}$$

then the system $(v_r(x))_{-\infty}^{\infty}$ is an orthonormal system of functions in $L^2 = L^2_{((-\infty,\infty),dx)}$. Further, Hardy remarked that this system is not complete in L^2 but in another subspace of L^2, namely in the space of all functions in L^2 whose Fourier transform is in L^2 and vanish (a.e.) outside the interval $(-\pi, \pi)$ is a complete system.

Following the interesting paper of Weston (1949), we now consider some subspaces of L^2 and further we indicate the connection with the communication theory.

We set

$$H'_2 = (f, f \in L^2, f \text{ vanish (a.e.) outside the interval } (-\pi, \pi)).$$

Obviously H_1 is isomorphic to the Hardy space H^2 corresponding to the space $L^2(-\pi, \pi)$.

Taking the Fourier transform of elements in H'_2, we obtain another space denoted by H_3.

The basic results of Hardy can be formulated by saying that the set of elements in L^2 whose Fourier transform vanishes (a.e.) outside the interval $(-\pi, \pi)$ is a complete inner product space having the system (v_r) as an orthonormal basis.

Thus if f is in H_3 then we can define (with respect the orthonormal basis (v_r)) the Fourier coefficients

$$a_r(f) = \int_{-\infty}^{\infty} f(x) v_r(x) \,\mathrm{d}x$$

and then

$$\|f\|^2 = \sum_{-\infty}^{\infty} |a_r|^2.$$

An interesting study of these spaces is contained in Weston's paper (1949).

We now present some of Weston's results.

THEOREM 3.7.1. *The mapping defined on L^2 by the formula*

$$(U_a)(f)(x) = |a|^{1/2} f(ax)$$

is a unitary operator; H_3 is an invariant subspace for U_a if $|a| \leqslant 1$.

THEOREM 3.7.2 (Weston, 1949). *The operator* id/dx *defined on H_3 is a bounded hermitian operator (whose matrix with respect to the orthonormal basis (v_r) of H_3 is*

$$a_{nm} = \begin{cases} (-1)^{n-m}, & m \neq n \\ 0, & m = n. \end{cases}$$

THEOREM 3.7.3 (Weston, 1949). *Let f be an element of H_3. A necessary and sufficient condition that f be orthogonal to all integral translations (of f) is that the modulus of its Fourier transform be constant (a.e.) in $(-\pi, \pi)$; in that case $\{f, \{f_r(,)\}\}$, f_r is the translate of f relative to the integer r), is a complete orthonormal system of H_3.*

An interesting application of these spaces was indicated by Weston (1949) to the theory of communication. The communication through a single channel system of a message consists essentially of the following facts:

(1) A physical magnitude is caused to vary in time, the variations representing messages (with respect to some code).
(2) The variations of the physical magnitude are transmitted to another place.
(3) The variations of the physical magnitude are received and interpreted.

The transmission system is a physical apparatus which realizes (2). The messages are represented by real-valued functions or a real-variable (generally the time) and it is assumed that f is such that the energy which enters the system is proportional to $|f(t)|^2$ and thus the total energy is

$$\int_{-\infty}^{\infty} |f(t)|^2 \, dt.$$

We consider the message as a point in L^2. Then we take the Fourier transform and get

$$f(t) = \int_{-\infty}^{\infty} F(u) e^{iut} \, du$$

with F in L^2.

It is also clear that the transmission system transmits f with distortion and an ideal transmission system is defined by the properties:

(1) $e^{iut} \to \begin{cases} e^{iu(t-\tau)} \\ 0 \end{cases},$

(2) a transformation which is linear,

for u in some interval or not.

We may assume that the interval is $[0, \pi]$. In this case the ideal system has the transformation mentioned above, the projection of L^2 onto H_3. The fact that the transmission may be without distortion means that the message is in H_3. In this case, it is a sum of cardinal series. For more information we refer to J. D. Weston's article (1949) and Higgins (1985).

CHAPTER 4

Metric Characterizations of Inner Product Spaces

4.0. Introduction

The problem of the kind of metrics on certain sets has attracted the attention of many mathematicians. Perhaps the oldest characterizations of metrics on certain sets of the projective space P^n is the famous result of the great Italian mathematician E. Beltrami, which is as follows.

THEOREM 4.0.1. (Beltrami, 1865; see Busemann, 1955). *Suppose that M is an open connected set in the projective space P^n and that the metric of M is a Riemannian metric with the property that the geodesics lie on projective lines. Then the metric of M is either Euclidean, elliptic or hyperbolic.*

For an excellent presentation of Beltrami's theorem we refer to the book by H. Busemann in 1955.

The first metric characterization of inner product spaces in the class of normed linear spaces was given by the great French mathematician Maurice Fréchet (1934) who proved that if X is a normed linear space with the norm

$$x \to \|x\|$$

and such that for any x, y, z in X,

$$\|x + y + z\|^2 + \|x\|^2 + \|y\|^2 + \|z\|^2 - \|x + y\|^2 - \|y + z\|^2 - \|x + z\|^2 = 0 \quad (+)$$

then there exists an inner product $\langle , \rangle$ on X such that for all x in X

$$\|x\|^2 = \langle x, x \rangle.$$

Jordan and von Neumann (1935) have remarked that if in the identity $(+)$ we take $y = -z$ then we obtain the identity

$$\|x + y\|^2 + \|x - y\|^2 = 2(\|x\|^2 + \|y\|^2) \qquad (++)$$

and they proved that this inequality, known as the 'parallelogram law', is as strong as the Fréchet identity.

The identity

$$\|x + y\|^2 + \|x - y\|^2 = 2(\|x\|^2 + \|y\|^2)$$

first appeared in the paper of Jordan *et al.* (1934); the importance of this identity in the theory of complete inner product spaces (or von Neumann spaces) was first recognized by the great mathematician Fr. Riesz in his 1934 paper. We note also that this identity in Euclidean geometry is exactly the well-known theorem of

Apollonius which asserts that the sum of the square of the lengths of the two diagonals of a parallelogram is equal to the sum of the squares of the lengths of the four sides.*

The problem of ascertaining the properties of a norm in a normed linear spaces – or more generally, the properties of the distance in metric spaces, which are necessary and sufficient to define an inner product, or the addition, multiplication and an inner product, respectively – has attracted the attention of many people, among them Aronszajn (1935), Belobrov (1968), Blumenthal (1953, 1955, 1968), Carlsson (1962, 1964), Ficken (1944), Freese (1968), Goƚab (1975), Goryachev (1970), Gurariĭ (1970), James (1945, 1947), Joichi (1966, 1968), Kasahara (1954), Krovonosov (1972), Kwapień (1972), Lorch (1948), Ohira (1952), Oman (1969, 1975, 1980), Lindenstrauss (1963), Nagumo (1936), Rubin (1953), Schoenberg (1952), Senechalle (1971, 1973), Schaeffer (1965), Sozonov (1970), Stamey (1957), and Yang Cong-ren (=Yang Ts'ung-sen).

In what follows we present some characterizations of inner product spaces in terms of addition, multiplication and the norm and, to some extent, characterizations of inner product spaces using only the distance concept. We note also that, as a consequence of some characterizations we give, it follows that the class of complete inner product spaces is the most appropriate setting for some operator theory (both linear and nonlinear), approximation theory, etc.

4.1. Inner Product Structures on Linear Spaces

Suppose that X is a linear space over K (R or C).

DEFINITION 4.1.1. An inner product $\langle , \rangle$ on X is a mapping on $X \times X$ with values in K such that the following conditions are satisfied:

(1) $\langle ax + by, z \rangle = a\langle x, z \rangle + b\langle y, z \rangle$,
(2) $\langle x, y \rangle = \overline{\langle y, x \rangle}$,
(3) $\langle x, x \rangle \geqslant 0, \quad x \in X$,
(4) $\langle x, x \rangle = 0$ iff $x = 0$.

If only conditions (1)–(3) are satisfied then we call $\langle , \rangle$ a pre-inner product on X.

We now give some simple properties of pre-inner products and of inner products on a linear space.

DEFINITION 4.1.2. The pair $(X, \langle , \rangle)$, where X is a linear space and $\langle , \rangle$ is an inner product, is said to be an inner product structure; if $\langle , \rangle$ is only a pre-inner product then we call it a pre-inner product structure.

PROPOSITION 4.1.3. *Let* $(X, \langle , \rangle)$ *be a pre-inner product structure. Then the function on* X *defined by the relation* $x \to \langle x, x \rangle^{1/2}$ *is a seminorm on* X.

Proof. Since the proof does not differ essentially from the proof given in 2.6.4 we omit it. □

* We shall shortly present another identity, also called the 'Apollonius identity', and another result which generalizes a theorem of Apollonius from Euclidean geometry (see 4.6.44 and 4.8).

COROLLARY 4.1.4. *If* $(X, \langle\,,\rangle)$ *is an inner product structure then* $x \to \langle x, x\rangle^{1/2}$ *is a norm on* X.

We now present some identities which have their own interest; the proof, being simple, is omitted.

PROPOSITION 4.1.5. *Let* $(X, \langle\,,\rangle)$ *be a pre-inner product structure and let* $p(x) = \langle x, x\rangle^{1/2}$. *Then the following relations hold:*

(1) $p(x+y)^2 + p(x-y)^2 = 2(p(x)^2 + p(y)^2)$ (*the parallelogram law for p (or the Apollonius theorem for p)*),
(2) $4\langle x, y\rangle = p(x+y)^2 + p(x-y)^2 + ip(x+iy)^2 - ip(x-iy)^2$.
(3) If $x_1, \ldots, x_n \in X$, $\bar{x} = \frac{1}{n}(x_1 + \cdots + x_n)$,

$$\sum_{i=1}^{n} p(x_i - \bar{x})^2 = \sum_{i=1}^{n} p(x_i)^2 - np(\bar{x})^2 \text{ (Apollonius' identity).}$$

4.2. Inner Product Structures and Complexification

Suppose now that X is a complex Banach space. Then we can associate a real Banach space denoted by X_r as follows: the elements of X_r are the elements of X and the addition and multiplication with real scalars is defined exactly as in X. The norm is exactly the norm for each element as an element of X. Then clearly X_r is a real Banach space and it appears of interest to know the connection between the inner product structures associated with the spaces X and X_r. As the reader may expect, we have the following folklore theorem.

THEOREM 4.2.1. *A Banach space* X *admits an inner product structure* $(X, \langle\,,\rangle)$ *with* $\|x\| = \langle x, x\rangle^{1/2}$ *for each* x *in* X *iff* X_r *admits an inner product structure* $(X_r, \langle\,,\rangle_r)$ *with* $\|x\| = \langle x, x\rangle$ *for each* x *in* X_r.

Proof. Suppose that X admits an inner product structure $(X, \langle\,,\rangle)$ with the above property. Then we can define an inner product on X_r by

$$\langle x, y\rangle_r = \operatorname{Re}\langle x, y\rangle. \qquad \square$$

It is obvious that this is indeed an inner product on X_r. Conversely, if $\langle\,,\rangle_r$ is such that $(X_r, \langle\,,\rangle_r)$ is an inner product structure with $\|x\| = \langle x, x\rangle^{1/2}$ then we can define an inner product on X by

$$\langle x, y\rangle = \langle x, y\rangle_r - i\langle ix, y\rangle_r.$$

Since

$$2\|x\|^2 + 2\langle ix, x\rangle_r = \langle x + ix, x + ix\rangle_r = 2\|x\|^2$$

we get that

$$\langle ix, x\rangle = 0$$

for all x in X. Thus

$$\langle x, x\rangle = \|x\|^2$$

and the theorem is proved since the other properties of the inner product are easy to justify.

REMARK 4.2.2. The result in Theorem 4.2.1 can be given, of course, for the case of seminorms; we leave the details to the reader.

In this book, taking the strong connection between the spaces X and X_r, we shall use the phrase 'we consider X to be a real Banach space' or 'X is a complex Banach space with its norm generated by an inner product iff X is a real Banach space with its norm generated by an inner product' and, of course, we mean the space X_r instead of X.

4.3. The Fréchet and Jordan–von Neumann Characterization of Inner Product Spaces

Let X be a normed linear space over K and consider on $X_{\times}X\setminus(0,0)$ the function

$$r(x,y)=\tfrac{1}{2}(\|x+y\|^2+\|x-y\|^2)/(\|x\|^2+\|y\|^2),$$

where $x\to\|x\|$ is the norm of x. Associated with the function r we define two numbers, m and M, as follows:

$$m=\inf(r(x,y),(x,y)\in X_{\times}X\setminus(0,0)),$$
$$M=\sup(r(x,y),(x,y)\in X_{\times}X\setminus(0,0)).$$

Obviously we have the following result.

LEMMA 4.3.1. *The following inequality holds,*

$$0\leqslant m\leqslant r(x,y)\leqslant M.$$

The following lemma gives more information about r and M,m.

LEMMA 4.3.2 (Jordan and von Neumann, 1935). *The following inequality holds,*

$$\tfrac{1}{2}\leqslant m\leqslant r(x,y)\leqslant M\leqslant 2.$$

The numbers m and M satisfy the relation $mM=1$.

Proof. First we remark that $r(x,y)>0$. Indeed, if for some (x,y) in $X_{\times}X\setminus(0,0)$ we have $r(x,y)=0$ then $\|x+y\|^2+\|x-y\|^2=0$. From these we obtain that $x=y$ and $x=-y$, which give that $x=y=0$. Thus we must have $r(x,y)>0$. But

$$r(x,y)=1/r(x+y,x-y)$$

and this implies obviously that $mM=1$.

Now

$$\begin{aligned}r(x,y)&=(\|x+y\|^2+\|x-y\|^2)/(2\|x\|^2+2\|y\|^2)\\&\leqslant(2\|x\|^2+2\|y\|^2)/(\|x\|^2+\|y\|^2)\leqslant 2\end{aligned}$$

and thus $M\leqslant 2$. From $mM=1$ we obtain that $m\geqslant\frac{1}{2}$. This completes the proof of the lemma. □

LEMMA 4.3.3. *If X is an inner product space then $r(x,y) = 1$ for all (x,y) in $X_\times X\setminus(0,0)$.*

Proof. Obvious. □

LEMMA 4.3.4 (Fréchet, 1935). *If X is an inner product space then for all x, y, z, in X the following relation holds:*

$$\|x+y+z\|^2 + \|x\|^2 + \|y\|^2 + \|z\|^2 - \|x+y\|^2 - \|x+z\|^2 - \|y+z\|^2 = 0.$$

Proof. Computing the norms using the inner product we easily obtain the relation. □

We are now ready to prove the converse of the results in Lemmas 4.3.4 and 4.3.3.

THEOREM 4.3.5 (Fréchet, 1935). *If X is a normed linear space and for any x, y, z in X*

$$\|x+y+x\|^2 + \|x\|^2 + \|y\|^2 + \|z\|^2 - \|x+y\|^2 - \|x+z\|^2 - \|y+z\|^2 = 0$$

then on X there exists an inner product $\langle\,,\rangle$ such that the norm generated by $\langle\,,\rangle$ coincides with the norm $\|\,,\|$.

THEOREM 4.3.6 (Jordan and von Neumann, 1935). *If X is a normed linear space such that for all x, y in X,*

$$\|x+y\|^2 + \|x-y\|^2 = 2(\|x\|^2 + \|y\|^2)$$

holds, then on X there exists an inner product $\langle\,,\rangle$ such that the norm generated by $\langle\,,\rangle$ is exactly the norm $\|\,,\|$.

Since Fréchet's condition reduces to that of Jordan–von Neumann if we set $z = -y$, it is clear that to prove Theorems 4.3.5 and 4.3.6 it suffices to prove Theorem 4.3.6.

In what follows we present two proofs of this theorem. The first constructs explicitly the inner product and the second is based on an interesting lemma of M. Fréchet about the characterization of polynomials in one variable.

The first proof of Theorem 4.3.6. For each x, y in X we set

$$\langle x,y\rangle = \tfrac{1}{4}(\|x+y\|^2 - \|x-y\|^2 + i\|x+iy\|^2 - i\|x-iy\|^2)$$

and we show that this is an inner product on X.

First we remark that

$$\langle x,y\rangle = \mathrm{Re}\langle x,y\rangle - i\,\mathrm{Re}\langle ix,y\rangle$$

and that

$$\mathrm{Re}\langle x,y\rangle = \tfrac{1}{4}(\|x+y\|^2 - \|x-y\|^2).$$

Let us consider x_1, x_2 and y in X. Then we have

$$\begin{aligned}
&\mathrm{Re}\langle (x_1+x_2),y\rangle + \mathrm{Re}\langle x_1-x_2,y\rangle\\
&\quad = \tfrac{1}{4}(\|x_1+x_2+y\|^2 - \|x_1+x_2-y\|^2 +\\
&\qquad + \|x_1-x_2+y\|^2 - \|x_1-x_2-y\|^2)\\
&\quad = \tfrac{1}{4}(2\|x_1+y\|^2 + 2\|x_2\|^2 - 2\|x_1-y\|^2 - 2\|x_2\|^2)\\
&\quad = \tfrac{1}{2}(\|x_1+y\|^2 - \|x_1-y\|^2) = 2\,\mathrm{Re}\langle x_1,y\rangle.
\end{aligned}$$

Since x_1, x_2 and y are arbitrary elements of X, taking $x_1 = x_2 = x$ we get the relation

$$\mathrm{Re}\langle 2x, y\rangle = 2\mathrm{Re}\langle x, y\rangle.$$

Changing now x_1 and x_2 in $\frac{1}{2}(x_1 + x_2)$ and $\frac{1}{2}(x_1 - x_2)$, respectively, we obtain the following relation

$$\mathrm{Re}\langle x_1, y\rangle + \mathrm{Re}\langle x_2, y\rangle = \mathrm{Re}\langle x_1 + x_2, y\rangle.$$

Changing now x_1 and x_2 in ix_1 and ix_2, respectively, we obtain the relation

$$\mathrm{Im}\langle x_1, y\rangle + \mathrm{Im}\langle x_2, y\rangle = \mathrm{Im}\langle x_1 + x_2, y\rangle.$$

From these relations we obtain that for all x_1, x_2 and y in X,

$$\langle x_1 + x_2, y\rangle = \langle x_1, y\rangle + \langle x_2, y\rangle$$

holds.

This obviously implies that for all integers m, n and all x, y in X the following relation holds:

$$\langle m/n\cdot x, y\rangle = m/n\cdot\langle x, y\rangle.$$

Indeed, clearly we have the relation

$$\langle m\cdot x, y\rangle = m\cdot\langle x, y\rangle$$

and applying this as follows,

$$\langle m/n\cdot x, y\rangle = \langle m\cdot(l/n\cdot x), y\rangle = m\cdot l/n\cdot\langle x, y\rangle$$

the assertion follows.

Thus for all rational numbers r and all x, y in X we have

$$\langle r\cdot x, y\rangle = r\langle x, y\rangle.$$

and this implies that for all real numbers a we have

$$\langle ax, y\rangle = a\cdot\langle x, y\rangle.$$

Since

$$\begin{aligned}\langle ix, y\rangle &= \tfrac{1}{4}(\|ix + y\|^2 - \|ix - y\|^2 + i\|ix + iy\|^2 - i\|ix - iy\|^2)\\ &= \tfrac{1}{4}(\|ix + y\|^2 - \|ix - y\|^2 + i\|x + y\|^2 - i\|x - y\|^2) = i\langle x, y\rangle\end{aligned}$$

using the above result we obtain that for all complex numbers a we have

$$\langle a\cdot x, y\rangle = a\langle x, y\rangle.$$

But

$$\begin{aligned}\overline{\langle y, x\rangle} &= \overline{\tfrac{1}{4}(\|x + y\|^2 - \|x - y\|^2 + i\|x + iy\|^2 - i\|x - iy\|^2)}\\ &= \tfrac{1}{4}(\|x + y\|^2 - \|x - y\|^2 - i\|ix + y\|^2 + i\|(y - ix)\|^2)\\ &= \tfrac{1}{4}(\|x + y\|^2 - \|x - y\|^2 - i\|ix + y\|^2 + i\|iy + x\|^2) = \langle x, y\rangle\end{aligned}$$

and thus $(x, y) \to \langle x, y \rangle$ satisfies all properties of the inner product. Since

$$\langle x, x \rangle = \|x\|^2$$

The first proof of Theorem 4.3.6 is complete.

The second proof of Theorem 4.3.6. In this proof we follow the presentation by Reznick (1978). First we need the following result about a characterization of polynomials among the continuous functions on R.

LEMMA 4.3.7 (Fréchet, 1909). *Let f be a continuous real-valued function on R. Suppose that for each r and s the following relation holds:*

$$\sum_{k=0}^{N} (-1)^{N-k}\binom{N}{k} f(r + ks) = 0. \qquad (*)$$

Then f is a polynomial of degree less than N.

Proof. We use the well-known fact that if (a_n) is a sequence of real numbers with the property that for all m the following relation holds:

$$\sum_{k=0}^{N} (-1)^{N-k}\binom{N}{k} a_{k+m} = 0$$

then there exists a polynomial $p(x)$ of degree less than N such that $p(n) = a_n$. □

If in the relation (*) we set $a_n = f(n)$, $s = 1$ and r is an arbitrary integer, we have that there exists a polynomial p (of degree less than or equal to N) such that $p(n) = a_n = f(n)$. We apply once again that $a_n = f(n/2)$, $s = \frac{1}{2}$ and r is an arbitrary integer. Then we obtain that there exists a polynomial (of degree less than or equal to N) such that $q(n) = f(n/2)$. This implies that the polynomials p and q satisfy the relation

$$p(n) = q(2n); q(x) = p(x/2)$$

for all n. Now we repeat this argument and get that for all n and m we have the relation

$$p(n \cdot 2^{-m}) = f(n \cdot 2^{-m}).$$

Since $(n \cdot 2^{-m})$ is a dense subset in R, the continuity of f implies that for all x in R, $f(x) = p(x)$ and the lemma is proved.

We are now in a position to give the second proof of Theorem 4.3.6. Since in X the 'parallelogram law' holds,

$$\|x + y\|^2 + \|x - y\|^2 = 2(\|x\|^2 + \|y\|^2)$$

we take $x = u + 2v$ and $y = v$. Then we have,

$$\|u + 3u\|^2 + \|u + v\|^2 = 2(\|u + 2v\|^2 + \|v\|^2)$$

and further, for $x = u + v$, $y = v$, this identity implies that

$$\|u + 2v\|^2 + \|u\|^2 = 2(\|u + v\|^2 + \|v\|^2)$$

and subtracting,

$$\sum_{k=0}^{3} (-1)^k \binom{3}{k} \|u + kv\|^2 = 0. \tag{**}$$

Since in this equality u and v are arbitrary elements of X, we can take $u = x + ry$ and $v = sy$ and (**) implies that

$$\sum_{k=0}^{3} (-1)^k \binom{3}{k} (x + (ry + ksy))^2 = 0. \tag{***}$$

If we consider on R the function defined by the formula

$$t \to \|x + ty\|^2 = f(t)$$

we have obviously that this is continuous and (***) shows that

$$\sum_{k=0}^{3} (-1)^k \binom{3}{k} f(r + ks) = 0.$$

From the lemma we have that f is a polynomial of degree 2, i.e.

$$f(t) = a(x, y) + tBb(x, y) + t^2 c(x, y).$$

We define on $X_\times X$ the function $(x, y) \to b(x, y)$ and we remark that this is a real-valued inner product on X. Indeed, if $x = x_1 + x_2$ then we have

$$\begin{aligned}
&\|x_1 + x_2 + ty\|^2 - \|x_1 + x_2 - ty\|^2 \\
&\quad = 2(\|x_1 + t/2 \cdot y\|^2 + \|x_2 + t/2 \cdot y\|^2 - \\
&\qquad - 2(\|x_1 - t/2 \cdot y\|^2 + \|x_2 - t/2 \cdot y\|^2) \\
&\quad = 2(a(x_1, y/2) + tb(x_1, y/2) + t^2 c(x_1, y/2)) + \\
&\qquad + 2(a(x_2, y/2) + tb(x_2, y/2) + t^2 c(x_2, y/2)) - \\
&\qquad - 2(a(x_1, y/2) - tb(x_1, y/2) + t^2 c(x_1, y/2)) - \\
&\qquad - 2(a(x_2, y/2) - tb(x_2, y/2) + t^2 c(x_2, y/2)) \\
&\quad = 4t(b(x_1, y/2) + b(x_2, y/2))
\end{aligned}$$

and thus

$$2b(x_1 + x_2, y) = 4(b(x_1, y/2) + b(x_2, y/2)).$$

Since

$$b(x, ay) = ab(x, y)$$

if $a > 0$ we obtain that

$$b(x_1 + x_2, y) = b(x_1, y) + b(x_2, y).$$

A similar computation shows that for all x, y in X,

$$b(x, y) = b(y, x)$$

and also

$$b(x, ay) = ab(x, y)$$

for $a < 0$ which implies that $b(x, y)$ defines a real inner product on X. Also it is clear that $b(x, x) = \|x\|^2$. If X is a linear space over $K = C$, the complex numbers, then a complex-valued inner product is defined by the formula

$$\langle x, y \rangle = \tfrac{1}{2}(b(x, y) + ib(x, iy)).$$

This completes the second proof of Theorem 4.3.6.

We leave the reader to prove that the Fréchet and Jordan–von Neumann characterizations may be generalized to the case where $\| = \|$ is replaced by $\| \leqslant \|$ or $\| \geqslant \|$; e.g. X has the norm generated by an inner product for all $x, y \in X$,

$$\|x + y\|^2 + \|x - y\|^2 \leqslant 2(\|x\|^2 + \|y\|^2).$$

See also Sections 4.12 and 4.13.

Interesting facts about the Apollonius theorem (the parallelogram law) and the inner product structures were stated by Fearnley-Sander and Symons (1974) and are related to some interesting remarks about quadratic forms contained in a paper by Gleason (1966). In what follows we present some of these statements.

Let us consider a linear space over a field K (where K is a field of arbitrary nature but with characteristic not equal to 2). Then, as is well known from linear algebra, a quadratic form Q on the linear space L over K is any function $Q: L \to K$ of the form

$$Q(x) = \sum_{\alpha, \beta \in I} k_{\alpha\beta} f_\alpha(x) f_\beta(x),$$

where $(f_\alpha)_{\alpha \in I}$ is a family of linear functionals dual to some basis $(x_\alpha)_{\alpha \in I}$ of L and $(K_{\alpha\beta})_{\alpha,\beta \in I}$ is a family of constants (in K). The right-hand side of the above equality is supposed to be finite for each x in L. Another way to define a quadratic form is using bilinear forms, i.e.

$$Q(x) = B(x, x),$$

where B is a bilinear form over L.

Since any quadratic form Q defines through the formula

$$S(x, y) = \tfrac{1}{4}(Q(x + y) - Q(x - y))$$

a symmetric bilinear form we see that there is a one-to-one correspondence between the set of all quadratic forms and the set of all bilinear forms.

If Q is a quadratic form then clearly we have the identity

$$Q(x + y) + Q(x - y) = 2(Q(x) + Q(y)) \quad \text{(the parallelogram law for } Q\text{)}.$$

Then the Jordan–von Neumann theorem for quadratic forms may be stated as follows.

THEOREM 4.3.8 (Gleason, 1966). *Suppose that the function Q on L with values in K satisfies the following identity:*

$$Q(x + y) + Q(x - y) = 2(Q(x) + Q(y)). \tag{+}$$

Then the mapping $S: L_\times L \to K$ defined by the relation

$$S(x, y) = \tfrac{1}{4}(Q(x + y) - Q(x - y))$$

is symmetric and additive (in each argument) and

$$S(x, x) = Q(x)$$

holds for all x in L.

Proof. For $x = y = 0$ we get from $(+)$ that $S(0) = 0$ and for $x = 0$ we obtain that

$$Q(y) = Q(-y).$$

This gives that S is a symmetric function. If $x = y$ then

$$Q(2x) = 4Q(x) \quad \text{and} \quad S(x, x) = Q(x).$$

As Q is symmetric, in order to prove that it is bilinear it suffices to prove that it is linear in the first argument. Suppose now that x, y, z are elements in L, then

$$\begin{aligned} 8S(x, z) + 8S(y, z) &= 2Q(x + z) - 2Q(x - z) - 2Q(y - z) + 2Q(y + z) \\ &= Q(x + y + 2z) + Q(x - y) - Q(x + y - 2z) - Q(x - y) \\ &= 4S(x + y, 2z) \end{aligned}$$

which implies that

$$2(S(x, z) + S(y, z)) = S(x + y, z). \qquad (+\,+)$$

We take $y = 0$ in $(+\,+)$ and thus

$$2S(x, z) = S(x, 2z)$$

since

$$S(0, z) = S(z, 0) = 0.$$

Thus $(+\,+)$ is in fact equivalent to

$$S(x, z) + S(y, z) = S(x + y, z)$$

and thus S is an additive function in the first argument. The assertion is proved. □

COROLLARY 4.3.9 (Gleason, 1966). *Let L be a linear space over K of dimension at least 2. Suppose that $Q: L \to K$ is a function with the property that for any two-dimensional subspace V of L, Q/V (the restriction of Q to V) is a quadratic form. Then Q is a quadratic form on L.*

In his interesting note, Gleason (1966) discusses some identities which are related to characterizations of quadratic forms. He gives an example to show that for a function $Q: L \to K$ satisfying the parallelogram law (i.e. the Apollonius theorem) and the relation

$$Q(ax) = a^2 Q(x), \quad a \in K, \quad x \in L$$

does not suffice to assure that Q is a quadratic form.

THEOREM 4.3.10 (Gleason, 1966). *Let $Q: L \to K$, K is assumed to be a field of characteristic not equal to 2 and having more than three elements. Suppose that Q*

satisfies the relation:

$$Q(x + ky) - Q(x - ky) = k(Q(x + y) - Q(x - y)). \tag{G$_1$}$$

Then Q is a quadratic form.

THEOREM 4.3.11 (Gleason, 1966). *Let $Q : L \to K$ where K is as in Theorem* 4.3.10. *Suppose that Q satisfies the relation:*

(1) $Q(0) = 0$; $\qquad$ (G)$_2$

(2) $Q(ky + x) - Q(ky - x) = k(Q(y + x) - Q(y - x))$. $\qquad$ (G)$_2$

Then Q is a quadratic form on L.

For a proof of these theorems we refer to Gleason's paper (1966) where the reader may find other interesting examples of quadratic forms.

We now prove a slight version of the Jordan–von Neumann theorem obtained by D. Fearnley-Sander and J. S. V. Symons, which essentially removes the condition of the triangle inequality in the conditions of the Jordan–von Newmann theorem.

THEOREM4.3.12 (Fearnley-Sander and Symons, 1974). *Let L be a linear space over K (R or C) and suppose that we have a function $\|,\|: L \to R^+$ such that the following relations hold:*

(1) $\|x\| = 0$ *iff* $x = 0$, $\|ax\| = |a|\|x\|$ *a in K.*

(2) *For any x, y in L the Apollonius theorem holds, i.e.*
$\|x + y\|^2 + \|x - y\|^2 = 2(\|x\|^2 + \|y\|^2)$.

Then there exists on L an inner product $\langle\, ,\rangle$ such that $\langle x, x\rangle = \|x\|^2$ holds for all x in L.

Proof. We may suppose that $K = R$, the extension to the complex case we leave to the reader. Then we define for any x, y in L

$$\langle x, y\rangle = \tfrac{1}{2}(\|x + y\|^2 - \|x\|^2 - \|y\|^2).$$

Then for all integers n we have

$$\langle nx, y\rangle = n\langle x, y\rangle$$

and from the Apollonius theorem we get that for all rational numbers r we have

$$\langle rx, y\rangle = r\langle x, y\rangle.$$

Since if

$$p(t) = \|x\|^2 t^2 + 2t\langle x, y\rangle + \|y\|^2$$

for t rational is $\|tx + y\|^2$, we get that p is positive for all t real. This obviously implies the Cauchy–Buniakowsky inequality

$$|\langle x, y\rangle| \leqslant \|x\|\,\|y\|.$$

Now for any real number t and any rational number r we have

$$|\langle tx, y\rangle - t\langle x, y\rangle| \leqslant |\langle (t - r)x, y\rangle| + |(r - t)|\,|\langle x, y\rangle|$$

and then, using the Cauchy–Buniakowsky inequality, the right-hand side can be made arbitrarily small for r sufficiently near to t. This obviously implies that

$$\langle tx, y\rangle = t\langle x, y\rangle$$

holds for all real numbers t. This clearly proves the theorem. □

REMARK 4.3.13. The conclusion of the theorem remains true if we suppose that the function $\|\,,\|: L \to R^+$ satisfies, instead of the condition $\|tx\| = |t|\,\|x\|$, the condition that $t \to tx$ is a continuous function at $t = 0$ (for each x of L).

As was noted by D. Fearnley-Sander and J. S. V. Symons, the condition cannot be omitted as the following example shows.

We consider R as a linear space over Q (the rationals) and if B is a Hamel basis for R then we may assume that $1, 2^{1/2}$ are in B. We set

$$\|x\|^2 = \sum_{i=1}^{n} |x_i|^2$$

if

$$x = x_1 b_{i_1} + \cdots + x_n b_{i_n}.$$

Obviously this satisfies the Apollonius theorem and $\|x\| \geqslant 0$ for all x in R. Also $\|x\| = 0$ if $x = 0$. But

$$\|2^{1/2}\| = 1 = \|1\|^2$$

and thus $\langle\,,\rangle$ is not homogeneous.

Very interesting and important results about quadratic forms related to inner product structures were obtained by Kurepa (1963, 1964, 1965), some of which are presented below.

As we remarked, if B is a bilinear form on the linear space X over the field C of complex numbers (i.e. B is linear in the first argument and antilinear in the second argument) then the function Q defined on X by

$$Q(x) = B(x, x)$$

has the properties:

(1) $Q(x + y) + Q(x - y) = 2(Q(x) + Q(y))$ (the Apollonius theorem for Q),
(2) $Q(zx) = |z|^2 Q(x)$

for all x, y in X and z in C.

DEFINITION 4.3.14. A function defined on the complex linear space X is said to be a quadratic functional if properties (1) and (2) above are satisfied.

I. Halperin (Kurepa, 1965) has raised the problem: as Q is a quadratic functional, does it have representation through a bilinear form on X? Kurepa's answer to this question is 'Yes' when X is over complex numbers and quaternions and possibly 'No' when X is over the reals, in which case condition (2) must be replaced by the condition

$$Q(rx) = r^2 x,$$

r being a real number.

The result was proved in a simple and ingenious way by Vrbová (1973) and generalized by Vukman (1984). Now the result is as follows.

THEOREM 4.3.15 (Kurepa, 1965). *Let X be a complex linear space and $Q: X \to C$ with the following properties:*

(1) $Q(x+y) + Q(x-y) = 2(Q(x) + Q(y))$,
(2) $Q(zx) = |z|^2 Q(x)$

for all x, y in X and z in C. Then there exists a function B: $X_\times X \to C$ defined by the formula

$$B(x, y) = 4^{-1}(Q(x+y) - Q(x-y) + iQ(x+iy) - iQ(x-iy))$$

which is a bilinear form and

$$Q(x, x) = B(x, x)$$

holds for all x in X.

The proof we give below is that of Vrbová (1973). First we prove a lemma about complex-valued functions on C.

LEMMA 4.3.16 (Vrbová, 1973). *Let $f: C \to C$ be a function satisfying the following properties:*

(1) $f(a+b) f(a) + f(b)$,
(2) $f(a) = -|a|^2 f(a^{-1}), \quad a \neq 0$

for all complex numbers a, b. Then f is of the form

$$f(z) = f(i) \operatorname{Im} z.$$

Proof. We prove first that the restriction of f to the interval $(-1, 1)$ is zero. Indeed, if r is a number in this interval we can choose s another number in this interval such that $r^2 + s^2 = 1$ and then we consider the complex number $z = r + is$. Then clearly

$$f(z) = f(r) + f(is) = -|z|^2 f(\bar{z}) = -f(r) + f(is).$$

Thus $f(r) = 0$. Now if the real number r is with $|r| > 1$ then r^{-1} is in the above interval and

$$f(r) = -r^2 f(r^{-1}) = 0.$$

Let us consider r in $(0, 1)$ and s in R with $r^2 + s^2 = s$ and define $z = r + is$. Then

$$f(z) = f(is) = -|z|^2 f(z^{-1}) = -sf(rs^{-1} - i) = sf(i).$$

If $s < 1$ then

$$f(is) = -s^2 f((is)^{-1}) = sf(i).$$

From the additivity property of f we get that

$$f(z) = -f(-z)$$

and thus

$$f(is) = sf(i)$$

which proves the lemma. □

Now the proof of Theorem 4.3.15 is as follows. We consider the function

$$f(x, y) = Q(x + y) - Q(x - y)$$

and we prove that f is additive in the first argument, i.e.

$$f(u + v, w) = f(u, w) + f(v, w)$$

for all u, v, w in X. Indeed,

$$\begin{aligned} f(u + v, w) &= Q(u + v + w) - Q(u + v - w) \\ &= 2Q(u + w) + 2Q(v) - w(u + w - v) - Q(u + v - w) \\ &= Q(u + w)/Q(v + w) + 2Q(v) - Q(u + w - v) - Q(u + v - w) \\ &= Q(u - w) + (2Q(u) + 2Q(w) - Q(u - w) + \\ &\quad + 2Q(v) - (2Q(u) + 2Q(v - w)) \\ &= f(u, w) + (2Q(w) + 2Q(v)) - 2Q(v - w) \\ &= f(u, w) + (Q(v + w) + Q(v - w) - 2Q(v - w)) = f(u, w) + f(v, w) \end{aligned}$$

and the assertion is proved.

If $u = v = 0$ then property (1) gives that

$$Q(0) = 0$$

and taking $u = -v$ we get the relation $Q(u) = Q(-u)$. From the form of f we conclude that

$$f(x, y) = f(y, x)$$

and f is additive in the second variable. Also for $z \neq 0$, z in C,

$$f(zx, y) = |z|^2 f(x \cdot z^{-1} y).$$

Now we remark that for all x, y in X we have the relations:

$$B(ix, y) = iB(x, y), B(x, iy) = -iB(x, y).$$

In order to obtain the structure of f we prove that

$$B(rx, y) = rB(x, y)$$

for all real numbers r.

We consider x, y arbitrary but fixed elements in X and define a function of a complex variable with values in C by the formula:

$$F(z) = B(zx, y) - B(x, zy).$$

Now for $z \neq 0$ we have,

$$\begin{aligned}4F(z) &= 4B(zx, y) - 4B(x, zy)\\ &= f(zx, y) + if(zx, iy) - f(x, zy) - if(x, izy)\\ &= f(zx, y) + if(zx, iy) - f(zy, x) - if(izy, x)\\ &= |z|^2(f(x, z^{-1}y) + if(x, iz^{-1}y) - f(y, z^{-1}x) - if(iy, z^{-1}x))\\ &= |z|^2(4B(x, z^{-1}y) - f(z^{-1}x, y) + if(z^{-1}x, iy))\\ &= |z^2|(4B(x, z^{-1}y) - 4B(z^{-1}x, y)) = -4|z|^2F(z^{-1})\end{aligned}$$

and thus

$$F(z) = z^2F(z^{-1}).$$

From Lemma 4.3.16 we conclude that

$$F(z) = F(i)\operatorname{Im} z.$$

Then for all real numbers r, $F(r) = 0$ and thus

$$B(rx, y) = B(x, ry).$$

Now if $z = ir$, r is a real number then

$$B(irx, y) - B(x, iry) = F(ir) = rF(i) = r(B(ix, y) - B(x, iy))$$

and thus

$$iB(rx, y) + iB(x, ry) = 2irB(x, y)$$

which implies that

$$2irB(x, y) = 2iB(rx, y)$$

or

$$rB(x, y) = B(rx, y)$$

and the theorem is proved. For other interesting related results we refer to Kurepa (1959, 1964, 1965) and Vukman (1984).

Since the parallelogram law plays a central role in the metric characterizations of inner product spaces in the class of normed spaces it was quite natural to propose other forms of such a law for some classes of normed (Banach) spaces. One of these is as follows.

DEFINITION 4.3.17 (Kay, 1967). Let X be a linear space over K and suppose that $x \to \|x\|$ is a norm on X. We say that the norm satisfies the weak parallelogram law (or that X is a weak parallelogram law space, in short a w.p. space) if for some r in $(0, 1]$ and all x, y in X the following inequality holds:

$$\|x + y\|^2 + r\|x - y\|^2 \leqslant 2(\|x\|^2 + \|y\|^2).$$

Then W. L. Bynum and J. H. Drew proved the following result.

THEOREM 4.3.18 (Bynum and Drew, 1972). *For p in $(1, 2]$ the space l^p of all* *sequences $x = (x_n)$ of all real numbers such that*

$$\|x\|^p = \sum_{n=1}^{\infty} |x_n|^p < \infty$$

is a w.p. space (and the largest possible constant r is $p - 1$).

Interesting characterizations of spaces which are w.p. spaces as well as new classes of spaces satisfying new kinds of parallelogram laws were given by Bynum (1976) Reznick (1978) and Jurgelas (1982). For an earlier result, see D. C. Kay and for connections with problems of geometry of metric spaces, see Kay and Haantjes (1947, 1951).

We have noted that if a space X has an inner product $\langle , \rangle$ and if we set

$$\|x\|^2 = \langle x, x \rangle$$

then the following identity holds:

$$4\langle x, y \rangle = \sum_{k=1}^{4} i^k \|x + i^k y\|^2.$$

Blumenson (1963) has extended this to the following result.

THEOREM 4.3.19 (Blumenson, 1963). *For every integer n there exists an integer N_n and complex numbers a_{jk}, b_k, $1 \leqslant j \leqslant 2n$, $1 \leqslant k \leqslant N_n$, such that for all complex numbers $z_1, \ldots, z_{2n}$ the following relation holds:*

$$z_1 \cdot \cdots \cdot z_n \cdot \bar{z}_{n+1} \cdot \cdots \cdot \bar{z}_{2n} = \sum_{k=1}^{N_n} b_k \left| \sum_{J=1_L}^{2n} a_{jk} z_j \right|^2.$$

This has led him to consider on the space $\Pi_{k=1}^{2n} X_k$, where $X_k = L^{2n}(\Omega, B, \mu)$, $k = 1, 2, \ldots, 2n$, the following function:

$$\langle f_1 \cdot f_2 \cdot \cdots \cdot f_n \cdot f_{n+1} \cdot \cdots \cdot f_{2n} \rangle = \int_\Omega \prod_{k=1}^{n} f_k(w) \bar{f}_{k+n}(w) \mathrm{d}\mu(w)$$

and we remark that the following properties hold:

$$\|f\|^{2n} = \langle f \cdot f \cdot \cdots \cdot f \rangle$$

$$f_k \to \langle f_1 \cdot \cdots \cdot f_{k-1} \cdot f_k \cdot f_{k+1} \cdot \cdots \cdot f_{2n} \rangle$$

is linear for $k = 1, 2, \ldots, n$ and conjugate linear for $k = n + 1, \ldots, 2n$,

$$\langle f_{2n} \cdot f_{2n-1} \cdot \cdots \cdot f_1 \rangle = \overline{\langle f_1 \cdot f_2 \cdot \cdots \cdot f_{2n} \rangle}.$$

DEFINITION 4.3.20 (Koehler, 1970). A Banach space X is said to be a G_{2n} space if there exists a mapping defined on the space

$$\prod_1^{2n} X_i, X_i = X, i = 1, 2, \ldots, 2n$$

and having all the three properties given above for $L^{2n}(\Omega, B, \mu)$.

For more details about such spaces we refer to Koehler (1970, 1972).

Another interesting identity characterization of inner product spaces was obtained by Ohira (1951) and is as follows.

THEOREM 4.3.21 (Ohira, 1951). *Let (X, d) be a metric space and suppose that for all p, q in X there exists a point r such that for all x in X the following identity holds:*

$$d(x,p)^2 + d(x,q)^2 \leqslant 2d(x,r)^2 + d(p,r)^2 + d(q,r)^2. \tag{O}$$

If X is complete and separable then X is isometric-isomorphic with a closed convex subset of a complete inner product space.

It is clear that this result contains the Jordan–von Neumann result.

We are interested now in the following problem: Suppose that X is a Banach space and for some subset P in X the parallelogram law holds. Under what conditions we can conclude that the parallelogram law holds for all points of X?

In what follows we give such a result obtained by Ellis (1953) and for this we need some notions from lattice theory. In what follows we present the necessary notions and the reader may want to consult the book of G. Birkhoff (*Lattice Theory* (3rd edn), 1967) for more information and details.

We recall that a poset is a set X in which we have defined a relation '$\leqslant$' such that for all x, y, z the following assertions hold:

(1) for all x in X, $x \leqslant x$ (Reflexivity)
(2) If $x \leqslant y$ and $y \leqslant z$ then $x = z$ (Transitivity)
(3) If $x \leqslant y$ and $y \leqslant x$ then $x = y$ (Antisymmetry)

It happens that $x \leqslant y$ and $x \neq y$ then we sometimes use the notation $x < y$ (and we say that x is less than y or that x is properly contained in y). We use also the notation $y > x$ for $x < y$. If X, Y are two posets and $f: X \to Y$ is a mapping then we say that f is isotone if whenever $x \leqslant y$ then $f(x) \leqslant f(y)$ (we say also that f is order-preserving). An isomorphism between two posets X, Y is an isotone mapping $f: X \to Y$ which has an isotone two-sided inverse, i.e. if $f(x) \leqslant f(y)$ then $x \leqslant y$.

A least element of a subset S of the poset X is an element m of S with the property that $m \leqslant x$ holds for all x in S. The greatest element M of S is an element M of S with the property that $x \leqslant M$ holds for all x in S.

A minimal element of the subset S of the poset X is an element u such that $u \leqslant x$ for no x in S holds. Dually we define the maximal elements.

It is obvious that a least element is a minimal element but the converse is not true. A similar statement holds for the greatest element and the maximal elements.

An upper bound for a subset S of the poset X is an element u of X containing every element s of S. The least upper bound (lub S) is an upper bound which is contained in every other upper bound of S. It is not difficult to see that if lub S exists then it is unique. Dually we define the notion of greatest lower bound of S (glb S) and similar results holds. The fundamental notion of lattice is defined as follows.

DEFINITION 4.3.22. A lattice is any poset X such that the following conditions hold:

(1) For any x, y in X the set (x, y) has a glb which is denoted by $x \wedge y$ (and called 'meet').

(2) For any x, y in X the set (x, y) has a lub which is denoted by $x \vee y$ (and called the 'join').

The lattice X is said to be complete if for each subset S of X, lub S and glb S exist.

If X is a complete lattice then it has obviously a least element (denoted O) and a greatest element (denoted I).

If X is a lattice and a, b elements of X then the set

$$\{x, a \leqslant x \leqslant b\} = [a, b]$$

is called the closed interval (with the endpoints a and b).

If S is a subset in a lattice X then we say that S is a convex subset if whenever a, b are in S then the interval $[a, b]$ is in S.

Suppose that X is a lattice; we say that X is distributive if either

$$x \wedge (y \vee z) = (x \wedge y) \vee (x \wedge z)$$

or

$$x \vee (y \wedge z) = (x \vee z) \wedge (x \vee z)$$

hold for all x, y, z.

The lattice X is said to be modular if $x \leqslant z$ then

$$x \vee (y \wedge z) = (x \vee y) \wedge z$$

holds for all x, y, z.

Let x be a lattice and $f: X \to R$ (R is the set of all real numbers). We say that f is a valuation if

$$f(x \vee y) + f(x \wedge y) = f(x) + f(y)$$

holds for all x, y in X.

We use the notation (for a valuation)

$$f(x) = v[x].$$

We say that a valuation v is isotone if whenever $x \leqslant y$ then

$$v[x] \leqslant v[y].$$

The valuation is said to be positive if whenever $x < y$ then

$$v[x] < v[y].$$

Let X be a lattice with a valuation v. Then we define

$$d(x, y) = v[x \wedge y] - v[x \wedge y].$$

The properties of the function d are given in the following proposition.

PROPOSITION 4.3.23. *The function d satisfies the following relations:*

(1) $d(x, x) = 0, d(x, y) \geqslant 0, d(x, y) = d(y, x)$,
(2) $d(x, z) \leqslant d(x, y) + d(y, z)$,
(3) $d(z \vee x, z \vee y) + d(z \wedge x, z \wedge y) \leqslant d(x, y)$

for all x, y, z in X.

Proof. The assertions in (1) are clear and thus we prove only (2) and (3). We begin with the proof of (3). We have,

$$d(z \vee x, z \vee y) + d(z \wedge x, z \wedge y)$$
$$= v[z \vee x \vee y] - v[(z \vee x) \wedge (z \vee y)] + v[(z \wedge x) \vee (z \wedge y)] - v[z \wedge x \wedge y])$$

and thus is at most

$$v[z \vee x \vee y] - v[z \vee (x \wedge y)] + v[z \wedge (x \vee y)] - v[z \wedge x \wedge y]$$

and using the valuation property we get further that this number is equal to

$$v[z] + v[x \vee y] - v[z] - v[x \wedge y] = d(x, y).$$

Thus (3) holds. For (2) we remark that

$$\begin{aligned} d(x, y) + d(y, z) &= d(x \vee y, y) + d(y, x \wedge y) + \\ &\quad + d(y \vee z, y) + d(y, y \wedge z) \\ &\geqslant d(x \vee y \vee z, y \vee z) + d(y \vee z, y) + \\ &\quad + d(y, x \wedge y) + d(x \wedge y, x \wedge y \wedge z) \end{aligned}$$

and using (3),

$$\geqslant d(x \vee y, y) + d(x \wedge y, x \wedge y \wedge z) \geqslant d(x \vee y \vee z, x \wedge y \wedge z)$$
$$\geqslant d(x \vee y, x \wedge y) = d(x, y).$$

From the properties of d given above we see that d is a pseudometric (i.e. all the properties of a metric hold except possibly the assertion $d(x, y) = 0$ does not necessarily means that $x = y$). But if the valuation is positive then d is clearly a metric on X.

DEFINITION 4.3.24. A lattice X is called a metric lattice if d defined above is a metric (on X).

Suppose now that G is a group which is a lattice too. In such a group G there is an important set, namely G^+, which consists of all positive elements of G, i.e. those elements x in G with $0 \leqslant x$.

It is obvious that the following properties hold:

(1) $G^+ \cap (-G^+) = (0)$,
(2) $G^+ + G^+ \subseteq G^+$,
(3) $(-a) + G^+ + a = G^+$ for all a in G.

The group G is said to be Archimedean if $na \leqslant b$ for all n, then $a = 0$.

Let us consider two elements in G^+. We say that they are disjoint if $a \wedge b = 0$ (and we use the notation $a \perp b$ for this fact). Let a be an arbitrary element in G. The absolute value of a is the element denoted by $|a|$ and defined by the formula

$$a \vee (-a)$$

The positive part of an element a of G is the element denoted by a^+ and defined as

$$a \vee 0.$$

The negative part of a denoted by a^- is defined by

$$a \wedge 0.$$

Obviously we have the relations:

$$a = a^+ + a^-$$
$$|a - b| = (a \vee b) - (a \vee b).$$

The notion of a Banach lattice is considered in the following definition.

DEFINITION 4.3.25. A Banach lattice X is a real Banach space which is also a lattice (and the operations '+' and '.' are compatible with the order, in the sense that $x \in X$, $y \in X$, $z \in X$, $y \leqslant z \to x + y \leqslant x + z$ and $x \geqslant 0$, $\lambda \geqslant 0 \to \lambda x \geqslant 0$) and the following property holds: if x, y are in X and $|x| \leqslant |y|$, then $\|x\| \leqslant \|y\|$.

An important class of Banach lattices is considered in the following definition.

DEFINITION 4.3.26 (Birkhoff, 1967). A Banach lattice X is said to be uniformly monotone (a UMB lattice) if for every $\varepsilon > 0$ there exists $\delta_\varepsilon > 0$ such that whenever $x \in X^+$, $\|x\| = 1$ and $\|x + y\| \leqslant 1 + \delta_\varepsilon$ then $\|y\| \leqslant \varepsilon$. Examples of such lattices are the spaces $L^p(\Omega, B, \mu)$, $1 < p < \infty$.

Another class of Banach lattices which is of interest to us is considered in the following definition.

DEFINITION 4.3.27. A complete inner product lattice is a Banach lattice which is also a complete inner product space.

In order to obtain a characterization of complete inner product lattices in the class of Banach lattices we need some result which are given as lemmas. The results which follow were obtained by Ellis (1953).

LEMMA 4.3.28. *Let X be a UMB-lattice and let $a < b$ be two arbitrary elements of X. Then $\|a\| < \|b\|$.*

Proof. Suppose that the assertion of the lemma is false. Thus we have $\|a\| = \|b\|$. Then

$$2\|a\| = \|2a\| = \|a\| + \|b\| \geqslant \|a + b\|.$$

But $a + b \geqslant 2a$ and thus

$$2\|a\| \leqslant \|a + b\| \leqslant 2\|a\| \quad \text{or} \quad \|a + b\| = 2\|a\|.$$

Now we use the fact that X is a UMB-lattice. Then for every $\varepsilon > 0$ there exists $\delta_\varepsilon > 0$ such that whenever $x, y \geqslant 0$, $\|x\| = 1$ and $\|x + y\| \leqslant 1 + \delta_\varepsilon$, then $\|y\| \leqslant \varepsilon$.

Now for δ_ε we have

$$1 + \delta_\varepsilon \geqslant \|1/\|a\| \cdot a\| = \tfrac{1}{2}(\|a\|^{-1})(\|a + b\|) = \|\|a\|^{-1}a + \tfrac{1}{2}\|a\|^{-1}(b - a)\|$$

which gives that

$$\|b - a\| \leqslant 2\|a\|\varepsilon$$

for all $\varepsilon > 0$. This clearly is impossible and thus the lemma is proved. □

Suppose now that X is a UMB-lattice which is a complete inner product space. Then we find an orthonormal set $(e_i)_{i\in I}$ which is a basis for X, i.e. for all x in X we have

$$x = \sum_{i\in I} \langle x, e_i \rangle e_i \quad \left(\text{and} \quad \|x\|^2 = \sum_{i\in I} |\langle x, e_i \rangle|^2\right).$$

Then we have the following result.

DEFINITION 4.3.29. For any $x, y \in X$ (X is as above) we have $x \leqslant y$ iff

$$\langle x, e_i \rangle \leqslant \langle y, e_i \rangle, i \in I.$$

Also the following result holds.

LEMMA 4.3.30. *If u, v are in X and we set $u \leqslant v$ iff $\langle u, e_i \rangle \leqslant \langle v, e_i \rangle$ for all i in I then $(X, \leqslant)$ is a Banach lattice.*

The following property is very important.

LEMMA 4.3.31. *$(X, \leqslant)$ is a complete inner product lattice.*

Proof. Since $\|x\|^2 = \Sigma_{i\in I} |\langle x, e_i \rangle|^2$, we have that $\|x\| = \|x\|$, and since $|x| \leqslant |y|$ implies that $\langle x, e_i \rangle \leqslant \langle y, e_i \rangle$ for all $i \in I$, we have that $|x| \leqslant |y|$ implies that $\|x\| \leqslant \|y\|$. □

LEMMA 4.3.32. *If X is a complete inner product lattice then it is a UMB-lattice.*

Proof. The assertion can be proved using the same device as to prove the fact that any complete inner product space is uniformly convex. □

The following result gives a structure result of complete inner product spaces with respect to order relations.

THEOREM 4.3.33. *Let X be a complete inner product space. Then there exists a partial order on X, say '$\leqslant$', such that $(X, \leqslant)$ is a UMB-lattice.*

Proof. We consider the following function on X,

$$v[x] = \|x\|^2$$

and we remark that

$$v[x] + v[y] = \tfrac{1}{2}(\|x + y\|^2 + \|x - y\|^2) = \tfrac{1}{2}(\|(x \vee y) + (x \wedge y)\|^2 + \\ + (x \vee y) - (x \wedge y)\|^2) = v[x \vee y] + v[x \wedge y]$$

which gives that v is a valuation. From the properties given above the result follows. □

LEMMA 4.3.34. *Let X be a Banach lattice such that for all x, y in X^+ the parallelogram law holds. Then X is a complete inner product lattic.*

Proof. We have for any x, y in X that,

$$2(\|x\|^2 + \|y\|^2) = 2(\||x|\|^2 + \||y|\|^2) = \||x| + |y|\|^2 + \||x| - |y|\|^2$$

But

$$|x| + |y| = |x + y| \vee |x - y|$$
$$||x| - |y|| = |x + y| \wedge |x - y|$$

and thus

$$\begin{aligned}2(\|x\|^2 + \|y\|^2) &= \||x+y| \vee |x-y|\|^2 + \||x+y| \wedge |x-y|\|^2 \\ &= \tfrac{1}{2}(\||x+y| + |x-y|\|^2 + \||x+y| - |x-y|\|^2) \\ &= \||x+y|\|^2 + \||x-y|\|^2 \\ &= \|x+y\|^2 + \|x-y\|^2.\end{aligned}$$

This proves the lemma. □

LEMMA 4.3.35. *Let X be a Banach lattice and suppose that X^+ is a metric lattice under*

$$v[x] = \|x\|^2.$$

Suppose further that whenever x, y are comparable elements (i.e. either $x \leqslant y$ or $y \leqslant x$ hold) then

$$\|x+y\|^2 + \|x-y\|^2 = 2(\|x\|^2 + \|y\|^2).$$

Then X is a complete inner product lattice.

Proof. By the result in the above lemma it suffices to prove that the parallelogram law is satisfied for elements in X^+. Indeed, if x, y are arbitrary elements in X^+ then $x \wedge y$ and $x \vee y$ are comparable elements. Since

$$x + y = (x \vee y) + (x \wedge y)$$

$$x - y = (x \vee y) - (x \wedge y)$$

$$\||x-y|\| = \|x-y\|$$

and using the parallelogram law for comparable elements, we get,

$$\|x+y\|^2 + \|x-y\|^2 = 2(\|x \vee y\|^2 + \|x \wedge y\|^2) = 2(\|x\|^2 + \|y\|^2)$$

and the lemma is proved. □

Summarizing, we have the following result.

THEOREM 4.3.36. *Let X be a real Banach space. Then the following assertions are equivalent:*

(1) *X is a complete inner product space.*
(2) *There exists a partial order on X such that X with that partial order is a Banach lattice and the parallelogram law holds for the elements of X^+.*
(3) *There exists a partial order on X such that X with that partial order is a Banach lattice and X^+ is a metric lattice under the valuation $v[x] = \|x\|^2$ and the parallelogram law holds for comparable elements in X^+.*
(4) *For all x, y in X,*

$$\|x+y\|^2 + \|x-y\|^2 = 2(\|x\|^2 + \|y\|^2).$$

We conjecture that the above results of D. Ellis may be extended using the Apollonius identity, i.e. in Theorem 4.3.36 condition (3) may be replaced by the following conditon:

(3′) *There exists a partial order on X such that X with this partial order is a*

Banach lattice and X^+ is a metric lattice under the valuation $v[x] = \|x\|^2$ and the Apollonius identity holds for comparable elements in X^+.

Also, condition (2) in the theorem may be replaced by the corresponding one using the Apollonius identity.

We shall now present a characterization of complete inner product spaces in the class of Banach spaces in terms of certain properties of the family of linear subspaces. Such a characterization has interesting and important applications in quantum mechanics, implying that certain structures arise essentially from complete inner product spaces.

We consider a lattice L and, for any s in L, we call the element s' the complement of s if:

(1) $s \wedge s' = 0$
(2) $s \vee s' = 1$.

DEFINITION 4.1.37. The lattice L is said to be modular if whenever x, y, a are in L with $z \leqslant x$ then

$$x \wedge (y \vee z) = (x \wedge y) \vee z$$

and is said to be distributive if for all x, y, z in L,

$$x \wedge (y \vee z) = (x \wedge y) \vee (x \wedge z)$$
$$x \vee (y \wedge z) = (x \vee y) \wedge (x \vee z).$$

If for any s in L there exists s', the complement of s, then we say that L is a complemented lattice; a complemented distributive lattice is called a Boolean algebra.

If L is a Boolean algebra and for any countable subset A of L, $\bigvee_{s \in A}$ and $\bigwedge_{s \in A}$ are in L then we say that L is a Boolean σ-algebra.

EXAMPLE 4.3.38. Let K be a division ring and X be a (left) linear space over K. Then the set L of all linear subspaces of X is a lattice which is not a Boolean algebra. This lattice is denoted by $L(X, K)$.

DEFINITION 4.7.39. Let L be a complemented modular lattice and let s be an element of L. We say that s is in the center of L if it has a unique complement in L. We call a lattice irreducible if the only elements in the center are 0 and 1. A chain in a lattice is any finite sequence

$$a_1 \leqslant a_2 \leqslant a_3 \leqslant \cdots \leqslant a_n, \quad n < \infty$$

and

$$a_{i-1} \neq a_i \neq a_{i+1}.$$

The number n is called the length of the chain.

DEFINITION 4.3.40. A lattice L is said to be of finite rank if for some integer $n = n(L)$ the length of every chain is at most n.

DEFINITION 4.3.41. A lattice L is said to be a geometry if:

(1) L is an irreducible complemented modular lattice,
(2) L is of finite rank.

DEFINITION 4.3.42. Let L and L_1 be two geometries of the same rank. A duality $\xi: L \to L_1$ is a mapping satisfying the following properties:

(1) ξ is injective and surjective.
(2) If $a \leqslant b$ ($a \neq b$) then $\xi(a) \leqslant \xi(b)$.

If $L_1 = L$ then we say that ξ is a duality on L.

EXAMPLE 4.3.43. $L(X, K)$ is a geometry and if we consider $L(X, \bar{K})$ where X is the set of all functions $f: X \to K$ such that

(1) $f(0) = 0$,
(2) $f(kx + y) = kf(x) + f(y)$

for any x, y in X and k in K, then for any M in $L(X, K)$ we set

$$M^0 = (f, f \in X, f(x) = 0, x \in M)$$

then the mapping $M \to M^0$ is a duality between the geometries $L(X, K)$ and $L(X, \bar{K})$. (Here $\bar{K}$ is the ring with the same elements as K but the operations '$\oplus$' and '$\odot$' are defined as $x \oplus y = x + y$, $x \odot y = yx$.)

DEFINITION 4.3.44. A polarity of $L(X, K)$ is a duality ξ with the property that $M \leqslant \xi(\xi(M))$ (and in fact iff $\xi(\xi(M)) = M$).

DEFINITION 4.3.45. An orthocomplementation in $L(X, K)$ is a polarity ξ with the property that if M is in X then $M \cap \xi(M) = 0$.

The basic result about orthocomplementation is the following theorem of G. Birkhoff and John von Neumann.

THEOREM 4.3.46 (Birkhoff and von Neumann, 1936). *Let X be a finite dimensional space* (dim $X = 3$) *over the division ring K and let ξ be an orthocomplementation of $L(X, K)$. Then there exists an involutive anti-isomorphism $j: K \to K$ and a nonsingular symmetric j-bilinear form $\langle\, , \rangle$ such that:*

(1) $\xi(M) = \{f, f(x) = 0, x \in M\}$,
(2) $\langle x, x\rangle = 0$ *iff* $x = 0$.

If we require that for some x_0 in X, $\langle x_0, x_0\rangle = 1$ then $\langle\, , \rangle$ and j are uniquely determined. More precisely, if $K = R$, Q (Q is the quaterion field) then there exists an inner product $\langle\, , \rangle$ on X such that the orthocomplementation induced by $\langle\, , \rangle$ coincides (up to a multiplication by a real number) with ξ. If $K = C$ (C is the set of complex numbers) and ξ is with the property that the anti-automorphism of C associated with ξ is the identity or the conjugation, then it is the conjugation and ξ is the orthocomplementation induced by an inner product on X (and ξ is determined up to a (real) constant).

The above theorem holds also in the case of infinite dimensional spaces and was proved by S. Kakutani and G. W. Mackey. We formulate the theorem now.

THEOREM 4.3.47 (Kakutani and Mackey, 1944, 1946). *Let X be a Banach space* *over K (R, C, Q) (R is the set of real numbers, C is the set of complex numbers, Q is the set of quaternions) and suppose that X is not finite dimensional. Suppose further that* $L_c(X)$, *of all closed subspaces of X, has the property that on* $L_c(X)$ *there exists an orthocomplementation. In this case there exists an inner product* $\langle , \rangle$ *on X (whose values are in K) such that:*

(1) $(X, \|\,,\|)$ *and* $(X, \|\,,\|_1)$, $\|\,,\|_1$ *is the norm induced by the inner product are isomorphic as Banach spaces;*
(2) *the orthocomplementation on* $L_c(X)$ *coincides with the orthocomplementation induced by the inner product, i.e.* $\xi(M) = M^{\perp}$.

For a proof of these theorems we refer to the original papers of G. Birkhoff and John von Neumann and S. Kakutani and G. W. Mackey or to the excellent presentation in V. S. Varadarajan (1968), Vol. 1, Theorems 4.4 and 7.1.

4.4. The Ficken Characterization of Inner Product Spaces

An interesting and useful characterization of inner product spaces was found by F. A. Ficken in 1946, and is as follows:

THEOREM 4.4.1 (Ficken, 1946). *A normed linear space X is an inner product space if and only if for all x, y in X with* $\|x\| = \|y\|$ *and all scalars a and b the following identity holds:*

$$\|ax + by\| = \|ay + bx\|.$$

Proof. First we remark that the condition is necessary. Indeed, if x, y are elements in X which has an inner product then for all scalars a and b we have

$$\|ax + by\|^2 = |a|^2\|x\|^2 + |b|^2\|y\|^2 + a\bar{b}\langle x, y\rangle + \bar{a}b\langle y, x\rangle = \|ay + bx\|^2. \qquad \square$$

Now we prove that the condition is sufficient. First we remark that from Ficken's condition it follows that if the elements x and y in X satisfy the relation

$$\|x\| = \|y\| = \tfrac{1}{2}\|x + y\|$$

then necessarily $x = y$. Indeed, we apply Ficken's condition to the elements $x + y$ and $-2y$; then we have

$$\|x + y - 2(2y)\| = \|2(x + y) - 2y\| = \|2x\| = 2\|x\|$$

which implies that

$$\|x - y - 2y\| = 2\|x\|.$$

An induction argument shows then that

$$\|(2n + 1)(x - y) - 2y\| = \|2x\| = 2\|x\| = \|2y\|. \qquad (*)$$

Indeed, we remark that the relation is true for $n = 0$. Suppose now that it is valid for all $i \leqslant n$ and we show that it holds for $i = n + 1$. For this we apply Ficken's condition to the elements $(2n + 1)(x - y) - 2y$ and $2y$ with $a = 2$ and $b = 2n + 3$.

Then we have,

$$\|2(2n+1)(x-y)-2y+(2n+3)(2y)\| \\ = \|(2n+3)(2n+1)(x-y)-2y+2(2y)\|$$

and this implies that

$$\|(2n+3)(x-y)-2y\| = \|2x\| = \|2y\|$$

holds and thus it holds for all n.

From (*) we get that

$$\|x-y\| \leqslant 2/(2n+1)\cdot(\|x\|+\|y\|) = 4/(2n+1)\cdot\|x\|$$

and for $n \to \infty$ this implies that $x - y = 0$. Thus the assertion is proved. From the assertion just obtained we have that if x, y are elements of X with

$$x/\|x\| = y/\|y\|$$

then

$$\|x+y\| < \|x\| + \|y\|. \tag{**}$$

In what follows we can suppose without loss of generality that $\|x\| \leqslant \|y\|$. Now we have

$$\|x+y\| = \big\|x + \|x\|\cdot y/\|y\| + (\|y\|-\|x\|)y/\|y\|\big\| \\ \leqslant \big\|x + \|x\|\cdot y/\|y\|\big\| + \|y\| - \|x\|.$$

If we consider the elements

$$x_1 = x, \qquad y_1 = \|x\|\cdot y/\|y\|$$

we have obviously $\|x_1\| = \|y_1\|$ and thus if $\|x_1 + y_1\| = 2\|x_1\|$ then necessarily $x_1 = y_1$, which is not possible (of course we suppose that x and y are linearly independent elements in X). Thus we must have the inequality

$$\|x_1 + y_1\| < 2\|x_1\|$$

and thus

$$\|x+y\| < 2\|x\| + (\|y\| - \|x\|) = \|x\| + \|y\|$$

which proves (**).

The following useful remark which is derived from Ficken's condition is: if x, y are arbitrary elements in X and r is in R^+ then there are no more than two distinct real numbers t such that

$$\|x + t(x-y)\| = r.$$

Indeed, suppose that there exist three distinct real numbers t_1, t_2 and t_3 such that

$$\|x + t_i(x-y)\| = r$$

holds. We define the elements x_i as follows:

$$x_1 = x + t_1(x-y), \qquad x_2 = x + t_2(x-y), \qquad x_3 = x + t_3(x-y) = ax_1 + (1-a)x_2$$

for some a in $(0, 1)$. As above, we suppose that x and y are linearly independent elements of X and thus we have

$$\|ax_1/\|x_2\|\,\| \neq \|(1-a)x_2/\|x_1\|\,\|$$

which implies that

$$\|x_3\| < ar + (1-a)r = r.$$

This contradiction proves our assertion.

Now we prove that from Ficken's condition it follows that in X the Jordan–von Neumann condition holds. Let x and y be arbitrary elements in X, both not equal to 0. Then we have,

$$\begin{aligned}\|x+y\| = \|x-y+2y\| &= \|\,\|2y\|/(\|x-y\|)\cdot(x-y) + (\|x-y\|)/\|y\|\cdot y\| \\ &= \|\,\|2y\|/\|x-y\|\cdot x + (\|x-y\|^2 - 2\|y\|^2)/(\|x-y\|\cdot\|y\|)\cdot y\|\end{aligned}$$

which may be written as:

$$\|y\|\,\|x+y\|\,\|x-y\| = \|2\|y\|^2 x + (\|x-y\|^2 - 2\|y\|^2)y\|. \qquad (*)$$

In a similar way we obtain the relations:

$$\|y\|\,\|x+y\|\,\|x-y\| = \|2\|y\|^2 x + (\|x-y\|^2 - 2\|x\|^2)y\| \qquad (**)$$

$$\|y\|\,\|x+y\|\,\|x-y\| = \|2\|y\|^2 x + (2\|y\|^2 - (\|x+y\|^2)y\| \qquad (***)$$

$$\|y\|\,\|x+y\|\,\|x-y\| = \|2\|y\|^2 x + (2\|x\|^2 - (\|x+y\|^2)y\|. \qquad (****)$$

We set

$$r = \|x+y\|\cdot\|x-y\|$$

and

$$x_1 = 2\|y\|^2 x, \qquad y_1 = 2\|y\|^2 x + y.$$

From relations (*), (**), (***) and (****) we see that the set

$$\begin{aligned}&(\|x-y\|^2 - 2\|y\|^2, \|x-y\|^2 - 2\|x\|^2, 2\|y\|^2 - \|x+y\|^2, \\ &\quad 2\|x\|^2 - \|x+y\|^2) = (a, b, c, d)\end{aligned}$$

contains in fact only two distinct elements. If $a = d$ or $b = c$ then we see that the Jordan–von Neumann condition is satisfied and thus we may suppose that $a \neq d$ and $b \neq c$. Now if we have $a = c$ and $b = d$ then the Jordan–von Neumann condition is again satisfied. Thus it remains to consider the case when $a = b$ and $c = d$. In this case we have that

$$\|x\| = \|y\| \text{ and } \|x+y\| = \|x-y\|.$$

If $a > 0$ then we have, applying the triangle inequality in (*), that

$$\|x+y\| < \|x-y\|.$$

If $a < 0$, similarly we have

$$\|x - y\| < \|x + y\|.$$

From these we see that we must have $a = b = 0$. Similarly we can prove that $c = d = 0$. In this case it is clear that the Jordan–von Neumann condition is satisfied for all x, y in X. Thus X is an inner product space.

Later we present extensions and localized versions of this result. Extensions and localized versions were obtained by Lorch (1948) and Oman (1969, 1975, 1980).

We close with a geometric interpretation of Ficken's result stated by J. A. Oman. Suppose that we have in the Euclidean plane R^2 an isosceles triangle, say ABC, with the sides AB and AC congruent. Let B' and C' be points on AB and AC respectively such that AB' and AC' are congruent. Then Ficken's condition is that $B'C$ and BC' are congruent. This, of course, is true and Ficken's theorem says that, in the plane, if a norm has the property that in such a triangle the above relations holds, then the norm is derived from an inner product.

In 1967 S. B. Stechkin, at the symposium on integral geometry at Petrozavodsk, asked the following question: Suppose that a Banach space X has the property that the Apollonius theorem holds in X (i.e. the geometric locus of the points for which the ratio of the distances to two fixed points is constant is a sphere). Is the norm necessarily generated by an inner product?

Danelich (1976) has proved, using Ficken's result, that the following assertion holds:

THEOREM 4.4.2. *Let X be a normed linear space in which the Apollonius theorem holds. Then the norm of X is generated by an inner product.*

For a proof we refer to I. A. Danelich's paper. For some extensions of Ficken's result, see Section 4.7.

It may be of interest to mention that many characterizations of inner product spaces are special cases of the following principle:

PRINCIPLE 4.4.3 (E. R. Lorch, J. A. Oman and, perhaps, many other people). *Let P be a property which holds in the Euclidean geometry. A normed linear space is an inner product space if and only if P holds in X.*

In what follows we present a number of results which illustrate this principle.

4.5. Closed Maximal Linear Subspaces and Inner Product Structures

Suppose that X is a Banach space and then (x_0) (the subspace generated by the element $x_0 \neq 0$) is a one-dimensional subspace of X. We find x^* in the dual space of X with the property that $x^*(x_0) \neq 0$ and if $N = (x, x^*(x) = 0)$ then $X = M \oplus N$.

Clearly N is a closed maximal linear subspace of X. It is also clear that if N is a closed linear maximal subspace of X then we find x^* in the dual of X such that

$$N = (x, x^*(x) = 0).$$

In their interesting note, Carlson and Hicks (1978) asked about conditions on M, N

which assure that the norm of X is generated by an inner product. In order to formulate their results we give some definitions.

DEFINITION. 4.5.1 (Carlson and Hicks, 1978). A Banach space X is said to satisfy condition (A) if for any one-dimensional subspace M of $X, M = (x_0)$ there exists a subspace N of X such that

$$M \oplus N = X \quad \text{and} \quad \|x + y\|^2 = \|x\|^2 + \|y\|^2,$$

where x is in M and y in N.

DEFINITION 4.5.2 (Carlson and Hicks, 1978). A Banach space X is said to satisfy condition (B) if for every closed maximal subspace N of X there exists a subspace $M = (x_0)$ such that

$$X = M \oplus N \quad \text{and} \quad \|x + y\|^2 = \|x\|^2 + \|y\|^2,$$

where x is in M and y in N.

Then we have the following result.

THEOREM 4.5.3 (Carlson and Hicks, 1978). *A Banach space X has its norm generated by an inner product iff X satisfies condition* (A).

Proof. It is clear that the condition is necessary. Suppose that it is satisfied. Let X, y be arbitrary element in X and first suppose that they are linearly independent. Take $M = (x)$ and N given by the condition of the theorem. Then since

$$y = ax + z,$$

where z is in N, we have

$$\begin{aligned}\|x + y\|^2 + \|x - y\|^2 &= \|x + ax + z\|^2 + \|x - ax - z\|^2 \\ &= \|(a + 1)x + z\|^2 + \|(1 - a)x - z\|^2 \\ &= (|a + 1|^2 + |a - 1|^2)\|x\|^2 + 2\|z\|^2 \\ &= 2((|a|^2 + 1)\|x\|^2 + \|z\|^2)\end{aligned}$$

which implies that

$$\|x + y\|^2 + \|x - y\|^2 = 2(\|x\|^2 + \|y\|^2).$$

If x, y are linearly dependent then the Apollonius theorem is trivially satisfied. Thus X is with the norm generated by an inner product according to Jordan–von Neumann's theorem. □

REMARK 4.5.4. Conditions (A) and (B) may also be formulated for the case of normed linear space. Then we have the following result.

THEOREM 4.5.5 (Carlson and Hicks, 1978). *A normed linear space X is a complete inner product space (i.e. has its norm generated by an inner product and is complete with respect to the norm derived from the inner product) iff conditions* (A) *and* (B) *are satisfied.*

4.6. Loewner's Ellipses. Ellipsoids

Many characterizations of inner product spaces are formulated using certain identities which hold for certain classes of elements in a normed linear space. In order to prove that a corresponding normed linear space has its norm generated by an inner product, one of the most useful methods used is to show that for any two-dimensional subspace the unit sphere is an ellipse, or that for any three-dimensional subspace the unit sphere is an ellipsoid. In such proofs some results about inscribed (circumscribed) ellipses for a centrally symmetric curve (a curve in the complex plane is said to be convex if the domain it encloses is a convex set) play a fundamental rôle as well as certain characteristic properties of ellipsoids in the three-dimensional spaces. In this section we present some results about such classes of ellipses (obtained by C. Loewner and J. A. Oman) and about ellipsoids (obtained by C. Loewner, G. Birkhoff, A. Marchaud, etc.).

DEFINITION 4.6.1. An inscribed (circumscribed) ellipse E in a closed convex symmetric curve C (in the plane) about the origin is said to be an inscribed (circumscribed) n-ellipse if $E \cap C$ (C is the curve) contains at least n independent points.

The following results of C. Loewner present some classes of 2-ellipses.

THEOREM 4.6.2 (C. Loewner). *If C is a symmetric closed convex curve in the plane, then there exists a unique ellipse of minimal area circumscribed about C.*

Proof. Suppose that the assertion is false. Then we find two ellipses which have the equations with respect to the principal axes as follows:

$$x^2 + y^2 = 1$$
$$x^2 \cdot a + y^2 \cdot b = 1.$$

Now since these are of equal area we obtain that $ab = 1$. From the hypothesis they contain the curve C and thus we have

$$x^2 + y^2 \leqslant 1$$
$$x^2 \cdot a + y^2 \cdot b \leqslant 1$$

for all $(x, y) \in C$. In this case for all (x, y) on C we have

$$(a + 1)\ x^2/2 + (b + 1)y^2 \leqslant 1$$

and since

$$(a + 1)x^2/2 + (b + 1)y^2/2 = 1$$

it is again an ellipse which contains C and has the area $2.1/((a + 1)(b + 1))^{1/2}$. From the minimality we obtain that $a = b$. Thus the theorem is proved. □

THEOREM 4.6.3 (C. Loewner). *If C is as in the Theorem 4.6.2 then the minimal circumscribed ellipse is a circumscribed 2-ellipse.*

Proof. The fact that the unique circumscribed ellipse of minimal area is a circumscribed 2-ellipse means that C touches C in at least four points which are symmetric pairwise.

Suppose that the minimal ellipse has the equation

$$x^2 + y^2 = 1$$

where (1,0) is the point of contact of C with E. We consider now the family of ellipses defined by the equation

$$(E_\varepsilon)\ x^2/(1 + \varepsilon) + y^2/(1 + \varepsilon) = 1$$

have the minimal area and C is not contained in (E_ε). Let p_ε be a point in C and not in (E_ε) and consider $\varepsilon = 1/n$. We may suppose without loss of generality that the sequence (P_n), $P_n = p_{1/n}$, is convergent, say, to a point P. Since C is a closed curve we must have $P \in C$. Since p_ε (for small ε) is within ε of E we see that P is on E. The y-coordinate of the points in $E \cap E_\varepsilon$ is $\pm 1/(2 + \varepsilon)^{1/2}$ and thus the absolute value is at least $1/2^{1/2}$. This implies that P is not either points (1,0) or (− 1,0). From the symmetry we have that the minimal ellipse circumscribed about C touches C in at least four points. Thus it is a circumscribed 2-ellipse. □

THEOREM 4.6.4 (C. Loewner). *If C is a symmetric closed curve about the origin then there exists a unique inscribed 2-ellipse of maximal area.*

Proof. Since the proof is similar to that for circumscribed ellipses, it is omitted. □

REMARK 4.6.5. The existence of inscribed (circumscribed) ellipses of maximal (minimal) area follows easily using a compactness argument.

Let us consider an ellipse with center O and let u^* be one of the points of E on the conjugate diameter to the diameter at u. Let uu^* be the shorter arc from u to u^* and, further, A_{u,u^*} is the union of uu^* and the reflection of uu^* in the origin.

DEFINITION 4.6.6 (Oman, 1975). A subset M of E is said to be spread with respect to u if it is not in the interior of A_{u,u^*} or in the interior of $A_{u,-u^*}$.

THEOREM 4.6.7 (Oman, 1975). *Let C be a symmetric closed curve about the origin (in the plane) and let w be a point inside M. Then there exists an ellipse $E(w)$ of maximal area inscribed in C, $E(w)$ with center at O, w is contained in $E(w)$. The contact of $E(w)$ is spread with respect to w.*

Proof. Since the existence of $E(w)$ among all ellipses with their center at O and containing w and having a maximal area is easy to prove, we show only the fact that the contact of $E(w)$ is spread with respect to w.

We suppose that the assertion is false and we obtain a contradiction, namely that the area is not maximal.

We may suppose without loss of generality that the equation of E is

$$x^2 + y^2 = 1$$

and that the point w is $(2^{1/2}/2,\ 2^{1/2}/2)$. Thus we have that $w^* = (-2^{1/2}/2, 2^{1/2}/2)$. Now $ww^* \cup (-(ww^*))$ contains the contact set, we find $\varepsilon > 0$ such that if the point $(x, y) \in C$ and $|y/x| \leqslant 1$ then $x^2 + y^2 \geqslant 1 + \varepsilon$. Then we consider the ellipse (E_ε) whose equation is

$$x^2/(1 + \varepsilon) + y^2 \cdot (1 + 2\varepsilon)/(1 + \varepsilon) = 1$$

which contains w and is inscribed in C. This has the area greater than of E. This contradiction proves the assertion about spread of contact of $E(w)$. □

REMARK 4.6.8. It should be noted that, in general, $E(w)$ is not unique with the above properties.

The following result gives information about the possibility of $E(w)$ being an n-ellipse.

THEOREM 4.6.9 (Oman, 1975). *Suppose that C is not an ellipse and that w is on C and not in the contact set of the Loewner inscribed ellipse in C with the maximal area. Then there exists r_0 in $(0, 1)$ such that $E(\gamma w)$ is an inscribed 2-ellipse for all r in $(r_0, 1)$.*

Proof. Suppose that the assertion of the theorem is false and thus there exists a sequence (r_n) in $(0_0, 1)$, $\lim r_n = 1$ such that $E(r_i w)$ are 1-ellipses. We may suppose that the contact points form a convergent sequence (if not we may change the notation for a corresponding subsequence) to a point, say, p on C such that p has a support line parallel to the line through the origin and w. This gives that some subsequence of the ellipses $E(r_n w)$ converges to an ellipse, say E, whose contact set contains p and w. We remark that p and w are then endpoints of conjugate diameters and then we must have that E is the Loewner ellipse. But this contradicts the property of w in the theorem. This completes the proof of the theorem. □

The following result gives information concerning the points of C which are in the contact set for some inscribed 2-ellipses.

THEOREM 4.6.10 (Oman, 1975). *Let C be a symmetric closed convex curve about the origin O in the plane, and let M be the set of all points of C which are in the contact set of some inscribed 2-ellipse. Then M is dense in C.*

Proof. Let us consider w in $C \setminus M$ and let (r_n) be a sequence of points in $(0, 1)$ such that $\lim r_n = 1$. By the above result $E(r_n w)$ are 2-ellipses. Then $(E(r_n w))$ is a convergent sequence of ellipses to an ellipse E which has w in the contact set, and this is possible unless all points of contact of $E(r_n w)$ converge to w. Thus, in any case, w would be a limit of a sequence of contact points of 2-ellipses. □

The following result deals with 2-ellipses with close points.

THEOREM 4.6.11 (Oman, 1975). *Let C be a curve as above but not an ellipse. Then given $\varepsilon > 0$ there exists at least two inscribed 2-ellipses such that the contact sets of these ellipses contain the points p and q, respectively, with the property that $|p - q| < \varepsilon$.*

Proof. If the set M in the proof of Theorem 4.6.10 is not C then the assertion follows. Thus we may suppose without loss of generality that $M = C$ (with M as above). Consider an arbitrarily inscribed 2-ellipse and let p, q be two contact points of this ellipse E such that the interior of the arc pq does not contain points of C. Then if w is any point of the arc of C between p and q then it is a contact point for some 2-ellipse, say E_1. In this case the second point of contact is also on the arc of C between p and q since $E \cap E_1$ contains at most four points. We can repeat this procedure to obtain an ellipse E^* which has the property stated property and using instead of E the ellipse E^* we obtain the ellipse with the property as stated in the theorem.

Of course we can ask if results similar to those given above hold for the case of circumscribed ellipses. Indeed, this is the case and the proofs are much like the above proofs with the difference that some additional hypotheses must be added. □

THEOREM 4.6.12 (Oman, 1975). *Let C be a closed, convex and symmetric curve about the origin (in the plane) and let w be a point outside C. Then there exists an ellipse $E(w)$ whose area is minimal among all ellipses with the center at the origin, circumscribed about C and containing w, and the contact set of $E(w)$ is spread with respect to w.*

Proof. Since all the proofs are almost exactly as for inscribed ellipses, except perhaps the last assertion, we prove it here.

We suppose that the assertion is false and then we get a contradiction, in the sense that we have an ellipse with no minimal area. We may suppose without loss of generality that the equation of the ellipse is

$$x^2 + y^2 = 1$$

and the point w is $(1, 0)$. If the assertion of the theorem is false, then there is $\varepsilon > 0$ such that the arc from the point $(\varepsilon, (1 - \varepsilon^2)^{1/2})$ to $w^* = (0, 1)$ has no points on the curve C and there exists δ in $(0, 1 - ((1 - \varepsilon)/(1 + \varepsilon))^{1/2}$ such that if (x, y) is on C and $y/x \leqslant 0$ or $y/x \geqslant (1 - \varepsilon^2)^{1/2}$, then $x^2 + y^2 \leqslant 1 - \delta$. In this case the ellipse

$$x^2 + bxy + cy^2 = 1,$$

where

$$b = 2((1 - \varepsilon)/(1 + \varepsilon))^{1/2} \cdot (1 - 1/(1 - \varepsilon)^2), \qquad c = (2/(1 - \delta)^2 - 1) + 1/\varepsilon)/(1 + 1/\varepsilon)$$

circumscribes the curve C, contains the point w and the area is smaller than the area of E. This contradiction proves the assertion. □

THEOREM 4.6.13 (Oman, 1975). *Let C be as in Theorem 4.6.12 and let w be a point on C but not in the contact set of the circumscribed Loewner ellipse attached to C. Then there is $r_0 > 1$ such that $E(rw)$ is a circumscribed 2-ellipse for all r in $(1, r_0)$.*

Proof. Since the proof is almost exactly as for Theorem 4.6.9, it is omitted. □

THEOREM 4.6.14 (Oman, 1975). *Let C be as above and let N be the set of all points of C which are in the contact set of some circumscribed 2-ellipse. Then if C is strictly convex then the set N is dense in C.*

Proof. The proof is the same as for inscribed ellipses with the exception that the contact points of $E(r_n w)$ may converge to points different from w but the limit ellipse may degenerate. But then C contains a segment. This is a contradiction and the proof of the theorem is complete. □

Using this we may prove, as for the case of inscribed ellipses, the following result.

THEOREM 4.6.15 (Oman, 1975). *Let C be as above and suppose moreover that it is a strictly convex curve and not an ellipse. Then given $\varepsilon > 0$ there exist at least two points on two circumscribed 2-ellipses, say p and q, with the property that $|p - q| < \varepsilon$.*

We now present certain characterizations of ellipsoids. In proving that certain convex bodies are solid ellipsoids the most frequently used method is like that used for ellipses, namely the minimal method (in a sense that the volume is minimal). Also, we mention certain extensions to the n-dimensional case.

THEOREM 4.6.16 (C. Loewner). *Let F be a closed, bounded set in R^3 and suppose that F has a positive volume. Then given a point p in R^3 there exists a unique solid ellipsoid E_p with center at p and with minimal volume and which contains F.*

Proof. A compactness argument shows immediately that such an ellipsoid exists. Suppose now that there are two such ellipsoids E_p^1 and E_p^2 with the above property. We must show that they are identical. Using an affine transformation we may suppose without loss of generality that E_p^1 is a closed ball of R^3. The affinity transforms E_p^2 into a solid ellipsoid E_p^{+2} and the set F into the set F_1. Using coordinates we have that the equation of the unit sphere of R^3 is

$$x^2 + y^2 + z^2 = r^2, \quad r > 0$$

and that the equation of E_p^{+2} is

$$x^2/b_1^2 + y^2/b_2^2 + z^2/b_3^2 \leqslant 1.$$

Now since the volumes are equal we must have that the solid ellipsoid

$$x^2/(r^2 + b_1^2) + y^2/(r^2 + b_2^2) + z^2/(r^2 + b_3^2) \leqslant 1$$

contains F_1 and its volume is not less than the volume of E_p^{+2}. This implies that

$$r^2 + b_i^2 = 2rb_i, \; i = 1,2,3$$

and this implies that the ellipsoids must be equal. □

THEOREM 4.6.17 (Birkhoff, 1935). *Let C be a compact convex body in R^3 with the bounding surface S satisfying the following property: there is a point O in $\mathring{C}$ (the interior of C) such that for any line L through O and $P \in S \cap L$, for any plane π through O with the property that its translate through P supports C, then for skew cylindrical coordinates (r, θ, z) with L the line $r = 0$ and π the plane $z = 0$, the equation of S is of the type $r = f(z)g(\theta)$. Then C is an ellipsoid.*

For a proof of this result we refer to G. D. Birkhoff's paper. For a related result, as well as for more information, we refer to an article by Robert (1982).

THEOREM 4.6.18 (Marchaud, 1959). *Let C be a compact convex body in R^3 with the bounding surface S satisfying the following property: for every direction d of R^3 there exists a corresponding plane π_d such that the cylinder in the direction d generated by the plane curve $S \cap \pi_d$ circumscribes C. Then C is an ellipsoid.*

For a proof we refer to A. Marchaud's paper (1959).

Related to Theorem 4.6.16 we mention now an interesting result obtained by Taylor (1947; see also Day, 1947).

THEOREM 4.6.19 (Taylor, 1947). *Let C be a symmetric convex body in R^n ($n \geqslant 2$). Then there exists a parallelipiped of minimal volume, π_n circumscribed on C and with the property that the midpoints of each face is on C.*

For a proof, as well for interesting applications to Banach space theory, see A. E. Taylor's paper.

We mention now two interesting characterizations of ellipsoids; for a proof we refer to H. Busemann's book (1955).

THEOREM 4.6.20. *Let C be a closed convex set with the bounding surface S having the following property: the locus of midpoints of each family of parallel chords is contained in a hyperplane. Then C is an ellipsoid.*

THEOREM 4.6.21. *Let C be a closed convex set with the bounding surface S. Suppose that there exists a point z in $\mathring{C}$ (the interior of C) such that for any points p, q in C we find an affine transformation T with the following properties:*

(1) $TC \subseteq C$,
(2) $Tp = q$,
(3) $Tz = z$.

The C is an ellipsoid.

Other interesting characterizations are in the papers of Auerbach (1940) and Auerbach *et al.* (1935).

4.7. Ellipses and Inner Product Spaces

We now give a result which is basic in many characterizations of inner product spaces in the class of normed linear spaces.

THEOREM 4.7.1 *Let X be a two-dimensional real normed space. Then the norm of X is generated by an inner product iff the unit sphere of $X, S_X = \{x, x \in X, \|x\| = 1\}$ is an ellipse.*

Proof. If X is an inner product space then using the set (e_1, e_2) with $\|e_i\| = 1$ and $\langle e_1, e_2 \rangle = 0$ we obtain immediately that S_X has the equation

$$u^2 + v^2 = 1$$

where $ue_1 + ve_2$ is an element of S_X. Thus the condition is necessary. Suppose now that the set S_X of X is an ellipse and we represent X as the set of all pairs (x, y) of real numbers. Then if we set

$$e_1 = (1, 0), \quad e_2 = (0, 1)$$

then any point of S_X is of the form $xe_1 + ye_2$ and S_X being an ellipse we must have

$$1 = \|xe_1 + ye_2\| = x^2/a^2 + y^2/b^2.$$

Then clearly we define the following inner product on X: if (x, y) and (x_1, y_1) are two arbitrary elements of X then we set

$$\langle (x, y), (x_1, y_1) \rangle = xx_1/a^2 + yy_1/b^2.$$

It is obvious that this satisfies the requirements of the theorem. □

From the Jordan–von Neumann theorem we have the following result.

COROLLARY 4.7.2. *A normed linear space is an inner product space (i.e. has its norm generated by an inner product) iff any two-dimensional subspace has the unit sphere as an ellipse.*

Thus in order to prove that a Banach space (or a normed linear space) has its norm generated by an inner product we must show that for any two-dimensional subspace the unit sphere has the equation of an ellipse.

4.8. The Integral Form of the Parallelogram Law

In a recent paper Penico and Stanojevič (1980) have considered an integral which to some extent, can be considered an integral analogue of the parallelogram law. More precisely, if X is a normed linear space with the norm $\|,\|$ then for each pair of elements of X, say x and y, we consider the following integral:

$$I(x, y) = \frac{1}{2\pi} \cdot \int_0^{2\pi} \|(\cos t)x + (\sin t)y\|^2 \, dt.$$

Then we have the following result.

THEOREM 4.8.1 (Penico and Stanojevič, 1980). *A normed linear space X has its norm generated by an inner product iff for all x, y in X, $\|x\| = 1 = \|y\|$ the following inequality holds:*

$$I(x, y) \leqslant 1.$$

Proof. It is easy to see that the condition is necessary. Suppose now that for all x, y with $\|x\| = 1 = \|y\|$, $I(x, y) \leqslant 1$. We consider now the subspace X_2 generated by x and y and then we denote by S_2 the subset in X_2 defined by

$$S_2 = (u, \|u\| = 1).$$

Then by the theorem of C. Loewner we find a Loewner ellipse which is minimal and encompasses S_2. Then we know that this meets the set S_2 in four distinct points, say $(x, -x, y, -y)$. The norm on X_2 which is generated by this ellipse, say $\|,\|_2$ has the following property: for each u in X_2,

$$\|u\|_2 \leqslant \|u\|. \tag{+}$$

Now clearly we have the following inequality:

$$1 = \frac{1}{2\pi} \cdot \int_0^{2\pi} \|(\cos t)x + (\sin t)y\|_2^2 \, dt \leqslant I(x, y) \leqslant 1.$$

Since (+) holds for all elements of X_2 the functions under the integral give the following inequality (taking into account the fact that these are continuous functions)

$$\|(\cos t)x + (\sin t)y\|_2 = \|(\cos t)x + (\sin t)y\|$$

and thus we must have an equality:

$$\|(\cos t)x + (\sin t)y\| = \|(\cos t)x + (\sin t)y\|_2.$$

But each point in $(v, \|v\|_2 = 1)$ is of the form $(\cos t)x + (\sin t)y$, and thus we have

$$\|v\|_2 = \|v\|.$$

Thus the norm of X_2 is generated by an inner product. This clearly implies that the norm of X is generated by an inner product since x, y were arbitrary in X with the property to be linearly independent and $\|x\| = 1 = \|y\|$.

Now if X is an arbitrary normed linear space then we can consider the problem of the best constants m and M with the property that for all x, y in X with $\|x\| = 1 = \|y\|$ then

$$m \leqslant I(x, y) \leqslant M. \qquad \square$$

The following result of A. J. Penico and Č. V. Stanojevič answers this question.

THEOREM 4.8.2 (Penico and Stanojevič, 1980). *For any normed linear space X the following inequality holds:*

$$2^{-1} + \pi^{-1} \leqslant I(x, y) \leqslant 2(2^{-1} + \pi^{-1})$$

and the above bounds are the best possible.

Proof. Since

$$I(x, y) = \pi^{-1} \int_0^\pi \|(\cos t)x + (\sin t)y\|^2 \, dt$$

we have

$$\pi \cdot I(x, y) = \int_0^\pi \|(\cos t)x + (\sin t)y\|^2 \, dt = \pi \cdot 2^{-1}(\|x\|^2 + \|y\|^2) +$$

$$+ \|x\| \, \|y\| \int_0^\pi |(\sin t)| \, dt = \pi \cdot 2^{-1}(\|x\|^2 + \|y\|^2) + 2\|x\| \, \|y\|.$$

Further,

$$\pi \cdot I(x, y) = \int_0^\pi \|(\cos t)x + (\sin t)y\|^2 \, dt = \int_0^{\pi/2} (\|(\cos t)x + (\sin t)y\|^2 +$$

$$+ \|(-\sin t)x + (\cos t)y\|^2) \, dt = \tfrac{1}{2} \int_{-\pi/4}^{\pi/4} (\|(\cos t - \sin t)x +$$

$$+ (\sin t + \cos t)y\|^2 + \|(\cos t + \sin t)x - (\cos t - \sin t)y\|^2) \, dt$$

$$= \tfrac{1}{2} \int_0^{\pi/4} (\|(\cos t - \sin t)x + (\sin t + \cos t)y\|^2 +$$

$$+ \|(\cos t - \sin t)x - (\cos t + \sin t)y\|^2 +$$

$$+ \|(\cos t + \sin t)x - (\cos t - \sin t)y\|^2 +$$

$$+ \|(\cos t + \sin t)x + (\cos t - \sin t)y\|^2) \, dt$$

$$\geqslant \tfrac{1}{4} \int_0^{\pi/4} (\|(\cos t - \sin t)x +$$

$$+ (\sin t + \cos t)y\| + \|(\cos t - \sin t)x - (\cos t + \sin t)y\|)^2 +$$

$$+ (\|(\cos t + \sin t)x + (\sin t - \cos t)y\| +$$

$$+ (\cos t + \sin t)x - (\cos t - \sin t)y\|)^2 \, dt$$

and this gives that

$$\pi \cdot I(x, y) = \int_0^{\pi/4} \|(\sin t + \cos t)\|^2 \mathrm{d}t(\|x\|^2 + \|y\|^2) = (\pi \cdot 4^{-1} + 2^{-1})\,(\|x\|^2 + \|y\|^2)$$

or

$$I(x, y) \geqslant 2^{-1} + \pi^{-1}$$

and the inequality is proved.

If we consider the space $X = l_2^2 = ((u, v), u, v \text{ in } R)$ with the norm

$$\|(u, v)\| = |u| + |v|$$

then

$$I((1,0),\ (0,1)) = \pi^{-1} \int_0^{\pi} (|\cos t| + |\sin t|)^2 \, \mathrm{d}t = 1 + 2\pi^{-1}.$$

Further,

$$I((2^{-1}, 2^{-1}),\ (2^{-1}, 2^{-1})) = \pi^{-1} \int_0^{\pi} 4^{-1}(|\cos t + \sin t| + |\cos t - \sin t|)^2 \, \mathrm{d}t$$
$$= 2^{-1} + \pi^{-1}$$

which proves that the constants in the inequality are the best possible. The following extension of Theorem 4.8.1 was obtained by Č. V. Stanojevič and A. M. Suchanek and the result is as follows.

THEOREM 4.8.3 (Stanojevič and Suchanek, 1981). *Let X be a normed linear space and G be an arbitrary compact group with the Haar measure. Then the norm of X is generated by an inner product iff for any character of G, $\neq 1$ and for all x, y in X the following relation holds:*

$$\int_G \|x + r(g)y\|^2 \, \mathrm{d}(g) = \|x\|^2 + \|y\|^2.$$

For a proof as well as for other examples we refer to the paper by Stanojevič and Suchanek. For some comments about the above results of A. J. Penico, Č. V. Stanojevič and A. M. Suchanek, see also Day (1981).

Suggested by the integral considered by A. J. Penico and Č. V. Stanojevič we can define the following integral $I_p(x, y)$ by the formula

$$I_p(x, y) = \left(\tfrac{1}{2}\pi \cdot \int_0^{2\pi} \|(\cos t)x + (\sin t)y\|^p \, \mathrm{d}t \right)^{1/p}$$

for p in $[1, \infty)$. It may be of interest to relate this integral to the structure of the normed linear space X.

4.9. Topological Innerproductability

Suppose that X is a topological linear space with the topology τ. One of the interesting problems is to decide whenever there exists an inner product on X,

$\langle\,,\rangle$ such that the topology defined by the distance defined by the inner product coincides with the topology τ. If X is a normed linear space this means we must decide whenever there exists an inner product on X which induces an equivalent norm on X. It is a well-known fact that on a normed linear space there exist many equivalent norms, some of which may be generated by an inner product. For example, if $X = R^n$ or C^n then the family of norms

$$\|x\|_p = (|x_1|^p + \cdots + |x_n|^p)^{1/p}, \quad 1 \leqslant p < \infty$$

are equivalent norms. However, only for $n = 2$ is the norm generated by an inner product.

We present below some results which answer the above problem. We also present some interesting facts about the topological character of complete inner product spaces. First we give a theorem.

THEOREM 4.9.1 (Lambert, 1968). *Let X be a topological linear space and suppose that there exists a Hamel basis H of X such that the set C, defined by*

$$C = \left(x, x = \sum_{i=1}^{n} a_i x_i, \sum_{i=1}^{n} |a_i|^2 < 1, \quad x_i \in H\right),$$

is open and bounded. Then there exists on X an inner product $\langle\,,\rangle$ such that the topology induced by the distance associated with the inner product is the topology of X.

Proof. First we note that C is a convex set. Indeed, if x, y are elements of C then we may suppose (adding terms with $a_i = 0$) that

$$x = \sum_{i=1}^{n} a_i x_i, \qquad y = \sum_{i=1}^{n} b_i x_i$$

and thus for t in $(0, 1)$ we have

$$tx + (1 - t)y = \sum_{i=1}^{n} (ta_i + (1 - t)b_i)x_i$$

which implies that

$$\left(\sum_{i=1}^{n} |ta_i + (1 - t)b_i|^2\right)^{1/2} \leqslant t\left(\sum_{i=1}^{n} |a_i|^2\right)^{1/2} + (1 - t)\left(\sum_{i=1}^{n} |b_i|^2\right)^{1/2} < 1. \qquad \square$$

It is obvious that C is a balanced set in X. But, by definition C is open and bounded and thus $(1/n \cdot C)$ is a countable basis at O since if N is an arbitrary neighborhood of the origin of X, we find a real number r such that $C \subseteq r \cdot N$. This gives that $r^{-1}C \subseteq N$ and then

$$r^{-1}C = \left(x, x = \sum_{i=1}^{n} a_i x_i, x_i \in H, \sum_{i=1}^{n} |a_i|^2 < r^{-2}\right).$$

We take x, y in X with $x \neq y$,

$$x - y = \sum_{i=1}^{n} a_i x_i, \quad x_i \in H,$$

and we set $a = \sum_{i=1}^{n} |a_i|^2$.

We consider r to have the property that $r^{-2} < a$. Then $r^{-1}C$ is a neighborhood of the origin of X and $x - y \notin r^{-1}C$. Clearly $r^{-1}C + y$ is a neighborhood of y and x is not in $r^{-1}C + y$. Since the argument is symmetric with respect to x, y we get that X is a T_1-space (i.e. a space in which the points are closed sets. (See Engelking, 1968, or Kelley, 1955, for results of T_1-spaces.)

Now since C is a bounded convex and balanced (with the origin in C) we can consider the Minkowski functional p_C of C. If $x = \Sigma_{i=1}^{n} a_i x_i$, $x_i \in H$ then $p_C(x) = a^{-1}$, where a is defined as above. Then

$$1 = ap_C(x) = p_C\left(\sum_{i=1}^{n} a_i x_i\right),$$

which gives that ax is in the boundary of C. Thus

$$p_C(x) = \sum_{i=1}^{n} |a_i|^2. \qquad \square$$

From this we get easily that p_C *satisfies the Apollonius parallelogram law. The other assertions of the theorem are clear.*

COROLLARY 4.9.2 (Lambert, 1968). *If X, H, C are as in the above theorem then H is an orthonormal basis for X.*

THEOREM 4.9.3 (Lambert, 1968). *Suppose that X is a complete inner product space. Then there exists a pre-inner product space $X_1 \subseteq X$ such that if H is its Hamel basis then the set*

$$C = \left(x, x = \sum_{i=1}^{n} a_i x_i,\ x_i \in H,\ \sum_{i=1}^{n} |a_i|^2 < 1\right)$$

is open and bounded (with respect to the relative topology of X_1).

Proof. Let K be a maximal orthonormal subset of X and let X_1 be the linear manifold generated by K. Then clearly H is a Hamel basis for X_1 and X_1 satisfies the requirements of the theorem. Let p_C and $\|\,,\|$ be the norm generated by the original inner product on X. Then a base at the origin of X is $(x, \|x\| < 1/n)$ and $(x, p_C(x) < 1/n)$ is a base at the origin in X_1. But for x in X_1,

$$x = \sum_{i=1}^{n} a_i x_i, \quad x_i \in H$$

we have

$$p_C(x) = \sum_{i=1}^{n} |a_i|^2$$

and thus $C = (x, p_C(x) < 1)$ is the (open) unit ball of X_1. $\square$

If U is an open neighborhood of the origin then for some $r > 0$ we have that

$$C \subseteq r\,U$$

and this proves the theorem.

The following result presents an existence theorem for inner product structures.

THEOREM 4.9.4 (Lambert, 1968). *Suppose that X is a complete topological space satisfying the following properties:*

(1) *X is a T_1-space.*
(2) *At the origin the topology of X has a countable basis.*

Then X is a complete inner product space iff there exists a dense set X_1 in X whose Hamel basis H is such that

$$C = \left(x, x = \sum_{i=1}^{n} a_i x_i,\ x_i \in H,\ \sum_{i=1}^{n} |a_i|^2 < 1 \right)$$

is an open and bounded subset.

Proof. The necessity is clear. Suppose now that X satisfies the condition. Then by the above results X is a metrizable space and Theorem 4.9.1 gives an inner product on X which induces the same topology as the original topology of X. The completion of X_1 is unique up to an isometry and X can be considered as the closure of X_1. The theorem is proved. □

We now present an example which illustrates the above results.

EXAMPLE 4.9.5 (Lambert, 1968). Let $X = L^p([0,1], dt)$, p in $(0,1)$, dt is the Lebesgue measure. We set for x in X,

$$\|x\| = \int_0^1 |x(t)|^p \, dt$$

and this is clearly a quasinorm:

(1) $\|x\| \geqslant 0, \qquad \|x\| = 0 \ \text{ iff } \ x = 0,$
(2) $\|ax\| = |a|^p \|x\|,$
(3) $\|x + y\| \leqslant \|x\| + \|y\|.$

The topology induced by the metric

$$d(x, y) = \|x - y\|$$

makes X a toplogical linear space.

Suppose now that the topology τ of this metric space is defined by an inner product on X. We mention an interesting property of X namely that the only nonempty convex set in X is exactly X. We consider now the set C defined as above. Then H is a maximal orthonormal family in X and the closure of C in X is also a convex set. The interior of the closure of C is an open and convex set which, by the above remark, must be X. Now if x_1, x_2 are in H then $x = 3x_1 + 2x_2$ is not in the interior of the closure of C and since $p_C(x) = 13$ we have a contradiction.

The problems related to the topological structure of complete inner product spaces have attracted the attention of many people. We mention, for example, the problem of studying the continuous images of complete inner product spaces which was posed by the great mathematician P. S. Aleksandrov. One of the earliest results on complete inner product spaces (from the dimension theory point of view) was the result of Witold Hurewicz in 1928 which asserts that a complete inner product space is not a countable union of finite-dimensional spaces. Another interesting result was observed by P. Erdös in 1940, namely, that the set of rational points of l^2 (i.e. the set of all points in l^2, $x = (r_i)$, r_i rationals) is one-dimensional. See also J. H. Roberts (1956). For more information, as well as for a detailed discussion of the above results, we refer to the excellent books by R. Engelking on general topology (1977) and on dimension theory (1978).

We now mention some results obtained concerning the Aleksandrov problem. The first result was obtained by R. Ju. Matskina (1951) and the next step was given in the papers by Huhunaisvili (1956). In order to formulate G. E. Huhunaisvili's result, we first give a definition.

DEFINITION 4.9.6. A set M is said to be linearly connected if for each pair of points of it there exists an arc joining them and lying in the set.

Then the result of G. E. Huhunaisvili is as follows.

THEOREM 4.9.7 (Huhunaisvili, 1956). *Let (X, d) be a metric space. In order that X be a continuous image of a complete inner product space it is necessary and sufficient that:*

(1) *X is a separable linear connected A-absolute set,*
(2) *X is of the form*

$$\bigcup_{n_i} \bigcap_{k=1}^{\infty} X_{n_1 \dots n_k}$$

with $X_{n_1 \dots n_k} \supset X_{n_1 \dots n_k n_{k+1}}$, $X_{n_1 \dots n_k}$ is linearly connected and the diameter converges uniformly to zero for $k \to \infty$.

An interesting and important result about the topological characterization of l^2 was obtained by Torunczyk (1981) and is as follows. For notation and terminology we refer to H. Torunczyk's paper and to the paper by Anderson *et al.* (1983).

THEOREM 4.9.8 (Torunczyk, 1981). *If X is a topologically complete separable metric AR then in order that X be homeomorphic to l^2 it is necessary and sufficient that for any map $f\colon \oplus_1^\infty Q_i \to X$ of the countable free union of Hilbert cubes into X is strongly approximable by maps $g\colon \oplus_1^\infty Q_i \to X$ for which the collection $(g(Q_i))$ is discrete.*

For equivalent formulations as well as for applications we refer to the papers by Torunczyk (1981) Anderson *et al.* (1983). For the interested reader we mention that a great number of results on topological problems of complete inner product spaces are in the monograph of Bessaga and Pelczyński (1975). See also Kramar (1981) for spaces in which the topology is generated by certain functions related to inner product structures.

4.10. Local Norm Characterizations of Inner Product Structures

We now present some important characterizations of inner product structures which are, to some extent, localizations of some characterizations given above. These results were obtained by Oman (1981).

We begin by formulating the general problem of localization of characterizations of inner product structures on normed linear spaces of Banach spaces.

THE LOCALIZATION PROBLEM 4.10.1. Suppose that X is a normed linear space with a norm $\|,\|$. The problem is to formulate the necessary and sufficient conditions for X to have a norm generated by an inner product on X using an order relation r ($=$, $\geqslant$, $\leqslant$) and certain relations related to r for elements in X which are 'near'. For example, we can consider a normed linear space X and we may ask if X has a norm generated by an inner product iff for some $\varepsilon > 0$ and $n \geqslant 2$ whenever $\|x\| = \|y\| = 1$ and $\|x - y\| < \varepsilon$ then there exist the real numbers a_i, b_i, c_i, $i = 1, 2, \ldots, n$ such that:

(1) $\sum_{i=1}^{n} a_i \|b_i x + c_i y\| \, r = 0,$

(2) $\sum_{i=1}^{n} a_i(a_i^2 + b_1^2) = \sum_{i=1}^{n} a_i b_i c_i = 0,$

(3) $a_i \neq 0, \quad i = 1, 2, 3, \ldots, n,$

(4) $b_i c_j \neq c_i b_j$ for $i \neq j.$

The formulation given above is to some extent too general to be of practical interest; in fact we give an example to show that in that it is generally false. But these are of interest since by strengthening one or other condition we may obtain a problem which has a positive solution.

We present two examples of spaces which we use later in connection with some variants of the above problem.

EXAMPLE 4.10.2. We consider the space R^2 with the unit sphere $S = (u, |u| = 1)$ to be the following set:

$$\left\{ (x, y), \begin{array}{lll} 4x^2 + (4/3)y^2 = 1 & \text{and} & |y| \geqslant 3|x|, \\ |x| + |y| = 1 & \text{and} & 3|x| \geqslant |y| \geqslant 3^{-1}|x| \\ 4y^2 + (4/3)x^2 = 1 & \text{and} & 3|y| \leqslant |x| \end{array} \right\}.$$

EXAMPLE 4.10.3. We consider again the space R^2 whose unit sphere is the set

$$\left\{ (x, y), \begin{array}{lll} x^2 + y^2 = 1 & \text{and} & xy \geqslant 0 \\ |x|^3 + |y|^3 = 1 & \text{and} & xy < 0 \end{array} \right\}.$$

We now define a symbol which we use to formulate necessary and sufficient conditions for the existence of inner product structures on a normed linear space (Banach space).

DEFINITION 4.10.4. If r is an order relation ($=$, $\geqslant$, $\leqslant$) then we say that the pair of elements of X, (x, y) is in the class $A(a, b, r)$ iff

$$b(1 - b)\|ax + (1 - a)y\|^2 + a(1 - a)\ \|bx + (1 - b)y\|^2 \, r \, ab(a + b - 2ab)\|x\|^2 + (1 - a)(1 - b)(a + b - 2ab)\ \|y\|^2$$

for a, b in $(0, 1)$.

We use $A(a,b,r)$ to formulate a problem like Problem 4.10.1.

PROBLEM 4.10.5. Let X be a normed linear space. Is the norm of X generated by an inner product iff for some set K in X the following condition is satisfied: for (x, y) *in* $K_x K$ there are (a, b) in $(0, 1)^2$ such that $(x, y) \in A(a, b, r)$ for some order relation R?

PROBLEM 4.10.6. Let X be a normed linear space. Is the norm of X generated by an inner product iff for some order relation r and some $\varepsilon > 0$ for all x, y in X with $\|x\| = \|y\| = 1$ and $\|x - y\| < \varepsilon$ we have that (x, y) is in $A(a, b, r)$ for some (a, b) in $(0, 1)^2$?

PROBLEM 4.10.7. Let X be a normed linear space. Is the norm of X generated by an inner product iff for some order relation r and some $\varepsilon > 0$ for all (x, y) with $\alpha(x, y) < \varepsilon$, $(x, y) \in A(a, b, r)$ for some (a, b) in $(0, 1)^2$? Here $\alpha(x, y)$ is the number $\|x/\|x\| - y/\|y\|\|$.

Now we use the spaces given in Examples 4.10.2 and 4.10.3 to discuss some of the problems considered above. First we consider the space in Example 4.10.2. Suppose that (u, v) are elements of the space with $\alpha(u, v) < \frac{1}{4}$. Then we have only three possibilities for u and v:

(1) $u/\|u\|$, $v/\|v\|$ lie on a flat spot of the sphere and at least one is not at the end of the flat spot;
(2) $u/\|u\|$, $v/\|v\|$ lie on an elliptical section of the sphere and at least one does not lie at the end of the elliptical section;
(3) one lies on a flat spot and one lies on an adjacent elliptical section.

Suppose now that either (1) or (2) holds. Thus for any (a, b) in $(0, 1)^2$ with $(bx - (1 - b)y)/\|(bx - (1 - b)y\|$ being on the same flat spot or elliptical section as $x/\|x\|$ and $y/\|y\|$, we have

$$(x, y) \in A(a, b, =).$$

Suppose that (3) holds. Suppose further that $x/\|x\|$ is on the flat spot and $y/\|y\|$ is on the adjacent elliptical section. If we choose (a, b) such that

(1) $(a, b) \in (0, 1)^2$,

(2) $(ax + (1 - a)y)/\|ax + (1 - a)y\|$, $(bx + (1 - b)y)/\|bx + (1 - b)y\|$

are on the same flat spot (elliptical section) as $x/\|x\|, y/\|y\|$ then we have

$$(x, y) \in A(a, b, =).$$

From the continuity of the norm we easily get that we may choose (a, b) such that (1) is valid and

$$(x, y) \in A(a, b, \geqslant).$$

We now consider Example 4.10.3 We consider the set K defined by the relation

$$K = ((t_1, t_2), t_i \geqslant 0).$$

Then for $(u \cdot v)$ in K and for all (a, b) in $(0,1)^2$ we have

$$(x, y) \in A(a, b, \leqslant).$$

We note that local conditions were formulated by Ficken (1944), Carlsson (1962), Kasahara (1954), Oman (1969, 1981) and many others. We begin with a technical lemma which has also an independent interest.

LEMMA 4.10.8. *Let X be a normed linear space. Then the norm of X is generated by an inner product iff there exists $\varepsilon > 0$ with the property that whenever (x, y) is a pair of elements of X with the properties:*

(1) $\|x\| = \|y\| = 1$,
(2) $\|x - y\| < \varepsilon$

then for some a in $(\frac{1}{2}, 1)$ and all b in $(0, 1)$ the following relation holds:

$$b(1-b)\|x + (1-a)y\|^2 - a(1-a)\|bx + (1-b)y\|^2 = -(b-a)(a+b-1).$$

Proof. Since the necessity is clear we prove only the sufficiency. Suppose indeed that X has the stated property. From the Jordan–von Neumann theorem we conlude that we may suppose without loss of generality that X is a two-dimensional space. Let (x, y) be arbitrary elements of X such that

$$\|x\| = \|y\| = 1,\ \|x - y\| < \varepsilon,\ \|ax + (1-a)y\| < 1, \text{ with } a \text{ arbitrary in } (0,1).$$

We consider the ellipse (X is identified with the plane) with its center at zero and x, y and $(bx + (1-b)y)/\|(bx + (1-b)y)\|$ lie on this ellipse. The ellipse determines a norm $\|,\|_2$ on X which is generated by an inner product. We now consider the cone determined by x, y in X, i.e. the set □

$$C(x, y) = (ax + by, a, b \geqslant 0).$$

We now consider the unit spheres corresponding to the two norms

$$S = (x, \|x\| = 1), \qquad E = (x, \|x\|_2 = 1).$$

Let a be in $(0,1)$. Then we have

$$\begin{aligned} b(1-b)\|ax + (1-a)y\|^2 &= a(1-a)\|bx + (1-b)y\|^2 - (b-a)(a+b-1) \\ &= a(1-a)\|bx + (1-b)y\|_2^2 - (b-a)(a+b-1) \\ &= b(1-b)\|ax + (1-a)y\|_2^2 \end{aligned}$$

which implies that

$$S \cap E \cap C(x, y) = S \cap C(x, y).$$

Let us consider x', y' in X such that

(1) $\|x'\| = 1 = \|y'\|$,
(2) $\|x' - y'\| < \varepsilon$,
(3) $\frac{1}{2}(x' + y') \in C(x, y)$,
(4) $y \notin C(x, y)$.

Then we obtain that

$$\|bx' + (1 - b)y'\| = \|bx + (1 - b)y\|_2 < 1.$$

since b is in $(\frac{1}{2}, 1)$. Then we find a new norm $\|, \|_3$ on X with the property that

(1) $\|x'\| = \|y'\|$,

(2) $\|bx' + (1 - b)y'\| = \|bx' + (1 - b)y'\|_3$.

If we set $E_3 = (u, \|u\|_3 = 1)$, then

$$E_3 \cap C(x', y') = S \cap C(x', y')$$

which can be proved as above. Thus

$$E \cap C(x', y) = S \cap C(x', y) = E_3 \cap C(x', y)$$

and since y' is not in $C(x, y)$,

$$C(x, y) \cap C(x', y') = C(x', y)$$

which gives that $E = E_3$. We then repeat the above procedure a finite number of times to get that $E = S$ holds, the lemma is proved. □

For the following lemma we need orthogonality in the Birkhoff sense $\perp_B$. From the definition we have the following known property.

PROPOSITION 4.10.9. *Let X be a normed linear space and let $x \perp_B y$. Then for all $a \neq 0 \neq b$ we have*

$$ax \perp_B by.$$

Proof. Obvious. □

LEMMA 4.10.10. *Let X be a normed linear space. Then the norm of X is generated by an inner product iff there exists an $\varepsilon > 0$ such that for any pair of elements (x, y) of X with*

(1) $\|x\| = \|y\| = 1$,

(2) $\|x - y\| < \varepsilon$

then $(x + y) \perp_B (x - y)$ holds.

We remark that the condition is clear necessary. In order to prove that the norm of a normed linear space having the stated property is generated by an inner product we first prove that the normed linear space is strictly convex.

Indeed, suppose that X is not strictly convex. Then we find two linearly independent elements x, y of X such that:

(1) $\|x\| = \|y\| = 1$, $\|x - y\| < \varepsilon$, $\|tx + (1 - t)y\| < 1$, t in $(\frac{1}{2}, 1)$,

(2) $\|sx + (1 - s)(tx + (1 - t)y)/\|tx + (1 - t)y\|\| = 1$, $s \in 0, 1$.

Then we have

$$(x + y) \perp_B (x - ((tx + (1 - t)y)/\|tx + (1 - t)y\|$$

and using (1) we see that $x - y$ and $x - ((tx + (1 - t)y))/\|tx + (1 - t)y\|$ are linearly independent. Since the set $S = (u, \ \|u\| = 1)$ can have only one support line at $(x + y)/\|x + y\|$; we have a contradiction. Thus X is a strictly convex space.

The next step is to show that if x, y in X have the properties

(1) $\|x\| = \|y\| = 1$,

(2) $\|x - y\| < \varepsilon$

then for all t in $(0, 1)$,

$$\|tx + (1 - t)y\| = \|ty + (1 - t)x\|.$$

Suppose that $tx + (1 - t)y = sx + (1 - s)y$ for $s \neq t$ both in $(0, 1)$. Then

$$(t + s)x + (2 - t - s)y \perp_B (x - y \quad \text{and} \quad x + y \perp_B (x - y).$$

But X is strictly convex by the above result and thus we must have $t + s = 2 - t - s$. Thus $s = 1 - t$ and the assertion is proved.

Suppose now that $\|x\| = \|y\|$ and $\|x - y\| < \varepsilon' = \min(1, \varepsilon)$ imply

$$t(1 - t)\|\tfrac{1}{2}(x - y)\|^2 - \tfrac{1}{4}\|tx + (1 - t)y\|^2 = -(t - \tfrac{1}{2})^2.$$

It is clear from the above results that we may consider only the case $0 \leqslant t \leqslant \frac{1}{2}$ and the relation we must prove is as follows:

If x, y are as above, then

$$\begin{aligned}\|ax + y\| &= \|a\|x + y\| \ (x + y)/\|x + y\| + (1 - a)y\| \\ &= \|(1 - a)x + y/\|x + y\| + a\|x + y\|y\|\end{aligned}$$

or

$$\|ax + y\|\|x + y\| = \|(1 - a)x + (1 - a - a\|x + y\|^2)y\|$$

and similarly

$$\begin{aligned}\|x + y\| &= \big\| \|x + ay\| \ x + ay/(\|x + ay\|) + (1 - a)y\big\| \\ &= \big\|(1 - a)(x + ay)/\|x + ay\| + \|x + ay\|y\big\|\end{aligned}$$

or

$$\|ax + y\|\|x + y\| = \|x + ay\|\|x + y\| = \|(1 - a)x(a - a^2 + \|x + ay\|^2)y\|$$

The function $g(t) = \|x + ty\|$ has the property that

$$g(-1) < g(0) = 1 < g(1)$$

and thus has a minimum at a negative t. This implies that

$$((1 - a)^2 + a\|x + y\|^2)/(1 - a) = (a - a + \|x + ay\|^2)/(1 - a)$$

or

$$a\|x + y\|^2 + (1 - a)^2 = \|x + ay\|^2.$$

Then by Lemma 4.10.8, the norm of the space is generated by an inner product.

Now we recall that Ficken (1944) has proved that a real normed linear space has a norm generated by an inner product iff for any pair (x, y) of elements of X with $\|x\| = \|y\|$ then for any real number t

$$\|tx + y\| = \|x + ty\|.$$

Lorch (1948) has greatly improved this result by showing that in order for X to have a norm generated by an inner product it is necessary and sufficient that for all pairs (x, y) of elements of X with $\|x\| = \|y\|$ for some r not equal to $0, \pm 1$, the following relation must be satisfied:

$$\|rx + y\| = \|x + ty\|.$$

In his Ph.D. Thesis, J. Oman has improved Lorch's result by obtaining a localized version as follows:

THEOREM 4.10.11 (Oman, 1969). *Let X be a real normed linear space. Then the norm of X is generated by an inner product iff for some $\varepsilon > 0$ and all $\|x\| = \|y\| = 1$, $\|x - y\| < \varepsilon$*

$$\|x + ry\| = \|rx + y\|$$

for some $r \neq 1$, $r > 0$.

Proof. Obviously the condition is necessary. Suppose that the normed linear space satisfies this condition. Then if we set $t = \min(1/(1 + r), r/(1 + r))$ then the hypothesis of the theorem is that there exists $\varepsilon > 0$ such that for all pairs $(x, y) \|x\| = \|y\| = 1$ and $\|x - y\| < \varepsilon$ there exists a in $(0, \frac{1}{2})$ such that

$$\|ax + (1 - a)y\| = \|(1 - a)x + ay\|.$$

Then by the result of Lemma 4.10.12 which we give immediately, for all t in $(0, 1)$

$$\|(tx + (1 - t)y)/\|tx + (1 - t)y\| - ((1 - t)x + ty)/\|(1 - t)x + ty\|\| < \varepsilon.$$

Now we can apply Lemma 4.10.10. In order to do this we show (using an induction argument) that

$$\|f(n)x + (1 - f(n))y\| = \|(1 - f(n))x + f(n)y\|$$

$(\|x\| = \|y\| = 1)$ and where $f(n)$ is a sequence which is defined in terms of x and y, recurrently.

If $\|x\| = \|y\| = 1$ and $\|x - y\| < \varepsilon$ then u is a scalar with the property that it is in $(0, \frac{1}{2})$ and $\|ux + (1 - u)y\| = \|(1 - u)x + uy\|$. Then we set $f(0)u$. For $n \geqslant 1$ we consider u_n be the scalar in $(0, \frac{1}{2})$ with the property that

$$\begin{aligned}&\|u_n(f(n - 1)x + (1 - f(n - 1))y + (1 - u_n))((1 - f(n - 1))x + f(n - 1)y)\| \\ &\quad = \|(1 - u_n))\,(f(n - 1)x + (1 - f(n - 1))y) + u_n((1 - f(n - 1)x + f(n - 1)y)\|\end{aligned}$$

and set

$$f(n) = (2u_n - 1)f(n - 1) + (1 - u_n).$$

If we prove that $f(n)$ is in $(0, 1)$ the induction implies that

$$\|f(n)x + (1 - f(n))y\| = \|(1 - f(n))x + f(n)y\|.$$

Now since

$$|\tfrac{1}{2} - f(n)| = |\tfrac{1}{2} - ((2u_n - 1)f(n-1) + (1 - u_n))|$$
$$= |\tfrac{1}{2} - f(n-1) + 2u_n - 1| < |\tfrac{1}{2} - f(n-1)|$$

we get that $(f(n))$ *converges to* $\frac{1}{2}$. Then we have that $(x + y)\perp_B(x - y)$ and Lemma 4.10.10 gives that the norm of X is generated by an inner product. □

LEMMA 4.10.12. *Let X be a normed linear space. Then for any x, y in X the following inequality holds:*

$$\alpha(z, w) \leqslant \alpha(x, y)$$

whenever z, w are in $C(x, y) = (ax + by, a, b \geqslant 0)$ (i.e. the cone generated by x and y).

Proof. We remark that we may assume that $\|x\| = \|y\| = \|z\| = \|w\| = 1$. Let $z = ax + by$ and $w = cx + dy$ with a, b, c, d positive numbers. Let $\|x - y\| = \varepsilon = \alpha(x, y)$. We may assume that $bc \geqslant ad$. □

Using the triangle inequality we have

$$(c + d)/\varepsilon = \|(c + d)/\varepsilon \cdot x\| = \|d(\varepsilon^{-1}x - \varepsilon^{-1}y + \varepsilon^{-1}(cx + dy)\| \leqslant d + \varepsilon^{-1}.$$

and

$$1 \leqslant \|cx + dy\| \leqslant c + d.$$

Thus

$$1 \leqslant c + d \leqslant d\varepsilon + 1.$$

We have to distinguish two cases:

CASE 1. $d \leqslant 1$. Then

$$\alpha(y, w) = \|y - (cx + dy)\| = \|(1 - d)(x - y) + (c + d - 1)x\|$$
$$\leqslant (1 - d)\varepsilon + (c + d - 1) = (1 - d)\varepsilon + d = \varepsilon.$$

CASE 2. $d \geqslant 1$. Then

$$\alpha(y, w) = \|y - (cx + dy)\| = \|c/d\,.\,x + (1 - 1/d)(cx + dy)\|$$
$$= c/d + 1 - 1/d = (c + d - 1)/d = (d\varepsilon)/d = \varepsilon.$$

Since $bc \geqslant ad$ we have then that z is in $C(y, w)$ and thus

$$\alpha(z, w) \leqslant \alpha(y, w) = \alpha(y, x)$$

and the lemma is proved. □

COROLLARY 4.10.13. *Let X be a normed linear space. Then the norm of X is generated by an inner product iff whenever $\|x\| = \|y\| = 1$ there exists $r \neq 1$, $r > 0$ (or $r \neq -1$, $r < 0$) such that*

$$\|rx + y\| = \|x + ry\|.$$

Proof. If $r > 0$ then the assertion follows from the above result. If $r < 0$ then $\|x\| = \|y\| = 1$ implies that $\|x\| = \|-y\| = 1$ and then

$$\|x - ry\| = \|rx - y\| = \|-rx + y\|$$

and the result follows from the first case. □

COROLLARY 4.10.14. *Let X be a normed linear space. Then the norm of X is generated by an inner product iff there exists $\varepsilon > 0$ and a in $(0, \frac{1}{2})$ such that for all pairs (x, y) of elements of X with $\|x\| = \|y\|$ and $\|x - y\| < \varepsilon$ the following relation holds:*

$$a(1-a)\|\tfrac{1}{2}\cdot x + \tfrac{1}{2}\cdot y\|^2 - \tfrac{1}{4}\cdot\|ax + (1-a)y\|^2 = -(\tfrac{1}{2} - a)^2.$$

Proof. The condition is clearly necessary. Suppose that X has the stated property. Let $\|x\| = \|y\|$ and $\|x - y\| < \varepsilon$. Then

$$a(1-a)\|\tfrac{1}{2}(x+y)\|^2 - \tfrac{1}{4}\cdot\|ax + (1-a)y\|^2 - (\tfrac{1}{2} - a)^2.$$

Applying this to the pair (y, x) we have

$$a(1-a)\|\tfrac{1}{2}(x+y)\| - \tfrac{1}{4}\|ay + (1-a)x\| = -(\tfrac{1}{2} - a)^2$$

and thus

$$\|ax + (1-a)y\| = \|ay + (1-a)x\|$$

and thus the norm of X must be generated by an inner product according to the above result. □

COROLLARY 4.10.15. *Let X be a normed linear space. Then the norm of X is generated by an inner product iff there exists $\varepsilon > 0$ and an order relation $r(=, \leqslant, \geqslant)$ such that for all pairs (x, y) in X with $\|x\| = \|y\| = 1$ for some $r > 1$ (r in $(0,1)$) the following relation holds:*

$$\|x + ry\| = \|y + rx\|.$$

Proof. It is clear that the condition is necessary. Suppose now that X has the property stated in the corollary. Let x, y be as follows: $\|x\| = \|y\|$, $\|x - y\| < \varepsilon$. Then we find a, a_1 in $(0,1)$ such that

$$\|x + ay\| r \|y + ax\|, \qquad \|x + a_1 y\| r \|y + a_1 x\|$$

where r is the order relation. We set

$$f(t) = \|x + ty\| - \|tx + y\|$$

which is a continuous function. If $f(a) \neq 0$ and $f(a_1) \neq 0$ then $a \neq a_1$. But $f(a)$ and $f(a_1)$ have opposite signs. This implies that for some s we have $f(s) = 0$. Then f is zero at some value of t which is not 1 and then the above theorem implies that the norm of X is generated by an inner product. □

The following is a complex version of an extension of Ficken's result.

THEOREM 4.10.16. *Let X be a complex normed linear space. Then the norm of X is generated by an inner product iff there exists an order relation r and a scalar*

$|a| \neq 0, 1$ *such that for all pairs of elements* (x, y) *with* $\|x\| = \|y\| = 1$ *the following relation holds:*

$$\|x + ay\| r \|y + ax\|.$$

Proof. Obviously the condition is necessary. Suppose now that X has the stated property. Then for $\|x\| = \|y\| = 1$ we have

$$\|x = \|\bar{a}/|a|y\| = 1 \quad \text{and} \quad \|x + a\cdot\bar{a}/|a|\cdot y\| r \|\bar{a}x + \bar{a}/|a|\cdot y\|$$

or

$$\|x + |a|y\| r \||a|x + y\|.$$

Since the roles of x and y are symmetric we obtain that

$$\|x + |a|y\| = \||a|x + y\|.$$

Then by Lorch's theorem 4.4.1, the norm of X is generated by an inner product. □

We denote by (L_1) the condition in the Lorch result, i.e. for some k not equal to 0 or 1 and for all x, y in X with $\|x\| = \|y\|$ the relation $\|x + ky\| = \|y + kx\|$ holds.

In his paper Lorch (1948) has considered other interesting conditions, which we list here.

(L_2) For all x, y, z in X with $\|x\| = \|y\|$ and $x + y + z = 0$ we have $\|x - z\| = \|y - z\|$.

(L_3) For all x, y, z, t in X with $\|x\| = \|y\|$, $\|z\| = \|t\|$ and $x + y + z + t = 0$ we have $\|x - t\| = \|y - z\|$ and $\|x - z\| = \|y - t\|$.

(L_4) If x, y, z are in X then $f(x, y, z) = \|x + y + z\|^2 + \|x + y - z\|^2 - \|x - y - z\|^2 - \|x - y + z\|^2$ does not depend on z,

(L_5) If x, y are in X and $\|x\| = \|y\|$ then for all real numbers $r \neq 0$ $\|rx + r^{-1}y\| \geqslant \|x + y\|$.

(L_6) If $x_1, \ldots, x_n$, $n \geqslant 3$, then *if* $x_1 + \cdots + x_n = 0$ we have the relation

$$\sum_{i \neq j} \|x_i - x_j\|^2 = 2n \sum_{i=1}^{n} \|x_i\|^2.$$

Then we have the following result.

THEOREM 4.10.17 (Lorch, 1948). *For a normed linear space* X *the following assertions are equivalent:*

(1) *The norm of* X *is generated by an inner product.*
(2) *For some* i *in* $[1, 6]$, *in* X, (L_i) *holds.*

For a proof we refer to E. R. Lorch's paper.

We now present an interesting theorem on the existence of inner product structures due to Rakestraw (1979).

THEOREM 4.10.18 (Rakestraw, 1979). *Let* X *be a normed linear space. Then the* *norm of X is generated by an inner product iff for some $n \geq 3$ and all $x_1, \dots, x_n$ in X and all real numbers $a_1, \dots, a_n$ with $a_1 + \cdots + a_n = 0$ the following relation holds:*

$$\left\| \sum_{i=1}^{n} a_i x_i \right\|^2 = - \sum_{1 \leq i < j \leq n} a_i a_j \|x_i - x_j\|^2.$$

Proof. Suppose that X has the stated property. Then for any $x(=x_1)$ and $y(=x_2)$ in X we have

$$\tfrac{1}{4}\|x+y\|^2 = \|\tfrac{1}{2}(x+y) + (-1)\cdot 0\|^2 = \tfrac{1}{2}\cdot\|x\|^2 + \tfrac{1}{2}\cdot\|y\|^2 - \tfrac{1}{4}\cdot\|x-y\|^2$$

or

$$\|x+y\|^2 + \|x-y\|^2 = 2(\|x\|^2 + \|y\|^2).$$

Then the Jordan–von Neumann theorem states that the norm of X is generated by an inner product.

Now we show the necessity of the above condition. Suppose first that $n = 3$. Then we have

$$\|a_1x_1 + a_2x_2 + a_3x_3\|^2 = a_1^2\|x_1 - x_3\|^2 + a_1a_2 \langle x_1 - x_3, x_2 - x_3\rangle + a_1a_2\langle x_2 - x_3, x_1 - x_3\rangle + a_2^2\|x_2 - x_3\|^2.$$

Then we replace $x_2 - x_3$ by $x_2 - x_3 + x_1 - x_2$ and $x_1 - x_3$ by $x_2 - x_3 + x_1 - x_2$ in the second and the third terms, and using the additivity of the inner product we obtain

$$\begin{aligned}\|a_1x_1 + a_2x_2 + a_3x_3\|^2 &= a_1\|x_1 - x_3\|^2 + a_1a_2\|x_1 - x_3\|^2 + \\ &\quad + a_1a_2\langle x_1 - x_3, x_2 - x_3\rangle + a_1a_2\|x_2 - x_3\|^2 + \\ &\quad + a_1a_2\langle x_2 - x_3, x_1 - x_2\rangle + a_2^2\|x_2 - x_3\|^2 \\ &= a_1(a_1 + a_2)\|x_1 - x_3\|^2 + a_2(a_1 + a_3)\|x_2 - x_3\|^2 + \\ &\quad + \langle x_2 - x_3, x_2 - x_1\rangle + \langle x_2 - x_3, x_1 - x_2\rangle \\ &= - a_1a_2\|x_1 - x_2\|^2 - \\ &\quad - a_1a_3\|x_1 - x_3\|^2 - a_2a_3\|x_2 - x_3\|^2\end{aligned}$$

and thus in this case the condition is satisfied.

Suppose that it is satisfied for all $i \leq n$ and we prove it for $n+1$. Let $x_1, \dots, x_{n+1}$ be arbitrary elements of X and $a_1, \dots, a_{n+1}$ be real numbers such that $a_1 + \cdots + a_{n+1} = 0$. Then if we set $a = a_1 + \cdots + a_{n-1}$ we have $a = -(a_n + a_{n+1})$ and then

$$\begin{aligned}\left\| \sum_{i=1}^{n+1} a_i x_i \right\|^2 &= a_{n+1}^2 \left\| \sum_{i=1}^{n} (a_i/ - a_{n+1})x_i - x_{n+1} \right\|^2 \\ &= a_{n+1}^2 \left\| \sum_{i=1}^{n-1} (a_i/ - a_{n+1})x_i + (a_n/ - a_{n+1})x_n - x_{n+1} \right\|^2 \\ &= a_{n+1}^2 \left\| (a/ - a_{n+1})\left(\sum_{i=1}^{n-1} (a_i/a)x_i \right) + (a_n/ - a_{n+1})x_n - x_{n+1} \right\|^2.\end{aligned}$$

But $(a/ - a_{n+1}) + (a_n/ - a_{n+1}) - 1 = 0$ and thus

$$\left\|\sum_{i=1}^{n+1} a_i x_i\right\|^2 = a_{n+1}^2\left(-\frac{aa_n}{a_{n+1}^2}\right)\left\|\sum_{i=1}^{n-1}(a_i/a)x_i - x_n\right\|^2 - $$

$$-(a/a_{n+1})\left\|\sum_{i=1}^{n-1}(a_i/a)x_i - x_{n+1}\right\|^2 - (a_n/a_{n+1})\|x_n - x_{n+1}\|^2.$$

By the induction hypothesis, we have further,

$$\left\|\sum_{i=1}^{n+1} a_i x_i\right\|^2 = (-aa_n)\left(\sum_{1\leqslant i<j\leqslant n}(a_i a_j)/a^2\|x_i - x_j\|^2\right) +$$

$$+(-aa_n)\left(\sum_{i=1}^{n-1}(a_i/a)\|x_i - x_{n+1}\|^2\right) +$$

$$+(aa_{n+1})\left(-\sum_{1\leqslant i<j\leqslant n-1} a_i a_j/a^2\|x_i - x_j\|^2\right) -$$

$$- a_n a_{n+1}\|x_n - x_{n+1}\|^2.$$

Simplifying we obtain the relation

$$\left\|\sum_{i=1}^{n+1} a_i x_i\right\|^2 = -\sum_{1\leqslant i<j\leqslant n} a_i a_j\|x_i - x_j\|^2$$

and the theorem is proved. □

In his interesting paper R. M. Rakestraw has proved (without using the inner product structure of X) that if a normed linear space has the property of the above theorem then

(1) the Jordan–von Neumann condition (the Apollonius parallelogram law) and
(2) the Ficken condition

hold for all x, y in $X, \|x\| = \|y\|$ and a, b in R, $\|ax + by\| = \|bx + ay\|$.

4.11. Other Norm Characterizations of Inner Product Structures

We have presented some characterizations of inner product structures on a normed linear space using some identities which must hold for some subsets of elements in the space. Such identities are the parallelogram law (the Apollonius theorem), the Fréchet identity, the Ficken identity, etc.

The activity in this area is very intensive and as result we have a lot of other similar results on the existence of inner product structures on normed linear spaces. Since we believe that these are of independent interest as well as of interest in applications we shall try to present some of them.

First we present two local theorems of Oman (1980).

THEOREM 4.11.1 (Oman, 1980). *Let X be a normed linear space. Then the norm of X is generated by an inner product iff there exists an ε in $(0, 2)$ with the property that whenever x, y are in X with $\|x\| = \|y\|$ and $\|x-y\| = \varepsilon$ then there exists a real* *number* $b \neq 0$ *with the property that, for all* $a \geqslant 0$,

$$a\|x + by\|^2 - \|bx + ay\|^2 = (a - b)(1 - ab).$$

Proof. The necessity. Let x, y in X with $\|x\| = \|y\|$ and a, b as in the theorem. Then we have

$$\begin{aligned} a\|x + by\|^2 - \|bx + ay\|^2 &= a(1 + b^2 + 2b\langle x, y\rangle - b(1a^2 + 2a\langle x, y\rangle \\ &= a + ab^2 - b - ba^2 = a - b + ab(b - a) \\ &= (a - b)(1 - ab). \end{aligned}$$

□

The sufficiency. In order to prove that the norm of X is generated by an inner product when X satisfies the condition of the theorem it suffices to consider only two-dimensional subspaces. Let us consider M to be any two-dimensional subspace of X and consider the unit sphere of $M, S_M = (x, x \in M, \|x\| = 1)$. Then by hypothesis there exists a real number $b \neq 0$ such that for all $a \geqslant 0$ the following relation holds:

$$a\|x + by\|^2 - b\|x + ay\|^2 = (a - b)(1 - ab).$$

There are several cases which we must consider

CASE 1. $\|x + by\|^2 = (1 - b)^2$. Then $\|x + ay\|^2 = (1 - a)^2$ for $a \geqslant 0$. Taking $a = 1$ we have $x + y = 0$. Thus x and y are linearly dependent. Since we suppose that x and y are linearly independent, this is a contradiction.

CASE 2. $\|x + by\|^2 = (1 + b)^2$. Then we have that for all $a \geqslant 0$, $x + ay^2 = (1 + a)^2$ and thus $S_M \cap C(x, y)$ is an open line segment joining x and y.

CASE 3. $\|x + by\|^2 \neq (1 + b)^2$ and $\|x + by\|^2 \neq (1 - b)^2$. If we consider in the plane determined by x and y the ellipse with center O and containing points x, y and $x + by/\|x + by\|$ then the norm associated with this ellipse (the Minkowski functional on M defined by the convex domain whose boundary is the above ellipse), say $|\,,|$ has the property that it is generated by an inner product and

$$|x| = |y| = 1, \qquad |x + by| = \|x + by\|.$$

If $a \geqslant 0$ then according to the hypothesis, we have,

$$\begin{aligned} b\|x + ay\|^2 &= a\|x + by\|^2 - (a - b)(1 - ab) \\ &= a|x + by| - (a - b)(1 - ab) = b|x + ay|^2 \end{aligned}$$

which gives that

$$S_M \cap C(x, y) = E(x, y) \cap C(x, y)$$

where $E(x, y)$ is the above ellipse which passes through x, y and $(x + by)/\|x + by\|$. Now we consider the family of open subset $(S_M \cap C(x, y);\ \|x\| = \|y\| = 1, \|x - y\| = \varepsilon)$ which is a covering of S_M. Since S_M is a compact set (M being two-dimensional) we find a finite subcovering of S_M, say $(S_M \cap C(x_i, y_i), i = 1,2,3, \ldots, n)$. We can suppose that this finite covering is minimal and that for all $i \leqslant n - 1$

$$C(x_i, y_i) \cap C(x_{i+1}, y_{i+1}) \neq \emptyset.$$

This means that for each $i \leqslant n - 1$, $C(x_i, y_i) \cap S_M$ and $C(x_{i+1}, y_{i+1}) \cap S_M$ are overlapping segments of portions of ellipses and if one is a portion of an ellipse then both are portions of the same ellipse or both are portions of the same line. Since at least one is a portion of an ellipse, S_M being a closed curve, the argument shows that C is in fact an ellipse. This proves that M is an inner product space and, being an arbitrary two-dimensional subspace of X, X must be an inner product space.

THEOREM 4.11.2 (Oman, 1969). *Let X be a normed linear space. Then X is an inner product space if and only if there exist a set K in X such that K is absorbing (i.e. for each $x \in X$ there exists a_x in R such that $a_x \cdot x \in K$) and for each x, y in K there exists λ, μ such that*

(1) *λ, μ are in $(0, 1)$,*
(2) *$(x, y) \in A(\lambda, \mu, r)$*

r being $=$, $\leqslant$, or $\geqslant$ and fixed for all pairs (x, y).

Proof. If X is an inner product space then the condition is obviously necessary. To prove that it is a sufficient condition, we may suppose without loss of generality that X_1 is a two-dimensional space in X, and we show that the unit sphere of X_1, S_{X_1}, is an ellipse. Then the assertion of the theorem follows. □

Suppose first that r means $\leqslant$. We consider now the Loewner ellipse with center in the origin of X and containing S_{X_1}. Since this Loewner ellipse is an 2-ellipse we can consider at least two points x and y (as well as the points $-x$ and $-y$) in S_{X_1} which are on the Loewner ellipse. If the Loewner ellipse does not coincide with S_{X_1} then there exist u, v on the ellipse and in S_{X_1} such that $(u, v) = S_{X_1} \cap C(u, v) \cap E$, E being the Loewner ellipse. Now we find α and β in R such that $\alpha u \in K$, $\beta v \in K$. Let us consider the Minkowski functional determined by the Loewner ellipse, say $|\,,|$. This is clearly induced by an inner product and $\|x\| \geqslant |x|$ for all x in X_1. According to the hypothesis, we have,

$$\begin{aligned}
&\lambda\mu(\lambda + \mu - 2\lambda\mu)\alpha^2 + (1 - \lambda)\cdot(1 - \mu)\cdot(\lambda + \mu - 2\lambda\mu)\beta^2 \\
&\quad \geqslant \mu(1 - \mu)\|\lambda\alpha u + (1 - \lambda)\beta v\|^2 + \lambda(1 - \lambda)\|\mu\alpha u - (1 - \mu)\beta v\|^2 \\
&\quad \geqslant \mu(1 - \mu)|\lambda\alpha u + (1 - \lambda)\beta v|^2 + \lambda(1 - \lambda)|\mu\alpha u - (1 - \mu)\beta v|^2 \\
&\quad = \lambda\mu(\alpha + \mu - 2\lambda\mu)\alpha^2 + (1 - \lambda)(1 - \mu)\cdot(\lambda + \mu - 2\lambda\mu)\beta^2
\end{aligned}$$

which implies that the equality holds. This implies further that

$$\begin{aligned}
&\pm(\lambda\alpha u + (1 - \lambda)\beta v)/\|\lambda\alpha u + (1 - \alpha)\beta v\| \\
&\pm(\mu\alpha u - (1 - \mu)\beta v/\|\mu\alpha u - (1 - \mu)\beta v\|
\end{aligned}$$

are in $S_{X_1} \cap E$, since at least one of these elements is in $C(u, v)$ and different from u and v. This is a contradiction, and thus S_{X_1} coincides with Loewner's ellipse and X_1 is an inner product space.

If r is $\|\geqslant\|$ then we consider Loewner's ellipse contained in S_{X_1} and, as above, we prove that S_{X_1} coincides with Loewner's ellipse and thus X_1 is an inner product space. The case of equality is obvious from the cases discussed. □

EXAMPLE 4.11.3. As an example of a set K satisfying the conditions of the theorem is any boundary of a closed, bounded set containing the origin in the interior.

COROLLARY 4.11.4 (Day, 1959). *A normed linear space X is an inner product space* *if and only if for each pair of elements* x, y *in* X *with* $\|x\| = \|y\| = 1$ *there exist* λ, μ *in* $(0, 1)$ *such that*

$$(\lambda + \mu - 2\lambda\mu)(\lambda\mu + (1 - \lambda)(1 - \mu)\, r\, \mu(1 - \mu)\|\lambda x + (1 - \lambda)$$
$$y\|^2 + \lambda(1 - \lambda)\|\mu x + (1 - \mu)y\|^2$$

where r is $\geqslant$, $\leqslant$ or $=$ and fixed for all pairs (x, y).

COROLLARY 4.11.5 (Kasahara, 1954). *A normed linear space* X *is an inner product space if and only if there exists a in* $(0, \frac{1}{2})$ *such that for any pair of elements* x, y *in* X *there exists* $b = b(x, y)$ *in* $[a, 1 - a]$ *such that the following inequality holds:*

$$b\|x\|^2 + (1 - b)\|y\|^2 \geqslant b(1 - b)\|x - y\|^2 + \|bx + (1 - b)y\|^2.$$

COROLLARY 4.11.6 (Schoenberg, 1952). *The normed linear space* X *is an inner product space if and only if in* X *Ptolemy's inequality holds: if* x, y, u, v *are arbitrary elements of* X *then*

$$\|y - x\| \cdot \|v - u\| + \|v - x\| \cdot \|u - y\| \geqslant \|u - x\| \cdot \|v - y\|.$$

We now present an interesting characterization of normed linear spaces having an inner product structure (i.e. the norm is generated by an inner product) obtained by L. M. Blumenthal (1968) and which is related to generalized ptolemaic inequality proved by Blumenthal: if p is in $[0, \frac{1}{2}]$ then for any (x, y, z) elements of a normed linear space the following inequality holds:

$$(I_p) \|x - y\|^p \|z\|^p + \|y - z\|^p \|x\|^p \geqslant \|x - z\|^p \|y\|^p.$$

THEOREM 4.11.7 (Blumenthal, 1968). *A normed linear space* X *has an inner product structure iff for any* x, y *in* X *for which* $\|x + y\| = \|x - y\|$ *holds we have* $\|x - y\|^2 = \|x\|^2 + \|y\|^2$.

Proof. Obviously the condition is necessary. Suppose now that it is satisfied in X. To prove that X has an inner product structure it suffices to show that whenever x, y are elements of norm 1 in X then

$$(+) \|x + y\|^2 + \|x - y\|^2 = 4$$

holds. Indeed, we consider the elements

$$u = \tfrac{1}{2}(x + y), \quad v = \tfrac{1}{2}(x - y)$$

and then we remark that $\|u + v\| = \|u - v\|$ and thus we must have the relation:

$$\|u\|^2 + \|v\|^2 = \|u - v\| = 1.$$

This relation is exactly the required equality $(+)$. □

In a recent paper M. S. Klamkin and A. Meir have introduced a new class of normed linear spaces called 'symmetric spaces'. The definition of this class is as follows.

DEFINITION 4.11.8 (Klamkin and Meir, 1982). A normed linear space X is said to be symmetric iff for any x, y in X, $\|x\| = \|y\| = 1$ the following equality holds for all real numbers r:

$$\|rx - y\| = \|x - ry\|.$$

In the same paper they assert that this class of normed linear space has an inner product structure. This assertion follows obviously from Lorch's theorem (see Theorem 4.4.1, 4.10.16) under very weak conditions. In fact the authors' proof of this result is false (see Istrăţescu, 1984).

For the proof of the following theorem we refer to Oman (1969).

THEOREM 4.11.9. *A normed linear space X is an inner product space if and only if there exists a set K with nonempty interior and a in* $(0, 1)$ *such that for any pair* (x, y) *of elements of K,* (x, y) *is in* $A(a, \frac{1}{2}, =)$.

THEOREM 4.11.10. *A normed linear space X is an inner product space if and only if there exists a set K in X with nonempty interior and such that for any pair* (x, y) *of elements of K there exists a in* $(0, 1)$ *with the property that* (x, y) *is in* $A(a, \frac{1}{2}, =)$.

THEOREM 4.11.11. *A normed linear space X is an inner product space if and only if there exists a convex cone K with interior nonempty and a in* $(0, 1)$ *such that for all pairs* (x, y) *of elements of K there is b in* $(0, 1)$ *such that* (x, y) *is in* $A(b, a, =)$.

Also we note that in Oman (1969) there are examples which show that '$=$' cannot be replaced by '$\leqslant$' or '$\geqslant$'.

We now give a characterization of inner product spaces using the so-called Clarkson angle measure between two points of the unit sphere of a Banach space.

DEFINITION 4.11.12. The Clarkson angle measure between the nonzero points x and y of a normed linear space X is a real number defined by the formula:

$$\big\| x/\|x\| - y/\|y\| \big\| = \alpha[x, y].$$

Using this we have the following result.

THEOREM 4.11.13 (Oman, 1969). *A normed linear space X is an inner product space if and only if there exists λ, μ in* $(0, 1)$ *and* $\varepsilon > 0$ *such that if* $\alpha[x, y] < \varepsilon$ *then* (x, y) *is in* $A(\lambda, \mu, r)$*, where r is either* $\leqslant$, $\geqslant$ *or* $=$.

Proof. If X is an inner product space it is not difficult to see that the condition of the theorem is satisfied; thus the necessity is clear. For the sufficiency we show that for any two-dimensional subspace X_1 or X the unit sphere is an ellipse and this obviously implies the assertion of the theorem. Suppose that r is $\leqslant$. Let S_{X_1} be the unit sphere of X_1 and also let E be Loewner's ellipse contained in S_{X_1} which we know to be a 2-ellipse. Let us consider the intersection of S_{X_1} with Loewner's ellipse E and let M be the intersection which we know to be not void (it contains at least four points). Let u and v be two points in this intersection with the property that $C(u, v)$, $M = (u, v)$. In this case there exist a and b in R^+ such that $x = au + bv$ and $y = (\mu/(1 - \mu)au - \lambda/(1 - \lambda)bv$ have the property that $\|x\| = 1$ and $\|x - v\| < \varepsilon/2$, $\|y/\|y\| - v\| < \varepsilon/2$. In this case $\alpha[x, y] < \varepsilon$ and the hypothesis of the theorem implies that ($|,|$ being the norm defined by the ellipse E)

$$\begin{aligned}&\mu(1-\mu)\|\lambda x+(1-\lambda)\,y\|^2+\lambda(1-\lambda)\|\mu x-(1-\mu)y\|^2\\&\quad\leqslant\lambda\mu(\lambda+\mu-2\lambda\mu)\|x\|^2+(1-\lambda)(1-\mu)(\lambda+\mu-2\lambda\mu\|y\|^2\\&\quad=\lambda\mu(\lambda+\mu-2\lambda\mu)|x|^2+(1-\lambda)(1-\mu)(\lambda+\mu-2\lambda\mu)|y|^2\\&\quad=\mu(1-\mu)|\lambda x+(1-\lambda)y|^2+\lambda(1-\lambda)|\mu x-(1-\mu)y|^2\\&\quad=\mu(1-\mu)\|\lambda x+(1-\lambda)y\|^2+\lambda(1-\lambda)\|\mu x-(1-\mu)y\|^2\end{aligned}$$

which shows that in fact we have an equality. This implies that x is in M and since x is also in $C(u, v)$, this contradiction shows that $M = S_{X_1}$. Thus X_1 is an inner product space and the theorem is proved since the cases $\geqslant$ and $=$ are similar. □

COROLLARY 4.11.14 (Oman, 1969). *A normed linear space X is an inner product space if and only if there exists a in* $(0,1)$, *a relation r and* $\varepsilon > 0$ *such that for all x, y in X with Clarkson angle measure less than* ε_1 *there is b in* $(0,1)$ *such that* (x,y) *is in* $A(a,b,r)$.

Proof. If X is an inner product space then it is easy to prove that the condition of the theorem is satisfied.

Suppose now that X has the property stated in the theorem. In order to prove that X is an inner product space we prove that every two-dimensional subspace X_1 of X is an inner product space. We suppose for simplicity that r means $\geqslant$, for other possibilities the treatment is similar. We now use the above result, namely that if x, y is in X with a Clarkson angle measure less than ε then (x, y) is in $A(\lambda, \mu, \geqslant)$ for all λ in $(0, 1)$. Then Theorem 4.11.13 implies the assertion of the theorem.

Suppose now that x, y are such that

$$\big\|x/\|x\| - y/\|y\|\big\| < \varepsilon.$$

We consider the points

$$x/\|x\|,\ y/\|y\|, \qquad (\mu x - (1-\mu)y)/\|\mu x - (1-\mu)y\|$$

and either there exists an ellipse through these points and center at origin or

$$\|\mu x - (1-\mu)y\| = \mu\|x\| + (1-\mu)\|y\|$$

or

$$\|\mu x - (1-\mu)y\| = |\mu\|x\| - (1-\mu)\|y\||.$$

Suppose that the last two cases happen and then for λ in $(0, 1)$, (x, y) is in $A(\lambda, \mu, \geqslant)$. Thus it remains to consider the possibility that through the above points passes an ellipse E and let $|,|$ be the norm induced by this ellipse E. Suppose that (x, y) is not in $A(\lambda, \mu, \geqslant)$ for all λ. Then there exist u, v in $C(x, y)$ such that

$$\|u\| = |u| = \|v\| = |v| = 1$$

and

$$\|au + bv\| < |au + bv|$$

if $ab > 0$. Let us consider $u = mx + ny$ and $v = px + qy$ with all m,n,p,q in R^+.

Define

$$k = \mu/(1-\mu)\cdot((1-\mu)m + \mu n)/(1-\mu)p + \mu q).$$

Since the Clarkson angle measure between x and y is less than ε there exists λ in $(0,1)$ such that

$$\begin{aligned}
&\lambda\mu(\lambda+\mu-2\lambda\mu)\|u\|^2 + (1-\lambda)(1-\mu)(\lambda+\mu-2\lambda\mu)|kv|^2\\
&\quad = \lambda\mu(\lambda+\mu-2\lambda\mu)\|u\|^2 + (1-\lambda)(1-\mu)(\lambda+\mu-2\lambda\mu)\|kv\|^2\\
&\quad = \mu(1-\mu)\|\lambda u + (1-\lambda)kv\|^2 + \lambda(1-\lambda)\|\mu u - (1-\mu)kv\|^2\\
&\quad < \mu(1-\mu)|\lambda u + (1-\lambda)kv|^2 + \lambda(1-\lambda)|\mu u - (1-\mu)kv|^2\\
&\quad = \lambda\mu(\lambda+\mu-2\lambda\mu)|u|^2 + (1-\lambda)(1-\mu)(\lambda+\mu-2\lambda\mu)|kv|^2
\end{aligned}$$

and this contradiction proves our assertion. □

If r is $\leqslant$ or $=$ then in a similar way we prove that all two-dimensional subspaces are inner product spaces.

Another localized characterization of inner product space in the same vein will be proved later using certain facts about orthogonality. We mention now some generalizations of the Jordan–von Neumann characterization. N. Aronszajn has remarked that Jordan–von Neumann admit the following reformulation:

THEOREM 4.11.15 (Aronszajn, 1935). *A normed linear space X is an inner product space if and only if for all x, y in X the following identity holds:*

$$\tfrac{1}{2}(\|x+y\|) = \tfrac{1}{2}(2\|x\|^2 + 2\|y\|^2 - \|x-y\|^2)^{1/2}.$$

This formula has the advantage that it may be formulated in terms of the triangle determined by the points O, x and y. Then we have the following result.

THEOREM 4.11.16 (Aronszajn, 1935). *A normed linear space X is an inner product space if and only if there exists a function defined on R^3 with values in R^+, say f, with the following properties:*

(1) $f(t, 0, t) = t$,
(2) $\tfrac{1}{2}\|(x+y)\| = f(\|x\|, \|y\|, \|x-y\|)$.

See also Senechalle (1968), Valentine and Andalafte (1973), and Section 4.15.

THEOREM 4.11.7 (Johnson, 1973). *A normed linear space X is an inner product space if and only if for some integer $k \geqslant 3$ and all x, y in X the following identity holds:*

$$\sum_{m=0}^{k} \binom{k}{m}(-1)^m \|x+my\|^2 = 0.$$

From his results on the characterization of inner product spaces using a form of the four-point property, R. W. Freese has obtained the following interesting characterization:

THEOREM 4.11.18 (Freese, 1968). *A Banach space is an inner product space iff the following property holds: if x, y are in X and $\|x-y\| = \|3x-y\|$ then*

$$\|y\|^2 = \|x-y\|^2 + 3\|x\|^2.$$

For a proof as well as for other related results see Section 4.13.

The following class of Banach spaces was introduced by Kantorovich (1935) and later identified by S. Kakutani as being in fact inner product spaces. Later we give a proof of this fact using Kakutani's characterization of inner product spaces.

DEFINITION 4.11.19 (Kantorovich, 1935). The normed linear space is said to be quasi-convex or a Kantorovich space if for any finite system of points $x_1, x_2, \ldots, x_n$ of X and x_0 in X there exists a point x_0^* of X in the real linear span of $(x_1, x_2, \ldots, x_n)$ such that

$$\|x_0^* - x_i\| \leqslant \|x_0 - x_i\|, \qquad i = 1, 2, \ldots, n.$$

Then we have the following result whose proof will be given later.

THEOREM 4.11.20 (Kakutani, 1939). *The normed linear space X is an inner product space if and only if X is a Kantorovich space.*

The following characterizations use a certain form of parallelogram inequality.

THEOREM 4.11.21 (Yang Cong-sen, 1963). *A normed linear space X is an inner product space if and only if the following inequalities hold:*

(1) *$p > 2$ and x, y arbitrary in X,*

$$2\|x\|^p + 2\|y\|^p \leqslant \|x + y\|^p + \|x - y\|^p$$

(2) *p in $(1, 2)$, $((1/p) + (1/q) = 1)$ and x, y arbitrary in X.*

An interesting characterization of inner product structures was obtained by D. Delbosco using a condition similar to that of P. Jordan and John von Neumann.

THEOREM 4.11.22 (Delbosco, 1975/76). *The norm of a Banach space X is generated by an inner product iff for all $\|x\| = \|y\| = 1$ and a in $[0, 4]$*

$$\|x + y\|^2 \geqslant a \quad \textit{and} \quad \|x - y\|^2 \geqslant 4 - a \rightarrow \|x + y\|^2 + \|x - y\|^2 = 4.$$

Proof. From the Jordan–von Neumann theorem it suffices to consider only two-dimensional subspaces. We suppose without loss of generality that X is two-dimensional. Then we can consider a Euclidean norm $\|\,,\|_2$ on X with the property that for all z in X, $\|z\|_2 \leqslant \|x\|$. The set $(z, \|z\|_2 = 1)$, by Loewner's theorem, has at least four points of contact with $(z, \|z\| = 1)$. Let us consider these points to be $(x, y, -x, -y)$. We may suppose that $y \neq \pm x$. Then we have,

$$\|x\|_2 = \|x\| = \|y\| = \|y\|_2 = 1$$

and thus

$$a = \|x + y\|_2^2 \leqslant \|x + y^2, \qquad 4 - a = \|x - y\|_2 \leqslant \|x - y\|^2$$

and thus we must have

$$\|x + y\|^2 + \|x - y\|^2 = 4.$$

This clearly implies that

$$\|x + y\|_2 = \|x + y\|, \qquad \|x - y\|_2 = \|x + y\|.$$

We can repeat the procedure and get a dense set of points which are common for the sets

$$(z, \|z\|_2 = 1), \qquad (z, \|z\| = 1).$$

This implies, by the density, that they are identical. Then the norm of X is generated by an inner product (i.e. it is $\|,\|_2$). Similarly, we can prove the following (to some extent, dual formulation) results. □

THEOREM 4.11.23 (Delbosco, 1975/76). *Let X be a Banach space. Then the norm of X is generated by an inner product iff for all x, y in X, $\|x\| = \|y\| = 1$ and any a in* $[0,4]$

$$\|x + y\|^2 \leqslant a \quad \textit{and} \quad \|x - y\|^2 \leqslant 4 - a \rightarrow \|x + y\|^2 + \|x - y\|^2 = 4.$$

REMARK 4.11.24. In the proof we use another result of Loewner, namely that if X is two-dimensional then there exists a Euclidean norm $\|,\|_2$ such that for all z in X, $\|z\|_2 \geqslant \|z\|$.

4.12. Orthogonality in Normed Linear Spaces and Characterizations of Inner Product Spaces

If X is a linear space over $K(R$ or $C)$ and if on X there exists an inner product denoted by $\langle\,,\rangle$ then on X we can define the notion of orthogonal elements by the relation: x is orthogonal to the element y if $\langle x, y\rangle = 0$.

We note the following properties of this notion. First we introduce the function $p\colon X \rightarrow R$ defined by the relation:

$$p(x) = \langle x, x\rangle^{1/2}.$$

Then the following properties hold:

(P_1) $p(ax) = |a|p(x)$,
(P_2) $p(x + y) \leqslant p(x) + p(y)$,
(P_3) If $\langle x, x\rangle = 0$ if and only if $x = 0$ then $p(x) = 0$ if and only if $x = 0$.

Then we have also the following characterization of orthogonal elements in X using the function p.

PROPOSITION 4.12.1. *An element x is orthogonal to an element y if and only if for all scalars k the following inequality holds:*

$$p(x + ky) \geqslant p(x)$$

PROPOSITION 4.12.2. *An element x is orthogonal to an element y if and only if for all scalars k*

$$p(x + ky) = p(x - ky).$$

PROPOSITION 4.12.3. *An element x is orthogonal to an element y if and only if*

$$p(x)^2 + p(y)^2 = p(x - y)^2.$$

The advantage of the formulation of the orthogonality property in Propositions

4.12.1–4.12.3 is that we use the function p which can be defined on more general spaces, since in fact the basic properties (P_1)–(P_3) are those of a norm. In what follows we present some possible definitions of orthogonality on normed linear space; if we require that the notion of orthogonality has properties like the orthogonality in inner product spaces then this forces the space to be an inner product space.

For the convenience of the reader, as well as for the motivation of some definitions given below, we list the basic properties of orthogonality defined using the inner products:

If $\langle , \rangle$ is the inner product of X and $x \perp y$ means $\langle x, y \rangle = 0$ then the following assertions hold:

(1) $x \perp y$ then $y \perp x$,
(2) $x \perp y$ then for all scalars t and s, $tx \perp sy$,
(3) $x \perp y$ and $x \perp z$ then $x \perp (y + z)$,
(4) if x, y are given then there exists a scalar k such that $x \perp (kx + y)$.

In what follows we give some definitions of orthogonality and use some of them to obtain a characterization of inner product spaces. In the last part, using certain differentiability properties of a norm, we consider an inner product which we call 'a generalized inner product'.

DEFINITION 4.12.4. Let X be a normed linear space and let $\perp$ be a function defined on a subset of $X \times X$. We call this 'an orthogonality'. We say that orthogonality is:

(1) symmetric, if $x \perp y$ then $y \perp x$;
(2) homogeneous if $x \perp y$ then for all scalars a and b, $ax \perp by$;
(3) additive on the right if $x \perp y$ and $x \perp z$ then $x \perp (y + z)$ and additive on the left if $x \perp y$ and $z \perp y$ then $(x + z) \perp y$;

and a crucial role is played by

(4) nontrivial, if x, y are in X then there exists a scalar k such that $x \perp (kx + y)$.

Now we formulate several orthogonality nations. As we show later only some of them are useful in characterizing the inner product spaces in the class of normed linear spaces. Historically the first definition of orthogonality in normed linear spaces was given by B. D. Roberts (1934), and is as follows.

DEFINITION 4.12.5 (Roberts, 1934). Let X be a normed linear space. We say that $x \perp_R y$, i.e. x is orthogonal to an element y in the sense of Roberts if for all scalars the following relation holds:

$$\|x + ky\| = \|x - ky\|.$$

As we remarked above, if X is an inner product space this orthogonality notion reduces to the natural orthogonality (i.e. that defined using the inner product). As we show later, following an example of R. C. James, this is not an appropriate definition for some Banach spaces.

DEFINITION 4.12.6 (*Isosceles orthogonality*). Let X be a normed linear space. An element x is said to be orthogonal to an element y, $x \perp_{\mathrm{I}} y$ if $\|x + y\| = \|x - y\|$.

DEFINITION 4.12.7 (*Pythagorean orthogonality*). Let X be a normed linear space. An element x is said to be orthogonal to an element y, $x \perp_{\mathrm{P}} y$ if $\|x\|^2 + \|y\|^2 = \|x - y\|^2$.

We remark that Definitions 4.12.6 and 4.12.7 formulate in terms of the norm the well-known properties of Euclidean geometry: two vectors are orthogonal if and only if their sum and their difference can be sides of an isosceles triangle and the property that two vectors are orthogonal if and only if there exists a right triangle with the two vectors as legs.

DEFINITION 4.12.8 (*Birkhoff orthogonality*). An element x of a normed linear space X is said to be orthogonal on an element $y \in X$, $x \perp_{\mathrm{B}} y$, if for all scalars k the following inequality holds:

$$\|x + ky\| \geqslant \|x\|.$$

DEFINITION 4.12.9 (Diminnie *et al.*, 1983). An element x of a normed linear space X is said to be orthogonal to an element y if

$$(1 + \alpha^2)\|x - y\|^2 = \|x - \alpha y\|^2 + \|\alpha x - y\|^2,$$

where α is a fixed real number $\neq 1$. We denote this orthogonality by $x \perp_{\alpha} y$.

DEFINITION 4.12.10 (Carlsson, 1961). An element x of a normed linear space X is said to be orthogonal to an element y of X if for some real numbers $(a_i, b_i, c_i)_1^n$ the following relations hold:

(1) $\sum_{i=1}^{n} a_i b_i^2 = \sum_{i=1}^{n} a_i c_i^2 = 0,$

(2) $\sum_{i=1}^{n} a_i b_i c_i = 1,$

(3) $\sum_{i=1}^{n} a_i \|b_i x + c_i y\|^2 = 0.$

In a recent note (1984) the author has remarked that we can introduce some weaker notions of orthogonality in normed linear space which have, however, the same power for the characterization of inner product spaces as the corresponding orthogonality notions which are stronger.

DEFINITION 4.12.11 (Istrăţescu, 1984). We say that an element x of a normed linear space X is isosceles lower semi-orthogonal to an element y if

$$\|x + y\| \leqslant \|x - y\|$$

holds. We use for this the notation $x \perp_{\mathrm{I},l} y$.

DEFINITION 4.12.12 (Istrăţescu, 1984). We say that an element x of a normed linear space X is isosceles upper semi-orthogonal to an element y if

$$\|x + y\| \geqslant \|x - y\|$$

holds. We use for this the notation $x \perp_{I,u} y$.

We can, of course define the Roberts, the Pythagorean, and the Birkhoff lower and upper semi-orthogonalities. We leave these to the interested reader to formulate.

DEFINITION 4.12.13. Given an orthogonality relation denoted by $\perp$ we say that this satisfies the 'rectangle inequality' if whenever $x \perp y$ then

$$\|x + y\|^2 + \|x - y\|^2 \, R \, 2(\|x\|^2 + \|y\|^2)$$

holds, where R means '$\geqslant$', '$\leqslant$' or '$=$'.

We now present some examples related to the above orthogonality notions. The first shows that in some Banach spaces the Roberts definition of orthogonality is not very useful.

EXAMPLE 4.12.14 (James, 1945). An example of a Banach space with the property that $x \perp_R y$ (i.e. orthogonality in the Roberts sense) if and only if either x or y is zero.

Indeed, for this purpose we consider the Banach space of all real-valued functions on $[0, 1]$ of the form $ax + bx^2$, where a, b are scalars and the norm on this space is the sup norm.

Let us suppose that f, g are two elements in this space which are orthogonal in the sense of Roberts. In this case, for all k we have the relation

$$\|f + kg\| = \|f - kg\|.$$

First, if f attains its norm at $t = 1$ (and supposing $f \neq 0$) for k sufficiently small the functions $|f - kg|$ and $|f + kg|$ attain their norm at a point near to $t = 1$. Suppose now that $g(1) \neq 0$. We can arrange k such that

$$|f(1) + kg(1)| > |f(1)|$$

and thus

$$\|f + kg\| > \|f\|.$$

In this case $\|f - kg\| < |f(1)|$ for t near to $t = 1$, which gives that for such k

$$\|f - kg\| < \|f\|.$$

Thus since $f \perp_R g$ this is not possible. Thus we must have $g(1) = 0$. This implies that g is of the form $a(x - x^2)$. The modulus of this function attains the maximum at $t = \frac{1}{2}$. In this case, arguing in a similar way as above, we must have $f(\frac{1}{2}) = 0$. Thus f is a of the form $b(x - 2x^2)$. Now f and g having the form given above, and being orthogonal in the sense of Roberts, we must have

$$\|f + 4g\| = \|f - 4g\|.$$

But

$$\|f + 4g\| = \|5x - 6x^2\| = 25/24$$
$$\|f - 4g\| = \|3x - 2x^2\| = 9/8$$

and thus f and g are orthogonal in the sense of Roberts if at least one is the zero function, when f or g attains the maximum value at $t = 1$. Now if f attains the maximum a only at $x = -a/2b$, which is the only other possible point where f attains its maximum, from the fact that f and g are orthogonal in the sense of Roberts we conclude that $g(-b/2a) = 0$. This implies that g is a multiple of $ax + 2bx^2$. Since f takes the maximum only at $t = -b/2a$, $f + kg$ attains its maximum near the point $x = -b/2a$ for k sufficiently small. Since

$$f + kg = a(1 + k)x + b(1 = 2k)x^2$$

if k is sufficiently small, f and g being orthogonal in the sense of Roberts, we must have

$$|a^2(1 + k)^2/(4b(1 + 2k)| = |a^2(1 - k)^2/(4b(1 - 2k)|$$

and thus

$$k = 0 \text{ or } k = \tfrac{1}{3}.$$

This implies that $f \perp_R g$ is impossible if either f or g attains its maximum in $(0, 1)$, except the case that either f or g is the zero function. Thus it remains to consider the case when f and g both attain their maximum in $(0, 1)$ and at $t = 1$. Then we can show, in a similar way, that f and g are multiples of $2(1 - 2^{1/2})x + x^2$. But then $f \perp_R g$ implies that at least one is the zero function.

Using some elements in the space of functions of the form $ax + bx^2$ we can show that the orthogonality notions introduced in Definitions 4.12.6 and 4.12.7 are independent.

EXAMPLE 4.12.15. (An example of elements orthogonal in the isosceles sense and not in the Pythagorean sense.) We consider in the above space the elements

$$f(x) = x; \qquad g(x) = x - x^2$$

and then we see immediately that they are isosceles orthogonal,

$$\|x + (x - x^2)\| = \|x - (x - x^2)\| = 1.$$

Since

$$\|x\|^2 + \|x - x^2\|^2 = 17/16 \quad \text{and} \quad \|x - (x - x^2)\| = 1$$

they are not orthogonal in the Pythagorean sense.

EXAMPLE 4.12.16. (An example of elements x, y orthogonal in the Pythagorean sense and not orthogonal in the isosceles sense.) Indeed, in the above same space we consider the elements

$$f(t) = t; \qquad g(t) = -2t + \tfrac{1}{4}\cdot(65)^{1/2} t^2.$$

Then we see that they are orthogonal in the Pythagorean sense, since

$$\|f\| = 1, \qquad \|g\| = (16/65), \qquad \|f - g\| = (81/65)^{1/2}.$$

But

$$\|f+g\| = (81/16 - \tfrac{1}{2}(65)^{1/2})^{1/2}$$

and thus they are not orthogonal in the isosceles sense.

We now present some properties of isosceles orthogonality and a characterization of inner product spaces using the isosceles orthogonality. We mention that all results were obtained by James (1945). First we note the following

PROPOSITION 4.12.17. *If X is a normed linear space and x, y are elements of X with* $x \perp_I y$ *they the following assertions hold*:

(a) If $|k| \geqslant 1$ *then* $\|x+ky\| \leqslant |k|\,\|x+y\|$ *and* $\|x+y\| \leqslant \|x+ky\|$.

(b) *If* $|k| \leqslant 1$ *then* $\|x+ky\| \leqslant \|x+y\|$ *and* $|k|\,\|x+y\| \leqslant \|x+ky\|$.

Proof. First we note that we have the identity (with respect to k),

$$x + ky = \tfrac{1}{2}(k+1)(x+y) - \tfrac{1}{2}(k-1)(x-y).$$

The triangle inequality implies then that the following inequality holds:

$$\|x+ky\| \leqslant \|\tfrac{1}{2}(k+1)(x+y)\| \mp \tfrac{1}{2}\|(k-1)(x-y)\|.$$

Suppose now that $x \perp_I y$ holds, i.e. $\|x-y\| = \|x+y\|$. Then, if $|k| \geqslant 1$ we obtain that

$$\|x+ky\| \leqslant |k|\,\|x+y\|.$$

Now since

$$\tfrac{1}{2}(k+1)(x+y) = x+ky+\tfrac{1}{2}(k-1)(x-y)$$

we obtain that

$$\tfrac{1}{2}|k+1|\,\|x+y\| \leqslant \|x+ky\| + \tfrac{1}{2}\|(k-1)(x-y)\|$$

and thus for $|k| \geqslant 1$

$$\|x+y\| \leqslant \|x+ky\|.$$

Using just the inequalities obtained we see that for $|k| \leqslant 1$ inequality (b) is satisfied. □

THEOREM 4.12.18. *Let x, y be elements of a normed linear space X and suppose that* $x \perp_I y$ *holds. Then for all* $|k| \geqslant 1$ *the following inequality holds:*

$$\|x+ky\| \geqslant \|x\|.$$

Proof. Since $2x = (x+y) + (x-y)$ we obtain, applying the triangle inequality, that

$$2\|x\| \leqslant \|x+y\| + \|x-y\| = 2\|x \pm y\|$$

since $x \perp_I y$. Thus

$$\|x\| \leqslant \|x+y\|$$

and from Proposition 4.12.17(a) the assertion follows. □

EXAMPLE 4.12.19. Theorem 4.12.18 cannot be extended to cover the case $|k| < 1$. Indeed, we consider X to be the Banach space R^2 with the norm

$$\|(x, y)\| = \max(|x|, |y|).$$

We consider $u = (1,0)$ and $v = (n - 1, n)$. Since

$$\|u + v\| = \|(n, n)\| = \|u - v\| = \|(2 - n, n)\| = n$$

for large n, u and v are isosceles orthogonal. Since

$$u - \tfrac{1}{2}n \cdot v = ((n + 1)/2n, \ -\tfrac{1}{2})$$

we have that

$$\|u - \tfrac{1}{2}n \cdot v\| = (n + 1)/2n$$

and we note that the sequence $((n + 1)/2n)$ converges to $\frac{1}{2}$. Thus if $\varepsilon > 0$ is given there exists N_ε such that for all $n \geqslant N_\varepsilon$,

$$\|u - \tfrac{1}{2}n \cdot v\| \leqslant \tfrac{1}{2} + \varepsilon.$$

But this means that $\|x + ky\| \geqslant \|x\|$ is impossible for $|k| < 1$.

THEOREM 4.12.20. *Let x be a nonzero element of a normed linear space X and suppose further that $x \perp_1 y$ holds. Then for all k the following inequality holds:*

$$\|x + ky\| > \tfrac{1}{2}\|x\|.$$

Proof. We remark that we have the identity (in k)

$$(k + 1)(x - ky) = (1 - k)(x + ky) + 2k(x - y)$$

and applying the triangle inequality we have that

$$|k + 1| \|x - ky\| \leqslant |1 - k| \|x + ky\| + 2|k| \|x - y\|.$$

Suppose first that for some k we have

$$\|x + ky\| = \tfrac{1}{2}\|x\|.$$

We may suppose without loss of generality that $k \geqslant 0$ (since, if this is not the case, we replace y by $-y$). From Theorem 4.12.18 we obtain that k is in $[0, 1)$.

Since

$$\|x - ky\| \geqslant \tfrac{1}{2} \cdot \|x\|$$

substituting in the inequality

$$|k + 1| \|x - ky\| \leqslant |1 - k| \|x + ky\| + 2|k| \|x - y\|$$

we obtain

$$3/2 \cdot (k + 1)\|x\| \leqslant \tfrac{1}{2}(1 - k)\|x\| + 2k\|x - y\|$$

or

$$(2k + 1)\|x\| \leqslant 2k\|x - y\|.$$

From Proposition 4.12.17(b)

$$k\|x - y\| \leqslant \|x + ky\|$$

and thus

$$k\|x - y\| \leqslant \tfrac{1}{2}\cdot\|x\|.$$

From this we have that

$$(2k + 1)\|x\| \leqslant \|x\|$$

which implies that $k = 0$. Since we suppose that for such k

$$\|x + ky\| = \tfrac{1}{2}\cdot\|x\|$$

we have a contradiction. Thus a scalar k with the property $\|x + ky\| = \tfrac{1}{2}\cdot \|x\|$ does not exist. For fixed x and y, $k \to \|x + ky\|$ is obviously a continuous function and thus

$$\|x + ky\| > \tfrac{1}{2}\cdot\|x\|$$

holds for all k. □

PROPOSITION 4.12.21. *The isosceles orthogonality is symmetric.*

Proof. Obvious

THEOREM 4.12.22. *The isosceles orthogonality is nontrivial.*

In order to prove this we need the following lemma.

LEMMA 4.12.23. *Let X be a normed linear space and let x, y be fixed elements in X. Then*

$$\lim_{n\to\infty} (\|(n + a)x + y\| - \|nx + y\|)$$

exists and is equal to $a\|x\|$.

Proof. We have the identity

$$n/(n + a) + a/(n + a) = 1$$

and thus

$$\|(n + a)x + y\| = \|nx + ny/(n + a)\| + a\|x + y/(n + a)\|$$

for n positive and sufficiently large. Further, we have that

$$\|(n + a)x + y\| - \|nx + y\| = (\|nx + ny/(n + a)\| - \|nx + y\|) + \\ + a\|x + y/(n + a)\|.$$

Since

$$|(\|nx + ny/(n + a)\| - \|nx + y\|)| \leqslant |a/(n + a)|\|y\|$$

and since $(|a/(n + a)|\|y\|)$ converges to zero,

$$\lim a\|x + y/(n + a)\| = a\|x\|$$

implies that the above limit exists and is equal to $a\|x\|$. The lemma is proved. □

Proof of Theorem 4.12.22. We consider the function

$$t \to f(t) = \|x + (tx + y)\| - \|x - (tx + y)\| = \|(t + 1)x + y\| - \|(t - 1)x + y\|$$

and we remark that, as in the above lemma, $\lim_{t\to\infty} f(t)$ exists and is equal to the limit we found in the lemma; similarly $\lim_{t\to-\infty} f(t)$ exists and is equal to that given by the above lemma. Since the first limit is equal to $2\|x\|$ and the second to $-2\|x\|$, f being continuous, we find a value k such that $f(k) = 0$. But this obviously implies the nontriviality of the isosceles orthogonality. Now we show that if the isosceles orthogonality has the other remaining properties from Definition 4.12.4 then the space is an inner product space and the orthogonality coincides with the natural orthogonality (i.e. that defined using an inner product). □

THEOREM 4.12.24. *If X is a normed linear space such that the isosceles orthogonality is homogeneous then the space X is an inner product space.*

Proof. For the proof we use the F. A. Ficken characterization of inner product spaces (Theorem 4.4.1). Let us suppose that x and y are elements of X with the property that $\|x\| = \|y\|$. Further, we remark that this may be written as

$$\|x + y + x - y\| = \|(x + y) - (x - y)\|$$

which means that $(x + y) \perp_I (x - y)$. Since we suppose the isosceles orthogonality to be homogeneous we have that

$$(a + 1)(x + y) \perp_I (a - 1)(x - y)$$

and thus

$$\|ax + y\| = \|x + ay\|.$$

Now if $a = m/n$ this implies that

$$\|m/n \cdot x + y\| = \|x + my/n\|$$

or

$$\|mx + ny\| = \|nx + my\|$$

for all m and n. This implies, by Ficken's theorem, that X is an inner product space. □

THEOREM 4.12.25. *A normed linear space X is an inner product space if and only if the isosceles orthogonality is additive.*

Proof. Since the condition is obviously necessary we prove only the sufficiency. For this purpose we show that, in fact, the isosceles orthogonality is homogeneous and thus the result follows from the theorem just proved.

Indeed, if x, y are elements in X such that $x \perp y$ then clearly $y \perp x$. From the additivity we have further that $nx \perp my$ for all integers n and m. Thus we have for all integers n and m

$$\|nx - my\| = \|nx + my\| \quad \text{or} \quad \|x - m/n \cdot y\| = \|x + m/n \cdot y\|.$$

Using the continuity of the norm we obtain that for all scalars k,

$$\|x - ky\| = \|x + ky\|.$$

This obviously implies that $x \perp ky$. From the symmetry we have that the isosceles orthogonality is homogeneous. The theorem is proved. □

Now we consider the case of Pythagorean orthogonality.

THEOREM 4.12.26. *Pythagorean orthogonality is not trival.*

Proof. Let x, y be arbitrary elements in X and consider the following function on R^+,

$$\begin{aligned} f(t) &= \|x\|^2 + \|tx + y\|^2 - \|x - (tx + y)\|^2 \\ &= \|x\|^2 + \|tx + y\|^2 - \|(t-1)x + y\|^2. \end{aligned}$$

Since we have the identity

$$(t-1)^2/t^2 + (2t-1)/t^2 = 1$$

the function f may be written as follows:

$$\begin{aligned} f(t) &= \|x\|^2 + (2t-1)/t^2\|tx + y\|^2 + \\ &\quad + (\|(t-1)x + (t-1)/t \cdot y\|^2 - \|(t-1)x + y\|^2) \\ &= \|x\|^2 + (2t-1)\|x + y/t\|^2 + \\ &\quad + (\|(t-1)x + ((t-1)/t)y\| - \|(t-1)x + y\|) \times \\ &\quad \times (\|(t-1)x + ((t-1)/t)y\| + \|(t-1)x + y\|) \geqslant \\ &\quad \geqslant \|x\|^2 + (2t-1)\|x + y/t\|^2 - \\ &\quad - |1/t| \cdot \|y\| \cdot (\|(t-1)x + ((t-1)/t) \cdot y\| + \|(t-1)x + y\|). \end{aligned}$$

Suppose now that $t \geqslant 1$ and since

$$\begin{aligned} \|x + y/t\|^2 &\geqslant (\|x\| - \|y/t\|)^2 \|(t-1)x + ((t-1)/t)y\| + \|(t-1)x + y\| \\ &\leqslant 2(t-1)\|x\| + (2t-1)/t \cdot \|y\| \end{aligned}$$

the function $f(t)$ satisfies the following inequality:

$$\begin{aligned} f(t) &\geqslant 2t\|x\|^2 + (2t-1)/t^2 \cdot \|y\|^2 - \\ &\quad - 2(2t-1)/t \cdot \|x\| \cdot \|y\| - \|y\| 2((t-1)/t)\|x\| + \\ &\quad + (2t-1)/t^2\|y\|) = 2t\|x\|^2 - 2 \cdot ((2t-2)/t)\|x\| \cdot \|y\| \\ &= 2\|x\|(t\|x\| - ((3t-2)/t)\|y\|). \end{aligned}$$

This implies that $f(t) \geqslant 0$ for t sufficiently large. We consider now the function g defined on R^+ by the formula:

$$g(t) = \|x\|^2 + \|tx - y\|^2 - \|(t+1)x - y\|^2.$$

Since we have the identity

$$(t+1)^2/t^2 - (2t+1)/t^2 = 1$$

we may write g as follows:

$$\begin{aligned} g(t) &= \|x\|^2 - (2t+1)/t^2 \cdot \|tx + y\|^2 + \\ &\quad + (\|(t+1)x - ((t+1)/t)y\|^2 - \|(t+1)x - y\|^2) \\ &= \|x\|^2 - (2t+1)\|x - y/t\|^2 + \\ &\quad + (\|(t+1)x - ((t+1)/t)y\| - \|(t+1)x - y\|) \times \\ &\quad \times (\|(t+1)x - ((t+1)/t)y\| + \\ &\quad + \|(t+1)x - y\| \leqslant \|x\|^2 - (2t+1)\|x - y/t\|^2 + \\ &\quad + \|y\|(\|(t+1)/t)x - ((t+1)/t^2 \cdot y\| + \\ &\quad + \|((t+1)/t)x - y/t\|). \end{aligned}$$

Thus we have that

$$\begin{aligned} g(t) &\leqslant -2t\|x\|^2 + ((2t+1)/t)\|x\| \cdot \|y\| - \\ &\quad - ((2t+1)/t^2 \cdot \|y\|^2 + \|y\|(2((2t+1)/t\|x\| + ((2t+1)/t^2 \cdot \|y\|) \\ &= -2\|x\|(t\|x\| - ((3t+2)/t)\|y\|). \end{aligned}$$

This implies that if t is sufficiently large $g(t) \leqslant 0$. The function

$$h(t) = \begin{cases} f(t), & t \geqslant 0 \\ g(-t) & t \leqslant 0 \end{cases}$$

is obviously continuous. For t positive and large h is positive and for t large and positive $h(-t)$ is negative. From the continuity it follows that for some t_0 we have $h(t_0) = 0$. But this means exactly that the Pythagorean orthogonality is nontrivial. □

EXAMPLE 4.12.27. (An example of two elements x and y with $x \perp_P y$ and not $x \perp_P (-y)$.) For this we consider the space of all pairs (x, y) with x, y in R and the norm

$$\|(x, y)\| = \max(|x|, |y|).$$

Let $x = (1, 0)$ and $y = (1 - 2^{1/2}, 1)$. Then

$$\|x\|^2 + \|y\|^2 = 1 + 1 = 2$$

and

$$\|x - y\|^2 = \|(2^{1/2}, 1)\| = 2.$$

Thus

$$x \perp_R y.$$

Since

$$\|x + y\|^2 = \|(2 - 2^{1/2}, 1)\| = 1.$$

the assertion is proved. □

Now we prove that if in a normed linear space the Pythagorean orthogonality is homogeneous, this forces the spaces to be inner product spaces.

THEOREM 4.12.28 *Let X be a normed linear space. Then X is an inner product space if and only if the Pythagorean orthogonality is homogeneous.*

Proof. Obviously the condition is necessary. Suppose now that X is a normed linear space with the Pythagorean orthogonality homogeneous. Let x and y be two arbitrary elements of X. From the nontriviality of the Pythagorean orthogonality, we find a scalar a such that $x \perp_P (ax + y)$. Since the orthogonality is supposed to be homogeneous, this implies that for all scalars k the following relation holds:

$$k\|x\|^2 + \|(ax + y\|^2 = \|kx - (ax + y)\|^2.$$

Now we consider $k = 1$ and $k = -1$ and thus we obtain that

$$\|x \pm y\|^2 = (a \pm 1)^2\|x\|^2 + \|ax + y\|^2.$$

This implies that

$$\|x + y\|^2 + \|x - y\|^2 = 2\|ax + y\|^2 + (2a^2 + 2)\|x\|^2$$

and from the homogeneity of the Pythagorean orthogonality we obtain that $ax \perp_P (ax + y)$ which implies that

$$\|ax + y\|^2 + a^2\|x\|^2 = \|y\|^2$$

and thus

$$\|x + y\|^2 + \|x - y\|^2 = 2(\|x\|^2 + \|y\|^2).$$

The Jordan–von Neumann characterization of inner product spaces implies that X is an inner product space. For author proof of the James result (1945), see also Perfect (1958). □

THEOREM 4.12.29. *The normed linear space X is an inner product space if and only if the Pythagorean orthogonality is additive on the right.*

Proof. It is easy to see that the condition is necessary. To prove sufficiency suppose that X is a normed linear space with the property that the Pythagorean orthogonality is additive. Let x, y be arbitrary elements of X. Since the Pythagorean orthogonality is not trivial there exists a scalar a such that $x \perp_P (ax + y)$. Now if we consider the elements x and $-y$ we find a scalar b such that $x \perp_P (bx - y)$. From the additivity we obtain that $x \perp_P (a + b)x$. This gives that $b = -a$. Thus we have $x \perp_P (-ax - y)$. From the additivity we obtain that $x \perp_P mu$, $x \perp_P (-mu)$, where $u = ax + y$. Now we remark that if U and V are elements in X and $U \perp_P V$ then $V \perp_P (-U)$. Applying the additivity property of the Pythagorean orthogonality we obtain that for all n and m the following relations hold:

$$n^2\|u\|^2 + m^2\|x\|^2 = \|nu - mx\|^2 \tag{*}$$

and

$$n^2\|u\|^2 + m^2\|x\|^2 = \|nu + mx\|^2. \tag{**}$$

From the continuity of the norm we obtain from these relations that for all scalars k the following relations hold:

$$\|u\|^2 + k^2\|x\|^2 = \|u + kx\|^2. \tag{***}$$

or

$$\|ax + y\|^2 + k^2\|x\|^2 = \|(a + k)x + y\|^2.$$

If $k = 1 - a$ we have that

$$\|x + y\|^2 = \|ax + y\|^2 + (1 - a)^2\|x\|^2$$

and if $k = 1 + 1$

$$\|x - y\|^2 = \|ax + y\|^2 + (1 + a)^2\|x\|^2$$

which implies that

$$\|x + y\|^2 + \|x - y\|^2 = 2(\|ax + y\|^2 + 2(1 + a^2)\|x\|^2.$$

For $k = a$ we have that

$$\|y\|^2 = \|ax + y\|^2 + a^2\|x\|^2$$

and thus

$$\|x + y\|^2 + \|x - y\|^2 = 2(\|x\|^2 + \|y\|^2).$$

Since x, y are arbitrary in X, by the Jordan–von Neumann characterization of inner product spaces, X is an inner product space. □

Now we consider the orthogonality in normed linear spaces in the sense of Birkhoff (1935) (Definition 4.12.8). In order to obtain characterizations of inner product spaces in terms of Birkhoff's orthogonality we need some preliminary results which have also an independent interest. For completeness we recall some definitions.

DEFINITION 4.12.30. Let X be a normed linear space. A hyperplane in X is any closed linear subspace M of X which is maximal (for the order relation defined by inclusion) or any translation of M, i.e. a set of the form $u + M = (u + m, m \in M)$.

DEFINITION 4.12.31. If C is a convex body in the normed linear space X then a hyperplane M is a supporting hyperplane for C if M does not contain an interior point of C and

$$d(M, C) = \inf(\|x - y\|, x \in M, y \in C) = 0.$$

The following result gives a characterization of Birkhoff's orthogonality using the elements of the dual space.

THEOREM 4.12.32 (James, 1947). *If X is a normed linear space then an element x is orthogonal to an element y if and only if there exists a linear functional f in X^* with the following properties:*

(1) $f(x) = \|f\| \cdot \|x\|$,
(2) $f(y) = 0$.

Proof. Let us first consider the case when $x \perp_B y$. Then for all scalars k we have $\|x + ky\| \geqslant \|x\|$. If we consider in X the subspace generated by x and y, denoted by

X_1, then we can define the following linear functional on X_1:

$$h(ax + by) = a\|x\|.$$

Since

$$h(ax + by) = ah(x + b/a \cdot y)$$

we obtain that

$$|h(ax + by)| \leqslant \|ax + by\|$$

because

$$\|ax\| \leqslant |a|\|x + b/a \cdot y\| = \|ax + by\|.$$

Thus h is continuous and by the extension theorems we can consider h defined on X and having the same norm (as on X_1). Clearly $h(y) = 0$ and thus the condition of the theorem is necessary. The sufficiency follows almost immediately. Indeed, we can suppose that if x, y are given elements and f is in X^* with the properties that $f(x) = \|f\| \cdot \|x\|$ and $f(y) = 0$ and also $\|f\| = 1$. Then clearly we have

$$f(x + ky) = \|x\| \leqslant \|x + ky\|$$

for all k. The theorem is proved. □

From this we have the following result.

COROLLARY 4.12.33 (James, 1947). *Birkhoff's orthogonality is not trivial.*

Proof. Indeed, let x, y be arbitrary elements of the normed linear space X. By the extension theorems we find a linear functional f in X^* with the following property:

$$f(x) = \|f\| \cdot \|x\|.$$

If we take $a = -f(y)/f(x)$ then we see that $f(ax + y) = 0$. Then an element x is orthogonal to an element $ax + y$ in Birkhoff's sense.

As we have seen, if X is a normed linear space and the isosceles orthogonality or the Pythagorean orthogonality is symmetric then a normed linear space is in fact an inner product space. Of course, we may ask if a similar assertion is true for the orthogonality in Birkhoff's sense. In this case it is possible that the space may have the property that the orthogonality in Birkhoff's sense is symmetric and the space is not an inner product space. This, however, may be the case only if the dimension of X is equal to 2. □

Oman (1969) gives constructions which produce all two-dimensional Banach spaces having Birkhoff's orthogonality symmetric. We give the following characterizations without proof. For details we refer to Oman's thesis (1969).

THEOREM 4.12.34 (Oman, 1969). *If X is a two-dimensional normed linear space then X has a symmetric Birkhoff orthogonality if and only if there exists a linear isometry $I: X \to X^*$ such that for all x in X, $(I(x))(x) = 0$.*

THEOREM 4.12.35 (Oman, 1969). *For a normed linear space X the following assertions are equivalent:*

(1) *X has the property that Birkhoff's orthogonality is symmetric,*

(2) *if* x, y *are in* X *and* $\|x\| = \|y\| = 1$ *with* $x \perp_B y$ *then* $(ax + by) \perp_B (cx + dy)$ *if and only if* $|ad - bc| = \|ax + by\| \cdot \|cx + dy\|$

(3) *if* x, y *are in* X, $\|x\| = \|y\| = 1$ *and* $(x + ay) \perp_B y$, $(y + bx) \perp_B x$ *then* $\|x + ay\| = \|y + bx\|$.

REMARK 4.12.36. In Theorem 4.12.35 no restriction about the dimension of X is assumed, so we have in fact a characterization of all normed linear space with the property that Birkhoff's orthogonality is symmetric. However, as we prove later, if the dimension of X is at least 3 then X is in fact an inner product space.

We now give an example of a two-dimensional Banach space having the property that Birkhoff's orthogonality is symmetric and is not an inner product space.

EXAMPLE 4.12.37. An example of a Banach space with nonsymmetric Birkhoff's orthogonality. We consider the space of all pairs of real numbers (x, y) and the norm is defined as follows. Let p be fixed in $(1, \infty)$. Let

$$\|(x, y)\|_p = \begin{cases} (|x|^p + |y|^p)^{1/p} & \operatorname{sgn} x = \operatorname{sgn} y, \\ (|x|^q + |y|^q)^{1/q} & \text{in the contrary case.} \end{cases}$$

We leave it to the reader to prove that Birkhoff's orthogonality is symmetric; obviously this space is not an inner product space.

For the characterizations which follow we use the characterization of inner product spaces in terms of linear operators defined on the space. The proof will be given in Chapter 8 since it is related more directly to the characterizations given there.

THEOREM 4.12.38 (Kakutani, 1939; Bohnenblust, 1942). *A normed linear space* X *is an inner product space if and only if for each two-dimensional subspace* X_1 *of* X *there exists a linear operator* $P: X \to X_1$ *with the following properties*:

(1) $P^2 = P$,

(2) $\|P\| = 1$.

We mention that our proof of this theorem in Chapter 8 is in some respects simpler that the ingenious proof of H. F. Bohnenblust which uses some beautiful and profound results from projective geometry; our proof is based on A. Grothendieck's theory of summing operators.

THEOREM 4.12.39 (Birkhoff, 1935; James, 1947). *Let* X *be a normed linear space of dimension at least* 3. *Then* X *is an inner product space if and only if* X *has the property that Birkhoff's orthogonality is symmetric.*

Proof. Let x, y be two linearly independent elements of X. Then by a procedure as in the proof of Theorem 4.12.32 there exists an element z in X which is orthogonal in Birkhoff's sense to all elements of the form $ax + by$, a, b being scalars. Let X_1 be the space generated by x, y and z. Then any element u in X_1 may be decomposed as follows:

$$P(u) + a_u z = u$$

and clearly P is a projection operator. Since z is orthogonal (in Birkhoff's sense) we obtain that $\|P\| = 1$ (considered as an operator on X_1). Then by the Kakutani–Bohnenblust theorem X_1 is an inner product space. This obviously implies that X is an inner product space since x and y are arbitrary points of X. □

The following result gives a characterization of inner product spaces using Theorem additivity on the left of orthogonality in Birkhoff's sense.

4.12.40 (James, 1947). *Let X be a normed linear space of dimension at least 3. Then X is an inner product space if and only if Birkhoff's orthogonality is additive on the left.*

Proof. The condition is obviously necessary. Suppose now that X is with Birkhoff's orthogonality additive on the left. Let x, y bet two arbitrary linearly independent elements of X. Then as in the proof of Theorem 4.12.32 we find two hyperplanes M_x and M_y such that x is orthogonal (in Birkhoff's sense) to all elements of M_x and a similar assertion is valid for y and M_y. From the additivity on the left we have then that for all scalars a and b we have

$$(ax+by) \perp_B m, \qquad m \in M_x \cap M_y.$$

Let z be an element in $M_x \cap M_y$ and then any element u in the space X_1 generated by x, y and z has a decomposition of the form

$$u = P(u) + a_z z$$

and from the orthogonality property of z we have that $\|P\| = 1$ and being a projection, by the Kakutani–Bohnenblust characterization X_1 is an inner product space. Since x, y are arbitrary in X this implies that X is an inner product space. □

Related to the above characterization we mention now some characterizations of inner product spaces obtained by M. M. Day and J. R. Holub. For proofs the reader is referred to the original papers.

THEOREM 4.12.41 (Day, 1962). *A normed linear space X is an inner product space (dimension of X at least 3) if and only if $x \perp_I y$ then $x \perp_B y$.*

THEOREM 4.12.42 (Holub, 1975). *A normed linear space X (dimension at least 3) is an inner product space if and only if $x \perp_B y$ then $x \perp_I y$.*

THEOREM 4.12.43 (Holub, 1975). *A normed linear space of dimension at least 3 is an inner product space if and only if for all x, y in X, $\|x\| = \|y\|$, then*

$$(x+y) \perp_B (x-y).$$

THEOREM 4.12.44 (Holub, 1975). *The normed linear space X is an inner product space if and only if whenever $(x_i)_1^n$, $n \geqslant 3$, is a linearly independent set of normalized elements (i.e. $\|x_i\| = 1$) and y is in the space generated by $(x_i)_1^n$ with $y \perp_B x_i$, $i = 1,2,3,\ldots,n$, then $y = 0$ or if $y \perp_I x_i$ $i = 1,2,3,\ldots,n$, then again $y = 0$.* Concerning the orthogonalities introduced in Definitions 4.12.11 and 4.12.6, as well as the others mentioned there, we now give some results obtained concerning their use in obtaining existence theorems for inner product structures.

THEOREM 4.12.45 (Istrăţescu, 1984). *Let X be a normed linear space. Then the norm of X is generated by an inner product iff the orthogonalities $\perp_{I,l}$ and $\perp_{I,u}$ (separately) are homogeneous.*

THEOREM 4.12.46 (Istrăţescu, 1984). *Let X be a normed linear space. Then the norm of X is generated by an inner product iff* $\perp_{P,l}$ *and* $\perp_{P,u}$ *(separately) are homogeneous.*

THEOREM 4.12.47 (Istrăţescu, 1984). *Let X be a normed linear space. Then the norm of X is generated by an inner product iff* $\perp_{I,l}$ *and* $\perp_{I,u}$ *(separately) have the additivity property.*

A similar result holds for Pythagorean and (α)-orthogonalities. For the proof, and for other related results, we refer to the author's paper (1984).

An interesting and important result about the existence of a norm generated by an inner product was formulated by R. C. James:

THEOREM 4.12.48 (James, 1947). *A normed linear space X has its norm generated by an inner product iff for all* x, y *in X with* $\|x\| = \|y\|$ *the following relation holds:*

$$\lim_{n \to \infty} (\|x + ny\| - \|y + nx\|) = 0.$$

In order to prove Theorem 4.12.48 we need some results, which are interesting in their own right.

LEMMA 4.12.49. *Let X be a Banach space and x in* $X \diagdown 0$, *y in X. Then*

$$\lim(\|nx + y\| = \|nx\|) = -a_1\|x\|,$$
$$\lim(\|nx\| - \|nx - y\|) = -b_1\|x\|,$$

where a_1, b_1 *are algebraically the smallest and the largest numbers a such that* $x \perp_B (ax + y)$.

Proof. First we remark that

$$\big| \|nx + y\| = \|nx\| \big| \leqslant \|y\|$$

and thus for $n \to \infty$, $\lim(\|nx + y\| - \|nx\|)$ exists. We set this limit as $-\tilde{a}_1\|x\|$. Let (n_i) be a subsequence of integers with the property that

$$\lim_{n_i \to \infty} (\|n_i x + y\| - \|n_i x\|) = -\tilde{a}_1\|x\|.$$

By Lemma 4.12.34, if $x \perp_B (ax + y)$ then

$$\lim(\|n_i x + (ax + y)\| - \|n_i x\| - a\|x\|) = \lim(\|n_i x + y\| - \|n_i x\|)$$

and since (taking into account that $x \perp_B (ax + y)$),

$$\|n_i x + y + ax\| \geqslant \|n_i x\|$$

we get that

$$\lim_{n_i \to \infty} (\|n_i x + y\| - \|n_i x\|) \geqslant a\|x\|.$$

If a_1 is the greatest lower bound of a with the property that $x \perp_B (ax + y)$ then we have

$$\lim_{n_i \to \infty} (\|n_i x + y\| - \|n_i x\|) = -a_1\|x\|.$$

Let $\varepsilon > 0$ be given, and we take N_ε such that

$$\big| \|n_i x + y\| - \|n_i x\| - \tilde{a}_1 \|x\| \big| < \varepsilon \|x\|$$

for all $n_i \geqslant N_\varepsilon$. Now if $n_i \geqslant |a_1|$ then the above inequality can be written as

$$\big| \|n_i x + y\| - \|(n_i - r)\tilde{a}_1 x\| \big| \leqslant \varepsilon \|x\|.$$

Then we find a real number r_i such that $|r_i| < \varepsilon$ and

$$\|n_i x + y\| = \|(n_i - \tilde{a}_1)x + r_i x\|$$

with

$$\|[(n_i - 2^{-1}\tilde{a}_1 + 2^{-1}r_i)x + 2^{-1}y] + [(2^{-1}\tilde{a}_1 - 2^{-1}r_i)x + 2^{-1}y]\|$$
$$= \|[(n_i - 2^{-1}\tilde{a}_1 + 2^{-1}r_i)x + 2^{-1}y] - [(2^{-1}\tilde{a}_1 - 2^{-1}r_i)x + 2^{-1}y]\|$$

gives that (for $\|k\| \geqslant 1$)

$$\|[(n_i - 2^{-1}\tilde{a}_1 + 2^{-1}r_i)x + 2^{-1}y] + k[(2^{-1}\tilde{a}_1 - 2^{-1}r_i)x + 2^{-1}y]\| \geqslant$$
$$\geqslant \|(n_i - 2^{-1}\tilde{a}_1 + 2^{-1}r_i)x + 2^{-1}y\|.$$

Replacing k by $2m$ and dividing the inequality just obtained by $|m|$ we obtain,

$$\|[((2n_i - \tilde{a}_1 + r_i)x + y)/2m] + (\tilde{a}_1 - r_i)x + y\| \geqslant \|[[(2n_i - \tilde{a}_1 + r_i)x + y)]/2m]$$

and since $|r_i| < \varepsilon$ this implies that

$$\lim r_i = 0.$$

This gives that if p is an arbitrary number with the property $m = n_i/p$ we have

$$\|px + (\tilde{a}_1 x + y)\| \geqslant \|px\|.$$

Now $-\tilde{a}_1\|x\|$ is a limit point of the sequence $\{\|nx + y\| - \|nx\|\}$ and this implies that $x \perp_B (a_1 x + y)$. But

$$-a\|x\| \geqslant -a_1\|x\| \quad \text{or} \quad a_1 \geqslant \tilde{a}_1$$

and thus the limit is equal to $-a_1\|x\|$.

Clearly $x \perp_B (a_1 x + y)$. Now if b_1 is the largest number a with the property that $x \perp_B (ax + y)$ then $-b_1$ is the smallest number a such that

$$x \perp_B (-ax - y)$$

and thus

$$-b_1\|x\| = \lim(\|nx\| - \|(nx - y)\|)$$

and the lemma is proved. □

In order to obtain some characterizations of inner product structures it is useful to use a function which is also important in other problems of the geometry of Banach spaces.

Let X be a Banach space and C be a convex set in X which contains the zero (element of X) as an internal point (i.e. for each x of X there exists $a_x > 0$ such that

$zx \in C$ whenever $|z| \leqslant a_x$). The point p of C is said to be internal if for each x of X there exists $a_x > 0$ such that $p + zx \in C$ whenever $|z| \leqslant a_x$.

The Minkowski functional of C, denoted by p_C, is defined by the relation

$$p_C(x) = \inf(a, a > 0, \qquad x \in aC).$$

The tangential functional $\tau(x, y)$ of C is defined by

$$\tau(x, y) = \lim_{t \to 0^+} t^{-1}(p_C(x + ty) - p_C(x)).$$

We must show that $\tau(,)$ is a well-defined function for any pair of elements x, y of X. First we note that the function on $(0, \infty)$ defined by the formula

$$t^{-1}(p_C(x + ty) - p_C(x))$$

is an increasing function. Indeed, if $t_1 \geqslant t_2$ then

$$\begin{aligned} p_C(t_1 x + t_2 t_1 y) &= p_C(t_1 x + t_2 x - t_2 x + t_1 t_2 y) \\ &\leqslant p_C((t_1 - t_2)x) + p_C(t_2(x + t_1 y)) \\ &= p_C((t_1 - t_2)x) + t_2 p_C(x + t_1 y), \end{aligned}$$

which implies that

$$t_1 (p_C(x + t_2 y) - p_C(x)) \leqslant t_2 (p_C(x + t_1 y) - p_C(x))$$

or

$$t_2^{-1}(p_C(x + t_2 y) - p_C(x)) \leqslant t_1^{-1}(p_C(x + t_1 y) - p_C(x))$$

and this clearly implies that the limit in the definition of τ exists. Thus $\tau(,)$ is well defined.

We remark that in the above proof no use is made of the Banach structure of the space and thus the proof is valid for linear spaces.

We now present the basic properties of the function τ which are important to us in what follows.

PROPOSITION 4.12.50. *If C is a set as above then the following assertions hold:*

(1) $\tau(x, y) \leqslant p_C(y)$.

(2) $\tau(x, y + z) \leqslant (x, y) + \tau(x; z)$.

(3) $\tau(ax, by) = b\tau(x, y)$, *$a, b$ in R^+.*

(4) $-\tau(x, -y) \leqslant \tau(x, y)$.

(5) $\tau(x, ax) = a\tau(x, x)$, $a \in R$.

(6) $\tau(x, 0) = 0, \tau(0, y) = p_C(y)$.

(7) *$\tau(,)$ is upper semicontinuous on $X_\times X$.*

Proof. Assertions (1) and (2) follow immediately from the definition. Assertions (3) and (5) are obvious and since

$$\tau(x, -y) + \tau(x, y) \geqslant \tau(x, 0) = 0$$

assertion (4) follows. The other assertions, (6) and (7), follow from the definition. □

REMARK 4.12.51. In the applications we use sets C which are the closed unit balls of Banach spaces. In this case it is easy to see that the Minkowski functional p_C is exactly the norm of the space.

From the definition of $\tau(\,,\,)$ and the definition of the orthogonality $\perp_B$ we have

PROPOSITION 4.12.52. *The following assertions hold:*

(1) $x \perp_B y$ *iff* $-\tau(x, -y) \leqslant 0 \leqslant \tau(x, y)$,

(2) $-\tau(x, -y) \leqslant a \leqslant \tau(x, y)$ *iff* $x \perp_B (x - ay/\|x\|)$.

We now present a result which shows the intimate connection between the properties of the function $\tau(\,,\,)$ and the geometric structure of the space.

THEOREM 4.12.53 (James, 1947). *Let X be a normed linear space of dimension at least 3. Then the following conditions are equivalent:*

(1) *The norm of X is generated by an inner product.*

(2) *Whenever $\|x\| = \|y\|$ and $\tau(x, y) = 0$ then $\tau(y, x) = 0$.*

Proof. Obviously (1) → (2). Suppose now that (2) holds and we must prove that (1) holds. The assertion follows from the following result which has also an independent interest.

THEOREM 4.12.54 (James, 1947). *The necessary and sufficient condition for $\tau(y, x) = 0$ for any nonzero elements x, y in X satisfying $\tau(x, y) = 0$ is that the orthogonality $\perp_B$ be symmetric and for each $x \neq 0$ and y in X there exists a unique a with the property that $x \perp_B (ax + y)$.*

Proof. If for all $x \neq 0$ of X and y in X there exists a unique number a such that $x \perp_B (ax + y)$ then from the properties of the function we get, as was noted at the properties of the function τ, that $x \perp_B y$ iff $\tau(x, y) = 0$. Thus if Birkhoff's orthogonality is symmetric we get that $\tau(y, x) = 0$ if $\tau(x, y) = 0$.

For the converse, suppose that x is $\neq 0$ and $x \perp_B y$. Let A, B be such that $x \perp_B (ax + y)$ iff $a \in [A, B]$, $A \leqslant 0 \leqslant B$. Then $a = 0$ is the smallest value of a such that $x \perp_B ((A + a'x + y)$ or $x \perp_B (ax + (Ax + y))$. Again, using the properties of the function τ, we get that

$$\tau(x, Ax + y) = 0.$$

Now if $Ax + y = 0$ then $(ax + y \perp_B x$ holds. Clearly a is the largest value such that $x \perp_B ((B + a)x + y)$ holds or $x \perp_B ((ax + (Bx + y))$.

Since

$$-\tau(x, -(Bx + y)) = 0$$

we have that $\tau(-Bx - y, x) = 0$ for $Bx + y \neq 0$. Thus $(Bx + y) \perp_B x$ and from the definition of A and B we have that $(Ax + y) \perp_B x$ and $(Bx + y) \perp_B x$. Then we get that $y \perp_B x$. This gives the symmetry of the $\perp_B$ orthogonality for nonzero elements with the property that $\tau(y, x) = 0$.

Suppose now that for $x \neq 0$ and y we have

$$x \perp_B (ax + y)$$

for a in $[A', B']$, $A' \neq B'$. From the symmetry we get that the relation $(ax + y) \perp_B x$ holds and $\|ax + y\|$ is minimum and constant for a in A', B'.

Let t be in $[A', B']$ and thus

$$\|tx + y + hx\| - \|tx + y\| = 0$$

for $A' < t \pm |h| < B'$. Thus

$$\tau(tx + y, x) = 0 = -\tau(x, -tx - y) \quad \text{or} \quad \tau(x, tx + y) = 0.$$

But this is impossible and the theorem is proved. □

LEMMA 4.12.55. *If the normed linear space X has the property that*

$$\lim_{n \to \infty} (\|nx + y)\| - \|x + ny\|) = 0$$

holds whenever $\|x\| = \|y\|$ *then the* $\perp_B$ *is symmetric, additive and unique and* $x \perp_B (ax + y)$, $y \perp_B (by + x)$ *imply that* $a = b$.

Proof. Let a_1, b_1 be the algebraically smallest and largest numbers a such that $x \perp_B (ax + y)$ and a'_1, b'_1 be the algebraically smallest and largest numbers a such that $y \perp_B (ay + x)$.

Suppose now that X satisfies the condition of the theorem and then we get that $a_1 = a'_1$, $b_1 = b'_1$. Then we have further that if $x \perp_B y$ then $y \perp_B x$. From the homogeneity of the orthogonality we may suppose without loss of generality that $\|x\| = \|y\|$. In this case the uniqueness follows. Indeed, in the opposite case, for some $\varepsilon > 0$ we have

$$x \perp_B (ax + y), \qquad (ax + y) \perp_B x$$

whenever $|a| < \varepsilon$.

But then for such a we have

$$\|ax + y\| = \|y\| = \|x\|$$

and

$$\lim(\|nx + (ax + y)\| - \|x + n(ax + y)\|) = 0,$$

since

$$\|x + n(ax + y)\| = \|(1 + na)x + ny\| = \|ny\|$$

for large n with $|(1 + na)|/n < \varepsilon$. But

$$\lim_{n \to \infty} (\|nx + (ax + y)\| - \|nx + y\|) = a\|x\|$$

and thus

$$\lim (\|nx + (ax + y)\| - \|x + n(ax + y)\|) = \lim(\|nx + y\| - \|ny\|) + a\|x\|$$

which is not zero for all $|a| < \varepsilon$. Thus the uniqueness is proved and we leave the interested reader to prove the additivity. □

Now the proof of Theorem 4.12.48 is as follows.

Suppose that x, y are both nonzero elements of X. We set

$$\langle x, y\rangle = -a\|x\|^2$$

when $x \perp_B (ax + y)$ and

$$\langle x, y\rangle = 0$$

if x or y (or both) are zero. Then we have that

$$\langle tx, y\rangle = t\langle x, y\rangle, \qquad \langle x, y\rangle = \langle y, x\rangle.$$

If

$x \perp_B (ax + y), x \perp_B (bx + z)$ then $x \perp_B ((a + b)x + (y + z))$ and

$$\langle x, y\rangle = -a\|x\|^2, \qquad \langle x, z\rangle = -b\|x\|^2$$
$$\langle x, y + z\rangle = -(a + b)\|x\|^2$$

which clearly gives the additivity.

Obviously we have

$$\langle x, x\rangle = x^2$$

and the theorem is proved since the condition of the theorem is obviously necessary. □

From this we get the following result.

THEOREM 4.12.56 (James, 1947). *The normed linear space has its norm generated by an inner product iff for all x, y in X the following relation holds:*

$$\tau(x, \|x\|y) = \tau(y, \|y\|x).$$

Proof. It is easy to see that the condition is necessary. Suppose now that it is satisfied by the normed linear space X. Then using the function τ we have

$$\lim_{t \to 0^+} (t^{-1}(\|x + ty\|x\|\| - \|x\|) - \|y + tx\|y\|\| - \|y\|)$$

$$= \lim_{t \to 0^+} t^{-1}(\|x\|y\| + ty\|x\|\| - \|y\|x\| + tx\|y\|\|)$$

$$= \lim_{t \to 0^+} t^{-1}(\|n(x\|y\|) + (y\|x\|)\| - \|(x\|y\|) + n(y\|x\|)\|)$$

and by the theorem it follows that the norm of X is generated by an inner product. □

A slight improvement was obtained by P. L. Papini, as follows.

THEOREM 4.12.57 (Papini, 1983). *The norm of a Banach space X is generated by an inner product iff for all x, y in $X, \|x\| = \|y\| = 1$*

$$|\tau(x, y)| = |\tau(y, x)|.$$

Proof. It is clear that the condition is necessary. Suppose now that the Banach space X has the property that the above relation holds. We may suppose that $\tau(x,y) = -\tau(y,x)$, $\|x\| = \|y\| = 1$. We may assume that $\tau(x,y) > 0$. From the definition of the function τ we have

$$(tx + y, x) > 1$$

for $t > 0$. We take $t_0 = -\tau(x,y)$ and thus

$$\tau(x, t_0x + y) = 0 \quad \text{or} \quad \tau(t_0x + y, x) = 0$$

which implies that

$$\|t_0x + y\| = \tau(t_0x + y, t_0x + y) = \tau(t_0x + y, y) \leqslant \|y\| = 1$$

which is a contradiction. Thus we must have

$$\tau(x,y) = \tau(y,x)$$

and this gives the assertion of the theorem. □

Jakstaš (1968) has given an interesting characterization of normed linear space whose norm is generated by an inner product using certain inequalities between the elements $\frac{1}{2}(x+y)$ and the elements $tx + (1-t)y$, t in $(0,1)$ and not equal to $\frac{1}{2}$. The characterization is as follows.

THEOREM 4.12.58 (Jakstaš, 1968). *Let X be a normed linear space of dimension at least 3. Then the norm of X is generated by an inner product iff for all x, y in X, $\|x\| = \|y\| = 1$, and all t in $(0,1)$, $t \neq \frac{1}{2}$, the following inequality holds:*

$$\|x + y\| < 2\|tx + (1-t)y\|. \tag{+}$$

Unfortunately, as was stated by P. K. Belobrov, Jakstaš' proof is not correct. A correct proof was supplied by Belobrov (1970). Goryachev (1970) has extended the result to include Banach spaces of dimension 2. Thus the theorem is as follows in its full generality.

THEOREM 4.12.59 (Jakstaš–Belobrov–Goryachev). *Let X be a normed linear space of dimension at least 2. Then the norm of X is generated by an inner product iff the inequality (+) holds.*

Proof. We present a proof, following Papini (1983), in which we suppose that the inequality (+) holds with '$\leqslant$' instead of '$<$'.

Let M be the segment $[x,y]$ and then we have, using the properties of the function τ that

$$x_0 \in P_Mx \quad \text{iff} \quad \tau(x - x_0, x_0 - y) \geqslant 0, \quad y \in M.$$

Since

$$\|x + y\| \leqslant 2\|(tx + (1-t)y)\|$$

holds for all x, y with $\|x\| = \|y\|$ we see that $x + y$ is in P_M0. Thus for any t in $(0,1)$ we have

$$\begin{aligned} 0 &= \tau((-(x+y)/2, (x+y)/2) - (tx + (1-t)y)) \\ &= \tau((x+y)/2, (t - \tfrac{1}{2})(x - y)). \end{aligned}$$

Now if t is in $(0, \frac{1}{2})$ then

$$\tau(x + y, x - y) \geqslant 0$$

and if t is in $(\frac{1}{2}, 1)$

$$\tau(x + y, -(x - y)) \geqslant 0$$

and thus

$$-\tau(x + y, y - x) \leqslant 0 \leqslant \tau(x - y, x - y)$$

which is $(x + y) \perp_B (x - y)$. But this implies that the norm of X is generated by an inner product. □

For another proof of this interesting result we refer to the original paper of P. K. Belobrov and to the interesting 1970 paper of N. I. Gurariĭ and Y. I. Sozonov.

For the proof of the following result we refer to Papini (1983).

THEOREM 4.12.60 *For a normed linear space X the following assertions are equivalent:*

(1) *X has its norm generated by an inner product.*
(2) *if $\|x\| = \|y\| = 1$ then $\tau(x, y) \geqslant 0$ implies $\tau(y, x) \geqslant 0$.*
(3) *if $\|x\| = \|y\| = 1$ then $-\tau(x, -y) \leqslant 0$ implies $\tau(y, -x) \leqslant 0$.*
(4) *if $\|x\| = \|y\| = 1$ then $\tau(x, y)\, \tau(y, x) \geqslant 0$.*
(5) *if $x, y \in X$, $x \neq 0$ then $\|y - \tau(x, y)\, x/\|x\|\| \leqslant \|y\|$.*
(6) *if $x, y \in X$, $x \neq 0$ then $\|y + \tau(x, -y) x/\|x\|\| \leqslant \|y\|$.*

Another interesting notion of orthogonality was proposed by Munroe (1946) and it may be of interest to know if this may be used to characterize the class of spaces with an inner product in the class of normed spaces.

Precupanu (1986) has considered the following interesting problem, namely to characterize the existence of an inner product which is in some sense compatible with the topology of the space. In order to present such a characterization we introduce several notions. All notions and results which follow are given by Precupanu (1968). Let X be a real linear space.

DEFINITION 4.12.61. The subset C of X is said to be H-convex if for any $t, s > 0$ and all x, y in C we find $t_1, s_1 > 0$ such that:

(1) $t_1^2 + s_1^2 \leqslant 2(t^2 + s^2)$;
(2) $tx + sy \in t_1 C$;
(3) $tx - sy \in s_1 C$.

This is equivalent to

PROPOSITION 11.12.62. *C is a H-convex subset of X iff for any $s, t > 0$ and any $x \in tC, y \in sC$ there exists t_1 and $s_1 > 0$ such that:*

(1) $t_1^2 + s_1^2 \leqslant 2(t^2 + s^2)$;
(2) $x + y \in t_1 C$;

(3) $x - y \in s_1 C$.

Now we consider a subset C in X which is H-convex. Then, under certain assumptions, we can consider the Minkowski functional of C. The properties of such a functional are given in

PROPOSITION 4.12.63. *Let C be a H-convex subset of X which we suppose to be absorbing. Then the Minkowski functional p_C of C defined by the formula*

$$p_C(x) = \inf(a > 0, x \in aC)$$

has the following properties:

(1) $p_C(x) \geqslant 0$,

(2) $p_C(rx) = |r| p_C(x)$,

(3) $\{x, p_C(x) < 1\} \subseteq C \subseteq \{x, p_C(x) = 1\}$,

(4) $p_C^2(x) + p_C^2(y) = \frac{1}{2}(p_C^2(x + y) + p_C^2(x - y))$,

(5) $p_C^2(x_1 + x_2 + y) - p_C^2(x + x_1 - y) = p_C^2(x_1 + y) + p_C^2(x_2 + y) - p_C^2(x_1 - y) - p_C^2(x_2 - y)$,

(6) $p_C^2(x + y) - p_C^2(x - y) \leqslant 4p_C(x)p_C(y)$,

(7) $p_C^2(rx + y) - p_C^2(rx - y) = r(p_C^2(x + y) - p_C^2(x - y))$,

(8) $p_C(x + y) \leqslant p_C(x) + p_C(y)$.

for all x_1, x_2, x, y in X and $r \in R$.

Proof. Since the proof is by computation and not difficult it is omitted.

COROLLARY 4.12.64. *Suppose that C is a subset as in Proposition 4.12.63. Then*

$$\langle x, y\rangle_C = 4^{-1}\ (p_C^2(x + y) + p_C^2(x - y))$$

has the following properties:

(1) $\langle x, x\rangle_C \geqslant 0$;

(2) $\langle rx, y\rangle_C = r\langle x, y\rangle_C$;

(3) $\langle x + y, z\rangle_C = \langle x, z\rangle_C + \langle y, z\rangle_C$;

(4) $\langle x, y\rangle_C = \langle y, x\rangle_C$.

REMARK 4.12.65. If F is a bilinear form on X which is symmetric, real and positive definite then the set

$$C_1 = \{x, f(x, x) \leqslant 1\}$$

is a H-convex set which is also absorbing.

Taking into account the above results, it follows that determining the quasi-inner products (i.e. functions like inner products without satisfying necessarily the condition that the value 0 is take only at $x = 0$) is equivalent with the problem of describing all the subsets of X which are H-convex subsets, absorbing in X.

It is also clear that we must consider that these subsets are different from the proper closed subsets of X which are linear subspaces.

DEFINITION 4.12.66. The function $F: X_{\times}X \to C$ is said to be of an inner product type if the properties

(1) $F(x,x) \geqslant 0$,

(2) $F(x,y) = \overline{F(y,x)}$,

(3) $F(ax+by),z) = aF(x,z) + bF(y,z)$

hold for all x,y,z in X and all a,b in C.

If F is of inner product type then Re F is a bilinear functional which is real symmetric and positive definite; conversely, if f is a bilinear functional on the real space X which is of the above type, then

$$F(x,y) = f(x,y) + if(ix,iy)$$

is a function of inner product type.

DEFINITION 4.12.67. A seminorm p on X is said to be of inner product type if for some function F of inner product type we have

$$p^2(x) = F(x,x).$$

If p is a norm then we say that p is a norm of inner product type.

THEOREM 7.12.68. *Let X be a linear topological Hausdorff space with a topology τ. Then a necessary and sufficient condition that τ be generated by the pseudometric defined by the seminorm of inner product type is that X must have a neighborhood of the origin which is bounded and H-convex.*

Proof. The necessity is obvious. Suppose now that C is a bounded H-convex neighborhood of the origin of X. Then the hull of C has the property that is H-convex and the Minkowski function p_C, is, according to the well-known theorem of Kolmogorov, a norm on X which generates the same topology. But P_C satisfies the Jordan–von Neuman theorem and thus is generated by an inner product. □

DEFINITION 4.12.69. Let X be a linear topological space with a topology τ. We say that τ is a topology of inner product type if on X there exists a norm of inner product type, $\langle , \rangle$ on X such that $(X, ||,||)$ is the topological space (X,τ).

A similar result to that given above can be proved for the case when the topology of a linear space X, say τ, is the topology generated by a family of seminorms of inner product type. Since this is not a very difficult task we leave it to the reader.

Another concept of orthogonality in Banach spaces was proposed by C. R. Diminnie (1983), and this was suggested by certain interesting considerations of E. Silverman (also the ideas of Silverman were at the origin of an interesting generalization of the notion of uniformly convex space considered by F. Sullivan).

Let X be a Banach space and X be its dual. If x,y are arbitrary elements in X then we set

$$||x,y|| = \sup_{\substack{||x^*|| \leqslant 1 \\ ||y^*|| = 1}} \left\{ \begin{vmatrix} x^*(x) & x^*(y) \\ y^*(x) & y^*(y) \end{vmatrix} \right\}.$$

The function on $X_x X$ defined above has the properties:

(1) $\|x, y\| \geqslant 0$,
(2) $\|x, y\| = 0$ iff $x = ay$ for some scalar a,
(3) $\|x, y\| = \|y, x\|$,
(4) $\|ax, y\| = |a| \, \|x, y\|$,
(5) $\|x + y, z\| \leqslant \|x, z\| + \|y, z\|$

for all x, y, z in X and a in K (R or C).

Suppose now that the norm of X is generated by an inner product, say $\langle \, , \rangle$. Then it is not difficult to see that

$$\|x, y\|^2 = \|x\|^2 - \|y\|^2 - \langle x, y \rangle$$

and this suggests the following definition.

DEFINITION 4.12.70 (Diminnie, 1983). In a Banach space X with the norm $\|, \|$ we say that the elements x, y of X are D-orthogonal if

$$\|x, y\|^2 = \|x\|^2 - \|y\|^2$$

and briefly we write this as $x \perp_D y$. This orthogonaly is denoted by $\perp_D$. From the above remark we see that for complete inner product spaces this reduces to the standard notion of orthogonality. Also we remark that it is not necessary to consider a Banach space; we can consider simply normed linear spaces.

DEFINITION 4.12.71 (Diminnie, 1983). We say that the orthogonality $\perp_D$ is unique if for each pair x, y of elements of $X, x \neq 0$, there exists a unique real number a such that

$$x \perp_D (ax + y).$$

Then the following results hold.

THEOREM 4.12.72 (Diminnie, 1983). *If $\perp_D$ is additive then it is unique.*

THEOREM 4.12.73 (Diminnie, 1983). *If $\perp_D$ is unique then $\perp_D$ is equivalent with the orthogonality in the sense of Birkhoff, $\perp_B$.*

THEOREM 4.12.74 (Diminnie, 1983). *For a Banach space X the following assertions are equivalent:*

(1) *X is a space whose norm $\|, \|$ is generated by an inner product $\langle \, , \rangle$,*
(2) *$\perp_B$ and $\perp_D$ are equivalent,*
(3) *$\perp_D$ is unique,*
(4) *$\perp_D$ is additive.*

In the same paper C. R. Diminnie gives some examples of two-dimensional spaces which show that, to some extent, the above results in Theorem 4.12. 74 are the best possible.

In what follows we give some characterizations of inner product spaces which in some sense may be considered as localized versions of known characterizations using orthogonalities. The result we give were obtained by Oman (1969, 1980).

THEOREM 4.12.75 (Oman, 1980). *The normed linear space is an inner product space if and only if there exists on* $\varepsilon > 0$ *such that whenever* $\|x\| = \|y\| = 1$ *and* $\|x - y\| < \varepsilon$ *then* $(x + y) \perp_B (x - y)$.

Proof. The condition is clearly necessary. Suppose now that X is a normed linear space with the property stated in the theorem. First we show that X is strictly convex, ie.e for any x, y linearly independent elements of $X, \|x + y\| < \|x\| + \|y\|$, when $\|x\| = \|y\|$.

Indeed, in the opposite case there exist x, y in X, $\|x\| = \|y\| = 1$ and t in $(\frac{1}{2}, 1)$ with the property that $\|x - y\|$ and $\|tx + (1 - t)y\| \in (0, 1)$, $\big\|sx + (1 - s)(tx + (1 - t)y)/\|tx + (1 - t)y\|\big\| = 1$ for all s in $[0, 1]$. Then the unit sphere has a unique support line at $(x + y)/\|x + y\|$ which is parallel to $x - (tx + (1 - t)y)/\|tx + (1 - t)y\|$. But the elements $x - y$ and $x - (tx + (1 - t)y)/\|tx + (1 - t)y\|$ are linearly independent elements and thus we have obtained a contradiction. Thus X is strictly convex. Now we show that if $\|x\| = \|y\| = 1$ and $\|x - y\| < \varepsilon$ then for all t in $[0, 1]$, $\|tx + (1 - t)y\| = \|ty + (1 - t)x\|$. Indeed, if for $t \neq s$, t, s in $[0, 1]$ we have

$$\|tx + (1 - t)y\| = \|sx + (1 - s)y\|$$

then

$$((t + s)x + (2 - t - s)y) \perp_B (x - y)$$

and

$$(x + y) \perp_B (x - y).$$

From the strict convexity of X we obtain that $t + s = 2 - t - s$ which is equivalent to $s + t = 1$ and our assertion is proved. Further, we show that if

$$\|x\| = \|y\| = 1 \quad \text{and} \quad \|x - y\| < \varepsilon_1, \quad \varepsilon_1 = \min(\varepsilon, 1)$$

then for all $a \geqslant 0$

$$a\|x + y\|^2 + (1 - a)^2 = \|x + ay\|^2.$$

We consider only the case when a is in $[0, 1]$. We remark that

$$\|x + y\| > 1 \quad \text{and} \quad \|(x + ay)/\|x + ay\| - y\| < \varepsilon_1.$$

Further, we have that,

$$\|x + ay\| = \|a\|x + y\|(x + y)/\|x + y\| + \\ + (1 - a)y\| = \|(1 - a)(x + y)/\|x + y\| + a\|x + y\|y\|$$

or

$$\|ax + y\|\|x + y\| = \|(1 - a)x + (1 - a + a\|x + y\|^2)y\|.$$

Similarly, we have

$$\|x + y\| = \big\|\,\|x + ay\|(x + ay)/\|x + ay\| + (1 - a)y\big\| \\ = \big\|(1 - a)(x + ay)/\|x + ay\| + \|x + ay\|y\big\|$$

or

$$\|ax + y\| \|x + y\| = \|x + ay\| \|x + y\|$$
$$= \|(1 - a)x + (a - a^2 + \|x + ay\|^2)y\|.$$

But

$a - a^2 + \|x + ay\|^2 \geqslant 0$ since a is in $(0, 1)$ and $1 - a + a\|x + y\|^2 \geqslant 0$ because $\|x + y\| > 1$. We consider now the function

$$h(t) = \|x + ty\|$$

which is clearly convex and $h(-1) < h(0) = 1 < h(1)$ and thus the minimum occurs at a negative value of t. Thus

$$(1 - a + a\|x + y\|^2)/(1 - a) = (a - a^2 + \|x + ay\|^2)/(1 - a)$$

which implies that

$$a\|x + y\|^2 + (1 - a)^2 = \|x + ay\|^2.$$

Since the left-hand side is symmetric in x and y we have that the following relation holds:

$$a\|x + y\|^2 + (1 - a)^2 = \|ax + y\|^2.$$

Now we see that the relations given above do hold for any $a \geqslant 0$. Indeed, if a is in $[0, 1]$ the assertions are clear. If $a \geqslant 1$, then if we apply the equalities for $1/a$ we have the desired relations. But then we see that X satisfies the conditions of Theorem 4.11.1 and thus it is an inner product space. □

THEOREM 4.12.76 (Oman, 1980). *A normed linear space X is an inner product space if and only if there exists $\varepsilon > 0$ such that whenever $\|x\| = \|y\| = 1$ and $\|x - y\| < \varepsilon$ then for some $r \neq 1$ and in R^+, $\|x + ry\| = \|rx + y\|$.*

Before we give the proof of this result we mention the following results as corollaries.

COROLLARY 4.12.77 (Lorch, 1948). *A normed linear space X is an inner product space if and only if for some real number $r \neq 0, r \neq \pm 1$ for all x, y in X with $\|x\| = \|y\| = 1$, $\|x + ry\| = \|rx + y\|$ holds.*

COROLLARY 4.12.78 (Ficken, 1944). *See Theorem* 4.4.1.

Proof of Theorem 4.12.76. We note that the condition is necessary. To prove sufficiency we may assume without loss of generality that X is a two-dimensional subspace. Also we may assume that r is in $(0, 1)$ (in the opposite case we may use $1/r$). Let x, y in X such that $\|x\| = \|y\| = 1$ and $\|x - y\| < \varepsilon$. Define

$$M = (a, a \in (0, 1), \|ax + y\| = \|x + ay\|).$$

We remark that M is bounded and closed and nonempty by hypothesis. Let m_s be the least upper bound of M. Suppose first that $m_s < 1$. Then

$$\|x + m_s y\| = \|m_s x + y\|$$

and

$$\big\|(x + m_s y)/\|x + m_s y\| - (m_s x + y)/\|m_s x + y\|\big\| < \varepsilon.$$

Then for some k in $(0, 1)$ we have

$$\big\|(x + m_s y)/\|x + m_s y\| + k(m_s x + y)/\|m_s x + y\|\big\|$$
$$= \big\|k(x + m_s y)/\|x + m_s y\| + (m_s x + y)/\|m_s x + y\|\big\|$$

and thus

$$\|x + (m_s + k)/(1 + km_s)\cdot y\| = \|(m_s + k)/(1 + km_s)\cdot x + y\|.$$

Since

$$m_s < (m_s + k)/(1 + km_s) < 1$$

we have a contradiction. This implies that $m_s = 1$ and thus $(x + y) \perp_B (x - y)$. Then X is an inner product. The theorem is proved. □

THEOREM 4.12.79 (Oman, 1980). *A normed linear space X is an inner product space if and only if there exists $r > 0, r \neq 1$ such that for any $\|x\| = \|y\|$ then $\|x + ry\| = \|x + ry\|$, or $r < 0, r \neq -1$ such that whenever $\|x\| = \|y\|$ then $\|x - ry\| = \|rx - y\|$.*

Proof. If $r > 0$ then the assertion is clear. If $r < 0$ then we apply the above result for $-r$ and the pair x and $-y$. □

THEOREM 4.12.80 (Oman, 1980). *A normed linear space X is an inner product space if and only if there exists an $\varepsilon > 0$ such that whenever $\|x\| = \|y\| = 1$ and $\|x - y\| < \varepsilon$ then for some $\bar{t} > 1$ or $\bar{s}$ in $(0, 1)$*

$$\|x + \bar{t}y\| \, r \, \|\bar{t}x + y\|, \qquad \|y + sx\| \, r \, \|\bar{s}y + x\|$$

where r means $=$, $\leqslant$ or $\geqslant$.

Proof. The condition is clearly necessary. To prove sufficiency we consider without loss of generality that X is a two-dimensional space and let

$$f(t) - \|x + ty\| - \|tx + y\|$$

which is obviously a continuous function. If $f(\bar{t})$ is not zero, then $f(\bar{t})$ and $f(\bar{s})$ have opposite signs and thus for some u between $\bar{t}$ and $\bar{s}$ we must have $f(u) = 0$. Thus we obtain that for some value not equal to 1 the hypotheses of Theorem 4.10.11 are satisfied and thus X is an inner product space. □

THEOREM 4.12.81 (Oman, 1980). *A normed linear space over C (the complex numbers) is an inner product space if and only if there exists a complex number t with $|t| \neq 0, 1$ with the property that for all $\|x\| = \|y\| = 1$, $\|tx + y\| = \|x + ty\|$.*

Proof. The condition is clearly necessary. To prove sufficiency we remark that if $\|x\| = \|y\| = 1$ and t has the above property then

$$\|x\| = \|\bar{t}/|t|y\| = 1$$

and then

$$\|x + t\cdot\bar{t}/|t|\cdot y\| r \|\bar{t}x + \bar{t}/|t|\cdot y\|$$

or

$$\|x + |t|y\| r \| |t|x + y\|$$

and the hypothesis of Theorem 4.12.79 is satisfied for $|t|$. Then X is an inner product space. □

We now present a localized form of a generalized form of a characterization of inner product spaces presented in Theorem 4.11.9 and Corollary 4.11.4. Since the proof is almost the same it is omitted.

THEOREM 4.12.82 (Oman, 1980). *A real normed linear space X is an inner product space if and only if there exists a subset K in X with the following properties:*

1. *If $x \in K$ then for all $t \in R$, $tx \in K$.*
2. *For each pair (x, y) of elements of K there exist the real numbers $a_n = a_n(x, y)$, $b_n = b_n(x, y)$ and $c_n = c_n(x, y)$ such that:*

(a) $a_n > 0$,

(b) $\sum_n a_n b_n c_n = 0$,

(c) $\sum_n a_n \|b_n x + c_n y\|^2 \, r \sum_n a_n (b_n^2 \|x\|^2 + c_n^2 \|y\|^2)$.

A localized form of the above result is given in the following theorem which extends a result of Senechalle (1971) and Oman (1969).

THEOREM 4.12.83 (Oman, 1980). *A real normed linear space X is an inner product space if and only if there exists $\varepsilon > 0$ and u in $(0, 1)$ such that whenever x, y are in X, $\|x\| = \|y\| = 1$ and $\|x - y\| < \varepsilon$ then we find the real numbers t, d_1, d_2, a_n, b_n and c_n with the following properties:*

(1) t $(0, 1)$,

(2) $d_i > 0$, $a_n \geqslant 0$,

(3) $\sum_n a_n b_n c_n = d_2 u(1 - u) - d_1 t(1 - t)$,

(4) $d_1 \|tx + (1 - t)y\|^2 + d_2 \|ux - (1 - u)y\|^2 +$

$$+ \sum_n a_n \|b_n x + c_n y\|^2 \, r \, d_1(t^2 + (1 - t)^2) +$$

$$+ d_2(u^2 + (1 - u)^2) + \sum_n a_n (b_n^2 + c_n^2)$$

where r means $\geqslant$ or $\leqslant$.

Proof. If X is an inner product space then it is not difficult to see that the condition is necessary. To prove the sufficiency part, suppose that r means $\geqslant$ (if r means $\leqslant$ the argument is similar). We suppose without loss of generality that X is a two-dimensional space. Let us consider the unit sphere S and let E be the ellipse inscribed in S with the contact points x and y such that $\|x - y\| < \varepsilon$. Let $|\,,\,|$ be the norm determined by the ellipse and thus for each m in the space we have $|m| \geqslant \|m\|$.

As we know the norm $|,|$ is determined by an inner product, and thus we have

$$d_1\|tx + (1-t)y\|^2 + d_2\|ux - (1-u)y\|^2 + \sum_n a_n\|b_n x + c_n y\|^2 \geqslant$$

$$\geqslant d_1(t^2 + (1-t)^2) + d_2(u^2 + (1-u)^2) + \sum_n a_n(b_n^2 + c_n^2)$$

$$= d_1|tx + (1-t)y|^2 + d_2|ux - (1-u)y|^2 + \sum_n a_n|b_n x + c_n y|^2 \geqslant$$

$$\geqslant d_1\|tx + (1-t)y\|^2 + d_2\|ux - (1-u)y\|^2 + \sum_n a_n\|b_n x + c_n y\|^2.$$

This clearly implies that

$$|tx + (1-t)y| = \|tx + (1-t)y\|$$
$$|ux - (1-u)y| = \|ux - (1-u)y\|$$
$$|b_n x + c_n y| = \|b_n x + c_n y\|.$$

We can repeat this procedure and then we get that the arc xy is in the contact set of E with S. If E is not S we have that the contact set consists of arcs and points having a distance between them greater than ε.

We have to distinguish three cases.

CASE 1. *u is in* $(0, \frac{1}{2})$. Let x, y be points in S such that x lies on an arc and y is the endpoint of the arc. Suppose that $x \to y$ and thus

$$w = -(ux - (1-u)y)/ux - (1-u)y$$

is in the contact set of E and $w \to y$ through points which are not on the arc. This contradicts the fact that on the arc there do not exist points that are contact points within ε of the arc.

CASE 2. $u = \frac{1}{2}$. We consider x, y and w as in Case 1. Then if $x \to y$, $w \to z$ which is parallel to a support line of S at y. Then E is the Loewner ellipse inscribed in S. By Theorem 4.6.11 we find a second ellipse having the same properties, and thus from the uniqueness of Loewner's ellipse we have a contradiction.

CASE 3. *u is in* $(\frac{1}{2}, 1)$. We can interchange the role of x and y and as in Case 1 we obtain a contradiction. Thus we must have $E = S$ and X is an inner product space. Since the arguments for $r =$ '$\leqslant$' work only for a strictly convex space we prove that if X satisfies the above property then it is indeed a strictly convex space.

Indeed, suppose that X has the property stated in the theorem and is not strictly convex. Then we find x, y in X with the following properties:

(1) $\|x\| = \|y\| = 1, \qquad \|x - y\| < \varepsilon,$
(2) $\|tx + (1-t)y\| = 1$ for all t in $(0, 1)$,
(3) $\|ux - (1-u)y\| > 1 - 2u$.

Since X satisfies the condition of the theorem we obtain that

$$\|bx + cy\|^2 = (b + c)^2$$

for all real numbers b and c. Then

$$d_1\|tx + (1-t)y\|^2 + d_2\|ux - (1-u)y\|^2 + \sum_n a_n\|b_n x + c_n y\|^2 >$$

$$> d_1 + d_2(1-2u)^2 + \sum_n a_n(b_n + c_n)^2 = d_1(t^2 + (1-t)^2) +$$

$$+ d_2(u^2 + (1-u)^2) + \sum_n a_n(b_n^2 + c_n^2)$$

and this contradicts the hypothesis. Thus X is strictly convex. □

REMARK 4.12.84 (Oman, 1980). There exists an example of a two-dimensional space satisfying the condition of Theorem 4.12.82 with u depending upon x and y and this space is not an inner product space. Thus the condition that u is independent of x and y is an essential condition.

For a related result see also Carlsson (1962).

As we know from the Jordan–von Neumann characterization of inner product spaces, a normed linear space is an inner product space if and only if all two-dimensional subspaces of that space are inner product spaces. The following result shows that a normed linear space is an inner product space if it has sufficiently many two-dimensional subspaces which are inner product spaces. The precise result is as follows:

THEOREM 4.12.85 (Oman, 1969, 1980). *A real normed linear space X in an inner product space if and only if there exist a hypersubspace H and an element x, $x \notin H$ such that:*

(1) *H is an inner product space,*
(2) *every two-dimensional subspace of X containing x is an inner product space.*

For the proof we need the following lemma. The proof is left to the reader.

LEMMA 4.12.89. *Let X and Y be two Banach spaces and consider the product spaces $X_\times Y$ with the norm*

$$\|(x, y)\| = (\|x\|^2 + \|y\|^2)^{1/2}.$$

Then $X_\times Y$ is an inner product space if and only if X and Y are inner product spaces.

Proof of Theorem 4.12.85. The condition is obviously necessary. Suppose now that X is a real normed linear space with the stated property. Let x and H be as given by the hypothesis of the theorem. We consider a hyperspace H_1 of X such that x is orthogonal in Birkhoff's sense to all elements of H_1. Then $X = H_1 \oplus V_x$ with $V_x = (rx, r \in R)$. We define on $H_1 \times V_x$ the following norm:

$$\|(x_1, rx)\| = (r\|x\|^2 + \|x_1\|^2)^{1/2}.$$

If we prove that $H_1 \times V_x$ is an inner product space then this obviously implies that X is an inner product space.

Let us consider h_1 in H_1 from the p hypothesis of the theorem that every two-dimensional subspace of X containing x is an inner product space, we obtain that

$$\|rx + h_1\|^2 = r^2\|x\|^2 + \|h_1\|^2.$$

This implies that in order to prove our assertion, it suffices to prove that H_1 is an inner product space. Let us consider u, v in H_1 and then for some real numbers p and q, $px + u$ is in H and $qx + v$ is also in H. Then we have,

$$\|(px + u) + (qx + v)\|^2 + \|(px + u) - (qx + v)\|^2 = 2(\|px + u\|^2 + \|(qx + v)\|^2)$$

and further

$$(p + q)^2\|x\|^2 + \|u + v\|^2 + (p - q)^2\|x\|^2 + \|u - v\|^2$$
$$= 2(p^2\|x\|^2 + \|u\|^2 + q^2\|x\|^2 + \|v\|^2)$$

which implies that

$$\|u + v\|^2 + \|(u - v)\|^2 = 2(\|u\|^2 + \|v\|^2)$$

i.e. the parallelogram law holds in H_1. This implies that X is an inner product space and the theorem is proved. □

COROLLARY 4.12.87. *A normed linear space X is an inner product space if and only if there exists a basis of the form $(x, x_i; i \in I)$ such that every two-dimensional subspace of X containing an x_i is an inner product space.*

We mention now the characterization of inner product spaces obtained by M. L. Gromov and A. Dvoretzky in connection with a problem of Banach (1932).

THEOREM 4.12.88 (Gromov, 1967). *Let X be an n-dimensional space and k be an integer in $(1, n)$. Suppose that X has the property that all the k-dimensional subspaces are isometric to each other. Then X is an inner product space if one of the following conditions hold:*

(1) *k is even,*
(2) *k odd, X is a real space and $n \geqslant 2k$,*
(3) *k odd, X real and $n \geqslant k + 2$.*

Dvoretzky (1960) has proved the same result under the hypotheses that k is finite, n may be infinite. For a proof of the above result we refer to the elegant papers of Gromov and Dvoretzky. The problem remains open for $k = 2p + 1$ and $n = k + 1$ in the case of real spaces and for $k = 2p$ and $n < 2k$ for complex spaces. It is worth mentioning the following result of D. A. Senechalle (unpublished manuscript).

THEOREM 4.12.89 (D. A. Senechalle). *A normed linear space X is an inner product space if and only if all two-dimensional subspaces are isometric.*

In what follows we show, following Tapia (1973), how we can define a 'generalized inner product' on normed linear spaces and a characterization of inner product spaces using this generalized inner product. Let X be a normed linear space.

DEFINITION 4.12.90. Let $f: X \to R$ be a real-valued function defined on X. The function f is said to have a first right-hand Gateaux derivative at $x_0 \in X$ in the direction $h \in X$ if

$$\lim_{t \to 0^+} (1/t(f(x_0 + th) - f(x_0)))$$

exists. In this case for the limit we use the notation $f'_+(x_0)(h)$.

Similarly, we define the first left-hand Gateaux derivative of f at x_0 in the direction h; it is denoted by $f'_-(x_0)(h)$.

The function f is said to be Gateaux differentiable at x_0 in the direction h if

$$f'_+(x_0) = f'_-(x_0)(h)$$

and this will be denoted by $f'(x_0)(h)$. In some papers the notation

$$f'_+(x_0)(h) = N_+(x_0, h)$$

is used (and a corresponding notation for $f'_-(x_0)(h)$). The basic properties of $f'_+(x_0)(h)$ for the case of the function $f(x) = \frac{1}{2} \cdot \|x\|$ were obtained by Ascoli (1932).

In what follows we are interested in the case of function f defined as follows:

$$f(x) = \tfrac{1}{2} \cdot \|x\|^2.$$

If we consider the function $g(t)$ on R^+ by the relation

$$g(t) = 1/t(\|x + th\| - \|x\|)$$

then we remark that it is nondecreasing and the values are in $[-\|h\|, \|h\|]$. Then $\lim g(t)$ for $t \to 0^+$ exists. From this we obtain that $f(x) = \frac{1}{2} \cdot \|x\|^2$ has the first right-hand Gateaux derivative at each x_0 and in any direction h. Indeed,

$$\tfrac{1}{2}(\|x + th\|^2 - \|x\|^2) = (\|x + th\| + \|x\|) \cdot (1/t(\|x + th\| - \|x\|))$$

and this obviously implies our assertion.

DEFINITION 4.12.91 (Tapia, 1973). The generalized inner product space on a normed linear space X is defined by the formula:

$$\langle x, y \rangle = f'_+(x)(y)$$

for the function

$$f(x) = \tfrac{1}{2} \cdot \|x\|^2.$$

The following proposition, whose proof, being simple, is omitted, gives some properties of the generalized inner product $\langle\, , \rangle$ defined above.

PROPOSITION 4.12.92. *The following properties hold:*

(1) $\langle x, x \rangle = \|x\|^2$,
(2) $|\langle x, y \rangle| \leqslant \|x\| \cdot \|y\|$,
(3) *if X is an inner product space then $\langle\, , \rangle$ coincides with the original inner product of X.*

The characterization of inner product spaces using the generalized inner product is given in the following theorem.

THEOREM 4.12.93 (Tapia, 1973). *The normed linear space X is an inner product space if and only if the following equivalent properties hold:*

(1) *the generalized inner product $\langle x, y \rangle$ is linear in the first variable, or*
(2) *$\langle x, y \rangle$ is symmetric (i.e. $\langle x, y \rangle = \langle y, x \rangle$ for all x, y in X).*

Proof. From the Jordan–von Neumann characterization of inner product spaces it suffices to prove the theorem for the case of two-dimensional subspaces. Of course, the conditions are necessary. Suppose now that the generalized inner product $\langle x, y \rangle$ is linear in the first variable. Now if (u, v) is a basis of the space, every element of the space is of the form $mu + nv$ with m, n scalars.

Let x, y be two arbitrary elements in the space and thus

$$x = au + bv, \qquad y = cu + dv.$$

Now the function $f(x) = \frac{1}{2} \cdot \|x\|^2$ is a convex function and thus it is differentiable at almost all unit elements of the space (which is two-dimensional). Thus x, y is linear in the second variable at almost all unit elements. Let us suppose that u, v are unit elements at which the function is differentiable and thus

$$\begin{aligned}\langle x, y \rangle &= a\langle u, cu + dv)\rangle + b\langle v, cu + dv \rangle \\ &= ac\langle u, u \rangle + ad\langle u, v \rangle + bc\langle v, u \rangle + bd\langle v, v \rangle\end{aligned}$$

and this is an inner product on X.

Suppose now that the generalized inner product is symmetric. Then x, y is linear in the first variable for almost all unit elements y and from the continuity in the second variable (which follows from 4.12.92(2)) $\langle x, y \rangle$ is linear in the first variable. Then from the above proved assertion, X is an inner product space. □

COROLLARY 4.12.94 (Tapia, 1973). *The normed linear space X is an inner product space if and only if the function $f(x) = \frac{1}{2} \cdot \|x\|^2$ is with the second right-hand Gateaux derivative at the origin, linear or symmetric, where the second right-hand Gateaux derivative of the function $h: X \to R$ at x in the directions u and v is the limit (if it exists)*

$$f''_+(x)(u, v) = \lim_{t \to 0^+} (1/t(f'_+(x + tu)(v) - f'_+(x)(v)).$$

Then we have that

$$\langle x, y \rangle = f''_+(0)(x, y).$$

REMARK 4.12.95. It may be of interest to know if there exist connections between the orthogonality in the sense of Roberts, the isosceles orthogonality, the Pythagorean orthogonality, the orthogonality in the Birkhoff sense and the orthogonality defined using the generalized inner product:

$$x \perp y \text{ if } \langle x, y \rangle = 0.$$

We shall now present some results obtained by Th. Precupanu in the case when the norm of space is supposed to have some smoothness properties.

First it is clear that if X is a normed linear space whose norm is generated by an inner product then it is Gateaux differentiable at each point x, $x \neq 0$, and the following relation holds:

$$\langle x, y\rangle = \|x\| f'_x(y).$$

Then we have the following results.

THEOREM 4.12.96 (Precupanu, 1978). *Let X be a normed (real) space. Then the norm of X is generated by an inner product iff one of the following equivalent properties hold:*

(1) $\|2f'_x(y)/\|x\|\cdot x - y\| = \|y\|$ *for all* x, y *in* X, $x \neq 0$,
(2) $\|2f'_x(y)/\|x\|\cdot x - y\| \leqslant \|y\|$ *for all* x, y *in* X, $x \neq 0$,
(3) $2f'_x(y) f'_z(x) \leqslant \frac{1}{2}(\|y\| + f'_z(y))\|x\|$ *for all* x, y, z *in* X, $x, z \neq 0$.

An improvement of this result was also obtained, as follows.

THEOREM 4.12.97 (Precupanu, 1978). *Let X be a normed linear space. Then the norm of X is generated by an inner product iff one of the equivalent conditions* (1), (2), (3) *of Theorem* 4.12.97 *are true whenever* (x, y) *is a pair for which* $f'_x(y)$ *exists.*

Using these results some very interesting metric characterizations of inner product structures were obtained:

THEOREM 4.12.98 (Precupanu, 1978). *Let X be a normed linear space. Then the norm of X is generated by an inner product if and only if one of the following inequalities hold:*

(1) $\big\| \|y\|^2 x + y \big\| \leqslant \|x + y\|$, $\|x\| = 1$ *and* $\|y\| \leqslant 1$,
(2) $\|tx + y\| \leqslant t^{-1}\|x + ty\|$, $\|x\| = \|y\| = 1$ *and* t *in* $(0, 1)$,
(3) $t\|tx + t^{-1}y\| \leqslant t^{-1}\|t^{-1}x + ty\|$, $\|x\| = \|y\| = 1$, t *in* $(0, 1)$,

For a proof we refer to Precupanu (1978). Concerning the differentiability properties of the norm and inner product structures the reader may consult the Ph.D. Thesis of Leduc (1968) and Leduc (1969). For other result concerning differentiability properties see also Chapter 8.

4.13. Approximation Theory and Characterizations of Inner Product Spaces

Let X be a normed linear space and let M be a nonempty subset of X. Let P_M be the function defined on X by the relation

$$x \to P_M x = (y, y \in M, d(x, M) = \|x - y\|),$$

where

$$d(x, M) = \inf(\|x - m\|, m \in M).$$

Clearly the values of P_M are in 2^M. The mapping P_M is called metric projection or best approximation operator, or proximity map, or Chebyshev operator, etc. In what follows we use the term metric projection.

DEFINITION 4.13.1. The set M is called an E-set (existence set) if $P_M x$ contains at least one element for each $x \in X$. The set M is called a U-set (uniqueness-set) if for each $x \in X$, $P_M x$ contains at most one element. A set M is an EU-set if for each $x \in X$, $P_M x$ contains exactly one element (i.e. the set M is an E-set and a U-set).

DEFINITION 4.13.2. A selection $s: X \to M$ for the metric projection P_M is any function s with the property that for each x, $s(x) \in P_M x$.

The following theorem lists some properties of P_M.

THEOREM 4.13.3. *Let X be a normed linear space and M be an E-set and $s: X \to M$ be a selection for P_M. Then the following assertions hold:*

(1) *M is closed.*
(2) *M convex $\to P_M x$ is convex for all $x \in X$.*
(3) *If M is a subspace of X then*

$$P_M(ax) = aP_M x$$

for all $x \in X$ and a scalar.
(4) *If M is a subspace then for each $y \in M$,*

$$P_M(x + y) = P_M x + y.$$

Proof. (1) Suppose that the assertion is not true. Then we can consider an element x in ClM-M and clearly $d(x, M) = 0$ but $\|x - m\| > 0$ for each $m \in M$. Then $P_M x = \emptyset$ which contradicts the fact that M is an E-set.

(2) Let y_1 and y_2 be two elements in M and $t \in (0, 1)$. Then

$$\begin{aligned}\|x - (ty_1 + (1 - t)y_2)\| &= \|t(x - y_1) + (1 - t)(x - y_2)\| \\ &\leqslant t\|x - y_1\| + (1 - t)\|x - y_2\| = d(x, M)\end{aligned}$$

which shows that $ty_1 + (1 - t)y_2$ is in $P_M x$.

(3) Let $x \in X$ and a be a fixed scalar. If $a = 0$ then

$$P_M 0 = (0) = aP_M x.$$

Suppose now that $a \neq 0$, then we get

$$\begin{aligned}d(ax, M) &= \inf(\|ax - m\|, m \in M) \\ &= |a| \inf(\|x - m/a\|, m \in M) = |a| d(x, M)\end{aligned}$$

and thus $y \in P_M(ax)$ if and only if

$$\|ax - y\| = d(ax, M) \leftrightarrow \|x - (1/a)y\| = d(x, M) \leftrightarrow (1/a)y \in P_M x$$

or $y \in aP_M x$ and the assertion is proved.

(4) If $y \in M$ and $x \in X$ then

$$\begin{aligned}x_0 \in P_M(x + y) &\leftrightarrow \|x + y - x_0\| = d(x + y, M) = d(x, M) \\ &\leftrightarrow x_0 - y \in P_M x \leftrightarrow x_0 \in P_M x + y.\end{aligned}$$

□

REMARK 4.13.4. There exists an example of a Banach space X and M a one-dimensional EU-subspace and P_M is nonlinear.

In fact X is the Banach space $C_{[0,1]}$ of all continuous functions on $[0, 1]$ with the sup norm and M is the subspace of all constant functions. For details, see Deutsch (1973).

The proof of the following result is simple and thus is omitted.

THEOREM 4.13.5. *Let X be a complete inner product space and let M be a closed subspace. Then P_M is a linear operator satisfying the relations:*

(1) $P_M^2 = P_M$,

(2) $P_M^* = P_M$.

In what follows we show that the property of P_M to be a linear operator characterizes the complete inner product spaces. First we prove the following result which has also independent interest.

THEOREM 4.13.6. *Let X be a normed linear space and M be an EU-subspace of X and let i: $X \to X/M$ be the canonical map (i.e. $x \to x + M$). Then the following assertions are equivalent:*

(1) *P_M is a linear operator.*

(2) *M^0 is a subspace, $u^0 = \{x, 0 \in P_M(x)\}$.*

(3) *M^0 is convex.*

(4) *i/M^0 is an isometry,*

where $M^0 = P_0^{-1}(0) = (x, x \in X, P_M x = 0)$.

Proof. First we remark that if M is an EU-subspace then $X = M \oplus M^0$, i.e. each $x \in X$ has a decomposition $x = u + v$ with $u \in M$ and $v \in M^0$. Indeed, for each $x \in X$ we set $u = P_M x$ and $v = x - u$.

Suppose now that P_M is linear. *Then for any elements p, q in M^0* and any scalars a, b we get,

$$P_M(ap + bq) = P_M(ap) + P_M(bq) = 0,$$

i.e. $ap + bq \in M^0$. Thus M^0 is a linear subspace.

It is trivial that (2) → (3). Suppose now that (3) holds and consider x, y in X. Then $x = m + m^0$, $y = m_1 + m_1^0$ with $m, m_1 \in M, m^0, m_1^0 \in M^0$. Further, we have

$$\begin{aligned} P_M(x + y) &= P_M((m + m_1) + (m^0 + m_1^0)) = m + m_1 + P_M(m^0 + m_1^0) \\ &= P_M(x) + P_M(y) + 2P_M(\tfrac{1}{2}(m^0 + m_1^0)) = P_M(x) + P_M(y) \end{aligned}$$

since $\frac{1}{2}(m^0 + m_1^0) \in M^0$, by the convexity. Thus P_M is additive and since the homogeneity property is true for any subspace we have that (3) → (1).

We show now that (1) → (4) and (4) → (1). Indeed, first we suppose that (1) holds and consider x, y in M^0. From the linearity of P_M we get,

$$\begin{aligned} \|i(x) - i(y)\| &= \|x + M - y - M\| = \|x - y + M\| \\ &= \|x - y - P_M(x - y)\| = \|x - y - P_M x + P_M y\| = \|x - y\| \end{aligned}$$

and thus i is an isometry.

Suppose now that (4) holds, i.e. that i is an isometry on M^0. Let x, y be two elements in X and since M is an EU-subspace we have

$$x = m + m^0, \quad y = m_1 + m_1^0, \quad (m, m_1) < M, \quad (m^0, m_1^0) < M^0$$

and thus

$$P_M x = m, \qquad P_M y = m_1.$$

Then we get

$$\begin{aligned} \|x - y - (P_M x - P_M y)\| &= \|m^0 - m_1^0\| \\ &= \|i(m_0) - i(m_1^0)\| = \|m_0 + M - (m_1^0 + M)\| = \\ &= \|x + M - (y + M)\| = \|x - y + M\| \\ &= \|x - y - P_M(x - y)\|. \end{aligned}$$

Since M is an EU-set we must have

$$P_M x - P_M y = P_M(x - y)$$

which gives that P_M *is additive.*

Since it is also homogenous we have that P_M is linear. Thus (4) → (1) and the theorem is proved. □

Now the characterization of complete inner product spaces is as follows:

THEOREM 4.13.7. *Let X be a Banach space of dimension at least 3. Then the following assertions are equivalent:*

(1) *X is a complete inner product space.*
(2) *Each closed convex subset of X is an EU-set with a nonexpansive metric projection* (i.e. $\|P_M x - P_M y\| \leqslant \|x - y\|$).
(3) *Each closed subspace M of X is an EU-subspace with a nonexpansive metric projection.*
(4) *Each closed subspace M of X is an EU-subspace and* $\|P_M x\| \leqslant \|x\|$, $x \in X$.
(5) *Each closed subspace M of X is an EU-subspace with a linear metric projection.*
(6) *For some (fixed) integer n, $1 \leqslant n \leqslant \dim X - 2$, each subspace of dimension n is an EU-subspace with a linear metric projection.*

Proof. By the above result we have (1) → (2) and obviously (2) → (3) as well as (3) → (4). Now we show that (4) → (1). For this it is sufficient to prove that every two-dimensional subspace is an inner product space.

Let M be a two-dimensional subspace of X and then, since it is an EU-subspace, P_M is linear and since $\|P_M\| \leqslant 1$ we have that $\|P_M\| = 1$.

The Kakutani–Bohnenblust theorem asserts that X is an inner product space. Thus (4) → (1). Obviously we have (1) → (4).

Clearly (5) → (6) and (1) → (5). We prove now that (6) → (1). Let n be an integer in $[1, \dim X - 2]$, and M be a subspace of X such that $\dim M = n$ and, of course, we suppose that it is an EU-subspace. Let M_2 be the space generated by two elements of X (we suppose them to be linearly independent) x_i', $i = 1, 2$, in X such that $\|x_1'\| = \|x_2'\| = 1$ and $x_i'(x_i) = 1$. Then the subspace $(x_i')_1^2$ is two-dimensional and

$$M_1 = (x, x_1'(x) = 0, x_2'(x) = 0)$$

is with codim $M_1 = 2$. This gives that dim $M = n$. Then M is an EU-subspace and thus P_M is linear and $I - P_M$ is also linear with $\|I - P_M\| = 1$. But

$$P^2 = (I - P_M)^2 = I - 2P_M + P_M^2 = I - 2P_M + P_M = I - P_M = P$$

and P is a projection. Let x be in M_1^0, which is the case if and only if $P_x = x$, which implies that M_1^0 is the range of P.

Since

$$\|x_1\| = d(x_1, M_1) = d(x_1, \ker X'_1) = x'_1(x_1)/\|x'_1\| = \|x_1\|$$

or

$$x_1 = d(x_1, M_1)$$

then x_1 is in M_1^0. Similarly, we prove that $x_2 \in M_1^0$. Since M_1^0 is a linear subspace we obtain that the subspace generated by x_1 and x_2 in M_1^0 is exactly M_1^0 because dim $M_1^0 =$ codim $M_1 = 2$. Thus P is the projection on the space generated by x_1 and x_2. Thus for any pair of linearly independent elements of X there exists a projection of norm 1 on that subspace. Then by the Kakutani–Bohnenblust characterization of inner product spaces, X is an inner product space.

Rudin and Smith (1961) have remarked that the Birkhoff–James characterization of inner product spaces of dimension at least 3, i.e. the normed linear space is an inner product space when the orthogonality in Birkhoff's sense is symmetric (Theorem 4.12.39), also follows from 4.13.7(6). Indeed, suppose that the orthogonality in Birkhoff's sense is symmetric. Let x, y be two linearly independent elements of the space X. If P_y is the projection on the one-dimensional subspace of X which contains the element y, then $x \perp_B y$ if and only if $P_y(x) = 0$. Suppose now that $x \perp_B y$ and we can suppose without loss of generality that $\|y\| = 1$. If Q is an $(n-1)$-dimensional plane containing the origin and such that $y + Q$ is a support hyperplane of the unit ball of X, then we have the relation

$$\|y + kx\| \geqslant \|y\| = 1$$

which gives that $y \perp_B x$. From the hypothesis we have that $x \perp_B y$. Then, by the above remark $P_y(x) = 0$. Thus for all x in Q, $P_y(x) = 0$ and thus the null space of P_y contains an $(n-1)$-dimensional plane. This implies that P_y is a linear mapping. But then X is an inner product. □

The result in Theorem 4.13.7(6) was given for $n = 2$ by James (1947), for $n = 1$ by Hirschfeld (1958b) and for arbirary n by Rudin and Smith (1961). It is clear that the method of proof given above may be applied to obtain the following result, as stated by Deutsch (1973).

THEOREM 4.13.8. *Let X be a Banach space of dimension $\geqslant 3$. Then X has an inner product structure iff for a fixed n, $2 \leqslant n \leqslant \dim X - 1$ and each subspace M of codimension n, M is an EU-set with a linear metric projection.*

In the same way we have the following,

THEOREM 4.13.9. *Let X be a Banach space of dimension $\geqslant 3$. Then X has an inner product structure iff for any linear subspace M of X, $1 \leqslant n \leqslant \dim X - 1$ the mapping P_M is contractive (i.e. $\|P_M u\| \leqslant \|u\|$).*

For $n = 1$ this result was obtained by Phelps (1957) and for arbitrary strictly convex spaces by Stiles (1965b). Other results related to these are noted in the book of Singer (1967, 1970). See also Papini (1975).

A characterization of inner product structures using properties of convex sets was noted by Phelps (1957, 1958) in connection with a very interesting characterization of inner product spaces of dimension 2 obtained by Motzkin (1935).

Let X be a normed linear space and let M be a set in X. We say that the set M is a nearest-point set if there exists a set N of X and element x_0 of X with the property that $M = P_N^{-1}(x_0)$. Then we have the following result.

THEOREM 4.13.10 (Phelps, 1957, 1958). *For a Banach space X and M a subset of X the following assertions hold:*

(1) $P_M^{-1}(x_0) = N_{x_0}$ *is a closed subset of X for each* x_0.
(2) *If the norm of X is generated by an inner product then* N_{x_0} *is a convex set.*
(3) *If for each M in X and each* x_0 *in* X, H_{x_0} *is a convex set, then the norm of X is generated by an inner product.*

It is worth noticing that sets of the form N_{x_0} were first considered by the great mathematician, Lipot Fejér, in 1922. A related characterization was obtained by Papini (1982). In order to formulate this we recall that we defined for any x, y in a normed linear space the number $\tau(x, y)$ by the formula

$$\tau(x, y) = \lim_{t \to 0^+} t^{-1}(\|x + ty\| - \|x\|)$$

which is well defined since

$$t \to t^{-1}(\|x + ty\| - \|x\|)$$

is a nondecreasing function.

DEFINITION 4.13.11. If M is a subset of X then the farthest points to x from M are

$$F_M(x) = \{u,\ u \in M, \|u - x\| \geqslant \|m - x\|,\ \forall m \in M\}.$$

Then it is easy to see that u_0 is in $F_M(x)$ iff

$$\tau(x - m, m - u_0) \geqslant 0$$

holds for all m in M.

Then we have

THEOREM 4.13.12 (Papini, 1982). *A Banach space X has its norm generated by an inner product iff for any convex and bounded set C in X,*

$$\tau(u - v, x - y) \geqslant 0$$

for all $u \in F_C(x), v \in F_C(y)$.

For the proof, as well as for other related results, we refer to Papini's paper (1982)

Related to the result in Theorem 4.13.9 we shall present in Chapter 8 an extension using nonlinear (nonexpensive) mappings. Another characterization of inner product structures was obtained by B. Beauzamy and B. Mory, using the so-called minimal points.

DEFINITION 4.13.13 (Beauzamy and Mory, 1977). Let C be a set in the normed linear space X. The point m is said to be minimal point for C if no poing x' in X, $x \neq x'$, has the property that

$$\|x' - c\| \leqslant \|x - c\|$$

holds for all c in C.

The set of all minimal points of the set C is denoted by Min C.

DEFINITION 4.13.14 (Beauzamy and Mory, 1977). The set C is said to be optimal iff $C = \text{Min } C$.

Then the characterization obtained by B. Beauzamy and B. Mory is as follows.

THEOREM 4.13.15 (Beauzamy and Mory, 1977). *Let X be a complete inner product space. Then for any C in X,*

$$\text{Min } C = \min (\text{conv } C) = \overline{\text{conv } C}.$$

If X is a strict convex, smooth and reflexive Banach space then the norm of X is generated by an inner product iff that closed unit ball of X is optimal.

Franchetti and Fiuri (1972) have introduced a map denoted by R_M and, using it, they have characterized the class of complete inner product spaces in the class of Banach spaces. We present here their results. For a proof we refer to the paper by Franchetti and Fiuri (1972) or Papini (1975) where the reader may find other interesting results.

DEFINITION 4.13.16 (Franchetti and Fiuri, 1972). Let X be a Banach space and let M be a subspace of X. The mapping R_M is defined by the following rule: for each x in X it associates the set of all y in M such that $\|m - y\| \leqslant \|x - m\|$, $m \in M$.

It is to be noted that when X is a complete inner product space then R_M is the orthogonal projection on M.

For the mapping R_M defined above we set $D(R_M)$ to be the set of all points of X such that $R_M x$ is nonempty. Then the characterization of complete inner product spaces is as follows.

THEOREM 4.13.17 (Franchetti and Fiuri, 1972). *Let X be a Banach of dimension $\geqslant 3$. Then the following assertions are equivalent:*

(1) *The norm of X is generated by an inner product.*
(2) *$D(R_M) = X$ for any subspace M and X is smooth.*
(3) *$D(R_M) = X$ for every hyperplane M.*
(4) *There exists a proper subspace X_0 in X such that for every subspace M of X, X-isomorphic to X_0, we have $R_M^0 \subset R_M x$ (R_M^0 is an arbitrary selection of R_M).*
(5) *X is strictly convex and there exists a proper subspace X_0 of X such that for every subspace M of X which is X-isomorphic to X_0, R_M^0 is defined on X and $I - R_M^0$ is a nonexpansive mapping.*

We recall that two subspaces of the Banach space X are said to be X-isomorphic if there exists an isomorphism T of X such that $TM = N$. This notion was also used earlier in the paper of Garkavi (1964). For some improvements and related result, see also Papini (1982, 1984). An interesting characterization of inner product structures using the best approximation elements was obtained by Belobrov (1968). The result is as follows.

THEOREM 4.13.18 (Belobrov, 1968). *Let X be a Banach space of dimension $\geqslant 3$. Then the norm of X is generated by an inner product iff for some subspace L_0 and for each subspace L of X which is X-isomorphic to L_0 the following condition is satisfied: if the element y in L has the best approximation x then $\|y\| \leqslant \|x\|$ holds.*

The proof of this theorem will follow from the following lemmas. First we give a definition.

DEFINITION 4.13.19. Let M be a subspace of the Banach space X. We saw that M is regularly closed if M is closed for the $\sigma(X^*, X)$-topology and T in $L(X)$ is said to be regularly continuous if it is continuous for the $\sigma(X^*, X)$-topology. The subspaces U and V of X are said to be regularly X^*-isomorphic if there exists a regular isomorphism T of X^* such that $TU = V$.

Then we have

LEMMA 4.13.20 (Garkavi, 1964). *The subspaces M, N of X are X-regular isomorphic iff M^+ and N^+ are regularly X-isomorphic.*

Proof. Let $T: M \to N$, T being an isomorphism. Consider $T^*: X^* - X^*$ which is again an isomorphism of X^*. If $x^* \in M^\perp$ then

$$T^*x^*(x) = x^*(Tx) = 0$$

for each x in M. Thus $T^*x^* \in M^\perp$. If y^* is in $M^\perp$ then for each x in M we have

$$T^{*-1}y^*(x) = y^*(T^{-1}x) = 0$$

which implies that $T^{*-1}y^* \in M^\perp$. Thus $M^\perp$ and $N^\perp$ and X-regular isomorphic and since the converse assertion may be proved in the same way, the lemma is proved. □

LEMMA 4.13.21 (Garkavi, 1964). *Let X be a Banach space and let M, N be two subspaces of X, one of them of finite dimension (index). Then M, N are X-isomorphic iff* $\dim M = \dim N$ *(the index equal).*

Proof. Let

$$M = \{x, f_i(x) = 0, \quad i = 1,2,\ldots,n, \quad \|f_i\| = 1\}$$
$$N = \{x, g_i(x) = 0, \quad i = 1,2,\ldots,n, \quad \|g_i\| = 1\}.$$

We may suppose that $(f_1,\ldots,f_m, g_1,\ldots,g_m)$ is a linearly independent set. We find $\{z_i\}_1^m$ in X such that

(1) $f_i(z_j) - g_i(z_j) = 0$,

(2) $g_i(z_k) = \delta_{ij}, \quad i,k = 1,2,\ldots,m$

and then we define

$$x \to Tx = x + \sum_{i=1}^{m} (f_i(x) - g_i(x))z_i$$

which obviously is an isomorphism of X. If M, N are X-isomorphic then we get that

$$\dim M^{\perp} = n, \quad \dim N^{\perp} = n$$

and the assertion of the lemma follows. The same holds for the dimension. □

Now we prove two lemmas from which Theorem 4.13.18 follows immediately.

LEMMA 4.13.22 (Belobrov, 1968). *Let X be a Banach space and let L_0 be a subspace of X. Suppose that for each subspace L of X which is X-isomorphic to L_0 the following condition is satisfied: for u in L the best approximation of v satisfies the inequality $\|u\| \leqslant \|v\|$. Then this condition is satisfied by arbitrary one-dimensional subspaces of X.*

Proof. Let x, x_0 in X be of norm 1 and linearly independent. Then for all t in R we have

$$\|x - tx_0\| \geqslant \|x - ax_0\| = d > 0.$$

We consider the element $z_0 = d^{-1}(x - ax_0)$ which clearly satisfies the inequality

$$\|z_0 - tx_0\| \geqslant 1$$

for all t in R.

If x_0 is not in L_0 then for some element $\bar{x}$ of L_0 we find an element f in X^* with the property that $f(x_0) = f(\bar{x}) = 1$. Then we can define the operator

$$T_0 u = u + f(u)(x_0 - \bar{x})$$

which is easily seen to be an isomorphism of X. Further, T transforms the space generated by $\bar{x}$ into the subspace generated by x_0, i.e. $T(\{t\bar{x}\}) = \{tx_0\}$.

We now consider the subspace $L_1 = T(L_0)$ and thus L_1, L_0 are X-isomorphic subspaces.

Since the distance of z_0 to (tx_0) is 1 and the norm of z_0 is 1 we find g in X with the properties that $\|g\| = 1$ and $g(x_0) = 1, f(z_0) = 0$. Further we consider the functional in X with the following properties: $\|h\| = 1$, $h/L_1 = 0$. By Lemma 4.13.21 we find an isomorphism $Tu = u + (h(u) - g(u))z$ which transforms the hyperplane $H_g = (u, g(u) = 0)$ onto the hyperplane $H_h = (u, g(u) = 0)$. The set $T(L_1)$ is a linear subspace of X and since the operator TT_0 is again an isomorphism which transforms L_0 onto $T(L_1)$, this subspace is X-isomorphic to L_0. But $x = ax_0 + dz_0$ and since $\|z - x\| \geqslant |g(z - x)|$ we get that $g(x) = d$. Thus the distance of x to $T(L_1)$ is equal to d. But ax_0 is in $L(x_0) = (tx_0)$ and this implies that $\|ax_0\| \leqslant \|x\|$. This proves the lemma. □

LEMMA 4.13.23 (Belobrov, 1968). *Let X be a Banach space of dimension $\geqslant 3$. Then the norm of X is generated by an inner product iff for each one-dimensional subspace L of X the following condition is satisfied: if $\inf_{u \in L} (\|x - u\|) = \|x - u\|$ then $\|y\| \leqslant \|x\|$.*

Proof. Clearly the condition is necessary. Suppose now that it is satisfied and take x, y in X, $\|x\| = \|y\| = 1$ and $\|x - ty\| \geqslant \|x\|$ for all t, i.e. $x \perp_B y$. Suppose that

$$\inf(\|y - tx\|) = \|y - x_0\| < 1.$$

Since

$$\inf(\|(x_0 - y) - ty)\| = \|(x_0 - y) - (-y)\|)$$

we get that $\|-y\| \leqslant \|x_0 - y\|$. This contradicts the fact that $\inf\|y - tx\|$ is $1 = \|y\|$. This clearly implies that $y \perp_B x$. Then since orthogonality in the sense of Birkhoff is symmetric we get that the norm of X is generated by an inner product.

Thus the lemma is proved and this together with Lemma 4.13.22, imply Theorem 4.13.24. □

From Lemma 4.13.23 we obtain another characterization of inner product structures, which is as follows.

THEOREM 4.13.24 (Belobrov, 1968). *Let X be a Banach space of dimension $\geqslant 3$. Then the norm of X is generated by an inner product iff the following condition is satisfied: if $f \perp_B g$ in X and y is the best approximation of x from $H_f = (u, f(u) = 0)$, x in $H_{f,c} = (u, f(u)c)$, then y is a common point of these hyperplanes.*

Proof. Let x_0 and y be two linearly independent elements of X with the norm equal to 1. Then we find φ in X, with the property that $\|\varphi\| = 1$, $\varphi(y) = 0$ and the distance between x and $H_\varphi = (u, \varphi(u) = 0)$, to be equal to the distance between x_0 and H_φ. We consider now the element g in X with the property that $|g(y)| = \|g\| = 1$. Then we see that g is orthogonal to H_φ. But $g(\bar{z}y) = g(x_0)$ or $\bar{z} = (x_0)$, where $\bar{z}$ is the number with the property that the distance between x_0 and (ty) is attained at $y^+ = zy$. Then we have clearly that $|\bar{z}| \leqslant 1$ and $\|y^+\| \leqslant 1$. By Lemma 4.11.23, the norm of X is generated by an inner product.

COROLLARY 4.11.25 (Belobrov, 1968; Singer, 1967, 1970). *Let X be a Banach space of dimension > 3. Then the norm of X is generated by an inner product iff for each subspace of finite index the following condition is satisfied: if y is the best approximation of x with respect to the subspace then $\|y\| \leqslant \|x\|$.*

Proof. This follows from the above results and Lemma 4.13.21 of A. L. Garkavi. □

COROLLARY 4.13.26 (Belobrov, 1968; Singer, 1967, 1970). *Let X be a Banach space of dimension $\geqslant 3$. Then the norm of X is generated by an inner product iff for each n-dimensional subspace (n is fixed in $(1, \dim X - 1)$)L_n the condition in Corollary 4.13.25 is satisfied.*

For this see also 4.13.7 and 4.13.9.

We now present a characterization of inner product structures related to a famous result of John von Neumann.

Suppose that X is a reflexive strict convex Banach space. Then for each M (closed linear subspace of X) the mapping P_M is well defined. Suppose that we have another closed linear subspace N of X. Take x arbitrary in X. The result of von Neumann (1949) is as follows.

THEOREM 4.13.27 (von Neumann, 1949). *Suppose that X is a complete inner product space and M, N are as above. Then for each x in X the following relation holds:*

$$\lim(P_M P_N \dots P_M P_N)(x) = P_{M \cap N}(x).$$

This result was also proved by Wiener (1955) and extended by Halperin (1962). It is quite natural to extend this problem to other Banach spaces. The following elegant results of Wilbur J. Stiles answer the problem.

THEOREM 4.13.28 (Wilbur, 1965). *Let X be a Banach space of dimension $\geqslant 3$. Then if X has the property that for each closed subspace M, P_M is single-valued and (if N is another arbitrary closed subspace) for any x in X,*

$$\lim(P_M P_N \dots P_M P_N)(x) = P_{M \cap N}(x)$$

then the norm of X is generated by an inner product.

THEOREM 4.13.29 (Wilbur, 1965). *Let X be a strictly convex Banach space of dimension $\geqslant 2$. Suppose that for any closed linear subspace R of X, P_R is a single-valued mapping and if M, N are two arbitrary closed subspaces for any x in X we have the relation*

$$\lim (P_M P_N \dots P_M P_N)(x) = P_{M \cap N}(x).$$

Then the norm of X is generated by an inner product.

REMARK 4.13.30 (Wilbur, 1965). In fact Theorem 4.13.29 is true if we restrict the class of subspaces M, N to be of dimension 2.

An interesting characterization of complete inner product spaces using the operator P_M was obtained by A. P. Bosznay (1978). In what follows we present these results.

For a Banach space X and an element x in X we set P_x as the mapping P_M corresponding to the subspace generated by an element x. Then we have

THEOREM 4.13.31 (Bosznay, 1978). *A real strictly convex Banach space X has its norm generated by an inner product iff for any pair of elements x, y of $X, \|x\| = \|y\| = 1$, the following relation holds:*

$$\|P_x(y)\| = \|P_y(x)\|.$$

Proof. Obviously the condition is necessary. Suppose now that the Banach space X is as in the theorem. We let x be an arbitrary element in X and let $y \neq 0$, $P_x(y) = 0$. Then clearly $P_y(x) = 0$. Also we have that $x + y \neq 0$. From the properties of the mapping P_M we get the following relations:

$$(1)\quad \|P_{x+y}(x)\| = \|x\| \|P_{(x+y)/\|x+y\|}(x/\|x\|)\|,$$

$$(2)\quad \|P_{x+y}(y)\| = \|y\| \|P_{(x+y)/\|x+y\|}(y/\|y\|)\|$$

and since

$$P_{x+y}(x) = x + y - P_{x+y}(y)$$

we get that

$$P_{x+y}(x) + P_{x+y}(y) = x + y$$

which, coupled with

$$P_x(x + y) = x + P_x(y) = x, \qquad P_y(x + y) = y + P_y(x) = y$$

implies that

$$\left\| P_x((x + y)/\|x + y\|)\right\| = \left\| P_{x/\|x\|}((x + y)/\|x + y\|)\right\| = \|x\|/\|x + y\|$$

$$\left\| P_y((x + y)/\|x + y\|)\right\| = \left\| P_{y/\|y\|}((x + y)/\|x + y\|)\right\| = \|y\|/\|x + y\|.$$

We now apply these results for the elements $x/\|x\|$ and $y/\|y\|$ and obtain that the following relations hold:

$$\left\| P_{x+y}(x)\right\| = \|x\|^2/\|x + y\|, \qquad \left\| P_{x+y}(y)\right\| = \|y\|^2/\|x + y\|.$$

But

$$P_{x+y}(x) + P_{x+y}(y) = x + y$$

and thus

$$\|x + y\|^2 = \|x\|^2 + \|y\|^2 \quad \text{or} \quad \|x + y\|^2 = \|x\|^2 - \|y\|^2.$$

Since $\|x + y\| \geqslant \max(\|x\|, \|y\|)$ the last relation is impossible and thus only the relation

$$\|x + y\|^2 = \|x\|^2 + \|y\|^2 \tag{+}$$

is correct.

Suppose now that x, y are two arbitrary linearly independent elements of X and consider z in $X, z \neq 0$, such that $P_{(x,y)}(z) = 0$. $\{x, y\}$ denotes the subspace generated by x and y. Then there exists a unique c in R such that $P_{(x,y)}(y - cz) = 0$. We apply the relation (+) and we get

$$\|y\|^2 = \|cz\|^2 + \|dx\|^2 (y = cz + dx)$$

$$\|x - y\|^2 = \|cz\|^2 + \|(1 - d)x\|^2,$$

$$\|x + y\|^2 = \|cz\|^2 + \|(1 - d)x\|^2$$

and thus

$$\|x + y\|^2 + \|x - y\|^2 = 2(\|x\|^2 + \|y\|^2)$$

which implies that the norm of X is generated by an inner product by the Jordan–von Neumann theorem.

We now give another result of A. P. Bosznay; for the proof we refer to the original paper.

THEOREM 4.13.32 (Bosznay, 1978). *Let X be a Banach space. Then the norm of X is generated by an inner product iff for each $x, x \neq 0$ there exist EU-spaces (sometimes* *called Chebyshev subspaces)* $\{X_i\}_1^\infty$ *of* X *such that the following conditions are true:*

(1) P_{X_i} *is a linear mapping.*

(2) $\bigcap X_i = \{x\}$.

THEOREM 4.13.33 (Bosznay, 1978). *Let* X *be a real Banach space of dimension* $\geqslant 3$. *Then the norm of* X *is generated by an inner product iff for every family of finite-dimensional subspaces* (X_i) *of* X *and every* $\varepsilon < 0$ *there exists a norm* $\|,\|_\varepsilon$ *on* X *such that:*

(1) $(1-\varepsilon)\|x\|_\varepsilon \leqslant \|x\| \leqslant (1+\varepsilon)\|x\|_\varepsilon$,

(2) X_i *as subspaces of* $(X, \|,\|_\varepsilon)$ *is an EU-set and* $P_i = P_{X_i}$ *is a linear mapping.*

A very interesting approximation theoretic characterization of inner product structures was discovered by Šmatkov (1971) in connection with some approximation problems of Deutsch and Morris (1969) (see also Holmes and Lambert, 1973). We present now Šmatkov's result. First we give

DEFINITION 4.13.34. Let X be a normed linear space and let M be a dense subset of X, with Γ a finite dimensional subspace of X^*. We say that the triple (X, M, Γ) has the SAIN* property if for every x in X and $\varepsilon > 0$ there exists y in M such that:

(1) $\|x\| = \|y\|$,

(2) $\|x - y\| < \varepsilon, \qquad x^*(x) = x^*(y), \quad x^* \in \Gamma.$

In their paper (1969), Deutsch and Morris have proved the following result.

PROPOSITION 4.13.35. *In order that* (X, M, Γ) *has the SAIN property it is necessary that all* x^* *in* Γ *attain their norm solely at the points of* M *or not at all.*

They raised the problem of whether this condition is in fact necessary and sufficient for a triple to have the SAIN property. Then Šmatkov proved the following result.

THEOREM 4.13.36 (Šmatkov, 1971). *Let* X *be a reflexive strictly convex Banach space. Then the norm of* X *is generated by an inner product iff the Deutsch–Morris condition is satisfied.*

REMARK 4.13.37. It may be of interest, in connection with Šmatkov's result, to have a characterization of normed linear spaces, or Banach spaces, which are isomorphic to spaces with an inner product, in terms of a property like SAIN.

Another approximation theoretic characterization of the existence of inner product structures on Banach spaces was stated by Amir and Deutsch (1983) and is as follows.

THEOREM 4.13.38. *Let* X *be a Banach space of dimension at least* 3. *Then each of the following conditions is necessary and sufficient for* X *to have its norm generated by*

*Simultaneous Approximation and Interpolation which is Norm-preserving (SAIN).

an inner product:

(1) *For any proximinal subspace V and any nonempty subset M of V*

$$P_M \circ P_V = P_M = P_V \circ P_M,$$

(2) *For every proximinal two-dimensional subspace V and any non-empty segment M of V*

$$P_M \circ P_V = P_V \circ P_M = P_M.$$

As is known in the theory of Banach spaces (approximation theory), certain classes of sequences associated with elements of the space are of great importance. In what follows we present an interesting characterization of the inner product structures on a Banach space using the so-called Aleksandrov diameters.

DEFINITION 4.13.39. Let X be a Banach space and M be an arbitrary subset of X. The n-dimensional Aleksandrov number of $M, d_n(M, X)$ is the number defined by the relation:

$$d_n(M, X) = \inf_{(K_n, T)} \sup_{x \in M} \|x - Tx\|$$

where the infimum is taken ever all pairs (K_n, T) in which K_n is an n-dimensional compact set in X and T is a continuous mapping from M into K_n. If M is symmetric with respect to the origin of X and the infimum is taken ever all pairs (K_n, T) in which K_n is an n-dimensional compact subset of X and T is a continuous odd mapping of M into K_n, then the corresponding number is called the odd n-dimensional Aleksandrov number of M, and is denoted by $\bar{d}_n(M, X)$.

Now the characterization of the existence of inner product structures on X is given in the following theorem.

THEOREM 4.13.40 (Stesin, 1974). *Suppose that X is a Banach space of dimension at least 3. If there exists an integer $n \geqslant 2$ such that for any n-dimensional ball of X its $(n-1)$-dimensional Aleksandrov number is equal to its radius, then the norm of X is generated by an inner product.*

For other interesting connections between the Aleksandrov numbers and Uryson diameters and Bernstein diameters we refer to the interesting papers of Stesin (1973, 1974).

4.14. Chebyshev Centers and Inner Product Structures

Let X be a real Banach space and M be a bounded subset in X.

DEFINITION 4.14.1. The point x_0 of X is said to be a Chebyshev center for the set M if

$$\sup_{x \in M} \{\|x - x_0\|\} = \inf_{y \in M} \sup_{x \in M} \|x - y\| = R_M.$$

The number R_M is called the Chebyshev radius of the set M. Geometrically, the Chebyshev center of a set is the center of a smallest ball containing the set.

The notion of a Chebyshev center is useful in many problems and the problem about the existence and uniqueness of Chebyshev centers if of importance. For some useful information about these problems we refer to the papers by Garkavi (1962, 1964) and the author's books on fixed point theory and strict convexity and complex strict convexity (1981, 1984). We mention, in particular, the beautiful result of A. L. Garkavi related to the uniqueness of Chebyshev centers, namely, the introduction of the class of spaces called Banach spaces which are uniformly convex in every direction, which is exactly the class of spaces in which the uniqueness holds.

In what follows we are interested in the applications of this notion to obtain criteria for the existence of an inner product on a Banach space which generates the norm. For this we need some results.

LEMMA 4.14.2 (Garkavi, 1962). *Let X be a complete inner product space and M be a bounded set in X. Then the Chebyshev center of M is a nonempty set which reduces to a single point which is in the closed convex hull of M.*

Proof. We present the proof for the fact that the Chebyshev center is in the set indicated in the lemma. Suppose that the assertion is false. Then we find a hyperplane H which separates (strongly) x_0 and M and $x_0 \notin H$. We may suppose, without loss of generality, that H is a subspace of X. Let P_H be the orthogonal projection of X onto H. Set $y_0 = P_H x_0$ and let y be the point of intersection of the segment z, x_0 with H, z being an arbitrary point of M. Then

$$P_H(y - x_0) = y - y_0$$

and thus

$$\|y - y_0\| = \|P_H(y - x_0)\| \leqslant \|y - x_0\|$$

which implies that

$$\|z - y_0\| = \|z - y + y - y_0\| \leqslant \|z - y\| + \|y - x_0\| = \|z - x_0\|$$

and thus for any point z of M we have

$$\|z - y_0\| \leqslant \|z - x_0\|$$

which contradicts the fact that x_0 is a Chebyshev center. The existence follows from the interesting result of Garkavi (1962). Suppose that the Banach space X has the property that there exists a projection of norm 1 from X onto X. Then every bounded set of X has a nonempty set of Chebyshev centers. The other property follows from the fact that X is a complete inner product space. □

Let us suppose that X is a strictly convex space of dimension 2, and M is a bounded arbitrary subset of X. Since for any one-dimensional subspace of X there exists a projection of norm 1 of X on that subspace, we have the following result.

THEOREM 4.14.3 (Garkavi, 1964). *Let M be a bounded subset in a two-dimensional subspace X. Then the Chebyshev center of M is a nonempty set and this is contained in* $\overline{\text{conv } M}$.

Proof. Let $\| , \|_2$ be the Euclidean norm on X. Then we set

$$\|x\|_n = \|x\| + n^{-1}\|x\|_2$$

which is obviously a norm on X and $(X, \|\,,\|_n)$ is strictly convex. By the above result the Chebyshev center of M exists, say x_0^n, and is in $\overline{\text{conv}}\, M$. But $\{x_0^n\}$ is a bounded sequence and we may consider a limit point of this sequence, say x_0. Then x_0 is in $\overline{\text{conv}}\, M$ and is a Chebyshev center for M. □

In his 1962 paper, V. Klee has proved the following result about the characterization of inner product structures.

THEOREM 4.14.4 (Klee, 1962). *Let X be a Banach space of dimension at least* 3. *Suppose that for any bounded set M of X the Chebyshev center is in the set* $\overline{\text{conv}}\, M$. *Then the norm of X is generated by an inner product.*

The following result is a powerful extension of Klee's result.

THEOREM 4.14.5 (Garkavi, 1964). *Let X be a Banach space of dimension at least* 3. *Suppose that for any set of the form $M = (x, y, z)$ of three points the Chebyshev center exists and lies in the plane generated by M. Then the norm of X is generated by an inner product.*

Proof. By M. Fréchet's theorem it suffices to consider only the case when dim $X = 3$. Let L be an arbitrary two-dimensional subspace of X and z_0 be arbitrary not in L. For each integer n we consider the sets:

$$D_n = (x, x \in L, \|x - z_0\| \leqslant n), \qquad \Gamma_n = (x, x \in L, \|x - z_0\| = n).$$

Suppose that (for $n \geqslant n_0$) these are nonempty. If $M = (y_1, y_2, y_3)$ are elements of Γ_n and, according to the hypothesis, the Chebyshev center exists and is in the plane determined by M, then if x' is in this Chebyshev center we have the following relations:

$$\|y_1 - x'\| \leqslant \|y_i - z_0\| = n, \qquad i = 1, 2, 3. \tag{$+$}$$

If

$$S(y_i) = \{x, x \in L, \|y_i - x\| \leqslant n\}$$

then, by $(+)$, $\cap_{i=1}^{3} S(y_i)$ is nonempty. Since $S(y)$, for y in Γ_n is a bounded closed set of a two-dimensional subspace,

$$S_n = \bigcap_{y \in \Gamma_n} S(y)$$

is nonempty. Take $x_n \in S_n$ and then for all y in Γ_n we have

$$\|y - x_n\| \leqslant \|y - z_0\|. \tag{$++$}$$

We consider now the set $D_n - \Gamma_n$ and if x is in this set (which is convex and its boundary is Γ_n), $x \notin L$, the segment $[x, x_n]$, $x_n \in L$, can be extended to the intersection with Γ_n since x is in the interior of $[x_n, y]$ for y in Γ_n. But

$$x_n - x = t(x_n - y), \quad t \in (0, 1)$$

and thus

$$z_0 - x = z_0 - x_n + x_n - x = z_0 - x_n + t(x_n - y).$$

Thus the inequality

$$\|x - x_n\| \leqslant \|x - z_0\|$$

is equivalent to the inequality

$$\|z_0 - x_n + t(x_n - y)\| \geqslant t\|x_n - y\|.$$

Suppose now that this last inequality is false. Then

$$\|z_0 - x_n + t(x_n - y)\| < t\|x_n - y\|$$

and thus

$$\begin{aligned}\|z_0 - y\| &= \|z_0 - x_n + t(x_n - y) + (1 - t)(x_n - y)\| \leqslant \\ &\leqslant \|z_0 - x_n + t(x_n - y)\| + (1 - t)\|x_n - y\| \\ &< t\|x_n - y\| + (1 - t)\|x_n - y\| = x_n - y\end{aligned}$$

which contradicts $(++)$. Thus the inequality is correct and thus for all x in $D_n - \Gamma_n$,

$$\|x - x_n\| \leqslant \|x - z_0\|.$$

We now consider the sequence of sets (D_n), $n \geqslant n_0$, and thus we find $x_n \in L$ such that

$$\|x - x_n\| \leqslant \|x - z_0\|$$

for all $x \in D_n$.

We consider a subsequence (x_{n_k}) which converges to an element u of L. Since

$$L = \bigcup D_n, \qquad n \geqslant n_0$$

we have

$$\|x - u\| \leqslant \|x - x_0\| \tag{$+++$}$$

for all x in L. Since for every z in X we have the decomposition

$$z = az_0 + x, \qquad x \in L$$

then we can define the mapping $Pz = au + x$ which is clearly a projection of X onto L. From $(+++)$ we obtain that

$$\|Pz\| = \|au + x\| = |a|\,\|u + x/a\| \leqslant |a|\,\|u + x/a\| \leqslant \|az_0 + x\| \leqslant \|z\|$$

and thus the norm of P is 1. Then by Kakutani's theorem, the norm of X is generated by an inner product. □

The proof of the theorem shows that the following version of Theorem 4.14.5 is true.

THEOREM 4.14.6 (Garkavi, 1964). *Let X be a Banach space of dimension at least* 3. *If each triple $(x_i)_1^3$ with $x_i = 1$ are in a ball whose radius is* 1 *and the center is in the plane determined by x_i, then the norm of X is generated by an inner product.*

The above result strengthens a theorem of S. Kakutani on Kantorovich's spaces:

THEOREM 4.14.7 (Kakutani, 1939). *Let X be a Banach space of dimension at least*

3 *and suppose that for each point* x_0 *of* X *and each triple* $(x_i)_1^3$ *of points in* X *there exists a point* u *such that*

$$\|x_i - u\| \leqslant \|x_i - x_0\|, \qquad i = 1, 2, 3.$$

Then the norm of X *is generated by an inner product.*

From this theorem of S. Kakutani we can obtain the result of Phelps (1958) about the characterization of inner product structures using the points considered by Féjer (1922). See Section 4.13.

We now mention another characterization of inner product structures using certain functions. Let X be a Banach space and for x and $\varepsilon > 0$ we define the set

$$S(x^*, \varepsilon) = (x, \|x\| = 1,\ x^*(x) \geqslant 1 - \varepsilon).$$

Then we have the following result.

THEOREM 4.14.8 (Garkavi, 1964). *A Banach space* X *has its norm generated by an inner product iff the Chebyshev radius* $R(x^*, \varepsilon)$ *of the set* $S(x^*, \varepsilon)$ *has the following properties:*

(1) $g(\varepsilon) = \sup_{\|x^*\| = 1} R(x^*, \varepsilon)$ *is for* $\varepsilon \to 0$ *strongly monotone,*

(2) $\lim g(\varepsilon) = 0$ for $\varepsilon \to 0$.

For a proof we refer to Garkavi's paper (1964). From the remarks by A. L. Garkavi in the above theorems, we can obtain the result of Comfort and Gordon (1961). In order to formulate the result of W. W. Comfort and H. Gordon we need some definitions.

DEFINITION 4.14.9. Let X be a normed linear space. We say that X has the n-spherical intersection property if whenever $(x_i)^n$ is a family of n elements in X,$(r_i)_1^n$ is a family of strictly positive numbers with

$$\bigcap_{i=1}^{n} \bar{B}_{r_i}(x_i) \neq \emptyset$$

then

$$\bigcap_{i=1}^{n} \bar{B}_{r_i}(x_i) \cap F_1(x_i) \neq \emptyset,$$

where $F_1(x_i)$ is the smallest flat set in X containing $(x_i)_1^n$. We recall that a set M in X is said to be flat if for some u in X the set

$$M - u = (x - u, x \in M)$$

is a linear subspace of X. Also, as usual,

$$\bar{B}_r(x) = (y, \|x - y\| \leqslant r).$$

Now if $n = 2$ then we say that X has a bi-spherical intersection property instead of a 2-spherical intersection property, and if $n = 3$ then we use the formulation that X

has a tri-spherical intersection property. We then have the following characterization of inner product structures.

THEOREM 4.14.10 (Comfort and Gordon, 1961). *A Banach space X has its norm generated by an inner product, iff it has a trispherical intersection property.*

For a proof we refer to the paper by Comfort–Gordon or the reader may prove the theorem using Theorem 4.14.7.

An interesting extension of Garkavi's results was obtained by Dan Amir and Z. Ziegler (1980) using the so-called 'relative Chebyshev centers'. We now present Amir–Ziegler's result.

If (X,d) is a metric space and A,G are subsets of X with A bounded then we set

$$r(x,A) = \inf(r > 0, A \subseteq B_r(x)),\ B_r(x) = (u, d(x,u) \leqslant r).$$

The relative Chebyshev radius of A is defined by

$$r(G,A) = \inf(r(x,a),\ x \in G).$$

The relative Chebyshev center of A is the set

$$Z(G,A) = (x, x \in G, r(x,A) = r(G,A)).$$

We remark that if $A = (y)$ then $Z(G,A)$ is the proximity map $P_G(y)$.

Rozema and Smith (1976) have proved that if X is a normed linear space whose norm is generated by an inner product then for any linear subspace X_1 of X and any points x,y of X,

$$Z(F,(x,y)) \cap P_F(x), P_F(y) \neq \emptyset,$$

where $[\,,]$ denotes the line segment.

Amir and Ziegler (1980) have proved the following result.

THEOREM 4.14.11 (Amir and Ziegler, 1980). *Let X be a normed linear space. Then the norm of X is generated by an inner product iff for any two-dimensional subspace X_1 of X and all x in X and y in $P_{X_1}(x)$ the following relation holds:*

$$Z(X_1, 0, x) \cap [0, y] \neq \emptyset.$$

For a proof as well as for other related results we refer to the paper by Amir and Ziegler.

For other results concerning Chebyshev centers and the proof of the following characterization of inner product structures we refer to Franchetti (1975).

Let X be a normed linear space and V be a closed subspace and a in X. If s is such that

$$d(a,V) = d < s$$

then the set

$$C_s = B_s(a) \cap V = (v, v \in V,\ \|v - a\| \leqslant s)$$

is called a hypercircle. The Chebyshev radius of C_s is denoted by r_s. The number

$l(a, V)$ is defined by

$$\inf(l_s), \qquad l_s = r_s - s.$$

Then we have the following result.

THEOREM 4.14.12 (Franchetti, 1975). *Let X be a reflexive Banach space,* dim $X \geqslant 3$. *Then the norm of X is generated by an inner product iff $l(a, V) = l \leqslant 0$ for every pair (a, V), with V a closed hyperplane, and a in X.*

Another interesting characterization of inner product spaces was formulated by J. A. Oman in terms of certain convex subsets of the normed linear space. In order to formulate it we need the following definitions.

DEFINITION 4.14.13. Let X be a normed linear space and let x, y be elements in X. Then $M(x, y)$ is the set

$$(a, \|x - ax\| = \|x - by\|, b \in K).$$

Then it is proved that for any $y \neq 0$, $M(x, y)$ exists and is a compact convex subset of K.

DEFINITION 4.14.14. Let X be a normed linear space and let x, y be elements of X. Then

$$L(x, y) = |a| \|y\|^2,$$

where a is in $M(x, y)$ and

$$|a| = \inf(m, m \in M(x, y))$$

and if $K = R$ we set

$$L(x, y) = a\|y\|^2 \quad \text{if} \quad y \neq 0 \quad \text{and} \quad 0 \text{ if } y = 0.$$

Here a is equal to

$$\mp \inf(\mp \tilde{a}, \tilde{a} \in M(x, y)).$$

We denote by L_C the function L for X as a complex space and L_R for X as a real space. Then we have the following characterization of inner product spaces.

THEOREM 4.14.15 (Oman, 1969). *The complex normed linear space X is an inner product space if and only if*

$$L_R(x, y) = L_R(y, x)$$

holds for all x, y in X.

4.15. On Some Norms on Two-Dimensional Spaces

Since many characterizations of the class of inner product spaces (or complete inner product spaces) in the class of normed spaces (or Banach spaces, respectively) are formulated using only two-dimensional subspaces, we now present some results about such spaces.

The result we give were obtained by D. A. Senechalle (1971). Let X be a two-dimensional linear space and f be a norm on X.

DEFINITION 4.15.1. The number $K(f)$ associated with f is defined by the formula:

$K(f) = \inf\{r > 0, g, g$ a norm of X, g is of inner product type and $rg \geqslant f, g \leqslant f\}$.
It is clear that $K(f) > 1$ if f is not of inner product type.

We say that a norm on a linear space is of inner product type if there exists an inner product such that for all x,

$$\|x\|^2 = \langle x, x \rangle.$$

Suppose now that x, y are two linearly independent elements of X, then we set

$$C(x, y) = (ax + by, a, b \geqslant 0),$$

$$W(x, y) = (ax + by,\ ab \geqslant 0).$$

DEFINITION 4.15.2. The quadruple of points in $X,(u, v, u', y')$, is said to be interlocking if:

(1) the elements are pairwise linearly independent;
(2) $C(v, v') \supset C(u, v) \supset C(u', v)$;
(3) for some norm $\|,\|$ on $X, \|u\| = \|v\| = \|u'\| = \|v'\|$.

Suppose that F is a function on X. Then we define

$$S(F) = (x, F(x) = 1)$$
$$U(F) = (x, 0 \leqslant F(x) \leqslant 1)$$

and we say that F is a subnorm on X if it is the restriction of a norm on X such that the domain, dom(F), has the following properties:

(1) dom(F) is closed;
(2) R dom(F) = dom(F).

DEFINITION 4.15.3. A function $H: X \to R$ is said to be a pre-norm if either H is of inner product type or $H = F$, F in X^{**}.
If H is a pre-norm then we set

$$E_H = (G, G \text{ is a pre-norm of inner product type dominating } H),$$
$$E^H = (G, G \text{ is a pre-norm of inner product type dominated by } H).$$

If f, g are in $E_H(E^H)$ then we define,

$$d(f, g) = \sup(g(x), 1/g(x);\ x \in S(f)),$$
$$d(f, O) = \inf(d(f, g): g \in O),$$

O being either E^H or E_H.
If $w = (u, v, u'v')$ is an interlocking quadruple then we set

$$K(w) = ((ab + cd)/(cd(a^2 + b^2) + ab(c^2 + d^2)))^{1/2},$$

where

$$u' = au + bv, \qquad v' = cu - dv.$$

The following properties are easy to prove.

PROPOSITION 4.15.4. (1) *There exists only one ordered pair* (r, c) *such that* $c = S(f)$ *for some pre-norm of inner product type,* c *contains* u', v' *and* rc *contains* u *and* v.

(2) $r = k(w)$.

(3) *For any subnorm* g *with* u, v, u', v' *in* $S(g)$ *and*

$$u' = (au + v/a)/ (g(au + v/a), v' = (bu - v/b)/ (f(bu - u/b))$$

then

$$k(w) = (g^2(au + v/b) + (g^2(bu - v/b))/(a^2 + a^{-2} + b^2 + b^{-2}))^{1/2}.$$

(4) *The quadruple* $w_1 = (u', v', u, v)$ *is also interlocking and*

$$k(w_1)k(w) = 1.$$

The properties of subnorms are given in the following theorem.

THEOREM 4.15.5. *Suppose that* h *is a subnorm on* X. *Then the following assertions hold*:

(1) E_h *and* E^h *have unique nearest elements, say* u *and* v, *to* h.
(2) $d(u, E^h) = d(u, E_h) = K(h), v = K(h)v$.
(3) *The set* $S(h) \cap S(u)$ *contains two linearly independent elements, say* x, y; $S(h) \cap S(v)$ *contains two linearly independent elements, say* x', y', *and* $W(x, y) \cap W(x', y') \neq \emptyset$.
(4) *There exist* (x, y, x', y') *such that*:
 (a) (x, y, x', y') *is an interlocking quadruple,*
 (b) $(x, y) \subset S(h)S(u), (x', y') \subset S(h)S(v)$
(5) $K(h) = k(x, y, x', y') = \sup(k(w), w$ *is an interlocking quadruple in* $S(u)$),
(6) $K(h) = \sup_{a,b>0} \{(h^2(ax + y/a) + h^2(bx - y/b))/(a^2 + a^{-2} + b^2 + b^{-2})\}^{1/2}$

$h(x) = h(y) = 1$, $ax + a^{-1}y \in \operatorname{dom} h$, $bx - b^{-1}y \in \operatorname{dom} h$.

Proof. From the definition of $d(h, E^h)$ it follows that we find a sequence of norms of inner product type in E^h converging pointwise to a pre-norm of inner product type, say f. Then clearly this gives

$$d(h, E^h) = K(f) = d(h, E_h).$$

If f is not a norm then $S(f)$ is of the form $G \cup (-G)$, where G is a line in X not containing the origin.

Suppose now that the assertion in (3) is not true. Then we find two points x, y in $S(f)$ such that

(1) $S(f) \cap S(h) \subset \mathring{W} \supset (x, y)$, $\{\mathring{M}$ denotes the interior of $M\}$,
(2) $\sup\{1/f(p), p \in S(f) \cap W(x, y)\} < d(h, E^h)$.

Now there exists a norm k of inner product type such that:

(1) $k(x) = k(y)$,
(2) $k(x + y) \cap k(x) + k(y)$,
(3) $\sup\{1/k(p), p \in S(h) \cap W(x, y)\} < d(h, E^h)$,
(4) $U(h) \cap U(k)$.

In this case k is in E^h and $d(h,k) < \sup\{1/h(p);\ p \in S(h)\} = d(h, E^h)$ which is a contradiction. Thus (3) holds.

Suppose now that E^h contains a pre-norm of inner product type, say k, such that

(1) $k \neq v$,
(2) $d(h, E^h) = d(k, h)$.

Set $m = (v^2 + k^2)^{1/2}$ and then m is in $E^h, d(h,m) = d(h, E^h)$. Thus m is a norm. We find a point x in $S(m) \cap \operatorname{dom} h$ with the property that

$$d(h,m) = h(x) \geqslant 1$$

since

$$h(x)/v(x) \leqslant d(h,k) = d(h,m) = h(x).$$

Similarly, we get that $k(x) = 1$ and since

$$m(x) = 1 = (v(x)^2 + k(x)^2)^{1/2}$$

we must have

$$v(x) = k(x) = 1.$$

Also $S(m)$ touches $S(h)$ at some point, say y, and thus $y = \pm x$ since

$$d(h,m) = f(x) \neq 1.$$

This implies that

$$S(h) \cap S(k) = \{\pm x, \pm y\}.$$

Since for some norm of inner product type, say $\|\,,\|$, we have

$$\|y\| = 1 = m(y), \|x + y\| = m(x + y), \|x\| > m(x)$$
$$d(h, \|\,,\|) < d(h, v),$$

which is clearly a contradiction. This gives that v is the unique nearest element of E^h to h. Then we set $u = d(h,v)v$ which is in E_h and $d(h,u) = d(h,v)$. Also, u is the nearest element of E_h to h. Thus (1) and (2) hold.

We show now that $S(h) \cap S(v)$ contains two linearly independent elements. Similarly we prove the assertion for $S(h)$ and $S(u)$.

Since, as we proved above, $W(x, y)$ contains $S(h)$ and $S(v)$ then it contains an element of $S(h)$, say t, with the property that

$$d(h, v) = 1/v(t).$$

This implies that t is in $S(v)$. Thus (3) is proved.

For (4) we remark that if h is of inner product type, then the assertion is clear. We suppose that h is not of inner product type. Then we find two linearly independent elements, say s and t in $S(h) \cap S(v)$ with the property that $W(s,t)$ contains $S(h) \cap S(v)$. From the property just proved in (3), and since $S(h) \cap S(v)$ being a symmetric set, we get that $C(s,t) \cap (S(h) \cap S(u)) \neq \emptyset$. Then we find a point t_1 in $S(h) \cap S(u) \cap C(s,t)$ such that $C(s,t_1)$ contains a point s_1 in the interior of $C(s,t_1)$ and with the property that for any r in $S(h) \cap S(v)$ and interior to $C(s,t_1)$, the set $C(s,r)$ is disjoint from the set $S(h) \cap S(u)$.

Suppose now that

$$W(s,t_1) \supset S(h) \cap S(v).$$

Then $C(s,t_1)$ contains two linearly independent elements, say p and q, p, q in $S(h) \cap S(u)$ and

$$W(p,q) \supseteq S(h) \cap S(u).$$

We find a point p_1 in $S(h) \cap S(u) \cap C(p,q)$; then p_1 is in the interior of $C(s,t_1)$ and $C(s,p_1)$ contains either p or q. This implies that $C(s,p_1) \cap (S(h) \cap S(u)$ is nonempty. This contradiction shows that $W(s,t_1)$ does not contain the set $S(h) \cap S(u)$ and thus for some point t_2 in $S(h) \cap S(u)$, $t_2 \notin S(h) \cap S(u)$ and $C(t_2,t) \supset C(s,t_1)$. Then clearly the quadruple $w = (s,t_1,s_1,t_2)$ satisfies the requirements in (4).

Now from the properties of the interlocking quadruples we get that there exists only one pair (r,c) such that $r > 0$ and $c = S(h)$ for some pre-norm inner product type. Then $r = k(w)$, where $w = (u,v,u',v')$ and the pair $(K(h),S(h))$ has the property that $k(w) = K(h)$. Let (m,n,m',n') be a quadruple which is interlocking. Let us consider a subnorm $\tilde{m}$ with the property that

$$S(\tilde{m}) = (\pm m, \pm n, \pm m', \pm n')$$

and thus $K(\tilde{m}) \leqslant K(h)$. Following a procedure as before, we prove that $k(m,n,m',n') = K(\tilde{m})$. Then we have that □

$$k(m,n,m',n') = K(\tilde{m}) = K(f) = k(u,v,u',v')$$

and

$$k(u,v,u',v') = \sup\{k(w), w \text{ is an interlocking quadruple}\}.$$

Thus (5) is proved. From the properties of interlocking quadruples and (4) we get (6). This completes the proof of the theorem. □

COROLLARY 4.15.6. *Let r be in $(1,2^{1/2}]$ and C be an ellipse in X with the center at* 0. *F denotes the sets of all subnorms p with the following properties:*

(1) $S(p) \subset [1,r]C$;
(2) *there exists an interlocking quadruple (x,y,x',y') of points of $S(p)$ such that $x, y \in rC$ and $x', y' \in C$.*

If F_1 is the set of all subnorms q with the property that $S(q) = TS(p)$ for some p in F and a reversible linear T, then F_1 is exactly the set

$$(h, h \text{ is a subnorm}, K(h) = r).$$

In order to state properties which are useful in establishing criteria for a norm to be of inner product type, we consider the following property $(D, \sim)$ defined as follows.

DEFINITION 4.15.7. If $\sim$ is one of the relations $\geqslant$, $\leqslant$ then we say that a subnorm p has the property $(D, \sim)$ if whenever $w(x, y, x', y')$ is an interlocking quadruple in $S(p)$ there exists an interlocking quadruple $w_1 = (x, y, x_1, y_1)$ in $S(p)$ such that

$$k(w_1) \sim 1.$$

DEFINITION 4.15.8. We call a subnorm of inner product type if $K(p) = 1$.

Then we have

THEOREM 4.15.9. *If a subnorm has the property $(D, \sim)$ then it is of inner product type.*

Proof. Let p be the subnorm having the property stated in the theorem. Suppose that the assertion is false. Consider u and v be the nearest members in E_p and E^p respectively. Then by the above theorem and since $S(p), S(u), S(v)$ are clearly closed subsets, we find an interlocking quadruple $w = (x, y, x', y')$ with the property that:

$$C(x, y) \cap S(p) \cap S(v) = \{x, y\}$$
$$C(x', y') \cap S(p) \cap S(u) = \{x', y'\}.$$

We now consider the quadruple $w_1 = (x, y, x_1, y_1)$ which exists, and then we have $k(w_1) > 1$.

If we set $w_2 = (x', y', x_1, y_1)$ then this is an interlocking quadruple in $S(p)$ and $k(w_2) < 1$. But this implies that p has not the property $(D, \sim)$. Thus we have a contradiction and the theorem is proved. □

Using the above results we obtain some interesting characterizations of norms of inner product type.

THEOREM 4.15.10. *Suppose that $\|,\|$ is a norm on L and that $\sim$ means $\geqslant$ or $\leqslant$. Then $\|,\|$ is a norm of inner product kind iff there exists $\varepsilon > 0$ such that if*

$$\|x\| = \|y\| = 1 \text{ and } \|x - y\| < \varepsilon$$

then for some a, b in $(0, \infty)$

$$\|ax + by\|^2 + ab\|x - y\|^2 \sim (a + b)^2.$$

Proof. Suppose that $\sim$ means $\geqslant$. Suppose, further, that the conclusion of the theorem is false and we get then a contradiction. Indeed, we consider u, v to be the nearest elements of $E_{\|,\|}$ and $E^{\|,\|}$ *to* $\|,\|$. Then we find an interlocking quadruple say (p, q, p', q') such that:

(1) $p, q \in S(u) \cap S(v)$,
(2) $p', q' \in S(u) \cap S(v)$.

Now if $v(p' - q') \leqslant 2^{1/2}$ then we set $p_1 = p'$ and $q_1 = q'$; otherwise we set $p_1 = p'$,

$q_1 = -q'$. In either case

$$p_1, q_1 \in S(\|\,,\|) \cap S(u), u(p_1 - q_1) \leqslant 2^{1/2}$$
$$S(\|\,,\|) \cap S(v) \cap C(p_1, q_1) \neq S(\|\,,\|) \cap C(p_1, q_1)$$

since $S(\|\,,\|) \cap C(p_1, q_1)$ contains either u or v.

Let $r = 1$ *and* $\|\,,\|_r$ be a norm of inner product type with the properties:

(1) $\|p_1\|_r = \|q_1\|_r = 1$,

(2) $\|p_1 + q_1\|_r = u((p_1 + q_1)/r)$.

Set

$$M = \sup\{r \geqslant 1, \|z\|_r \geqslant 1 \text{ for some } z \text{ in } S(\|\,,\|_r) \cap C(p_1, q_1\}$$

and

$$S_r = (z, z \in S(\|\,,\|) \cap C(p_1, q_1), \|z\|_r \geqslant 1.$$

Further, if

$$l(r) = \sup\{\|x - y\|, x, y \text{ are in the same component of } S_r\}$$

then if $r \in (1, M]$, we consider the points p_r, q_r such that

$$\|p_r - q_r\| = l(r).$$

Thus the arc $C(p_r, q_r)$ is a component of S_r.

We remark that p_r, q_r are in $S(\|\,,\|_r)$ and that the function $r \to l(r)$ is nonincreasing on $(1, M]$. Then either

(1) $l(M) > 0$,

or

(2) $\lim_{r \to M} l(r) = 0$.

If the first holds, then S_M contains an arc of $S(\|\,,\|_M)$. If the second holds, then for some r in $(1, M]$ we have the relation $\|p_r - q_r\| < \varepsilon$. Thus in both cases, for some r in $(1, M]$ and two elements x, y in $C(p_1, q_1) \cap S(\|\,,\|) \cap S(\|\,,\|_r)$ we have the relation $\|x - y\| < \varepsilon$, and if a, b are in R^+,

$$\|ax + by\| \leqslant \|ax + by\|_r.$$

Now $u(p_1 - q_1) \leqslant 2^{1/2}$ and p_1, q_1 are in $S(\|\,,\|)$, $x - y$ is in the interior of $W(p_1, -q_1)$ which gives that

$$\|x - y\| \leqslant \|x - y\|_r.$$

Thus if a, b are in R^+ we have

$$\|ax + by\|^2 + ab\|x - y\|^2 < \|ax + by\|_r^2 + ab\|x - y\|_r^2 = (a + b)^2.$$

This is a contradiction and thus $\|\,,\|$ is of inner product type.

If $\sim$ means $\leqslant$ then similar arguments give the assertion (of course we must take $S(\|\,,\|) \cap S(u)$ and M is taken with the inequality '$<$' as well as S_r). □

THEOREM 4.15.11. *Suppose that* $\|,\|$ *is a norm on* X *with the following property: if* $x \in X$, $\|x\| = 1$ *then there exists* $a,b > 0$ *(not necessarily independent of* x*) and a sequence* (x_n) *in* $\{u, \|u\| = 1\} \setminus \{x\}$ *such that*

$$\lim_{n \to \infty} ((a+b)^2 - (\|ax + bx_n\|)^2/(ab\|x - x_n\|^2)) = 1.$$

Then $\|,\|$ *is a norm of inner product type.*

Proof. Suppose that the assertion is false. Then we consider the sets $E_{\|,\|}$ and $E^{\|,\|}$ and u,v to be the nearest elements in these sets to $\|,\|$, respectively. We may assume without loss of generality that X is a plane, $S(u), S(v)$ are circles with radius 1 and greater than 1, respectively, $(1,0)$ is in $S\ (\|,\|) \cap S(u)$ and $C((1,0), 1,1)) \cap S(\|,\|) \cap S(v) \neq \emptyset$.

Now if $s \to r(s)$ is a function then we consider such functions such that $r(s) = (r(s)\cos s, r(s)\sin s)$ is a point in $S(\|,\|)$, $\theta = -a\tan(r'(s)/r(s)), \theta_- = -a\tan(r'_-(s)/r(s)), r'_-(s)$ being the left-hand derivative of the function r. If R is the radius of the circle $S(v)$ then we set $k = 1 - R^2$ and then clearly this is a number in $(0, \frac{1}{2}]$.

We consider the function $s \to \theta_-(s) + s + \pi/2$ if $(\theta'_-)(s) = -k(s)$. Then $\theta_-(s) + s$ is nondecreasing and for s in $\mathrm{dom}(\theta'_-)$ we have the inequality $(\theta'_-)(s) + 1 \geqslant 1 - k$. But $(1,0) \in S(\|,\|) \cap S(u)$ and thus $\theta_-(0) = 0$. Then for every s, $\theta_-(s) \geqslant (1-k)s$ or $\theta_-(s) \geqslant -ks$. Thus we have

$$a\tan(r'_-(s)/r(s)) \leqslant ks, (\ln r(s))'_- \leqslant \tan(ks)$$

and

$$r(s) \leqslant \exp\left(\int_0^s \tan(ks)\, ds\right) = \cos(ks)^{-1/k}.$$

Now for some s_0 in $[0, \pi/4)$ we have $r(s_0) = R = (1-k)^{-1/2}$. Then

$$(1-k)^{-1/2} = \cos(ks_0)^{-1/k} \quad \text{and} \quad (1-k)^{k/2} = \cos(ks_0) \geqslant \cos(k \cdot \pi/4).$$

If we set

$$f(t) = (\ln(1-t)t/2 - \ln(\cos(t \cdot \pi/4)), t \in [0, \tfrac{1}{2}].$$

we get that

$$f(k) \geqslant 0. \tag{+}$$

Then we have that $f(0) = f'(0) = 0$ and $f''(t) \leqslant 0$ for t in $(0, \frac{1}{2}]$. Then we must have the inequality $f(t) < 0$ for t in $(0, \frac{1}{2}]$. Thus $(+)$ holds. This implies, further, that for some s_1 we have $\theta'_-(s_1) = -k$. By a rotation we may assume without loss of generality that $s_1 = 0$.

Now we find $a,b \geqslant 0$ and a sequence of numbers (a_n), $\lim a_n = 0$, $a_n \neq 0$, with the property that

$$\lim_{n \to \infty} ((a+b)^2 - \|ap(0) + bp(a_n)\|^2)/(ab\|p(0) - p(a_n)\|^2) \geqslant 1.$$

We may assume without loss of generality that $a + b = 1$ (if this not the case then we replace a by $a/(a+b)$, b by $b/(a+b)$). Then if we consider a norm in the plane given by the formula

$$|(s,t)| = (s^2 + t^2)^{1/2}$$

then for any point P of the plane, sgn $P = P/\|P\|, P \neq 0$. For s in $[-\pi/2, \pi/2]$ we have

$$|\operatorname{sgn}(ap(0) + bp(s)| = ra\tan(br(s)\sin s/(ar(0) + br(s)\cos s)))$$

and

$$\lim_{n\to\infty} (1 - \|ap(0) + bp(a_n)\|^2)/(ab\|p(0) - p(a_n)\|^2) =$$

$$= \lim_{n\to\infty} |\operatorname{sgn}(p(0) - p(a_n)|^2/|(\operatorname{sgn}(ap(0) + bp(a_n))|^2 g(a_n)/h(a_n),$$

where

$$g(s) = r^2(a\,tan(br(s)\sin s/(ar(0) + br(s)\cos s))$$
$$-(a^2r^2(0) + 2abr(0)r(s)\cos s + b^2r^2(s))$$

and

$$h(s) = ab(r^2(0) - 2r(0)r(s)\cos s + r^2(s)).$$

The function

$$a\tan(br(s)\sin s/(ar(0) + br(s)\cos s))$$

is increasing on an open subinterval containing 0. We conclude that g is a left and right differentiable function. The function h clearly has the property to be left and right differentiable.

If g'' and h'' are the derivatives of g'_-, h'_- respectively, then since

$$g(0) = h(0) = g'(0) = h'(0) \neq h''(0)$$

we have the following properties:

(1) $I \subset \operatorname{dom} g \cap \operatorname{dom} g'_- \cap \operatorname{dom} g'_+ \cap \operatorname{dom} h \cap \operatorname{dom} h'_- \cap \operatorname{dom} h'_+$, I being the above open interval containing 0,

(2) $0 \in \operatorname{dom} g' \cap \operatorname{dom} g'' \cap \operatorname{dom} h' \cap \operatorname{dom} h''$.

Then we have the relation

$$\lim g(s)/h(s) = g''(0)/h''(0)$$

and thus

$$\lim g(a_n)/h(a_n) = g''(0)/h''(0) = R^{-2}.$$

Since

$$g''(0)/h''(0) = 1 + \theta'(0) \quad and \quad \theta'_-(0) \geqslant R^{-2} - 1$$

we have a contradiction. Thus $\|\,,\|$ is of inner product type. □

From this result we derive the following sufficient conditions for a norm to be of inner product type.

COROLLARY 4.15.12. *Let* $\|,\|$ *be a norm on* X *and suppose that for each* x *in* X, $\|x\| = 1$, *there exist* $a,b > 0$ *with the property that for any* $\varepsilon < 0$ *there exists an element* y *in* X, $\|y\| = 1$ *with the property that* $\|x - y\| < \varepsilon$ *and*

$$\|ax + by^2 + ab\|x - y\|^2 = (a + b)^2.$$

Then, $\|,\|$ *is a norm of inner product type.*

COROLLARY 4.15.13. *Let* $\|,\|$ *be a norm on* X *and suppose that for each* x *in* X, $\|x\| = 1$, *there exist* $a,b > 0$ *with the property that*

$$\lim_{y \to x} ((a + b)^2 - \|ax + by\|^2/(ab\|x - y\|))$$

exists and is in $[1,00]$. *Then* $\|,\|$ *is a norm of inner product type.*

COROLLARY 4.15.14. *Is* δ_X *is the modulus of convexity of* X *and for some sequence of positive numbers* (a_n), $\lim a_n = 0$ *we have*

$$\lim \delta_X(a_n) = 8^{-1}.$$

Then $\|,\|$ *is a norm of inner product type.*

REMARK 4.15.15. Corollary 4.15.14 was obtained independently by Kirčev and Troianskiĭ (1975) with another interesting characterization. See Section 4.17 for the exact formulation of Kirčev–Troianskiĭ results.

The following result is useful to obtain some general sufficient conditions for a norm to be of inner product type.

THEOREM 4.15.16. *Let* X *be a two-dimensional space and* $\|,\|$ *be a norm on* X. *For* $a, b, c, d > 0$, $|c - d| < \varepsilon < c + d$ *we set*

$$W = ((cd\|ax + by\|^2 + ab\|cx - dy\|^2)^2/cd(a^2 + b^2) + ab(c^2 + d^2):$$
$$\|x\| = \|y\| = 1, \|cx - dy\| = \varepsilon).$$

Then $1 \in W$ *and* $a \in W$, $a > 1$ *iff some* $b \in W$, $b < 1$.

Proof. We may suppose without loss of generality that X is a plane and consider as above the function $r = r(s)$ such that

$$p(s) = (r(s)\cos s, r(s) \sin s) \in S(\|,\|).$$

For each s we have

$$\|cp(s) - dp(s)\| = |c - d|, \|cp(s) - dp(s + \pi)\| = c + d$$

and we find $h(s)$ in $[s, s + \pi]$ such that

$$\|cp(h(s)) - dp(h(s))\| = \varepsilon.$$

We set

$$f(s) = r(h(s)) \cos h(s),$$
$$g(s) = r(h(s)) \sin h(s)$$

and

$$q(s) = (f(s), g(s)).$$

Then f, g are obviously continuous functions and $q(s) \in S(\|\,.\,\|)$.

Let a, b be two real numbers and consider the curve $ap(s) + bq(s)$ which enclosed an area $A(a,b)$. Then we have,

$$(a^2 + b^2)A(1,0) - A(a,b) = (a^2 + ab)A(1,0) + (b^2 + ab)A(0,1) - A(a,b)$$

$$= (a^2 + ab)\int_0^{2\pi} r(s) \cos s d(r(s) \sin s) + (b^2 + ab)\int_0^{2\pi} f(s) dg(s) -$$

$$- \int_0^{2\pi} (ar(s) \cos s \; br(s) \sin s)\; d(ar(s) \sin s + brg(s))$$

$$= ab \int_0^{2\pi} (r(s) \cos s - f(s)) d(r(s) \sin s - g(s)) = abA(1, -1).$$

Since for all s

$$\|cr(h(s)) \cos h(s) - dr(h(s)) \sin h(s)\| = \varepsilon$$

we get that

$$A(c, -d) = \varepsilon^2 A(1,0)$$

and thus

$$(a^2 + b^2)A(1,0) - A(a,b) = (a,b)A(1, -1)$$
$$(c - d)^2 A(1,0) - \varepsilon^2 A(1,0) = -cd\, A(1, -1).$$

This gives that

$$cd\, A(a,0) + ab\varepsilon^2 = cd(a^2 + b^2) + ab(c^2 + d^2).$$

If $t(s)$ is a function with the property that

$$(t(s) \cos s, t(s) \sin s)$$

is a point of the curve $ap(\,,) + bq(\,,)$, then since

$$cd \cdot \int_0^{2\pi} t^2(s)\, ds \Big/ \left(\int_0^{2\pi} r^2(s)\, ds \right) + ab^2 = cd(a^2 + b^2) + ab(c^2 + d^2)$$

there exists s_0 with the property that

$$cdt^2(s_0)/r^2(s_0) + ab\,\varepsilon^2 < cd(a^2 + b^2)ab(c^2 + d^2)$$

and there exists s_1 such that

$$abt(s_1)/r^2(s_1)^2 + ab\,\varepsilon > cd(a^2 + b^2) + ab(c^2 + d^2).$$

Then clearly the existence of s_1 with the above property implies that 1 is in W. The existence of $\tilde{s}_0$, where we have equality giving the result since

$$\tilde{t}(\tilde{s}_0)^2/r^2(\tilde{s}_0) = \|ap(s_2) + bq(s_2)\|^2 \text{ for some } s_2$$

COROLLARY 4.15.17. *Let X be a two-dimensional space and $\|,\|$ be a norm on X. Suppose that there exist $a,b > 0$ a subset M in $[0,2]$, 0 is an accumulation point, a function on M with real values with the property that*

$$F(\|x-y\|) = ax + by$$

for x, y in $S(\|,\|), \|x-y\| \in M$. Then $\|,\|$ is a norm of inner product type.

Proof. The function

$$F(\|x-y\|) = ((a+b)^2 - ab\|x-y\|^2)^{1/2}$$

for $\|x-y\| \in M$. The conditions of Theorem 4.15.11 are satisfied and thus the assertion is proved. □

REMARK 4.15.18. This extends an earlier result of Senechalle (1968).

4.16. Parameters Associated with Normed Linear Spaces and Inner Product Structures

Associated with a normed linear space there are several quantities which are useful in the study of the structure of normed linear spaces or of mappings defined on them. In what follows we intend to present some parameters associated with normed linear spaces which are useful in obtaining existence theorems of the inner product structure on normed linear spaces. In Section 4.17 two parameters are also considered with applications to inner product structures.

We consider a real normed linear space and $\perp$ to be an orthogonality defined on X.

DEFINITION 4.16.1. If $\perp$ is an orthogonality on X then we define the numbers $\mu\perp(X)$, $K\perp(X)$ and $A\perp(X)$ by the following formulas:

(1) $K\perp(X) = \sup(\|y\|/\|x-y\|,\ y \neq 0,\ x \perp y)$,

(2) $\mu\perp(X) = \sup(\mu(x,y){:}x \perp y, \|x\| = \|y\| = 1)$,

where

$\mu(x,y) = \sup(\|x\| + \|ty\|)/\|x+ty\|$

at least one of (x, y) is different from zero,

(3) $A\perp(X) = \frac{1}{2}\sup(\|x+y\| + \|x-y\|, \|x\| = \|y\| = 1$ and $x \perp y)$.

REMARK 4.16.2. For the case of Birkhoff's orthogonality ($\perp_B$) the number $K\perp(X)$ is discussed in Baronti (1981), Thele (1974); the parameter $\mu\perp(X)$ was introduced for Birkhoff's orthogonality by Joly (1969) and $A\perp_B(X)$ was defined by Baronti (1981). We define now other parameters.

DEFINITION 4.16.3 (Smith, 1981). Let X be a real Banach space and let M be a proper closed subspace of X. We set

$$\|P_M\| = \sup(\|y\|, y \in P_M(x), \|x\| \leqslant 1)$$

and then the 'metric projection bound' of X is the number defined by:

$$\mathrm{MPB}(X) = \sup(\|P_M\|, M \text{ is a proximinal subspace of } X).$$

Another related parameter was defined by Baronti (1981):

DEFINITION 4.16.4 (Baronti, 1981). Let X be a Banach space. Then the number $\mathrm{MPB}'(X)$ is defined by:

$$\mathrm{MPB}'(X) = \sup(\|P_M\|, M \text{ is a one-dimensional subspace of } X).$$

We now present some parameters which measure certain deviations.

DEFINITION 4.16.5 (Fitzpatrick and Reznick, 1983). The skewness of a normed linear space X is the number $s(X)$ defined by:

$$s(X) = \sup\left(\lim_{t\to 0} t^{-1}(\|x + ty\| - \|y + tx\|)\right).$$

DEFINITION 4.16.6 (Istrăţescu, 1984). Let X be a normed linear space with the norm Gateaux differentiable. If we set

$$n(x, y) = \lim_{t\to 0} t^{-1}(\|x + ty\| - \|x\|)$$

then the Gateaux skewness of X is the number defined by:

$$s_{\mathrm{G}}(X) = \sup_{\|x\| = \|y\| = 1} (n(x, y) - n(y, x)).$$

DEFINITION 4.16.7 (Istrăţescu, 1984). The projection skewness is the number defined by:

$$s_{\mathrm{P}}(X) = \sup_{\|x\| = \|y\| = 1} (\|P_{[x]}(y) - P_{[y]}(x)\|),$$

where $[z]$ denotes the space generated by the element z.

DEFINITION 4.16.8 (Istrăţescu, 1984). The metric skewness of X is the number defined by:

$$s_{\mathrm{m}}(X) = \sup_{\|x\| = \|y\| = 1} (d(x,[y]) - d(y,[x])).$$

DEFINITION 4.16.9 (Istrăţescu, 1984). The Ficken skewness of X is the number defined by:

$$s_{\mathrm{F}}(X) = \sup_{\|x\| = \|y\| = 1} \left(\min_{t\in R} \|x - ty\| - \|y - tx\|\right).$$

DEFINITION 4.16.10 (Istrăţescu, 1984). The Apollonius skewness of X is the number defined by:

$$s_{\mathrm{A}}^{n}(X) = \sup_{\|x_i\| = 1} \left(\sum_{i=1}^{n} \|x_i - n^{-1}(x_1 + \cdots + x_n)\|^2 - \sum_{i=1}^{n} \|x_i\|^2 - n^{-1}\left\|\sum_{i=1}^{n} x_i\right\|^2\right),$$

where n is a fixed integer $n \geqslant 2$.

DEFINITION 4.16.11. Let X be a normed linear space. The radial projection R is the mapping defined on X by the formula:

$$Rx = \begin{cases} x & \|x\| \leqslant 1 \\ x/\|x\|, & \|x\| \geqslant 1. \end{cases}$$

DEFINITON 4.16.12. The Lipschitz constant of a normed linear space X is the number $K(X)$ defined by:

$$\sup_{x \neq y} (\|Rx - Ry\|/\|x - y\|).$$

From this definition we get immediately that $1 \leqslant K(X) \leqslant 2$. It is easy to see that if the norm of X is generated by an inner product then $K(X) = 1$.

We begin with a result about $K(X)$.

THEOREM 4.16.13 (Thele, 1974). *For every normed linear space X the following relation holds*:

$$K(X) = \sup(\|y\|/\|y - tx\|, y \neq 0, x \perp_B y, t \in R).$$

To prove the theorem we need the following result.

LEMMA 4.16.14 (Thele, 1974). *For any linear normed space X the following relations hold*:

$$K(X) = \sup\Big(\big\|x - y/\|y\|\big\|/\|x - y\|, \|x\| \leqslant 1 < \|y\| \Big)$$

$$= \sup(\|x - y\|/\|x - ty\|, \|x\| \leqslant \|y\| \leqslant 1, t > 1).$$

Proof. Suppose that $x \perp_B y$ and t is in R. Then

$$\|(t/s)x\| \leqslant \|(t/s)x - y\|$$

for all s in R and using $(t/s)x - y$ instead of y we have

$$\begin{aligned} K(X) &\leqslant \sup(\|y\|/\|y - s(t/s)x - y\|, x \perp_B y, y \neq 0, t \in R, s > 0) \\ &= \sup(\|y\|/\|(1 + s)y - tx)\|, x \perp_B y, y \neq 0, t \in R) \\ &= \sup(\|y\|/\|y - tx\|, x \perp_B y, y \neq 0, t \in R). \end{aligned}$$

But

$$\|(y - sv) - (s + 1)y\| = s\|v + y\| \leqslant s\|v\| = \|(y - sv) - y\|$$

the minimum of the function

$$t - \|(y - sv) - ty\|$$

is attained at some point $t_0 > 1$. Thus $x \perp_B y$ if

$$x = (y - sv) - t_0 y$$

which implies that

$$\|y\|/\|y - sv\| = \|y\|/\|t_0 y + x\| \leqslant \|y\|/\|y_0 + t_0^{-1} x\|$$

and thus

$$K(X) \leqslant \sup(\|y\|/\|y - tx\|, x \perp_{\mathrm{B}} y, y \neq 0, t \in R)$$

which gives the required equality. □

COROLLARY 4.16.15 (de Figueiredo *et al.*, 1967). *Let X be a normed linear space. Then the norm of X is generated by an inner product iff $K(X) = 1$.*

Proof. As we remarked, the condition is necessary. Suppose now that X has the property that $K(X) = 1$. Then if $x \perp_{\mathrm{B}} y$, $y \neq 0$, we have

$$\|y\|/\|y - tx\| \leqslant 1$$

for t in R and thus $y \perp_{\mathrm{B}} y$. Thus the orthogonality in Birkhoff's sense is symmetric and thus the norm of X must be generated by an inner product ($\dim X \geqslant 3$). □

REMARK 4.16.16. In Chapter 8 we shall present some related results for the case of nonexpansive mappings.

We now mention an interesting result on the behavior of the parameter $K(X)$ obtained by C. Franchetti.

THEOREM 4.16.17 (Franchetti, 1980). *Let X be a real normed linear space. Then $K(X) = K(X^*)$.*

J. L. Joly has proved the following result.

THEOREM 4.16.18 (Joly, 1969). *Let X be a normed linear space of dimension at least 3. Then the norm of H is generated by an inner product iff*

$$\mu \perp_{\mathrm{B}}(X) = 2^{1/2}.$$

This result was extended by Benitez and Del Rio (1977) to include spaces of dimension 2. O. P. Kapoor and S. B. Mathur have proved the following result.

THEOREM 4.16.19 (Kapoor and Mathur, 1981). *Let X be a normed linear space. Then the following assertions are equivalent*:

(1) *X has its norm generated by an inner product.*
(2) *For any pair of elements (x, y) of X with $\|x\| = \|y\| = 1$* and $x \perp_{\mathrm{B}} y$ then $x \perp_{\mathrm{I}} y$ ($\perp_{\mathrm{I}}$ *indicates isosceles orthogonality*).
(3) $\mu \perp_{\mathrm{B}}(X) = 2^{1/2}$.

In presenting results about the parameter $\mu \perp (X)$ we shall follow the elegant paper of Baronti (1981). First we give a theorem.

THEOREM 4.16.20. *Let X be a normed linear space. Then*

$$\mu \perp_{\mathrm{B}}(X) \geqslant 2^{1/2}.$$

Proof. Let (x, y) be a pair of elements of X with $\|x\| = \|y\| = 1$ and $x \perp_{\mathrm{B}} y$. If $\|x - y\| \leqslant 2^{1/2}$ then the assertion of the theorem follows.

Suppose now that the contrary inequality holds. Then we remark that x and

$z = x - y$ are linearly independent elements of X. Then

$$rz + x = r(x - y) + x = (r + 1)x - ry$$

and

$$((r + 1)x - ry) \perp_B (x - y)$$

for all r in R. Now if $r = 1$ then we obtain that

$$y \perp_B (x - y)$$

and thus

$$\mu \perp_B(X) \geqslant (\|y\| + \|x - y\|)/\|x\| \geqslant 2^{1/2} + 1 \geqslant 2^{1/2}.$$

Now if we set $r_1 = -r/(r + 1)$ then we have that

$$(x + r_1 y) \perp_B (x - y)$$

and thus

$$\mu \perp_B(X) \geqslant (1 + |r_1|)/\|x + r_1 y\|$$

which gives the required inequality if we prove that

$$(1 + |r_1|)/\|x + r_1 y\| \geqslant 2^{1/2}.$$

Suppose that we have

$$(1 + |r_1|)/(\|x + r_1 y\|) < 2^{1/2}.$$

But we have $\|x - y\| \geqslant 2^{1/2}$ and $(x + r_1 y) \perp_B (y - x)$ and thus

$$\begin{aligned}\mu \perp_B(X) &\geqslant (\|x + r_1 y\| + \|y - x\|)/\|x + r_1 y + y - x\| \\ &\geqslant \|x + r_1 y\|/(1 + |r_1|) + 2^{1/2}/|1 + r_1| \\ &\geqslant 2^{-1/2} + 2^{1/2}(1 + |r_1|)^{-1}.\end{aligned}$$

The last number is greater than $2^{1/2}$ for r_1 in $[-3, 1]$. Since $x + r_1 y \perp_B (x - y)$ we have also

$$\begin{aligned}\mu \perp_B(X) &\geqslant (\|x + r_1 y\| + |r_1|\,\|x - y\|)/\|x + r_1 y + r_1 x - r_1 y\| \\ &\geqslant (\|x + r_1 y\|/(1 + |r_1|) + r_1 \cdot 2^{1/2}(1 + |r_1|)^{-1} \\ &\geqslant 2^{-1/2} + r_1 \cdot 2^{1/2} \cdot (1 + |r_1|)^{-1}.\end{aligned}$$

Then the last number is greater than $2^{1/2}$ for $r_1 \geqslant 1$ and for $r_1 \leqslant -3$. This completes the proof of the inequality. □

LEMMA 4.16.21 (Baronti, 1981). *Let X be a normed linear space and let (x, y) be a pair of elements of X with $x \neq \pm y$ and $\|x\| = \|y\|$. For t in R we set*

$$f(t) = (1 + |t|)/\|x + ty\|.$$

Then the following assertions hold:

(1) $\max(f(t), t \geqslant 0)$ *exists,*
(2) *if* $f(t_0) = \max(f(t), t \geqslant 0)$ *then this is the case iff* $(x - t_0 y) \perp_B (x - y)$,

(3) $\min(f(t), t \leqslant 0)$ *exists*, $f(t_0) = \max\{f(t), t \leqslant 0\}$ *iff* $(x + t_0 y) \perp_B (x + y)$.

Proof. We consider the functions

$$(x, y) = \lim_{t \to 0^+} t^{-1}(\|x + ty\| - \|x\|)$$

which we used earlier. First we remark that assertion (1) is obvious. Now if t_0 is as in (2) then for t_0 we have $(x + t_0 y) \perp_B (x - y)$ when $x - (t_0/(1 + t_0))(x - y) \perp_B (x - y)$. Then we have

$$\|x - t_0/(1 + t_0)(x - y)\| \leqslant \|x - (t_0/(1 + t_0) + u)(x - y)|, \quad \text{for } u \geqslant 0.$$

If we take

$$u = t/(1 + t_0) - s/(1 + s), \quad s \geqslant 0,$$

then

$$\|x - t_0/(1 + t_0)(x - y)\| \leqslant \|x - s/(1 + s)(x - y)\|$$

and thus

$$f(s) \leqslant f(t_0).$$

Conversely, suppose that $f(t_0) \geqslant f(t)$ for $t \geqslant 0$. Then $g(t) = \|x + ty\|$ is a convex function and there exists the right derivative

$$g'_+(t) = \tau(x + ty, y)$$

and the left derivative

$$g'_-(t) = -\tau(x - ty, -y)$$

for $t > 0$. But then we have

$$\|x + t_0 y\| - (1 + t_0)\tau(x + t_0 y, y) \leqslant 0 \leqslant \|x + t_0 y + (1 + t_0)\tau(x - ty, -y)$$

and taking into account that

$$\tau(x, y) = -\tau(x, -y)$$

we get that $(x + t_0 y) \perp_B (x - y)$.

In the above considerations we supposed that $t > 0$. If $t = 0$ a simple calculation shows the truth of the assertion. Similarly we can prove the rest of the assertions. □

Following Baronti (1981) we consider the following properties which a Banach space may or may not have.

(P_1) If $\|x\| = \|y\| = 1$, $x \perp_B y$ then $\|x + y\| = 2^{1/2}$,
(P'_1) If $\|x\| = \|y\| = 1$, $x \perp_B y$ then $\|x + y\| \leqslant 2^{1/2}$,
(P_2) $\mu \perp_B (X) = 2^{1/2}$,
(P_3) If $\|x\| = \|y\| = 1$ then $(x + y) \perp_B (x - y)$,
(P_4) If $\|x\| = \|y\| = 1$ then $x \perp_B y$ implies that $\|x + y\| = \|x - y\|$.

Concerning these properties we have the following result.

THEOREM 4.16.22 (Baronti, 1981). *For any Banach space X the following assertions hold*:

$$(\mathrm{P}_1) \longleftrightarrow (\mathrm{P}'_1) \longleftrightarrow (\mathrm{P}_2) \begin{cases} \to (\mathrm{P}_3) \\ \to (\mathrm{P}_4) \end{cases}.$$

Proof. Suppose first that (P'_1) holds. If $\|x\| = \|y\| = 1$ and $x \perp_{\mathrm{B}} y$ then

$$\mu \perp^{+}_{\mathrm{B}}(x, y) = \sup_{t \geqslant 0} (\|x\| + \|ty\|)/(\|x + ty\|) = (1 + t_0)/(\|x + t_0 y\|).$$

By the above lemma $(x + t_0 y) \perp_{\mathrm{B}} (x - y)$ and thus

$$2^{1/2} \geqslant \big\| (x + t_0 y)/\|x + t_0 y\| \pm (x - y)/\|x - y\| \big\|$$

which implies further that

$$2\|x + t_0 y\| \geqslant \big\| x(\|x - y\|) \pm \|x + t_0 y\|) + y(t_0 \|x - y\| \mp \|x + t_0 y\|) \big\|.$$

Then we have

$$\begin{aligned} 2\|x + t_0 y\| &\geqslant t_0 \|x - y\| + \|x + t_0 y\| - \|x - y\| - \|x + t_0 y\| \\ &= t_0 \|x - y\| + \|x + t_0 y\| - \|x + t_0 y\| + \|x - y\| = (1 + t_0)\|x - y\| \end{aligned}$$

which gives that

$$2/\|x - y\| \geqslant (1 + t_0)/\|x + t_0 y\| = \mu^+ \perp_{\mathrm{B}} (x, y) \geqslant 2/\|x + y\|.$$

Thus we must have

$$\|x + y\| \geqslant \|x - y\|$$

and similarly

$$\|x - y\| \geqslant \|x + y\|$$

and thus the equality.
Thus we have

$$\|x + y\| = \|x - y\| = 2(\mu^+ \perp_{\mathrm{B}} (x, y))^{-1}$$

and applying the above Lemma we get that

$$(x + y) \perp_{\mathrm{B}} (x - y) \text{ and } (x - y) \perp_{\mathrm{B}} (x + y).$$

Thus we have

$$(\mathrm{P}'_1) \longleftrightarrow (\mathrm{P}_3), (\mathrm{P}'_1) \to (\mathrm{P}_4).$$

Obviously $(\mathrm{P}_1) \to (\mathrm{P}'_1)$. Suppose now that (P_4) holds and if $\|x\| = \|y\| = 1, x \perp_{\mathrm{B}} y$ then $\|x + y\| = \|x - y\| = c \leqslant 2^{1/2}$. We apply (P'_1) to the pair

$$((x + y)/c, (x - y)/c)$$

and thus

$$\|x + y + x - y\| \leqslant 2^{1/2} c.$$

Then $c \geqslant 2^{1/2}$. Thus we must have $c = 2^{1/2}$ and thus (P_1) holds. Suppose now that

(P_1) holds and $\|x\| = \|y\| = 1$, $x \perp_B y$. From (P'_1) we get that

$$\mu^+ \perp_B (x, y) = 2^{1/2} = \mu^- \perp_B (x, y)$$

where

$$\mu^- \perp_B (x, y) = \sup_{t \leqslant 0} (\|x\| + \|ty\|)/(\|x + ty\|).$$

Then clearly $(P_1) \to (P_2)$.

If (P_2) holds then for $\|x\| = \|y\| = 1$, $x \perp_B y$ we have, using the fact that $\mu \perp_B (X) = 2^{1/2}$, arguing as in the proof of the Lemma 4.16.21 that

$$(x + y) \perp_B (x - y), \|x + y\| = 2^{1/2}.$$

Thus

$$(P_2) \to (P_3).$$

Similarly we prove that $(P_2) \to (P_4)$. □

In his 1981 paper **M.** Baronti introduces another useful property which may be or may be not true in a Banach space – property (P'_3), which is as follows: *for* $\|x\| = \|y\| = 1$, $x \perp_B y$ *there exists* $t \geqslant 0$ *such that*

$$(x + ty) \perp_B (x - y).$$

Then we have the following result on Banach spaces for which $\perp_B$ is symmetric.

THEOREM 4.16.23 (Baronti, 1981). *If X is a Banach space in which* (P'_3) *holds then* $\perp_B$ *is a symmetric orthogonality.*

Thus if dim X is at least 3 and a Banach space has one of the properties (P_1), (P'_1), (P_2), (P_3), (P'_3), then necessarily the norm of X is generated by an inner product.

Another interesting result obtained by M. Baronti is:

THEOREM 4.16.24 (Baronti, 1981). *For any Banach space X the following relations hold*:

$$K(X) = \mathrm{MPB}(X) = \mathrm{MPB}'(X).$$

Another proof of this result was obtained by Kapoor and Mathur (1983).

Related to the parameters considered above we mention another interesting parameter introduced by Franchetti (1983).

Let X be a real Banach space and for each closed subspace X_1 of X we set

$$\lambda(X_1, X) = \inf(\|P\|, P \text{ is a projection of } X \text{ onto } X_1).$$

Then clearly

$$1 \leqslant \lambda(X_1, X) \leqslant 2$$

and Franchetti has defined the parameter $F(X)$ as follows:

$$F(X) = \sup(\lambda(X_1, X), X_1 \text{ is a hyperplane in } X).$$

Then clearly we have the inequality

$$1 \leqslant F(X) \leqslant 2$$

and Franchetti mentions that using the Garkavi's theorem (Section 4.14) the following assertion holds: the norm of X is generated by an inner product iff $F(X) = 1$.

An interesting characterisation of inner product structures was obtained by Gulevič (1982) and is related to a measure of nonconvexity of a set and the result of Routledge (1952).

4.17. The Modulus of Convexity and the Modulus of Smoothness and Inner Product Spaces

Let X be a Banach space of dimension at least 2. The modulus of convexity of X is defined as follows:

$$\delta_X(t) = \tfrac{1}{2} \inf_{\substack{\|x\| = \|y\| = 1 \\ \|x - y\| = t}} (2 - \|x + y\|).$$

The modulus of smoothness of X is defined as follows:

$$\rho_X(s) = \sup(\|x + y\| + \|x - y\| - 1, \|x\| = 1, \|y\| = s).$$

For a detailed exposition of these functions we refer to Istrăţescu (1984). If X is an inner product space then a computation shows that the following assertions hold.

LEMMA 4.17.1. *If X is an inner product space then*

$$\delta_X(t) = 1 - (1 - t^2/4)^{1/2}$$

and

$$\rho_X(s) = (1 + s^2)^{1/2} - 1.$$

Then we have the following result.

THEOREM 4.17.2 (Lindenstrauss, 1963). *A normed linear space X is an inner product space if and only if the modulus of convexity is*

$$1 - (1 - t^2/4)^{1/2}$$

or, equivalently, the modulus of smoothness is

$$(1 + s^2)^{1/2} - 1.$$

Proof. Suppose first that X is a normed linear space whose modulus of smoothness is $(1 + s^2)^{1/2} - 1$. Then for x, y in X the following inequality holds:

$$\|x + y\| + \|x - y\| \leqslant 2(\|x\|^2 + \|y\|^2)^{1/2}$$

and thus by Theorem 4.11.22, X is an inner product space.

Suppose now that X is a normed linear space with the modulus of convexity given by the formula

$$1 - (1 - t^2/4)^{1/2}.$$

Then by the Lindenstrauss duality theorem (see Istrăţescu, 1984, Th. 2.7.3) we conclude that the modulus of smoothness of X^* is given by the formula

$$(1+s^2)^{1/2}-1.$$

Then X^* is an inner product space and thus X is an inner product space. □

We now present an interesting characterization of normed linear spaces in which the norms are generated by an inner product using the behavior of the modulus of convexity and the modulus of smoothness. The results we give were obtained by Kirčev and Troianskiĭ (1975).

THEOREM 4.17.3 (Kirčev and Troianskiĭ, 1975). *For a normed linear space X the following conditions are equivalent:*

(1) *the norm of X is generated by an inner product,*

(2) $\lim_{t\to 0} t^{-2}\delta_X(t) = 1/8,$

(3) $\lim_{s\to 0} s^{-2}\rho_X(s) = 1/2.$

For the proof we need some results, which are given as lemmas.

LEMMA 4.17.4. *Suppose that f, g are convex-real-valued functions on the real axis and*

$$F(t,s,u) = f(t)/g^2(s)((1+us)f((t+u)/(1+us)) + (1-us)f((t-u)/(1-us)) - 2f(t))$$

for $t, s \geqslant 0$ and $u < s^{-1}$.

Suppose further that the following conditions are satisfied:

(1) $f(t) \geqslant 1 = f(0) = g(0) \leqslant g(s),$
(2) $f(t) = tg(t^{-1}),\ t \neq 0$
(3) $\lim_{u\to 0} u^{-2}F(t,s,u) = 1$ *for* $s = 0$, *uniformly in* t.

Then

$$f(t) = (1+t^2)^{1/2}.$$

Proof. Since f is a convex function, then from conditions (1) and (3) we obtain that the first derivative of f is absolutely continuous. Suppose that t is a point where the second derivative exists for f and then condition (3) implies that

$$|(1-st)/(g(s)|^2 f(t)f''(t) \leqslant 1. \tag{$+'$}$$

Let us consider the functions

$$h(t) = t - f(t)/f'(t), \qquad k(t) = h(t)^{-1}.$$

Now if we take in condition (3), $s = h(t)$, $l(t) = f^2(t)$, we get

$$h'(t)/l(h(t)) \leqslant l(t)^{-1} \quad \text{a.e.}(dt).$$

From conditions (1) and (2) we have that on the intervals $(-\infty, 0)$ and $(0,\infty)$ the function $h(t)$ increases monotonously and

$$\lim_{t \to \pm\infty} l(t)/t^2 = 1, \tag{$+$}$$

$$\lim_{t \to \pm\infty} h(t) = 0, \tag{$++$}$$

$$\lim_{t \to \pm 0} h(t) = \mp \infty \tag{$+++$}$$

$$th(t) < 0, \quad t \neq 0. \tag{$++++$}$$

Suppose now that $v > w$, $wv > 0$. The last inequality implies that $h(u)h(w) \geqslant 0$. Now h is absolutely continuous in every subinterval (finite) which does not contain 0. From $(+')$ we get that the primitive function of k, say G, satisfies the following inequalities:

$$G(h(h(u))) - G(h(h(w))) \leqslant G(h(u)) - G(h(w)) \leqslant G(v) - G(w). \tag{$+_1$}$$

This inequality implies that if

$$v > w > 0, z < -w < 0, \tag{$+_2$}$$

$$G(h(-w)) - G(h(h(z))) \leqslant G(-w) - G(z) \tag{$+_3$}$$

$$G(h(h(u))) + G(h(h(-w))) - G(h(h(z))) - G(h(h(w))) \\ \leqslant G(v) + G(-w) - G(w) - G(z). \tag{$+_4$}$$

From $(+)$ for $v \to \infty$ we get, using $(+_1)$, that

$$-G(h(h(w))) \leqslant -G(w), G(h(h(-w))) \leqslant G(-w). \tag{$+_\sim$}$$

If $w \to 0$ in $(+_4)$, $z \to -\infty$ in $(+_\sim)$, we have,

$$-G(h(h(z))) \leqslant -G(z), \quad G(h(h(-v))) \leqslant G(-v).$$

From the inequalities just obtained we have that for $z \neq 0$,

$$G(h(h(z))) = G(z).$$

From $(+_1)$ and the above equality we have that for $vw > 0$

$$G(h(u)) - G(h(w)) = G(v) - G(w)$$

and thus a.e.,

$$h'(t)/l(h(t)) = l(t)^{-1}.$$

But

$$G(h(h(t))) = G(t)$$

implies that

$$l'(t)/l(t) = -l'(h(t))/l(h(t))$$

and multiplying this with h', followed by an integration, we get

$$l'(t)^2 l(v)/(l(t)l'(v)^2) = l(h(u))/l(h(t)). \tag{$+_5$}$$

Condition (2) in lemma 4.17.4, combined with $(+_4)$, gives in the above relation that

$$\lim_{v \to 0} l(h(u))l'(h(v))^2/h(v) = 4.$$

Then $(+_5)$ implies that for all $t \neq 0$,

$$l'(t)^2 l(h(u))/l(t) = 4$$

and from

$$h'(t)/l(h(t)) = l(t)^{-1}$$

we get that for almost all t,

$$l'(t)^2 h'(t) = 4 \quad \text{or} \quad 2l''(t)l(t) - l'(t)^2 - 4 = 0.$$

Since $l(0) = 1$, $l'(0) = 0$ we must have $l(t) = (1 + t^2)$ and the lemma is proved. □

Now the proof of the theorem is as follows. It is obvious that (1) → (2), and (1) → (3). Now suppose that (3) holds. It is clear that we can suppose that X is a two-dimensional space and we choose in X an Auerbach basis which exists in every (finite) dimensional space. Then if (e_1, e_2) is the basis, we set

$$f(t) = e_1 + te_2, g(t) = te_1 + e_2.$$

We remark that these are functions satisfying the condition of Lemma 4.17.4 since, in this case,

$$F(t, s, u) = u^2 r^{-2}\left(\left\| x/\|x\| + y/\|y\| \right\| + \left\| x/\|x\| - y/\|y\| \right\| - 2 \right),$$

where

$$r = u\|y\|/\|x\|, x = e_1 + te_2, y = te_1 + e_2.$$

Then, by the lemma we have

$$\|e_1 + te_2\| = (1 + t^2)^{1/2}$$

which implies that the norm of the space is generated by an inner product.

Suppose now that X satisfies condition (2). Then the dual space X^* satisfies the relation

$$\lim_{s \to 0} s^{-2} \rho_{X^*}(s) = \tfrac{1}{2},$$

which gives that X is a space whose norm is generated by an inner product. This clearly implies that the norm of X is generated by an inner product. □

REMARK 4.17.5. The equivalence between conditions (1) and (2) was independently proved by Senechalle (1972). See Section 4.15 for details.

4.18. Spaces Isomorphic to Inner Product Spaces

In this section we present some results concerning the problem of a geometric characterization of normed spaces which are isomorphic to inner product spaces.

We limit ourselves to characterizations using finite dimensional subspaces; for characterizations involving the lattice structure of subspaces we refer to the paper by Kakutani and Mackey (1944).

First we give a result of J. Lindenstrauss which characterizes the spaces isomorphic to inner product spaces using the family of all separable subspaces.

THEOREM 4.18.1 (Lindenstrauss, 1963). *Let X be a Banach space with the property that all separable subspaces of X are isomorphic to inner product spaces. Then X is isomorphic to an inner product space.*

Proof. Let X_1 be an arbitrary separable subspace of X and set

$$k(X_1) = \inf(k,(\exists) \text{ an inner product on } X_1, \langle\, , \rangle \text{ such that}$$
$$\|x_1\| \leqslant \langle x_1, x_1\rangle^{1/2} \leqslant k\|x_1\|, x_1 \in X_1).$$

First we show that there exists a constant k such that for all X_1, separable subspace in X, $k(X_1) \leqslant k$. Indeed, in the contrary case for some sequence of separable subspaces (X_1^n) the inequality $k(X_1^n) \geqslant n$ holds. Since the subspace of X containing all $X_1^n, n = 1, 2, 3, \ldots$, is obviously separable it is clearly not isomorphic to an inner product space, a fact which contradicts our hypothesis. Thus the constant k exists.

Of course, we may suppose that X is not a separable space. Suppose first that X has a dense subset of cardinality $\aleph_1$. Then X is of the form $X = \cup_{\alpha\in\Omega} X_\alpha$, where Ω is the set of all ordinal numbers of cardinality smaller than the first uncountable ordinal, and we have $X_\alpha \subset X_\beta$ for $\alpha < \beta$ and X_α is separable for all α in Ω. Let us consider $\|\,,\|_\alpha$ be a norm on X_α which is generated by an inner product and the following inequality is satisfied:

$$\|x\| \leqslant \|x\|_\alpha \leqslant k\|x\|, \quad x \in X.$$

We can consider on the set Ω the Stone–Cech compactification and let

$$s \in \bigcap_{i\in\Omega} \mathrm{Cl}(\beta, \beta \geqslant \alpha)$$

where the closure is taken in the Stone–Cech compactification of Ω. For each x in X we set

$$f_x(\alpha) = \begin{cases} 0, & x \notin X_\alpha \\ \|x\|_\alpha, & x \in X_\alpha \end{cases}$$

and let $\hat{f}_x$ be the continuous extension of f_x to the Stone–Cech compactification of Ω. Then we set

$$\|x\|_* = \hat{f}_x(s)$$

for x in X.

Now we remark that

$$x \to \|x\|_*$$

is a norm on X which satisfies the Jordan–von Neumann condition:

$$f_{x+y}^2(\alpha) + f_{x-y}^2(\alpha) = 2(f_x^2(\alpha) + f_y^2(\alpha)).$$

Also

$$||x|| \leqslant ||x||_* \leqslant k||x||$$

holds, which proves the theorem in this case. If we use an argument based on transfinite induction then we have that on X there exists an inner product which generates a norm equivalent to the original norm of X. Thus the theorem is proved. □

We now present the elegant characterization of Banach spaces which are isomorphic to inner product spaces obtained by Dvoretzky (1967). For this we need the following definition.

DEFINITION 4.18.2. Let $\mathscr{F}$ be a family of finite dimensional Banach spaces. We say that $\mathscr{F}$ is equi-isomorphic if there exists a finite positive constant k with the property that for any two elements M, N of $\mathscr{F}$ there exists $T: M \to N$ which is injective, $T \in L(M, N)$ and $T^{-1} \in L(N, M)$ with

$$\max(||T||, ||T^{-1}||) \leqslant k.$$

The following characterization fo Banach spaces isomorphic with inner product spaces, in terms of equi-isomorphic families of subspaces was obtained by A. Dvoretzky.

THEOREM 4.18.3 (Dvoretzky, 1967). *A Banach space X is isomorphic to an inner product space if and only if the family of finite dimensional subspaces are equi-isomorphic.*

The above condition means that a Banach space X is isomorphic to an inner product space if and only if there exists a constant $k < \infty$ such that for any two subspaces M and N of the same finite dimension there exists a linear operator $T: M \to N$ which is an injective and surjective linear mapping and $\max(||T||, ||T^{-1}||) \leqslant k$.

Proof. The condition is necessary. If X is a Banach space which is isomorphic to an inner product space, say H, then let $T: X \to H$. Let $k = ||T|| \cdot ||T^{-1}||$. If M and N are subspaces in X of the same dimension then $M_T = TM, N_T = TN$ are subspaces in H of the same dimension. Then these are isometric subspaces. This obviously implies that the condition of the theorem is necessary. For the sufficiency part we need the following result, stated as a lemma.

LEMMA 4.18.4 (Dvoretzky, 1967). *If X is a Banach space such that there exists $k < \infty$ with the property that for all finite dimensional subspaces M of X there exists an injective and surjective linear operator T of M onto a Euclidean space with $||T|| \leqslant k$, $||T^{-1}|| \leqslant k$ then X is isomorphic to an inner product space.*

Proof. Let X be a Banach space as in the lemma. Suppose first that X is separable. We consider (x_i) be a countable dense linearly independent subset of elements of x. We consider in a separable complete inner product space H an orthonormal basis, say (e_i). For each fixed n let H_n be the space generated by $e_1, \ldots, e_n$. Then, according to our hypothesis there exists a linear operator T_n defined on the space generated by $x_1, \ldots, x_n$ onto the space H_n satisfying the inequality

$\|T_n\| \leqslant k$, and $\|T_n^{-1}\| \leqslant k$. Let $y_i^{(n)} = T_n x_i, i = 1,2,3,\ldots,n$. Since $(x_1,\ldots,x_n)$ are linearly independent we have that $(y_i^{(n)})_1^n$ is a linearly independent set. Then there exists an isometry of H_n onto itself such that $I_n y_i^{(n)} \in H_i$, for $i = 1,2,3,\ldots,n$. We set $S_n = T_n^{-1}.I_n^{-1}$ which is clearly a linear operator and sends H_n onto the space generated by $x_1,\ldots,x_n$. This has also the property that sends H_k in the subspace generated by $x_1,\ldots,x_k$ for all $k = 1,2,3,\ldots,n$. Since

$$\|S_n\| = \|T_n^{-1}\cdot I_n\| = T_n^{-1}$$

$$\|S_n^{-1}\| = \|I_n \cdot T_n\| = \|T_n\|$$

we have that

$$\max(\|S_n\|, \|S_n^{-1}\|) \leqslant k.$$

Let us denote $z_i^{(n)} = S_n e_i$, $i = 1, 2, 3,\ldots,n$ and then $z_i^{(n)}$ are in B_i for all $i = 1, 2, \ldots, n$. Thus we obtain that

$$\|z_i^{(n)}\| \leqslant k$$

$$\|z_i^{(n)} - x\| \geqslant 1/k, \qquad x \in (x_1,\ldots,x_k) = \text{the span of } x_1,\ldots,x_k.$$

since

$$\|S_n^{-1}(z_i^{(n)} - x)\| = \|e_i - S_n^{-1}x\| = \|e_i\| = 1.$$

Now $z_1^{(n)}$ is a multiple of x_1 then there exists an increasing sequence of integers (n_1) such that $z_1^{n_1} \to z_1$. Since $z_2^{(n)} \in (x_1, x_2)$ we find a sequence of integers (n_2) such that for some z_2 in (x_1, x_2), $z_2^{n_2} \to z_2$. We can continue this procedure and we have a sequence of points (z_k) in $(x_1,\ldots,x_k)$ such that for some sequence of integers (n_k), $z_k^{(n_k)} \to z_k$. Of course we have that (n_k) is a subsequence of the sequence (n_{k-1}). Then by the diagonal procedure we obtain a sequence of integers (n_i) such that

$$\lim S_{n_i} e_k = z_k, \quad k = 1, 2, 3,\ldots.$$

We now show that we can construct a linear operator from the union of all H_i into X by defining T by the relations:

$$Te_i = z_k, \quad k = 1, 2, 3,\ldots.$$

Indeed, if u, v are arbitrary elements in the union of all H_i we have that

$$\|Tu - Tv\| = \lim_{i\to\infty} \|S_{n_i}(u - v)\| \leqslant k\|u - v\|.$$

Thus the map T so defined may be extended by continuity to the linear map from H to X. The extension is denoted again by T. Then clearly the inequality $\|Tu\| \leqslant k\|u\|$ is satisfied for all u in H. Since on every $(x_1,\ldots,x_k)$, $S_{n_i}^{-1} \to T^{-1}$, T^{-1} extends to the closure of (x_i), i.e. X. The extension satisfies the inequality $\|T^{-1}\| = k$. Thus the existence of an isomorphism betwen X and H is proved, and thus in this case the lemma is proved. If X is not separable the step considered above and the Lindenstrauss theorem given above complete the proof. □

Proof of the sufficiency of Theorem 4.18.3. First we remark that we may suppose that X is infinite dimensional since otherwise there is nothing to prove. Then by a result of Dvoretzky (1950, 1961) for each integer k there exists a subspace of dimension k which is arbitrarily near in the following sense

$$\Delta(M, N) = \inf \log \max(\|T\|, \|T^{-1}\|)$$

where the infimum is taken over all one-to-one linear operators on the space M or on the space N (both are of the same finite dimension), to a Euclidean space. If X is as in the theorem then for some k_1 with $\Delta(M, N) = k_1$ and we can choose N such that $\Delta(N, E) \leqslant 1$, E being a Euclidean space. Then by the triangle inequality we have that

$$\Delta(M, E) \leqslant 1 + k_1$$

and this implies that X satisfies the conditions of the lemma. Thus X is isomorphic to an inner product space. □

COROLLARY 4.18.5 (Dvoretzky, 1967). *A Banach space X is isomorphic to an inner product space if and only if all its quotient spaces relative to subspaces of finite* co-dimension are equi-isomorphic.

We now mention the following interesting characterization of Banach spaces isomorphic to inner product spaces obtained by H. Fakhoury (unpublished manuscript).

THEOREM 4.18.6 (H. Fakhoury). *For a real Banach space X the following assertions are equivalent*:

(1) *X is isomorphic to an inner product space.*
(2) *For any subspace M of X and any compact operator $T: M \to c_0(M)$ there exists an extension to X which is also compact.*
(3) *For any subspace M of X and any compact operator T of M in a closed linear subspace N of $c_0(N)$ there exists an extension to X which is again compact.*
(4) *There exists t in $[1, \infty)$ such that for any subspace M of X and any operator of M into X_1 of finite dimension and with the closed unit ball having a finite number of extreme points, there exists an extension of the operator whose norm satisfies the inequality $\|\bar{T}\| \leqslant t\|T\|$, T being the operator, $\bar{T}$ the extension.*

There are a number of very interesting characterizations of Banach spaces isomorphic to inner product spaces using L_p spaces. We mention here only the characterization obtained by Grothendieck (1956).

THEOREM 4.18.7 (Grothendieck, 1956). *A Banach space X is isomorphic to an inner product space if and only if X is isomorphic to a subspace of an L_1-space and to a quotient of an L_∞-space.*

In the previous section we used the modulus of convexity and the modulus of smoothness to characterize the class of normed linear spaces for which the norm is generated by an inner product. We show now how we can use the modulus of convexity and the modulus of smoothness to obtain a characterization of Banach spaces which are isomorphic to complete inner product spaces. First we give the following result.

Let X be a Banach space and suppose that for some $p > 0$ and all $t > 0$

$$\delta_X(t) \geqslant pt^2 \tag{+}$$

and that

$$\rho_X(s) \leqslant qs^2 \tag{++}$$

for some $q > 0$ and s in $[0, 1]$. Here δ_X and δ_X are the modulus of convexity and the modulus of smoothness of X, respectively.

PROPOSITION 4.18.8 (Lindenstrauss, 1963). *Suppose that X is a separable Banach space with an unconditional basis. If the modulus of convexity and the modulus of smoothness of X satisfy the inequalities $(+)$ and $(++)$ then X is isomorphic to l^2.*

Proof. For this we use the theorem of Kadets (1956 – see Istrăţescu, 1984, for a proof) which gives that the convergence of $\Sigma\, a_n e_n$ implies that

$$\Sigma\, |a_n|^2 < \infty.$$

Conversely, if $\Sigma |a_n|^2 < \infty$ then by a theorem of Lindenstrauss (1963 – see the same reference for a proof) we find a sequence of numbers (t_n), $t_n = \pm 1$ such that

$$\Sigma\, t_n a_n e_n$$

converges. Thus the mapping

$$\Sigma a_n e_n \to (a_1, a_2, \ldots)$$

gives the required isomorphism. For the proof of the theorem we need also the following lemma.

LEMMA 4.18.9 (Lindenstrauss, 1963). *Suppose that a Banach space X has two equivalent norms $\|\,,\|_1$ and $\|\,,\|_2$. Suppose that with respect to $\|\,,\|_1$, the modulus of convexity satisfies $(+)$ and with respect to $\|\,,\|_2$, the modulus of convexity and the modulus of smoothness satisfy $(+)$ and $(++)$ respectively. Then the norm*

$$\|x\|_3 = \|x\|_1 + \|x\|_2$$

is a norm which satisfies both the condition $(+)$ and $(++)$.

Proof. Let us consider m and M as positive numbers such that

$$\|x\|_1 \leqslant m\|x\|_2, \qquad \|x\|_2 \leqslant M\|x\|_1$$

for all x in X. Let us consider x, y in X with $\|x\|_3 = \|y\|_3 = 1$ and $\|x - y\|_3 = \varepsilon$. We may assume without loss of generality that $\|x\|_2 \geqslant \|y\|_2$. Set

$$k = \|y\|_2/\|x\|_2 \quad \text{and} \quad z = kx.$$

Then

$$\|z - y\|_3 \geqslant \varepsilon/2$$

since

$$\|z - y\|_3 \geqslant \|y\|_3 - \|z\|_3 = 1 - k = \varepsilon/2.$$

for $k \leqslant 1 - \varepsilon/2$; if $k \geqslant 1 - \varepsilon/2$ then

$$\|z - y\|_3 \geqslant \|y - x\|_3 - \|x - z\|_3 = \varepsilon - 1 + k \geqslant \varepsilon/2.$$

Thus we have

$$\|z - y\|_3 \geqslant \varepsilon/(2a + 2).$$

But $(+)$ is satisfied and thus

$$\|z\|_2 = \|y\|_2$$

which implies that

$$\|y + z\|_2 \leqslant \|x - z\|_2 + \|z + y\|_2 \leqslant (1 - k)\|x\|_2 + 2\|y\|_2 (1 - k\varepsilon^2(2m + 1)\|y\|_2)^{-2}) = \leqslant \|x\|_2 + \|y\|_2 - k\varepsilon^2/(2(m + 1)^2).$$

Thus

$$\|x + y\|_3 \leqslant 2 - k\varepsilon^2/2(m + 1)^2)$$

and thus $(+)$ is satisfied.

For $(+\,+)$ suppose that $\|x\|_3 = 1$ and $\|y\|_3 = s$. Then

$$\|x\|_1 \geqslant (M + 1)^{-1}, \qquad \|x\|_2 \geqslant (m + 1)^{-1}$$

which imply that

$$\|x + y\|_3 + \|x - y\|_3 = \|x + y\|_1 + \|x - y\|_1 + \|x + y\|_2 + \|x - y\|_2 \leqslant$$
$$\leqslant 2\|x\|_1(1 + K_1\|y\|_1^2\|x\|_1^{-2}) + 2\|x\|_2(1 + K_2\|y\|_2^2\|x\|_2^{-2}) \leqslant$$
$$\leqslant 2 + 2(M + 1)K_1 + (m + 1)K_2 s^2$$

and thus $(+\,+)$ is satisfied. □

LEMMA 4.18.10 (Lindenstrauss, 1963). *Let X be a Banach space with a norm $\|\,,\|$. Then X has property $(+)$ iff X_1 has property $(+\,+)$, and conversely.*

Proof. This follows immediately from the Lindenstrauss result about the modulus of convexity of X and the modulus of smoothness of X (see also Istrăţescu, 1984, for more information on these matters).

COROLLARY 4.18.11. *Let X be a Banach space with a norm $\|\,,\|$. Then X has properties $(+)$ and $(+\,+)$ iff X_1 has properties $(+)$ and $(+\,+)$.*

COROLLARY 4.18.12 (Lindenstrauss, 1963). *Let X be a Banach space with the norm $\|\,,\|$. Suppose that $(+)$ and $(+\,+)$ hold. Then for any given $\varepsilon > 0$ there exists an equivalent norm $\|\,,\|_\varepsilon$ on X satisfying $(+)$ and $(+\,+)$ and being arbitrarily close to the original norm (i.e. $\|x\| \leqslant \|x\|_\varepsilon \leqslant (1 + \varepsilon)\|x\|$).*

LEMMA 4.18.13 (Lindenstrauss, 1963). *Let X be a separable Banach space with an unconditional basis and $(+)$ and $(+\,+)$ hold. Then X is isomorphic to l^2.*

Proof. We use for the proof the Kadets (=M. I. Kadec) result about unconditional convergence (see the author's book, 1984). If (e_i) is the unconditional basis of X with $\|e_i\| = 1$ then, by Kadets result, if $\Sigma a_i e_i$ converges then $\Sigma|a_i|^2 < \infty$.

Conversely, if $\Sigma|a_n|^2 < \infty$, then by the Lindenstrauss result (see also Istrăţescu (1984)). We find $\varepsilon_i = \pm 1$ such that $\Sigma a_i \varepsilon_i e_i$ converges. But the basis is unconditional and thus $\Sigma a_i e_i$ is convergent. Then the mapping $\Sigma a_i e_i \to (a_1, a_2, \ldots)$ satisfies the assertions of the lemma.

From these lemmas we have the following result.

THEOREM 4.18.14 (Lindenstrauss, 1963). *Let X be a Banach space with a norm* $\|,\|$. *Then X is isomorphic to a complete inner product space iff the following assertions hold*:

(1) *If* X_1 *is an arbitrary separable subspace of X then* X_1 *has an unconditional basis.*
(2) *For any* $\varepsilon > 0$ *there exists, arbitrarily close to the norm* $\|,\|$, *an equivalent norm* $\|,\|_\varepsilon$ *satisfying* $(+)$ *and* $(++)$.

4.19. Inner Product Spaces and Classes of Metric Spaces

As the reader knows, a Fréchet metric space was defined using as a primitive concept the notion of 'distance' while in a normed space or Banach space, distance is a derived notion, defined using three primitive notions; for any two elements x, y of a normed space the distance between x and y is the number defined by

$$d(x, y) = \|x - y\|,$$

where $\|,\|$ is the norm of the space.

Thus it appears that in the case of normed spaces or Banach spaces the distance has properties which are not present for all metric space.

The metric characterization program consists in the determination of those distinctive properties of normed linear spaces (or Banach spaces) with the property that the norm is generated by an inner product, i.e. on the space there exists an inner product $\langle , \rangle$ such that for all x in the space the following relation holds:

$$\langle x, x\rangle = \|x\|^2.$$

In other words, the theme of the metric characterization program is to determine the necessary and sufficient properties of the distance function of the metric space in order to assure that it is possible to define the usual notions of a normed linear space (addition, multiplication and the norm with the known properties). Then these properties (expressed in terms of distance) are interpreted as a set of metric postulates for defining normed space (Banach spaces or spaces with the norm generated by an inner product).

Among the contributors in this aspect of the theory of normed linear space we mention Andalafte (1964, 1982), Blanc (1938), Blumenthal (1935, 1936, 1943, 1944, 1950, 1953, 1955, 1964, 1974), Busemann (1941, 1955), Diminnie (1982), Fréchet (1958), Freese (1961, 1982), Loveland (1978), Martin (1976), Menger (1928, 1931, 1932), Stamey (1957), Valentine (1971, 1972, 1973, 1978, 1980), Wayment (1972, 1973, 1978, 1980) and Wilson (1932).

In this section we present some characterizations of complete inner product spaces in the sense mentioned above. For a detailed account we refer to the excellent presentations in the books of Blumenthal (1953) and Blumenthal and Menger (1970) – where the reader may find many other interesting results – as well as to the original papers of the authors mentioned above.

Let (X, d_X) and (Y, d_Y) be two metric spaces.

DEFINITION 4.19.1. The point c of the metric space X is said to be between the points a and b of X if:

(1) $a \neq c \neq b$
(2) $d_X(a, c) + d_X(c, b) = d_X(a, b)$.

We write this briefly as acb.

DEFINITION 4.19.2. The pair of points of X, (a, b) is said to be congruent to the pair (a_1, b_1) of Y if

$$d_X(a, b) = d_Y(a_1, b_1).$$

If M is any subset in X and N is a subset in Y then we say that M is congruent to N if there exists a one-to-one map of M onto N

$$f: M \to N$$

such that for all m_1, m_2 in M,

$$d_Y(f(m_1), f(m_2)) = d_X(m_1, m_2).$$

DEFINITION 4.19.3 (K. Menger). A subset C of the metric space X is said to be (metrically) convex if for any pair a, b of points of C there exists at least one point c of C such that acb holds.

The following notion plays a fundamental role in the formulation of metric characterizations of Banach spaces and of complete inner product spaces.

DEFINITION 4.19.4. If (X, d) is a metric space and p, q are two points of X then a subset M of X is a metric segment with the endpoints p and q if M is congruent with a line segment of length $d(p, q)$ and p, q are in M.

DEFINITION 4.19.5. A metric space (X, d) is said to be (metrically) convex if for each points p, q in the space, $p \neq q$ there exists at least one point c in X such that pcq holds.

DEFINITION 4.19.6. The metric space (X, d) is said to be metrically externally convex if whenever p, q are distinct points of X there exists at least one point r of X such that pqr holds.

DEFINITION 4.19.7. A subset L of the metric space (X, d) is said to be a metric line if it is congruent with R (the set of all real numbers with the usual metric).

DEFINITION 4.19.8. A triple of points (p_1, p_2, p_3) of the metric space X is said to be collinear if there exists a mapping

$$f: (p_1, p_2, p_3) \to R$$

such that for all $i, j = 1, 2, 3$ the following relations hold:

$$|f(p_i) - f(p_j)| = d(p_i, p_j).$$

DEFINITION 4.19.9. Let p, q be two distinct points of the metric space (X, d). We say that the point r of X is the middle point of p and q if

$$d(p, r) = d(r, q) = \tfrac{1}{2}d(p, q).$$

DEFINITION 4.19.10. A plane π in the metric space (X, d) is the closure of the set of all points of X which are collinear to a fixed point p and a point q situated on a fixed line L of X.

REMARK 4.19.11. In the above definition we suppose that the point p is not situated on the line L.

DEFINITION 4.19.12. Let M be a subset in the metric space (X, d). The point x is said to be a foot point of $p \in X$ if for all $m \in M$

$$d(x, p) \leqslant d(m, p).$$

REMARK 4.19.13. It is not difficult to see that if M is compact then foot points exists.

As mentioned above the basic problem of the metric characterization program is to determine those metric properties which are necessary and sufficient to distinguish the members of a class of metric spaces from all others of a bigger class. One (and a very important) approach, due to K. Menger, was through the imbedding problem.

In order to formulate some characterizations of complete inner product spaces in terms of the distance we need such notions as Euclidean k-point property, weak Euclidean four-point property, etc. Some facts about the three-point property were stated by W. A. Wilson in 1932 and he initiated the study of metric spaces having a Euclidean four-point property. The very useful notion of weak Euclidean four-point property was introduced by L. M. Blumenthal in 1933, as well as the notion of feeble Euclidean four-point property.

DEFINITION 4.19.14. The metric space (X, d) is said to have a Euclidean k-point property if each k-tuple of its points is congruent to a subset of the Euclidean space $R^2 (= R_\times R)$.

DEFINITION 4.19.15. Let (X, d) be a metric space. We say that the quadruple (p_1, p_2, p_3, p_4) is in the class

(1) C_1 iff it contains a linear triple;
(2) C_2 iff p_3 is between p_2 and p_4 and $d(p_2, p_3) = d(p_3, p_4)$;
(3) C_3 iff p_2, p_3, p_4 are colinear and $d(p_1, p_2) = d(p_1, p_4)$;
(4) C_4 iff p_3 is between p_2 and p_4 and $d(p_2, p_3) = d(p_3, p_4)$, $d(p_1, p_2) = d(p_1, p_4)$;
(5) C_5 iff p_3 is between p_2 and p_4 and $d(p_2, p_3) = 2d(p_3, p_4)$, $d(p_1, p_2) = d(p_1, p_3)$.

DEFINITION 4.19.16. The metric space (X, d) is said to have a weak, feeble, isosceles weak, isosceles feeble, external isosceles feeble Euclidean four-point property provided every quadruple (p_1, p_2, p_3, p_4) of points of X of class C_1, C_2, C_3, C_4, C_5, respectively, is congruent to a set in the Euclidean space R^2.

DEFINITION 4.19.17. A metric space (X, d) is said to have a queasy Euclidean four-point property if for each pair (p, q) of points of X there exists a point r of X such that prq holds and for any s in X the quadruple (p, q, r, s) is congruent to a set in the Euclidean space R^2.

The following result, proved for the case of compact spaces by K. Menger and for arbitrary spaces by N. Aronszajn, is very important in characterizations of inner product spaces in the class of metric spaces.

THEOREM 4.19.18 (K. Menger and N. Aronszajn). *Let (X, d) be a complete metric space which is (metrically) convex. Then for any two points p, q in the space there exists a segment in the space whose endpoints are p and q.*

Proof. Let p and q be given and consider the set M_1 defined by

$$M_1 = \{p, q\}$$

and the real-valued function on M_1, f_1, defined by

$$f_1(p) = 0, \qquad f_1(q) = d(p, q).$$

We construct, by induction, a sequence of sets in X and a sequence of functions defined on these subsets as follows. Suppose that $(f_1, \ldots, f_n)$ are functions defined of the subset $M_1, \ldots, M_n$ such that f_i is an extension of $f_{i-1}(f_0 = f_1)$ and the ranges of f_n are in $[0, d(p, q)]$ and, further, each f_i is an isometry. We consider the set $\mathscr{F}$ of all such functions. Consider now I^n be the set of 2^n closed intervals obtained by dividing $[0, d(p, q)]$ in equal nonoverlapping intervals of length $d \cdot 2^{-n}$. For any f in $\mathscr{F}$ we denote by $N_n(f)$ the number of all subintervals of I^n whose intersection with $R(f)$ (the range of f) is nonempty. Obviously, $N_n(f) \leqslant 2^n$ and for each f in F we may choose f_{n+1} in $\mathscr{F}$ such that

$$N_n(f_{n+1}) \geqslant N_n(f).$$

Thus we have a sequence (f_n) which clearly defines a function f whose domain is $\cup D(f_n)$ and the range is $\cup R(f_n) \subseteq [0, d(p, q)]$. This function is also an isometry and clearly is an extension of each of the functions f_n. We show that it admits an extension to the closure of $\cup D(f_n)$. Consider p_1 be a point in the closure of the set $\cup D(f_n)$ and not in the set. Then we find a sequence (p_n) in $\cup D(f_n)$ such that

$$\lim p_n = p_1.$$

Obviously $(f(p_n))$ is convergent and let $s = \lim f(p_n)$. Clearly s is in $[0, d(p, q)]$ and we remark that s is uniquely determined, in the sense that it is independent of the sequence we choose (and whose limit is p_1). We extend f by defining $f(p_1) = s$. We denote by $\tilde{f}$ this extension and we remark that it is an isometry.

We now show that $R(\tilde{f}) = [0, d(p, q)]$. This, of course, completes the proof of the theorem since then $D(\tilde{f})$ is the required segment. Suppose now that f has not the above property, i.e. $R(\tilde{f})$ is not $[0, d(p, q)]$. Since this range is a closed subset, the complement in $[0, d(p, q)]$ is an open set and thus we may suppose that there exists an interval (a, b) with the following properties:

(1) It a subset of $[0, d(p, q)]$.
(2) a, b are in the range of f.
(3) Any point of (a, b) is not in the range of $\tilde{f}$.

Thus we find r, t in X with the property that

$$\tilde{f}(r) = a, \tilde{f}(t) = b$$

and thus

$$d(r, t) = b - a.$$

Now X is supposed to be convex and thus we find s in X with the property that

$$d(r, s) + d(s, t) = d(r, t).$$

This gives that we find c in (a, b) with the properties that

(1) $d(t, s) = b - c$,
(2) $d(r, s) = c - a$.

Since c is not in the range of $\tilde{f}$, s is not in the domain of $\tilde{f}$. We consider the set $D(f) \cup \{s\} = D_1$ and a function h on this set defined by $h(s) = c$, $h(u) = \tilde{f}(u)$ whenever u is in $D(\tilde{f})$. We remark that h is an isometry. Indeed, to prove this it suffices to prove that

$$d(x, s) = |h(x) - h(s)| = |c - \tilde{f}(x)|.$$

We consider the set

$$(a = \tilde{f}(r), b = \tilde{f}(t), \tilde{f}(x))$$

and since this is a set of real numbers one of them is between the other two. The function $\tilde{f}$ is an isometric mapping and thus we get that one of the points r, x, t is between the other two. Since c is not in (a, c) we must have xrt or rtx.

If xrt holds then rtx implies that xrs holds, i.e.

$$d(x, r) + d(r, s) = d(x, s)$$

or

$$d(x, s) = a - \tilde{f}(x) + c - a = c - \tilde{f}(x).$$

In the case rtx holds and rst we have stx and thus

$$d(x, s) = c - b + \tilde{f}(x) - c = \tilde{f}(x) - b.$$

Thus h is an isometry and clearly it is an extension of f_n for all n. We take n sufficiently large such that $d \cdot 2^{-n} < \min\{c - a, b - c\}$ and then at least one of the closed intervals of I^n contains c and is contained in (c, b). This gives that $R(h)$ has a point in common with at least on interval of I^n whose intersection with $R(f_{n+1})$ is empty. But h is an extension of each f_n and thus $R(f_{n+1})$ is in the range of h. This gives that the range of h has a common point with all those intervals of I^n that have a nonempty intersection with $R(f_{n+1})$. Thus

$$N_n(h) \geqslant N_n(f_{n+1}) + 1$$

and since h is in f_n this contradicts the definition of f_{n+1}. This completes the proof of the theorem.

THEOREM 4.19.19 (K. Menger). *If (X, d) is a complete metric convex space, a metric segment with the endpoints p and q is characterized in the family of all closed subsets of X and points between p and q as those subsets which have for each pair of points exactly one middle point.*

Proof. Let p, q be the segment with the endpoints p and q. It is congruent with the segment $[0, d(p, q)]$. Suppose now that F is a closed set in X satisfying the properties stated in the theorem. Then clearly F is a closed convex metric space which clear contains the metric segment p, q. Suppose now that there exists a point in F, say r, which is not in $[p, q]$. If we have prq then we get the existence of a metric segment $[p, q]^+$ of F containing r. Clearly this is not the segment $[p, q]$. Now we find, traversing $[p, q]^+$ from r to p, a first point, say s, of $[p, q]$ and traversing $[p, q]$ from r we find a first point of $[p, s]$, say t.

Then the subsegments of $[p, s]$ and $[p, s]^+$ with the endpoints s and t have only those points in common. But each of these contains a middle point of s and t, and this is clearly a contradiction. This proves the theorem. □

THEOREM 4.19.20. *Let (X, d) be a complete externally convex metric space and let p, q be arbitrary points (distinc) in X. Then they are on a metric line of X.*

Proof. From the above result we find a segment p, q in X with the endpoints p and q. We consider now in X the set:

$$L(p, q) = (x, x \in X, pqx \text{ holds}).$$

We show first that this is an unbounded set in X. Indeed, in the contrary case we find a finite positive number r_0 such that

$$\text{lub}(d(p, x) = r_0.$$

We can consider p_1 in $L(p, q)$ such that

$$d(p_1, p) > r_0 - 1, \qquad p \in L(p, p_1)$$

Then p is in $L(p, p_1)$ and we find r_1 such that

$$\text{lub}(d(x, p) = r_1, \quad x \in L(p, p_1)$$

and we can take p_2 in $L(p, p_1)$ such that

$$d(p, p_2) > r_1 - \tfrac{1}{2}.$$

Continuing in this way we get a sequence of points $(p_n) \subset X$ and a sequence of real numbers (r_n) such that:

(1) $r_{n-1} = \text{lub}(d(p, s); s \in L(p, p_{n-1}))$,

(2) $p_n \in L(p, p_{n-1})$, $\qquad r_n = d(p, p_n) = r_{n-1} - n^{-1}$,

(3) $r_0 \geqslant r_1 \geqslant r_2 \geqslant \cdots \geqslant r_n \geqslant r_{n+1} \geqslant \cdots$,

(4) $L(p, q) \supseteq L(p, p_1) \supseteq L(p, p_2) \supseteq \cdots$.

Then it is clear that $\lim r_n$ exists,

$$\lim r_n = r = d(p, q) \quad \text{and} \quad \lim d(p_n, p) = r.$$

Now pp_np_{n+1} holds ($n = 1, 2, 3, \ldots$) and then we get that for all k pp_np_{n+k} holds. Thus

$$d(p, p_n) + d(p_n, p_{n+k}) = d(p, p_{n+k}) \tag{+}$$

which gives for $n, k \to \infty$,

$$\lim d(p_n, p_{n+k}) = 0$$

and thus (p_n) is a Cauchy sequence. Thus we can consider the element $\bar{p} = \lim p_n$. Then relation (+) gives that

$$\lim d(p, p_n) = d(p, \bar{p}).$$

Now in (+) for n fixed and for $k \to \infty$ we have

$$d(p, p_n) + d(\bar{p}, p_n) = d(p, \bar{p})$$

and since $\bar{p} \neq p$, $pp_n\bar{p}$ holds, for $n = 1, 2, 3, \ldots$. If s is a point of X such that $p\bar{p}s$ holds then we choose n sufficiently large such that

$$r_n < r + 2^{-1} d(\bar{p}, s). \tag{++}$$

But $pp_n\bar{p}$ and $p\bar{p}s$ imply that pp_ns holds and thus

$$s \in L(p, p_n)$$

which gives that $d(p, \bar{p}) \leqslant r_n$. But $d(p, \bar{p}) + d(\bar{p}, s) \leqslant r_n$, or

$$r + d(\bar{p}, p) < r$$

which contradicts (++). Thus $L(p, q)$ is an unbounded set.

Now the proof of the theorem is as follows. We suppose that $p \neq q$. We consider the (metric) segment $S_0 = [p, q]$, i.e. the metric segment of X with the endpoints p and q. Further we consider another segment of X, containing S_0 and determined by a point q_1 such that $d(q_1, p) > 2d(p, q)$. We can continue in this way and we get a sequence of segments of X, (S_n) with the following properties:

(1) $S_n \subseteq S_{n+1}$,

(2) $\operatorname{diam} S_n \to \infty$.

Then $L = \cup_n S_n$ is clearly congruent with R and this proves the theorem. □

It appears of interest to determine when the metric space (X, d) has the property that each pair of distinct points of X lie on an unique metric line. In order to formulate a necessary and sufficient condition for a metric space to have that property we need the following notion.

DEFINITION 4.19.21. The metric space (X, d) is said to have the two-triple property if, for each of its quadruples of pairwise points, linearity of any two triples implies the linearity of the remaining two triples.

Then we have

THEOREM 4.19.22. *A metric space (X, d) which is complete, convex, externally convex has unique (metric) lines through any two points iff it has the two-triple property.*

Proof. The condition is necessary. Suppose now that X is a metric space as in the theorem and we must show that pair of points (p, q), $p \neq q$ lies exactly on one metric line. Suppose that a, b, c d is a quadruple such that a, b, c and a, b, d are linear triples. Then the line $L(a, b)$ contains c and d and hence the triples a, c, d and b, c, d are linear triples.

Now we show that the condition is sufficient, i.e. that if X has the two-triple property then any two distinct points are on a unique (metric) line.

Suppose that a, b are two distinct points of X and that the assertion is false. Then a, b are on two distinct (metric) lines. Let us denote these by $L(a, b)$ and $L_1(a, b)$. Then we find a point c on $L(a, b)$ which is not in $L_1(a, b)$. We have to consider two cases. □

CASE 1. *c is on the metric segment $[a, b]$ determined on $L(a, b)$ by the points a, b.* Then obvious $a \neq c \neq b$. We choose c' on the segment $[a, b_1]$ on $L_1(a, b)$ (determined by a and b) and such that

$$d(a, c') = d(a, c), \qquad d(b, c') = d(b, c).$$

Then the quadruple (a, b, c, c') has the points pairwise distinct. Since acb and $ac'b$ hold, then by the two-triple property of X the other two-triples are linear triples. Thus (a, c, c') and (b, c, c') are linear triples. But $d(a, c') = d(a, c)$, $d(b, c') = d(b, c)$ hold and thus acc', $ac'c$, $bc'c$, bcc' are not possible. Thus we have cac' and cbc'.

If d' is in a, b_1 with the property that $ad'c'$ holds, since $ac'b$ holds we get that $d'c'b$ holds as well as $ad'b$. The quadruple (b, c, c', d') has the triples (b, c, c'), (b, c', d') linear. Then cbb' and $bc'd'$ hold. Thus (c', c, d') and (b, c, d') are linear triples. But $cc'd'$ and $c'cd$ are not possible since we have with cac' a contradiction with $ad'c'$. Thus $c'd'c$ holds. From $bd'c$ and cad' we get bad' which contradicts $ad'b$ and also abd' and $bc'd'$ imply $cc'd'$ which contradicts $c'd'c$. From these we conclude that Case 1 is impossible.

CASE 2. *c is not in $[a, b]$.* We may suppose without loss of generality that abc holds. If c' denotes the point of $L_1(a, b)$ such that abc' holds and $d(a, c) = d(a, c')$ then $d(b, c) = d(b, c')$. The quadruple (a, b, c, c') is of pairwise distinct points. From the two-triple property, (a, c, c') and (b, c, c') are linear triples. Then cac' and cbc' hold. Thus we get

$$d(a, b) + d(b, c) = d(a, c), d(a, b) + d(b, c') = d(a, c')$$

and since cac' and cbc' hold we get that $d(a, b) = 0$. But $a \neq b$ and this contradiction proves the assertion. □

We have also

THEOREM 4.19.23. *If X is a metric space which has a weak Euclidean four-point property then X has a two-triple property.*

Proof. Obvious.

COROLLARY 4.19.24. *If X is a complete, convex, externally convex metric space with a Euclidean four-point property then any two distinct points of X lie on a unique (metric) line.*

THEOREM 4.19.25. *Let X be a complete, convex, externally convex metric space having a Euclidean four-point property. Suppose that $a, b, c, b \neq c$ are points in X and that (a, b, c) is congruent with the triple (a', b', c') of R^2. Then this congruence may be extended to the congruence between the sets*

$$a \cup L(b, c), \quad a' \cup L(b', c'). \tag{+}$$

Proof. It is clear that the congruence between (b, c) and (b', c') has a unique extension to the lines determined by (b, c), (b', c') respectively. The assertion of the theorem is clear if a is on the line $L(b, c)$. Thus we suppose that a is not on this line. Let p, q be arbitrary elements of $a \cup L(b, c)$ and p', q' be the corresponding elements in $a' \cup L(b', c')$. We have to consider two cases.

CASE 1. *p, q are on $L(b, c)$.* Then $d(p, q) = d(p', q')$.

CASE 2. $p = a, p' = a'$. Then the quadruple (a, b, c, q) is congruent with the quadruple $(\bar{a}, \bar{b}, \bar{c}, \bar{q})$ in R^2. But $(\bar{a}, \bar{b}, \bar{c})$ is congruent to (a, b, c) which is congruent to (a', b', c'). Since $(b', c', \bar{q})$ is congruent to (b, c, q) which is congruent to (b', c', q') we must have $q' = \bar{q}$. Then (a, b, c, q) is congruent to (a', b', c', q') and $d(a, q) = d(a', q')$. This proves the theorem. □

In what follows we denote by $\mathcal{M}$ the class of all complete, convex, externally convex metric spaces. Then we have,

THEOREM 4.19.26. *Let $X \in \mathcal{M}$ and suppose that (a, b, c) are pairwise distinct elements of X and that this triple is congruent to the triple (a', b', c') of R^2. Then this congruence has a unique extension to a congruence between the sets*

$$L(a, b) \cup L(b, c), \qquad L(a', b') \cup L(b', c'). \tag{++}$$

Proof. Using the congruence between the lines we can define a one-to-one correspondence between the above sets, such that the pairs

$$(a, b), \qquad (a', b'); \qquad (b, c), \qquad (b', c')$$

are congruent.

Let p, q be two distinct points in $L(a, b) \cup L(b, c)$ and p', q' be the corresponding points in $L(a', b') \cup L(b', c')$. We may suppose without loss of generality that p is on $L(a, b)$ and q is on $L(b, c)$. From Theorem 4.19.25 we get that the quadruple (a, b, c, p) is congruent to the quadruple (a', b', c', p'). But the quadruple (a, b, c, q) is congruent to a quadruple $(\bar{a}, \bar{b}, \bar{p}, \bar{q})$ of R^2 and since $(\bar{a}, \bar{b}, \bar{q})$ is congruent to (a', b', q') we get that (a, b, p, q) is congruent to $(a', b', \bar{p}, q')$ where $\bar{p}$ is on $L(a', b')$. Since (a', b', p') is congruent to (a, b, p) which is congruent to $(a', b', \bar{p})$ we must have $p' = \bar{p}$. Then

(a, b, p, q) is congruent to (a', b', p', q') and $d(p, q) = d(p', q')$ which proves the assertion of the theorem.

THEOREM 4.19.27. *Let $X \in \mathcal{M}$ and suppose that p_0 is a point of X and L is a (metric) line of X. Then there exists a unique foot point p of p_0 on L.*

Proof. We recall that p on L is said to be a foot point for p_0 if for all x on L the following inequality holds:

$$d(p, p_0) \leqslant d(p_0, x).$$

We remark that if p_0 is on L then the foot point p is exactly p_0. Thus we may assume that p is not on L. We take two distinct points a, b on L and let p'_0, a', b' be points in the Euclidean plane R^2 such that the triple (p_0, a, b) is congruent to (p'_0, a', b'). Then $p_0 \cup L(a, b)$ is congruent to $p'_0 \cup L(a', b')$.

We consider q to be a point which is nearest (on the line $L(a', b')$) from the point p'_0. Then if $\bar{p}$ is the corresponding point on the line $L(a, b)$ we get

$$d(p_0, \bar{q}) = d(p'_0, q) \leqslant d(p'_0, x') = d(p_0, x)$$

for all x on $L(a, b)$.

This proves the existence and the uniqueness follows from the uniqueness in the Euclidean plane. □

THEOREM 4.19.28. *Let $X \in \mathcal{M}$ and p be a point in X and let L be a (metric) line in X. Then the set $p \cup L$ is congruent to the set $q \cup R$ (R is the set of all real numbers with the usual metric), q is a point in R^2 iff*

$$d(p, L) = d(q, R).$$

Proof. The condition is obviously necessary. Suppose now that it is satisfied. Then if $d(p, L) = d(q, R) = 0$ the assertion is clear. We suppose thus that $d(p, L) \neq 0$. Let us consider p_1 as the foot point of p on L and q_1 as the foot point of q on R. Then

$$d(p, p_1) = d(q, q_1).$$

If x is an arbitrary point of L we choose x_1 on R such that

$$d(x, p) = d(x_1, p_1).$$

Now for the triple (p, x, p_1) we find the points r, s in R^2 such that (p, x, p_1) is congruent to (r, x_1, q_1), which is congruent to (s, x_1, q_1). By Theorem 4.19.27 we can consider these congruences extended to $p \cup L(x, p_1)$, $r \cup L(x_1, q_1)$ and the last is congruent to $s \cup L(x_1, q_1)$. Then q_1 is the foot point for r and s and since

$$d(r, q_1) = d(p, p_1) = d(q_1, q)$$

we get that r or s is p_1. Then this implies that the set $p \cup L$ is congruent to $q \cup R$. The theorem is proved. □

We now define certain subsets of a space X in $\mathcal{M}$ which play a fundamental role in the metric characterization of complete inner product spaces.

DEFINITION 4.19.29. Let X be in $\mathcal{M}$. We say that a subset M of X is:

(1) a zero-dimensional subspace if M reduces to a single point;

(2) a k-dimensional subspace if
   (a) it is the closure in X of all points in X on a line with a point in a $(k-1)$-dimensional subspace, not belongint to it,
   (b) if $p_0, \ldots, p_k$ are not in a $(k-1)$-dimensional subspace then there is one and only one k-dimensional subspace of X in which they lie,
   (c) this k-dimensional subspace is linear in the sense that with any two points contains the line determined by the points,
   (d) the subspace is congruent with the Euclidean space R^k.

By an induction method we can easily prove the following extension of Theorem 4.19.28: *if X is in $\mathscr{M}$ then the set $p \cup L_n$, L_n is an n-dimensional subspace of X and is congruent with the set $q \cup R^n$ iff*

$$d(p, L_n) = d(q, R^n).$$

The following result gives the isometric structure of subspaces of an n-dimensional subspace of X.

THEOREM 4.19.30. *Let X be in $\mathscr{M}$. Then for any n, the k-dimensional subspaces of L_n are congruent to R^k.*

Proof. It is clear that it suffices to prove the assertion for $k+1$ with $k > 0$. Suppose that L_{k+1} is generated by L_k and the point p_0. We write this as $L_{k+1} = (p_0, L_k)$. We consider R^k to be the k-dimensional subspace of R^{k+1} and consider p_0' in R^{k+1} such that

$$d(p_0, L_k) = d(p_0', R^k).$$

Let x be an arbitrary point in R^{k+1} $(x \neq p_0)$ and p be a point in L_k such that (p_0, p, x) is a collinear triple. If p' is the point in R^k which corresponds to p through the congruence of the sets $p_0 \cup L_k$, $p_0' \cup R^k$ then $L_1(p_0, p) \cup L(p_0', p')$ are congruent. Then to x there corresponds an element x'. Now we show that $L_{n+1} \cup p_0$ may be mapped into a subset M of R^{k+1} which is obtained from R^{k+1} by deleting a k-dimensional subspace through p_0' and which is parallel to R^k.

Let x, x', y, y' be two pairs of corresponding points and consider the triple (p_0, q, y) is congruent to the triple (p_0', q', y') (and all the above triple are collinear) and $q \cup L_k$, $q' \cup R^k$ are congruent sets. But (p_0, p, q) and (p_0', p', q') are congruent triples and thus we get that $L_1(p_0, p) \cup L_1(p_0, q)$ is congruent to $L(p_0', p') \cup L(p_0', q')$. This implies that $d(x, y) = d(x', y')$ and the above mapping is a congruence.

But X and R^{k+1} are complete metric spaces and the above mapping, being an isometry, extends uniquely to the closures. This clearly gives the assertion of the theorem. □

From this we also get the following result.

COROLLARY 4.19.31. *For each $k = 1, 2, 3, \ldots, L_k$ is a linear space.*

The following result describes the geometry of $n+1$ points in a space which is a member of $\mathscr{M}$.

THEOREM 4.19.32. *Let $X \in \mathscr{M}$ and n be a positive integer. Then for any points*

$p_0, \ldots, p_n$ in X which are not in any $(n-1)$-dimensional subspace of X there exists a unique n-dimensional subspace of X containing them.

Proof. We proceed by induction. Clearly the assertion of the theorem is true for $n = 1$. Suppose that the assertion of the theorem is valid for all $k \leqslant n$ and we prove it for $n + 1$. We consider the points $p_0, \ldots, p_{n+1}$ and then we consider the set $(p_1, \ldots, p_{n+1} = N)$ which is not in any L_{k-1} for $k \leqslant n$. Indeed, in the contrary case the subspace generated by p_0 and the set N is an L_k subspace. This is a contradiction and thus the assertion is true. From the induction hypothesis we get that $p_1, \ldots, p_{n+1}$ is contained in a unique L_n-subspace of X. Then clearly p_0 and this subspace generate a subspace L_{k+1} which contains $(p_0, \ldots, p_{n+1})$.

Now this gives the existence. The uniqueness is proved as follows. Suppose that there are two $(n + 1)$-dimensional subspaces of X which contain the above points, say L_{n+1} and L^1_{n+1}. From the linearity we get that they contain a unique subspace L_n which contains the points $(p_1, \ldots, p_{n+1})$ (by the induction hypothesis). Then we get that L^1_{n+1} contains L_{n+1}. But L^1_{n+1} contains a point, say r, which is not in L_{n+1}. Then L^1_{n+1} contains a subspace generated by L_{n+1} and r, i.e. an $(n + 2)$-dimensional subspace. This is clearly impossible since

$$L^1_{n+1} \subseteq R^{n+1}, \quad L_{n+2} \subseteq R^{n+2} \text{ ('}\subseteq\text{' means 'congruence')}$$

and

$$L_{n+2} \subseteq L^1_{n+1}$$

gives that R^{n+1} is congruent to R^{n+2}. This is false and thus the uniqueness is proved. □

The above results give then the following basic result.

THEOREM 4.19.33. *If $X \in \mathcal{M}$ then X has a Euclidean n-property for all $n = 1, 2, 3, 4, \ldots$.*

Proof. Let $p_0, \ldots, p_n$ be n points of X. By the above theorem either they are contained in some k-dimensional subspace of X $(k = n - 1)$ or they are in a unique n-dimensional subspace of X. Thus it is clear that they are congruent with some points $(p'_0, \ldots, p'_{n+1})$ of R^n. □

REMARK 4.19.34. Theorem 4.19.33 was proved by W. A. Wilson in 1932 with the hypothesis that the metric space X has a Euclidean four-point property; the extension, using a weak Euclidean four-point property, was obtained by L. Blumenthal in 1955. Above we followed essentially Blumenthal.

COROLLARY 4.19.35. *Let X be a Banach space which is separable. We suppose further that X, with a metric induced by the norm, has a weak Euclidean four-point property. Then X is congruent with a Banach space whose norm is generated by an inner product.*

Interesting metric characterizations of Euclidean space were obtained by G. Berg in (1973). We mention them here; for a proof we refer to the original paper.

THEOREM 4.19.36 (Berg, 1973). *Let n be an integer in $[1, 3]$. The topological space*

X is the Euclidean space R^3 in the class of locally compact n-dimensional spaces iff:

(1) *X is a metric space such that for any two points in X they lie on a unique segment;*
(2) *if two segments have an endpoint and other point in common then one is contained in another;*
(3) *each segment in X can be extended, at either end, to a segment strictly containing it.*

THEOREM 4.19.37 (Berg, 1973). *Let $n > 3$. Then the topological space X is R^n iff it satisfies* (1), (2), (3) *of Theorem* 4.19.36 *in the class of locally compact n-dimensional spaces which contain a homeomorph of an n-ball.*

For other interesting and related results, as well as for references about these matters, we refer to Berg's paper.

As mentioned above, many properties were introduced in order to obtain more general criteria for a metric space to be congruent with a complete inner product space (and also with a Banach space, which although very interesting, forms no part of our principal object in this book). Some of these properties are mentioned above; later we present some new ones. Here we give some results about characterizations obtained using the properties mentioned above.

THEOREM 4.19.38 (Blumenthal, 1955). *In the class $\mathcal{M}$ the following assertions are equivalent:*

(1) *X has a weak Euclidean four-point property.*
(2) *X has a feeble Euclidean four-point property.*

We now prove another interesting property of spaces in the class $\mathcal{M}$.

THEOREM 4.19.39 (Freese, 1968). *For X in $\mathcal{M}$ the following assertions are equivalent:*

(1) *X has a weak Euclidean four-point property.*
(2) *X has an externally isosceles feeble Euclidean four-point property.*

Proof. Since the implication (1) → (2) is clear we prove only the implication (2) → (1). This will be given as a consequence of some lemmas which are interesting in their own right.

LEMMA 4.19.40. *Let $X \in \mathcal{M}$ and suppose that X has an externally isosceles feeble Euclidean four-point property. Then any two distinct points of X are the endpoints of a unique (metric) segment.*

Proof. Suppose that the assertion is false, and thus we have two distinct (metric) segments S_1 and S_2 having the endpoints p and q. Let r be a point in S_1 and not in S_2. When we traverse S_1 from r to p we meet a first point u of S_2, and traversing S_1 from r to q we meet a first point of S_2, say v. Then we consider the segments in S_1 and S_2 having only the endpoints u and v in common. Further, we consider the midpoints of these segments, say m_1 and m_2 and then we consider the segment $[m_1, m_2]$ which has m_1 and m_2 as endpoints. Then either we find x in $[m_1, m_2]$ such that at least one of the numbers $d(u, x)$, $d(v, x)$ is greater than $\frac{1}{2}d(u, v)$ or else

$$d(u, x) = d(v, x) = \tfrac{1}{2}d(v, u).$$

In the latter case if s is in $[m_1, m_2]$ with the property that

$$d(m_1, s) = 2d(s, m_2)$$

we get that the quadruple (u, m_1, s, m_2) is congruent with a quadruple in the plane, and thus $d(u, s) < d(u, m_2)$, which is a contradiction. Thus we find x with the property

$$d(u, x) > \tfrac{1}{2}d(u, v)$$

or

$$d(v, x) > \tfrac{1}{2}d(u, v).$$

Then we find t in $[m_1, m_2]$ such that $d(u, t)$ is maximal and

$$d(u, t) > \tfrac{1}{2}d(u, v).$$

But $d(u, t) > d(u, m_1) = d(u, m_2)$ and thus we find the firsts points of $[m_1, m_2]$, say w_1, w_2, such that $d(u, w_1) = d(u, w_2) = d(u, m_1)$. The points are to be found by traversing $[m_1, m_2]$ in each direction from t. From the continuity property we find y, z in $[m_1, m_2]$ such that

$$d(u, y) = d(u, z) > d(u, m_1)$$
$$d(y, z) = 2d(z, w_1).$$

Then the quadruple (w_1, y, z, u) is congruent to a quadruple in the plane, and this implies that $d(u, w_1) > d(u, y)$. This is a contradiction and the lemma is proved. □

LEMMA 4.19.41. *Let $X \in \mathcal{M}$ satisfying an externally isosceles feeble Euclidean four-point property. Then the metric segments have unique prolongations.*

Proof. It is clear that we must prove only the uniqueness. Suppose that we have a segment with the endpoints p, q, $[p, q]$ and that this has two distinct prolongations. Then by the continuity property we find r, s on these prolongations such that for some x, q^+ in p, q the following relations hold:

(1) $r \neq s$,
(2) $d(s, q^+) = d(s, r)$,
(3) $d(q^+, r) = 2d(q^+, x)$,
(4) q^+sr and xqr hold.

The quadruple (x, q^+, r, s) is congruent to a quadruple in the Euclidean plane (x', q', r', s') with s in the prolongation of $[x, q^+]$. This gives that (x', q', s') is a collinear triple, which is a contradiction. This proves the lemma. □

COROLLARY 4.19.42. *Let X be as in Lemma* 4.19.41. *Then each two points of X are contained in a unique line of X.*

LEMMA 4.19.43. *Let X be as in Lemma* 4.19.41. *Then for any (metric) line L of X and any point p of X, $p \notin L$, there exists exactly one foot point p_1 of p on L.*

Proof. Since the existence is easy to prove we give the proof for the uniqueness. Suppose that in fact we have two points p_1 and p_2 which are foot points for p. If all points between p_1 and p_2 are foot points (of p) then if q is any of them the quadruple (p, p_1, p_2, q) is congruent to a quadruple in the plane if $d(p_1, q) = 2d(p_2, q)$. This gives that $d(p, p_1) > d(p, p_2)$ which contradicts the fact that p_i, $i = 1, 2$ are foot points of p.

Now if not all points between p_1 and p_2 are foot points of p let r be such that:

(1) r is in p_1, p_2,
(2) $d(p, r) > d(p, p_1)$,
(3) $d(p, r) \geqslant d(p, x)$ for all x in $[p_1, p_2]$.

Suppose now that x is in $[p_1, r]$ and let $y = y(x)$ be the point in $[x, p_2]$ such that $d(x, y) = 2d(y, p_2)$. Then we find x^+, y^+ such that

(1) $d(p, x^+) = d(p, y^+)$,
(2) $d(x^+, y^+) = 2d(y^+, p_2)$.

Since X has an externally isosceles feeble Euclidean four-point property we get that $d(p, p_2) > d(p, y^+)$ which contradicts the definition of p_2. This gives the assertion of the lemma. □

LEMMA 4.19.44. *Let X be as in Lemma* 4.19.41 *and let L be a (metric) line in X with $p \notin L$, p in X. If $\bar{p}$ is the foot point of p on L and q,r are on L and such that $rq\bar{p}$ holds then $d(p, q) < d(p, r)$.*

Proof. Suppose that the assertion is false. Then either for some $u \in [q, r]$, $d(p, u) - d(q, r) \neq 0$ or $d(p, u) = d(q, r)$.

If $d(q, r) \neq d(p, u)$ we find s, r^+, r' in $[\bar{p}, r]$ such that $d(p, r) = d(p, r') < d(p, s)$. If $[r, s]$ and $[s, \bar{p}]$ are traversed from s to the other endpoint we encounter the first points such that the distance from p is $d(p, r')$, with s in $[r', r^+]$ and $d(p, s) \geqslant d(p, r)$, r in $[r', r^+]$. Then we find x^+, y^+ in $[r', r^+]$ such that $d(x^+, y^+) = 2d(y^+, r^+)$ and $d(p, x^+) = d(\bar{p}, y^+)$. Then the quadruple $(\bar{p}, x^+, y^+, r^+)$ is congruent to a quadruple in the plane and this implies that $d(\bar{p}, r^+) > d(p, y^+)$, which contradicts the definition of r^+.

Suppose now that $d(p, u) = d(p, r)$ holds for all u in $[q, r]$. If z^+ is the point in q, r with $d(q, z^+) = 2d(z^+, r)$ then we have that the quadruple (p, q, r, z^+) is congruent to a quadruple in the plane, which gives that $d(p, r) > d(p, z^+)$. This is a contradiction and the lemma is proved. □

LEMMA 4.19.45. *Let X be as in Lemma* 4.19.41. *Let L be a (metric) line in X and p be a point of X not on L. Suppose that q, r are on L and that $d(p,q) = d(p,r)$. Then the midpoint of q, r is the foot point of p on L.*

Proof. By the result in the above lemma, if p_1 is the foot point of p on L then it lies between q and r. If m denotes the midpoint of q and r and $m \neq p_1$ we may assume that p_1 lies between q and m. We find an integer n and $\varepsilon > 0$ such that

$$d(m, q) - d(m, p_1) < 2^n\varepsilon - \varepsilon,$$
$$2^n\varepsilon < d(m, q) + d(m, p_1) - \varepsilon.$$

Since we find s, t such that p lies between s and t with

$$d(p,s) = d(p,t), \qquad d(s,t) = 2\varepsilon$$

the midpoint of s and t has a distance from p_1 less than ε. Since $d(p, s) = d(p, t)$, applying n-times the externally isosceles feeble Euclidean four-point property, we get that there exist u and v on L such that $d(p, u(=d(p, v))$ with q between p_1 and u, v between p_1 and r. This contradicts the assertion of Lemma 4.19.44, i.e. $d(p, q) = d(p, r)$. This proves the lemma. □

THEOREM 4.19.46. *Let X be as in Lemma* 4.13.41. *Then X has an isosceles feeble Euclidean four-point property.*

Proof. We consider the (metric) line $L(q, r)$, $p \notin L(q, r) = L$, and take p_1 as its foot point on L. Let (p', q', r') be a triple of point in the plane which is congruent to the triple (p, q, r) and consider p'_1 as the foot point of p' on $L(q', r')$. If m is the midpoint of the points q and r and if we consider the segments $[m, r_{n-1}]$, $[m, q_{n-1}]$, $q_0 = q$, $r_0 = r$ and the midpoints r_n, q_n of these segments, then from the continuity of the distance function (and the monotonicity of $x \to d(p, x)$) we get that $d(p, r_n) = d(p, q_n)$ holds for all n. Applying many times the externally isosceles feeble Euclidean four-point property we get that $d(p',r'_n) = d(p,r_n) = d(p,q_n)$ holds and r'_n is the point in the plane corresponding to r_n under the correspondence between $L(q, r)$ and $L(q', r')$. Since $\lim r_n = m$ we get that $\lim r'_n = m'$ and thus $\lim d(p, r_n) = d(p, m)$, $\lim d(p', r'_n) = d(p', m')$. Then from $d(p, r_n) = d(p', r'_n)$ we get that $d(p, p_1) = d(p', p'_1)$, which proves the theorem. □

THEOREM 4.19.47. *Let X be as in Lemma* 4.19.41. *Then X has an isosceles weak Euclidean four-point property.*

Proof. Indeed, if (p, q, r) is a triple of points in X such that $d(p, q) = d(q, r)$ and s is a point on $L(q, r)$, then, by the above theorem, we may assume that $s \neq p_1$, the foot point of p on $L = L(q, r)$. If (p', q', r') is the triple in R^2 congruent to (p, q, r) we get that $d(p, p_1) = d(p', p'_1)$, p'_1 being the foot point of p' on $L(q', r')$. If we consider the correspondence between L and $L(q',r')$ then to s corresponds s', and if we take t on $L(q, r)$ such that $d(p, s) = d(p, t)$ then the quadruple (p, p_1, s, t) is congruent to the quadruple $(\bar{p}, \bar{p}_1, \bar{s}, \bar{t})$ in R^2, and thus

$$d(\bar{p}'_1, \bar{s}) = d(p'_1, s') \qquad d(\bar{p}, \bar{p}_1) = d(p', p'_1)$$

which, combined with Pythagoras' theorem (in R^2), gives that $d(p,s) = d(\bar{p},\bar{s}) = d(p',s')$. Then the quadruple (p,q,r,s) is congruent to the quadruple (p',q',r',s') of R^2 and the theorem is proved. □

Now the proof of the implication (2) → (1) is as follows. Suppose that (p_1, p_2, p_3, p_4) is a quadruple in X and that (p_2, p_3, p_4) lie on a line L and p_1 is not on L. We may assume that $p_2 \neq p$, the foot point of p_1 on L. Then we find r on L such that $d(p_1, r) = d(p_1, p_2)$ and by the above theorem (p_1, p_2, r, p_4) is congruent to a quadruple of R^2 and the linearity of (p_2, r, p_3, p_4) gives the fact that (p_1, p_2, r, p_3, p_4) is congruent to a subset of R^2. This proves the theorem. □

Now this obviously implies the following result.

THEOREM 4.19.48. *If X is as in Lemma 4.19.41 then H has an inner product structure.*

Proof. This follows from Blumenthal's theorem given above. □

REMARK 4.19.49. In proving Freese's theorem we followed the exposition in Freese's paper quoted above.

REMARK 4.19.50. Suppose now that X is a Banach space and has the property that the space X with the metric induced by the norm has an externally isosceles feeble four-point property. Then this is equivalent with the assertion: *if x, y are elements in X such that $\|x-y\| = \|3x-y\|$ then we have the equality* $\|y\|^2 = \|x-y\|^2 + 3\|x\|^2$. We shall call this the 'Freese metric condition'; then clearly we have the following result.

THEOREM 4.19.51 (Freese, 1968). *A Banach space X has its norm generated by an inner product iff X satisfies the Freese metric condition.*

We now present some new four-point properties which are important in obtaining sufficient conditions for a space in the class $\mathcal{M}$ to be with an inner product structure.

DEFINITION 4.19.52 (Valentine and Wayment, 1972). A metric space X is said to have a Euclidean externally symmetric isosceles queasy four-point property if for each pair of distinct points p and r there exists a number $\lambda = \lambda(p, r)$ in $[1, k]$, k fixed (but arbitrary), such that if q is a point with the property that r lies between p and q and $d(q, r) = \lambda d(p, r)$ then for each point s in X with $d(p, s) = d(r, s)$ the quadruple (p, q, r, s) is congruent to a quadruple in R^2.

We remark that the above theorem of R. W. Freese refers to the case when $k = \frac{3}{2}$ and, as stated by J. E. Valentine and S. G. Wayment, his proof can be easily adapted to obtain the following results:

THEOREM 4.19.53 (Valentine and Wayment, 1972). *Suppose that X in $\mathcal{M}$ has a Euclidean externally symmetric isosceles queasy four-point property (for some k). Then X has an inner product structure.*

To formulate the other result we need the notion given in

DEFINITION 4.19.54 (Valentine and Wayment, 1972). A metric space X is said to have a Euclidean k-externally isosceles feeble four-point property iff $k > 0$ and (p_1, p_2, p_3, p_4) is a quadruple of points in X, p_2 between p_1 and $p_3, d(p_4, p_1) = d(p_2, p_4)$, $kd(p_1, p_2) = d(p_1, p_3)$ is congruent to a quadruple of the Euclidean plane (R^2).

A quadruple as in this definition is denoted by J. E. Valentine and S. G. Wayment to be of type $C_{4,k}$. Thus the space X is said to have a Euclidean k-isosceles feeble four-point property if any quadruple of $C_{4,k}$ type in X is congruent to a quadruple of R^2. Then we have.

THEOREM 4.19.55 (Valentine and Wayment, 1972). *A space X in $\mathcal{M}$ has an inner product structure iff it has a Euclidean k-externally isosceles feeble four-point property.*

For other related characterizations of inner product structures we refer to the paper of Valentine and Wayment (1972). Another interesting characterization of inner product structures was obtained by E. Z. Andalafte, J. E. Valentine and S. G. Wayment using the so-called 'median bisector property (MBP) and the 'median intersection property' (MIP). These are defined as follows.

DEFINITION 4.19.56 (Andalafte, *et al.*, 1979). A metric space X is said to have a median intersection property iff $(p, q, r) \subset X$, not collinear, and if (p', q', r') are the midpoints of (q, r), (r, p) and (p, q) respectively, then any three segments $[p, p']$, $[q, q']$ and $[r, r']$ contain a common point which satisfies the relations

$$d(p, t)/d(p', t) = d(q, t)/d(q', t) = d(r, t)/d(r', t) = 2,$$

where t is the common point.

DEFINITION 4.19.57 (Andalafte *et al.*, 1979). A metric space X is said to have a median bisector property (MBP) iff (p, q, r) is a nonlinear triple and p', q', r' are as in Definition 4.13.56, then any midpoint s of p, p' is a midpoint of q', r' and, conversely, any midpoint of q' and r' is a midpoint of p and p'.

DEFINITION 4.19.58 (Lorch, 1948). A metric space X is said to have an I_2 property iff (p, q, r) is a nonlinear triple, in which q', r' are the midpoints of p and r, and p and q respectively, then

$$d(p, q) = d(r, p)$$

iff

$$d(q, q') = d(r, r').$$

Then we have

THEOREM 4.19.59 (Andalafte *et al.*, 1979). *Let X be in $\mathcal{M}$. Then X is a real inner product space iff X satisfies either the MIP or MBP and I_2 holds.*

Another characterization of the inner product spaces in the class $\mathcal{M}$ was obtained by Valentine and Wayment (1975), which consists in localizing the properties considered for the characterization of inner product spaces in the class $\mathcal{M}$, in the sense that it is required that the corresponding property holds in a spherical neighborhood of each point of the space (of course, the radius of the sphere depends upon the point).

DEFINITION 4.19.60. Let X be a metric space and P be a property which may or may not be satisfied by the points of X. We say that P holds locally if for each point p of X there exists $\varepsilon_p > 0$ such that in the subset $S_p(\varepsilon_p)$ the property holds. For example, as property P we consider the Euclidean, weak, feeble or queasy four-point property.

Another way to improve certain characterizations of inner product spaces was considered by Loveland and Valentine (1978) by defining two new classes of four-point properties:

(1) the Banach weak, Banach feeble, Banach queasy, four-point property,

(2) Banach, Banach weak, Banach weak queasy, Banach feeble, four-point property at p (p is a point in X).

The precise formulation is as follows.

DEFINITION 4.19. 61 (Loveland and Valentine, 1979). A metric space X is said to have a Banach four-point property if each quadruple (p_1, p_2, p_3, p_4) of distinct points containing a linear triple (p_2, p_3, p_4) such that p_3 is between p_2 and p_4, and if $(m_i)_1^4$ are the midpoints of (p_1, p_2), (p_2, p_3), (p_3, p_4) and (p_4, p_1) respectively, $d(m_1, m_2) = d(m_3, m_4)$, $d(m_1, m_4) = d(m_3, m_2)$ hold.

DEFINITION 4.19.62 (Loveland and Valentine, 1979). A metric space X is said to have a Banach weak feeble four-point property if, for each quadruple (p_1, p_2, p_3, p_4) of distinct points of X containing a triple (p_2, p_3, p_4) such that p_3 is the midpoint of p_2 and p_4, and if $(m_i)_1^4$ are as in the above definition, $d(m_1, m_2) = d(m_3, m_4)$ and $d(m_1, m_4) = d(m_2, m_3)$ hold.

DEFINITION 4.19.63 (Loveland and Valentine, 1979). A metric space X is said to have a Banach queasy four-point property if, for each pair of points (p_2, p_4) of X, there exists a point p_3 between p_2 and p_4 such that for any point p_1 of X, and if $(m_i)_1^4$ are as above, $d(m_1, m_2) = d(m_3, m_4)$, $d(m_1, m_4) = d(m_2, m_3)$.

Then we have clearly the following assertion.

PROPOSITION 4.19.64. *For any metric space X the following implications hold: X has a Banach weak four-point property $\to X$ has a Banach feeble four-point property $\to X$ has a queasy four-point property.*

The above properties can be formulated as being properties which hold at some point of X.

DEFINITION 4.19.65 (Loveland and Valentine, 1979). A metric space X is said to have a Banach four-point property at p_1 if, for any three distinct points (p_2, p_3, p_4) in $X - \{p_1\}$, and if $(m_i)_1^4$ are as above, $d(m_1, m_2) = d(m_3, m_4)$, $d(m_1, m_4) = d(m_2, m_3)$.

The following example shows that none of the four-point properties which hold at some point p of X is sufficient to assure the inner product structure of a metric space in the class $\mathcal{M}$.

EXAMPLE 4.19.66 (Loveland and Valentine, 1978). We consider a space X which consists of the union of three metric rays $R(p_1, p_2)$, $R(p_1, p_3)$, $R(p_1, p_4)$ whose pairwise intersections are each (p_1) and whose pairwise unions are metric lines. This space is called 'a convex, externally convex triped'. If (p_2, p_3, p_4) are such that $d(p_1, p_2) = d(p_1, p_3) = d(p_1, p_4)$ and $(m_i)_1^4$ are defined as above, then $m_2 = m_3 = p_1$ and $d(m_2, m_3) = 0 \neq d(m_1, m_4)$. This gives that X is a space with a Banach weak, feeble and queasy four-point property not holding at p_1.

In order to remove this deficiency of the above properties L. D. Loveland and J. E. Valentine have considered a property known as 'the Young postulate at p'. In order to define this we give the following definitions.

DEFINITION 4.19.67 (Andalafte and Blumenthal, 1964). A metric space X is said to satisfy the Young postulate if, for every three points (p_1, p_2, p_3) of X, if (m_1, m_2) are the midpoints of (p_1, p_2) and (p_2, p_2) respectively, then $2d(m_1, m_2) = d(p_2, p_3)$.

DEFINITION 4.19.68 (Loveland and Valentine, 1978). We say that a metric space X satisfies the Young postulate at p if, for each pair of points (q, r) of X, if m_1, m_2 are the midpoints of p, q and q, r respectively, then

$$2d(m_1, m_2) = d(q, r).$$

REMARK 4.19.69 (Loveland and Valentine, 1978). The convex externally convex triped considered in Example 4.19.65 satisfies the Young postulate at p_1.

Then we have the following characterizations of an inner product structure in metric spaces.

THEOREM 4.19.70 (Loveland and Valentine, 1978). *Suppose that X is in $\mathcal{M}$. If X has one of the following properties – local Euclidean weak, feeble or queasy four-point property at p – and if X satisfies the Young postulate at p, then X has an inner product structure.*

THEOREM 4.19.71 (Loveland and Valentine, 1978). *Let X be in $\mathcal{M}$ and suppose that X has a Euclidean feeble four-point property and, at some point p, satisfies a local Euclidean feeble four-point property. Then X has an inner product structure.*

THEOREM 4.19.72 (Loveland and Valentine, 1978). *Suppose that X is in $\mathcal{M}$ and that, at some point p, X satisfies a Euclidean feeble four-point property and at the same point a local Euclidean feeble four-point property. Then X has an inner product structure.*

REMARK 4.19.73 (Loveland and Valentine, 1978). There exists an example of a complete metric space (the closed Euclidean half plane P and the interior C of a Euclidean half-circle including the open diameter of C and the boundary of P identified under an isometry; this example is given in Freese (1961) which has, at each point of P, a local Euclidean feeble four-point property (and clearly does not have an inner product structure).

The importance of local properties is shown by the following result.

THEOREM 4.19.74 (Valentine and Wayment, 1975). *Let X be in $\mathcal{M}$ and suppose that X has either a local Euclidean weak, feeble or queasy four-point property. Then X has an inner product structure.*

In other words, the above result shows that in the class $\mathcal{M}$ all the above local properties are equivalent with the existence of an inner product structure on the space and thus are equivalent properties.

For a proof of the above results, as well as other interesting facts about the properties considered above, we refer to the quoted papers.

The four-point properties considered above (or some of them) require that if the triple (p, q, r) is congruent to a triple (p', q', r') of R^2 then we can extend this congruence to the quadruples (p, q, r, s) and (p', q', r', s'), where s (and s') are suitable points (related to some extent to the triple). Valentine and Andalafte (1973) have defined the so-called 'intrinsic four-point properties' using this idea, which may be of interest when studying the properties of a metric space using the fact that a certain congruence between triples may be extended to certain quadruples. The precise definitions are as follows.

DEFINITION 4.19.75 (Andalafte and Valentine, 1973). A metric space X is said to have an intrinsic weak four-point property if for each triple (p_1, p_2, p_3) of points of X which is congruent to the triple (p'_1, p'_2, p'_3) of X we can extend this congruence to a congruence between $p_1 \cup L(p_1, p_2)$ and $p'_1 \cup L(p'_1, p'_2)$.

DEFINITION 4.19.76 (Andalafte and Valentine, 1973). A metric space X is said to have an intrinsic feeble four-point property if for each triple (p_1, p_2, p_3) of X which is congruent to a triple (p'_1, p'_2, p'_3) of points of X we can extend this to a congruence between (p_1, p_2, p_3, m) and (p'_1, p'_2, p'_3, m'), where $m(m')$ is the midpoint of $(p_2, p_3) \cup ((p'_2, p'_3))$.

DEFINITION 4.19.77 (Andalafte and Valentine, 1973). A metric space X is said to have an intrinsic queasy four-point property if for each pair of points (p_2, p_4) of X there exists p_3 between p_2, p_4 such that for any point p_1 of X with the property that the triple (p_1, p_2, p_4) is congruent to (p'_1, p'_2, p'_4) of X there is p'_3 in X such that $(p_i)_1^4$ is congruent to $(p'_i)_1^4$.

It is to be noted that complete inner product spaces and hyperbolic spaces have a metric feeble and intrinsic queasy four-point property.

The following results show that, to some extent, these are only some of the spaces that have such properties. Another space with a metric feeble and intrinsic queasy four-point property is the space given in Example 4.19.67 (the convex, externally convex tripod).

THEOREM 4.19.78 (Andalafte and Valentine, 1973). *Let X be a finitely compact (a metric space is said to be finitely compact if any bounded set in the space is relatively compact) convex, externally convex metric space having a two-triple property. Then H has either an inner product structure or X has a hyperbolic structure provided the intrinsic queasy four-point property holds for X.*

From this result we have

COROLLARY 4.19.79 (Andalafte and Valentine, 1973). *If X is in $\mathcal{M}$ and has a two-triple property then it has either an inner product structure or a hyperbolic structure provided the intrinsic weak (intrinsic feeble) four-point property holds in X.*

For a proof of these results we refer to the paper by Andalafte and Valentine (1973).

COROLLARY 4.19.80 (Lorch, 1948). *Let X be a real Banach space. Then the norm of X is generated by an inner product (i.e. X has an inner product structure) iff whenever three sides of a triangle are equal respectively to three sides of an another triangle then their corresponding medians are equal.*

Proof. The property assumed in the corollary is an intrinsic feeble four-point property and any complete real inner product space has this property. In order to prove that if the property holds then the Banach space has an inner product structure, it suffices to consider only two-dimensional subspaces which are obviously finitely compact spaces. Let X_1 be such a space. Then we remark that X_1 has the property that each two distinct points lie on a unique line of X_1. Indeed, suppose that p, q are two distinct points in X_1. First we show that they are the endpoints of a unique segment in X_1. If we suppose that this assertion is false, we thus have (at least) two segments with endpoints p and q, say p, q and p, q_1. Let m, m_1 be the midpoints of these segments and consider the triples (p, m, q) and (p, m_1, q) which are clearly congruent. Then we may extend this to the congruence between the quadruples (p, m, q, m_1) and (p, m, q, m). This gives that p, q lie on a unique metric segment. Now if L, L_1 are two (distinct) metric lines on which p, q lie we consider the segments with endpoints p, q on these lines. They are identical by the above remark. Suppose now that the line L_1 has the property that some r on L is not on L_1 and some s on L_1 is not on L. We may assume without loss of generality that L_1 is the algebraic line through p and q. We consider now the algebraic segment with the endpoints p and r, p, r_a. Then $q \notin p, r_a$ and this contradicts the above established property that p, r are the endpoints of a unique segment. Thus X_1 has a two-triple property. Then X_1 has, by the above result, either an inner product structure or a hyperbolic structure. But it is clear that a Banach space does not have a hyperbolic structure. Thus X_1 has its norm generated by an inner product and since it is arbitrary in X, by the Jordan–von Neumann theorem X has an inner product structure. □

Other interesting four-point properties were introduced by Valentine (1980) in order to improve certain characterizations of spaces having an inner product structure using the four-point properties at some points.

DEFINITION 4.19.81 (Valentine, 1980). A metric space X is said to have a Banach–Euclidean four-point property at p_1 provided for each triple (p_2, p_3, p_4) if m_1, m_2, m_3, m_4 are the midpoints of (p_1, p_2), (p_2, p_3), (p_3, p_4) and (p_4, p_1), respectively. Then the following relations hold:

(1) $d(m_1, m_2) = d(m_3, m_4)$, $d(m_1, m_4) = d(m_2, m_4)$.
(2) The quadruple $(m_i)_1^4$ is congruent to a quadruple in R^2.

DEFINITION 4.19.82 (Valentine, 1980). A metric space X is said to have a Banach–Euclidean feeble four-point property if $(p_i)_1^3$ is a triple in X and p_4 is the midpoint of (p_2, p_3), and if $(m_i)_1^4$ are the midpoints of (p_1, p_3), (p_3, p_4), (p_3, p_2) and (p_2, p_1) respectively, then:

(1) $d(m_1, m_2) = d(m_3, m_4)$, $d(m_1, m_4) = d(m_2, m_4)$,
(2) the quadruple $(m_i)_1^4$ is congruent to a quadruple in R^2.

Then we have the following result.

THEOREM 4.19.83 (Valentine, 1980). *Let X be in $\mathcal{M}$. Suppose that X has either*

(1) *a Banach–Euclidean four-point property at some point p_1 of X, or*
(2) *a Banach–Euclidean feeble four-point property. Then X has an inner product structure.*

For the proof see Valentine (1980).

REMARK 4.19.84. It may be of interest to know if the properties defined above by J. E. Valentine may have a useful local version or an intrinsic version in the sense we used in the definitions given above.

As we know, one of the most interesting and useful characterizations of complete inner product structures in the class of Banach spaces was given by F. A. Ficken (see, especially Sections 4.4 and 4.10). There are a number of interesting formulations of Ficken's condition in metric spaces. We present some of them below.

DEFINITION 4.19.85 (Valentine and Wayment, 1972). A metric space X is said to satisfy a strong Ficken postulate if for any points p, p' in X and L, L' lines in X, p_1, p'_1 being the foot points of p, p' on L, L', respectively, then if $d(p, p_1) = d(p', p'_1)$ we have that for any points q, r on $L(p, p_1)$ and any pair (q', r') on $L(p', p'_1)$ respectively and for which $d(p, q) = d(p', q')$, $d(p_1, r) = d(p', r')$ then $d(q, r) = d(q', r')$.

DEFINITION 4.19.86. A metric space X is said to have a congruence extension property (CEP) if for any two congruent triples $(p_i)_1^3$, $(p'_i)_1^3$ there exists a congruence between

$$p_1 \cup L(p_2, p_3) \quad \text{and} \quad p'_1 \cup L(p'_2, p'_3).$$

Then we have the following result of J. E. Valentine and S. G. Wayment.

THEOREM 4.19.87 (Valentine and Wayment, 1972). *If X is a finitely compact Banach space then X has an inner product structure iff either:*

(1) *the strong Ficken postulate holds in X, or*
(2) *the CEP holds.*

Other very interesting four-point properties were defined and used to obtain characterizations of inner product structures by Freese *et al.* (1982) and Freese and Andalafte (1982). The basic idea in defining these new four-point properties is to take a quadruple containing a linear triple and some three distances from the six distances among the four points that are required to be equal. For the very interesting results about these four-point properties we refer to the above papers.

We now mention another interesting and important approach to the problem of characterizing inner product structure proposed by Blumenthal (1950, 1953), in terms of the primitive notion introduced in the following definition.

DEFINITION 4.19.88 (Blumenthal, 1950, 1953). Let X be a (nonempty) set. We say that on X the function $\langle\,,\rangle$ defined on $X \times X$ with real values is a quasi-inner product iff the following conditions are satisfied:

(1) For all x, y in X, $\langle x, y\rangle = \langle y, x\rangle$.
(2) For each x in X, $\langle x, x\rangle \geqslant 0$.
(3) If $\langle x, y\rangle = \langle x, x\rangle = \langle y, y\rangle$ then $x = y$.

For any points $(x_1, \ldots, x_n)$ in X we consider the determinant

$$G(x_1, \ldots, x_n) = \det(|\langle x_i, x_j\rangle|)_{i,j=1,\ldots,n}.$$

The properties we require for these determinants are as follows:

(S_1) $G(x_1, x_2) \geqslant 0$,

(S_2) $G(x_1, x_2) = 0 \to G(x_1, x_2, x_3) \geqslant 0$,

(S_3) $G(x_1, x_2, x_3) = 0 \to G(x_1, x_2, x_3, x_4) \geqslant 0$.

Also the following axioms are required for some special elements of X.

(E_1) *There exists an element denoted by* 0 *in* X *such that* $\langle 0, x\rangle = 0$ *for all* x *in* X.

(E_2) *For* x *in* X *and* r *in* R *there exists* y *in* X *such that* $G(x, y) = 0$ *and* $\langle x, y\rangle = r\langle x, x\rangle$.

(E_2) *For* x, z *in* X *with* $G(x, z) = 0$ *there exists* y *in* X *such that* $B(x, y, z) = G(x, y, z) = 0$, $G(x, y) = G(y, z)$, *where* $B(x_1, \ldots, x_n)$ *is the determinant obtained from* $G(x_1, \ldots, x_n)$ *by bordering it with a row and a column of* 1's *and whose intersection is the set with the element* (0), $0 \in R$.

Then we have the following result of L. M. Blumenthal.

THEOREM 4.19.89 (Blumenthal, 1950, 1953). *A set* X *which has a quasi-inner product satisfying the axioms* (S_1)–(S_3) *and* (E_1)–(E_3) *has an inner product structure.*

4.20. Other Metric Characterizations of Inner Product Spaces

As we know, if the norm of a Banach space X is to be generated by an inner product (i.e. to be an inner product space) it is necessary and sufficient that all of its two-dimensional subspaces have their norms generated by an inner product.

Thus if we are interested in obtaining necessary and sufficient conditions for a Banach space to have its norm generated by an inner product it suffices to consider only spaces of low dimension. Among the very interesting characterizations of Euclidean (and related classes) of spaces are those using motions, angular measures, or certain properties of spheres, etc. In this section we present some of these characterizations without proof; the reader interested in these characterizations may want to consult the excellent book of H. Busemann for proofs as well as for other related results. For notions and terms not defined here we refer to Busemann (1955).

Let (X, d) be a metric space. A motion on X is any function $f: X \to X$ which is a surjective isometry (i.e. f is an isometry and $f(X) = X$).

DEFINITION 4.20.1 (Busemann, 1955). The complete metric (X, d) is said to be a G-space if the following assertions hold:

(1) In X the Bolzano theorem holds (i.e. if M is a bounded subset in X then it has at least one accumulation point).

(2) For any points x, z in X, $x \neq z$, there exists y in X such that

$$d(x, y) + d(y, z) = d(x, z)$$

(i.e. xyz holds).

(3) For any p in X there exists $r_p > 0$ such that for any points x, y in $S_{r_p}(p) = \{u, d(p, u) < r_p\}$ there exists z such that xyz holds.

(4) If xyz_1 and xyz_2 hold and $d(y, z_1) = d(y, z_2)$ then $z_1 = z_2$.

Then we have.

THEOREM 4.20.2 (Busemann, 1955). *A straight Riemann space having convex spheres and satisfying the parallel axiom is Euclidean.*

THEOREM 4.20.3 (Busemann, 1955). *If X is a Minkowski G-space,* $\dim X = 2$, *and with the property that the perpendicularity between lines is symmetric, then the space is Euclidean.*

THEOREM 4.20.4 (Beltrami). *If X is a Desarguesian and Riemannian G-space then it is either Euclidean, hyperbolic or elliptical.*

THEOREM 4.20.5 (Busemann, 1955). *Let X be a G-space and suppose that for each u, v in X, $B(u, v) = (x, d(u, x) = d(v, x))$ contains with any points m, n at least one segment with the endpoints m and n respectively. Then X is either Euclidean, hyperbolic or spherical of* $\dim \geqslant 1$.

THEOREM 4.20.6 (Busemann, 1955). *Suppose that X is a straight space with the property that any two spheres about the centers p and q permits rotation. Then the space X is either Euclidean or hyperbolic.*

The following result is a local version of the famous problem of Helmholtz–Lie, the origin of which lies in Riemann's lecture (1854). Very profound contributions to this problem were made by Weyl (1923), Kolmogorov (1930) and Tits (1952).

THEOREM 4.20.7. *Suppose that X is a G-space with the property that for every p in X there exists a neighborhood of p of the form $S_{r_p}(p)$ such that for any points a, a', b, b' in $S_{r_p}(p)$ and with $d(a, p) = d(p, a')$, $d(b, p) = d(p, b')$, $d(a, b) = d(a', b')$ and an isometry of $S_{r_p}(p)$ onto itself with p fixed and which takes a into a' and b into b', then the space is either Euclidean, hyperbolic or spherical.*

Another result, which is mainly a solution of the Helmholtz–Lie problem, is as follows.

THEOREM 4.20.8. *Suppose that X is a G-space with the property that for any points p, p', q, r with $d(p, q) = d(p, p')$, $d(p, r) = d(p', r)$ there exists a motion which leave q, r fixed and takes p into p'. Then X is either Euclidean, hyperbolic or spherical.*

DEFINITION 4.20.9. We say that on the G-space X, a notion of triangle area is defined, and there exists a function with positive values on the space $X \times X \times X$ which has the following properties:

(1) $f(x, y, z)$ is symmetric in the variables x, y, z.
(2) $f(x, y, z) = 0$ iff x, y, z lie on a segment.
(3) If x, y, z, t are in X and ytz holds, then $f(x, y, t) + f(x, t, z) = f(x, y, z)$.

THEOREM 4.20.10. *Suppose that X is a G-space with the property that for each p in X there exists a neighborhood V_p of p of the form $S_{r_p}(p)$, $r_p > 0$, in which a notion of triangle area is defined and such that it has the same value for isometric triples. Then the space X is either Euclidean, hyperbolic or spherical.*

THEOREM 4.20.11 (Wang, 1951). *Suppose that X is a G-space with finite odd dimension (or two) and with the property that there exists a pairwise transitive group of motions on X. Then X is either Euclidean, hyperbolic, spherical or elliptical.*

4.21. Angles and Complete Inner Product Spaces

Karl Menger, in his studies on geometrical aspects of metric space theory, has considered the notion of 'angles' and indicated applications to the characterizations of complete inner product spaces (of finite or infinite dimension; in many papers the class of finite dimensional inner product spaces also appears under the name 'Euclidean spaces' or 'unitary spaces').

In this section we present some results on the angles and some applications to the characterization of complete inner product spaces.

DEFINITION 4.21.1 (Menger, 1931). Let X be a metric space. We say that on X we have defined a notion of angle if for each point a of X and for each pair of points b, c of X (both different from a) there is a defined real number denoted by bac and called the angle (with vertex a) and such that the following properties hold:

$$bac = cab = \begin{cases} 0 & \text{if } d(a,b) + d(b,c) = d(a,c) \text{ or } d(a,c) + d(b,c) = d(a,b) \\ >0 & \text{otherwise} \\ \pi & \text{if } d(b,a) + d(a,c) = d(b,c). \end{cases}$$

DEFINITION 4.21.2 (Menger, 1931). Let X, Y be two metric spaces and let $f: X \to Y$ be a one-to-one mapping (continuous). F is said to be a conformal mapping if it preserves the angles.

We recall that in a metric space X, a segment is a set which is congruent (i.e. it is a one-to-one isometry of the set on a segment in R^1), a ray is a set which is congruent with a half-line, and a line is a set which is congruent to R^1.

DEFINITION 4.21.3 (Wilson, 1932). Let X be a metric space and a, b, c be points of X. The angle bac is defined by

$$bac = \arccos((d(a,b)^2 + d(a,c)^2 - d(b,c)^2)/(2d(a,b)d(a,c))).$$

It is easy to see (from the triangle inequality) that the above definition is correct. We now define the angle between two rays in X which have the property that they have a common initial point.

DEFINITION 4.21.4 (Wilson, 1932). Let X be a metric space and r_1, r_2 be two rays in X with a common initial point, say a. Let b on r_1 and c on r_2 arbitrary (and both not equal to a). The angle between the rays r_1 and r_2 (if it exists) is defined by

$$\text{angle}(r_1, r_2) = \lim_{\substack{b \to a \\ c \to a}} bac.$$

The problem which appears is about the existence of such angles between rays. Some interesting examples concerning this problem are given in the papers of Wilson (1931).

Following Valentine and Wayment (1971), we present a characterization of complete inner product spaces using properties of angles.

Let X be a real Banach space with a norm $\| , \|$. If a, b are two distinct points of X then the set

$$L(a,b) = \{x, x = ra + (1 - r)b : r \in R^1\}$$

is called the algebraic line (determined by a and b). If we set

$$ra + (1 - r)b \to (1 - r)\|a - b\|$$

then we get that $L(a, b)$ is congruent to R^1 (i.e. it is a metric line). On a Banach space X we suppose that the following property holds:

PROPERTY 4.21.5 (Valentine and Wayment, 1971). *For each point a of X and each ray r_1 and r_2 with common initial point a, $\lim_{b\to a, c\to a}$ bac exists. (Here b is on r_1 and c on r_2.)*

Then we have the following result.

THEOREM 4.21.6 (Valentine and Wayment, 1971). *Suppose that a Banach space X has the property mentioned in* 4.21.5. *Let a be a point of X. Then if r_1 and r_2 are two rays with common initial point a, then*

$$\text{angle}(r_1, r_2) = \arccos((\|a - b\|^2 + \|a - c\|^2 - \|b - c\|^2)/(2\|a - b\|\,\|a - c\|)),$$

where $b \in r_1$, $c \in r_2$.

Proof. Since lim bac exists (when $b \to a$, $c \to a$) we have,

$$\begin{aligned}\lim_{\substack{b\to a\\ c\to a}} bac &= \lim_{t\to 1} ((\|a - (1 - t)b - ta\|^2 + \|a - (1 - t)c - ta\|^2 - \\ &\quad - \|(1 - t)b + ta - (1 - t)c - ta\|^2)/ \\ &\quad (2(\|a - (1 - t)b - ta\|\,\|a - (1 - t)c - ta\|)) \\ &= \arccos \lim_{t\to 1} ((1 - t)^2\|a - b\|^2 + \\ &\quad + (1 - t)^1\|a - c\|^2 - (1 - t)^2\|b - c\|^2).\end{aligned}$$

$((1 - t)^2\|a - b\|\,\|a - c\|)^{-1} = bac.$

THEOREM 4.21.7 (Valentine and Wayment, 1971). *Let X be a Banach space as in Theorem* 4.21.6 *and (a_1, a_2, a_3, a_4) be an arbitrary quadruple of elements of X such that a_2, a_3, a_4 are on an algebraic line. Then there exists a quadruple $(b_1, \ldots, b_4)$ in the Euclidean plane such that $(a_1, \ldots, a_4)$ and $(b_1, \ldots, b_4)$ are congruent quadruples.*

Proof. Since a_2, a_3, a_4 are on an algebraic line then one of them is between the other two. Suppose that a_3 is between a_2 and a_4. Then we find in the Euclidean plane the points b_1, b_2, b_4 such that (a_1, a_2, a_4) is congruent to the (b_1, b_2, b_4). If b_3 is the point in the Euclidean plane which is between b_2 and b_4 and such that

$$\|b_2 - b_4\| = \|a_2 - a_4\|, \qquad \|b_3 - b_4\| = \|a_3 - a_4\|$$

and

$$\|a_1 - a_2\| = \|b_1 - b_2\|, \qquad \|a_1 - a_4\| = \|b_1 - b_4\|,$$
$$\|a_2 - a_3\| = \|b_2 - b_3\|, \qquad \|b_3 - b_4\| = \|a_3 - a_4\|.$$

To prove the assertion of the theorem it suffices to show that

$$\|a_1 - a_3\| = \|b_1 - b_3\|.$$

From Theorem 4.21.6 we get that ($r(,)$ is the ray determined by the points $(,)$),

$$\begin{aligned}
&\cos(r(a_2, a_1), r(a_2, a_4)) \\
&\quad = (\|a_1 - b_2\|^2 + \|a_2 - a_3\|^2 - \|a_1 - a_3\|^2)/(2\|a_1 - a_2\| \cdot \|a_2 - a_3\|) \\
&\quad = (\|a_1 - a_2\|^2 + \|a_2 - a_1\|^2 - \|a_1 - a_4\|^2)/(2\|a_1 - a_2\| \, \|a_2 - a_4\|) \\
&\quad = (\|b_1 - b_2\|^2 + \|b_2 - b_4\|^2 - \|b_1 - b_4\|^2)/(2\|b_1 - b_2\| \, \|b_2 - b_4\|) \\
&\quad = (\|b_1 - b_2\|^2 + \|b_2 - b_3\|^2 - \|b_1 - b_3\|^2)/(2\|b_1 - b_2\| \, \|b_2 - b_3\|)
\end{aligned}$$

which implies that

$$\|a_1 - a_3\| = \|b_1 - b_3\|$$

and the theorem is proved. □

THEOREM 4.21.8 (Valentine and Wayment, 1971). *If a, b, c are elements of X (which is supposed as in Theorem* 4.21.6) *and are on a line with b between a and c then*

$$\text{angle}(r(b, a), r(b, c)) = \pi.$$

Proof. Suppose that $a_n \neq b \neq c_n$ are sequences on the rays $r(b, a)$ and $r(b, c)$ respectively, then

$$\|a_n - c_n\| = \|a_n - b\| + \|b - c_n\|$$

and then

$$\lim_{\substack{a_n \to b \\ c_n \to b}} \text{angle}\,(a_n, b, c_n) = -1$$

which gives the assertion of the theorem. □

THEOREM 4.21.9 (Valentine and Wayment, 1971). *Suppose that a, b are points of X with $a \neq b$ (X is as in Theorem* 4.21.6). *Then a, b determine a uniquely algebraic (and metric) line.*

Proof. First we show that a, b are the endpoints of exactly one segment. It is clear that a, b are the endpoints of an algebraic segment which is also a metric segment. Suppose now that the assertion of the theorem is false. Then we find another metric segment, say I_1, which has the endpoints a and b and is different from the algebraic segment with endpoints a and b which is denoted by $I(a, b)$.

Let us consider a point in I_1 and not in I. We choose a point e such that

$$\|b - e\| = \|b - d\|,$$

d being the point in I_1 and not in I. Now consider c to be a point such that b is between a and c. In this case b is between d and c. If $r(b, d)$ and $r(b, c)$ are the rays determined by (b, d) and (b, c), respectively, we have,

$$\text{angle}(r(b, d), r(b, c)) = \pi.$$

Now if we consider the ray $r(b, c)$ then, according to Theorem 4.20.8,

$$\text{angle}(r(b, e), r(b, d)) + \text{angle}(r(b, d), r(b, c)) = \pi.$$

Then

$$\text{angle}(r(b, e), r(b, d)) = 0$$

which implies that

$$(\|b - d\|^2 + \|b - e\|^2 - \|e - d\|^2)/(2\|b - e\| \, \|b - d\|) = 1$$

or

$$\|b - d\|^2 + \|b - e\|^2 = \|e - d\|^2$$

and

$$\|b - d\| = \|b - e\| + \|e - d\|$$

or

$$\|b - d\| + \|d - e\| = \|b - e\|.$$

Then $\|e - d\| = 0$, which implies that $d = e$. This is a contradiction and the uniqueness of the segment is proved. □

Now we consider the algebraic line determined by a and b which is clearly a metric line. The segment determined by a and b has the property that it is not prolongable to another line. This follows from the above argument with the exception that we choose d so that d is between a and c and c is in the algebraic line determined by a and b such that b is between a and c.

COROLLARY 4.21.10 (Valentine and Wayment, 1971). *If X is a Banach space as in Theorem* 4.21.6 *and if a, b, c are linear then they are on an algebraic line of X.*

From Theorems 4.21.7 and 4.21.9 we get the following.

THEOREM 4.21.11 (Valentine and Wayment, 1971). *Let X be a Banach space as in Theorem* 4.21.6. *Then any quadruple of points $(a_1, \ldots, a_4)$ of X which contains a linear triple, is congruently embeddable in the Euclidean plane.*

REMARK 4.21.12. Theorem 4.21.11 asserts exactly that a Banach space X (which is as in Theorem 4.21.6) has a 'weak Euclidean four-point property'. (A metric space (Y, d) is said to have a weak Euclidean four-point property if for each quadruple of pairwise distinct points of X, and containing a linear triple, it is congruently embeddable in the Euclidean plane.)

Now using a result of L. M. Blumenthal (4.19.35) that every convex complete metric space which is externally convex with a weak Euclidean four-point property is a complete inner product space, we have the following result.

THEOREM 4.21.13 (Valentine and Wayment, 1971). *Let X be a Banach space as in Theorem* 4.21.6. *Then the norm of X is generated by an inner product iff* $\lim_{b\to a, c\to a}$ *bac, where b is on the ray r_1 and c on the ray r_2, both having a common initial point* ($=a$) *for each triple of points and each pair of rays.*

Another interesting characterization of Banach spaces having their norm generated by an inner product was obtained by J. J. Schaeffer in 1965. In order to formulate the results obtained by J. J. Schaeffer we recall the notion of 'Clarkson angles'.

DEFINITION 4.21.14. Let X be a Banach space and x in X, $x \neq 0$. Then $\operatorname{sgn} x = x/\|x\|$. The 'Clarkson angle' between the points x and y of X is then defined as

$$a(x, y) = \|\operatorname{sgn} x - \operatorname{sgn} y\|.$$

We prove in Chapter 8 that

$$\|x - y\| \geqslant \tfrac{1}{2}a(x, y) \max(\|x\|, \|y\|).$$

Suppose now that X is a complete inner product space. Then

$$\|x - y\| \geqslant ka(x, y)\max(\|x\|, \|y\|),$$

where k is close to 1 when $a(x, y)$ is small.

The main result of J. J. Schaeffer is that such a property is in fact characteristic for a Banach space to have the norm generated by an inner product. Below we give Schaeffer's results; for a proof we refer to the original paper. Note that $\perp_B$ denotes orthogonality in the sense of Birkhoff.

THEOREM 4.21.15 (Schaeffer, 1965). *Let X be a Banach space of dimension not less than* 2. *Then the following assertions are equivalent:*

(1) $\mu = \lim\limits_{a(x,y)\to 0} (\|x - y\|)/(a(x, y) \max(\|x\|, \|y\|)) = 1.$

(2) $\lim\limits_{t\to 0^+} t/a(u, u + tv) = 1, \qquad \forall u, v, \qquad \|u\| = \|v\| = 1, \quad u \pm v \neq 0.$

(3) $\perp_B$ is symmetric.

If $\dim X = 3$ *then either* (1), (2) *or* (3) *implies that the norm of X is generated by an inner product.*

In order to formulate the next result we need the following notion.

DEFINITION 4.21.16. A curve in $X\setminus 0$ is the image of a continuous function $f\colon [\alpha, \beta] \to X\setminus 0$, $[\alpha, \beta]$ is a compact interval. Further, for any curve Γ we set

$$d(\Gamma) = \min(\|f(s)\|, s \in [\alpha\ \beta]).$$

If Γ is a curve then the function $s \to \operatorname{sgn} f(s)$ is called the radial projection of the curve and is denoted by $\operatorname{sgn} \Gamma$. Then we have

THEOREM 4.21.17 (Schaeffer, 1965). *For any real Banach space X of dimension not less than* 2, *conditions* (1), (2), (3) *of Theorem* 4.21.15 *are equivalent to each of the* *following conditions:*

(4) *Every rectifiable curve* Γ *in* $X\setminus 0$ *has the property that* $l(\operatorname{sgn}\Gamma) = l(\Gamma)/d(\Gamma)$, ($l(\ )$ denotes the length of ( )).

(5) *For every rectifiable curve* Γ *in* X *that contains no interior points (i.e.* $d(\Gamma) \geqslant 1$) *the following inequality holds:* $l(\operatorname{sgn}\Gamma) \leqslant l(\Gamma)$.

If $\dim X \geqslant 3$ *then* X *has either property* (4) *or* (5) *iff the norm of* X *is generated by an inner product.*

Let X be a normed linear space.

DEFINITION 4.21.18 (Martin and Valentine, 1976). Let k be a fixed number in $[-1, 1]$. We say that the law of cosines is k-signum homogeneous whenever x, y are $\neq 0$ elements of X satisfying the relation

$$(\|x\|^2 + \|y\|^2 - \|x - y\|^2) = 2k\|x\|\,\|y\| \tag{*}$$

then for all real numbers a and b the following equality holds:

$$\|ax\|^2 + \|by\|^2 - \|ax - by\|^2 = 2k\|ax\|\,\|by\|\operatorname{sgn}(ab). \tag{**}$$

The following lemma shows the existence of pairs of elements satisfying (*).

LEMMA 4.21.19 (Martin and Valentine, 1976). *Let* X *be a normed linear space of dimension greater than* 1. *Then for any fixed* x *in* X *and any* k *fixed in* $(-1, 1)$ *there exists an element* y *satisfying the relation* (*).

Proof. Since X has dimension at least 2, given x there exists another element u in X such that the set (x, u) is a linear independent set of elements in X. We define the following function on $[0, \pi]$ by the formula:

$$f(t) = (\|x\|^2 + \|x\cos t + u\sin t\|^2 - \|x(1 - \cos t) - u\sin t\|^2) \times$$
$$\times 1/(2\|x\|\,\|x\cos t + u\sin t\|.$$

From the form of the function f we see that it is continuous and $f(0) = 1, f(\pi) = -1$ and thus given k in $(-1, 1)$ there exists t_0 in $[0, \pi]$ such that $f(t_0) = k$. If we set $y = x\cos t_0 + u\sin t_0$ then this element satisfies the assertion of the lemma.

The following result shows that if the law of cosines is k-signum homogeneous for any fixed k in $(-1, 1)$ then this forces X to be an inner product space.

THEOREM 4.21.20 (Martin and Valentine, 1976). *A normed linear space* X *is an inner product space if and only if the law of cosines is* k*-signum homogeneous for each fixed* k *in* $(-1, 1)$, *i.e. if* k *is in* $(-1, 1)$ *and* x, y *are elements of* X ($\neq 0$) *satisfying* (*), *then* (**) *is satisfied for all real numbers* a *and* b, *and* X *is an inner product space.*

Proof. We show that the Jordan–von Neumann condition is satisfied in X. Indeed, if x is given then for $y = tx$ the Jordan–von Neumann condition is obviously satisfied. Thus we suppose without loss of generality that x and y are linearly independent elements in X. Given y we use Lemma 4.21.19 to obtain an element u in the space generated by x and y, such that

$$\|y\|^2 + \|u\|^2 - \|y - u\|^2 = 2k\|y\|\,\|u\|.$$

We now use the hypothesis that the cosine law is k-signum homogeneous and thus

$$||ay||^2 + ||bu||^2 - ||ay - bu||^2 = 2k||ay||\,||bu||\text{sgn}(ab)$$

Now as x is in the space generated by y and u there exist scalars m and n such that $x = my + nu$ and thus

$$||x||^2 = ||nu||^2 + ||my||^2 - 2k||my||\,||nu||\text{sgn}(mn) \quad (*)$$

$$||x - y||^2 = ||nu||^2 + ||(m-1)y||^2 - 2k||nu||\,||(m-1)y||\text{sgn}(n(m-1)) \quad (**)$$

$$||x + y||^2 = ||nu||^2 + ||(m+1)y||^2 - 2k||nu||\,||(m+1)y||\text{sgn}(n(m+1)). \quad (***)$$

We obtain $||nu||$ from (*) and use it in (**) and (***). We add the results obtained and this easily implies that the Jordan–von Neumann condition is satisfied. Thus X is an inner product space.

The following result is a localized version of 4.21.20.

THEOREM 4.21.21 (Martin and Valentine, 1976). *A normed linear space X is an inner product space if and only if for some fixed k in $(-1, 1)$ and for each pair of elements x, y in X $(\neq 0)$ with*

$$||x||^2 + ||y||^2 - ||x - y||^2 = 2k||x||\,||y||$$

the following relation holds:

$$\lim_{t,s\to 0} ((||tx||^2 + ||sy||^2 - ||tx - sy||^2)/(2||tx||\,||sy||) - k\,\text{sgn}(st)) = 0$$

Proof. We apply the above relation for fixed s and t and the pair (mt, ms) for $m \to 0$. Then we get that

$$\lim_{m\to 0} ((||mtx||^2 + ||msy||^2 - ||mtx - msy||^2)/(2||mtx||\,||msy||) - k\,\text{sgn}(m^2ts)) = 0$$

and using the fact that t and s are fixed,

$$\lim_{m\to 0} ((||mtx||^2 + ||msy||^2 - ||mtx - msy||^2)/(2||mtx||\,||msy||))$$
$$= ((||tx|| + ||sy||^2 - ||tx - sy||^2)/(2||tx||\,||sy||))$$

which implies that the law of cosines is k-homogeneous. Then Theorem 4.21.20 gives us the result.

REMARK 4.21.22. Since in the Jordan–von Neumann condition we may replace = by $\geqslant$ or $\leqslant$ it would be of interest to know if Theorems 4.21.20 and 4.21.21 may be extended (using a suitable definition of k-signum homogeneity for the law of cosine).

CHAPTER 5

Banach Algebras

5.0. Introduction

The notion of Banach algebra was first considered by M. Nagumo in 1936 under the name 'metric rings' but the true development began with the fundamental papers of I. M. Gelfand (1939, 1940, 1941) and of members of his school, among whom we mention M. A. Naimark, G. E. Shilov, V. A. Ditkin and D. A. Raikov. One of the spectacular success of Gelfand's Banach algebra theory was a simple proof of a famous result of N. Wiener: if f is a function on $[-\pi, \pi]$ and

$$f(t) = \sum_{-\infty}^{\infty} a_n \mathrm{e}^{int}, \quad \sum_{-\infty}^{\infty} |a_n| < \infty$$

and $f(t) \neq 0$ for all t, then

$$g(t) = (f(t))^{-1} = \sum_{-\infty}^{\infty} b_n \mathrm{e}^{int}, \quad \sum_{-\infty}^{\infty} |b_n| < \infty.$$

The results and methods of Banach algebras have many applications, among which we mention the applications in operator theory, approximation theory, and harmonic analysis on locally compact abelian groups.

5.1. Definition of Banach Algebras and Some Examples

DEFINITION 5.1.1. A Banach algebra A over $K(R$ or $C)$ is a Banach space over K such that there exists a mapping of $A_\times A$ with values in A satisfying the following conditions:

$(x, y) \to xy$

(1) $x(y + z) = xy + xz, \qquad (x + y)z = xz + yz,$

(2) $x(yz) = (xy)z,$

(3) $\lambda(xy) = (\lambda x)y,$

(4) $\lambda(\mu x) = (\lambda\mu)x,$

(5) $1x = x,$

(6) $\|xy\| \leqslant \|x\| \, \|y\|.$

for all x, y, z in A and λ, μ in K.

If $K = R$ then A is said to be a real Banach algebra and if $K = C$, A is called a complex Banach algebra.

DEFINITION 5.1.2. A Banach algebra A over K is said to be:

(1) with identity if there exists an element $e \in A$ such that $ex = xe = x$,
(2) commutative if $xy = yx$

for all x, y in A.

In what follows we suppose that $A \neq (0)$.

DEFINITION 5.1.3. A subalgebra A_1 of a Banach algebra A over K is a subspace of A which is closed with respect to multiplication $(x, y) \rightarrow xy$, i.e. for all x, y in A_1, xy is again in A_1.

DEFINITION 5.1.4. (1) A left ideal I_l of a Banach algebra A is a subalgebra of A with the property that for all $x \in A$, $y \in I_l$, $xy \in I_l$.

(2) A right ideal I_r of a Banach algebra A is a subalgebra of A such that for any $x \in A$ and $y \in I_r$, $yx \in I_r$.

(3) A two-sided ideal I of a Banach algebra A is a subalgebra of A with a left and right ideal.

(4) If I_1 and I_2 are two ideals (of the same type – i.e. left ideal, right ideal, two-sided ideal), then we say that $I_1 \leqslant I_2$ if I_1 is included in I_2.

(5) An ideal is maximal if there is no ideal (of the same type) in which it is contained.

EXAMPLE 5.1.5. Let S be a compact Hausdorff space and $C_K(S)$ be the set of all continuous functions defined on S with values in K. If we define for any f, g in $C_K(S)$ and $k \in K$,

(1) $(f + g)(s) = f(s) + g(s)$
(2) $(kf)(s) = kf(s)$
(3) $(fg)(s) = f(s)g(s)$
(4) $f \rightarrow \|f\| = \sup(|f(s)|, s \in S)$

then $C_K(S)$ is a commutative Banach algebra with identity. Of course, the identity is the function e, defined by

$$e(s) = 1 \text{ for all } s \in S.$$

Let s_0 be a fixed point in S and consider the set M_{s_0} where

$$M_{s_0} = (f, f \in C_K(S), f(s_0) = 0).$$

Obviously M_{s_0} is an ideal in $C_K(S)$. More generally, if F is any closed subset in S then the set M_F,

$$M_F = (f, f \in C_K(S), f(s) = 0, s \in F)$$

is an ideal in $C_K(S)$.

EXAMPLE 5.1.6. Let $D = (z, |z| < 1)$ and $\Gamma = (z, |z| = 1)$. We consider the set A of all complex-valued continuous functions on $D \cup \Gamma$ which are analytic in D. With the norm and the multiplication fg defined in the usual way, A is a commutative Banach algebra with identity. An example of an ideal in A is as follows: let z_0 be a fixed point in $D \cup \Gamma$ and let M_{z_0} be the set

$$M_{z_0} = (f, f \in A, f(z_0) = 0).$$

EXAMPLE 5.1.7. Let S be a locally compact space and let f be a continuous k-valued function defined on S. We say that f is null at infinity if for each $\varepsilon > 0$ there exists a compact subset of S, K_ε such that $|f(s)| < \varepsilon$ for all s in $S - K_\varepsilon$.

We denote by $C_0(S)$ the set of all continuous k-valued functions on S which are null at infinity. With the multiplication fg defined in the usual way and the norm

$$f \to f = \sup(|f(s)|, s \in S),$$

this is a commutative Banach algebra without identity. Indeed, if the identity is e, then e is a function which has the property that for all $s \in S$, $e(s) = 1$. But this function is obviously not in $C_0(S)$.

EXAMPLE 5.1.8. Let us consider the set W(after Wiener) of all sequences of complex numbers $x = (x_n)_{\infty}^{\infty}$ such that

$$x \to \|x\| = \sum_{\infty}^{\infty} |x_n| < \infty.$$

This is obviously a normed linear space and it is not difficult to prove that it is, in fact, a Banach space. Now we show that W can be organized as a Banach algebra. Suppose that we have another element $y = (y_n)_{\infty}^{\infty}$ and we define the element z, the product of x and y, to be a sequence with terms

$$z_n = \sum_{\infty}^{\infty} x_{n-m} y_m.$$

We now show that the product is well defined. Indeed, we have,

$$\sum_{-\infty}^{\infty} |z_n| = \left| \sum_{\infty}^{\infty} \sum_{\infty}^{\infty} x_{n-m} y_m \right| \leqslant \sum_{\infty}^{\infty} \sum_{\infty}^{\infty} |x_{n-m}| \, |y_m| \leqslant \|x\| \, \|y\|.$$

Thus $z = (z_n)_{\infty}^{\infty}$ is in W and clearly we have $z = xy = yx$, i.e. W is a commutative Banach algebra with identity

$$e = \left(\ldots, \underset{-1}{0}, \underset{0}{1}, \underset{1}{0}, \underset{2}{0}, \underset{3}{0}, \underset{4}{0}, \ldots \right).$$

REMARK 5.1.9. An interesting generalization of the Banach algebra W is the algebra W_α. If $\alpha = (a_i)_{-\infty}^{\infty}$ is a sequence of positive numbers such that:

(1) $a_i \geqslant 1$,
(2) $a_{1+j} \leqslant a_i a_j$

for all i, j, then W_α denotes the set of all sequences of complex numbers $x = (x_n)_{\infty}^{\infty}$ such that

$$x \to \|x\|_\alpha = \sum_{-\infty}^{\infty} a_n |x_n| < \infty.$$

With the operations defined exactly as for the Banach algebra W, W_α is a commutative Banach algebra with identity.

REMARK 5.1.10. Some interesting subalgebras of W and W_α are the algebras W^+ and W_α^+ and consist in the elements in W, respectively, with $x_i = 0$ for all $i = -1, -2, -3, \ldots$

EXAMPLE 5.1.11. Let $I = [0, 1]$ and consider the Banach space $L^1(I, \mathrm{d}x)$, i.e. the space of all functions (classes of functions) on I such that

$$f \to \|f\| = \int_0^1 |f(s)| \, \mathrm{d}s < \infty.$$

A multiplication of elements of $L^1(I, \mathrm{d}x)$ may be defined using the convolution: if f, g are in L^1 then

$$(f \otimes g)(x) = \int_0^x f(x - s) g(s) \, \mathrm{d}s$$

is the product of f and g. With this multiplication $L^1(I, \mathrm{d}x)$ is a commutative Banach algebra without identity.

EXAMPLE 5.1.12. Let H be a complete inner product space over C and let $L(H)$ be the set of all bounded linear operators on H with the norm

$$T \to \|T\| = \sup(\|Tx\|, \|x\| \leqslant 1).$$

Then with the natural laws of addition and scalar multiplication $L(H)$ is a Banach space. If we set for any T, S in $L(H)$,

$$(TS)(x) = T(Sx)$$

then $L(H)$ is a Banach algebra with identity I, $Ix = x$ for all $x \in H$. An ideal in $L(H)$ is the set K of all compact operators and it is easy to see that this is a two-sided ideal. Later, we give other examples of ideals in $L(H)$ (see Chapter 7).

5.2. Ideals in Banach Algebras

We now consider a Banach algebra over K and denote by M_l the set of all left ideals of A.

DEFINITION 5.2.1. Suppose that A has an identity and consider an arbitrary element $x \in A$. We say that x has an inverse y if the following relations are satisfied:

$$xy = yx = e.$$

We denote the inverse of x (if it exists) by x^{-1}. It is easy to see that the inverse is unique. First we remark that if a Banach algebra has an identity then this is unique. Indeed, suppose that A has at least two identities, say e and f. Then we have

$$ef = fe = e \quad \text{and} \quad fe = ef = f$$

which implies that $e = f$. Now if x has two inverses, say x_1^{-1} and x_2^{-1}, then

$$x_1^{-1} = x_1^{-1}e = x_1^{-1}x x_2^{-1} = x_2^{-1}$$

and the assertion is proved.

Now if $I_l \in M_l$ and $e \in I_l$, then $A = I_l$. In what follows we consider $I_l \neq A$.

PROPOSITION 5.2.2. *Any element $x \in I_l$ is not invertible (i.e. the inverse of x does not exist).*

Proof. Indeed, otherwise we would clearly get that $e \in I_l$, which implies that $I_l = A$. This contradition proves the assertion. □

Let $I_l \in M_l$ and we suppose that in M_l are only closed ideals. Then we can consider the Banach space A/I_l which can be organized as a Banach algebra if we set, for any two equivalence classes, say $\hat{x}$ and $\hat{y}$,

$$\hat{x}\hat{y} = (\hat{xy}),$$

where $x \in \hat{x}$ and $y \in \hat{y}$. Then it is easy to see that this is a well-defined multiplication (i.e. it is independent of $x \in \hat{x}$ and $y \in \hat{y}$ chosen) and if A has an identity then A/I_l has also an identity, $\hat{e}$. Thus we get that A/I_l is a Banach algebra.

The following result is an important property of Banach algebras with identity.

PROPOSITION 5.2.3. *If A is a Banach algebra with identity then M_l is an inductive set.*

Proof. Let P be a totally ordered subset of M_l. We define

$$I_P = (x, x \in A, \text{ there exists } I_l \in P \text{ such that } x \in I_l).$$

We show that this is in M_l. For this it suffices to show that if x, y are in I_P then $x + y$ is in I_P and that for any $z \in A$, zx is in I_P. From the definition of I_P there exist $I_{l,x}$ and $I_{l,y}$ such that $x \in I_{l,x}$ and $y \in I_{l,y}$. Since P is totally ordered we have either $I_{l,x} \leqslant I_{l,y}$ or $I_{l,y} \leqslant I_{l,x}$. In both cases $x + y$ is clearly in I_P. Also, zx is in I_P and I_P contains all elements of P, and since e is not in any member of P it is not in I_P, which shows that $I_P \neq A$. If I_P is not closed, the closure satisfies the assertion.

COROLLARY 5.2.4. *If I_l is in M_l then there exists a maximal left ideal containing I_l.*

Proof. In this case we can consider the family of elements in M_r containing I_l and Zorn's lemma gives the assertion. □

DEFINITION 5.2.5. A Banach algebra which is a field is called a Banach field.

The following proposition characterizes the Banach fields among Banach algebras with identity.

PROPOSITION 5.2.6. *A Banach algebra with identity is a field if and only if the set M_l is equal to ((0)).*

Proof. First suppose that A is a field. Then any $x \neq 0$ has an inverse, which implies that the set M_1 reduces to the ideal (0). We note that we considered in M_l only ideals different from A.

Conversely, if M_l has the above form we show that A is a field. Take x an arbitrary nonzero element of A. Then by the above result it is clear that x is contained in a maximal left ideal, say $I_{x,l}$. From our assumption this is necessarily A. Since, if e is the identity of A we must have an element $y \in A$ such that $yx = e$. Of course, $y \neq 0$, which implies that there exists another element z of A such that $zy = e$. Then we have

$$xy = exy = (zy)xy = z(yx)y = zy = e$$

which gives that x is invertible and the proposition is proved.

PROPOSITION 5.2.7. *Let A be Banach algebra with identity and I be a closed two-sided ideal in A. The Banach algebra A/I is a field if and only if I is a maximal two-sided ideal of A.*

Proof. Let us consider the canonical mapping $\pi: A \to A/I$. Then it is easy to see that if i_l is a left ideal in A/I then $\pi^{-1}(i_l)$ is a left deal in A and this ideal contains I. From the maximality of I we see that any left ideal i_l is (0). This implies by the above result that A/I is a field. Conversely, if A/I is a field then we show that I is a maximal ideal. Indeed, in the opposite case, I is contained in a maximal two-sided ideal I_m and we can form A/I_m. This is a left ideal in A/I which must be zero. This is a contradiction and the proposition is proved. □

REMARK 5.2.8. We give later, following I. M. Gelfand and S. Mazur, the structure of complex Banach fields. The structure theorem is the famous theorem of Gelfand–Mazur that every such a field is isomorphic and isometric with the field C. This fact is of fundamental importance in the study of commutative complex Banach algebras.

The following result gives information regarding the position of noninvertible elements of a commutative Banach algebra with identity.

PROPOSITION 5.2.9. *If x is a noninvertible element of a commutative Banach algebra with identity then there exists a maximal ideal containing x.*

Proof. We consider the set

$$m_x = (xy, y \in A)$$

which is obviously an ideal. Since x is not invertible we get that $m_x \neq A$. The maximal ideal which contains m_x clearly satisfies the proposition.

5.3. The Spectrum of an Element in a Complex Banach Algebra with Identity

In what follows we suppose that A is a complex Banach algebra with identity e. Since $e \neq 0$ we get that $\|e\| \leqslant \|e\| \|e\|$, and thus $\|e\| \geqslant 1$.

DEFINITION 5.3.1. Let A and B be two Banach algebras over C and having identities. A continuous mapping $f: A \to B$ satisfying the relations

(1) $f(ax + by) = af(x) + bf(y)$
(2) $f(xy) = f(x)f(y)$

for all x, y in A and a, b in C is said to be a homomorphism. If $f^{-1}(0)$ is (0) and $f(A) = B$ then f is called an isomorphism.

THEOREM 5.3.2. *Every Banach algebra A with identity e is isomorphic with a Banach algebra B with identity f and $\|f\| = 1$.*

Proof. For every $x \in A$ we consider the mapping $T_x: A \to A$ defined by the relation

$$T_x(y) = xy.$$

It is clear that the mapping $x \to T_x$ is a homomorphism of A into $L(A)$. We consider now in $L(A)$ the subalgebra $(T_x, x \in A)$. Clearly this has the identity $f = T_e$. Obviously A and $B = (T_x, x \in A)$ are isomorphic.

Now we define the norm of B as the norm induced by the norm of $L(A)$, i.e. we set

$$|||x||| = \|T_x\|.$$

Since

$$\|T_x\| = \sup_{\|y\| \leqslant 1} \|xy\| \geqslant \sup_{\|y\| \leqslant \|e\|} \|xy\| \cdot 1/\|e\| \geqslant 1/\|e\| \, \|x\|$$

we get

$$1/\|e\| \cdot \|x\| \leqslant |||x||| \leqslant \|x\|.$$

Since it is obvious that $f = T_e$ has $|||f||| = 1$ the assertion is proved.

In the rest of the book, when a Banach algebra has an identity we shall assume that the norm of the identity is 1.

DEFINITION 5.3.3. A complex number is said to be in the resolvent set of the element x of a complex Banach algebra with identity A if $x - ze = x - z$ has an inverse. The resolvent set of x is denoted by $\rho(x)$.

The following theorem gives some information about the resolvent set $\rho(x)$ of an element x.

THEOREM 5.3.4. *Let A be a complex Banach algebra with identity e. Then the following assertions hold*:

(1) *If $x \in A$ and $\|x - e\| < 1$ then x^{-1} exists.*
(2) *If z is a complex number such that $|z| > \|x\|$ then $x - z$ has an inverse.*
(3) *For any $x \in A$, $\rho(x)$ is an open set in the complex plane.*

Proof. (1) We consider the series (formally)

$$\sum_{n=0}^{\infty} (e - x)^n,$$

where for any $y \in A$, $y^{\circ} = e$. Since $q = \|e - x\| < 1$ the above series converges and let s be the sum. Since we have $x = e - (e - x)$ we get that

$$xs = \sum_{i=0}^{\infty} (e - x)^n - \sum_{i=1}^{\infty} (e - x)^n = e = sx$$

and this proves assertion (1).

For (2) we remark that

$$x - ze = z(x/z - 1)$$

and assertion (1) applies.

For (3) we consider $x \in A$ arbitrary and let $z \in \rho(x)$. Then $(x - z)^{-1}$ exists as the inverse of $x - z$. We now consider u in the complex plane such that

$$|z - u| < 1/\|(x - z)^{-1}\|$$

which is thus in a neighborhood of z. We now show that $x - u$ has an inverse. For this purpose consider again a series like that at (1):

$$(x - z)^{-1}\left(e + \sum_{i=1}^{\infty} (z - u)^n (x - z)^{-n}\right)$$

(if $a \in A$ has an inverse a^{-1} then $a^{-n} = (a^{-1})^{-n}$). Since u is chosen in the neighborhood of z with the above property, we see immediately that the series converges. Let s be the sum. Then a similar computation as in (1) shows that s is the inverse of $x - u$. This proves (3).

The following results contain further information about invertible elements.

COROLLARY 5.3.5. *If A is a complex Banach algebra with identity then*

$$A^{-1} = (x, x \in A,\ x \text{ invertible})$$

is an open set in A.

Proof. From Theorem 5.3.4 it follows that there is a neighborhood of e (the identity of A) of the form

$$(y, y \in A,\ \|y - e\| < 1)$$

such that any element of this neighborhood $V(e)$ is in A^{-1}. Let us now suppose that $x \in A^{-1}$. Then there exists a neighborhood $V(x)$ of x such that any element $y \in V(x)$ is such that $yx^{-1} \in V(e)$. Indeed, since we have the identity

$$xv - x_0 v_0 = (x - x_0)(v - v_0) + x_0(v - v_0) + v_0(x - x_0)$$

the existence of the above neighborhood follows. Now it is clear, from Theorem 5.3.2(1), that $y \in A^{-1}$ and thus the assertion is proved.

PROPOSITION 5.3.6. *The mapping (defined on A^{-1}) $x \to x^{-1}$ is continuous.*

Proof. Let (x_n) be a sequence in A^{-1} and suppose that $x = \lim x_n$ is in A^{-1}. Since $x^{-1}x_n \to e$ (which follows easily from $x_n \to x$) for $\varepsilon > 0$ there exists N_ε such that for all $n \geqslant N_\varepsilon$,

$$\|x^{-1}x_n - e\| < 1.$$

If we write (for $n \geqslant N_\varepsilon$) (formally)

$$x^{-1}x_n = e + (e - x^{-1}x_n) + \cdots + (e - x^{-1}x_n)^n + \cdots$$

then we see that

$$\|e + (e - x^{-1}x_n) + \cdots + (e - x^{-1}x_n)^n + \cdots\|$$

$$\leqslant \sum_{i=0}^{N} \|(e - x^{-1}x_n)^i\| + \sum_{i=N+1}^{\infty} \varepsilon^i < \infty.$$

Further, if we suppose that for the same N_ε we have

$$\|x - x_n\| \leqslant \varepsilon/(2\|x^{-1}\|)$$

then

$$\|e - x^{-1}x_n\| \leqslant \sum_{i=1}^{\infty} \|(e - x^{-1}x_n)^i\| \leqslant \varepsilon\left(\sum \tfrac{1}{2}^i \varepsilon\right) \leqslant \varepsilon.$$

Thus

$$\lim x_n^{-1} = x^{-1}$$

and the assertion is proved. □

DEFINITION 5.3.7. For any element x of a complex Banach algebra with identity A, the complement $C\rho(x)$ of the resolvent set $\rho(x)$ is called the spectrum of x and is denoted by $\sigma(x)$. The number $r_x = \sup(|z|, z \in \sigma(x))$ is called the spectral radius of x. The element x is called quasi-nilpotent if $r_x = 0$.

We have a basic problem about the spectrum. Is this set nonempty?

THEOREM 5.3.8. *For any $x \in A$, $\sigma(x)$ is a compact nonempty set.*

Proof. Since $\rho(x)$ is open, $\sigma(x)$ is a closed set. It is compact because any point z with $|z| > \|x\|$ is in the resolvent set. Thus the spectrum (if it is nonempty) is contained in $(z, |z| \leqslant \|x\|)$.

The nonemptiness of the set $\sigma(x)$ may be proved in different ways. There exists a very elegant proof by Tornheim which uses only the continuity of $z \to (x - z)^{-1}$. In what follows we present a proof using some results concerning Banach-algebra-valued holomorphic functions. The proofs of these results are essentially the same as for complex-valued functions.

DEFINITION 5.3.9. Let D be a domain in the complex plane and $f: D \to A$ be a function with values in the Banach algebra A with identity. f is said to be holomorphic in D if for each $z_0 \in D$,

$$\lim_{z \to z_0} (f(z) - f(z_0))/(z - z_0)$$

exists. The limit is denoted by $f'(z_0)$.

Sometimes we use the word 'analytic' to denote a holomorphic function. If $f: D \to A$ is any holomorphic function with values in A then for any fixed z_0 in D the function f has the form

$$(*) \quad f(z) = \sum_0^{\infty} a_n(z - z_0)^n$$

with a_n in A. A holomorphic function of the form (*) is called a power series. If $\Sigma_0^\infty a_n(z-z_0)^n$ is a power series then this converges in a circle with the radius given by the Cauchy-Hadamard formula

$$R^{-1} = \limsup \|a_n\|^{1/n}.$$

It is important and useful for this class of functions that the Liouville theorem is true: if $f: C \to A$ is a bounded holomorphic then it is necessarily a constant.

Another notion of holomorphy can be defined using the elements of the dual space A^*. If $f: D \to A$ is a function then we say that f is weakly holomorphic if for each $a^* \in A^*$, $a^*(f)$ is analytic. A famous result of N. Dunford asserts that the weak holomorphic function are exactly the holomorphic functions (strong holomorphic functions) defined above.

Now we are ready to prove the nonemptiness of the spectrum for any element of a Banach algebra with identity. Suppose, on the other hand, that $\sigma(x)$ is empty. Then the function $z \to (x-z)^{-1}$ is holomorphic on C. Indeed, first we show that

$$z \to R_z(x) = (x-z)^{-1}$$

is holomorphic on $\rho(x)$. First we note that the following identity holds:

$$R_z(x) - R_{z_1}(x) = -(z-z_1)R_z(x)R_{z_1}(x).$$

Indeed, we have,

$$R_z^{-1}(z)R_{z_1}(x) = (x-z)(x-z_1)^{-1} = (x-z_1+(z_1-z))(x-z_1)^{-1}$$
$$= e + (z_1-z)(x-z_1)^{-1} = e + (z_1-z)R_{z_1}(x)$$

and thus

$$R_{z_1}(x) = R_z(x) + (z_1-z)R_z(x)R_{z_1}(x)$$

which is equivalent with out identity. From the continuity of $R_z(x)$ we get that

$$\lim (R_z(x) - R_{z_0}(x))/(z-z_0) = -R_{z_0}^2(x)$$

and thus $z \to R_z(x)$ is a holomorphic mapping on $\rho(x)$.

Now we remark that we have the relation

$$\lim_{z\to\infty} (e - 1/zx) = e, \qquad \lim_{z\to\infty} (e-1/z)^{-1} = e.$$

Since

$$(x-z)^{-1} = -1/z(e-1/zx)^{-1}$$

we get that if $\sigma(x)$ is an empty set, $z \to R_z(x)$ is a bounded holomorphic function. Then, by Liouville's theorem, it is a constant, say a. But

$$\lim_{z\to\infty} (x-z)^{-1} = \lim 1/z(e-x/z)^{-1} = 0,$$

i.e. the constant a is the null element of A. This implies that $x - ze = 0$ which is a contradiction. Thus $\sigma(x)$ is nonempty and the theorem is proved.

COROLLARY 5.3.10 (S. Mazur and I. M. Gelfand). *If A is a Banach field then it is isomorphic and isometric with C.*

Proof. Let $x \in A$ and $x \neq 0$. We consider the mapping

$$f_x(z) = x - z$$

and if this mapping is not zero then $R_z(x)$ exists for all z, which is impossible. Thus there exists z_x such that $x - z_x = 0$.

The mapping $x \to z_x$ obviously satisfies the corollary.

5.4. The Gelfand Representation. The Representation and Structure of Commutative Banach Algebras

Let A be a commutative Banach algebra over C and suppose that A has an identity. In this section we denote the identity by 1. We now consider an important class of functionals on A, the so-called multiplicative functionals.

DEFINITION 5.4.1. A function $f: A \to C$ is called a multiplicative functional if the properties

(1) $f(ax + by) = af(x) + bf(y)$,

(2) $f(xy) = f(x)f(y)$

are satisfied for all x, y in A and a, b in C.

We consider only nontrivial multiplicative functionals, i.e. satisfying the property to be nonidentically zero.

PROPOSITION 5.4.2. *If f is a multiplicative functional on A then the following assertions hold:*

(1) $f(1) = 1$.

(2) *If x is in A^{-1} then* $f(x^{-1}) = (f(x))^{-1}$.

(3) *f is in A^*.*

Proof. The first two assertions follow from the definition of multiplicative functionals in an obvious way. We now prove (3). Let z in C and if x is in A such that $|z| > \|x\|$, then $z - x$ is in A^{-1}. Then $f(x) \neq z$ and since x is arbitrary this gives that $|f(x)| \leqslant \|x\|$. Thus f is in A^* and since $f(1) = 1, \|f\| = 1$.

We denote by M_A the set of all multiplicative functionals on A. The following result gives, in part, a justification for this notation.

PROPOSITION 5.4.3. *If f is in M_A then*

$$m_f = (x, x \in A,\ f(x) = 0)$$

is a maximal ideal of A.

Proof. The fact that m_f is an ideal of A is obvious. Now we show that m_f is maximal. Indeed, if y is in A and not in m_f then the decomposition

$$x = (x - f(x)/f(y) \cdot y) + f(x)/f(y) \cdot y = x_1 + (f(x)/f(y))y$$

with x_1 in m_f clearly implies the maximality.

PROPOSITION 5.4.4. *For each maximal ideal m of A there exists a multiplicative functional f_m such that*

$$m = (x, f_m(x) = 0) = m_{f_m}.$$

Proof. We consider the quotient algebra A/m and since m is a maximal ideal of A, A/m is a Banach field and by the Mazur–Gelfand result A/m is isomorphic with C. Thus for each $x \in A$ we can consider the corresponding class in A/m which we identify, by the above result about A/m, with a complex number λ_x. Since the spectrum of any element is contained in a circle with radius $\|x\|$ we get that

$$|\lambda_x| \leqslant \|x\|$$

holds for all $x \in A$. Now it is clear that we can define the multiplicative functional m_f by the relation

$$m_f(x) = \lambda_x = \hat{x}(m)$$

and clearly this satisfies the proposition.

From this proposition we see that there exists a one-to-one correspondence between M_A and the set of all maximal ideals of A. But M_A is a subset in A^* or, more precisely, of the set $(a^*, \|a^*\| \leqslant 1)$. Using this fact we now define a topology on M_A such that this will be a compact Hausdorff space. For this we consider the product space

$$S = \prod_{x \in A} D_x, \quad D_x = (z, |z| \leqslant \|x\|)$$

and we note that for each $x \in A$, D_x is a compact space with the topology induced by the topology of the complex plane. We can now consider on S the product topology of Tychonov and then S is a compact space. Now we define for each maximal ideal space m of A the mapping $m \to \varphi(m) \in S$ by the relation

$$\varphi(m) = (\hat{x}(m))_{x \in A}.$$

Propositions 5.4.3 and 5.4.4 show that φ is one-to-one. Now we prove that ($\varphi(m)$, m a maximal ideal of A) is a closed subset of S. Indeed, let $(\hat{x}(m_d))_{x \in A}$ be a convergent net to an element, say $(z_x)_{x \in A}$ of S. Since for each $x \in A$ we have

$$\lim \hat{x}(m_d) = z_x$$

we get that for all complex numbers $c \in C$ and any $x \in A$,

$$z_{cx} = cz_x.$$

In the same way for all x_1 and x_2 in A,

$$z_{x_1 + x_2} = z_{x_1} + z_{x_2}$$

$$z_{x_1 x_2} = z_{x_1} z_{x_2}.$$

Now we have that $|z_x| \leqslant \|x\|$ and thus $x \to z_x = \psi(x)$ is a multiplicative functional. Then there exists a maximal ideal, say m_0, such that $\hat{x}(m_0) = \psi(x)$. This gives that ($\varphi(m)$, m a maximal ideal of A) is a closed subset of S. Since S is compact we get that ($\varphi(m)$, m a maximal ideal of A) is a compact subset of S. Now we consider a topology on M_A by taking the set of the form $\varphi^{-1}(G), G$ open in $(\varphi(m), m$ a maximal ideal of $A))$ and this clearly gives a compact space. This topology is called the Gelfand topology on M_A. Clearly the functions $m \to \hat{x}(m)$ are continuous.

Thus for any commutative Banach algebra with identity A we can consider the family of functions $(\hat{x}(m))_{x\in A}$ defined on M_A and each $\hat{x}(\,,)$ is continuous. We denote by $\hat{A}$ the family $(\hat{x}(m))_{x\in A}$.

Summarizing we have the following result.

THEOREM 5.4.5. *For any commutative Banach algebra with identity there exists a compact Hausdorff space M_A and an algebra of continuous functions on M_A such that A is homomorphic to $\hat{A}$.*

An interesting and important problem arises about the possibility that this homomorphism is an isomorphism or an isometry. In what follows we give some results concerning these questions.

We recall that an element of a Banach algebra with identity is called quasinilpotent if the spectrum reduced to (0).

DEFINITION 5.4.6. A commutative Banach algebra with identity is called semisimple if the set of quasinilpotent elements reduces to (0).

THEOREM 5.4.7. *A necessary and sufficient condition for a commutative Banach algebra with identity to have $A \xrightarrow{\text{isomorphism}} \hat{A}$ (defined above) is that A be semisimple.*

Proof. This follows from the fact that the kernel of the above homomorphism is the set of all quasinilpotent elements of A(i.e. the radical of A).

Let A be a commutative Banach algebra with identity and consider an element $x \in A$. Connected with x we can consider three numbers defined as follows:

$$r_x = \lim \|x^n\|^{1/n}, \qquad \|x\|_{M_A} = \sup_{m\in M_A} (|x(m)|, \qquad r_{\sigma,x} = \sup(|z|, z \in \sigma(x)).$$

Then we have

THEOREM 5.4.8. *For each $x \in A$,*

$$r_x = \|x\|_{M_A} = r_{\sigma,x}.$$

Proof. Since $x - z$ is invertible if and only if for all maximal ideals $m, \hat{x}(m) - z$ is not zero we get that

$$\|x\|_{M_A} = r_{\sigma,x}.$$

Since for each m

$$|\hat{x}(m)| \leqslant \|x^n\|^{1/n} = a_n^{1/n}$$

we get that $\|x\|_{M_A} \leqslant r_x$. But $x - z$ is invertible if and only if z is not a value of $\hat{x}(m), m$ being a maximal ideal of A. This implies that $\|x\|_{M_A} = r_x$.

REMARK 5.4.9. The limit $\lim \|x^n\|^{1/n}$ exists because the sequence a_m satisfies the relation $a_{n+m} \leqslant a_n a_m$

We now consider an important class of Banach algebras for which the isomorphism of Theorem 5.4.7 is an isometry. For this we need some new notions which are considered below.

DEFINITION 5.4.10. Let A be a complex Banach algebra. We say that A is an involutive Banach algebra if there exists a mapping '*' defined on A with values in A such that the properties

(1) $(ax + by) = \bar{a} + x^* + \bar{b} + y^*$,

(2) $(xy)^* = y^* x^*$,

(3) $(x^*)^* = x$,

(4) $\|x^* x\| = \|x\|^2$

are satisfied for all x, y in A and any complex numbers a, b in C.

DEFINITION 5.4.11. If A is an involutive Banach algebra then an involutive subalgebra A_1 of A is any subalgebra of A such that it is invariant for '*', i.e. if $x \in A_1$ then $x^* \in A_1$.

PROPOSITION 5.4.12. *Let A be an involutive Banach algebra and $(A_i)_{i \in I}$ be a family of involutive subalgebras (closed). Then $\cap_i A_i$ is an involutive subalgebra.*

Proof. Obvious.

DEFINITION 5.4.13. Let A be an involutive Banach algebra and $x \in A$. The involutive Banach algebra generated by x is the smallest closed involutive subalgebra of A containing x and the identity of A.

PROPOSITION 5.4.14. *If A is an involutive Banach algebra with unit then the following assertions hold:*

(1) $0^* = 0$,
(2) $1^* = 1$.

Proof. The first assertion is obvious. The proof of the second is given by

$$1^* = 1 \cdot 1^* = 1^{**} \cdot 1^* = (1 \cdot 1^*)^* = (1^*)^* = 1.$$

DEFINITION 5.4.15. Let A be an involutive Banach algebra. An element $x \in A$ is called hermitian if $x^* = x$.

PROPOSITION 5.4.16. *Every element x of an involutive Banach algebra has a unique decomposition*

$$x = x_1 + ix_2$$

with $x_i, i = 1, 2$, hermitian elements.

Proof. Let $x_1 = \frac{1}{2}(x + x^*), x_2 = \frac{1}{2}i(x^* - x^*)$. Then $x = x_1 + ix_2$ satisfies the proposition. Suppose that we have two decompositions,

$$x = x_1 + ix_2 = y_1 + iy_2$$

with $x_i, y_i, i = 1, 2$ hermitian elements. This obviously reduces the problem to the following: if $0 = z_1 + iz_2$, $z_i, i = 1, 2$ hermitian elements, then $z_1 = z_2 = 0$.

Indeed, if $0 = z_1 + iz_2$, $z_i, i = 1,2$, hermitian elements, the element (iz_2) is a hermitian element which implies that $(iz_2)^* = iz_2$, i.e. $-iz_2 = iz_2$ which implies that $2iz_2 = 0$. This gives that $z_2 = 0$ and thus also $z_1 = 0$. The uniqueness is thus proved.

DEFINITION 5.4.17. An element x of an involutive Banach algebra A is called normal if $x^* x = xx^{**}$.

Then it is easy to prove the following result.

THEOREM 5.4.18. *An element x is normal if and only if $x = x_1 + ix_2$ with $x_1 x_2 = x_2 x_1$.*

DEFINITION 5.4.19. An element of an involutive Banach algebra with identity is positive if it is hermitian and the spectrum is in R^+. An element u is called unitary if $u^* u = uu^{**} = 1$.

Now we give some examples of involutive Banach algebras.

EXAMPLE 5.4.20. Let S be a compact Hausdorff space and let $C(S)$ be the Banach algebra of all complex continuous functions on S with the sup norm,

$$f \to \|f\| = \sup(|f(s)|, s \in S).$$

If we define '*' using the complex conjugation,

$$(f^*)(s) = (\overline{f(s)})$$

then it is obvious that $C(S)$ is a commutative involutive Banach algebra with identity.

EXAMPLE 5.4.21. Let H be a complete inner product space and let $L(H)$ be the Banach algebra $L(H)$ of all continuous linear operators on H. If we set as '*' the mapping which, for $T \in L(H)$, gives T^*, the adjoint of T, then this defines on $L(H)$ an involution and $L(H)$ is an involutive (noncommutative) Banach algebra with identity.

EXAMPLE 5.4.22. Let S be a locally compact space and let $C_0(S)$ be the Banach algebra of all continuous complex-valued functions defined on S which are null at infinity. With the mapping '*' defined by

$$(f^*)(s) = (\overline{f(s)}) \quad \text{(the bar denotes the complex conjugate)}$$

this is a commutative involutive Banach algebra without identity.

REMARK 5.4.23. As an example of a locally compact space which is noncompact we note R (the set of real numbers) or, more generally, R^n, C^n.

The following theorem will be of fundamental importance in the theorems about the structure of involutive commutative Banach algebras with identity.

THEOREM 5.4.24. *If A is an involutive commutative Banach algebra with identity then for each $x \in A$,*

$$r_x = \|x\|.$$

(in other words, each element is normaloid).

Proof. First we show that for each integer n,

$$\|x^{2^n}\| = \|x\|^{2^n}.$$

From the formula (known as the Beurling–Gelfand formula for the spectral radius)

$r_x = \lim \|x^n\|^{1/n}$, the assertion follows. First we remark that we have

$$\|x^* x^2\| = \|x^* x\|^2$$

and further

$$\|(x^* x)^{2^n}\| = \|x^* x\|^{2^n}.$$

But $\|x^* x\| = \|x\|^2$ and thus

$$\|x\|^2 = \|x^* x\|^{2^n/2^n} = \|x^{*2^n} x^{2^n}\|^{1/2^n} = \|x^{*2^n}\|^{1/2^n}\|x^{2^n}\|^{1/2^n}$$

which gives that

$$\|x\|^2 = r_x r_{x^*}.$$

Since $r_x = r_{x^*}$ we get that $r_x = \|x\|$ and the assertion is proved.

THEOREM 5.4.25 (Arens, 1947). *If A is as in Theorem 5.4.24 and x is a hermitian element, then $\sigma(x) \subset R$.*

Proof. Suppose that the assertion is not true. Then there exists in $\sigma(x)$ a point z with $\operatorname{Im} z \neq 0$. Consider now the element x_t, for t real, defined by the formula

$$x_t = x + it$$

which has in the spectrum the point $z + it$ and $z^* - it$ is in the spectrum of x_t^*. The element $x_t x_t^*$ is hermitian and if $z = u + iv$ then

$$u^2 + (v + t)^2 = |z + it|^2 \leqslant \|x_t x_t^*\| = x^2 + t^2$$

or

$$u^2 + v^2 + 2ut \leqslant \|x\|^2.$$

But t is arbitrary and this is then a contradiction which proves the theorem.

Now the basic result of this section is the following structure theorem for involutive commutative Banach algebras with identity.

THEOREM 5.4.26 (I. M. Gelfand–M. A. Naimark). *If A is a commutative, involutive Banach algebra with identity then it is isomorphic and isometric with an algebra of continuous functions on a compact Hausdorff space.*

Proof. Let us consider the algebra A. Since each $x \in A$ has the form $x = x_1 + ix_2$ *with* $x_i, i = 1, 2$, hermitian elements we have that

$$x = x_1 + ix_2 \quad \text{and} \quad x^* = x_1 - ix_2.$$

Since the functions $x_i, i = 1, 2$, are real-valued we get that the involution '*' of A is transformed in the natural involution of the algebra $C(M_A)$, i.e. the complex conjugation. Now A has the property that separates the points of M_A (i.e. for each pair m_1, m_2 of distinct point of M_A there exists an x such that $x(m_1)$ is not equal to $x(m_2)$). This implies that A is dense in $C(M_A)$ by the well-known Stone–Weierstrass theorem. Since we have

$$\|x\|_{M_A} = \sup_{m \in M_A} |x(m)| = r_x = \|x\|$$

we have that $x \to \hat{x}$ is an isometry. The density implies that $A = C(M_A)$. The theorem is proved.

Now we use this theorem to obtain a structure theorem for involutive, commutative Banach algebras without identity. For this purpose we remark that any such algebra can be embedded isometrically and isomorphically into an algebra of the same type but with identity. Indeed, we consider for this purpose the family of all pairs (z, x), where $z \in C$ and $x \in A$. We define the corresponding operations by

(1) $(z_1, x) + (z_2, y) = (z_1 + z_2, x + y)$,

(2) $(z_1, x)(z_2, y) = (z_1 z_2, z_2 x + z_1 y + xy)$

(3) $a(z, x) = (az, zx)$

and the norm of a pair (z, x) by

$$(z, x) = \sup_{u \in A, \|u\| = 1} \|zu + ux\|.$$

It is not difficult to see that with these operations the family of all pairs $\{(z, x)\}$ is a Banach algebra with identity (the pair (1,0)). Clearly A may be identified with the subfamily $\{(0, x), x \in A\} = A^{\sim}$. From the form of $A^{\sim}$ it is clear that it is an ideal in $\{(z, x)\}$.

REMARK 5.4.27. If A is a noncommutative Banach algebra this construction works also and we obtain a noncommutative algebra with identity.

We now show that in fact $A^{\sim}$ is a maximal ideal of $\{(z, x)\}$. Indeed, if there exists an ideal m_1 containing $A, m_1 \neq A$ then there exists an element in m_1 of the form (z, x) with $z \neq 0$. Then $(z, x) - (0, x)$ is in m_1 which gives that $(1, 0)$ is in m_1. This is a contradiction and the assertion is proved.

Now, for reader's convenience we prove that $\sup_{a \in A, \|a\| = 1} \|za + ax\|$ defines a norm on $\{(z, x)\}$. The properties

(1) $\|(z_1 + z_2, x + y)\| \leqslant \|(z_1, x)\| + \|(z_2, y)\|$

(2) $\|a(z, x)\| = |a| \|(z, x)\|$

are obvious. Now if $\|(z, x)\| = 0$ we show that $z = 0$ and $x = 0$. We suppose, of course, that (z, x) is not the pair $(0, 0)$.

Case 1. $(z, x) = (z, 0)$. Then clearly $z = 0$.

Case 2. $(z, x) \neq (0, 0)$ but $(z, x) = (0, x)$. Then $x^* \neq 0$ and from the hypothesis we get that $\|x^* x\| = \|x\|^2 = 0$. Thus $x = 0$.

Case 3. $(z, x) \neq (0.0)$, $z \neq 0$, $x \neq 0$. Then we get that

$$zu + xu = 0$$

for all $u \in A$. We can consider the element $e = -x/z$ which is in A. Then the above relation gives that $u = eu$ for all $u \in A$. This implies that e is an identity for A, which is a contradiction. Thus we must have $z = 0$ and $x = 0$.

If A has an involution '*' then on the algebra $\{(z, x)\}$ we can define an involution by the formula

$$(z, x) \to (z, x)^* = (z^*, x^*).$$

We now show that this defines (also in the noncommutative case) on $\{(z, x)\}$ an involution such that $\{(z, x)\}$ is an involutive Banach algebra with identity. For this purpose we remark that all properties of an involution are clear, except possibly the property

$$\|(z, x)^*(z, x)\| = \|(z, x)\|^2.$$

Here is the proof of this property. Of course we can suppose without loss of generality that $\|(z, x)\| \neq 0$.

Let $\varepsilon > 0$ and take p, q in A, $\|p\| = \|q\| = 1$ with the property

$$\varepsilon\|(z, x)\| \leqslant \|zp + px\|$$
$$\varepsilon\|(z^*, x^*)\| \leqslant \|z^*q + qx^*\|.$$

Then we have

$$\varepsilon^3\|(z, x)\|\,\|(z^*, x^*)\| \leqslant \varepsilon\|z^*q + qx^*\|\,\|zp + px\|$$
$$= \varepsilon\|zq^* + q^*x\|\,\|zp + px\| \leqslant \varepsilon\|(z, x)\|\,\|zp + px\|$$

and further

$$\varepsilon^3\|(z, x)\|\,\|(z^*, x^*)\| \leqslant \|zp + px\|^2 = \|(z^*p^* + p^*x^*)(zp + px)\|.$$

Also,

$$\|zp + px\|\,\|z^*p^* + p^*x^*\| = \|(z^*p^* + p^*x^*)(zp + px)\|$$
$$= \|z^*zp^*p + z^*p^*px + zp^*x^*p + p^*x^*px\|$$
$$\leqslant \|(z^*, x^*)(z, x)\|$$

because $\|p^*p\| = 1 = \|p\| = \|p^*\|$.

Since the inequality in the converse sense is easy to prove, we get the equality

$$\|(z, x)^*(z, x)\| = \|(z, x)\|^2.$$

We now indicate briefly how we can prove that $\{(z, x)\}$ is complete.

LEMMA 5.4.28. *Let $\{(z_n, x_n)\}$ be a sequence in $\{(z, x)\}$ and suppose that converges to* $(0, 0)$. *Then the following assertions hold*:

(1) $\lim z_n = 0$.

(2) *lim* $x_n = 0$.

Proof. If $\lim z_n = 0$ then from the inequality

$$\|(z_n, x_n)\| \geqslant \|z_n x_n^* + x_n^* x_n\|/\|x_n^*\| \geqslant \|x_n\| - |z_n|$$

we obtain that $\lim \|x_n\| = 0$ and this implies that $\lim x_n = 0$. Now if $\lim z_n = 0$ is not true then there exists $\varepsilon > 0$ and a subsequence (z_{n_k}) such that $|z_{n_k}| \geqslant \varepsilon$. We can suppose that just (z_n) has this property. Since

$$\|(1,x_n/z_n) - (1,x_m/z_m)\| = \|(0,x_n/z_n - x_m/z_m)\| \geqslant \|x_n/z_n - x_m/z_m\|$$

and

$$\|(1, x_n/z_n) - (1, x_m/z_m)\| \leqslant 1/\varepsilon(\|(z_n, x_n)\| + \|(z_m, x_m)\|)$$

we see that (x_n/z_m) is a Cauchy sequence. Let $u = \lim x_n/z_n$. Then we have for all $y \in A$,

$$\|y + uy\| = \lim \|y + x_n y/z_n\| = \lim 1/|z_n| \|z_n y + x_n y\| \leqslant$$
$$\leqslant \|y\|/\varepsilon \lim \|(z_n, x_n)\| = 0.$$

Thus $y = -uy$ for all $y \in A$ which means that $-u$ is an identity for A. This contradiction proves the lemma.

Now we are ready to show the completeness of $\{(z,x)\}$. From the lemma we see that if $\{(z_n, x_n)\}$ is a Cauchy sequence then the double sequence

$$((z_n - z_m, x_n - x_m))$$

converges to zero. Then (z_n) and (x_n) are Cauchy sequences and let

$$z = \lim z_n, \ x = \lim x_n.$$

Thus $((z_n, x_n))$ converges to (z,x).

Now we can give the structure theorem for commutative, involutive Banach algebras without identity.

THEOREM 5.4.29. *Let A be a commutative, involutive and without identity Banach algebra. Then A is isomorph-isometrical with a Banach algebra of functions which are null at infinity on a locally compact Hausdorff space.*

Proof. We consider the Banach algebra $\{(z,x)\}$ as above. We know that $((0,x), x \in A)$ is a maximal ideal space of $\{(z,x)\}$. By the structure theorem for commutative, involutive Banach algebras with identity $\{(z,x)\}$ is isomorphic and isometric with an algebra of continuous complex-valued functions on the compact Hausdorff space $M_{\{(z,x)\}}$. Then we get that A is isomorphic and isometric with the family of all functions on $M_{\{(z,x)\}}$ which are zero at the maximal ideal $((0,x), x \in A)$. The space $M_{\{(z,x)\}} - ((0,x, x \in A)$ with the induced topology is clearly locally compact, which proves the theorem.

5.5. The Representations of B*-Algebras with Identity

We now give the theorem of Gelfand–Naimark about the representation of B*-algebras as closed selfadjoint subalgebras of operators on a complete inner product space (over C).

We now consider A to be a B*-algebra, i.e. a Banach algebra as in Definition

5.4.10. We remark that from the axiom

$$\|x^* x\| = \|x\|^2$$

we easily obtain that

$$\|x^* x\| = \|x^*\|\|x\|$$

as we now show. More complicated is the converse assertion, i.e. if there exists on A an involution '*' satisfying properties (1)–(3) of Definition 5.4.10 and

$$\|x^* x\| = \|x\|\|x^*\|,$$

then necessarily it satisfies axion (4) of Definition 5.4.10. (This and other information on B*-algebras can be found in almost all books on the subject.)

Let us suppose that A is a B*-algebra and take $x \in A$ arbitrary. Then we have that

$$\|x\|^2 = \|x^* x\| \leqslant \|x\|\|x^*\|$$

and thus if $x \neq 0$ we obtain the inequality

$$\|x\| \leqslant \|x^*\|$$

and applying this inequality for x^* we get that in fact $\|x^*\| = \|x\|$. Suppose now that $x \in A$ is hermitian, i.e. $x = x^*$. Then we get that any power of x, x^n, is a hermitian element and thus

$$\|x^2\| = \|x\|^2 \quad \text{and} \quad \|x^{2n}\| = \|x\|^2.$$

This obviously implies that for any hermitian element x $r_x = \|x\|$. We recall that we have proved that for any hermitian element, $x, \sigma(x) \subset R$.

DEFINITION 5.5.1. A state φ of a Banach algebra A with identity and a B*-algebra is any element of A^* satisfying the property

$$\|\varphi\| = \varphi(e) = 1.$$

Concerning the states and hermitian elements we can prove the following result.

THEOREM 5.5.2. *Let A be a B^*-algebra with identity. Then the following hold*:

(1) *If h is a hermitian element and $r \in \sigma(h)$ then there exists a state of A, φ, such that $\varphi(h) = r$,*
(2) *h is positive if and only if for any state φ of $A, \varphi(h) \geqslant 0$,*
(3) *If h, k are positive elements then $ah + bk$ is positive whenever $a \geqslant 0, b \geqslant 0$,*
(4) *If h is hermitian then $h = h_1 - h_2$ with $h_i, i = 1, 2$, positive elements and $h_1 h_2 = 0$,*
(5) *h is positive if and only if $h = a^* a$ for some $a \in A$.*

Proof. (1) Let A_h be the commutative B*-algebra generated by h and e. Then A_h is isomorphic and isometric with an algebra of continuous complex-valued functions on a compact Hausdorff space M_{A_h} which, in this case, may be identified with $\sigma(h)$. On A_h we have, for any polynomial $p(h)(p(t) = a_i t^i)$, $a_i \in R$)),

$$f(p(h)) = p(r)$$

and this clearly is a state of A_h. The extension to A is a state of A. Clearly this satisfies the assertion.

(2) If h is positive, then we have that for any state of $A, \varphi, \varphi/A_h$ is positive and thus the value at h is positive. Then we get a functional induced by a state on A_h has a norm 1 and the value at e(the identity of A) is 1. This easily implies that the functional is positive. This gives that $\sigma(h) \subset R^+$. Conversely, if ($\varphi(h) \geqslant 0$, φ a state of A), by (1) any point in the spectrum of h is in R^+ and this gives our assertion.

(3) This follows from (2), directly.

(4) This follows from the fact that any continuous function is the difference of two positive functions $f = f_1 - f_2$ and $f_1 f_2 = 0$,

(5) Suppose first that h is positive. Then the function $\hat{h}$ on $\sigma(h)$ is positive and thus $\hat{h}^{1/2}$ exists and is continuous. Then let $x^{1/2}$ be an element in A_h with $\hat{x}^{1/2} = \hat{h}^{1/2}$. Then clearly we have $x = x^{1/2} \cdot x^{1/2}$. Conversely, if $h = a^* a$ for some $a \in A$ then we show that h is positive. Obviously h is hermitian. We now show that it is positive. We have the decomposition

$$h = h_1 - h_2$$

with h_i hermitian elements. Clearly we have the property $h_1 h_2 = h_2 h_1 = 0$. We note that the h_i are also positive and we show now that $h_2 = 0$. Since

$$(ah_2)^*(ah_2) = h_2 a^* a h_2 = h_2(h_1 - h_2)h_2 = -h_2^3$$

the positivity of h_2 implies that

$$(ah_2)^*(ah_2) = 0.$$

But $ah_2 = u + iv$ with u, v hermitian elements. Then we get further

$$(ah_2)^*(ah_2) = 2(u^2 + v^2) = 0.$$

But clearly $u^2 + v^2 = 0$ and thus we must have

$$u^2 = v^2 = 0.$$

This implies that $h_2 = 0$, and proves (5).

We give now a result which is very important in the proof of the Gelfand–Naimark theorem.

PROPOSITION 5.5.3. *Let A be a B^*-algebra with identity e. Let φ be a state of A. Then the following assertions hold*:

(1) $\varphi(x^*x) \geqslant 0, \quad x \in A$.

(2) $\varphi(x) = \varphi(x^*)^*, \quad x \in A$.

(3) *if $\langle\ ,\ \rangle_\varphi$ is defined by $\langle x, y\rangle_\varphi = \varphi(y^* x)$ then $\langle\ ,\ \rangle_\varphi$ is a function satisfying all the properties of an inner product except (possibly) the property $\langle x, x\rangle_\varphi = 0$ if and only if $x = 0$.*

(4) $\langle xz, y\rangle_\varphi = \langle z, x^* y\rangle_\varphi$.

Proof. Properties (1) and (2) follow from the results just obtained and property (3) follows from the definition of an involution. For (4) we remark that

$$\langle xz, y\rangle_\varphi = \varphi(y^* xz) = \varphi((x^* y)^* z) = \langle z, x^* y\rangle_\rho.$$

PROPOSITION 5.5.4. *Let A be a B^*-algebra with identity e and φ be a state of A. Then the following assertions hold:*

(1) *$I_\rho = (x, \langle x, x\rangle_\rho = 0)$ is a left (closed) ideal of A.*

(2) *$H_\varphi = A/I_\varphi$ is an inner product space with an inner product $\langle\, ,\rangle_\varphi$.*

(3) *If for $a \in A$ we set $T_a^\varphi : H_\varphi - H_\varphi$ defined by*

$$T_a(x) = ax + I_\varphi$$

then $\|T_a^\varphi\| \leqslant \|a\|$, $\langle a, a\rangle_\varphi \leqslant \|T_a^\varphi\|^2$.

Proof. Since $\langle\, ,\rangle_\varphi$ is almost an inner product it is easy to see that the Cauchy–Buniakowsky inequality holds:

$$|\langle x, y\rangle_\varphi|^2 \leqslant \langle x, x\rangle_\varphi \langle y, y\rangle_\varphi.$$

This implies that I_φ is a linear space and also the inequality

$$\langle yx, yx\rangle_\varphi = \langle x, y^* yx\rangle_\varphi \leqslant (\langle x, x\rangle_\varphi \langle y, y\rangle_\varphi)^{1/2}$$

gives that I_φ is a closed left ideal of A. Clearly H_φ is an inner product space by the above remarks. Since I_φ is a closed left ideal of A, the operator T_a^φ is well defined. Now we have,

$$\|ax\|_\varphi^2 = \varphi(x^* a^* ax)$$

and

$$\|a\|^2 \|x\|_\varphi^2 = \varphi(\|a\|^2 x^* x)$$

from which we get

$$\|a\|^2 \|x\|_\varphi^2 - \|ax\|_\varphi^2 = \varphi(x^*(\|a\|^2 - a^* a)x).$$

Since

$$\|a\|^2 - a^* a \geqslant 0$$

we get

$$\varphi(x^*(a^2 - a^* a)x) \geqslant 0$$

and thus

$$\|ax\|_\varphi^2 \leqslant \|a\|^2 \|x\|_\varphi^2,$$

i.e.

$$\|T_a^\varphi\| \leqslant \|a\|.$$

Since

$$\|ae\|_\varphi^2 = \varphi(a^* a) = \|a\|_\varphi^2$$

we get that

$$\|T_a^\varphi\| \geqslant \|a\|_\varphi.$$

Thus the proposition is proved.

Now for each state φ of A we denote by $\tilde{H}_\varphi$ the completion of H_φ and since $\|T_a^\varphi\| \leqslant \|a\|, T_a^\varphi$ induces an operator on $\tilde{H}_\varphi$, denoted by $\tilde{T}_a^\varphi$, with the same norm. Then $a \to T_a^\varphi$ is an *-homomorphism of A into $L(\tilde{H}_\varphi)$. Also we have $(\tilde{T}_{a^*})^* = T_a^\varphi$ for each $a \in A$. Now the Gelfand–Naimark representation theorem for **B***-algebras is as follows.

THEOREM 5.5.5 (Gelfand–Naimark). *Let A be a B^*-algebra with identity. Then A is *-isomorphical and isometric with a self-adjoint (i.e. an algebra with the property that whenever x is in the algebra then x^* is also in the algebra) of operators on a complete inner product space.*

Proof. Let S be the family of all states of A and for each φ in S we consider the space $\tilde{H}_\varphi$ constructed above. We now consider the space of all elements $h = (h\varphi)_{\varphi \in S}$ with the properties:

(1) h_φ is in $\tilde{H}_\varphi$,

(2) $\sum_S \|h_\varphi\|_\varphi^2 < \infty$.

We denote the family of such elements by H and it is clear that it is a complete inner product space with an inner product

$$\langle h,r\rangle = \sum \langle h_\varphi, r_\varphi\rangle_\varphi.$$

For each $a \in A$ we define an operator on H by the formula

$$T_a(h) = (\tilde{T}_a^\varphi h_\varphi)_{\varphi \in S}.$$

Clearly we have $\|T_a\| \leqslant \|a\|$. Also $T_{a^*} = (T_a)^*$ and thus $a \to T_a$ is the *-homomorphism of A into $L(H)$. Let a be hermitian in A and thus there exists $\varphi \in S$ such that $\varphi(a^2) = \|a\|^2$. This implies that for hermitian elements of $A, a \to T_a$ is an isometry. Now we show that $a \to T_a$ is an isometry for all $a \in A$. Indeed, if a is arbitrary in A the element $a^* a$ is hermitian. Thus we have, by the above remark,

$$\|a^* a\| = T_{a^* a}\| = \|T_{a^*} T_a\| = \|a^*\|\|a\|$$

which implies that $\|T_a\| = \|a\|$. The theorem is proved.

5.6. Approximate Identities in Banach Algebras

The presence of an identity in a Banach algebra is of great importance, and if the algebra has no identity then we have seen that we can construct a new Banach algebra containing an ideal which is isomorphic and isometric with the original algebra. This new algebra has an identity. But there are problems which cannot be treated fully even with the help of the new algebra which possesses an identity. There are some cases in which the absence of an identity can be replaced successfully by a so-called 'approximate identity'.

DEFINITION 5.6.1. Let A be a Banach algebra and let $(e_i)_{i\in I}$ be a net of elements of A. This net is said to be an approximate identity for A if for each x in A the following relations hold:

$$\lim \|xe_i - x\| = 0 \tag{*}$$

$$\lim \|e_i x - x\| = 0. \tag{**}$$

If there exists a constant $M < \infty$ such that $\|e_i\| \leqslant M$ for all i in I then we say that A has a bounded approximate identity. Also, if only (*) holds then we say that A has a right approximate identity; if on the contrary, only (**) holds then we say that A has a left approximate identity.

Not every Banach algebra without identity has an approximate identity. We show, in what follows that some Banach algebras without identity have an approximate identity.

DEFINITION 5.6.2. A summability kernel on R (the reals) is a family of functions (k_s), s being a discrete or continuous parameter such that the following conditions are satisfied:

(1) For each s,

$$\int_{-\infty}^{\infty} |k_s(t)\,\mathrm{d}t| = 1,$$

(2)

$$\|k_s\|_{L^1(R,dx)} = 0(1) \text{ as } s \to \infty,$$

(3)

$$\lim_{s\to\infty} \int_{|t|\geqslant d} |k_s(t)\,\mathrm{d}t| = 0 \text{ for all } d > 0.$$

An example of such a summability kernel is the so-called Fejér kernel on R defined by

$$k_s(t) = sk(st),\ s > 0.$$

with

$$k(t) = \tfrac{1}{2}((\sin t/2)/(t/2))^2 = \tfrac{1}{2}\int_{-1}^{1} (1 - |s|)\mathrm{e}^{ist}\,\mathrm{d}s.$$

Then, using the results about Fejér's kernel given in Section 3.4 we can prove that the following result holds.

PROPOSITION 5.6.3. *If (k_s) is a summability kernel on R then (k_s) is an approximate identity for $L^1(R, \mathrm{d}s)$ with the multiplications given by the convolution.*

We now show that a certain class of Banach algebra having an involution has an approximate identity.

Let A be a Banach algebra with involution and $\bar{A}$ be the Banach algebra obtained from A by adjoining an identity. Then A has a natural involution. We denote by $\tilde{\sigma}(a)$ the spectrum of a as an element of $\bar{A}$.

DEFINITION 5.6.4. An element of a is called positive if a is hermitian (i.e. $a^* = a$) and $\tilde{\sigma}(a)$ is a subset of R^+ (in which case we say that the spectrum of a is positive, and we then call a a positive element of A.

DEFINITION 5.6.5. We say that two hermitian elements of A, h and k, are in the relation $h \leqslant k$ if $k - h$ is positive.

RELATION 5.6.6. This relation coincides with the ordinary order relation in the set of all bounded hermitian operators defined on an inner product space (of course complete). We now consider A be a B*-algebra. If $p(z)$ is a complex polynomial then for any element a of A we can define (when A has an identity, if A is without identity we consider a polynomial with $p(0) = 0$) the element $p(a)$ by the usual formula. Now if f is a real-valued function defined on some interval $[0, T]$ of R then there exists a sequence of polynomials of the form $p_n(t) = tg_n(t)$ with q_n a polynomial, converging uniformly (on $[0, T]$) to f. Then for any normal element N of the B*-algebra $A, p_n(N)$ converges to an element of A which we denote by $f(N)$. It is not difficult to see that $f(N)$ is independent of the sequence of polynomials converging to f. For further details we refer the reader to the author's book on linear operators (1981), where the results are presented for hermitian operators but the case of B*-elements presents no difficulties.

Let $\varepsilon > 0$ and let h be a positive element of the B*-algebra A (without unit). We consider on R the function

$$f(t) = t/(t + \varepsilon) = t(t + \varepsilon)^{-1}.$$

Then we have the following result.

PROPOSITION 5.6.7. *Let h_1 and h_2 be two hermitian positive elements of the B*-algebra A (without identity). Then the following assertions hold*:

(1) *If $h_1 \leqslant h_2$ then $f_\varepsilon(h_1) \leqslant f_\varepsilon(h_2)$.*

(2) $\lim \| f_\varepsilon(h_1)x - x \| = 0$

for all x in the closed right ideal of A generated by h_1.

(3) $\lim \| yf_\varepsilon(h_1) - y \| = 0$

for all y in the closed left ideal of A generated by h_1.

Proof. If $\bar{A}$ denotes the B*-algebra obtained from A by adjunction of an identity and if $h_1 \leqslant h_2$ then clearly

$$(h_1 + \varepsilon) \leqslant (h_2 + \varepsilon)$$

and thus

$$(h_2 + \varepsilon)^{-1/2}(h_1 + \varepsilon)(h_2 + \varepsilon)^{-1/2} \leqslant 1$$

(1 denotes the identity of $\bar{A}$. This implies that

$$(h_1 + \varepsilon)^{-1} \geqslant (h_2 + \varepsilon)^{-1}.$$

This gives further that

$$f_\varepsilon(h_1) = (h_1 + \varepsilon)^{-1} h = 1 - \varepsilon(h_1 + \varepsilon)^{-1} \leqslant 1 - \varepsilon(h_2 + \varepsilon)^{-1} = h_2(h_2 + \varepsilon)^{-1} = f_\varepsilon(h_2).$$

But

$\|f_\varepsilon(t)\| \leqslant 1$ and thus, from the normality of $f_\varepsilon(h_1)$ we have that

$\|f_\varepsilon(h_1)\| \leqslant 1.$

We define

$$m_{h_1} = (x, x \in A, \lim \|f_\varepsilon(h_1)x - x\| = 0)$$

which is obviously a right ideal of A. Since

$$h_1 - f_\varepsilon(h_1)_1 = h_1 - (h_1 + \varepsilon)^{-1} h_1^2 = \varepsilon(h_1 + \varepsilon)^{-1} h_1 = \varepsilon f_\varepsilon(h_1)$$

we obtain that h_1 is in m_{h_1}.

Let x be in $\bar{m}_{h_1}$. From the definition of $\bar{m}_{h_1}$ there exists a sequence (x_n) in m_{h_1} with the property that given $\delta > 0$ there exists an integer N_ε such that if $n \geqslant N_\varepsilon$

$$\|x - x_n\| \leqslant \delta.$$

Let δ_1 such that for $0 < \varepsilon < \delta_1$

$$\|f_\varepsilon(h_1)x_n - x_n\| \leqslant \delta_1.$$

Then for ε in $(0, \delta_1)$ we have,

$$\|f_\varepsilon(h_1)x - x\| = \|f_\varepsilon(h_1)x - f_\varepsilon(h_1)x_n + f_\varepsilon(h_1)x_n - x_n + x_n - x\| \leqslant 2\delta + \delta_1.$$

This obviously implies that $m_{h_1} = \bar{m}_{h_1}$. The other assertions may be proved similarly.

We are now ready to prove the existence of an approximate identity in each B*-algebra. This approximate identity is constructed using the partial order '$\leqslant$' defined on the set of all positive elements of A.

Let $B_0(1) = (x, x \in A, \|x\| < 1)$ be the open unit ball of A. If m is a left (right) ideal of A then m is a left (right) ideal of $\bar{A}$.

PROPOSITION 5.6.8. *For any left (right) ideal of a B*-algebra A the set* $B_0(1) \cap m_+$, *where* $m_+ = m \cap H^+(A), H^+(A) =$ *(a, a positive) is an approximate identity for the closure* $\bar{m}$ *of m.*

We shall use the following notation: $B_0(1) \cap m_+ = (u_u, u \in B_0(1) \cap m_+, B_0(1) \cap m_+$ is considered also as an index set).

Proof. To prove the assertion of the proposition it suffices to show that $B_0(1) \cap m_+$ is a directed set for the (partial) order relation '$\leqslant$' and that the conditions of the approximate identity are satisfied.

If p, q are in $B_0(1) \cap m_+$ then we set

$$P = (1 - p)^{-1} p, \qquad Q = (1 - q)^{-1} q$$

and thus

$$p = (1 + P)^{-1} P, \qquad q = (1 + Q)^{-1} Q.$$

Then $r = p + q$, $R(1 + r)^{-1}r$ are in m_+ and also $R \geqslant p \cdot R \geqslant q$. But

$$||R|| = ||r||/(1 + ||r||) < 1$$

which implies that R is in $B_0(1) \cap m_+$. This shows that $B_0(1) \cap m_+$ is a directed set for the (partial) order '$\leqslant$'.

We now show that the conditions of the approximate identity are satisfied. Indeed, let $\varepsilon > 0$. If x is in m then x^*x is in m_+ and

$$u_\varepsilon = (x^*x + \varepsilon)^{-1}x^*x \in m \cap B_0(1).$$

Also,

$$||x(1 - u_\varepsilon)^{1/2}|| = ||x(x^*x + \varepsilon)^{-1/2}|| = \varepsilon||(x^*x + \varepsilon)^{-1/2}x^*x(x^*x^*_{+\varepsilon})^{-1/2}||^{1/2} =$$
$$= \varepsilon||(x^*x + \varepsilon)^{-1}x^*x|| \leqslant \varepsilon.$$

Let $y \geqslant u_\varepsilon$, $y \in m_+ \cap B_0^{(1)}$. Then we have,

$$||x - xy|| = ||x(1 - y)|| \leqslant ||x(1 - y)^{1/2}|| ||(1 - y)^{1/2}|| \leqslant ||x(1 - y)x^*||^{1/2}$$
$$= r^{1/2}_{x(1-y)x^*} \leqslant r^{1/2}_{x(1-u_\varepsilon)x^*} = ||x(1 - u_\varepsilon)x^*||^{1/2} =$$
$$= ||x(1 - u_\varepsilon)^{1/2}|| \leqslant \varepsilon.$$

We now define

$$n = (x, x \in A, \quad \lim_{u \in m_+ \cap B_0(1)} \quad xu = x)$$

which is clearly a right ideal of A. We note that m is contained in n and we show that n is closed (i.e. $n = \bar{n}$).

Let x be an element of $\bar{n}$ and thus if $\varepsilon > 0$ is given there exists y in n such that $||x - y|| < \varepsilon$. Further, by the above results, we find $\bar{u}$ in $m_+ \cap B_0(1)$ show that if $u \geqslant \bar{u}$, u in $m_+ \cap B_0(1)$,

$$||y - yu|| \leqslant \varepsilon.$$

Then we obtain the following relations

$$||x - xu|| = ||x - y + y - yu + yu - xu|| \leqslant 3\varepsilon.$$

This obviously implies that n is closed. This clearly gives the assertion of the proposition. From there we obtain the following important result.

THEOREM 5.6.9. *Every B^*-algebra A has an approximate identity. If the B^*-algebra is a separable Banach space then the approximate identity may be a sequence.*

Proof. The first assertion was proved in Proposition 5.6.8. Suppose now that A is a B*-algebra which is a separable Banach space, and thus there exists a sequence (u_n) which is dense in $H^+(A) \cap B_0(1)$. We define a sequence in $B_0(1) \cap H^+(A)$ as follows; if $\bar{u}_n$ was defined, then $\bar{u}_{n+1}$ is an element of $B_0(1) \cap H^+(A)$ with the following properties:

(1) $\bar{u}_{n+1} \geqslant u_i, \quad i = 1, 2, 3, \ldots, n,$

(2) $\bar{u}_{n+1} \geqslant \bar{u}_n.$

That such an element exists follows clearly from Proposition 5.6.8. Now, since for every normal element of a B*-algebra the spectral radius r_a of a coincides with the norm of a, it follows that, on the set of positive elements of A, to take the square root is a continuous function. Thus if $\varepsilon > 0$ is given for x in A there exists N_ε such that for all $n \geqslant N_\varepsilon$,

$$\|x(1 - \bar{u}_n)^{1/2}\| \leqslant \varepsilon$$

and this implies that

$$\|x - xu_n\| \leqslant \varepsilon.$$

Thus (u_n) is a right approximate identity for A. But (u_n^*) (which is equal to (u_n)) is then a left approximate identity for A and the theorem is proved.

The above result permits us to prove the following important result which, for the case of the B*-algebra of all bounded linear operators on a complete inner product space, was proved by Calkin (1941); for the general B*-algebras this result was proved by Kaplansky (1951).

THEOREM 5.6.10. (Calkin–Kaplansky). *Let A be a B*-algebra and m be a self adjoint closed ideal of A. Then A/m is a B*-algebra.*

Concerning the terminology used in the formulation of the above result we mention that a subset of a Banach algebra with an involution is called selfadjoint whenever x is in M. Then x^* is also in M.

Proof. Let $(u_i)_{i \in I}$ be an approximate identity for m. Then we have that for all x in m the following relations hold:

$$\lim \|u_i x - x\| = \|\lim \|x^* u_i - x^*\| = 0.$$

Thus $x^* u_i \in m$ for all i in I. Now we know that the involution of A induces in A/m an involution '*'. To prove that A/m with this involution is a B*-algebra it suffices to prove that for all x in A,

$$\|\hat{x}^* \hat{x}\| = \|\hat{x}\|^2,$$

where $\hat{x}$ denotes the class of x in A/m.

For this we remark that it suffices to prove the inequality

$$\|\hat{x}^* \hat{x}\| \geqslant \|\hat{x}\|^2$$

(since the contrary inequality is obvious). Let y be an element of m. Then we have

$$\lim \|y - yu_i\| = 0$$

and thus

$\lim \sup\|x - xu_i\| = \lim \sup\|x - xu_i + xu_i - yu_i - y\| = \lim\sup \|(x + y)(1 - u_i)\| \leqslant \|x + y\|$. Since by definition of the norm in A/m,

$$\|\hat{x}\| = \inf(\|x + y\|, y \in m) \geqslant \lim\sup(\|x - xu_i\|)$$
$$\geqslant \lim\inf(\|x - xu_i\|) \geqslant \inf(\|x + y\|, y \in m) = \|\hat{x}\|$$

and thus we obtain the following useful formula:

$$\|\hat{x}\| = \lim\|x - xu_i\|.$$

But A is supposed to be a B*-algebra and thus,

$$\|\hat{x}\|^2 = \lim \|x - xu_i\|^2 = \lim \|(x - xu_i)^*(x - xu_i)\| = \lim \|(1 - u_i)x^*x(1 - u_i)\|$$

$$= \lim \|(1 - u_i)x^*x(1 - u_i) + (1 - u_i)\| y(1 - u_i)\| = \lim (\|1 - u_i)(x^*x + y)$$
$$(1 - u_i)\| \leqslant \|x^*x + y\|$$

for y in m. But then

$$\|\hat{x}\|^2 = \inf_{y \in m}(\|x^*x + y\|) = \|\hat{x}^*\hat{x}\|$$

and thus A/m is a B*-algebra.

COROLLARY 5.6.11. (Calkin, 1941). *Let X be a complete inner product space. Let $L(X)$ be the Banach algebra of all bounded linear operators on X and $K(X)$ be let the ideal of compact operators in $L(X)$. Then $L(X)/K(X)$ is a B*-algebra with respect to the involution induced by the natural involution of $L(X)$.*

NOTE 5.6.12. The B*-algebra $L(X)/K(X)$ is now called the Calkin algebra of X.

5.7. Classes of Elements in Banach Algebras

We now present some classes of elements in Banach algebras; for motivation and proofs we refer to the author's book on linear operator theory (1981).

Let A be a complex Banach algebra with identity 1.

DEFINITION 5.7.1. (Vidav, 1956). An element h in A is said to be hermitian if one of the following equivalent assertions hold:

(1) For all real numbers $t, \|e^{ith}\| = 1$,

(2) $\|1 + ith\| = 1 + 0(t)$ for $t \to 0$.

DEFINITION 5.7.2. An element h in A is said to be positive if h is hermitian and the spectrum is in R^+.

DEFINITION 5.7.3. An element a in A is said to be decomposable if

$a = h + ik$, where k, h are hermitian elements.

If a is a decomposable element of A then by $\bar{a}$ we denote the element $h - ik$ when $a = h + ik$. Sometimes we use the notation a for $\bar{a}$.

DEFINITION 5.7.4. A decomposable element n is said to be normal if, when $n = h + ik$, $hk = kh$.

DEFINITION 5.7.5. The decomposable element n is said to be normal if when $n = h + ik$, all elements $h_{p,q} = h^p k^q$ (where p, q are positive intergers) are hermitian elements.

REMARK 5.7.6. There are examples of Banach algebras where the normality defined in 5.7.4 and 5.7.5 are not equivalent. In this respect it is worth mentioning an example of a Banach algebra with an element h hermitian and h^2 not hermitian, and that there exist Banach algebras with the property that they have only trivial hermitian elements, i.e. only the elements 1 and -1 (as well as real multiples of such elements by real numbers).

We have the following important result.

THEOREM 5.7.7 *Let A be a complex Banach algebra with identity. Then if h,k are hermitian elements, the element $i(hk-kh)$ is hermitian. Suppose now that A is a Banach algebra with an involution. Then when A is a B*-algebra, the above notion of hermitianity coincides with the usual one, i.e. h is hermitian iff $h = h^*$.*

In B*-algebras there are several classes of elements which are important, and in fact they are the analogue of corresponding classes when A is $L(X)$ the Banach algebra of all bounded linear operators on a complex complete inner product space H. Below we present some of them.

DEFINITION 5.7.8. An element a of a B*-algebra A is said to be hyponormal if $a^*a - aa^* \geqslant 0$.

DEFINITION 5.7.9. An element of a complex Banach algebra A is said to be normaloid if $\|a\| = r_a$ (r_a is the spectral radius of a).

DEFINITION 5.7.10. An element of a complex Banach algebra A with identity is said to be transloid if for all complex numbers $z, a + z$ is a normaloid operator.

DEFINITION 5.7.11. An element a of a complex Banach algebra with identity is said to have the G_1-property if

$$\|(a - z)^{-1}\| \leqslant d(z, \sigma(a))^{-1}$$

for any z in the resolvent set of a.

Suppose now that A is a B*-algebra and for each a in A we define the numerical range by the formula

$$W(a) = (\varphi(a),\ \varphi \text{ is a state of } A).$$

Then it is not difficult to see that this is a convex set of the complex plane. An element of A, h is hermitian iff the numerical range is a subset of the reals.

DEFINITION 5.7.12. We say that the element a of a B*-algebra A is convexoid if

$$\operatorname{conv}\sigma(a) = \operatorname{Cl} W(a) \quad \text{(Cl means the closure)}.$$

DEFINITION 5.7.13. An element a of a B*-algebra A is said to be in the class (W − N) if

$$|a|^2 \geqslant (\operatorname{Re} a)^2,$$

where $|a|$ is the positive square root of a^*a.

REMARK 5.7.14. This class was introduced by the author in 1977 (quoted, for example, in Campbell, *Pacific J. Math* 1978) and some results about this class were given in a joint paper (1979).

DEFINITION 5.7.15 (Istrăţescu, 1980). An element a of a B*-algebra A is said to be in the class trans-(W–N) if for all complex numbers z the element $a - z$ is in the class (W–N).

DEFINITION 5.7.16. The element a of the B*-algebra A is said to be in the class θ if $a^* a$ commutes with $a + a^*$.

For the reader interested in results about the classes of elements in Definitions 5.7.14 and 5.7.15 we refer to the author's papers of 1980 and 1985.

CHAPTER 6

Bounded and Unbounded Linear Operators

6.0. Introduction

Let H be a complex inner product space and let $L(H)$ be the set of all bounded linear operators on H. In this chapter we present some basic facts about the set $L(H)$ as well as about the set of unbounded operators on H. We also present some classes of operators whose structure is better understood, among which we mention the class of hermitian, unitary and normal operators, selfadjoint and essentially selfadjoint operators, and the class of unbounded normal operators. Several examples illustrate the exposed results. Also, since there has recently been a growing interest in new classes of unbounded operators, we also present some classes of unbounded operators related to bounded normal operators, namely the classes of quasinormal operators, subnormal operators, hyponormal operators, operators of class (N), W–N operators, θ operators, etc. For more detail on the classes of bounded operators we refer to the author's book (1981).

6.1. Classes of Bounded Linear Operators on Complete Inner Product Spaces

Let H and $L(H)$ be as above.

PROPOSITION 6.1.1. *A linear operator $T: H \to H$ is bounded if and only if it is continuous.*

Proof. If T is bounded then it is clearly continuous. For the converse suppose that T is not bounded. Then there exists a sequence of unit vectors (x_n) such that $\|Tx_n\| \geqslant n$. Now we consider the sequence (y_n) with $y_n = x_n/n$. Obviously (y_n) converges to zero. Since $\|Ty_n\| \geqslant 1$ this contradicts the continuity of T at $x = 0$. The proposition is proved. □

The proof of the following result is left to the reader.

PROPOSITION 6.1.2. *$L(H)$ is a Banach algebra with identity. The norm on $L(H)$ is*

$$T \to \|T\| = \sup(\|Tx\|, \|x\| \leqslant 1).$$

We now show how we can consider another useful norm on $L(H)$, the so-called 'numerical radius' which is equivalent with the norm considered above. The numerical norm of an operator $T \in L(H)$ is defined as follows:

$$T \to w(T) = \sup(|z|, z \in W(T)),$$

where $W(T)$ is the numerical range of T, i.e. the set

$$(\langle Tx, x\rangle, \|x\| = 1).$$

THEOREM 6.1.3. *For any $T \in L(H)$ the following inequality holds:*

$$\tfrac{1}{2}\|T\| \leqslant w(T) \leqslant \|T\|.$$

Proof. We consider s and t real numbers, $t \neq 0$. Then we have the following identity

$$\|Tx\|^2 + e^{2is}\langle T^2x, x\rangle = \tfrac{1}{2}\langle (t\,e^{2is}T^2x + t^{-1}e^{is}Tx), (t\,e^{is}Tx + t^{-1}x)\rangle - \\ - \tfrac{1}{2}\langle (t\,e^{2is}T^2x - t^{-1}e^{is}Tx), (t\,e^{is}Tx - t^{-1}x)\rangle.$$

Since for all $y \in H$,

$$|\langle Ty, y\rangle| \leqslant w(T)\|y\|^2$$

the above identity implies

$$\big|\|Tx\|^2 + e^{2is}\langle T^2x, x\rangle\big| \leqslant \tfrac{1}{2}w(T)(\|t\,e^{is}Tx + t^{-1}x\|^2 + \|t\,e^{is}Tx - t^{-1}x\|^2) \leqslant \\ \leqslant w(T)(t^2\|Tx\|^2 + t^{-2}\|x\|^2).$$

Suppose now that $Tx \neq 0$ and take s such that $|\langle T^2x, x\rangle| = e^{2is}\langle T^2x, x\rangle$ and t such that $t^2\|Tx\| = \|x\|$. Then we get

$$\|Tx\|^2 + |\langle T^2x, x\rangle| \leqslant 2w(T)\|Tx\|\,\|x\|.$$

This implies that

$$\|Tx\|^2 \leqslant 2w(T)\|Tx\|\,\|x\|$$

or

$$\|Tx\| \leqslant 2w(T)\|x\|.$$

Thus we have

$$\|T\| \leqslant 2w(T).$$

But clearly

$$w(T) \leqslant \|T\|$$

and thus the theorem is proved. □

REMARK 6.1.4. The inequality

$$\|Tx\|^2 + |\langle T^2x, x\rangle| \leqslant 2w(T)\|Tx\|\,\|x\|$$

may be written in the form

$$(\|Tx\| - w(T))^2 + |\langle T^2x, x\rangle| \leqslant (w(T))^2$$

which implies that

$$w(T) \leqslant (w(T^2))^{1/2}.$$

From this fact we obtain obviously that for all n,

$$w(T^{2^n}) \leqslant (w(T))^{2^n}$$

REMARK 6.1.5. There exist operators $T \in L(H)$ for which

$$\tfrac{1}{2}\|T\| = w(T).$$

We now consider several classes of bounded operators on a Hilbert space.

If X is a complete inner product space then we know that the spaces X and X^* are very strongly related and in fact we know that there exists a mapping $\sigma\colon X^* \to X$ such that if x^* is an element of X^* then it is of the form

$$x^*(x) = \langle x, x_0 \rangle$$

and

$$\sigma(x^*) = x_0.$$

This mapping is linear, norm-preserving and thus one-to-one. Thus there exists the mapping σ^{-1}. For any T in $L(X)$ we can consider $\bar{T} = \sigma T \sigma^{-1}$. Then this is an operator in $L(X)$ and has the property that for all x, y in X the following relation holds:

$$\langle Tx, y \rangle = \langle x, \bar{T}y \rangle.$$

Instead of the notation $\bar{T}$ we use the standard notation $\bar{T} = T^*$ and whenever X is a Banach space whose norm is generated by an inner product, the adjoint of T in $L(X)$ is understood to be the operator T^* defined above. Then it is easy to see that the following properties hold:

(1) $(T + S)^* = T^* + S^*$,
(2) $(aT)^* = \bar{a}T^*$,
(3) $(TS)^* = S^*T^*$,
(4) $I^* = I$,
(5) $T^{**} = (T^*)^* = T$,
(6) $\|T^*\| = \|T\|$,
(7) If $T^{-1} \in L(X)$ then $T^{*^{-1}}$ exists and is in $L(X)$, and $(T^{-1})^* = T^{*^{-1}}$,
(8) If $T^{*^{-1}}$ exists and is in $L(X)$ then T^{-1} exists and $(T^{-1})^* = (T^*)^{-1}$.

We also note that in this case the operator T^* defined above for each T in $L(X)$ is sometimes called the inner product adjoint of T.

DEFINITION 6.1.6. An operator $T \in L(H)$ is called Hermitian if $T = T^*$.

The following theorem gives equivalent formulations of the hermitianity of an element $T \in L(H)$.

THEOREM 6.1.7. *For an operator $T \in L(H)$ the following assertions are equivalent:*

(1) *T is Hermitian.*
(2) *The function on H, $x \to \langle Tx, x \rangle$, is real valued.*
(3) *For any real t, $\|e^{itT}\| = 1$.*
(4) *For $t \in R^+$, $t \to 0$, $\|I + itT\| = 1 + 0(t)$.*

For a proof we refer to the author's book (1981).

We now consider an important subclass of Hermitian operators.

DEFINITION 6.1.8. A Hermitian operator $T \in L(H)$ is called positive if for all $x \in H$, $\langle Tx, x \rangle \geqslant 0$. We write this as $T \geqslant 0$.

It is obvious that the set of all positive elements in $L(H)$ is a cone. Now using the cone of positive elements we can define a partial order in the set of all Hermitian operators: if R, S are two Hermitian operators then we say that $R \geqslant S$ if $R - S \geqslant 0$.

It is easy to see that:

(1) for any $T \in L(H)$, the operator T^*T and TT^* are positive;
(2) if T is Hermitian then any power T^n is Hermitian and any power T^{2n} is positive;
(3) if T, S are positive operators then $aT + bS$ is positive for all $a, b \in R^+$.

LEMMA 6.1.9. (The Cauchy–Buniakowsky generalized inequality). *If A is a positive operator then for all x, y in H,*

$$|\langle Ax, y \rangle|^2 \leqslant \langle Ax, y \rangle \langle Ay, y \rangle.$$

Proof. Since the proof does not differ essentially from the proof of Cauchy–Buniakowsky inequality (the case $A = I$) we omit it.

DEFINITION 6.1.10. Let $(T_n)_1^\infty$ be a sequence of elements of $L(H)$. We say that the sequence converges strongly if for each $x \in H$, $\lim T_n x$ exists. (Sometimes we say that (T_n) converges pointwise.)

LEMMA 6.1.11. *Let (T_n) be a sequence of positive operators and suppose that the sequence is monotone, i.e. the inequalities $T_n \leqslant T_{n+1}$ hold for all n. Then (T_n) converges strongly to a Hermitian operator in $L(H)$.*

Proof. First we remark that we can suppose, without loss of generality, that the sequence satisfies the properties

(1) $T_n \leqslant T_{n+1}$ for all n,
(2) $T_n = I$ for all n,

and thus for all $n - m \geqslant 0$, $T_n - T_m$ is positive. Then, by the Cauchy–Buniakowsky generalized inequality we get

$$\begin{aligned}\|(T_n - T_m)x\|^4 &= \langle (T_n - T_m)x, (T_n - T_m)x \rangle^2 \\ &\leqslant \langle (T_n - T_m)x, x \rangle \langle (T_n - T_m)^2 x, (T_n - T_m)x \rangle \\ &\leqslant \langle (T_n - T_m)x, x \rangle \|x\|^2.\end{aligned}$$

But the sequence (T_n) is monotone and thus $(\langle T_n x, x \rangle)$ is monotone. This implies that $(\langle T_n x, x \rangle)$ is convergent. From this we get that $(T_n x)$ is a Cauchy sequence and thus $\lim T_n x$ exists. We define a linear mapping on H by the formula

$$Tx = \lim T_n x$$

and clearly T is an element of $L(H)$.

We now prove a basic result concerning positive operators. □

THEOREM 6.1.12. *Let T be a positive element of $L(H)$. Then there exists a unique operator $S \in L(H)$ such that:*

(1) $S^2 = T$,
(2) $S \geqslant 0$.

Proof. We remark that we can suppose without loss of generality that $0 \leqslant T \leqslant I$. We now define a sequence of operators in $L(H)$ by

$$S_{n+1} = S_n + \tfrac{1}{2}(S - S_n^2)$$

and $S_0 = 0$. Any element of the sequence (S_n) is a Hermitian operator and we now show that $0 \leqslant S_n \leqslant I$. For $n = 0$ this is clearly true. Suppose now that the assertion is true for all $i \leqslant n$ and we now prove that it is true for $n + 1$. Indeed, we have,

$$I - S_{n+1} = \tfrac{1}{2}(I - S_n)^2 + \tfrac{1}{2}(I - S)$$

and this clearly implies our assertion. Thus $0 \leqslant S_n \leqslant I$ holds for all n. We remark that S_n commutes with each operator commuting with T and thus

$$S_{n+1} - S_n = \tfrac{1}{2}(I - S_{n-1}) + (I - S_n)(S_n - S_{n-1})$$

is positive. Thus the sequence (S_n) is monotone and by the above lemma it converges strongly to an operator $S \in L(H)$. Obviously S is Hermitian. From the form of S_{n+1} we get that $S^2 = T$. Thus the existence of S such that $S^2 = T$ and $S \geqslant 0$ is proved. We now prove the uniqueness.

Suppose that this is not true and thus there exists another S_1 such that $S_1^2 = T^2$ and $S_1 \geqslant 0$. Now as S and S_1 are positive operators there exist positive operators $\bar{S}$ and $\bar{S}_1$ such that

$$\bar{S}^2 = S, \quad \bar{S}_1^2 = S_1.$$

We remark that T commutes with S_1 since

$$TS_1 = S_1^2 S_1 = S_1^3 = S_1 \cdot S_1^2 = S_1 T$$

and thus S and S_1 are commuting operators.

Let $x \in H$ and set $y = (S - S_1)x$. Then we have

$$\begin{aligned} \|\bar{S}_1 y\|^2 + \|\bar{S}y\|^2 &= \langle \bar{S}_1^2 y, y\rangle + \langle \bar{S}^2 y, y\rangle = \langle (S + S_1)y, y\rangle \\ &= \langle (S + S_1)(S - S_1)x, y\rangle = \langle (S^2 - S_1^2)x, y\rangle = 0. \end{aligned}$$

Thus $\bar{S}y = \bar{S}_1 y = 0$, which implies that $Sy = S_1 y = 0$. Thus

$$\|Sx - S_1 x\|^2 = \langle (S - S_1)x, (S - S_1)x\rangle = \langle (S - S_1)y, x\rangle = 0$$

which gives that $Sx = S_1 x$ and the uniqueness is proved. □

An important and useful class of positive operators is the class of projections.

DEFINITION 6.1.13. An element $P \in L(H)$ is said to be a projection if the following properties are satisfied:

(1) P is Hermitian.
(2) $P^2 = P$.

PROPOSITION 6.1.14. *Every projection is positive.*

Proof. Let x be arbitrary in H. Then we have

$$\langle Px, x\rangle = \langle P^2x, x\rangle = \langle Px, Px\rangle \geqslant 0. \qquad \square$$

PROPOSITION 6.1.15. *If P is a projection then $Q = I - P$ is a projection.*
Proof. Obvious. □

PROPOSITION 6.1.16. *There exists a one-to-one correspondence between projections and closed subspaces of H.*
Proof. Let P be a projection and consider the space

$$m_P = PH = (Px, x \in H).$$

Then clearly m_P is a closed linear subspace of H and for each $x \in m_P$,

$$Px = x.$$

If for each closed subspace m of H we set

$$x = u + v$$

with $u \in m$ and $v \in m^{\perp}$ then the mapping $x \to u$ is a linear operator and, as is easy to prove, a projection. The proposition is proved. □

We remark that the mapping $x \to v$ is again a projection, corresponding to the orthogonal subspace.

The following proposition gives a connection between projection and orthogonal subspaces.

PROPOSITION 6.1.17. *Let P, R be two projections. Then the subspaces PH and RH are orthogonal if and only if $PR = RP = 0$.*
Proof. Let $PH = m$, $RH = n$ and suppose first that $m \perp n$. Then for any x, y in H we have

$$\langle PRx, y\rangle = \langle Rx, Py\rangle = 0 \qquad \langle RPx, y\rangle = \langle Px, Ry\rangle = 0$$

and this implies that $PR = RP = 0$.
Conversely, if $PR = RP = 0$ then for any $x \in m$, $y \in n$ we have

$$\langle x, y\rangle = \langle Px, Ry\rangle = \langle RPx, y\rangle = 0$$

and the proposition is proved. □

PROPOSITION 6.1.18. *If P, R are projections then $S = P + R$ is a projection if and only if $PR = RP = 0$.*
Proof. If $PR = RP = 0$ then clearly S is a projection. Conversely, if S is a projection, then

$$S^2 = P + R + PR + RP = S$$

which implies that

$$PR + RP = 0.$$

Now let x be arbitrary in H and thus if

$$a + ib = \langle PRx, x\rangle$$

we get

$$\langle PRx, x\rangle = \langle -RPx, x\rangle = -\langle x, PRx\rangle = -\langle PRx, x\rangle^* = -(a - ib) = -a + ib]$$

which implies that $a = 0$. Since $PR + RP = 0$ we get that

$$\langle PRx, x\rangle = -\langle PRPx, x\rangle$$

which shows that $\langle PRx, x\rangle$ is a real number. Then $b = 0$. Thus for all $x \in H$, $\langle PRx, x\rangle = 0$ which implies that $PR = 0$. Since $RP = (PR)^*$, $RP = 0$ and the proposition is proved. □

The following proposition gives a condition under which the product of two projections is again a projection.

PROPOSITION 6.1.19. *If P, R are two projections then $S = PR$ is a projection if and only if $PR = RP$.*

Proof. If $PR = RP$ then clearly S is a projection. Conversely, if S is a projection then

$$S = S^* = RP = PR$$

and the assertion is proved. □

PROPOSITION 6.1.20. *Lf P, R are two projections then $S = P - R$ is a projection if and only if $PR = R$.*

Proof. If $P - R$ is a projection then

$$I - (P - R) = I - P + R$$

is a projection and by 6.1.18 this is a projection if and only if

$$(I - P)R = R(I - P) = 0$$

or

$$R = PR = RP.$$

Conversely, if $PR = R$ then $P - R$ is a projection as a computation shows. □

PROPOSITION 6.1.21. *If P, R are two projections then the following assertions are equivalent:*

(1) $RP = R$.
(2) $\|Rx\| \leqslant \|Px\|$, $x \in H$.
(3) $R \leqslant P$.

Proof. (1) → (2). Let x in H arbitray. Then if (1) holds we have $Rx = RPx$ and thus

$$\|Rx\| = \|R \cdot Px\| \leqslant \|Px\|$$

since for any projection, the norm is 1.

(2) → (3). Let $x \in H$ and then

$$\langle (P - R)x, x \rangle = \langle Px, x \rangle - \langle Rx, x \rangle = \|Px\|^2 - \|Rx\|^2 \geqslant 0.$$

(3) → (1). Let $x \in H$ and thus we have

$$\langle Px, x \rangle \geqslant \langle Rx, x \rangle$$

which implies that $\|Rx\|^2 \leqslant \|Px\|^2$ and thus if $y = Rx$ we get

$$\|y\|^2 = \|Ry\|^2 \leqslant \|Py\|^2 \leqslant \|y\|^2$$

which gives that $\|y\| = \|Py\|$. Further we have,

$$\|(I - P)y\|^2 = \|y\|^2 + \|Py\|^2 - \langle Py, y \rangle - \langle y, Py \rangle = \|y\|^2 - \|Py\|^2 = 0$$

and thus $y = Py$. Therefore, y is in the range of P. Thus, if $y = Ry$ then $y = Py$ and $Ry = Rpy$ or $R = RP$. The proposition is proved. □

PROPOSITION 6.1.22. *Let $(P_i)_{i \in I}$ be a family of projections and let I a directed set such that whenever $i \leqslant j$ then $P_i \leqslant P_j$. Suppose that $(P_i)_{i \in I}$ converges strongly (i.e. pointwise) to P. Then P is a projection.*

Proof. It is clear that P is Hermitian and since, for each $x \in H$,

$$\langle Px, x \rangle = \lim \langle P_i x, x \rangle = \lim \|P_i x\|^2 = \|Px\|^2$$

we get that P is a projection. □

PROPOSITION 6.1.23. *Let $(P_i)_{i \in I}$ be a family of projections and let I be a directed set for $\geqslant$, such that if $i \geqslant j$ then $P_i \geqslant P_j$. Suppose that $(P_i)_{i \in I}$ converges strongly. Then* $\lim P_i$ *exists and is a projection.*

Proof. Similar to the proof of 6.1.22. □

NOTE 6.1.24. It is not difficult to compute the space $(x, Px = x)$ for the projections in 6.1.22 and 6.1.23 in terms of subspaces m_i with $m_i = (x, P_i x = x)$. This is left to the reader.

COROLLARY 6.1.25. *Let $(P_i)_{i \in I}$ be a family of projections with the property that for $i, j \in I$, $i \neq j$, $P_i P_j = 0$. Then $(P_i x)_{i \in I}$ is summable and $Px = \Sigma_{i \in I} P_i x$ is a projection.*

Proof. If we consider the family $P_F I$ of all finite parts of I then with the partial order defined using the inclusion (i.e. if A, B are two finite parts then $A \leqslant B$ if A is a subset of B) it is a directed set. Then clearly

$$\lim P_J x = \lim \sum_{i \in J} P_i x, \quad J \in P_F I,$$

exists and defines a projection. □

We now present an inequality which follows from the Cauchy–Buniakowsky inequality. First we give the following result.

LEMMA 6.1.26 (Istrăţescu, 1983). *Let A, B be in $L(X)$, where X is a complete inner product space and A, B satisfy the conditions*

(1) $A \geqslant 0$,
(2) $AB = e^{it} B^* A$

where t is a real number and B^* is the adjoint of B. Then for all x in X the following inequality holds:

$$|\langle ABx, x\rangle| \leqslant r_B\langle Ax, x\rangle,$$

where r_T is the spectral radius of T.

Proof. We show that for all integers n the following inequality holds:

$$|\langle ABx, x\rangle|^{2^n} \leqslant |\langle AB^{2^n}x, x\rangle|\langle Ax, x\rangle^{2^n-1}.$$

Indeed, for $n = 1$, applying the Cauchy–Buniakowsky inequality for the positive operator A, we get that

$$\begin{aligned}|\langle ABx, x\rangle|^2 &\leqslant \langle ABx, Bx\rangle\langle Ax, x\rangle = \langle B^*ABx, x\rangle\langle Ax, x\rangle\\ &= e^{-it}\langle AB^2x, x\rangle\langle Ax, x\rangle\end{aligned}$$

or

$$|\langle ABx, x\rangle^2 \leqslant |\langle AB^2x, x\rangle|\langle Ax, x\rangle,$$

which is what we need.

Now we remark that for all integers k the following relations hold:

$$AB^k = e^{ikt}B^{*k}A.$$

Then, if we suppose that for all $i \leqslant n$ the above inequality holds, we have

$$\begin{aligned}(|\langle ABx, x\rangle|^{2^n})^2 &= (|\langle AB^{2^n}x, x\rangle|\langle Ax, x\rangle^{2^n-1})^2\\ &= |\langle AB^{2^n}x, B^{2^n}x\rangle|\langle Ax, x\rangle\langle Ax, x\rangle^{2^{n+1}-1}\\ &= |\langle AB^{2^{n+1}}x, x\rangle|\langle Ax, x\rangle^{2^{n+1}-1},\end{aligned}$$

which shows that the required inequality holds for $n + 1$. Thus we may suppose that it holds for all n and then

$$|\langle ABx, x\rangle| \leqslant \|A\|\|B^{2^{n+1}}\|\|x\|^{2\,1/2^{n+1}}\langle Ax, x\rangle^{(2^{n+1}-1)/2^{n+1}}$$

which for $n \to \infty$ implies the inequality of the lemma. □

REMARK 6.1.27. It is clear from the proof that the above inequality is valid if we suppose that A, B satisfy the conditions

(1) $A \geqslant 0$,
(2) $AB = SB^*A$

where S is an invertible operator which commutes with B^*, B and A.

From the above inequality we obtain the following inequalities.

REID'S INEQUALITY 6.1.28. *Suppose that A, B are in L(X) and $A \geqslant 0$ and $AB = B^*A$. Then*

$$|\langle ABx, x\rangle| \leqslant \|B\|\langle Ax, x\rangle.$$

Reid's inequality was improved by Halmos (1967) as follows: *A, B are as in Reid's inequality and then*

$$|\langle ABx, x\rangle| \leqslant r_B\langle Ax, x\rangle.$$

6.2. Normal, Unitary and Partial Isometry Operators

We now consider other important classes of bounded operators. The most important class is considered in the following definition.

DEFINITION 6.2.1. An element $T \in L(H)$ is said to be normal if $T^*T = TT^*$.

PROPOSITION 6.2.2. *An element $T \in L(H)$ is normal if and only if $T = T_1 + iT_2$ with T_i, $i = 1, 2$ Hermitian operators and $T_1T_2 = T_2T_1$.*

Proof. Since the proof is simple it is omitted. □

PROPOSITION 6.2.3. *An operator $T \in L(H)$ is normal if and only if for all $x \in H$, $\|Tx\| = \|T^*x\|$.*

Proof. It is obvious that the condition is necessary. Suppose now that for each $x \in H$, $\|Tx\| = \|T^*x\|$. This implies that

$$\|Tx\|^2 = \|T^*x\|^2 \quad \text{or} \quad \langle T^*Tx, x\rangle = \langle x, TT^*x\rangle.$$

This gives that

$$\langle T^*Tx, x\rangle - \langle TT^*x, x\rangle = 0 = \langle (T^*T - TT^*)x, x\rangle$$

and thus, necessarily, $T^*T - TT^* = 0$ and the proposition is proved. □

The following proposition gives further properties of normal operators.

PROPOSITION 6.2.4. (1) *The set of all normal operators $T \in L(H)$ is closed for a strong convergence.*

(2) *If $T \in L(H)$ is normal then $aT + bI = aT + b$ is normal.*

(3) *If T is normal then $\|T^2\| = \|T\|^2$.*

Proof. (1) Let (T_n) be a sequence of normal operators converging strongly to an operator $T \in L(H)$. Then for each $x \in H$ we have

$$\|Tx\|^2 = \lim\|T_nx\|^2 = \lim\|T_n^*x\|^2 = \|T^*x\|^2$$

and thus T is normal.

(2) Obvious.

(3) Since for each operator $S \in L(H)$, $\|S\|^2 = \|S^*S\|$, we show that $\|T^*T\| = \|T^2\|$. From this it is clear that our assertion follows. Let us consider $x \in H$ and from the normality of T we get

$$\begin{aligned}\|T^2x\|^2 &= \langle T^2x, T^2x\rangle = \langle T^{*2}T^2x, x\rangle = \langle (T^*T)^2x, x\rangle \\ &= \langle T^*Tx, T^*Tx\rangle = T^*Tx\|^2\end{aligned}$$

and thus $\|T^2x\| = \|T^*Tx\|$, which implies $\|T^2\| = \|T^*T\|$ and the assertion is proved. □

We now consider an important subclass of normal operators, namely the class of unitary operators.

DEFINITION 6.2.5. An operator $U \in L(H)$ is said to be unitary if

$$U^*U = UU^* = I.$$

From the definition it is clear that any unitary operator is normal. We now give some characterizations of unitary operators.

PROPOSITION 6.2.6. *An operator $T \in L(H)$ is unitary if and only if*

(1) $TH = H$,
(2) $\langle Tx, Ty \rangle = \langle x, y \rangle$.

Proof. If T is unitary then it is clear that properties (1) and (2) are satisfied. Suppose now that (1) and (2) are satisfied for $T \in L(H)$.

From (2) we get that $T^*T = I$ and

$$\langle T^*x, T^*y \rangle = \langle x, TT^*y \rangle.$$

But $x = Tx_1$ and $y = Ty_1$ and thus

$$\langle T^*x, T^*y \rangle = \langle T^*Tx_1, T^*Ty_1 \rangle = \langle x_1, y_1 \rangle = \langle Tx_1, Ty_1 \rangle = \langle x, y \rangle$$

which implies that $TT^* = I$ and the proposition is proved. □

In a similar way we can prove the following proposition.

PROPOSITION 6.2.7. *$T \in L(H)$ is unitary if and only if T^{-1} exists and is unitary.*

DEFINITION 6.2.8. A unitary operator is called a symmetry if its square is the identity.

It is clear that any symmetry is hermitian.

PROPOSITION 6.2.9. *There is a one-to-one correspondence between projections and symmetries.*

Proof. For this it is sufficient to remark that if U is a symmetry then $P = \frac{1}{2}(U + I)$ is a projection and, conversely, if P is a projection then $U = 2P - I$ is a symmetry. □

The following proposition gives examples of unitary operators.

PROPOSITION 6.2.10. *Let $(e_i)_{i \in I}$ and $(f_i)_{i \in I}$ be two bases of H. If $x = \Sigma \langle x, e_i \rangle e_i$ then*

$$Ux = \sum \langle x, e_i \rangle f_i$$

is a unitary operator. Conversely, if U is a unitary operator then it is of the above form for some bases.

Proof. It is clear that U defined in this way is unitary. Now if U is unitary we consider a basis $(e_i)_{i \in I}$ of H. Then $(Ue_i)_{i \in I}$ is an orthonormal basis for H. Set $f_i = Ue_i$ and this satisfies the proposition. □

PROPOSITION 6.2.11. (1) *The set of unitary operators is closed in $L(H)$.*

(2) *The set of unitary operators is closed for the transformation $(U_1, U_2) \to U_1U_2$ (or U_2U_1) and with this 'multiplication' the set of of unitary operators is a group.*

Proof. Since the proof of the assertions is simple it is omitted. □

PROPOSITION 6.2.12. *Let $T \in L(H)$ be an invertible operator. Then the operator $T_1 = T(T^*T)^{-1/2}$ ($(T^*T)^{1/2}$ is the square root of T^*T) is unitary.*

Proof. Computing $T_1 T_1^*$ and $T_1^* T_1$ we see immediately that they are equal to I. □

DEFINITION 6.2.13. An operator $T \in L(H)$ is called an isometry if for all x, y in H,

$$\langle Tx, Ty \rangle = \langle x, y \rangle.$$

Then we have clearly the following assertion.

PROPOSITION 6.2.14. *An operator $T \in L(H)$ is an isometry if and only if $T^*T = I$.* Now we can prove the following characterization of isometries.

PROPOSITION 6.2.15. *An element $T \in L(H)$ is an isometry if and only if for all $x \in H$, $\|Tx\| = \|x\|$.*

Proof. Clearly the condition is necessary. Suppose now that the condition holds and thus from the formula

$$4\langle x, y \rangle = \|x + y\|^2 - \|x - y\|^2 + i\|x + iy\|^2 - i\|x - iy\|^2$$

we get

$$\begin{aligned} 4\langle Tx, Ty \rangle &= \|T(x + y)\|^2 - \|(T(x - y))\|^2 + i\|T(x + iy)\|^2 - i\|T(x - iy)\|^2 \\ &= 4\langle x, y \rangle \end{aligned}$$

i.e. T is an isometry. □

DEFINITION 6.2.16. An operator $T \in L(H)$ is said to be a partial isometry if there exists a closed linear subspace m of H such that:

(1) $\langle Tx, Ty \rangle = \langle x, y \rangle$ for all x, y in m,
(2) $Ty = 0$ for all $y \in m^{\perp}$.

The space m is called the initial variety of T and Tm is called the final variety of T.

PROPOSITION 6.2.17. *If T is a partial isometry then T^* is also a partial isometry.*

Proof. This is left to the reader. □

EXAMPLE 6.2.18. Let H be a separable Hilbert space with the basis $(e_i)_0^{\infty}$. The operator $S: H \to H$ defined by the relations $Se_i = e_{i+1}$, $i = 0, 1, 2, \ldots$, is an isometry. We recall that this is the so-called 'shift' operator. It is easy to see that T is an isometry and T^* is a partial isometry.

As is well known any complex number z has the decompositions

$$z = x + iy, \qquad z = r\,e^{it},$$

where x, y and r, t are real numbers. Corresponding to the decomposition $z = x + iy$ we have the decomposition $T = T_1 + iT_2$ of an operator $T \in L(H)$ with T_i, $i = 1, 2$ hermitian operators.

We now give a decomposition corresponding to the decomposition $z = r\,e^{it}$. First we consider the case of normal operators.

THEOREM 6.2.19. *Let $T \in L(H)$ be a normal operator. Then T may be factorized to $T = UR$, where U is a unitary operator, R is a positive operator, and $UR = RU$.*

Proof. We consider the operator T^*T which is positive (and equal to TT^*); thus its square root exists and will be denoted by R. Thus we have $R^2 = T^*T$. In this case for any $x \in H$,

$$\|Rx\|^2 = \langle Rx, Rx \rangle = \langle R^2x, x \rangle = \langle T^*Tx, x \rangle = \|Tx\|^2 = \|T^*x\|^2.$$

Let us consider the subspace $m = (y, y = Rx, x \in H)$ and set $\bar{m}$ be the closure of this subspace. Then we have

$$m^\perp = (z, z \perp m) = \{z, Tz = 0\} = \{z, T^*z = 0\}.$$

Now we define the operator U as follows: for any $y = Rx$ we set $Uy = Tx$ and we have

$$Uy = Tx = Rx.$$

Then $Um \to m$ and is isometric. Thus we can extend U uniquely to $\bar{m}$ and the extension is again isometric (we denote the extension by the same letter U). Now we define $U: H \to H$ to be the identity operator on $m^\perp$. The operator U is clearly unitary and from the definition we have $Tx = URx$.

We now show that we have $Tx = RUx$. Of course, this is trivially true for any $x \in m^\perp$. Thus it remains to prove the assertion for $x \in \bar{m}$. Every y in $\bar{m}$ is of the form $y = \lim y_n$ with y_n in m. Thus $y_n = Rx_n$ for some x_n in H. Then we get,

$$Ty = \lim TRx_n = R \lim Tx_n = R \lim URx_n = RU \lim Rx_n = RUy$$

and the theorem is proved. □

Now we consider the case of arbitrary elements of $L(H)$.

THEOREM 6.2.20. *Let $T \leftarrow L(H)$. Then T may be factorized to $T = VR$ where R is a positive operator and V is a partial isometry.*

Proof. We consider the operator T^*T which is obviously positive and thus has a (unique) square root which we denote by R. We now consider the subspace

$$M = \overline{(Rx, x \in H)}$$

and $M^\perp = N$ is the orthogonal complement of M. We define an operator $V: M \to H$ as follows: if $y = Rx$ then we set $Vy = Tx$ and we remark that

$$\|Vy\|^2 = \langle Tx, Tx \rangle = \langle T^*Tx, x \rangle = \langle R^2x, x \rangle = \|Rx\|^2 = \|y\|^2$$

and thus since the elements y of the form Rx are dense in M, V may be extended to M and this operator is denoted again by V, and clearly this is an isometry.

We remark that since $\|Vy\| = \|Rx\|$ the operator V is well defined in the sense that if $y = Rx = Rx_1$ then $Tx = Tx_1$ since, in this case, we have that $R^2(x - x_1) = 0$, which gives

$$\langle R^2x - R^2x_1, x - x_1 \rangle = 0 = \langle (T^*T)(x - x_1) \rangle = \|Tx - Tx_1\|^2$$

and thus $Tx = Tx_1$. We can extend the operator V on the entire space H by defining it to be equal to the 0 operator on $M^\perp$.

We remark that if $y = Rx$ then from the definition of the operator V we have

$$Vy = Tx = VRx$$

and thus the required decomposition holds for such elements. Now, if $y = \lim Rx_n$ then we have, from the continuity of T, that

$$Ty = \lim TRx_n \quad \text{and} \quad Vy = \lim Tx_n,$$

which implies that

$$VRy = VR(\lim Rx_n) = V(\lim R^2 x_n) = \lim R^2 x_n = \lim TRx_n = T \lim Rx_n = Ty$$

and thus d, the decomposition, holds for all elements in M.

If y is in $M^{\perp}$ then clearly $Ty = VRy$.

We note that if $T^{-1} \in L(H)$ then the operator $T_1 = T^* R^{-1/2}$ is unitary. □

Since the (partial) isometries are intimately connected with the structure of a general bounded linear operator on H we now give some other properties of this class of operators.

PROPOSITION 6.2.21. *Let $T: H \to H$ be an isometric operator and $N = (TH)^{\perp}$. Then the subspaces $T^k N$ are pairwise orthogonal* ($k = 0, 1, 2, 3, \ldots$).

Proof. Since T is supposed to be an isometry it is clear that $T^k N$ are closed subspaces. Let m, n be two integers and we may suppose without loss of generality that $n > m$. Let $x \in T^n N$, $y \in T^m N$ and thus for some u and v in N we have that $x = T^n u$, $y = T^m v$. Thus

$$\langle x, y \rangle = \langle T^n u, T^m v \rangle = \langle T^{n-m} u, v \rangle = 0$$

since $n - m \geqslant 1$. This proves the assertion. □

PROPOSITION 6.2.22. *Let $T: H \to H$ be an isometric operator and $N = (TH)^{\perp}$. Let M be the orthogonal complement of the space containing all the space $T^k N$. Then the following assertions hold:*

(1) $H = M \oplus N \oplus TN \oplus T^2 N \oplus \cdots \oplus T^k N \oplus \cdots$

(2) *M is an invariant subspace of T and T/M is a unitary operator on M.*

(3) *Every element m in M has the property that it is in the range of T^k for all k.*

Proof. Let $m \in M$ and then it is in the range of T, i.e. for some n_1 we have that $m = Tn_1$ and we show that $n_1 \in M$. Indeed, if n is arbitrary in N we have

$$\langle T^k n, n_1 \rangle = \langle T^{k+1} n, Tn_1 \rangle = \langle T^{k+1} n, m \rangle = 0$$

and thus we must have that $n_1 \in M$. The argument applies now to n_1 and thus m is in the range of T^2, etc.

Thus (3) is proved. Assertion (1) follows from the definition of M and assertion (2) follows from the fact that T/M is surjective.

DEFINITION 6.2.23. An isometry T on H is called a pure isometry if $\cap_0^{\infty} T^n H = (0)$. The multiplicity of an isometry T is the dimension of $(TH)^{\perp}$.

The following structure theorem is known as the Wold decomposition of an isometry.

THE WOLD DECOMPOSITION 6.2.24. *Let $T: H \to H$ be an isometry. Then there exists a closed linear subspace M of H with the following properties:*

(1) *M reduces T (i.e. M is invariant under T and T^*).*
(2) *T/M is a pure isometry.*
(3) *$T/M^\perp$ is unitary.*

Proof. We consider the subspace $M = \cap_{n=0}^{\infty} T^n H$ and it is obvious that it is an invariant subspace of T. Let $x \in M$ and then for each $m \in M$ we have $\langle x, m\rangle = 0 = \langle Tx, Tm\rangle$ which implies that M reduces T. Obviously T/M is an isometry and since T/M is onto we see that T/M is unitary. Also $T/M^\perp$ is an isometry and clearly $\cap_0^\infty T^n M^\perp = (0)$. □

We leave the reader to identify the Wold decomposition from the decomposition given in Proposition 6.2.22.

We now show how we can represent the elements of $L(H)$ using matrices. We recall that a matrix is a set of complex numbers

$$(a_{\alpha\beta})_{\alpha\in I, \beta\in J},$$

where I, J are two sets of indices. If $I = J$ then we say that we have a square matrix. Let $M_{I,J}$ be the family of all matrices having I and J as sets of indices. We can organize $M_{I,J}$ as a complex linear space by defining the sum and the multiplication with scalars, as follows:

(1) $(a_{\alpha\beta}) + (b_{\alpha\beta}) = (a_{\alpha\beta} + b_{\alpha\beta})$,
(2) $z(a_{\alpha\beta}) = (za_{\alpha\beta})$,

It is easy to see that $M_{I,J}$ is a complex linear space. In what follows we consider only square matrices.

DEFINITION 6.2.25. A square matrix $(a_{\alpha\beta})_{\alpha,\beta\in I}$ is called bounded if there exists a constant $K < \infty$ such that

$$\sum_{\alpha\in I} \left| \sum_{\beta\in J} a_{\alpha\beta} c_\beta \right|^2 \leqslant K^2 \sum_{\beta\in J} |c_\beta|^2$$

for any finite set J and where (c_β) are complex numbers.

It is easy to see that the family of all bounded matrices can be organized as a complex space (a subspace of $M_{I,I} = M_I$). We denote this subspace by B_I.

From the definition of bounded matrices it follows that if $M = (a_{\alpha\beta})_{\alpha,\beta\in I}$ is in B_I then

$$\sum_{(\alpha,\beta)} |a_{\alpha\beta}|^2 < \infty.$$

To prove this we take

$$c_\beta = \begin{cases} 0, & \beta \neq \gamma \\ 1, & \beta = \gamma \end{cases}$$

and from the definition of boundedness we get

$$\sum_{\alpha\in I} |a_{\alpha\gamma}|^2 < \infty$$

and, further, for any finite part J of I,

$$\sum_{\alpha\in I}\left|\sum_{\beta\in J} a_{\alpha\beta}\, a^*_{\beta\gamma}\right|^2 \leqslant K^2 \sum_{\beta\in J} |a_{\beta\gamma}|^2 \leqslant K^2 \sum_{\beta\in I} |a_{\beta\gamma}|^2$$

which implies that

$$\sum_{(\alpha,\beta)} |a_{\alpha\beta}|^2 < \infty.$$

Now if M is in B_I then the smallest constant K satisfying the definition of boundedness is called the norm of M and is denoted by $\|M\|$.

Now if H is a complete inner product space we consider a basis $(e_i)_{i\in I}$ of H and if $T\in L(H)$ then we can consider the complex numbers $a_{\alpha\beta} = \langle Te_\beta, e_\alpha\rangle$ where

$$Te_\beta = \sum_\alpha \langle Te_\beta, e_\alpha\rangle\, e_\alpha.$$

We consider now the matrix $M(T)$ with the elements $\{a_{\alpha\beta}\}$, i.e.

$$M(T) = (a_{\alpha\beta})_{\alpha,\beta\in I}.$$

The mapping

$$\varphi\colon T \to M(T)$$

has obviously the following properties:

(1) $\varphi(T+S) = \varphi(T) + \varphi(S)$.
(2) $\varphi(zT) = z\varphi(T)$.
(3) $\varphi(O) = O$.†
(4) $\varphi(I) = I$.‡

Now we show that in fact φ is an isometry. Indeed, if $x\in H$ then for any finite part J of I

$$x = \sum_\alpha \langle x, e_\alpha\rangle e_\alpha, \qquad x_J = \sum_{\alpha\in J} \langle x, e_\alpha\rangle e_\alpha$$

and thus

$$\|Tx_J\|^2 = \sum_{\alpha\in I} |\langle Tx_J, e_\alpha\rangle|^2 = \sum_{\alpha\in I}\left|\sum_J \langle x, e_\beta\rangle\langle Te_\beta, e_\alpha\rangle\right|^2 = \sum_{\alpha\in I}\left|\sum_{\beta\in J} \alpha_{\alpha\beta}\langle x, e_\beta\rangle\right|^2$$

which implies that

$$\|Tx_J\| \leqslant K\|x_J\|$$

if and only if

$$\sum_{\alpha\in I}\left|\sum_{\beta\in J} a_{\alpha\beta}\langle x, e_\beta\rangle\right|^2 \leqslant K^2 \sum_{\beta\in J} |\langle x, e_\beta\rangle|^2.$$

But $(x_J, x\in H)$ is a dense set in H and this implies that $\|M(T)\| = \|T\|$. Suppose now

† The matrix O has $a_{\alpha\beta} = 0$ for all α, β in I.
‡ The matrix I has $a_{\alpha\beta} = \delta_{\alpha\beta}$ (the Kronecker symbol).

that T, S are in $L(H)$ and

$$\varphi(T) = (a_{\alpha\beta})_{\alpha,\beta\in I}$$
$$\varphi(S) = (b_{\alpha\beta})_{a,\beta\in I}$$

then $R = TS$ has the matrix

$$(R) = (c_{\alpha\beta})_{\alpha,\beta\in I},$$

where

$$c_{\alpha\beta} = \sum_{\alpha\in I} a_{\alpha\beta} b_{\alpha\beta}.$$

Also if $T\in L(H)$ has the matrix $M(T) = (a_{\alpha\beta})_{\alpha,\beta\in I}$ then the adjoint T^* of T has the matrix $(a^1_{\alpha\beta})$, where $a^1_{\alpha\beta} = a^*_{\beta\alpha}$.

Of course the property for an element $T\in L(H)$ to be hermitian, unitary or, more generally, a normal operator can be characterized in terms of the corresponding matrices. We leave this for the interested reader.

6.3. Semispectral and Spectral Families of Radon Measures

We now present some results about semispectral and spectral families of Radon measures which will be used to obtain a spectral decomposition for some classes of operators.

DEFINITION 6.3.1. Let H be a complete innerproduct space (over C) and S be a locally compact space. We say that a family $\mathscr{F} = (\mu_{x,y})_{x,y\in H}$ of Radon measures† on S is a semispectral family if the following conditions are satisfied:

(1) For each $y\in H$ fixed, $x \to \mu_{x,y}$ is a linear map.
(2) For each $x, y\in H$, $\mu_{x,y} = \mu_{y,x}{}^* \, (\mu^*(f) = (\mu(f^*))^*)$,
(3) For each $x, y\in H$, $|\mu_{x,y}| \leqslant \|x\|\|y\|$.

If $\mathscr{F} = (\mu_{x,y})_{x,y\in H}$ is a semispectral family then we say that each $\mu_{x,y}$ is a semispectral measure.

We now define for each complex-valued continuous function with compact support f,

$$[x,y]_f = \int_S f(s)\, d\mu_{x,y}(s)$$

and we remark that the following properties hold:

(1) $[x_1 + x_2, y]_f = [x_1, y]_f + [x_2, y]_f$.
(2) $[ax, by]_f = ab^*[x,y]_f$.
(3) $|[x,y]_f| \leqslant \|f\|_s \|x\|\|y\|$.
(4) $[x,y]_f = [y,x]^*_f$.

We now need the following result.

† See the Appendix (p. 824) for the notion of a Radon measure.

LEMMA 6.3.2. *Let H be as above, let H_1 and H_2 be two subspaces which are dense in H and let $\langle\, , \rangle^{\sim}$ be a mapping of $H_1 \times H_2$ with values in C such that:*

(1) $\langle x_1 + x_2, y\rangle^{\sim} = \langle x_1, y\rangle^{\sim} + \langle x_2, y\rangle^{\sim}$,
(2) $(ax, by\rangle^{\sim} = ab^* \langle y, x\rangle^{\sim *}$,
(3) $\langle x, y_1 + y_2\rangle^{\sim} = \langle x, y_1\rangle^{\sim} + \langle x, y_2\rangle^{\sim}$
(4) *for each $x \in H_1$ there exists K_x such that for all $y \in H_2$,*

$$|\langle x, y\rangle^{\sim}| \leqslant K_x \| y \|.$$

Then the mapping

$$\langle x, y\rangle^{\sim} \to C$$

may be extended to a mapping of $H_1 \times H$ such that properties (1)–(3) *hold for this mapping and there exists $T: H_1 \to H$ a linear mapping such that*

$$\langle x, y\rangle^{\sim} = \langle Tx, y\rangle.$$

Proof. For each $x \in H_1$ fixed we consider the linear functional

$$F_x(y) = \langle x, y\rangle^{\sim *}$$

and from property (4) we see that it is bounded. Thus, being in H^* there exists y_1 such that

$$F_x(y) = \langle y, y_1\rangle.$$

We define

$$x \to Tx = y_1$$

and it is easy to see that this satisfies all the required properties. Now we remark that if $\mathscr{F} = (\mu_{x,y})_{x,y \in H}$ is a semispectral family and for f a continuous complex-valued function with compact support the mapping

$$[x, y]_f = \int_S f(s)\, d\mu_{x,y}(s)$$

satisfies all the above properties ((2) when f is real-valued). Then this implies that there exists $U_f \in L(H)$ such that

$$[x, y]_f = \langle U_f x, y\rangle. \qquad \square$$

Now is we denote by $K(S)$ the space of all continuous complex-valued functions with compact support defined on S, we can consider the mapping $f \to U_f$ of $K(S)$ into $L(H)$.

The following proposition gives some important properties of this mapping.

PROPOSITION 6.3.3. *The mapping*

$$f \to U_f$$

from $K(S)$ into $L(H)$ has the following properties:

(1) $U_{f+g} = U_f + U_g, U_{af} = aU_f$,

(2) $U_{f^*} = (U_f)^*$,
(3) $T \in L(H)$ *commutes with all* U_f *if and only if for all* x, y *in* H,

$$\mu_{Tx,y} = \mu_{x,T^*y}.$$

Proof. Property (1) is clear. For (2) let $f \in K(S)$ and x, y in H. Then we have

$$\langle U_{f^*}x, y\rangle = \int_S f^*(s) d\mu_{x,y}(s) = \left(\int_S f(s)\, d\mu^*_{x,y}(s)\right)^* = \langle U_f y, x\rangle^* = \langle x, U_f y\rangle$$

which implies the assertion.

Now for (3) suppose first that T commutes with all U_f. Then we have

$$\int_S f(s)\, d\mu_{Tx,y}(s) = \langle U_f(Tx), y\rangle = \langle T(U_f)x, y\rangle = \langle U_f x, T^*y\rangle$$
$$= \int_S f(s)\, d\mu_{x,T^*y}(s)$$

and this gives clearly that

$$\mu_{Tx,y} = \mu_{x,T^*y}.$$

Conversely, if this relation holds, then we have

$$\langle U_f(Tx), y\rangle = \int_S f(s)\, d\mu_{Tx,y}(s) = \int_S f(s)\, d\mu_{x,T^*y}(s)$$
$$= \langle U_f x, T^* y\rangle = \langle T(U_f x), y\rangle$$

and x, y being arbitrary in H we get the relation

$$TU_f = U_f T.$$ □

Now we consider an important class of semispectral family of Radon measures, the so-called spectral families of Radon measures.

DEFINITION 6.3.4. Let μ be a Radon measure on S and $g \in K(S)$. We define the product $g \cdot \mu$ as the Radon measure

$$g \cdot (f) = \int_S f(s) g(s)\, \mathrm{d}\mu(s).$$

DEFINITION 6.3.5. A semispectral family of Radon measures $\mathscr{F} = (\mu_{x,y})_{x,y \in H}$ on a locally compact space S is called a spectral family if the relation

$$U_{fg} = U_f U_g$$

holds for each f, g in $K(S$,

The following proposition gives a characterization of spectral families in terms of measures $\mu_{x,y}$.

PROPOSITION 6.3.6. *A family* $\mathscr{F} = (\mu_{x,y})_{x,y \in H}$ *is a spectral family of Radon mea-*

sures if the following properties hold:

(1) $\mathscr{F}$ *is a semispectral family of Radon measures.*
(2) *For each* $g \in K(S)$, $g \cdot \mu_{x,y} = \mu_{U_g x, y}$.

Proof. Suppose first that $\mathscr{F}$ is a spectral family of Radon measures. Consider f, g in $K(S)$ and thus

$$\int_S f(s)g(s)\,\mathrm{d}\mu_{x,y}(s) = \int_S f(s)\,\mathrm{d}\mu_{U_g x, y}(s),$$

which shows that (1) and (2) are satisfied. Conversely, if (1) and (2) hold, then for any f, g in $K(S)$ we have

$$\langle U_{fg}x, xy\rangle = \int_S f(s)g(s)\,\mathrm{d}\mu_{x,y}(s) = \int_S f(s)\,\mathrm{d}\mu_{U_g x, y}(s) = \langle U_f U_g x, y\rangle$$

and thus $U_{fg} = U_f U_g$.

COROLLARY 6.3.7. *Let* $\mathscr{F} = (\mu_{x,y})_{x,y \in H}$ *be a spectral family. Then for all* $x \in H$, $\mu_{x,x} \geqslant 0$.

Proof. Let f in $K(S)$ and thus

$$\int_S f(s)f^*(s)\,\mathrm{d}\mu_{x,x} = \langle U_{ff^*}x, x\rangle = \langle U_{f^*}x, U_f^* x\rangle \geqslant 0. \qquad \square$$

COROLLARY 6.3.8. *Let* $\mathscr{F} = (\mu_{x,y})_{x,y \in H}$ *be a spectral family. Then for any* x, y H *and* $g \in K(S)$,

$$\mu_{U_g x, y} = \mu_{x, U_g^* y} = \mu_{x, U_{\bar{g}}^* y}.$$

Proof. We have

$$\langle U_f U_g x, y\rangle = \langle U_g U_f x, y\rangle = \langle U_f x, U_g^* y\rangle = \langle U_f x, U_{g^*} y\rangle$$

and these imply our assertion. $\square$

We now consider the problem of integrability with respect to a spectral family. For this we need the following notion.

DEFINITION 6.3.9. Let $\mathscr{F} = (\mu_{x,y})_{x,y \in H}$ be a spectral family of Radon measures. We consider the superior integral of $\mu_{x,x}$ for a function f and this will be denoted by $\mu_{x,x}^*(f)$. A positive function f is said to be $\mathscr{F}$-negligible if for each $x \in H$, $\mu_{x,x}^*(f) = 0$. If f is a complex-valued function on S we say that f is $\mathscr{F}$-negligible if $|f|$ is $\mathscr{F}$-negligible.

Then we have the following proposition.

PROPOSITION 6.3.10. (1) *The set of all* $\mathscr{F}$*-negligible functions is a complex linear space.*
(2) *If* $f \geqslant 0$ *is* $\mathscr{F}$*-negligible and* $0 \leqslant g \leqslant f$, *then* g *is also* $\mathscr{F}$*-negligible.*
(3) *If* f *is* $\mathscr{F}$*-negligible then for each* x, y *in* H f *is* $\mu_{x,y}$*-negligible.*

Proof. Assertions (1) and (2) are obvious. Assertion (3) follows from the following identity:

$$\mu_{x,y}\tfrac{1}{4}\,(\mu_{x+y,x+y} - \mu_{x-y,x-y} + i\mu_{x+iy,x+iy} - i\mu_{x-iy,x-iy}).$$

DEFINITION 6.3.11. A set M in S is said to be negligible if

$$\varphi_M(s) = \begin{cases} 1, & s \in M \\ 0, & s \notin M \end{cases}$$

is $\mathscr{F}$-negligible.

PROPOSITION 6.3.12. (1) *If $A \subseteq B$ and B is $\mathscr{F}$-negligible then A is $\mathscr{F}$-negligible.*

(2) *If $(A_n)_1^\infty$ is a sequence of $\mathscr{F}$-negligible subsets then $\cup_n A_n$ and $\cap_n A_n$ are negligible subsets.*

Proof. Since the proof is simple it is omitted. □

We now introduce the measurability with respect to a spectral family.

DEFINITION 6.3.13. A complex-valued function f defined on S is said to be $\mathscr{F}$-measurable if it is $\mu_{x,y}$-measurable for each $\mu_{x,y}$ in $\mathscr{F}$. The identity mentioned above shows that it is sufficient to consider only measures of the form $\mu_{x,x}$, $x \in H$.

DEFINITION 6.3.14. A set $M \subset S$ is $\mathscr{F}$-measurable if φ_M is $\mathscr{F}$-measurable where

$$\varphi_M(s) = \begin{cases} 1, & s \in M \\ 0, & s \notin M \end{cases}.$$

It is clear that if (A_n) is a sequence of measurable subsets then $\cup_n A_n$ and $\cap_n A_n$ are again $\mathscr{F}$-measurable subsets and if M is $\mathscr{F}$-measurable then C_M is also $\mathscr{F}$-measurable.

We now consider an important Banach space related to a spectral family of Radon measures, namely the space $L^\infty(\mathscr{F})$. We consider the family of all complex-valued functions on S such that $\|f\|_\infty = \inf(a > 0, M_a = (s, |f(s)| > a)$ is $\mathscr{F}$-negligible$) < \infty$.

It is clear that if f, g are two $\mathscr{F}$-measurable functions and $(s, f(s) \neq g(s))$ is $\mathscr{F}$-negligible then $\|f\|_\infty = \|g\|_\infty$ (if at least one is finite).

If we define an equivalence in the class of $\mathscr{F}$-measurable functions as "two $\mathscr{F}$-measurable functions f and g are in the same class if $(s, f(s) = g(s))$ is $\mathscr{F}$-negligible", then it is not difficult to see that this satisfies all properties of an equivalence relation. We denote by $L^\infty(\mathscr{F}) = L^\infty$ the set of all equivalence classes of functions such that if $f \in \mathring{f}$ then

$$\|f\|_\infty = \|\mathring{f}\|_\infty < \infty.$$

Clearly the function defined on L by $\mathring{f} \to \|\mathring{f}\|_\infty$ is a norm. We now show that if the class of equivalence $\mathring{f}$ contains a continuous function f then

$$\|f\|_\infty = \|\mathring{f}\|_\infty = \sup\{|f(s)|, s \in S).$$

Indeed, let us consider

$$M = (s, |f(s)| > \|f\|_\infty)$$

and from the continuity of f it is clear that M is open. Also it is $\mathscr{F}$-negligible and thus M is negligible for each measure $\mu_{x,x}$ for $x \in H$. Let Sup $\mu_{x,x}$ be the support of the measure $\mu_{x,x}$ and thus we have,

$$M \cap \operatorname{Sup} \mu_{x,x} = \emptyset.$$

This implies that

$$\|f\|_\infty \geqslant \sup(|f(s)|, s \in \bigcup_x \operatorname{Sup} \mu_{x,x}.$$

Let $N = S - \cup_x \operatorname{Sup} \mu_{x,x}$. Then it is contained in each C Sup $\mu_{x,x}$ and thus is $\mathscr{F}$-negligible. If $a < \|f\|_\infty$ then $(s, |f(s)| > a)$ is not negligible and we find in this set a point of $\cup$ Sup $\mu_{x,x}$ which implies that

$$\|f\|_\infty \leqslant \sup(|f(s)|, s \in \bigcup \operatorname{Sup} \mu_{x,x})$$

which implies that

$$\|f\|_\infty = \sup(|f(s)|, s \in \bigcup_x \operatorname{Sup} \mu_{x,x}).$$

COROLLARY 6.3.15. *For any* $f \in \mathring{f} \in L^\infty$ *we have*

$$\int_S f(s)\, \mathrm{d}\mu_{x,y}(s) \leqslant \|f\|_\infty \mu_{x,y}.$$

We now prove an important fact about the mapping $f \to U_f$ considered before for elements of $K(S)$. First we give the following:

LEMMA 6.3.16. If $f \in K(S)$ then $\|f\|_\infty = \|U_f\|_\infty$.

Proof. From the inequality $\|U_f\|_\infty \leqslant \|f\|_\infty$

we have to show only the converse. Let $\varepsilon > 0$ and set

$$M_\varepsilon = (s, |f(s)| \geqslant \varepsilon + \|f\|_\infty)$$

and we show that this is an $\mathscr{F}$-negligible set. Then it is clear that this implies our assertion. To prove that M_ε is $\mathscr{F}$-negligible we consider g an arbitrary function in $K(S \cap M)$. Then we show that

$$\int_S |g|^2\, \mathrm{d}\mu_{x,x}(s) = 0$$

for all $x \in H$. Now we have

$$\|U_{fg}x\|^2 = \|U_f(U_g x)\| \leqslant \|U_f\|^2\ U_g x\|^2 = \|U_f\|^2 \int_S |g|^2\, \mathrm{d}\mu_{x,x}(s)$$

and

$$\|U_{fg}x\|^2 = \int_S fg(s)^2\, \mathrm{d}\mu_{x,x}(s) \geqslant (\varepsilon + \|f\|_\infty)^2 \int_S |g|^2\, \mathrm{d}\mu_{x,x}(s)$$

which gives that

$$\int_S |g|^2\, \mathrm{d}\mu_{x,x}(s) = 0.$$

Thus M_ϵ is $\mathscr{F}$-negligible and then $\|f\|_\infty = \|U_f\|_\infty$. This completes the proof. □

Now for each $\mathring{f} \in L^\infty$ we set (for simplicity we write f only)

$$[x, y]_f = \int_S f(s)\, \mathrm{d}\mu_{x,y}(s)$$

and it is clear that this depends only upon the equivalence class. The mapping $[x, y]_f$ satisfies all properties for the similar mapping considered for functions from $K(S)$. Then we get that there exists an operator $U_f \in L(H)$ such that

$$[x, y]_f = \langle U_f x, y \rangle.$$

Thus we can consider the mapping $f \to U_f$ defined on L^∞ with values in $L(H)$ and we note that the mapping has the following properties:

(1) $U_{f+g} = U_f + U_g$.
(2) $U_{af} = aU_f$.
(3) $U_{fg} = U_f U_g$.
(4) $U_{f^*} = (U_f)^*$.
(5) $f \to \langle U_f x, y \rangle$ is continuous for each x and y in H.

We note that any mapping with properties (1)–(5) is said to be a representation. Thus we have a representation of L^∞ in $L(H)$.

We now show that the above representation has a property similar to that of 6.3.3(3).

THEOREM 6.3.17. *An element $T \in L(H)$ commutes with all $U_f, f \in L^\infty$ if and only if for all x, y in H,*

$$\mu_{Tx,y} = \mu_{x,T^*y}.$$

Proof. Let us consider first the case when $TU_f = U_f T$. Then we have

$$\langle TU_f x, y \rangle = \langle U_f Tx, y \rangle = \int_S f(s)\, \mathrm{d}_{Tx,y}(s) = \langle U_f x, T^*y \rangle$$

$$= \int_S f(s)\, \mathrm{d}\mu_{x,T^*y}(s)$$

which gives that

$$\mu_{Tx,y} = \mu_{x,T^*y}.$$

Conversely, if this relation holds then we have

$$\langle TU_f x, y \rangle = \langle U_f x, T^*y \rangle = \int_S f(s)\, \mathrm{d}\mu_{x,T^*y} = \int_S f(s)\, \mathrm{d}\mu_{Tx,y} = \langle U_f Tx, y \rangle.$$

Thus T commutes with each U_f and the assertion is proved. □

Consider now M be a measurable set (with respect to $\mathscr{F}$) and the operator corresponding to φ_M, $U_{\varphi_M} = P_M$.

Since $\varphi_M^2 = \varphi_M$ we get that $P_M = P_M^2$ and since φ_M is real-valued we get that in fact P_M is a projection. Also we have $P_{\emptyset} = 0$, where $\emptyset$ is the empty set. The following proposition gives some properties of the projections obtained in this way.

PROPOSITION 6.3.18. *Let $(M_n)_1^\infty$ be a sequence of $\mathscr{F}$-measurable subsets of S and suppose that:*

(1) $M_1 \supseteq M_2 \supseteq M_3 \supseteq M_4 \supseteq \cdots \supseteq M_n \supseteq M_{n+1} \supseteq \cdots$; *or*
(2) $M_1 \cap M_j = \emptyset$ *for* $i \neq j$.

Then if (1) *holds*

$$P\left(\bigcap_n M_n\right) = \inf_n (P_{M_n})$$

and if (2) *holds*

$$P\left(\bigcup_n M_n\right) = \sum_n P_{M_n}.$$

Proof. Suppose that (1) holds. Then for each $x \in H$ and any integer n we have

$$\|P_{M_n} x\|^2 = \int_S \varphi_{M_n}(s)\, \mathrm{d}\mu_{x,x}(s) \geqslant \int_S \varphi_{M_{n+1}}(s)\, \mathrm{d}\mu_{x,x} = \|P_{M_{n+1}} x\|^2$$

which gives that

$$P_{M_n} \geqslant P_{M_{n+1}}.$$

Further, we have for each $x \in H$,

$$\|P_{M_n} x - P_{M_m} x\|^2 = \int_S |\varphi_{M_n} - \varphi_{M_m}|^2 (s)\, \mathrm{d}\mu_{x,x}$$

and since

$$\lim_{n \to \infty} \varphi_{M_n} = {}_{\cap_m M_m}$$

we get that

$$\inf_n P_{M_n} = P_{\cap_n M_n}.$$

Now if (2) holds then

$$\varphi_{\cup M_m} = \lim_{n \to \infty} \varphi_{\cup_{i=1}^n M_i}$$

and thus for each $x \in H$,

$$\lim_{n \to \infty} \|P_{\cup_m M_m} x - P_{\cup_{i=1}^n M_i} x\| = 0$$

which implies that

$$P_{\cup_m M_m} = \sum_n P_{M_n}.$$

The following theorem gives a condition for an element $T \in L(H)$ to commute with U_f in terms of the projections P_M.

THEOREM 6.3.19. *The element $T \in L(H)$ commutes with U_f, for each f, if and only if for each compact set $M \subset S$,*

$$TP_M = P_M T.$$

Proof. The condition is necessary. This is clear from the above definition of P_M. Suppose now that this condition is satisfied. Since for any compact subset M of S,

$$\mu_{Tx,y}(M) = \mu_{x,T^*y}(M)$$

this implies, using the regularity, that

$$\mu_{Tx,y} = \mu_{x,T^*y}$$

and thus T commutes with U_f according to 6.3.17.

Let us consider M be a $\mathscr{F}$-measurable set in S and thus

$$\langle P_M x, y\rangle = \int_S \varphi_M(s)\,\mathrm{d}\mu_{x,y}(s) = \mu_{x,y}(M)$$

and thus

$$\langle U_f x, y\rangle = \int_S f(s)\,\mathrm{d}\langle P(s)x, y\rangle.$$

Now we consider a decomposition of S as follows:

$$\Delta = (M_i)_{i\in I}, \qquad \bigcup_i M_i = S$$

and where each M_i is $\mathscr{F}$-measurable. In what follows we consider only the case when I is a finite set.

Associated with the above decomposition we consider the sum,

$$S(f, \Delta)x = \sum_i f(s_i) P_{M_i} x.$$

Now on the set of all (finite) decompositions of S we consider a partial order (by inclusion) and thus we get a directed set. Thus we can consider

$$\lim_\Delta S(f, \Delta)x$$

and show that it exists. Let y arbitrary in H and thus

$$\langle (U_f x - S(f, \Delta)x), y\rangle = \int_S \Big(f(s) - \sum_i f(s_i)\varphi_{M_i}(s)\Big)\,\mathrm{d}\mu_{x,y}(s)$$

and if our decomposition is sufficiently fine, i.e.

$$\Big| f(s) - \sum_i f(s_i)\varphi_{M_i}(s)\Big| < \varepsilon$$

then

$$\lim_\Delta \| U_f x - S(f, \Delta)x\| = 0. \qquad (*)$$

In what follows we write U_f as $\int_S f(s)\,\mathrm{d}P(s)$ and this notation is justified by relation (*).

We now consider a method to construct spectral families. For this purpose we need the notion of 'the spectral decomposition of unity' which is now defined.

DEFINITION 6.3.20. A mapping of R with values in the set of projections on the space H is said to be a spectral decomposition of unity if the following conditions are satisfied:

(1) If $s \leqslant t$ then $P_s \leqslant P_t$.
(2) For each $t \in R$, $P_{t-0} = P_t$.
(3) For each $x \in H$, $\lim_{t \to -\infty} P_t x = 0$.
(4) For each $x \in H$, $\lim_{t \to \infty} P_t x = x$.

We now define two projections $P_{-\infty}$ by the relation $P_{-\infty} x = 0$ and $P_\infty x = x$. Then conditions (3) and (4) in Definition 6.3.20 assert that

$$\lim_{t \to -\infty} P_t x = P_{-\infty} x, \qquad \lim_{t - \infty} P_t x = P_\infty x.$$

From condition (1) of Definition 6.3.20 we get that

$$P_t P_s = P_{\min(t,s)}.$$

Let us consider x, y in H, and we define a function on R by the relation

$$\mu_{x,y}(t) = \langle P_t x, y \rangle$$

and clearly the following properties hold:

(1) $\mu_{x,x}$ is monotone increasing.
(2) $\mu_{x,x}$ is left continuous.
(3) $\mu_{x,x}$ is bounded.
(4) $\mu_{x,y}$ has a bounded variation.

We denote by $\tilde{\mu}_{x,y}(s)$ the Radon measure associated with the function $\mu_{x,y'}$

$$\tilde{\mu}_{x,y}(f) = \int_R f(s)\,\mathrm{d}\mu_{x,y}(s)$$

and the above properties of the functions $\mu_{x,y}$ imply that the following assertion is true.

PROPOSITION 6.3.21. (1) *The family* $(\tilde{\mu}_{x,y})_{x,y \in H}$ *is a semispectral family of Radon measures.*

(2) *The family* $(\tilde{\mu}_{x,y})_{x,y \in H}$ *is spectral.*

Proof. Property (1) is clear from the above discussion. We prove only (2). Let f, g be two continuous functions with compact support. Then

$$\langle U_f(U_g)x, y \rangle = \int_R f(s)\,\mathrm{d}\langle P_s(U_g x), y \rangle$$

and since

$$\langle P_s(U_g x), y\rangle = \langle U_g x, P_s y\rangle = \int_R g(t)\,\mathrm{d}\langle P_t x, P_s y\rangle = \int_{(-\infty,t)} g(t)\,\mathrm{d}\langle P_s x, y\rangle$$

we get that

$$\langle U_f(U_g x), y\rangle = \int_R f(s)g(s)\,\mathrm{d}\langle P_s x, y\rangle = \langle U_{fg}x, y\rangle.$$

Since x, y are arbitrary, we get that

$$U_{fg} = U_f U_g$$

and the spectrality of the family is proved. □

PROPOSITION 6.3.22. *Let $(P_t)_{t\in R}$ be a spectral decomposition of unity. Then the following assertions hold:*

(1) $P_t H \subseteq P_s H$ *for* $t \subseteq s$.
(2) $P_s - P_t$ *is a projection for* $t \leqslant s$.

Proof. Since the proof is simple it is omitted. □

PROPOSITION 6.3.23. *Let $(P_t)_{t\in R}$ be a decomposition of unity. Then for each $t \in R$,*

$$\lim_{s\downarrow t} P_s = P_{t+0}$$

and P_{t+0} is a projection.

Proof. Let $t \in R$ be fixed and (s_n) be a monotone decreasing sequence converging to t, i.e.

(1) $s_1 \geqslant s_2 \geqslant s_3 \geqslant \cdots \geqslant s_n \geqslant s_{n+1} \geqslant \cdots$
(2) $\lim s_n = t$.

Then we have that $(Q_n)_1^\infty$, where $Q_n = P_{s_n} - P_{s_{n+1}}$ is a sequence of projections with $Q_n Q_m = 0$, $n \neq m$.

Let $n > m$ and $x \in H$ fixed. Then

$$\|(P_{s_n} - P_{s_m})x\|^2 = \sum_m \|(P_{s_{i+1}} - P_{s_i})x\|^2$$

which implies, for $n \to \infty$ and $m = 1$,

$$\sum_1^\infty \|(P_{s_{i+1}} - P_{s_i})x\|^2 \leqslant \|x\|^2.$$

This gives that for any $\varepsilon > 0$ there exists N_ε such that for all $n, n \geqslant N_\varepsilon$ the following inequality holds:

$$\|(P_{s_n} - P_{s_m})x\|^2 \leqslant \varepsilon.$$

This means that the sequence $(P_{s_n}x)$ is a Cauchy sequence. Let $\lim P_{s_n}x = P_{t+0}x$. To show that the notation is justified we must show that $P_{t+0}x$ is independent of the sequence (s_n). Indeed, if (s'_n) is another sequence satisfying the conditions:

(1) $s'_1 \geqslant s'_2 \geqslant s'_3 \geqslant \cdots \geqslant s'_n \geqslant s'_{n+1} \geqslant \cdots$

(2) $\lim s'_n = t$

then the sequence $(P_{s'_n} x)$ converges to $P_{t+0} x$.

Since for each n there exist $N = N(n)$ and $M = M(n)$ such that

$$s_N \leqslant s'_n \leqslant s_M$$

we get that

$$(P_{s_n} - P_{s'_n})P_{s_N} = P_{s_N} - P_{s_N} = 0$$

$$(P_{s_n} - P_{s'_n})P_{s_M} = P_{s_n} - P_{s'_n}$$

and thus for $n \geqslant N_{\varepsilon'}$

$$\|(P_{s_n} - P_{s'_n})x\| = \|(P_{s_n} - P_{s'_n})(P_{s_M} - P_{s_N})x\| \leqslant \|(P_{s_M} - P_{s_N})x\| \leqslant \varepsilon.$$

This clearly implies that

$$\lim P_{s_n} x = \lim P_{s'_n} x$$

and thus our notation is justified.

Now we show that P_{t+0} is a projection. Let x, y be arbitrary elements of H and thus

$$\langle P_{t+0}x, y\rangle = \lim_{s \downarrow t} \langle P_s x, y\rangle = \lim_{s \downarrow t} \langle x, P_s y\rangle = \langle x, P_{t+0} y\rangle$$

and thus P_{t+0} is hermitian.

Let $r > t$, then

$$P_{t+0}x = \lim_{s \downarrow t} P_s x = \lim_{s \downarrow t} P_r P_s = P_r \lim_{s \downarrow t} P_s x = P_r P_{t+0}$$

which implies that

$$P_{t+0}^2 = \lim_{r \downarrow t} P_r P_{t+0} = P_{t+0}$$

and thus P_{t+0} is a projection.

We now consider a spectral decomposition of unity with several dimensions; the spectral decomposition of unity considered in Definition 6.3.20 is sometimes called 'the spectral decomposition of unity with one dimension'.

DEFINITION 6.3.24. A mapping of R^n $(n \geqslant 2)$ with values in the set of projections of a space H is called a decomposition of unity if the following properties are satisfied (we denote the value of the mapping at $(t_1, \ldots, t_n)$ with $P_{t_1 \ldots t_n}$):

(1) If $t_1 \leqslant t'_1, \ldots, t_n \leqslant t'_n$, then $P_{t_1 \ldots t_n} \leqslant P_{t'_1 \ldots t'_n}$,

(2) For each $(t_1, \ldots, t_n) \in R^n$,

$$P_{t_1 - 0 \ldots t_n - 0} = P_{t_1 \ldots t_n},$$

(3) For each $x \in H$,

$$\lim_{t_1 \to -\infty, \ldots t_n \to \infty} P_{t_1 \ldots t_n} x = 0.$$

(4) For each $x \in H$,

$$\lim_{t_1 \to \infty, \dots t_n \to \infty} P_{t_1 \dots t_n} x = x.$$

We now define two projections, $P_{-\infty \cdots -\infty}$, $P_{\infty \cdots \infty}$ by the formulas:

$$P_{-\infty \cdots -\infty} x = 0, \qquad P_{\infty \cdots \infty} x = x$$

and condition (3) and (4) in Definition 6.3.24 tell us that

$$\lim_{t_1 \to -\infty, \dots, t_n \to \infty} P_{t_1 \dots t_n} x = P_{-\infty \cdots -\infty} x$$

$$\lim_{t_1 \to \infty \dots t_n \to \infty} P_{t_1 \dots t_n} x = P_{\infty \cdots \infty} x.$$

For each x, y in H we define the following function on R^n:

$$\mu_{x,y}(t_1, \dots, t_n) = \langle P_{t_1 \dots t_n} x, y \rangle$$

and from Definition 6.3.24 we get that the following properties hold:

(1) $\mu_{x,x}$ is increasing in each t_i.

(2) $\mu_{x,y}$ has a bounded variation.

(3) $\lim\limits_{t_i \to -\infty} \mu_{x,y}(t_1, \dots, t_n) = 0,$

(4) $\lim\limits_{t_i \to \infty} \mu_{x,y}(t_1, \dots, t_n) = x, y.$

As in the case of spectral decompositions with one dimension we define, by the use of Radon measure associated with $\mu_{x,y}(t_1, \dots, t_n)$,

$$\tilde{\mu}_{x,y}(f) = \int_{R^n} f(s)\, \mathrm{d}\mu_{x,y}(s), \quad s = (s_1, \dots, s_n)$$

and we can prove similarly the following result:

PROPOSITION 6.3.25 *The family $(\tilde{\mu}_{x,y})_{x,y \in H}$ is a spectral family of Radon measures on R^n.*

REMARK 6.3.26. If we have $n = 2$ then we write $P_{t,s} = P_z$ with $z = t + is$.

We now describe another way to construct families of Radon measures.

DEFINITION 6.3.27. Let S be a locally compact Hausdorff space and B_S be the σ-algebra generated by the compact subsets of S.

DEFINITION 6.3.28. A mapping on B_S with values in the set of projections on a space H, $M \to P_M$, is said to be a spectral measure if the following conditions are satisfied:

(1) $P_{M \cap N} = P_M P_N$.

(2) If $(M_n)_1^\infty \subset B_S$, $\quad M_i \cap M_j = \emptyset$, $\quad i \neq j$, $\quad$ then $P_{\cup M_n} = \sum\limits_n P_{M_n}$.

(3) $\sup\limits_{M \in B_S} P_M = I$.

DEFINITION 6.3.29. A spectral measure $M \to P_M$ is said to be regular if and only if, for any $M \in B_S$,

$$P_M = \inf_{M \subset G} P_G, \quad G \text{ open}.$$

As in the case of real measures the reader can verify the following.

THEOREM 6.3.30. *A spectral measure is regular if and only if for any $M \subset B_S$,*

$$P_M = \sup_{F \subset M} P_F, F \text{ closed}.$$

Suppose now that $M \to P_M$ is a spectral measure and for each x, y in H we set

$$\mu_{x,y}(M) = \langle P_M x, y \rangle$$

and consider the family $\mathscr{F} = (\mu_{x,y})_{x,y \in H}$.

PROPOSITION 6.3.31. *The spectral measure $M \to P_M$ is regular if and only if $\mu_{x,x}$ is regular.*

Proof. We leave this to reader to verify. □

THEOREM 6.3.32. *The family $\mathscr{F} = (\mu_{x,y})_{x,y \in H}$ is a spectral family.*

Proof. Let f, g in $K(S)$ and then we have

$$\langle U_f(U_g x), y \rangle = \int_S f(s)\, \mathrm{d}\langle P(s) U_g x, y \rangle$$

and since

$$\langle P_M U_g x, y \rangle = \langle U_g x, P_M y \rangle = \int_S g(s)\, \mathrm{d}\langle P(s)x, P_M y \rangle$$

$$= \int_S g(s)\varphi_M(s)\, \mathrm{d}\langle P(s)x, y \rangle$$

we get that

$$\langle U_f\, U_g x, y \rangle = \int_S f(s)g(s)\, \mathrm{d}\langle P(s)x, y \rangle = \langle U_{fg} x, y \rangle$$

and as x, y is arbitrary this implies that $U_{fg} = U_f U_g$. Thus the family $\mathscr{F} = (\mu_{x,y})_{x,y \in H}$ is a spectral family. □

We now give some important applications of the above results for the case of some classes of operators on a complete inner product space H over C. We now consider a commutative closed subalgebra A of $L(H)$ and we suppose that A is selfadjoint (i.e. if $T \in A$ then also $T^* \in A$). We suppose that A is without identity and then by the structure theorem for commutative B*-algebras without identity, A is *-isomorphic and isometric with some algebra $C_0(S)$, where S is a locally compact space. For each $a \in A$ we set f_a as the function which is null at infinity, which corresponds to the element a in $C_0(S)$, through the above isomorphism. Then it is clear that the following properties hold:

(1) $f_{a+b} = f_a + f_b$,

(2) $f_{a^*} = f_a^*$,

(3) $f_{za} = zf_a$, $z \in C$,

(4) $\|a\| = \|f\|_S = \sup(|f_a(s)|, s \in S)$.

If $a \geqslant 0$ (i.e. $a = a^*$ and in the algebra $\{(z, a)\}$, $\sigma(a) \geqslant 0$) then $f_a \geqslant 0$ and $f^{1/2}$ exists and is in $C_0(S)$. Let $a^{1/2}$ be the element in A which has the property that $f_{a^{1/2}} = f_a^{1/2}$.

If φ is in A^* then $\psi(f_a) = \varphi(a)$ is a continuous functional on $C_0(S)$; if φ is positive (i.e. if $\varphi(b * b) \geqslant 0$ for all $b \in A$) then ψ is also positive on $C_0(S)$. Then by the Riesz–Kakutani representation theorem there exists a measure μ_φ on S such that

$$\varphi(a) = \int_S f_a(s)\, \mathrm{d}\mu_\varphi(s).$$

Now if x, y are elements of H we can define on A the following functional

$$\varphi_{x,y}(T) = \langle Tx, y\rangle$$

which is obviously continuous

$$|\varphi_{x,y}(T)| \leqslant \|T\|\,\|x\|\,\|y\|$$

and clearly positive. Thus for any $T \in A$ there exists, for each pair x, y in H, a measure $\mu_{x,y}^T$ such that

$$\varphi_{x,y}(T) = \langle Tx, y\rangle = \int_S f_T(s)\, \mathrm{d}\mu_{x,y}^T(s).$$

PROPOSITION 6.3.33. *The family* $\mathscr{F} = (\mu_{x,y}^T)_{x,y \in H}$ *is a spectral family of Radon measures.*

Proof. It is clear that $\mathscr{F}$ is a semispectral family. We show that $\mathscr{F}$ in fact is a spectral family. Let f, g be two functions in $C_0(S)$ and T_f, T_g be the operators in A such that

$$T_f \to f, \qquad T_g \to g$$

through the above isomorphism. Then we have

$$\int_S f(s)g(s)\, \mathrm{d}\mu_{x,y}(s) = \langle T_{fg}x, y\rangle = \langle T_f(T_g x), y\rangle$$

which implies that

$$U_{fg} = U_f U_g$$

and the spectrality is proved. □

It follows that we find a spectral measure $M \to P_M$ defined on the σ-algebra generated by the compact subsets of S and with values in the set of projections on H such that, for any $T \in A$,

$$\langle Tx, y\rangle = \int_S f_T(s)\, \mathrm{d}\mu_{x,y}(s) = \int_S f_T(s)\, \mathrm{d}\langle P(s)x, y\rangle.$$

Suppose now that A is a subalgebra of $L(H)$ with the following properties:

(1) A is commutative and selfadjoint.
(2) There exists an operator $T \in L(H)$ such that (T, I) generates A.

In this case A is isomorphic and isometric with the algebra of continuous functions on a compact spaces S and, as above, for each $a \in A$, f_a denotes the corresponding function for the element a. Then the mapping $a \to f_a$ satisfies the properties which are the same as for the case of algebra without identity. The important fact is that the compact space S may be identified now with the spectrum of T, $\sigma(T)$.

Thus A and $C(\sigma(T))$ are isomorphic and isometric algebras and we note that the function on $\sigma(T)$ which corresponds to T is $f_T(z) = z$.

We now consider some special cases.

(1) The operator T is hermitian. Then $\sigma(T)$ is in R and then A is isomorphic and isometric with an algebra generated by 1 and t.
(2) The operator T is unitary. Then $\sigma(T) \subseteq (z, |z| = 1)$ and A is isomorphic and isometric with the algebra generated by (1 and z).
(3) T is normal. Then A is isomorphic and isometric with the algebra of continuous functions of a closed subset contained in $(z, |z| \leqslant \|T\|)$.

If $M \to P_M$ is the spectral measure associated with the algebra generated by T and I then the spectral measure associated with the operator T is defined as follows:

$$M \to P_{M \cap \sigma(T)} = E_M$$

and thus for any x, y in H,

$$\langle Tx, y \rangle = \int_{\sigma(T)} z \, \mathrm{d}\langle E(z)x, y \rangle$$

and

$$T = \int_{\sigma(T)} z \, \mathrm{d}E_z.$$

It is clear that

$$T^* = \int_{\sigma(T)} z^* \, \mathrm{d}E_z$$

and thus for any polynomial $p(z, z^*) = \Sigma a_{ij} z^i z^{*j}$, $p(T, T^*) = \Sigma a_{ij} T^i T^{*j}$,

$$p(T, T^*) = \int_{\sigma(T)} p(z, z^*) \, \mathrm{d}E_z.$$

Then we have

$$\|p(T, T^*)\| = \sup |p(z, z^*)| = \|p(z, z^*)\|_{\sigma(T)}$$

and we have a homomorphism of the algebra of polynomials in z, z^* into A which is an isometry. This can be extended to continuous functions, since $(p(z, z^*)$ is dense in

$C(\sigma(T))$. This gives further that

$$\sigma(p(T, T^*)) = p(\sigma(T), \sigma(T)^*).$$

and

$$\sigma(f(T)) = f(\sigma(T))$$

for any continuous function on $\sigma(T)$.

REMARK 6.3.34. The relation

$$\sigma(p(T, T^*)) = p(\sigma(T), \sigma(T)^*),$$

can be proved in an elementary way and leads to an elementary proof of the spectral theorem for normal operators. See Istrăţescu (1981).

6.4. Unbounded Operators

Let us consider the space $L^2_{[0,1]}$. In $L^2_{[0,1]}$ we consider the subspace of all functions f such that $f' = df/dx$ is in $L^2_{[0,1]}$ and we denote this subspace by C^2. Then we have that

$$d/dx: C^2 \to L^2_{[0,1]},$$

defined by $d/dx(f) = f'$, is a linear mapping. C^2 as a subspace of L^2 is clearly not closed but it is dense in L^2. The above linear mapping d/dx is not bounded in the sense that there exists no $K > 0$ such that for all $f \in C^2$, $\|d/dx\, f\|_{L^2} \leqslant K\|f\|_{L^2}$.

Indeed, it is clear that C^2 contains the function $f_n(s) = s^n$, $n = 0, 1, 2, \ldots$. Then we have,

$$\|d/dx\, f_n\|^2_{L^2} = n^2 \int_0^1 x^{2n-2}\, dx = n^2/(2n+3)$$

which obviously implies the nonexistence of K satisfying the above inequality.

In what follows we are interested in mappings defined on linear subspaces of Banach or complete inner product spaces. Among the most interesting and important classes of mappings which we consider are the class of hermitian operators, the class of selfadjoint and essentially selfadjoint operators, and normal operators. Also an important role is played by the class of closed operators.

DEFINITION 6.4.1. Let H be a complete inner product space and let $T: D_T \to H$ be a map. We say that T is a linear operator if the following conditions are satisfied:

(1) D_T is a linear subspace of H.

(2) For all x, y in D_T and any a, b in C, $T(ax + by) = aTx + bTy$.

REMARK 6.4.2. As in the case of bounded operators we also use the notation Tx for $T(x)$. The subspace D_T is called the domain of T and $T(D_T)$ is the range of T, sometimes denoted by $R(T)$ or Δ_T.

DEFINITION 6.4.3. If T, S are two linear operator then we say that S is an extension of T if the following conditions are satisfied:

(1) $D_S \supset D_T$.
(2) For all $x \in D_T$, $Tx = Sx$.

Now we introduce certain operations in the set of all linear operators.

DEFINITION 6.4.4. Let T and S be two linear operators. We define the sum of T and S as the operator denoted by $T + S$, which is defined as follows: the domain is $D_T \cap D_S$ and for each $x \in D_T \cap D_S$

$$(T + S)x = Tx + Sx.$$

The multiplication with a scalar is defined as in the case of bounded linear operators.

DEFINITION 6.4.5. The product of two linear operators T and S is the operator denoted by TS, defined as follows: $D_{TS} = (S^{-1}(D_T))$ and for each $x \in D_{TS}$, $(TS)x = T(Sx)$. Here

$$S^{-1}(M) = (x, x \text{ with the property that } x \in D_S \text{ and } Sx \in D_T).$$

DEFINITION 6.4.6. The inverse of a linear operator T is the operator denoted by T^{-1}, defined as $D_T - 1 = R(T)$, and satisfying the conditions

(1) $T^{-1}(Tx) = x$
(2) $T(T^{-1}y) = y$

for all $x \in D_T$ and all $y \in R(T)$.

DEFINITION 6.4.7. If T is a linear operator with the domain D_T and V_1 is a linear subspace of D_T then we define a linear operator $T_1 : V_1 \to H$ by the relation $T_1 x = Tx$. The operator T_1 is called the restriction of T at V_1 and sometimes we use the notation $T_1 = T/V_1$. Also, we denote by O the operator defined on H by $Ox = 0$ and by I the operator defined on H by $Ix = x$ for all $x \in H$.

We now define an important notion for a pair of linear operators.

DEFINITION 6.4.8. Let T and S be two linear operator with $D_T = H$. We say that T and S are permutable operators if $ST$is an extension of TS. If $D_S = H$ this notion reduces to the definition of commuting operators.

PROPOSITION 6.4.9. *If T and S are two linear operators then the following assertions hold*:

(1) $T + S = S + T$.
(2) *If R is another linear operator then* $(T + S) + R = T + (S + T)$.
(3) $(TS)R = T(SR)$.
(4) *If T is an extension of S and R is an extension of Q then TR is and extension of SQ.*
(5) $(T + S)R = TR + SR$.
(6) *If T is permutable with R and S and $D_T = H$ then T is permutable with $R + S$.*

Proof. The verification of these assertions is left to the reader. □

As we know, one of the most important sets associated with a bounded linear operator (or more generally to an element of a Banach algebra with identity) is the spectrum. We now define for unbounded operators the analogue set.

DEFINITION 6.4.10. Let $T: D_T \to H$ be a linear operator. We say that the complex number z is in the resolvent set $\rho(T)$ of T if $T_z = T - z = T - zI$ has an inverse which is an element of $L(H)$.

The spectrum of T is the set $\sigma(T)$ defined by the relation:

$$\sigma(T) = C_{\rho(T)}.$$

There are certain subsets in $\sigma(T)$ which in some problems are very important. We now consider these subsets.

DEFINITION 6.4.11. $\sigma_c(T)$ (the continuous spectrum of T) is the subset of all z in $\sigma(T)$ with the property that T_z has an inverse which is not bounded and the range of T_z is dense in X.

DEFINITION 6.4.12. $\sigma_r(T)$ (the residual spectrum of T) is the subset f all z in $\sigma(T)$ with the property that T_z has an inverse and the range of T_z is not dense in X.

DEFINITION 6.4.13. $\sigma_p(T)$ (the point spectrum of T) is the subset of all z in $\sigma(T)$ for which T_z does not have an inverse.

From the definition we see that the following relation holds:

$$\sigma(T) = \sigma_p(T) \cup \sigma_c(T) \cup \sigma_r(T).$$

The elements of $\sigma_p(T)$ are called eigenvalues of T and the subspace

$$E_z = (x, Tx = zx)$$

is called the eigenspace associated with the eigenvalue z.

DEFINITION 6.4.14. An eigenvalue z of T is said to be a normal eigenvalue if

$$E_z(T) = E_{\bar{z}}(T^*).,$$

REMARK 6.4.15. It is obvious that for normal operators every eigenvalue is a normal eigenvalue; if T is a hyponormal operator then obviously we have the relation:

$$E_{\bar{z}}(T^*) \subseteq E_z(T).$$

There are simple examples to show that the inclusion may be strict. We leave the reader to prove that if T is normal then $\sigma_r(T)$ is an empty set.

6.5. Closed and Closable Operators

The important and useful class of closed operators is introduced below and also several properties are proved.

DEFINITION 6.5.1. A linear operator $T: D_T \to H$ is called closed whenever $x_n \to x$, $x_n \in D_T$ and $y = \lim Tx_n$, then $x \in D_T$ and $Tx = y$.

We have the following result.

PROPOSITION 6.5.2. *Every bounded linear operator $T \in L(H)$ is closed.*

Proof. Let $x_n \to x$ and $y = \lim Tx_n$. Since T is continuous we have

$$y = \lim Tx_n = T(\lim x_n) = Tx$$

and the assertion is proved. □

REMARK 6.5.3. An operator may be closed and not continuous. We give later an example of such an operator which will be a differential operator.

The following result shows that veritable closed operators (i.e. not continuous) are defined on a strictly smaller part of the space H.

THEOREM 6.5.4. *If $T: H \to H$ is a closed linear operator then it is continuous.*

Proof. This is the so-called 'closed graph theorem' and for the proof we refer the reader to Chapter 2. □

Let H be a complete inner product space and let T be a linear operator. The following result gives a characterization of closed operators in terms of properties of the space D_T.

THEOREM 6.5.5. *The operator T is closed if and only if D_T with the inner product*

$$[x, y] = \langle x, y \rangle + \langle Tx, Ty \rangle$$

is a complete inner product space.

Proof. Suppose first that T is closed and to prove our assertion it suffices to show that if (x_n) is a Cauchy sequence with respect to the norm

$$x \to [x, x]^{1/2} \tag{*}$$

then (x_n) converges to an element $x \in D_T$.

From the form of the inner product $[\,,]$ we see that (x_n) and (Tx_n) are Cauchy sequences and thus we can consider $x = \lim x_n$ and $y = \lim Tx_n$. Now the closedness of T asserts that $y = Tx$ and this implies that (x_n) converges with respect to the norm (*) to x. Thus D_T is complete.

Conversely, if D_T is complete with respect to the inner product $[\,,]$ then we show that T is closed. Let us consider (x_n) in D_T such that $x = \lim x_n$ and $y = \lim Tx_n$. Then (x_n) is a Cauchy sequence with respect to the norm (*) and since D_T is complete our sequence converges to an element x in D_T and clearly $Tx = y = \lim Tx_n$. The theorem is proved. □

DEFINITION 6.5.6. Let T be a linear operator. We say that T is closable if T has an extension which is a closed operator.

The following result gives information about the set of closed extensions of a closable operator.

THEOREM 6.5.7. *Let T be a closable operator. Then in the family of all closed extensions of T there exists a minimal one, in the sense that any other closed extension C_T of T is an extension of this minimal closed extension, denoted by $\bar{T}$.*

Proof. Let S be a closed extension of T. Then we know from Theorem 6.5.5 that

D_S is a complete inner space with an inner product

$$[x, y] = \langle x, y \rangle + \langle Sx, Sy \rangle.$$

Now S is an extension of T, and thus for all $x \in D_T$ we have $Tx = Sx$. Now we define a linear operator $\bar{T}$ on the closure of D_T in D_S by the formula $\bar{T}x = Sx$ for all x in $\bar{D}_T$ (the closure of D_T in D_S). Then clearly $\bar{T}$ is an extension of T and Theorem 6.5.5 gives that $\bar{T}$ is closed. Now to prove the theorem we show the minimality property. Let us consider for this purpose that R is another extension of T (of course closed). Then for all $x \in D_T$, $Rx = \bar{T}x$. Let us consider $x \in \bar{D}_T$. Then there exists (x_n) in D_T such that

$$\lim(\langle x_n - x, x_n - x \rangle + \langle \bar{T}x_n - \bar{T}x, \bar{T}x_n - \bar{T}x \rangle) = 0.$$

Then we get that $x_n \to x$ and

$$\lim \bar{T}x_n = \lim Tx_n = \lim Rx_n = \bar{T}x = Rx$$

which gives that $x \in D_R$. Thus R is a closed extension of $\bar{T}$ and the minimality is proved. □

We now define the notion of an adjoint of a linear operator. In what follows we suppose that D_T, the domain of the operator, is dense in H.

DEFINITION 6.5.8. Let T be a linear operator with D_T dense. If we set

$$D_{T^*} = (y, y \in H, \text{ there exists } y^* \in H \text{ such that for all } x \in D_T, \langle Tx, y \rangle = \langle x, y^* \rangle)$$

then $T^*: D_{T^*} \to H$, defined by the relation $T^*y = y^*$, is a linear operator, called the adjoint of T.

First we remark that from the density of D_T we get immediately that T^* is well defined, i.e. y^* is unique. Also it is easy to verify that T^* is just a linear operator on D_{T^*}.

THEOREM 6.5.9. *Let T be a linear operator with D_T dense in H. Then the adjoint T^* of T has the properties*

(1) *T^* is a closed operator,*
(2) *if S is an extension of T then T^* is an extension of S^*,*
(3) *if T is closable then $T^* = (\bar{T})^*$, where $\bar{T}$ is the operator defined in the proof of Theorem 6.5.7.*

Proof. (1) Let (y_n) be a sequence of elements of D_{T^*} such that

(1) $y_n \to y$,
(2) $T^*y_n \to z$.

In this case, for any $x \in D_T$ we have

$$\langle x, z \rangle = \lim\langle x, T^*y_n \rangle = \lim\langle Tx, y_n \rangle = \langle Tx, y \rangle$$

which implies that $y \in D_{T^*}$ and $z = T^*y$. Thus T^* is a closed operator.

(2) If $x \in D_T$ and $y \in D_{S^*}$ then we have

$$\langle Tx, y \rangle = \langle Sx, y \rangle = \langle x, S^*y \rangle$$

which gives that $y \in D_{T^*}$ and that $T^*y = S^*y$ and thus (2) is proved.

(3) Since T is supposed closable, $\bar{T}$, as defined in the proof of Theorem 6.5.7, is closed and is an extension of T. In this case we get that T^* is an extension of $(\bar{T})^*$. We show now that $(\bar{T})^*$ is an extension of T^*. Indeed, if $x \in D_{\bar{T}}$ then there exists $(x_n) \subset D_T$ such that

(1) $\lim x_n = x$,
(2) $\lim Tx_n = \bar{T}x$.

Now, since for each n and $y \in D_{T^*}$,

$$\langle Tx_n, y \rangle = \langle x_n, T^*y \rangle$$

for $n \to \infty$, we get that

$$\langle \bar{T}x, y \rangle = \langle x, T^*y \rangle$$

and thus $y \in D_{\bar{T}^*}$, $T^*y = (\bar{T})^*y$. But y is arbitrary in D_{T^*} and thus the assertion is proved. □

We now give other properties of the adjoint of operators.

PROPOSITION 6.5.10. *Let T, S be two linear operators with dense domains. Then the following assertions hold:*

(1) *If T is an extension of S then S^* is an extension of T^*.*
(2) *If D_{T+S} is dense then $(T+S)^*$ is an extension of $T^* + S^*$.*
(3) *If D_{TS} is dense then $(TS)^*$ is an extension of S^*T^*.*
(4) *For any $z \neq 0$, $(zT)^* = z^*T^*$, $z \in C$.*
(5) *If D_{T^*} is dense then T^{**} is an extension of T.*
(6) *If $R(T)$ is dense and T^{-1} exists then $(T^*)^{-1} = (T^{-1})^*$.*
(7) *For any $S \in L(H)$ which commutes with T, S^* commutes with T^* (i.e. they are permutable).*
(8) *If T^{-1} and $(T^*)^{-1}$ exist then $R(T)$ is dense in H and (6) holds.*

Proof. (1) Let $y \in D_{T^*}$ and thus for each $x \in D_S$ we have

$$\langle Tx, y \rangle = \langle Sx, y \rangle = \langle x, T^*y \rangle$$

which gives that $y \in D_{S^*}$ and $S^*y = T^*y$ and (1) is proved.

(2) Let $y \in D_{T^*+S^*}$ and $x \in D_{T+S}$ and thus

$$\langle (T+S)x, y \rangle = \langle Tx, y \rangle + \langle Sx, y \rangle = \langle x, T^*y \rangle + \langle x, S^*y \rangle$$

which implies that $y \in D_{(T+S)^*}$ and $(T+S)^*y = T^*y + S^*y$ which gives (2).

(3) Let $x \in D_{TS}$ and $y \in D_{S^*T^*}$. From the definition of the product of two linear operators we have that $Sx \in D_T$ and $T^*y \in D_{S^*}$. Further, we get that

$$\langle (TS)x, y \rangle = \langle T(Sx), y \rangle = \langle Sx, T^*y \rangle$$

and thus $T^*y \in D_{S^*}$. Now,

$$\langle (TS)x, y \rangle = \langle Sx, T^*y \rangle = \langle x, S^*(T^*y) \rangle$$

which implies that $y \in D_{(TS)^*}$ and $(TS)^*y = S^*(T^*y)$ which proves (3).

(4) Since we define for each $z \neq 0$, $(zT)x = zTx$ we have clear $D_T = D_{zT}$ and if

$y \in D_{T^*}$ then we have

$$\langle (zT)x, y \rangle = \langle zTx, y \rangle = z\langle Tx, y \rangle = z\langle x, T^*y \rangle = \langle x, z^*T^*y \rangle$$

and (4) is proved.

(5) Let $x \in D_T$ and $y \in D_{T^*}$ and thus $\langle y, Tx \rangle = \langle T^*y, x \rangle$ which implies that $x \in D_{T^{**}}$ and $T^{**}x = Tx$. Thus we have (5).

(6) Since $D_T - 1$ is dense $(T^{-1})^*$ exists and let now $x \in D_T$ and y in $D_{(T^{-1})^*}$. Then we get

$$\langle x, y \rangle = \langle T^{-1}(Tx), y \rangle = \langle Tx, (T^{-1})^*y \rangle$$

which gives that $y \in D_{T^*}$ and

$$T^*((T^{-1})^*y) = y. \qquad (*)$$

Now if $x \in R(T)$ and $y \in D_{T^*}$ we have

$$\langle x, y \rangle = \langle T(T^{-1})x, y \rangle = \langle T^{-1}x, T^*y \rangle$$

which implies that $y \in D_{(T^{-1})^*}$ and

$$(T^{-1})^*(T^*(y)) = y. \qquad (**)$$

From (*) and (**) we get assertion (6).

(7) Since $S \in L(H)$, $D_S = D_{S^*} = H$ and thus $D_{S^*T^*} = D_{T^*}$. Let $x \in D_T$ and $y \in D_{S^*T^*}$ and thus

$$\langle Tx, S^*y \rangle = \langle STx, y \rangle = \langle TSx, y \rangle = \langle Sx, T^*y \rangle = \langle x, S^*(T^*y) \rangle$$

which gives that $S^*y \in D_{T^*}$. Further,

$$\langle x, T^*(S^*y) \rangle = \langle x, S^*(T^*y) \rangle$$

and thus

$$T^*(S^*y) = S^*(T^*y)$$

for all $y \in D_{S^*T^*}$ and (7) is proved.

(8) To prove that $R(T)$ is dense we suppose that this is not so and we get a contradiction. Indeed, if $R(T)$ is not dense then there exists a nonzero element of H which is orthogonal to $R(T)$. Let z be such an element and thus we have for all $x \in D_T$,

$$\langle Tx, z \rangle = 0.$$

Now if $y \in D_{T^*}$ then $\langle Tx, y \rangle = \langle x, T^*y \rangle$ and with the above relation gives

$$\langle Tx, y + z \rangle = \langle x, T^*y \rangle$$

i.e. $T^*(y + z) = T^*y$, which contradicts the existence of $(T^*)^{-1}$. Thus $R(T)$ is dense and the assertion follows. Clearly (6) holds. □

6.6. The Graph of Linear Operators and Some Applications

Let H be a complete inner product space and $T: D_T \to H$ be a linear operator.

DEFINITION 6.6.1. The graph of the linear operator T is the set $((x, Tx), x \in D_T)$ in $H \oplus H$. We denote the graph of T by G_T. It is clear that G_T is a linear subspace.

The following proposition determines the subspaces in $H \oplus H$ which are the graphs of operators.

PROPOSITION 6.6.2. *A linear subspace V of $H \oplus H$ is the graph of a linear operator if and only if $(0, y) \in V$; then $y = 0$.*

Proof. Suppose that V is the graph of a linear operator. Then $(0, 0)$ is in V and $(0, y) \in V$. Then $(0, y) = (0, T0) = 0, 0)$. This implies $y = 0$.

Conversely, if V has the above property then we define a linear operator T as follows: D_T is the projection of V on the first coordinate and for $(x, y) \in V$ we set $y = Tx$. Then clearly T is linear and the above property of V implies that T is well defined. Indeed, if $Tx = y = y'$ then we have that (x, y) is in V and that (x, y') is in V. Since V is a linear subspace we get that $(0, y - y')$ is in V. This implies that $y = y'$ and thus T is well defined. □

The following proposition gives a characterization of closed operators using the graph.

PROPOSITION 6.6.3. *An operator $T: D_T \to H$ is closed if and only if G_T is closed in $H \oplus H$.*

Proof. Suppose first that T is closed. Then we show that G_T is closed. Let us consider a sequence in G_T which converges to an element of $H \oplus H$. Let (x_n, y_n) be the sequence in G_T converging to (x, y). Then clearly $x = \lim x_n$ and $y = \lim y_n = \lim Tx_n$. Since T is closed, $y = Tx$. Thus G_T is closed. Similarly we prove the converse. □

We now consider an arbitrary linear operator $T: D_T \to H$ and we consider the set G_T and let $\bar{G}_T$ be the closure of G_T in $H \oplus H$. The following proposition gives information about $\bar{G}_T$ and its relation to closable operators.

PROPOSITION 6.6.4. *The set $\bar{G}_T$ is a graph of an operator if and only if T is closable.*

Proof. If $\bar{G}_T$ is the graph of an operator then clearly this is closable and thus the condition is sufficient. The necessity follows from Theorem 6.5.5. □

REMARK 6.6.5. If $\bar{G}_T$ is the graph of an operator then it is exactly the graph of the operator $\bar{T}$ considered in Theorem 6.5.7.

We now give some results concerning the graph and the adjoint for linear operators. For this purpose we consider the mapping $U: H \oplus H \to H \oplus H$ defined by the relation

$$U(x, y) = (iy, -ix)$$

and we remark that the following properties hold:

(1) $U^2 = I$,
(2) U is unitary.

THEOREM 6.6.6. *Let $T: D_T \to H$ be a linear operator and suppose that T^* exists.*

Then

$$G_{T^*} = \overline{(UG_T)}^{\perp}.$$

Proof. Let $x \in D_T$ and since T^* exists we have

$$\langle Tx, y \rangle = \langle x, T^*y \rangle$$

and thus

$$\langle ((iTx), -ix, (y, T^*y)) \rangle = \langle iTx, y \rangle + \langle -ix, T^*y \rangle = i\langle Tx, y \rangle - i\langle x, T^*y \rangle.$$

This implies that $(y, T^*y) \in (UG_T)^{\perp}$ since x is arbitrary. This clearly implies that

$$G_{T^*} \subset (UG_T)^{\perp}. \tag{*}$$

Now if (*) holds then

$$\langle Tx, y \rangle = \langle x, z \rangle$$

for all $x \in D_T$ and thus

$$(y, z) = (y, T^*y) \in G_{T^*}$$

or

$$(UG_T)^{\perp} \subset G_{T^*} \tag{**}$$

and from (*) and (**) our assertion follows. □

We note that since $(UG_T)^{\perp}$ is a closed set we get that if T^* exists then it is a closed operator. Thus we have a new proof of Theorem 6.5.9(1).

PROPOSITION 6.6.7. *Let V be a linear subspace of $H \oplus H$ and suppose that $(U(V))^{\perp}$ is a graph. Then the projection on the first coordinate is a dense set in H.*

Proof. Suppose that the assertion is not true and thus there exists z_0 which is orthogonal to each element of the projection, i.e.

$$\langle x, z_0 \rangle = 0 \quad \text{for all} \quad (x, y) \in (U(V))^{\perp}.$$

Let (x_1, y_1) in $H \oplus H$ and thus we have

$$\begin{aligned}\langle (x, y + z_0), (iy_1, -ix_1) \rangle &= \langle (x, y), (iy_1, -ix_1) \rangle + \langle (0, z_0), (iy_1, -ix_1) \rangle \\ &= \langle 0, iy_1 \rangle - \langle z_0, ix_1 \rangle = 0,\end{aligned}$$

which implies that $(x, y + z_0) \in (U(v))^{\perp}$. Since $(U(V))^{\perp}$ is a graph, this is a contradiction because $y \neq y + z_0$.

PROPOSITION 6.6.8. *Let T be a closed linear operator with D_T a dense subset. Then T^* exists and $T^{**} = T$.*

Proof. By the above result we have

$$G_{T\cdot} = (U(G_T))^{\perp}$$

and thus

$$U(G_{T^*}) = (G_T)^{\perp}$$

or

$$G_T = (U(G_{T^*})).$$

This implies that D_{T^*} is dense. Since

$$G_{T^{**}} = (U(G_{T^*}))^{\perp}$$

we get that $T^{**} = T$. □

6.7. Hermitian, Selfadjoint and Essentially Selfadjoint Operators

First we consider the notion of Hermitian and selfadjoint operators.

DEFINITION 6.7.1. A linear operator $T: D_T \to H$ is called Hermitian if the following conditions are satisfied:

(1) D_T is dense in H.
(2) For each x, y in D_T, $\langle Tx, y\rangle = \langle x, Ty\rangle$.

The following proposition gives a characterization of Hermitian operators in terms of the adjoint.

PROPOSITION 6.7.2. *The linear operator $T: D_T \to H$ is hermitian if and only if T^* is an extension of T.*

Proof. Let T be Hermitian. Then we see that any point $x \in D_T$ is in D_{T^*} and $Tx = T^*x$. Thus T^* is an extension of T. Conversely, if T^* is an extension of T then for any $x \in D_T$, $y \in D_T$ we have

$$\langle Tx, y\rangle = \langle T^*x, y\rangle = \langle x, Ty\rangle$$

and thus T is hermitian. □

We now define the important notion of selfadjoint operator.

DEFINITION 6.7.3. A linear operator $T: D_T \to H$ is called selfadjoint if the following conditions are satisfied:

(1) D_T is dense in H.
(2) $T = T^*$.

From the above definition we have.

THEOREM 6.7.4. *Let T be a linear operator. Then*

(1) *if T is selfadjoint then T is Hermitian,*
(2) *if T is Hermitian and $D_T = H$ then T is bounded.*

Proof. The first assertion is obvious. The second assertion is known as the 'Hellinger–Toeplitz theorem' and may be proved as follows. Suppose that the assertion is false and thus there exists a sequence (x_n) of unit vectors such that $\|Tx_n\| \geqslant n$. We consider now for each n the functional

$$F_y(x) = \langle Tx, y\rangle$$

with $y = Tx_n$. Then clearly $\|F_y\| \leqslant \|y\|$, and since $f(x) = \sup |F_y(x)|$ is a subadditive functional, there exists by the Gelfand lemma (Lemma 2.5.6), a constant $K > 0$ such that

$$|f(x)| \leqslant K\|x\|.$$

This gives that $\|Tx_n\| \leqslant K$ and this contradiction proves the assertion. □

DEFINITION 6.7.5. A Hermitian operator T is called maximal if any Hermitian extension of T is exactly T. A selfadjoint operator T is maximal if any selfadjoint extension of T is exactly T.

THEOREM 6.7.6. *A linear operator $T: D_T \to H$ is selfadjoint if and only if the following conditions are satisfied:*

(1) T is closed.
(2) T^* is hermitian.

Proof. From the properties of the adjoint and the definition of Hermitian and selfadjoint operators, we see that conditions (1) and (2) are necessary. Suppose now that T satisfies (1) and (2). Since T^* is an extension of T and the adjoint of T^* is T we get that T is an extension of T^*. This implies that $T = T^*$. □

THEOREM 6.7.7. *Let T be a selfadjoint operator and $S \in L(H)$ be a Hermitian operator. Then $T + S$ is selfadjoint.*

Proof. Since $D_{T+S} = D_S$ and $S = (S + T) - T$ we get that S^* is an extension of $(S + T)^* - T^*$ and thus $S^* + T^*$ is an extension of $(S + T)^*$. This gives that $(S + T)^* = S^* + T^*$ and further $(S + T)^* = S + T$. □

THEOREM 6.7.8. *Let T be a linear operator such that T^{-1} exists. Then T is selfadjoint if and only if T^{-1} is selfadjoint.*

Proof. Since

$$(T^{-1})^{-1} = T$$

to prove the assertion it is sufficient to show that if T is selfadjoint then T^{-1} is selfadjoint.

Since $T = T^*$ we get that $(T^*)^{-1}$ exists and since

$$(T^{-1})^* = (T^*)^{-1}$$

we get that

$$(T^{-1})^* = T^{-1}$$

and the theorem is proved. □

The following theorem gives an useful criteria for a Hermitian operator to be selfadjoint.

THEOREM 6.7.9. *Let T be a Hermitian operator and suppose that for a complex number z, $T - z$ and $T - z^*$ have ranges equal to H. Then T is selfadjoint.*

Proof. Since T is Hermitian we have that T^* is an extension of T. To prove that $T = T^*$ it is sufficient to show that D_{T^*} is a subset of D_T. Let us consider $y \in D_{T^*}$ and thus, according to the definition of the adjoint, for all $x \in D_T$

$$\langle Tx - zx, y \rangle = \langle Tx, y \rangle - z\langle x, y \rangle = \langle x, T^*y \rangle - \langle x, z^*y \rangle = \langle x, T^*y - z^*y \rangle$$

and since the range of $(T - z^*)$ is H there exists $u \in D_T$ such that $(T - z^*)u = T^*y - z^*y$ and thus

$$\langle Tx - zx, y \rangle = \langle x, (T - z^*)u \rangle = \langle Tx, u \rangle - \langle zx, u \rangle = \langle Tx - zx, u \rangle.$$

Now the range of $(T - z)$ is H and thus

$$\langle Tx - zx, u - y \rangle = 0$$

implies that $y = u$. But $u \in D_T$ and thus $y \in D_T$. The theorem is proved. □

COROLLARY 6.7.10. *Let T be a Hermitian operator. Then T is selfadjoint if and only if for any complex number z, $\operatorname{Im} z \neq 0$, the range of $T - z$ is H.*

Proof. From 6.7.7 we know that

$$(T - zI)^* = T^* - z^*I = T^* - z^*$$

and now we prove that:

(1) $H = \bar{R}(T - z) \oplus N(T^* - z^*)$,
(2) $\|(T - z)x\| \geqslant |\operatorname{Im} z| \, \|x\|, \quad x \in D_T$.

Indeed, if $x \in D_T$ and $y \in D_{T^*}$ then we have

$$\langle Tx - zx, y \rangle = \langle x, T^*y \rangle - z\langle x, y \rangle = \langle x(T^* - z^*)y \rangle$$

and thus $y \in D_{(T-z)^*}$ and (x being arbitrary in D_T)

$$(T - zI)^*y = T^*y - z^*y.$$

If $x \in D_T$ and $y \in D_{(T-zI)^*}$ then

$$\begin{aligned}\langle Tx, y \rangle &= \langle Tx - zx, y \rangle + z\langle x, y \rangle = \langle x, (T - z)^* \rangle + z\langle x, y \rangle \\ &= \langle x, ((T - z)^*y + zy) \rangle\end{aligned}$$

which gives that $y \in D_{T^*}$.

Now if $y \in N(T^* - z^*)$ then for any $x \in D_T$ we have

$$\langle Tx - zx, y \rangle = \langle x, (T - z^*)^*y \rangle = 0$$

and thus $y \in \bar{R}(T - z)$. Also, if $y \in \bar{R}(T - z)^-$, then for $x \in D_T$ we get

$$\langle Tx - zx, y \rangle = 0$$

which implies that $y \in D_{(T-z)^*}$ and $T^*y - z^*y = (T - z)^*y$, i.e. $y \in N((T - z)^*)$ and (1) is proved.

Let $x \in D_T$ and then

$$\|Tx - zx\|^2 = \langle Tx, Tx \rangle + (\operatorname{Re} z)^2\|x\|^2 - 2\operatorname{Re} z\langle Tx, x \rangle + (\operatorname{Im} z)^2\|x\|^2.$$

But

$$2|\operatorname{Re} z\langle Tx, y \rangle| \leqslant \|Tx\|^2 + (\operatorname{Re} z)^2\|x\|^2$$

which implies that

$$\|Tx - zx\|^2 \geqslant (\operatorname{Im} z)^2\|x\|^2$$

which is equivalent to (2).

Now, if $\operatorname{Im} z \neq 0$ the range of $T - z$ is H and by Theorem 6.7.9, T is selfadjoint. Conversely, if T is selfadjoint we show that for each z with $\operatorname{Im} z \neq 0$, the range of $T - z$ is H.

First, let $y \in N(T^* - z^*)$ and thus

$$\operatorname{Im} z^*\|y\|^2 = \langle \operatorname{Im} z^* y, y\rangle = \langle \operatorname{Im} T^* y, y\rangle = \operatorname{Im}\langle Ty, y\rangle$$

and as Ty, y is a real number we get that $y = 0$. From the above decomposition of H we have that

$$H = \bar{R}(T - z) \tag{*}$$

and then $N(T - z) = (0)$. Thus $(T - z)^{-1}$ exists and its domain $D_{(T-z)^{-1}}$ is $R(T - z)$.

Let $y \in H$ and not in $R(T - z)$. By (*) there exists a sequence (y_n) in $R(T - z)$ such that $y = \lim y_n$. Let $x_n = (T - z)^{-1} y_n$ and since

$$\|(T - z)^{-1} y_n\| = \|x_n\| \leqslant 1/|\operatorname{Im} z| \cdot \|y_n\|$$

we obtain that (x_n) is a Cauchy sequence. Let $x = \lim x_n$ and thus we get

$$Tx_n = (T - z)x_n + zx_n = y_n + zx_n \to y + zx.$$

But $T = T^*$ is a closed operator and thus $x \in D_T$ and $Tx = y + zx$, i.e. $y = Tx - zx$ which shows that $R(T - z)$ is closed and the corollary is proved. □

THEOREM 6.7.11. *Let T be a selfadjoint operator and S be a Hermitian operator satisfying the conditions*

(1) $D_T \subset D_S$,
(2) *For some* $d \in [0, 1)$ *and* $c \geqslant 0$, $\quad \|Sx\| \leqslant d\|Tx\| + c\|x\|, x \in D_T$.

Then the operator $T + S$ with $D_{T+S} = D_T$ is selfadjoint.

Proof. Obviously $T + S$ is Hermitian and to prove that it is selfadjoint we show that for some nonzero real number r the range of $T + S - ir$ is H. Then by Theorem 6.7.9, $T + S$ is selfadjoint.

Indeed, we have the following inequality: for r a real number and $y \in H$,

$$\|(T - ir)^{-1} y\| \leqslant 1/|r| \cdot \|y\|.$$

Now for all $x \in D_T$,

$$\|Tx - irx\|^2 = \|Tx\|^2 + r^2\|x\|^2$$

and if $y = (T - ir)x$ we have that for all $y \in H$,

$$\|T(T - ir)^{-1} y\| \leqslant \|y\|.$$

Now let y be fixed in H and consider the following equation

$$z + S(T - ir)^{-1} z = y \tag{*}$$

and since for all $u \in H$,

$$\|S(T - ir)^{-1}u\| \leqslant d\|T(T - ir)^{-1}u\| + c\|(T - ir)^{-1}u\|$$

or

$$\|S(T - ir)^{-1}u\| \leqslant d\|u\| + c/d\|u\|.$$

This shows that $S(T - ir)^{-1} \in L(H)$ and if $r > r_0$, $S(T - ir)^{-1}$ is a strict contraction. In this case, equation (*) has a unique solution z and if $x = (T - ir)^{-1}z$ then $x \in D_T$ and

$$(T + S - ir)x = Tx - irx + Sx = z + S(T - ir)^{-1}z = y$$

which implies that $R(T + S - ir) = H$. Then $T + S$ is selfadjoint. □

For some decomposition theorems the following results will be of importance.

THEOREM 6.7.12. *Let T be a linear operator which is closed. Then the following assertions hold:*

(1) *$(I + T^*T)^{-1}$ exists and is a positive Hermitian contraction in $L(H)$.*
(2) *$T(I + T^*T)^{-1}$ is in $L(H)$ and is a contraction operator.*

Proof. Let z in H arbitrary and since

$$H \oplus H = G_T \oplus U(G_{T^*})$$

the element $(z, 0)$ can be written uniquely as

$$(z, 0) = (x, Tx) + i(T^*y, -y) \tag{*}$$

with $x \in D_T$ and $y \in D_{T^*}$. This implies that

$$z = x + iT^*y \quad \text{and} \quad 0 = Tx - iy$$

and thus

$$z = (I + T^*T)x.$$

We show now that x is unique. Indeed, if there exists another x' such that

$$z = (I + T^*T)x'$$

then

$$(x, -iTx) \neq (x', -iTx')$$

and satisfy (*), which is not possible. Thus we must have $x = x'$. This shows that $(I + T^*T)^{-1}$ exists and since

$$\|(I + T^*T)^{-1}z\| = \|x\| \leqslant \|z\|$$

$(I + T^*T)^{-1}$ is a contraction in $L(H)$. Now we show that it is Hermitian and positive.

Indeed, when u, v are in H, if $R = (I + T^*T)^{-1}$ then

$$\begin{aligned}\langle Ru, v\rangle &= \langle Ru, (I + T^*T)Rv\rangle = \langle Ru, Rv\rangle + \langle Ru, T^*TRv\rangle \\ &= \langle Ru, Rv\rangle + \langle TRu, TRv\rangle = \langle Ru, Rv\rangle + \langle T^*TRu, Rv\rangle \\ &= \langle (I + T^*T)Ru, Rv\rangle = \langle u, Rv\rangle\end{aligned}$$

and thus R is Hermitian. Since

$$\langle Ru, u\rangle = \langle Ru, Ru\rangle + \langle TRu, TRu\rangle$$

we get that R is positive. Since

$$D_{T(I+T^*T)^{-1}} = H \quad \text{and} \quad \|T(I + T^*T)^{-1}z\| = \|Tx\| \leqslant \|z\|$$

the operator $T(I + T^*T)^{-1}$ is a contraction in $L(H)$.

REMARK 6.7.13. If T is a closed linear operator then similarly we can prove the results for the operators $(I + TT^*)^{-1}$ and $T(I + TT^*)^{-1}$.

We now consider the class of essentially selfadjoint operators.

DEFINITION 6.7.14. A linear operator $T: D_T \to H$ is said to be essentially self-adjoint if the following conditions are satisfied:

(1) T is Hermitian.

(2) The closure $\bar{T}$ of T is a selfadjoint operator.

It is clear that any selfadjoint operator is essentially selfadjoint; we give later an example of a differential operator which is essentially selfadjoint but not selfadjoint.

6.8. Some Examples of Selfadjoint and Essentially Selfadjoint Operators

EXAMPLE 6.8.1. We consider the space $L^2(R, \mathrm{d}x)$ of all functions (classes of functions) such that with respect to the Lebesgue measure $\mathrm{d}x$ on R,

$$\int_R |f|^2 \,\mathrm{d}x < \infty$$

with the inner product

$$\langle f, g\rangle = \int_R f(x)g^*(x)\,\mathrm{d}x$$

and then $L^2(R, \mathrm{d}x)$ is complete.

Define T as follows:

$$D_T = (f, f \in L^2(R, \mathrm{d}x), xf(x) \in L^2(R, \mathrm{d}x))$$

and

$$T: D_T \to L^2(R, \mathrm{d}x)$$

by the formula

$$((Tf)x) = xf(x).$$

It is known and not difficult to prove that D_T is dense in $L^2(R, \mathrm{d}x)$. We now show that T is Hermitian. Indeed, we have for f, g in D_T,

$$\langle Tf, g\rangle = \int_R xf(x)g^*(x)\mathrm{d}x = \int_R f(x)(xg(x))^*\mathrm{d}x = \langle f, Tg\rangle$$

and thus T is Hermitian.

We now show that T is in fact selfadjoint. For this it is sufficient to show that if for some fixed z, $\operatorname{Im} z \neq 0$ the equation

$$(T - z)f = g$$

has a solution for each fixed g in $L^2(R, \mathrm{d}x)$. Indeed, it is clear that for each z, $\operatorname{Im} z \neq 0$ the function

$$x \to g(x)/(x - z) = f(x)$$

is in $L^2(R, \mathrm{d}x)$. Since clearly we have the relation

$$(T - z)f = g$$

then according to Theorem 6.7.9 or 6.7.10, T is selfadjoint.

REMARK 6.8.2. If we consider the space $L^2(R^n, \mathrm{d}x)$, $\mathrm{d}x$ is the Lebesgue measure on R^n, then the operators T_i, $i = 1, 2, \ldots, n$, defined as

$$D_{T_i} = (f, f \in L^2(R^n, \mathrm{d}x), x_i f(x) \in L^2(R^n, \mathrm{d}x))$$
$$(T_i f)(x) = x_i f(x)$$

are selfadjoint. In a similar manner we can prove that the operators

$$T_{i_1 \ldots i_k}$$

defined as

$$D_{T_{i_1 \ldots i_k}} = (f, f \in L^2(R^n, \mathrm{d}x), x_{i_1} \ldots x_{i_k} f(x) \in L^2(R^n, \mathrm{d}x))$$

$$(T_{i_1 \ldots i_k} f)(x) = x_{i_1} \ldots x_{i_k} f(x)$$

are selfadjoint.

EXAMPLE 6.8.3. For this example we need the so-called Sobolev space $W_2^i(R^n, \mathrm{d}x)$, $i = 1, 2$. These are defined as follows. Let $k = a_1 + \cdots + a_n$ and define $W_p^k(R^n, \mathrm{d}x)$ as the space of all functions in $L^p(R^n, \mathrm{d}x)$ (classes of functions) having the derivatives

$$(\partial^{a_1 + \cdots + a_n}/\partial x_1^{a_1} \ldots \partial x_n^{a_n}) f = D^a f, \quad a = (a_1, \ldots, a_n)$$

in $L^p(R^n, \mathrm{d}x)$ and such that

$$f \to \| f \|_{W_p^k(R^n, \mathrm{d}x)} = \left(\sum_{a_1 + \cdots + a_n = k} \| D^a f \|^p_{L^p(R^n, \mathrm{d}x)} \right)^{1/p} < \infty.$$

It is not difficult to see that $C_0(R^n)$ is dense in $W_p^k(R^n, \mathrm{d}x)$ (i.e. $f C_0^\infty(R^n)$ if it is of C^∞ class and for some $M_f > 0, f\chi_{\complement[-M_f, M_f]} = 0$). We consider the space $W_2^2(R^n, \mathrm{d}x)$ which is an inner product space and the operator Δ (the Laplacian) in $L^2(R^n, \mathrm{d}x)$:

$$D_\Delta = W_2^2(R^n, \mathrm{d}x).$$

and

$$T: D_\Delta \to L^2(R^n, \mathrm{d}x)$$

defined by

$$Tf = -\Delta f.$$

Then clearly D_Δ is dense in $L^2(R^n, \mathrm{d}x)$ and we show that T is Hermitian. Indeed, if f, g are in $C_0(R^n)$ then we get

$$\langle -\Delta f, g\rangle = \int_{R^n} (-\Delta f) g^* \,\mathrm{d}x = \int_{R^n} \sum_{l=1}^{n} \partial f/\partial x^l \cdot \partial g^*/\partial x^l$$

$$= \int_{R^n} f(-\Delta g^*)\,\mathrm{d}x = \langle f, -\Delta g\rangle$$

and since $C_0^\infty(R^n)$ is dense in $W_2^i (i = 1, 2)$ we get that T is Hermitian. Now we show that in fact T is selfadjoint. Let z be an arbitrary complex number with $\operatorname{Im} z \neq 0$. The Fourier transform of g, $\hat{g} \in L^2(R^n, \mathrm{d}x)$ and clearly

$$1/(\|y\|^2 - z) \cdot \hat{g}(y) \in L^2(R^n, \mathrm{d}x).$$

Since the inverse transform of $1/(\|y\|^2 - z)\hat{g}(y)$ is in $L^2(R^n, \mathrm{d}x)$ as well as the inverse transform of

$$(y^{a_1 + \cdots + a_n}/(\|y\|^2 - z))\hat{g}(y) \in L^2(R^n, \mathrm{d}x)$$

(for $a_1 + \cdots + a_n \leqslant 2$), this implies that $D^a f$ are in $L^2(R^n, \mathrm{d}x)$ and since

$$Tf - zf = -\Delta f - zf = \text{Inv. Transf. } (1/(|y|^2 - z)\hat{g}(y)) = g,$$

and according to Theorem 6.7.9 or 6.7.10, T is selfadjoint.

EXAMPLE 6.8.4. We now consider the operator T defined on $C_0^\infty(R^n)$ and we show that it is essentially selfadjoint.

We note that in the above example we proved that $-\Delta$ is Hermitian. We now show that the closure is selfadjoint. Indeed, we consider on $W_2^2(R^n, \mathrm{d}x)$ the norm

$$f \to \|f\|_\Delta = (\|f\|^2 + \|Tf\|^2)^{1/2}$$

then this is equivalent with the norm of $W_2^2(R^n, \mathrm{d}x)$.

Let f be an arbitrary element of $W_2^2(R^n, \mathrm{d}x)$ and thus

$$\|f\|_\Delta \leqslant \|f\|_{W_2^2(R^n, \mathrm{d}x)} = \left(\sum_a \|D^a f\|^2_{L^2(R^n, \mathrm{d}x)}\right)^{1/2}$$

and if we suppose that f is in $C_0^\infty(R^n)$ we get

$$\|Tf\|^2 = \sum_{l,k} \int_{R^n} ((\partial^2 f/\partial x_l^2)(\partial^2 f^*/\partial x_k^2))$$

$$= -\sum_{l,k} \int_{R^n} ((\partial^3 f/(\partial x_l^2 \partial x_k))(\partial f^*/\partial x_k))\mathrm{d}x$$

$$= \sum_{l,k} ((\partial^2 f/(\partial x_i \partial x_k))(\partial^2 f^*/(\partial x_j \partial x_k)))\,\mathrm{d}x$$

and since $C_0^\infty(R^n)$ is dense in $W_2^2(R^n, \mathrm{d}x)$ the above formulas hold for any f in $W_2^2(R^n, \mathrm{d}x)$. But

$$\langle Tf, f\rangle + \|f\|^2 + \|Tf\|^2 \geqslant \sum_a \|D^a f\|^2_{L^2(R^n,\mathrm{d}x)} = \|f\|^2_{W_2^2(R^n,\mathrm{d}x)}$$

and thus for $f \in W_2^2(R^n, \mathrm{d}x)$,

$$\begin{aligned}\|f\|^2_{W_2^2(R^n,\mathrm{d}x)} &= \langle Tf, f\rangle + \|f\|^2 + \|Tf\|^2 \\ &\leqslant \|Tf\|\,\|f\| + \|f\|^2 + \|Tf\|^2 \\ &\leqslant M(\|f\|^2 + \|Tf\|^2) = M\|f\|^2_\Delta.\end{aligned}$$

This means that $C_0^\infty(R^n)$ is a dense subset of D_T and thus the closure of $-\Delta$ is exactly our operator on D_T. Thus $-\Delta$, with $C_0^\infty(R^n)$ as domain, is essentially selfadjoint.

6.9. Selfadjoint Extensions

Since the selfadjoint operators play a fundamental role in applications in physics, quantum mechanics, partial differential and integrodifferential equations it appears quite natural to ask about the existence of selfadjoint extension for Hermitian operators. Among the main contributors to the solution of this problem we mention John von Neumann (1929), M. H. Stone (1932), M. G. Krein (1947), and M. G. Krein and M. A. Krasnoselskiĭ (1947).

In this section we present some results about this problem. The reader interested in the subject may look at the very interesting papers of John von Neumann (1929), M. H Stone (1932), M. G. Krein and M. A. Krasnoselskiĭ, M. S. Birman (1956).

We begin with two preliminary results. Let X be a Banach space and let P, Q be two projections. We consider the subspaces PX and QX. Then we have the following result.

PROPOSITION 6.9.1. *If* $\|P - Q\| < 1$ *then the subspaces* PX *and* QX *have the same dimension.*

Proof. Since

$$\|P - PQ\| = \|P^2 - PQ\| = \|P(P - Q)\| \leqslant \|P\|\,\|P - Q\| < 1$$

and

$$\|Q - QP\| < 1 \tag{*}$$

we can consider PQ as an element of a Banach space $L(PX)$ and thus we see that the inequality $\|P - PQ\| < 1$ holds, and since P is the identity operator of $L(PX)$ we see that PQ has an inverse in $L(PX)$ and this, as is easy to see, is given by the formula

$$P + \sum_{n=1}^{\infty} (P - PQ)^n.$$

Then we obtain that P applies the Banach space X on a subspace whose dimension is at least dim QX, and from (*), exactly in the same way, we obtain that Q applies X onto a subspace whose dimension is at least dim PX. Thus we have dim $PX = \dim QX$. □

We now consider a complete inner product space X and let M, N be two closed linear subspaces of X.

DEFINITION 6.9.2. The openness between subspaces M and N is defined as follows:

$$\theta(M, N) = \sup (\|(P_M - P_N)x\|/\|x\|, x \in X, x \neq 0).$$

Here P_M, P_N are the orthogonal projections associated with subspaces M and N, i.e. for each $x \in M$, $y \in N$, $P_M x = x, P_N y = y$ and on the orthogonal complements are zero.

PROPOSITION 6.9.3. *For any closed subspaces M, N in X we have*

$$\theta(M, N) = \max(\sup\|(I - P_M)x\|, \sup\|((I - P_N)y\|, x \in M, y \in N, \|x\| = \|y\| = 1)).$$

Proof. From the definition of the openness between M and N we have

$$\begin{aligned}\theta(M, N) &= \sup(\|(P_M - P_N)x\|/\|x\|, x \neq 0)\\ &= \sup(1/\|x\|(\|P_M(I - P_N)x\|^2 + \|(I - P_M)P_N x\|^2)^{1/2})\end{aligned}$$

and thus

$$\begin{aligned}\theta(M, N) &= \sup_{x\in N} (1/\|x\|(\|P_M(I - P_N)x\|^2 + \|(I - P_M)P_N x\|^2)^{1/2})\\ &\geqslant \sup_{x\in N}\|(I - P_M)x\|/\|x\| = m_1\end{aligned}$$

and similarly we obtain the relation

$$\theta(M, N) \geqslant \sup_{x\in M} \|(I - P_N)x\|/\|x\| = m_2$$

which implies that

$$\theta(M, N) \geqslant \max(m_1, m_2).$$

To prove the assertion of the proposition we must show that the inequality in the contrary sense holds. For this we remark that

$$\|(I - P_M)P_N x\|^2 \leqslant m_2^2\|P_N x\|^2$$

and also

$$\begin{aligned}\|P_M(I - P_N)x\|^2 &= \langle P_M(I - P_N)x, P_M(I - P_N)x\rangle = \langle P_M(I - P_N)x, (I - P_N)x\rangle \\ &= \langle P_M(I - P_N)x, (I - P_N)^2x\rangle \\ &= \langle (I - P_N)P_M(I - P_N)x, (I - P_N)x\rangle \\ &\leqslant \|(I - P_N)P_M(I - P_N)x\| \, \|(I - P_N)x\|\end{aligned}$$

and using the definition of m_1 we have,

$$\|P_M(I - P_N)x\|^2 \leqslant m_1\|P_M(I - P_N)x\| \, \|(I - P_N)x\|$$

and thus

$$\|P_M(I - P_N)x\| \leqslant m_1\|(I - P_N)x\|.$$

Since

$$\begin{aligned}\|(I - P_M)P_Nx\|^2 + \|P_M(I - P_N)x\|^2 &\leqslant m_2^2\|P_Nx\|^2 + m_1^2\|(I - P_N)x\|^2 \\ &\leqslant \max(m_1^2, m_2^2)(\|P_Nx\|^2 + \|(I - P_N)x\|^2) \\ &\leqslant \max(m_1^2, m_2^2)\|x\|^2\end{aligned}$$

which implies that

$$\theta(M, N) \leqslant \max(m_1, m_2)$$

and the proposition is proved. □

REMARK 6.9.4. The openness between two subspaces can be defined for not necessarily closed linear subspaces, and formulas like those given above hold.

Let X be as above, a complete inner product space, and let T be a linear operator on X with the domain D_T.

DEFINITION 6.9.5. A point z in C (the set of complex numbers) is said to be a regular point for T if there exists $k > 0$ such that for all x in D_T,

$$\|(T - zI)x\| \geqslant k\|x\|.$$

LEMMA 6.9.6. *The set of all regular points of an operator is an open set.*

Proof. Indeed, z and z_1 are two points in C such that z is a regular point, then if z_1 is such that $|z - z_1| \leqslant k(z)/2$, then we have

$$\begin{aligned}\|(T - z_1)x\| &= \|(T - z)x + (z - z_1)x\| = \|(T - z)x\| - \|(z - z_1)x\| \\ &\geqslant (k(z) - (|z - z_1|))\|x\| \geqslant k(z)/2 \cdot \|x\|\end{aligned}$$

and the assertion is proved. □

6.9.7. *Let T be a Hermitian operator. Then every point $z = x_1 + iy$, $y \neq 0$ is a regular point of T.*

Proof. Indeed, if z is as above then we have,

$$\|(T - z)x\|^2 \geqslant |y|^2\|x\|^2$$

and we can take $k = |y|$.

It is clear that if T is a Hermitian operator then the subsets $(z, \operatorname{Im} z < 0)$ and $(z, \operatorname{Im} z > 0)$ are two connected components of the set of regular points of T. □

The following result gives an information concerning the connected components of the set of regular points of an operator.

THEOREM 6.9.8 (Krein and Krasnoselskiĭ, 1947). *Let T be a linear operator with the domain C_T and let M be a connected component of the set of regular points of T. Then the function defined on M by*

$$z \to \dim((T - z)X)^{\perp}$$

is a constant function.

Proof. Let z and z_0 be two points in M and let P_z, P_{z_0} be the projections on the subspaces $(T - z)X^{\perp}$, $((T - z_0)X)^{\perp}$ respectively. We show that if z is sufficiently near to z_0 then $\|P_z - P_{z_0}\| < 1$ and the assertion follows from the connectedness of M. For this we choose $d = \frac{1}{3} \cdot k(z_0)$ and since

$$k(z_0)\|x\| \leqslant \|(T - z_0)x\| = \|(T - z)x + (z - z_0)x\| \leqslant \|(T - z)x\| + |z - z_0| \, \|x\|$$

for $|z - z_0| \leqslant d$ we obtain that

$$\|(T - z)x\| \geqslant \tfrac{2}{3}k(z_0)\|x\|.$$

For $|z - z_0| \leqslant d$ and $\|u\| = 1$, $u \in (T - z)X$, we have

$$\begin{aligned} \|(I - P_{z_0})u\| &= \sup_{x \in D_T} |\langle u, (T - z_0)x\rangle| / \|(T - z_0)x\| \\ &= \sup_{x \in D_T} |\langle u, (T - z)x + (z - z_0)x\rangle| / (\|T - z_0)x\| \\ &\leqslant \sup_{x \in D_T} |z - z_0| |\langle u, x\rangle| / \|(T - z_0)x\| \leqslant \tfrac{1}{2} \end{aligned}$$

and, similarly, for v in $(T - z_0)$, $\|v\| = 1$ and z as above,

$$\|(I - P_z)\| \leqslant \tfrac{1}{2}.$$

This implies, using the formula for the openness between two subspaces, that

$$\|P_z - P_{z_0}\| \leqslant \tfrac{1}{2}.$$

And this implies that

$$\dim P_z X = \dim P_{z_0} X.$$

The theorem is proved. □

DEFINITION 6.9.9. Let T be a Hermitian operator with the domain D_T. The defect index of T is the ordered pair of numbers (m, n) defined as follows:

$$m = \dim(T - z)X, \quad \operatorname{Im} z < 0,$$
$$n = \dim(T - z)X, \quad \operatorname{Im} z > 0.$$

DEFINITION 6.9.10. Let T be an isometric operator on X. The defect index of T is the ordered pair (m, n) of numbers defined as follows:

$$m = \dim(T - z)X, \quad |z| > 1,$$
$$n = \dim(T - z)X, \quad |z| < 1.$$

REMARK 6.9.11. The numbers m, n in the above definition are called defect numbers and may be finite of infinite.

From the results obtained above we have the following consequences.

COROLLARY 6.9.12. *If T is a hermitian operator on X and T has a real regular point then the defect numbers are equal.*

COROLLARY 6.9.13. *Let T be a hermitian operator with the defect index (m, n). Then for all complex numbers z, $\operatorname{Im} z \neq 0$, the following assertions hold:*

(1) *If $\operatorname{Im} z > 0$ then z is an eigenvalue for T^* of multiplicity m.*
(2) *If $\operatorname{Im} z < 0$ then z is an eigenvalue for T^* of multiplicity n.*

COROLLARY 6.9.14. *If T is a hermitian operator and T_1 is a hermitian operator in $L(X)$ then the defect index of $T + T_1$ is the same as the defect index of T.*

For the proof of the last corollary it is useful to consider for z, $\operatorname{Im} z \neq 0$, the following family of operators T_t with t in $[0, 1]$,

$$T_t = T - z + tT_1.$$

Let us consider T to be a closed hermitian operator and z to be a complex number with $\operatorname{Im} z \neq 0$. For x in D_T we set

$$y = (T - zI)x, \qquad y_1 = (T - \bar{z}I)x$$

and then we remark that

$$X_1 = (y, x \in D_T), \qquad X_2 = (y_1, x \in D_T)$$

are closed subspaces in X.

We define $V: X_2 \to X_1$ by the formula $Vy_1 = y$.

LEMMA 6.9.15. *The operator V is an isometry.*

Proof. Computation. □

DEFINITION 6.9.16. The operator V defined above is called the Cayley transformation of the hermitian operator T.

We now take $z = i$, $i^2 = -1$, and then we have

$$(T + i)x = y_1, \qquad (T - i)x = y$$

and since $Vy_1 = y$ we obtain that

$$Tx = \tfrac{1}{2}(y_1 + Vy_1) = \tfrac{1}{2}(I + V)y_1, \qquad x = \tfrac{1}{2}i\cdot(I - V)y_1$$

which implies that

$$Tx = i(I + V)(I - V)^{-1}x.$$

Also, we have that

$$Vy = (T - zI)(T - \bar{z}I)^{-1}y$$

and this implies the following result.

LEMMA 6.9.17. *The defect index of T is equal to the defect index of V.*

Another property of the operator V is given in

PROPOSITION 6.9.18. *Suppose that V has the property that the range of $V - I$ is dense in X. Then the operator*

$$Tx = (\bar{z}V - zI)(V - I)^{-1}x$$

for $x \in D_{V-I}$ is a hermitian operator.

Proof. Since $(V - I)X$ is dense in X the operator $(V - I)^{-1}$ exists, since in the contrary case, 1 is an eigenvalue of V and thus for some nonzero u we have $Vu = u$, which implies that

$$\langle Vx - x, u\rangle = \langle Vx, u\rangle - \langle x, u\rangle = \langle Vx, Vu\rangle - \langle x, u\rangle = 0$$

for all x in D_{V-I}. This implies that $u = 0$, which is a contradiction. Thus we can define T by the formula in the theorem and we remark that the domain D_T is dense in X. We now prove that T is symmetric.

Let us consider u, v in $R(V - I)$ and set

$$u = (V - I)p \quad \text{and} \quad v = (V - I)q$$

for p, q in D_V. Then we have

$$Tu = (\bar{z}V - z)p = \bar{z}Vp - zp \quad \text{and} \quad Tv = \bar{z}Vq - zq$$

and then

$$\langle Tu, v\rangle = \langle \bar{z}Vp - zp, Vq - q\rangle = \langle p, q\rangle(z + \bar{z}) - \bar{z}\langle Vq, p\rangle - z\langle p, Vq\rangle$$

But

$$\langle u, Tv\rangle = \langle Vp - p, \bar{z}Vq - zq\rangle = \langle p, q\rangle(z + \bar{z}) - z\langle p, Vq\rangle - \bar{z}\langle Vq, p\rangle$$

and thus

$$\langle Tu, v\rangle = \langle u, Tv\rangle$$

and the assertion of the proposition is proved. □

THEOREM 6.9.19. *Let T be a hermitian operator and V be the Cayley transform of T. Then T is selfadjoint if and only if V is a unitary operator.*

Proof. Suppose first that T is a hermitian operator with the property that $T = T^*$. Since

$$V = (T - iI)(T + iI)^{-1}$$

the unitarity of V follows.

Conversely, if V is unitary, $V^* = V^{-1}$ and this implies that the hermitian operator T is selfadjoint. For the relation between the decomposition of unity associated with

T and V we refer to the book of N. I. Achiezer and I. M. Glazman. □

The above results show a very strong connection between the hermitian operators and isometric operators, and we may ask about the connections concerning possible extensions of operators. In what follows some results in this direction are given. First we give

DEFINITION 6.9.20. A hermitian operator T is called maximal if there are no extensions of T which are hermitian operators.

DEFINITION 6.9.21. An isometric operator is called maximal if there is no isometric extension of the operator.

Then we have the following result.

THEOREM 6.9.22. *Let T be a closed Hermitian operator. Then the following assertions hold:*

(1) *T is a maximal Hermitian operator if and only if V(the Cayley transform of T) is a maximal isometric operator or, equivalently, one of the defect numbers of T is* $(0, 0)$.
(2) *T is a selfadjoint operator if and only if V is an isometric unitary operator or, equivalently, the defect index of T is* $(0, 0)$.

Proof. The assertions follows from the above results and the considerations below. If V is the Cayley transform of the Hermitian operator T and D_V is the domain of V and $R(V)$ the range of V, and if M, N are two closed subspaces in the orthogonal complement of D_V and of $R(V)$, respectively, M, N of the same dimension, then we consider the subspaces $D_V \oplus M$ and $R(V) \oplus N$. Let

$$V_1 x = \begin{cases} Vx & \text{if } x \in D_V \\ \bar{V}x & \text{if } x \in M \end{cases}$$

where $\bar{V}$ is an arbitrary isometry from M to N. It is clear that V_1 is an extension of V. If T is given, then using V_1 we can obtain extensions of T and the assertions follows. □

COROLLARY 6.9.23. *If T is a Hermitian operator with the defect index (m, n), then T has an extension which is a maximal hermitian operator. If $m \neq n$ then there is no selfadjoint extension of T. If $m = n < \infty$ then each maximal Hermitian operator is selfadjoint; if $m = n$ then there exists at least one extension of T which is a maximal Hermitian operator and at least one selfadjoint extension of T.*

Basic results for the extensions of Hermitian operators were obtained by John von Neumann. In what follows, after the exposition of Achiezer and Glazman, we expose some of von Neumann's results.

THEOREM 6.9.24 (von Neumann, 1929). *Let T be a Hermitian operator and let z be a complex number with* $\operatorname{Im} z \neq 0$. *Then D_{T^*} has the following decomposition*

$$D_{T^*} = D_T \oplus D_{\bar{z}} \oplus D_z$$

where for any a, D_a is the subspace $(R(T - aI))^\perp$, and every extension of T is of the *form*

$$\bar{T}x = Tx_0 + zg_z + \bar{z}V'g_z$$

with g_z *in* $M_{\bar{z}}$ $D_{\bar{z}}$, V' *is a unitary operator of* $M_{\bar{z}}$ *on a subspace* N_z *in* D_z.

Proof. Let x be in D_{T^*} and then we have that

$$T^*x - zx = u + v$$

with u in $R(T - zI)$, v in $(R(T - zi)^{\perp}) = D_z$. Then

$$u = Tx_0 - zx_0 \quad \text{and} \quad v = (\bar{z} - z)g_{\bar{z}},$$

since each element in $((R(T - zI)^{\perp})$ is an element corresponding to the eigenvalue $\bar{z}$.

Further,

$$T^*(x - x_0 - g_{\bar{z}}) = z(x - x_0 - g_{\bar{z}})$$

and thus

$$x - x_0 - g_{\bar{z}} \subset R(T - \bar{z})^{-} = D_{\bar{z}}.$$

This gives that

$$x - x_0 - g_{\bar{z}} = g_z$$

and the required decomposition is proved. The uniqueness is easy to prove and thus it is omitted.

We prove now the decomposition for the extension. Let z be a complex number with $\operatorname{Im} z \neq 0$. We consider two subspaces U_z in $D_{\bar{z}}$, V_z in D_z with the same dimension and let V' be an isometric operator $V': U_z - V_z$. Let

$$\bar{V}x = \begin{cases} Vx, & x \in D_T \\ V'x, & x \in U_z \end{cases},$$

which is obviously an extension of the isometric operator V, and let $\bar{T}$ be the associated operator corresponding to $\bar{V}$. Since

$$\begin{aligned} D_{\bar{T}} &= (\bar{V} - I)D_{\bar{V}} = (\bar{V} - I)(D_V \oplus U_z) = (V - I)D_V \oplus (V' - I)U_z \\ &= D_T \oplus (V' - I)U_z, \end{aligned}$$

then, if we take $-V'$, we obtain that

$$D_{\bar{T}} = D_T \oplus (V' + I)U_z.$$

But $\bar{T} \subset T^*$ and thus if $x = x_0 + g_z + V'g_z$ we have

$$\bar{T}x = Tx + zg_z + \bar{z}V'g_z$$

and the proof of the assertion is complete. □

COROLLARY 6.9.25. *If* $\bar{T}$ *is a hermitian extension of the hermitian operator* T *and* $(g_z) = D_{\bar{z}}$, $(g_{\bar{z}}) = D_z$ *then* $\bar{T}$ *is a selfadjoint operator.*

6.10. Extensions of Semibounded Linear Operators

If T is a hermitian operator then we know that it has some special properties and we may ask if there exist selfadjoint extensions which preserve these properties. In the case of a semiboundedness property this problem was first considered by John von Neumann and has also conjectured a number of results which have attracted the attention of many people, among whom we mention M. Krein, M. A. Krasnoselskiĭ, and M. H. Stone.

Below we present some results related to semibounded operators and their extensions.

DEFINITION. 6.10.1. Let T be a Hermitian operator. Then we say that T is bounded from below if there exist a constant c such that, for all x in D_T,

$$\langle Tx, x \rangle \geqslant c\|x\|^2.$$

DEFINITION 6.10.2. The Hermitian operator T is said to be bounded from above if there exists a constant c such that for all x in D_T,

$$\langle Tx, x \rangle \leqslant c\|x\|^2.$$

DEFINITION 6.10.3. The largest (respectively, upper bound) of all c with the property in Definition 6.10.1 (respectively Definition 6.10.2, the smallest) is called the lower bound (upper bound) of T.

DEFINITION 6.10.4. A Hermitian operator T is called semibounded if it is bounded from below or from above.

6.10.5. The core of an operator with a domain D_T is any subset D in D_T such that $(x, Tx)_{x\in D}$ is dense in G_T.

Let T be a lower bounded Hermitian operator. Suppose first that the lower bound is zero. Then for any positive real number a we have that

$$\|Tx + ax\|^2 = \|Tx\|^2 + \langle 2Tx, x \rangle + |a|^2\|x\|^2 \geqslant a^2\|x\|^2$$

and thus T has a real regular point. This implies that the following corollary holds.

COROLLARY 6.10.6. *Every Hermitian operator which is lower bounded has a defect index of the form* (m, m) *and thus has a selfadjoint extension.*

Concerning the extensions of lower bounded Hermitian operators, we first prove the following important theorem of John von Neumann.

THEOREM 6.10.7 (von Neumann, 1929). *If* T *is a lower bounded Hermitian operator with a lower bound* m, *then there exists for each* $m' < m$ *a selfadjoint extension of* T *which has a lower bound* m'.

Proof. If we consider the operator

$$T_1 = (1/(m' - m))((T - m')/(m' - m))$$

then

$$\langle T_1 x, x \rangle \geqslant \|x\|^2, \qquad x \in D_T = D_{T_1}$$

and thus we may assume, without loss of generality, that $m = 1$.

If this is the case then $T^{-1}: R(T) \to X$ exists and is well defined if $N = (R(T)^{\perp})$ is the space with the property that $T^*x = 0$ for all x in N.

We define $\bar{T}: D_{\bar{T}} \to X$ by the formula

$$\bar{T}x = Tx_T$$

where $x = x_T + n_N, n_N \in N$. Then we see that $\bar{T}$ is a hermitian operator which is obviously an extension of T. Since

$$\langle \bar{T}x, x \rangle = \langle Tx_T, x_T + n_N \rangle = \langle Tx_T, x_T \rangle \geqslant \|x_T\|^2$$

the lower bound for $\bar{T}$ is zero.

We now show that $\bar{T}$ is a selfadjoint operator. For this it suffices to remark that the range of $\bar{T} - iI$ is X. Then Theorem 6.7.9 says that $\bar{T}$ is selfadjoint. Thus von Neumann's theorem is proved in the particular case when $m = 1$ and $m' = 0$.

From the above consideration of the operator T_1 we can conclude that the assertion of the theorem holds. □

John von Neumann has conjectured that a strong form of his result holds, namely that the following is true.

THEOREM 6.10.8 (Stone, 1932). *Let T be a Hermitian operator which is lower bounded with a lower bound m. Then there exists a selfadjoint extension $\bar{T}$ of T with the same lower bound.*

In what follows we present two proofs of this important result. The first is a more general result of Krein (1947) in which a description of all selfadjoint extensions of a lower bounded Hermitian operator is obtained, and the second is obtained by Friederichs (1934).

First we give the fundamental idea of Krein's method in the theory of extensions of Hermitian operators. Let us remark that we can translate the operator T such that the lower bound $m = 0$ and thus we may suppose without loss of generality that $T \geqslant 0$. Then $z = 1$ is a regular point of T and we can consider

$$S_T = (I - T)(I + T)^{-1}$$

with $D_S = (I + T)D_T$. The application $T \to S_T$ is a one-to-one map in the set of positive Hermitian operators to the set of all Hermitian operators S_T with $\|S_T\| \leqslant 1$. Here

$$\|S_T\| = \sup\{\|S_T x\|, x \in D_{S_T}, \|x\| = 1).$$

We now consider S a linear Hermitian operator with the domain D_S not necessarily dense in X(the Hermitianity means the fact that for all x, y in D_S the following relation holds:

$$\langle Sx, y \rangle = \langle x, Sy \rangle.$$

The operator is said to be bounded if

$$\|S\| = \sup(\|Sx\|, x \in D_S, \|x\| \leqslant 1)$$

is a finite number.

Then we can prove the following result.

PROPOSITION 6.10.9. *If S is a bounded Hermitian operator with a domain D_S*

which is not dense in X then there exists an extension $\bar{S}$ of S with the following properties:

(1) $\bar{S}$ *is hermitian.*
(2) $\|S\| = \|\bar{S}\| = \sup(\|\bar{S}x\|, x \in X, \|x\| \leqslant 1)$.

Proof. We can suppose without loss of generality that $\|S\| = 1$ and that D_S is closed. Let P_S be the orthogonal projection of X onto D_S and set

$$S_1 = P_S S, \qquad S_2 = (I - P_S)S, \qquad D_S = D_{S_1} = D_{S_2}.$$

For x in D_S, y in D_S we have

$$\langle Sx, y\rangle = \langle x, Sy\rangle = \langle P_S x, Sy\rangle = \langle x, P_S Sy\rangle = \langle x, S_1 y\rangle$$

or

$$\langle Sx, y\rangle = \langle x, S_1 y\rangle$$

and we remark that in the last equality x may be in X. Thus for y fixed in D_S, $x \to \langle x, S_1 y\rangle$ is a continuous linear functional on D_S and by Riesz's representation theorem there exists z in D_S such that

$$\langle x, S_1 y\rangle = \langle x, z\rangle$$

and $\|z\| \leqslant \|y\|$. We define the value of an extension $\bar{S}_1$ of S_1 at y to be the element z from the above relation. From the definition it follows that $\|\bar{S}_1\| \leqslant 1$ and the range of $\bar{S}_1$ is in D_S.

Now an extension of S_2 may be defined as follows. We consider the bilinear form

$$[x, y] = \langle x, y\rangle\langle \bar{S}_1 x, \bar{S}_1 y\rangle$$

which has the properties of an inner product except that $[x, x] = 0$ does not imply that $x = 0$. We consider the subspace $N = (x, [x, x] = 0)$ and we remark that it is a closed subspace. We can consider the space X/N whose completion is denoted by X_1. This space X_1 is clearly an inner product space. We remark that the subspace N is an invariant subspace for S_2 and then S_2 induces on the subspace D_S/N an operator, and this further induces in the completion an operator which is denoted by $\mathbf{S}_2$. If P is the orthogonal projection of X_1 on the completion of D_S/N then we have that

$$\mathbf{S}_2\mathbf{x} = S_2 P\mathbf{x} \qquad \mathbf{x} \in X_1$$

and

$$\|\mathbf{S}_2\mathbf{x}\|^2 = \|S_2 Px\|^2 = [x, x].$$

Then we can define an extension of S_2 by the formula

$$\bar{S}_2 x = \bar{\mathbf{S}}_2\mathbf{x}$$

where $\bar{\mathbf{S}}_2$ is the above extension of S_2 and $\mathbf{x}$ is the class of x. The definition has sense since $\bar{\mathbf{S}}_2$ applies X_1 on $D_S^{\perp}$ and by the fact that if x, y are in D_S, $x - y \in N$, then $S_2 x = S_2 y$.

But

$$\|\mathbf{S}_2 x\|^2 = [x, x] = \langle x, x\rangle - \langle \bar{S}_1 x, \bar{S}_1 x\rangle$$

and we define an operator on X by the formula

$$x \to \bar{S}_1 x + \mathbf{S}_2 x = Rx$$

which is clearly an extension of S. We remark that $S \subset R^*$. Indeed, if x in D_S and y in X we have

$$\langle R^*x, y\rangle = \langle x, Ry\rangle = \langle x, \bar{S}_1 y + \mathbf{S}_2 y\rangle = \langle x, \bar{S}_1 y\rangle = \langle Sx, y\rangle$$

and thus

$$R^*x = Sx$$

for all x in D_S. Also we have that $\|R^*\| = 1$ and then we consider the operator

$$\bar{S} = \tfrac{1}{2}(R + R^*)$$

which represents a selfadjoint extension of S.

PROPOSITION 6.10.10. *Let S be a Hermitian operator which is positive. Let X_1 be a subspace of X and let F_S be the family of all selfadjoint operators T satisfying the conditions*

(1) $T \leqslant S$,
(2) $Tx_1 = 0,\ x_1 \in X_1$.

Then there exists in F_S a maximal operator (with respect to the order relation '='), say $\bar{S}$, which is given by the formula

$$\bar{S} = S^{1/2} P S^{1/2},$$

where P is the orthogonal projection of X onto $(S^{1/2}x_1, x_1 \in X_1)$.

Proof (I. M. Gelfand). Let R be an element of F_S and then we have $\langle Rx, x_1\rangle = 0$ for all x in X and x_1 in X_1. Further, we have for x in X and x_1 in X_1,

$$\langle Rx, x\rangle = \langle R(x - x_1), x - x_1\rangle$$

which gives that

$$\langle Rx, x\rangle = \langle S(x - x_1), x - x_1\rangle = \|S^{1/2}x - S^{1/2}x_1\|^2$$

and thus

$$\langle Rx, x\rangle \leqslant \inf_{x_1 \in X_1} (\|S^{1/2}x - S^{1/2}x_1\|).$$

Then we have

$$\langle Rx, x\rangle \leqslant \|Ps^{1/2}x\|^{1/2} = \langle S^{1/2}PS^{1/2}x, x\rangle = \langle \bar{S}x, x\rangle$$

and the proposition is proved. □

PROPOSITION 6.10.11. *Let S be an operator as in the above proposition (with $\|S\| \leqslant 1$) and $\bar{S}_0$ be a fixed selfadjoint extension preserving the norm of S. Let $\bar{S}$ be an arbitrary selfadjoint extension of S (preserving the norm) and let $C = \bar{S} - \bar{S}_0$. Then C*

is an operator satisfying the following properties:

(1) $-(I+\bar{S}_0) \leqslant C \leqslant (I-\bar{S}_0)$.
(2) *if* x *is in* D_S *then* $Cx = 0$.

Conversely, if C *is an operator satisfying the above two conditions then* $\bar{S}$ *is a selfadjoint extension of* S *(preserving the norm).*

Proof. Clear. □

The following theorem describes the selfadjoint extensions of operators of S. Let P_1 and P_2 be the orthogonal projections on the subspaces $((I+\bar{S}_0)D_S)^{\perp}$ and $((I-\bar{S}_0)D_S)^{\perp}$ respectively and let $\bar{C}_1$, $\bar{C}_2$ be the operators defined by the following formulas:

$$\bar{C}_1 = (I+\bar{S}_0)^{1/2} P_1 (I+\bar{S}_0)^{1/2}, \qquad \bar{C}_2 = (I-\bar{S}_0)^{1/2} P_2 (I-\bar{S}_0)^{1/2}.$$

Then we have the following result.

THEOREM 6.10.12. *The selfadjoint operator* T *is an extension preserving the norm of* S *if and only if*

$$S' \leqslant T \leqslant S'', \tag{*}$$

where

$$S' = \bar{S}_0 - \bar{C}_1, \qquad S'' = \bar{S}_0 + \bar{C}_2.$$

Proof. The conditions are clearly necessary by the above results. Suppose now that T is a norm preserving the selfadjoint extension of S, $\|T\| = \|S\|$. Let $R = T - S'$ and then we have

$$0 \leqslant R \leqslant S'' - S'$$

and since $S \subset S'$, $S \subset S''$ we obtain that for all x in D_S the following inequality holds:

$$0 \leqslant \langle Rx, x\rangle = \langle (S''-S')x, x\rangle = 0.$$

This implies that, for all x in D_S, $Rx = 0$ since $R \geqslant 0$. This implies that $Tx = S's$ for all x in D_S and thus T is an extension of S.

Suppose now that S is a Hermitian operator with $\|S\| \leqslant 1$ and consider the inner product

$$\langle x, y\rangle_{\bar{S}} = \langle x, y\rangle + \langle \bar{S}x, y\rangle$$

with $\bar{S}$ a selfadjoint extension of S. Then we have the following theorem.

THEOREM 6.10.13. *Let* T *be a selfadjoint extension preserving the norm of* S $(\|S\| = \|T\|)$. *Then* T *coincides with* S' *of* (*) *if* D_S *is a dense linear subspace in* X *with the norm*

$$x \to \|x\|_S = \langle x, x\rangle_T^{1/2}.$$

Proof. First we remark that if $T = \bar{S}_0 = S'$ is equivalent to the fact that $\bar{C}_1 = 0$, from the definition of $\bar{C}_1$ we have that, for x in X,

$$\langle \bar{C}_1 x, x\rangle = \|P(I+\bar{S}_0)^{1/2}x\|^2$$

and, further,

$$\begin{aligned}\|P_1(I+\bar{S}_0)x\|^2 &= \inf_{y\in D_S}(\|(I+\bar{S}_0)^{1/2}x-(I+\bar{S}_0)^{1/2}y\|^2)\\ &= \inf_{y\in D_S}(\|x-y\|^2\bar{S}_0)\end{aligned}$$

and clearly this implies the assertion of the theorem.

From the above facts it is clear that given S we can associate the extensions S' and S'' and then the following notations are justified by the results in Theorem 6.10.12:

$$\bar{S}_{\min} = S' \leqslant T \leqslant S'' = \bar{S}_{\max}. \tag{**}$$

We are now interested in the extensions of lower bounded Hermitian operators. Let T be a closed lower bounded operator whose lower bound is zero (i.e. $T = 0$). Then the point $z = -1$ is a regular point of T and thus the equation $Tx + x = y$ has a solution for all x in D_T and y in $(I + T)D_T$.

Let us consider $D_S = (I + T)D_T$ and S is an operator defined on D_S by the formulas:

$$\left.\begin{aligned} y &= x + Tx\\ Sy &= x - Tx\end{aligned}\right\} \tag{*}$$

Then we can show that S is a Hermitian operator and the norm is equal to 1 (and the domain is not dense). Indeed, if $y_1 = x_1 + Tx_1 \in D_S$, $y_2 = x_2 + Tx_2 \in D_S$ then from the definition of S and the hermitianity of T we obtain that

$$\langle Sy_1, y_2\rangle = \langle y_1, Sy_2\rangle$$

and since

$$\begin{aligned}\|Sy\|^2 &= \langle x - Tx, x - Tx\rangle = \|x\|^2 - 2\langle x, Tx\rangle + \|Tx\|^2\\ &\leqslant \|x\|^2 + 2\langle x, Tx\rangle + \|Tx\|^2 \leqslant \|y\|^2\end{aligned}$$

the norm of S is $\leqslant 1$.

Also from the definition of S we obtain that the following formulas hold:

$$\left.\begin{aligned} x &= \tfrac{1}{2}(I + S)y\\ Tx &= \tfrac{1}{2}(I - S)y\end{aligned}\right\} \tag{**}$$

which shows the existence of a one-to-one correspondence between the set of Hermitian operators T and the family of operators defined by the formulas in (*) and (**) which may be written as follows:

$$\begin{aligned} S &= (I - T)(I + T)^{-1}\\ T &= (I - S)(I + S)^{-1}.\end{aligned}$$

Now we know that every Hermitian operator with a lower bound zero has a selfadjoint extension. The following theorem gives a more precise result.

THEOREM 6.10.14. *Let T be a Hermitian operator whose lower bound is zero. Then T has a selfadjoint extension with a lower bound zero.*

Proof. We consider

$$S = (I - T)(I + T)^{-1}$$

and we remark that this has a norm equal to 1 (in the sense defined above). S has a selfadjoint extension which is with the same norm. Let us denote this by $\bar{S}$. First we can show that -1 is not an eigenvalue for $\bar{S}$. Since D_T is dense in X and $(I + S)D_S = D_T$ this gives the assertion about -1. The operator

$$\bar{T} = (I - \bar{S})(I + \bar{S})^{-1}$$

satisfies the conditions of the theorem.

If F_S denotes the family of all selfadjoint norm-preserving extensions of S and F_T is the family of all selfadjoint extensions of T, and the connection between T and S is given by the formula

$$S = (I - T)(I + T)^{-1}$$

then

$$F_T = (\bar{T} = (I - \bar{S})(I + \bar{S})^{-1}, \bar{S} \in F_S).$$

Since in F_S we have two special extensions, this gives that

$$T_m = (I - \bar{S}_{\min})(I + \bar{S}_{\min})^{-1}$$
$$T_M = (I - \bar{S}_{\max})(I + \bar{S}_{\max})^{-1}$$

have the property that every element of F_T is between T_m and T_M. The selfadjoint extension T_m of T is the so-called Friederichs' extension of the operator T. Now we give the Krein characterization of T_m. For this purpose we consider on D_T the inner product

$$\langle x, y \rangle_T = \langle x, y \rangle + \langle Tx, Ty \rangle. \qquad \square$$

DEFINITION 6.10.15. A sequence (x_n) D_T is said to be T-convergent if

(1) (x_n) converges in X to an element x,
(2) $\lim_{n,m \to \infty} \langle T(x_n - x_m), T(x_n - x_m) \rangle = 0$.

The closure of D_T with respect to this convergence is denoted by $D_{[T]}$.

LEMMA 6.10.16 (Friederichs, 1934). *Let (x_n), (y_n) be two sequences in D_T which are T-convergent sequences to x and y respectively. Then*

$$\lim \langle x_n, Ty_n \rangle = \lim \langle Tx_n, y_n \rangle$$

exists and is independent of the sequences which are T-convergent (i.e. depends only on x and y). If $\bar{T} \in F_T$ we have the relations:

(1) $D_T \subset D_{(\bar{T})^{1/2}}$
(2) $\lim \langle Tx_n, y_n \rangle = \langle (\bar{T})^{1/2} y, (\bar{T})^{1/2} x \rangle$, x, y in $D_{[T]}$.

Proof. Property (1) follows from the definitions of T and $\bar{T}$. To show that $\lim Tx_n$, y_n exists and is independent of the choice of sequences we consider $(\bar{x}_n)$ to be another T-convergent sequence to x and set $u_n = x_n - \bar{x}_n$ for all $n = 1, 2, 3, \ldots$. Then clearly

$\lim u_n = 0$ and also

$$\lim_{n,m\to\infty} (\|u_n - u_m\|^2 + \|T(u_n - u_m)\|^2)^{1/2} = 0. \tag{*}$$

Let

$$a = \lim(\|u_n\|^2 + \|Tu_n\|^2)^{1/2}.$$

But

$$\begin{aligned}
&|\langle u_n, u_m\rangle_T - \langle u_p, u_q\rangle_T| \\
&\quad \leqslant |\langle u_n, u_m\rangle_T - \langle u_n, u_q\rangle_T + \langle u_n, u_q\rangle_T - \langle u_p, u_q\rangle_T| \\
&\quad \leqslant \|u_n\|_T \|u_m - u_q\|_T + \|u_n - u_p\|_T \|u_q\|_T.
\end{aligned}$$

From (*) it follows that if $\varepsilon > 0$ is given then we find N_ε such that if $n, m, p, q \geqslant N_\varepsilon$ then the left-hand side of (*) is less than ε. Then this implies that if $n, m, p, q \geqslant N_\varepsilon$ then

$$|\langle u_n, u_m\rangle_T - \langle u_p, u_q\rangle_T| \leqslant \varepsilon.$$

Since

$$\begin{aligned}
a^2 &= \lim_{n,m\to\infty} \langle u_n, u_m\rangle_T = \lim\langle u_n, u_m\rangle + \langle Tu_n, u_m\rangle \\
&= \lim(\langle Tu_n, 0\rangle + \langle u_n, 0\rangle) = 0
\end{aligned}$$

and thus $a = 0$, i.e. $\lim\|u_n\|_T = 0$.

If $\varepsilon > 0$ is given then there exists N_ε such that if $n \geqslant N_\varepsilon$

$$|\|x_n\|_T - \|\bar{x}_n\|_T| \leqslant \varepsilon.$$

This clearly implies that

$$\lim \|x_n\|_T = \lim \|\bar{x}_n\|_T.$$

Now if $(\bar{y}_n)$ is another T-convergent sequence to y, in a similar way we prove that

$$\lim \|y_n\|_T = \lim\|\bar{y}_n\|_T.$$

Since for any inner product the following relation holds

$$4\langle x, y\rangle = \|x + y\|^2 - \|x - y\|^2 = i\|x + iy\|^2 - i\|x - iy\|^2$$

the assertion of the lemma follows.

From the T-convergence we obtain that if (x_n) is a T-convergent sequence then

$$\lim_{n,m\to\infty} T^{1/2}(x_n - x_m) = 0$$

and $(\bar{T})^{1/2}$ being a closed operator, for $x_n \to x$ we obtain that x is in $D_{(\bar{T})^{1/2}}$ and $\lim(\bar{T})^{1/2}x_n = (\bar{T})^{1/2}x$. Similarly, $\lim(\bar{T})^{1/2}y_n = (\bar{T})^{1/2}y$ which clearly implies the assertion of the lemma.

THEOREM 6.10.17 (M. G. Krein, 1947). *Let T be a Hermitian operator which is positive. Then among the selfadjoint extensions of T bounded from below there exists a unique extension $\bar{T}$ with $D_{\bar{T}}$ in $D_{[T]}$ and this extension coincides with $\bar{T}_m$ and $D_{\bar{T}_m} = D_{[T]}$.*

Proof. Let $\bar{T}$ be in F_T and then we consider

$$\bar{T} = (I - \bar{S})(I + \bar{S})^{-1}$$

with $\bar{S}$ in F_S, where

$$S = (I - T)(I + T)^{-1}.$$

If $D_{\bar{T}}$ is in $D_{[T]}$ then D_T is dense in $D_{\bar{T}}$ with respect to $\langle , \rangle_{\bar{T}}$.

Let x in $D_{\bar{T}}$ and then if $x + \bar{T}x = y$ then $y + \bar{S}y = 2x$. Thus

$$2(\langle x, x\rangle + \langle \bar{T}x, x\rangle) = \langle y, y\rangle + \langle \bar{S}y, y\rangle$$

or

$$2\|x\|_{\bar{T}}^2 = \|y\|_{\bar{S}}^2.$$

But D_T is dense in $D_{\bar{T}}$ with respect to $\langle , \rangle_T$ and thus $(I + T)D_T$ is dense in $D_{\bar{S}} = (I + \bar{T})D_{\bar{T}} = X$ with respect to $\langle , \rangle_S$. Then from Theorem 6.10.13 we have that $\bar{S} = \bar{S}_{\min}$ and thus $\bar{T} = \bar{T}_{\min}$.

Conversely, if $\bar{T} = \bar{T}_{\min}$ then $\bar{S} = \bar{S}_{\min}$ and the domain of S is dense in x by Theorem 6.10.13 with respect to $\langle , \rangle_{\bar{T}}$. This implies that D_T is dense in $D_{\bar{T}}$ with respect to $\langle , \rangle_{\bar{T}}$. Then we have that $D_{\bar{T}} \subset D_{[T]}$.

But $D_{\bar{T}} \supset D_T$ and thus $D_{[T]} = D_{[\bar{T}]}$.

It remains to prove that $D_{\bar{T}} \subset D_{[T]}$ implies that $\bar{T} = \bar{T}_{min}$. Let $a > 0$ such that $\bar{T} + aI > 0$. Then by the result just proved, and the fact that

$$D_{\bar{T}+aI} = D_{\bar{T}} \subset D_{[T]} = D_{[T+aI]},$$

we obtain the relation

$$\bar{T} + aI = \overline{(T + aI)}_{\min}$$

and thus for $\bar{T} = \bar{T}_{\min}$ we obtain the relation

$$\bar{T}_{\min} + aI = \overline{(T + aI)}_{\min}$$

which implies that

$$\bar{T} = \bar{T}_{\min}.$$

The theorem is proved. □

For a variant of Friederichs' construction in a Banach space setting, we refer to Birman (1956).

6.11. Unbounded Normal Operators and Some Related Classes of Operators

Let H be a complete inner product space. In this section we consider a class of operators that is related to the class of bounded normal operators. Some related classes are also considered. We mention that the theory of these classes is not much advanced comparatively to the theory for some classes of bounded linear operators related to normal bounded linear operators.

DEFINITION 6.11.1. A linear operator T is said to be normal if the following conditions are satisfied:

(1) T is closed.
(2) D_T is dense in H.
(3) $T^*T = TT^*$.

REMARK 6.11.2. It is obvious that any selfadjoint operator is normal.

The following result gives an important property of normal operators.

PROPOSITION 6.11.3. *If T is a normal operator and S is another normal operator, and if S is an extension of T then necessarily $T = S$.*

Proof. Since S is an extension of T, T^* is an extension of S^* and thus $D_{T^*} = D_T$, $D_S = D_{S^*}$ implies that $D_T = D_S$. Thus $T = S$ and the proposition is proved. □

A characterization of normal operators is given in,

THEOREM 6.11.4. *The operator T is normal if and only if T^* exists and $D_T = D_{T^*}$ is dense in H and for all x in D_T, $\|Tx\| = \|T^*x\|$.*

Proof. It is obvious that the condition of the theorem is necessary. Suppose now that T satisfies the conditions of the theorem. First we prove that T is closed. Let us consider (x_n) be a sequence in D_T such that $x = \lim x_n$ and $y = \lim Tx_n$. We remark that if u, v are in D_T then

$$\|Tu \pm Tv\| = \|T^*u \pm T^*v\|$$

which implies that (T^*x_n) is a convergent sequence. Let $y_1 = \lim T^*x_n$. We know that the adjoint is closed and thus $y_1 = T^*x$. Further, we have that

$$\|T^*x_n - y_1\| = \|T^*x_n - T^*x\| = \|Tx_n - Tx\|$$

and thus $y = Tx$. Thus T is closed.

We must now show that $T^*T = TT^*$. For this we recall that we have the identity

$$4\langle x, y\rangle = \|x + y\|^2 - \|x - y\|^2 + i\|x + iy\|^2 - i\|x - iy\|^2$$

and thus for all x, y in D_T,

$$\langle Tx, Ty\rangle = \langle T^*x, T^*y\rangle.$$

Let us consider y to be in D_{TT^*} and thus $T^*y \in D_T = D_{T^*}$ which implies that for all x in D_T,

$$\langle Tx, Ty\rangle = \langle T^*x, T^*y\rangle = \langle x, TT^*y\rangle = \langle x, T^{**}(T^*y)\rangle = \langle x, T(T^*y)\rangle$$

and thus $Ty \in D_{T^*}$. But

$$\langle Tx, Ty\rangle = \langle x, T^*(Ty)\rangle$$

which gives that $y \in D_{T^*T}$. Since x is arbitrary in D_T, T^*T is an extension of T^*T. Similarly, we obtain that T^*T is an extension of TT^* and thus $TT^* = T^*T$. The theorem is proved. □

We define now the class of hyponormal operators.

DEFINITION 6.11.5. The operator T is called hyponormal if the following conditions are satisfied:

(1) T is closed.
(2) $D_T = D_{T^*}$, D_T is dense in H.
(3) $T^*T - TT^* \geqslant 0$.

The following result gives a characterization of hyponormal operators.

THEOREM 6.11.6. *The operator T is hyponormal if and only if the following conditions are satisfied:*

(1) T^* exists, $D_T = D_{T^*}$ and D_T is dense in H.
(2) $\|Tx\| \geqslant \|T^*x\|$.

Proof. It is clear that the conditions of the theorem are necessary. Suppose now that T is an operator which satisfies the conditions of the theorem. Then we note that for each u, v in D_T we have the inequality

$$\|Tu \pm Tv\| \geqslant \|T^*u \pm T^*v\|.$$

Using this inequality we obtain that T is a closed operator. Indeed, if $(x_n) \in D_T$ has the properties that $x = \lim x_n$ and $y = \lim Tx_n$ then from the above inequality we obtain that (T^*x_n) is a Cauchy sequence. Thus we can consider $y_1 = \lim T^*x_n$. Then we have that $y_1 = T^*x$ since T^* is a closed operator. Now as in the case of normal operators we can prove that $y = Tx$. Thus the closedness of T is proved. □

Let $x \in D_{TT^*}$ and thus T^*x is in $D_T = D_{T^*}$. This implies that

$$\langle T^*x, T^*x \rangle = \langle x, T(T^*x) \rangle \leqslant \langle Tx, Tx \rangle = \langle x, T^*(Tx) \rangle$$

and thus

$$T^*T - TT^* \geqslant 0.$$

6.11.7. *If T^* is hyponormal then we say that T is co-hyponormal.*

PROPOSITION 6.11.8. *If T is hyponormal then for any complex number z, $T_z = T - zI = T - z$ is a hyponormal operator.*

We leave the reader to prove the following result.

PROPOSITION 6.11.9. *If T is hyponormal and $(T - z)^{-1}$ exists then it is a hyponormal operator.*

THEOREM 6.11.10. *Let T be a hyponormal operator. If x is in $\cap_n D_T n$ then for each integer m the following inequality holds:*

$$\|T^m x\| \geqslant \|Tx\|^m \|x\|^{l-m}.$$

Proof. We may suppose without loss of generality that $\|x\| = 1$ and thus from

$$\langle Tx, Tx \rangle = \langle x, T^*(Tx) \rangle \leqslant \|x\| \, \|T^*(Tx)\| \leqslant \|x\| \, \|T^2 x\|$$

we see that the inequality is valid for $m = 2$. Then an induction argument (taking into consideration the domains) like in the case of bounded operators gives the assertion. □

For details in the case of bounded operators we refer to Istrăţescu (1981).

REMARK 6.11.11. This result for the case of unbounded operators has applications in the theory of some operators connected with particle theory in quantum mechanics. The above result suggests the consideration of some classes of operators related to similar classes of bounded linear operators.

DEFINITION 6.11.12. The operator T is called of class (N, k), $k \geqslant 2$, if the following conditions are satisfied:

(1) D_{T^i}, $i \leqslant k$, are dense in H.
(2) T^i, $i \leqslant k$, is a closed operator.
(3) For each $x \in D_{T^i}$, $i \leqslant k$ and $\|x\| = 1$, the following inequality holds:

$$\|Tx\|^k \leqslant \|T^k x\|.$$

DEFINITION 6.11.13. The linear operator T is called subnormal if there exists a complete inner product space K containing H as a closed subspace with the properties:

(1) $D_T \subset H$, D_T is dense in H,
(2) There exists a normal operator N on K such that $D_T \subset D_N$ and for each x in D_T, $Tx = Nx$.

We have the following result.

6.11.14. *Every subnormal operator is hyponormal.*

Proof. Since N is an extension of T we get that T^* is an extension of N^* and thus if x is in D_{TT^*} we have that $T^*x \in D_T \subset D_N = D_{N^*}$ which gives further,

$$\|T^*x\|^2 = \langle x, TT^*x \rangle = \langle x, NT^*x \rangle = \langle N^*x, T^*x \rangle \leqslant \|Nx\| \|T^*x\|.$$

Thus

$$\|T^*x\| = \|Tx\|$$

and thus T is hyponormal.

It remains to show that T is closed. Let $(x_n) \subset D_T$ with the properties:

(1) $\lim x_n = x$,
(2) $\lim Tx_n = y$.

Since N is closed we obtain that $x \in D_N$ and $y = Nx$. We must show that $y = Tx$. Let u be arbitrary in D_T and thus

$$\begin{aligned}\langle u, y - Tx \rangle &= \langle u, y \rangle - \langle u, Tx \rangle = \lim \langle u, Nx_n \rangle - \langle T^*u, x \rangle \\ &= \lim(\langle N^*u, x_n \rangle - \langle T^*u, x \rangle) \\ &= \lim(\langle N^*u, x_n - x \rangle + \langle N^*u, x \rangle - \langle T^*u, x \rangle) = 0\end{aligned}$$

since T^* is an extension of N^*. Then T is closed and the theorem is proved. □

Let H be a separable complete inner product space and (e_n) be an orthonormal basis for H. We consider a sequence (a_m) called the weights of complex numbers and we define the associated shift operator on H by the formula:

$$se_m = a_m e_{m+1}.$$

If T is an operator on H which is injective, i.e. if for some $x \in D_T$, $Tx = 0$ then $x = 0$, we consider the following sequence of weights (a_m) defined by

$$a_m = \|T^m x\| / \|T^{m-1} x\|.$$

The associated shift operator is denoted by S_x. Our problem concerning the operator T and the family of shifts is about the characterization of the appartenence to a class of operators through the appartenence of each shift S_x. This seems to be the case at least for subnormal operators.

We leave the reader to formulate the definition of quasinormal operators and of subnormal operators. We now mention some generalizations of the notion of hyponormal operators. From the result in Theorem 6.11.10 we see that any hyponormal operator is of class (N, k) for all k. There are simple examples, even in the case of bounded operators, which show that the class considered in Definition 6.11.12 is larger than the class of hyponormal operators. See, for details, Istrăţescu (1981). We now present two notions which are related to hyponormality.

DEFINITION 6.11.15. An operator $T: D_T \to H$ is said to be M-hyponormal if for some $M > 0$ and all x in $D_T = D_{T^*}$ the following inequality holds:

$$\|(T^* - \bar{z})x\| \leqslant \|(T - z)x\|.$$

DEFINITION 6.11.16. The operator $T: D_T \to H$ is said to be dominant if the following properties hold:

(1) $D_T = D_{T^*}$.
(2) For each z in $\sigma(T)$ there exists $M_z > 0$ such that for all x in D_T the following inequality holds:

$$\|(T^* - \bar{z})x\| \leqslant \|(T - z)x\|.$$

One of the interesting problems connected with these classes of operators seems to be the existence of a generalized Fuglede theorem. Of course, it is of interest to extend many of the known properties from the class of bounded hyponormal operators of these classes.

Concerning the generalization of hyponormal operators we mention an interesting approach considered by Friedland (1982) and which consists in the consideration of certain real-valued functions defined on the reals with a bounded operator. For the case when the operator is in a definite class of operators, then these functions have some very interesting properties.

Let us consider H to be a complete inner product space, T to be a bounded linear operator, and x to be a fixed element of H, $x \neq 0$. Then we define the following function:

$$g_{x,T}(t) = \log\|e^{tT} x\|$$

for t in R.

It is not difficult to see that if T is normal then the family $\{g_{x,T}\}$ of functions is a family of convex functions. Of course, we may ask if, for operators in other classes, this may be the case, particularly for hyponormals.

An interesting characterization of normal operators using the functions g_x was obtained by S. Friedland and is as follows:

THEOREM 6.11.17 (Friedland, 1982). *The operator $T \in CL(H)$ is normal iff*

$$d^2/dt^2(g_{x,T}(t))(0) \geqslant 0$$

for all $x \neq 0$ in H when $\sigma(\operatorname{Re} T) = \sigma_p(\operatorname{Re} T)$.

For the proof we need some results, which have their own interest too.

LEMMA 6.11.18. *The function $g_{x,T}$ is convex on R if the operator T is hyponormal.*

Proof. Since

$$d^2/dt^2(\log e^{tT}x)(0) = \tfrac{1}{2}\langle x,x\rangle^{-2}(\langle (T^2 + T^{*2} + 2T^*T)x, x\rangle \cdot \langle x,x\rangle - \langle (T + T^*)x, x\rangle^2) \qquad (+)$$

the positivity is equivalent to

and applying the Cauchy–Buniakowsky inequality we get

$$\langle (T + T^*)x, x\rangle^2 \leqslant \langle (T + T^*)^2 x, x\rangle \cdot \langle x, x\rangle. \qquad (++)$$

But

$$(T + T^*)^2 = T^2 + T^{*2} + 2T^*T - (T^*T - TT^*)$$

and the hyponormality of T implies the assertion of the lemma. In fact we have proved that $g_{x,T}$ is convex iff

$$d^2/dt^2(g_{x,T}(t))(0) \geqslant 0. \qquad \square$$

For another equivalent formulation we need the following lemma.

LEMMA 6.11.19. *Let U, V, W be positive Hermitian elements in $L(H)$ and then we have the inequality*

(1) $\langle Ux, x\rangle^2 \leqslant \langle Vx, x\rangle\langle Wx, x\rangle$

iff for all positive t

(2) $2U \leqslant t^{-1}V + tW$.

Proof. If (1) holds then the arithmetic geometric inequality implies the validity of (2). Suppose that (2) holds. If $\langle Vx, x\rangle = 0$ then for $t \to 0$ we conclude that $\langle Ux, x\rangle = 0$, i.e. (1) holds. Thus we may assume without loss of generality that

$$\langle Vx, x\rangle \neq 0, \qquad \langle Wx, x\rangle \neq 0.$$

Then for t equal to $(\langle Vx, x\rangle/\langle Wx, x\rangle)^{1/2}$ we get inequality (1).

LEMMA 6.11.20. *The inequality (++) is equivalent to the inequality*

$$i/2(QP - PQ) \leqslant (P - t)^2$$

for all real t. Here $T = P + iQ$, P, Q are hermitian operators.

Proof. We remark that the inequality $(+)$ is invariant under translations for T, i.e. it is valied for all $T + zI$, z a complex number. In this case we may assume that $P_T \geqslant 0$. Then if $(+)$ holds, we have

$$\langle Px, x\rangle^2 \leqslant (\langle P^2 + i/2(PQ - QP) \times x, x\rangle)\langle x, x\rangle.$$

From Lemma 6.11.19 this is equivalent with the required inequality for $t \geqslant 0$. But the required inequality trivially holds for $t \geqslant 0$ since $P \geqslant 0$. Since the above inequality (in P, Q is invariant under translations we have the assertion of the lemma. □

The following lemma is a key for the proof of a more general result which we prove later.

LEMMA 6.11.21. *Let P, Q be Hermitian operators such that $(+\,+)$ holds and suppose that for some $u \neq 0$ in H we have $Pu = tu$. Then $PQu = tQu$.*

Proof. Let $y = u + sx$, where s is a complex number and $\langle u, x\rangle = 0$. Then we have $(q = i/2(PQ - QP))$,

$$\langle qu, u\rangle = \langle (P - t)^2 u, x\rangle = 0$$

and then $(+\,+)$ implies that

$$2\,\mathrm{Re}(\bar{s}\langle qu, x\rangle) + |s|^2\langle qx, x\rangle \leqslant |s|^2(\langle P - t)^2 x, x\rangle).$$

But s is arbitrary and thus $\langle qu, x\rangle = 0$ if $\langle u, x\rangle = 0$. Then $qu = ru$ for some r. Since $\langle qu, u\rangle = 0$ we must have $r = 0$. But this is the required inequality.

Now the proof of Theorem 6.11.17 is as follows. Since P has only a point spectrum, H is the sum of invariant eigenspaces for P. Then the above lemma implies the assertion of the theorem. □

Using a rather delicate result, S. Friedland has obtained the following improvement of Theorem 6.11.17.

THEOREM 6.11.22. *The operator $T \in L(H)$ is normal iff $\{g_{x,T}, g_{x,T}\}$ are with the second derivative at $t = 0$ in R^+.*

We mention that the assertion can be proved using a somewhat nonstandard theory which was presented by Berberian (1962) (W. A. Luxemburg, *General Theory of Monads*, 1969). The construction is exposed in the author's book (1981).

Using that result we may consider without loss of generality that the operator T is defined on a complete inner product space (in general nonseparable) and if $T = P + iQ$ then $\sigma_p(P) = \sigma(P)$. Since

$$\mathrm{d}^2/\mathrm{d}t^2(\log \|\mathrm{e}^{tT} x\|)(0) \geqslant 0$$

is equivalent with $(+\,+)$ then the assertion follows from Lemma 6.11.21. Thus we have the result in Theorem 6.11.22.

Now we are ready to prove the following conjecture of Friedland (1982):

CONJECTURE. *The relation*

$$\mathrm{d}^2/\mathrm{d}t^2(\log \|\mathrm{e}^{tT} x\|)(0) \geqslant 0$$

iff T is hyponormal.

Proof. Since

$$d^2/dt^2(\log \|e^{tT} x\|) \geqslant (0) = 0$$

for all $x \neq 0$, is equivalent with the inequality $(++)$ we may consider that $T = P + iQ$ has the property that $(++)$ holds.

Now we may suppose, as mentioned above, that P has the property that $\sigma(P) = \sigma_p(P)$. Then clearly for each x an eigenelement of P we have

$$\langle i/2(PQ - QP)x, x\rangle \leqslant 0$$

according to Lemma 6.11.21. Clearly this relation holds for all elements in the space generated by the eigenelements of P. On the orthogonal complement of this space, the relation $PQ = QP$ is satisfied and thus T must be a hyponormal operator. □

Suggested by the above results, we may define the notion of hyponormal operator on Banach spaces or unbounded operators as follows.

DEFINITION 6.11.23. Let X be a complex Banach space and T in $L(X)$. We say that T is hyponormal if for all $x \neq 0$ the functions $g_{x,T}(t)$ have the property that

$$d^2/dt^2(g_{x,T}(t))(0) \geqslant 0.$$

DEFINITION 6.11.24. The operator $T: D_T \to X$ is said to be hyponormal if for all $x \neq 0$, $x \in D_T n$, $n = 1, 2, 3, \ldots$, the functions $g_{x,T}$ have the property

$$d^2/dt^2(g_{x,T}(t))(0) \geqslant 0.$$

For more detail about these we refer to the author's papers: 'Generalizing normality for operators on Banach spaces: hyponormality.' Another possibility of defining a generalization of normal operators was suggested by the class of W–N operators introduced by the author in 1977.

DEFINITION 6.11.25. The operator $T: D_T \to H$ is said to be a W–N operator if for all x in $D_{T^*T} \cap D_{(T+T^*)^2}$ the following inequality holds:

$$\langle (4T^*T - (T + T^*)^2)x, x\rangle \geqslant 0.$$

6.12. Some Decomposition Theorems

In what follows H denotes a complete inner product space. Let T be a linear operator defined on the domain D_T.

DEFINITION 6.12.1. A closed linear subspace H_1 of H reduces T if the following relation holds: if P is the orthogonal projection of H onto H_1 then the operator $T \circ P$ is an extension of the operator $P \circ T$. We say sometimes that H_1 is a reducing subspace for T.

The fact that H_1 reduces T is obviously equivalent to the following assertion.

PROPOSITION 6.12.2. *Let T be a linear operator with the domain D_T. The closed subspace H_1 reduces T if and only if the following conditions are satisfied:*

(1) *$x \in D_T$ then $Px \in D_T$,*
(2) *$TPx = PTx$, $x \in D_T$.*

6.12.3. If T is a linear operator with a nontrivial reducing subspace then T is called reducible; otherwise T is irreducible.

PROPOSITION 6.12.4. *Let T be a linear operator with the domain D_T and let H_1 be a closed linear subspace of H. Then the following assertions hold:*

(1) *H_1 reduces T if and only if $H_1^\perp$ reduces T.*
(2) *If H is a closed subspace of H such that $H_1 \cap H_2$ and the H_i reduce T, then $H_2 \cap H_1^\perp$ reduces T.*

Proof. If $P_{H_1^\perp}$ denotes the orthogonal projection onto $H_1^\perp$ then, if H_1 reduces T we have

$$TP_{H_1^\perp} = T(I - P_{H_1})$$

and $T - TP_{H_1}$ is an extension of $T(I - P_{H_1})$. Since

$$T - TP_{H_1} = (I - P_{H_1})T = P_{H_1^\perp}T$$

we obtain that $H_1^\perp$ reduces T. Thus the first assertion is proved. For the second assertion we note that, according to the result just proved, $H_1^\perp$ and H_2 are reducing subspaces of T. Then since

$$\begin{aligned} T(P_{H_2 \bigcirc P_{H_1^\perp}}) &= (TP_{H_2})P_{H_1^\perp} \supset P_{H_2}(P_{H_1^\perp}T) = (P_{H_2}P_{H_1^\perp})T \\ &= P_{H_2 \cap (H_1^\perp)}T \end{aligned}$$

the assertion follows.

REMARK 6.12.5. It is not difficult to prove the following decomposition of a linear operator connected with reducing subspaces: if T is a linear operator with the domain D_T and H_1 is a closed subspace which reduces T and P_{H_1} and $P_{H_1^\perp}$ are the orthogonal projections on the subspaces H_1 and $H_1^\perp$ respectively, then

$$T = TP_{H_1} + TP_{H_1^\perp}.$$

The following result is connected with families of reducing subspaces.

THEOREM 6.12.6. *Let T be a linear operator with the domain D_T and suppose that the following conditions hold:*

(1) *T is closed.*
(2) *There exists a family $(H_i)_{i\in I}$ of reducing subspaces of T which are pairwise orthogonal.*

Then the subspace

$$H_1 = \sum_{i\in I} \oplus H_i$$

is a reducing subspace for T.

Proof. First we remark that the assertion is true if the set I is finite. Indeed, in this case we have that the family is $(H_1, \ldots, H_n)$ and thus since are pairwise orthogonal we have that

$$P_{H_1\oplus\cdots\oplus H_n}T = \left(\sum_{i=1}^{n} P_{H_i}\right)T = \sum_{i=1}^{n} P_{H_i}T.$$

Since $\Sigma_{i=1}^{n} TP_{H_i}$ is an extension of $\Sigma_{i=1}^{n} P_{H_i}T$ and

$$T\left(\sum_{i=1}^{n} P_{H_i}\right) = \sum_{i=1}^{n} TP_{H_i}$$

we obtain that $TP_{H_1\oplus\cdots\oplus H_n}$ is an extension of $P_{H_1+\cdots\oplus H_n}T$ and thus the assertion of the theorem is true when the index set of the family if finite. We now consider an arbitrary family with an index set I. Let $P_F(I)$ be the family of all finite subsets of I. We define a partial order in $P_F(I)$ and we can define the notion of summability. Then we have that for x in D_T,

$$P_{\Sigma\oplus H_i}Tx = \lim_{P_F(I)} P\left(\sum_{i\in J, J\in P_F(I)} P_{H_i}\right)Tx$$

$$= \lim_{P_F(I)} T\left(\sum_{i\in J, J\in P_F(I)} P_{H_i}x\right).$$

Since

$$P_{\Sigma_{i\in I}\oplus H_i}x = \lim_{P_F(I)} P_{\Sigma_{i\in J, J\in P_F(I)}H_i}$$

and

$$\lim T(P_{\Sigma_{i\in J, J\in P_F(I)}\oplus H_i}x) = P_{\Sigma_{i\in J, J\in P_F(I)}H_i}Tx$$

which implies, T being closed, that

$$P_{\Sigma_{i\in I}\oplus H_i}Tx = TP_{\Sigma_{i\in I}\oplus H_i}x.$$

Thus $TP_{\Sigma_{i\in I}\oplus H_i}$ is an extension of $P_{\Sigma_{i\in I}\oplus H_i}T$. This gives that $\Sigma_{i\in I}\oplus H_i$ is a reducing subspace and the theorem is proved. □

The following result relates the reducing subspaces to the adjoint of operators.

THEOREM 6.12.7. *Let T be a linear operator with the domain D_T, $\bar{D}_T = H$, and suppose that H_1 reduces T. Then*

$$(TP_{H_1})^* = T^*P_{H_1}.$$

Here P_{H_1} is the orthogonal projection of H onto H_1.

Proof. Since H_1 is reducing we obtain that $P_{H_1}T$ is an extension of TP_{H_1}. Then $T^*P_{H_1}$ is an extension of $(TP_{H_1})^*$. Indeed, by the general properties of the adjoint, $(P_{H_1}T)^*$ is an extension of (TP_{H_1}) and since $(P_{H_1}T)^* = T^*P_{H_1}$ the assertion follows. Further, $(TP_{H_1})^* = (P_{H_1}TP_{H_1})^*$ and since the last operator is an extension of $P_{H_1}T^*P_{H_1}$ which is $T^*P_{H_1}$ it follows that $T^*P_{H_1}$ is an extension of $(TP_{H_1})^*$ which is also an extension of $T^*P_{H_1}$. This implies that $T^*P_{H_1} = (TP_{H_1})^*$. □

The following result gives a decomposition of closed operators with respect to certain families of reducing subspaces.

THEOREM 6.12.8. *Let T be a linear operator on H with the domain $\bar{D}_T = H$ such that the following assertions hold:*

(1) *T is closed.*
(2) *There exists a family $(H_i)_{i\in I}$ of pairwise orthogonal reducing subspaces for T and $H = \Sigma_{i\in I} \oplus H_i$.*

Then $x \in D_T$ if and only if for each i in I, $P_{H_i}x = P_i x \in D_T$ and the family $(TP_i x)_{i\in I}$ is summable; if x is in D_T then the following decomposition holds:

$$Tx = \sum_{i\in I} \oplus TP_i x.$$

Proof. Let x be in D_T and since for each i, H_i reduces T, $P_i = P_{H_i}$ has the property that $P_i x \in D_T$ and $TP_i x = P_i Tx$ and thus we get that $(TP_i x)_{i\in I}$ is summable.

Conversely, if for each i, $P_i x \in D_T$ since $H = \Sigma_{i\in I} \oplus H_i$ we have that $x = \Sigma_{i\in I} \oplus P_i x$ and as $(TP_i x)_{i\in I}$ is summable and T is closed we obtain that x is in D_T and $Tx = \Sigma_{i\in I} \oplus TP_i x$, and the theorem is proved. □

From this theorem we have the following result.

COROLLARY 6.12.9. *Let T be a linear operator such that there exists a family $(H_i)_{i\in I}$ of subspaces of H satisfying condition* (2) *of Theorem* 6.12.8. *Suppose that D_T is dense in H. Then x is in D_T if and only if for each $i \in I$, $P_i x \in D_T$ and $(T^* P_i x)_{i\in I}$* is summable.

Proof. The operator T^* is closed and if H_i reduces T then H_i reduces T^* and the assertion follows from Theorem 6.12.8.

We consider in what follows some applications to the decomposition of normal operators.

THEOREM 6.12.10. *Let T be a normal operator on H. Then there exists a family of subspaces $(H_i)_{i\in I}$ with the following properties:*

(1) *$(H_i)_{i\in I}$ is a family of pairwise orthogonal reducing subspaces of T.*
(2) *$H = \Sigma_{i\in I} \oplus H_i$.*
(3) *For each $i \in I$, TP_i is normal.*
(4) *$x \in D_T$ if and only if for each $i \in I$, $P_i x = P_{H_i} x$ is in D_T and $(TP_i x)_{i\in I}$ is summable.*

Proof. Since the operator T is normal the operator $A = (I + T^*T)^{-1}$ is hermitian as well as the operator $B = (I + TT^*)^{-1}$. As we know these are bounded operators. Then for any x, y in H we have that

$$\langle Ax, y\rangle = \int_{\sigma(A)} t \,\mathrm{d}\langle E_t x, y\rangle; \qquad (0, 1] \supset \sigma(A),$$

where (E_t) is a spectral measure for A. Since A is invertible, $E((0)) = 0$. Let $(1/n)_1^\infty$ and denote I_n as the subset

$$I_n = (1/(n + 1), 1/n].$$

Clearly

$$E(I_n)E(I_{n+1}) = 0 \quad \text{and} \quad \Sigma_n E(I_n) = I$$

since $\lim 1/n = 0$. We set $H_n = E(I_n)H$ for $n = 1, 2, \ldots$, and we remark that (H_n) is a denumerable family of pairwise orthogonal subspaces of H with the property that $H = \Sigma \oplus H_m$.

We consider now the family of functions $(u_n)_1^\infty$ defined as follows:

$$u_n(t) = \begin{cases} 0, & t \in I_n \\ 1, & t \notin I_n \end{cases}$$

The operator corresponding to u_n is denoted by U_n and $P_n = E(I_n)$. Since the operator TB commutes with A and $TP_n = (T(I + T^*T)^{-1} U_n$ we see that TP_n is in $L(H)$ and TP_n is an extension of $P_n T$. This gives that H_n reduces T and then TP_n is normal (the reader may prove the general fact that for any normal operator R and any reducing subspace H_1 of R the operator RP_{H_1} is normal). The assertions of the theorem are proved. □

REMARK 6.12.11. We can obtain other families of pairwise reducing subspaces if we choose, instead a sequence (a_n), a net $(a_\tau)_\tau$ converging to zero. We leave the details to the reader to perform this.

The following result is in some sense a converse of the result obtained above.

THEOREM 6.12.12. *Let T be an operator defined on H with the domain D_T and suppose that there exists a family $(H_i)_{i \in I}$ of pairwise orthogonal subspaces such that the following conditions are satisfied:*

(1) $H = \Sigma_{i \in I} \oplus H_i$.
(2) *For each i in I, H_i reduces T.*
(3) *$TP_i \in L(H)$ for each i in I, TP_i is normal.*
(4) *x is in D_T if and only if $(T(P_i x))_{i \in I}$ is summable.*
(5) *For each x in D_T the following formula holds:*

$$Tx = \sum_{i \in I} T(P_i x).$$

Then T is normal.

Proof. We remark that D_T is dense in H and thus we can consider the adjoint T^* of T. Then for each i in I we have the relation

$$(TP_i)^* = T^* P_i$$

and thus if x is in D_T the family $(T(P_i x))_{i \in I}$ being summable and

$$\| T(P_i x) \| = \| T^*(P_i x) \| \tag{*}$$

we obtain that $(T^*(P_i x))_{i \in I}$ is summable and then x is in D_{T^*}. In a similar way we obtain that if x is in D_{T^*} then x is in D_T and thus $D_T = D_{T^*}$. From (*) we obtain that T is normal and the theorem is proved.

CHAPTER 7

Ideals of Operators on Complete Inner Product Spaces and on Banach Spaces

7.0. Introduction

It is well known that if the composition of operators is taken as multiplication then the space of all bounded linear operators on a Banach space becomes an algebra. The study of this algebra and its subalgebras when the underlying Banach space is a complete inner product space was initiated in the monumental work of F. J. Murray and John von Neumann in 1941. J. W. Calkin's studies of the two-sided ideals in that algebra, and their related congruences, has supplied the foundation for much of the work on such ideals. A fundamental contribution to the study of operators on Banach spaces was made in the papers of A. Grothendieck where the facts about absolutely summing and p-absolutely summing operators were obtained.

In this chapter we present an introduction to the theory of classes of ideals of operators on complete inner product spaces and on Banach spaces. The reader requiring more information on this important subject may consult the expository paper of Retherford (1975) and the survey paper of Kisliakov (1977), both of which also contain extensive bibliographies.

We first present some results about a very important class of ideals on the case of complete inner product spaces, the so-called Schatten–von Neumann ideals, C_p. Further, we present some basic results about the class of absolutely summing and absolutely p-summing operators of Grothendieck. We also present a proof of the basic Grothendieck inequality, the so-called 'Grothendieck fundamental theorem of metric theory of tensor products'. The chapter ends with the indication of some applications.

7.1. Some Terminology and Notations

Concerning the ideal (left, right, two-sided) our terminology is standard (see chapter 5). Also, for x^* in X^*, X is a Banach space, we use the notation $x^* \otimes y$ to denote the operator defined on X with values in the Banach space Y:

$$x^* \otimes y(x) = x^*(x)y.$$

If p, q are numbers in $[1, \infty]$ then we use the notation $l_p \hat{\otimes} l_q$ to denote the closure of finite rank operators $T: l_p \to l_1$ with respect to the norm

$$T \to \|T\| = \inf \sum_{i=1}^{n} \|x_i\| \, \|x_i^*\|,$$

where 'inf' is taken over all representations of T of the form $Tx = x_1^*(x)x_1 + \cdots + +x_n^*(x)x_n$. Also the Banach space X is said to have an approximation property if every compact operator T in $L(Y, X)$, where Y is another Banach space, can be approximated uniformly in $L(Y, X)$ with finite rank operators; a Banach space X is said to have a bounded approximation property if there is a constant $k > 0$ such that if X_1 is a finite dimensional subspace of X then there exists a finite rank operator T on X with the properties:

(1) $\|T\| \leqslant k$,
(2) $Tx = x,\ x \in X_1$.

If $k = 1$ then we say that X has a metric approximation property. For another approximation property (compact approximation property) we refer to Lebow and Schechter (1971). Other notations and terminology will be given at appropriate places.

7.2. Ideals of Operators on Complete Inner Product Spaces

Let X be a complete inner product space and let $L(X)$ be the set of all bounded linear operators on X. Suppose now that $\dim X < \infty$. In this case it is known that $L(X)$ has no proper (nonzero) two-sided ideals. Indeed, in this case $L(X)$ is isomorphic to the algebra of all $n \times n$ matrices and if K is a (nonzero) two-sided ideal in $L(X)$ then it is not difficult to show that K contains any $n \times n$ matrix with a single entry 1 and all others equal to 0. This clearly gives that $K = L(X)$.

If $\dim X$ is not finite then the above argument does not work. But simple modifications give the following result.

PROPOSITION 7.2.1. *If K is a two-sided ideal in $L(X)$ then K contains all operators of $L(X)$ of finite rank.*

We recall that a bounded operator T in $L(X)$ is said to be compact if whenever (x_n) is a sequence with $\|x_n\| \leqslant 1$ then (Tx_n) has a convergent subsequence (in X). The following result gives some important properties of compact operators.

THEOREM 7.2.2. *For an element T in $L(X)$ the following assertions are equivalent:*

(1) *T is compact.*
(2) *T maps bounded sets in X in relatively compact subsets in X.*
(3) *T maps weakly convergent sequences in X in convergent sequences in X (i.e. strong convergent sequences).*
(4) *For any orthonormal system (h_n) in X, $\lim\langle Th_n, h_n\rangle \to 0$.*
(5) *There exists a sequence (T_n) in $L(X)$, each T_n finite rank operator, with the property that*

$$\lim\|T - T_n\| = 0.$$

Proof. (1) → (2). Obvious. Suppose now that (2) holds and we prove (3). If (x_n) is a weak convergent sequence in X then we may suppose without loss of generality that (x_n) converges weakly to zero. We must show that (Tx_n) has the property that

$\|Tx_n\| \to 0$. Otherwise, for some $\varepsilon > 0$ and a sequence of integers (n_k),

$$\|Tx_{n_k}\| \geqslant \varepsilon.$$

From the weak convergence of (x_n) to zero it follows that for any x in X,

$$\lim \langle Tx_n, x\rangle = 0. \tag{*}$$

Since (2) holds we may assume that for some y in X, $\lim Tx_{n_k} = y$. Then (*) implies that $\|y\| \geqslant \varepsilon^2$. But

$$\lim \langle Tx_n, y\rangle = 0$$

which is clear a contradiction. Thus (2) → (3).

Suppose now that (3) holds and we prove (4). Suppose that assertion (4) is false. Then for some orthonormal family (h_n) and some $\varepsilon > 0$,

$$|\langle Th_n, h_n\rangle| \geqslant \varepsilon.$$

Since (h_n) is an orthonormal family of elements we may suppose that (h_n) converges weakly to an element h. From property (3) we obtain that $\lim \|Th_n - Th\| = 0$. Thus if $\varepsilon > 0$ is given, there exists an integer N_ε such that if $N_\varepsilon \leqslant n$ then

$$\|Th_n - Th\| \leqslant \varepsilon/2.$$

Then we have that

$$|\langle Th_n, h_n\rangle - \langle Th, h_n\rangle| \leqslant \varepsilon/2. \tag{**}$$

But for any x in X we have the inequality

$$\|x\|^2 \geqslant \sum_n |\langle x, h_n\rangle|^2$$

which implies that

$$\|Th\|^2 \geqslant \sum |\langle Th, h_n\rangle|^2.$$

From (**) we obtain that $\Sigma|\langle Th, h_n\rangle|^2$ is not a convergent series, and this is a contradiction. Thus (4) holds.

(4) → (5). For fixed n we consider the family F_n of all orthonormal subsets in X such that if F is in F_n then

$$|\langle Tx, x\rangle \geqslant (4n)^{-1}$$

for x in F. With the order relation by inclusion we can easily prove that may Zorn's lemma can be applied to F_n. Thus we can find a maximal family F_m (which is, of course, finite) and if X_{F_m} is the finite dimensional subspace of X generated by F_m then we have the inequality

$$|\langle Tu, u\rangle| \leqslant (4n)^{-1}$$

for all u in $X_{F_m}^\perp$, $\|u\| = 1$. This gives further that

$$|\langle Tu, u\rangle| \leqslant n^{-1}$$

for all u in $X_{F_m}^{\perp}$, $\|u\| \leqslant 2$. But

$$4\langle Tm, n\rangle = \langle T(m+n), m+n\rangle - \langle T(m-n), m-n\rangle + i\langle T(m+in), m+in\rangle - i\langle T(m-in), m-in\rangle$$

and thus for all u, v in $X_{F_m}^{\perp}$, $\|u\| \leqslant 1$, $\|v\| \leqslant 1$,

$$|\langle Tu, v\rangle| \leqslant n^{-1}.$$

Let P_m be the orthogonal projection of X onto X_{F_m} and thus for any x, y in X,

$$|\langle (I - P_m)T(I - P_m)x, y\rangle| \leqslant n^{-1}$$

if $\|x\| \leqslant 1$, $\|y\| \leqslant 1$. (The above inequality is applied to $u = (I - P_m)x$, $v - (I - P_m)y$.) This clearly implies the inequality

$$\|(I - P_m)T(I - P_m)\| \leqslant n^{-1}.$$

We now define the operator

$$T_n = P_m T + T P_m - P_m T P_m$$

which has a finite rank and then, from the above inequality, we have that $\|T - T_n\| \leqslant n^{-1}$, which shows that (T_n) converges in $L(X)$ to T. The fact that the set of all compact operators is closed implies that $(5) \to (1)$ is true. The theorem is proved. □

REMARK 7.2.3. The characterization of compact operators given above is not valid for arbitrary Banach spaces. This is the case of properties (3) and (5).

DEFINITION 7.2.4. Let T be an element of $L(X)$ which is positive or $\Sigma_{i\in I}|\langle Te_i, e_i\rangle| < \infty$ for some orthonormal basis $(e_i)_{i\in I}$ of X. Then $\Sigma_{i\in I}\langle Te_i, e_i\rangle$ is called the trace of the operator T and is denoted by Tr T.

The definition is justified by.

LEMMA 7.2.5. *If T is an operator as in Definition 7.2.4 then*

$$\sum_{i\in I}\langle Te_i, e_i\rangle$$

is independent of the orthonormal basis $(e_i)_{i\in I}$ of X.

Proof. Let $(f_i)_{i\in I}$ be another orthonormal basis of X. Then we have

$$\begin{aligned}\sum_{i\in I}\langle Te_i, e_i\rangle &= \sum_{i\in I} T\Big(\sum_{j\in I}\langle e_i, f_j\rangle f_j, \sum_{j\in I}\langle e_i, f_j\rangle f_j\Big)\\ &= \sum_{i\in I}\sum_{j\in I}\langle Tf_j, f_j\rangle|\langle e_i, f_j\rangle|^2\\ &= \sum_{j\in I}\langle Tf_j, f_j\rangle\sum_{i\in I}|\langle e_i, f_j\rangle|^2 = \sum_{j\in I}\langle Tf_j, f_j\rangle\end{aligned}$$

and the assertion is proved. □

We now consider in $L(X)$ an operator T; then T^*T is positive and has a unique square root denoted by $(T^*T)^{1/2}$. Suppose now that T is compact and then $(T^*T)^{1/2}$ is a compact operator. Then the spectrum of $(T^*T)^{1/2}$ can be arranged, since it is at most denumerable with a possible accumulation point 0, in a decreasing sequence (according to the corresponding multiplicity):

$$\mu_1(T) \geqslant \mu_2(T) \geqslant \cdots = \mu_n(T) \geqslant \cdots, \qquad \lim \mu_n(T) = 0.$$

DEFINITION 7.2.6 (Schatten and von Neumann, 1948). A compact operator $T \in L(X)$ is said to be in the C_p class $(0 < p < \infty)$ if

$$\|T\|_p = \sum_{i=1}^{\infty} \mu_i(T)^p)^{1/p} < \infty.$$

For $p = \infty$ we set $\|T\|_\infty = \|T\|$.

Since the spectrum of T^*T and TT^* are the same, except possibly the point $z = 0$, we obtain that if T is in C_p then T^* is also in C_p (i.e. C_p is selfadjoint).

We prove in what follows some properties of operators in C_p-classes and we suppose that $1 \leqslant p < \infty$.

THEOREM 7.2.7. *T in $L(X)$ is in C_p if and only if for each orthonormal system $(h_i)_{i \in J}$, J a set with card $J \leqslant$ card I, where $(e_i)_{i \in I}$ is a basis of X, the following holds:*

$$\sum_{i \in J} |\langle Th_i, h_i \rangle|^p < \infty.$$

Proof. Suppose that T is in C_p and thus (T^*T) is a compact operator. Let $(\mu_i(T))$ be the elements in $((T^*T)^{1/2})$. Then there exists an orthonormal system (e_i) such that for all i,

$$(T^*T)^{1/2} e_i = \mu_i(T) e_i.$$

Then we can complete (e_i) with an orthonormal family (f_j) such that $(e_i) \cup (f_j)$ is a basis of X. Of course $(T^*T)^{1/2} f_j = 0$. Then clearly, if $\{h_s\} = \{e_i\} \cup \{f_j\}$, we have that

$$\sum_s |\langle (T^*T)^{1/2} h_s, h_s \rangle|^p = \sum_i \mu_i(T)^p < \infty.$$

If $(\bar{e}_i)$ is another orthonormal basis of X we have that

$$\langle (T^*T)^{1/2} \bar{e}_i, \bar{e}_i \rangle = \sum_n \mu_i(T) |\langle \bar{e}_i, e_n \rangle|^2$$

and if q is the conjugate index of p we obtain further the following inequalities:

$$|\langle (T^*T)^{1/2} \bar{e}_i, \bar{e}_i \rangle| = \sum_n \mu_n(T) |\langle \bar{e}_i, e_n \rangle|^{2/p} |\langle \bar{e}_i, e_n \rangle|^{2/q} \leqslant$$

$$= \left(\sum_{n=1}^{\infty} \mu_n(T)^p |\langle \bar{e}_i, e_n \rangle|^2 \right)^{1/p} \left(\sum_{n=1}^{\infty} |\langle \bar{e}_i, e_n \rangle|^2 \right)^{1/q}$$

But $\Sigma_{n=1}^{\infty} |\langle \bar{e}_i, e_n \rangle|^2 \leqslant 1$ and thus

$$\sum_s |\langle T\bar{e}_s, \bar{e}_s \rangle|^p = \sum_i \sum_n \mu_n(T)^p |\langle e_i, e_n \rangle| = \sum_n \mu_n(T)^p \sum_i |\langle \bar{e}_i, e_s \rangle|^2$$

$$= \sum_n \mu_n(T)^p \|e_n\|^2 = \sum_n \mu_n(T)^p < \infty.$$

Conversely, if T is a compact operator and $(T^*T)^{1/2}$ has the eigenvalues $(\mu_n(T))$ then from the condition it follows that T is in C_p.

COROLLARY 7.2.8. *C_p is a linear space in $L(X)$.*

Another very important property of the class C_p is given in the following theorem.

THEOREM 7.2.9. *T in $L(X)$ is in C_p if and only if there exists a sequence of finite rank operators (T_n)*, rank $T \leqslant n$ *such that*

$$\sum_{n=1}^{\infty} \|T - T_n\|^p < \infty.$$

Proof. For the proof we consider the family F_p of all operators in $L(X)$ with the property that if T is in F_p then there exists a sequence of finite rank operators (T_n), rank $T_n = n$ with the property that

$$\sum \|T - T_n\|^p < \infty.$$

First we show that F_p is a linear subspace of $L(X)$. Indeed, if T, S are in F_p and for S there exists the sequence of finite rank operators (S_n) with the property that

$$\sum \|S - S_n\|^p < \infty$$

then we consider the following sequence of finite rank operators (R_n) with

$$R_1 = 0, \qquad R_{2n} = R_{2n-1} = T_n + S_n$$

and thus if $R = T + S$ then clearly we have

$$\sum_n \|R - R_n\|^p < \infty.$$

Since if T is in F_p then for all complex numbers z, zT is obviously in F_p and we obtain that F_p is a linear subspace.

From the fact that the adjoint has a norm equal to the original operator we see that if T is in F_p then T^* is in F_p. Thus F_p is a selfadjoint subset of $L(X)$. Of course, this is valid for C_p also and the theorem is proved if we show that the selfadjoint parts of F_p and C_p coincide. Let us consider T be a Hermitian operator which is compact. Then T is of the form

$$Tx = \sum_{n=1}^{\infty} a_n \langle x, e_n \rangle e_n,$$

where (a_n) are eigenvalues and (e_n) is an orthonormal subset in X. Further, we know that we can arrange (a_n) as

$$a_1 \geqslant a_2 \geqslant a_3 \geqslant \cdots \geqslant a_n \geqslant a_{n+1} \geqslant \cdots,$$

where each a_i appears in this sequence according to its multiplicity. Also if the sequence is finite we denote the missing terms by 0 so that in all cases we have $\lim a_n = 0$.

For each n we put

$$T_n x = \sum_{i=1}^{n} a_i \langle x, e_i \rangle e_i$$

and then $\|T - T_n\| \leqslant a_{n+1}$. If T is in C_p then

$$\sum_n \|T - T_n\|^p \leqslant \sum_n a_n^p < \infty$$

and thus T is in F_p.

Suppose now that T is in F_p and let $\mu(T_n)$, $\dim R(T) \leqslant n$ and $\Sigma_n \|T - T_n\|^p < \infty$. Then each T_n is of the form

$$T_n x = \sum_{i=1}^{n} \langle x, e_i \rangle f_i.$$

If $k_1, \ldots, k_{n+1}$ are complex numbers satisfying the relations ,

$$\langle (k_1 e_1 + \cdots + k_n e_n + k_{n+1} e_{n+1}), e_i \rangle = 0, \quad i = 1, 2, 3, \ldots, n$$

then we have that

$$\begin{aligned} \|(T - T_n)(k_1 e_1 + \cdots + k_{n+1} e_{n+1})\| &= \|T(k_1 e_1 + \cdots + k_{n+1} e_{n+1})\| \\ &= \|a_1 k_1 e_1 + \cdots + a_{n+1} k_{n+1} e_{n+1}\| \\ &= a_1^2 k_1^2 + \cdots + a_{n+1}^2 k_{n+1}^2 \\ &\geqslant a_{n+1} \left(\sum_{i=1}^{n+1} k_i^2 \right)^{1/2} \\ &= \|k_1 e_1 + \cdots + k_{n+1} e_{n+1}\| \end{aligned}$$

which gives that $\|T - T_n\| \geqslant a_{n+1}$. But

$$\sum_i a_i^p = \sum_i \|(T - T_i)\|^p < \infty$$

and thus T is in C_p, and the theorem is proved. □

COROLLARY 7.2.10. *The set C_p is a two-sided ideal in $L(X)$.*

The above characterization of C_p permit us to consider a more general class of operators. This may be done in the following way. Let (a_n) be a sequence of positive numbers such that the sequence (a_n) is unbounded.

DEFINITION 7.2.11 (Istrăţescu, 1983). The class $C_p((a_n))$ is the set of all elements T in $L(X)$ such that there exists a sequence of finite rank operators (T_n), rank $T_n \leqslant n$ with the property that

$$\sum a_n^{1/p} \|T - T_n\|^p < \infty.$$

REMARK 7.2.12. A related class of operators was considered by Markus (1964). A further generalization of this class to arbitrary semisimple Banach algebras may be given as follows. Call an element of the semisimple Banach algebra A, finite if $(axa, a \in A)$ is a finite dimensional subset of A, the rank of x is the dimension of the set.

DEFINITION 7.2.13. An element is in the C_p of A if there exists a sequence (x_n) of elements of finite rank, rank $x_n \leqslant n$, such that

$$\sum \|x - x_n\|^p < \infty.$$

It may be of interest to compare these classes with the classes C_p of H*-algebras considered by Smith (1972) and C_p-classes of the dual of B*-algebras defined by Pan-Ken Wong (1974). An interesting generalization of C_p classes to the so called ces-p classes was obtained by Constantin (1973). For results about the C_p, and $C_p((a_n))$ classes, see Istrăţescu (1983).

7.3. The Banach Spaces C_p

In what follows we define on the spaces C_p a norm such that C_p with this norm will be a Banach space. For this purpose we need some preliminary results which have their own interest.

LEMMA 7.3.1 (McCarthy, 1967). *Let T be a positive element of $L(X)$. Then the following inequalities hold:*

$$\langle T^r x, x\rangle \leqslant \langle Tx, x\rangle^r \|x\|^{2(1-r)}, \quad 0\langle r\rangle \leqslant 1$$
$$\langle T^r x, x\rangle \geqslant \langle Tx, x\rangle^r \|x\|^{2(1-r)} \quad 1 \leqslant r < \infty.$$

Proof. Let us first consider the case when $r = 1$ and then if (E_s) is the decomposition of unity of T we have that

$$\langle Tx, x\rangle = \int_0^\infty s\,\mathrm{d}\langle E_s x, x\rangle = \left(\int_0^\infty s^r\,\mathrm{d}\langle E_s x, x\rangle\right)^{1/r}$$
$$\left(\int_0^\infty 1\,\mathrm{d}\langle E_s x, x\rangle\right)^{(r-1)/r} = \langle T^r x, x\rangle \|x\|^{2(r-1)/r}.$$

For r in $(0, 1)$ we apply the inequality just proved for the number $1/r$. □

COROLLARY 7.3.2 (McCarthy, 1967). *Let T be in $L(X)$. Then the following assertions hold:*

(1) $0 < p \leqslant 2\|T\|^p \leqslant \inf \Sigma\|Te_i\|^p$,
(2) $2 \leqslant p < \infty$, $\|T\|_p^p \geqslant \sup \Sigma\|Te_i\|^p$,

where the inf and sup are taken over all orthonormal bases of X.

Proof. Let $r = p/2$ and thus if (e_i) is an arbitrary orthonormal basis of X (for $p \leqslant 2$)

$$\langle (T^*T)^{1/2} e_i, e_i\rangle \leqslant \langle (T^*T)e_i, e_i\rangle^{p/2} = \|Te_i\|^p$$

and $(p \geqslant 2)$

$$\langle (T^*T)^{p/2} x, x\rangle \geqslant \langle (T^*T)e_i, e_i\rangle^{p/2} = \|Te_i\|^p$$

and summing we obtain the assertions of the corollary.

COROLLARY 7.3.3 (McCarthy, 1967). *For any T, S in C_p we have*

$$\|T + S\|_p \leqslant \|T\|_p + \|S\|_p$$

and thus the function defined on the linear space C_p

$$T \to \|T\|_p$$

is a norm.

Proof. Since the assertions follows from the above results, the proof is omitted. □

THEOREM 7.3.4. $(C_p, \|\,,\|_p)$ *is a Banach space.*

Proof. Let (T_n) be a Cauchy sequence in C_p. Since the norm of S in C_p is $\geqslant$ than the norm of S as an element of $L(X)$ we obtain that (T_n) converges in $L(X)$ to an element T of $L(X)$. Also, since C_p is a selfadjoint set in $L(X)$ we may suppose without loss of generality that T_n and T are positive operators. Also we have that $T_n^p \to T^p$, and then if (e_i) is an arbitrary orthonormal basis of X we have

$$\sum_{i\in J} \|T^p e_i\|^2 = \lim_{n\to\infty} \sum_{i\in J} \|T_n^p e_i\|^2$$

$$= \liminf \sum_{i\in J} \|T_n^p\|_2$$

for any finite part J of I. Thus J, being an arbitrary finite subset of I,

$$\|T^p\|_2^2 = \sup \sum_{J\in I} \|T_i^p e_{i_J}\|^2 \leqslant \liminf \|T_n^p\|_2^2.$$

If $r = p/4$ we obtain that

$$\|T\|_p^p = \|T^p/4\|_2^2 = \liminf \|T_n^{p/4}\|_2^2 = \liminf \|T_n\|_p^p$$

and we remark that this implies that T is in C_p and since

$$\|T - T_n\|_p \leqslant \liminf_{m\to\infty} \|T_n - T_m\|$$

we obtain that

$$\lim \|T - T_n\|_p = 0$$

which gives that $(C_p, \|\,,\|_p)$ is a Banach space. □

For further details concerning the Banach structure of C_p we refer to McCarthy (1967).

7.4. Ideal Sets and Ideals of Compact Operators

A fundamental result concerning the ideals of compact operators on a complete inner product space X was established by J. W. Calkin, who essentially has proved that between the family of ideals of compact operators on X and a certain family of subsets in c_0 there is a one-to-one correspondence.

In order to establish this we need some preliminary results which have also importance in many problems concerning eigenvalues of compact Hermitian operators.

LEMMA 7.4.1. *If* $T \in L(X)$ *and* T *is a compact Hermitian operator then there exists an orthonormal family* (e_n) *of elements of* X *such that*

$$Tx = \sum_{n=1}^{\infty} a_n\langle e, x\rangle e_n, \qquad \lim a_n = 0,$$

where (a_n) *is the set of all eigenvalues of* T *which may be arranged in a sequence*

$$a_1 \geqslant a_2 \geqslant a_3 \geqslant \cdots \geqslant a_n \geqslant a_{n+1} \geqslant \cdots$$

(where, if the set of eigenvalues of T *is finite, the missing terms are* 0) *and the number of times that* a_n *appears is equal to its multiplicity.*

We have already used this simple result, but for completeness, we give a proof.

Proof. Let a be an element of $\sigma(T)$ such that $|a| = T$. For the proof of this property, as well as for other properties of operators for which the spectral radius is equal to its norm, we refer to the author's book (1981). Then it follows that a is an eigenvalue of T. Let $M_a = (x, Tx = ax)$, and this reduces T. On $M_a^{\perp}$, $T/M^{\perp}$ is again a compact Hermitian operator and if we continue the above procedure we shall, at the end, have the decomposition of the theorem. □

COROLLARY 7.4.2. *Let* T *be an arbitrary compact element of* $L(X)$. *Then there exist two orthonormal families* (e_n) *and* (f_n) *of elements of* X *such that*

$$Tx = \sum_{n=1}^{\infty} a_n\langle x, e_n\rangle f_n, \qquad \lim a_n = 0,$$

where (a_n) *is the family of all eigenvalues of* $(T^*T)^{1/2}$ *which may be arranged in a sequence*

$$a_1 \geqslant a_2 \geqslant a_3 \geqslant \cdots \geqslant a_n \geqslant a_{n+1} \geqslant \cdots$$

(where, if the family is finite, the missing terms of the sequence are 0) *and each eigenvalue appears according to its multiplicity in the sequence.*

Proof. Since $T = WR$, where $R = (T^*T)^{1/2}$ and W is a partial isometry, we have that

$$Rx = \sum_{n=1}^{\infty} a_n\langle x, e_n\rangle e_n$$

and thus

$$Tx = WRx = \sum_{n=1}^{\infty} a_n\langle x, e_n\rangle We_n = \sum_{n=1}^{\infty} a_n\langle x, e_n\rangle f_n$$

and the assertion is proved. From this form of compact operator T we see that

$$a_n = \max_{\|x\| = 1, \langle x, l_i\rangle = 0, i = 1, 2, 3, \ldots, n} \langle Tx, x\rangle$$

We now present a formula of R. Courant which gives an expression of a_n using the class of subspaces in X.

THEOREM 7.4.3. *The eigenvalue* a_n *of the compact Hermitian operator* $T \in L(X)$ *is given by the formula*

$$a_n = \min_{\substack{M \text{ is} \\ (n-1)\text{-dimensional}}} \left(\max_{\|x\|=1, x \in M^{\perp}} \langle Tx, x \rangle \right).$$

Proof. Let M_n be the subspace generated by $e_1, \ldots, e_n$ and $x \in M_{n-1}^{\perp}$ then we have that

$$Tx, x = \sum_{i \geq n} a_i \langle x, e_i \rangle^2 \geq a_n \|x\|^2.$$

But $\langle T_i, e_i \rangle = a_i$ and thus

$$\max_{x \in M_{n-1}^{\perp}} \langle Tx, x \rangle \leq a_n.$$

If M is an arbitrary $(n-1)$-dimensional subspace of X then there exists x_0 in M_n, $\|x_0\| = 1$ which is in $M^{\perp}$. Then we have that

$$Tx_0, x_0 = \sum_{n=1}^{n} a_i |\langle x_0, e_i \rangle|^2 \geq a_n \sum_{i=1}^{n} |\langle x_0, e_i \rangle|^2 = a_n.$$

Thus the converse inequality is proved and we must then have an equality. The theorem is proved. □

COROLLARY 7.4.4. *Let $T \in L(X)$ be a compact operator and $R \in L(X)$. If (a_i) and (b_i) are the sequence in the decomposition of T and RT then the following inequality holds:*

$$a_n \leq \|R\| b_n, \; n = 1, 2, 3, \ldots .$$

Proof. Since a_n^2 is an eigenvalue for T^*T we have that

$$a_n^2 = \min(\max \langle T^*Tx, x \rangle) = \min(\max \|Tx\|^2)$$

and thus

$$a_n = \min(\max \|Tx\|).$$

Similarly, for the operator RT,

$$b_n = \min(\max \|RTx\|),$$

which obviously implies the assertion of the corollary. □

COROLLARY 7.4.5. *Let $T, S \in L(X)$ be compact operators and let (a_n), (b_n) be the associated sequence in the way described above. Let (c_n) be the corresponding sequence for the operator $T + S$. Then the following inequalities hold:*

$$c_{2i} \leq c_{2i-1} \leq 2(a_i + b_i).$$

Proof. Since

$$(T^* + S^*)(T + S) + (T^* - S^*)(T - S) = 2(T^*T + S^*S)$$

then

$$(T + S)^*(T + S) \leq 2(T^*T + S^*S).$$

Let us consider, for each integer n, $e_1, \ldots, e_n$ to be the eigenvectors of $(T^*T)^{1/2}$ with the corresponding eigenvalues $a_1, \ldots, a_{n-1}$. Then $e_1, \ldots, e_{n-1}$ are eigenvectors for T^*T with the eigenvalues $a_1^2, \ldots, a_{n-1}^2$ and

$$a_n^2 = \max_{\|x\|=1, \langle x, e_i\rangle = 0, i=1,\ldots,n-1} (\langle T^*T)x, x\rangle).$$

Similarly, for the operator S, if $f_1, \ldots, f_{n-1}$ are the eigenvectors of S, then

$$b_n^2 = \max_{\|x\|=1, \langle x, f_i\rangle = 0, i=1,2,\ldots,n-1} (\langle (S^*S)x, x\rangle)$$

By Courant's theorem given above, remarking that $(e_1, \ldots, e_{n-1}, f_1, \ldots, f_{n-1})$ is a $(2n-2)$-dimensional subspace of X, we have,

$$b_{2n-1}^2 = \max_{\|x\|=1, \langle x, e_i\rangle = 0, \langle x, f_i\rangle = 0, i=1,2,\ldots,n-1} \langle (T+S)^*(T+S)x, x\rangle$$

$$\leqslant 2 \max_{\|x\|=1, \langle x, e_i\rangle = 0} \langle (T^*Tx), x\rangle + 2 \max_{\|x\|=1, \langle x, f_i\rangle = 0} \langle (S^*Sx), x\rangle$$

$$\leqslant 2 \max_{\|x\|=1 \langle x, e_i\rangle = 0} \|Tx\|^2 + 2 \max_{\|x\|=1, \langle x, f_i\rangle = 0} \|Sx\|^2 \leqslant 2(a_n^2 + b_n^2)$$

and the inequality of the corollary obviously follows from this.

We are now in a position to describe a very useful connection between the family of ideals of compact operators on a complete inner product space X and certain families of sets in the Banach space c_0 of all sequences converging to zero. First we mention an important and interesting theorem of Calkin (1941) which gives information concerning the special position of the two-sided ideal of compact operators.

THEOREM 7.4.6 (Calkin, 1941). *Let X be a separable complete inner product space and let K be a two-sided ideal in $L(X)$. If K contains the ideal of compact operators then K is either the ideal of compact operators or $L(X)$.*

For a proof of this result we refer to R. Schatten's monograph (1960) and for an extension of this theorem to certain Banach spaces we refer to the monographs of I. C. Gohberg and M. G. Krein (1965, 1969).

Let $T \in L(X)$ be a compact operator and let (a_n) be the sequence of eigenvalues of $(T^*T)^{1/2}$ (where, if this sequence is finite, the missing terms of the sequence are 0) and

$$a_1 \geqslant a_2 \geqslant a_3 \geqslant \cdots \geqslant a_n \geqslant a_{n+1} \geqslant \cdots, \qquad \lim a_n = 0$$

and each a_n appears in the sequence according to its multiplicity.

DEFINITION 7.4.7. The sequence (a_n) is called the characteristic sequence of the operator T.

Let K be a two-sided ideal in $L(X)$ and then we remark that K is a selfadjoint set in $L(X)$. Indeed, each T in $L(X)$ has the form $T = WR$, where $R = (T^*T)^{1/2}$ and W is a partial isometry. Clearly, $R = W^*T$ and thus R is in K. But $T^* = RW^*$ and thus T^* is in K and the assertion is proved. □

DEFINITION 7.4.8. Let K be a two-sided ideal of compact operators in $L(X)$. The characteristic set of the ideal K is the set of all characteristic sequences of elements of K.

EXAMPLE 7.4.9. If K is the ideal of finite rank operators then the characteristic set of K is the family of all nonincreasing sequences of nonnegative real numbers having only a finite number of nonzero terms.

THEOREM 7.4.10. *A family J of sequences (a_n), $a_i \geqslant 0$, with the properties*

(1) *(a_n) is nonincreasing,*
(2) $\lim a_n = 0$

is a characteristic set of an ideal of compact operators if and only if the following properties hold:

(3) $(a_1, a_2, a_3, \ldots) \in J$ *then* $(a_1, a_1, a_2, a_2, \ldots, \ldots) \in J$,
(4) $(a_n) \in J$, $(b_n) \in J$ *then* $(a_n + b_n) \in J$,
(5) $(a_n) \in J$, $b = (b_n)$, $0 \leqslant b_n \leqslant a_n, n = 1, 2, 3, \ldots,$ *then* $b \in J$.

DEFINITION 7.4.11. A set $J \subset C_0$ with the above properties is called an 'ideal set'.

Proof. If J is the characteristic set of the ideal K then K is exactly the family of all operators in $L(X)$ of the form

$$Tx = \sum_{n=1}^{\infty} a_n \langle x, e_n \rangle f_n,$$

where (e_n) and (f_n) are orthonormal families of elements in X. We fix an orthonormal set in X, say (h_n), and then if $a = (a_n)$, $b = (b_n)$ are elements in J then we consider the operators

$$Tx = \sum a_n \langle x, h_{2n-1} \rangle h_{2n-1}$$
$$Sx = \sum a_n \langle x, h_{2n} \rangle h_{2n}$$

and we remark that T, S are in K and thus $T + S$ is in K. This implies that (3) holds. With the same orthonormal set we consider the operators

$$Tx = \sum a_n \langle x, h_n \rangle h_n, \; Sx = \sum b_n \langle x, h_n \rangle h_n$$

which are in K and then, from the fact that $T + S$ is in K, we obtain that (4) holds. Suppose now that $(a_n) \in J$ and b_n is as in (5). We consider the subset M of all n with $b_i \neq 0$. Then $0 < b_i/a_i \leqslant 1$ for all i in M. We now define the operator

$$T_1 x = \sum_{i \in M} b_i/a_i \langle x, h_i \rangle h_i$$

and composed with T defined as above we obtain that (5) holds.

Conversely, suppose that J is a family of sequences in c_0 satisfying the conditions of the theorem and then we consider the set K_J of all operators in $L(X)$ of the form

$$\sum_i a_i \langle x, e_i \rangle f_i$$

with $a = (a_n) \in J$ and (e_i), (f_j) as two orthonormal families of elements of X. We show that K_J is an ideal.

Let

$$Tx = \sum a_n \langle x, e_n \rangle f_n, \qquad Sx = \sum b_n \langle x, \bar{e}_n \rangle \bar{f}_n,$$

where $(\bar{e}_n)$, $(\bar{f}_n)$ are another pair of orthonormal subsets of X. Let (c_n) be the characteristic sequences of $(T + S)^*(T + S)$ (T, S are obviously compact operators). Then the following inequalities hold:

$$c_{2i} \leqslant c_{2i-1} \leqslant 2(a_i + b_i)$$

and thus from the fact that $(2a_n)$ and $(2b_n)$ are in J, we obtain that (c_n) is in J. But this implies that $T + S$ is in K_J.

Let T be in K_J and $RL(X)$ arbitrary. Then we can consider the characteristic sequence of T and RT which we denote by (a_i) and (b_i) respectively. Since we have the inequality

$$|b_i| \leqslant \|R\| \, |a|_i$$

we can choose an integer m with the property that $m = R$. Since $(ma_n) \in J$ we obtain that $b = (b_i)$ is in J and thus RT is in J_K. Thus J_K is a left ideal of $L(X)$. But if T is in $L(X)$ then the characteristic sequence for T and T^* is the same and thus K_J is a right-ideal. Thus K_J is a two-sided ideal and the theorem is proved. □

For some interesting results concerning the ideals of compact operators and some subspaces of sequences in c_0 we refer to Holub (1973, 1974, 1975), where the result of Calkin (1942), about the maximality of the ideal of compact operators (for the case of separable complete inner product spaces), and that of Mitiagin (1964), about the existence of intermediate ideals, are properly extended.

We now intend to present an extension due to Istrăţescu to operators on Banach spaces. For this purpose we need the notion of an approximate number of an operator.

Let X, Y be two Banach spaces and let $L(X, Y)$ be the space of all bounded operators defined on X with values in Y. $L_n(X, Y)$ denotes the space of all operators defined on X with values in Y whose range has a dimension less than or equal to n. All operators considered here are, of course, linear.

DEFINITION 7.4.12. For any T in $L(X, Y)$ the nth approximate number $a_n(T)$ is defined as

$$a_n(T) = \inf(\|T - T_n\|, T_n \in L_n(X, Y)).$$

Obviously we have the relations

$$a_0(T) \geqslant \|T\| \geqslant a_1(T) \geqslant a_2(T) \geqslant \cdots \geqslant a_n(T) \geqslant a_{n+1}(T) \geqslant \cdots.$$

We now list some properties of the approximation numbers whose proof, being simple, is omitted.

PROPOSITION 7.4.13. *The following assertions hold:*

(1) *T is of finite rank if and only if $a_n(T) = 0$ for some n.*
(2) $a_{n+m}(T) \leqslant a_n(T) + a_m(T)$.

(3) *For any T, S in $L(X, Y)$ and any integer n,*

$$a_n(T - S) \leqslant \|T - S\|.$$

(4) *For any scalar k, $a_n(kT) = |k| a_n(T)$.*
(5) *If $T \in L(X, Y)$, $S \in L(Y, Z)$ then*

$$a_{n+m}(ST) \leqslant a_n(S) a_m(T).$$

Now if T is a compact operator in $L(X)$, X a complete inner product space, then we can show that the set of approximation numbers coincides with the characteristic set of the operator T. This permits us to consider some classes of mappings related to C_p as well as to ask about the existence of a theory related to the ideal set in complete inner product spaces.

DEFINITION 7.4.14 (Istrăţescu, 1983). Let $a = (a_n)$ be a sequence of strictly positive numbers. Then $C_{a,p}$ is the set of all operators T in $L(X, Y)$ such that

$$\sum_{n=1}^{\infty} a_n^{p-1} a_n(T)^{\bar{p}}, \bar{p} = g(p).$$

converges. If $a_n = 1$ this reduces to a certain class of operators related to C_p (see J. R. Retherford for references); if $a_n = n$ this is related to the class of p-factorable operators considered by Hutton (1975) (in fact, this contained Hutton's class for $a_n = n$).

For the definition below, we recall the following classical result of S. Bernstein: *if (x_n) is a sequence of linearly independent elements of a Banach space, then, given arbitrary positive monotonically decreasing null sequence $a = (a_n)$, there exists x_0 in X such that for all n,*

$$a_n = \mathrm{d}(x_0, (x_1, \ldots, x_n)).$$

DEFINITION 7.4.15 (Hutton *et al.*, 1976). The Banach spaces X and Y are said to be a Bernstein pair if for any positive monotonic null sequence (b_n) there exists an operator T in $L(X, Y)$ and a constant $M > 0$ such that

$$b_n \leqslant a_n(T) \leqslant M b_n$$

for all $n = 1, 2, 3, \ldots$.

DEFINITION 7.4.16. Let $a = (a_n)$, $b = (b_n)$ be two monotonic null sequences. We say that $a \sim b$ if there exists a constant $M > 0$ such that for all n,

$$a_n \leqslant a_n \leqslant b_n M.$$

It is not difficult to see that the following is true.

ASSERTION 7.4.17. On the set of all monotonic null sequences '$\sim$' defines an equivalence relation.

We denote the equivalence classes of the set of all monotonic null sequences with respect to this equivalence relation by $\bar{c}_{m,0}$. We remark that $\bar{c}_{m,0}$ is a cone.

We are now ready to define a concept of an ideal set which may be applied to certain classes of Banach spaces.

DEFINITION 7.4.18 (Istrăţescu, 1983). A subset $\bar{J}$ in $\tilde{c}_{m,0}$ is called an ideal set if the following properties hold:

(1) If $\bar{a} \in \bar{J}$, $\bar{b} \in \bar{J}$ then $\bar{a} + \bar{b} \in J$.
(2) If $\bar{a} \in \bar{J}$, $a = (a_1, a_2, \ldots) \in \bar{a}$, then $(a_1, a_1, a_2, a_2, \ldots) \in \bar{a}$.
(3) If $\bar{a} \in \bar{J}$, $a = (a_n)$ and $\bar{b} \in \tilde{c}_{m,0}$, $\bar{b}$ contains $b = (b_n)$ with $|b_n| \leqslant |a_n|$ for all n, then $\bar{b} \in J$.

Let us consider an ideal K in $L(X)$, X is now a Banach space, and consider for T in K the sequence of approximation numbers $(a_n(T)$.

DEFINITION 7.4.19 (Istrăţescu, 1983). We say that a Banach space X is a Bernstein space if (X, X) is a Bernstein pair.

Following the conjecture of Hutton *et al.* (1976) that any pair of Banach spaces is a Bernstein pair, we conjecture that every Banach space is a Bernstein space.

In the rest of this we assume that X is a Bernstein Banach space.

DEFINITION 7.4.20. For any Banach space X, $K_{a,0}$ denotes the set of all elements in $L(X)$ such that the sequence of approximation numbers of the respective elements are in c_0 (the space of all sequences converging to zero).

DEFINITION 7.4.21. Let T be in $K_{a,0}$. Then we define the characteristic set of T to be the equivalence class of $(a_n(T))$.

Then we can prove the following result.

THEOREM 7.4.22. *Let K be a two-sided ideal of operators in $K_{a,0}$ with the property that if $s \in \mathscr{L}(x)$ and $\alpha_n(S) \leqslant a_n(T)$ for some $T \in K$ then $S \in K$. If K_J denotes the family of all characteristic sets of elements in K then this is an ideal set in the sense of Definition 7.4.18.*

Proof. Let $\bar{a}$, $\bar{b}$ in K_J and a, b in $\bar{a}$, $\bar{b}$ respectively. Then since X is a Bernstein Banach space there exists T, S in $K_{a,0}$ such that

$$a_n \leqslant a_n(T) \leqslant a_n M, \qquad b_n \leqslant a_n(S) \leqslant b_n N,$$

where M, N are two constants depending only on the sequences (a_n) and (b_n) and T and S respectively. The sequence $(a_n + b_n)$ obviously satisfies the inequality

$$a_n + b_n \leqslant a_n(T) + a_n(S) \leqslant (M + N)(a_n + b_n)$$

and since X is a Bernstein Banach space there exists an operator R in $K_{a,0}$ such that

$$(a_n + b_n) \leqslant a_n(R) \leqslant (M + N)(a_n + b_n)$$

and thus $(a_n + b_n)$ has the property that it is in the characteristic set (equivalence class) of an operator in $K_{a,0}$.

Let $\bar{a} \in K_J$ and $b = (b_n)$ with the property that for some $b \in \bar{b}$, $b_n \leqslant a_n$, $n = 1, 2, \ldots$, and thus since X is a Bernstein space there exists $T \in L(X)$ with the property that $a_n \leqslant a_n(T) \leqslant Ma_n$. Again from the fact that X is a Bernstein space we find S in $L(X)$ such that $b_n \leqslant a_n(S) \leqslant Nb_n$. Now we note that since $a \in \bar{a}$ we may suppose that T is in K. But $b_n(S) \leqslant Nb_n \leqslant Na_n \leqslant NMa_n(T)$ and thus S is in K. Thus $\bar{b} \in K_J$ and then K_J is an ideal set of an ideal and the theorem is proved.

For more detail about ideal sets for operators in Banach spaces we refer to Istrăţescu (1983).

7.5. Banach Ideals. Classes of Summing Operators

We now present some terminology that is widely used concerning classes of ideals of operators.

DEFINITION 7.5.1. A class of bounded linear operators is called an ideal if this class is defined for each pair X, Y of Banach spaces (which, in this case, is denoted by $C(X, Y)$) such that the following properties hold:

(1) For any $x^* \in X^*$ and y in Y, the operator

$$x \to x^*(x)y = x^* \otimes y(x) \text{ is in } C(X, Y).$$

(2) $C(X, Y)$ is a linear subspace of $L(X, Y)$.
(3) If $T \in L(X, Y)$, $S \in C(Y, Z)$, $R \in L(Z, V)$ then $RST \in C(X, V)$.

EXAMPLES 7.5.2. (1) The set of all finite rank operators is clearly an ideal of operators in this sense.

(2) The set of all compact operators is also an ideal of operators in this sense.

We now define the corresponding notion of 'ideal norm'.

DEFINITION 7.5.3. If $\mathscr{A}$ is an ideal of operators in the sense of the above definition then a function α defined on the members of this ideal of operators is called an ideal norm if the following conditions are satisfied:

(1) For any x^* in X^*, y in Y, $\alpha(x^* \otimes y) = \|x^*\| \|y\|$.
(2) For any T, S in $C(X, Y)$, $\alpha(T + S) \leqslant \alpha(T) + \alpha(S)$.
(3) T, S, R as above and then $\alpha(RST) = \|T\| \alpha(S) \| T\|$.

The notion of a Banach ideal is defined as follows.

DEFINITION 7.5.4. An ideal $\mathscr{A}$ is said to be a Banach ideal if for any pair of Banach spaces, $(C(X, Y), \alpha)$ is a Banach space.

There are also some interesting associated ideals, defined as follows.

DEFINITION 7.5.5. If $\mathscr{A}$ is an ideal then the dual ideal $(\mathscr{A}', \alpha')$ is the set of all operators T in $L(X, Y)$ such that T^* is in $C(Y^*, X^*)$. We set $\alpha'(T) = \alpha(T^*)$.

DEFINITION 7.5.6. If $\mathscr{A}$ is an ideal then the conjugate ideal $(\mathscr{A}^{\Delta}, \alpha^{\Delta})$ is defined as the set of all operators T in $L(X, Y)$ such that for some $c > 0$ and any operator in the set of finite rank operators from X in Y (denoted by $\mathscr{F}(X, Y)$, S), we have

$$|\text{Trace } ST| \leqslant c\alpha(S).$$

The norm of T is defined as the inf of all c satisfying the above inequality.

DEFINITION 7.5.7. If $\mathscr{A}$ is an ideal then the adjoint ideal $\mathscr{A}^* A^*(X, Y)$ is the set of all elements T in $L(X, Y)$ such that there exists $c > 0$ with the property that for all finite dimensional Banach spaces V, V_1 and for all $R \in L(V, X)$, S in $L(Y, V_1)$,

$P \in C(V_1, V)$, the following inequality holds:

$$|\operatorname{trace} RPLT| \leqslant c\|R\| \, \|L\| \alpha(P).$$

We now define some very important classes of ideals defined more or less explicitly by A. Grothendieck. For this purpose we need some notations.

Let (x_n) be a sequence of elements in a Banach space X. We then use the following notations:

$$\varepsilon_p((x_n)) = \sup\left(\sum_{n=1} |x^*(x_i)|^p\right)^{1/p}, \|x^*\| \leqslant 1, \qquad 1 \leqslant p < \infty,$$

$$\varepsilon_\infty((x_n)) = \sup\left(\sup_n |x^*(x_i)|, \|x^*\| \leqslant 1\right)$$

$$\alpha_p((x_n)) = \left(\sum_{n=1} \|x_n\|^p\right)^{1/p}, \qquad 1 \leqslant p < \infty$$

$$\alpha_\infty((x_n)) = \sup_n \|x_n\|$$

$$\sigma_p((x_n)) = \sup\left(\left|\sum_n x_n^*(x_n)\right|, \varepsilon_q((x_n^*)) \leqslant 1\right), \qquad 1/p + 1/q = 1.$$

We are now ready to define the ideals of operators that are of great importance in some applications.

Many of the following definitions are implicitly, or explicitly, in the works of Grothendieck.

DEFINITION 7.5.8 (A. Grothendieck, 1953). The operator $T \in L(X, Y)$ is called nuclear if T has the representation

$$Tx = \sum_{n=1}^{\infty} x_n^*(x) y_n$$

with

$$\sum_n \|x_n^*\| \, \|y_n\| < \infty.$$

DEFINITION 7.5.9 (A. Grothendieck, 1953). The operator $T \in L(X, Y)$ is called absolutely p-summing if there exists $C > 0$ such that for all finite sequences $(x_1, \ldots, x_n)$ of elements of X,

$$\alpha_p((Tx_i)) = C\varepsilon_p((x_i)).$$

If $p = 1$ then we say simply that T is absolutely summing. The norm $\Pi_p(T)$ of an operator which is absolutely summing is defined as the inf of all C satisfying the above inequality. For the set of all absolutely p-summing operators we use the notation $(\Pi_p(X, Y), \Pi_p)$.

DEFINITION 7.5.10 (A. Grothendieck). The operator T in $L(X, Y)$ is said to be p-integral if there exists a measure space (Ω, B, P), $P(\Omega) = 1$ such that if $j: L_\infty \to L_p$

(with respect to the above measure space) is the canonical injection of L_∞ into L_p and i the canonical embedding of Y into Y^{**} the following diagram is commutative:

$$\begin{array}{ccccc} X & \xrightarrow{T} & Y & \xrightarrow{i} & Y^{**} \\ {\scriptstyle P}\downarrow & & & & \uparrow{\scriptstyle Q} \\ L & & \xrightarrow{j} & & L_p \end{array}$$

with $P \in L(X, L_\infty)$ and $Q \in L(L_p, Y^{**})$. The norm of T is defined as

$$\inf \|P\| \, \|Q\| = i_p(T).$$

where the inf is taken over all such possible representations. The set of all p-integral operators on X with values in Y is denoted by (I_p, i_p).

DEFINITION 7.5.11. The operator $T \in L(X, Y)$ is said to be p-nuclear if there exists (x_n^*), (y_n) such that

(1) $\alpha_p((x_n^*)) < \infty$, $\varepsilon_q((y_n)) < \infty$, $\quad 1/p + 1/q = 1$

and if $p = 1$, $\lim x_n^* = 0$. The norm of T is defined as

$$\nu_p(T) = \inf \alpha_p((x_n^*))\varepsilon_q((y_n))$$

where the inf is taken over all possible representations of T with the above properties.

The set of all p-nuclear operators is denoted by $(N_p, \nu_p(\ ))$.

DEFINITION 7.5.12 (Cohen, 1973). The operator $T \in L(X, Y)$ is called strongly p-summing if and only if there exists $C > 0$ such that for all finite sequences $(x_1, \ldots, x_n)$ of elements of X the following inequality holds:

$$\sigma_p((Tx_i)) \leqslant C\alpha_p((x_i))$$

and the norm of T,

$$d_p(T) = \inf C$$

where the inf is taken over all finite subsets $(x_1, \ldots, x_n)$ of elements of X. The set of all strongly p-summing operators on X with values in Y is denoted by (D_p, d_p).

Another class of operators which may be considered on a Banach space is as follows. Let X, Y be two Banach spaces and s be a continuous seminorm on X.

DEFINITION 7.5.13 (Istrăţescu, 1983). The operator $T \in L(X, Y)$ is said to be absolutely summing with respect to s if for all x in X there exists a constant $C > 0$ such that

$$\|Tx\| \leqslant Cs(x).$$

The norm of T, $\Pi_s(T)$ is the inf of all C satisfying the above inequality. The class of all absolutely summing operators with respect to s is denoted by (Π_s, ν_s).

We show later that for an appropriate choice of s this reduces to the class of absolutely p-summing operators of A. Grothendieck (in fact the above definition was suggested by some results of A. Grothendieck).

Below we give some properties of absolutely p-summing operators. First we remark that the following assertions are true; the proof, being simple, is omitted.

PROPOSITION 7.5.14. *For all Banach spaces X, Y and all fixed p, $(\Pi_p(X,Y), \Pi_p)$ is a linear space.*

7.5.15. *For each fixed p the family of all absolutely p-summing operators is an ideal.*

EXAMPLE 7.5.16. The following example of an absolutely p-summing operator is of importance in the theory of absolutely p-summing operators and in the applications of these operators.

We consider a compact Hausdorff space K and consider $C(K)$ to be the Banach algebra of all continuous functions on X with the sup norm. We suppose that on K we have a positive measure of the Borel subsets of K and for fixed p we consider the Banach space $L^p(K, B, \mu)$. We suppose that $\mu(K) < \infty$. We use the notations $X = C(K)$ and $Y = L^p = L^p(K, B(K), \mu)$.

We can consider the operator $T: X \to Y$ defined as follows:

$$(Tf)(t) = f(t) \quad \text{a.e. } (\mu).$$

It is obvious that T is in $L(X, Y)$. We show that in fact $T \in \Pi_p(X, Y)$. Indeed, if $f_1, \ldots, f_n$ are continuous functions, then

$$\begin{aligned}
\sum_{i=1}^{n} \|Tf_i\|^p &= \sum_{i=1}^{n} \int_K |f_i(t)|^p \, d\mu(t) \\
&\leqslant \sup_{T\in K} \left(\sum_{i=1}^{n} |f_i(t)|^p \right) \cdot \|\mu\| \cdot \\
&\leqslant \|\mu\| \cdot \sup_{t\in K} \left(\sum_{i=1}^{n} |\delta_t(f_i)|^p \right) \\
&\leqslant \|\mu\| \sup_{x^* \in X^*, \|x^*\| \leqslant 1} \sum_{i=1}^{n} |x^*(f_i)|^p
\end{aligned}$$

where δ_t is the Dirac measure $\delta_t(f) = f(t)$.

From the above inequality it follows that T is an absolutely p-summing operator and that $\Pi_p(T) \leqslant \|\mu\|^{1/p}$. If we consider the identity function $1_K(t) = 1$ for all t in K then we obtain that $\Pi_p(T) = \|\mu\|^{1/p}$.

The following characterization of absolutely p-summing operators, using an integral inequality, is very useful in some problems.

THEOREM 7.5.17 (see S. V. Kisliakov (1977) for historical details). *For an operator $T \in L(X, Y)$ the following assertions are equivalent:*

(1) $T \in \Pi_p(X, Y)$.
(2) *There exists a positive measure μ on*

$$K = (x^*, x^* \in X^*, \|x^*\| \leqslant 1)$$

(with (X^*, X)-topology) such that for all x in X,

$$Tx = \left(\int_K |x^*(x)|^p \, d\mu(x^*)\right)^{1/p} \cdot C, \quad C > 0.$$

Proof. Suppose that T is an absolutely p-summing operator and consider the subset K as in (2). Then we know that K is a compact set. For each system of elements $(x_1, \ldots, x_n)$, $n < \infty$, we define the following function on K:

$$G_{(x_1,\ldots,x_n)}(x^*) = \sum_{i=1}^{n} \|Tx_i\|^p - \Pi_p(T) \sum_{i=1}^{n} |x^*(x_i)|^p$$

and we remark that the family of such function is a cone in $C(K)$. Since T is supposed to be absolutely p-summing, each function $G_{(x_1,\ldots,x_n)}$ is nonpositive and thus this cone does not intersect the cone of all strictly positive functions on K. According to the separation theorem for convex subsets there exists a continuous functional x_0^* such that

$$x^*(m) > 0 > x^*(G_{(x_1,\ldots,x_n)}),$$

where m is a strictly positive function. From the Riesz representation of continuous linear functionals on $C(K)$ there exists a measure μ such that

$$\int_K m(x^*) \, d\mu(x^*) > 0 > \int_K G_{(x_1,\ldots,x_n)}(x^*) \, d\mu(x^*).$$

This gives clearly that (1) → (2).

Suppose now that (2) holds. If $x_1, \ldots, x_n$ is a finite subset of elements in X then we have that

$$\sum_{i=1}^{n} \|\mathrm{T}x_i\|^p \leqslant C \sum_{i=1}^{n} \int_K |x^*(x_i)|^p \, d\mu(x^*)$$

$$\leqslant C_1 \sup_{\|x^*\| \leqslant 1} \sum_{i=1}^{n} |x^*(x_i)|^p$$

and the proof of the theorem is complete. □

REMARK 7.5.18. In the inequality

$$\|Tx\| \leqslant C\left(\int_K |x^*((x))| \, d\mu(x^*)\right)^{1/p}$$

we may assume without loss of generality that μ is a probability measure (i.e. $\mu(K) = 1$). Also we may assume that μ is concentrated on the subset of extreme points of K and that for any operator T in $\Pi_p(X, Y)$ we have the inequality

$$\Pi_p(T) \leqslant \|T\|.$$

An interesting theorem was obtained by Pelczyński (1969) in the case of some special Banach spaces.

Let X be a closed subspace of the Banach space $C(G)$, where G is a compact abelian group with the Haar measure dg. We suppose further that X is invariant for translations (i.e. if $f \in X$, then for any $g \in G$, f_g (with $f_g(u) = f(u+g)$) is in X). Let Y be a Banach space such that there exists a continuous homomorphism from G into the group of isometries of Y, say p, such that

$$((p(g))T)(u) = T(u_g),$$

where $T \in L(X, Y)$. If T is in $\Pi_p(X, Y)$ then the following result holds.

THEOREM 7.5.19 (Pelczyński, 1969). *For T as above the following inequality holds:*

$$\|Tx\| \leqslant \Pi_p(T)\left(\int_G |x(g)|^p \, dg\right)^{1/p}, \quad x \in X.$$

For a proof we refer to Pelczyński's interesting paper, where other results can be obtained. For the following result (essentially a reformulation of Theorem 7.5.17) we need the following notion.

DEFINITION 7.5.20. Let K be a compact Hausdorff space and let $C(K)$ be the Banach space of all continuous functions on K with the sup norm. Let μ be a probability measure on the Borel subsets of K and the spaces $L^p = L^p(K, B, \mu)$.

The operator $I_{\mu,p}\colon C(K) \to L^p$, defined by the formula $I_{\mu,p} f(t) = f(t)$, a.e. (μ), will be called the $I_{\mu,p}$-operator. Then we have.

THEOREM 7.5.21 (The factorization theorem). *Let X, Y be two Banach spaces. Then $T \in L(X, Y)$ is an absolutely p-summing operator if and only if the following diagram is commutative:*

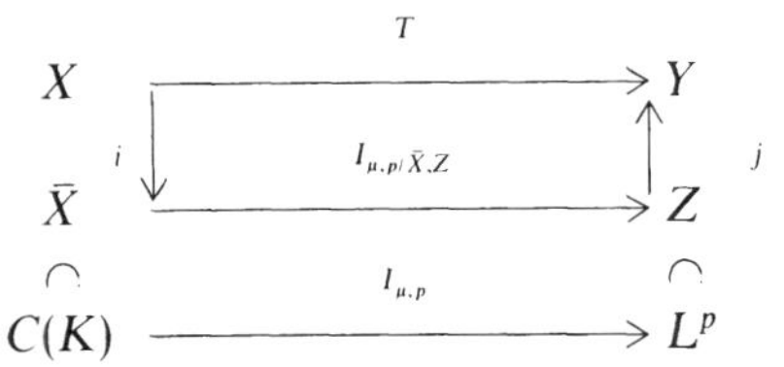

where K is the subset in X^* equal to the closed unit ball, i is the mapping defined by the relation: $ix(x^*) = x^*(x)$ for all $x \in X$, $\bar{X}$ is the closure of the range of i, Z is the closure of the range of the $I_{\mu,p}$-operator in L^p. The fact that T is absolutely p-summing implies that the operator j defined on the closure of the range of the $I_{\mu,p}$-operator by the formula $j(I_{\mu,p}, \bar{x}) = T(i^{-1}, \bar{x})$ $\bar{x} \in \bar{X}$ is continuous in the L^p-norm and thus has an extension to a continuous operator on Z (denoted again by j). Then clearly we have the following representation for T:

$$T = j_0(I_{\mu,p}/\bar{X}, Z)_0 i$$

where the notation for $I_{\mu,p}$ denotes a restriction operator to the spaces indicated in the formula.

Conversely, if there is a such factorization for T (i.e. a formula for T) then clearly T is an absolutely p-summing operator.

THEOREM 7.5.22. *Let T be an operator in $\Pi_p(X, Y)$, where X, Y are arbitrary*

Banach spaces. Then the following assertions hold (p in $[1, \infty)$):

(1) *T is weakly compact.*
(2) *T is compact.*

Proof. From Theorem 7.5.21, Tadmits a factorization throught a reflexive Banach space L^p since we may suppose $p > 1$, and this clearly implies assertion (1). The second assertion follows from the inequality

$$\|Tx\| \leqslant C\left(\int_K |x^*(x)|^p \, \mathrm{d}(x^*)^p\right)^{1/p}$$

and the Lebesgue dominated convergence theorem. □

As an interesting application of the results about absolutely p-summing operators which we presented above we give a proof of the famous result of Dvoretzky and Rogers (1950), known as the Dvoretzky–Rogers theorem. To this end we recall the notion of an unconditionally convergent series in Banach space.

DEFINITION 7.5.23. The series $\Sigma_{n=1}^{\infty} x_n$ of elements of a Banach space X is said to be unconditionally convergent if, when $a = (a_n)$ is a bounded sequence of scalars, the series $\Sigma_{n=1}^{\infty} a_n x_n$ converges.

If X is a finite dimensional space then a series $\Sigma_{n=1}^{\infty} x_n$ is unconditionally convergent if and only if $\Sigma_{n=1}^{\infty} \|x_n\|$ converges. The Dvoretzky–Rogers theorem says essentially that the above result characterizes the finite dimensional spaces.

THEOREM 7.5.24 (Dvoretzky and Rogers, 1950). *If X is an infinite dimensional Banach space then for each sequence (a_n) of positive real numbers with $a = (a_n)$ in l^2 there exists an unconditionally convergent series $\Sigma_{n=1}^{\infty} x_n$ such that $\|x_n\| = a_n$ for all n.*

We present an essentially equivalent assertion to the Dvoretzky–Rogers theorem. First we give the following result.

THEOREM 7.5.25. *Let X be an infinite dimensional Banach space. Then the set of all sequences (x_n) with $c_p((x_n))$ is not equal to the set of all sequences (y_n) with the property that $\alpha_p((y_n))$.*

Here (x_n) and (y_n) are sequences of elements in X.

Proof. Suppose that the contrary assertion is true. In this case we obtain that the closed unit ball of X is a compact subset of X and then, by the Riesz theorem, X must be a finite dimensional space. □

COROLLARY 7.5.26. *If X is an infinite dimensional Banach space then there exists a series Σx_n which is unconditionally convergent but it is not an absolutely convergent series (i.e. $\Sigma\|x_n\|$ does not converge).*

Proof. We consider the space of all unconditionally convergent series of elements of a Banach space X. We denote this space by $S_{\text{u.c.}}(X)$ and we consider the space of all series Σx_n with

$$\sup_{\|x^*\| \leqslant 1}\left(\sum |x^*(x_n)|^p\right)^{1/p} < \infty$$

By Baire's theorem there exists $C > 0$ such that

$$\left(\sum_{n=1}^{\infty} |x^*(x_n)|^p\right)^{1/p} \leqslant C\|x^*\|.$$

The smallest constant C with the above property is the norm of the sequence (x_n); this norm is denoted by $s_p(x_n)$. Thus, this is a Banach space and the space of all unconditionally convergent series is a closed subspace of the above space, in fact a proper subspace. Then the above result (for $p = 1$) and these consideration clearly imply the assertion of the corollary. □

We remark that the proof of the Dvoretzky–Rogers theorem given in the above corollary is an existential proof. □

REMARK 7.5.27. Since for each p and any probability measure μ on $K = (x^*, x^* \in X^*, \|x^*\| \leqslant 1)$ the function

$$x \to \left(\int_K |x^*(x)|^p \, d\mu(x^*)\right)^{1/p}$$

is a seminorm, we see that the class of absolutely p-summing operators is a special case of operators considered in Definition 7.5.13. Related to Definition 7.5.13 we may ask if we can define the dual ideal of operators considered there. For this purpose we mention that it is useful to define the so-called dual seminorm of the seminorm s. This is defined as follows.

DEFINITION 7.5.28 (Istrăţescu, 1983). Let s be a continuous seminorm on a Banach space X. The dual seminorm of s, denoted by s', is the seminorm defined on X^* by the relation

$$s'(x^*) = \sup_{x \in X} |x^*(x) - s(x)|.$$

Sometimes we call the seminorm s', defined above, the Fenchel conjugate seminorm of s. Using this seminorm we may define a class of operators which represents, in some sense, the analogue of $\Pi_q(X, Y)$ for the case of $\Pi_p(X, Y)$. For more details about this we refer to the author's 1983 paper.

Another interesting class of operators related to absolutely p-summing operators was considered by Mitiagin and Pelczyński (1966), known as absolutely (p, q)-summing operators.

DEFINITION 7.5.29 (Mitiagin and Pelczyński, 1966). The operator $T \in L(X, Y)$ is said to be absolutely (p, q)-summing if there exists a constant $C > 0$ such that for all finite sequences $(x_1, \ldots, x_n)$ in X the following inequality holds:

$$\alpha_p((Tx_i)) = C\varepsilon_q((x_i)).$$

Here p, q are in $[1, \infty)$.

The smallest constant C with the above property is called the norm of T and is denoted by $\Pi_{p,q}(T)$. Many papers present extensions of the class of absolutely p-summing operators using, instead of the spaces l^p, other classes of sequence spaces.

The following problem is interesting and difficult: There exists a pair of Banach spaces X and Y such that for all p, q we have the equality $\Pi_p(X, Y) = \Pi_q(X, Y)$. In order to obtain some results about this problem we need Grothendieck's fundamental theorem of the metric theory of tensor products.

7.6. Grothendieck's Fundamental Theorem

We present a proof of Grothendieck's theorem; this follows the presentation by J. Lindenstrauss and A. Pełczyński of Grothendieck's proof.

THEOREM 7.6.1 (Grothendieck, 1955). *Let $A = (a_{ij})_{i,j=1}^n$ be a real matrix and let M be a positive constant such that for all real numbers $t_1, \ldots, t_n$; $s_1, \ldots, s_n$ in $[-1, 1]$ the following inequality holds:*

$$\left| \sum_{i,j=1}^{n} a_{ij} t_i s_j \right| \leqslant M.$$

Then for any $x_1, \ldots, x_n$; $y_1, \ldots, y_n$ in a real complete inner product space X the following inequality holds:

$$\left| \sum_{i,j=1}^{n} \right| a_{ij} \langle x_i, y_j \rangle \leqslant M \cdot K_G \sup_i \|x_i\| \cdot \sup_i \|y_i\|.$$

The number K_G is called Grothendieck's constant and satisfies the inequality

$$K_G \leqslant \sinh \pi/2.$$

REMARK 7.6.2. It is clear that we have a complex variant and the constant of Grothendieck in this case is $2K_G$.

REMARK 7.6.3. A better majorant for the Grothendieck constant K_G was obtained by Rietz (1974).

For the proof of Grothendieck's theorem we need the following lemma.

LEMMA 7.6.4. *Let R^n with the usual Euclidean norm (i.e. the norm derived from an inner product) and $S = S^{n-1} = (x, x \in R^n, \|x\| = 1)$. Then for any x, y in S^{n-1} we have the relation*

$$\int_{S^{n-1}} \operatorname{sgn}\langle x, m \rangle \operatorname{sgn}\langle y, m \rangle \, \mathrm{d}\mu(m) = 1 - \tfrac{2}{\pi}\theta(x, y)$$

where θ is the real number in $[0, \pi]$ with $\cos\theta = \langle x, y \rangle$. Here $\mathrm{d}\mu$ is the Lebesgue measure on R^n multiplied with a constant such that $\mathrm{d}\mu(S^{n-1}) = 1$.

Proof. We consider a basis in R^n such that $x = (1, 0, 0, \ldots, 0)$ and $y = (\cos\theta, \sin\theta, 0, \ldots, 0)$ and for the computation we use the polar coordinates $u = (u_1, \ldots, u_{n-1})$ and thus the integral

$$\int_{S^{n-1}} g(f) \, \mathrm{d}\mu(t) = |S^{n-1}| \int_{I^{n-1}} g(f(u)) J(u) \, \mathrm{d}u,$$

where

$$f(u) = (f^1(u), \dots, f^n(u)), \qquad f^1(u) = \prod_{i=1}^{n-1} \sin u_i$$

$$f^k(u) = \cos u_{k-1} \prod_{i=k}^{n-1} \sin u_i, \qquad k = 2, 3, \dots, n$$

$$I^{n-1} = (u, 0 \leqslant u_1 < 2\pi : 0 \leqslant u_i \leqslant \pi, i = 2, 3, \dots, n-1)$$

$$J(u) = \prod_{i=2}^{n-1} (\sin u_i)^{i-1}, \qquad |S^{n-1}| = \int_{I^{n-1}} J(u)\,\mathrm{d}u = 2\pi \prod_{i=2}^{n-1} \int_0^\pi (\sin u_i)^{i-1}\,\mathrm{d}u_i.$$

If we set

$$h(m) = \langle x, m\rangle\langle y, m\rangle = m^1(m^1 \cos\theta + m^2 \sin\theta) \text{ then}$$

$$h(m(u)) = \left(\prod_{i=2}^{n-1} \sin u_i\right)^2 \sin u_1(\sin u_1 \cos\theta + \cos u_1 \sin\theta).$$

Then if $g(m) = \operatorname{sgn} h(m)$ then we obtain that $g(m(u)) = \operatorname{sgn}(\sin u_1 \sin(u_1 + \theta)) = p(u_1, \theta)$ which implies that p is 1 on $(0, \pi - \theta)$ and on $(\pi, 2\pi - \theta)$, and equal to -1 on $(\pi - \theta, \pi)$ and $(2\pi - \theta, 2\pi)$. This gives further that the above integral is

$$\int_{S^{n-1}} \operatorname{sgn}\langle x, m\rangle \operatorname{sgn}\langle y, m\rangle\,\mathrm{d}\mu(m) = |S^{n-1}| \int_{I^{n-1}} p(u_1, \theta) J(u)\,\mathrm{d}u$$

$$= S^{n-1} \int_0^{2\pi} p(u_1, \theta)\,\mathrm{d}u_1 \times$$

$$\times \prod_{i=2}^{n-1} \int_0^\pi (\sin u_i)^{i-1}\,\mathrm{d}u_i = 1 - 2\theta/\pi$$

and the relation is proved.

The proof of Grothendieck's theorem is as follows. First we remark that if the matrix (a_{ij}) satisfies the conditions of the theorem then for any real numbers a_i, b_j the matrix $(a_{ij}a_ib_j)$ satisfies again the conditions of the theorem (of course, with another constant M). Thus, taking into account the homogeneity of the inequality we may suppose without loss of generality that $\|x_i\| = \|y_i\| = 1$, $i = 1, 2, 3, \dots, n$.

The subspace generated by x_i and y_i may be considered to be in R^{2n} and thus for each m in S^{2n-1} we can consider the numbers

$t_i(m) = \operatorname{sgn}\langle mx_i, s_j(m)\rangle = \operatorname{sgn}\langle m, y_j\rangle$. Then for any $p_1, \dots, p_n/; q_1, \dots, q_n$

real numbers we have the inequality

$$-M \leqslant \sum_{i,j=1}^n a_{ij} t_i(m) s_j(m), \qquad p_i q_j \leqslant M.$$

Integrating this inequality with respect to the measure $\mathrm{d}\mu(m)$ on S^{2n-1}, and using the above result, we have that

$$-\pi/2\cdot M \leqslant \sum_{i,j=1}^{n} a_{ij}(\pi/2-\theta(x_i,y_j)p_iq_j \leqslant \pi/2\cdot M.$$

This implies that the matrix $(a_{ij}(\pi/2-\theta(x_i,y_j))$ satisfies the conditions of the theorem where the constant M is $\pi/2\cdot M$. We repeat this procedure and thus for each integer n we have the inequality

$$-(\pi/2)^n M > \sum_{i,j=1}^{n} a_{ij}(\pi/2-\theta(x_i,y_j))^n \leqslant (\pi/2)^n M.$$

This implies that

$$\begin{aligned}\langle x_i, y_j\rangle &= \sin(\pi/2-\theta(x_i,y_j))\\ &= \sum_{p=0}^{\infty}(-1)^p(\pi/2-(x_i,y_j))^{2p+1})/(2p+1)!\end{aligned}$$

and thus

$$i, \sum_{j=1}^{n} a_{ij}\langle x_i, y_j\rangle \leqslant M\sum_{p=0}^{\infty}(\pi/2)^{2p+1}/(2p+1)! = M\sinh\pi/2.$$

The proof of Grothendieck's theorem is finished. □

7.7. On the Coincidence of Classes of Absolutely p-Summing Operators

In order to formulate some of the fundamental theorems of Grothendieck about the coincidence of classes of absolutely p-summing operators we need some notions.

DEFINITION 7.7.1. Let X, Y be two Banach spaces. Then the Banach–Mazur distance between X and Y, denoted by $d(X,Y)$, is given by the formula:

$$d(X,Y) = \inf\|T\|\,\|T^{-1}\|$$

where the inf is taken over all isomorphisms between X and Y; if X and Y are not isomorphic then by definition, $d(X,Y)=\infty$.

DEFINITION 7.7.2 (Lindenstrauss, 1969; Morrell and Retherford, 1971, 1972). A Banach space X is said to be an $L_{p,\lambda}$-space if for each n, X contains subspaces X_n which are uniformly complemented (i.e. there are continuous projections P_n, $P_nX = X_n$, $\|P_n\| \leqslant M$, $n = 1, 2, 3, \ldots$, M being a positive constant) with the property that $d(X_n, l_n^p) = \lambda$. A Banach space X is said to be an L_p-space if it is an $L_{p,\lambda}$-space for some $\lambda \geqslant 1$.

For a detailed discussion of these spaces, as well as for other related classes of Banach spaces, we refer to the paper by Morrel and Retherford (1972). Then we have the following fundamental theorems, proved by A. Grothendieck.

THEOREM 7.7.3 (A. Grothendieck). *Let X be an L_1-space and let Y be an L_2-space. Then $L(X,Y) = \Pi_1(X,Y)$.*

THEOREM 7.7.4 (A. Grothendieck). *Let X be an L_∞-space and let Y be an L_p-space, $1 \leqslant p \leqslant 2$. Then $L(X,Y) = \Pi_2(X,Y)$.*

In what follows we prove only Theorem 7.7.3; for a proof of Theorem 7.7.4 we refer to the elegant paper by Lindenstrauss and Pełczyński (1968).

Proof of Theorem 7.7.3. From the definition of L_p-spaces, to prove the theorem it suffices to prove that for each n and m and any operator $L(l_m^1, l_n^2)$ the following inequality holds:

$$\Pi_1(T) \leqslant K_G \| T \|.$$

(l_n^p means the space R^n with the norm $x = (|x_1^p| + \cdots + |x_n^p|)^{1/p}$). This means that we have to prove that for any $x_1, \ldots, x_k$ in $l_m^1 (k < \infty)$ and any x^* in $(l_m^1)^* = l_m^\infty$ the inequality

$$\sum_{i=1}^{k} \| Tx_i \| \leqslant K_G \| T \|$$

holds if

$$\sum_{i=1}^{k} |x^*(x_i)| \leqslant \|x^*\|.$$

If the sequence (x_i) satisfies the above inequality we consider the matrix (a_{ij}) defined using the elements x_i:

$$x_i = (a_{i1}, \ldots, a_{ij}, a_{ij+1}, \ldots, a_{ik})$$

and then the relation

$$\sum_{i=1}^{k} |x^*(x_i)| \leqslant \|x^*\|$$

is

$$\sum_{i=1} \left| \sum_{j=1}^{k} a_{ij} t_j \right| \leqslant 1, \ |t_j| \leqslant 1$$

and thus it follows that if $|s_i| \leqslant 1$ then

$$\left| \sum_{i,j=1}^{k} a_{ij} t_i s_j \right| \leqslant 1$$

for all $|t_i| \leqslant 1$, $|s_j| \leqslant 1$. From Grothendieck's theorem (7.6.1) applied to the matrix (a_{ij}) we obtain the assertion as follows. Let $Tx_i = u_i$, v_i be any element in l_n^2 such that $\langle u_i, v_i \rangle = \|u_i\|$, $i = 1, 2, \ldots, k$, and $\|v_i\| = i = 1, 2, 3, \ldots, n$. Then we remark that

$$\sum_{i=1}^{k} \| Tx_i \| = \sum_{i=1}^{k} \|u_i\| = \sum_{i=1}^{k} \langle u_i, v_i \rangle = \sum_{i,j=1}^{k} a_{ij} \langle v_i, Te_j \rangle .$$

Here (e_j) are the elements of l_m^1 defined by the relations:

$$e_j = \Big(0, 0, \ldots, \underset{j-1}{0}, \underset{j}{1}, \underset{j+1}{0}, \ldots, 0\Big).$$

Since the matrix (a_{ij}) satisfies the conditions of Grothendieck's theorem (Theorem 7.6.1), we obtain that

$$\sum_{i=1}^{k} \|Tx_i\| \leqslant K_G \sup_i \|v_i\| \sup_j \|Te_j\| = K_G \|T\|$$

and the proof is complete. □

REMARK 7.7.5. We mention now the very interesting fact that from Theorem 7.7.3 we can obtain Theorem 7.6.1 and thus they are equivalent. For another interesting proof of Theorem 7.7.3 we refer to Pelczyński's paper (1976).

COROLLARY 7.7.6. *Let X be a Banach space which is an L_1-space. Then for any Banach space Y, $\pi_1(X, Y) = \Pi_2(X, Y)$.*

Proof. We have to prove that every absolutely 2-summing operator $T \in L(X, Y)$ is in $\Pi_1(X, Y)$.

Since T is an absolutely 2-summing operator it has a factorization through L_2 and thus by Theorem 7.7.3 it is absolutely summing (=absolutely 1-summing).

The result in the corollary gives rise the following problem: suppose that p is fixed in $(1, \infty)$ and then we may ask about the class of Banach spaces X with the property that for any Banach space Y, $\Pi_1(X, Y) = \Pi_p(X, Y)$. Connected with this problem we mention the set I_X associated with a Banach space X and which consists of all number q such that for all p in $(0, q)$ the following relation holds:

$$\Pi_p(X, Y) = \Pi_q(X, Y)$$

for all Banach spaces Y. Then I_X satisfies the relation

$$(0, 1) \subset I_X \subset (0, 2)$$

and I_X is always an interval which is either open or coincides with $(0, 2]$.

For more information about I_X we refer to the excellent expository paper of Kisliakov (1977) which also contains some new results and a bibliography.

Concerning the coincidence of classes $\Pi_p(X, Y)$ we mention the following result of Kisliakov *et al.* (1977, 1972).

THEOREM 7.7.7. *Let p be in* $(1, 2]$ *and let X be a Banach space. Then the following conditions are equivalent:*

(1) $\Pi_1(X, Y) = \Pi_p(X, Y)$ *for all Banach spaces Y.*
(2) $\Pi_q(L_\infty, X) = L(L_\infty, X)$.
(3) *There exists a positive constant* $c > 0$ *such that for any finite system of elements of X,* $x_1, \ldots, x_n$ *and for any* $x_1^*, \ldots, x^*$ *in* X^* *the following inequality holds:*

$$\sum_{j=1}^{n} \left(\sum_{i=1}^{n} |x_i^*(x_j)|^p \right)^{1/p} \leqslant c \left(\sum_{i=1}^{m} \|x_i^*\|^p \right)^{1/p} \sup_{|a_j| \leqslant 1} \left\| \sum_{j=1}^{n} a_j x_j \right\| .$$

From this theorem we have Grothendieck's theorem which asserts that for p in $(1, 2]$ the following relation is true:

$$L(L_\infty, L_p) = \Pi_2(L_\infty, L_p).$$

For a proof of Theorem 7.7.7 we refer to Kisliakov's paper.

7.8. Types, Cotypes and Rademacher Averages in Banach Spaces

Let us consider on the interval [0, 1] the sequence of Rademacher functions $(r_n(t))$ where

$$r_n(t) = \operatorname{sgn} \sin 2^{n-1} \cdot 2\pi t.$$

Then there exists the classical inequalities of A. Khinchine, which essentially asserts on the linear space generated by these functions that for all p in $(0, \infty)$ the norms are equivalent; the analytical form of these inequalities is as follows: for any p in $(0, \infty)$ and any scalars k_i there exists constants A_p and B_p such that

$$A_p\left(\int_0^1 \left|\sum_{n=1}^{k} k_i r_i(t)\right|^2 dt\right)^{1/2} \leqslant \left(\int_0^1 \left|\sum_{n=1}^{k} k_i r_i(t)\right|^p dt\right)^{1/p}$$
$$\leqslant B_p\left(\int_0^1 \left|\sum_{n=1}^{k} k_i r_i(t)\right|^2 dt\right)^{1/2}.$$

For a proof we refer to A. Zygmund's book on trigonometrical series. J. P. Kahane has proved that there is a variant of the Khinchine inequalities for k_i elements in a Banach space: if $x_1, \ldots, x_k$ are elements of a Banach space X then there exists a constant $C_p > 0$ such that

$$C_p^{-1}\left(\int_0^1 \left\|\sum_{i=1}^{n} x_i r_i(t)\right\|^2 dt\right)^{1/2} \leqslant \left(\int_0^1 \left\|\sum_{i=1}^{n} x_i r_i(t)\right\|^p dt\right)^{1/p}$$
$$\leqslant C_p\left(\int_0^1 \left\|\sum_{i=1}^{n} x_i r_i(t)\right\|^2 dt\right)^{1/2}.$$

Quite naturally there follows the problem about the existence of an inequality of the type: for some constants A_p and C_p and all $x_1, \ldots, x_n$ in a Banach space X,

$$A_p\left(\sum_{i=1}^{n} \|x_i\|^2\right)^{1/2} \leqslant \left(\int_0^1 \left\|\sum_{i=1}^{n} x_i r_i(t)\right\|^p dt\right)^{1/p}$$
$$\leqslant B_p\left(\sum_{i=1}^{n} \|x_i\|^2\right)^{1/2}.$$

Here the constants A_p and B_p depend only upon p and the Banach spaces X. Kwapien (1972) has proved that if a Banach space admits such an inequality then it is isomorphic to a complete inner product space (see Chapter 9). As suggested by the Khinchine inequalities, some interesting classes of Banach spaces were considered.

DEFINITION 7.8.1. The pth Rademacher average of a finite set of elements $x_1, \ldots, x_n$ of a Banach space X is the number $R^p_{(x_1, \ldots, x_n)}$ defined as:

$$R^p_{(x_1, \ldots, x_n)} = \left(\int_0^1 \left\|\sum_{i=1}^{n} x_i r_i(t)\right\|^p dt\right)^{1/p}.$$

For p in $(0, 2]$ we set

$$p^* = \begin{cases} p & \text{if } p < 2 \\ \infty & \text{if } p = 2. \end{cases}$$

DEFINITION 7.8.2. The operator $T \in L(X, Y)$ is said to be of p-Rademacher type if there is an r in $(0, \infty)$ and a constant $c > 0$ such that for all finite sets of elements $x_1, \ldots, x_n$ in X the following inequality holds:

$$\mathrm{R}^r_{(Tx_1,\ldots,Tx_n)} \leqslant c\alpha_p((x_n)).$$

DEFINITION 7.8.3. A sequence of functions (f_n) on $[0, 1]$ is said to be stable of order p if there exists a constant $c_{p,n} > 0$ such that

$$\int_0^1 \mathrm{e}^{if_n(t)}\,\mathrm{d}t = \mathrm{e}^{-(c_{p,n}|\lambda|^{1/p})}.$$

DEFINITION 7.8.4. The operator $T \in L(X, Y)$ is said to be p-stable if for some constant $c > 0$ and r in $(0, p^*)$ for all finite sets $x_1, \ldots, x_n$ of elements of X the following inequality holds:

$$\left(\int_0^1 \left\|\sum_{i=1}^n f_i(t)Tx_i\right\|^r \mathrm{d}t\right)^{1/r} \leqslant c\alpha_p((x_i)),$$

where (f_n) is a stable sequence of order p on $[0, 1]$.

DEFINITION 7.8.5. A Banach space X is said to be of p-Rademacher type if the identity operator is of p-Rademacher type; a Banach space X is said to be p-stable if the identity operator is p-stable.

Notions related to the above are those of cotype and infratype defined below.

DEFINITION 7.8.6. A Banach space X is said to be of cotype p-Rademacher (p in $[2, \infty)$) if there exists a constant $c_p > 0$ such that for all finite sets of elements of X, $x_1, \ldots, x_n$, the following inequality holds:

$$\alpha_p((x_i)) \leqslant c_p R^2_{(x_1,\ldots,x_n)}.$$

DEFINITION 7.8.7. A Banach space X is said to be of infratype p (p in $(0, 2]$) if for some constant $c > 0$ and all finite sets $x_1, \ldots, x_n$ of elements of X the following inequality holds:

$$\inf_{t_i = \pm 1} \left\|\sum_{i=1}^n t_i x_i\right\| \leqslant c\alpha_p((x_i)).$$

REMARK 7.8.8. If X is a complete inner product space then X is of type and cotype 2. The space L^p is of type 2 for $2 \leqslant p < \infty$ and cotype 2 for $1 \leqslant p \leqslant 2$.

As illustration of the use of the above notions we give the following result.

THEOREM 7.8.9. *Let X be a Banach space which is of cotype 2-Rademacher. Then for any Banach space Y and any $p \geqslant 2$, $\Pi_p(X, Y) = \Pi_2(X, Y)$.*

Proof. Let T be in $\Pi_p(X, Y)$ and μ be the measure on K such that for all x in X

the following inequality holds:

$$\|Tx\| \leqslant \left(\int_K |x^*(x)|^p \, d\mu(x^*)\right)^{1/p}$$

(the constant is included in the measure, i.e. we take $\mu = c\mu$, without loss of generality). Then if $x_1, \ldots, x_n$ is a finite set of elements of X, we have

$$\begin{aligned}
\sum_{i=1}^{n} \|Tx_i\|^2 &\leqslant c \int_0^1 \left\| \sum_{i=1}^{n} r_i(t) Tx_i \right\|^2 dt \\
&\leqslant c \int_0^1 \left\| T\left(\sum_{i=1}^{n} r_i(t) x_i \right) \right\|^2 dt \\
&\leqslant c \int_0^1 \left(\int_K \left| x^*\left(\sum_{i=1}^{n} r_i(t) x_i \right) \right|^p d\mu(x^*) \right)^{2/p} dt \\
&\leqslant c \left(\int_0^1 \int_K \left| \sum_{i=1}^{n} r_i(t) x^*(x_i) \right|^p d\mu(x^*) \, dt \right)^{2/p}
\end{aligned}$$

Now we can change the order of integration, and using the Khinchine's inequality we obtain that

$$\begin{aligned}
\sum_{i=1}^{n} \|Tx_i\|^2 &\leqslant c_1 \left(\int_K \left(\sum_{i=1}^{n} |x^*(x_i)|^2 \right)^{p/2} d\mu(x^*) \right)^{2/p} \\
&\leqslant c_1 \|\mu\|^{2/p} \times \sup\left(\sum_{i=1}^{n} |x^*(x_i)|^2; \|x^*\| \leqslant 1 \right)
\end{aligned}$$

and thus T is in $\Pi_2(X, Y)$ and since the converse inclusion holds for all spaces the assertion is proved. □

COROLLARY 7.8.10. *If X is a Banach space of cotype 2-Rademacher then*

$$L(L_\infty, X) = \Pi_2(L_\infty, X).$$

COROLLARY 7.8.11. *If X is a complete inner product space then the following equality holds: for all $p \geqslant 2$.*

$$\Pi_p(X) = \Pi_2(X).$$

Proof. Since every inner product space is of type and cotype 2-Rademacher the assertion follows from Theorem 7.8.9. □

REMARK 7.8.12. The assertion in Corollary 7.8.11 is valid for all p. See S. V. Kisliakov's paper for an indication of the proof.

For more details and information concerning types and cotypes of Banach spaces see Pisier (1973), Retherford (1975), Kisliakov (1977), Muštari (1980) and the references quoted.

We now mention some classes of operators on complete inner product spaces that have received a much attention in recent years. Let $1 < r$, $s < \infty$ and we consider the Lorentz space $l_{r,s}$ which consists of all sequences $a = (a_i)$ of real numbers with the following property: if $\bar{a}$ represent the decreasing rearrangement of a, $\bar{a} = (\bar{a}_i)$ then a is in $l_{r,s}$ if

$$\|a\|_{r,s} = \left(\sum_{i=1}^{\infty} i^{s/r-1-s}\left(\sum_{k=1}^{\infty} \bar{a}_k\right)^s\right)^{1/s} < \infty.$$

From the definition it follows that $l_{p,p} = l^p$. The definition of the class of operators in $\gamma_{r,s}$ is exactly as in the cases of the Schatten–von Neumann classes c_p.

DEFINITION 7.8.13. The compact operator $T \in L(X)$, X is a complete inner product space is in $\gamma_{r,s}(x)$ if the characteristic sequence of T is in $l_{r,s}$.

We now give one of the basic result about this class of operators. For a proof we refer to Kwapien (1968), Bennett *et al.* (1975) and Mitiagin (1977).

THEOREM 7.8.14. *Concerning the classes* $\Pi_p(X)$, $\Pi_{p,q}(X)$, $\gamma_{r,s}(X)$, *X is a complete inner product space, the following assertions hold:*

(1) $p = q \to \Pi_{p,q}(X) = c_2$,
(2) $q^{-1} - p^{-1} \geqslant \frac{1}{2} \to L(X) = \Pi_{p,q}(X)$,
(3) $q^{-1} - p^{-1} < \frac{1}{2}$, $\quad q < 2 \to \Pi_{p,q}(X) = c_r$ *with* $r^{-1} = p^{-1} - q^{-1} + \frac{1}{2}$,
(4) $2 < q < p \to \Pi_{p,q}(X) = \gamma_{2p/q,p}$.

We now mention without proof several very interesting applications of absolutely p-summing operators. In the next chapter we present a simple proof using absolutely p-summing operators of a characterization of complete inner product spaces obtained by S. Kakutani and H. F. Bohnenblust.

If T^k is the k-dimensional torus (T is the torus), then $A(T^k)$ denotes the subspace in $C(T^k)$ (the space of all continuous functions on T^k with the sup norm) generated by $(e^{in_1t_1} \cdots e^{in_kt_k})$, $n_i \geqslant 0$, the closure of $A(T^k)$ in L^p (with respect to the Haar measure (the normalized Lebesgue measure on T^k) is denoted by $H^p(T^k)$ and is the so-called Hardy space. Then using the theory of absolutely p-summing operators and related classes considered in this chapter, the following result of Henkin (1967, 1968) holds: the spaces $A(T)$ and $A(T^2)$ are not isomorphic B. S. Mitiagin has proved that the classical result of S. Bernstein concerning absolutely convergent Fourier series for functions in the space Lip_a, $a > \frac{1}{2}$ is the best possible with respect to the papameter a, i.e. for $a = \frac{1}{2}$ there exists a function in $\mathrm{Lip}_{1/2}$ with an absolutely divergent Fourier series. Another interesting application, obtained by Lindenstrauss and Pelczyński (1968), proved that l^1 and c_0 have a unique unconditional basis.

In closing we mention the following characterization of Banach spaces which are isomorphic to inner product spaces obtained by Grothendieck (1955).

THEOREM 7.8.15. *If X is a Banach space which is isomorphic to a subspace of an L_1-space and to a quotient space of an L_∞-space, then X is isomorphic to an inner product space.*

Concerning this result it is not known if it holds with 'isometric' instead 'isomorphic'.

CHAPTER 8

Operator Characterizations of Inner Product Spaces

8.0. Introduction

As we have seen, if on a normed linear space X there exist certain relations among finite sets of elements in X then this forces the space to be an inner product space. Sometimes this fact may be interpreted to state that certain functions defined on the space X have some very special properties. In the first part of this chapter we present further characterizations of inner product spaces along these lines. Our approach is that we first present certain characterizations using nonlinear mappings. We begin with a characterization of inner product spaces obtained by I. J. Schoenberg using a class of functions related to negative definite functions. Secondly, we present certain inequalities on Banach spaces, including the characterization of inner product spaces using a sharp form of an inequality. The case of nonexpansive mappings and their connections with the characterization of inner product spaces, as well as with the m-accretive and maximal accretive mappings, is discussed further. Here we present some of the classical results of, for example, K. Kirszbraun and F. Valentine and also some recent ones by B. Grünbaum, S. Schönbeck, etc.

In the second part we emphasize the case of linear mappings. First we present the characterization of inner product spaces obtained by S. Kakutani and H. F. Bohnenblust in terms of the possible extensions of linear operators and their projections. Related to these we present the characterizations of E. Berkson, N. J. Kalton and G. V. Wood using hermitian projections. Further, we present the beautiful characterization of inner product spaces using the notion of spectral sets obtained by John von Neumann.

A number of other related characterization are also included, among them we mention some characterizations using smooth functions on Banach spaces and the Mielnik probability spaces (which are of interest in the foundation of quantum mechanics).

8.1. 0-Negative Definite Functions and Inner Product Spaces

In this section we present a characterization of inner product spaces in the class of normed linear space using certain classes of functions which we call 0-negative definite functions.

DEFINITION 8.1.1. A function $f: X \to R$ is said to be a 0-negative definite function on the normed linear space X if for all finite set of points $x_1, \ldots, x_n$ and all systems of real numbers $r_1, \ldots, r_n$ with $r_1 + r_2 + \cdots + r_n = 0$ the following inequality holds:

$$\sum_{i,j=1}^{n} f(x_i - x_j) r_i r_j \leqslant 0.$$

Now we have the following characterization of inner product spaces.

THEOREM 8.1.2 (Schoenberg, 1938). *A normed linear space X is an inner product space if and only if the function $f(x) = \|x\|^2$ is a 0-negative definite function.*

Proof. It is not difficult to see that the condition is necessary. Suppose now that X is a normed linear space with the above property. Then we choose $n = 4$ and set

$$x_1 = x, \qquad x_2 = y, \qquad x_3 = -y, \qquad x_4 = 0, \qquad r_1 = 1 - 2a, \qquad r_2 = r_3 = a,$$
$$r_4 = -1,$$

where x, y are arbitrary elements of X and a is in $(0, \frac{1}{2})$. Then we have the inequality

$$(1 - 2a)\|x - y\|^2 + (1 - 2a)\,a\|x + y\|^2 \leqslant (1 - 2a)\|x\|^2 + 2a(1 - 2a)\|y\|^2$$

or

$$\|x + y\|^2 + \|x - y\|^2 \leqslant 2(\|x\|^2 + \|y\|^2). \tag{*}$$

If we set in the above inequality

$$x = \tfrac{1}{2}(x_1 + y_1), \qquad y = \tfrac{1}{2}(x_1 - y_1)$$

then

$$2(\|x_1\|^2 + \|y_1\|^2) \leqslant \|x_1 + y_1\|^2 + \|x_1 - y_1\|^2 \tag{**}$$

and these inequalities imply that for all u, v in X,

$$\|u + v\|^2 + \|u - v\|^2 = 2(\|u\|^2 + \|v\|^2)$$

and, according to the Jordan–von Neumann characterization of inner product spaces, X is an inner product space.

The approach in Fréchet's paper on the characterization of inner product structures on normed linear spaces (Banach spaces) has suggested to Schoenberg (1935) an interesting solution of a problem considered by Menger (1931): Suppose that $A = (a_{ij})$ is a matrix with $a_{ij} = a_{ji}$, $i \neq j$, $a_{ij} > 0$. The problem is to find the necessary and sufficient conditions such that a_{ij} are the lengths of the edges of an n-simplex, and more generally to find the necessary and sufficient conditions such that a_{ij} are the lengths of the edges of an n-simplex in a Euclidean space of dimension p (p in $[1, n]$) and not in a Euclidean space of dimension $p - 1$. The solution obtained by I. J. Schoenberg is as follows.

THEOREM 8.1.3 (Schoenberg, 1935). *The necessary and sufficient conditions such that a_{ij} be the lengths of the edges of an n-simplex in a Euclidean space of dimension p*

and not in a Euclidean space of dimension p − 1 is that the function

$$f(t_1, t_2, \ldots, t_n) = \sum_{i=1}^{n} a_{0i}^2 t_i^2 + \sum_{\substack{i,j=1 \\ i<j}} (a_{0i}^2 + a_{0j}^2 - a_{ij}^2) t_i t_j$$

$$= \frac{1}{2} \sum_{i,j=1} (a_{0i}^2 + a_{0j}^2 - a_{ij}^2) t_i t_j$$

be positive and of rank p.

For a proof as well as for extensions we refer to Schoenberg (1935).

Let X be a topological space and

$f: X_\times X \to C$(C the complex numbers).

We say that f is positive definite if

(1) f is continuous,

(2) $f(t, s) = \overline{f(s, t)}$,

(3) $f(t, t) = f(s, s)$.

(4) For any integer n and any $t_1, \ldots, t_n$ in X and any complex numbers $c_1, \ldots, c_n$,

$$\sum_{i,j} f(t_i, t_j) c_i \bar{c}_j \geqslant 0. \qquad (+)$$

The set of positive definite functions on X is denoted by $P(X)$. The set of all functions f satisfying (1)–(4) when $c_1 + \cdots + c_n = 0$ is denoted by $P_0(S)$.

Suppose now that on X we have a metric d.

DEFINITION 8.1.4 (Bochner, 1941). We say that d is an inner product distance (metric) on X if there is a mapping $T: X \to H$, H being a complete inner product space, such that

$$\|T(t) - T(s)\| = d(t, s)$$

for all t, s in X.

Then we have the following result of Menger and Schoenberg (1938).

THEOREM 8.1.5 (Menger and Schoenberg, 1938). *The metric d on X is an inner product metric iff the function $-d$ is in $P_0(X)$.*

In his paper, Bochner (1941) considers other types of positive definiteness. We refer to Faraut and Harallah (1974) for the case of homogeneous space as well as to the papers quoted there and to Chapter 10.

8.2. Some Inequalities and a Characterization of Inner Product Spaces

We now prove certain inequalities which are of importance in the study of a class of nonlinear mappings, called the nonexpansive mappings, which have received great interest in recent years.

THEOREM 8.2.1 (Massera and Schaeffer, 1958). *For any $x \neq 0 \neq y$ in the normed* *linear space X the following inequality holds:*

$$\big\| \|y\| x - \|x\| y \big\| \max(\|x\|, \|y\|) \leqslant 2\|x\| \cdot \|y\| \, \|x - y\|.$$

Proof. First we remark that this is equivalent to the inequality

$$\big\| x/\|x\| - y/\|y\| \big\| \cdot \tfrac{1}{2} \cdot \max(\|x\|, \|y\|) \leqslant \|x - y\|.$$

For the proof we may assume that $\|x\| \leqslant \|y\|$ (since the above inequality is symmetric in x and y). Let $a = \big\| x/\|x\| - y/\|y\| \big\|$. If $\|x\| \leqslant (1 - a/2)\|y\|$ then we have

$$\|y\| \leqslant \|y - x\| + \|x\| = \|y - x\| + \|x\| \leqslant \|y - x\| + (1 - a/2)\|y\|$$

or

$$a/2 \cdot \|y\| \leqslant \|x - y\|$$

and in this case the required inequality is satisfied.

Now if $\|x\| \leqslant (1 - a/2)\|y\|$ is not true then

$$(1 - a/2)\|y\| \leqslant x$$

and thus

$$(1 - a/2)\|y\| \leqslant \|x\| \leqslant \|y\|$$

which gives further that

$$a\|y\| = \big\| y - \|y\| \cdot x/\|x\| \big\| = \big\| y - x - (\|y\| - \|x\|)x/\|x\| \big\| \leqslant \|x - y\| + a/2 \cdot \|y\|$$

and then the inequality is satisfied in this case also.

COROLLARY 8.2.2. *For any $x \neq 0 \neq y$ in X the following inequality holds:*

$$4\|x\| \cdot \|y\| \, \|x - y\| \geqslant (\|x\| + \|y\|) \big\| \|x\| y - \|y\| x \big\|.$$

The proof of the following inequality is left to the reader.

LEMMA 8.2.3. *Let X be an inner product space. Then for any $x \neq 0 \neq y$ in X the following inequality holds:*

$$2\|x\| \|y\| \|x - y\| \geqslant (\|x\| + \|y\|) \big\| x\|y\| - y\|x\| \big\|. \tag{*}$$

In answer to a question by Dunkl and Williams (1964) of whether the inequality (*) characterizes inner product spaces, Kirk and Smiley (1964) have proved the following result.

THEOREM 8.2.4 (Kirk and Smiley, 1964). *If X is an inner product space in which the inequality (*) holds, then X is an inner product space.*

Proof. Let x, y be arbitrary elements in X, $x \neq y$. Let $x_1 = kx$, $y_1 = y$ and thus by (*) we have

$$2k\|x\|^2 \|kx - y\| \geqslant (k + 1)\|x\|^2 k \|x - y\|$$

if $k \geqslant 0$ and

$$2|k| \, \|x\|^2 \|kx - y\| = (|k| + 1)\|x\|^2 |k| \cdot \|x - y\|$$

if $k \leqslant 0$. Then we have that

$$\|kx - y\| \geqslant (|k| + 1)/2 \cdot \|x - y\|$$

for all k.

If we take $-y$ then we have the inequality

$$\|kx + y\| = (|k| + 1)/2 \cdot \|x + y\|.$$

We now show that for all s in R we have the inequality $s \neq 0$:

$$\|s(kx + y) + s^{-1}(x + ky)\| \geqslant \|(ax + y) + (x + ay)\|$$
$$= \|(a + 1)x + (a + 1)y\| = |a + 1| \, \|x + y\|.$$

Indeed,

$$\|s(ax + y) + s^{-1}(x + ky)\| = \|(sk + s^{-1})x + (s + s^{-1}k)y\|$$
$$= |sk + s^{-1}| \, \|x + (sk + s^{-1})^{-1}(s + s^{-1}k)y\|$$
$$\geqslant (|sk + s^{-1}|) \cdot \tfrac{1}{2} \cdot (1 + |sk + s^{-1}|^{-1}|s + s^{-1}k| \, \|x + y\|).$$

We have thus reduced the problem to prove the inequality

$$|sk + s^{-1}| + |s + s^{-1}k| \geqslant 2|k + 1|.$$

We can suppose that k is positive and then the inequality is obvious. Thus for fixed k we have that for all $s \neq 0$

$$\|s(kx + y)\| + \|s^{-1}(x + ky)\| \geqslant (k + 1) \, k + 1 \|x + y\|.$$

But this implies that for such k

$$\|kx + y\| = \|x + ky\|.$$

Now k being positive (and arbitrary in R^+), from 4.8.3 X is an inner product space.

We remark that for all x, y in X with $\|x\| = \|y\|$ the following inequality holds:

$$\|kx + k^{-1}y\| \geqslant \|x + y\|,$$

for all $k \neq 0$. Indeed we apply the inequality (*) to the pair kx and $-k^{-1}y$ and the relation follows.

Now if the inequality:

$$\|ku + k^{-1}u\| \geqslant \|u + v\|$$

holds for u, v in X and all $k \neq 0$, then $\|u\| = \|v\|$. (This is what was important for our proof.) First, if $u + v = 0$ then $u = -v$ and the assertion follows. If $u + v \neq 0$ we can suppose that $u \neq 0$ and then $v \neq 0$. Let $k_0 = \|v\|/\|u\|$ and $v = k_0 v_0$. Then clearly $\|v_0\| = \|u\|$. Then we have, by the above inequality, that

$$\|u + v\| = \|u + k_0 v_0\| = k^{1/2}\|k^{1/2}x + k^{1/2}v_0\| \geqslant k^{1/2}\|u + v_0\|$$

and this is possible if and only if $k^{1/2} = 1$. Then $\|u\| = \|v\|$ and the assertion is proved. □

REMARK 8.2.5. In the above theorem the dimension of X is at least 3. Another way to look at the behavior of the Clarkson angle measure between two elements of a normed linear space is discussed in the interesting paper of Schaeffer (1965).

8.3. Nonexpansive Mappings and the Extension Problem

Let (X, d) and (Y, d^{-}) be two metric spaces and let $T: M \to N$ be a mapping defined on the subset M of X and with values in the subset N of Y.

DEFINITION 8.3.1. The mapping T is said to be a Lipschitzian mapping with the Lipschitz constant k if for all x, y in M the following inequality holds:

$$d^{-}(Tx, Ty) \leqslant kd(x, y).$$

If $k = 1$ then we say that the mapping T is nonexpansive; if k is in $(0, 1)$ then we say that T is a strict contraction mapping.

An interesting and important class of nonexpansive mappings is considered in the following definition.

DEFINITION 8.3.2. Let (X, d) be a metric space and let M be a subset of X. A mapping $R: X \to X$ which is continuous and for all $m \in M$, $Rm = m$, is called a retraction of X onto M.

DEFINITION 8.3.3. Let (X, d) be a metric space and let M be a subset in X. A mapping which is continuous and is the identity on M, i.e. $R: X \to X$ is continuous and $Rx = x$ for all x in M, is said to be a nonexpansive retraction of X onto M if R is a nonexpansive mapping.

LEMMA 8.3.4. *If the set M in the metric space X has a retraction of X onto M then M is a closed subset of X.*

Proof. Let (x_n) be a sequence of elements of M converging to an element x of X. Since $Rx_n = x_n$ for all n the continuity of R implies that $Rx = x$. Thus M is closed. □

COROLLARY 8.3.5. *If M is a subset in a metric space X and has a nonexpansive retraction, then M is closed.*

The following lemma considers the case of strictly convex spaces.

LEMMA 8.3.6. *Let X be a strictly convex Banach space and let M be a subset of X having a nonexpansive retraction. Then M is a convex subset of X.*

Proof. Let u, v be two arbitrary elements of M and t in $(0, 1)$. Let $x = tu + (1 - t)v$. Then we have

$$\|u - Rx\| \leqslant \|u - x\| = \|tu + (1 - t)u - tu - (1 - t)v\| = (1 - t)\|u - v\|,$$
$$\|v - Rx\| \leqslant \|v - x\| = \|tv + (1 - t)v - tu - (1 - t)v\| \leqslant t\|u - v\|.$$

Then the strict convexity of X implies that $Rx = x$ and the assertion is proved. □

From Chapter 4 we know that the proximity map defined for subsets in a complete inner product space is nonexpansive. Another important example of a mapping on a Banach space is the so-called radial retraction on the balls centered at the origin.

DEFINITION 8.3.7. The radial retraction on the ball of radius r and with the center at the origin is the mapping R_r defined as follows:

$$R_r x = \begin{cases} rx/\|x\| & \text{if } \|x\| \geqslant r \\ x & \text{if } \|x\| \leqslant r. \end{cases}$$

In what follows we restrict ourselves to the case $r = 1$ (see Section 4.16).

THEOREM 8.3.8. *The normed linear space X is an inner product space* (dim $X \geqslant 3$) *if and only if the radial retraction on the unit ball is a nonexpansive mapping.*

Proof (Istrăţescu, 1982). The condition is necessary. We now show that a normed linear space X with the above property is an inner product space by proving that in X the condition of W. A. Kirk and M. F. Smiley is satisfied (Theorem 8.2.4). Let x, y be arbitrary elements of X and we may suppose without loss of generality that $\|x\| \leqslant \|y\|$. Then the elements $x/\|x\|$ and $y/\|x\|$ are not in the interior of the unit ball and thus, by the nonexpansiveness of the radial retraction on the unit ball, we must have the inequality

$$\|x/\|x\| - y/\|y\|\| \leqslant 1/\|x\| \cdot \|x - y\|$$

or

$$\|x - y\| \geqslant \|x\| \|x/\|x\| - y/\|y\|\|. \tag{*}$$

(Of course, we suppose without loss of generality that $x \neq 0 \neq y$.)

We now consider the elements $x_1 = \|x\|^a/\|y\|^b \cdot x$ and $y_1 = \|x\|^2/\|y\|^b \cdot y$

and thus

$$\|x_1\| = \|x\|^{1+a}/\|y\|^b, \|y_1\| = \|x\|^a/\|y\|^{1-b}$$

and for $b \geqslant 2$ we have

$$\|x_1\| \leqslant \|y_1\|.$$

Then we can apply the above inequality and

$$\begin{aligned} \|x/\|x\| - y/\|y\|\| &\leqslant \|x\|^{1+a}/\|y\|^b \cdot \|x\|^a/\|y\|^b \cdot \|x - y\| \\ &= \|x\|^{1+2a}/\|y\|^{2b} \cdot \|x - y\| \end{aligned}$$

Then for $2b = 2a + 2$ we obtain the inequality

$$\|x/\|x\| - y/\|y\|\| \leqslant 1/\|y\| \cdot \|x - y\| \tag{**}$$

and with (*) this implies that

$$2\|x - y\| \geqslant (\|x\| + \|y\|) \|x/\|x\| - y/\|y\|\|.$$

Then the result of Kirk and Smiley implies that X is an inner product space. □

REMARK 8.3.9. A radial retraction mapping is not expansive even in dimension 2. Let us consider the space R^2 of all pairs of real numbers with the norm

$$\|(x, y)\| = |x| + |y|.$$

Then if $u = (1, 0)$, $v = (1, \frac{1}{2})$ we have clearly that a radial retraction on the unit ball has the values

$$Ru = u, \qquad Rv = (\tfrac{2}{3}, \tfrac{1}{3})$$
$$\|Ru - Rv\| = \tfrac{2}{3}$$

and $\|u - v\| = \frac{1}{2}$. Thus R is not nonexpansive.

The following result shows a connection between the nonexpansive retraction on closed convex sets and the Lipschitz mappings. First we give a definition.

DEFINITION 8.3.10 A Banach space X is said to have an extension property for Lipschitzian mappings in closed convex sets (in short an (EC)-property) if for all closed convex sets C and all Lipschitzian mappings $T: C \to X$ there exists a Lipschitzian mapping (with the same Lipschitz constant) $\bar{T}: X \to X$ such that for all x in C, $Tx = \bar{T}x$.

Then we have the following result.

LEMMA 8.3.11. *A Banach space X has an (EC)-property if and only if X has a nonexpansive retraction on every closed convex set of X.*

Proof. Suppose first that X has an (EC)-property and let C be a closed convex set in X. Then we can consider the identity mapping $I: C \to C$ and by our hypothesis there exists an extension of I to X with a Lipschitz constant equal to 1. Then this clearly is a retraction of X onto C.

Conversely, suppose that X is a Banach space with the stated property and that C is a closed and convex subset of X. Then $T: C \to X$ is a Lipschitzian mapping. If R_C is the retraction of X onto C (which is nonexpansive) then $T_0 R_C$ satisfies the requirements of the lemma. □

We consider in what follows the problem of extending a nonexpansive mapping defined on a subset of a Banach space with values in another Banach space (not necessarily distinct from the original one) to the whole space. For a proper formulation of the problem, as well as of the results, we need some definitions.

DEFINITION 8.3.12. A Banach space X is said to have the 'property (E)' if for any nonexpansive mapping $T: M \to X$, M being a subset of X, there exists a nonexpansive mapping $\bar{T}: X \to X$ such that for all x in M, $Tx = \bar{T}x$.

DEFINITION 8.3.13. A pair of Banach spaces (X, Y) is said to have a nonexpansive (isometric) extension property if $T: M \to Y$ is a nonexpansive (isometric) mapping defined on a subset M of X; then there exists a nonexpansive (isometric) mapping $\bar{T}: X \to Y$ such that for all x in M, $Tx = \bar{T}x$.

We say that the pair has property (E) only in cases concerning nonexpansive mappings.

Concerning the property (E) considered in Definition 8.3.12 and 8.3.13, the basic problem can be formulated as follows.

PROBLEM 8.3.14. Characterize the Banach spaces and the pairs of Banach spaces that have property (E).

A basic notion in the study of the extension of nonexpansive mappings is the following notion concerning families of closed balls. In order to formulate this, we recall that $B(x, r)$ is the set $(y, \|x - y\| \leqslant r)$.

DEFINITION 8.3.15. A Banach space X is said to have the Kirszbraun property for the cardinal γ (in short, 'property (K, γ)') if for each set of points $(x_i)_{i \in I}$, I is an index set of cardinality γ, such that $\|y_i - y_j\| \leqslant \|x_i - x_j\|$ for all i, j in I, then $\cap_{i \in I} B(x_i, r_i) \neq \emptyset$ implies that $\cap_{i \in I} B(y_i, r_i) \neq \emptyset$.

We say that X has the Kirszbraun property (in short, 'property (K)') if it has property (K, γ) for all cardinal numbers γ.

DEFINITION 8.3.16. A pair of Banach spaces (X, Y) is said to have the Kirszbraun property for cardinal γ (in short, 'property (K, γ)') if for each set of points $(x_i)_{i \in I}$, where I is an index set with cardinality γ, such that $\|y_i - y_j\| \leqslant \|x_i - x_j\|$, x_i in X, y_i in Y for all i in I, we have that $\cap_{i \in I} B(x_i, r_i) \neq \emptyset$ implies that $\cap_{i \in I} B(y_i, r_i) \neq \emptyset$.

We say that the pair (X, Y) has the Kirszbraun property (in short, 'property (K)') if it has property (K, γ) for all cardinal γ.

DEFINITION 8.3.17. A Banach space X is said to have a 'binary intersection property' if every family of mutually intersecting balls of X has a nonempty intersection.

We now present some examples in order to show that a Banach space having the property (E) or a pair of Banach spaces having the property (E) must be, in some sense, a special Banach space, or pair of Banach spaces.

EXAMPLE 8.3.18 (Schönbeck, 1966). Let us consider $R^2 = R_{\times} R$. We consider in R^2 a regular hexagon C_H and let C be the circle inscribed in C_H. From these we obtain, by Minkowski's functional, two norms on R^2 denoted by $\|,\|_{C_H}$ and $\|,\|_C$ respectively.

Since the unit balls of R^2 with respect to the two norms are the hexagon and the inscribed circle, we consider two points which are consecutive points of contact of the circle with the hexagon. Let us denote these by x_1 and x_2. We consider the subset of R^2 as a Banach space with the norm $\|,\|_{C_H}$, $M = (0, x_1, x_2)$, and the same set in R^2 as a Banach space with the norm $\|,\|_C$. We define the following mapping on M, $T: M \to R^2$, $T0 = 0$, $Tx_1 = x_2$, $Tx_2 = x_1$. Since

$$\|x_1\|_{C_H} = \|x_1\|_C = \|x_2\|_{C_H} = \|x_2\|_C = \|x_1 - x_2\|_{C_H} = \|x_1 - x_2\|_C = 1$$

we see that T is a nonexpansive mapping. We show that T does not possess an extension on the whole R^2. Indeed, suppose that such an extension exists. Then at the point $x = \frac{1}{3}(x_1 + x_2)$ we have the value $Tx = y$ which satisfies the inequalities

$$\|y\|_C \leqslant \tfrac{1}{2}, \qquad \|y - x_1\|_C \leqslant \tfrac{1}{2}, \qquad \|y - x_2\| \leqslant \tfrac{1}{2}$$

and

$$\|x\|_{C_H} = \|x - x_1\|_{C_H} = \|x - x_2\|_{C_H} = \tfrac{1}{2}.$$

Then clearly such a y does not exist.

EXAMPLE 8.3.19 (Schwartz, 1969). We consider the space R^n of all elements x, $x = (x_1, \dots, x_n)$ with $x_i \in R$ and the norm on R^n is defined by

$$x \to \|x\|_p = (|x_1|^p + \cdots + |x_n|^p)^{1/p},$$

where p is in $(1, \infty)$ and $p \neq 2$. Then the pair (R^n, R^n) does not have the property (K) when $n > 1$.

We remark that from the Clarkson inequalities it follows that $(R^n, \|\,,\|_p)$ is a uniformly convex space. (For the reader's convenience we recall that a Banach space X with a norm $\|\,,\|$ is said to be uniformly convex if, whenever $\varepsilon > 0$, there exists $\delta(\varepsilon) > 0$ such that for all $x, y, \|x\| = \|y\| = 1$ and $\|x - y\| \geqslant \varepsilon$ then $\|x + y\| \leqslant 2(1 - \delta(\varepsilon))$.) Let us consider the points $x_1 = (0, 0, \dots, 0)$, $x_2 = (0, 1, 0, \dots, 0)$, $x_3 = (1, 0, 0, \dots, 0)$ and $r_i = r = 2^{(1-p)/p}$. Since we have

$$\|x_1 - x_2\| = \|x_1 - x_3\| = 1, \qquad \|x_2 - x_3\| = 2^{1/p}$$

we see that

$$\bigcap_{i=1}^{3} B(x_i, r_i) = (\tfrac{1}{2}, \tfrac{1}{2}, 0, 0, \dots, 0).$$

We now consider the points in R^n defined as follows:

$$y_1 = x_1, y_2 = ((1 - 2^{1-p})^{1/p}, 2^{(1-p)/p}, 0, 0, \dots, 0)$$

$$y_3 = ((1 - 2^{1-p})^{1/p}, -2^{(1-p)/p}, 0, \dots, 0).$$

We now consider the balls $B(y_i, r_i)$. Since

$$\|y_1 - y_2\| = \|y_1 - y_3\| = 1, \qquad \|y_2 - y_3\| = 2^{1/p}$$

we see that the condition in the definition of the property $(K, 3)$ is satisfied. But we have that

$$\bigcap_{i=1}^{3} B(y_i, r_i) = \emptyset.$$

Indeed, we see that $B(y_2, r_2) \cap B(y_3, r_3)$ is nonempty and equal to $(((1 - 2^{1-p})^{1/p}, 0, 0, \dots, 0))$ by the uniform convexity of the space R^n (with respect to the norm $\|\,,\|_p$). But $((1 - 2^{1-p})^{1/p}, 0, \dots, 0)$ is not in $B(y_1, r)$. Here we used the fact that $p \neq 2$. If $p \neq 2$ and p in $(1, 2)$ then we can reverse the role of the x_i's with the y_i's and the radius $r = (1 - 2^{1-p})^{1/p}$.

In what follows we now prove the basic result of Kirszbraun (1934) which tell us that (R^n, R^n) has property (K); this result was extended to the case of complete inner product spaces by Valentine (1945).

THEOREM 8.3.20 (Kirszbraun, 1934). *The pair (R^n, R^n), R^n considered as an inner product space, has property (K).*

For the proof we need the following interesting result of E. Helly (see Danzer *et al.*, (1963) for a proof as well as for related results).

THEOREM 8.3.21 (Helly, 1921). *Let X be a finite dimensional space of dimension n.*

If $(C_i)_{i \in I}$, *where* I *is an arbitrary index-set, is a family of convex subsets of* X *then* $\cap_{i \in I} C_i \neq 0$ *if and only if any subfamily of* $(C_i)_{i \in I}$ with $n+1$ sets *has a nonempty intersection.*

Now the proof of Theorem 8.3.20 is as follows. From Helly's theorem we have that the pair (R^n, R^n) has property (K) if and only if it has the property $(K, n+1)$. This is equivalent to the extension property (K, γ): we say that a Banach space (or a pair of Banach spaces) has property (E, γ) if for any subset M in X with cardinality less than or equal to γ, and $T: M \to X$ $(T: M \to Y)$ has an extension to a subset of the form $M \cup (x_0)$, $x_0 \notin M$ and the extension is nonexpansive. Equivalence will be proved in the next theorem.

Suppose now that $x_1, \ldots, x_{n+1}$ are $n+1$ points in R^n and there exists a nonexpansive mapping defined for each x_i with values in R^n. Let us consider x_0 to be an arbitrary point in R^n different from $x_1, \ldots, x_{n+1}$. Then we show that there exists y_0 in R^n with the property that

$$\|Tx_i - y\| \leqslant \|x_i - x_0\| \tag{*}$$

holds for all $i = 1, 2, \ldots, n+1$.

We consider the following function defined on R^n by the formula

$$f(t) = \max(\|Tx_i - t\| / \|x_i - x_0\|).$$

We remark that f is continuous and $f(t) \to \infty$ as $t \to \infty$. Then f, being positive, has a minimum value at some point, say y_0,

$$m = f(y_0) = \min\,(f(t), t \in R^n).$$

If $m = 0$ then $y_0 = Tx_i$ for all i and then (*) is true. Thus we may assume that $m > 0$. We may assume (reordering if necessary) that for some m the following relations hold:

$$\|y_0 - Tx_i\| = m\|x_0 - x_i\|, \qquad i = 1, 2, 3, \ldots, m$$
$$\|y_0 - Tx_i\| < m\|x_0 - x_i\|, \qquad i = m+1, \ldots, n+1.$$

We show now that y_0 is in the convex hull of $Tx_1, \ldots, Tx_m$. If not, then by the separation theorem and by the Fréchet–Riesz representation of linear and continuous functionals, there exists z in R^n such that

$$\langle z, y_0 \rangle < \langle z\, Tx_i \rangle, \qquad i = 1, 2, 3, \ldots, m.$$

In this case if ε is sufficiently small $f(y_0 + \varepsilon t) < f(y_0)$. This contradicts the definition of y_0. Thus y_0 is in the convex hull of the elements $Tx_1, \ldots, Tx_m$. Then we have that

$$y_0 = a_1 Tx_1 + \cdots + a_m Tx_m, \qquad a_i \text{ in } [0, 1], \qquad a_1 + \cdots + a_m = 1.$$

We now show that $m \leqslant 1$ and this clearly implies that if we set $Tx_0 = y_0$ the extension of T at x_0 gives us a nonexpansive mapping.

Now we know that T is nonexpansive on the set $x_1, \ldots, x_{n+1}$. Then we have

$$\|Tx_i - Tx_j\|^2 = \|Tx_i - y_0 - (Tx_j - y_0)\|^2 \leqslant \|(x_i - x_0 - (x_j - x_0)\|^2$$

or

$$(m^2 - 1)(\|x_i - x_0\|^2 + \|x_j - x_0\|^2) \leqslant 2(\langle Tx_i - y_0, Tx_j - y_0\rangle \\ - \langle x_i - x_0), (x_j - x_0\rangle$$

for all $i = 1, 2, 3, \ldots, m$.

We multiply the last inequality with $a_i a_j$ and, adding up, we have that the following inequality holds:

$$2(m^2 - 1)\sum_{i=1}^{m} a_i\|x_i - x_o\|^2 = 0.$$

Since x_0 does not coincide with any of x_i this inequality implies that $m \leqslant 1$.

COROLLARY 8.3.22 (Valentine, 1945). *For any complete inner product space X the pair (X, X) has property (K).*

Proof. We can suppose without loss of generality that X is of infinite dimension. Let $((B(x_i, r_i))_{i\in I}$, be a family of closed balls in X and $((B(y_i, r_i))_{i\in I}$ be another family of closed balls in X. Suppose that $\cap_{i\in I} B(x_i, r_i)$ is nonempty and that for all i, j in I, $\|x_i - x_j\| \leqslant \|y_i - y_j\|$. Since the set $B(x, r)$ is weakly compact from the characterization of compactness in a given topology, we have to show that for any finite subset J of I, $\cap_{i\in J} B(y_i, r_i) \neq \emptyset$. If J is given then we consider the finite dimensional subspace generated by x_i with $i \in J$, and by y_i with $i \in J$. First we remark that $\cap_{i\in J} B(x_i, r_i) \cap V_J \neq \emptyset$, where V_J is the subspace generated by x_i, y_i, with i in j. Let P_J be the orthogonal projection of X onto V_J and if x is in $\cap_{i\in I} B(x_i, r_i)$ then Px satisfies the relation $\|Px - x_i\| \leqslant \|x - x_i\| \leqslant r_i$ for $i \in J$. This implies that Px is in $\cap_{i\in J} B(x_i, r_i)$. Then by the above remark we have that Px is in $V_J \cap_{i\in J} B(x_i, r_i)$. The Kirszbraun theorem gives us that $V_J \cap B(y_i, r_i) \neq \emptyset$. This clearly implies the assertion of the theorem.

REMARK 8.3.23. Using the fact that two complete inner product spaces with bases of the same cardinality are isomorphic and isometric, we have from Corollary 8.3.22 that for any complete inner product spaces X, Y the pair (X, Y) has property (K).

Now we prove the equivalence between property (K, γ) and property (E, γ) which, of course, implies equivalence between property (K) and property (E).

THEOREM 8.3.24. *The pair of Banach spaces (X, Y) has property (K, γ) if and only if (X, Y) has property (E, γ).*

Proof. First suppose that the pair (X, Y) has property (E, γ). Let x_i, y_i be points in X and Y respectively with $i \in I$, card $I = \gamma$. We suppose that $\|y_i - y_j\| \geqslant \|x_i - x_j\|$. We consider the set $D = (x_i)_{i\in I}$. We define a mapping on D with values in Y by the relation $Tx_i = y_i$ for each $i \in I$. Then from the above relation between the x_i's and y_i's the mapping is nonexpansive. Then since the pair has property (E, γ) there exists an extension of T to a subset of X including D and a point, say x_0, not in D. Thus we find a point Tx_0 in Y such that for all i

$$\|Tx_0 - Tx_i\| \leqslant \|x_0 - x_i\|.$$

This implies that $\|Tx_0 - y_i\| \leqslant r_i$ for all i. Then clearly this implies that Tx_0 is in $\cap_{i\in I} B(y_i, r_i)$. Thus the pair (X, Y) has property (K, γ).

Conversely, suppose that the pair (X, Y) has property (K, γ). Let D be a subset in X with card $D = \gamma$. If $T: D \to Y$ is a nonexpansive mapping we show that T admits a nonexpansive extension to a set of the form $D \cup (x_0)$, where x_0 is an arbitrary point in X and not in D. Let x be a point in D and consider the number $r_x = \|x - x_0\|$. Since for all y, x in D

$$\|Tx - Ty\| \leqslant \|x - y\|$$

and $\cap_{x \in D} B(x, r_x)$ obviously contains x_0 and is nonempty, then since the pair (X, Y) has property (K, γ) we conclude that $\cap_{x \in D} B(Tx, r_x)$ is nonempty. Let us denote by y_0 a point in this intersection. We define an extension of T to the set $D\,(x_0)$ by defining $\bar{T}$ equal to T at each point of D and at x_0 the value is y_0. Then clearly this is the desired extension of T. □

REMARK 8.3.25. The equivalence between property (K) and property (E) follows from the above result by using a well-known argument based on Zorn's lemma. We leave the details to the reader.

REMARK 8.3.26. With appropriate definitions the above results can be extended to metric spaces, or at least to some classes of metric spaces.

We now give some results about property (K) (or property (K, γ)) and some intersection properties, which are defined below, extending the notion of binary intersection considered above.

DEFINITION 8.3.27. Let (X, d) be a metric space and let γ be a cardinal. We say that the metric space X has a $(\gamma, 2)$ intersection property if for each set of balls $B(x_i, r_i)$, $i \in I$, card $I = \gamma$ such that $B(x_i, r_i) \cap B(x_j, r_j) \neq \emptyset$ for all i, j in I, then $\cap_{i \in I} B(x_i, r_i) \neq \emptyset$.

A metric space X is said to have a finite-2 intersection property if it has a $(\gamma, 2)$ intersection property for all finite γ. A metric space X is said to have a binary intersection property if property $(\gamma, 2)$ holds for all cardinal numbers γ.

THEOREM 8.3.28. *If X is a metric space with a $(\gamma, 2)$ intersection property then it has property (K, γ).*

Proof. If $(x_i)_{i \in I}$, $(y_i)_{i \in I}$ are two families of elements of X, where I is an index set of cardinality, then if $\cap_{I \in I} B(x_i, r_i)$ is nonempty we show that $\cap_{I E I} B(y_i, r_i)$ is nonempty when $d(y_i, y_j) \geqslant d(x_i, x_j)$ for all i, j in I.

First we remark that, given two balls $B(y_i, r_i)$ and $B(y_j, r_j)$, their intersection is not empty. Indeed, if x is any point in $B(x_i, r_i) \cap B(x_j, r_j)$ then

$$d(y_i, y_j) \leqslant d(x_i, x_j) \leqslant d(x_i, x) + d(x, x_j) \leqslant r_i + r_j.$$

Now using the binary intersection property we conclude that $B(y_i, r_i) \neq \emptyset$.

COROLLARY. 8.3.29. *Let X be a metric space having a binary intersection property. Then X has property (K).*

For more information about intersection properties we refer to Lindenstrauss (1964), Kelley (1952), Goodner (1950) and Klee (1963).

If we require the space to have an extension property for isometries then the following example shows that we must restrict ourselves to the finite dimensional case.

EXAMPLE 8.3.30. Let l^2 be the space of all sequence of real numbers $x = (x_i)_1^\infty$ with the property that

$$\|x\|^2 = \sum_{i=1}^{\infty} |x_i|^2 < \infty.$$

We consider the set D in l^2 of all elements x with $x_1 = 0$. On D we define the mapping $T\colon D \to l^2$ by the relation

$$Tx = (x_2, x_3, \ldots)$$

when $x = (x_1, x_2, \ldots)$. It is clear that T coincides with the adjoint of the shift operator. Then T is an isometry and cannot be extended to an isometry $\tilde{T}$ to the element $(1, 0, 0, \ldots)$.

Concerning finite dimensional spaces we have the following result.

LEMMA. 8.3.31. *Let V be a finite dimensional inner product space and let D be a subset in V. Then any isometric mapping $T\colon D \to V$ can be extended to an isometry of the linear subspace generated by D.*

Proof. Let a be a fixed element in D. We define an isometry on $D - a = (x - a, x \in D)$ by the formula

$$T_1 x = T(x + a) - Ta.$$

We remark that 0 is in $D - a$ and that $T_1 0 = 0$. Now we define T_1 on the space generated by the elements in $D - a$ by the formula

$$T_1(a_1 x_1 + \cdots + a_n x_n) = a_1 T_1 x_1 + \cdots + a_n T_1 x_n \cdot a_i \in R.$$

Then it is easy to see that T_1 so defined is a linear isometry from the linear space generated by $D - a$ to a linear subspace generated by $T_1(D - a)$. Since T_1 is an isometry, the dimensions of the domain and the range are the same. On the orthogonal complements, which have the same dimension, we can extend P_1 in an obvious way. Thus we have an extension of the original isometry on the whole space, which can be obtained in an obvious way from the extension of T_1.

As we have seen, in general, pairs of Banach spaces do not have the extension property.

We show now that, in a special case, if the extension property holds for the pair (X, X) then X is necessarily a complete inner product space. First we give a result of Thompson *et al.* (1967) concerning nonexpansive mappings on strictly convex spaces. For the reader's convenience we recall that a Banach space is said to be strictly convex if, whenever x, y are elements of X and $\|x\| = \|y\|$, then $\|x + y\| < \|x\| + \|y\|$.

PROPOSITION 8.3.32. (Thompson *et al.*, 1967). *Let X be a strictly convex Banach space and $T\colon X \to X$ be a nonexpansive mapping. Suppose that there exists a subset D in X such that $T_D\colon D \to X$ is an isometry. Then for any $x_1, \ldots, x_m$ in D and any a_i in*

$(0,1)$ *such that* $a_1 + \cdots + a_m = 1$ *we have*

$$T(a_1 x_1 + \cdots + a_m x_m) = a_1 Tx_1 + \cdots + a_m Tx_m.$$

Here T_D *is the restriction of* T *to the subset* D.

Proof. Let x, y be two elements of D and let t be in $(0,1)$. Set $z = tx + (1-t)y$ and thus

$$\|Tx - Tz\| \leqslant \|x - z\|, \qquad \|Ty - Tz\| \leqslant \|y - z\|.$$

Then we have

$$\|x - z\| = (1-t)\|x - y\| = \|x - y\| - \|z - y\| \leqslant \|Tx - Ty\| - \|Tz - Ty\| \leqslant \|Tx - Tz\| \leqslant \|x - z\|$$

which implies that equality holds throughout. Thus we have

$$\|Tx - Tz\| = \|x - z\|, \qquad \|Tz - Ty\| = \|z - y\|$$

which imply that

$$\|Tx - Tz\| + \|Tz - Ty\| = \|x - z\| + \|z - y\| = \|x - y\| = \|Tx - Ty\|.$$

Since X is strictly convex we must have

$$Tx - Tz = (1-t)/t \cdot (Tz - Ty).$$

Thus if D consists of two elements the assertion of the proposition is proved.

We now use an induction argument to complete the proof. Suppose that the assertion of the proposition is true if D consists of i elements with $i \leqslant n$. We prove that the assertion of the proposition is true if D consists of $n+1$ elements.

Let $x_1, \ldots, x_n, x_{n+1}$ be elements in D and let $z = a_1 x_1 + \cdots + a_n x_n + a_{n+1} x_{n+1}$ with a_i in $(0,1)$ and $a_1 + \cdots + a_n + a_{n+1} = 1$. We consider the following elements in X:

$$u = a_1 x_1 + \cdots + a_{n-1} x_{n-1} + (1 - a_1 - a_2 - \cdots - a_{n-1}) x_n$$
$$v = a_1 x_1 + \cdots + a_{n-1} x_{n-1} + (1 - a_1 - a_2 - \cdots - a_{n-1}) x_{n+1}.$$

These are obviously convex combinations of elements of D and by the induction hypothesis we have:

$$Tu = a_1 Tx_1 + \cdots + a_{n-1} Tx_{n-1} + (1 - a_1 - \cdots - a_{n-1}) Tx_n$$
$$Tv = a_1 Tx_1 + \cdots + a_{n-1} Tx_{n-1} + (1 - a_1 - \cdots - a_{n-1}) Tx_{n+1}.$$

Also since for $n = 2$ the assertion of the proposition gives that

$$T(tu + (1-t)v) = tTu + (1-t)Tv \tag{*}$$

for all t in $(0,1)$. If we set $t = a_n/(1 - a_1 - \cdots - a_{n-1})$ then (*) is exactly the assertion of the proposition for $n+1$. Thus the proposition is proved. □

THEOREM 8.3.33 (Thompson *et al.*, 1967). *Let* X *be a strictly convex space. Then the pair* (X, X) *has property* (E) *if and only if* X *is an inner product space.*

Proof. If X is an inner product space then the pair (X, X) has property (E).

Suppose now that X is not an inner product space. Then by the E. R. Lorch characterization of inner product spaces there exists a pair of elements x, y in X such that for some $c \neq 0$ and $1, \|x\| = \|y\|$,

$$\|x - cy\| > \|y - cx\|. \tag{*}$$

Then we consider the set $D = (0, x, y)$ and define the mapping $T: D \to X$ by the relations

$$T0 = 0, \qquad Tx = y, \qquad Ty = x.$$

It is clear that T is nonexpansive on D. Suppose now that T has an extension $\bar{T}$ on the whole X which is again a nonexpansive mapping. It is easy to see that we can suppose in (*) that c is not in $(0, 1)$. Now the point $c^{-1}y$ is in X and as X is strictly convex and $c^{-1}y = c^{-1}y + (1 - c^{-1})0$ we have

$$\bar{T}(c^{-1}y) = c^{-1}Ty + (1 - c^{-1})T0 = c^{-1}x.$$

The nonexpansiveness of $\bar{T}$ then implies the relation

$$\|\bar{T}(c^{-1}y - \bar{T}x\| = \|c^{-1}x - y\| = c^{-1}\|x - cy\| \leqslant c^{-1}\|cy - x\| = \|y - c^{-1}x\|$$

which is a contradiction. Thus T does not have a nonexpansive extension on the whole X.

The case of arbitrary Banach spaces was considered by Grünbaum (1958) and Schönbeck (1967). The result of Grünbaum says that a two-dimensional Banach space has property $(K, 3)$ if and only if it is an inner product space or if the unit sphere is a parallelogram and the extension obtained by Schönbeck says that a finite dimensional Banach space has property $(K, 4)$ (equivalently property (K)) if and only if X is an inner product space or if the unit sphere is a parallelotop. Concerning the general situation of a pair of Banach spaces having an extension property, Schönbeck has proved that if a pair of Banach spaces (X, Y) has an extension property and Y is a Banach space such that $\dim Y \geqslant 2$ and Y is strictly convex, then X, Y are inner product spaces. In what follows we expose these results.

THEOREM 8.3.34 (Schönbeck, 1966, 1967). *Let (X, Y) be a pair of Banach space and suppose that Y is strictly convex. Then (X, Y) has property (E) if and only if X, Y are inner product spaces* ($\dim Y \geqslant 2$).

Proof. The necessity is clear from the results given above. In order to prove that the condition is sufficient we need some auxiliary results, which are also interesting. □

THEOREM 8.3.35 (Oman, 1969). *Let X be a normed linear space of dimension 2. Then X has a symmetric orthogonality in the sense of Birkhoff if and only if there exists a linear isometry $I: X \to X^*$ such that $(I(x))(x) = 0$, for all x in X.*

Proof. First we prove that the condition is necessary. Suppose that X has a symmetric orthogonality in the sense of Birkhoff. Let us consider x, y to be elements in X with $\|x\| = \|y\| = 1$ and $x \perp_B y$. We define the following mapping on X, I,

$$(I(ax + by))(cx + dy) = ad - bc.$$

Then it is clear that $(I(x))(x) = 0$. Suppose now that $(ax + by) \perp_B (cx + dy)$ holds.

Then there exists a linear functional on X, say $\bar{f}$, such that $\bar{f}(ax + by) = \|\bar{f}\|\|ax + by\|$. From the fact that X is two-dimensional, there exists only one functional (up to a constant) such that $f(cx + dy) = 0$ and thus

$$|(I(cx + dy))(ax + by)| = \|I(cx + dy)\|\|ax + by\|.$$

But $(cx + dy) \perp_B (ax + by)$ and thus

$$|(I(ax + by))(cx + dy)| = \|I(ax + by)\|\|cx + dy\|.$$

If we identify X with X^{**} the above relations imply $I(ax + by) \perp_B I(cx + dy)$ and $I(cx + dy) \perp_B I(ax + by)$. Then there exists a constant k such that

$$\|I(ax + by)\| = k\|ax + by\|.$$

Since

$$|(I(x))(ax + by)| = |b| \leqslant \|ax + by\|$$

we have that $\|I(x)\| \leqslant 1$ and since $|(I(x))(y)| = 1 = \|y\|$ we obtain that $\|I(x)\| = 1$. Then $k = 1$.

Thus the condition is necessary. Suppose now that there is an isometry $I: X \to X^*$ such that $(I(x)(x) = 0$ for all x in X. Suppose now that $x \perp_B y$ and from the two-dimensionality of X we obtain that

$$|(I(y))(x)| = \|I(y)\|\|x\| = \|x\|\|y\|.$$

But $(I(x + y))(x + y) = 0$ and thus

$$(I(x))(y) + (I(y))(x) = 0.$$

Hence

$$|(I(x))(y)| = |(I(y))(x)| = \|x\|\|y\|$$

which is the fact that $y \perp_B x$. □

THEOREM 8.3.36 (Oman, 1969). *If X is a normed linear space then the following assertions are equivalent*:

(1) *The orthogonality in Birkhoff's sense is symmetric.*

(2) *If x, y are elements of X with $\|x\| = \|y\| = 1$ and $x \perp_B y$ then $(ax + by) \perp_B (cx + dy)$ if and only if $|ad - bc| = \|ax + by\| \cdot \|cx + dy\|$.*

(3) *If x, y are in X, $\|x\| = \|y\| = 1$, $x + ay \perp_B y$, $y + bx \perp_B x$ then $\|x + ay\| = \|y + bx\|$.*

Proof. We have to consider only the case when $\dim X = 2$ since in the contrary case X is an inner product space by the results in Chapter 4.

(1)→(2) I being an isometry, the assertion follows.

(2)→(3) If x, y are as in (3) then

$$x = \|x + ay\| \cdot (x + ay)/\|x + ay\| - ay$$
$$y + bx = b \cdot \|x + ay\| \cdot (x + ay)/\|(x + ay)\| + (1 - ab)y.$$

We apply assertion (2) to the pair $(x + ay)/\|x + ay\|$ and y and thus

$$\|y + bx\| = \|x\|\|y + bx\| = \|x + ay\|(1 - ab) + ab\|x + ay\| = \|x + ay\|.$$

(3)→(1) Let x, y in X with $\|x\| = \|y\|, x \perp_B y$ and $y + bx \perp_B x$. Then

$$\|x\| = \|y + bx\| = 1 = \|y\|,$$

which shows that $y \perp_B x$. □

LEMMA 8.3.37 (Schönbeck, 1967). *Let (X, Y) be a pair of Banach spaces with Y strictly convex. If the pair (X, Y) has property (E) then if $x_i \in X$, $y_i \in Y$ $i = 1, 2$ and $\|x_1\| = \|y_1\|$, $\|x_2\| = \|y_2\|$, $\|x_1 - x_2\| = \|y_1 - y_2\|$ then for all real numbers a, b $\|ay_1 + by_2\| \leqslant \|ax_1 + bx_2\|$ holds.*

Proof. First we consider the case when a, b are positive numbers and $a + b = 1$. We consider the mapping T defined on the set $D = (0, x_1, x_2)$ with values in $(0, y_1, y_2)$ by the relations

$$T0 = 0 \qquad \text{and} \qquad Tx_i = y_i, \quad i = 1, 2.$$

Then clearly T is a nonexpansive mapping and there exists a nonexpansive extension of T denoted by $\bar{T}$. Consider the element $x = ax_1 + bx_2$ and thus for some y in Y we have $\bar{T}x = y$ and from the nonexpansiveness of $\bar{T}$ we have that the following relations hold:

$$\|y\| \leqslant \|x\| = \|ax_1 + bx_2\|$$
$$\|y - y_1\| \leqslant \|x - x_1\| = b\|x_1 - x_2\|$$
$$\|y - y_2\| \leqslant \|x - x_2\| = a\|x_1 - x_2\|.$$

But

$$\|y_1 - y_2\| = \|y_1 - y + y - y_2\| \leqslant (a + b)\|x_1 - x_2\| = \|y_1 - y_2\|$$

and thus, by the strict convexity of Y, we must have

$$y_1 - y = (\|y - y_1\|/\|y - y_2\|(y - y_2) = b/a\cdot(y - y_2)$$

and this gives that

$$y = ay_1 + by_2.$$

If a, b are not as here, then we can consider multiples of a and b and the mappings defined as follows:

$$0 \to 0$$
$$-x_1 \to -y_1$$
$$-x_1 + x_2 \to -y_1 + y_2$$

and

$$0 \to 0$$
$$-x_2 \to -y_2$$
$$-x_2 + x_1 \to -y_2 + y_1.$$

□

LEMMA 8.3.38 (Schönbeck, 1967; Wells and Williams, 1975). *Let X be a two-*

dimensional normed linear space and let $\|,\|$ *be a norm on* X. *Then*

(1) *there exists* x, y *in* X *with* $\|x\| = \|y\| = 1$ *and* $x \perp_B y$, $y \perp_B x$,
(2) *if* $x \in X$ *then there exists* $y \neq 0$ *in* X *such that* $x \perp_B y$,
(3) *there exists* x, y *in* X *such that* $\|x\| = \|y\| = 1$ *and for all* $a \neq 0$, $\|x + ay\| > 1$,
(4) *if* x, y *are in* X *and* (r_n) *is a sequence of positive numbers converging to zero then there exists the sequences* (x_n) *and* (y_n) *in* X *with the following properties*:

(1′) $\|x_n\| = \|y_n\| = 1$

(2′) $\lim x_n = x, \quad \lim y_n = x, \quad x_n = (1 - r_n) y_n$.

Proof. We consider, without loss of generality, that $X = R_\times R$ and let S be the unit sphere of X. We define the papping $f: S_\times S \to R$ as follows: $f(u, v)$ is equal to the area of the parallelogram determined by u and v. Since $S_\times S$ is a compact subset there exists a point (u_0, v_0) where f attains its maximum. Then the straight line through u_0 parallel to v_0 cannot intersect the interior of C. Thus $u_0 \perp_B v_0$. Since the role of u_0 and v_0 is symmetric we have also $v_0 \perp_B u_0$.

The assertion in (2) was proved in arbitrary normed linear spaces in Chapter 4. We now consider a disc of least radius containing the set S, and for some point of contact of S with the boundary of the disc, say x, the point y which lies on the boundary of the disc and parallel to the tangent to S at x satisfies the relation $\|x + ay\| > 1$. Thus (3) is proved.

For (4) we consider the function $g(t) = \|x + ty\|$ which is either strictly increasing for $t \geqslant 0$ or there exists $t_0 > 0$ such that $g(t) = 1$ for $0 \leqslant t \leqslant t_0$.

If the first situation does not arise then if t_1 is in $0, t_0$ we have

$$1 \leqslant \|x + t_1 y\| = \|x + t_0 y\| = r.$$

Let a be in $(0, 1]$ with the property that

$$(1 - a)/(1 + (r - 1)a) = t_1/t_0.$$

Then we have clearly the relation

$$\|arx + (1 - a)(x + t_0 y)\| \leqslant ar + (1 - a)r = r$$

and

$$\|arx + (1 - a)(x + t_0 y)\| = \|(1 + (r - 1)a)\|x + t_1 y\| = (1 + (r - 1)a)r$$

which implies that $1 + (r - 1)a \leqslant 1$. This gives that $r = 1$ and thus $g(t) = 1$ for t in $[0, t_0]$. This gives that the second situation occurs and in this case we set $x_n = x$ for all n and $y_n = (x + r_n y)/ x + r_n y$. Clearly $\|x_n\| = \|y_n\| = 1$ and $\lim x_n = \lim y_n = x$. If we set $r_n \leqslant t_1$ then $y_n - x_n = r_n y_n$.

Now if the first situation occurs we consider the function

$$h(t) = g^{-1}(\|x - ty\|)$$

defined on R^+. Then the function

$$f(t) = (h(t)y + ty)/\|x - ty\| = (h(t) + t)/\|x - ty\|$$

is continuous, $f(0) = 0$ and $\lim_{t \to \infty} \inf f(t) = 2$. Then there exists a sequence of positive numbers, say (t_n), converging to zero such that $u(t_n) = r_n$, for all n sufficiently large. We set

$$x_n = (x - t_n y)/\|x - t_n y\|, \quad y_n = (x + h(t_n)y)/\|x - t_n y\|$$

and then we have that

$$\lim x_n = \lim y_n = x, \quad \|x_n\| = 1$$

and

$$\begin{aligned}\|y_n\| &= \|(x + h(t_n)y)\|/\|x - t_n y\| \\ &= \|(x + g^{-1})\|(\|x - t_n y\|y)/\|x - t_n y\| = 1.\end{aligned}$$

Also

$$y_n - x_n = f(t_n) = r_n y$$

for large n. This completes the proof of the lemma. □

LEMMA 8.3.39 (Schönbeck, 1967). *Let (X, Y) be a pair of Banach spaces satisfying the conditions of Lemma* 8.3.37. *Then if $x_i \in X, y_i \in Y, i = 1,2$, are unit elements such that $x_1 \perp_B x_2, y_1 \perp_B y_2$ then for all real numbers a, b we have the inequality*

$$\|ay_1 + by_2\| \geqslant \|ax_1 + bx_2\|$$

and the orthogonality in Birkhoff's sence is symmetric in both spaces.

Proof. From the above lemma there exist the sequences $(x_n^1), (x_n^2)$, (y_n^1) and (y_n^2) in the spaces generated by x_1, x_2 and y_1, y_2 respectively such that

(1) $\lim x_n^1 = \lim x_n^2 = x_1, \quad \lim y_n^1 = \lim y_n^2 = y_1,$

(2) $\|x_n^1\| = \|x_n^2\| = \|y_n^1\| = \|y_n^2\| = 1$ for all n,

(3) $x_n - x_n = 1/n \cdot x_2, \qquad y_n^2 - y_n^1 = 1/n \cdot y_2$ for large n.

Then Lemma 8.3.37 asserts that for all real numbers a and b we have the inequality

$$\|ay_n^1 + by_n^2\| \geqslant \|ax_n^1 + bx_n^2\| \quad \text{for all large } n.$$

This obviously implies our relation of the lemma.

To prove that the orthogonality in Birkhoff's sense is symmetric in both spaces we take x_1, x_2 in X with $x_1 \perp_B x_2$ and $\|x_1\| = \|x_2\| = 1$. Then there exists y_1 and y_2 in Y such that $y_1 \perp_B y_2$ and $y_2 \perp_B y_1$, $\|y_1\| = \|y_2\| = 1$. Then we have the relations:

$$\|x_2 + tx_1\| = \|y_2 + ty_1\| \geqslant \|y_2\| = \|x_2\|$$

which gives that $x_2 \perp_B x_1$. Thus the orthogonality in Birkhoff's sense is symmetric in X. To prove that this is true in Y take y_1, y_2 in Y with $y_1 \perp_B y_2$ and $\|y_1\| = \|y_2\|$. Then there exists x_1, x_2 in X with the properties

(1) $\|x_1\| = \|x_2\| = 1$

(2) $x_1 \perp_B x_2, \quad x_2 \perp_B x_1,$

(3) $\|x_1 + tx_2\| \geqslant 1$ for all $t \neq 0$.

Then

$$\|x_2 + t(ax_1 + bx_2)\| = \|tax_1 + (1 + tb)x_2\| \geqslant \|tay_1 + (1 + tb)y_2\|$$
$$= \|y_2 + t(ay_1 + by_2)\| = \|y_2\| = 1.$$

This gives that $x_2 \perp_B (ax_1 + bx_2)$ and from the fact that the orthogonality in Birkhoff's sense is symmetric in X we obtain that

$$\|ax_1 + bx_2\| = \|ax_1 + bx_2 + (-bx_2)\| = |a|.$$

Since

$$\|ax_1 + bx_2\| = |a|\|x_1 + ba^{-1}x_2\| > |a|$$

for $b \neq 0$ we must have $b = 0$. Then this implies that $y_2 \perp_B y_1$ and the orthogonality in Birkhoff's sense is symmetric in Y.

Now we are in a position to prove Theorem 8.3.34. Let us consider x_i in X and y_i in Y such that

(1) $\|x_i\| = \|y_i\| = 1, \qquad i = 1,2,$

(2) $x_1 \perp_B x_2, y_1 \perp_B y_2$.

We can identify the subspaces generated by x_1 and x_2 and y_1 and y_2 with $R_\times R$ and then on $R_\times R$ we have two norms $\|,\|_1, \|,\|_2$ induced by the norms of X and Y restricted to the above subspaces. Then by Lemma 8.3.39 the orthogonality in Birkhoff's sense is symmetric with respect to the norms $\|,\|_1$ and $\|,\|_2$. Also we have

$$\|(a,b)\|_2 \leqslant \|(a,b)\|_1$$

and

$$\|(a,b)\|_2 = |(ac - bd)|/\|(c,d)\|_2$$
$$\geqslant |ac - bd|/\|(c,d)\|_1 = \|(a,b)\|_1$$

and thus the equality of norms $\|,\|_1$ and $\|,\|_2$.

In the above proof we have considered only (c,d) which are orthogonal in Birkhoff's sense to (a,b). We can interchange y_1 and y_2 and y_1 and $-y_2$ and then for all a,b in R

$$\|(a,b)\|_1 = \|(b,a)\|_1 = \|(a,-b)\|_1.$$

Let us consider the point (c,d) such that $\|(a,b)\|_1 = \|(c,d)\|_1$. Let m,n be real numbers and then applying Lemma 8.3.37 we obtain that

$$\|m(a,b) + n(c,d)\|_1 = \|m(c,d) + n(a,b)\|_2 = \|n(c,d) + m(a,b)\|_1$$

and from the symmetry we have that

$$\|m(a,b) + n(c,d)\|_1 = \|m(c,d) + n(a,b)\|_1.$$

Since m,n are arbitrary by Ficken's characterization of inner product spaces $\|,\|_1$ is generated by an inner product. Then clearly $\|,\|_2$ is generated by an inner product. Thus we see any two-dimensional subspace of X and Y is an inner product space which implies that X and Y are inner product spaces. □

REMARK 8.3.40. It may be of interest of know if a proof of Theorem 8.3.34 may be given using the ideals of operators as we characterize inner product spaces in terms of absolutely summing operators.

We now mention some further interesting results about the problems considered above.

THEOREM 8.3.41 (Grünbaum, 1958; Schönbeck, 1967; Wells and Williams, 1975). *Let X be a separable Banach space. Then X has an extension property iff the norm of X is generated by an inner product or X has a binary intersection property.*

This result was obtained for $\dim X = 2$ by B. Grünbaum, for the case when $\dim X < \infty$ by S. Schönbeck and in the form given above by J. H. Wells and L. R. Williams. R. C. O'Brien (1977) has removed the restriction on the separability on X. For a proof of these results and the improvement, we refer to the book by Wells and Williams and the paper by O'Brien.

8.4. Fixed Point Sets for Nonexpansive Mappings and Inner Product Structures

In this section we present a characterization of the fact that the norm of a Banach space is generated by an inner product using fixed point sets for nonexpansive mappings.

If X is a Banach space and C is a closed subset in X then a mapping $T: C \to X$ is said to be nonexpansive if for all x, y in C,

$$\|Tx - Ty\| \leqslant \|x - y\|.$$

The fixed point set of T is the subset of C defined by the relation

$$F(T) = \{x, x \in C, Tx = x\}.$$

For more information about fixed point results for nonexpansive mappings we refer to Istrăţescu (1981b).

In order to formulate the first result we need:

DEFINITION 8.4.1. We say that a Banach space has a fixed point property for nonexpansive mappings if whenever bounded, closed and convex subset C of X and any nonexpansive mapping $T: C \to C$ the set $F(T)$ is nonempty.

Then we have the following result of Belluce and Kirk (1967).

THEOREM 8.4.2 (Belluce and Kirk, 1967). *Let X be a reflexive Banach space with a fixed point property for nonexpansive mappings. Then the following assertions are equivalent:*

(1) *The norm of X is generated by an inner product.*
(2) *Every closed, convex set in X is exactly the convex closure of $F(T)$ for some T, nonexpansive.*

Proof. If the norm of X is generated by an inner product then every closed, convex set is the convex closure of a fixed point set of a nonexpansive mapping. For this it suffices to consider the nearest point map associated with the set.

Conversely, suppose that a Banach space X has the stated property. Let M be a set whose convex closure is the set K (M is closed). Let x be in $C(M, X) = (x, x \in M$, $r_x(M) = r(M, X)$ where $r_x(U) = \sup(\|x - y\|, y \in U)$ and $r(U, V) = \inf(r_x(U), x \in V)$ and thus for each z in M we have

$$\|Tx - z\| \leqslant \|Tx - Tz\| \leqslant \|x - z\| \leqslant r_x(M)$$

which gives further that

$$r_{Tx}(M) \leqslant r_x(M) = r(M, X).$$

Thus Tx is in $C(M, X)$. But X has a fixed point property for nonexpansive mappings and thus

$$C(M, X) \cap M \neq \emptyset.$$

But

$$C(M, X) = C(K, X)$$

and thus

$$C(K, X) \cap K \neq \emptyset$$

for each bounded, closed convex subset K of X. Then by Garkavi's theorem (Ch. 4.14) the norm of X is generated by an inner product. Related to these is the problem about nonexpansive retractions. We recall that if C is a closed convex subset of a Banach space X then a nonexpansive retraction is a nonexpansive mapping $R: X \to C$ with the property that $Rx = x$ if x is in C. Then Burck Jr. (1974) has proved that if the dimension of X is at least 3 then the norm of X is generated by an inner product iff there exists a bounded smooth nonexpansive retract of X with a nonempty interior. For explanation and a proof we refer to Burck (1974).

We now mention two interesting results about retractions and inner product structures. For a proof we refer to Wells and Williams (1975).

THEOREM 8.4.3 (Wells and Williams, 1975). *Let X be a real finite dimensional normed space which is strictly convex.*

Suppose that at each point of the unit ball there exists a unique support hyperplane. Then on X there exists a nonexpansive retraction on the unit ball iff the norm of X is generated by an inner product.

THEOREM 8.4.4 (Grünbaum, 1958). *Let X be a Banach space. Then for every subset S of X and any nonexpansive mapping $T: S \to X$ there exists an extension to X with the range in the closed convex hull of $T(S)$ iff X is a space with the norm generated by an inner product or X is two-dimensional and has the binary intersection property.*

We now present a result about the existence of inner product structures on normed linear spaces using the class of monotone operators. The result was obtained by Cernes (1974). In order to formulate this result we need some notions about certain classes of mappings on Banach spaces (or subsets of Banach spaces).

DEFINITION 8.4.5. Let X be a Banach space and let X^* be the dual space of X. A duality mapping is a mapping $J: X \to P(X^*)$ defined by

$$J(x) = (x^*, x^* \in X^*, x^*(x) = \|x\|^2).$$

DEFINITION 8.4.6. A mapping $A: D_A \to P(X)$ defined on a subset of a Banach space D_A and taking the values in the collection of all subsets of X, is called accretive if for all $a > 0$ and all x, y in D_A the inequality

$$\|(x + au) - (y + av)\| \geqslant \|x - y\|$$

holds for all $u \in Ax$ and $v \in Ay$. For simplicity, we write this as

$$\|(x + aAx) - (y + aAy)\| \geqslant \|x - y\|.$$

It is not difficult to prove that a mapping TA is accretive if and only if $\langle u - v, x^* \rangle \geqslant 0$ for all x, y in D_A, all $u \in Ax, v \in Ay$ and some x^* in $J(x - y)$. For simplicity, we write this as

$$\langle Ax - Ay, x^* \rangle \geqslant 0.$$

If X is an inner product space then we call also an accretive mapping, monotone. If $A: D_A \to P(X)$ then the graph of $A, G(A)$ is the subset of $X_\times X$, defined as follows:

$$G(A) = ((x, y), x \in D_A, y \in Ax).$$

DEFINITION 8.4.7. A subset G in $X_\times X$ is called accretive if for all pairs $(x, u), (y, v)$ in G

$$\|(x + au) - (y + av)\| \geqslant \|x - y\|$$

holds for all $a > 0$.
Then a mapping A is accretive if and only if its graph is an accretive subset.

DEFINITION 8.4.8. A mapping A is said to be maximal accretive if there is no accretive set G in $X_\times X$ containing properly the graph of A.

DEFINITION 8.4.9. A mapping $A: D_A \to P(X)$ is called m-accretive if the following conditions hold:

(1) A is accretive.
(2) The range of $I + aA, R(I + aA)$ is X for all $a > 0$.

(The range $R(I + aA)$ is the set $(x + au, x \in Au,\ x \in D_A, u \in Ax)$.

There is a strong connection between monotone mappings, accretive mappings and nonexpansive mappings. For such a connection we refer to books in fixed point theory (e.g. Istrăţescu, 1981b.)

We now mention some results of A. Cernes (for a proof see his Ph. Thesis).

THEOREM 8.4.10 (Cernes, 1972). *If X has dimension 2 and m-accretiveness and maximal accretiveness are equivalent, then X is either an inner product space or the unit sphere is a parallelogram.*

THEOREM 8.4.11 (Cernes, 1972). *Let X be a uniformly convex space of infinite* *dimension. Suppose that X is not an inner product space. Then there exist maximal accretive mappings which are not m-accretive.*

8.5. Support Mappings and Inner Product Structures

In this section we present some results about support mappings, the smoothness of the norm of a Banach space and inner product structures. We recall that if X is a Banach space with a dual X^* then a support mapping is any function g on $X - 0$ with values in $X^* - 0$ and with the following properties:

(1) If $\|x\| = 1$ then $\|g(x)\| = 1 = 1 = g(x)(x)$,
(2) For $t \geqslant 0$, $g(tx) = tg(x)$.

In what follows we adhere to the usual convention and denote a support mapping by $x \to f_x$. The interested reader can find more information about support mappings in Istrăţescu (1984). The first result we give is given by B. J. Pettis.

LEMMA 8.5.1 (Pettis, 1939). *Let X be a Banach space with a dual X^*. Suppose that X^* is strictly convex. Then for each $x \neq 0$ there exists a unique element f_x in X with the properties $f_x(x) = 1$, $\|f_x\| = \|x\|^{-1}$.*

Proof. We remark that existence clearly follows from the Hahn–Banach theorem. Suppose now that we have at least two such functionals. Let f_x be one and x^* be another. Then

$$u = \tfrac{1}{2}(f_x + x^*)$$

satisfies the same properties, which contradicts the strict convexity of X^*. Thus uniqueness is guaranteed.

LEMMA 8.5.2. *Let X be a Banach space with a dual X. Suppose that X is uniformly convex. Then the mapping $x \to f_x$ (which is obviously a support mapping) is continuous.*

Proof. We recall that the uniform convexity of a Banach space Z can be formulated as follows: for each $\varepsilon > 0$ there exists $\delta > 0$ such that whenever $\|z_1 + z_2\| \geqslant 2 - \delta$ then $\|z_1 - z_2\| < \varepsilon$. Now using this (equivalent formulation of) definition of uniform convexity we consider $\varepsilon > 0$ and δ exists from the definition of uniform convexity of X. If $\|x - y\| < \delta$ then

$$\|f_x + f_y\| + \|x - y\| \geqslant 2$$

and thus

$$\|f_x + f_y\| \geqslant 2 - \delta.$$

But X is uniformly continuous and thus we must have $\|f_x - f_y\|$. Thus continuity is proved. □

COROLLARY 8.5.3. *Let X be a Banach space with X^* a uniformly convex space. Then*

$$[x, y] = \begin{cases} \|x\|^2 f_x(y), & x \neq 0 \\ 0, & x = 0 \end{cases}$$

is a continuous linear functional for each x in X and for each y is a continuous function of x.

REMARK 8.5.4. It is clear that the above map is a semi-inner product in the sense of Lumer (1961).

A very useful result for what follows is the theorem of S. Mazur.

THEOREM 8.5.5 (Mazur, 1933). *Let X be a Banach space with X^* strictly convex. Then for all $\|x\| = \|y\| = 1$ we have*

$$f_x(y) = \mathrm{d}/\mathrm{d}t(x + ty)_{t=0}.$$

Proof. We remark that if $x = \pm y$ then the assertion is clear since

$$\mathrm{d}/\mathrm{d}t(\|x + ty\|) = \pm 1.$$

Suppose now that x, y are linearly independent. Then we consider the space generated by x and by x, y. We define the functional on the space generated by x by $g(tx) = t$. We can extend this to the subspace generated by x and y. Since X is strictly convex the extension is unique. From the usual proof of the Hahn–Banach theorem we get

$$\sup_t(-\|tx + y\| - t) = f_x(y) = \inf_t(\|tx + y\|).$$

If $a > b$ then

$$-\|ax + y\| + \|bx + y\| \leqslant \|(b - a)x\| = a - b.$$

Thus

$$-\|ax + y\| - a \leqslant \|bx + y\| - b$$

and thus

$$f_x(y) = \lim_{a \to -\infty}(-\|ax + y\| - \mathrm{a}) = \lim_{a \to 0-}(-\|a^{-1}x + y\| - a^{-1})$$
$$= \lim_{t \to 0-} t^{-1}(\|x + ty\| - \|x\|)$$

and the assertion of the theorem is proved. □

We can now formulate the characterization of the inner product structures on X.

THEOREM 8.5.6 (Ritt, 1955). *Let X be a Banach space whose dual X^* is uniformly convex. Then the following assertions are equivalent*:

(1) *If $\|x\| = \|y\| = 1$ then $f_x(y) = f_y(x)$.*
(2) *$[x, y] = [y, x]$ for all x, y in X.*
(3) *$[,]$ is an inner product space on X.*

Proof. Suppose that (1) holds. Then since for each $t \neq 0$, $f_{tx} = t^{-1}f_x$, we have

$$[x, y] = \|x\|\|y\|f_{x/\|x\|}(y/\|y\|)$$
$$= \|x\|\|y\|f_{y/\|y\|}(x/\|x\|) = [y, x].$$

Thus (1) → (2).

Obviously (2) → (1). Suppose now that (2) holds. Then

$$[x + y, z] = [x, z] + [y, z], \tag{1'}$$
$$[x, x] = \|x\|^2, \tag{2'}$$
$$[x, y] = [y, x], \tag{3'}$$
$$[tx, y] = [t[x, y]. \tag{4'}$$

Thus (2) → (3). Obviously (3) → (2). The theorem is proved. □

We present now a result of P. M. Miličić which requires a continuity condition for the function $\tau(x, y)$ (considered earlier, see Section 4.12). For this purpose we consider two functions which are related to $\tau(x, y)$, and are defined as.

$$\tilde{\tau}_{\pm}(x, y) = \|y\| \lim_{t \to 0\pm} t^{-1}(\|y + tx\| - \|y\|).$$

Then we have the following useful result.

THEOREM 8.5.7 (Miličić, 1973). *Suppose that a Banach space X has the property that*

$$\lim_{s \to 0} \tilde{\tau}_{+}(x, y + sx) = \tilde{\tau}_{+}(x, y).$$

Then

$$\tilde{\tau}_{+}(x, y) = \tau_{-}(x, y).$$

Proof. First we remark that

$$\tilde{\tau}_{+}(x, y) = \tau(y, x),$$

where τ is the function used earlier. We note that the functions $\tilde{\tau}_{\pm}(x, y)$ have the following properties:

(1) $\tilde{\tau}_{\pm}(tx, y) = \tilde{\tau}_{\pm}(x, ty) = t\tilde{\tau}_{\pm}(x, y), \quad t \geqslant 0,$
(2) $\tilde{\tau}_{\pm}(tx, y) = \tilde{\tau}_{\pm}(x, ty) = t\tilde{\tau}_{\mp}(x, y), \quad t \leqslant 0,$
(3) $\tilde{\tau}_{\pm}(x, x) = \|x\|^2,$
(4) $|\tilde{\tau}_{\pm}(x, y)| \leqslant \|x\| \|y\|,$
(5) $|\tilde{\tau}_{\pm}(x + y, z)| \lesseqgtr \tilde{\tau}_{\pm}(x, z) + \tilde{\tau}_{\pm}(y, z),$
(6) If $[u, v] = \frac{1}{2}(\tilde{\tau}_{+}(u, v) + \tilde{\tau}_{-}(u, v))$ then
$[iu, v] = -[u, iv].$

Since the proof of these relations follows from the properties of the function τ they are omitted.

Now the proof of the assertion of the theorem is as follows. Suppose first that $x \neq 0 \neq y$ and set $u = x/\|x\|, v = y/\|y\|$. Then

$$\lim_{s \to 0^+} (x, y + sx) = \|x\| \|y\| \lim_{s \to 0} \tau_{+}(u, v + \|x\|/\|y\| su) =$$
$$= \|x\| y\| \tilde{\tau}_{+}(u, v) = \tilde{\tau}_{+}(x, y).$$

From the hypothesis we conclude then that

$$\lim_{s\to 0} \tilde{\tau}_+(x, y + sx) = \tilde{\tau}_+(x, y).$$

If in (5) we set $y = -x$ then we get

$$\tilde{\tau}_+(0, y) \leqslant \tilde{\tau}_+(x, y) + \tilde{\tau}_+(-x, y)$$

and since $\tau_+(0, u) = 0$ we obtain that

$$\tilde{\tau}_+(x, y) \geqslant \tilde{\tau}_-(x, y).$$

Using the properties of $\tilde{\tau}_\pm$ we have, for $t < 0$,

$$\begin{aligned} t^{-1}(\|x + ty\| - \|x\|) &= (\|x + ty\|^2 + \|x\|\|x + ty\|)/(t\|x + ty\|) \geqslant \\ &\geqslant (\tilde{\tau}_-(x + ty, x + ty) - \tilde{\tau}_-(x, x + ty))/(t\|x + ty\|) \\ &= (\tilde{\tau}_-(x, x + ty) + \tilde{\tau}_-(ty, x + ty) - \tilde{\tau}_-(x, x + ty) \\ &\quad \times 1/(t\|x + ty\|) \geqslant \tau_-(ty, x + ty)/(t\|x + ty\|) \\ &\geqslant \tilde{\tau}_+(y, x + ty)/\|x + ty\| \end{aligned}$$

or, for $t \leqslant 0$,

$$(\|x + ty\| - \|x\|)/t \geqslant \tilde{\tau}_+(y, x + ty)/\|x + ty\|.$$

This gives that

$$\lim_{t\to 0^-} (\|x + ty\| - \|x\|)t^{-1} = \lim_{t\to 0^-} \tau_+(y, x + ty)/\|x + ty\|$$

and thus

$$\tau_-(x, y) \geqslant \tau_+(x, y)$$

which implies equality and the theorem is proved. □

We now show that two interesting results mentioned in Chapter 4, namely F. Ficken's theorem and R. C. James' theorem, follow from the above theorem.

COROLLARY 8.5.8 (Ficken, 1944). *Let X be a Banach space and suppose that for all real numbers r and all $\|x\| = \|y\|$ in X we have the relation*

$$\|x + ry\| = \|rx + y\|.$$

Then the norm of X is generated by an inner product.

COROLLARY 8.5.9. (James, 1947). *Let X be a Banach space over the reals. Suppose that for all x, y in X the following relations hold*:

$$\|x\| \lim_{s\to 0+} (\|x + sy\|)s^{-1} = \|y\| \lim_{s\to 0+} (\|y + sx\| - \|y\|)s^{-1}.$$

Then the norm of X is generated by an inner product.

It is obvious that Corollary 8.5.8 follows immediately from Corollary 8.5.9. We may suppose that X is over the complex numbers. Then we have clearly that

$$\tilde{\tau}_\pm(x, y) = \tilde{\tau}_\pm(y, x)$$

and since

$$|\tilde{\tau}_+(x,y) - \tilde{\tau}_+(x_0,y)| = \|y\| \left| \lim_{s\to 0+} (\|y+sx\| - \|y\|)s^{-1} - \lim_{s\to 0+} (\|y+sx_0\| - \|y\|)s^{-1} \right|$$

$$= \|y\| \left| \lim_{s\to 0+} (\|y+sx\| - \|y+sx_0\|)s^{-1} \right|$$

$$\leqslant \lim_{s\to 0+} (\|y+sx-y-sx_0\|)s^{-1}$$

$$= \|y\|\|x-x_0\|.$$

Thus $\tilde{\tau}_+(x,y)$ is continuous in x and the above relation implies that τ_+ satisfies the condition of Theorem 8.5.7. Then we have

$$[u,v] = [v,u]$$

for all u,v in X and

$$[u+u_1,v] = [u,v] + [u_1,v].$$

Using property (6) we see that $[\,,\,]$ is an inner product on X which generates the norm of X. The corollary is proved. □

For some interesting results about the case of generalized semi-inner products and functions like $\tilde{\tau}(x,y)$ we refer to P. M. Miličić's 1973 paper.

8.6. Smooth Functions on Banach Spaces and Inner Product Structures

In some problems it is very useful that real-valued functions can be uniformly approximated by infinitely differentiable functions. This is well known in the case of spaces of finite dimension; in spaces of infinite dimension the situation is quite complicated. For the case of spaces $C_{[0,1]}$ and l^1, Kurzweil (1954) has proved that real-valued Fréchet differentiable functions with a bounded nonempty support do not exist. Restrepo (1964, 1965) and M. I. Kadets have proved that if the Banach space X has the property that dens $X^* >$ dens X (if Z is a Banach space, dens Z is the minimum cardinal number of a dense subset of Z) then on X there does not exist a Fréchet-smooth norm and Leduc (1970) has proved, under the same assumption, that no continuously Fréchet differentiable function with bounded nonempty support exists on X.

Leach and Whitfield (1972) have proved that if dens $X^* >$ dens X then on X there exists an equivalent norm $\|\,,\|_1$ which has a uniformly discontinuous differential and that if a Banach space X has an equivalent norm which has a uniformly discontinuous differential then there do not exist any Fréchet differentiable functions with a bounded nonempty support.

After Leach and Whitfield, a norm p on a Banach space X is said to have a uniformly discontinuous differential p' if there exists $\varepsilon > 0$ such that for any x in X and $\eta > 0$ there exist u, x_1, x_2 in X with the property that $p(x-x_1) < \eta$, $p(x-x_2) < \eta$, $p(u) = 1$ and $(p'x_2 - p'x_1)u \geqslant \varepsilon$.

For information about smooth functions on Banach manifolds we refer to the 1966 paper of R. Bonic and J. Frampton. In that paper they mention a problem which has attracted much attention – the characterization of complete inner product spaces in terms of smooth functions – and they forward the conjecture that it is likely that if a Banach space X and its dual X^* (away from zero) have C^2-smooth norms then the norm of the space must be generated by an inner product.

The following result was obtained almost simultaneously in similar forms by Rao (1967), Bonic and Reis (1966) and Sundaresan (1966).

THEOREM 8.6.1. *Let X be a Banach space and suppose that away from zero the norms of X and of X^* are of class C^2. Then the norm of X is generated by an inner product.*

Another interesting characterization of the existence of an inner product on a normed linear space generating a norm which coincides with the original norm was obtained by Rao (1972) and consists in the use of such conditions as

$$\sup_{\|x\|=\|y\|=1} ((\|x+ty\| - \|y+tx\|)t^{-2}) < \infty.$$

Veits (1971) has obtained an interesting and important characterization in terms of the continuity of the second derivative of the function $x \to \|x\|^2$. A very interesting and important point of view was taken by Meshkov (1978) by considering the so-called Hölder smoothness. In the same paper V. Z. Meshkov develops some of the basic techniques used later by Makarov (1979) and by Fabian *et al.* (1983) who have proved some results about the existence of equivalent norms generated by inner product on Banach spaces, using a class of points called 'Lipschitz smooth points'. In what follows we present some of the results mentioned above about equivalent norms generated by inner products. We begin with a result.

THEOREM 8.6.2 (Rao, 1972). *Let X be a normed linear space. Then the norm of X is generated by an inner product iff:*

(1) away from zero either X or X^* have a twice differentiable norm and (respectively) one of the following conditions is satisfied:

(2) $$\sup_{t>0,\|x\|=\|y\|=1} ((\|x+ty\| - \|y+tx\|t^{-2}) < \infty,$$

(2′) $$\sup_{t>0,\|x^*\|=\|y^*\|=1} ((\|x^*+ty^*\| - \|y^*+tx^*\|)t^{-2}) < \infty.$$

Proof. If the norm of X is generated by an inner product then it is clear that the above conditions are satisfied.

Suppose now that X has the stated property, and we consider that the norm of X is twice differentiable and satisfies (2). Then for each point $x_0, \|x_0\| = 1$ we have, by Taylor's formula,

$$\|x_0 + ty\| = \|x_0\| + td_{x_0}(y_0) + t^2/2!d^2_{x_0}(y_0, y_0) + O(t^2)$$

where d_{x_0} and $d^2_{x_0}$ are Fréchet differentials of the norm of X at x_0. Also we have, similarly,

$$\|y_0 + tx_0\| = \|y_0\| + td_{y_0}(x_0) + t^2/2!d^2_{y_0}(x_0, x_0) + O(t^2)$$

and thus

$$\lim_{t\to 0} |((\|x_0 + ty_0\| - \|y_0 + tx_0\|)t^{-2}) - (d_{x_0}(y_0) - d_{y_0}(x_0))/t|$$
$$= \tfrac{1}{2}d^2_{x_0}(y_0, y_0) - d^2_{y_0}(x_0, x_0).$$

But (2) holds and thus we must have

$$d_{x_0}(y_0) = d_{y_0}(x_0).$$

But

$$d_{x_0}(x_0) = \|x_0\|^2$$

and thus $\langle x, y\rangle = d_x(y)$ is an inner product which satisfies the requirements of the theorem.

REMARK 8.6.3. It is possible to weaken the conditions in the theorem by an examination of the proof; for example, the full force of Fréchet differentiability was not essentially used and thus it suffices to have the Gateaux differential, as well as the second Gateaux differential, etc. See also Section 4.12.

We now present a result of B. E. Veits in which some smoothness conditions near to the origin are used.

THEOREM 8.6.4 (Veits, 1971). *Let X be a Banach space. Then a necessary and sufficient condition for X to have an equivalent norm generated by an inner product is that the function $x \to \|x\|^2$ must have a second (Fréchet) derivative uniformly continuous in a neighborhood of zero.*

For the proof we need some results which are also of independent interest. Let X be a Banach space and suppose that the norm

$$x \to \|x\| = p(x)$$

is Fréchet differentiable at each point away from zero. Then X_x is the closed subspace $d_{p_x}^{-1}(0)$.

If $x \to f(x)$ is a function defined on X which has a second Fréchet derivative $d^2_{f_x}$ at the point $x(\neq 0)$, then we say that f is a continuous (resp. uniformly continuous) derivative if

$$x \to d^2_{f_x}$$

is a continuous (resp. uniformly continuous) function. Then we have the following result.

THEOREM 8.6.5 (Sundaresan, 1967). *Let X be a Banach space. Then a necessary and sufficient condition for the existence on X of an equivalent norm generated by an inner product is that either of the following conditions hold:*

(1) $\displaystyle\inf_{h\in X_x, \|h\|=1} d^2_{f_x}(h,h) > 0,$

(2) X_x *with the norm* $u \to \|u\|_1^2 = d^2_{f_x}(u,u)$ *is a Banach space.*

Proof. First we prove that (1)→(2). Suppose that (1) holds and let

$0 < a = \inf d^2_{f_x}(u, u)$. Then

$$\|y\| \leqslant \|y\|_1 \leqslant (\|d^2_{f_x}\|^{1/2})\|y\|$$

and thus X_x with the norm $\| , \|_1$ is a Banach space. Conversely, suppose that (2) holds. Then the identity mapping $(X_x, \| , \|)$ and $(X, \| , \|_1)$ is continuous and thus we find $m, M > 0$, such that

$$m\|y\| \leqslant \|y\|_1 \leqslant M\|y\|$$

which implies that

$$d^2_{f_x}(y, y) \leqslant \|y\|_1^2 \leqslant m\|y\|^2$$

and thus (1) holds.

Now in order to prove the theorem we set

$$\|y\|_2^2 = d^2_{f_x}(y, y) + k^2\|x\|^2,$$

where

$$y = u + kx, \quad u \text{ in } X_x$$

is the unique decomposition of y. Then since the norm $\| , \|_2$ is generated by an inner product, the assertion of the theorem follows.

COROLLARY 8.6.6 (Sundaresan, 1967; Rao, 1967; Bonic and Reis, 1966). *Let X be a Banach space. Then on X there exists an equivalent norm generated by an inner product iff the norms of X and of X^* are twice differentiable (away from zero).*

Proof (Sundaresan, 1967). If X has a norm generated by an inner product then clearly the above conditions are satisfied. Conversely, suppose that X satisfies the above conditions. Then we consider $x, x^*, \|x\| = 1 = \|x\|$, the derivatives T_x and T_{d_x} of the norms

$$x \to \|x\|, \; x^* \to \|x^*\|.$$

at x and d_x (d_x being the derivative of the norm of X at x), taking into account that

$$d_{x^*}(d_x) = \pi x/\|x\|.$$

Here π is the canonical map of X into X^{**}. Let x be in X_x and since the norm of X is twice differentiable we have

$$\begin{aligned} d_{x^*}(d_{x+ty}) &= d_{x^*}(d_x) + T_{d_x}(d_{x+ty} - d_x) + \theta_x(t) \\ &= \pi x + T_{d_x}(tT_x(y)) + \varphi_x(t) + \theta_x(t) \\ &= \pi x + tT_{d_x}(T_x(y)) + T_{d_x}(\varphi_x(t)) + \theta_x(t) \end{aligned}$$

and

$$\lim_{t\to 0} \varphi_x(t)/t = \lim_{t\to 0} \theta_x(t)/t = 0$$

because

$$\lim_{t\to 0} \|t^{-1}(d_{x+ty} - d_x)\| = \|T_x(y)\|.$$

But

$$\begin{aligned} T_{d_x}(T_x(y)) &= \lim_{t\to 0} t^{-1}(\pi(x+ty)/\|x+ty\| - \pi x) \\ &= \lim_{t\to 0} t^{-1}(\pi(x+ty) - \|x+ty\|\pi x)/(\|x+ty\|) \\ &= \lim_{t\to 0} ((\pi x + t\pi y - (1 + td_x(y) + t^2 T_x(y,y) + \\ &\quad + \varepsilon(t)\pi x)/(t\|x+ty\|)) = \pi y. \end{aligned}$$

But π is an isometry and thus $\|T_{d_x}(T_x(y))\| = \|y\|$ and thus for y in $X_x \cap (u, \|u\| = 1)$ we have

$$\|T_x(y)\| \geqslant \|T_{d_x}\| - 1,$$

which gives the assertion of the corollary by the above theorem. Later we shall present a proof by Meshkov (1978) as an illustration of his methods.

Now the proof of Veits' theorem is as follows. Since

$$\|x+h\|^2 = \langle x+h, x+h\rangle = \|x\|^2 + \langle x, h\rangle + \langle h, x\rangle + \|h\|^2,$$

it follows that if the norm of X is generated by an inner product, the condition is necessary. Suppose now that the Banach space X has the property that the norm satisfies the condition of Veits.

Applying Taylor's formula (Dieudonné, 1960) we have

$$\begin{aligned} \|x+h\|^2 &= \|x\|^2 + 2\|x\| df_x(h) + (df_x(h))^2 + \\ &\quad + \|x\|^2 d^2 f_x(h,h) + w(x,h) \end{aligned}$$

with

$$\lim_{h\to 0} \|w(x,h)\|\,\|h\|^{-2} = 0.$$

Here df_x, d^2f_x are the Fréchet derivatives of the norm (of X at x). Since we suppose d^2f_x to be uniformly continuous in a neighborhood of zero, for $\varepsilon = \frac{1}{2}$ we find $\delta > 0$ such that for all x in $B_0(r)$ and all h in $B_0(\delta)$,

$$\|w(x,h)\| \leqslant 2^{-1}\|h\|^2.$$

Thus for h in $B_0(\delta) \cap X_{dx}$, where

$$X_{dx} = (u, d_{f_x}(\mathrm{u}) = 0)$$

we have

$$\begin{aligned} \|x+h\|^2 &= \|x\|^2 + \|x\| d^2_{f_x}(h,h) + w(x,h) \\ &\leqslant \|x\|^2 + \|x\| d^2 f_x(h,h) + \tfrac{1}{2}\|h\|^2 \end{aligned}$$

or

$$(\|h\| - \|x\|)^2 \leqslant \|x\|^2 + \|x\| d^2 f_x(h,h) + 2^{-1}\|h\|^2.$$

Let $r_1 = \min(r, 8^{-1}\delta, 1)$ and thus for x in $B_0(r_1)$, h in $B_0(\delta) \cap X_{dx}$,

$$\tfrac{1}{2}\|h\|^2 - \|h\|\,\|x\| \leqslant d^2 f_x(h,h)$$

and

$$\tfrac{1}{2}\|h\|^2 - 4^{-1}\delta\|h\| \leqslant d^2f_x(h, h).$$

In this case if

$$h \in X_{dx}, \|h\| = \delta$$

then

$$4^{-1}\delta^2 \leqslant d^2f_x(h, h)$$

and if h is in X_{dx}, $\|h\| = 1$ then x satisfies the inequality

$$4^{-1}\delta^2 \leqslant d^2f_x(h, h) \leqslant \delta^2 d^2f_x(h, h).$$

In this case for h in X_{dx} and $\|h\| = 1$ we have

$$d^2f_x(h, h) \geqslant 4.$$

Suppose now that $\|x\| \geqslant r_1/2$ and thus $xr_1/2\|x\| \in B_0(r_1)$, which gives that

$$d^2f_{r_1x/2\|x\|}(h, h) \geqslant 4^{-1}$$

and since, for $t \neq 0$,

$$d^2f_{tx}(h, h) = |t^{-1}|d^2f_x(h, h)$$

we get that for h in X_{dx} and $\|h\| = 1$

$$d^2f_x(h, h) = 2\|x\|/r_1 d^2f_{xr_1/2\|x\|}(h, h)$$
$$\geqslant \tfrac{1}{2}\|x\|r_1 \geqslant 4^{-1}$$

Thus from Theorem 8.6.5 the assertion of the theorem follows.

We shall now present the basic result of Meshkov (1978). For this we need some preliminary notions. We assume in what follows that X is a real Banach space and that differentiability means Fréchet's differentiability.

DEFINITION 8.6.7. Let $f: X \to R$ and $p \geqslant 1$ and let k be the greatest integer less than p. We say that f is an H^p-function if f is k-times continuously differentiable and for each x in X there exists a neighborhood O_x of x such that for all y, z in O_x,

$$\|d_y^k f - d_z^k f\| \leqslant C\|y - z\|^{p-k},$$

where C is a constant.

DEFINITION 8.6.8. We say that a Banach space X is H^p-smooth if there exists $f: X \to R$ which is H^p and $f(x) = 0$ for $\|x\| \geqslant 1$.

DEFINITION 8.6.9. We say that a Banach space X is C^r-smooth if on X there exists a C^r-function which has the property that $f(x) = 0$ for $\|x\| \geqslant 1$.

Now the result of V. Z. Meshkov is the following theorem.

THEOREM 8.6.10 (Meshkov, 1978). *Let X be a (real) Banach space. Then on X there exists an equivalent norm generated by an inner product iff X and X^* are C^2-smooth spaces.*

In order to prove this theorem we need auxiliary results, some of which are proved while others are only stated, and for the proof we refer to the original papers.

LEMMA 8.6.11 (Davis *et al.*, 1968). *Let X be a Banach space and let X_1 be a finite dimensional subspace of X. Then there exists for any $\varepsilon > 0$ a subspace X_2 of X of finite codimension and such that:*

(1) $X_1 \cap X_2 = (0)$,
(2) *The projection of $X_1 \oplus X_2$ onto X_1 has a norm less that $1 + \varepsilon$.*

Proof. Let $a > 0$. Since the set

$$(x, x \in X_1, \|x\| = 1)$$

is compact we find an a-net say $(u_1, \ldots, u_p)$. We take $x_1^*, \ldots, x_p^*$ in X with the properties

(1) $\|x_i^*\| = 1, \quad i = 1, 2, \ldots, p,$
(2) $x_i^*(u_i) = 1, \quad i = 1, 2, \ldots, p.$

We set

$$X_2 = (x, x \in X, x_i^*(x) = 0, i = 1, 2, \ldots, p).$$

Then clearly the codimension of X_2 is finite and for each x in X_1, $\|x\| = 1$, we have, for a suitable i,

$$\begin{aligned}\|x - y\| &\geqslant \|u_i - y\| - \|x - u\| > |x_i^*(u_i - y)| - a \\ &= |x_i^*(u_i)| - a = 1 - a\end{aligned}$$

and thus

$$d(X_1, X_2) = \inf(\|x - y\|, x \in X_1, \|x\| = 1, y \in X_2) > 1 - a.$$

In order to obtain the assertion of the lemma we choose $a = \varepsilon/(1 + \varepsilon)$.

LEMMA 8.6.12 (Meshkov, 1978). *Let X be a Banach space and let X_1 be a finite dimensional subspace of X. Then there exists a subspace X_2 of X^* of finite codimension such that for each y_2^* in X_2 there exists x in X such that*

$$y_2(x) \geqslant \tfrac{1}{3}\|y_2\|\,\|x\|.$$

Proof. Let $V = X_1^{\perp}$, and applying the above lemma with $\varepsilon = \frac{1}{2}$ we find a subspace of finite codimension in X^*, say Y, such that:

(1) $V \cap Y = 0$,
(2) $\|P_Y\| \leqslant \frac{5}{2}$.

Let y be in $Y$and let $\bar{x}$ be in X with $\|\bar{x}\| = 1$ and $\bar{x}(y) = \|y\|$. Such an element exists by the Hahn–Banach theorem. Then the functional $\bar{x}P_Y$ is defined on $Y + V$, and is zero on V. We can extend this to X^*. Applying Helly's theorem we find x in X with the properties

$$v(x) = 0,\ v \text{ in } V,\ y(x) = \|y\|,\ \|x\| \leqslant 3$$

Then clearly

$$y(x) \geqslant \tfrac{1}{3}\|y\|\,\|x\|.$$

LEMMA 8.6.13 (Meshkov, 1978). *Let X be a Banach space and let X_1 be a subspace of finite codimension in X^*. Then there exists a subspace U_1 of X of finite codimension such that for any x in U_1 there exists y in X_1 such that $y(x) \geqslant \tfrac{1}{3}\|x\|\,\|y\|(y \neq 0)$.*

Proof. If π is the canonical map of X into X then we apply the above lemma for the subspace

$$\pi^{-1}(\tilde{X} \cap \pi(X)),\ \tilde{X} \subset X^{**}\ (\tilde{X} \text{ given by Lemma 8.6.12}).$$

THEOREM 8.6.14 (Bessaga and Pelczyński, 1958). *Let X be a Banach space with the property that it does not contain a subspace isomorphic to c_0. Then every weakly absolutely convergent series in X converges unconditionally.*

THEOREM 8.6.15 (Bessaga and Pelczyński, 1958). *Let X be a Banach space with its dual X^*. Suppose that X^* contains a subspace isomorphic to c_0. Then X contains a complemented subspace isomorphic to c_0.*

For the reader's convenience we recall that a series $\Sigma\, x_n$ in a Banach space X is said to be weakly absolutely convergent if for each x^* in X^*, $\Sigma x^*(x_n)$ is absolutely convergent. The series Σx_n is said to be unconditionally convergent if for any sequence $\{\Sigma_n\}$, $\Sigma\varepsilon_n x_n$ converges, where $\varepsilon_n = \pm 1$.

LEMMA 8.6.16 (Meshkov, 1978). *Let X be a Banach space which is also H^p-smooth and X is H^q-smooth, where p, q are in $(1, \infty)$. Suppose that (x_n) is a sequence of points either of X or of X^* and with the property that the partial sums $\Sigma\, \varepsilon_i x_i$, are bounded. Then for any sequence (ε_n) the series $\Sigma\, \varepsilon_n x_n$ converges and the set $\{\Sigma_{n=1}^{p} \varepsilon_n x_n\}$ is compact. For this we need the following result of R. Bonic and J. Frampton.*

THEOREM 8.6.17 (Bonic and Frampton, 1966). *Let X be a Banach space which is H^p-smooth. Then X does not contain a subspace isomorphic to l^1.*

In this theorem p is in $(1, \infty)$.

The proof of Lemma 8.6.16 is as follows. Suppose that X contains a subspace isomorphic to c_0. Then we conclude that the dual X contains a subspace isomorphic to l^1. Suppose now that X contains a subspace isomorphic to c_0, then from Theorem 8.6.15 we conclude that X has a subspace isomorphic to l^1. Theorem 8.6.17 gives that neither X nor X^* has a subspace isomorphic to l^1. The compactness of the set follows from Theorem 8.6.14.

We now present an interesting result which assures the existence of an equivalent norm on a Banach space and which is generated by an inner product.

THEOREM 8.6.18 (Figiel and Meshkov, 1978). *Let X be a Banach space and suppose that there exist $Q_1, \ldots, Q_m$, $m < \infty$, bounded symmetric bilinear forms on X such that for all x in X the following inequality holds:*

$$\|x\|^2 \leqslant \sup(Q_i(x, x), i = 1, 2, \ldots, m).$$

Then on X there exists an equivalent norm which is generated by an inner product.

Proof. Let (r_i) be the sequence of Rademacher functions and let $(x_i)_1^n$ be a finite set of elements of X. Then, using the Hölder inequality, we get

$$\left(\int_0^1\left(\left\|\sum_1^n x_ir_i(t)\right\|^2\right)\mathrm{d}t\right)^2 \leqslant \int_0^1\left\|\sum_{i=1}^n x_ir_i(t)\right\|^4 \mathrm{d}t$$

$$\leqslant \int_0^1\left(\sum_{J=1}^m Q_j\left(\sum_{i=1}^n x_ir_i(t), \sum_{i=1}^n x_ir_i(t)\right)\right)\mathrm{d}t$$

and since

$$\int_0^1 Q_j^2\left(\sum_{i=1}^n x_ir_i(t), \sum_{i=1}^n x_ir_i(t)\right)\mathrm{d}t = \sum_{m,i,k,l=1}^n Q_j(x_i, x_k)Q_j(x_l, x_m)\int_0^1 r_i(t)r_k(t)r_m(t)\,\mathrm{d}t$$

$$= \left(\sum_{i=1}^n Q_j(x_i, x_i)\right)^2 + 2\sum_{\substack{k,l=1\\ k\neq l}}^n Q_j^2(x_k, x_l).$$

Further,

$$\int_0^1 Q_j^2\left(\sum_{i=1}^n x_ir_i(t), \sum_{i=1}^n x_ir_i(t)\right)\mathrm{d}t \leqslant 3\|Q_j\|^2\left(\sum_{i=1}^n \|x_i\|^2\right)^2.$$

We set $M = \sup\|Q_j\|$, $j = 1, 2, \dots, m$, and thus

$$\int_0^1\left\|\sum_{i=1}^n x_ir_i\right\|^2 \mathrm{d}t \leqslant (3m)^{1/2}M\left(\sum_{i=1}^n \|x_i\|^2\right)^2.$$

We remark now that

$$\sum_{i=1}^n \|x_i\|^2 \leqslant \sum_{i=1}^n\left(\sum_{j=1}^m |Q_j(x_i, x_i)|\right)$$

$$= m \max_j \sum_{i=1}^n |Q_j(x_i, x_i)| = m\sum_{i=1}^n |Q_{j_0}(x_i, x_i)|$$

where j_0 is in $[1, m]$.

If we set

$$A = (i, Q_{j_0}(x_i, x_i) \geqslant 0), B = [1, m] - A$$

then

$$\sum_{i\in A} Q_{j_0}(x_i, x_i) \geqslant -\sum_{i\in B} Q_{j_0}(x_i, x_i)$$

and thus

$$\sum_{i=1}^n \|x_i\|^2 \leqslant 2m\sum_{i\in A} Q_{j_0}(x_i, x_i).$$

We write

$$\{x_1, x_2, \dots, x_p\} = \{x_i, i\in A\}$$

changing the notation if necessary. Then

$$\sum_{i=1}^{n} \|x_i\|^2 \leqslant 2m\left(\sum_{i=1}^{k} Q_{j_0}(x_i, x_i)\right)$$

$$= 2m\int_0^1 Q_{j_0}\left(\sum_{i=1}^{k} x_i r_i(t), \sum_{i=1}^{k} x_i r_i(t)\right)dt$$

$$\leqslant 2mM \int_0^1 \left\| \sum_{i=1}^{k} x_i r_i(t) \right\|^2 dt$$

$$= 2mM \sum_{\varepsilon_i = \pm 1} 1/2^k \left\| \sum_{i=1}^{k} \varepsilon_i x_i \right\|^2$$

$$\leqslant 2mM \sum_{\varepsilon_i = \pm 1} 1/2^n \left\| \sum_{i=1} \varepsilon_i x_i \right\|^2$$

$$= 2mM \int_0^1 \left\| \sum_{i=1}^{n} x_i r_i(t) \right\|^2 dt$$

since for any u, v in X,

$$\|u + v\|^2 + \|u - v\|^2 \geqslant 2\|u\|^2.$$

Now if

$$\sum_{i\in A} Q_{j_0}(x_i, x_i) \leqslant -\sum_{i\in B} Q_{j_0}(x_i, x_i)$$

then

$$\sum_{i=1}^{n} \|x_i\|^2 \leqslant 2m \sum_{i\in B} (-Q_{j_0}(x_i, x_i))$$

and using the same procedure as above we get

$$\sum_{i=1}^{n} \|x_i\|^2 \leqslant 2mM \int_0^1 \left\| \sum_{i=1}^{n} x_i r_i(t) \right\|^2 dt$$

and then S. Kwapień's theorem [9.2.1] asserts the existence of an equivalent norm on X which is generated by an inner product.

COROLLARY 8.6.19 (Meshkov, 1978). *Let X be a Banach space and suppose that on X there does not exist an equivalent norm which is generated by an inner product. Then if Y is a subspace of finite codimension in X and $(Q_i)_1^m$ are bounded symmetric bilinear forms on X then for any $\varepsilon > 0$ there exists u in X, $\|u\| = 1$ such that for all $p = 1, \ldots, m$, $|Q_p(u, u)| < \varepsilon$.*

Proof. Obvious from Theorem 8.6.18. □

The following result is mentioned, for example, in S. Lang's book and may be proved using the case of real-valued functions and the Hahn–Banach theorem.

THEOREM 8.6.20. *Let X, Y be two Banach spaces and let U be an open set in X. Let x, y be elements of U and such that the segment $x + ty$, $0 \leqslant t \leqslant 1$, lies in U. Suppose that $f\colon U \to Y$ is a C^p-map. Let us denote by $y^{(p)}$ the element $(y, y, \ldots, y)$. Then $d^p_{f_{x+ty}}$ is continuous in t and the following formula holds:*

$$f(x + ty) = f(x) + df_x(y) + \cdots + d^{p-1}f_x(y^{(p-1)})/(p-1)! +$$

$$+ \int_0^1 (1-t)^{p-1}/(p-1)!\,d^p_{f_{x+ty}}(y^{(p)})\,dt.$$

COROLLARY 8.6.21. *Let X be a Banach space and let K be a compact set in X. Suppose that f is a C^2-function on X. Then*

$$f(x + h) = f(x) + df_x(h) + \tfrac{1}{2}d^2f_x(h, h) + r(f)(x, h)$$

with

$$\lim_{\|h\| \to 0} r(f)(x, h)\|h\|^{-2} = 0.$$

COROLLARY 8.6.22. *Let X be a Banach space and let K be a compact set. If f is a H^p-function on a neighborhood of K then there exists $K > 0$, $\delta > 0$ such that for $\|h\| < \delta$,*

$$f(x + h) = f(x) + df_x(h) + \tfrac{1}{2}!\,d^2f_x(h, h) + \cdots + 1/k!\,d^kf_x((h^{(k)}) + r(f)(x, h)$$

with

$$|r(f)(x, h)| \leqslant M\|h\|^p.$$

Proof of Theorem 8.6.10. Since X is supposed to be C^2-smooth, and by the definition of C^2-smoothness we find a nonconstant C^2-function, say g, which is zero for $\|x\| \geqslant 1$. Since g is nonconstant we can choose a point in $\|x\| < 1$ such that $g(u) = a \neq 0$. We now consider C^∞ on the real axis, say f, and with the property that $f(a) = 0$, $f(0) = 2$.

We now consider the function

$$\varphi(x) = f(g(2x + u)).$$

We construct similarly on X a function ψ such that $\psi(0) = 0$, $\psi(x^*) = 2$ when $\|x^*\| \geqslant 1$.

We define on $X \times X^*$ the following function (with real values)

$$\theta(x, x) = \varphi(x) + \varphi(x^*).$$

Then

(1) $\theta(0, 0) = 0$,
(2) $\theta(x, x^*) = 2$ when $\max(\|x\|, \|x^*\|) \geqslant 2$.

Suppose now that X has the property that on X there exists no equivalent norm which is generated by an inner product. Then we prove the following property of the function $\theta(x, x)$: if K and K^* are compact sets in X, respectively in X^*, and $F \subset K$, $F^* \subset K^*$ are finite sets, then $Q(F, F^*, H, G)$ is the set of all pairs (h, g) with the property that $\|h\| \leqslant \frac{1}{2}$, $\|g\| \leqslant \frac{1}{2}$ and

$$|\theta(x \pm h, x \pm g) - \theta(x, x)| \leqslant g(h)$$

for all x in F, x in F and where H, G are subspaces of finite codimension in X and X respectively. Let further

$$a(F, F, H, G) = \sup(g(h), (h, g) \in Q(F, F, H, G)).$$

Then there exists a constant denoted by $k(K, K^*)$ such that

$$a(F, F, H, G) \geqslant k(K, K) > 0.$$

Proof. We consider the following set

$$G_1 = G \cap \ker d\psi_{x^*}.$$

Then, according to Lemma 8.6.12 we find a subspace of finite codimension, say V of X, with the property that for every x in V we find $g \neq 0$ in G_1 such that

$$g(x) \geqslant \tfrac{1}{3}\|g\|\,\|x\|.$$

We now consider the set $H_1 = H \cap V \cap \ker df_\varphi$, and set

$$M = \sup(\|d^2 f_x\|, x^* \in K^*).$$

Since we may assume that $M \geqslant 1$ this is supposed to be the case in what follows. According to Corollary 8.6.21 we find $\delta > 0$ (and we may suppose this to be in $(0, \frac{1}{2})$ such that if $\|h\| \leqslant \delta$ and $\|g\| \leqslant \delta$, then

$$r(\varphi)(x, h) \leqslant 100^{-1} M^{-1}\|h\|^2, \qquad r(\varphi)(x^*, g) \leqslant 100^{-1} M^{-1}\|g\|^2$$

when x is in K and x^* in K^*.

From Corollary 8.6.19 we find h in H_1 such that $\|h\| = \delta$ and

$$\|d^2\varphi_x(h, h)| \leqslant 100^{-1} M^{-1}\delta^2$$

for all x in F. Then we choose g in G_1 with the properties that

$$\|g\| = (10\,M)^{-1}\delta, \qquad g(h) \geqslant \tfrac{1}{3}\|h\|\,\|g\| > (30\,M)^{-1}\delta^2.$$

Thus for x in F and x^* in F^* we have

$$\begin{aligned}|\theta(x \pm h, x \pm g) - \theta(x, x)| = {} & |\tfrac{1}{2}d^2\varphi_x(h, h) \\ & + \tfrac{1}{2}d^2\psi_{x^*}(g, g) + r(\varphi)(x, \pm h) + \\ & + r(\varphi)(x^*, \pm g)| \leqslant (200\,M)^{-1}\delta^2 \\ & + (200\,M)^{-1}\delta^2 + (100\,M)^{-1}\delta^2 + (100\,M)^{-1}\delta^2 \\ & \leqslant 3 \cdot 100\,M)^{-1}\delta^2 < (30\,M)^{-1}\delta^2 < g(h).\end{aligned}$$

If we set

$$k(K, K^*) = (30\,M)^{-1}\delta^2$$

then the property stated above is proved. We shall prove theorem 8.6.10 by obtaining a contradiction if we suppose that on X there does not exist an equivalent norm generated by an inner product. Indeed, we use Lemma 8.6.16 and the property proved above to construct inductively two sequences of elements (h_n) and (g_n) in X and X respectively.

For $n = 0$ we set $h = 0$ and $g = 0$. Suppose now that h_i and g_i are defined when $i \leqslant n$ and we shall define the elements h_{n+1} and g_{n+1}. Let

$$F_n\left(\sum_{i=0}^{n} \varepsilon_i h_i, \varepsilon_i = \pm 1, i = 0, 1, \ldots, n\right),$$

$$F_n = \left(\sum_{i=0}^{n} \varepsilon_i g_i, \varepsilon_i = \pm 1, i = 0, 1, 2, \ldots, n\right).$$

We set further

$$H_n = \ker \bigcap_0^n g_i, \qquad G_n = \bigcap_{i=0}^{n} (x^*, x^*(h_i) = 0),$$

$$Q_{n+1} = Q(F_n, F_n, H_n, G_n)$$

$$a_{n+1} = a(F_n, F_n, H_n, G_n).$$

We choose the pair (h_{n+1}, g_{n+1}) in Q_{n+1} with the property that $g_{n+1}(h_{n+1}) = (a_{n+1})/2$. Then the following properties are satisfied:

(1) $g_k(h_n) = 0$, $k \neq n$,
(2) $g_k(h_k) = 4^{-1}$.

We have also that the series $g_k(h_k)$ has a sum equal to 2. Indeed, suppose that the assertion is false and thus $\Sigma\, g_k(h_k) < 2$. Then

$$\theta\left(\sum_{i=0}^{p} \varepsilon_i h_i, \sum_{i=0}^{p} \varepsilon_i g_i\right) \leqslant \sum_{i=0}^{p-1}$$

$$\theta\left(\sum_{i=0}^{p+1} \varepsilon_i h_i, \sum_{i=0}^{p+1} \varepsilon_i g_i\right) -$$

$$-\, \theta\left(\sum_{i=0}^{n} \varepsilon_i h_i, \sum_{i=0}^{n} \varepsilon_i g_i\right)$$

$$\leqslant \sum_{i=1}^{p} g_i(h_i) < 2.$$

This implies that

$$\left\| \sum_{i=0}^{p} \varepsilon_i h_i \right\| < 1, \qquad \left\| \sum_{i=0}^{p} \varepsilon_i g_i \right\| < 1$$

for all p and all $\varepsilon_i = \pm 1$. According to Lemma 8.6.16 we find the compact set K and K^* in X and X^* respectively and F_n and F_n for all n. Then by the above property we have $a_n = k(K, K^*)$. But $g_n(h_n) \geqslant a_n/2 \geqslant k(K, K^*)/2 > 0$ for all n. This is a contradiction with the convergence of the series $g_n(h_n)$.

Suppose now that p has the property that

$$1 < \sum_{i=0}^{p} g_i(h_i) < 22.$$

According to the definition of the sequences (h_n) and (g_n) we have

$$\theta\left(\sum_{i=0}^{p} h_i, \sum_{i=0}^{p} g_i\right) \leqslant \sum_{i=0}^{p} g_i(h_i) < 2$$

and thus

$$\sum_{i=0}^{p} \|h_i\| < 1 \quad \text{and} \quad \sum_{i=0}^{p} \|g_i\| < 1 \quad \text{while} \quad \left(\sum_{i=0}^{p} g_i\right)\left(\sum_{i=0}^{p} h_i\right) > 1.$$

This contradiction proves the theorem. □

Related to the above result an interesting theorem was obtained by Fabian *et al.* (1983). Let X be a Banach space. A function $f: X \to R$ is said to be a bump function if f has a bounded nonempty support.

DEFINITION 8.6.23 (Fabian *et al.*, 1983). Let X be a Banach space. X is said to be locally uniformly smooth if on X there exists a bump function with df_x locally uniformly continuous. If df_x is locally Lipschitz then we say that X is locally Lipschitz smooth.

A Banach space X is said to be uniformly smooth if the above local smoothness is a global property; X is Lipschitz smooth if the above locally Lipschitz smooth property is global.

DEFINITION 8.6.24 (Fabian *et al.*, 1983). The point x_0, $\|x_0\| = 1$, is said to be a smooth Lipschitz point of X if when f_x denotes the support functional of $B_x(0)$ at x there exists a constant K such that for all h in X the following inequality holds:

$$\|x + h\| - \|x\| - f_x(h) \leqslant K\|h\|^2.$$

Then we have the following result.

THEOREM 8.6.25 (Fabian *et al.*, 1983). *Let X be a Banach space with the property that x_0 is a Lipschitz smooth point for X and the norm of X is twice Gateaux differentiable on the unit sphere of X. Then on X there exists an equivalent norm which is generated by an inner product.*

For a proof, as well as for other interesting related results, we refer to the paper by Fabian *et al.* (1983).

We remark that we can formulate S. O. Carlsson's condition on the existence of an inner product structure on a normed linear space (Banach space) as follows:

$$\|x\|^2 = \sum_{i=1}^{n} a_i\|b_i x + c_i y\|^2,$$

where a_i, b_i, c_i satisfy certain conditions or, more generally,

$$\|x\|^2 = G((\|b_i x + c_i y\|)_1^n),$$

where G is a function on R^n satisfying certain smoothness conditions. In his 1973 paper, D. A. Senechalle has considered such functions and has proved, under some assumptions on G, that the normed linear space (the Banach space) has an equivalent norm which is generated by an inner product, i.e. X is isomorphic to a (complete) inner product space. In what follows we present Senechalle's results. First we recall that a function f defined on an open set of X with values in Y (X, Y are Banach spaces or normed linear spaces) is said to be twice-differentiable at x_0 of the open set if there exist the mappings T in $L(X, Y)$ and R a bounded bilinear mapping such that

$$\lim_{\substack{\|t\|\to 0,\\ x+t\in G}} (f(x+t) - f(x) - Tt - \tfrac{1}{2}R(t,t))/\|t\|^2 = 0$$

(here G is the open set on which f is defined). Then we have the following result.

THEOREM 8.6.26 (Senechalle, 1973). *Let $f: X \to R$ be a twice-differentiable mapping at $x = 0$ and suppose that for $0 < a \leqslant b$ we have*

$$a^2\|x\|^2 \leqslant \lim_{r\to 0} f(rx)/r^2 \leqslant b^2\|x\|^2, \quad x \in X.$$

Then $I(X) = \inf(\|R\|, \|R^{-1}\|$, R is an isomorphism of X onto a complete inner product space) $\leqslant b/a$.

Proof. Suppose that T and R are linear (respectively bilinear) operators which exist since f is twice-differentiable at $x = 0$ and

$$\lim(f(x) - f(0) - T(0) - \tfrac{1}{2}R(0,0)) = 0$$

which gives that f is continuous at $x = 0$. Since $\lim f(rx)/r^2$ exists we must have $f(0) = 0$. Now

$$\begin{aligned} 0 &= \lim_{r\to 0} (f(rx) - rTx - \tfrac{1}{2}r^2R(x,x)/r^2) \\ &= \lim_{r\to 0} (f(rx)/r^2 - \tfrac{1}{2}R(x,x)) - \lim_{r\to 0} Tx/r. \end{aligned}$$

This clearly implies that $Tx = 0$ or $T = 0$. Thus

$$\lim_{r\to 0} f(rx)/r^2 = \tfrac{1}{2}R(x,x).$$

Now we define

$$\langle x, y\rangle = \tfrac{1}{4}(R(x,y) + R(y,x))$$

which is clearly an inner product on X and

$$a^2\|x\|^2 \leqslant \langle x, x\rangle \leqslant b^2\|x\|^2$$

gives the estimation of the theorem.

From the above theorem we obtain.

COROLLARY 8.6.27 (Senechalle, 1973). *Suppose that X is a Banach space with the property that $x \to \|x\|^2$ is twice-differentiable at $x \neq 0$. Then the norm of X is generated by an inner product.*

Another sufficient condition for a Banach space to be isomorphic to a complete inner product space is given in the following result.

THEOREM 8.6.28 (Senechalle, 1973). *Let X be a Banach space and suppose that for some y in X, $\|y\| = 1$, $x \to \|x\|$ is there twice-differentiable and that there exist real numbers b_i, $c_i \neq 0$, $i = 1, 2, \dots, n$ and a function defined on a subset M of R^n with values in R such that*

$$a^2\|x\|^2 = \lim_{t \to 0} G((\|b_i tx + c_i y\|)/t^2 \leqslant b^2\|x\|^2$$

for some $0 < a \leqslant b$ and all x in X. Then

$$I(X) \leqslant b/a.$$

Proof. We consider a Banach space $Y = \underbrace{X \times X \times \dots \times X}_{n-\text{times}}$ and set $f: R \to Y$, $g: Y \to R^n$, where

$$f(t) = (b_i tx + c_i y), \qquad g((u_i)) = (\|u_i\|).$$

Then f is twice-differentiable at 0 and g is twice-differentiable at $f(0)$.
Then we can apply Theorem 8.6.28 and since

$$a^2\|x\|^2 \leqslant \lim_{t \to 0} G_0 g_0 f(tx)/t^2 \leqslant b^2\|x\|^2$$

for all x in X we obtain that $I(X) = b/a$.

COROLLARY 8.6.29 (Senechalle, 1973). *Let X be a Banach space and suppose that at some y, $\|y\| = 1$ the norm $x \to \|x\|$ is twice-differentiable and that for some numbers $0 < a \leqslant b$*

$$\begin{aligned} a^2\|x\|^2 &\leqslant \lim_{t \to 0} (\|tx + y\|^2 + \|tx - y\|^2 - 2\|y\|^2)/2t^2 \\ &\leqslant b^2\|x\|^2 \end{aligned}$$

holds for all x in X. Then $I(X) \leqslant b/a$.

Proof. We have only to consider the function

$$G(t_1, t_2, t_3) = \tfrac{1}{2}(t_1^2 + t_2^2 - 2t_3^2)$$

and since this is twice-differentiable at $(1 \cdot 1, 1)$ the assertion follows from the above theorem.

REMARK 8.6.30. It is clear that the above corollary is an improvement of the Jordan–von Neumann theorem (the Apollonius parallelogram law).

COROLLARY 8.6.31 (Senechalle, 1973). *Let X be a two-dimensional Banach space with the property that for some y, $\|y\| = 1$, $x \to \|x\|$ is twice-differentiable there and for some pairwise linearly independent points $((b_i, c_i))$, $i = 0, 1, \dots, n$, there exists a function of R^n which is twice-differentiable at*

$$(|b_i c_0 - c_0 b_i|/(b_0^2 + c_0^2)^{1/2})$$

and such that for some numbers $0 < a \leqslant b$,

$$a^2\|b_0 p + c_0 q\|^2 \leqslant G((\|b_i p + c_i q\|)) \leqslant b^2\|b_0 p + c_0 q\|^2$$

holds for all p, q *in* X. *Then* $I(X) \leqslant b/a$.

Proof. We set

$$b_i' = (b_0 b_i + c_0 c_i)/(b_0^2 + c_0^2)^{1/2}, \quad i = 0, 1, 2, \ldots, n$$

$$c_i' = (b_0 c_i - c_0 b_i)/(b_0^2 + c_0^2)^{1/2}$$

and then $c_i' \neq 0$ with $(b_0', c_0') = (1, 0)$. If x is arbitrary in X then for

$$p = (b_0 x - c_0 y)/(b_0^2 + c_0^2)^{1/2}$$

$$q = (c_0 x + b_0 y)/(b_0^2 + c_0^2)^{1/2}$$

we have

$$a^2\|x\|^2 \leqslant a^2\|b_0 p + c_0 q\|^2 \leqslant G((\|b_i p + c_i q\|))$$

$$G((\|b'x + c'y\|)) \leqslant b^2\|b_0 p + c_0 q\|^2 \leqslant b^2\|x\|^2.$$

Then from the above theorem the assertion of the corollary follows.

COROLLARY 8.6.32 (Carlsson, 1962). *Let* X *be a Banach space with the property that for some* a_i, b_i, c_i *such that:*

(1) $a_i \neq 0$,
(2) $\Sigma_{i=1} a_i b_i^2 = \Sigma a_i b_i c_i = \Sigma a_i c_i^2 = 0$,
(3) $\Sigma a_i \|b_i x + c_i y\| \geqslant 0$, x, y *in* X,
(4) (b_i, c_i) *and* (b_j, c_j) *are linearly independent for* $i \neq j$.

Then the norm of X *is generated by an inner product.*

Proof. We apply the above corollary to X_1, X_1 being an arbitrary two-dimensional subspace of X and to the function

$$G(t_1, \ldots, t_n) = -a_0^{-1}(a_1 t_1^2 + \cdots + a_n t_n^2).$$

Then we obtain that the norm of X_1 is generated by an inner product and thus, by the Jordan-von Neumann theorem, the norm of X is generated by an inner product. □

In his interesting paper, Senechalle (1973) made some remarks about the necessity of twice-differentiability of the function G. He has remarked (Senechalle, 1971) that for $n = 3$ the twice-differentiability may be omitted and that for other values of n this is not the case (Senechalle, 1973).

8.7. Classes of Functions on Banach Spaces and Inner Product Structures

We have presented in Section 8.6 some important results concerning the existence of some functions on Banach spaces and inner product structures. In this section we present another class of results in which some properties of certain functions determine the existence of an inner product structure on the Banach space. In this respect it is worth mentioning the result of M. Nagumo in which certain properties are assumed to hold which are inspired from the properties of the center of gravity in Euclidan three-dimensional space. Another interesting result, related to some important new notions in quantum mechanics, was obtained by Č. V. Stanojevič and his students, who have introduced and extensively used the notion of a Mielnik space. Another result we mention is that of W. J. Stilles and L. Dor, which is connected with some of the properties of inversion. We also present the very interesting results of P. Enflo which characterize the existence of an equivalent norm generated by an inner product in terms of certain homeomorphisms; and, finally, we include an interesting result of L. N. Krovonsov.

Let X be a complex inner product space and let $x_1, \ldots, x_n$ be n-points in the space. The center of gravity of these points is the element

$$1/n(x_1 + \cdots + x_n).$$

It is easy to see that this point has the following property: if we consider the function on X defined by the formula

$$f(x) = \sum_{i=1}^{n} \|x - x_i\|^2$$

then this has a minimum exactly at the center of gravity of the points $x_1, \ldots, x_n$.

Indeed this follows from the fact that the Apollonius identity holds in X, i.e.

$$\sum_{i=1}^{n} \|u - x_i\|^2 = \sum_{i=1}^{n} \|x_i - m\|^2 + n\|u - m\|^2,$$

where

$$m = 1/n(x_1 + \cdots + x_n).$$

Now we show that, to some extent, this is a characteristic property of spaces having their norm generated by an inner product. This result was obtained by M. Nagumo.

THEOREM 8.7.1. (Nagumo, 1936). *Let f be a continuous positive function on* $[0, \infty)$ *with the properties that* $f(0) = 0, f(1) = 1$. *A Banach space X is a space with an inner product structure iff for any integer n and any points $x_1, \ldots, x_n$ in X the function*

$$x \to g(x) = \sum_{i=1}^{n} f(\|x - x_i\|)$$

has a minimum at the point $m = 1/n(x_1 + \cdots + x_n)$ *and then* $f(t) = t^2$.

Proof. The condition is clearly necessary. Suppose now that it is satisfied. We consider the function

$$g(x) = f(\|x - a\|),$$

where a is a fixed element of X. First we show that g is a convex function of X and that if x lies on a line the function $t \to g(tx)$ on R is differentiable. (Of course, here x is fixed.)

Let us consider x_0 to be an arbitrary point of X and a' to be an element of X such that

$$x_0 = \tfrac{1}{2}(a + a').$$

We now consider the system of points (a, a') which has at the center of gravity the element x_0. The function

$$x \to g(x - a) + g(x - a')$$

has a minimum at $x = x_0$.

We set $x = x_0 + y$ and then

$$x - a = x_0 + y - a, \qquad x - a' = x_0 + y - (2x_0 - a) = -(x_0 - y - a)$$

which gives that (for $y \neq 0$)

$$f(\|x_0 + y - a\|) + f(\|x_0 - y - a\|) > 2f(\|x_0 - a\|).$$

Since x_0 and y are arbitrary, this implies that

$$x - f(x - a)$$

is a convex function.

We now consider a line in X,

$$x(t) = x_0 + ty, \qquad y \neq 0, \quad t \in R$$

and set

$$g(t) = f(\|x(t) - a\|).$$

By the above result $t \to g(t)$ is a convex function of t. This clearly implies that g has left and right derivatives $D_-g(t)$, $D_+g(t)$ and these are, in general, not equal. We prove now that they are in fact equal. We may suppose that we have $t = 0$ and there

$$D_-g(0) < D_+g(0).$$

If a' is a point in X with the property that $x_0 = \frac{1}{2}(a + a')$ then

$$g(-t) = f(\|x(t) - a\|)$$

and

$$D_\pm f(x(t) - a')_{t=0} = D_\pm g(-t)_{t=0} = -D_\pm g(t)_{t=0}.$$

We consider now the point

$$1/(2n + 1)(2nx_0 + \bar{a})$$

which is clearly the center of gravity for the points $(\underbrace{a, \ldots, a, a', \ldots, a'}_{n\text{-times}}, \bar{a})$ which lie on our line.

Now if n is sufficiently large, this gives that

$$n(D_+(g(t) - D_-g(t))_{t=0}) > \max(-D_+\bar{g}(t)_{t=0}, D_-\bar{g}(t)_{t=0})$$

with

$$\bar{g}(t) = f(\|x(t) - \bar{a}\|).$$

Since

$$D_+(nf(\|x(t) - a\|) + nf(\|x(t) - a'\|) + f(\|x(t) - \bar{a}\|)_{t=0} > 0$$

and

$$D_-(nf(\|x(t) - a\|) + nf(\|x(t) - a'\|) + f(\|x(t) - \bar{a}\|)_{t=0} < 0$$

and using the convexity (just proved) we get further that

$$h(t) = nf(\|x(t) - a\|) + f(\|x(t) - a'\|) + f(\|x(t) - \bar{a}\|)$$

has a minimum at x_0 which is clearly a contradiction with out assumption since the center of gravity is $1/(2n+1)(2nx_0 + \bar{a})$. This contradiction implies the differentiability of our function.

Now we prove that the following parallelogram law holds for f,

$$f(\|x+y\|) + f(\|x-y\|) = 2(f(\|x\|) + f(\|y\|)). \qquad (++)$$

Let $x(t) = a + tb$ be an arbitrary line in X and let $x(t_0)$ be an arbitrary point on the line. We consider the system of points $(x, -x, 3x(t_0))$ and thus the function

$$u \to f(\|-x-u\|) + f(\|x-u\|) + f(\|3x(t_0) - u\|)$$

has a minimum at $u = x(t_0)$. Thus for $x(t) = a + tb$ we get

$$\mathrm{d}/\mathrm{d}t(f(\|x(t) + x\|) + f(\|x - x(t)\|) + f(\|3x(t_0) - x(t)\|))_{t=t_0} = 0. \qquad (+++)$$

If we consider the system

$$(0, 0, 3x(t_0))$$

then the center of gravity is $x(t_0)$ and thus we have

$$\mathrm{d}/\mathrm{d}t(2f(\|x(t)\|) + f(\|3x(t_0) - x(t)\|))_{t=t_0} = 0 \qquad (++++)$$

and from these relation the following relation follows:

$$\mathrm{d}/\mathrm{d}t(f(\|x + x(t)\|) + f(\|x - x(t)\|) - 2f(\|x(t)\|))_{t=t_0} = 0. \qquad (\mathrm{v})$$

But $x(t)$ is arbitrary and thus

$$f(\|x + x(t)\|) + f(\|x - x(t)\|) - 2f(\|x(t)\|)$$

is a constant function. Thus

$$f(\|x+y\|) + f(\|x-y\|) - f(\|y\|) = 2f(\|x\|) - 2f(0)$$

and thus the parallelogram law for f is proved.

Now we prove that $f(t) = t^2$. But this follows from the proof of the Jordan–von Neumann theorem. The case of complex spaces follows in a well-known manner from the case of real spaces.

Recently the author has obtained the following result which may be considered an extension of Nagumo's result given above.

THEOREM 8.7.2 (Istrăţescu, 1984). *Let X be a Banach space and suppose that for some fixed n ($\geqslant 2$) and any points $x_1, \ldots, x_n$ the function*

$$g(x) = \sum_{i=1}^{n} f(\|x_i - x\|)$$

attains the minimum at $\bar{x} = 1/n(x_1 + \cdots + x_n)$ *and where* f *is a continuous function on* $[0, \infty)$ *with values in the same interval and* $f(0) = 0, f(1) = 1$. *Then the norm of* X *is generated by an inner product and* f *is in fact the function* $f(t) = t^2$.

We now consider a characterization of inner product spaces using Mielnik's probability spaces.

DEFINITION 8.7.3. A. Mielnik space (S, p) is a pair where S is a nonempty set and p is a function defined on $S_{\times}S$ with real values satisfying the following axioms:

(1) $p(a, b) = 1$ if and only if $a = b$.
(2) $p(a, b) = p(b, a)$ for all $a, b \in S$.
If a, b are in S then we say that a is orthogonal to b if $p(a, b) = 0$. A maximal set of elements of S which are mutually orthogonal is a basis of S.
(3) For any a in S and any basis $(b_\alpha)_{\alpha \in I}$, I an index set, $\Sigma_{\alpha \in I}\, p(a, b_\alpha) = 1$.

The elements of S are viewed as possible states of a physical system and $p(a, b)$ is interpreted as the probability that a system is in state b after being in state a.

If X is a complete inner product space we may define S as the set of all equivalence classes of the unit sphere of X via the following equivalence relation: $x\,R\,y$ if $y = zx$ with z a complex number with $|z| = 1$. Then if x, y are two elements of S and u, v two elements in $\hat{x}$ and $\hat{y}$, respectively, then $p(x, y) = |\langle u, v\rangle|^2$ (which, as is easy to see, depends upon only the equivalence class).

In papers by Č. V. Stanojevič, A. R. Blass and S. J. Guccione, Jr some interesting results concerning Mielnik probability spaces and inner product spaces were obtained. Some of these are given below.

8.7.4 (Stanojevič, 1970). *Let* X *be a normed linear space and let* $S = (x, x \in X, \|x\| = 1)$ *be the unit sphere of* X. *Then* X *is an inner product space if and only if* (S, p) *is a Mielnik probability space, where*

$$p(a, b) = \tfrac{1}{4}(\|a + b\|)^2.$$

Proof. If X is an inner product space then it is easy to see that the conditions of the theorem are satisfied.

Suppose now that X is a normed linear space with the properties stated in the theorem. From the definition of orthogonality in a Mielnik space we see that for p defined as above, for any x in S the only element orthogonal to x is $-x$. Since for each b in S and a in S we have

$$\tfrac{1}{4}(\|a + b\|^2 + \|a - b\|^2) = 1$$

and this clearly implies, by the results in Chapter 4, that X is an inner product space. The result may easily be improved if, in the axioms of Mielnik's probability spaces, axiom (1) is weakened to axiom (1′) below:

(1′) For any a, b in S, $p(a, b)$ is in $[0, 1]$ and $p(a, b) = 1$ implies that $a = b$.

Then we have the following result.

THEOREM 8.7.5 (Stanojevič, 1973). *Let* (S, p) *be a Mielnik probability space where the axiom is* (1′) *instead of* (1). *If* S *is the unit sphere of a normed linear space and* p *satisfies the inequality*

$$p(a, b) \geqslant \tfrac{1}{4}(\|a + b\|^2)$$

then if the dimension of X *is* 2 (*the cardinal of a basis of* S *as a Mielnik space*), X *is an inner product space.*

For the following characterization we need the following class of real-valued functions.

DEFINITION 8.7.6. F is the set of all continuous real-valued functions defined on $[0, 2]$ with values in $[0, 1]$ such that the following properties are satisfied:

(1) $f(0) = 0$, $f(2) = 1$,
(2) f is increasing

if $f \in F$. The class G is the set of all functions with the domain and range $[0, 2]$ such that the following properties hold:

(1) $g(0) = 2$, $g(2) = 0$,
(2) g is continuous and decreasing.

THEOREM 8.7.7 (Stanojevič, 1973). *A real normed linear space* X *is an inner product space if and only if there exists* f *in* F *such that for all* x, y *in the unit sphere of* X *the following relation holds:*

$$f(\|x + y\|) + f(\|x - y\|) = 1.$$

THEOREM 8.7.8 (Stanojevič, 1973). *Let* X *be a real normed linear space and let* S *be the unit* p *sphere of* X. *For* f *in* F *we set* $p(x, y) = f(\|x + y\|)$ *for* x, y *in* S. *Then* X *is an inner product space if and only if for some* f *in* F, (S, p) *is a Mielnik space of dimension* 2 (*as a Mielnik space*).

REMARK 8.7.9. In Stanojevič (1973) it is pointed out that if the dimension of X is 3 (as a Mielnik space) then X is not necessarily an inner product space when (S, p) with $p(x, y) = f(\|x + y\|)$ is a Mielnik space.

Let X be a real normed linear space. The inversion of X in the unit sphere $S_X = (x, \|x\| = 1)$ of X is the mapping defined on $X - 0$ by the formula

$$x \to x/\|x\|^2 = I(x)$$

This is clearly a mapping defined on $X \setminus 0$ with values in $X \setminus 0$. Then Stiles (1969) has proved the following results.

THEOREM 8.7.10 (Stiles, 1969). *Let* X *be a two-dimensional space and suppose that the inversive image of some hyperplane is a sphere then the norm of* X *is generated by an inner product.*

THEOREM 8.7.11 (Stiles, 1969). *If* X *is a normed linear space with the property that the inversive image of every supporting hyperplane has a center of symmetry then the norm of* X *is generated by an inner product.*

Recently L. E. Dor has proved related results inspired by a classical result of J. L. Walsh.

THEOREM 8.7.12 (Dor, 1980). *Let X be a normed linear space and let D be a closed bounded subset in X which is not reduced to a single point. Suppose that for any u in $X \setminus D$, the image through I of the set $D - u$ is convex. Then X has the norm generated by an inner product and D is the intersection of X with a closed unit ball in the completion of X.*

NOTE 8.7.13. The classical result of Walsh (1922) asserts that when X is the plane then D is a disc. In the same paper L. E. Dor has obtained the following extension of the above result. This is given in

THEOREM 8.7.14 (Dor, 1980). *Let X be a normed linear space and $a > 1$. We set*

$$I_a: X \setminus 0 \to X$$

$$I_a(x) = x/\|x\|^a.$$

Then if D is a bounded, closed subset of X which does not reduce to a single point, and for any u not in D, $I_a(D - u)$ is convex, then the norm of X is generated by an inner product and D is the intersection of X with a closed ball in the completion of X and $a = 2$.

In order to formulate Enflo's result we need the following notion.

DEFINITION 8.7.15. Let X be a topological linear space. We say that X is a locally bounded linear space if the zero of the space has a bounded neighborhood.

Then we have the following results.

THEOREM 8.7.16 (Enflo, 1970). *Let X be a Banach space with the property that X is uniformly homeomorphic to a complete inner product space, then on X there exists an equivalent norm generated by an inner product.*

THEOREM 8.7.17 (Enflo, 1970). *Let X be a locally bounded linear space which is uniformly homeomorphic to a complete inner product space. Then on X there exists a norm which is generated by an inner product and generates the same topology.*

For other related results, and for proofs, we refer to Enflo's papers.

Now we mention the result of L. V. Krovonosov.

THEOREM 8.7.18 (Krovonosov, 1972). *Let X be a real linear space of dimension at least 2. Suppose that for some abstract set Y we have a function $h: X \to Y$ satisfying the properties*

(1) $h(x) = h(-x)$,
(2) *for each $x \neq 0$ in X and for each two-dimensional subspace X_2 of X containing x the set $(u, h(u + a) = h(u - a))$ is a one-dimensional subspace not containing a.*

Then there exists an injective mapping $f: h(X) \to [0, \infty)$ such that $x \to f(h(x))$ is a norm on X which is generated by an inner product.

8.8. Linear Operators and Inner Product Structures

One of the basic theorem of linear functional analysis, the Hahn–Banach–Bohnenblust–Sobczyc–Sukhomlinov theorem, asserts that if X is a Banach space over $K(R, C, Q$, the real, the complex numbers or the quaternions) then every bounded linear functional on a closed subspace X_1 of $X, f: X_1 \to K$, has a norm-preserving extension, i.e. there exists F in the dual of X with the following properties:

(1) $F/X_1 = f$,
(2) $\|F\| = \sup\{|F(x)|, x \in X, \|x\| = 1\} = \|f\| = \|f\|_{X_1}$
$= \sup\{|f(x)|, x \in X_1, \|x\| = 1\}$.

Then we may ask if the above theorem may be extended to the case of bounded linear operators, $T: X_1 \to Y$, Y being another Banach space over K which is linear and bounded, i.e. if there exists $\bar{T}: X \to Y$ in $L(X, Y)$ with the properties

(1) $\bar{T}/X_1 = T$,
(2) $\|\bar{T}\| = \sup\{\|Tx\|, x \in X, \|x\| = 1\} = \|T\|$
$= \sup\{\|Tx\|, x \in X_1, \|x\| = 1\}$

and secondly, we may ask if it is possible to extend all elements of the dual of X_1 to elements in X with the property that if u denoted the application $u(f) = F, f$ in X_1 and F in X then we require that

(1) u is linear, i.e. $u(f + g) = u(f) + u(g)$,
(2) u is an isometry, i.e. for all f, $\|u(f)\| = \|f\|$.

Concerning the first problem, special results were obtained by Murray (1937) and Bohnenblust (1938) and for the second by Kantorovič (1935).

Definitive results concerning the above problems were obtained by Kakutani (1939) which has proved that the above problems have an affirmative answer iff X is a space whose norm is generated by an inner product. The result from which this answer follows is contained in the following theorem.

THEOREM 8.8.1 (Kakutani, 1939; Bohnenblust, 1942). *Let X be a Banach space of dimension at least* 3. *Then the norm of X is generated by an inner product iff for any two-dimensional subspace X_1 of X there exists a projection $P: X \to X_1$ of norm* 1.

REMARK 8.8.2. The above theorem was proved by S. Kakutani for real spaces and by F. H. Bohnenblust for complex spaces. We mention that the ingenious proof of F. H. Bohnenblust uses some basic facts from projective geometry. In what follows we prefer a proof which uses some properties of compact operators and Loewner ellipses, i.e. the existence of Euclidean norms on X_1 which majorizes or are smaller than the original norm (restricted to X_1).

Our proof was inspired by a proof given by Cohen (1972) for the case of absolutely 2-summing operators. See Theorem 8.8.8.

Proof of Theorem 8.8.1. Take X_1 a two-dimensional subspace of X and there a basis (e_1, e_2). Then the projection P of X onto X_1 is the operator given by the formula

$$Px = x'_1(x)e_1 + x'_2(x)e_2$$

where x'_1, x'_2 are bounded linear functionals on X satisfying the relations

$$x'_1(e_1) = x'_2(e_2) = 1, \qquad x'_1(e_2) = x'_2(e_1) = 0.$$

We can consider on X_1 a new norm which is Euclidean (i.e. generated by an inner product), say $\| \, , \|_2$, and such that for all z in X_1 the following inequality holds:

$$\|z\|_2 \geqslant \|z\|.$$

Now we remark that the condition is necessary. Suppose now that it is satisfied for X. Then the operator P that we choose is a projection with a norm 1. Then we get the following inequalities:

$$1 \leqslant \|P\| = \sup\{|x^*(P(x)|, \|x\| \leqslant 1, \|x^*\| \leqslant 1\}$$
$$= \sup_{\substack{\|x\| \leqslant 1 \\ \|x^*\| \leqslant 1}} (|x_1^*(x)|^2 + |x_2^*(x)|^2)^{1/2} \times (|x^*(e_1)^2 + |x^*(e_2)|^2)^{1/2}.$$

Also,

$$1 = \sup_{\|x\| \leqslant 1} \|Px\| \leqslant \sup_{\|x\| \leqslant 1} \|P(x)\|_2 \leqslant \sup\{|x'(x)|^2 + |x'(x)|^2\}^{1/2}$$

since we may suppose that the elements (e_1, e_2) for an orthonormal basis of X_1 with respect to the norm $\| \, , \|_2$.

If x', y' are elements of X_1 which are also of norm 1 then, by the extension theorems, we can consider these as elements of X (with the same norm) and thus supposing (without loss of generality) that they are linearly independent, we obtain that for the elements

$$u' = \tfrac{1}{2}(x' + y'), \; v' = \tfrac{1}{2}(x' - y').$$

which is a basis for X_1,

$$\|u'\|^2 + \|v'\|^2 \geqslant 1$$

and thus

$$\|x' + y'\|^2 + \|x' - y'\|^2 \geqslant 4.$$

This implies that X_1 is an inner product space and the theorem is proved. □

We leave the reader to prove the following version of Theorem 8.8.1.

THEOREM 8.8.3. *Let X be a Banach space of dimension at least* 3. *Then the norm of X is generated by an inner product iff for any closed linear subspace X_1 of codimension* 2 *there exists a projection $P\colon X \to X_1$ of norm* 1.

We now mention briefly the solution of the above-mentioned problems. Suppose that the projection P exists and thus if we set for any f,

$$F(x) = f(Px)$$

the mapping $f \to F$ satisfies the requirements. In order to see the necessity we may suppose without loss of generality that the space has a dimension equal to 3 and that X_1 is a two-dimensional subspace which we identify with the plane (ξ, η). Then any element of the dual is of the form

$$f(x) = a\xi + b\eta, \qquad a, b \text{ constants}$$

and

$$x = (\xi, \eta, 0).$$

Then we define

$$F(x) = a\xi + b\eta + cv,$$

where $x = (\xi, \eta, v)$ and c is a constant depending upon a and b only, i.e. $c = c(a, b)$. Now since the mapping $f \to F$ is linear we must have the relations:

(1) $c(a + a_1, b + b_1) = c(a, b) + c(a_1, b_1)$,
(2) $c(\lambda(a, b)) = \lambda c(a, b)$.

Thus if we set $A = c(1, 0)$, $B = c(0, 1)$ we have

$$c(a, b) = aA + bB$$

and thus

$$F(x) = a\xi + b\eta + v(aA + bB) = a(\xi + vA) + b(\eta + vB).$$

If we define for $x = (\xi, \eta, v)$

$$Px = (\xi + vA, \eta + vB, 0)$$

then we obtain a projection of X onto X_1. To prove the assertion of the theorem it remains to show that the norm of this projection is 1. Let x_0 be an arbitrary but fixed element in X,

$$x_0 = (\xi_0, \eta_0, v_0)$$

and consider f with the properties

(1) $f(x) = a\xi + b\eta$,

(2) $\|f\|_{X_1} = 1$,

(3) $|f(Px_0) = \|Px_0\|$.

If F is the extension of f then

$$F(x) = f(Px)$$

and

$$\|Px_0\| = f(Px_0) = Fx_0 \leqslant \|F\|_X \|x_0\|$$

which gives that $\|Px_0\| \leqslant \|x_0\|$. Since x_0 is assumed arbitrary, the norm of P is 1. Since the answer to the first problem is quite clear (see Corollary 8.8.7 below) we now consider the spaces introduced by Kantorovič (1935) in order to solve the second problem. In Chapter 4 we defined the Kantorovič spaces (quasiconvex spaces) as those Banach spaces X with the property that for any finite set of points $x_0, X_1, \ldots, x_n$ in X there exists a point $\bar{x}$ in X of the form $a_1x_1 + \cdots + a_nx_n$ such that

$$\|x_i - \bar{x}\| \leqslant \|x_i - x_0\|.$$

Then we have

COROLLARY 8.8.4 (Kakutani, 1939). *A Banach space X (dimension of $X \geqslant 3$) is a Kantorovič space iff the norm of X is generated by an inner product.*

Proof. The condition is necessary. Conversely, suppose that X is a Kantorovič space. We may suppose without loss of generality that the dimension of X is equal to 3.

Let X_1 be a two-dimensional subspace of X, and we show that there exists a projection of norm 1 of X onto X_1. For this we consider $x_1, \ldots, x_n$ to be arbitrary points in X_1 and x_0 to be an arbitrary point in $X \rightarrow X_1$. Then there exists a point $\bar{x}_0$ in X_1 such that

$$\|x_i - \bar{x}_0\| \leqslant \|x_i - x_0\|, \quad i = 1, 2, 3, \ldots, n.$$

Let y be arbitrary in X_1 and $K(y) = (x, x \in X_1, \|y - x\| \leqslant \|x - x_0\|)$ then $\Pi_{i=1}^n K(x_i) \neq 0$. Since $x_1, \ldots, x_n$ are arbitrary, Helly's theorem implies that $\Pi_{x \in X_1} K(x) \neq 0$.

Let u_0 be an arbitrary point in the intersection, then any point of X has the unique representation

$$x = x_1 + ax_0,$$

where x_1 is in X_1 and a is a scalar. The mapping

$$Px = x_1 + au_0$$

is a projection of X onto X_1 and we show that the norm is 1. Indeed, we have,

$$\|Px\| = \|x_1 + au_0\| = \|a\| \, \|-x_1/a - u_0\| \leqslant |a| \, \|-x_1/a - x_0\| = \|x_1 + ax_0\| = \|x\|$$

and thus the norm of P is 1. The assertion follows since X_1 is arbitrary in X.

COROLLARY 8.8.5 (J. Gil de Lamadrid, 1962). *Let X be a Banach space. Then on X there exists a norm equivalent with the original norm and generated by an inner product iff there exists $K > 0$ such that for every closed linear subspace X_1 of X there exists a projection with the property that the multiplicative semigroup generated by these projections (P_{X_1}) is bounded by K.*

Proof. It is obvious that the condition is necessary. Suppose now that X satisfies the condition stated above. Then we define a new norm on X by the formula

$$\|x\|_1 = \sup \|P_{X_1} x\|,$$

where X_1 varies on the set of all above projections. It is easy to see that this norm is equivalent to the norm $\|\,,\|$ of X. Now if X_1 is an arbitrary closed linear subspace of X then the projection $P\colon X \rightarrow X_1$ has the norm with respect to $\|\,,\|_1$ equal to 1. Then the Kakutani–Bohnenblust theorem asserts that $\|\,,\|_1$ is generated by an inner product. The corollary is proved. □

COROLLARY 8.8.6 (Kakutani, 1939). *Let X be a Banach space such that for any closed linear subspace X_1 of X and any bounded linear operator $T: X_1 \to Y$, Y being another Banach space, has an extension with the same norm, then it is necessary and sufficient that the norm of X be generated by an inner product.*

Proof. If we take $X = Y$ then we see that for any two-dimensional subspace the identity operator may be extended to the whole space and this means that there exists a projection of norm 1. But this implies that X is an inner product space. Conversely, if X is an inner product space then it is easy to see (using the orthogonal decomposition of X) that an extension exists. □

REMARK 8.8.7. We remind the reader that we have used Theorem 8.8.1 several times in Chapter 4.

For an interesting condition assuring the existence of projections of norm 1 we refer to the paper by Solomiak (1957). We now present a characterization of inner product spaces using the ideal of absolutely 2-summing operators.

THEOREM 8.8.8 (Cohen, 1970). *Let X be a Banach space. Then X is an inner product space if and only if for all Banach spaces Y and for all absolutely 2-summing operators $T: X \to Y$ the adjoint $T^*: Y^* \to X^*$ is absolutely 2-summing and the following inequality is satisfied:*

$$\Pi_2(T^*) = \Pi_2(T).$$

Proof. The condition is necessary. Suppose that X is an inner product space and let $T: X \to Y$ be an absolutely 2-summing operator defined on X with values in the Banach space Y. From the structure theorem of absolutely summing operators in Chapter 7 we find a compact Hausdorff space S and a positive Radon measure μ with $\mu(S) = 1$, such that the following diagram is commutative and $\|P\| = 1$, $\|Q\| = \Pi_2(T)$:

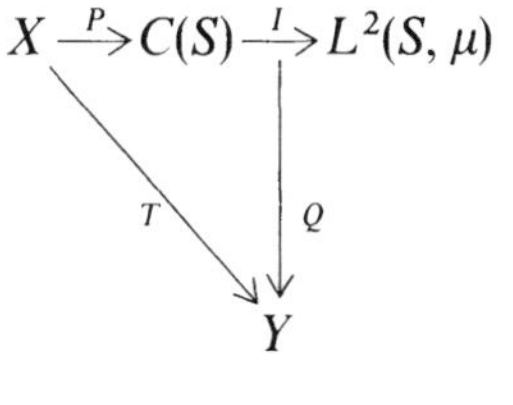

(Here I means the embedding operator.)

Since the class of absolutely 2-summing operators on complete inner product spaces coincides with the class of Hilbert–Schmidt operators. As is well known, the class of Hilbert–Schmidt operators is self adjoint, and the Hilbert–Schmidt norm is the same. Then the following relations hold:

$$\Pi_2((IP)^*) = \Pi_2(IP),$$

and since $T^* = (IP)^*Q^*$ we obtain that T^* is absolutely 2-summing and

$$\Pi_2(T^*) = \Pi_2((IP)^*Q^*) \leqslant \Pi_2((IP)^*)\|Q^*\| = \Pi_2((IP)^*)\|Q\| = \Pi_2(IP)\|Q\|$$
$$\leqslant \Pi_2(I)\|P\|\,\|Q\| = \|Q\| = \Pi_2(T).$$

Thus the necessity is proved. □

Suppose now that X has the property stated in the theorem. We consider the space X_2 of dimension 2 in an inner product space U. Let (e_1, e_2) be any orthonormal basis of X_2 and define the following mapping from X into X_2:

$$Tx = x_1'(x)e_1 + x_2'(x)e_2.$$

Obviously T is a compact operator and the range being finite dimensional (the dimension of the range is 2) it is absolutely 2-summing. Then the adjoint of T is absolutely 2-summing and

$$\Pi_2(T) \geqslant \Pi_2(T^*).$$

But

$$\Pi_2(T) = (\|x_1'\|^2 + \|x_2'\|^2)^{1/2} \sup_{\|u\| \leqslant 1}(|\langle u, e_1 \rangle|^2 + |\langle u, e_2 \rangle|^2)^{1/2} \leqslant (\|x_1'\|^2 + \|x_2'\|^2)^{1/2}.$$

and

$$\begin{aligned}(\|T^*e_1\|^2 + \|T^*e_2\|^2)^{1/2} &\leqslant \Pi_2(T^*) \sup_{\|u\| \leqslant 1}(|\langle u, e_1 \rangle|^2 + |\langle u, e_2 \rangle|^2)^{1/2} \\ &\leqslant \Pi_2(T^*) \leqslant \Pi_2(T).\end{aligned}$$

Thus, since $T^*e_i = x_i'$, $i = 1, 2$, we have that

$$(\|x_1'\|^2 + \|x_2'\|^2)^{1/2} \leqslant \Pi_2(T)$$

and thus

$$\|x_1'\|^2 + \|x_2'\|^2 = \Pi_2(T)^2.$$

If we set

$$y_1^* = (x_1' + x_2')/2^{1/2}, \qquad y_2^* = (x_1' - x_2')/2^{1/2}$$
$$u_1 = (e_1 + e_2)/2^{1/2}, \qquad u_2 = (e_1 - e_2)/2^{1/2}$$

and using the same computation as above we get that

$$\Pi_2(T) = (\|y_1^*\|^2 + \|y_2^*\|^2)^{1/2} = ((\|x_1' + x_2'\|^2 + \|x_1' - x_2'\|^2)^{1/2}/2^{1/2}.$$

Then we have that

$$\|x_1' + x_2'\|^2 + \|x_1' - x_2'\|^2 = 2(\|x_1'\|^2 + \|x_2'\|^2).$$

But x_1', x_2' may be arbitrary elements in the dual of X. This implies that X^* is an inner product space, and this, of course, implies that X is an inner product space.

In connection with the above characterization we mention the following result of S. Kwapień which characterizes the class of Banach spaces isomorphic to spaces in the class of inner product spaces. For a proof we refer to Kwapień's paper.

RESULT 8.8.9 (Kwapień, 1970). *Let X be a Banach space. Then X is isomorphic to an inner product space if and only if for any bounded linear operator T in $\Pi_2(X, l^2)$ the adjoint T^* is in $\Pi_2(l^2, X^*)$.*

We now mention some very interesting characterizations of the existence of norms or equivalent norms generated by inner products. For a proof we refer to the elegant paper of Lindenstrauss and Pełczyński (1968).

THEOREM 8.8.10 (Lindenstrauss and Pełczyński, 1968). *Let X, Y be two infinite dimensional Banach spaces with the following properties:*

(1) *X has an unconditional basis.*
(2) *$L(X, Y)$ consist only from absolutely summing operators. Then Y is isomorphic to an inner product space.*

THEOREM 8.8.11 (Grothendieck, 1953). *A Banach space X has an equivalent norm generated by an inner product if and only if X is isomorphic to a subspace of an $\mathscr{L}_1$-space and to a quotient of an $\mathscr{L}_\infty$-space.*

See also Kisliakov (1979) for a related interesting result.

THEOREM 8.8.12 (Lindenstrauss and Pełczyński, 1968). *Let X be a Banach space such that X and X^* are isomorphic to a subspace of $\mathscr{L}_1$-spaces or they are quotients of $\mathscr{L}_\infty$-spaces. Then X has an equivalent norm generated by an inner product.*

THEOREM 8.8.13 (Lindenstrauss and Pełczyński, 1968). *If a Banach space X is an L_2-space then X has an equivalent norm generated by an inner product.*

We mention now a characterization of spaces isomorphic to inner product spaces which is, in some sense, related to the famous Grothendieck theorem of the metric theory of tensor products.

THEOREM 8.8.14 (Lindenstrauss and Pełczyński, 1968). *Let X be a Banach space such that there exists a constant K with the following property: if $(a_{ij})_i$, $i = 1, 2, \ldots, n$, is a finite real-valued matrix for which*

$$\left| \sum_{i,j=1,2,\ldots,n}^{n} a_{ij} t_i s_j \right| \leqslant 1$$

for all $|t_i| \leqslant 1$ and $|s_j| \leqslant 1$, then for any choice of $(x_i)_1^n$ in X and $(x_i^*)_1^n$ in X^*

$$\left| \sum_{i,j=1,2,\ldots,n} a_{ij} x_j^*(x_i) \right| \leqslant K \sup\|x_i\| \sup\|x_j^*\|.$$

Then X has an equivalent norm generated by an inner product.

We now mention two results in which some classes of linear operators are used to characterize inner product structures.

THEOREM 8.8.15 (Joichi, 1966). *Let X be a normed linear space. Then on X there exists an equivalent norm generated by an inner product iff for some constant $K \geqslant 1$ for each finite dimensional subspace X_1 of X there exists a linear mapping $T_1: X_1 \to H$, H is a complete inner product space such that the following inequality is satisfied:*

$$K^{-1}\|x\| \leqslant \|T_1 x\| \leqslant K\|x\|$$

for all x in X_1.

The following result is related to the famous Weyl inequalities and to the class of nuclear operators of A. Grothendieck.

THEOREM 8.8.16 (Johnson *et al.*, 1979). *Let X be a Banach space. Then on X there exists an equivalent norm generated by an inner product iff for each nuclear operator on X the eigenvalues are absolutely summable.*

Concerning the existence of an equivalent norm generated by an inner product, there was a conjecture related to complemented subspaces. We recall that if X is a Banach space then a closed subspace X_1 of X is said to be complemented if there exists a closed subspace X_2 of X with the property that for each x in X,

$$x = x_1 + x_2,$$

where the x_i are in X_i, $i = 1, 2$. We write this as $X = X_1 \mathring{\oplus} X_2$.

For each complemented subspace X_1 of X we can associate the projection P_{X_1} defined by the formula

$$P_{X_1}x = x_1,$$

where x has the above decomposition. The fact that in Banach spaces there exist noncomplemented subspaces was remarked by Banach and Mazur (1933). The conjecture was as follows.

CONJECTURE 8.8.17. *If X is a Banach space with the property that every closed linear subspace is complemented then on X there exists an equivalent norm which is generated by an inner product.*

The conjecture was positively settled by J. Lindenstrauss and L. Tzafriri in 1971. For an excellent exposition of a proof of the conjecture we refer to the paper by Kadets and Mityagin (1973).

8.9. Algebraic Characterizations of Inner Product Structures

In this section we present conditions for the existence of inner product structures on Banach spaces using certain properties of the Banach algebra of all bounded linear operators on the Banach space. The first result concerning the connection between Banach spaces X and Y and Banach algebras $L(X)$ and $L(Y)$ (if Z is a Banach space the $L(Z)$ denotes the set of all bounded linear operators on Z) was obtained by Eidelheit (1939) who proved that $L(X)$ and $L(Y)$ are isomorphic iff X and Y are isomorphic. The next result was a characterization of inner product structures on Banach spaces obtained by Kakutani and Mackey (1944, 1946) which is as follows. Let X be a Banach space and suppose that on $L(X)$ we have a mapping $*$: $L(X) \to L(X)$ called involution and which is supposed to satisfy the conditions

(1) $(T + S)^* = T^* + S^*$,
(2) $(T^*)^* = T$,
(3) $(TS)^* = S^*T^*$,
(4) $T^*T = 0$ iff $T = 0$.

Then on X there exists an equivalent norm which is generated by an inner product.

A new proof of the results of Eidelheit and Kakutani and Mackey was given by Kawada (1943) who has obtained other related results.

Of course we may ask under what conditions on $L(X)$ satisfying the conditions (1)–(4) is the new equivalent norm in fact the original one? The first result in this direction was obtained by Prijatelj (1964) who proved the following result: If X is a Banach space and $L(X)$ is a B^*-algebra then the norm of X is generated by an inner product.

The result of Prijatelj has inspired many authors. Vukman (1980) proved the following result: If X is a complex Banach space and there exists a strictly irreducible Banach *-algebra $\mathscr{A}$ in $L(X)$ and a minimal left ideal such that whenever l is in L and $\|l^*l\| = \|l\|^2$, then the norm of X is generated by an inner product. The author (1984) has obtained the following result: On $L(X)$ there exists an involution such that conditions (1)–(3) are satisfied and whenever P is a hermitian projection then

$$\|P\|^2 = \|P\|^2.$$

Then the norm of X is generated by an inner product. A related result to Prijatelj's was obtained by Vukman (1982) and is as follows: If $L_F(X)$ denotes the subset of all elements in $L(X)$ with finite dimensional range, and there we have an involution satisfying conditions (1)–(3) and one of the following conditions is fulfilled, then on X there exists an equivalent norm generated by an inner product and the adjoint with respect to the inner product coincides with the given involution. The conditions are:

(1) there exists a minimal left ideal L in $L_F(X)$ with the property that whenever l is in L and $l^*l = 0$ then $l = 0$,
(2) for each one-dimensional subspace X_1 of X there exists an operator T in L such that $T = T$, $T^2 \neq 0$, range $T = X_1$,
(3) if P is in $L_F(X)$ is a one-dimensional projection then $P^*P = 0$ iff $P = 0$.

Vukman (1983) has proved that if X is a real or complex Banach space and there exists L in $L(X)$ which contains I and all bounded linear operators with finite dimensional range, and with an involution such that each unitary operator U in L has norm 1, then on X there exists an equivalent norm generated by an inner product and the involution coincides with the adjoint with respect to the inner product for each T in L.

Also Vukman (1983) has obtained an extension of the Kakutani–Mackey results for bornological spaces. We shall present some of these results below. First we present Vukman's result in the case of bornological spaces. For the reader's convenience we recall that if S is a set then a bornology on S is a family of subsets B of S with the following properties:

(1) B is a covering of S, i.e. $S = \cup b$, $b \in B$,
(2) B is hereditary under the inclusion, i.e. if b is in B then any subset of b is in B,
(3) B is stable under finite union. The elements of B are the bounded sets of S.

If X is a linear space then a bornology B on X is said to be compatible with the linear structure of X (or to be a linear bornology) if B is stable under addition, homothetic transformations and the formation of circled hulls. A bornological space is a pair (X, B), where X is a linear space and B is a linear bornology on X. We say briefly that X is a bornological space. A separated bornological space is one in which (0) is the only subspace of X which is bounded.

THEOREM 8.9.1 (Vukman, 1983). *Let X be a real or complex linear bornological barrelled Hausdorff locally convex space. Suppose that $L_F(X)$, the space of all continuous linear operators on X with finite dimensional range, has an involution $T \to T^*$ such that on some minimal left ideal L in $L_F(X)$, if $T^*T = 0$ then $T = 0$. Then on X there exists an inner product $\langle \, , \rangle$ such that the topological space $(X, \| \, , \|)$, which is the norm generated by the inner product, coincides with the original topological space and for each T in $L_F(X)$ and all x, y in X,*

$$\langle Tx, y \rangle = \langle x, T^*y \rangle.$$

For the proof we need the following lemma.

LEMMA 8.9.2. (Rickart, 1960). *Let A be an arbitrary *-algebra with the property that $T^*T = 0$, then $T = 0$. Then in A very minimal left ideal is of the form Ap, where p is a unique hermitian idempotent.*

Proof. We may suppose without loss of generality that $L$is not (0). We consider a nonzero element x in $L$and define h*h. Then clearly h is not zero and, since $L$is assumed minimal, $L = Lh$ holds. Also it is clear that the left annihilator of h in L reduces to (0). From the relation $L = Lh$ we have that for some e in L, $h = eh$. Also $e^2h = e(eh) = eh$ and thus $(e^2 - e)h = 0$ which gives that $e^2 = e$.

If we consider the element $p = e^*e$ then p is in L and $ph = h$ (and as above) $p^2 = p$. Clearly p is nonzero and $L = Ap$.

Now the uniqueness is as follows. Suppose that we have another hermitian idempotent q with the property that $L = Aq$. Then $qp = p$, $pq = q$ and we get obviously that $p = q$. The lemma is proved. □

Now the proof of Theorem 8.9.1 is as follows. According to the lemma we have $L = L_F(X)p$, where p is a hermitian idempotent. Now since $L$is supposed to be minimal, the range of the element p is one-dimensional and thus isomorphic to R or C.

For a, b in L we consider the element b^*a which is clearly of the form $z_{a,b}p$ where $z_{a,b}$ is a number. We define the application of $L_\times L$ in R (or C) by $\langle a, b \rangle = z_{a,b}$. Then from the definition it follows that the above mapping is linear in the first variable. Now, since

$$a^*b = (b^*a)^* = (z_{a,b}p)^* = \bar{z}_{a,b}p$$

we have that

$$\langle a, b \rangle = \overline{\langle b, a \rangle}$$

and from $a^*a = 0$ iff $a = 0$ we obtain that the above mapping is an inner product.

Now if $A \to T_A$, $T_A(a) = Aa$, then

$$\langle T_A a, b \rangle_p = b^*Aa = (A^*b)^*a = \langle a, T_{A^*}b \rangle p$$

and thus $T_{A^*} = (T_A)^*$, where $(\ )^*$ means the adjoint with respect to the inner product $\langle\, ,\rangle$.

Now the idempotent p is of the form $x_0 \oplus x_0^*$, where x_0 is in X and x_0^* is in X^*, $x_0^*(x_0) = 1$. We remark that there is an isomorphism between X and the set of all operators of the form $x \oplus x_0^*$, and that this set coincides with a minimal left ideal. Then we set for each x, y in X,

$$\langle x, y\rangle = \langle x \oplus x_0^*, y \oplus x_0^*\rangle$$

and this is clearly an inner product on X. Now we remark that for all x, y in X the relation

$$\langle Tx, y\rangle = \langle x, T^*y\rangle$$

holds, which follows from the relations

(1) $Tx \oplus x_0^* = T(x \oplus x_0^*)$,
(2) $\langle Tx, y\rangle p = \langle (Tx) \oplus x_0^*,\ y \oplus x_0^*\rangle p = \langle T(x \oplus x_0^*),\ y \oplus x_0^*\rangle p = (y \oplus x_0^*)^*$ $(T(x \oplus x_0^*))$,
(3) $\langle x \oplus x_0^*,\ T^*(y \oplus x_0^*)\rangle p = (T^*(y \oplus x_0)^*(x \oplus x_0^*) = (y \oplus x_0^*)^*\ (T(x \oplus x_0^*))$.
(4) $\langle x, T^*y\rangle p = \langle x \otimes x_0^*, (T^*y) \otimes x_0^*\rangle p$.

We now show that for each y in X we find a continuous seminorm p on X such that for all x in X,

$$|\langle x, y\rangle| \leqslant p(x).$$

We consider a, b in $L_F(X)$ and for each continuous seminorm $\bar{p}$ we find another continuous seminorm $\bar{q}$ with the property that

$$\bar{p}(ab(x)) \leqslant q(b(x))$$

holds for all x in X.

We now consider the hermitian idempotent p and thus we find a continuous seminorm p_1, $p_1(x_0) \neq 0$. From the definition of the inner product we have

$$\begin{aligned}|\langle x, y\rangle| p_1(x_0) &= p_1(\langle x, y\rangle p) = p_1((y \oplus x_0^*)^*(x \oplus x_0^*)(x_0))\\ &\leqslant ((x \oplus x_0^*)(x_0)) = q(x)\end{aligned}$$

and thus

$$|\langle x, y\rangle| \leqslant p_1(x_0)^{-1} q(x)$$

which proves the assertion. Thus for each fixed y, $x \to \langle x, y\rangle$ is a continuous linear functional on X. Now we prove that every x in X is of the form $\langle\, , y\rangle$ for some y in X. This means that the Fréchet–Riesz representation holds in X.

Let x be arbitrary in X. We may suppose without loss of generality that x is not identically zero. We consider an element u in X with the property that u, $u = 1$. Then

$$x^*(x) = \langle (u \oplus x^*)x, u\rangle = \langle x, (u \oplus x^*)^* u\rangle$$

and thus the assertion is proved.

Now we prove that the topology induced by the inner product is the same as the original topology of X. For this it suffices to show that the bounded sets in the respective topologies are the same.

We now consider M to be a bounded set in the original topology of X and consider the family of functionals $(F_m)_{m\in M}$ with $F_m(x) = \langle x, m\rangle$.

Since we find a seminorm p_x with the property that

$$|F_m(x)| = |\langle x, m\rangle| \leqslant p_x(m),$$

and the fact that M is bounded, this implies that

$$\sup_{m\in M}|F_m(x)| < \infty.$$

The space X is assumed barrelled and thus we find a continuous seminorm p_1 with the property that $F_m(x) \leqslant p_1(x)$ for all x in X and all m in M. This clearly means that M is bounded with respect to the norm induced by the inner product.

Conversely, if M is bounded with respect to the norm induced by the inner product, i.e.

$$\langle m, m\rangle \leqslant C_M, \quad m\in M,$$

where C_M is a fixed constant, and if x is arbitrary in X, then by the result just obtained concerning the Fréchet–Riesz representation we get that

$$x^*(x) = \langle x, m_x\rangle, \quad m_x\in X$$

and applying the Cauchy–Buniakowsky inequality we obtain that x^* is bounded on M. Thus M is bounded with respect to the original topology of X.

Thus the topology induced by the inner product coincides with the original topology of X and the theorem is proved. □

REMARK 8.9.3. For the case of Banach spaces the above result is an improvement on the Kakutani–Mackey result. This improvement was obtained by Vukman (1982).

We now present results related to the problem when the new norm constructed is in fact the original one. As mentioned above, the first result in this direction was obtained by N. Prijatelj in 1964. The author has improved Prijatelj's result in the sense that it is required that axiom (4) holds only for the class of hermitian idempotent elements. We mention also that the author's proof was inspired by the elegant paper of Prijatelj.

THEOREM 8.9.4 (Istrăţescu, 1984). *Let X be a complex Banach space and suppose that on $L(X)$ there exists an involution '*' such that the following conditions are satisfied:*

(1) $(T + S)^* = T^* + S^*$,
(2) $(T^*)^* = T$,
(3) $(TS)^* = S^*T^*$,
(4) *If P is a hermitian projection (i.e. $P = P^* = P^2$) then* $\|P\| = \|P\|^2$.

Then on X there exists an inner product with the property that

$$\|x\|^2 = \langle x, x\rangle$$

and the involution coincides with the 'adjoint' with respect to the inner product, i.e. for all x, y *in* X,

$$\langle Tx, y \rangle = \langle x, T^*y \rangle.$$

Proof. Let us consider a, b to be two arbitrary linearly independent elements in X which generate a two-dimensional (closed) linear subspace of X which we denote by X_2. We consider the element f in X^* with the following properties:

(1) $f(a) = \|x\|$,
(2) $\|f\| = 1$.

Using f we define an operator P_1 by the formula

$$P_1 x = f(x)a/\|a\|.$$

It is not difficult to see that this operator is hermitian and $P_1 = P_1^2$. Since

$$b = P_1 b + (I - P_1)b$$

we conclude that $(I - P_1)b \neq 0$ since a and b are linearly independent. We choose g in X with the properties

(1) $g(b) = \|b\|$,
(2) $\|g\| = 1$.

We then consider the following operator on X, defined by

$$P_2 x = g(x)/\|(I - P_1)b\| \cdot (I - P_1)b.$$

This is again a hermitian projection and we have clearly the relations:

$$P_1 P_2 = P_2 P_1 = 0.$$

Then the operator, defined by $P = P_1 + P_2$, is a hermitian operator with the norm 1. Then the Kakutani–Bohnenblust theorem asserts that the norm of X is generated by an inner product since X_2 was an arbitrary subspace of dimension 2 in X.

We now present a proof of the Kakutani–Mackey theorem using some ideas from Prijatelj's elegant paper (1964). Let us consider an element $a \neq 0$ in a Banach space X and let f be in X^* with the properties

(1) $f(a) = \|a\|$,
(2) $\|f\| = 1$.

We consider on $L(X)$ the following functional g, defined by

$$g(T) = f(Ta)$$

and it is clear that this is a positive functional (i.e. $g(I) = 1 = \|g\|$). Using this functional we define the following bilinear form on $L(X)$ (we remark that the positivity of g implies that $g(T) = (g(T^*)^*)$,

$$\langle T, S \rangle = g(S^*T)$$

which has all the properties of an inner product except the fact that $\langle T, T \rangle = 0$ does not necessarily imply that $T = 0$. Suppose now that the algebra has the following additional property: for any u, v in, $\|u\| = \|v\|$ there exists a unitary operator U (i.e. $U^*U = UU^* = I$) such that $Uu = v$.

We now consider an operator, defined on $L(X)$ with values in X, which is defined as follows:

$$W(T) = Ta.$$

Then it is clear that $\|W\| = \|a\|$ and if $N(W)$ is the null space of W then $N(W)$ is a left ideal in $L(X)$. Another left ideal in $L(X)$ is defined as follows:

$$\{R, R \in L(X), \langle R, R \rangle = 0\} = N.$$

Then clearly we have the inclusion $N(W) \subset N$ and we show that we have in fact an equality. Now since we suppose that $L(X)$ has the additional property with respect to unitary operators, we consider an aribtrary element x of X, $x \neq 0$. Then we find an unitary operator U with the property that

$$x/\|x\| = Ua.$$

Since $\|x\|U$ is in $L(X)$ we get that

$$W(\|x\|U) = x.$$

Now in order to prove the above equality we must prove the inclusion in the converse sense, and for this it suffices to prove that $C_{N(W)} \subset C_N$. Suppose that T is in $C_{N(W)}$ and then

$$W(T) = Ta = x \neq 0.$$

We find a unitary operator U in $L(X)$ with the property that

$$x = \|x\|Ua \quad \text{or} \quad Ta = \|x\|Ua$$

which gives that

$$(T - \|x\|U)a = 0.$$

This implies that

$$(T - \|x\|U) \in N(W) \subset N.$$

Now if T, S are in $L(X)$ and $T - S$ is in N then

$$g(T^*T) = g(S^*S).$$

Further, this means that

$$\begin{aligned} |g(T^*T) - g(S^*S)| &= |g(T^*(T - S)) + g((T - S)^*S)| \\ &\leqslant |g(T^*(S - S))| + |g((T - S)^*S)|. \end{aligned}$$

and applying the Cauchy–Buniakowsky inequality (for positive functionals) we get that

$$g(T^*(T - S)) = 0, \qquad g((T - S)^*S) = 0.$$

Since $(T - \|x\|U)$ is in N we have that

$$g(T^*T) = g(\|x\|U^*\|x\|U) = \|x\|^2 g(U^*U) = g(I)\|x\|^2.$$

Thus if $x \neq 0$, $g(T^*T) \neq 0$ and thus we have that T is in C_N, which implies the inclusion $N \subset N(W)$ and thus the equality follows.

We now form the quotient space $L(X)/N(W) = L(X)/N$ and we remark that this is isomorphic to X, for each x in X the corresponding element in $L(X)/N$ is the equivalence class determined by the condition $W(T) = |x|$.

Using this isomorphism we can define an inner product on X as follows: if x, y are arbitrary elements in X and $[x]$, $[y]$ denotes the corresponding elements in $L(X)/N$ then we set

$$\langle x, y \rangle = g(u, v)$$

with u arbitrary in $[x]$ and v arbitrary in $[y]$.

From the Cauchy–Buniakowsky inequality (for positive functionals) we see immediately that $\langle\ \rangle$ is well defined in the sense that it does not depend on u, v in the equivalence classes.

Of fundamental importance of the above inner product is the property

$$\langle x, x \rangle = \|x\|^2.$$

Indeed, suppose that $W(T) = x$ and thus

$$\langle x, x \rangle = g(T^*T) = \|x\|^2$$

and the assertion follows.

In order to complete the proof of the Kakutani–Mackey theorem we must show that $L(X)$ satisfies the additional property formulated above.

Let us consider a two-dimensional subspace X_2 of X generated by the linearly independent elements e_1 and e_2. We may suppose without loss of generality that (e_1, e_2) satisfies the properties

(1) $\|e_i\| = 1$, $i = 1, 2$,
(2) for some fixed f in X^*, $f(e_1) = 1$, $f(e_2) = 0$.

Suppose now that x is an arbitrary element of X_2, the subspace generated by e_1 and e_2. We may suppose also that the norm of x is 1. Then

$$x = Ae_1 + Be_2$$

where A, B are scalars.

We must prove that there exists a unitary operator U with the property that $Ue_1 = x$. Now, every operator on X_2 represented by a 2×2 matrix

$$\begin{pmatrix} a_{11} & a_{12} \\ a_{21} & a_{22} \end{pmatrix}$$

and the adjoint by the matrix

$$\begin{pmatrix} \tilde{a}_{11} & \tilde{a}_{12} \\ \tilde{a}_{21} & \tilde{a}_{22} \end{pmatrix}.$$

Since property (1) holds the elements a_{ij} are linear combinations of the conjugate, i.e.

$$a_{ij} = \sum_k b_{ij}^{ik} \bar{a}_{kj}.$$

Further, since properties (2) and (3) hold, taking into account that the identity operator is Hermitian and that the operator $P_1 x = f(x)e_1$ is Hermitian and a projection as well as the fact that for any operator T, T^*T is Hermitian, we get that the coefficients satisfy the following relations:

$$1 = b_{11}^{12} + b_{11}^{22}, \qquad 0 = b_{12}^{11} + b_{21}^{22},$$

$$0 = b_{21}^{11} + b_{21}^{22}, \qquad 1 = b_{22}^{11} + b_{22}^{22},$$

$$1 = b_{11}^{11}, \qquad 0 = b_{12}^{11}, \qquad 0 = b_{21}^{11}, \qquad 0 = b_{22}^{11},$$

$$b_{11}^{21} = b_{12}^{12} = b_{11}^{12} = b_{11}^{21} = b_{22}^{12} = b_{22}^{21} = 0,$$

$$b_{12}^{21} = b_{12}^{21} = r, \quad r \text{ a real number},$$

$$b_{21}^{12} = r^{-1}.$$

Thus we have

$$a_{11} = \bar{a}_{11}, \qquad a_{12} = \bar{a}_{21}, \qquad a_{21} = \bar{a}_{12}, \qquad a_{22} = \bar{a}_{22}$$

(where the bar denotes the conjugate).

Now we use the fact that for unitary operators U the relations must agree again, and thus the following relations must hold:

$$a_{11}\bar{a}_{11} + ra_{21}\bar{a}_{21} = 1 = a_{11}\bar{a}_{11} + r^{-1}a_{12}\bar{a}_{12},$$
$$a_{12}\bar{a}_{11} + ra_{22}\bar{a}_{21} = 0 = ra_{11}\bar{a}_{21} + a_{12}\bar{a}_{22},$$
$$r^{-1}a_{11}\bar{a}_{12} + a_{21}\bar{a}_{22} = 0 = a_{21}\bar{a}_{11} + r^{-1}a_{22}\bar{a}_{12}$$
$$r^{-1}a_{12}\bar{a}_{12} + a_{22}\bar{a}_{22} = 1 = ra_{21}\bar{a}_{21} + a_{22}\bar{a}_{22}.$$

In this case we must have

$$a_{22}\bar{a}_{22} = a_{11}\bar{a}_{11}, \qquad a_{12}\bar{a}_{12} = r(1 - a_{11}\bar{a}_{11}) = r^{-1}(1 - a_{11}\bar{a}_{11}).$$

Thus if we set

$$a_{11} = r\,\mathrm{e}^{i\theta}$$

we have that

$$a_{22} = -a\,\mathrm{e}^{i(\theta - s + t)}, \qquad a_{12} = ((1 - a^2)r)^{1/2}\,\mathrm{e}^{i(\theta + t)}$$

$$a_{21} = ((1 - a^2)r^{-1})^{1/2}\,\mathrm{e}^{i(\theta - s)}$$

where θ, s, t are real numbers (parameters). Thus the matrix representing a unitary operator has the form:

$$\begin{pmatrix} a\,\mathrm{e}^{i\theta} & ((1 - a^2)r)^{1/2}\mathrm{e}^{i(\theta + s)} \\ ((1 - a^2)r^{-1})^{1/2}\,\mathrm{e}^{i(\theta - t)} & -a\,\mathrm{e}^{i(\theta - s - t)} \end{pmatrix}$$

Now if x is an arbitrary element of X_2 (of norm 1) we must show that for some unitary operator U, $Ue_1 = x$.

$$a\,\mathrm{e}^{i\theta}\,e_1 + (1 - a^2)^{1/2}\,\mathrm{e}^{i(\theta - t)} = Ae_1 + Be_2 = x.$$

Now if

$$A = a_1\,\mathrm{e}^{i\theta_1}, B = a_2\,\mathrm{e}^{i\theta_2}$$

from

$$A/B = (a_1/a_2)\cdot e^{i(\theta_1-\theta_2)} = (1-a^2)^{1/2}a^{-1}\,e^{-it}$$

we conclude that

$$\theta_1 - \theta_2 = t, \quad a = (a_1^2/(a_1^2 + a_2^2))$$

and thus

$$x = C(a\,e^{i\theta_1}e_1 + (1-a^2)^{1/2}\,e^{i(\theta_1-\theta_2)}e_2.$$

From the fact that the norm of x is 1 we conclude that $a = a_1$, $a_2 = (1-a^2)^{1/2}$. Thus the matrix of the unitary operator is

$$\begin{pmatrix} A\,e^{i\theta_1} & (1-A^2)^{1/2}\,e^{i(\theta+s)} \\ (1-A^2)^{1/2}\,e^{i(\theta-t)} & -A\,e^{i(\theta+t-s)} \end{pmatrix}.$$

Thus the additional property holds for our algebra and this completes the proof of the theorem. □

8.10. Hermitian Decomposition of a Banach Space and Inner Product Spaces

As we know, if X is a Banach space and on each two-dimensional subspace there exists a projection of norm 1 then the norm of X is in fact generated by an inner product. Thus we see that in order for a Banach space to have its norm generated by an inner product it is necessary and sufficient that every two-dimensional subspace be the range of a projection of norm 1. If X is an inner product space then it is clear that the projection is Hermitian. As we know, a successful generalization of the notion of Hermitianity for the case of elements of complex Banach algebras was proposed by Vidav (1956). Later G. Lumer used his 'semi-inner products' on Banach spaces to characterize Hermitianity which, in some sense, is similar to the case of complete inner product spaces.

Using certain classes of Hermitian elements, E. Berkson, N. J. Kalton and G. V. Wood were able to characterize inner product spaces and at the same time they pointed out certain parts in a Banach space which are like inner product spaces. In this section give some results on these facts. For the reader's convenience we recall that an element h of a complex Banach algebra is said to be Hermitian if, for all t in R, the element e^{ith} (defined in the usual way) has norm 1; equivalently if

$$\|1 + aih\| = 1 + o(a)$$

for a in R and $a \to 0$. (Here we assume that the complex Banach algebra has an identity. For further information on Hermitian elements in a Banach algebra we refer to the author's book on linear operators (1981).

DEFINITION 8.10.1. A Hermitian projection on a Banach space X is any element P in $L(X)$ which is Hermitian (an element of $L(X)$) and $P^2 = P$.

DEFINITION 8.10.2 (Kalton and Wood, 1976). A (Hermitian) splitting of a Banach space X is a family $(P_\alpha, \alpha \in I)$ of Hermitian projections with the property that $P_\alpha \cdot P_\beta = 0$ for $\alpha \neq \beta$. The extent of the splitting $(P_\alpha, \alpha \in I)$ is the smallest closed linear subspace containing all $P_\alpha X$. We say that $(P_\alpha, \alpha \in I)$ is a Hermitian decomposition of X if $(P_\alpha, \alpha \in I)$ is a (Hermitian) splitting of X whose extent is X.

DEFINITION 8.10.3 (Berkson, 1972). We say that a closed linear subspace X_1 of the Banach space X is 'split' if there exists a Hermitian projection P in $L(X)$ whose range is X_1.

REMARK 8.10.4. It is not difficult to see that the projection in Definition 8.10.3 is unique.

LEMMA 8.10.5. Let $(P_\alpha, \alpha \in I)$ be a (Hermitian) splitting of a Banach space X and let X_1 be its extent. Then for each x in X_1, $(P_\alpha x, \alpha \in I)$ is summable with the sum equal to x.

Proof. Let us consider the set of all finite parts of I, $P_F(I)$ and let $J \in P_F(I)$. Let us consider further the element $P_J = \Sigma_{\alpha \in J} P_\alpha$. Then from the splitting property we see that P_J is a Hermitian projection. This gives that the norm of P_J is 1. Then $(P_J, J \in P_F(I))$ is an equicontinuous family of mappings and thus the set $(x, P_J x \to x)$ is a closed linear subspace of X and since for each α in I, $P_\alpha X$ is in this subspace, the assertion follows.

The following proposition shows a connection between the property of a space to be split and a Hermitian decomposition.

PROPOSITION 8.10.6 (Kalton and Wood, 1976). *Let X be a Banach space and $(P_\alpha, \alpha \in I)$ be a splitting of X whose extent is X_1. Then the following assertions are equivalent:*

(1) *For $x \in X$, $\sum_{\alpha \in I}^{P} \alpha I$ converges.*
(2) *X_1 is split.*
(3) *There exists a Hermitian decomposition of X containing $(P_\alpha, \alpha \in I)$.*

Proof. (1)→(2). We define for each x in X, Px by the formula

$$Px = \sum P_\alpha x.$$

Then clearly P is a Hermitian operator which is also projection and the range is X_1.

(2)→(3). Let P be the projection of X onto X_1 and we consider the system of projections $(P_\alpha, \alpha \in I) \cup (I - P)$. Then clearly we have (3).

(3)→(1). Suppose that $P_\alpha, \alpha \in I) \cup (P_\beta, \beta \in L)$ is a Hermitian decomposition of X. Then for each x in X, $\Sigma_\gamma Q_\gamma x$ is convergent, $\{Q_\gamma \zeta = (P_\alpha, \alpha \in I) \cup (P_\beta, \beta \in L)$. This implies that $\Sigma_\alpha P_\alpha x$ is convergent. The proposition is proved. □

In Kalton and Wood (1976) it is proved that if X contains no subspace isomorphic to c_0 (the space of all convergent sequences, converging to zero) then (1), (2), (3) are implied by this fact.

The following result is concerned with linear operators related to diagonal linear operators in the theory of sequence spaces.

THEOREM 8.10.7 (Kalton and Wood, 1976). *Let X be a Banach space and $(P_\alpha, \alpha \in I)$ be a Hermitian decomposition of X. Let T be in $L(X)$. Then the following* *assertions hold:*

(1) $\Sigma_{\alpha\in I}P_\alpha TP_\alpha$ *converges in the strong topology to an operator* $D\in L(X)$ *with* $\|D\|\leqslant\|T\|$.

(2) *If T is Hermitian then D is Hermitian.*

Proof. First we show that the assertions hold for I a finite set, say $I=(1,2,3,\ldots,n)$. We consider the operator S_p defined by the relation:

$$S_p=\sum_{j=1}^{n-p}\sum_{k=1}^{n-p}P_jTP_k+\sum_{j+n-p+1}^{n}P_jTP_j$$

for p in $[0,n]$ (with the usual convention that the summation over the empty set is zero). Then we show that $\|S_p\|\leqslant\|T\|$ and when T is Hermitian then S_p is Hermitian. Since $S_0=T$ the assertion is trivially true in this case. Suppose that it is true for $p=q$ with $q\geqslant 0$. We consider now the operator $I-2P_{n-q}$ and we remark that $R^2=(I-2P_{n-q})^2=I$. This implies that the inverse of $I-2P_{n-q}$ is $I-2P_{n-q}$ and the fact that it is an isometry. Then we have

$$\|(I-2P_{n-q})S_q(I-2P_{n-q})\|\leqslant\|T\|$$

and thus

$$\|(I-2P_{n-q})S_q(I-2P_{n-q})+S_q\|\leqslant 2\|T\|$$

or

$$\|S_q-P_{n-q}S_q-S_qP_{n-q}+2P_{n-q}S_qP_{n-q}\|\leqslant\|T\|.$$

But

$$S_q-P_{n-q}S_q-S_qP_{n-q}+2P_{n-q}S_qP_{n-q}=S_{q+1}$$

and since if S_q is Hermitian then

$$e^{it(I-2P_{n-q})S_q(I-2P_{n-q})}=(I-2P_{n-q})e^{itS_q}(I-2P_{n-q})$$

which implies that S_{q+1} is Hermitian.

From the induction we have that S_n is Hermitian and $\|S_n\|\leqslant\|T\|$.

We now consider the case when I is not finite. For $\varepsilon>0$ and x in X there exists a finite subset of $I, J_{x,\varepsilon}$ such that for any J with $J\cap J_{x,\varepsilon}=\emptyset$ the following inequality holds:

$$\left\|\sum_{\alpha\in J}Px\right\|\leqslant\varepsilon.$$

If J is a finite subset with $J\cap J_{x,\varepsilon}=\emptyset$ then

$$\left\|\sum_{\alpha\in J}P_\alpha TP_\alpha+QTQ\right\|\leqslant\|T\|,$$

where

$$Q=I-\sum_{\alpha\in J}P_\alpha$$

and thus

$$\left\| \sum_{\alpha\in J}(P_\alpha TP_\alpha + QTQ)\left(\sum_{\alpha\in J} P_\alpha x\right)\right\| \leqslant \|T\|.$$

Then we have that for any finite subset J with $J \cap J_{x}, = \emptyset$,

$$\left\| \sum_{\alpha\in J} P_\alpha TP_\alpha x \right\| \leqslant \varepsilon\|T\|$$

which implies that

$$\sum_{\alpha\in I} PTP$$

converges to an operator D in the strong operator topology.

If J is a finite subset in I we set $Q_J = I - \Sigma_{\alpha\varepsilon J} P_\alpha$ and thus for x in X and $J \supseteq J_{x,\varepsilon}$ we have that

$$\|Q_J TQ_J x\| = \left\| \left(\sum_{\alpha\in J} P_\alpha TP_\alpha + Q_J TQ_J\right) Q_J x \right\| \leqslant \|T\|\|Q_J x\| \leqslant \varepsilon\|T\|$$

and thus $(Q_J TQ_J)$ converges to zero in the strong operator topology and thus

$$D = \lim_{J\in P_F(I)} \left(\sum_{\alpha\in J} P_\alpha TP_\alpha + Q_J TQ_J\right)$$

which implies that

$$\|D\| \leqslant \|T\|$$

and the fact that if T is Hermitian then D is Hermitian.

We now define the notion of a Hermitian element of a Banach space which plays a fundamental role in the following characterizations of inner product spaces.

DEFINITION 8.10.8. Let X be a complex Banach space and let x be an element in X. We say that x is a Hermitian element if there exists a Hermitian projection P_x in $L(X)$ such that the range of P_x is the linear span of x.

The set of all Hermitian elements of a Banach space X is denoted by $h(X)$ and the linear closure of $h(X)$ is denoted by $\hat{h}(X)$.

Using the extension theorems we can easily prove that on X there exists a mapping with values in X^* such that (if this mapping is denoted by $x \to x^*$) the following conditions are satisfied:

(1) $\|x\| = \|x^*\|$.

(2) $x^*(x) = \|x\|^2$.

(3) If $x \in h(X)$ then $\|x\|^2 P_x = x^*(\,,)x$.

We give now some properties of $h(X)$.

PROPOSITION 8.10.9. *$h(X)$ is a closed subset of X.*

Proof. Let (x_n) be a sequence of elements of $h(X)$ and suppose that $\lim x_n = x$. We suppose without loss of generality that $x \neq 0$. Thus $\inf \|x_n\| = m > 0$. We consider the following elements of X^* defined by

$$g_n = \|x_n\|^{-2} x_n^*$$

which gives that

$$y \to g_n(y)x_n \tag{*}$$

is a Hermitian projection. Also we have that

$$\|g_n\| = \|x_n\|^{-1} \leqslant m^{-1}.$$

We can suppose without loss of generality that (g_n) converges in the *-weak topology to an element, say g. Clearly g is in X^* and

$$|1 - g_n(x)| = |g_n(x) - g_n(x_n)| = |g_n(x - x_n)| \leqslant m^{-1}\|x - x_n\|$$

which gives that $g(x) = 1$ and thus

$$y \to g(y)x \tag{**}$$

is a nonzero projection. Let z be arbitrary in X and let t be arbitrary in R. Then we have

$$\|z + (e^{it} - 1)g_n(z)(x_n)\| = \|z\|$$

from the hermitianity of (*). For $n \to \infty$ we obtain that

$$\|z + (e^{it} - 1)g(z)x\| = \|z\|$$

and thus (**) is a Hermitian operator. This completes the proof. □

DEFINITION 8.10.10. If $x \to x^*$ is the mapping defined above then we set $[x, y] = y^*(x)$.

REMARK 8.10.11. It is easy to see that this satisfies all the properties of a semi-inner product in Lumer's sense.

PROPOSITION 8.10.12. *For any x, y in $h(X)$ with $[x, y] = 0$, it follows that $[y, x] = 0$.*

Proof. From $[x, y] = 0$ we have that $Pyx = 0$ and thus $P_y P_x = 0$. Then the assertion follows from the following general result. □

PROPOSITION 8.10.13. *Let P, Q be two Hermitian projections and suppose that $PQ = 0$. Then $QP = 0$. Since the spectrum of PQ is the same as the spectrum of PQ except eventually the point zero, and from the fact that $i(PQ - QP)$ is hermitian, we must have $QP = 0$.*

PROPOSITION 8.10.14. *If x, y are in $h(X)$ and $[x, y] = 0$ then $y \perp_B x$.*

Proof. Indeed, since $y^*(x) = 0$ we have that

$$y^*(y + kx) = y^*(y)$$

and this clearly gives our assertion if we consider y^* normalized.

REMARK 8.10.15. The result in Proposition 8.10.14 will be used to give an alternate proof of characterization of inner product spaces in terms of Hermitian elements when the dimension of the space is at least 3.

The following notion defines, in the setting of Banach spaces, the corresponding notion of orthonormal systems from the case of inner product spaces.

DEFINITION 8.10.16 (Berkson, 1972; Kalton and Wood, 1976). Let X be a complex Banach space and let $(e_\alpha, \alpha \in I)$ be a family of elements of X. Then we say that $(e_\alpha, \alpha \in I)$ is an orthonormal system if the following conditions are satisfied:

(1) $e_\alpha \in h(X)$ for each α in I,

(2) $[e_\alpha, e_\beta] = \delta_{\alpha\beta}$.

The extent of an orthonormal system is the closed linear span of $(e_\alpha, \alpha \in I)$. We say that the orthonormal system is complete if the extent of $(e_\alpha, \alpha \in I)$ is X.

DEFINITION 8.10.17. We say that two closed subspaces of X, M and N, are mutually orthogonal if there exists two Hermitian projections P_M and P_N whose ranges are M and N respectively and $P_M P_N = 0$.

For the characterization of inner product spaces in terms of $h(X)$ we need the following technical propositions.

PROPOSITION 8.10.18. *Let us consider the space* $X = C^2$ *with a norm* $\|,\|$ *such that* $\|(1,0)\| = \|(0,1)\| = 1$. *Let*

$$T = \begin{pmatrix} 0 & a \\ b & 0 \end{pmatrix}$$

be the matrix of a Hermitian operator on X. *Then* $a = \bar{b}$.

Proof. We suppose that $T \neq 0$. Then e^{itT} is an isometry for all real numbers t. This implies that $T^2 \neq 0$ and thus $ab \neq 0$. Since T is Hermitian, T^2 has only positive numbers in the spectrum. Then we must have $ab = c^2$ for some $c > 0$ (since $ab \neq 0$). But

$$e^{itT} = \begin{pmatrix} \cos ct & iac^{-1} \sin ct \\ ibc^{-1} \sin ct & \cos ct \end{pmatrix}.$$

We take $t_0 = /2c$ and thus

$$e^{it_0 T} = \begin{pmatrix} 0 & iac^{-1} \\ ibc^{-1} & 0 \end{pmatrix}$$

But this is an isometry and from $\|(1,0)\| = \|(0,1)\| = 1$ we obtain that $c = |a| = |b|$. But this clearly implies the assertion of the proposition.

PROPOSITION 8.10.19. *Let* X *be a Banach space and let* (e_1, e_2) *be an orthonormal system in* X. *Suppose that for some* T *in* $L(X)$, T *Hermitian, we have* $[Te_1, e_2] \neq 0$. *Then whenever* $A = (a_{jk})_{j,k=1}^2$ *is a Hermitian matrix the operator*

$$y \to \sum_{j,k=1}^{2} a_{jk} e_j^*(y) e_k = Ry$$

is Hermitian and for and a, b in C

$$\|ae_1 + be_2\| = |a|^2 + |b|^2.$$

Proof. Let

$$m = [Te_1, e_2], \qquad n = [Te_2, e_1]$$

and then the operator

$$y \to me_1^*(y)e_2 + ne_2^*(y)e_1$$

is Hermitian (since it is of the form $PTQ + QTP$ where P, Q are Hermitian projections, T Hermitian and $PQ = 0$). From the above proposition (if we consider the restriction to the invariant subspace generated by e_1 and e_2) we have that $m = \bar{n}$. If we now consider the operator

$$y \to i(me_1^*(y)e_2 - ne_2^*(y)e_1)$$

then this is a Hermitian operator (being of the form $i(PTQ - QTP)$ with P, Q Hermitian projection, T Hermitian and $PQ = 0$).

Since the set of Hermitian operators on a Banach space is a closed real subspace we obtain that

$$y \to e_1^*(y)e_2, \qquad y \to e_2^*(y)e_1$$

are Hermitian operators. This clearly implies the first part of the proposition. If a, b are complex numbers such that $|a^2 + |b|^2 = 1$ then the operator with the matrix

$$U = \begin{pmatrix} a & \bar{b} \\ b & \bar{a} \end{pmatrix}$$

is unitary and thus there exists a Hermitian operator whose matrix is (a_{jk}), say A, such that $U = e^{iA}$. Then

$$y \to \sum_{J,k} a_{jk} e_j^*(y)e_k$$

is Hermitian and thus

$$y \to ae_1^*(y)e_1 + \bar{a}e_2^*(y)e_2 + be_1^*(y)e_2 - \bar{b}e_2^*(y)e_1$$

is an isometry. Applying this to $y = e_1$ we obtain that

$$\|ae_1 + be_2\| = 1.$$

If a, b are arbitrary then we apply the result obtained for $a/(|a|^2 + |b|^2)$, $b/(|a|^2 + |b|^2)$.

PROPOSITION 8.10.20. *Let X be a Banach space and let (e_1, e_2) be an orthonormal system in X. Suppose that for some a, b with $ab \neq 0$, $ae_1 + be_2$ is a Hermitian element of X. Then there exists T is L(X), T Hermitian, with $[Te_1, e_2] \neq 0$ and Proposition* 8.10.19 *holds.*

Proof. We consider the linear projection P on the subspace generated by $ae_1 + be_2$. Then P has a nontrivial diagonal matrix when it is restricted to the line determined by e_1 and e_2. This implies that either $[Pe_1, e_2] \neq 0$ or $[Pe_2, e_1] \neq 0$. But then both are nonzero and the linear span of e_1 and e_2 is in $h(X)$. □

THEOREM 8.10.21 (Berkson, 1972). *Let X be a Banach space. Then X is an inner product space if and only if $h(X) = X$.*

Proof. The condition is clearly necessary. Suppose now that X is a Banach space with the property that $h(X) = X$. Let (e_1, e_2) be a normalized basis of a two-dimensional subspace of X with the property that $[e_2, e_1] = 0$. Then from Proposition 8.10.20 for any complex numbers a, b we have

$$\|ae_1 + be_2\| = |a|^2 + |b|^2$$

and thus we see that the Jordan–von Neumann condition is satisfied. This implies that X is an inner product space. □

REMARK 8.10.22 (V. I. Istrăţescu). Another proof of Theorem 8.10.21 for Banach spaces having dimension at least 3 follows from Proposition 8.10.14. Indeed, if $h(X) = X$ then we see that the orthogonality in Birkhoff's sense is symmetric. From the results in Chapter 4 (Theorem 4.12.34) we have that X is an inner product space.

We mention now two related characterizations of inner product spaces. For a proof we refer to N. J. Kalton and G. V. Wood paper (1976).

First we give the following.

DEFINITION 8.10.23 (Rolewicz, 1972). Let X be a Banach space with the norm $\|,\|$. We say that X is convex transitive if whenever x_0 is in X, $\|x_0\| = 1$ then the convex closed cover of $(Ix_0, I \in G_I)$, G_I the group of isometries of X, is the unit ball of X.

It is known (Rolewicz, 1972) that a Banach space $L^p_{[0,1]}$, $p \in (1, \infty)$ is convex transitive and $C_{[0,1]}$ is not convex transitive.

THEOREM 8.10.24 (Kalton and Wood, 1976). *Let X be a Banach space which is convex transitive. If $h(X) \neq (0)$ then X is an inner product space.*

REMARK 8.10.25. It is clear from the definition of convex transitivity that this is an intrinsic property of the norm of a Banach space.

For the following characterization we need the notion of maximal norm.

DEFINITION 8.10.26 (Rolewicz, 1972). Let X be a Banach space and let $\mathrm{GL}(X)$ be the group of all bounded invertible operators on X. Then if we set

$$\|x\|_* = \sup(\|Tx\|,\ T \in \mathrm{GL}(X))$$

then this is a new norm, equivalent with the original norm of X and each T in $\mathrm{GL}(X)$, with respect to this new norm, is obviously an isometry. We say that a subgroup of $\mathrm{GL}(X)$ is maximal if it is not contained in any larger subgroup of $\mathrm{GL}(X)$. We say that a norm is maximal if the subgroup of isometries is maximal.

From the above definition of the norm $\|,\|_*$ we have the following obvious result.

PROPOSITION 8.10.27. *If G is a maximal subgroup then there exists an equivalent norm on X such that this is a maximal norm.*

THEOREM 8.10.28 (Rolewicz, 1972). *Let X be a finite dimensional Banach space with the norm $\|,\|$. Then $\|,\|$ is a maximal norm if and only if the norm is Euclidean.*

A result of A. Pełczyński and S. Rolewicz states that in c_0 and l^p, L^p $(1 \leqslant p < \infty)$, the usual norms are maximal. For some compact spaces $C(K)$ has no maximal norms.

REMARK 8.10.29. Related to the above characterization of a finite dimensional inner product space we may ask under what additional assumption (related, for example, to $h(X)$) is a Banach space whose norm is maximal an inner product space or an isomorph to an inner product space?

For other interesting results on transitive norms we refer to Chapter 9 of S. Rolewicz's excellent book (1972).

8.11. Classes of Hermitian Elements and Inner Product Structures

We have remarked that the existence of some classes of elements in $L(X)$ implies in fact that the norm of X, or an equivalent norm, is generated by an inner product. In this section we present further results of this type in which the class of Hermitian elements is involved. We remark that the Kakutani–Bohnenblust theorem may be formulated as follows.

THEOREM 8.11.1 *Let X be a Banach space of dimension at least* 3. *Then the norm of X is generated by an inner product iff every two-dimensional subspace is the range of a Hermitian projection.*

We have the following result, which is well known.

LEMMA 8.11.2. *Let P be a Banach space of dimension at least* 3. *Then the operator P is Hermitian iff I–P is Hermitian.*

Proof. This follows immediately considering the numerical range of P. We leave the details to the reader.

From this, in conjunction with the Kakutani–Bohnenblust theorem, we get the following result.

PROPOSITION 8.11.3. *Let X be a Banach space of dimension at least* 3. *Then the norm of X is generated by an inner product iff every one-dimensional subspace of X is the range of a Hermitian idempotent.*

Also, we have the following result by P. L. Papini:

THEOREM 8.11.4 (Papini, 1974). *Let X be a Banach space. Then the norm of X is generated by an inner product iff every one-dimensional subspace is the range of an adjoint abelian projection.*

Before we give the proof we present some notions which may be unknown to the reader.

DEFINITION 8.11.5 (Stampfli, 1969). The operator T in $L(X)$ is said to be adjoint abelian if for some inner product $[\,,]$ on X we have

$$[Tx, y] = [x, Ty]$$

for all x, y in X.

DEFINITION 8.11.6. An operator T in $L(X)$ is said to be a numerical contraction if the numerical radius of T,

$$w(T) = \sup(|a|, a \in W(T), W(T) = \{[Tx, x], \|x\| = 1\}) = 1.$$

Then we have the following result.

PROPOSITION 8.11.7. *For any idempotent P in $L(X)$ the following conditions are equivalent:*

(1) $\|P\| = 1$.
(2) $w(P) = 1$.

Since the proof is not difficult we leave it to the reader.

Proof of Theorem 8.11.4. It is clear that the condition is necessary. Suppose now that X satisfies the condition. If $\dim X \geqslant 3$ then the assertion follows from Theorem 8.11.1. If $\dim X = 2$ then, if X is an adjoint abelian projection on $[x]$,

$$X = [x] \oplus [y], \qquad [y] = (I - P)X$$

and this gives the inner product on X.

For other interesting results in which the set of all hermitian elements and the set of all adjoint abelian elements coincide we refer to Fleming and Jamison (1974, Th. 3.6).

Now we present a result in which the existence of many hermitian elements forces the Banach algebra to be a B*-algebra.

THEOREM 8.11.8. *Let A be a complex Banach algebra and suppose that $A = H(A) + iH(A)$, $H(A)$ is the set of all hermitian elements of A and the left regular representation of $A(a \to T_a, T_a(x) = ax)$ is faithful. Then A is a pre-B*-algebra with the norm*

$$a \to \|R_a\|,$$

where R is the left representation of A.

COROLLARY 8.11.9. *Let X be a complex Banach space and let $L(X)$ be the Banach algebra of all bounded linear operators on X. If $A = H(A) + iH(A)$ then the norm of X is generated by an inner product.*

For a proof of these result we refer to J. Duncan and F. Bonsall's (1973) book, vol. 2.

8.12. A Variational Characterization of Inner Product Structures

Let X be a complex complete inner product space and let $L(X)$ be the set of all bounded linear operators on X. By $\langle\, , \rangle$ we denote the inner product on X. For any closed and bounded subset in the plane we denote by $r(\ )$ the radius of the minimal (unique) closed disk

$$D_r(p) = (q, \|p - q\| \leqslant r)$$

such that this contains the set ( ). For any T in $L(X)$ we set

$$n_*(T) = \inf_{z\in C}(\|T - z\|^2)$$

and for each x in $X, \gamma(T, x)$ denotes the number defined by the formula:

$$\gamma(T,x) = \|Tx\|^2 - |\langle Tx, x\rangle|^2.$$

From a simple computation we conclude that for all z in C,

$$\gamma(T - z, x) = \gamma(T, x).$$

We define

$$\gamma(T) = \sup(\gamma(T, x), \|x\| = 1).$$

Zeev Nehari (see Seidman, 1969) has remarked the following inequality holds for all bounded operators on X:

$$\gamma(T) \leqslant n_*(T)$$

and has raised the question: for which bounded linear operators on X do we have in fact an equality?

It is not difficult to see that equality holds whenever T is a normal operator. Seidman (1969) has proved that equality holds for a class of operators called 'normal-like', i.e. operators T for which

$$n_*(N) = r^2(\sigma(T)).$$

It is clear also that

$$\gamma(T) = n_*(T) \tag{*}$$

may be formulated as a 'minimax principle',

$$\sup_{\substack{x\in X\\ \|x\|=1}} \inf_z \|(T - z)x\| = \inf_z \sup_{\substack{x\in X\\ \|x\|=1}} \|(T - z)x\|. \tag{**}$$

which follows from the fact that

$$\inf_{z\in C}\|(T - z)x\|^2 = \|Tx\|^2 - |\langle Tx, x\rangle|^2 = \gamma(T, x).$$

Then we have the following result of E. Asplund and V. Pták.

THEOREM 8.12.1 (Asplund and Ptak, 1971). *Let X be a Banach space. Then the following assertions are equivalent*:

(1) *The norm of X is generated by an inner product.*
(2) (**) *holds for all bounded linear operators on X.*

In what follows we present a proof of the implication (1)→(2), and for (2)→(1) we refer to Asplund and Ptak (1971). In order to prove the assertion it suffices to prove the inequality

$$n_*(T) \leqslant \gamma(T).$$

We show that there exists a point in the complex plane, say u_T, with the property that

$$n_*(T) = \|T - u_T\|^2.$$

For this we show that there exists a sequence of elements (x_n) in X such that:

(1) $\|x_n\| = 1, n = 1, 2, 3, \ldots$

(2) $\langle (T - u_T)x_n, x_n \rangle \to 0.$

(3) $\|(T - u_T)x_n\| \to \|T - u_T\|.$

Suppose that such a sequence does not exist. We may assume without loss of generality that for all sequences satisfying (1) and (3) the sequence in (2) is away from zero, more preci3ely in $\operatorname{Re} z \geqslant t$ for some $t > 0$.

Let $M = (x, \|x\| = 1$ and $\operatorname{Re}\langle Tx, x\rangle \leqslant t/2)$ and $m = \sup(\|(T - u_T)x\|, x \in M)$ which is clear strictly smaller than $\|T - u_T\|$. If we set

$$c = \min\ (t/2,\ \|T - u_T\| - m/2)$$

then for the operator $T - u_T - c$ we have

$$\|(T - u_T - c)x\| \leqslant \|Tx - u_T x\| + c < \|T - u_T\|$$

for x in M. Now if

$$(T - u_T)x = (a + ibx) + y,$$

where x is in M and $\langle x, y\rangle = 0$, we get

$$\|(T - u_T - c)x\|^2 = (a - c)^2 + b^2 + \|y\|^2 = \|(T - u_T)x\|^2 + \\ + (c - 2ac) < \|T - u_T\|^2$$

and thus

$$\|T - u_T - c\| < \|T - u_T\|$$

which is a contradiction. Thus such a sequence exists and clearly we have

$$\lim \|(T - u_T)x_n\|^2 = \lim \sup(\|Tx_n\|^2 - |u_T|^2) = \gamma(T).$$

Thus

$$n_*(T) \leqslant \gamma(T)$$

if we prove the existence of u_T with the above property. But this follows from the fact that $z \to \|T - z\|^2$ is a strictly convex function.

8.13. Von Neumann Spectral Sets and a Characterization of Inner Product Spaces

Let X be a complex Banach space and let $L(X)$ be the set of all bounded linear operators on X. If T is in $L(X)$ then for any bounded rational function with poles outside the spectrum $\sigma(T)$ of T we can define the operator $f(T)$. We may ask about the connection between the norm of $f(T)$ and the norm of $\|f\|$ as an element of $C(\sigma(T))$. In a well-known paper von Neumann (1951) has introduced the notion of spectral set of an operator which permits the expression of some beautiful and important properties of certain classes of operators defined on complete inner product spaces.

DEFINITION 8.13.1. (von Neumann, 1951). A closed set F in the plane is said to be a spectral set for T in $L(X)$ if the following conditions are fulfilled:

(1) F contains $\sigma(T)$.
(2) For any rational function with poles outside F the operator $f(T)$ satisfies the inequality

$$\|f(T)\| \leqslant \sup(|f(z)|, z \in F) = \|f\|_F.$$

One of the basic result of John von Neumann is as follows:

THEOREM 8.13.2 (von Neumann, 1951). *Let X be a complete inner product space and let T be in $L(X)$ with $\|T\| \leqslant 1$. Then $\bar{D}_1 = (z, |z| \leqslant 1)$ is a spectral set for T.*

There are several proofs of this result. The original proof of John von Neumann uses certain results of I. Schur. Another uses the dilation theory, and will be given in Chapter 10.

Of course we may ask if a similar result holds for operators defined on Banach spaces. The following result found by John von Neumann shows that the property stated in Theorem 8.13.2 is a characteristic property of inner product spaces.

THEOREM 8.13.3 (John von Neumann). *If X is a Banach space such that the set $\bar{D}_1$ is a spectral set for any operator T in $L(X)$ with $\|T\| \leqslant 1$, then X is an inner product space.*

Proof. (Fusias, 1957). Let x_0 be an arbitrary element of X and then, by the extension theorems, there exists an element x_0^* in X^* with the property that $\|x_0\|\|x_0^*\| < 1$. We define the following operator on X:

$$Tx = x_0^*(x)x_0$$

and this clearly is a contradiction. Then by the hypothesis of the theorem, $\bar{D}_1$ is a spectral set for T. The function

$$r_a(z) = (z + a)/(1 + \bar{a}z), \quad |a| < 1$$

is clearly a rational function and then $\|r_a(T)\| \leqslant 1$ because $\|r_a\|_{\bar{D}_1} \leqslant 1$. This implies that for all x in X the following inequality holds:

$$\|(T + aI)x\| \leqslant \|(I + \bar{a}T)x\|.$$

From the form of T we have further that for all x in X the following relation holds:

$$\|x_0^*(x)x_0 + ax\| \leqslant \|x + \bar{a}x_0^*(x)x_0\|.$$

Now if x, y are arbitrary elements in X we can assume without loss of generality that $\|x\| = \|y\| > 0$. Then we find x_0^* in X^* such that $\|x_0^*\| = \|x\|^{-1}$, $x_0^*(x) = 1$ and for $y = x_0$ the above inequality shows that

$$\|y + ax\| \leqslant \|x + \bar{a}y\|$$

and a continuity argument shows that the above inequality holds also in the case

$|a| = 1$. Suppose now that $\|x\| = \|y\|$ and from the symmetric role of x and y and a and $\bar{a}$ we conclude that for all $|a| = 1$ and all x, y with $\|x\| = \|y\|$ the following inequality holds:

$$\|x + \bar{a}y\| \leqslant \|y + ax\|.$$

Thus if $|a| \leqslant 1$ we must have the equality:

$$\|x + ay\| = \|y + ax\|.$$

If $|a| > 1$ then $b = a^{-1}$ with $|b| < 1$ and then, applying the result just obtained, we have that

$$\|x + ay\| = \|y + ax\|.$$

Thus

$$\|x + ay\| = \|y + ax\| \tag{*}$$

holds for all x, y in X with $\|x\| = \|y\|$ and all a. This implies by the results in Chapter 4, Section 4.4 (Ficken's theorem (4.4.1)), that X is an inner product space.

8.14. A Series-Immersed Isomorphic Characterization of Complete Inner Product Spaces

An interesting characterization of Banach spaces isomorphic to complete inner product spaces was obtained by Kalton and Ruckle (1973) and Ruckle (1977, 1983) using the notion of a 'series immersed in a Banach space' defined as follows.

DEFINITION 8.14.1. A Banach space X is said to be series immersed in the Banach space Y if the following assertion holds: if Σx_n is an absolutely divergent series in X (i.e. $\Sigma\|x_n\| = \infty$) then there exists T in $L(X, Y)$ such that $\Sigma T x_n$ diverges absolutely.

An equivalent formulation can be given using absolutely p-divergent series ($1 \leqslant p \leqslant \infty$) (see Ruckle, 1977).

For a proof of the following result we refer to Kalton and Ruckle (1973).

THEOREM 8.14.2 (Kalton and Ruckle, 1973). *A Banach space X is not isomorphic to a subspace of a space $L^p(\mu)$ for any measure μ($1 \leqslant p \leqslant \infty$) iff there exists in X a series Σx_n such that*

(1) $\sum \|x_n\| = \infty$,
(2) *for each T in $L(X, l^p)$, we have $\sum \|T x_n\|^p < \infty$.*

COROLLARY 8.14.3 (Kalton and Ruckle, 1973). *A Banach space X is not isomorphic to a complete inner product space iff there exists a series $\Sigma x_n, x_n \in X$, such that*

(1) $\sum \|x_n\| = \infty$,
(2) *for each T in $L(X, H)$, H is a complete inner product space, $\Sigma \|T x_n\| < \infty$ holds.*

For conditions on a Banach space X to be series immersed in another Banach space

Y, as well as for applications to the famous Dvoretzky–Rogers theorem, we refer to the papers by N. J. Kalton and W. H. Ruckle mentioned above.

8.15. A Symmetric-Invariant Characterization of $L^2_{[0,1]}$.

An interesting characterization of $L^2_{[0,1]}$ in the class of symmetric spaces was obtained by Potepun (1974). In this section we present his interesting result as well as some related results.

THEOREM 8.15.1 (Potepun, 1974). *Let X be a symmetric Banach space of measurable functions on* $[0,1]$. *Suppose that $I: X \to L^2_{[0,1]}$ is an isomorphism. Then* $X = L^2_{[0,1]}$.

Proof. We consider the following inner product on X,

$$\langle x, y\rangle = \langle Ix, Iy\rangle_{L^2_{[0,1]}}$$

and

$$x \to \|x\|_1 = \langle x, x\rangle^{1/2}.$$

Since I is supposed to be an isomorphism we have the following inequalities:

$$m^{-1}\|x\|_1^2 \leqslant \|x\|^2 \leqslant m\|x\|_1^2$$

for some $m > 0$.

Let $(e_i)_1^N$ be a decomposition of $[0,1]$ such that:

(1) $e_i \cap e_j = \emptyset, \quad i \neq j$,

(2) $\mu(e_i) = 1/N$.

We now consider the subspace of all functions of the form

$$x = \sum_{i=1}^N a_i \chi_{e_i}(t) \qquad \chi_e(t) = \begin{cases} 1, & t \in e \\ 0, & t \notin e \end{cases}$$

which is clearly finite dimensional. It is clear also that

$$\langle x, x\rangle = \sum_{i,j=1}^N a_{ij} a_i a_j.$$

Now if $\varepsilon = (\varepsilon_1, \ldots, \varepsilon_N)$, $\varepsilon_i = \pm 1$, then card $(\varepsilon_i) = 2^N$ and if we set

$$x_\varepsilon(t) = \sum_{i=1}^N \varepsilon_i a_i \chi_{e_i}(t)$$

we have

$$\|x_\varepsilon\|_1^2 = \langle x_\varepsilon, x_\varepsilon\rangle = \sum_{i,j=1}^N \varepsilon_i \varepsilon_j a_{ij} a_i a_j.$$

But

$$\sum_\varepsilon \|x_\varepsilon\|_1^2 = \sum_\varepsilon \sum_{i,j=1}^N \varepsilon_i \varepsilon_j a_{ij} a_i a_j = 2^N \sum_{i=1}^N a_{ii} a_i^2$$

and since the norm $\|,\|$ is monotone, we get that

$$2^N m^{-1}\|x\|^2 \leqslant \sum \|x_\varepsilon\|_1^2 \leqslant 2^N m \sum_{i=1}^{N} a_{ii} a_i^2 = 2^N m\|x\|^2$$

or

$$m^{-1}\|x\|^2 \leqslant \sum_{i=1}^{N} a_{ii}a_i^2 \leqslant m\|x\|^2.$$

We now consider the function

$$f(t) = \|\chi_{[0,t]}\|_X$$

(which is called the fundamental function of the space X); then for

$$a_k = \begin{cases} 0, & k \neq i \\ 1, & k = i \end{cases}$$

we obtain,

$$m^{-1} f^2(1/N) \leqslant a_{ii} \leqslant mf^2(1/N). \tag{*}$$

Further, if $a_k = 1$ for all k then we get

$$m^{-1} f^2(1) \leqslant \sum_{i=1}^{N} a_{ii} \leqslant mf^2(1). \tag{++}$$

Using the inequalities in $(+)$ and $(++)$ we get that

$$\|x\|_1^2 \leqslant m \sum_{i=1}^{N} a_{ii} a_i^2 \leqslant m^2 f^2(1/N)\cdot N \sum_{i=1}^{N} (a_i^2/N)$$

$$\leqslant m^4 f^2(1) \int_0^1 x^2(t)\,dt$$

and similarly

$$\|x\|_1^2 \geqslant M^4 f^2(1) \int_0^1 x^2(t)\,dt$$

which means that

$$m^{-2} f^2(1) \int_0^1 x^2(t)\,dt \leqslant \|x\| \leqslant m^2 \int_0^1 x^2(t)\,dt$$

or

$$m^{-2} f^2(1)\|x\|_{L^2_{[0,1]}} \leqslant \|x\| \leqslant m^2 f^2(1) \int_0^1 x^2(t)\,dt. \tag{+++}$$

Thus $(+++)$ holds for all decompositions of $[0,1]$ in parts of equal measure. Now since X is separable the set of elements

$$\sum_{i=1}^{N} a_i \chi_{e_i}(t) \tag{++++}$$

is dense in X as well as in $L^2_{[0,1]}$. Thus if X is an arbitrary element of $L^2_{[0,1]}$ then

$$x = \lim x_n$$

with x_n of the form given in $(++++)$. The convergence is in $L^2_{[0,1]}$ as well as in X. Thus x is in X. Conversely, if x is arbitrary in X then $x = \lim x_n$, x_n of the form given in $(++++)$ and the convergence is in $L^2_{[0,1]}$ as well as in X. This clearly means that $X = L^2_{[0,1]}$.

REMARK 8.15.2. The arguments given in the proof of the above theorem show that in fact the following slightly general result holds.

Let ν be a continuous finite measure on [0,1] and X be a space of measurable functions on [0,1]. We say that X is ν-symmetric if X is normally embedded in $S_{\nu[0,1]}$ and for each f in X every ν-measurable function g with $\nu\{s, f(s) - g(s)\} = 0$ is in X.

THEOREM 8.15.3 (Potepun, 1974). *If X is a ν-symmetric Banach space of functions and if I is an isomorphism of X onto $L^2_{[0,1]}$ then X is $L^2_{[0,1]}$.*

NOTE 8.15.4. Theorem 8.15.1 extends the following earlier result of Semenov (1964): if X is a symmetric complete inner product space of functions on [0, 1] then $X = L^2_{[0,1]}$.

CHAPTER 9

Probability Theory and Inner Product Structures

9.0. Introduction

In recent years the study of extensions of classical probability theory to more general settings has known a very intensive activity. The earliest contribution in this direction seems to be that of Kolmogorov (1935) who has defined Fourier transform for any measure on a Banach space. Further contributions were made by Mourier (1954), LeCam (1957), Sazonov (1958) and especially Prohorov (1957). For a detailed bibliography we refer to Woyczyński (1978) where the reader may find an excellent account of many of the recent results in this area.

In this chapter we are interested in the possible connections between certain probabilistic properties and the inner product structures on a Banach space. We present results which show that when a Banach space has some properties – for example, certain martingale results hold, or when the central limit theorem is valid, etc. – then the norm (or an equivalent norm) is generated by an inner product.

9.1. Probabilities on Banach Spaces

Let S be a metric space and let $\mathscr{B}$ be the σ-algebra of Borel sets of S, i.e. the σ-algebra generated by the open subsets of S. A probability measure μ on S is a Borel measure on $\mathscr{B}$ which has the property that $\mu(S) = 1$. We use the notation μ as a p.m. We say that μ is tight on a subalgebra of subset B_1 of if

$$\mu(A) = \sup(\mu(K), K \subseteq A, K \in B_1 \quad \text{and} \quad K \text{ compact})$$

for all A in B_1.

$\mathscr{P}(S)$ denotes the set of all p.m. on S.

The weak topology on $\mathscr{P}(S)$ is the weak *-topology and its neighborhood base are the sets

$$O(\mu,(f_i)_1^n,(\varepsilon_i)_1^n) = \left\{ \nu, \nu \in \mathscr{P}(S), \qquad \left| \int_S f_i \,\mathrm{d}\mu - \int_S f_i \,\mathrm{d}\nu \right| < \varepsilon_i, i = 1, \ldots, n \right\},$$

where $(f_i)_1^n$ are continuous real-valued functions on S and $(\varepsilon_i)_1^n$ are positive numbers.

If μ is the weak limit of the p.m. (μ_n) then we use the notation

$$\mu = \lim{}_w \mu_n \quad \text{or} \quad \mu_n \xrightarrow{w} \mu.$$

From just the definition of the weak topology, $\mu_n \xrightarrow{w} \mu$ iff for each f in $C(S)$ (the space of all continuous real-valued functions on S),

$$\lim \int_S f \,\mathrm{d}\mu_n = \int_S f \,\mathrm{d}\mu.$$

If $(\Omega, \mathscr{B}, \mu)$ is a probability space and X is a separable Banach space then a function $f: \Omega \to X$ is said to be weakly measurable if for each x^* in X^* the function $x^*(f)$ defined by

$$(x^*(f))(w) = x^*(f(w))$$

is measurable.

The function $f: \Omega \to X$ is said to be strongly measurable if for any Borel set M in $X, f^{-1}(M)$ is in $\mathscr{B}$.

Then we have the following.

LEMMA 9.1.1. *If X is a separable Banach space then a function $f: \Omega \to X$ is weakly measurable iff f is strongly measurable.*

DEFINITION 9.1.2. Any (weakly) measurable function $f: \Omega \to X$, is said to be a X-valued random variable.

For X-valued random variables there are many (nonequivalent) notions of integrals. In what follows we present the Bochner integral and the Gelfand–Pettis integral.

DEFINITION 9.1.3. Let $f: \Omega \to X$ be a X-valued random variable. We say that f is Bochner integrable if there exists a sequence of simple functions (f_n) which are X-valued random variables and

(1) $f_n \to f$ a.s.

(2) $\lim\limits_{n \to \infty} \int \| f_n - f \| \,\mathrm{d}\mu = 0.$

Then we set

$$\int_A f \,\mathrm{d}\mu = \lim \int_A f_n \,\mathrm{d}\mu$$

and where

$$f_n = \sum_{i=1}^{p} a_i \chi_{A_i}, \quad a_i \in X, \quad A_i \text{ are measurable and}$$

$$\int_A f_n \,\mathrm{d}\mu = \sum_{i=1}^{p} a_i \mu(A_i \cap A).$$

We use also the probabilistic terminology (see 'weak topology' above) for the integral of f, i.e. we call $\int_\Omega f \,\mathrm{d}\mu$ the expected value of f,

$$Ef = \int f \,\mathrm{d}\mu.$$

The following lemma holds.

LEMMA 9.1.4. *Let $f:\Omega \to X$ be a X-valued random variable. Then f is Bochner integrable (i.e. f has a Bochner integral) iff $f_1:\Omega \to R$, defined by*

$$f_1(w) = \|f(w)\|,$$

is integrable. Then we have the inequality

$$\|Ef\| \leqslant E\|f\| \cdot \|f\|(w) = \|f(w)\|$$

Suppose now that $f:\Omega \to X$ is a X-valued random variable and has the property that for each x^* in X^* the random variable $x^*(f)$ is integrable with respect to the p.m. μ. Then we can prove that there exists x^{**} in X^{**} such that for all x^* in X^* the following equality holds:

$$x^{**}(x^*) = \int_\Omega x^*(f(w))\,\mathrm{d}\mu(w).$$

Now the notion of the Gelfand–Pettis integral is defined as follows.

DEFINITION 9.15. Let $f:\Omega \to X$ be a X-valued random variable. We say that f is Gelfand–Pettis integrable iff there exists an element x_f in X such that for all x^* in X^* we have the equality:

$$x^*(x_f) = \int_\Omega x^*(f(w))\,\mathrm{d}\mu(w).$$

PROPOSITION 9.1.6. *Let $f:\Omega \to X$ be a X-valued random variable. If f is Bochner integrable then f is Gelfand–Pettis integrable and both integrals are the same.*

REMARK 9.1.7. There are X-valued random variables which are Gelfand–Pettis integrable and not Bochner integrable.

Let us suppose that $f:\Omega \to X$ is a X-valued random variable which is Gelfand–Pettis integrable. Then we can define 'covariance' as follows.

DEFINITION 9.1.8. If $f:\Omega \to X$ is a X-valued Gelfand–Pettis integrable random variable then the 'covariance' is the bilinear form mapping defined on X^* by the formula:

$$(\operatorname{cov} f)(x^*, y^*) = E(x^*(f - Ef)y^*(f - Ef)).$$

Then we have the following result.

PROPOSITION 9.1.9. *The* cov *f is a continuous bilinear mapping.* We can define a general notion of covariance as follows.

DEFINITION 9.1.10. Let X be a Banach space. A covariance C on X is a function

$C: X^* {}_\times X^* \to R$ with the following properties:

(1) $C(x^*, x^*) = 0$.
(2) $C(x^*, y^*) = C(y^*, x^*)$, $\quad x^*, \quad y^*$ in X^*.
(3) $C(x^*, \ )$ is linear for each x^* in X^*.

If C is a covariance on X then we can consider the subspace

$$X_1^* = \{x^*, C(x^*, x^*) = 0\}.$$

We now define the Fourier transform of a measure μ on a Banach space as follows.

DEFINITION 9.1.11 (Kolmogorov, 1935). Let X be a real Banach space and let μ be a measure on X. Then the Fourier transform of μ is the function on X^* defined by the formula:

$$F(\mu)(x^*) = \int_X e^{ix^*(x)}\, d\mu(x).$$

F is called the characteristic functional of the measure. In the finite dimensional space there are necessary and sufficient conditions for a function on X to be the characteristic functional of a measure. In the case of infinite dimensional spaces, interesting contributions to elucidate the problem about the necessary and sufficient conditions for a function on X^* to be the Fourier transform of a measure were made by Mourier (1954), Getoor (1957), LeCam (1957), Sazonov (1958) and many other authors. For an excellent account we refer to Mandrekar (1977).

Sazonov (1958) has presented a necessary and sufficient condition for a function on a complete inner product space to be the characteristic functional of a measure P and satisfying the condition

$$\int \|x\|^2\, dP(x) < \infty.$$

In Section 9.4 we shall mention the beautiful result of D. K. Muštari concerning the following problem: let X be a real separable Banach space with the dual X^*, and let f be a function on X with the following properties:

(1) $f(0) = 1$,
(2) f is positive definite.

The problem is: What are the necessary and sufficient conditions such that the continuity of f in some topology will be the necessary and sufficient for X to be the characteristic functional of a measure on X?

As mentioned above the case of complete inner product spaces has attracted the attention of many mathematicians and the first paper which is concerned with measures on the smallest σ-algebra containing the open subsets of the space was a paper by C. Loewner. In that paper Loewner is interested in properties of measures which are similar (in some sense) to the Lebesgue measure on finite dimensional spaces. It is noted that such a measure does not exist. Indeed, the basic properties of the Lebesgue measure connected with the linear structure of the finite dimensional spaces are the invariance with respect to translations and rotations. If μ is a measure on the smallest σ-algebra containing the open subsets of the complete inner product space X and has the properties to be invariant to translations and rotations, then we conclude that it is the measure identically zero. Indeed Loewner remarked that if for some ball with radius a, $\mu(B_a(x))$ is strictly positive then for any b with $b < a/2^{1/3}$ we have the inequality

$$nf(b) \leqslant f(a), \qquad f(s) = \mu(B_s(x)),$$

where n is an arbitrary integer. The above inequality follows from the fact that in a complete inner product space of infinite dimensions we find arbitrary number of pairwise disjoints balls of radius b.

Now since n is arbitrary the above inequality implies the assertion. For more details, as well as for proofs, we refer to the interesting paper of Loewner (1931).

Another interesting aspect of the measure theory on complete inner product spaces was (and is) related to problems in quantum mechanics. The problem there was to obtain information about the measures defined on the family of all closed subspaces of a complete inner product space. An interesting and fundamental result of Gleason (1957) asserts that in the case of separable spaces on the family of closed subspaces the measure is induced by a positive nuclear operator. Šerstnev (1974) has shown that Gleason's result is also valid for nonseparable spaces. For more detail we refer to Section 16.16.

We now present some specific results for measures in complete inner product spaces. The reader is referred to the excellent book of V. Sazonov for a complete treatment.

Let X be a complete inner product space and let μ be a measure on X. Suppose now that L is a finite dimensional subspace of X and then we can consider B_L (the Borel sets of L) and the family of subsets in X denoted by $P_L^{-1}B_L$, where P_L is the projection of X onto L, and

$$(P_L^{-1}M_L), \qquad M_L \in B_L, \qquad P_L^{-1}M_L = (x, x \in X, P_L x \in M_L).$$

Such a set

$$P_L^{-1}M_L$$

is said to be a cylindrical set with base in L. It is obvious that $(P_L^{-1}M_L, M_L \in B_L)$ is a σ-algebra.

If f is a function which is measurable with respect to the σ-algebra of cylindrical subsets, then this is called a cylindrical function.

Associated with the measure μ on (X, B) are the measures μ_L, L being a finite dimensional subspace in X, with

$$\mu_L(\mathrm{M}_L) \sim \mu(P_L^{-1}(M_L))$$

for each M_L in B_L.

The measures $(\mu_L, L$ finite dimensional in $X)$ are sufficient to compute the integrals of cylindrical functions since if h is a bounded B_L-measurable function then we have clearly the relation

$$\int_L h(x)\,\mathrm{d}\mu_L(x) = \int_X h(Px)\,\mathrm{d}\mu(x).$$

Suppose now that L is a subspace of another finite dimensional subspace L_1 of X and then we note that, concerning the measures (μ_L), the following relation holds:

$$\mu_L(M_L) \sim \mu_{L_1}(L_1 \cap P_L^{-1}(M_L)). \tag{*}$$

The conditions in (*) satisfied by the measures $(\mu_L$, L finite dimensional subspace in $X)$ are called 'compatibility conditions' and any family of measures defined on the family of finite dimensional subspaces of X satisfying the relations (*) is said to be a compatible family of finite dimensional distributions.

The characteristic functional of the measure μ_L, L is a finite dimensional subspace of X, is given by the formula:

$$\varphi_L(z) = \int e^{i\langle z, x\rangle}\, d\mu_L(x) \tag{**}$$

for each z in L.

Of course this coincides with $\varphi(z)$ when z is in L. Conversely, if (μ_L, L is finite dimensional in X) is a compatible family of distributions then there exists a function $\varphi(z)$ on X such that (**) holds. If (μ_L) is the family of compatible distributions then we set

$$\varphi_L(z) = \int e^{i\langle z, x\rangle}\, d\mu_L(x)$$

and if L_1 is a subspace containing the subspace L (both L and L_1 are finite dimensional) then for z in L we have,

$$\begin{aligned}\varphi_{L_1}(z) &= \int e^{i\langle z, x\rangle}\, d\mu_{L_1}(x) = \int e^{i\langle P_L z, x\rangle}\, d\mu_{L_1}(x)\\ &= \int e^{i\langle z, P_L x\rangle}\, d\mu_{L_1}(x) = \int e^{i\langle z, x\rangle}\, d\mu_L(x) = \varphi_L(z).\end{aligned} \tag{***}$$

Thus we can define $\varphi(z)$ correctly using the above formulas.

Now, conversely, if (μ_L, L finite dimensional subspace in X) is a family of measures such that $(\varphi_L(z))$ satisfies the relation (***), then (μ_L) is a family of compatible distributions.

Let z be arbitrary in X and let L_z denote the subspace of X generated by z (which is obviously one-dimensional and set

$$\varphi(z) = \varphi_{L_z}(z).$$

Let L be a finite dimensional subspace in X and for each z in L we consider the subspace L_z (generated by z). Then clearly L_z is in L and

$$\varphi_L(z) = \varphi_{L_z}(z) = \varphi(z) \tag{****}$$

and, conversely, if (z) is a function on X satisfying the property (****) for each z in L and $L \subset L_1$,

$$\begin{aligned}\int e^{i\langle z, x\rangle}\, d\mu_L(x) &= \int e^{i\langle z, x\rangle}\, d\mu_{L_1}(x) = \int e^{i\langle z, P_L x\rangle}\, d\mu_{L_1}(x)\\ &= \int e^{i\langle z, x\rangle}\, d\mu_{L_1}(P_L^{-1} x).\end{aligned}$$

This gives the fact that the measures μ_L and $\mu_{L_1}(P_L^{-1} x)$ have the same characteristic functional and this clearly implies that they must coincide.

We now consider the problem of moments of a measure defined on $(X, \mathscr{B})$.

DEFINITION 9.1.12. Let μ be a measure on $(X, \mathscr{B})$ and then we say that μ has a moment of order k if the integral

$$\int \langle x, z_1 \rangle \cdot \langle x, z_2 \rangle \cdots \langle x, z_k \rangle \, \mathrm{d}\mu(x)$$

exists. This is denoted by $m_k(z_1, \ldots, z_k)$. In the above integral z_i are arbitrary points in X.

From the definition it is clear that a necessary and sufficient condition for the existence of the moments of order k is the existence of the integral $\int |\langle x, z \rangle|^k \, \mathrm{d}\mu(x)$.

If μ is a measure on $(X, \mathscr{B})$ which has the moments of order k then we see that the function

$$(z_1, \ldots, z_k) \to m_k(z_1, \ldots, z_k)$$

is obviously symmetric in all variables z_i and clearly it is continuous and is, in fact, a continuous k-linear form.

DEFINITION 9.1.13. For a measure μ which has $m_1(z)$ we call the element a the mean value of μ if $m_1(z) = \langle z, a \rangle$ (which exists by the Fréchet–Riesz theorem). The number

$$m_2(z_1, z_2) - m_1(z_1)m_1(z_2)$$

is called the correlation of μ.

Now from the continuity and the fact that the correlation is bilinear, it follows that there exists a unique bounded linear operator $B \in L(X)$ such that

$$m_2(z_1, z_2) - m_1(z_1)m_1(z_2) = \langle Bz_1, z_2 \rangle \qquad (*****)$$

for all z_1 and z_2 in X. This permit us to give the following definition.

DEFINITION 9.1.14. The correlation operator of the measure μ on $(X, \mathscr{B})$ is the unique bounded linear operator on X satisfying relation (*****).

The correlation operator B has very interesting properties. One of these is given in the following proposition.

PROPOSITION 9.1.15. *The correlation operator B on a complete inner point space X is nuclear if and only if the following condition is satisfied*:

$$\int \|x\|^2 \, \mathrm{d}\mu(x) < \infty.$$

In this case if (a_i) are the eigenvalues of B the following relation holds:

$$\sum_{i=1}^{\infty} a_i = \int \|x\|^2 \, \mathrm{d}\mu(x) - \|a\|^2$$

where a is the mean value of μ.

Proof. If (e_i) are the eigenvectors of B then the following formula holds:

$$\sum_{i=1}^{n} \langle Be_i, e_i \rangle = \sum_{i=1}^{n} |\langle x, e_i \rangle|^2 \, \mathrm{d}\mu(x) - \sum_{i=1}^{n} \langle a, e_i \rangle$$

and for $n \to \infty$ this gives the assertion. □

The following important theorem of Minlos (1959) and Sazonov (1958) gives the necessary and sufficient condition for a function of X to be the characteristic functional of a measure on $(X, \mathscr{B})$.

THEOREM 9.1.16 (Minlos–Sazonov). *The necessary and sufficient condition for a complex-valued function φ on X to be the characteristic functional of a measure on (X, B) is that for any $\varepsilon > 0$ there exists a nuclear (positive operator)K_ε such that if $\langle K_\varepsilon x, x\rangle \leqslant 1$ then* $\mathrm{Re}(\varphi(x) - \varphi(0)) \leqslant \varepsilon$.

In what follows we prove only the necessity of the condition of the theorem. For the sufficiency part we refer to Sazonov (1959).

Suppose that μ is a measure on $(X, \mathscr{B})$ and let φ be its characteristic function. In this case we have obviously the formula:

$$\mathrm{Re}\,\varphi(z) - \mathrm{Re}\,\varphi(0) = \int (1 - \cos z, x)\,\mathrm{d}(x) \leqslant \int_{\|x\| \leqslant c} 2\sin\langle x, z\rangle^2/2\mathrm{d}\mu(x) +$$

$$+ 2\int_{\|x\| \geqslant c} \mathrm{d}\mu(x) = \tfrac{1}{2}\int_{\|x\| \leqslant c} |\langle x, z\rangle|^2\,\mathrm{d}\mu(x) + 2\mu(x, \|x\| \geqslant c).$$

Since $\int_{\|x\| \leqslant c} |\langle x, z\rangle|^2\,\mathrm{d}\mu(x)$ is for each fixed c a quadratic functional with respect to z, and thus there exists a bounded linear operator (which is a positive operator on X) such that

$$\langle B_c z, z\rangle = \tfrac{1}{2}\int_{\|x\| \leqslant c} |\langle x, z\rangle|^2\,\mathrm{d}\mu(x)$$

and since

$$\int_{\|x\| \leqslant c} \|x\|^2\,\mathrm{d}\mu(x) \leqslant \mu(X)c^2$$

we can choose c such that

$$\mu(x, \|x\| > c) < \varepsilon/4$$

and thus if we set $K_\varepsilon = \varepsilon^{-1}B_c$ then the following relation holds:

$$\mathrm{Re}\,\varphi(z) - \mathrm{Re}\,\varphi(0) < \varepsilon/2 + \tfrac{1}{2}\langle B_c z, z\rangle \leqslant \varepsilon/2 + \varepsilon/2 + \langle K_\varepsilon z, z\rangle = \varepsilon.$$

Thus the condition of the theorem is necessary.

The above theorem shows that there exists a strong connection between the properties of the characteristic functionals (or correspondingly, measures on X) and the class of positive nuclear operators. It may be of some interest to know if other classes of operators may be connected with objects defined on $(X, \mathscr{B})$.

We now consider a special and important class of measures defined on $(X, \mathscr{B})$, the so-called Gaussian measures. These are defined as follows.

DEFINITION 9.1.17. The measure μ defined on $(X, \mathscr{B})$ is said to be a Gaussian measure if for each fixed z in X the function

$$f_z(x) = \langle x, z \rangle$$

is a Gaussian random variable.

Suppose that μ is a Gaussian measure on $(X, \mathscr{B})$. We consider the mean value of μ,

$$a_z = m_1(z) = \int \langle z, x \rangle \, \mathrm{d}\mu(x) = \langle z, a \rangle$$

for some fixed a in X and

$$b_z = m_2(z, z) - \|a_z\|^2 = \int \langle z, x \rangle^2 \, \mathrm{d}\mu(x) - \|a_z\|^2 = \langle Bz, z \rangle$$

where B is the correlation operator. Since we suppose that μ is a Gaussian measure we have that

$$\varphi(z) = \int \mathrm{e}^{i\langle z, x \rangle} \, \mathrm{d}\mu(x) = \mathrm{e}^{(i\langle z, a \rangle - 1/2\langle Bz, z \rangle)}, \quad B \geqslant 0$$

and thus we see that in this case the characteristic functional of the measure μ is of special form and clearly satisfies the relation

$$\varphi(tz) = \mathrm{e}^{it\langle a, z \rangle - 1/2t^2\langle Bz, z \rangle}, \; B \text{ nuclear and } B \geqslant 0 \tag{R*}$$

which is, in some sense, like to the familiar characteristic function of a Gaussian random variable.

Conversely, if μ is a measure on $(X, \mathscr{B})$ with the property that the characteristic functional φ satisfies the relation (R*) then μ is a Gaussian measure on X.

In recent years, by considering random variables with values in inner product spaces, several classical theorems of probability theory were extended. We mention at this point the extension of the celebrated Kolmogorov three-series theorem.

9.2. Bernoulli and Gaussian Random Independent Variables and Inner Product Structures

We now present some very interesting and important characterizations of Banach spaces having an equivalent norm which is generated by an inner product. The results were obtained by Kwapień (1972). We shall follow mainly the Kwapień Exposé (No. VIII).

We recall that a random variable ξ defined on the probability space $(\Omega, \mathscr{B}, P)$ is said to be a Bernoulli random variable if

$$P(w, \xi(w) = 1) = P(w, \xi(w) = -1) = 1/2.$$

The random variable η is said to be Gaussian if

$$P(w, \eta(w) < t) = (2\pi)^{-1} \int_{-\infty}^{t} \mathrm{e}^{-s^2/2} \, \mathrm{d}s.$$

Then we have the following result of S. Kwapień.

THEOREM 9.2.1 (Kwapień, 1972). *For a Banach space X the following assertions are equivalent:*

(1) *X has the property that some equivalent norm on X is generated by an inner product.*

(2) *For some $K > 0$ and all finite sequences $(x_i)^n$ of elements of X and any sequence (ξ_i) of independent identically distributed Bernoulli random variables,*

$$K^{-1}\sum_{i=1}^{n}\|x_i\|^2 \leqslant E\left(\left\|\sum_{i=1}^{n}\xi_i x_i\right\|^2\right) \leqslant K\sum_{i=1}^{n}\|x_i\|^2,$$

(3) *For some K_1 0 and all finite sequences $(x_i)_1^n$ of elements of X and any sequence (η_i) of Gaussian independent random variables,*

$$K_1^{-1}\sum_{i=1}^{n}\|x_i\|^2 \leqslant E\left(\left\|\sum_{i=1}^{n}\eta_i x_i\right\|^2\right) \leqslant K_1\sum_{i=1}^{n}\|x_i\|^2,$$

(4) *If the norm of the operator $T_n\colon l_2^n \to l_2^n$ with the matrix $A = (a_{ij})_1^n$ is T_n then*

$$\sum_{i=1}^{n}\left(\left\|\sum_{j=1}^{n}a_{ij}x_j\right\|^2\right) \leqslant C^2\|T_n\|^2\sum_{i=1}^{n}\|x_i\|^2,$$

where $(x_i)_1^n$ is arbitrary in X and C is a constant.

Proof. Since

$$E\left(\left\|\sum_{i=1}^{n}\xi_i x_i\right\|^2\right) = \sum_{i=1}^{n}\|x_i\|^2$$

when the norm of X is generated by an inner product then clearly any equivalent norm satisfies a similar inequality and thus (1) → (2).

Suppose now that (2) holds. For m fixed we set

$$u_i^m = n^{-1/2}\sum_{k=0}^{m-1}\xi_{im+k}.$$

Then the Moivre–Laplace theorem asserts that the common distribution of the random variables (u_i^m) converges to the common distribution of $(\eta_1, \ldots, \eta_n)$ when $m \to \infty$. Thus for any bounded continuous funcion $h\colon R^n \to R$ we have the relation:

$$\lim E(h(u_1^m, \ldots, u_n^m)) = Eh(\eta_1, \ldots, \eta_n) \qquad (+)$$

(i.e. weak convergence).

The following facts improve this assertion, as follows. □

LEMMA 9.2.2 (Kwapień, 1972). *For any continuous function $h\colon R^n \to R$ with the property that*

$$\lim_{(|t_1| + \cdots + |t_n|) \to \infty} h(t_1, \ldots, t_n)e^{-(|t_1| + \cdots + |t_n|)} = 0 \qquad (++)$$

the relation $(+)$ holds.

Proof. We consider the set of all functions $h\colon R^n \to R$ satisfying the condition

$(++)$ and which are continuous. If we set

$$\|h\| = \sup |h(t_1, \ldots, t_n)| e^{-(|t_1| + \cdots + |t_n|)}$$

then this is a Banach space. We set

$$F_m(h) = E(h(u_1^m, \ldots, u_n^m))$$

for h in the space. Then clearly this is a functional on the space and if we set

$$F(h) = E(h(\eta_1, \ldots, \eta_n))$$

$$= (2\pi)^{-n/2} \int_{-\infty}^{\infty} \cdots \int_{-\infty}^{\infty} h(t_1, \ldots, t_n) e^{-} \, dt_1 \ldots dt_n$$

then the Moivre–Laplace theorem quoted above implies that (F_m) converges to F for each bounded continuous function in our space. But it is not difficult to see that the set of all such functions is dense in the space and if we note that (F_m) is uniformly bounded then the assertion of the lemma follows.

Since the random variables (ξ_i) are independent we have

$$\|F_m\| \leqslant E\left(\exp \sum_{i=1}^{n} |u_i^m|\right) \leqslant (E \exp|u_i^m|)^n \leqslant (E \exp|u_1^m| + E \exp(-|u_1^m|))^n$$

$$\leqslant ((E \exp \eta_1 \cdot m^{-1/2})^m + (E \exp(-\eta_1 \cdot m^{-1/2}))^m)^n$$

$$= 2^n (2^{-1}(\exp m^{-1/2} + \exp(-m^{-1/2})))^{mn} \leqslant 2(2e^{1/2})^n$$

and thus

$$(F_m)$$

is uniformly bounded. The lemma is proved. □

Now if $(x_i)_1^n$ is a finite family of elements of X we set

$$h(t_1, \ldots, t_n) = \|t_1 x_1 + \cdots + t_n x_n\|$$

and since we suppose that (2) holds we get that

$$K^{-1} \sum_{i=1}^{n} \|x_i\|^2 = E(\|u_1^m x_1 + \cdots + u_n^m x_n\|)^2 \leqslant K \sum_{i=1}^{\infty} \|x_i\|^2$$

and (3) follows from lemma 9.2.2.

Suppose now that (3) holds. Then the closed unit ball of the space $L(l_2^n, l_2^n)$ is a compact convex set. Then by the Krein–Milman theorem on extreme points we get that any element of this set is a convex combination of extreme points. But is not difficult to see that an extreme point of the above set is a matrix which represent an operator being an isometry. Thus, in order to prove (4) it suffices to consider only T_n which are isometries. Then according to (3) we have

$$\sum_{i=1}^{n} \left\| \sum_{j=1}^{n} a_{ij} x_j \right\|^2 \leqslant mE\left(\left\| \sum_{i=1}^{n} \sum_{j=1}^{n} a_{ij} x_j \right) \eta_i \right\|^2$$

$$\leqslant mE\left(\left\| \sum_{j=1}^{n} x_j \left(\sum_{i=1}^{n} a_{ij} \eta_i \right) \right\|^2\right) \leqslant mE \left\| \sum_{i=1}^{n} x_i \bar{\eta}_i \right\|^2,$$

where

$$\bar{\eta}_i = \sum_{j=1}^{n} a_{ji}\eta_j.$$

Now (a_{ij}) represents an isometry and the random variables $(\bar{\eta}_i)$ are with the same distributions as (ξ_i). This gives that

$$mE\left(\left\|\sum_{j=1}^{n} x_j\bar{\eta}_j\right\|^2\right) \leqslant mE\left\|\sum_{j=1}^{n} x_j\eta_j\right\|^2 \leqslant m^2 \sum_{i=1}^{n} \|x_i\|^2.$$

The case of complex matrices follows easily from the case of real matrices, and thus we omit it. Thus (4) holds.

Suppose now that (4) holds. Let us consider the Banach space $L(l_I^1, X)$, where I is a set of sufficiently great cardinality. Then we find T in $L(l_I^1, X)$ which is a surjection. For each i in the set I we set

$$e_i = (\delta_{ij}), \quad \delta_{ij} \text{ is the Kronecker symbol}$$

and

$$x_i = Te_i, \quad i \in I.$$

We show that T is a 2-absolutely summing operator. For this it suffices to show that for some S in $L(l_I^2, l_I^1)$ we have the inequality:

$$\sum_{i \in I} \|TSe_i\|^2 < \infty.$$

If we set

$$\Delta(a_i) = (c_i a_i), \quad (c_i) \text{ in } l_I^2$$

then

$$\|\Delta\|^2 = \|(c_i)\|^2$$

then by the factorization theorem of A. Grothendieck we have

$$S = \Delta R$$

with R in $L(l_I^2, l_I^2)$. But

$$\sum_{i \in I} \|TRe_i\|^2 = \sum_{i \in I} \sum_{j \in I} \|a_{ji} c_j x_j\|^2$$

$$\leqslant K^2 R^2 \sum_{j \in I} \|c_j x_j\|^2 \leqslant K^2 \|R\|^2 \|\Delta\|^2 \|T\|^2$$

by applying (4) and the inequality $\|x_i\| \leqslant T$. But every 2-absolutely summing operator factorizes through a complete inner product space H. Since T is a surjection we have

$$T = P\bar{R}, \qquad P \in L(H, X), \qquad \bar{R} \in L(l_I^1, H)$$

and thus X has an equivalent norm which is generated by an inner product on X.

We can formulate the above result using the notion of type and cotype which are now recalled.

DEFINITION 9.2.3. Let X, Y be Banach spaces. An element T in $L(X, Y)$ is said to be of type p if for some $K > 0$ and for all finite families of independent random variables in L^p with $Ef_i = 0$, $i = 1, 2, \ldots, n$, the following inequality holds:

$$E\left(\left\|T \sum_{i=1}^{n} f_i\right\|\right)^p \leqslant K \sum_{i=1}^{n} E\|f_i\|^p.$$

DEFINITION 9.2.4. A Banach space X is said to be of type p if the identity operator is of type p.

DEFINITION 9.2.5. Let $p = 2$. A separable Banach space X is said to be of cotype p if, with (f_i) as in Definition 9.2.3, the following inequality holds:

$$E\left\|\sum_{i=1}^{n} f_i\right\|^p \geqslant K \sum_{l=1}^{n} E\|f_i\|p.$$

The notion of space of cotype p (for $p = 2$) was introduced by Mouchtari (1973). See Woyczyński (1978) for historical informations.

THEOREM 9.2.6 (Kwapień, 1972). *A Banach space X has an equivalent norm which is generated by an inner product iff it is of type and cotype* 2.

We now present another necessary and sufficient condition for a Banach space X to have an equivalent norm generated by an inner product in terms of certain inequalities using bases of the space L^2 $([0, 1], dx)$, dx being the Lebesgue measure on $[0, 1]$.

THEOREM 9.2.7 (Kwapień, 1972). *Let X be a Banach space. Then on X there exists an equivalent norm generated by an inner product iff (e_i) being an orthonormal system which is complete in L^2 we find a constant $K > 0$ such that for all finite systems of elements in X, $(x_1, \ldots, x_n)$ the following inequalities hold:*

$$K^{-1} \sum_{i=1}^{n} \|x_i\|^2 \leqslant \int_0^1 \left\|\sum_1^n e_i x_i\right\|^2 dt \leqslant K \sum_{i=1}^{n} \|x_i\|^2. \tag{$+$}$$

Proof. It is easy to see that the above condition is necessary. Suppose now that the Banach space X satisfies the above condition. Let (f_i) be an orthonormal system in L^2 and thus for each $k \neq 1$, $\langle f_k, f_i \rangle = 0$, $i \neq k$, and for each finite system of elements $(x_1, \ldots, x_n)$ of X we have the inequalities:

$$K^{-1} \sum_{i=1}^{n} \|x_i\|^2 \leqslant \int_0^1 \left\|\sum_1^n f_i x_i\right\|^2 dt \leqslant K \sum_{i=1}^{n} \|x_i\|^2, \tag{$++$}$$

where K is the same constant as in $(+)$.

Indeed,

$$\int_0^1 \left\| \sum_{i=1}^n f_i x_i \right\|^2 \mathrm{d}t \leqslant \int_0^1 \left\| \sum_{i=1}^n x_i \sum_{j=1}^\infty \langle f_i, e_j \rangle e_j \right\|^2 \mathrm{d}t$$

$$= \int_0^1 \left\| \sum_{j=1}^\infty \left(\sum_{i=1}^n \langle f_i, e_j \rangle x_i \right) e_j(t) \right\|^2 \mathrm{d}t$$

and

$$\sum_{i=1}^n \|x_i\|^2 = \sum_{i=1}^n \|x_i\|^2 \sum_{j=1}^\infty |\langle f_i, e_j \rangle|^2 \leqslant \sum_{j=1}^\infty \left(\left\| \sum_{i=1}^n \langle f_i, e_j \rangle x_i \right\|^2 \right)$$

which gives our inequalities.

Suppose now that (ξ_i) is a sequence of Bernoulli independent identically distributed random variables. Then by the gliding hump procedure we get for each $\varepsilon > 0$ an increasing sequence (n_k) of integers and an orthonormal system (f_i) with the above property and

$$\int_0^1 |\xi_{n_k}(t) - f_k(t)|^2 \,\mathrm{d}t \leqslant \varepsilon 2^{-k}.$$

But

$$K^{-1} \sum_{i=1}^n \|x_i\|^2 = \int_0^1 \left\| \sum_{i=1}^n x_i \xi_{n_i}(t) \right\|^2 \mathrm{d}t \leqslant K \sum_{i=1}^n \|x_i\|^2$$

and we can apply Theorem 9.2.1 and the assertion of the theorem follows. □

REMARK 9.2.8. In Theorem 9.2.1 the proof of the implication (3) → (4) holds only for real spaces but it is not difficult to modify it for the case of complex spaces.

We now present a characterization of inner product structures on a Banach space using some properties of Fourier transform. We mention that this characterization has its origin in an interesting and important paper by Peetre (1969).

Let X be a complex Banach space and let $\mathrm{d}x$ be the Lebesgue measure on R. If (A_i) is a finite family of disjoint measurable subsets of R then a function of the type

$$g(t) = \sum_{i=1}^n a_i \chi_{A_i} \qquad \chi_A(t) = \begin{cases} 1, & t \in A \\ 0, & t \notin A \end{cases} \qquad a_i \in X$$

is said to be a simple function. The set of all simple functions is a linear space. We define a norm on this space by the formula

$$\|g\|^2 = \int_R \|g(t)\|^2 \,\mathrm{d}t$$

(with the usual identification of functions which are equal a.e. $\mathrm{d}x$). This space is denoted by $L_R^{2,S}(X)$ and the completion is denoted by $L_R^2(X)$.

We now define a map

$$\mathscr{F}: L_R^{2,S}(X) \to L_R^2(X)$$

by the formula

$$(\mathscr{F}f)(t) = (2\pi)^{-1} \int_{-\infty}^{\infty} e^{-ist} f(s)\, ds$$

and the 'inverse' by the formula

$$(\mathscr{F}^{-1}f)(t) = (2\pi)^{-1} \int_{-\infty}^{\infty} e^{ist} f(s)\, ds.$$

We remark that from the definition of $\mathscr{F}$ and of $\mathscr{F}^{-1}$ these are linear maps. For the study of these mappings we need the following result given as lemma.

LEMMA 9.2.9 (Kwapień, 1972). *If we set*

$$h_n = \sum_{k=-n}^{n} x_k/a^{1/2} \cdot \chi_{[ka,(k+1)a]} \qquad a > 0, \quad x_i \in X, \quad n > 0$$

then the following assertions hold:

(1) $\|h_n\|^2 = \sum_{-n}^{n} \|x_k\|^2$,

(2) $\|\mathscr{F}h_n\|^2 = \int_0^1 \left\| \sum_{k=-n}^{n} e^{-2ikt} x_k \right\|^2 dt.$

Proof. Since the proof of the first assertion is easy we omit it. For (2) we remark that

$$(\mathscr{F}h_n)(t) = (2\pi)^{-1} \int_{-\infty}^{\infty} \sum_{k=-n}^{n} x_k/a^{1/2}\, \chi_{[ka,(k+1)a]}(s)\, e^{-ist}\, ds$$

$$= (2\pi a)^{-1/2} \sum_{k=-n}^{n} x_k \int_{ka}^{(k+1)a} e^{ist}\, ds = (a/2\pi)^{1/2}\, (\sin at/2)/(at/2) \times$$

$$\times (e^{-iat/2}) \sum_{k=-n}^{n} x_k e^{-ikat}$$

and after the change of variable $u = at/2$ we obtain further,

$$\|\mathscr{F}h_n\|^2 = \int_{-\infty}^{\infty} ((\sin \pi u)/(\pi u))^2 \left\| \sum_{k=-n}^{n} x_k e^{-2iku} \right\|^2 du$$

$$= \sum_{p=-\infty}^{\infty} \int_p^{p+1} ((\sin \pi u)/(\pi u))^2 \left\| \sum_{k=-n}^{n} x_k e^{-2iku} \right\|^2 du$$

$$= \int_0^1 \sum_{p=-\infty}^{\infty} ((\sin \pi u)/(\pi(u+p)))^2 \left\| \sum_{k=-n}^{n} x_k e^{-2iku} \right\|^2 du.$$

But

$$\sum_{p=-\infty}^{\infty} ((\sin \pi u)/(\pi(u+p)))^2 = 1$$

for all u in R and thus

$$\|\mathscr{F}h_n\|^2 = \int_0^1 \left\| \sum_{k=-n}^{n} x_k e^{-2iku} \right\|^2 du$$

which is the assertion (2). □

Then we have the following important result.

THEOREM 9.2.10 (Kwapień, 1972). *Let X be a Banach space. Then on X there exists an equivalent norm generated by an inner product iff one of the following equivalent conditions hold:*

(1) *There exists $K > 0$ such that for all finite families of elements $(x_1, \ldots, x_n)$ of elements of X the following inequality holds:*

$$\int_0^1 \left\| \sum x_k e^{2ikt} \right\|^2 dt \leqslant K \sum_{i=-n}^{n} \|x_k\|^2.$$

(2) *For some $l < 0$*

$$l^{-1} \cdot \sum_{k=-n}^{n} \|x_k\|^2 \leqslant \int_0^1 \left\| \sum_{k=-n}^{n} x_k e^{2ikt} \right\|^2 dt.$$

(3) *The Fourier transform $\mathscr{F}$ is bounded as a map between $L_R^{2,S}(X)$ and $L_R^2(X)$.*

Proof. If on X there exists a norm generated by an inner product which is equivalent to the original one, then clearly we have that (1) holds. Suppose now that (1) holds, then the simple functions are dense in $L_R^2(X)$ and we find $K > 0$ such that for all h in $L_R^{2,S}(X)$,

$$\|\mathscr{F}h\| \leqslant K\|h\|.$$

Thus (3) holds. We remark that the same estimate holds for $\mathscr{F}^{-1}$. But $\mathscr{F}$ and $\mathscr{F}^{-1}$ are either bounded or unbounded simultaneously. Thus we get that (3) → (2). Now if (2) is satisfied then (1) is satisfied, and by Theorem 9.2.6 on X there exists an equivalent norm generated by an inner product. □

REMARK 9.2.11 (Pełczyński and Kwapień, 1972). A. Pełczyński has remarked that using a result of Landau and Shepp (1971) the following result holds: on a Banach space X there exists an equivalent norm generated by an inner product iff for some p, q in $(0, \infty)$ and all finite families of elements $(x_1, \ldots, x_n)$ in X and any sequence of independent identically distributed Gaussian random variables (r_i) the following inequalities hold:

$$K^{-1}\left(E\left(\left\| \sum_{i=1}^{n} r_i(w)x_i \right\|^p\right)^{1/p}\right) \leqslant \left(\sum_{i=1}^{n} \|x_i\|^2\right)^{1/2}$$

$$\leqslant K\left(E\left\| \sum_{i=1}^{n} r_i(w)x_i \right\|^q\right)^{1/q}.$$

REMARK 9.2.12. The above results are in intimate connection with the classical

Khinchine inequality: for any p in $(1, \infty)$ there are constants A_p and B_p such that

$$B_p\left(\sum_i |a_i|^2\right)^{1/2} = \left(\int_0^1 \left|\sum_{i=1}^n a_i r_i(t)\right|^p \mathrm{d}t\right)^{1/p} = A_p\left(\sum_{i=1}^n |a_i|^2\right)^{1/2}$$

for all scalars $a_1, \ldots, a_n$ and where (r_i) is the Rademacher orthogonal system (we recall that r_i is defined by the formula

$$r_i(t) = \operatorname{sign}\sin(2^{(i-1)}2\pi t,\ t \text{ in } (0, 1).$$

The natural problem which appears with this inequality is: For what Banach spaces does the inequality still remain true if the scalars a_i are elements of the respective Banach space and, of course, if the absolute value is replaced by the norm? We may suppose that A_p and B_p are connected with the Banach space. The Kwapień result given above says that if a Banach space has the property that the Khinchine inequality holds, then on X there is an equivalent norm which is generated by an inner product.

Kahane (1968) has extended the Khinchine inequality in a form valid for all Banach spaces: for every p in $(1, \infty)$ there exist universal constants a_p and b_p such that for every Banach space X the following inequality holds:

$$b_p\left(\int_0^1 \left\|\sum_{j=1} a_j r_j(t)\right\|^2 \mathrm{d}t\right)^{1/2} \leqslant \left(\int_0^1 \left\|\sum a_j r_j\right\|^p \mathrm{d}t\right)^{1/p}$$
$$\leqslant a_p\left(\int_0^1 \left\|\sum a_j r_j(t)\right\|^2 \mathrm{d}t\right)^{1/2}$$

for all elements $a_1, \ldots, a_n$ in X.

For more interesting results related to the Khinchine inequality we refer to Young (1976) and Szarek (1976) as well as to the papers quoted there.

9.3. Biconvex Functions and a Characterization of Complete Inner Product Spaces

Motivated by certain problems about martingales, Burkholder (1980), (1981) has investigated some classes of Banach spaces using some functions defined on the closed unit ball of Banach spaces. In this section we present Burkholder's characterization of complete inner product spaces using symmetric biconvex functions. For applications to probability theory we refer to Burkholder's papers quoted above.

DEFINITION 9.3.1. Let X be a Banach space. We say that X is ξ-convex if there exists a function

$$\xi: X \times X \to R$$

such that:

(1) ξ is symmetric (i.e. $\xi(x, y) = \xi(y, x)$),
(2) ξ is biconvex (i.e. both $\xi(x, \cdot)$ and $\xi(\cdot, y)$ are convex functions),

(3) whenever x, y are in X and $\|x\| \leqslant 1 \leqslant \|y\|$ then $\xi(x, y) \leqslant \|x + y\|$.

(Here $\|,\|$ denotes the norm of X).

REMARK 9.3.2. As examples of spaces which are ξ-convex we may quote the function space L^p, $p \in (1, \infty)$.

Then we have the following result.

THEOREM 9.3.3 (Burkholder, 1981). *A Banach space X has a norm generated by an inner product iff X is ξ-convex and*

$$\xi(0, 0) = 1.$$

Proof. Suppose first that X is a ξ-convex Banach space with $\xi(0, 0) = 1$. We remark that we can suppose without loss of generality that ξ has the property

$$\xi(x, y) = \xi(-x, -y).$$

Indeed, the function

$$\xi_1(x, y) = \tfrac{1}{2}(\xi(x, y) + \xi(-x, -y))$$

has this property and, of course, the properties of ξ.

Now we prove that for all x in X and all a, b in R with $\max(\|ax\|, \|bx\|) \leqslant 1$ the following relation holds:

$$\xi(ax, bx) = 1 + ab\|x\|^2.$$

First we remark that the relation is true if $x = 0$ and thus we suppose that $x \neq 0$. Of course we may suppose that $\|x\| = 1$ and that a, b satisfy the inequality $|a| \leqslant 1$, $\|b\| \leqslant 1$.

Since ξ is supposed to be a symmetric function we have,

$$\xi(x, 0) = \xi(0, x) \leqslant \|0 + x\| = 1$$
$$\xi(-x, 0) = \xi(0, -x) \leqslant \|0 - x\| = 1.$$

But ξ is biconvex and thus

$$\xi(ax, 0) = 1 \tag{+}$$

for all a, $|a| \leqslant 1$. For a, $|a| \leqslant 1$, using the property of (3) we get that

$$\xi(ax, \pm x) \leqslant \|ax \pm x\| = 1 \pm a. \tag{++}$$

Using again the biconvexity of ξ we have that

$$\xi(ax, by) = ((1 - b)(1 - a) + (1 + b)(1 + a))/2 = 1 + ab$$

which is equivalent to our relation.

Now we define

$$F(x, y) = \|x + y\|^2 + \|x - y\|^2 - 2(\|x\|^2 + \|y\|^2),$$
$$G(x, y) = 2\xi(x, y) + 2\xi(x, -y) - 4$$

and we show that if

$$\max(\|x\|, \|y\|) \leqslant 1$$

then

$$F(x, y) \leqslant G(x, y). \qquad (+++)$$

We consider the elements

$$u = \tfrac{1}{2}(x + y), \qquad v = \tfrac{1}{2}(x - y)$$

and then we have that

$$4\xi(u, v) = 2\xi(x, u) + 2\xi(y, u) \leqslant \xi(x, x) + \xi(y, y) + 2\xi(x, y).$$

Using the relations derived above we get that

$$4\|u\|^2 \leqslant \|x\|^2 + \|y\|^2 + 2\xi(x, y) - 2$$
$$4\|v\|^2 \leqslant \|x\|^2 + \|y\|^2 + 2\xi(x, -y) - 2.$$

Adding these we have

$$4(4^{-1}\|x + y\|^2 + 4^{-1}\|x - y\|^2) \leqslant 2(\|x\|^2 + \|y\|^2) + 2(\xi(x, y) + \xi(x, -y)) - 4$$

which is equivalent to $(+++)$. But

$$2(\xi(x, y) + \xi(x, -y)) \geqslant 4\xi(x, 0) = 2(\xi(x, 0) + \xi(-x, 0)) \geqslant 4\xi(0, 0) = 4$$

we get that the function G is nonnegative. Further we remark that G has the properties:

(1) $G(\pm x, \pm y) = G(x, y)$,
(2) $G(x, 0) = G(0, x) = 0$,
(3) $G(x, x) = G(x, 0) = 0$.

Suppose now that x, y are in the closed unit ball of X and that a, b are in $[0, 1]$. Then we have,

$$G(ax, by) = (1 - a)G(0, by) + aG(x, by) = aG(x, by) \leqslant abG(x, y).$$

This clearly implies that if $a < b$ for all x, y in X with $\max(\|ax\|, \|by\|) \leqslant 1$ the following relation holds:

$$G(ax, ay)a^{-2} = G(ab^{-1}bx, ab^{-1}by)a^{-2} \leqslant G(bx, by)b^{-2}.$$

This implies that

$$\lim_{a \downarrow 0} G(ax, ay)a^{-2}$$

exists and we denote this by $H(x, y)$. From the properties of the function G and the definition of H we obtain the following assertions:

(1) H is a nonnegative function.
(2) H is a symmetric function.
(3) H is biconvex.

Now we remark that

$$F(x, y) = F(ax, ay)a^{-2} \leqslant G(ax, ay)a^{-2} \leqslant G(x, y)$$

which gives that for all x, y in X with $\max(\|x\|, \|y\|) = 1$ the following inequality holds:

$$F(x, y) \leqslant H(x, y) \leqslant G(x, y). \qquad (++++)$$

Now we prove that $H(x, y) = 0$. If this is true, then F is nonnegative on a neighborhood of the origin of $X_{\times}X$ and hence mall of $X_{\times}X$, since, if as above

$$2u = x + y, \qquad 2v = x - y$$

then

$$F(x, y) = -2F(u, v).$$

Now we remark that for all $a > 0$ and x, y in X,

$$H(ax, ay) = a^2 H(x, y).$$

Indeed, we have

$$H(ax, ay) = \lim_{b \downarrow 0} G(abx, aby)b^{-2} = a^{-2}G(x, y).$$

Further we remark that for $\max(\|x\|, \|y\|) = 1$ and a in $[0, 1]$ the following relations hold:

(1) $H(\pm x, \pm y) = H(x, y)$.
(2) $H(x, x) = 0$.
(3) $H(ax, y) \leqslant aH(x, y)$.
(4) $H(ax, y) \geqslant aH(x, y)$.

Since the first three properties are clear we present a proof for (4). If a is in $(0, 1)$ then

$$aH(x, y) = aH(x, a^{-1}ay) \leqslant a^2 H(x, a^{-1}y) = H(ax, y)$$

which is our assertion.

In order to prove that $H(x, y) = 0$ we distinguish two cases.

CASE 1. The elements (u, v) are linearly dependent (over R),

$$2u = x + y, \qquad 2v = x - y.$$

Then $u = aw$, $v = bw$ for some w in X. We consider $d > 0$ such that $|ad| \leqslant 1$, $|bd| \leqslant 1$, and $|dw| \leqslant 1$. Then we have clearly that

$$d^2 H(u, v) \leqslant H(|a|d(dw), |b|d(dw)) \leqslant |a|\,|b|d^2 H(dw, dw)$$

and since $H(dw, dw) = 0$ we must have $(d \neq 0)$ $H(u, v) = 0$.

CASE 2. The elements u and (and thus x, y) are linearly independent (over R). In this case we consider the two-dimensional subspace X_2 of X generated by x, y. We consider the set

$$S_{X_2} = \{z, z \in X_2, \|z\| = 1\}.$$

Then by Loewner's results we find a new norm on X_2 which is Euclidean (i.e. generated by an inner product) and satisfies the following inequality:

$$\|z\| \leqslant \|z\|_2,$$

where $\|\,,\|_2$ is the Euclidean norm of X_2.

From Loewner's result that

$$S_{X_2}^{+} = \{z, z \in X_2, \|z\|_2 = 1\}$$

has at least four point of contact with S_{X_2}, say $(t, s, -t, -s)$, we get that for the points (t, s) we have the relation $H(t, s) = 0$. Indeed, for $k > 0$ we have

$$\begin{aligned} kH(t, s) &\leqslant H(kt, s) \leqslant G(kt, s) \leqslant 2(\|kt + s\| + \|kt - s\|) - 4 \\ &= 2(\|kt + s\|_2 + \|kt - s\|_2) - 4 \leqslant 2(\|kt + s\|_2^2 + \|kt - s\|_2^2)^{1/2} \cdot 2^{1/2} - 4 \\ &= 2.2^{1/2}(2k^2 + 2)^{1/2} - 4 \leqslant 2k^2. \end{aligned}$$

Since k is arbitrary in $(0, 1)$ we obtain that $H(t, s) = 0$.

We now consider the points

$$u_1 = \tfrac{1}{2}(t + s), \qquad v_1 = \tfrac{1}{2}(t - s)$$

and then, using the symmetry and biconvexity of H we get that

$$H(u_1, v_1) = 0.$$

We can continue in this way and then we get that for a dense (countable) subset M of the set $S_{X_2}^{+}$, for all m_1, m_2 in M we have $H(m_1, m_2) = 0$. But this clear implies that H is identically zero.

Now we prove that the condition of the theorem is necessary. Suppose that a Banach space X has its norm generated by an inner product, say $\langle\,,\rangle$. Then for any x, y in X we set

$$\xi(x, y) = 1 + \mathrm{Re}\langle x, y\rangle.$$

It is obvious that $\xi(0, 0) = 1$ and that ξ is symmetric and biconvex. We now show that the condition

$$\xi(x, y) \leqslant \|x + y\|$$

is satisfied whenever $\|x\| \leqslant 1 \leqslant \|y\|$. Since

$$|1 + \mathrm{Re}\langle x, y\rangle|^2 \leqslant 1 + 2\,\mathrm{Re}\langle x, y\rangle + \|x\|^2\,\|y\|^2 \leqslant \|x + y\|^2 + (1 - \|x\|^2)(1 - \|y\|^2)$$

the assertion follows.

We present very briefly a characterization of the existence of an equivalent norm on a Banach space which is generated by an inner product in terms of certain properties of martingales. For more detail as well as for other related results we refer to Burkholder (1981).

Suppose that $(\Omega, \mathscr{B}, p)$ is a probability space and that $f = (f_i)_1^\infty$ is a X-valued martingale with the difference sequence $(d_i)_1^\infty$, $f_n = \Sigma_{k=1}^n d_k^k$, $d_k^f: \Omega \to X$ is strongly measurable to $\mathscr{B}_k$, $(\mathscr{B}_k)_0^\infty$ is a nondecreasing sequence of sub-σ-fields of $\mathscr{B}$.

We are interested in the class of Banach spaces for which, given two X-valued martingales F, G with difference sequences satisfying the inequalities

$$\|d_k^F(w)\| \leqslant \|d_k^G(w)\|,$$

there exists p in $(1, \infty)$ and a positive constant K_p (universal) such that

$$\|G\|_p \leqslant K_p \|F\|_p. \tag{+}$$

Then we have the following result.

THEOREM 9.3.4 (Burkholder, 1981). *K_p exists iff on X there exists an equivalent norm which is generated by an inner product.*

Proof. It was proved in Burkholder (1981) that if the relation (+) holds for some p in $(1, \infty)$ then it holds for all p in the interval $(1, \infty)$. Thus we may suppose that $p = 2$. If (x_n) is a sequence of elements of norm 1 then we set

$$d_n = r_n x_n, \qquad e_n = r_n \|x_n\| z, \qquad z \in X, \quad \|z\| = 1,$$

where (r_n) is a sequence of independent Bernoulli random variables,

$$P(r_n = \pm) = \tfrac{1}{2}.$$

Then (d_n) and (e_n) are X-valued martingales with difference sequences equal. Since (+) holds we obtain that the following inequalities hold:

$$\begin{matrix} c^{-1} \\ c \end{matrix} \left(\int_0^1 \left\| \sum_{i=1}^n r_i x_i \right\|^2 dp \right)^{1/2} \begin{matrix} \leqslant \\ \geqslant \end{matrix} \int_0^1 \left(\left\| \sum_{i=1}^n r_i \|x_i\| z \right\|^2 dp \right)^{1/2}$$

$$= \int_0^1 \left(\left\| \sum_{i=1}^n r_i \|x_i\| \right\|^2 dp \right)^{1/2} = \left(\sum_{i=1}^n \|x_i\|^2 \right)^{1/2}$$

and thus S. Kwapień's result gives the assertion of the theorem, since if X has a norm equivalent to the original norm, and this norm is generated by an inner product, then the inequality (+) holds.

9.4. Other Probabilistic Characterizations of Inner Product Structures

We mention now some very interesting and important results concerning the connection between certain probabilistic properties and the existence of an inner product structure on a Banach space. First we give a result by D. K. H. Muštari.

THEOREM 9.4.1 (Muštari, 1976). *Let X be a real separable Banach space and let f be a function on X^* with the following properties:*

(1) $f(0) = 1$.
(2) *f is positive definite.*

If there exists a topology τ on X^* such that:

(1) *τ is generated by a family of bilinear forms,*
(2) *the continuity of f in τ is necessary and sufficient for f to be the characteristic functional of a measure μ on X,*

then on X there exists an equivalent norm which is generated by an inner product.

For a proof of this result we refer to Muštari (1976) or to Mandrekar (1978) where a simpler proof is presented.

An interesting connection between certain facts on martingales and inner product structures was discovered by Dang Hung Thang and Nguen Duy Tien. In order to present the result we need some notations.

If (X_n) is a X-valued martingale (associated with the σ-algebras (a_n)) then we define

$$B_0(r) = \|X_0\|^r,$$

$$B_n(r) = \|X_0\|^r + \sum_{i=1}^{n} E(\|X_i - X_{i-1}\|)^r/a_{i-1}$$

$$B_\infty(r) = \|X_0\|^r + \sum_{i=1}^{\infty} (E(\|X_i - X_{i-1}\|)^r/a_{i-1})$$

$$V(r) = \|X_0\|^r + \sum_{i=1}^{\infty} \|X_i - X_{i-1}\|^r.$$

Then we have the following result.

THEOREM 9.4.2 (Dang Hung Thang and Hguen Duy Tien, 1981). *On a Banach space X there exists an equivalent norm generated by an inner product iff for each X-valued martingale (X_n) with*

$$E(\sup\|X_i - X_{i-1}\|^2) < \infty$$

then

$$(w, B_\infty(2) < \infty) = (w, V(2) < \infty) = (w, X_n \to \text{a.s.})$$

$(\to, X_n(w)$ converges).

Jain (1977) and Aldoux (1976) have obtained a very interesting result concerning the structure of Banach spaces in which a central limit theorem golds.

If X is a Banach space and U is a X-random variable and the (X_n) are independent identically distributed X-random variables and are distributed as U then we say that U satisfies the central limit theorem (CLT) if

$$\lim n^{-1/2} \sum_{i=1}^{n} X_n$$

converges (weakly) (to a Gaussian random variable). We say, in short, that $n^{-1/2}\Sigma_{i=1}^n X_n$ converges.

Then the result of Jain (the necessary part) and Aldoux (the sufficient part) is as follows.

THEOREM 9.4.3 (N. Jain–D. Aldoux). *Let X be a Banach space. Then on X there exists an equivalent norm generated by an inner product iff U satisfies* CLT *iff* $E(\|U\|^2) < \infty$, $E(X) = 0$.

For a proof we refer to Aldoux and Jain's papers.

Concerning the function space $L^2 = L^2(\Omega, \mathscr{B}, P)$ some results were obtained by G. I. Baidak, M. Š. Braverman and Ju. I. Petunin which characterize the space L^2 in some classes of function spaces by certain properties of some parameters which are very useful in statistics.

Let $(\Omega, \mathscr{B}, P)$ be a measure space and let $x(\,,\,)$ be a random variable (i.e. a measurable function on Ω (and as usual we identify functions which are equal a.e. (μ)). We suppose that we have a linear space X which consists of classes of functions (equally a.e. define the classes and we speak simply of functions). We suppose that X is a Banach space with respect to a certain norm and has the property that whenever x is in X and y is a measurable function with the property that

$$y(w) = x(w) \quad \text{a.e. } (\mu)$$

then y is in X.

For each x in X we define $m(x)$ to be the number

$$x(w)\,\mathrm{d}P(w)$$

(when it exists) and we set further

$$\delta(x) = \|x - m(x)\|_X$$

$(\|\,,\|_X$ means the norm of X). Further, we suppose that the function

$$e(w) = 1\,(\text{a.e.}(\mu))$$

is in X and

$$\|e\|_X = 1.$$

Then we have the following characterizations of the space L^2.

THEOREM 9.4.4 (Baidak *et al.*, 1983). *A Banach space X coincides with L^2 iff for each x in X the following relation holds:*

$$\|x + e\|_X^2 = \|x\|_X^2 + 1$$

whenever $m(x) = 0$.

THEOREM 9.4.5 (Baidak *et al.*, 1983). *A Banach space X coincides with L^2 iff for any two uncorrelated elements x, y we have the relation:*

$$\|x + y - m(x) - m(y)\|_X^2 = \|x - m(x)\|_X^2 + \|y - m(y)\|_X^2.$$

REMARK 9.4.6. The elements x, y in X are said to be uncorrelated if

$$m((x - m(x))(y - m(y))) = 0.$$

We mention that the paper by Baidak *et al.* contains other interesting related results.

CHAPTER 10

Positive Definite Functions, Functions of Positive Type and Inner Product Structures

10.0. Introduction

Positive definite functions seem to be a class of functions encountered in many branches of mathematics; we mention, especially, probability theory where the best known place is that offered by the famous Bochner theorem of characteristic functions. By now there is a great deal of information on this class of functions due essentially to D. A. Raikov, A. Weil, R. Godement, etc.

In this chapter we present some results about this class of functions, as well as some applications. The applications we give are concerned with complex function theory and operator theory. Concerning the applications in complex function theory we present the solution of the Nevanlinna problem about the necessary and sufficient conditions for a complex-valued function defined on a subset in $(z, |z| < 1)$ to admit an extension to an analytic function in $(z, |z| < 1)$ and having the real part nonnegative. The solution we present is that obtained by Korańyi *et al.* (1957). Further, we present the characterization of Loewner's monotone matrix functions obtained by Korańyi (1956). We mention that these Loewner functions have importance in quantum mechanics (Wigner and von Neumann, 1954). Some applications to the dilation theory of operators are also presented. We mention that in the next chapter we present further results about the dilation theory of operators using reproducing kernels. Of course, the class of positive definite functions is intimately related to reproducing kernels, but we prefer to present a separate treatment of the theory of reproducing kernels.

10.1. Positive Definite Functions. Definitions and Some Examples

Let us consider an abelian group which we suppose to be locally compact with multiplicative notation and the neutral element e.

We recall that a character of the locally compact abelian group G is any continuous function χ on G satisfying the condition $|\chi(g)| = 1$ for all g in G. The set of all characters of G is called the 'dual group' and is denoted by $\hat{G}$. If we define on G the topology which has as a neighborhood basis the sets of the form

$$(\chi, |\chi(g) - \chi_0(g)| < \varepsilon, g \in K),$$

where $\varepsilon > 0$ is a positive number and K is a compact subset in G, then $\hat{G}$ is a locally compact abelian group.

For more details we refer to Rudin (1962) and Naimark (1960).

DEFINITION 10.1.1. Let G be an Abelian group and let X be a complete inner product space. A function $f: G \to L(X)$ is said to be positive definite if for any finite subset $(g_1, \ldots, g_n)$ of G and $(c_1, \ldots, c_n)$ complex numbers for any $x \in X$ the following inequality holds:

$$\sum_{i,j=1}^{n} \langle f(g_i^{-1} g_j)x, x \rangle c_j \bar{c}_i \geqslant 0.$$

DEFINITION 10.1.2. A function $g: G \to L(X)$ is said to be of positive type if for any finite subset $(g_i)_1^n$ of G and $(x_i)_1^n$ in X the following inequality holds:

$$\sum_{i,j=1}^{n} \langle g(g_i^{-1} g_j)x_j, x_i \rangle \geqslant 0.$$

REMARK 10.1.3. The above definition can be formulated for any group; for the case of groups considered here we have the following important result (the validity for other classes of groups is not known):

THEOREM 10.1.4. *For G an Abelian locally compact group the classes of functions considered in Definitions* 10.1.1 *and* 10.1.2 *coincide.*

In fact we have a more general result.

RESULT 10.1.5. *Let G be an arbitrary Abelian group. Then the class of positive definite functions and the class of functions of positive type coincide.*

In Section 10.2 we present a proof of this result since we need more information about complex-valued positive definite functions. The following result gives some information about the realizations of positive definite functions. First we need the following definition.

DEFINITION 10.1.6. Let G be a group (with multiplicative notation, with the neutral element e). A unitary representation of G is any mapping $g \to U_g$ defined on the group G with values of unitary operators satisfying the following conditions:

(1) $U_{gg_1} = U_g U_{g_1}$,
(2) $U_e = I$.

Then we have the following result.

PROPOSITION 10.1.7. *If $g \to U_g$ is a unitary representation of the group G, then for any $x \in X$, X is the complete inner product space on which the unitary operators act, the function $g \to U_g x, x$ is a complex-valued positive definite function.*

Proof. Indeed if $(g_i)_1^n$ is a finite set in G and $(c_i)_i^n$ are arbitrary complex numbers then we obtain that

$$\left\langle \sum_{i,j=1}^{n} U_{g_i g_j}^{-1} x, x \right\rangle c_j \bar{c}_i = \left\langle \sum_{j,i=1}^{n} U_{g_i}^{-1} U_{g_j} x, x \right\rangle c_j \bar{c}_i$$
$$= \left\langle \sum_{i=1}^{n} c_i U_{g_i} x, \sum_{i=1}^{n} c_i U_{g_i} x \right\rangle \geqslant 0$$

and the assertion is proved. □

The converse assertion is given in

PROPOSITION 10.1.8. *If $g \to f(g)$ is a nonidentically zero positive definite function on the group G then there exists a unitary representation $g \to U_g$ of the group G such that $f(g) = \langle U_g x, x \rangle$ for some x in an inner product space X and where U_g acts.*

Proof. The following construction is in fact the prototype of many constructions in the proofs in this chapter.

We consider the space L of all complex-valued functions $\xi(g)$ on G which assumes only a finite number of nonzero values. If $\eta(g)$ is another function in this space then we set

$$\langle \xi, \eta \rangle = \sum_{g,h \in G} f(h^{-1}g)\xi(g)\bar{\eta}(h).$$

Then this is clearly a bilinear function which has all the properties of an inner product but the property $\langle \xi, \xi \rangle = 0$ only if $\xi = 0$. We now consider in the above subspace of functions the following subspace

$$N = (\xi, \langle \xi, \xi \rangle = 0)$$

and the quotient subspace L/N, and this is now a space with an inner product. We can complete this, and thus we obtain a complete inner product space X. Now for each g_0 in G we define a mapping on L by the formula

$$T_{g_0}\xi(h) = \xi(h^{-1}g_0)$$

and obviously this is an element of L. Further, N is an invariant subspace of T_{g_0}. Indeed, if η is another element of L we have

$$\langle T_{g_0}\xi, T_{g_0}\eta \rangle = \sum_{g,h \in G} f(h^{-1}g)\xi(g_0^{-1}g)\eta(g_0^{-1}h)$$
$$= \sum_{g,h \in G} f(h^{-1}g_0^{-1}g_0 g)\xi(g)\bar{\eta}(h) = \langle \xi, \eta \rangle.$$

Thus N is an invariant subspace for T_{g_0}, and T_{g_0} preserves the inner product and thus induces on L/N an isometric operator. This may be extended to an isometric operator on X. This is denoted by U_{g_0}. Since it is obvious the $U_{g_0^{-1}}$ is the inverse of U_{g_0} we obtain that in fact it is a unitary operator. Obviously $g \to U_g$ is a unitary representation of the group G. To prove that for some x in X

$$f(g) = \langle U_g x, x \rangle$$

we consider on the group G the function in L defined by the formula

$$\xi_0(g) = \begin{cases} 1, & g = e \\ 0, & g \neq e \end{cases}$$

and we remark that this is a function in L. Let ξ be the class of this function in L/N and then

$$\langle U_{g_0}\xi_0, \xi_0 \rangle = \langle T_{g_0}\xi_0, \xi_0 \rangle = \sum_{g,h \in G} f(h^{-1}g)\xi_0(g_0^{-1}g)\bar{\xi}_0(h) = f(g_0).$$

This clearly gives the assertion of the proposition.

REMARK 10.1.9. If G is assumed to be a topological group and f is assumed to be a continuous complex-valued positive definite function, then the representation $g \to U_g$ is also continuous.

Let G be a locally compact Abelian group with the Haar measure dx. We consider the space $L^1(G) = L^1(G, \mathscr{B}, \mathrm{d}x)$. This Banach space may be organized as a Banach algebra if we define the multiplication by the following formula: if f, g are two arbitrary elements in L^1 then we set

$$(f \otimes g)(y) = \int_G f(x - y)g(y)\,\mathrm{d}y,$$

which is an element of L^1, and the following inequality holds:

$$\|f \otimes g\|_{L^1} \leqslant \|f\|_{L^1}\|g\|_{L^1}.$$

Since $f \otimes g = g \otimes f$ we see that $L^1(G)$ is a commutative Banach algebra. This Banach algebra has identity if and only if G is a discrete group. For a proof of this fact we refer to Naimark (1960, p. 372) as well as the fact that $L^1(G)$ has an approximate identity (Naimark, 1960, p. 373).

DEFINITION 10.1.10. Let G be a locally compact Abelian group (nondiscrete) and let $L^1(G)$ be the Banach algebra considered above with the multiplication given by '$\otimes$' (and called the convolution). The group ring of a locally compact Abelian group G is the Banach algebra obtained from $L^1(G)$ by adjoining the identity. This group ring is denoted by $R(G)$.

If we define for f in $L^1(G)$,

$$f^*(g) = \bar{f}(g^{-1})$$

then

$$\|f^*\|_{L^1} = \|f\|_{L^1}$$

and $f \to f^*$ is an involution on $L^1(G)$. Then this induces on $R(G)$ an involution in a natural way. Now we are interested in the form of maximal ideals of the commutative Banach algebra with identity $R(G)$. Of course, one ideal maximal is $L^1(G)$. This ideal is denoted by m_0. If m is another maximal ideal of $R(G)$ then we find f in $L^1(G)$ such that $\hat{f}(m) \neq 0$. Then clearly we may assume, without loss of generality, that $\hat{f}(m) = 1$.

We now define the function

$$f_{g_0}(g) = f(g_0^{-1}g)$$

and set

$$\chi(g_0) = \hat{f}_{g_0}(m).$$

Then it is not difficult to see that the following relations hold:

$$\chi(e) = 1$$

$$|\chi(g_0)| \leqslant \|f_{g_0}\|_{L^1(G)}$$

$$f_{g_0^1} \otimes f_{g_0^2} = f \otimes f_{g_0^1 \cdot g_0^2}$$

The last relation is a consequence of the fact that G is an Abelian group. Then we have that

$$\chi(g_0^1 \cdot g_0^2) = \chi(g_0^1)\chi(g_0^2)$$

which gives that

$$|\chi(g)| = 1$$

and since

$$|\chi(g) - \chi(g_0)| \leqslant \|f_g - f_{g_0}\|_{L^1(G)}$$

we obtain that $\chi(\,,)$ is uniformly continuous on G. We see that for any maximal ideal $m \neq m_0$ we can associate a character of the group G. We show now that conversely, given the characters, we can associate maximal ideals. This is equivalent to indicating a construction of a Gelfand transform of a Banach algebra $R(G)$. Indeed, if f, g are in $L^1(G)$ then

$$(f \otimes g)(y) = \int_G f(x)g(x^{-1}y)\,\mathrm{d}x = \int_G f(x)g_x(y)\,\mathrm{d}x$$

and since $f \to \hat{f}(m)$ is continuous, we obtain that (for the ideal m we suppose, without loss of generality, that $\hat{g}(m) = 1$)

$$\hat{f}(m) = \hat{f}(m)\hat{g}(m) = \int f(x)g_x(m)\,\mathrm{d}x = \int f(x)\chi(x)\,\mathrm{d}x.$$

Since every element of $R(G)$ is of the form $ze + f$, where z is a complex number, we have that the Gelfand transform of this element is $z + \int_G f(x)\chi(x)\mathrm{d}x$.

A computation shows that the converse is also true, i.e. the formula

$$ze + f \to z + \int f(x)\chi(x)\,\mathrm{d}x \qquad (*)$$

defines a homomorphism of the Banach algebra $R(G)$ into the set of complex numbers. This gives in fact that the following result holds.

PROPOSITION 10.1.11. *There is a one-to-one correspondence between the maximal ideals of $R(G)$ which are not $L^1(G)$ and the dual group of G and this correspondence is given by the formula* (*).

We remark that a character of the group G may be obtained using the above correspondence as follows: We take a function f in $L^1(G)$ with the property that $\hat{f}(m) \neq 0$. Then we set

$$\chi(y) = \hat{f}_y(m)/\hat{f}(m).$$

and it is clear that this is a character of the group G.

Let G be a group and let $F_{\text{p.d.}}$ be the set of all positive definite complex-valued functions. We define a partial order on $F_{\text{p.d.}}$ as follows: if f, g are two positive definite complex-valued functions then we say that $f \leqslant g$ if $g - f$ is in $F_{\text{p.d.}}$. We say that a positive definite complex-valued function is elementary if every positive definite function dominated by that function is a multiple of that function. If f is a normalized positive functional on $R(G)$ (i.e. $f(e) = 1$ and $f(x^*x) = 0$) then it is of the form

$$f(ze + x) = z + \int_G x(g)\chi(g)\,dg \quad (dg \text{ is the Haar measure})$$

and if we suppose that f is elementary, then from the commutativity of $R(G)$, we obtain that for some ideal m we have

$$f(ze + x) = z + x(m)$$

which, comparing the representations of f, gives that

$$\varphi(g) = \chi(g)$$

i.e. every elementary positive definite function on a locally compact Abelian group is a character.

If a linear functional on $R(G)$ is of the form

$$f(ze + x) = z \cdot k, \quad k \text{ is a constant,}$$

then we call such a functional 'degenerate'. A positive functional will be called regular if it does not dominate the degenerate functional.

Let us consider a representation $x \to U_x$ of the Banach algebra $R(G)$ into the Banach algebra of all bounded linear operators on a complete inner product space X. Then clearly

$$N = (v, v \in X, A_x v = 0, x \in L^1(G))$$

is an invariant subspace for the representation (i.e. if v_0 is in this space then $A_x v_0$ is again in this space). We remark that on this invariant subspace the representation $x \to A_x$ is of the form zI, where $x = ze + a$, $a \in L^1(G)$. Then we may delete from X the subspace N and we can consider the representation only on $N^\perp$ and in this case we say that the representation does not contain the degenerate representation (which is the representation $(ze + a) \to zI$). Then we have the following result.

THEOREM 10.1.12. *For each representation $x \to A_x$ of the group ring $R(G)$ which does not contain the degenerate representation there exists a unitary representation $g - U_g$ of the group G and, conversely, to each measurable unitary representation $g \to U_g$ of the group G there corresponds a representation $x \to A_x$ of the group ring $R(G)$ and these representations are connected by the formula*

$$A_{ze+x} = zI + \int_G x(g) U_g \, dg.$$

We recall that the measurability of the representation $g \to U_g$ means that for any $x \in X$, $g \to \langle U_g x, x \rangle$ is a measurable function. For a proof we refer to Naimark (1960, pp. 377–380).

If $x \to A_x$ is a representation then a vector $v_0 \in X$ is said to be cyclic if $(A_x v_0)$ is dense in X. The representation is said to be a cyclic representation if it has a cyclic vector. If $(A_x^i)_{i \in I}$, I is an index set and each A^i is a representation of the Banach algebra A into the Banach algebra of all bounded linear operators on the complete inner product space X^i, then we can define $l_I^2(X^i) = X$ and if

$$\|A_x^i\| \leqslant M_x, \quad \forall i \in I$$

then

$$x \to (A_x^i v_i), \quad v = (v_i) \in l_I^2(X^i)$$

is a bounded linear operator on $l_I^2(X^i)$ which is denoted by A_x. Then $x \to A_x$ is a representation of A into $L(l_I^2(X^i))$ and is called the direct sum of the representations A_x^i,

$$A_x = \sum_i \oplus A_x^i.$$

It is not difficult to see that any representation is a direct sum of cyclic representations.

Let $g \to U_g$ be a unitary representation of the locally compact Abelian group G which is constructed with a positive definite function f defined on G. This unitary representation gives rise to a representation of the group ring $R(G)$, $x \to A_x$ not containing the degenerate representation. For $x \to A_x$ we have the regular positive functional on $R(G)$ by the formula

$$f(ze + x) = z\langle u, u\rangle + \langle A_x u, u\rangle = z\varphi(e) + \int_G x(g)\langle U_g u, u\rangle \, dg$$
$$= z\varphi(e) + \int_G x(g)\varphi(g) \, dg.$$

Conversely, if f is a regular positive functional on $R(G)$ then for this we have a representation $x \to A_x$ of $R(G)$ which does not contain the degenerate representation and has the property that

$$f(ze + x) = z\langle u, u\rangle + \langle A_x u, u\rangle,$$

where u is a cyclic vector of the representation. If $g - U_g$ is the continuous unitary representation associated with $x \to A_x$, then

$$A_x = \int_G x(g) U_g \, dg$$

and thus

$$f(ze + x) = z\varphi(e) + \int_G x(g)\langle U_g u, u\rangle \, dg = z\varphi(e) + \int_G x(g)\varphi(g) \, dg$$

with

$$\varphi(g) = \langle U_g u, u\rangle.$$

Thus we have the following result.

THEOREM 10.1.13. *Let G be a locally compact Abelian group G. Then there exists a one-to-one correspondence between the regular positive functionals of the ring group R(G) of G and the set of all continuous positive definite functions of G. This is given by the formula*

$$f(ze + x) = z\varphi(e) + \int_G x(g)\varphi(g)\,\mathrm{d}g,$$

where f is a positive regular functional and φ is a continuous positive definite function.

Now we know that every positive functional on $R(G)$ is of the form:

$$f(ze + x) = \int_M x(m)\,\mathrm{d}\mu(m),$$

where M is the set of all maximal ideals of $R(G)$ and μ is a positive Borel measure on M. In M we have the ideal $L^1(G)$ and we show that the support of μ is $M - L^1(G) = M - m_0$.

Indeed, if $\mu(m_0) > 0$ then

$$f(x) = \int_M x(m)\,\mathrm{d}\mu(m) = \int_{m_0 \cup (M - m_0)} x(m)\,\mathrm{d}\mu(m) = f(m_0)\mu(m_0) + \int_{M - m_0} x(m)\,\mathrm{d}\mu(m)$$

which implies that the degenerate functional $h(ze + x) = z$ is dominated by f, a fact which contradicts that the property of f is regular. Thus

$$f(x) = \int_{M - m} x(m)\,\mathrm{d}\mu(m)$$

and since $M - m_0$ is in one-to-one correspondence with the maximal ideals of $L^1(G)$, which is in one-to-one correspondence with the dual group (the characters of G), we can consider the measure supported on the group of characters, $\hat{G}$ of G. For $x = ze + a$, a in $L^1(G)$, we have

$$x(m) = z + \int_G a(g)\chi(g)\mathrm{d}g$$

and thus f is of the form

$$z(e) + \int_G a(g)\varphi(g)\,\mathrm{d}g = z\varphi(\hat{G}) + \int_{\hat{G}} \left(\int_G a(g)\chi(g)\,\mathrm{d}g \right) \mathrm{d}\mu(\chi)$$

which implies that

$$\mu(\hat{G}) = \varphi(e)$$

and

$$\int_G a(g)\varphi(g)\,\mathrm{d}g = \int_{\hat{G}} \left(\int_G a(g)\chi(g)\,\mathrm{d}g \right) \mathrm{d}\mu(\chi).$$

We can interchange the order of integration in the right-hand side of the last expression and since $a \in L^1(G)$ is arbitrary and $\chi(g)$ are continuous we conclude that the following equality must hold:

$$\varphi(g) = \int_{\hat{G}} \chi(g)\,\mathrm{d}\mu(\chi).$$

The converse assertion can be easily verified, i.e. each function on G of this form is positive definite. Thus we have the following result.

THEOREM 10.1.14 (Weil, 1938; Raikov, 1940). *If φ is a continuous positive definite function on G then it is of the form*

$$\varphi(g) = \int_{\hat{G}} \chi(g)\, d\mu(\chi), \tag{*}$$

where μ is a positive measure on $\hat{G}$ with $\mu(\hat{G}) = \varphi(e)$. Conversely, if φ is a function of the form (*) with μ satisfying $\mu(\hat{G}) = \varphi(e)$ then it is a continuous positive definite function on G.

10.1.15. If $G = R$ then $\hat{G} = R$ and the Weil–Raikov theorem reduces to the Bochner representations theorem since the characters of G may be identified with the functions

$$\chi_s(t) = e^{its}.$$

Thus every continuous positive definite function on R is of the form

$$\varphi(s) = \int_R e^{its}\, d\mu(t),$$

where $\mu(\,,\,)$ is a positive finite measure on R.

If we consider the group of integers $(0, \pm 1, \pm 2, \ldots, \pm n, \ldots)$ then this is an Abelian group and the dual group is the segment $[0, 2\pi]$ with 0 and 2π identified. Then we have that any positive definite function on $Z = (\pm n)$ is of the form

$$f(n) = \int_0^{2\pi} e^{ins}\, d\mu(s)$$

for some positive measure on $[0, 2\pi]$.

10.2. The Coincidence of Classes of Positive Definite Functions and Functions of Positive Type on Locally Compact Abelian Groups

First we remark that if f is positive definite then it is obviously of positive type. Thus to prove the assertion we must prove that if f is of positive type then it is a positive definite function.

Since f is of positive type, for each x in X the function $g \to \langle f(g)x, x\rangle$ is positive definite and thus by the Weil–Raikov theorem there exists a positive measure on the dual group $\hat{G}$ of G (we consider on G the discrete topology and then T is a compact Abelian group) such that

$$\langle f(g)x, x\rangle = \int_{\hat{G}} \chi(g)\, d\mu_x(\chi) = \int_G g(\hat{g})\, d\mu_x(\hat{g}) \quad (\chi(g) = g(\hat{g})).$$

We consider now on $\hat{G}$ the algebra of continuous functions generated by the characters of G. This is a subalgebra of $C(\hat{G})$ containing the constant (the character

$\hat{e}$) and is selfadjoint (the involution being the complex conjugation). If

$$f(\hat{g}) = c_1\chi_{g_1}(\hat{g}) + \cdots + c_n\chi_{g_n}(\hat{g}) = \sum_{i=1}^{n} c_i g_i(\hat{g}),$$

where c_i are complex numbers and χ_i are characters, we let

$$Tf = c_1 T_{g_1} + \cdots + c_n T_{g_n}$$

and thus

$$\left\langle \left(\sum_{i=1}^{n} c_i T_{g_i} \right) x, x \right\rangle = \int_{\hat{G}} f \, d\mu_x.$$

This gives that the operator Tf is well defined and depends only from f and from the representation of f used. Thus we may extend it to all f in the subalgebra generated by the characters of G and thus

$$\langle Tfx, x \rangle = \int_{\hat{G}} f \, d\mu_x.$$

The mapping $f \to Tf$ is obviously linear and positive (i.e. if $f \geqslant 0$ then $Tf \geqslant 0$). Since $T_1 = T_e$ we have the inequality

$$\| Tf \| \leqslant 2 \| f \|_\infty \| T_e \|. \tag{**}$$

Thus since, according to the Stone approximation theorem, the subalgebra generated by the characters is dense in $C(\hat{G})$, from (**) we obtain that $f \to Tf$ may be extended to all f in $C(\hat{G})$ and the positiveness property is preserved. Then by an operator version of the representation theorem of Riesz–Markov, we find an operator-valued measure F such that

$$Tf = \int_{\hat{G}} f(g) \, dF(g).$$

Since

$$T(\chi(g)) = \int_{\hat{G}} \chi(g) \, dF(\chi)$$

we have that

$$T_g = \int_{\hat{G}} \chi(g) \, dF(\chi)$$

and this implies that $s \to T_s$ is a positive definite function. The assertion is proved. □

COROLLARY 10.2.1 (Naimark's moment theorem). *Let G be a locally compact Abelian group and G be its character group. Let $g \to T_g$ be a weakly continuous operator-valued function on G (the values are bounded linear operators on a complete inner product space X) with $T_e = I$. Then the following assertions are equivalent:*

(1) *$g \to T_g$ is of positive type.*

(2) *$g \to T_g$ is a positive definite function.*
(3) *For some regular weakly Borel operator-valued measure μ on $\hat{G}$ the following representation holds,*

$$T_g = \int_{\hat{G}} \chi(g)\, \mathrm{d}\mu(\chi).$$

Proof. The equivalence of (1) and (2) was proved before. Let us suppose that (2) holds. For each x in X we consider the function

$$g \to \langle T_g x, x \rangle$$

which has the representation

$$\langle T_g x, x \rangle = \int_{\hat{G}} \chi(g)\, \mathrm{d}\mu_x(\chi)$$

where μ_x is a measure on $\hat{G}$. If x, y are arbitrary elements in $\hat{G}$ then we define the following Borel measure on X,

$$\mu_{x,y} = \tfrac{1}{4}(\mu_{x+y} - \mu_{x-y} + i\mu_{x+iy} - i\mu_{x-iy})$$

and then for all x, y in X we have that $g \to \langle T_g x, y \rangle$ has the representation

$$\langle T_g x, y \rangle = \int_{\hat{G}} \chi(g)\, \mathrm{d}\mu_{x,y}(\chi).$$

An approximation argument shows that $\mu_{x,y}$ is the unique measure with the above property. Further, for any Borel set E in $\hat{G}$, we have

$$|\mu_{x,x}(E)| = \mu_x(E) \leqslant \int_{\hat{G}} \chi(g)\, \mathrm{d}\mu_x(\chi) = \langle T_e x, x \rangle \leqslant \|x\|^2 \cdot \|T_e\|.$$

This implies that for each Borel set E in G there exists a bounded linear operator $\mu(E)$ such that

$$\langle \mu(E)x, y \rangle = \mu_{x,y}(E)$$

and thus

$$\langle \mu(E)x, x \rangle = \mu_x(E)$$

which gives that $E \to \mu(E)$ is a weakly Borel measure on G and the following representation formula holds:

$$\langle T_g x, x \rangle = \int_{\hat{G}} \chi(g) \mathrm{d}\mu_x(\chi) = \left\langle \int_{\hat{G}} \chi(g)\, \mathrm{d}\mu(\chi)x, x \right\rangle.$$

This gives that (3) holds. Conversely, if (3) holds it is easy to prove that (2) holds by a computation.

REMARK 10.2.2. From the proof of Corollary 10.2.1 it follows that the condition $T_e = I$ is unnecessary.

COROLLARY 10.2.3 (Stone–Godement–Ambrose). *Let G be a locally compact*

Abelian group and let $g \to U_g$ be a unitary representation which is weakly continuous. Then there exists a regular weakly Borel spectral measure μ such that

$$U_g = \int_{\hat{G}} \chi(g)\, \mathrm{d}\mu(\chi).$$

We recall that an operator-valued measure is called a spectral measure if the values are projection operators and the value for the entire space G is the identity operator I.

REMARK 10.2.4. The first theorem of this type was proved by Stone (1932) for the case $G = R$.

REMARK 10.2.5. Another result proved by Naimark (1943) was as follows: if $g \to T_g$ is a weakly continuous positive definite function on the locally compact Abelian group G with value that are bounded linear operators on a complete inner product space X with $T_e = I$ then there exists a complete inner product space X_1 containing X as a closed subspace and a unitary representation $g \to U_g$ such that for all g in G,

$$T_g = PU_g,$$

where P is the orthogonal projection of X_1 onto X.

We prove this result in a more general formulation later, but here we present a proof which uses the construction mentioned in the proof of 10.1.8. We consider the space $L_0(G)$ of all functions defined on G with values in the complete inner product space X which vanishes on G except on a finite subset. If ξ and η are two elements in $L_0(G)$ then we set

$$\langle \xi, \eta \rangle = \sum_{g,h} \langle T_{h^{-1}g}\xi(g), \eta(h) \rangle$$

which has all the properties of an inner product but the property that $\langle \xi, \xi \rangle = 0$ if and only if $\xi = 0$. We remark that for the function

$$\xi_x(e) = x, \qquad \xi_x(g) = 0, \quad g \neq e$$

we have

$$\langle \xi_x, \xi_y \rangle = \langle T_e x, y \rangle.$$

We consider the subspace N in $L_0(G)$ defined as

$$N = (\xi, \langle \xi, \xi \rangle = 0)$$

and we remark that this is an invariant subspace for all mappings defined on $L_0(G)$ by the formula

$$U_{g_0}(\xi)(g) = \xi(g_0^{-1}g).$$

Then the quotient space $L_0(G)/N$ is a space with an inner product. The mappings U_g defined above preserves the inner product $\langle \xi, \eta \rangle$ defined above and thus induces an isometric operator on the completion of $L_0(G)/N$. We denote the induced operator again by U_g. Since G is an Abelian group, U_g is unitary and this gives the relation

$$\langle U_s \xi_x, \xi_y \rangle = \langle T_s x, y \rangle$$

which implies that

$$T_s = PU_s,$$

where P is the orthogonal projection of the completion of $L_0(G)/N$ onto the subspace of all functions of the form $(\xi_x, x \in X)$ which may be considered as the space X (identified).

We now note the following theorem of Fr. Riesz which has many interesting applications.

THEOREM 10.2.6 (Fr. Riesz). *Let $D = (z, |z| < 1)$ be the open unit disc in the complex plane and let $f: D \to L(X)$ be a function on D with values in the set of all bounded linear operators on a complete inner product space such that $f(0) = I$ and* Re *$f(z)$ is a positive Hermitian operator. Then there exists a complete inner product space X_1 containing X as a closed subspace and a unitary operator $U: X_1 \to X_1$ with the property that*

$$f(z) = P(I + zU)(I - zU)^{-1},$$

where P is the orthogonal projection of X_1 onto X.

Proof. Since f is an operator-valued analytic function then it has the form

$$f(z) = T_0 + T_1 z + \cdots + T_n z^n + \cdots$$

where T_i are in $L(X)$. Of course we have that $f(0) = T_0 = I$ and we consider on the set of integers, which is an Abelian group (and compact if we consider the discrete topology), the following function with values in $L(X)$:

$$n \to h(n) = \begin{cases} \frac{1}{2} T_n & \text{if } n \geqslant 1 \\ \frac{1}{2} T^*_{-n} & \text{if } n \leqslant -1. \end{cases}$$

We now show that h is positive definite. Let n be fixed and consider the function

$$m(t) = \sum_{-n}^{n} x_k \mathrm{e}^{ikt}, \quad x_i \in X$$

and thus

$$\tfrac{1}{2}\pi \int_0^{2\pi} \left\langle \mathrm{Re} \langle f(r\,\mathrm{e}^{it}) m(t), m(t) \rangle \, \mathrm{d}t = \sum_{p,q} r^{|p-q|} \langle T_{p-q} x_p, x_q \rangle \right\rangle$$

which implies that h is positive definite. Thus there exists a complete inner product space X_1 containing X as a closed subspace and a unitary representation

$$n \to U_n = (U_1)^n = U^n, \quad U_1 = U$$

such that

$$h(n) = PU_n = PU^n,$$

where P is the orthogonal projection of X_1 onto X.

Since $f(z)$ is of the form

$$f(z) = I + T_1 z + T_2 z^2 + \cdots = I + 2(h(l)z + h(2)z^2 + \cdots)$$
$$= I + 2P \sum_{i=1}^{\infty} (U^n z^n) = P(I + zU)(I - zU)^{-1},$$

the theorem is proved. □

REMARK 10.2.7. It is easy to see that the converse assertion is also true, namely that any function of the form

$$f(z) = P(I + zU)(I - zU)^{-1}$$

has the property that $f(0) = I$ and $\operatorname{Re} f(z)$ is a positive Hermitian operator.

NOTE 10.2.8. Of course, in applications it is of great importance to find useful positive definite functions or operator-valued functions on D with the real part a positive Hermitian operator.

PROPOSITION 10.2.9. *If X is a complete inner product space and $T \in L(X)$ is an operator with $\|T\| \leqslant 1$ (i.e. T is a contraction) then the function on D $(=(z, |z| < 1))$ defined by the formula*

$$f(z) = (I - zT)^{-1}(I + zT)$$

satisfies the conditions of Fr Riesz's theorem.

Proof. Obviously $f(0) = I$ and since for elements of the form $u = (I - zT)x$, $x \in X$, we have that

$$\langle \operatorname{Re} f(z)x, x \rangle = \operatorname{Re}\langle (I + zT)u, (I - zT)u \rangle = \|u\|^2 - |z|^2 \|Ty\|^2,$$

the assertion is proved. □

COROLLARY 10.2.10. *If T is any contraction operator on a complete inner product space X then there exists a complete inner product space X_1 containing X as a closed subspace and a unitary operator $U: X_1 \to X_1$ such that $T^n = PU^n$, where P is the orthogonal projection of X_1 onto X.*

Proof. This follows from the above result if we compare the above expressions after expanding in both sides. □

Another useful function satisfying the conditions of Fr. Riesz's theorem was noticed by C. Berger, and is given in the following lemma.

LEMMA 10.2.11 (Berger, 1965). *Let T be in $L(X)$, X is a complete inner product space, such that*

$$w(T) = \sup\{|\langle Tx, x \rangle|, \|x\| = 1\} \leqslant 1$$

(i.e. T is a numerical contraction). Then the function $f(z)$ defined on D by the formula

$$f(z) = (I - zT)^{-1}$$

satisfies the conditions of Fr. Riesz's theorem.

Proof. Obviously $f(0) = I$ and if x is arbitrary in X then for $u = (I - zT)^{-1}x$ we have

$$\langle \operatorname{Re} f(z)x, x\rangle = \operatorname{Re}\langle u, (I - zT)u\rangle = \|u\|^2 - \operatorname{Re}\langle zTu, u\rangle \geqslant 0$$

and the assertion is proved. □

COROLLARY 10.2.12 (Berger, 1965). *Let T be an operator as in Lemma 10.2.11. Then there exists a complete inner product space X_1 containing X as a closed subspace and a unitary operator $U: X_1 \to X_1$ such that $T^n = 2PU^n$ for all $n = 1, 2, 3, \ldots$.*

Proof. The assertion follows from the above result if we compare the terms in the expansions in both sides of the representation formula for f. □

We give now another result of M. A. Naimark on the dilation of the so-called set functions with values in the set of all bounded operators on a complete inner product space.

THEOREM 10.2.13 (Naimark, 1943). *Let $(S, \mathscr{B})$ be a pair where S is a set, $\mathscr{B}$ is a ring of subsets of S and suppose that on $\mathscr{B}$ there is defined a function $F: \mathscr{B} \to L(X)$ such that the following conditions are satisfied:*

(1) $F(\emptyset) = 0$, $F(S) = I$,
(2) *For any N in $\mathscr{B}$, $F(N)$ is a positive operator.*
(3) *If $M \cap N = \emptyset$ then $F(M \cup N) = F(M) + F(N)$.*

Then there exists a complete inner product space X_1 containing X as a closed subspace and a function $\mu: \mathscr{B} \to L(X_1)$ whose values are projections and

(1) $\mu(\emptyset) = 0$, $\quad \mu(S) = I$,
(2) *if M, N are disjoint subsets in $\mathscr{B}$, then $\mu(M \cup N) = \mu(M) + \mu(N)$.*

Proof. We consider the space L_0 of all measurable simple functions defined on S and with values in X. If ξ, η are two elements of L_0 we find a partition of S in measurable subsets $S_1, \ldots, S_m$ such that on each S_i, ξ and η are constant. Then we set

$$\langle \xi, \eta\rangle = \sum_{i=1}^{m} F(S_i)x_i, y_i$$

with x_i, y_i the values of ξ, η on S_i respectively. Then this is a function having all the properties of an inner product but $\langle \xi, \xi\rangle = 0$ if and only if $\xi = 0$. We can consider the subspace N as above and L_0/N may be completed to a complete inner product space X_1. The subspace X may be considered as a subspace of X_1 if we define for each $M \in \mathscr{B}$, a function equal to x on M and 0 on $S - M$. Then we can define the projections in $L(X_1)$ which have the property stated in the theorem.

For a related notion of dilation we refer to Loynes (1965).

10.3. Completely Positive Maps. Stinespring's Theorem

M. A. Naimark's result given above about operator-valued functions on a ring of subsets suggests a problem about mappings defined on B^*-algebras and with values in the algebra $L(X)$, X being a complete inner product space. For the precise formulation of the problem and the result of W. F. Stinespring we consider a B^*-algebra A. If B is another B^*-algebra then a mapping (linear) $T: A \to B$ is said to be positive if whenever $a \in A$ is positive then Ta is positive.

DEFINITION 10.3.1. The mapping $T: A \to B$ is called completely positive if for each n, the mapping defined on the $M_n(A) = (a_{ij})_{i,j=1}^n$ with values in $M_n(B)$ by the formula $T_n:(a_{ij}) \to (Ta_{ij})$, $a_{ij} \in A$, is positive. Then we have the following result.

THEOREM 10.3.2 (Stinespring, 1955). *Let A be a B^*-algebra with identity and let T be a linear mapping $T: A \to B$, where B is another B^*-algebra contained in $L(X)$, X being a complete inner product space. Then a necessary and sufficient condition for T to be of the form*

$$Ta = S^*\rho(a)S,$$

*where S is a bounded linear operator defined on a complete inner product space X_1 with values in X and ρ is a *-representation of A into $L(X)$, is that T must be completely positive.*

Proof. Suppose first that T is of the above form and consider for each n the element (a_{ij}) in $M_n(A)$. Then (Ta_{ij}) is an operator in $L(X \oplus \cdots \oplus X)$, the direct sum of X with itself n-times. Since ρ is a *-representation, the matrix (Ta_{ij}) is a positive operator on $X \oplus X \oplus \cdots \oplus X$ (n-times) and thus if $(x_1, \ldots, x_n)$ is given we have that

$$\sum_{i,j} \langle Ta_{ij}x_i, x_j \rangle = \sum_{i,j} \langle \rho(a_{ij})Sx_j, Sx_i \rangle \geqslant 0.$$

Thus the condition is necessary.

Conversely, suppose that T is a completely positive linear map and then we consider the tensor product of $A \otimes X$. If

$$\xi = \sum_i^n a_i \otimes x_i, \qquad \eta = \sum_j^m b_j \otimes y_j$$

are two arbitrary elements in $A \otimes X$ then we set

$$\langle \xi, \eta \rangle = \sum_{i,j} \langle T(b_j^* a_i)x_i, y_j \rangle$$

and since T is supposed to be completely positive we have that the function $\langle\, , \rangle$, defined above, has all properties of an inner product but the property $\langle \xi, \xi \rangle = 0$ if and only if $\xi = 0$.

We can now consider the following mapping defined on $A \otimes X$ by the formula: $a \in A$

$$\left(\sum_i a_i \otimes x_i\right) \to \sum_i (aa_i) \otimes x_i$$

and this gives clearly a mapping ρ_1 defined on A with values in the space of all linear operators on $A \otimes X$. We show now that we have the inequality:

$$\langle \rho_1(a)\xi, \rho_1(a)\xi \rangle \leqslant \|a\|^2 \langle \xi, \xi \rangle.$$

Indeed, suppose that the assertion is not true and thus there exists ξ in $A \otimes X$ with the property that $\langle \xi, \xi \rangle \leqslant 1$ and $\|a\| < 1$ with $\langle \rho_1(a)\xi, \rho_1(a)\xi \rangle > 1$. From the Cauchy–Buniakowsky inequality we have that

$$\langle \rho_1(a^*a)\xi, \rho_1(a^*a) \rangle > 1$$

and similarly for all k

$$\langle \rho_1(a^*a)^{2^k}\xi, \xi \rangle > 1.$$

Now we remark that T, being completely positive, is also positive and thus if h is a Hermitian element then clearly we have the relation

$$-\|h\|e \leqslant h \leqslant \|h\|e \quad (e \text{ is the identity of } A)$$

and the positivity of T implies that

$$-\|h\|Te \leqslant Th \leqslant \|h\|Te$$

which implies the inequality

$$Th \leqslant \|h\|Te.$$

Thus T is uniformly continuous on the set of all Hermitian elements and since A is a B*-algebra this implies clear that T is uniformly continuous on A. Since we have that $\|a\| < 1$ this implies obviously that $(a^*a)^{2^k} \to 0$ (in norm) and thus

$$\lim \langle \rho_1(a^*a)^{2^k}\xi, \xi \rangle = \lim \sum_{i,j} \langle T(b_j^*(a^*a)^{2^k}b_i x_i, x_i \rangle = 0$$

and this is a contradiction. Thus the above inequality holds. We now consider the space $N = (\xi, \langle \xi, \xi \rangle = 0)$ and this is obviously a linear subspace in $A \otimes X$ and the quotient space $(A \otimes X)/N$ is a space with an inner product which can be completed; we denote the completion by X_1. For each a in A we can define a bounded linear operator on X_1 as follows. If a is in A then we set

$$\rho_{(a)}\left(\sum_i a_i \otimes x_i\right) = \sum_i (aa_i) \otimes x_i$$

which is a bounded linear operator on $A \otimes X$. We remark that N is an invariant subspace for each $\rho_{(a)}$ and thus $\rho_{(a)}$ induces an operator on $A \otimes X/N$ which also induces an operator on X_1 (which is denoted again by $\rho_{(a)}$). We now define the following operator on X with values in X_1 as follows: this operator, denoted by S, is the operator induced by the operator defined by the formula

$$x \to e \otimes x + N$$

and we remark that $\|Sx\|^2 \leqslant \|Te\| \langle x, x \rangle$. Then this is a bounded linear operator on X with values in X_1. But

$$\langle S^*(a)Sx, y \rangle = \langle \rho(a)Sx, Sy \rangle = \langle \rho_1(a)(e \otimes x), (e \otimes y) \rangle = \langle a \otimes x, e \otimes y \rangle = \langle Tax, y \rangle$$

and this shows that

$$Ta = S^*(a)S.$$

REMARK 10.3.3. In Stinespring (1955) there is an example which shows that positivity implies necessarily the complete positivity.

The following results present examples of completely positive mappings.

PROPOSITION 10.3.4 (Stinespring, 1955). *Let A be a commutative B^*-algebra and let $f: A \to L(X)$ be a linear positive mapping. Then f is completely positive.*

Proof. We know that A is isomorphic and isometric with the algebra of all continuous complex-valued functions vanishing at infinity on a locally compact Hausdorff space S, and in what follows we use this algebra of functions instead of A. If (a_{ij}) are complex-valued functions on S (vanishing at ∞) then we must show that

$$\sum_{i,j} \langle f(a_{ij})x_j, x_i \rangle \geqslant 0$$

for any $x_1, \ldots, x_n$ in X. X is supposed, of course, to be a complete inner product space and $L(X)$ is the set of all bounded linear operators on X. Since for each i

$$\langle f(a)x_i, x_i \rangle$$

is a continuous linear functional, by the Riesz–Markov representation theorem there exists a Borel measure m such that

$$\langle f(a)x_i, x_i \rangle = \int_S a(s)\, \mathrm{d}m(s)$$

and thus by the Lebesgue–Nikodym theorem there exist functions h_{ij} such that

$$\langle f(a)x_j, x_i \rangle = \int_S a(s)h_{ij}(s)\, \mathrm{d}m(s).$$

We now consider the matrix $(h_{ij}(s))$ which is nonnegative in S (a.e. (m)). This implies that $(a_{ij}(s)h_{ij}(s))$ is a positive matrix on S (a.e. (m)). The theorem is proved. □

PROPOSITION 10.3.5. *Let A be a B^*-algebra and let f be a positive linear functional on A. Then f is completely positive.*

Proof. This follows from the above result since $B = C$ is commutative. □

REMARK 10.3.6. Concerning these results we note that in fact the following result holds: let A, B be two B*-algebras and suppose that $f: A \to B$ is a positive linear map. Then if either A or B is commutative, f is completely positive.

We note also that completely positive maps play an important role in some recent works in tensor products of B*-algebras. For this we refer to Lance (1973) and to the references quoted there.

10.4. The Nevanlinna Problem

Let $D = (z, |z| < 1)$ be the open unit disc in the complex plane and let E be a subset in D. Suppose that $f: E \to C$ is a complex-valued function. The Nevanlinna problem is as follows:

PROBLEM 10.4.1. Under what conditions does there exist a function $g: D \to C$ with the following properties:

(1) $g/E = f$.

(2) g is analytic in D,
(3) $\operatorname{Re} g(z) \geqslant 0$?

In his classical paper of 1920, R. Nevanlinna found the answer in terms of 'positive definiteness' as follows:

THEOREM 10.4.2. *A necessary and sufficient condition for the Nevanlinna problem to have a solution is that for any integer n and any points $z_1, \ldots, z_n$ in E and any complex numbers $c_1, \ldots, c_n$ the following relation holds:*

$$\sum_{i,j=1}^{n} k(z_i, z_j) c_i \bar{c}_j \geqslant 0,$$

where

$$k(u, v) = (1 - u\bar{v})(f(u) + \bar{f}(v))$$

is defined on $E_{\times} E$.

An interesting solution of the Nevanlinna problem was given by Krein and Rechtman (1938) by reducing it to a moment problem. Another interesting solution of the Nevanlinna problem was found by Koranýi *et al.* (1957) using, in an ingenious way, some simple facts from the theory of operators on complete inner product spaces. Also, using the same device they obtained an extension to operator-valued functions of the Nevanlinna problem. In this section we follow essentially A. Koranýi *et al.*'s solution of the Nevanlinna problem as exposed in N. Ahiezer's book. First we prove that the condition is necessary.

Indeed, suppose that g exists and thus by Poisson's formula we have that

$$g(z) = i \operatorname{Im} g(0) + \tfrac{1}{2} \int_0^{2\pi} (\operatorname{Re} g(r\,\mathrm{e}^{it})(r\,\mathrm{e}^{it} + z)/(r\,\mathrm{e}^{it} - z)\mathrm{d}t$$

and thus if $z_1, \ldots, z_n$ are given points in E and $c_1, \ldots, c_n$ are complex numbers then

$$\sum_{i,j}^{n} k(z_i, z_j) c_i \bar{c}_j = \sum_{i,j=1}^{n} 1/(1 - z_i \bar{z}_j) \cdot (f(z_i) + \bar{f}(z_j)) c_i \bar{c}_j$$

$$= \frac{1}{2} \int_0^{2\pi} \Bigg(\sum_{i,j=1}^{n} (\operatorname{Re} g(r\,\mathrm{e}^{it}))(r\,\mathrm{e}^{it} + z_i)/(r\,\mathrm{e}^{it} - z_i) +$$

$$+ (r\,\mathrm{e}^{-it} + \bar{z}_j)/(r\,\mathrm{e}^{-it} - \bar{z}_j)) c_i \bar{c}_j \Bigg) \mathrm{d}t$$

$$= \frac{1}{2} \int_0^{2\pi} \operatorname{Re} g(r\,\mathrm{e}^{it}) \Bigg(\sum_{i,j=1}^{n} ((r\,\mathrm{e}^{it} + z_i)/(r\,\mathrm{e}^{it} - z_i) +$$

$$+ (r\,\mathrm{e}^{-it} + \bar{z}_j)/(r\,\mathrm{e}^{-it} - \bar{z}_j)) c_i \bar{c}_j \Bigg) \mathrm{d}t \geqslant 0$$

and thus the necessity is proved. □

We now prove that the condition is sufficient. First we suppose that the point

$z = 0$ is in E. We consider the space L_0 of all functions defined on E with complex values, and vanishing except only a finite subset of E.

If ξ, η are two functions in L_0 then we set

$$\langle \xi, \eta \rangle = \sum_{s,t \in E} k(s,t)\xi(s)\bar{\eta}(t).$$

We remark that $\langle \xi, \eta \rangle$ is well defined since the sum is a finite sum. Using the property of $k(s,t)$, the function $\langle \, , \rangle$ has obviously all the properties of an inner product, but the property $\langle \xi, \xi \rangle = 0$ if and only if $\xi = 0$. We now consider the subspace

$$N = (\xi, \langle \xi, \xi \rangle = 0)$$

and then L_0/N is an inner product space whose completion we denote by X. For each s in E we consider the function

$$e_s(u) = \begin{cases} 1, & s = u \\ 0, & s \neq u \end{cases}$$

which is obviously in L_0. From the definition of the form $\langle \, , \rangle$ we see that

$$\langle e_s, e_t \rangle = k(s,t)$$

for all s, t in E.

On $E - 0$ we define the function $e_{1,s} = 1/s \cdot e_s = s^{-1}e_s$ and then we remark that the following relation holds:

$$\langle e_{1,s}, e_{1,t} \rangle = \langle e_s, e_t \rangle.$$

Indeed,

$$\begin{aligned} \langle e_{1,s}, e_{1,t} \rangle &= s^{-1}t^{-1}\langle e_s, e_t \rangle - \langle e_s, e_0 \rangle - \langle e_0, e_t \rangle + \langle e_0, e_0 \rangle \\ &= s^{-1}t^{-1}(k(s,t) - k(s,0) - k(0,t) + k(0,0)) \\ &= s^{-1}t^{-1}((f(s) + \bar{f}(t))/(1 - s\bar{t})) - \\ &\quad -(f(s) + \bar{f}(0)) - (f(0) + \bar{f}(t)) + (f(0) + \bar{f}(0)) = \\ &= (f(s) + \bar{f}(t))/(1 - s\bar{t}) = k(s,t) \end{aligned}$$

and the assertion is proved.

We consider now the subspace generated by the functions $(e_s)_{s \in E - 0}$ and on this subspace we define the mapping with values in L_0 by the formula:

$$Ve_s = e_{1,s}$$

and the above result shows that V is an isometric operator. This implies that V may be extended on the completion to an isometric operator. From the definition of V we see that the following relation holds:

$$(I - sV)e_s = e_s - se_{1,s} = e_0.$$

Since N is an invariant subspace of V we see that on the completion X of L_0/N, V induces an isometric operator (which is denoted again by V) and since every isometric operator can be extended to a unitary operator defined on a space X_1.

We consider the space $X \oplus X$ and we may identify X with the subspace of all elements of the form $(x, 0)$ with $x \in X$; and we now consider the subspace M generated by the functions e_s, $s \neq 0$ and M_1 is the image of M through V;
then any $x \in X$ has the following decompositions:

$$x = x_M + x_M^\perp = x_{M_1} + x_{M_1}^\perp.$$

The unitary operator which is an extension of V is defined as follows:

$$U(x, y) = (Vx_M + g_{M_1^\perp}, V^{-1}y_{M_1} + x_M).$$

Then we have that

$$(I - sU)e_s = e_0$$

(e_s is identified with the element $(e_s, 0)$) for all s in $E - 0$. We remark that this relation is true also for $s = 0$. Further, we note that

$$\begin{aligned}\langle \tfrac{1}{2}(I + sU)(I - sU)^{-1}e_0, e_0\rangle &= \langle (I - sU)^{-1}e_0, e_0\rangle - \tfrac{1}{2}\langle e_0, e_0\rangle \\ &= \langle e_s, e_0\rangle - \tfrac{1}{2}\langle e_0, e_0\rangle = k(s, 0) - \tfrac{1}{2}k(0, 0) \\ &= f(s) + \bar{f}(0) - \tfrac{1}{2}(f(0) + \bar{f}(0)) \\ &= f(s) - i\,\mathrm{Im}\, f(0).\end{aligned}$$

Thus f has the following representation

$$f(s) = i\,\mathrm{Im}\, f(0) + \tfrac{1}{2}\langle (I + sU)(I - sU)^{-1}e_0, e_0\rangle$$

and we can define $g(s)$ by the formula

$$g(s) = i\,\mathrm{Im}\, f(0) + \tfrac{1}{2}\langle (I + sU)(I - sU)^{-1}e_0, e_0\rangle$$

which represent an analytic function in D which has $\mathrm{Re}\, g(s) \geqslant 0$ since

$$\mathrm{Re}\ g(s) = \langle \mathrm{Re}\, \tfrac{1}{2}(I + sU)u, (I - sU)u\rangle = \tfrac{1}{2}(1 - |s|^2\|u\|^2)$$

where $u = (I - sU)^{-1}e_0$.

Thus Nevanlinna's problem has a solution which may be represented by the above formula if the set E does not contain the point $z = 0$. If E does not contain $z = 0$ then we consider a fixed point in E, say s_0, and let

$$s \to (s - s_0)/(1 - s\bar{s}_0) = u$$

and E_1 is the image of E through this transformation. Obviously E_1 is in D and the function

$$h(u) = f(s)$$

satisfies the relation: if $u_1, \ldots, u_n$ are arbitrary points in E_1 and $c_1, \ldots, c_n$ are arbitrary complex numbers then

$$\sum_{i,j=1}^{n} (h(u_i) + \bar{h}(u_j))/(1 - u_i\bar{u}_j)c_i\bar{c}_j = \sum_{i,j=1}^{n} (f(s_i) + \bar{f}(s_j))/(1 - s_i\bar{s}_j)b_i\bar{b}_j$$

with

$$u_i = (s_i - s_0)/(1 - s_1\bar{s}_0), \qquad b_i = (1 - s_i\bar{s}_0)/(1 - s_0\bar{s}_0)^{1/2} \cdot c_i,$$

then the function h and the set E_1 satisfy the conditions of the above result and h admits an extension, say h_1. Then the function

$$s \to h_1((s - s_0)/(1 - s\bar{s}_0))$$

gives the solution of the Nevanlinna problem without restriction on the set F.

As mentioned in the beginning in their paper, A. Koranýi *et al.* have solved the Nevanlinna problem in the case when the function f has values in the space $L(X)$ of all bounded linear operators defined on a complete inner product space X.

10.4.3 (The operator-valued Nevanlinna problem). Let X be a complete inner product space, let E be a subset in $D = (z, |z| < 1)$, and let $f: E \to L(X)$ be a function. The Nevanlinna problem is: What are the necessary and sufficient conditions for the existence of a function $g: D \to L(X)$ such that

(1) $g/E = f$,
(2) g is analytic in D,
(3) $\operatorname{Re} g(s)$ is a positive operator?

The answer is given in terms of positive definiteness of a function defined using the function f. Let us consider, as above, the operator-valued function on $E_\times E$ defined by the formula:

$$k(s, t) = (1 - s\bar{t})(f(s) + f^*(t)),$$

where $f^*(t)$ is the adjoint of $f(t)$.

Concerning the operator-valued function k we consider two types of positive definiteness:

DEFINITION 10.4.4. The function k is said to be of positive type if for all $s_1, \ldots, s_n$ in E and all $x_1, \ldots, x_n$ in X the following relation holds:

$$\sum_{i,j=1}^{n} \langle k(s_i, s_j)x_i, x_j \rangle \geqslant 0.$$

DEFINITION 10.4.5. The function k is said to be of weak positive type if for all $s_1, \ldots, s_n$ in E and all complex numbers $c_1, \ldots, c_n$ the operator

$$\sum_{i,j=1}^{n} k(s_i, s_j)c_i\bar{c}_j$$

is positive.

REMARK 10.4.6. It is easy to see that if k is of positive type then k is of weak positive type.

Then we have the following result.

THEOREM 10.4.7. *If the function k is of positive type then the Nevanlinna problem has a solution.*

Proof. We consider the space L_0 of all functions defined on E with values in X and vanishing on E except of a finite subset of E. If ξ, η are two elements of L_0 then we define

$$\langle \xi, \eta \rangle = \sum_{s,t \in E} \langle k(s,t)\xi(s), \eta(t) \rangle$$

and as in the case considered above for complex-valued functions, this function has all the properties of an inner product but the property $\langle \xi, \xi \rangle = 0$ if and only if $\xi = 0$. We consider the subspace $N = (\xi, \langle \xi, \xi \rangle = 0)$; then L_0/N is a space with an inner product which may be completed. The completion is denoted by $\bar{X}$.

We suppose that E contains the point $z = 0$ and that $f(0) = I$. For each x in X we consider the function defined on E with values in X by the formula

$$e_x(s) = xe_s, \qquad e_s(t) = \begin{cases} 1, & s = t \\ 0, & s \neq t \end{cases}$$

We may identify the space X with the space of all functions e_x.

For $s \in E - 0$ we consider the mapping V defined by

$$V(xe_s) = s^{-1}(xe_s - xe_0).$$

As above, we note that the following relation holds:

$$\langle V(xe_s), ye_t \rangle = \langle xe_s, ye_t \rangle.$$

This implies that V is an isometric operator defined on the subspace M generated by the function xe_s. Let M_1 be the image of M through V. Again, as above, we remark that for all s in E the following relation holds:

$$(I - sV)(xe_s) = x.$$

As above we can extend the operator V to a unitary operator U defined on a complete inner product space containing X as a closed subspace and such that

$$(I - sU)(xe_s) = x,$$

which gives that

$$xe_s = (I - sU)^{-1}x$$

and remarking that

$$P(xe_s) = k(s, 0)x$$

is a projection operator of $\bar{X}$ onto X (the identified subspace of X in $\bar{X}$). Then we have that

$$\begin{aligned} P(\tfrac{1}{2}(I + sU)(I - sU)^{-1})x &= P((I - sU)^{-1} - \tfrac{1}{2}I)x = P(xe_s) - \tfrac{1}{2}(xe_0) \\ &= K(s, 0)x - \tfrac{1}{2}k(0, 0)x \\ &= (f(s) + f^*(0))x - \tfrac{1}{2}(f(0) + f^*(0))x = f(s)x. \end{aligned}$$

Then Theorem 10.4.7 is proved. □

Suppose now that the function k is of weak positive type. Then for each x in X the function

$$f_x(s) = \langle f(s)x, x \rangle$$

has an extension to D and this is represented by an integral formula, such as

$$f_x(s) = \int_0^{2\pi} (1 + s\,e^{it})(1 - s\,e^{it})\,d\mu_x(t),$$

where μ_x is a measure on $[0, 2\pi]$.

The measure μ_x is not necessarily unique but there are some cases when this is the case, for example when the set E has an accumulation point in D. Then the mapping $t \to \mu_x([0, t])$ gives an operator P_t which satisfies the conditions of the Naimark theorem and thus there exists a complete inner product space $\bar{X}$ and a spectral measue (E_t) on $\bar{X}$ such that $P_t = PE_t$, where P is the orthogonal projection of $\bar{X}$ onto X. Then the unitary operator

$$U = \int_0^{2\pi} e^{it}\,dE_t$$

may be used to obtain the integral representation of the Nevanlinna problem as follows:

$$g(s) = \tfrac{1}{2}P((I + sU)(I - sU))^{-1}.$$

REMARK 10.4.8. It appears of some interest to know if the result on the Nevanlinna problem for operator-valued functions may be extended to functions with values in a B*-algebra.

10.5. The Monotone Functions of C. Loewner

The class of monotone matrix functions of arbitrarily high order were introduced by C. Loewner in 1934 in connection with his studies of real-valued functions which are analytic in some domain and are continuous in a complex domain with the property that they are analytic in the upper half-plane and have a nonnegative imaginary part. A detailed study of monotone functions of Loewner is given in the paper of Bendat and Sherman (1955) where there are also some integral representations.

The interest in monotone functions of Loewner has many sources: among them we mention that they (or a subclass) appear in the quantum theory of collision (later we shall give the exact formulation of functions which appear there) and there exist connections with electrical network theory (see Richards, 1947, and the literature quoted there) and, as mentioned in Bendat–Sherman's paper, Loewner has noted a connection with the special theory of relativity.

Let X be a complete inner product space and let $L(X)$ be the set of all bounded linear operators on X.

DEFINITION 10.5.1. A function $f\colon (a, b) \to R$ is said to be monotone if for each bounded Hermitian operators A, B whose spectrum lie in a, b the relation $A \leqslant B$ implies the relation $f(A) \leqslant f(B)$ whenever both $f(A)$ and $f(B)$ are defined. Then f is called 'monotone increasing'. Similarly we define monotone decreasing functions.

For the proof of the integral representation we need the following characterization whose proof is in Loewner's paper (1934).

THEOREM 10.5.2 (Loewner, 1934). *Let $f:(a,b)\to R$ be a continuously derivable function. Then a necessary and sufficient condition for f to be a monotone function is that for all n*

$$\sum_{i,j=1}^{n} k(t_i,t_j)c_i\bar{c}_j \geqslant 0$$

for all $t_1,\dots,t_n$ in (a,b) and all complex numbers $c_1,\dots,c_n$ and where

$$k(t,t)=f'(t),\qquad k(s,t)=(f(s)-f(t))/(s-t)\cdot(s\neq t).$$

If $C_{(a,b),P}$ denotes the set of all functions satisfying the hypothesis of Loewner's theorem, then Loewner has proved that this is identical with the class of all functions which are analytic in (a,b) and can be extended analytically onto the entire upper half-plane to an analytic function with a nonnegative imaginary part. An elegant proof of this fact is given in Koranýi (1955).

Now we give an ingenious proof of A. Koranýi for the integral representation of functions in $C_{(a,b),P}$.

We note that we may consider, without loss of generality, that $(a,b)=(-1,1)$. Then we have the following result.

THEOREM 10.5.3 (Bendat and Sherman, 1955). *The function f is in $C_{(-1,1),P}$ if and only if there exists a measure μ on $(-1,1)$ such that*

$$f(s)=f(0)+\int_{-1}^{1} s/(1-st)\,\mathrm{d}\mu(t).$$

Proof (Korányi, 1956). We consider the set L_0 of all functions defined on $(-1,1)$ and which are of the form:

$$a(t)=\sum_{i=1}^{n} a_i k(t,s_i)$$

with $s_i\in(-1,1)$.

If b is another function in L_0,

$$b(t)=\sum_{j=1}^{m} b_i k(t,t_j)$$

then the 'inner product of a and b' is defined by the formula:

$$\langle a,b\rangle=\sum_{i,j=1}^{n,m}\sum k(t_j,s_i)a_i\bar{b}_j.$$

Then this has all the properties of an inner product since

$$\langle a,b\rangle=\sum_{j=1}^{m} a(t_j)b_j=\sum_{i=1}^{n}\bar{b}_i(s_i)a_i.$$

We denote the completion of L by X. If $t\neq 0$ then we define for each element of the form $k(s,t)=k_t(s)$ the mapping

$$T_1k_t(s)=t^{-1}(k(s,t)-k(s,0)).$$

Then if $k_t(s)$ and $k_p(s)$ are two elements with $t \neq 0 \neq p$, we have

$$\begin{aligned}\langle T_1 k_t(s), k_p(s)\rangle &= t^{-1}\langle (k(s,t) - k(s,0)), k(s,p)\rangle = t^{-1}(k(p,t) - k(p,0)) \\ &= t^{-1}((f(t) - f(p))/(t-p) - (f(t) - f(0))/t) = p^{-1}(k(p,t) - k(0,t)) \\ &= k(s,t), p^{-1}(k(s,p) - k(s,0)) = \langle k(s,t), T_1 k(s,p)\rangle.\end{aligned}$$

We remark that this holds also if $t = p$. Thus T may be extended by linearity on the subspace generated by the elements k_t.

Since

$$\lim_{t\to 0}\|k(s,t) - k(s,0)\|^2 = \lim_{t\to 0}(f'(t) - 2(f(t) - f(0))/t + f'(0)) = 0$$

we see that $k(s,0)$ is a limit of elements of the form k_t, which gives that T_1 is densely defined. Of course, we restrict ourselves to a consideration of the subspace generated by the elements k_t and $k(,\ \ 0)$. Also we remark that if

$$\sum_{i=1}^{n} a_i k_{t_i} = 0$$

then

$$\sum_{i=1}^{n} a_i T_1 k_{t_i} = 0.$$

From the above relation we see that T_1 is a symmetric operator and we see that it has a selfadjoint extension, say S. We show that for all $|t| < 1$ the operator $(I - tS)^{-1}$ exists. Since $I - tS$ is selfadjoint it suffices to show that the range of $I - tS$ is dense. But

$$(I - tS)k(s,p) = k(s,p) - t/p(k(s,p) - k(s,0)) = (1 - t/p)k(s,p) + \frac{t}{p}(k(s,0))$$

which implies that

$$k(s,0) = (I - tS)k(s,t)$$

$$k(s,p) = (1 - t/p)^{-1}\left((I - tS)k(s,p) - \frac{t}{p}(k(s,0))\right), \quad t \neq p$$

and thus $k(s,p) = k_p(s)$ is in the range of $I - tS$ if $p \neq t$. Since

$$\lim_{p\to t} k(s,p) = k(s,t)$$

and thus $k(s,t)$ is in the range of $I - tS$. For $t \neq 0$ we have

$$\begin{aligned}(f(t) - f(0))/t = k(0,t) &= \langle k(s,t), k(s,0)\rangle \\ &= \langle (I - tS)^{-1}(I - tS)k(s,t), k(s,0)\rangle \\ &= \langle (I - tS)^{-1}k(s,0), k(s,0)\rangle\end{aligned}$$

which implies that

$$f(t) = f(0) + \langle (t(I - tS)^{-1}k(s,0), k(s,0))\rangle.$$

Since S is a selfadjoint operator we have that

$$f(t) = f(0) + \int_{-\infty}^{\infty} t(1 - ts)\,\mathrm{d}\langle E_s k(u,0), k(u,0)\rangle = f(0) + \int_{-\infty}^{\infty} t(1 - st)^{-1}\,\mathrm{d}\mu(s).$$

To complete the proof of the theorem we must show that the support of the measure μ is contained in $(-1, 1)$. If $|s| < 1$ then we have

$$\int_{-\infty}^{\infty} (1 - st)^{-1}\,\mathrm{d}\mu(t) = \|(I - sA)^{-1}k(u,0)\|^2 = \|k(u,s)\|^2 = k(s,s) = f'(s)$$

which implies that for $|s| > 1$,

$$\int_{-\infty}^{\infty} (s - t)^{-2}\,\mathrm{d}\mu(t) = s^{-2}f'(s^{-1})$$

and thus if $|s| \geqslant m > 1$,

$$\int_{-\infty}^{\infty} (s - t)^{-2}\,\mathrm{d}\mu(t) = \max_{u \leqslant 1/m} f'(u) = K_m.$$

Let $m \leqslant M < N$ and take $s = \frac{1}{2}(M + N)$. Then

$$K_m \geqslant \int_M^N (s - t)^{-2}\,\mathrm{d}\mu(t) \geqslant ((N - M)/2)^{-2} \int_M^N \mathrm{d}\mu(t) = 4(N - M)^{-2}(\mu((M, N)))$$

and

$$\mu((M,N))/(N - M) \leqslant (N - M)/n \cdot K_m.$$

From these we obtain that

$$F(t) = \mu((-\infty, t))$$

is differentiable and for $t > m$, $F'_{(t)} = 0$ which implies that F is constant for $s > m$. Since m is arbitrary in $(1, \infty)$ this implies that F is constant on $(1, \infty)$. A similar argument shows that F is constant on $(-\infty, -1)$. Thus the measure μ has the support in $(-1, 1)$.

Conversely, if f has a representation of the above form then

$$k(s,t) = \int_{-1}^{1} (1 - su)^{-1}(1 - tu)^{-1}\,\mathrm{d}\mu(u)$$

and thus if $t_1, \ldots, t_n$ are arbitrary elements in $(-1, 1)$ and $c_1, \ldots, c_n$ are arbitrary complex numbers we have that

$$\sum_{i,j=1}^{n} k(t_i, t_j)c_i\bar{c}_j = \sum_{i,j=1}^{n} \int_{-1}^{1} (1 - ut_i)^{-1}(1 - t_j u)^{-1}\,\mathrm{d}\mu(u)c_i\bar{c}_j$$

$$= \int_{-1}^{1} \left(\sum_{i=1}^{n} c_i/(1 - t_i u)\right)^2 \mathrm{d}\mu(t) \geqslant 0.$$

The theorem is proved. □

10.5.4. *If f has the integral representation*

$$f(t) = f(0) + \int_{-1}^{1} (1 - st)^{-1}\, \mathrm{d}\mu(s)$$

then this implies that f can be extended to an analytic function in the upper half-plane and with an imaginary part nonnegative.

This class of function is related to the class of functions with the so-called property (P) studied by E. P. Wigner and John von Neumann: the function f has the property (P) if is defined on $t > 0$ and differentiable at all points of R^+ except a set of isolated points t_i, and the matrix $(a_{ij})_1^n$ is positive semidefinite and nonsingular, where

$$a_{ii} = f'(s_i), a_{ij} = (f(s_i) - f(s_j))/(s_i - s_j)$$

and $s_1, \ldots, s_n$ are arbitrary points not equal to any of t_i.

For the results about operator-valued version of Wigner–von Neumann results we refer to papers by Narcovich (1974), and Allen and Narcovich (1976). It may be of interest to see how the approach of A. Koranýi may be used to obtain some of the results of Allen and Narcovich.

For the connections with the quantum theory of collision we refer to the excellent papers of Wigner (1946, 1964) and Wigner and von Neumann (1954); and for connections with the theory of nuclear reactions, see Lane and Thomas (1958).

CHAPTER 11

Reproducing Kernels and Inner Product Spaces. Applications

11.0. Introduction

The theory of complete inner product spaces with reproducing kernel functions has received a growing interest partly because of its own beauty and partly because of numerous applications in the theory of functions of one or several variables and in the theory of partial differential equations. It seems worthy of notice that, perhaps, the space in which the existence of a reproducing kernel was first recognized was the space of harmonic functions on a complex domain and which are elements of L^2. This fact was stated by S. Zaremba in 1908. The theory of reproducing kernel functions was decisively influenced by the works of S. Bergman in which the so-called 'Bergman kernel' was introduced and studied intensively. The abstract theory of reproducing kernel functions was greatly influenced by the works of Aronszajn (1943, 1950). For more information about the applications of the theory of reproducing kernels we refer to Bergman (1950), Bergman and Schiffer (1951), Epstein (1965) as well as to the references quoted in our bibliography.

In this chapter we present an introduction to the theory of the reproducing kernel function and we give also some applications.

11.1. Reproducing Kernels. Basic Properties

Let X be a complete inner product space which consists of functions defined on a set S. The functions are supposed to be complex-valued.

The notion of a reproducing kernel for X is considered in the following definition.

DEFINITION 11.1.1. The function $K: S_{\times} S \to C$ is said to be a reproducing kernel for the space X if the following conditions are satisfied:

(1) For each $s \in S$ the function $K(\ ,s) \in X$.
(2) For each $f \in X$ and all $s \in S$ the following representation holds:

$$f(s) = \langle f, K(\ ,s)\rangle$$

where $\langle\ ,\ \rangle$ denotes the inner product on X.

Then we have the following.

PROPOSITION 11.1.2. *If K is a reproducing kernel for X then the following* *assertions hold:*

(1) $K(s, s) \geqslant 0$.
(2) $K(s, t) = \bar{K}(t, s)$.
(3) $|K(s, t)|^2 \leqslant K(s, s) \cdot K(t, t)$.

Proof. Since for each $s \in S$, $K(\ ,s)$ is in X and

$$K(s, s) = \langle K(\ ,s), K(\ ,s) \rangle,$$

property (1) follows. For (2) we consider t, s arbitrary points in S and we have, using the property of K, that

$$\langle K(.,s), K(.,t) \rangle = K(s, t) = \overline{\langle K(.,t), K(.,s) \rangle} = \bar{K}(t, s).$$

Thus (2) is proved. Assertion (3) is an obvious consequence of the Cauchy–Buniakowsky inequality. □

The following property is one of the most important property of a reproducing kernel function.

THEOREM 11.1.3. *If K is a reproducing kernel for X then K is 'positive definite', i.e. for any finite set $s_1, \ldots, s_n$ in S and any complex numbers $c_1, \ldots, c_n$ the following inequality holds:*

$$\sum_{i,j=1}^{n} K(s_j, s_i) c_i \bar{c}_j \geqslant 0.$$

Proof. We remark that

$$\sum_{i,j=1}^{n} K(s_j, s_i) c_i \bar{c}_j = \left\langle \sum_{i=1}^{n} c_i K(\ ,s_i), \sum_{j=1}^{n} c_j K(\ ,s_j) \right\rangle = 0$$

and the assertion follows. □

If X is a complete inner product space of functions on a set S and X has a reproducing kernel K then, quite naturally, we have the problem: Does X have many such functions? The following result answers this question.

THEOREM 11.1.4. *If X is a complete inner product space with a reproducing kernel K then X has only one reproducing kernel.*

Proof. Suppose that X has reproducing kernels K and K_1. Then we have that for all s in S, $K(.,s)$ and $K_1(.,s)$ are elements of X as well as $K(s,.)$, $K_1(s,.)$ (this follows from Proposition 11.1.2(2)). Then using the 'reproducing property' of the functions K and K_1 we obtain

$$\begin{aligned} K(s, t) &= \langle (K(.,t))(s) \rangle = \langle K(.,t), K_1(.,s) \rangle \\ &= \overline{\langle K_l(.,s), K(.,t) \rangle} = \overline{K_1(t, s)} = K_1(s, t). \end{aligned}$$ □

If X is a complete inner product space whose elements are functions defined on a set S we may ask if X has a reproducing kernel. The following theorem gives the complete answer to this question.

THEOREM 11.1.5. *Let X be a complete inner product space of functions on a set S.*

Then a necessary and sufficient condition for X to have a reproducing kernel is that for each s in S the functional on X given by the evaluation at s be a continuous functional, i.e. for each s in S, $f \to f(s)$ is continuous.

Proof. Suppose first that X is a complete inner product space with a reproducing kernel function K. Then for each s in S we have the representation

$$f(s) = \langle f, K(.,s) \rangle$$

and thus by the Cauchy–Buniakowsky inequality we have that

$$|f(s)| \leqslant \|f\| \cdot \|K(.,s)\|$$

and thus the continuity of $s \to f(s)$ follows.

Conversely, suppose that X is an inner product space whose elements are functions on the set S and for each s in S the functional $s \to f(s)$ is continuous. Then by the Fréchet–Riesz representation theorem there exists a function g_s in X such that

$$f(s) = \langle f, g_s \rangle.$$

We define a function of $S_\times S$ with complex values by the formula

$$K(t,s) = g_s(t)$$

and this obviously is a reproducing kernel for X. □

COROLLARY 11.1.6. *Let X be a complete inner product space and let X_1 be a closed linear subspace of X. If X has a reproducing kernel then X_1 has also a reproducing kernel.*

Proof. In this case if s is an arbitrary point in S the functional

$$f \to f(s)$$

is obviously continuous and thus the assertion follows from the above theorem.

COROLLARY 11.1.7. *Let X_1 be a closed linear subspace of a complete inner product space with reproducing kernel K. Then if K_1 is the reproducing kernel of X_1 the orthogonal projection of X onto X_1 is given by the formula:*

$$(P_{X_1} f)(s) = \langle f, K_1(\,,s) \rangle.$$

Proof. Let f be arbitrary in X and let $f = f_{X_1} + f_{X_1^\perp}$ be the orthogonal decomposition of f with $f_{X_1} \in X_l$ and $f_{X_1^\perp} \in X_1^\perp$. Then we have

$$f_{X_1}(s) = \langle f_{X_1}, K_1(.,s) \rangle = \langle f_{X_1} + f_{X_1^\perp}, K_1(.,s) \rangle = \langle f, K_1(.,s) \rangle$$

and the assertion is proved. □

COROLLARY 11.1.8. *If for any closed subspace of a complete inner product space K_M denotes the reproducing kernel of the respective linear subspace M then for any closed linear subspace of X the following relation holds:*

$$K_M + K_M^\perp = K,$$

where K is the reproducing kernel of X.

11.2. Linear Functionals and Linear Operators on Spaces with Reproducing Kernels

Suppose now that X is a complete inner product space with a reproducing kernel K. We are interested in the form of functionals and in the form of linear operators in terms of the reproducing kernel K. We mention that we consider only continuous linear functional or linear continuous operators.

Let F be a continuous linear functional on X. Then by the Fréchet–Riesz representation theorem there exists a unique element g in X with the property that

$$F(f) = \langle f, g \rangle.$$

Our problem is: How can we compute g using the reproducing kernel K? Concerning this, we can prove the following result.

PROPOSITION 11.2.1. *If F and g are as above then*

$$g(s) = \bar{F}(K(\ ,s)).$$

Proof. Indeed, if $s \in S$ then we have

$$g(s) = \langle g, K(.,s) > = \langle g(t), K(t,s) \rangle = \langle \overline{K(t,s), g(t)} \rangle = \bar{F}(K(\ ,s))$$

and the assertion is proved. □

The following result extends the above proposition for the case of bounded linear operators.

PROPOSITION 11.2.2. *Let X be a complete inner product space with a reproducing kernel K. If $T \in L(X)$ then T has the representation:*

$$(Tf)(s) = \langle f, G_s \rangle,$$

where $G_s \in X$, $\quad G_s = T^*K(.,s)$, $\quad T^*$ is the adjoint of T.

Proof. Indeed, if s is in S then

$$(Tf)(s) = \langle f(t), T^*K(.,s) \rangle$$

and the assertion follows. □

REMARK 11.2.3. Sometimes the family of functions $(G_s)_{s \in S}$ is called the kernel of the operator T.

Now suppose that we have two complete inner product spaces X and Y with reproducing kernels K_X and K_Y, respectively. Then we have the following result.

THEOREM 11.2.4 (Chalmers, 1975). *Let $T: X \to Y$ be a bounded linear operator such that the range of T is dense in Y. Then the reproducing kernel K_Y is obtained from K_X throughout formula:*

$$K_Y(m,n) = T_t^{*-1}(T_s K_x(s,t))_n)_m.$$

(The subscript s means that T is operating on $K_X(s,t)$ as a function of s, the subscript n indicates that for each fixed t the function $T_s(K_X(s,t))$ in Y is evaluated at n, etc.)

Proof. Let n be a fixed point in the set on which the functions of Y are defined, then $n \to (Tf)(n)$ is a continuous linear functional on Y. Then there exists a function g_n in X such that

$$(Tf)(n) = \langle f, g_n \rangle$$

and thus, if we set $h(s) = K_X(s, t)$, we have the relation

$$T_s(K_X(s, t))_n = \langle K_X(s, t), g_n(t) \rangle = \overline{\langle g_n(t), K_X(t, s) \rangle} = \bar{g}_n(t). \qquad (*)$$

Since Y has a reproducing kernel, we have the relation

$$(Tf)(n) = \langle Tf, K_Y(., n) \rangle = \langle (Tf)(m), K_Y(m, n) \rangle = \langle f(s), T_m^*(K_Y(m, n))_t \rangle \qquad (**)$$

and with (*) implies that

$$T_m^*(K_Y(m, n))_t = (T_s(K_X(s, t))_n).$$

Since the range of T is dense in Y, the adjoint of T, T^* has an inverse defined on $(R(T^*))$ and then the last formula implies our result. □

REMARK 11.2.5. Later we shall give some applications of this result which were obtained by B. L. Chalmers to the reproducing kernel of some spaces of analytic functions.

For the following result we need the notion of continuous linear selection, which is defined below.

DEFINITION 11.2.6. Let X, Y be normed linear spaces and let $T: X \to Y$ be a linear operator defined on X with values in Y. Then $R(T)$ is isomorphic to $X/N(T)$ with $N(T) = (x, Tx = o)$. If S is a linear mapping of $X/N(T)$ in X with the property that $S(x + N(T)) \in x + N(T)$ and is a continuous linear mapping, then we say that S is a continuous linear selection on $X/N(T)$.

THEOREM 11.2.7 (Chalmers, 1975). *Let X, Y be two complete inner product spaces with reproducing kernels K_X and K_Y respectively. Let $T: X \to Y$ be a bounded linear mapping. Let S be any continuous linear selection on $X/N(T)$ and for each $x \in X$ let $x^* = S(x + N(T))$. Then $\{x^*\}$ can be organized as an inner product space and has a reproducing kernel, and also $R(T)$ can be organized as an inner product space with a reproducing kernel.*

Proof. First we note that $R(S) = \{x^*\}$ is an inner product space if we set

$$\langle x^*, y^* \rangle = \langle Tx, Ty \rangle.$$

It is clear that we get an inner product and we must show that it is a complete space. Let (x_n^*) be a Cauchy sequence with respect to the norm induced by the inner product defined above. Since $x_n^* = S(x_n + N(T))$ then we see that (x_n) is a Cauchy sequence in X, and this obviously implies that (x_n^*) converges to an element x^* of the space by the continuity of S. Now on $R(T)$ we define an inner product by the above formula, and since the operator between the spaces is an isometry the assertion follows from Theorem 11.2.4. □

11.3. Some Properties of Reproducing Kernels

Let us consider a complete inner product space X of functions on a set S and suppose that X has a reproducing kernel K. Let S be the set on which the functions in X are defined.

From the definition the element $K(\cdot, s)$ is for each $s \in S$ in the space X. Of course, we may ask if the function defined on S by the formula $s \to K(s, s)$ is an element of X. The following example of O. Lehtö shows that, generally, this is not the case.

EXAMPLE 11.3.1 (Lehtö, 1950). We consider a monotone sequence of real numbers (a_n) satisfying the conditions:

(1) $\lim a_n = o$.
(2) $a_n \leqslant \frac{1}{2}$.

We now consider a sequence of positive numbers (b_n) with the properties:

(1) (b_n) is monotone.
(2) $\lim b_n = 1$.
(3) $b_n \geqslant \frac{1}{2}$.

We consider a sequence of functions defined on $(-1, 1)$ as follows:

$$\varphi_n(t) = \begin{cases} 0, & -1 < t < a_{n+1} \\ 2(t - a_{n+1})/(a_n - a_{n+1}), & a_{n+1} \leqslant t \leqslant (a_n + a_{n+1})/2 \\ -2(t - a_n)/(a_n - a_{n+1}), & (a_n + a_{n+1})/2 \leqslant t \leqslant a_n \\ 0, & a_n \leqslant t \leqslant b_n \\ 2\,k_n(t - b_n), & b_n < t < (b_n + b_{n+1})/2 \\ -2k_n(t - b_{n+1})/(b_{n+1} - b_n), & (b_n + b_{n+1})/2 \leqslant t \leqslant b_{n+1} \\ 0, & b_{n+1} \leqslant t < 1. \end{cases}$$

Here k_n are constants determined by the relations

$$\int_{-1}^{1} |\varphi_n(t)|^2 \, dt = 1, \quad n = 1, 2, 3, \ldots \tag{*}$$

and we note that from the definition of the functions $\varphi_n(t)$ they are continuous on $(-1, 1)$ and since

$$\int_{-1}^{1} \varphi_n(t)\varphi_m(t) \, dt = 0, \quad n \neq m \tag{**}$$

we see that this system of functions is an orthonormal system on $(-1, 1)$.

We consider the space X of all functions on $(-1, 1)$ which are of the form

$$a = \sum_{n=1}^{\infty} a_n \varphi_n(t)$$

with (a_n) in l^2. If

$$b = \sum_{n=1}^{\infty} b_n \varphi_n(t)$$

is another function in the space we set

$$\langle a, b\rangle = \sum_{n=1}^{\infty} a_n\bar{b}_n$$

then X is a complete inner product space.

We now show that X has a reproducing kernel. Indeed, we consider for this purpose the following function defined on $(-1, 1)^2$:

$$K(s, t) = \sum_{n=1}^{\infty} \varphi_n(s)\bar{\varphi}_n(t).$$

From the relations (*) and (**) we see that K is indeed a reproducing kernel function of X. We consider now the function $s \to K(s, s)$ defined on $(-1, 1)$. Let $\varepsilon > 0$ sufficiently small and then we remark that $K(s, s)$ is uniformly bounded on each interval of the form $(-1, 1 - \varepsilon)$, and this implies that all elements in X are continuous functions on $(-1, 1)$. Now we show that $K(s, s)$ is not continuous at $s = 0$ and this implies that $s \to K(s, s)$ is not in X.

First we remark that $K(0, 0) = 0$ and that

$$K(\tfrac{1}{2}(a_n + a_{n+1})/2), \tfrac{1}{2}(a_n + a_{n+1})) = \varphi_n(\tfrac{1}{2}(a_n + a_{n+1}))^2 = 1$$

for all n. This implies that

$$\lim_{s\to 0} K(s, s) = 1$$

and the assertion is proved. □

Let us consider X to be a complete inner product space and suppose that X has a reproducing kernel K. Suppose that S is the set on which the functions in X are defined. Then we have the following result, which gives the form of the reproducing kernel K in terms of a basis of X.

THEOREM 11.3.2. *Let $(e_i)_{i\in I}$, I an index set, be a basis of X. Then*

$$K(s, t) = \sum_{i\in I} e_i(s)\,\bar{e}_i(t).$$

Proof. This follows immediately from the definition of K and the fact that $(e_i)_{i\in I}$ is a basis of X. □

REMARK 11.3.3. If X is a complete inner product space of functions defined on a set S and $(e_i)_{i\in I}$ is a basis of X satisfying the property that $\Sigma_{i\in I} |e_i(t)|^2 < \infty$, then

$$K(s, t) = \sum_{I\in I} e_i(s)\bar{e}_i(t)$$

is a reproducing kernel for X.

Let X be a complete inner product space of functions defined on the set S and suppose that X has a reproducing kernel K. We say that on a subset S_1 of S, $K(s, s)$ is bounded if for some $M < \infty$ and all $s \in S_1$ the following inequality holds:

$$K(s, s) \leqslant M.$$

Then we have the following result.

PROPOSITION 11.3.4. *Let* (f_n) *be a convergent sequence of elements of* X *converging (in* X*) to* f*. Then* (f_n) *converges uniformly to* f *on each subset of* S *on which* $K(s, s)$ *is bounded.*

Proof. Indeed, for each s in S we have the following relation:

$$|f_n(s) - f(s)| \leqslant |\langle f_n - f, K(., s)\rangle| \leqslant \| f_n - f\| \, K(s, s)^{1/2}.$$

Now this formula obviously implies our assertion. □

REMARK 11.3.5. The above formula shows that for any sequence (f_n) in X converging in X to f, (f_n) convergence on each compact subset of S.

THEOREM 11.3.6 (Chalmers, 1969). *Let* X *be a complete inner product space with the reproducing kernel* K*. Let* $(x_n^*)_1^\infty$ *be a sequence of linearly independent elements in* X^**. For each* n *we set*

$$X_n = (x, x \in X, x_i^*(x) = 0, i = 1, 2, 3, \ldots, n).$$

If K_n *denotes the reproducing kernel of* X_n *then*

$$K_\infty(s, t) = \lim K_n(s, t)$$

is the reproducing kernel for the space $X_\infty = (x, x \in X, x_i^*(x) = 0, i = 1, 2, 3, \ldots)$ *and the convergence in the above limit is pointwise and uniformly on each subset of* S *on which* $s \to K(s, s)$ *is bounded.*

Proof. For each functional x_i^* in our sequence there exists, by Riesz's representation theorem, an element g_i in X such that

$$x_i^*(x) = x, g_i$$

and $\|x_i^*\| = \|g_i\|$, holds. □

This system of elements $(g_i)_1^\infty$ can be transformed in an orthogonal system, say $(e_i)_1^\infty$. Then we have that

$$K(s, t) = K_\infty(s, t) + \sum_{n=1}^{\infty} e_n(s)\bar{e}_n(t)$$

and for each n

$$K_n(s, t) = K_\infty(s, t) + \sum_{i=n+1}^{\infty} e_i(s)\bar{e}_i(t).$$

Also,

$$\|K_n(s, t) - K_\infty(s, t)\|^2 = \|K_n(s, t)\|^2 - \langle K_n(s, t), K_\infty(s, t)\rangle - \langle K_\infty(s, t), K_n(s, t)\rangle + \\ + \|K_\infty(s, t)\|^2 = \sum_{i=n+1}^{\infty} |e_i(t)|^2 \to 0.$$

Since

$$|K_n(s, t) - K_\infty(s, t)| \leqslant K(s, s) \, |K_n(t, t) - K_\infty(t, t)|$$

we have that the convergence of (K_n) to K_∞ is uniform on each subset of S on which $s \to K(s, s)$ is bounded. The theorem is proved. □

REMARK 11.3.7. This result has applications in the the reproducing kernels for some spaces of analytic functions on bounded domains in C^n which are pseudoconformal.

11.4. Functional Completion of a Space of Functions. The Existence of Complete Inner Product Spaces with Reproducing Kernels

Suppose that we have a normed linear space which is not complete. Then, using a procedure like that to obtain the real numbers from the rationals, we obtain a Banach space which contains an isomorphic copy of the original space and which is dense in the newly constructed Banach space. The procedure gives no information about the concrete form of the elements of the Banach space. In what follows we are interested in a completion of a normed linear space whose elements are functions defined on a set S which, by completion, have a concrete representation as functions. The following result of N. Aronszajn gives the necessary and sufficient conditions for this.

THEOREM 11.4.1 (Aronszajn, 1950). *Let X be a normed space of functions defined on a set* S. *Then* X *has a functional completion with elements as functions on S if and only if the following conditions are satisfied:*

(1) *for each* s *in* S, $s \to f(s)$ *is a continuous linear functional on* X.
(2) *If* (f_n) *is a sequence in* X *and* $\lim f_n(s) = 0$ *for each* s *in* S *then* $\lim \|f_n\| = 0$.

Proof. It is clear that the conditions are necessary. Suppose now that X is a normed linear space of functions on S satisfying conditions (1) and (2). □

Let (f_n) be a Cauchy sequence in X and thus for each s in S the sequence $(f_n(s))$ is a Cauchy sequence and we can define a function f on S by the formula

$$f(s) = \lim f_n(s).$$

Now we consider the space X_1 of all such functions and obviously we have that X is a subspace of X_1. It remains to define a norm on X_1 and to show that, with this norm, this is a Banach space (and for each f in X the two norms coincide).

If f is defined as above then we set

$$f \to \|f\| = \lim \|f_n\|$$

and we remark that this is a number which is independent of the Cauchy sequence which defines f. Indeed, if (g_n) is another Cauchy sequence which defines f then clearly

$$\lim (f_n(s) - g_n(s)) = 0$$

for each s. Now since condition (2) holds this implies that

$$\lim \|f_n - g_n\| = 0$$

and thus

$$\|f\| = \lim \|f_n\| = \lim \|g_n\|.$$

It is clear from the definition of X_1 that X is a dense linear subspace of X_1, and thus the theorem is proved.

REMARK 11.4.2. The functional completion X_1 of X is obviously unique.

The following example shows that not necessarily every normed linear space of functions has a functional completion in the sense of Theorem 11.4.1.

EXAMPLE 11.4.3. We consider for this the set $S = ((1 - n^{-1}, n = 1, 2, 3, \ldots)$ and on S we consider the functions

$$u_n(z) = (a_n - z)/(1 - a_n z)$$

with

$$a_n = 1 - n^{-2}.$$

We consider the so-called Blaschke product associated with (a_n),

$$p(z) = \prod_{n=1}^{\infty} u_n(z)$$

and we note that the following properties hold:

(1) $p(a_n) = 0, n = 1, 2, 3, \ldots$

(2) $|p(z)| < 1$ for $|z| \leqslant 1$

(3) $\lim_{n \to \infty} \iint_{|z|<1} |p(z) - p_n(z)|^2 \, \mathrm{d}x \, \mathrm{d}y = 0,$

where (p_n) is some sequence of polynomials.

We now consider the space X of all polynomials on the set S and we define an inner product on X by the formula: if f, g are two polynomials then

$$\langle f, g \rangle = \iint_{|z|<1} f(z)\bar{g}(z) \, \mathrm{d}x \, \mathrm{d}y.$$

We show that X does not admit a functional completion in the sense of Theorem 11.4.1. Indeed, suppose that it has a functional completion and thus the Blaschke product considered above is in the functional completion and thus has a strictly positive norm. But $p(a_n) = 0$, and this is a contradiction.

COROLLARY 11.4.4. *Let us suppose that X is a complete inner product space of functions defined on a separable Hausdorff space S (i.e. in S there exists a countable subset which is dense in S). If $(s_i)_1^\infty = S_1$ is the countable dense subset then if X has a reproducing kernel K then the set $((K(., s_i))_1^\infty$ is a dense subset in X.*

Proof. If h is a function in X which is in the orthogonal complement of the subspace generated by $((K, s_i))$ then we obtain easy that $h(s) = 0$ for all s in S. We have remarked (11.1.3) that if K is a reproducing kernel of a complete inner product space X then it is 'positive definite' in the sense that for any finite subset of S, $s_1, \ldots, s_n$ and any complex numbers $c_1, \ldots, c_n$

$$\sum_{i,j=1}^{n} K(s_j, s_i) c_i \bar{c}_j \geqslant 0$$

holds.

We now show that in fact also the converse assertion holds. □

THEOREM 11.4.5. *Let S be a nonempty set and suppose that $K: S_{\times} S \to C$ is a function which for any finite number of points $s_1, \ldots, s_n$ of S and any complex numbers $c_1, \ldots, c_n$ the following inequality holds:*

$$\sum_{i,j=1}^{n} K(s_j, s_i) c_i c_j \geqslant 0.$$

Then there exists a complete inner product space of functions defined on S with a reproducing kernel equal to K.

Proof. We consider the space L of all functions of the form

$$s \to \sum_{i=1}^{m} a_i K(s, s_i) = a(s)$$

and if

$$s \to \sum_{j=1}^{n} b_j K(s, s_j) = b(s)$$

is another function then we define

$$\langle a, b \rangle = \sum_{i,j}^{n} K(s_j, s_i) a_i \bar{b}_j$$

and we remark that this has all the properties of an inner product but the property $\langle a, b \rangle = 0$ if and only if $a = 0$. We construct a completion of this space as follows. If (f_n) is a Cauchy sequence, then we easily obtain that for each s in S the sequence $(f_n(s))$ is a Cauchy sequence, and this permits us to define a function f by the relation

$$f(s) = \lim f_n(s). \qquad (*)$$

We consider the space of all such functions and we see that it is a linear space. If we define

$$\|f\| = \lim \|f_n\|$$

then we see that this is a normed linear space whose norm is generated by an inner product. From (*) it is easy to see that this space is complete and the theorem is proved.

11.5. Some Examples of Complete Inner Product Spaces with Reproducing Kernels

In what follows we present some examples of complete inner product spaces having a reproducing kernel.

EXAMPLE 11.5.1. Let X be the space of all complex-valued functions on $[0, 1]$ which are absolutely continuous satisfying the following properties:

(1) For some k, $f(0) = kf(1)$, k is independent of the function f, $k \neq 1$.

(2) $\int_0^1 |f'(t)|^2 \, dt < \infty.$

If, for two arbitrary elements f, g in X, we set

$$\langle f, g \rangle = \int_0^1 f'(t)\bar{g}'(t) \, dt$$

then X is a complete inner product space.

Now we remark that every f in X has the following integral representation:

$$f(s) = \int_0^s f'(t) \, dt + k(1-k)^{-1} \int_0^1 f'(t) \, dt.$$

Now if we set

$$K(s,t) = \min(s,t) + s\bar{k}(1-k)^{-1} + tk(1-k)^{-1} + |k(1-k)^{-1}|^2$$

then K is a reproducing kernel for X. Further, we remark that K has the properties:

(1) $K(s, o) = \bar{k}(1-k)(s + k(1-k)^{-1}) = \bar{k}K(s,1)$,

(2) $|K(0,1)|^2 = K(0,0)\, K(1,1)$.

EXAMPLE 11.5.2. We consider a complete separable inner product space X and let $(e_i)_1^\infty$ be a basis of X. We now consider a function defined on $(1, 2, 3, \ldots)^2$ by the formula:

$$K(s,t) = \delta_{s,t} = \begin{cases} 1, & s = t \\ 0, & s \neq t \end{cases}$$

and we remark that this is a reproducing kernel for X. Of course we can consider any element x of X as a function on $(1, 2, 3, \ldots)$.

We remark that this example gives another reproducing kernel for which the function $s \to K(s,s)$ is not in X.

REMARK 11.5.3. As we know, we can consider for any arbitrary set I the complete inner product space l_I^2. Generally speaking, these spaces do not have a reproducing kernel. For an interesting proof of this fact for $I = [0, 1]$ we refer to Aronszajn (1943).

EXAMPLE 11.5.4. We now consider the space $H_{\bar{D}}^2$ which is the space in which the existence of a reproducing kernel was first noted by S. Zaremba in 1908. Here D is a domain in the complex plane and $H_{\bar{D}}^2$ denotes the set of all harmonic functions in D and which are in $L_{\bar{D}}^2(dx\,dy)$. If f, g are two elements in $H_{\bar{D}}^2$ then we define the inner product by the formula

$$\langle f, g \rangle = \iint_{\bar{D}} f\bar{g} \, dx \, dy.$$

Then it is not difficult to see that this is a closed subspace of $L_{\bar{D}}^2$ and thus it is a complete inner product space. Now we show that $H_{\bar{D}}^2$ has a reproducing kernel. For this according to Theorem 11.1.5 it suffices to show that, for each z_0 in $D, f \to f(z_0)$

is a continuous functional. For this we remark that if z_0 is in D then there exists $r > 0$ such that the disc centered at z_0 and with radius r is contained in D,

$$\bar{D}(z_0, r) = (z, |z - z_0| \leqslant r) \subset D.$$

Then since f is a harmonic function in D we obtain that

$$f(z_0) = 1/r^2 \iint_{\bar{D}(z_0, r)} f(x, y)\, \mathrm{d}x\, \mathrm{d}y$$

and thus by the Cauchy–Buniakowsky's inequality we have that

$$|f(z_0)| \leqslant (\pi r)^{-2} \| f \| (\pi r^2)^{1/2} = \| f \| (\pi r^2)^{-1/2} \cdot m$$

and the continuity of $f \to f(z_0)$ follows. Thus $X = H_D^2$ is an inner product space with a reproducing kernel.

EXAMPLE 11.5.5. Let D be a domain in the complex plane whose boundary G consists of a finite number of smooth Jordan curves. We consider the space H_G^2 of all analytic functions in G with the property that

$$\iint_G |f(z)|^2\, \mathrm{d}x\, \mathrm{d}y < \infty.$$

If f, g are two elements in H_G^2 then we set

$$\langle f, g \rangle = \iint_G f(z) \bar{g}(z)\, \mathrm{d}x\, \mathrm{d}y$$

and this is clearly an inner product. We show that H_G^2 is a closed linear subspace of $L^2(G, \mathrm{d}x\, \mathrm{d}y)$. Indeed, if $(f_n) \subset H_G^2$ converges to f in $L^2(G, \mathrm{d}x\, \mathrm{d}y)$ then we have that for each z in G there exists $r > 0$ such that the disc $\bar{D}(z_0, r)$ is contained in G. Then we have that the following inequality holds:

$$|f_n(z_0) - f_m(\tau_0)| \leqslant (\pi r^2)^{-1} \iint_G |f_n(z) - f_m(z)|^2\, \mathrm{d}x\, \mathrm{d}y, \quad z = x + iy$$

and using the Cauchy–Buniakowsky inequality we obtain that $(f_n(z_0))$ is a Cauchy sequence. This permits us to define the function

$$g(z) = \lim f_n(z)$$

which is clearly analytic in G (being the limit of a sequence of functions which are analytic in G and converge uniformly to g on each $\bar{D}(z_0, r)$ for r sufficiently small). Then f coincides with g a.e. $(\mathrm{d}x\, \mathrm{d}y)$ and thus H_G^2 is a closed subspace in $L^2 = L^2(G, \mathrm{d}x\, \mathrm{d}y)$.

Now if z_0 is arbitrary in G then for r sufficiently small we have (as above) that the following inequality holds:

$$|f(z_0)| \leqslant m(\pi r^2)^{-1/2} \| f \|$$

and thus the continuity of the linear functional $f \to f(z_0$ follows. Then this implies that H_G^2 has a reproducing kernel denoted by K_G.

DEFINITION 11.5.6. For G as above the function K_G is called the Bergman kernel of G and sometimes we use the notation $K_G(z, u)$.

REMARK 11.5.7. If $G = (z, |z| < 1)$ then the Bergman kernel is the function

$$K_G(z, \bar{u}) = \pi^{-1}(1 - z\bar{u})^{-2}.$$

REMARK 11.5.8. Suppose that G is a domain in the complex plane such that we can apply the Riemann mapping theorem, and if h is the function which realizes the conformal application of G into $(z, |z| < 1)$, then H_G^2 considered above for this domain has a reproducing kernel given by the formula

$$K_G(z, \bar{u}) = h(z)\bar{h}'(u)/(\pi(1 - h(z)\bar{h}(u))^2).$$

EXAMPLE 11.5.9. Let μ be a positive measure on R and let $\mu(R) < \infty$. Then we can define a function

$$f(s) = \int_R e^{its}\,\mathrm{d}\mu(t)$$

which has the following properties:

(1) f is continuous and bounded on R.
(2) $f(s) = \bar{f}(-s)$
(3) f is a positive definite function on R.

We now consider the function defined on R by the formula

$$K(s, t) = f(s - t)$$

and from the general theory this is a reproducing kernel for a complete inner product space of functions on R. This space is, in fact, the space of all functions on R of the form

$$F(t) = \int_R \mathrm{e}^{-ist}\,h(s)\,\mathrm{d}\mu(s),$$

where

$$\int_R |h(s)|^2\,\mathrm{d}\mu(s) < \infty.$$

If G is another function

$$G(t) = \int_R \mathrm{e}^{-ist}\,k(s)\,\mathrm{d}\mu(s), \qquad \int_R |k(s)|^2\,\mathrm{d}\mu(s) < \infty$$

then we define

$$\langle F, G\rangle = \int_R h(s)\bar{k}(s)\,\mathrm{d}\mu(s)$$

which clearly defines an inner product on our space.

We leave the reader to extend these constructions to the case when R is changed to a locally compact abelian group and f is a continuous positive definite function on the group.

EXAMPLE 11.5.10 (The space H^2 of Hardy). Let us consider the space of all analytic functions in $(z, |z| < 1)$ and such that

$$\lim_{r \to 1} \frac{1}{2\pi} \int_{-\pi}^{\pi} |f(r\,e^{it})|^2\, dt = \|f\|_2^2 < \infty.$$

Then this can be organized as a linear space and if we set, for f, g two arbitrary elements in H^2,

$$\langle f, g \rangle = \lim_{r \to 1} \frac{1}{2\pi} \int_{-\pi}^{\pi} f(r\,e^{it}) \bar{g}(r\,e^{it})\, dt$$

then this defines an inner product on H^2.

It is not difficult to see that for any z_0, $|z_o| < 1$ the functional $f \to f(z_0)$ is continuous on H^2. We obtain that H^2 is a space with a reproducing kernel. Since for any $z = r\,e^{it}$

$$f(z) = 1/(2\pi) \int_{-\pi}^{\pi} P_r(t - s) f^*(e^{is})\, ds$$

we obtain that the reproducing kernel of H^2 is exactly the Poisson kernel

$$K(z, \bar{u}) = 1/(2\pi)\, \mathrm{Re}(1 + z\bar{u})/(1 - zu).$$

11.5.11. We shall use this space when we consider the subspaces of the so-called 'shift operator'.

We present some remarks which seem to be of interest in the structure theory of the maximal ideal space of a commutative Banach algebra of functions on a space X. Here X is a compact Hausdorff space and A is a closed subalgebra of $C(X)$ which separates the points of X and contains the constants. Let M_A be the maximal ideal space of A. For each m in M_A let μ_m be the positive Baire measure on X which represents the m, i.e.

$$\hat{f}(m) = \int_X f(s)\, d\mu_m(s).$$

The Hardy spaces H_m^p (for the representing measure μ_m) are the closure of A in $L^p(X, \mu_m)$. For $X = [-\pi, \pi]$, with $-\pi$ and π identified, this reduces for $p = 2$ and $m = 0$ to the space considered in Example 11.5.10. This suggest us the following conjecture. Suppose that M contains a Gleason part which does not reduce to a single point (we recall that in M_A we can define the following equivalence relation: $m_1 \sim m_2$ if $\|m_1 - m_2\| < 2$; a Gleason part is an equivalence class in M_A); then for each m in this part H_m^2 has a reproducing kernel. If, perhaps, the answer to the general case may be 'no', a positive answer is to be expected at least in the case of Dirichlet algebras of Gleason or the log-modular algebras of Hoffman (1962).

11.5.12. *The G. Szegö reproducing kernel.* We consider a domain G in the complex plane whose boundary Γ consists of an analytic curve and G is supposed to be connected. We consider the space H_Γ^2 of all functionals analytic in G and with the property that for such space every element f has the property that

$$\int_\Gamma |f(s)|^2 \, ds < \infty.$$

Obviously, if we set for any f, g in H_Γ^2,

$$\langle f, g \rangle = \int_\Gamma f(s)\bar{g}(s) \, ds$$

then this is a complete inner point space. Now we remember that the Riemann mapping theorem applies to G and thus there exists a function $z = z(w)$ which is the inverse mapping of the conformal representation of G onto $(z, |z| < 1)$. Since in the Hardy space H_Γ^2 (see Example 11.5.10) the system $(z^n)_0^\infty$ is a complete orthonormal system, it follows that in our space the system

$$((z'(w))^{1/2}/(2^{1/2}\pi^{1/2})\cdot(z(w))^{n-1}) \qquad (*)$$

is a complete orthonormal system. Now we know that for an arbitrary complete inner product space with a reproducing kernel we have the formula

$$K(s, t) = \sum_{n=0}^{\infty} w_n(t)\bar{w}_n(s)$$

where (w_n) is a complete orthonormal system of elements of the space. Using this fact and the system of functions in (*) we obtain that the reproducing kernel of the space H_Γ^2 is given by

$$K(s, t) = (2\pi)^{-1}(z'(t))^{1/2}(\bar{z}'(s))^{1/2}.$$

This reproducing kernel is called the Szegö reproducing kernel for H_Γ^2.

11.5.13. *The reproducing kernel functions for spaces of analytic functions in several variables.* The notion of analytic function may be defined for functions of several complex variables, and it is not difficult to see that many theorems from the case of one variable extend to the case of several variables. But there are new phenomena even in the case of two complex variables. We now mention two of the most striking differences from the theory of analytic functions of one and several variables. As is well known in the case of functions of one complex variable, if we have a domain D in the complex plane, then we find an analytic function defined on D which is also analytic on D and all the points of the boundary of D are singular points of the function.

In the case of several variables (also for $n = 2$) there are domains such that any function which is analytic on that domain is analytic in a larger domain.

The second difference we mention refers to the Riemann 'mapping theorem'. This famous results asserts that any two bounded, simply connected domains can be mapped conformally one onto the other. This correspondence is unique if we fix a pair of corresponding points and the corresponding directions are chosen at these points.

In the case of several complex variables there are pairs of domains D_1 and D_2

with the property that there are no analytic mappings

$$w_1 = f_1(z_1, \ldots, z_n), \ldots, w_n = f_n(z_1, \ldots, z_n)$$

defined on D_1 and with values in D_2 and being one-to-one (i.e. $(z_1, \ldots, z_n) \to (w_1, \ldots, w_n)$ is one-to-one). Such an example of domains is as follows $(n = 2)$:

$$D_1 = ((z_1, z_2), |z_i| < 1, i = 1, 2); \qquad D_2 = ((w_1, w_2), |w_1^2| + |w_2^2| < 1).$$

For the theory of analytic functions of several variables we refer to the excellent books of Hervé (1963), Malgrange (1958) and Gunning and Rossi (1965).

To consider certain classes of analytic functions we find it convenient to introduce the notion of a 'Reinhardt domain'. This is given in the following definition.

DEFINITION 11.5.14. A domain D in C^n is a Reinhardt domain if whenever $z = (z_1, \ldots, z_n)$ is in D then $(e^{it} 1z_1, \ldots, e^{it} nz_n)$ is in D for all t_i in R, $i = 1, 2, \ldots, n$.

In what follows we restrict ourselves to the case $n = 2$ but the results can easily be extended to arbitrary n.

For each (n, m) we consider the functions

$$z_1^m z_2^n = \omega_{mn}(z_1, z_2)$$

which are analytic in the Reinhardt domain D (in C^2).

We consider the numbers

$$a_{mn}^2 = \int_D |z_1^{2m}||z_2^{2n}| \, d\omega$$

and we define the following space.

DEFINITION 11.5.15. The space H_D^2 is the set of all functions defined on D with complex values and such that if

$$f(z_1, z_2) = \sum_{m,n} c_{mn} z_1^m z_2^n$$

then $\Sigma_{m,n} |a_{mn} c_{mn}|^2 < \infty$. Then it is not difficult to see that the space H_D^2 admits a scalar product and that in fact H_D^2 is a complete inner product space.

REMARK 11.5.16. In the above sum we consider only for $m \geqslant 0$, $n \geqslant 0$. If we consider translates of a Reinhardt domain then it is possible to remove the fact that m, n are integers in $(0, \infty)$. In this case the system of functions $(z_1^m z_2^n/(a_{mn}))$ is an orthonormal system for the space H_D^2. Using a two-dimensional Cauchy formula we can easily prove that H_D^2 has a reproducing kernel function. Since the above system is orthonormal, we obtain that this is given by the formula:

$$K(z_1, z_2; w_1, w_2) = \sum_{m,n} \bar{\omega}_{mn}(z_1, z_2) \, \bar{\omega}_{mn}(w_1, w_2).$$

If D is $((z_1, z_2), |z_i| < 1, i = 1, 2)$ then the reproducing kernel is given by the formula:

$$(\pi)^{-2}(1 - \bar{z}_1 w_1)^{-2}(1 - \bar{z}_2 w_2)^{-2}.$$

If D is $((z_1, z_2), |z_1^2| + |z_2^2| < 1)$ then the reproducing kernel of H_D^2 is given by the formula:

$$2\pi^{-2}r^2(r^2 - \bar{z}_1 w_1 - \bar{z}_2 w_2)^{-3}.$$

The reader may try to consider the case of the analogue of the Szegö spaces in several complex variables and to find the corresponding reproducing kernels. For the reproducing kernel functions of domains which are pseudoconformal to another domain (i.e. there exists a biholomorphic mapping between the two domains) we refer to Chalmers (1969), Morita (1956) and Maschler (1956), as well as for results concerning the boundary behavior of the reproducing kernel (the Bergman kernel function).

For the kernel function associated with some spaces of harmonic functions and the connections with Green's function we refer to the book of Bergman and Schiffer (1953), Nehari (1949, 1952). For the connection between the so-called Schwarz lemma and the Szegö reproducing kernel functions we refer to Garabedian (1948, 1949).

11.6. Operations on the Reproducing Kernel Functions (the Sum, Products and Limits of Reproducing Kernels)

In this section we are interested in certain operations defined on the family of reproducing kernels. These operations are as follows: the sum, the product and the limit of sequences of reproducing kernels.

First we consider the problem for the sum of two reproducing kernel functions: if K_1 and K_2 are the reproducing kernels for two complete inner product spaces of functions defined on the same set S then, of course, we can define the function $K = K_1 + K_2$ on $S'_\times S$. The problem is to determine if this function is again a reproducing kernel and, if the answer is 'yes', to determine the corresponding complete space whose kernel function it is in terms of the spaces associated with K_1 and K_2. Let us denote the spaces associated with K_1 and K_2 by H_1 and H_2 respectively. We consider the space $H = H_1 \cup H_2$ of all pairs $h = (h_1, h_2)$, $h_i \in H_i$, $i = 1, 2$. An inner product on H may be defined, as we know, by the formula:

$$\langle h, h^* \rangle = \langle h_1, h_1^* \rangle + \langle h_2, h_2^* \rangle, \quad h^* = (h_1^*, h_2^*).$$

Associated with the spaces H_1 and H_2 is the space $V = H_1 \cap H_2$ (which may reduce to (0)) and related to this we consider the subspace M in H of the elements of the form $(f, -f), f$ in V. It is not difficult to see that M is a closed subspace of H. Then clearly we can consider the orthogonal decomposition

$$H = M \oplus M^{\perp}.$$

Thus every element $h = (h_1, h_2)$ in H has the decomposition

$$h = (f, -f) + (h'_1, h'_2).$$

We now consider the set

$$F = (h_1 + h_2, h_i \in H_i)$$

which is obviously a linear subspace and we show that there is a one-to-one correspondence between F and $M^{\perp}$.

First we define the mapping on $M^{\perp}$ with values in F by the formula:

$$T(h'_1, h'_2) = h'_1 + h'_2.$$

First we remark that this is 'onto' and thus to prove our assertion it suffices to show that $T(h'_1, h'_2) = 0$ if and only if $h'_1 = h'_2 = 0$.

Suppose that this is not the case and thus we find two elements h'_1 and h'_2 such that $h'_1(s) + h'_2(s) = 0$ for all s in S and the functions h'_1 and h'_2 are not identically zero. Thus we must have

$$h'_2(s) = -h'_1(s) \quad \text{or} \quad h'_2 = -h'_1.$$

This implies that $(h'_1, -h'_1)$ is in $M^{\perp}$ and obviously the above element is in M. Thus, this contradicts the property of the decomposition $H = M \oplus M^{\perp}$. The map T is thus one-to-one.

Now we can define an inner product on F by the formula:

$$\langle f, g\rangle = \langle f'_1, g'_1\rangle + \langle f'_2, g'_2\rangle,$$

where

$$f = T(f'_1, f'_2), \qquad g = T(g'_1, g'_2).$$

Then we have the following property.

PROPOSITION 11.6.1. *F with the above inner product is a complete inner product space.*

Proof. Since the proof is simple it is omitted.

As we see, the elements of F are functions defined on S and we now show that F is a space with a reproducing kernel.

THEOREM 11.6.2. *A complete inner product space F has a reproducing kernel which is equal to* $K_1 + K_2$.

Proof. Let t be a fixed element in S and thus, from the definition of K_i, we have that $(K_1(\,,\,,t),\ K_2(\,,\,,t))$ is an element of H. Then we can consider the element $K_1(\,,\,,t) + K_2(\,,\,,t)$ which has a decomposition of the form

$$K_1(\,,\,,t) + K_2(\,,\,,t) = T(K'_1(\,,\,,t), K'_2(\,,\,,t))$$

and thus for each f in F,

$$f = T(f'_1, f'_2)$$

we have the following relations:

$$\begin{aligned}
f(s) &= f'_1(s) + f'_2(s) = \langle f'_1(t), K_1(t,s)\rangle + \langle f'_2(t), K_2(t,s\rangle \\
&= \langle (f'_1, f'_2), (K_1(\,,\,,s), K_2(\,,\,,s))\rangle \\
&= \langle (f'_1, f'_2), (K'_1(\,,\,,s), K'_2(\,,\,,s))\rangle \\
&\quad + \langle (f'_1, f'_2), (K_1(,\,,s) - K'_1(,\,,s)), (K_2(,\,,s) - K'_2(,\,.s))\rangle \\
&= \langle (f'_1, f'_2), (K'_1(\,,\,,s), K'_2(\,,\,,s))\rangle = \langle f, K_1(\,,\,,s) + K_2(\,,\,,s)\rangle.
\end{aligned}$$

This proves the assertion of the theorem.

From the above relations we have also the following result.

COROLLARY 11.6.3. *For any element of F, f the following formula holds:*

$$\| f \| = \min(\| f_1 \|^2 + \| f_2 \|^2, f = f_1 + f_2, f_i \in H_i, i = 1, 2),$$

where $f \to \| f \|$ is the norm generated by the inner product defined on F by the above formula.

We now consider the case of the product of reproducing kernel functions. Suppose now that the complete inner product spaces H_i, $i = 1, 2$, are formed of functions defined on the sets S_1 and S_2 and the reproducing kernel functions are K_i, $i = 1, 2$. We can consider the set $S = S_1 {}_\times S_2$ and on $S {}_\times S$ the function

$$K(s_1, t_1, s_2, t_2) = K_1(s_1, t_1) K_2(s_2, t_2)$$

and we may ask if this function is a reproducing kernel of a complete inner product space, and, if the answer is 'yes', to determine the form of the space, with respect to the elements of the spaces H_i.

THEOREM 11.6.4 (Aronszajn, 1950). *The function K is a reproducing kernel function.*

Proof. We consider the set L of all functions of the form

$$(s, t) \to \sum_{i=1}^{n} f_i(s) g_i(t) = F(s, t)$$

where f_i and g_i are functions defined on S_1 and S_2 respectively. If G is another element of this set,

$$G(s, t) = \sum_{j=1}^{m} h_j(s) k_j(t)$$

then we set

$$\langle F, G \rangle = \sum_{i=1}^{n} \sum_{j=1}^{m} \langle f_i, h_j \rangle \langle g_i, k_j \rangle$$

and we remark that $\langle F, F \rangle \geqslant 0$. For this we consider F to be of the form

$$F = \sum_{i=1}^{n} \sum_{j=1}^{m} a_{ij} u_i(s) v_j(t),$$

where (u_i), (v_j) are bases for the spaces H_1 and H_2 respectively. In this case we have that

$$\langle F, F \rangle = \sum_{i=1}^{n} \sum_{j=1}^{m} |a_{ij}|^2$$

and thus

$$\langle F, F \rangle = 0 \quad \text{if and only if } F = 0.$$

Now we remark that all the properties of an inner product are satisfied by the function $\langle F, G \rangle$. We remark that the family of functions of the form

$$\sum_{i,j}^{n}\sum^{m} a_{ij}\, u_i(s) v_j(t)$$

is a linear subspace of the space of all functions on S of the form

$$\sum_{i,j=1}^{\infty} a_{ij}\, u_i(s) v_j(t), \qquad \sum_{i,j=1}^{\infty} |a_{ij}|^2 < \infty.$$

If g is another element of this space then we set

$$\langle f, g \rangle = \sum_{n,m}^{\infty} a_{ij} \bar{b}_{ij}$$

and this defines an inner product which is clearly a complete inner product space. Also, it is clear that $K = K_1 K_2$ is the reproducing kernel of this space. The fact that this space has a reproducing kernel follows from the inequality

$$|f(s,t)| \leqslant (K_1(s,s)\, K_2(t,t))^{1/2} \|f\|.$$

The above result of Aronszajn is connected with a classical result of Schur (1914) which asserts that the product (in the sense of Schur) of two positive matrices is again a positive matrice.

For the reader's convenience we recall that if $A = (a_{ij})$, $B = (b_{ij})$ are two $n_\times n$ matrices then the Schur product of A and B is the matrix $C = (c_{ij})$ with $c_{ij} = a_{ij} b_{ij}$. Of course, this product may be defined for infinite matrices as well as the positivity – which is equivalent to the fact that the operator whose matrix is the given matrix (with respect to an orthonormal basis, is a positive hermitian operator) is valid for infinite matrices.

The above construction suggests the following method of obtaining new complete inner product spaces from given spaces. Suppose that H is a complete inner product space of functions on the set S with a reproducing kernel K. Let us consider S_1 to be a proper subset in S and

$$F_{S_1} = (f/S, f \in H),$$

i.e. the set of all functions which are restrictions of functions in our space H. Obviously this is a linear space. We defined a norm (which is easy to see is generated by an inner product) by the formula

$$\| f/S_1 \| = \inf(\|g\|, g/S_1 = f/S_1).$$

The following proposition gives the basic property of the space F_{S_1}.

THEOREM 11.6.5. *The space F_{S_1} is a complete inner product space with a reproducing kernel.*

Proof. First we consider the subspace M of H consisting of all functions in H which have the restriction to the subset S_1 equal to zero. It is easy to see that this is a closed linear subspace of H. Then we decompose the space H as follows:

$$H = M \oplus M^\perp.$$

Since for each s, $K(, , s)$ is in H we have the decomposition

$$K(, , s) = K_M(, , s) + K_{M^\perp}(, , s).$$

This gives that for all s in S_1,

$$K_M(s_1, s) = 0$$

and then for this s_1, $K(s_1, t) = K_M(s_1, t)$. If f, g are in H then they have the decomposition

$$f = f_M + f_{M^\perp}, g = g_M + g_{M^\perp}$$

and clearly they have the same restriction to the set S_1 if and only if $g_{M^\perp} = f_{M^\perp}$. Then we can define the inner product of the restrictions as the inner product of $f_{M^\perp}$ and $g_{M^\perp}$. This obviously gives our assertion.

We are now interested in the problem concerning sequences of reproducing kernel functions. Suppose that we have a set S and a sequence of subsets (S_n) with the properties

(1) $S_1 \subset S_2 \subset S_3 \subset \cdots \subset S_n \subset S_{n+1} \subset \cdots$

(2) $S = \bigcup_n S_n$

and for each n there is a complete inner product space H_n with a reproducing kernel K_n such that the following assertions hold:

(1′) If $f_i \in H_i$ then for all $j < i, f_i/S_j$ is in H_j,

(2′) For all pairs of indices i,j as in (1′)

$$\|f_i/S_j\| H_j \leqslant \|f_i\| H_i.$$

Then we can consider the sequence of the functions K_n, $(K_n(s, t))$ and suppose that we can define in some way $\lim K_n$ which may be a function on $S_\times S$. Then it is possible that this function may be a reproducing kernel for a complete inner product space of functions defined on S and if 'yes', to describe the elements of this space. We show that, following N. Aronszajn, we can associate to the sequence (K_n) a function which is a reproducing kernel for a space of functions which is completely determined by the sequence of spaces (H_n).

First we define the function $K(s, t)$ as follows. Let $\varepsilon > 0$ and N_ε such that (s, t) is in S_n for $n \geqslant N_\varepsilon$. Then we say that $K(s, t)$ is the limit of the sequence $(K_i(s, t))$ if for all $n \geqslant N$,

$$|K(s, t) - K_n(s, t)| \leqslant \varepsilon.$$

It is clear that in this way we obtain a function defined on $S_\times S$. We write this as

$$K(s, t) = \lim K_n(s, t). \qquad (*)$$

Now we show that this is a reproducing kernel for a complete inner product space of functions on S.

We define the space H as follows:

$$H = (f, f\colon S \to C, f/S_i \text{ is in } H_i \text{ for each } i \text{ and } \lim \|f/S_i\|_{H_i} < \infty).$$

If we define for f, g in H,

$$\langle f, g\rangle = \lim_i \langle f/S_i, g/S_i\rangle_{H_i}$$

then clearly H is an inner product space, and from the form of the sequence (S_i) it easily follows that H is complete.

Now we have the following result.

THEOREM 11.6.6 (Aronszajn, 1950). *The function K is a reproducing kernel for H defined above.*

Proof. Let us consider $i > j$ and then for t fixed in S and $K_{ij} = K_i - K_j$, with $j < k < i$,

$$\|K_i(s, t)/S_k - K_j(s, t)/S_k\|^2 = \|K_i(s, t) - K_j(s, t)/S_k\|^2 \leqslant K_i(t, t) - K_j(t, t) \qquad (**)$$

and from the properties of the norms in the spaces H_i (property (2′) above) we see that $(K_n(t, t))$ is a monotone sequence and thus convergent. The above inequality shows then that $(K_i(s, t)/S_k)$ is a convergent sequence in H_k and this clearly implies the pointwise convergence of $(K_n(s, t)/S_k)$. Thus we can set

$$h_t^k(s) = \lim K_i(s, t)/S_k$$

and from the property of the norms in the spaces H_i we see that we can define a function on $S_\times S$ using these functions, $h_t(s)$ will be the value of $h_t^k(s)$ at s. Then $K(s, t)$ will be the values of h_t at s. □

We must show that K is a reproducing kernel for H defined above. Indeed, if f is in H then for each n, using the fact that H_n has a reproducing kernel for the function K_n, we have

$$f(t) = \langle f/S_n)(t)\rangle = \langle f/S_n(s), K_n(s, t)\rangle_{H_n} = \langle f/S_n(s), h_t^n(s)\rangle_{H_n} + \\ + \langle f/S_n(s), K_n(s, t) - h_t^n(s)\rangle_{H_n}$$

which for $n \to \infty$ converges to $f(s)$, $K(s, t)$ and thus the assertion of the theorem is proved.

REMARK 11.6.7. An interesting application of this result was noted by Aronszajn (1950) concerning the reproducing kernel (the Bergman function) of the domain

$$S = (z, |\mathrm{Im}\, z| \leqslant c, c \text{ is a constant}).$$

In this case we can choose the subsets S_n as follows:

$$S_n = (z, z = x + iy, x^2/a_n^2 + y^2/c^2 \leqslant 1, a_n > 0, \lim a_n = \infty),$$

and the reproducing kernel for S_n can be determined explicitly and we have thus a way to determine K for S.

11.7. Interpolation, Extremal and Minimal Problems and Reproducing Kernels

As is known, if D is a domain in the complex plane and (z_i) is a sequence of points of

D and

$$f_n(z) = (z - z_1)\cdots(z - z_{n-1})$$
$$f_1(z) = 1$$

then if the analytic function in D satisfies some additional requirements, it can be developed as a Newton interpolation series

$$f(z) = \sum_{n=1}^{\infty} a_n f_n(z), \qquad a_1 = f(z_1), \qquad a_n = \sum_{j=1}^{n} f(z_j)\Big/\left(\sum_{k=1}^{n} (z_j - z_k)\right), \quad n \geqslant 2.$$

It is also well known that the Newton interpolation series does not converge in general in all of D. In his paper, Bergman (1947) has introduced an interpolation series for certain spaces of analytic functions in a domain D which converges in all of D. The representation of the function takes the form

$$f(z) = \sum_{n=1}^{\infty} a_n h_n(z)$$

where (h_n) is a complete orthonormal system of elements of the space and the coefficients a_n are determined as for the Newton interpolation series.

These formulas look at the function f using the numbers $(f(z_i))$ where z_i are points in D. Of course we can consider these as the values of some functionals defined on the space (in our case these are the evaluation at the points z_i) and this point of view was largely considered by Chalmers (1970, 1971). In what follows we present some of Chalmers' results. These are related to the problem of interpolation of bounded analytic functions, which is concerned with the existence of a sequence of points in the unit disk (z_n), with the property that given an arbitrary bounded sequence of complex numbers (w_n), there is a bounded analytic function in the unit disk having the property that

$$f(z_i) = w_i, i = 1, 2, 3, \ldots$$

and any sequence (z_i) with this property is called an 'interpolation sequence'. We mention that the problem about interpolation sequences was posed by R. C. Buck and that the basic results about such sequences were obtained by Carleson (1958), Hayman (1958) and Newman (1959). A beautiful proof of results about the interpolation sequences, as well as new results, for the case of Hardy spaces H^p, was given by Shapiro and Shields (1961) using some Banach space techniques. Independently of the problem of interpolation sequences, some related results were obtained by Naftalevich (1956) and Kabaila (1958). See also Hoffman (1962).

In order to specialize the general results about interpolation, we give here the basic definitions concerning classes of sequences in the unit disk.

DEFINITION 11.7.1. A sequence (z_i) in the unit disk is called a Carleson sequence if for some $\delta > 0$

$$\prod_{i \neq j} |(z_i - z_j)|/|(1 - \bar{z}_j z_i)| \geqslant \delta j = 1, 2, 3, \ldots.$$

DEFINITION 11.7.2. The sequence (z_i) in the unit disk is said to satisfy condition

(H) if, for all functions f in the Hardy space H^1,

$$\sum_n |f(z_n)|(1 - |z_n|) < \infty. \tag{H}$$

DEFINITION 11.7.3. The sequence (z_i) in the unit disk is said to be an exponential sequence if for some r in $(0, 1)$,

$$(1 - |z_{j+1}|)/(1 - |z_i|) \leqslant r, \quad j = 1, 2, 3, \ldots.$$

DEFINITION 11.7.4. The sequence (z_i) in the unit disk is called radial if all z_i lie on one radius.

From the definition it is clear that any exponential sequence is a Carleson sequence and that a radial Carleson sequence is an exponential sequence. Let us consider a complete inner product space H with an inner product $\langle\, , \rangle$ and suppose that H is a space with a reproducing kernel $K(s, t)$; the set on which the elements of H are defined is denoted by S. If x^* is a bounded linear functional on H then there exists a unique element f_{x^*} of H such that

$$x^*(x) = \langle x, f_{x^*} \rangle.$$

If $(x_1^*, x_2^*, x_3^*, \ldots)$ is a sequence of elements of H^* (the dual of H), we consider the elements $(f_{x_1^*}, f_{x_2^*}, \ldots)$ and we suppose that the set (x_n^*) is linearly independent.

Associated with the above sequence of elements of H^* we consider the following sequence of determinants $(D_n)_0^\infty \cdot D_0$ is by definition equal to 1, D_n is the determinant of the matrix $\overline{(f_{x_i^*}, f_{x_j^*})}_{1 \leqslant i,j \leqslant n} = Q_n$. The first problem we consider is:

PROBLEM 11.7.5. Does there exist an element x of H such that given n linearly independent elements of H^*, $x_1^*, \ldots, x_n^*$ and n complex numbers, $a_1, \ldots, a_n$ such that

(1) $x_i^*(x) = a_i, i = 1, 2, 3, 3, \ldots, n$

(2) x has a minimal norm among the elements of H satisfying 1?

The following theorem gives the answer to this problem.

THEOREM 11.7.6. *Problem* 11.7.5 *has a unique solution which admits the representation* $f_{x^*} = -D_a(0)/D_n$, *where* D_a *is the (formal) determinant of the matrix*

$$\begin{pmatrix} 0 & X_n^* \\ A_n^T & Q_n \end{pmatrix}$$

where $X_n^* = (f_{x_1^*}, \ldots, f_{x_n^*})$, $A_n = (a_1, \ldots, a_n)$ *and* Q_n *is the matrix in the determinant* D_n.

Proof. It is clear that f_{x^*} is an element satisfying (1) and, from the form of it, it follows that f_{x^*} is in the subspace generated by $f_{x_1^*}, \ldots, f_{x_n^*}$. Thus

$$f_{x^*} = \sum_{i=1}^{n} c_i f_{x_i^*}$$

and suppose now that we have another solution, say g. We can then consider the element $u = g - f_{x^*}$ and clearly $x_1^*(u) = 0$ for $i = 1, 2, 3, \ldots, n$. This implies that

$$\langle u, f_{x_i^*} \rangle = 0, \qquad i = 1, 2, 3, \ldots, n$$

and then

$$\|g\|^2 = \|f_{x^*}\|^2 + \|u\|^2$$

and from the fact that condition (2) is satisfied, f_{x^*} and g have equal norms. Thus u must be the zero element.

We now consider the case when the sequence (x_i^*) is infinite. Then we have the following result.

THEOREM 11.7.7 (Chalmers, 1970). *Let (x_i^*) be an infinite set of elements in H^* which is linearly independent. Let $a = (a_i)$ be a sequence of complex numbers. Then there exists a solution of the following problem: x in H with the property that*

(1) $x_1^*(x) = a_i, \quad i = 1, 2, 3, \ldots,$

(2) *x has a minimal norm among the elements satisfying* 1, *if and only if*

$$\lim \|f_{x^*}\| = M < \infty$$

where $f_{x_n^*}$ is the solution of Problem 11.7.5 *for the truncated sequence $(a_1, \ldots, a_n^n)$ and correspondingly $(x_1^*, \ldots, x_n^*)$.*

Proof. Suppose first that $\lim \|f_{x_n^*}\| = M < \infty$. In this case we have clearly the relation

$$\|f_{x_n^*} - f_{x_m^*}\|^2 = \|f_{x_n^*}\|^2 + \|f_{x_m^*}\|^2 - 2\,\mathrm{Re}\langle f_{x_n^*}, f_{x_m^*}\rangle$$

and suppose that $m \geqslant n$. Then $\langle f_{x_n^*}, f_{x_m^*}\rangle = \|f_{x_n^*}\|^2$ since $\langle f_{x_n^*}, f_{x_m^*}\rangle$ is obtained by replacing the first row in the determinant of $f_{x_n^*}$ by $0\ \bar{a}_1 \ldots \bar{a}_n$. Thus we have the relation

$$\|f_{x_n^*} - f_{x_m^*}\|^2 = \|f_{x_n^*}\|^2 - \|f_{x_m^*}\|^2$$

which implies that $(f_{x_n^*})$ is a Cauchy sequence. Let

$$f_{x^*} = \lim f_{x_n^*}$$

and we show that this is indeed the required solution of our problem. It is clear that for all $i, x_i^*(f_{x^*}) = a_i$ and that $\|f_{x^*}\| = \lim \|f_{x_n^*}\|$. Now if g is another solution of our problem, let $u = f_{x^*} - g$. Then for all $i, x_1^*(u) = 0^n$ and thus

$$\langle u, f_{x^*}\rangle = \lim \langle u, f_{x_n^*}\rangle = 0$$

which implies that

$$\|f_{x^*}\|^2 + \|u\|^2 = \|g\|^2.$$

Since f and g have equal norms we must have $u = 0$. Thus the solution is unique. Now if the problem has a solution then we have that

$$\|f_{x_1^*}| \leqslant \|f_{x_2^*}\| \leqslant \cdots \leqslant \|f_{x_n^*}\| \leqslant \|f_{x_{n+1}^*}\| \leqslant \cdots \leqslant \|f_{x^*}\|,$$

where f_{x^*} is the solution of our problem. The argument given above shows that f_{x^*} is equal to $\lim f_{x_n^*}$ and $\|f_{x^*}\| = \lim \|f_{x_n^*}\|$. The theorem is proved.

For the interpolation series an useful result is the following theorem.

THEOREM 11.7.8 (Chalmers, 1970). *Suppose that* (x_i^*) *is a linearly independent subset of elements in* H *then the following sequence of functions* (h_n) *is an orthonormal system, where*

$$h_n = (|Q_{(n-1)\times(n-1)}|\cdot|Q_{n\times n}|)^{1/2} Q_{(n-1)\times n}/(x_1^*(K(,,z)\ldots x_n^*(K(,,z)),$$

where $Q_{ij} = \overline{x_i^*(x_j^*(K(w,z)))}$ *where, first,* x_j^* *is applied to* $(K(,,z))$ *and, next,* x_i^* *to* $x_j^*(K(,,,z))$ *as a function of* z, *and it is an orthonormal system in* H.

Proof. Let n be fixed and consider the solution of Problem 11.7.5 with the data $x_i^*(f_{x_n^*}) = \delta_{nm}$ (the Kronecker symbol), $m = n$. Then the solution is given by the formula in Theorem 11.7.6 where $f_{x_1^*} = x_i^*(K(,,z))$, $i = 1,2,3,\ldots,n$. Also a is $(0,\ldots,0,1)$ and since $\langle f_{x_n^*}, f_{x_m^*}\rangle = \delta_{nm} D_{n-1}/D_n$ the orthormality of the system follows.

COROLLARY 11.7.9 (Chalmers, 1970). *If* (x_i^*) *is a complete set in* H *then* (h_n) *is a basis of* H.

This follows from the form of the solution of the minimum problem for each n.

COROLLARY 11.7.10. *If* (x_i^*) *is as in 11.7.9 then every function* f *in* H *has the following development*

$$f(z) = \sum a_n h_n(z).$$

Proof. Obvious. □

For concrete examples of systems for which the above theorems apply, we refer to the interesting by Bergman (1947). We now consider an interesting class of systems of elements in H^*.

DEFINITION 11.7.11. A sequence (x_n^*) of elements of H^* is called an M-interpolating sequence for the set M in the space of all sequence of complex numbers, if the mapping on H defined by

$$T(f) = (x_i^*(f))$$

is onto M.

For the proof of the following interesting result, we refer to Chalmers (1970).

THEOREM 11.7.12. *For the sequence* (x_i^*) *of* H^* *the following conditions are equivalent*:

(1) (x_n^*) *is an* l^2*-interpolating sequences,*
(2) *the matrix* $Q = (a_{ij})$ *is bounded below and above by strictly positive constants and where the entries in the matrix are given by*

$$a_{ij} = x_i^* x_j^*(K(w,z)).$$

11.7.13. Another interesting characterization for the case of the Hardy space H^2 of the unit disk and the functionals being evaluations at points, was given by H. S. Shapiro and A. L. Shields using Carleson sequences (in fact the sequence of points of evaluations forming a Carleson sequence is the characterization). For a related interesting result see also the very interesting paper of Bari (1951).

For the interpolation problem in generally complete inner product spaces H with a reproducing kernel, we present first certain classes of sequences of elements in the dual space which represents, to some extent, the analogue of the classes of sequences in the unit disk considered above. We mention that these classes of sequences of functions were defined by Chalmers (1971).

DEFINITION 11.7.14. Let x^* be a nonzero element in H^*. The element $x^*/\|x^*\|$ is called x^*-normalized and is denoted by $\hat{x}^*$. Also, $f_{x^*}/\|f_{x^*}\| = \hat{f}_{x^*}$.

DEFINITION 11.7.15. A sequence (x_n^*) of elements in H^* is said to be a Carleson δ-sequence if the distance of $\hat{f}_{x_i^*}$ to the subspace generated by $(f_{x_j^*})_{j \neq i}$ is at least $\delta > 0$ for all i.

DEFINITION 11.7.16. The sequence (x_n^*) is said to be an exponential (r, c)-sequence if

$$\|x_n^*\|/\|x_{n+1}^*\| \leqslant r < 1 \quad \text{and} \quad |< x_n^*, x_m^* >| \leqslant c\|x_n^*\|$$

for some constant c and all $n, m = 1, 2, 3, \ldots$.

EXAMPLE 11.7.17. The following example shows that, the above notions are generalizations of the notions defined for points in the unit disk. Consider for this the Hardy space H^2 corresponding to the unit desk, and for a sequence of points in the unit disk, say (z_n), we can define the following sequence of elements in $(H^2)^*$ by the formula:

$$x_n^*(f) = f(z_n).$$

In this case the Fréchet–Riesz representation theorem is as follows:

$$f(z_n) = x_n^*(f) = \langle f, 1/(1 - z\bar{z}_i)\rangle$$

and this gives that

$$\|x_n^*\|/\|x_{n+1}^*\| = (1 - |z_{n+1}|^2)^{1/2}/(1 - |z_n|^2)^{1/2}$$

$$|\langle x_n^*, x_m^*\rangle| = 1/(|1 - z_n\bar{z}_n|) \leqslant 1/(1 - |z_n|) \leqslant 2/(1 - |z_n|^2) = 2\|x_n^*\|^2$$

for all n, m. This clearly shows that the notions defined above are proper generalizations of the notions introduced for sequences in the unit disk.

The following result gives a characterization of Carleson δ-sequences.

THEOREM 11.7.18 (Chalmers, 1971). *The sequence (x_n^*) in H^* is a Carleson δ-sequence if and only if there exist functions in H, (x_n) with the property that*

$$\hat{f}_{x_n^*}(x_m) = \delta_{nm} = \begin{cases} 1, & n = m \\ 0, & n \neq m. \end{cases}$$

Proof. First we remark that the condition is necessary. Indeed, if we take $x_n = e_n/\hat{x}_n^*(e_n)$, where e_n represents the element of H which appears in the element of H^* which is defined by the component of $\hat{x}_n^*$ orthogonal to $(\hat{x}_i^*)_{i \neq n}$.

For the sufficiency part, suppose now that the condition of the theorem is satisfied for the sequence (x_n^*). In this case we have

$$\delta^{-1}\left\| x_n^* - \sum_{i\neq n} a_i x_i^* \right\| \geqslant \left| \left(x_n^* - \sum_{i\neq n} a_i x_i^* \right)(x_n) \right| = \hat{x}_n^*(x_n) = 1$$

and the theorem is proved. □

DEFINITION 11.7.19 (Chalmers, 1971). If (x_n^*) is a sequence in H^* and T is defined by $T(x) = (x_n^*(x))$, then we say that (x_n^*) is a normalized interpolation sequence if l^2 is contained in the range of T.

Then we have the following result.

THEOREM 11.7.20 (Chalmers, 1971). *A necessary condition for (x_n^*) to be a normalized interpolation sequence is that (x_n^*) be a Carleson δ-sequence for some $\delta > 0$.*

Proof. We consider the subspace of H defined by

$$H_1 = (x, x \in H, T(x) = 0).$$

Then the mapping T induces a mapping $\hat{T}$ from H/H_1 onto a subset of the space of all bounded sequences which contain l^2. Then we can apply the closed graph theorem and obtain that, for some constant $M > 0$, the following inequality holds:

$$\|\hat{x}\| \leqslant M \|\hat{T}\hat{x}\|$$

for all $\hat{x}$ in H/H_1. Then Theorem 11.7.18 asserts that (x_n^*) is a Carleson l/M-sequence. The theorem is proved.

For the proof of the following results we refer to a paper B. L. Chalmers.

THEOREM 11.7.21 (Chalmers, 1971). *If (x_n^*) is a sequence in H^* then a sufficient condition for the range of T to be in l^2 is that the sequence (x_n^*) be an exponential (r, x)-sequence, for some r, $c > 0$.*

THEOREM 11.7.22. *If (x_n^*) is a sequence in H^* then a sufficient condition for the range of T to contain l^2 is that (x_n^*) be an exponential (r, c)-sequence for some $r, c > 0$.*

For the precise inequalities which must be satisfied by r, c in the above results we refer to the above-mentioned paper by Chalmers, where there is also a necessary and sufficient condition for a class of sequences (x_n^*) such that the range of T be equal to l^2.

11.8. Conformal Mappings and Reproducing Kernel Functions

In this section we are interested in the relations between the kernel functions of a space of analytic functions defined on regions of the complex plane and the conformal properties of that region. For the reader's convenience we repeat some definitions from the theory of conformal mappings.

DEFINITION 11.8.1. For each complex number $z = x + iy$, x, y are real numbers and $z \neq 0$, we have a direction from the origin defined by the point

$$\text{Dir } z = z|z|$$

on the unit circle.

DEFINITION 11.8.2. Let Ω be a region in the complex plane and let z_0 be a point in Ω. Suppose further that $f: \Omega \to C$ is a function and for some deleted neighborhood $D'(z_0, r)$, $f(z) \neq f(z_0)$, $z \in D'(z_0, r)$ holds. We say that f preserves angles at z_0

$$\lim e^{-i\theta} \text{Dir}(f(z_0 + r\, e^{i\theta}) - f(z_0)), \quad r > 0$$

exists and is independent of θ.

Let Ω be a region in the complex plane and let $f: \Omega \to C$ be a function. Then the following result is well known (for a proof see books on complex function theory of one variable): if at some z_0 of $\Omega, f'(z_0)$ exists and $f'(z_0) \neq 0$ then f preserves angles at z_0. Also the converse assertion holds: if the differential of f exists and is not 0 at $z = z_0$ and if f preserves angles at $z = z_0$ then $f'(z_0)$ exists and is not equal to 0.

This gives a reason to call the analytic functions with nonvanishing derivative 'conformal mappings'.

DEFINITION 11.8.3. Two regions Ω_1 and Ω_2 are called conformally equivalent if there exists f in $H(\Omega_1)$ with the properties:

(1) f is one-to-one in Ω_1,

(2) $f(\Omega_1) = \Omega_2$.

It is not difficult to see that 'conformally equivalence' defines an equivalence relation on regions. Also, it is worth mentioning that if Ω_1 and Ω_2 are two conformally equivalent regions then the ring of analytic functions $H(\Omega_1)$ and $H(\Omega_2)$ are isomorphic. Thus it is possible to get information about $H(\Omega_1)$ knowing that Ω_1 and Ω_2 are conformally equivalent regions from the properties of the ring $H(\Omega_2)$. The most important case is that when Ω_2 is the unit disk and this is based on the famous Riemann mapping theorem.

In what follows we present some applications of the inner product spaces to some problems of conformal mappings, and these are some extremal problems. In our presentation we follow Stone (1960) and Hille (1972). The application of inner product space methods to conformal mappings is made possible, as is well known, by considering the following extremal problem:

PROBLEM 11.8.4 (An extremal problem). Let Ω be a region in the complex plane. We are interested in the problem of determining the set of analytic functiosn in Ω which minimize the integral

$$\iint_\Omega |f'(z)|^2 \, dx \, dy$$

subject to

$$1 - |f'(z_0)| = 0$$

for some fixed point z_0 of Ω.

In order to eliminate the case when the above integral is ∞ for every competing function we shall assume in what follows that Ω has a finite area. If we consider the function

$$h(z) = 1 - f'(z)$$

then the above problem may be formulated as follows.

PROBLEM 11.8.5 (An equivalent formulation of Problem 11.8.4). Let Ω be as above. The problem is to determine all analytic functions h in Ω which minimize the integral

$$\iint_{\Omega} |1 - h|^2 \, dx \, dy$$

subject to the condition that $h(z_0) = 0$ for some fixed point in Ω.

If we consider the complete inner product space L^2 $(\Omega, dx\,dy)$ then this is equivalent to the problem of determining the distance of the function 1 (i.e. the function which has the value 1 at each point of Ω) to some subspace of $L^2 = L^2(\Omega, dx\,dy)$ determined by the auxiliary condition

$$h(z_0) = 0$$

for some fixed z_0 of Ω.

For this we need the following known result.

THEOREM 11.8.6. *Let H be a complete inner product space and C be a closed and convex subset of H. Then in C there exists a unique element of minimal norm, i.e. there exists a unique element x_0 with the property that*

$$\|x_0\| = \inf(\|x\|, x \in C).$$

Proof. Let $d = \inf(\|x\|, x \in C)$. It is obvious that we may suppose without loss of generality that $d > 0$. In this case there exists a sequence of points in C with the property that

$$d = \lim \|x_n\|.$$

Now for any n, m we have that $\frac{1}{2}(x_n + x_m) \in C$ and thus

$$d \leqslant \tfrac{1}{2}(\|x_n + x_m\|).$$

This obviously implies that

$$\lim_{n,m \to \infty} \|x_n + x_m\| = 2d. \tag{*}$$

Now we have the identity

$$\|u + v\|^2 + \|u - v\|^2 = 2\|u\|^2 + 2\|v\|^2$$

which implies that

$$\|x_n - x_m\|^2 = 2(\|x_n\|^2 + \|x_m\|^2) - \|x_n + x_m\|^2.$$

The relation (*) implies that (x_n) is a Cauchy sequence and thus we can consider

$x_0 = \lim x_n$. It is clear that x_0 satisfies the relation

$$\|x_0\| = d.$$

Since every complete inner product space is strictly convex we obtain easily that x_0 is the unique point of C with property (**).

REMARK 11.8.7. The above result holds for some classes of Banach spaces including uniformly convex spaces. For a proof we refer to V. I. Istrăţescu ('Strict convexity and complex strict convexity', 1984). We mentioned this result earlier.

In order to distinguish a closed linear subspace in $L^2(\Omega, \mathrm{d}x\mathrm{d}y)$ which is appropriate to our extremal problem we must consider some connections between the uniform convergence on compact subsets of Ω and the weak topology on $H^2 = L^2 \cap H(\Omega)$. As in well known, the topology of uniform convergence on compact parts of Ω is defined with the aid of the family of seminorms $(v_K)_K$, K compact in Ω and where

$$v_K(f) = \sup(|f(z)|, z \in K).$$

Now if f is in $H(\Omega)$ and z is in Ω, $d = \inf\{|z - u|, u \in \partial\Omega, \partial\Omega$ being the boundary of Ω, the Cauchy formula implies clearly h the following inequality:

$$|f^{(k)}(z)| \leqslant ((k+2)\|f\|)/((2d^{k+1})\pi^{1/2}) \tag{**}$$

for all $k = 0, 1, 2, 3, \ldots$. Indeed, if we set $w = r\,e^{i\theta}$, $0 < r \leqslant R < d$ and $C(R, z)$ being the circular disk with center at z and radius R, from the Cauchy integral formula we get:

$$r^{k+1}|f^{(k)}(z)| = r^{k+1}\left|1/(2\pi)i \cdot \int_{|w|=r} f(z+w)\mathrm{d}w/w^{k+1}\right|$$

$$\leqslant \frac{1}{2}\int_0^{2\pi} |f(z+w)|r\,\mathrm{d}\theta.$$

This implies that

$$R^{k+2}/(k+2)\cdot|f^{(k)}(z)| \leqslant (2\pi)^{-1}\int_0^R\int_0^{2\pi} |f(z+w)|\,r\,\mathrm{d}\theta\,\mathrm{d}r$$

$$= 1/(2\pi)\iint_{C(R,z)} |f(z)|\,\mathrm{d}x\,\mathrm{d}y$$

$$= (2\pi)^{-1}\left(\iint_{C(R,z)} |f(z)|^2\,\mathrm{d}x\,\mathrm{d}y\right)^{1/2}\cdot\left(\iint_{C(R,z)} 1\cdot\mathrm{d}x\,\mathrm{d}y\right)^{1/2}$$

$$= R(2\pi^{1/2})^{-1}\|f\|$$

and this clearly gives the required inequality.

COROLLARY 11.8.8. *For any f in $H(\Omega) \cap L^2$ the following inequalities hold:*

$$v_K(f^{(k)}) \leqslant (k+2)(2\pi^{1/2}d^{k+1})\|f\|, \quad k = 0, 1, 2, 3, \ldots.$$

COROLLARY 11.8.9. *If Γ is any rectifiable curve in Ω then*

$$\left|\int_\Gamma f^{(k)}(z)\,dz\right| \leqslant l(\Gamma)v_\Gamma(f^{(k)}) \leqslant (k+2)/(2\pi^{1/2}d^{k+1})\cdot\|f\|$$

where $l(\Gamma)$ is the length of Γ and $\tilde{\Gamma}$ means the compact set in Ω bordered by Γ. Then we can prove the following useful theorem.

THEOREM 11.8.10. *$L^2 \cap H(\Omega)$ is a closed subset in L^2.*

Proof. Indeed, suppose that (f_n) is a sequence in $L^2 \cap H(\Omega)$. Then from Corollary 11.8.8 we see that (f_n) is a convergent sequence for the topology of uniform convergence on compact subsets of Ω. This implies that there exists a function-f/in $H(\Omega)$ which is the limit of this sequence (in the topology of uniform convergence on compact subsets of Ω). But then it is easy to see that (f_n) converges in L^2 to f. Thus $H^2(\Omega) = L^2 \cap H(\Omega)$ is a closed subspace in L^2 which is called the Hardy space of Ω.

We now mention two very important properties of the space H^2.

COROLLARY 11.8.11. *For any fixed z_0 in Ω the functionals on H^2*

$$f \to f^{(k)}(z_0) \quad \textit{and} \quad f \to \int_\Gamma f^{(k)}(z)\,dz,$$

where Γ is a rectifiable curve in Ω, are linear continuous functional for each $k = 0$, 1, 2, 3, ...

DEFINITION 11.8.12. We say that a function f in H^2 is with single-valued integrals in Ω if for any closed rectifiable curve Γ in Ω,

$$\int_\Gamma f(z)\,dz = 0.$$

REMARK 11.8.13. In the above definition it is sufficient to require the condition satisfied only for a certain closed rectifiable curve.

Then we have the following result.

THEOREM 11.8.14. *The set of all elements in $H^2 = H^2(\Omega)$ which have single-valued integrals is a closed linear subspace of H^2.*

COROLLARY 11.8.15. *The weak topology and the topology of uniform convergence on compact subsets of Ω coincide on bounded subsets of $H^2 = H^2(\Omega)$.*

Proof. First we recall a result which is easy to prove, namely that the unit ball (or, more generally, any bounded closed set) is weakly compact. Thus to prove our assertion it suffices to prove that any bounded closed ball has the property that the uniform convergence topology on compact subsets is weaker that the weak topology. Indeed, in this case the identity operator is continuous and closed when we consider the weak topology to the set when we consider the topology of uniform convergence on compact subsets. Let us consider a compact subset K of Ω and let ε be a positive number. We choose $0 < \varepsilon < d$, $\eta + 3a/(\pi^{1/2}d^2)$ and we choose a covering of K with open circular disks with centers at $z_1, \ldots, z_n$ $(n < \infty)$ and a common radius δ. This is possible since K is supposed to be compact, and in fact we can construct an open cover of K with open circular disks at each point with radius δ. Then the compactness of K implies that we can choose $(z_i)_1^n$ as above. Then the inequalities

$$\|f\| \leqslant 2a, |f(z_i)| < \eta, \quad i = 1, 2, 3, \ldots, n$$

imply that

$$v_K(f) < \varepsilon.$$

Indeed, if z is in K then we find an index, say j, with the property that z is in an open circular disk with center at z_j, and this implies that

$$\begin{aligned} |f(z)| &= |f(z) - f(z_j) + f(z_j)| \leqslant |f(z) - f(z_j)| + |f(z_j)| \\ &\leqslant \left| \int_{z_j}^{z} f(u)\,\mathrm{d}u \right| + |f(z_j)| < 3\delta/(2\pi^{1/2} d^2) \cdot \|f\| + \eta \\ &\leqslant 3\delta a/(\pi^{1/2} d^2) + \eta < \varepsilon. \end{aligned}$$

In this case, if $\|f\| \leqslant a$, $\|f_0\| \leqslant a$ then $\|f - f_0\| \leqslant 2a$ and if we set $f - f_0$ in place of f, we conclude that for any f which is in the real neighborhood of f_0 determined by the inequalities

$$|f(z_j) - f_0(z_j)| < \eta, \quad j = 1, 2, 3, \ldots, n$$

then $v_K(f - f_0) < \varepsilon$. This implies that the topology of uniform convergence is weakly on that sphere than the weak topology. Thus, on the bounded sets the two topologies coincide.

For the solution of the above extremal problem we must consider the subclass of analytic functions on Ω which are univalent. It is well known that if we add to this family the constants (the functions which are constant on Ω) then we obtain a subset which is closed in the topology of uniform convergence on compact subsets of Ω.

From the corollary above we know that the linear functionals $f \to f(z)_0$ and $f \to f'(z)_0$ are continuous for the topology of uniform convergence on the compact subsets of Ω and thus the subset of functions g with the property that $g'(z)_0 = 1$, for some fixed z_0 in Ω, is closed in the topology of uniform convergence on compact subsets of Ω. From the above result we get that this is a weakly compact subset of H^2.

We note also the following observation which is very important for the solution of the extremal problem.

OBSERVATION 11.8.16. C is the set of all univalent function in H^2 with the property that $f'(z)_0 = 1$. Then this is a convex closed subset of H^2.

From Theorem 11.8.6 we obtain the following result.

THEOREM 11.8.17. *The extremal problem* 11.8.4 *has a unique solution.*

In what follows we denote the solution by h_{z_0}, where z_0 is the fixed point in Ω

from Problem 11.8.4. In this case the minimum of the integral

$$\inf \iint_\Omega |1 - m(z)|^2 \, dx \, dy = \|h_{z_0}\|^2 = \langle 1 - h_{z_0}, 1 \rangle.$$

Now if f is any function in H^2 then

$$g = f - f(z_0)$$

is a function which satisfies the property

$$\langle f - f_{z_0}, h_{z_0} \rangle = 0$$

and thus

$$\langle f, h_{z_0} \rangle / \|h_{z_0}\|^2 = f(z_0). \qquad (+)$$

This suggest the consideration of the function on $\Omega_\times \Omega$ defined by

$$K_1(z, t) = K_{\Omega,1}(z, t) = h_t(z)/\|h_t\|^2$$

and thus $(+)$ may be written as

$$\iint_\Omega \bar{K}_1(z, t) f(z) \, dx \, dy = f(t) \qquad (++)$$

for all f in H^2. (The bar denotes the complex conjugate.)

The space $L^2 = L^2(\Omega, dx\,dy)$ is separable and thus every closed subspace has this property. Thus H^2 is separable and we can consider an orthonormal basis of H^2, say $(e_i(z))_1^\infty$. From $(++)$ we conclude that the Fourier coefficients of K as a function of z are $(\bar{e}_n(t))$. Then from the property of bases in a complete inner product space we obtain the relation

$$K_1(z, t) = \sum_{n=1}^{\infty} e_n(z) \bar{e}_n(t)$$

where the right side is absolutely convergent for all (z, t) in $\Omega_\times \Omega$, stronly convergent in H^2 for each fixed t.

The function

$$K(z, t) = \bar{K}_1(z, t)$$

is, as we know from the case of some domains, the so-called Bergman kernel function of Ω. Now the solution of the extremal problem is given in terms of the Bergman kernel function by the formula:

$$h_{z_0} = 1 - K(z, z_0)/K(z_0, z_0)$$

and the minimum of the integral is equal to

$$\|h_{z_0}\| = (1/K(z_0, z_0))^{1/2}.$$

Another important result which follows from the result obtained above is the following inequality.

THEOREM 11.8.18. *Suppose that f is an analytic function on Ω and consider the* *area integral*

$$A_f = \iint_\Omega |f'|^2 \,\mathrm{d}x\,\mathrm{d}y.$$

Then the following inequality holds:

$$|f'(z)| \leqslant (A_f / h_z^2)^{1/2}$$

with equality holding if and only if $f'(z) = f'(z_0)h_{z_0}$.

Proof. We may consider that f does not have the property that $f'(z) = 0$ or $A_f = \infty$. Now to prove the assertion of the theorem we apply the result about the extremal problem obtained above to the function

$$g(z) = 1 - f'(z)/f'(z_0)$$

and thus from the minimum property we have the inequality

$$\|1 - g\| \geqslant \|h_{z_0}\|$$

and since the equality in this inequality holds if and only if $1 - g = h_{z_0}$, the inequality

$$\|A_f\|/|f'(z_0)| \geqslant \|h_{z_0}\|$$

which implies the assertion of the theorem.

Suppose now that Ω is the open circular disk with center at $z = 0$ and radius R. Then the above inequality implies that for any analytic function f the following inequality holds:

$$A_f \geqslant \pi R^2$$

and the equality holds if and only if $f(z) = z$.

In order to obtain this assertion we remark that an orthonormal system for H^2 of the open circular disk with center at $z = 0$ and radius R is

$$\left\{\frac{n^{1/2} z^{n-1}}{\pi^{1/2} R^n}\right\}_1^\infty.$$

Then

$$K(z, t) = R^2/(R^2 - z\bar{t})$$

$$\|h_{z_0}\| = (R^2 - |z_0|^2)^2/R^2$$

$$h_{z_0}(z) = (R^2 - |z_0|^2)^2/(R^2 - z\bar{z}_0)^2.$$

The above inequality then implies the inequality

$$A_f \geqslant \pi R^2$$

and the equality holds if and only if $f(z) = z$.

It is a well-known fact in the theory of conformal mappings that many of the basic theorems can be proved using the Bergman kernel. In what follows we show how some of them can be obtained from the results given above. Among them we give the proof of the Riemann mapping theorem.

We shall consider only finitely connected domains Ω in the complex plane. The first observation concerning such domains is that, if the boundary Ω is sufficiently smooth then it is conformally equivalent to a parallel slit plane (i.e. a domain which is obtained from the extended complex plane by removing from it a finite number of parallel line segments). In this conformal equivalence an arbitrary point z_0 of Ω corresponds to the point at ∞. P. Garabedian and M. Schiffer have considered this problem when the set Ω has the property that the boundary $\partial\Omega$ has two tangent circles of fixed radius bounding open disks, one of which is in the interior of Ω and the other in the exterior of Ω. In this case the domain Ω is simply connected and conformally equivalent to a parallel slit plane having a single slit (which may be the line segment with the endpoints at -1 and 1). But this is conformally equivalent to the unit disk in which the point at ∞ and the center of the disk are corresponding points. This implies that Ω is conformally equivalent to the unit disk in which the arbitrary fixed point z_0 and the center of the unit disk are corresponding points. A multiple of this conformal mapping is a solution of the extremal problem 11.8.4, and this maps Ω on the open disk with center at $z = 0$ and radius R. Thus the Riemann mapping theorem for this class of domains follows from the result of Garabedian and Schiffer.

First we remark that if Ω is any simply connected domain and h is a competing function of the extremal problem 11.8.4, and, further, if h maps Ω on an open circular disk with the center at $z = 0$ and radius R, then the solution of the extremal problems is given by the function h_{z_0}.

From the inequality proved above for the integral area A_f it follows that when the domain Ω is a circular disk then the solution of the extremal problem is the identity mapping.

Suppose now that Ω is a simply connected domain and we may suppose without loss of generality that Ω is not the entire complex plane (because the analytic functions which are in our space are only the constant functions). Since Ω is simply connected the boundary $\partial\Omega$ is not empty and does not reduce to a point. Then it is a well-known fact that Ω is conformally equivalent to a domain with finite area. In this case the above extremal problems have a unique solution in terms of the functions h_{z_0} and K.

In order to obtain this solution we construct an exhaustion of Ω consisting of domains (Ω_n) with a sufficiently smooth boundary (and, of course, all containing the point z_0). Then we obtain the solutions of the extremal problems for each domain Ω_n, say $h^n_{z_0}$ and $1 - h^n_{z_0}$, since $h^n_{z_0}$ maps Ω_n on the open circular disk with center at $z = 0$ and radius R_n with

$$\pi R_n^2 = \|1 - h^n_{z_0}\|^2.$$

Since $(h^n_{z_0})$ is a Cauchy sequence in L^2, it converges in this space to a function h which is in H^2 (which follows from the connection between the weak topology and the topology of uniform convergence on compact subsets given above).

Then it is clear that

$$\pi R^2 = \|1 - h\|^2,$$

where $R = \lim R_n$. The function h maps Ω on the open circular disk $(z, |z| < R)$ since

$|(h^n_{z_0})z)| \leqslant R$ implies that $|h(z)| \leqslant R$. We must show that we can identify $h(\Omega)$ with the circular disk $(z, |z| < R)$. For this purpose we consider the function

$$k_n(z) = (R/R_n)h(h^n_{z_0})^{-1}(R_n z/R),$$

where $(h^n_{z_0})^{-1}$ is the inverse function. Then k_n is a competing function for the disk and has an area integral equal to

$$(R^2/R_n^2)\iint_{\Omega n} |h'|^2 \, dx \, dy.$$

In this case $(1 - k_n')$ is a minimizing sequence for the extremal problem in the case of the open circular disk. Then this converges in L^2 to the minimizing function for the open circular disk. Then (k_n) converges in the topology of uniform convergence on compact subsets of Ω. We now consider an arbitrary point of the disk, say u_0, and let r be sufficiently small so that $(z, |z - u_0| = 2r) = C_{r,u_0}$ is in the disk. Now the equation $k_n(z) = z$ has a finite number of roots, which is given by the formula

$$\frac{1}{2\pi i}\int_{C_{r,u_0}} k_n'(z)/(k_n(z) - z_1)\, dz = (2\pi i)^{-1}\int_{C_{r,u_0}} dz/(z - z_1) = 1$$

in the closed disk $(z, |z - z_1| \leqslant r)$. Then the set $(R/R_n)h(\Omega)$ contains the disk

$$(z, |z - u_0| \leqslant r)$$

and thus $h(\Omega)$ contains the disk

$$(z, ||z - u_0 R_n|| \leqslant R_n r/R)$$

if n is sufficiently large, which implies that u_0 is in $h(\Omega)$.

This gives the following result asserted by B. Riemann.

THEOREM 11.8.19 (The Riemann conformal mapping theorem). *Suppose that Ω is a simply connected domain properly contained in the complex plane. Then Ω is conformally equivalent with a circular disk with the following property: a prescribed point z_0 of Ω and the center of the disk are corresponding points through the mapping.*

From the proof we see that this mapping is uniquely determined by the derivative number $h'(z_0)$.

COROLLARY 11.8.20. *If Ω is a finitely connected domain in a complex plane then it is conformally equivalent to a bounded domain in the complex plane which is bounded by a finite number of isolated points and a finite number of simple, closed analytic curves.*

For results about the conformally invariant length, applications to the principle of hyperbolic measure and connections with the H. Poincaré model of the Lobačevskiĭ non-Euclidean geometry, we refer to Hille's expository paper (1972).

11.9. Invariant Subspaces for Generalized Translations and Reproducing Kernels

If l^2 is the space of all sequences of complex numbers $x = (x_n)$ with $\|x\|^2 = \Sigma_{\Omega}^{\infty} |x_n|^2 < \infty$ then, as we know, this can be organized as a complete inner product space. An interesting operator defined on this space is the so-called 'shift' which is defined as follows: if $x = (x_n)_1^{\infty}$ then $Sx = (0, x_1, x_2, \ldots)$. Of course, S is an isometric operator with I-SS^* one-dimensional.

In a well-known paper (1948), A. Beurling posed and resolved the problem of describing all invariant subspaces of the operator S. This is realized in fact in the space H^2 of Hardy for the unit disk and an exposition of Beurling's theorem is given in Section 17.2. Here we are interested in a more general problem, namely the problem of describing certain classes of spaces of analytic functions on certain subsets of the complex plane which will be described axiomatically in what follows, and which contain the Hardy space H^2 and the Paley–Wiener space of the upper half plane which was considered by Lax (1959).

We mention that this interesting approach to the problem of invariant subspaces for the shift operator, or operators like the shift, was considered in the interesting paper of H. S. Shapiro in 1964, and in what follows we give an account of Shapiro's results. First we consider a class of complete inner product spaces H defined as follows:

DEFINITION 11.9.1. The elements of H are functions f defined on an open subset Ω of the complex plane whose boundary is Γ and such that the following properties hold:

(1) Every f in H is analytic in Ω.
(2) For each z in Ω, the functional evaluation at z, i.e. $f \to f(z)$, is a bounded linear functional on H.
(3) There is a family of subsets of Γ, called negligible sets, which have the following properties:

   (a) an enumerable union of negligible sets is negligible;
   (b) Γ is not negligible. A property is said to be nearly everywhere if it is enjoyed by all points of Γ except possibly of a negligible set;
   (c) for every function f in H there is a function $f(t)$ defined on Γ and nonvanishing nearly everywhere (we denoted the function with the same letter); this function is called the boundary function;
   (d) for nearly every point t of Γ there is a sequence of points (z_n) in Ω such that $\lim f(z_n) = f(t)$ for every f in H holds nearly everywhere; the points (z_n) are called the admissible sequence belonging to the point t of Γ);
   (e) the norm of f in H is uniquely determined by the values of the boundary function nearly everywhere.

(4) There is a class of functions, called "multipliers", M, which have the following properties:

   (a) every element of M is an analytic function on Ω,
   (b) if $m \in M$ and $f \in H$ then mf is in H,

(c) if f_i, g_i are in H and m is in M and

$$\langle f_1, mg_1\rangle = \langle f_2, mg_2\rangle, \qquad \langle mf_1, g_1\rangle = \langle mf_2, g_2\rangle$$

then

$$f_1(t)\bar{g}_1(t) - f_2(t)\bar{g}_2(t) = 0$$

nearly everywhere on Γ.

Then we have clearly the following result.

PROPOSITION 11.9.2. *H is a space with a reproducing kernel.*

Proof. This follows from Axiom (2) and Theorem 11.1.5. □

We now mention some space which satisfy the above axioms. The first example is that of the Hardy space of the unit disk. Here Γ is the unit circle and the negligible subsets are the subsets of measure zero. The boundary functions are the non-tangential limits. The multipliers are the bounded analytic functions. The same with the spaces E_2 of V. Smirnov (a detailed account of these spaces is given in I. Privalov's book (1950). Another important space in our class is the space of Paley–Wiener of analytic functions in the upper half-plane whose boundary functions are in L^2 $(0, \infty)$. If we consider the Fourier transform for any f in $L^2(0, \infty)$

$$F(z) = \int_0^\infty f(t)\, e^{itz}\, dt$$

then $F(z)$ is in the Paley–Wiener space, and conversely. See Rudin (1966, Ch. 19) for details.

Now the generalized form of Beurling's theorem for the class of complete inner product spaces satisfying the above axioms obtained by H. S. Shapiro is as follows.

THEOREM 11.9.3 (Shapiro, 1964). *Let H be a complete inner product space satisfying the above axioms. Let $\mathcal{F}$ be the family of all closed subspaces of H with the following property; if H_1 is in $\mathcal{F}$ then for all f in H_1 and all m in M, $mf \in H_1$. Then H_1 is of the form $H_1 = w_H(\,,)H$, where w_{H_1} an analytic function in Ω and $|w_{H_1}(t)| = 1$ nearly everywhere on Γ. This function w_{H_1} is uniquely determined except of a factor of modulus 1.*

Before the proof of the theorem we remark that any subspace of H of the form wH, where w is like w_{H_1}, is a subspace with the property that for all f in that space and m in M, mf is in the subspace.

The fact that w_{H_1} has modulus equal to 1 nearly everwhere means that for nearly every point t in Γ and for every admissible sequence (z_n) belonging to t, $\lim w_{H_1}(z_n)$ exists and has a modulus equal to 1.

Now the proof of the theorem is as follows.

From Proposition 11.9.2 we know that H has a reproducing kernel and thus we denote it by $K_\xi\,(z)$. We have the relation: if u, v are in Ω then

$$k_u(v)K_u(t)\bar{K}_v(t) = K_u(v)\, k_v(t)\bar{K}_v(t) \tag{+}$$

nearly everwhere on Γ. Indeed, for this we consider the following functions:

$$f_1(z) = k_u(v)K_u(z), \qquad f_2(z) = K_u(v)k_u(z)$$
$$g_1(z) = K_v(z), \qquad g_2(z) = k_v(z)$$

and the above relation is proved and we show, according to Axiom (4c) that the following relations hold:

$$k_u(v)\langle K_v, mK_v\rangle = K_u(\xi)\langle k_u, mk_\xi\rangle \qquad (*)$$

$$k_u(v)\langle mK_u, K_v\rangle = K_u(v)\langle mk_u, k_v\rangle \qquad (**)$$

for all m in M. Since mk_v and mk_u are in H for all m and K is a reproducing kernel for H, relations (*) and (**) are equivalent to relations

$$k_u(v)\bar{m}(u)\bar{K}_v(u) = K_u(v)\bar{m}(u)\bar{k}_v(u) \qquad (***)$$

$$k_u(v)m(v)K_u(v) = K_u(v)m(v)k_u(v) \qquad (****)$$

But these are clearly from the reproducing kernel property of K.

We now consider the meromorphic function

$$n_u(z) = k_u(z)/K_u(z)$$

which implies that relation (+) is of the form

$$n_u(v) = n(t)\bar{n}_v(t)$$

which for $u = v$ implies that

$$|n_u(t)|^2 = n_u(u) = k_u(u)/K_u(u)$$

nearly everwhere on Γ.

Now since in any complete inner product space with a reproducing kernel the following inequality holds

$$|f(s)|^2 \leqslant K_s(t)\|f\|^2$$

we obtain, for $f(z) = k_u(z)$, that

$$k_u(u) \leqslant K_u(u).$$

This clearly implies that nearly everywhere $|n_u(t)| = 1$ on Γ. Thus for all t in the complement of the negligible set S_1 we have that for all $u_1|n_u(t)| \leqslant 1$ and $|n_v(t)| \leqslant 1$, $v \in C_{S_1}$,

$$n_u(v) = n_u(t)\bar{n}_v(t),$$

Thus if t is in the above set we obtain that

$$|n_u(v)| \leqslant 1, \qquad v \in C_{S_1}.$$

Since n_u is a meromorphic function in Ω which is bounded on the complement of a negligible set it follows that it is a bounded analytic function on Ω.

We now consider u,v in the complement of the above negligible set, and then we remark that the following relation holds:

$$n_u(v) = n_u(t)\bar{n}_v(t).$$

Now, according to (3d) we can choose a point t in Γ and an admissible sequence (z_n) belonging to t with the property that

$$\lim n_u(z_n) = n_u(t) \qquad (++)$$

exists and is finite and also

$$n_u(v) = \lim n_u(z_n)\bar{n}_v(z_n). \qquad (+++)$$

We now consider the sequence of functions $(h_n(z))$ defined by the relation

$$h_n(z) = k_{z_n}(z)/K_{z_n}(z)$$

(and we remark that if n is sufficiently large, say greater than n_0, $K_z(z) \neq 0$ and thus in the above definition of h_n we may suppose that $n \geqslant n_0$). From the above relations we have that

$$|h_n(z)| \leqslant 1, \quad n \geqslant n_0$$

and from $(++)$ and $(+++)$ it follows that

$$\lim h_n(u) = \bar{n}_u(t)$$

$$\lim h_n(u)h_n(v) = h_u(v).$$

The sequence of functions (h_n) is uniformly bounded by 1 and converges pointwise nearly everywhere. This implies that (h_n) coverges uniformly on compact subsets of Ω to a function h satisfying the relation

$$n_u(v) = \bar{h}(u)h(v).$$

Since this holds nearly everywhere, from the continuity property these hold for all u, v in Ω. Now if we set $u = v$ we have that

$$|n_u(u)| = |h(u)|^2$$

and

$$|n_u(v)|^2 = |h(u)|^2\,|h(v)|^2.$$

Now we fix u and v run through an admissible sequence (z_n) belonging to a boundary point t. Then we obtain that $h(v) \to 1$ and we say that

$$|h(t)| = 1$$

nearly everywhere on Γ.

We denote the closed subspace H_2 of H which is of the form hH, where h is the above function. Since

$$k_v(z) = K_v(z)\bar{h}(v)h(z)$$

is in H_1 for each v. Now if $g = hf$ is in H_2 then we have,

$$g, k_v = h(z)\,f(z), \qquad K_v(z)\bar{h}(v)h(z) = h(z)\,hf, \qquad hK_v = g(v)$$

which implies that $k_v(\,,)$ is a reproducing kernel for H_1. But a subspace of H is uniquely determined by its reproducing kernel, and this implies that H_1 is equal to H_2. Of course we have $h = w_{H_1}$. The theorem is proved.

We now present, following H. S. Shapiro, two interesting applications of the above structure theorem of the subspaces invariant by the multiplication with elements of the multipliers set M to some well-known extremal problems in the theory of bounded analytic functions.

We suppose Ω to be a simple connected open set in the complex plane (so that the Riemann mapping theorem applies to Ω to obtain a conformal mapping of Ω into the unit disk) and we consider the set $H(\Omega)$ of all bounded analytic functions on Ω (see Rudin, 1966; Hoffman, 1962). The set $H(\Omega)$ is a Banach space under the norm

$$f \to \|f\|_\infty = \sup(|f(z)|, z \in \Omega).$$

The extremal problems are as follows:

PROBLEM 11.9.4. What is the value of

$$\max(|f(v)|, f(u) = 0, \|f\|_\infty \equiv 1)?$$

PROBLEM 11.9.5. What is the value of

$$\max(|f'(u)|, f \in H(\Omega), f(u) = 0)?$$

In order to obtain a solution to these problems we suppose that Ω satisfies the following two additional properties with respect to the space H defined axiomatically above.

First we suppose that for each m in $H(\Omega)$, the operator $f \to mf$ is a bounded linear operator whose norm is less than or equal to $\|m\|_\infty$ and, secondly, that no z in Ω is a common zero of all f in H.

It is not difficult to see that in fact the norm of the operator $f \to mf$ is equal to $\|m\|_\infty$. In order to solve these problems we consider the subspace of H consisting of all f in H with the property that $f(u) = 0$. We denote this space by H_u. Obviously H_u is a closed linear subspace of H and for each m in $H(\Omega) = M$ (the set of multipliers), mf is in H_u whenever f is in H_u. Then by the above theorem there exists a unique function (part of a factor of modulus 1) $w_u = w_{H_u}$ such that $H_u = w_u H$. We show that this function w_u is the unique extremal function for Problems 11.9.4 and 11.9.5.

Indeed, suppose that g is another function which is extremal for Problem 11.9.4, and thus

$$\begin{aligned}|g(v)|^2 K_v(v)^2 &= |\langle g(z)K_v(z), k_v(z)\rangle|^2 = |w_u(v)|^2 |\langle g(z)K_v(z), w_u(z)K_v(z)\rangle|^2 \\ &\leqslant |w_u(v)|^2 K_v(v)^2\end{aligned}$$

which gives our assertion.

For problem 11.9.5 we consider f to be an arbitrary element in H, and thus

$$f(v) = \langle f(z), k_v(z)\rangle$$

which implies that

$$f'(v) = \langle f(z), (\partial/\partial v)k_v(z)\rangle$$

and since

$$k_v(z) = \langle K_v(z) w_u(v) \bar{w}_u(z)$$

we have that

$$\partial k_v(z)/\partial v(z) = \bar{K}_v(z)\bar{w}'_u(v)w_u(z) + (\partial K_v(z)/\partial v(z))\hat{w}_u(v)w_u(z).$$

Then we have further that, for

$$f(z) = g(z)K_u(z)$$

(and taking $u = v$), the following inequalities hold:

$$|g'(u)|^2 K_u(u)^2 = |w_u'(u)|^2 |\langle g(z)K_u(z), w_u(z)K_u(z)\rangle|^2 \leqslant |w_u'(u)|^2 K_u(u)^2$$

and, as above, the assertion follows.

Now the explicit value of the maximum can be obtained by computing the value of $w_u'(u)^2$ which is given by the formula

$$\begin{vmatrix} \partial^2 K(s,t)/\partial s\,\partial t & \partial K(s,u)/\partial s \\ \partial K(u,t)/\partial t & K(u,u) \end{vmatrix} \Big/ K(u,u)^2$$

for $s = t = u$.

REMARK 11.9.6. It is an interesting and important property of the function w_u that it is the mapping which applies conformally Ω on the unit disk.

11.10. Spline Functions and Inner Product Spaces with Reproducing Kernels

A polynomial spline function results from splicing polynomial arcs in such a way that the resulting function is sufficiently smooth (we give below a precise definition). Spline functions have many applications in interpolation theory, quadrature formulas, approximation theory, integration of differential equations, etc. For an excellent exposition we refer to the book of Ahlberg *et al.* (1967) and to more recent book of Schumacher (1981). It is worth mentioning that splines appeared early in the study of extremal problems; for example, in a paper of Favard (1937) and in a paper of Achiezer and Krein (1937) in which they are interested in the problem of the functions which are farthest in the norm of $L^{\infty}(-\pi,\pi)$ from the trigonometric polynomials of degree n and have the properties that they

(1) are 2-periodic functions, and,
(2) are m-fold integrals of functions in the unit ball of $L^{\infty}(-\pi,\pi)$.

The connection of spline theory with complete inner product spaces has its origin in a discovery of Holladay (1957) that the cubic interpolating spline function is an extremal of the functional:

$$f \to \int_a^b (f''(t))^2 \, dt.$$

Several interesting minimum properties of (odd) spline functions were discovered by De Boor (1963) and De Boor and Lynch (1966). The basic observation is that the spline functions are representers of appropriate bounded linear functionals on appropriate complete inner product space.

If H is a complete f inner product space whose elements are functions, and if x^* is a bounded linear functional on H then by the Riesz–Fréchet representation theorem there exists a function in H, f_{x^*} such that for all x in H,

$$x^*(x) = \langle x, f_{x^*} \rangle.$$

The function f_{x^*} is then called a representer.

We mention that the notion of spline functions was extended by Atteia (1965, 1966) to abstract inner product spaces. In what follows we present some results about spline functions that were given by De Boor and Lynch (1966). We begin with the definition of spline functions.

DEFINITION 11.10.1. A spline function $S(x)$ of degree $(k \geqslant 0)$ is, by definition, a function on the real axis with the property that it reduces to a polynomial on each of the $n+1$ intervals in which the knots $(x_i)_1^n$ divide the real axis and where we suppose that

$$x_1 < \cdots < x_{n-1} < x_n.$$

It is not difficult to see that the function S depends linearly on $n+k$ parameters, and in fact that most general form of S is given by

$$S(x) = \sum_{i=0}^{k} a_i x^i + \sum_{j=1}^{n} b_j (x - x_j)_+^k,$$

where a_i, $i = 0, \ldots, k$, and b_j, $j = 1, 2, \ldots, n$, are real numbers and

$$(x)_+^m = \begin{cases} xm, & \text{if } \quad x > 0 \\ 0, & \text{if } \quad x < 0. \end{cases}$$

In what follows we suppose that $k = 2m - 1$, with $1 \leqslant m \leqslant n$. We consider the family S_0 of polynomially spline functions of odd degree k with the knots $x_1 < x_2 < \cdots < x_{n-1} < x_n$ with the property that it reduces to polynomials of degree at most $m - 1$ in each of the intervals $(-\infty, x_1)$, (x_n, ∞). Given the knots (x_i), $1 \leqslant i \leqslant n$, we consider $[a, b]$ to be a finite interval which contains the points x_1, $i = 1, 2, 3, \ldots, n$, and we consider further a space of functions defined on a, b as follows:

$F^m((a, b)) = (f, f\colon [a, b] \to R, f \in C^{(m-1)}[a, b],\ f^{(m-1)}$ is absolutely continuous and $f^{(m)} \in L^2_{[a,b]})$.

This space will play a fundamental role in our considerations. The basic property of the space $F^m((a, b))$ is given in the following theorem.

THEOREM 11.10.2 (De Boor and Lynch, 1966). *The set $F^m((a, b))$ can be organized as a complete inner product space with the inner product defined by the following formula: if f, g are two arbitrary elements in $F^m = F^m((a, b))$ then*

$$\langle f, g \rangle = \sum_{i=1}^{n} f(x_i) g(x_i) + \int_a^b f^{(m)}(t) g^{(m)}(t)\, dt.$$

Proof. It is clear that F^m may be organized as a linear space where the addition and multiplication by scalars are defined in the usual way. Now we suppose that (f_n) is a Cauchy sequence in F^m. Then it is obvious that for each i, $i = 1, 2, 3, \ldots, n$, the sequence $(f_p(x_i))$ is a Cauchy sequence in R and thus converges to a real number r_i. Since L^2 is a complete space it follows that the Cauchy sequence $(f_i^{(m)})$ converges to a function in L^2, say h. Then h is integrable on $[a, b]$ and this gives rise a function of C^{m-1} whose m-derivative coincides a.e. with h. Since this function depends on some constants, these may be determined such that the values at x_i are to be taken equal to r_i. Then (f_p) converges in F^m to this function, and the theorem is proved. □

REMARK 11.10.3. We can consider a more general space than F^m, namely a space which may be useful in considering spline functions with an infinite (denumerable) set of knots. Indeed, suppose that (t_i) is a denumerable subset in R; we consider, as above, the interval $[a, b]$ which contains all points t_i and F^m is defined as the set of elements of C^{m-1} with the property that $f^{(m)}$ is in $L^2_{[a,b]}$ and $\Sigma_{i=1}^{\infty} f(t_i)^2 < \infty$. Then F^m may be organized as a complete inner product space with the inner product given by the formula: if f, g are two elements in F^m then

$$\langle f, g\rangle = \sum_{i=1}^{\infty} f(t_i)g(t_i) + \int_a^b f^{(m)}(t)g^{(m)}(t)\,dt.$$

The proof is essentially the same as above.

Another possibility is to consider weighted space which may be also useful when the condition $\Sigma_{i=1}^{\infty} |f(t_i)|^2 < \infty$ is very difficult to satisfy for functions which are of interest to us. For this purpose we consider a sequence of strictly positive numbers (p_i) and a function $p(t)$ on a, b. The space $F^m(p, (p_i))$ is then defined similarly and for two arbitrary elements of the space, say f, g, the inner product is defined by the formula

$$\langle f, g\rangle = \sum_{i=1}^{\infty} p_i f(t_i)g(t_i) + \int_a^b p(t)f^{(m)}(t)g^{(m)}(t)\,dt.$$

The proof of the completeness of the space is essentially as in the case of the space F^m.

Now we return to the case of the space F^m. A crucial property of the space F^m is as follows.

THEOREM 11.10.4. *The space F^m is a reproducing kernel space and the reproducing kernel is given by the formula:*

$$K(s, t) = \sum_{i=1}^{m} c_i(s)c_i(t) + (-1)^m\left((s-t)_+^{2m-1} + \sum_{i=1}^{m}\sum_{j=1}^{m} (x_i - x_j)_+^{2m-1}\, c_i(s)c_i(t)\right)$$

$$- \sum_{i=1}^{m} ((s - x_i)_+^{2m-1}\, c_i(t) + (x_i - t)_+^{2m-1}\, c_i(s))/(2m-1)!,$$

where

$$c_i(u) = \sum_{j=1, j\neq i}^{m} (u - x_i)/(x_i - x_j), \quad i = 1, 2, 3, \ldots, m.$$

Proof. The fact that this space has a reproducing kernel follows from Theorem

11.1.4. The fact that the above function is a reproducing kernel for the space gives a new proof for the existence of a reproducing kernel.

The following lemma, when applied to F^m, gives an interesting property of spline functions.

LEMMA 11.10.5. *Let H be a complete inner product space and (x_i^*), $i \in I$ be a set in H^*. Let $(f_{x_i^*})$ be the elements in H with the property that $x_i^*(x) = \langle x, f_{x_i^*} \rangle$, $i \in I$. If P is the orthogonal projection of H onto the subspace generated by $(f_{x_i^*})$ then for each x in H,*

$$x_i^*(Px) = x_i^*(x), \quad i \in I.$$

Proof. First we remark that the elements Px and $(I - P)x$ are orthogonal and thus

$$\|x\|^2 = \|x - Px\|^2 + \|Px\|^2. \qquad \square$$

Also, P is a hermitian operator and thus for any x, y in H we have,

$$\langle Px, y \rangle = \langle Px, Py \rangle = \langle x, Py \rangle$$

and thus for $y = f_{x_i^*}$, $i \in I$, we obtain the assertion.

This obviously implies the following.

COROLLARY 11.10.6. *Given x in H, Px is the unique element in the space generated by representers $(f_{x_i^*})$ which interpolates with respect to (x_i^*), i.e.*

$$x_i^*(x) = x^{*i}(Px),$$

for all i in I.

In our concrete example I is a finite set, and thus we obtain the existence of a spline which interpolates f at the knots $x_1, \ldots, x_n$ and is the best approximation with respect to the norm

$$g \to \|g\| = \left(\sum_{i=1}^{n} |g(x_i)|^2 + \int_a^b |g^{(m)}(t)|^2 \, dt \right)^{1/2}.$$

This obviously implies the following minimum property of spline functions:

COROLLARY 11.10.7. *For any f in F^m and any s in S_0 the following inequality holds:*

$$\int_a^b ((f^{(m)} - s^{(m)})(t))^2 \, dt \geqslant \int_a^b ((f^{(m)}(t) - (Pf)^{(m)}(t))^2 \, dt)^{1/2}.$$

We now consider the subspace $H_1 = PH$ and we remark that an element u of H is in H_1 if and only if $x_i^*(u) = x_i^*(x)$, and from the relation

$$\|x\|^2 = \|x - Px\|^2 + \|Px\|^2$$

we obtain that for all any u in H_1 we have

$$\|Px\| \leqslant \|u\|$$

and the equality holds if and only if $u = Px$. Thus we have the following result.

COROLLARY 11.10.8. *An element x in H is in H_1 if and only if x is the element of minimal norm in the set of elements y of H which interpolate x with respect to* (x_i^*).

This, when applied to the space F^m, implies the following minimum property of spline functions:

COROLLARY 11.10.9. *For any f in F^m the following inequality holds:*

$$\int_a^b (f^m(x))^2\,dx \geqslant \int_a^b ((Pf)^m(x))^2\,dx.$$

There is another minimum property of spline functions related to be approximation of linear functionals in $(F^m)^*$ through linear functionals of the form

$$\sum_0^{m-1} \int_{[a,b]} f^{(i)}(t)\,d\mu_i(t),$$

where (μ_i) are bounded measures on $[a, b]$ and $f^{(i)}$ are integrable with respect to these measures.

Generally, the problem of the approximation of linear functionals in the dual space of F^m may be formulated as follows: we have, as above, the set in H^*, (x_i^*) and we are interested in the distance of a linear functional x^* in H to the subspace (closed) generated by (x_i^*). Of course, from the Riesz–Fréchet representation theorem it suffices to find a representer for the functional which realizes the minimum distance. Then we can prove the following result.

THEOREM 11.10.10. *The representer of the best approximation of x^* through the elements of the closed subspace generated by (x_i^*) is Pf_{x^*}.*

Proof. Since the best approximant exists, let us denote it by u^* and then

$$u^*(x) = \langle x, Pf_{x^*}\rangle = \langle Px, f_{x^*}\rangle = L(Px)$$

and the assertion is proved. □

This clearly implies the following result.

RESULT 11.10.11. *For any given x in H the values of the best approximation to x^* at x is equal to the values of x^* at the best approximation of x (through the elements of the space generated by representers).*

The third minimum property of spline functions is related to the so-called 'best approximation of x^*' in the sense of Sard (1963).

We consider the space F^m and let π_{m-1} be the set of all polynomials of degree at most $m-1$, and if f is in π_{m-1} then we find $a_1, \ldots, a_n$ with the property that $Rf = (x^* - \Sigma_{i=1}^m a_i x_i^*)(f) = 0$ (since $m = n$). We consider further the set

$$M = \left(x_1^* = \sum_{i=1}^m a_i x_i^*,\ x_1^*(x) = x^*(x), x \in \pi_{m-1}\right)$$

which is clearly nonempty. Peano's theorem asserts that

$$R(f) = (x^* - x_1^*)(f) = \int_a^b K(s) f^{(m)}(s)\,\mathrm{d}s, \quad x_1^* \in M,$$

where

$$K(s) = R(u - s)_+^{m-1}/(m-1)!.$$

Since M is a convex and closed set, we define x^* to be the unique element in M which minimizes the following integral:

$$\int_a^b (K(s))^2\,\mathrm{d}s$$

and this is exactly the fact that x^* minimizes $\|x^* - x_1^*\|$ on M. Thus, this is identical with the functional whose representer was given above. We than have the following property, given in

COROLLARY 11.10.12. *If x^* is a linear functional on F^m of the form*

$$x^*(f) = \sum_{i=1}^{m-1} f^{(i)}(s)\,\mathrm{d}\mu_i(s)$$

then the best approximation of $x_1^*(f)$ for all f in F^m in the sense of Sard is given by operating x^* on Pf, i.e. $x_1^* = x^*P$.

There is another way to look at the space F^m. Indeed, if we consider the differential operator in normal form

$$\mathrm{D}^m = \mathrm{d}^m/\mathrm{d}x^m + \sum_{i=0}^{m-1} a_i(x)\,\mathrm{d}^i/\mathrm{d}x^i, \tag{*}$$

where (a_i) are functions sufficiently that are smooth on the interval $[a, b]$ and $x_1^*, \ldots, x_m^*$ are m linear functionals on F^m. We can define, for any two elements in F^m, the number

$$\langle f, g\rangle = \sum_{i=1}^{m} x_i^*(f) x_i^*(g) + \int_a^b (\mathrm{D}^m f)(t)(\mathrm{D}^m g)(t)\,\mathrm{d}t \tag{+}$$

and this defines an inner product on F^m.

In fact, the following result is valid. For a proof we refer to De Boor and Lynch (1966).

THEOREM 11.10.13. *Let D^m be any mth-order ordinary linear differential operator in the normal form (*) and let $(a_i)_1^{m-1}$ be a subset of $C_{[a,b]}$. Let us suppose further that the kernel of D^m in $D_{[a,b]}^m$ has dimension m and that $x_1^*, \ldots, x_m^*$ are linear functionals of the form*

$$\sum_0^{m-1} f^{(i)}(t)\,\mathrm{d}\mu_i(t).$$

and are linearly independent in the kernel of D^m. Then F^m is a complete inner product *space with the inner product defined as in (+) and has a reproducing kernel given by the formula:*

$$K(s, t) = \sum_{i=1}^{m} c_i(s)c_i(t) + \int_a^b G(s, u)G(t, u)\,du, \quad s, t \in [a, b]$$

where (c_i) *is the dual basis to* (x_i^*) *in the kernel of* D^m, *and* G *is Green's function of the differential equation* $\mathrm{D}^m f(t) = e(t)$ *with* $x_i^*(f) = 0$, $i = 1, 2, \ldots, m$.

Further results, as well as some extensions of the extremal properties of spline functions to general m-order linear differential operators (in normal form), are given in the paper of C. De Boor and R. E. Lynch (the case considered above was of the operator $\mathrm{D}^m = \mathrm{d}^m/\mathrm{d}x^m$).

11.11. Dilation Theory and Reproducing Kernels

In Chapter 10 we presented some results about the dilation of certain classes of operators and we noted an interesting application about the dilation of a single contraction operator. In an appendix to the book on functional analysis by Fr. Riesz *et al.* (1953) the following extension of the fundamental theorem of M. A. Naimark is proved.

THEOREM 11.11.1. *Let* Γ *be a *-semigroup with identity* ε *and let* $(T_s)_{s\in\Gamma}$ *be a family of bounded linear operators on a complete inner product space* H (H *is a complex space) satisfying the following properties:*

(1) $T_\varepsilon = I$.

(2) $(T_s)_{s\in\Gamma}$ *is completely admissible, i.e. for all finite sums*

$$\sum_{i=1}^{n}\sum_{j=1}^{n} \langle T_{s_j^* a^* a s_i} x_i, x_j\rangle \leqslant M_a^2 \sum_{i=1}^{n}\sum_{j=1}^{n} \langle T_{s_j^* s_i} x_i, x_j\rangle$$

for all a *in* Γ *and where* M_a *is a positive constant.*

(3) $s \to T_s$ *is of positive type.*

Then there exists a representation of Γ *in* $L(H)$, $(D_s)_{s\in\Gamma}$, *where* K *is a complete inner product space containing* H *as a closed subspace and such that*

$$T_s = PD_s, s \in \Gamma$$

with P *the orthogonal projection of* K *onto* H.

We recall that Γ is a *-semigroup if there exists a mapping $*: \Gamma \to \Gamma$ with the following properties:

(1) $s^{**} = s$ for all s in Γ.

(2) $(st)^* = t^* s^*$.

(3) $\varepsilon^* = \varepsilon$ (ε is the identity of Γ).

A function $f: \Gamma \to L(H)$, H is a complete inner product space, is said to be of positive type if for all $s_1, \ldots, s_n$ in Γ ($n < \infty$) and all $x_1, \ldots, x_n$ in H the following inequality holds:

$$\sum_{i,j=1}^{n} \langle T_{s_j^* s_i} x_i, x_j \rangle \geqslant 0.$$

In recent years several interesting extensions of Naimark's result have appeared, some of which have been in connection with deep problems of probability theory (see Chapter 10). In what follows we are interested in the dilation theorem that is related to the reproducing kernel theory. Such an approach was proposed by Faulkner and Shonkwiler (1978). We assume that Γ is a *-semigroup that does not necessarily have an identity but has an element, say ε, with the property that $\varepsilon^* = \varepsilon$.

In order to state precisely the content of the dilation theorem we give some definitions. First we consider a complex complete inner product space H with H_0 a dense linear subspace of H. We denote by $L(H_0)$ the set of all linear operators (possibly unbounded) whose domain contains H_0.

DEFINITION 11.11.2. A reproducing kernel complete inner product space in a complete inner product space $\mathscr{H}$ of functions defined on the set $\Gamma, f\colon \Gamma \to H$, such that there is a function $K\colon \Gamma^2 \to L(H_0)$ with the property that for each s in $\mathscr{H}$ and each x in H_0, $K(.,s)x \in \mathscr{H}$ and for g in $\mathscr{H}$ the "reproducing property" g holds:

$$\langle g(s), x \rangle_H = \langle g, K(,, s)x \rangle.$$

The function K is called the reproducing kernel of $\mathscr{H}$.

DEFINITION 11.11.13. The reproducing kernel 'splits' if for all s, t in Q and x in H_0 the following relation holds:

$$K(s,t)x = K(s,\varepsilon)K(\varepsilon,t)x,$$

where ε is the distinguished element of Γ with the property that $\varepsilon^* = \varepsilon$.

DEFINITION 11.11.4. We say that the kernel $\tilde{K}$ is a dilation of the kernel K, $\tilde{K}(s,t) \in L(\tilde{H}_0)$ for all s, t in Γ, if H_0 is a dense linear subspace in the space $\tilde{H}_0$, $\tilde{H}_0$ being a dense linear subspace of a superspace $\tilde{H}$ containing H as a closed subspace and $K(s,t) \in L(H_0)$, $D_{\tilde{K}(s,t)} \supset D_{K(s,t)}$ for all s, t in Γ; and for each x in $D_{K(\varepsilon,s)}$ the following relation holds:

$$K(\varepsilon, s)x = \tilde{P}\tilde{K}(\varepsilon, s)x \qquad (*)$$

where $\tilde{P}$ is the orthogonal projection of H onto H.

The space H is called the dilation space and for simplicity we write relation (*) as follows:

$$K = \tilde{P}\tilde{K}.$$

REMARK 11.11.5. The function $\tilde{K}$ is not necessarily a reproducing kernel function. From (*) it is clear that for all y in H and x in $D_{K(\varepsilon,s)}$ the following relation holds:

$$K(\varepsilon, s)x, y = K(\varepsilon, s)x, y.$$

DEFINITION 11.11.6. We say that the kernel K conjugates if for each s, t in Γ

there exists the adjoint $K^*(s,t)$ of $K(s,t)$ and

$$K^*(s,t) \supset K(s,t).$$

REMARK 11.11.7. If K is a reproducing kernel such that for all s, t in Γ, $D_{K(s,t)}$ is dense, then K conjugates.

The following proposition gives a property of kernels whicn conjugate and split.

PROPOSITION 11.11.8. *Suppose that the kernel K has the following properties:*

(1) *$D_{K(s,t)} = H_0$ is dense in H for all s, t in Γ.*

(2) *K conjugates and splits.*

Then K is of positive type on H_0.

Proof. Let n be a fixed arbitrary integer and let $s_1, \ldots, s_n$ be elements of Q and $x_1, \ldots, x_n$ in H_0. Then we have,

$$\sum_{i,j=1}^{n} \langle K(s_j, s_i)x_i, x_j\rangle = \sum_{i,j=1}^{n} \langle K(s_j, \varepsilon)K(\varepsilon, s_i)x_i, x_j\rangle$$

$$= \sum_{i,j=1}^{n} \langle K(\varepsilon, s_i)x_i, K(\varepsilon, s_j)x_j\rangle =$$

$$= \sum_{i=1}^{n} \|K(\varepsilon, s_i)x_i\|^2 \geqslant 0$$

and the assertion is proved. □

We note also the following proposition, whose proof is obvious.

PROPOSITION 11.11.9. *Suppose that the kernel K satisfies the following properties:*

(1) *K splits,*

(2) *For each s in Q, $K(s, \varepsilon)$ and $K(\varepsilon, s)$ conjugates.*

Then K conjugates.

The setting of the dilation theorem of G. Faulkner and R. Shonkwiler may be described as follows: we have a measure space (W, B_W, μ), μ is a real measure, and the functions

$$c: \Gamma^2{}_{\times} W \to C, \qquad g: \Gamma^2{}_{\times} W \to \Gamma$$

and K is defined as a function on Γ^2 with values in $L(H_0)$ by the following integral:

$$K(s,t) = \int_W c(s^*, t, w)T(g(s^*, t, w))\,\mathrm{d}\mu(w),$$

where $T: \Gamma \to L(H)$ and T has the property that the above integral in the definition of K is absolutely convergent.

In their paper, G. Faulkner and R. Shonkwiler consider four cases represented by different choice of Γ and the measure space as well as the functions c and g. In what follows we consider only two of them, namely the case of the semigroup kernel and the case of the cosine kernel. For the other two cases we refer to the original paper.

CASE 11.11.10 (The semigroup case). Here we consider Γ to be a *-semigroup with no necessary identity but a distinguished element ε with the property that $\varepsilon^* = \varepsilon$, the measure space (W, B_W, μ) will be $W = ((1))$, $\mu(W) = 1$, $c = 1$ and

$$g(s, t, 1) = st.$$

In this case

$$K(s, t) = T(s^*t)$$

and all operators are bounded linear operators on a complete inner product space $H = Y_0$.

CASE 11.11.11 (The cosine case). Here Γ will be the set of real numbers with the semigroup operation, the usual addition, and * is defined by $(s)^* = -s$. The measure space (W, B_W, μ) is as follows: $W = ((-1, 1))$ and the measure μ is defined as follows,

$$\mu((1)) = \tfrac{1}{2}, \qquad \mu((-1)) = \tfrac{1}{2}.$$

The function $c = 1$ and g is defined by the following relations:

$$g(s, t, 1) = t - s, \qquad g(s, t, -1) = s + t.$$

In this case

$$K(s, t) = \tfrac{1}{2}(T(s + t) + T(s - t)).$$

We make an additional assumption, in this case, that the function T has the property that $T(s) = T(-s)$. Of course, here we consider only bounded linear operators on a complete inner product space $H = H_0$. Now the dilation theorem is as follows.

THEOREM 11.11.12 (The kernel dilation theorem: Faulkner and Shonkwiler, 1978). *Suppose that K is as in Cases* 11.11.10 *and* 11.11.11 *and that the following two properties hold:*

(1) $K(\varepsilon, \varepsilon) = I$.

(2) K *is of positive type.*

Then there exists a dilation space H containing H_0 as a closed subspace and a dilated kernel K in H which conjugates and splits.

Proof. We consider the space H_0 of all H-valued functions $u(s)$, $s \in \Gamma$, of the following form:

$$u(s) = \sum_{i=1}^{n} K(s, s_i)x_i, \quad s_i \in \Gamma, x_i \in H, \quad n < \infty \text{ but not fixed.}$$

Then if v is another element of H_0,

$$v(s) = \sum_{j=1}^{m} K(s, t_j)y_j, \quad t_j \in \Gamma, \quad y_j \in H$$

we define

$$\langle u, v\rangle = \sum_{i=1}^{n}\sum_{j=1}^{m} K(t_j, s_i)x_i, \qquad y_j = \sum_{j=1}^{n} u(t_j), \qquad y_i = \sum_{i=1}^{n} x_i, \quad v(s_i). \qquad \square$$

We remark that u, v is independent of the representation of u and v but depends only from the values of u and v, and $(u, v) \to \langle u, v\rangle$ is positive, sesquilinear and symmetric. We show now that $\langle u, u\rangle = 0$ if and only if $u = 0$. Indeed, if $\langle u, u\rangle = 0$ then, using the Cauchy–Buniakowsky inequality we get

$$|\langle u(s), y\rangle|^2 = |\langle u, K(\ ,s)y\rangle|^2 \leqslant \langle u, u\rangle\langle K(\ ,s)y, K(\ ,s)y\rangle = 0.$$

Since this holds for any y in H we must have $u(s) = 0$, and since s is arbitrary the assertion is proved. Thus $\langle u, v\rangle$ is an inner product space on H_0. In general, H_0 is not a complete space and thus we can consider the completion which we denote by $\tilde{H}$. We now show that $\tilde{H}$ can be realized as a space of functions on Γ. Indeed, if (u_n) is a Cauchy sequence in H_0 which converges to the ideal element $((u_n))$ then we can consider, first, s to be a fixed point in Γ and for y in H_0 we have

$$F_0^s(y) = \lim\langle y, u_n(s)\rangle = \lim\langle K(\ ,s)y, u_n(s)\rangle = \langle K(\ ,s)y, ((u_n))\rangle.$$

Then F_0 is a bounded linear functional on H_0 and thus has a unique extension to a bounded linear on H. Then there exists a unique element $u(s)$ in H with the property that

$$\langle y, u(s)\rangle = F_0^s(y) = \lim\langle y, u_n(s)\rangle, \quad y \in H.$$

If (u_n^1) is another Cauchy sequence converging to $((\mu_n))$ then

$$\lim\langle K(.\,, s)y, u_n^1(s)\rangle = \langle K(.\,, s)y, \qquad ((u_n^1))\rangle = \langle K(.\,, s)y, ((u_n))\rangle$$

and thus $u(s)$ is independent of the Cauchy sequence associated with the ideal element of $\tilde{H}$.

Thus, for each ideal element $((u_n))$, we can choose a function $s \to u(s)$ on Γ and this clearly can be organized as an inner product space which is isometric to $\tilde{H}$. We have that K is a reproducing kernel for this space. Indeed,

$$\langle ((u_n)), k(.\,, s)y\rangle = \langle u(\,,), K(.\,, s)y\rangle = \langle u(s), y\rangle$$

and the assertion is proved.

If we define $x \to K(\cdot\,\varepsilon)x$, then H may be embedded in H if we identify x with $K(.,.,\varepsilon)x$ which is possible since

$$\langle K(t,\varepsilon)x, K(t,\varepsilon)\,y\rangle = \langle K(\varepsilon,\varepsilon)x, y\rangle = \langle x, y\rangle.$$

Now the orthogonal projection of H onto $\tilde{H}_0$ is given by the formula:

$$PK(s,t)x = K(s,\varepsilon)K(\varepsilon,t)x.$$

We remark also that for $K(s,\varepsilon)y \in H$ we have the relation

$$\langle K(s,t)x, K(s,\varepsilon)K(\varepsilon,t)x, K(s,\varepsilon)\,y\rangle = \langle K(\varepsilon,t)x, y - K(\varepsilon,\varepsilon)K(\varepsilon,t)\,y\rangle = 0.$$

The dilated kernel of K is defined as follows:

$$\tilde{K}(\varepsilon,t)K(u,v)x = \int_W c(t,v,w)K(u,g(t,v,w))x \, \mathrm{d}\mu(w)$$

and

$$K(s,t)K(u,v)x = \int_W c(s^*,t,w)K(u,g(s^*,t,w))\,K(u,v)\,\mathrm{d}\mu(w).$$

First we remark that $K(s,t)$ is well defined if we suppose that K conjugates. Indeed, if $h(p)$ is a representation of zero,

$$h(p) = \sum_{i=1}^{n} K(p,s_i)x_i$$

then $k(p) = K(p,m)y$ gives that

$$\langle K(s,t)h,k\rangle = \sum_{i=1}^{n} \langle K(s,t)K(p,s_i)x_i, K(p,q)\,y\rangle = \sum_{i=1}^{n} \langle K(p,s_i)x_i, K^*(s,t)K(p,q)\,y\rangle$$
$$= \langle h(p), K(t,s)K(p,q)\,y\rangle = 0.$$

This relation holds for all k in $\tilde{H}_0$ and thus $\tilde{K}(s,t)h(\ ,\ ) = 0$ which implies our assertion.

Let us now suppose that we have a semigroup kernel. Then

$$\tilde{K}(\varepsilon,t)K(u,v)x = K(u,tv)x$$

and

$$\tilde{K}(s,t)K(u,v)x = \tilde{K}(\varepsilon,s^*t)K(u,v)x.$$

But

$$\tilde{K}(s,\varepsilon)K(u,v)x = K(u,s^*v)x$$

and thus the definition of $\tilde{K}(s,t)$ is correct.

Now the fact that the kernel conjugates, splitting follows from the relations

$$\tilde{K}^*(\varepsilon,t) \supset K(t,), \qquad \tilde{K}(\varepsilon,s) \subset \tilde{K}^*(s,\varepsilon);$$

and these relations follow from the relations

$$\langle \tilde{K}(\varepsilon,t)K(u,v)x, K(u,p)\,y\rangle = \langle K(u,v)x, \tilde{K}(t,\varepsilon)K(u,p)\,y\rangle$$

with $K(\varepsilon,\mathrm{t}) = T(t)$.

Thus, since

$$\begin{aligned}\tilde{P}\tilde{K}(s,t)K(u,\varepsilon)x &= \tilde{P}(\tilde{K}(\varepsilon,s^*t)K(u,\varepsilon)x)\\ &= \tilde{P}K(u,s^*t\varepsilon)x = K(u,\varepsilon)K(\varepsilon,s^*t)x\\ &= K(u,\varepsilon)T(s^*t)x = K(u,\varepsilon)K(s,t)x\end{aligned}$$

the projection property is established.

Let us now suppose that we have a cosine kernel. Then the dilated kernel is

$$K(0,t)K(u,v) = \tfrac{1}{2}K(u,-t+v)x + \tfrac{1}{2}K(u,t+v)$$

and

$$K(s,t)K(u,v)x = \tfrac{1}{2}K(0,s+t)K(u,v)x + \tfrac{1}{2}K(0,-s+t)K(u,v)x$$

with

$$K(s,0)K(u,v)x = \tfrac{1}{2}K(u,s+v)x + \tfrac{1}{2}K(u,s+v)x.$$

Since $K(0,t) = T(t)$ the splitting and conjugacy properties hold, and from

$$\begin{aligned} PK(0,t)K(u,0)x &= \tfrac{1}{2}P(K(u,-t)x + K(u,t)x) \\ &= \tfrac{1}{2}K(u,0)K(0,-t)x + \tfrac{1}{2}K(u,0)K(0,-t)x \\ &= K(u,0)K(0,t)x \end{aligned}$$

the projection property of the dilated kernel is proved.

For cosine representations of abelian *-semigroups of operators we refer to Faulkner and Shonkwiler (1978).

11.12. Some Applications of Reproducing Kernels

As mentioned above, the reproducing kernel theory has important and useful applications; in this section we present other applications. First we mention the work of E. Parzen who used the spaces with reproducing kernels in an ingenious way in several problems of time series analysis (especially to the locally minimum variance of unbiased estimations, to probability densities of normal random functions, etc.). We now present very briefly some aspects of these and for further details we refer to the work of E. Parzen.

We consider a stochastic process $(x_t, t \in T)$ (T is in the most useful case R or $(0, \pm 1, \ldots)$) with the property that for each t in T, $E(|x_t|^2) < \infty$. We define $L^2(x_t)$ as the smallest closed linear subspace of $L^2 = L^2(\Omega, \mathcal{B}, P)$ containing all the functions x_t.

The covariance function of the stochastic process (x_t) is the function R defined by the relation

$$R(s,t) = E(x_t x_s).$$

Then we have the following result.

LEMMA 11.12.1. *The function R is positive definite.*

DEFINITION 11.12.2. The process (x_t) is said to be strongly continuous if

$$\lim_{s \to t} E(|x_t - x_s|^2) = 0,$$

where T is suppose to be with a metric d.

We suppose now that T is R.

DEFINITION 11.12.3. The stochastic process (x_t) has a strong derivative at t_0 if

$$\lim_{h \to 0} E(h^{-1}(x_{t_0+h} - x_{t_0}), h^{-1}(x_{t_0+h} - x_{t_0}))$$

exists.

We now mention a famous result obtained by Mercer: suppose that K is a continuous symmetric ($K(s,t) = K(t,s)$) and nonnegative function on I^2,I is a compact interval. Then there exists a sequence of normalized eigenfunctions (φ_n) and a sequence (λ_n) of positive numbers such that

$$(1)\ \int_I K(s,t)\varphi_n(t)\mathrm{d}s = \lambda_n \varphi_n(t),$$

$$(2)\ \int_I \varphi_n(t)\,\varphi_m(t)\,\mathrm{d}t = \delta_{nm},$$

$$(3)\ K(s,t) = \sum_{i=1}^{\infty} \varphi_i(t)\,\varphi_i(s)$$

where the series in (3) converges uniformly and absolutely on I. Let (x_t) be a stochastic process with the property that R is a kernel, as in Mercer's theorem. Then we consider the space of all real-valued sequences $a = (a_n)$ with the property that

$$\sum_{n=1}^{\infty} |a_n^2| \lambda_n < \infty$$

and if $b = (b_n)$ is another sequence then we set

$$\langle a, b \rangle = \sum_{n=1}^{\infty} a_n b_n \lambda_n$$

and as this clearly defines an inner product, the space is then a complete inner product space, denoted by H.

It is also clear that for each t,

$$\varphi(t) = (\varphi_n(t))$$

is in H and

$$R(s,t) = \langle \varphi(s), \varphi(t) \rangle.$$

We now consider the space $L^2(\varphi_n)$ and then we set

$$\varphi(t) \in L^2(\varphi_n) \to \chi_t \in L^2(x_t)$$

which is an isomorphism (in fact an isometric isomorphism). We now consider for each n the element in H defined by

$$e_n = (\underbrace{0, \ldots, 0}_{n\text{-places}}, 1, 0, \ldots)$$

and ξ_n is the image of this element through the above isomorphism. Then clearly

$$E(\xi_n \xi_m) = \chi_m \delta_{nm}.$$

We now consider the space $L^2(\xi_n)$. We define the stochastic integral

$$\int_I g(s)x_s\,\mathrm{d}s$$

for functions of the form

$$g(s) = \sum_{n=1}^{\infty} \lambda_n a_n \varphi_n(s)$$

by the formula

$$\int_I g(s)x_s\,\mathrm{d}s = \sum_{n=1}^{\infty} a_n \xi_n.$$

It is obvious that the set of all functions g of the above form is a linear space and we can introduce an inner product on this space as follows. If h is another function in the space

$$h(s) = \sum \lambda_n b_n \varphi_n(s)$$

then we set

$$\langle g,h \rangle = \sum \lambda_n a_n b_n.$$

It is not difficult to see that this is indeed an inner product. The space with this inner product is denoted by $H(R)$ and has two interesting properties, namely:

(1) for each t in $I, R(\ ,t)$ is in $H(R)$,
(2) for each f in $H(R), f, R(,s) = f(s)$.

This means in fact that $H(R)$ is a space with a reproducing kernel.

We now mention the basic problem of statistical inference. This is the problem of least squares linear prediction.

Let (x_t) be a stochastic process and let z be a random variable, which we assume to be in L^2, and the problem is to find a random variable in $L^2(x_t)$ whose distance from z is smallest, i.e. to predict the values of z be observing the values of the process (x_t).

The existence of such a random variable follows from the general results on the orthogonal projection in complete inner product spaces. In the case of an inner product space with a reproducing kernel the projection has a neat formula. We refer to Parzen (1969) where the reader may find other interesting results about the applications of $H(R)$ considered above, as well as other spaces.

We now mention another aspect of applications in which the point of view of a statistician and that of numerical analyst are to some extent unified. Generally speaking, working with numerical data we are looking for a function of which we know only some values, but we wish to interpolate. This problem has two aspects: we must

(1) construct an estimation for the unknown value; and
(2) determine how accurate the estimated value is likely to be.

In numerical analysis the problem is treated by considering that the function is algebraic, and the problem is then reduced to determine certain parameters. In statistics, on the other hand, we consider a joint probability distribution and then try to determine the conditional distribution for the unknown value.

In 1972 F. M. Larkin proposed a method in which both approaches are considered simultaneously by regarding the interpolating function as a random variable in some L^2. His paper was inspired by the seminal work of Sul'din (1959, 1960) in which the minimum variance estimation of values of linear functionals on a Wiener space was considered. For details we refer to Larkin (1972) and the paper by Kuelbs *et al.* (1972).

CHAPTER 12

Inner Product Modules

12.0. Introduction

The notion of inner products with values in C*-algebras (commutative C*-algebras with identity) was first considered by I. Kaplansky in 1951 and the results were applied to obtain a solution of some problems in operator theory, namely, the structure of derivations on AW*-algebras. Further, the idea introduced by Kaplansky has attracted the attention of many people who have developed the theory of inner products with values in B*-algebras for modules over B*-algebras and with values in spaces with an involution (the so-called VH-spaces).

12.1. Inner Product Modules. Definition and Some Examples. Bounded Module Maps

Let B be a B*-algebra and let $\mathcal{M}$ be a module over B.

DEFINITION 12.1.1. By a right module $\mathcal{M}$ over B we mean an additive abelian group $\mathcal{M}$ together with a function

$$m: B_{\times}\mathcal{M} \to \mathcal{M}$$

from the Cartesian product $B_{\times}\mathcal{M}$ into $\mathcal{M}$, that satisfies the following three axioms:

$$\begin{aligned} m(x_1 b_1 + b_2) &= m(x, b_1) + m(x, b_2) \\ m(x_1 + x_2, b) &= m(x_1, b) + m(x_2, b) \end{aligned} \qquad (\mathrm{m}_1)$$

holds for all b, b_1, b_2 in B and x, x_1, x_2 in $\mathcal{M}$,

For arbitrary b_1, b_2 in B and any $x \in \mathcal{M}$,

$$m(m(x, b_1), b_2) = m(x, b_1 b_2), \qquad (\mathrm{m}_2)$$

For every element $x \in \mathcal{M}$,

$$m(x, 1) = x \qquad (\mathrm{m}_3)$$

where 1 is the identity of B.

The function m is called the scalar multiplication of the (right) module $\mathcal{M}$. For each $b \in B$ and each $x \in \mathcal{M}$ the element $m(b, x)$ is called the scalar product of x by b and will be denoted by xl. Then we note that the above axioms may be written as follows:

$$x(b_1 + b_2) = xb_1 + xb_2 \tag{m$_1$}$$

$$(x_1 + x_2)b = x_1b + x_2b$$

$$(xb_1)b_2 = x(b_1b_2) \tag{m$_2$}$$

$$x \cdot 1 = x = x1 \tag{m$_3$}$$

DEFINITION 12.1.2. A submodule $\mathcal{M}_1$ of the module $\mathcal{M}$ over B is any nonempty set of $\mathcal{M}$ which is itself a module over B relative to the addition and the scalar multiplication of the module $\mathcal{M}$. This means that the submodule $\mathcal{M}_1$ is a subgroup of $\mathcal{M}$ which is invariant under the scalar multiplication of the module $\mathcal{M}$.

We give now some examples.

EXAMPLE 12.1.3. If B_1 is any subalgebra of B containing the identity of B then m: $B_\times B_1 \to B$ defined by

$$m(x, b_1) = xb_1$$

satisfies all the above axioms. Then we get that any B*-algebra is a module over any of its subalgebra containing the identity.

EXAMPLE 12.1.4. Let S be a compact Hausdorff space and let B be a B*-algebra with identity denoted by 1. Let $\mathscr{C}(S)$ be the set of all continuous B-valued functions defined on S. Then $\mathscr{C}(S)$ is a module over B: the addition is defined in the usual way,

$$(f + g)(s) = f(s) + g(s)$$

and the multiplication m: $\mathscr{C}(S)_\times B \to \mathscr{C}(S)$ by the relation

$$m(b, f)(s) = f(s)b.$$

Then it is not difficult to see that all the axioms of the (right) module over B are satisfied.

REMARK 12.1.5. Analogously we can define the left modules over B. We note that in the case of a commutative B*-algebra the notions of right and left modules over B coincide with each other.

DEFINITION 12.1.6. If $\mathcal{M}$ and $\mathcal{N}$ are two (right) modules over B then any function $f: \mathcal{M} \to \mathcal{N}$ which is a homomorphism of the additive abelian group of $\mathcal{M}$ into the additive abelian group of $\mathcal{N}$ and preserves the scalar multiplication is called a homomorphism.

Thus if f is a homomorphism we have

$$f(x + y) = f(x) + f(y)$$

$$f(xb) = f(x)b$$

and this is clearly necessary and sufficient for f to be a homomorphism. The homomorphism is called monomorphism if it is injective and is said to be an epimorphism if it is surjective. A bijective (i.e. monomorphism and epimorphism) homomorphism is called an isomorphism.

If $\mathcal{M} = \mathcal{N}$ then the homomorphism is called endomorphism and any isomorphism is called an automorphism.

We now consider the notion of a pre-inner product module over a B*-algebra.

DEFINITION 12.1.7. If $\mathcal{M}$ is a right module over a B*-algebra B with the identity denoted by 1, then a B-inner product is any conjugate bilinear function $\langle\, ,\rangle$ defined on $\mathcal{M}_\times\mathcal{M}$ with values in B such that the properties

(1) $\langle x, x\rangle \geqslant 0, \quad \langle x, x\rangle = 0$ if and only if $x = 0$,
(2) $\langle x, y\rangle = \langle y, x\rangle^*$,
(3) $\langle xb, y\rangle = \langle x, y\rangle b$

hold for all x, y in $\mathcal{M}$ and $b \in B$.

Any module $\mathcal{M}$ with an inner product with the above properties is called a pre-inner product module.

EXAMPLE 12.1.8. If B is any B*-algebra with identity and B_1 is any B*-subalgebra of B containing the identity of B then B_1 can be considered as a right module over B and an inner product on B_1 can be defined by the formula

$$[x, y] = y^*x$$

and it is easy to see that all the above axioms are satisfied.

Let $\mathcal{M}$ be a pre-inner product B-module and we define for each $x \in \mathcal{M}$ the number $\|x\|_{\mathcal{M}} = \|\langle x, x\rangle\|^{1/2}$. Then we have the following result.

PROPOSITION 12.1.9. *The function* $x \to \|x\|_{\mathcal{M}}$ *is a norm on* $\mathcal{M}$ *and satisfies the following properties:*

(1) $\|xb\|_{\mathcal{M}} \leqslant \|x\|_{\mathcal{M}}\|b\|, \quad x \in \mathcal{M}, \quad b \in B,$
(2) $\langle y, x\rangle\langle x, y\rangle \leqslant \|y\|^2_{\mathcal{M}}\langle x, x\rangle, \quad x, y \in \mathcal{M},$
(3) $\|\langle x, y\rangle\| \leqslant \|x\|_{\mathcal{M}}\|y\|_{\mathcal{M}}, \quad x, y \in \mathcal{M}.$

Proof. If f is any positive functional on B then $x \to f(\langle x, x\rangle)^{1/2}$ is a pseudonorm, and since □

$$\|x\|_{\mathcal{M}} = \sup(f(\langle x, x\rangle)^{1/2}, f \text{ a state of } B)$$

we get that $x \to \|x\|_{\mathcal{M}}$ is a pseudonorm on $\mathcal{M}$. From property (1) of the inner product on $\mathcal{M}$ we get that it is a norm.

Now, if $x \in \mathcal{M}$ and $b \in B$, we have

$$\|xb\|^2_{\mathcal{M}} = \|\langle xb, xb\rangle\| = \|b^*\langle x, x\rangle b\| \leqslant \|b\|^2\langle x, x\rangle$$

and this proves (1).

To prove (2) we take x, y in $\mathcal{M}$ and f a positive linear functional on B. Since for each positive linear functional the Cauchy–Buniakowsky inequality holds, we get,

$$\begin{aligned} f(\langle y, x\rangle\langle x, y\rangle) &= f(\langle y\langle x, y\rangle, x\rangle) \leqslant f(\langle y\cdot\langle x, y\rangle, y\cdot\langle x, y\rangle\rangle)^{1/2} f(\langle x, x\rangle)^{1/2} \\ &\leqslant f(\langle y, x\rangle\langle y, y\rangle\langle x, y\rangle)^{1/2} f(\langle x, x\rangle)^{1/2} \\ &\leqslant \|\langle y, y\rangle\|^{1/2} f(\langle y, x\rangle\langle x, y\rangle)^{1/2} f(\langle x, x\rangle)^{1/2} \end{aligned}$$

and thus

$$f(\langle y,x\rangle(x,y\rangle) \leqslant \|y\|_{\mathcal{M}}^2 f(\langle <x,x\rangle)^{1/2}.$$

But f is an arbitrary positive linear functional on B and this implies (2).

For (3) we note that since

$$f(\langle y,x\rangle\langle x,y\rangle) \leqslant \|y\|_{\mathcal{M}}^2 f(\langle x,x\rangle)^{1/2}$$

is true for any positive linear functional on B, taking the supremum over all states of B, assertion (3) follows. □

DEFINITION 12.1.10. A pre-inner product module $\mathcal{M}$ is said to be an inner product module (or a Kaplansky module) if it is complete under the norm $x \to \|x\|_{\mathcal{M}}$.

Following the standard procedure of completion we get that any pre-inner product module may be densely embedded into a complete inner product module.

Now we consider a class of operators representing, in some sense, the analogue of the set of all bounded linear operators on a complete inner product space. Let $\mathscr{a}(\mathcal{M})$ be the set of all bounded linear operators on $\mathcal{M}$ such that, for each $T \in \mathscr{a}(\mathcal{M})$ there exists an element $T^* \in \mathscr{a}(\mathcal{M})$ with the property that for all x, y in $\mathcal{M}$,

$$\langle Tx, y\rangle = \langle x, T^*y\rangle.$$

LEMMA 12.1.11. *$\mathscr{a}(\mathcal{M})$ is a B^*-algebra under the involution defined via $*: T \to T^*$.*

Proof. It is easy to see that the element T^* with the above property is unique and that $\mathscr{a}(\mathcal{M})$ is closed in $L(\mathcal{M})$. Also for each $T \in \mathscr{a}(\mathcal{M})$ we have,

$$\|T^*T\|_{\mathcal{M}} = \|T\|_{\mathcal{M}}^2.$$

The following property shows a connection of $\mathscr{a}(\mathcal{M})$ with the module structure of $\mathcal{M}$. □

LEMMA 12.1.12. *Every $T \in \mathscr{a}(\mathcal{M})$ satisfies the property*

$$T(xb) = (Tx)b$$

for all $x \in \mathcal{M}$ and $b \in B$.

Proof. Let y be arbitrary in $\mathcal{M}$ and thus

$$\langle T(xb), y\rangle = \langle xb, T^*y\rangle = \langle x, T^*y\rangle b = \langle Tx, y\rangle b = (Tx)b, y\rangle$$

and the assertion is proved. □

EXAMPLE 12.1.13. We show that in general not all elements of $L(\mathcal{M})$ are in $\mathscr{a}(\mathcal{M})$, even in complete inner product modules.

Let B be an arbitrary B*-algebra with identity and let J be a closed right ideal of B with the property that no elements of $J^* = (j^*, j \in J)$ acts as a left multiplicative identity on J. We consider the right module $J \cap B$ with the inner product defined as follows:

$$\langle (j_1, b_1), (j_2, b_2)\rangle = j_2^* j_1 + b_2^* b_1$$

for all j_1, j_2 in J and b_1, b_2 in B. Since for all $(j, b) \in \mathcal{M}$ we have

$$\|(j, b)\|_{\mathcal{M}} = \|(j^*j + b^*b)^{1/2}\|$$

$$\max(\|j\|, \|b\|) \leqslant \|(j, b)\|_{\mathcal{M}} \leqslant (\|j\|^2 + \|b\|^2)^{1/2}$$

we obtain that $\mathcal{M}$ is complete with respect to $\|,\|_{\mathcal{M}}$.

Define now $T: \mathcal{M} \to \mathcal{M}$ by the formula

$$T(j, b) = (0, b)$$

which is clearly a module map. We show that it is not in $\mathscr{a}(\mathcal{M})$.

Suppose that the contrary assertion is true, and thus let T^* be the element satisfying the relation,

$$\langle Tx, y \rangle = \langle x, T^*y \rangle$$

for all x, y in $\mathcal{M}$. (The element T^* with this property is called the adjoint of T.) Then $T^*(0, 1) = (g^-, b^-)$ and thus for any (j, b) in we have

$$j = \langle T(j, b), (0, 1) \rangle = \langle (J, b), (g^-, b^-) \rangle$$

which implies that $b^- = 0$. But $(g^-)^* \, j = j$ for all $j \in J$. Since j^* is the J^* this contradicts the defining property of J. Thus Tconsidered above is not in $\mathscr{a}(\mathcal{M})$. We note that $\mathscr{a}(\mathcal{M})$ contains all operators of the form $Tx = xb$ for some b in B. As an algebra, B and J can be as follows: B is the algebra of all continuous functions on $[0, 1]$ and J is the ideal of all functions in B which vanish at $t = 0$.

LEMMA 12.1.14. *If $T \in \mathscr{a}(\mathcal{M})$ then for all $x \in \mathcal{M}$,*

$$\langle Tx, Tx \rangle \leqslant \|T\|^2_{\mathcal{M}} \langle x, x \rangle.$$

Proof. Let x be an arbitrary element of $\mathcal{M}$ and let f be a positive linear functional on B. Then we get, using the Cauchy–Buniakowsky inequality for positive functionals,

$$\begin{aligned} f(\langle Tx, Tx \rangle) &= f(\langle T^*Tx, x \rangle) \leqslant f(\langle T^*Tx, T^*Tx \rangle)^{1/2} f(\langle x, x \rangle)^{1/2} \\ &\leqslant f(\langle (T^*T)^2 x, (T^*T)^2 x \rangle)^{1/4} f(\langle x, x \rangle)^{1/2 + 1/4} \leqslant \cdots \\ &\leqslant f(\langle (T^*T)^n_{x}, (T^*{}_{-n}^{T})^n_x \rangle)^{2^{-n}} f(\langle x, x \rangle)^{1/2 + \cdots + 1/2^n} \\ &\leqslant (\|f\| \|x\|^2)^2 \|T\|^2_{\mathcal{M}} f(\langle x, x \rangle)^{1/2 + \cdots + 1/2^n} \end{aligned}$$

and for $n \to \infty$ we get

$$f(\langle Tx, Tx \rangle) \leqslant \|T\|^2_{\mathcal{M}} f(\langle x, x \rangle)$$

which implies the assertion.

We now give a characterization of module maps, i.e. if $\mathcal{M}$ and $\mathcal{N}$ are two right modules over B and A respectively, $T: \mathcal{M} \to \mathcal{N}$ is a module homomorphism. The characterization gives another formulation in terms of boundedness.

THEOREM 12.1.15. *For any linear map $T: \mathcal{M} \to \mathcal{N}$ the following conditions are equivalent:*

(1) *T is bounded and $T(xb) = (Tx)b$,*

(2) *there is* $k \geqslant 0$ *such that* $\langle Tx, Tx\rangle_A \leqslant k\langle x, x\rangle_B$

for all x *in* $\mathcal{M}$ *and* $b \in B$.

For the proof we need the following lemma.

LEMMA 12.1.16. *Let* A *and* B *be two* B^**-algebras and we suppose that* B *is a closed* $*$*-subalgebra of* A *and that the identity of* A *is also the identity of* B. *If* $\tau: B \to A$ *is a linear map such that for some* $k \geqslant 0$ *and all* $x \in B$,

$$\tau(x)^*\tau(x) \leqslant kx^*x$$

then $\tau(x) = (1)x$. *Here* 1 *denotes the identity element of* A(*and* B).

Proof. Let x in B and thus $x^*\tau(1)^*\tau(1)x \leqslant \|\tau(1)\|^2\, x^*x = kx^*x$ which implies that

$$-(x^*\tau(1)^*\tau(x) + \tau(x)^*\tau(1)x) \leqslant \tau(x)^*\tau(x) + x^*\tau(1)^*\tau(1)x \leqslant 2kx^*x.$$

Then we get

$$(\tau(x) - \tau(1)x)^*(\tau(x) - \tau(1)x) \leqslant 2kx^*x - (x^*\tau(1)^*\tau(x) + \tau(x)^*\tau(1)x) \leqslant 4kx^*x$$

and we set $\tau_1: B \to A$ by $\tau_1(x) = (2k^{1/2})^{-1}\,(\tau(x) - \tau(1)x)$. We remark that $\tau_1(1) = 0$ and $\tau_1(x)^*\tau_1(x) \leqslant x^*x$ and we show that this implies that $\tau_1 = 0$. Since A is a B*-algebra we may assume that $A = L(H)$ for some complete inner product space H and B is a subalgebra of $L(H)$ with the identity operator in B. Then from our property of τ_1 we have for all $T \in L(H)$, $\tau_1(T)^*\tau_1(T) \leqslant T^*T$. This gives that τ_1 extends to a linear map, denoted by the same notation, from B'' which is the strong closure of B in $L(H)$ and the above inequality remains true for T in B''. Let P be any projection in B'' and thus $\tau_1(P)^*\tau_1(P) \leqslant P$ and also $\tau_1(I-P)^*\tau_1(I-P) \leqslant I - P$. These relations gave that $\tau_1(P) = 0$. Since B'' is a W*-algebra, it is equal to the linear closure of the projections in B'' and thus $\tau_1 = 0$ which is the assertion of the lemma.

Now the proof of Theorem 12.1.15 is as follows. Suppose first that (1) holds, i.e. T is a linear and bounded mapping with the property that $T(xb) = (Tx)b$. We may suppose without loss of generality that $\|T\| \leqslant 1$. Let x be an arbitrary point in $\mathcal{M}$ and consider the sequence $((\langle x, x\rangle_B + n^{-1})^{-1/2}) = (u_n)$. Then the sequence $(x \cdot u_n)$ satisfies the following property:

$$\langle x \cdot u_n, x \cdot u_n\rangle = ((\langle x, x\rangle_B \langle x, x\rangle_B + n^{-1})^{-1} \leqslant 1.$$

Then, since T is contractive,

$$\langle T(x \cdot u_n),\ T(x \cdot u_n)\rangle \leqslant 1$$

which implies that for all n,

$$\langle T(x \cdot u_n), T(x \cdot u_n)\rangle_A \leqslant 1.$$

Now

$$\langle T(x \cdot u_n), T(x \cdot u_n)\rangle \leqslant u_n \langle Tx, Tx\rangle_A u_n$$

which implies that

$$\langle Tx, Tx\rangle_A \leqslant u_n^{-2} = \langle x, x\rangle_B + n^{-1}$$

and for $n \to \infty$ this gives

$$\langle Tx, Tx \rangle_A \leqslant \langle x, x \rangle_B.$$

Suppose now that (2) holds. We suppose that the constant $k = 1$ and thus

$$\langle Tx, Tx \rangle_A \leqslant \langle x, x \rangle_B.$$

This implies that T is a contraction mapping. Let $x \in \mathcal{M}$ and $y \in \mathcal{N}$ and set $\tau: B \to A$ by the relation

$$\tau(b) = \langle T(x \cdot b), y \rangle_A.$$

From the properties of the norms of the inner product modules, we get

$$\begin{aligned}\tau(b)^*\tau(b) &= \langle y, T(x \cdot b)_A T(x \cdot b), y \rangle_A \\ &\leqslant \|y\|^2 \langle T(x \cdot b), T(x \cdot b) \rangle_A \leqslant \|y\|^2 \langle x \cdot b, x \cdot b \rangle_B \\ &\leqslant \|y\|^2 b^* \langle x, x \rangle_B b \leqslant \|y\|^2 \|x\|^2_{\mathcal{M}} b^* b\end{aligned}$$

and thus τ satisfies the conditions of Lemma 12.1.16. Then $\tau(b) = (1)b$ which implies that

$$\langle T(x \cdot b), y \rangle_A = \langle Tx, y \rangle_A b = \langle (Tx) \cdot b, y \rangle_A$$

and this is equivalent to (1) since x, y are arbitrary.

We now present some results obtained by G. G. Kasparov concerning inner product modules.

DEFINITION 12.1.17 (Kasparov, 1980a). Let $(X_i)_{i \in I}$ be a family of inner product modules over the B*-algebra A (with identity). The direct sum $\Sigma_i \oplus X_i$ is the completion of an algebraic direct sum with respect to a norm generated by an A-valued inner product

$$[x, y] = \sum_i [x_i, y_i],$$

where

$$x = \sum_i \oplus x_i, \; y = \sum_i \oplus y_i$$

with x_i, y_i in X_i.

For each finite integer n we set

$$\sum_{i \leqslant n} \oplus X_i = X^n, \quad X_i = X, i = 1, 2, 3, \ldots, n.$$

DEFINITION 12.1.18. Let A be an arbitrary Banach algebra. The double centralizer algebra $M(A)$ is the set of all pairs of mappings $T_i: A \to A$ with the following property: for each x, y in A we have the relation

$$xT_1 y = T_2(x) \cdot y.$$

If A has an involution then we can define an involution on $M(A)$ by the relation:

$$\bar{T}_i(a) = (T_i a)^*, \quad i = 1, 2.$$

Let (X_i), $i = 1, 2$, be complete inner product modules over the B*-algebra A with identity. Let x in X_1 and y in X_2. For each z in X_2 we set

$$\theta_{x,y}(z) = x \cdot [y, z]$$

and clearly this is in (X_2, X_1) and the adjoint is

$$\theta^*_{x,y} = \theta_{y,x}.$$

Further, if $u \in X_3$, X_3 being another inner product module over A, and T is in $\mathscr{a}(X_2, X_1)$, S is in $\mathscr{a}(X_3, X_2)$ we have

$$T_0\theta_{u,v} = \theta_{Tu,v}, \qquad \theta_{x,y\,0}S = \theta_{x,S^*y}.$$

DEFINITION 12.1.19 (Kasparov, 1980a). The closure of the linear span of $(\theta_{x,y}\colon x \in X_1, y \in X_2)$ in (X_2, X_1) is denoted by $K(X_2, X_1)$. Also we use the notation $K(X) = K(X, X)$.

Then we have the following result.

THEOREM 12.1.20 (Kasparov, 1980a). If X is an arbitrary complete inner product module over the B*-algebra A with identity then there exists an isomorphism between $\mathscr{A}(X)$ and $M(K(X))$ given by $T \to (T_1, T_2)$, where

$$T_1(\theta_{x,y}) = \theta_{Tx,y}, \; T_2(\theta_{x,y}) = \theta_{x,T^*y}.$$

Proof. For each S in the linear span of $(\theta_{x,y})$ and each T in (X) we have clearly the relations:

$$T_1 = TS, \quad T_2 = ST$$

and thus

$$\|T_i\| \leqslant \|T\|\,\|S\|, \quad i = 1, 2,$$

which implies that T_i can be extended to $K(X)$ and (T_1, T_2) is in $M(K(X))$. Further, if

$$T_1(\theta_{x,y}) = 0$$

for all x, y in X then $\theta_{Tx,Tx} = 0$ which implies that $T = 0$. Thus the above mapping is injective. To prove surjectivity we take (T_1, T_2) in $M(K(X))$ and for each x in X we set

$$Tx = \lim_{t\to 0^+} T_1(\theta_{x,x})(x)([x, x] + t^{-1},$$

$$Tx = \lim_{t\to 0^+} T_2(\theta_{x,x})(x)([x, x] + t)^{-1}.$$

Since

$$(T_1(S)^*)T_1(S) \leqslant \|T_1\|^2 S^*S$$
$$T_2(S)(T_2(S))^* \leqslant \|T_2\|^2 SS^*$$

for all S in $K(X)$ we obtain that T and T^* are well defined. But for each x in X

$$\lim \theta_{x,x}(x)([x, x] + t)^{-1} = x$$

and then

$$\begin{aligned}[x, T^*y] &= \lim_{t\to 0^+} [x, (T_2(\theta_{y,y}))(y)([y,y]+t)^{-1} \\ &= \lim_{t\to 0^+} [T_2(\theta_{y,y})\theta_{x,x}([x,x]+t)^{-1}, ([y,y]+t)^{-1}] \\ &= \lim_{t\to 0^+} [\theta_{y,y}\cdot T_1(\theta_{x,x})(x)([x,x]+t)^{-1}, y([y,y]+t)^{-1}] \\ &= \lim_{t\to 0^+} [T_1(\theta_{x,x})(x)([x,x]+t)^{-1}, \theta_{y,y}([y,y]+t)^{-1}] = [Tx, y].\end{aligned}$$

Then T is in $\mathscr{A}(X)$. If we set for z in X, $w = T_1(\theta_{x,y}(z)) - \theta_{Tx,y}(z)$ we get that $T_1(\theta_{x,y}) = \theta_{Tx,y}$ and similarly we get the relation $T_2(\theta_{x,y}) = \theta_{x,Ty}$ and the theorem is proved.

In order to present the second result of Kasparov we need some notions which may be known to the reader. We present them briefly.

Let G be a compact second countable group and we suppose that this acts as a group of automorphisms on A; we suppose further that the action of G is continuous in the sense that $G_\times A \to A$ defined by the relation $(g, a) \to g(a)$ is norm continuous. A linear map on A is said to be equivariant if it preserves the G-actions and the real involution, i.e. we have on g the involution $g \to \bar{g}$ and $g(a)^* = \bar{g}(a)$ for all a in A. We consider (E_i) to be a countable family of finite dimensional spaces with the Euclidean norm and with a unitary G-action such that every finite dimensional unitary representation of G occurs an infinite number of times in (E_i). The inner product on E_i is assumed to be linear in the second variable and antilinear in the first variable. We consider $E_i \otimes A$ with the inner product

$$[x \otimes a_1, y \otimes a_2] = [x, y]a_1^* a_2$$

and then the direct sum

$$\sum_i \oplus E_i \otimes A$$

which is denoted by K_A.

DEFINITION 12.1.21 (Kasparov, 1980a). A set of elements (x_i), $i \in I$, is said to be a system of generators for the complete inner product module X if the set of all finite sums

$$\sum x_{i_k} a_k, \quad x_{i_k} \in (x_i), \quad a_k \in A$$

is dense in X.

We say that the module is finite dimensional if in X we have a finite system of generators; the module X is said to be countable generated if in X we have a countable system of generators.

Then we have the following result.

THEOREM 12.1.22 (Kasparov, 1980a). *Suppose that A is a B^*-algebra with a continuous action and X is a countable generated complete inner product module over*

A. Then we have the relation

$$X \oplus K_A \simeq K_A.$$

In the (1980b) paper of G. G. Kasparov, the reader may find other interesting results about inner product modules, among which we mention a beautiful extension of Stinespring's theorem of (1955). Other results related to fields of inner product spaces can be found in Takahashi (1979). For rigged inner product modules we refer to the author's note 'Rigged Kaplansky modules' (forthcoming).

12.2. Some Representation Theorems

We now consider $\mathcal{M}$ be a pre-inner product module over a B*-algebra with identity 1. For each state of B, φ (i.e. φ is a positive linear functional on B with $\|\varphi\| = \varphi(1) = 1$) we define

$$\langle x, y\rangle_\varphi = \varphi(\langle x, y\rangle)$$

where $\langle\, ,\rangle$ is the B-inner product of the module $\mathcal{M}$. Then it is clear that the following properties hold:

(1) $\langle x, y\rangle_\varphi \geqslant 0$,
(2) $\langle x, y\rangle_\varphi = \langle y, x\rangle_{\varphi^*}$,
(3) $N_\varphi = (x, x \in \mathcal{M}, \langle x, x\rangle_\varphi = 0)$ is a (right) submodule of $\mathcal{M}$.

From this it is clear that we can consider the quotient space $\mathcal{M}/N_\varphi$ which is a right module over B; we can consider on $\mathcal{M}/N_\varphi$ an inner product defined by the formula:

$$\langle m_1 + N, m_2 + N\rangle = \langle m_1, m_2\rangle_\varphi.$$

Then with this inner product $\mathcal{M}/N_\varphi$ is an inner product space and we consider the completion of this which will be denoted by $\tilde{H}_\varphi$. Of course the completion is again a right module over B. If S denotes the set of all states of B, we consider the direct sum of all spaces $(\tilde{H}_\varphi)_{\varphi\in S}$, i.e. we consider the set of all elements h of the form $(h_\varphi)_{\varphi\in S}$, $h_\varphi \in \tilde{H}_\varphi$ with the property that $\Sigma_{\varphi\in S}\|h_\varphi\|^2 < \infty$ with the inner product

$$\langle h, k\rangle = \sum_{\varphi\in S} \langle h_\varphi, k_\varphi\rangle.$$

Since all the members of this direct sum are right modules over B, the direct sum may be organized in a natural way as a right module over B.

We denote the module constructed above by $\Sigma_{\varphi\in S} \oplus \tilde{H}_\varphi$. Then from the above considerations we have the following result.

THEOREM 12.2.1. *For every pre-inner product module over B there exists an embedding into a module complete inner product module over B,* $\Sigma_{\varphi\in S} \oplus \tilde{H}_\varphi$.

We now show how we can use this complete inner product space which is a module over B to represent certain classes of linear operators on the module $\mathcal{M}$.

Let T be a linear operator on $\mathcal{M}$ and we suppose that it is a module map. Then we define in each H_φ an operator which is also a module map by the relation: first we set for each $T\tilde{m}_\varphi$ in $\tilde{H}_\varphi$,

$$\tilde{T}\tilde{m} = [\tilde{T}\tilde{m}], \quad \tilde{m} \ni m$$

and we see immediately that this is well defined, in the sense that the value is independent of the representor of the class $\tilde{m}$. It is now clear that on $\Sigma_S \oplus \tilde{H}_\varphi$ we can define the operator T by the formula: if $h = (h_\varphi)_{\varphi \in S}$ then

$$\tilde{T}h = (\tilde{T}\tilde{h}_\varphi)_{\varphi \in S}.$$

Then we have the following result.

THEOREM 12.2.2. *The (right) module of all module maps on the module $\mathcal{M}$ is isomorphic to a (right) module over B of linear operators on the complete inner product space $\Sigma_S \oplus \tilde{H}_\varphi$.*

THEOREM 12.2.3. *The (right) module of all bounded module maps on $\mathcal{M}$ is topologically isomorphic (as a module) with a module of bounded linear maps on $\Sigma_S \oplus \tilde{H}_\varphi$.*

Proof. All the properties are clear except the fact that this is a topological isomorphism. But this follows from Theorem 12.1.15 since for any bounded module map T,

$$\|T\| = \inf(k^{1/2}, \langle Tx, Tx\rangle_A \leqslant k\langle x, x\rangle_B).$$

COROLLARY 12.2.4. *The set of all bounded linear operators having an adjoint with respect to the B-valued inner product is topologically isomorphic to a C*-algebra of operators on $\Sigma_{\varphi \in S} \oplus \tilde{H}_\varphi$.*

The results of Section 12.2 are taken from the author's note (1984).

12.3. Dilation Theory and Inner Product Modules

We have presented in other chapters some results about dilation theory. Now we are interested in a possible dilation theory in the setting of inner product modules. The result we give were obtained by Itoh (1980). We wish to know if the results on dilation theory are extendable to Banach modules, or, more precisely, if there are analogues of the dilation theorems proved by Stroescu (1971, 1972, 1973, 1974) for the case of Banach modules. Some results concerning this question are given in the author's note 'Remarks on dilation theory of Banach modules' (forthcoming).

In what follows $\mathcal{M}$ is a right module over the B*-algebra B and $\alpha(\mathcal{M})$ denotes the set of all bounded module maps which have an adjoint with respect to the B-valued inner product.

We now consider a *-semigroup Γ with unit ε and involution $*: \Gamma \to \Gamma$ satisfying the following (usual) conditions:

(1) $a \in \Gamma, \quad a^{**} = a$,
(2) $(ab)^* = b^*a^*$
(3) $\varepsilon^* = \varepsilon$.

$C(\Gamma, \mathcal{M})$ is the set of all functions defined on Γ with values in $\mathcal{M}$ with the property that the values $f(s)$ are 0 except for a finite number of elements in Γ.

DEFINITION 12.3.1. A mapping $p: \Gamma \to \mathcal{a}(\mathcal{M})$ is said to be positive definite (PD) if for any f in $C(\Gamma, \mathcal{M})$ the following relation holds:

$$\sum_{a,b\varepsilon} p(b^* a) f(a), f(b) \geqslant 0$$

REMARK 12.3.2. If p is (PD) then for all a in Γ, $p(a^*) = (p(a))^*$.

DEFINITION 12.3.3. Let $p: \Gamma \to \mathcal{a}(\mathcal{M})$ be a (PD)-map. We say that p satisfies the boundedness condition (BC) if there exists a function $h: \Gamma \to [0, \infty)$ with the property that for each a in Γ and each f in $C(\Gamma, \mathcal{M})$ the following inequality holds:

$$\sum_{b,c} p(c^* a^* ab) f(b), f(c) \leqslant h(a) \sum_{b,c} p(c^* b) f(b), f(c).$$

REMARK 12.3.4. If the semigroup Γ is a group then we can take $h(a) = 1$, i.e. on groups every (PD)-map satisfies the boundedness condition (BC).

We need the following result about certain classes of operators on $\mathcal{M}$. First we give the following definition.

DEFINITION 12.3.5. The element T in $\mathcal{a}(\mathcal{M})$ is called hermitian if $T = T^*$.

We have obviously the following result.

PROPOSITION 12.3.6. *Every element T in $\mathcal{a}(\mathcal{M})$ admits a unique decomposition*

$$T = T_1 + iT_2,$$

where T_i, $i = 1, 2$, are hermitian elements of $\mathcal{a}(\mathcal{M})$.

DEFINITION 12.3.7. The element $T \in \mathcal{a}(\mathcal{M})$ is said to be positive if and only if for each x in $\mathcal{M}$, $[Tx, x] \geqslant 0$.

From Lemma 12.2.11 we know that $\mathcal{a}(\mathcal{M})$ is a B*-algebra. Then in a B*-algebra we have a definition of Hermitian elements, positiveness, normality, etc. The following proposition shows that for $\mathcal{a}(\mathcal{M})$ the notion of positiveness is the same in both cases.

PROPOSITION 12.3.8. *The operator T in $\mathcal{a}(\mathcal{M})$ is positive in the sense of Definition 12.3.7 if and only if it is a positive element of $\mathcal{a}(\mathcal{M})$ as a B^*-algebra.*

Proof. Suppose first that T is a hermitian operator which is positive in the sense that for all x in $\mathcal{M}$, $[Tx, x] \geqslant 0$. Now we show that T is positive as an element of the Banach B*-algebra $\mathcal{a}(\mathcal{M})$. Now, in the case of B*-algebras, using the states of B, it is easy to see that $T \in B$ hermitian is positive if and only if $\|T\|e - T$ has the property that

$$\big\| \|T\|e - T \big\| \leqslant \|T\|. \tag{*}$$

Now if

$$0 \leqslant [Tx, x] \leqslant \|T\|[x, x]$$

then

$$0 \leqslant [(\|T\|e - T)x, x] \leqslant \|T\|[x, x].$$

Applying the states of B we get that (*) holds for our operator and thus T is positive in the B*-algebra $\mathscr{A}(\mathscr{M})$.

Conversely, if T is a positive element in $\mathscr{A}(\mathscr{M})$ then $T = S^*S$, where S is an element in $\mathscr{A}(\mathscr{M})$. This implies clearly that T is positive in the sense of 12.3.7.

The above proposition was proved by Loynes (1965) for the case of VH-spaces and by Paschke (1973) for the case of inner product modules.

PROPOSITION 12.3.8 (Itoh, 1980). *Let $p: \Gamma \to \mathscr{A}(\mathscr{M})$ be PD. Then for every f in $C(\Gamma, \mathscr{A}(\mathscr{M}))$ the following inequality holds*:

$$\sum_{a,b} f(b)^* p(b^*a) f(a) \geqslant 0. \tag{*}$$

If the B-algebra B has an identity e and in $\mathscr{M}$ there exists u with the property that $\langle u,u \rangle = e$ then the converse holds.*

Proof. Suppose first that p is PD. Then for each x in $\mathscr{M}$ and f in $C(\Gamma, \mathscr{A}(\mathscr{M}))$ we have

$$\sum_{a,b} [f(b)^* p(b^*a) f(a)x, x] \geqslant 0.$$

From Proposition 12.3.8 we obtain the inequality (*).

Conversely, suppose that (*) holds as well as the additional conditions. In this case for any f in $C(\Gamma, \mathscr{A}(\mathscr{M}))$ we define the element g_f in $C(\Gamma, \mathscr{A}(\mathscr{M})$ by the relation

$$f(a) \cdot \langle u, z \rangle = g_f(a)(u)$$

for all u in $\mathscr{M}$. In this case

$$\sum_{a,b} [g_f(b)^* p(b^*a) g_f(a)x, x] = 0$$

and then

$$\sum_{a,b} [p(b^*a) f(a), f(b)] = 0.$$

PROPOSITION 12.3.9. *Let Γ be a commutative B*-algebra and let $p: \Gamma \to \mathscr{A}(\mathscr{M})$ be a linear and positive mapping. Then p is PD.*

Proof. Obvious. □

Concerning the BC conditions we note that interesting conditions were found recently and for an interesting discussion we refer to P. Masani's papers (1978).

PROPOSITION 12.3.10 (Itoh, 1980). *Let Γ and $\mathscr{M}$ be as above. Let $p: \Gamma \to \mathscr{A}(\mathscr{M})$ be PD. Then the following assertions are equivalent:*

(1) *p is BC.*
(2) *For some function $h: \Gamma \to [0, \infty)$ and each a, b in Γ, and each x in $\mathscr{M}$, the following inequality holds:*

$$[p(b^*a^*ab)x, x] \leqslant h(a)[p(b^*b)x, x],$$

(3) *There exists a function $k: \Gamma \to [0, \infty)$ and a strictly positive constant K such that*

for every a, b in Γ *the following inequality holds:*

$$\|p(a)\|_X \leqslant K \cdot k(a), \qquad k(ab) \leqslant k(a)k(b).$$

Proof. First we remark that (1) → (2). Suppose now that (2) holds. Then we set

$$t(a) = \inf(h(a),\ h \text{ satisfying condition}(2)).$$

Then t also satisfies condition (2) and clearly

$$t(ab) \leqslant t(a)t(b).$$

Suppose now that φ is any positive linear form on the B*-algebra B. Then we have

$$\left|\sum_{a,b} \langle p(b^*a)f(a), g(b)\rangle_\varphi\right|^2 = \left(\sum_{a,b} \langle p(b^*a)f(a), f(b)\rangle_\varphi\right) \times$$

$$\times \left(\sum_{a,b} \langle p(b^*a)g(a), g(b)\rangle_\varphi\right)$$

by the Cauchy–Buniakowsky inequality for positive linear forms on B*-algebras. Thus for each u in $a(m)$ and a in Γ

$$[p(a)u, p(a)u]_\varphi^2 \leqslant t(a)[p(\varepsilon)u, u]_\varphi[p(\varepsilon)p(a)u, p(a)u]_\varphi$$

and

$$[p(\varepsilon)x, x]_\varphi \leqslant \|p(\varepsilon)\|_X[x, x]_\varphi.$$

Applying Lemma 12.1.14 we have further that

$$[p(\varepsilon)p(a)u, p(a)u]_\varphi \leqslant \|p(\varepsilon)\|_X[p(a)u, p(a)u]_\varphi.$$

Since φ is an arbitrary positive linear form on B we have

$$[p(a)u, p(a)u] \leqslant t(a)\|p(\varepsilon)\|_X^2[u, u]$$

and then we can take as the constant K the number $\|p(\varepsilon)\|_X$ and as a function on k the function $a \to t(a)^{1/2}$. Then we remark that condition (3) is satisfied.

Now suppose that (3) is satisfied. We consider the element $b = a^*a$ in Γ and then

$$k(b^{2^n}) \leqslant (k(b))^{2^n}$$

for all $n = 1, 2, 3, \ldots$. Let φ be a positive linear form on B and let f be arbitrary in $C(\Gamma, \mathcal{M})$. Then we have

$$0 \leqslant \sum_{a,c}[p(c^*b^{2^n}a)f(a), f(c)]_\varphi \leqslant K \cdot k(b)^{2^n}\|p\|\sum_{c,a} k(c^*)\|f(c)\|_X\|f(a)\|_X.$$

Then

$$\liminf_n \left(\sum_{a,c}[p(c^*b^{2^n}a)f(c), f(a)]_\varphi\right)^{2-n} \leqslant k(b).$$

If we set

$$h(b) = \sum_{a,c}[p(c^*ba)f(a), f(c)]_\varphi, \quad r = h(\varepsilon)$$

from the fact that p is PD we obtain that

$$h(b^{2^n}) \leqslant rh(b^{2n+1})$$

for $n = 0, 1, 2, 3, \ldots$, and an induction argument shows that for all n the following inequality holds:

$$h(b) \leqslant r^{1-2^{-n}} h(b^{2^n})^{2^{-n}}.$$

This implies that

$$h(b) \leqslant rk(b).$$

To see that the BC condition is satisfied we take as a function h in the definition of BC the function $k(a^*a)$.

Now we are ready to prove a dilation theorem. First we need the notion of a representation of Γ.

DEFINITION 12.3.11. Let Γ and $\mathscr{M}$ be as above. Then a representation of Γ in $\mathscr{M}$ is a mapping P of Γ in $\mathscr{A}(\mathscr{M})$ satisfying the following properties:

(1) $P(a)^* = P(a^*)$,
(2) $P(ab) = P(a)P(b)$,
(3) $P(\varepsilon) = I_{\mathscr{M}}$.

Here $I_{\mathscr{M}}$ means the mapping defined on $\mathscr{M}$ by the relation $I_{\mathscr{M}}(u) = u$ for all u in $\mathscr{M}$.

DEFINITION 12.3.12. Let $\mathscr{N}$ be another module over B which is supposed to be an inner product module with the values of the inner product on B. A module map $T: \mathscr{M} \to \mathscr{N}$ is said to be an isometric embedding if for all x, y in $\mathscr{M}$,

$$[Tx, Ty] = [x, y].$$

If T is surjective then we say that $\mathscr{M}$ is isomorphic to $\mathscr{N}$.

12.3.13. *Let $p: \Gamma \to \mathscr{A}(\mathscr{M})$ be a PD mapping satisfying the BC condition. Then the following assertions hold:*

(1) *There exists a super inner product module $\mathscr{N}$, a bounded linear module map $r: \mathscr{M} \to \mathscr{N}$ and a bounded linear module map $q: \mathscr{N} \to \mathscr{M}$ such that $qr = p(\varepsilon)$ and Γ admits a representation in $\mathscr{N}$, say P, with the property that $p(a) = qP(a)r$ for all a in Γ.*
(2) *$(P(a)r(u), a \in \Gamma, u \in \mathscr{M})$ is a dense submodule in $\mathscr{N}$.*
(3) *If $p(\varepsilon) = I_{\mathscr{M}}$ then r is an isometric embedding and identifying $\mathscr{M}$ with $r(\mathscr{N})$ we can say that q is a projection.*

Proof. The set $C(\Gamma, \mathscr{M})$ is obviously a right module over B. We now show that we can define a B-valued inner product (or something like an inner product with values in B). Indeed, if f, g are two elements in $C(\Gamma, \mathscr{M})$ then using the fact that p is PD we set

$$[f, g] = \sum_{a,b} [p(b^*a)f(a), f(b)]$$

and we remark that the values are in B and the following properties hold:

(1) $f, f \geqslant 0$.
(2) $[f, g] = [g, f]^*$.
(3) If b is in B then $[f \cdot b, g] = [f, g]b$.

We now consider the set

$$N = (f, f \in C(\Gamma, \mathscr{M}), [f, f] = 0).$$

Since an inequality like Cauchy–Buniakowsky's holds for the linear positive forms over B, we see that N is a B-submodule. Then we can form $C(\Gamma, \mathscr{M})/N$ which is again a module and the bilinear form (with values in B), $[f,g]$ induces on $C(\Gamma, \mathscr{M})/N$ a bilinear conjugate form such that this is an inner product module.

Let δ_b be a function defined on Γ by the formula

$$\delta_b(a) = 1 \text{ if } a = b \text{ and } 0 \text{ in the contrary case.}$$

If f is in $C(\Gamma, \mathscr{M})$ then we have that $f = \Sigma\delta_{a_i}(\quad)m_i$, where $m_i \in \mathscr{M}$. Let a be in Γ and we consider the mapping $p_a: C(\Gamma, \mathscr{M}) \to C(\Gamma, \mathscr{M})$ defined as follows: $p_a f = \Sigma\delta_{aa_i} m_i$.

We remark that N is an invariant submodule for each p_a. This implies that we can define a mapping on $C(\Gamma, \mathscr{M})/N$ by the formula

$$p_0(a)(f + N) = p_a f + N.$$

Now we can consider the completion of $C(\Gamma, \mathscr{M})/N$ with respect to the norm

$$\|x\|_X = \text{norm of } [x, x] \text{ in } B$$

and it is easy to see that this is an inner product module which is complete.

In this case the mapping $p_0(a)$ can be extended uniquely to an element of $\mathscr{a}(\mathscr{N})$, being the completion of $C(\Gamma, \mathscr{M})/N$. We denote the extension of $p_0(a)$ by $P(a)$ and it is clear that $a \to P(a)$ is a representation of Γ on $\mathscr{N}$.

Now we set $r_1: \mathscr{M} \to C(\Gamma, \mathscr{M})$ and $q_1: C(\Gamma, \mathscr{M}) \to \mathscr{M}$, defined by the formulas:

$$r_1 u = \delta_\varepsilon u, \qquad q_1 f = \sum_a p(a) f(a).$$

Now if f is in N and φ is a positive linear form on B, then for all g in $C(\Gamma, \mathscr{M})$ we have

$$([g,f]_\varphi)^2 \leqslant ([g,g]_\varphi)([f,f]_\varphi)$$

which implies, since φ is arbitrary, that $[g, f] = 0$. This gives that

$$[q_1 f, q_1 f] = [r_1(q_1 f), f] = 0$$

and thus the mapping $q_0: C(\Gamma, \mathscr{M})/N \to \mathscr{M}$, defined by the formula

$$q_0(h + N) = q_l f$$

is well defined. Further, we set $r: \mathscr{M} \to \mathscr{N}$ by $rx = r_1 x + N$.

Then

$$[rx, rx] = [p(\varepsilon)x, x] \leqslant \|p(\varepsilon)\|_X [x, x]$$

which shows that r is a bounded mapping with a norm less than $\|p(\varepsilon)\|_X^{1/2}$.

From the definition of r it is obvious that this is a module map, and since for any f in $C(\Gamma, \mathcal{M})$ and x in $\mathcal{M}$,

$$[q_0(f+N), q_0(f+N)] = [rq_0(f+N), f+N]$$

we have

$$\|q_0\|_X \leqslant \|p(\varepsilon)\|_X^{1/2}.$$

Then q_0 is a bounded module map and this has an extension to a mapping defined on $\mathcal{N}$ with values in $\mathcal{M}$. Also, for each x in m and y in n we have the relations

$$[rx, y] = [x, qy], \qquad qP(a)rx = p(a)x.$$

The minimality condition is obvious.

Now if $p(\varepsilon) = I_{\mathcal{M}}$ then for each x, y in $\mathcal{M}$ we have

$$[rx, ry] = [p(\varepsilon)x, y] = [x, y]$$

and $q = q^* = q^2$.

The above theorem was given by Itoh (1980). We remark that this theorem may be applied to the case of C*-algebras with identity since, if we set for any x, y in $\mathcal{M}$ the algebra,

$$[x, y] = y^*x$$

then the algebra is an inner product module. The set of all bounded module maps having an adjoint is (L_x, x in the algebra) and where

$$L_x(y) = xy.$$

For further remarks concerning the dilation theorems for inner product modules we refer to Itoh (1980).

12.4. Von Neumann Algebra Module

The intial motivation of I. Kaplansky to introduce the inner product modules was that they offer a powerful tool in some delicate problems about algebras of operators on complete inner product spaces or other classes of algebras. For example, we mention his proof that every derivation of an AW* of type I is an inner derivation (i.e. it is of the form $ax - xa$ for some fixed element a in the respective AW*-algebra).

Another interesting application was made by Takemoto (1973) in the theory of representation of von Neumann algebras as continuous fields of von Neumann algebras. Thus it may be of interest to consider a class of subalgebras in the algebra $\mathcal{A}(\mathcal{M})$ of a complete inner product module $\mathcal{M}$, where $\mathcal{A}(\mathcal{M})$ is the set of all bounded module maps having an adjoint.

DEFINITION 12.4.1. Let (T_n) be a sequence in $\mathcal{A}(\mathcal{M})$. We say that (T_n) converges strongly to T in $\mathcal{A}(\mathcal{M})$ if, for each x in $\mathcal{M}$,

$$\lim \|T_n x - Tx\|_X = 0.$$

The strong topology on a complete inner product module $\mathcal{M}$ is defined by the family of seminorms

$$p_x(T) = \|Tx\|_X.$$

It is clear that this is a locally convex topology.

DEFINITION 12.4.2. A von Neumann algebra module $\mathscr{a}$ is a *-algebra of operators in $\mathscr{a}(\mathscr{M})$, $\mathscr{a}$ is a complete inner product module which is closed in the strong topology on $\mathscr{a}$.

As usual, for any subset S in $\mathscr{a}(\mathscr{M})$ the commutant is defined as the set

$$S' = (t, t \in \mathscr{a}(\mathscr{M}), ts = st \text{ for all } s \in S).$$

It is easy to see that S' is an algebra and contains I.

We now define another class of module algebras.

DEFINITION 12.4.3. An AW*-algebra module is a B*-subalgebra of $\mathscr{a}(\mathscr{M})$, where $\mathscr{M}$ is a complete inner product module satisfying the following conditions:

(1) In the set of projections, each collection of orthogonal projections has a least upper bound.
(2) Each maximal commutative selfadjoint subalgebra is generated by its projections.

We mention that AW*-algebras for the case when $B = C$, the set of complex numbers, were introduced by I. Kaplansky. For a study of these algebras we refer to the author's note 'Von Neumann algebra module and Kaplansky module AW*-algebras' (forthcoming).

12.5. The AW*-Modules of Kaplansky

As mentioned above, one of the first successes of the inner product module theory of I. Kaplansky was the result about the structure of derivations of AW*-algebras of type I. Another interesting application of Kaplansky modules was obtained by Azoff (1977) on the selfadjointness of a von Neumann algebra generated by a weakly closed algebra of operators acting on a seperable complete inner product space. In this section, following Kaplansky (1953), we present some results about the notion of AW*-modules.

DEFINITION 12.5.1 (Kaplansky, 1953). An inner product module $\mathscr{M}$ over a Banach algebra B which is supposed to be an AW*-algebra is said to be an AW*-module if the following conditions are satisfied:

(1) For any family $(e_i)_{i \in I}$, I is an index set of orthogonal projections with lub $e_i = e$ and for any x in $\mathscr{M}$ with the property that $e_i x = 0$, $i \in I$, then $ex = 0$.
(2) For any family $(e_i)_{i \in I}$, I is an index set of orthogonal projections with lub $e_i = 1$, is the identity of B, and any bounded sequence in $\mathscr{M}$, (x_i), there exists x in $\mathscr{M}$ with the property that $e_i x = e_i x_i$ for all $i \in I$.
(3) B is a commutative algebra.

From the properties of commutative AW*-algebras it is known that any such algebra is *-isomorphic to an algebra $C(S)$ of continuous complex-valued functions on a Stone space – i.e. a compact space with the property that the closure of open sets are open. This class of topological spaces may also be considered as not necessarily compact.

Now the element x in condition (2) is clearly unique and this justifies the notation

$$x = \sum_{i \in I} e_i x_i.$$

PROPOSITION 12.5.2. *Let B be a commutative AW^*-algebra and $(e_i)_{i \in I}$ be a family of orthogonal projections with* lub $e_i = 1$. *Suppose that $\mathscr{M}$ is an AW^*-module over B. Then for each y in $\mathscr{M}$ and* (x_i) a bounded subset in $\mathscr{M}$ the following relation holds:

$$\left[\sum_{i \in I} e_i x_i, y\right] = \sum_{i \in I} e_i [x_i, y].$$

Proof. Since (x_i) is a bounded family of elements in $\mathscr{M}$, it follows that $([x_i, y])$ is again a bounded family of elements in B. Then we see that the right side of the above inequality is well defined. To prove the equality above, it suffices to show that it holds after multiplication with e_i, i arbitrary in I. In this case the right side is equal to

$$e_i[x_i, y]$$

and the left side is equal to

$$e_i\left[\sum_{j \in I} e_j x_j, y\right] = e_i \sum_{j \in I} [e_j x_j, y] = [e_i x_i, y] = e_i[x_i, y]$$

and thus the proposition is proved. □

PROPOSITION 12.5.3. *Let $\mathscr{M}$ be an AW^*-module over B. Let S be a set in $\mathscr{M}$ and*

$$S^\perp = (b \in B,\ bx = 0 \textit{ for all } x \in S).$$

Then $S^\perp$ is a direct summand of B.

Proof. From the definition it is easy to see that $S^\perp$ is an ideal of B which is obviously closed. Since any closed ideal of an AW*-algebra is again an AW*-algebra a maximal family of orthogonal projections in $S^\perp$. Then their lub e_i is in I and if $e = \text{lub}\, e_i$ we show that $S^\perp = eB$.

For this it suffices to prove that

$$(1 - e)S^\perp = 0.$$

If $(1 - e)S^\perp$ is not zero then it contains a nonzero projection and this contradicts the fact that (e_i) is a maximal family of orthogonal projections in $S^\perp$. Thus the proposition is proved. □

DEFINITION 12.5.4. We say that an AW*-module is faithful if $\mathscr{M}^\perp$ is 0.

In what follows we suppose without loss of generality that our AW*-modules are faithful.

PROPOSITION 12.5.5. *Let x be the nonzero element of an AW^*-module $\mathscr{M}$ over B.*

Then for some nonzero projection p in B and an element a in B we have

$$p = a[x, x]^{1/2}.$$

Proof. Since x is supposed to be a nonzero element in $\mathcal{M}$, $[x, x]$ is a nonzero element of B and thus the root $[x, x]^{1/2}$ is also nonzero. This implies that some nonzero function representing $[x, x]^{1/2}$ is bounded away from zero in pB. The inverse of $[x, x]^{1/2}$ in pB is the required element a.

PROPOSITION 12.5.6. *If $\mathcal{M}$ is a faithful AW*-module then there exists x in $\mathcal{M}$ with the property that*

$$[x, x] = 1.$$

Proof. We consider (x_i) to be a maximal collection of elements in $\mathcal{M}$ with the property that $[x_i, x_i]^{1/2} = p_i$, (p_i) is an orthogonal family of projections. First we show that such a family exists. From Proposition 12.5.5 we can construct x_1 and p_1 and then we apply Zorn's lemma to get the above fact. Let lub $p_i = e$. Suppose that e is not 1 and then $(1 - e) \neq 0$. Applying Proposition 12.5.5 we find an element y and, added to the family (x_i), we obtain a contradiction with the maximality. Thus we have $1 = e$. Since $[x_i, x_i] = 1$ we may consider

$$x = \sum_{i \in I} p_i x_i.$$

In this case we obtain that $[x, x] - 1$ annihilates each p_i and thus must be 0. Thus x satisfies the proposition. □

DEFINITION 12.5.7. An orthonormal set in an AW*-module over B is any set (x_i) with the property that

$$[x_i, x_j] = \delta_{ij} \cdot 1.$$

From the above results we see that the family of orthonormal sets is not void.

DEFINITION 12.5.8. If S is a subset in $\mathcal{M}$ then the orthogonal complement S^{-} is the subset in $\mathcal{M}$ with the property that

$$[s, s^{\perp}] = 0$$

for all $s \in S$, $s^{\perp} \in S^{\perp}$.

DEFINITION 12.5.9. If $\mathcal{M}$ is an AW*-module then an orthonormal basis of $\mathcal{M}$ is an orthonormal set (e_i) whose orthogonal complement reduces to 0.

DEFINITION 12.5.10. An AW*-module is said to be homogeneous if it has an orthonormal basis.

We define now the notion of an AW*-submodule.

DEFINITION 12.5.11. Let $\mathcal{M}$ be an AW*-module over B. An AW*-submodule $\mathcal{N}$ is a set satisfying the following conditions:

(1) $\mathcal{N}$ is a submodule of $\mathcal{M}$.
(2) $\mathcal{N}$ is closed in the strong topology of $\mathcal{M}$.

(3) If (x_i) is a bounded set in $\mathcal{N}$ and (p_i) is an orthogonal family of projections in B with lub $p_i = 1$ then $\Sigma_{i \in I} p_i x_i$ is in $\mathcal{N}$.

COROLLARY 12.5.12. *Let $\mathcal{M}$ be an AW*-module over B and let S be a subset in $\mathcal{M}$. Then $S^\perp$, the orthogonal complement of S, is an AW*-submodule of $\mathcal{M}$.*

Proof. Properties (1) and (2) are clear. To prove (3) let x_i and p_i be as above and suppose that $[x_i, y] = 0$. We must show that

$$\left[\left(\sum_{i \in I} p_i x_i\right), y\right] = 0$$

From Proposition 12.5.2 we see that the left side is equal to

$$\sum_{i \in I} p_i [x_i, y] = 0.$$

Thus the corollary is proved. □

COROLLARY 12.5.13. *Let $\mathcal{M}$ and $\mathcal{N}$ be two AW*-modules over B and suppose that $T: \mathcal{M} \to \mathcal{N}$ is a bounded module map. Then the kernel* ker T *is an AW*-submodule of $\mathcal{M}$.*

Proof. It is clear that conditions (1) and (2) of Definition 12.5.11 are satisfied. To prove that condition (3) is also satisfied, we consider x_i and p_i as above with $x_i \in \ker T$. We can consider the element

$$u = T \sum_{i \in I} p_i x_i$$

and multiplying with p_i we obtain

$$p_i u = p_i T x_i = 0$$

for all $i \in I$. Then $u = 0$. □

COROLLARY 12.5.14. *Let $\mathcal{M}$ and $\mathcal{N}$ be two AW*-modules over B and consider $\mathcal{M} \otimes \mathcal{N}$ which is clearly an AW*-module with the usual B-valued inner product. Then each of the modules $\mathcal{M}$, $\mathcal{N}$ is an AW*-submodule of $\mathcal{M} \otimes \mathcal{N}$.*

The following theorem gives the structure of AW*-modules in terms of homogeneous AW*-modules.

THEOREM 12.5.15. *Let $\mathcal{M}$ be a faithful AW*-module over B. Then there exists a family $(p_i)_{i \in I}$ of orthogonal projections in B with* lub $p_i = 1$ *and such that for each $i \in I$, $p_i H$ is a homogeneous AW*-module over $p_i B$.*

Proof. Let (x_i) be a maximal orthonormal family of elements in $\mathcal{M}$. Let S be the orthogonal complement of this family. Since $\mathcal{M}$ is supposed to be faithful and S is an AW*-submodule (by Corollary 12.5.12) we find a projection e with the property that $eS = 0$. Then we show that (ex_i) is an orthonormal basis of eH with respect to the AW*-algebra eB. Indeed, if y is an element orthogonal to all elements ex_i then we have

$$0 = [ex_i, y] = e[x_i, y] = [x_i, ey]$$

and thus $ey = y$. But $eS = 0$ and thus $y = 0$. Thus eH is a homogeneous AW*-

submodule. In $(1 - e)\mathcal{M}$ we can repeat the procedure, and continuing in this way we obtain the above decomposition of $\mathcal{M}$.

Suppose now that B is a commutative AW*-algebra. Then it is known that B is a conditionally complete lattice, i.e. every bounded set of B has a least upper bound. Suppose that we have a family of positive elements $(a_i)_{i\in I}$ in B. Suppose further that for each finite subset J of I,

$$\sum_{i\in J} a_i \leqslant a,$$

where a is an element of B, and is a least upper bound for all such finite sets. We say that $\Sigma_{i\in I} a_i$ converges and contains the sum of this least upper bound a. Further, we define the convergence of arbitrary $\Sigma_{i\in I} a_i$ if $\Sigma_{i\in I}|a_i|$ converges. Then we have.

PROPOSITION 12.5.16. *Suppose that $(a_i)_{i\in E}$, $(b_i)_{i\in I}$ are two families of elements of B with the same index set I. Suppose further that $\Sigma\,|a_i|^2$ and $\Sigma\,|b_i|^2$ converge. Then $\Sigma\, a_i b_i$ converge and*

$$\left|\sum a_i b_i\right|^2 \leqslant \left(\sum_{i\in I} |a_i|^2\right)\left(\sum_{i\in I} |b_i|^2\right). \tag{*}$$

Proof. First we remark that (*) holds for finite sums and we note that on the right-hand side we can consider infinite sums, which implies the assertion of the inequality.

We now give a method of constructing homogeneous AW*-modules. Let I be an index set and suppose that for each i in I we have an AW*-module $\mathcal{M}_i$ over B. We define the set $\mathcal{M}$ as the set of all $x = (x_i)_{i\in I}$ with x_i in $\mathcal{M}_i$ for each i in I and $\Sigma\,|x_i|^2$ converges.

It is not difficult to see that $\mathcal{M}$ can be organized as a module over B and if y is another element in $\mathcal{M}$ then we define

$$[x, y] = \sum x_i y_i$$

which defines an inner product (with values in B) such that $\mathcal{M}$ is an inner product module. We show that in fact $\mathcal{M}$ is an AW*-module over B.

Let $x = (x_i)$ be an element in $\mathcal{M}$ and let (p_s) be a family of projections in B with lub $p_s = e$. Suppose that for each s, $p_s x = 0$. Since $p_s x_i = 0$ we have $ex_i = 0$. This implies that $ex = 0$. Let $(x(s))$ be a family of elements in $\mathcal{M}$, $x(s) = (x_i(s))$ with $\Sigma_{i\in I}|x_i(s)|^2$ bounded by a constant K. Let (p_j) be a family of orthogonal projections in B with lub $p_j = 1$, $j \in I$. Let s be fixed and we have that

$$|x_i(s)|^2 \leqslant K$$

and since $\mathcal{M}_i$ is an AW*-module, there exists $y_i = \Sigma\, p_s x_i(s)$, $p_s y_i = p_i x_i(s)$ for all s.

Now for any finite subset J of I,

$$\sum_{i\in J} |y_i|^2 = \sum_{i\in J}\sum_{s} p_s|x_i(s)|^2. \tag{**}$$

If we multiply by p_t, the right-hand side becomes $\Sigma_{i\in J}\, p_t|x_i(t)|^2$ which is bounded by K. Since t is arbitrary the left-hand side of (**) is also bounded by K. This implies that $y = (y_i)$ is in $\mathcal{M}$ and also $p_s y = p_s x_s$ and y is the element satisfying condition (3) of an AW*-module.

We now show that $\mathcal{M}$ is a complete AW*-module. Indeed, if (x_n) is a sequence of elements in $\mathcal{M}$, which is a Cauchy sequence, we show that this converges. Indeed, $x_p = (x_i(p))$ and for each fixed i, $(x_i(p))$ is a Cauchy sequence in $\mathcal{M}_i$ and thus converges to an element, say y_i. Since for some constant K, $x_p^2 = K$ we show that $\Sigma |y_i|^2 \leqslant K$. Let J be a finite subset in I and then

$$\sum_{i\in J} |y_i|^2 = \lim_{m\to\infty} \sum_{i\in J} |x_i(m)|^2$$

and since

$$\sum_{i\in J} |x_i(m)|^2 \leqslant |x_m|^2 \leqslant K$$

we have that

$$\sum_{i\in J} |y_i|^2 \leqslant K.$$

This implies that $y = (y_i)$ is in $\mathcal{M}$. Now we show that our original sequence converges to y. Let $\varepsilon > 0$ and N_ε such that for all p, $q \geqslant N_\varepsilon$ we have

$$\|x_p - x_q\| \leqslant \varepsilon.$$

Let J be a finite subset in I and then

$$\sum_{i\in J} |x_i(p) - y_i|^2 = \lim_{q\to\infty} \sum_{i\in J} |x_i(p) - x_i(q)|^2 \leqslant \varepsilon$$

and thus

$$\sum_{i\in I} |x_i(p) - y|^2 \leqslant \varepsilon.$$

Thus $\|x_p - y\| \leqslant \varepsilon$ if $p \geqslant N_\varepsilon$ which implies the convergence of (x_p) to y.

Suppose now that each $\mathcal{M}_i$ is isomorphic to B. Since B is a homogeneous AW*-module the AW*-module constructed above has an orthonormal basis, i.e. it is homogeneous. The cardinal of the orthonormal basis of the AW*-module so constructed is equal to card I.

The following result is the analogue of the Fréchet–Riesz representation theorem for continuous linear functionals on complete inner product spaces.

12.5.17. *Let $\mathcal{M}$ be an AW*-module and let $f: \mathcal{M} \to B$ be a module homomorphism which is continuous. Then there exists a unique element x in $\mathcal{M}$ such that*

$$f(y) = [y, x].$$

Proof. Let N_f be the kernel of f, i.e. the set

$$N_f = (u, u \in \mathcal{M}, f(u) = 0)$$

and we remark that this is an AW*-submodule. Let $N_f^\perp$ be the orthogonal complement of N_f. Then it is not difficult to see that

$$\mathcal{M} = N_f \oplus N_f^\perp.$$

Let eB be the annihilator of $\mathcal{M}$ in B. Then we have

$$ef(y) = f(ey) = 0.$$

This implies that the range of f is in $(1 - e)B$. But $(1 - e)B$ is again an AW*-algebra and we can consider, without loss of generality, f on $\mathcal{M}$ as a module over $(1 - e)B$. From Proposition 12.5.6 we find an element u with the property that $[u, u] = 1$ and we show that the orthogonal complement of the $(1 - e)B$-module generated by z is 0. Indeed, in the contrary case we find a nonzero element, say v, with $[u, v] = 0$. According to Proposition 12.5.5 we find a nonzero projection e_1 with the property that $e_1|v|$ is a regular element in $e_1 B$.

Since $|e_1 u| = e_1$ we obtain that $f(e_1 u) \neq 0$. For some projection $e_2(e_2 e_1 = e_2)$ the element $f(e_2 u)$ is regular in $e_2 B$. Now for some b we have

$$bf(e_2 u) = e_2 f(z)$$

whence

$$f(be_2 u - e_2 v) = 0, \quad be_2 u = e_2 v$$

and thus

$$[be_2 u, e_2 v] = [e_2 v]^2 = 0.$$

Since $e_2 v$ cannot be zero this is a contradiction and thus the orthogonal complement of the module generated by u is 0. If we set (as in the case of complete inner product spaces) $x = f(u)^* u$ then this is the representer for f, i.e.

$$f(y) = [y, x].$$

Since the uniqueness can be proved as above, it is omitted. □

From this we have

12.5.18. *Let $\mathcal{M}$ be an AW*-module and let $T: \mathcal{M} \to \mathcal{M}$ be a module map. Then T is continuous if and only if T has an adjoint.*

Proof. Suppose first that T is continuous. Then for each x fixed in $\mathcal{M}$, $y \to [Ty, x]$ is a continuous module map. Then, by the above representation theorem, this is of the form $[Ty, x] = [y, u]$ for some u in $\mathcal{M}$. If we set $x \to u = T^* x$, then this is obviously an adjoint for T.

Suppose now that T has an adjoint, say T^*. For the continuity we need only the boundedness of T on the unit sphere. But the uniform boundedness principle gives the desired assertion.

12.6. Classes of Kaplansky's Inner Product Modules

In a series of papers P. P. Saworotnow, J. C. Friedel, G. R. Giellis and J. F. Smith have developed a theory of inner product modules when the algebra A (over which the module is considered) is a H*-algebra. (For definition and some properties of such algebras, see Ch. 14). Also, Saworotnow (1980) has presented an interesting generalization of the LVH-spaces of Loynes (1965) in the sense that the values of the inner product are in an H*-algebra but the module structure is not required. In this section we present results concerning such inner product modules as well as some related results.

Let A be a proper H*-algebra with an inner product $\langle\, , \rangle$.

DEFINITION 12.6.1. The trace class $\tau(A)$ of A is the set $(xy; x, y \in A)$.

DEFINITION 12.6.2. A projection in A is any nonzero hermitian idempotent. A projection is called primitive if p is not of the form $r + s$, where r, s are orthogonal projections.

DEFINITION 12.6.3. A projection base for A is any maximal family of mutually orthogonal projections.

Saworotnow and Friedel (1970) have proved that on $\tau(A)$ we can define an algebra norm if we set

$$\tau(a) = \mathrm{tr}[a],$$

where $[a]$ denotes the unique positive element satisfying the relation

$$[a]^2 = a^*a$$

and

$$\mathrm{tr}(xy) = \mathrm{tr}\langle x, y^* \rangle.$$

Then (A, τ) is a Banach *-algebra.

Now we define the inner product modules considered by P. P. Saworotnow.

DEFINITION 12.6.4 (Saworotnow, 1968). A right inner product module $\mathcal{M}$ is a complex linear space such that the following conditions are satisfied:

(1) $\mathcal{M}$ is a right module over the proper H*-algebra A,

(2) There exists on $\mathcal{M}$ an inner product $[\, ,]$ which is $\tau(A)$-valued and such that for each u, v, w in $\mathcal{M}$ and $a \in A$, $z \in C$:

(a) $[u + v, w] = [u, w] + [v, w]$,
(b) $[u, v] = [v, u]^*$,
(c) $[u, va] = [u, v]a$,
(d) $[u, zv] = z[u, v]$,
(e) $[u, v] = [a]^2, \quad [a] \geqslant 0$,
(f) $\mathcal{M}$ is complete with respect to the norm induced by $[u, v] = \mathrm{tr}[v, u]$, $|\mathrm{tr}[u, v]|^2 = \tau(u, u) \geqslant (v, v)$.

DEFINITION 12.6.5 (Saworotnow, 1968). A mapping $T: \mathcal{M} \to \mathcal{M}$ is said to be A-linear if

(1) $T(u + v) = Tu + Tv$,
(2) $T(au) = (Tu)a$

for all u, v in $\mathcal{M}$ and a in A.

DEFINITION 12.6.6 (Saworotnow, 1968). The adjoint of an A-linear operator (boundedness is defined exactly as for Kaplansky modules) is an operator denoted by T^* and satisfying the following relation: for all u, v in $\mathcal{M}$,

$$[Tu, v] = [u, T^*v]$$

holds.

Then we have the following result.

THEOREM 12.6.7 (Saworotnow, 1968). *If* $T: \mathcal{M} \to \mathcal{M}$ *is a bounded linear operator then* T *has an adjoint which is also bounded and* A*-linear.*

We mention now some characterizations of modules introduced by P. P. Saworotnow.

THEOREM 12.6.8 (Saworotnow, 1976). *Let* A *be a proper* H^**-algebra with the property that*

$$(\|p\|,\ p \text{ is a minimal projection in } A)$$

has a finite upper bound. Suppose now that $\mathcal{M}$ *is a right module over* A *satisfying the following properties:*

(1) $\mathcal{M}$ *is a complete inner product space.*
(2) *For all* $m \in \mathcal{M}$, $a \in A$ *and* $z \in C$,
$z(ma) = m(za) = (zm)a$.
(3) *For some* $k > 0$ *and all* $m \in \mathcal{M}$, $a \in A$
$\|ma\| \leqslant k\|m\|\,|a|$,
(4) $ma = 0$ *for all* $a \in A$ *implies* $m = 0$.

Then $\mathcal{M}$ *is a module over* A *in the sense of Definition* 12.6.4.

Another characterization of such modules is as follows.

THEOREM 12.6.9 (Giellis, 1972). *Let* A *be a proper* H^**-algebra and let* $\mathcal{M}$ *be a right module over* A*. Suppose that* $\mathcal{M}$ *has the following properties:*

(1) $\mathcal{M}$ *is a complete inner product space with respect to the inner product* $[\,,]$.
(2) *For all* $m \in \mathcal{M}$, $a \in A$ *and* $z \in C$,
$z(ma) = m(za) = (zm)a$.
(3) $[ma, n] = [m, na^*]$,
(4) $ma = 0$ *for all* $a \in A$ *implies that* $m = 0$.

Then $\mathcal{M}$ *is a module over* A *as in Definition* 12.6.4.

For a proof we refer to the paper by Saworotnow (1976). For a characterization related to VH-spaces see Saworotnow (1980). Interesting structure theorems for such modules are in the paper by Saworotnow and Smith (1974). For applications to stationary processes and inner product valued processes see Saworotnow (1979).

Another class of inner product modules considered (independently) by many authors represents the analogue of inner product spaces with a reproducing kernel. We refer for this to the papers by Gilbert and Hile (1977), Kakihara (1981, 1982), and Delange and Bracks (1977). We present below some results about these types of modules, that were considered by Kakihara (1981).

Suppose that $\mathcal{M}$ is an $L(H)$-module where H is a complete inner product space and let, further, N_H be the set of all elements in $L(H)$ with trace. We suppose that our module has an inner product with values in N_A and we set

$$[x, y] = \operatorname{tr}\langle x, y\rangle$$

and a norm on $\mathcal{M}$ is defined by the formula:

$$\|x\| = [x, x]^{1/2}.$$

We consider only modules which are complete with respect to this norm.

DEFINITION 12.6.10 (Kakihara, 1981). Let Ω be a set and let $f: \Omega^2 \to N_A$ be a function. We say that f is a positive definite kernel (p.d.k.) if for every finite set of points in Ω, $(w_1, \ldots, w_n)$ and every set of elements $(a_1, \ldots, a_n)$ in N_H,

$$\sum_i a_i f(w_i, w_j) a_j \geqslant 0.$$

We remark that the construction of a pre-inner product space when f is a complex-valued function can be used to construct an inner product module associated with the p.d.k. function f. As usual we consider the set $F(\Omega)$ of all functions h defined on Ω with the property that only on a finite subset in Ω is the value of h not zero. We form the tensor product $F(\Omega) \otimes L(H)$ and for each w in Ω we set for a in $L(H)$,

$$w \otimes a = \delta_w \otimes a,$$

where

$$\delta_w(t) = \begin{cases} 1, & \text{if } w = t \\ 0, & \text{otherwise.} \end{cases}$$

We then consider the left $L(H)$-module $\mathcal{M}$ generated by $\Sigma w_i \otimes a_i$, $a_i \in L(H)$ with

$$a \cdot \left(\sum w_i \otimes a_i \right) = \sum w_i \otimes a a_i.$$

An inner product with values in N_H may be defined by the formula:

$$h, g = \sum_{i,j} a_i f(w_i^1, w_j^2) b_j,$$

where

$$h = \sum w_i^1 \otimes a_i, \qquad g = \sum w_j^2 \otimes b_j.$$

If we consider the submodule

$$N_f = (h, [h, h] = 0)$$

then $\mathcal{M}/N_f$ is the required module.

We now define the notion of a reproducing kernel for inner product modules.

DEFINITION 12.6.11 (Kakihara, 1981). Let $\mathcal{M}$ be an $L(H)$-module which consists of functions on a set Ω with values in N_H. We call a function on Ω^2 a reproducing kernel for the module $\mathcal{M}$ if the following conditions are satisfied:

(1) If w is in Ω then $(w,\ )\in\Omega$,
(2) If w is in Ω and m is in $\mathcal{M}$ then

$$m(w) = [m(s), \Gamma(s, w)].$$

Then we have the following result.

THEOREM 12.6.12 (Kakihara, 1981). *If Γ is a N_H-valued p.d.k. on the set Ω then there exists a unique module $\mathcal{M}_\Gamma$ of N_H-valued functions having Γ as a reproducing kernel.*

Proof. We consider the set $\mathcal{M}_\Gamma^0$ of all functions on the set Ω with values in N_H which are of the form

$$f(s) = \sum_{i=1}^{m} a_i\Gamma(w_i, s), \qquad a_i\in L(H), \quad w_i\in\Omega.$$

Then this is clearly a left $L(H)$-module with

$$af = \sum_i aa_i\Gamma(w_i,\ ).$$

If g is another element in this module then we set

$$[f, q]_1 = \sum_{i,j} a_i\Gamma(w_i^1, w_j^2)b_j^*$$

(where the w's are clearly those for f and g, respectively). From the definition it is clear that

$$h(w) = [h(\cdot), \Gamma(w,\ )].$$

But if τ is the trace norm, then

$$\tau(h(w)) \leqslant \tau(h, h)^{1/2}\cdot\tau(\Gamma(w, w))^{1/2}$$

and then our module is a pre-inner product module. If (h_n) is a Cauchy sequence the above inequality implies that for some h with values in N_H we have the equality $\lim \tau(h_n(w) - h(w)) = 0$. We denote by $\mathcal{M}_\Gamma$ the set of all such functions h obtained from Cauchy sequences. If we set

$$[h, k] = \lim[h_n, k_n],$$

where k is defined by the Cauchy sequence (k_n), then the above formula defines an inner product and $\mathcal{M}_\Gamma$ is complete with respect to this and the existence is proved. The uniqueness up to module isomorphism we leave to the interested reader.

DEFINITION 12.6.13 (Kakihara, 1981). Let $\mathcal{M}$ be a complete inner product module and let G be a locally compact abelian group. A stochastic $\mathcal{M}$-valued process is a mapping on G with values in $\mathcal{M}$, $t\to x(t)$. We consider associated with this process, which is denoted by $(x(t))_{t\in G}$, the covariance function

$$\Gamma(s, t) = [x(t), x(s)].$$

For any stochastic process $(x(t))_{t\in G}$, the time domain is the submodule generated by $(x(t))_{t\in G}$ and is denoted by $H(\bar{x})$. The observation space with respect to a subset F in F is the submodule generated by $(x(t))_{t\in F}$.

Now if $\mathcal{M}$, $\mathcal{N}$ are two complete inner product modules over the same algebra then $U: \mathcal{M} \to \mathcal{N}$ is said to be a unitary operator if

(1) $U(\mathcal{M}) = \mathcal{N}$,
(2) for any x, y in $\mathcal{M}$, $[Ux, Uy] = [x, y]$.

Two stochastic processes $(x(t))_{t\in G}$ and $(y(t))_{t\in G}$ are said to be equivalent if there exists a unitary operator between $H(\bar{x})$ and $H(\bar{y})$.

PROPOSITION 12.6.14 (Kakihara, 1981). *If $(x(t))_{t\in G}$ is a stochastic process then the covariance function Γ is a p.d.k. function.*

Proof. Clear. □

DEFINITION 12.6.15. The stochastic process $(x(t))_{t\in G}$ is said to be stationary if $\Gamma(s, t)$ depends only on $s - t$.

DEFINITION 12.6.16. For each s in G the operator U_s is defined by the formula:

$$U_s(f(t)) = f(s + t) = f_s(t).$$

Then we have

PROPOSITION 12.6.17. *The process $(x(t))_{t\in G}$ is stationary iff U_s on $H(\bar{x})$ is unitary for each s in G.*

Proof. Since the proof is simple it is omitted. □

We now define the notion of channel.

DEFINITION 12.6.18 (Kakihara, 1981). Let G be a locally compact group and let $\mathcal{M}$ be a Kaplansky module. The triple $(G, \nu, \mathcal{M})$ is said to be a channel with input G and output $\mathcal{M}$. Here ν is a real-valued function on $G_\times B_{\mathcal{M}}$ with the properties:

(1) for every s in G, $\nu(s, \cdot)$ is a probability measure on $\mathcal{M}$ such that

$$\int_{\mathcal{M}} \|x\| \nu(s, \mathrm{d}x) < \infty,$$

(2) for every B in $B_{\mathcal{M}}$, $\nu(\cdot, B)$ is $B_{\mathcal{M}}$-measurable.

If $(G, \nu, \mathcal{M})$ is a channel then we can associate a stochastic process as the following proposition shows.

PROPOSITION 12.6.19 (Kakihara, 1981). *If $(G, \nu, \mathcal{M})$ is a channel then there exists a unique stochastic process $(x(t))_{t\in G}$ such that*

$$[x(t), y] = \int_{\mathcal{M}} \langle x, y \rangle \, \mathrm{d}\nu(t, \mathrm{d}x)$$

for each y in $\mathcal{M}$.

Proof. For fixed s in G we set

$$g_t(y) = \int_{\mathcal{M}} \langle x, y \rangle \, \mathrm{d}\nu(t, \mathrm{d}x)$$

and then we have the inequality:

$$|g_t(y)| \leqslant \int_{\mathcal{M}} |\langle x, y \rangle| \, \mathrm{d}\nu(t, \mathrm{d}x) \leqslant \|y\|_{\mathcal{M}} \int_{\mathcal{M}} \|x\| \nu(t, \mathrm{d}x)$$

and then the Fréchet–Riesz representation theorem gives the existence of $(x(t))_{t \in G}$ with the required properties.

For other results about Kaplansky modules related to some classes of stochastic processes, such as harmonizable stochastic processes, weakly harmonizable stochastic processes, connections with harmonic analysis, we refer to the papers by Ozawa (1980), Kakihara (1980, 1981, 1983, 1984), and Kakihara and Terasaki (1979).

CHAPTER 13

Quaternionic Complete Inner Product Spaces

13.0. Introduction

In this chapter we present an introduction to the theory of inner product spaces over the field of quaternions. As is known, the discovery of quaternions is attributed to the great Irish mathematician R. Hamilton*. In recent years the interest in spaces over the quaternions has increased due in part to their possible application in quantum mechanics. We shall present some facts concerning linear spaces over Q and linear operators on such spaces (Q will be used to denote the set of quaternions). Other topics are very briefly mentioned.

13.1. The Quaternions

Let $1, i, j, k$ be the standard basis of the space R^4 over the reals.

DEFINITION 13.1.1. A quaternionic product on R^4 is a bilinear mapping $R^4{}_\times R^4 \to R^4$ with unity 1 defined by the formulas:

(1) $i^2 = j^2 = k^2 = -1$,

(2) $ij = k = -ji, jk = i = -kj, ki = j = -ik$.

PROPOSITION 13.1.2. *A quaternionic product is associative.*

Proof. Since the proof is simple we omit it.

REMARK 13.1.3. A quaternionic product on R^4 is not commutative.

DEFINITION 13.1.4. A linear space R^4 with a quaternionic product is called the algebra of quaternions.

DEFINITION 13.1.5. Every quaternion q is uniquely decomposed as

$$q = q_1 + q_2,$$

where q_1 is in R and q_2 in R^3. The element q_1 is called the real part of q and q_2 is called the pure part of q.

PROPOSITION 13.1.6. *The centre of the algebra of quaternions is R. R is identified with the real part of quaternions.*

Proof. If q is in R then clearly it is in the centre of the algebra of quaternions.

*As is mentioned by C. Truesdell (*Bull. Amer. Math. Soc.* 9 (1983), 306), 'Quaternions themselves were first discovered, applied and published by Rodrigues, Poisson's former pupil, in 1840'.

Conversely, if q is a quaternion in the centre then q is of the form

$$q = a + bi + cj + dk$$

where a, b, c, d are real numbers. Since q is in the centre we get immediately that $b = c = d = 0$. □

DEFINITION 13.1.7. The conjugate of the quaternion $q = a + bi + cj + dk$ is the quaternion denoted by $\bar{q}$ and defined by the formula

$$\bar{q} = a - bi - cj - dk.$$

Then we have

PROPOSITION 13.1.8. *The conjugation $q \to \bar{q}$ on the algebra Q of the quaternions has the following properties*:

(1) $\overline{(q + p)} = \bar{p} + \bar{q}$.

(2) $r \in R$, $\overline{(rq)} = r\bar{q}$.

(3) $(\bar{\bar{q}}) = q$.

(4) $\overline{(pq)} = \bar{q}\bar{p}$.

Proof. Clear. □

On the quaternion algebra Q we define a quaternionic-valued inner product as follows.

DEFINITION 13.1.9. An inner product on the algebra of quaternions is defined as follows: if q_1, q_2 are two arbitrary quaternions then we set

$$\langle q_1, q_2 \rangle = \tfrac{1}{2}(q_1\bar{q}_2 + \bar{q}_2 q_1).$$

PROPOSITION 13.1.10. *The inner product defined above satisfies the following properties*:

(1) $\langle q, q \rangle \geqslant 0$, $\langle q, q \rangle = 0$ *iff* $q = 0$,

(2) $\langle q_1, q_2 \rangle = 0$ *iff* q_1 *and* q_2 *are anticommuting quaternions*,

(3) $\langle\, ,\, \rangle$ *is linear in the first variable*,

(4) $\langle q_1, q_2 \rangle = \overline{\langle q_2, q_1 \rangle}$.

Proof. Since the proofs are simple we omit them. □

DEFINITION 13.1.11. The norm of a quaternion q is the number defined by the formula

$$|q| = \langle q, q \rangle^{1/2}.$$

Then we have the following results.

PROPOSITION 13.1.12. *For the algebra Q of quaternions the following assertions* *hold*:

(1) *If* $q \neq 0$ *then* q *is invertible and*
$$q^{-1} = |q|^{-2}\bar{q}.$$
(2) *If* q_1, q_2 *are two arbitrary quaternions then*
$$|q_1 q_2| = |q_1| \cdot |q_2|.$$

Proof. If q is a nonzero quaternion then we set

$$p = |q|^{-2}\bar{q}$$

and we remark that

$$pq = qs = 1.$$

This gives the first assertion. For the second assertion we remark that

$$|q_1 q_2|^2 = \overline{(q_1 q_2)}(q_1 q_2) = \bar{q}_2 \bar{q}_1 q_1 q_2 = \bar{q}_2 |q_1|^2 q_2 = |q_1|^2 |q_2|^2$$

and clearly the second assertion follows.

DEFINITION 13.1.13. A quaternion q with the property that $|q| = 1$ is called a unit quaternion.

For more information about quaternions and their geometrical applications we refer to Qualding (1979).

13.2. Linear Spaces over Quaternions

Since the algebra of quaternions is not commutative we must distinguish two types of linear spaces over Q, namely the left linear spaces and the right linear spaces.

DEFINITION 13.2.1. A left linear space X over Q is a linear space over Q with the map $Q_{\times} X \to X$ defined by $(q, x) \to qx$ such that all axioms of a linear space hold and that

$$p(qx) = (pq)x$$

for any x in X and p, q in Q.

DEFINITION 13.2.2. A right linear space over Q is a linear space over Q with the map $X_{\times} Q \to X$ such that all axioms of a linear space hold and that

$$(xp)q = x(pq)$$

holds for all x in X and p, q in Q.

Since the theory of left linear spaces is quite similar to the theory of right linear spaces, in what follows we consider only left linear spaces.

DEFINITION 13.2.3. Let X, Y be two left linear spaces over Q. We say that the mapping $T: X \to Y$ is a linear operator on X with values in Y if the conditions

(1) $T(x + y) = Tx + Ty$ (we use the notion $Tx = T(x)$),

(2) $T(rx) = rTx$

are satisfied for all x, y in X and r in R.

REMARK 13.2.4. A simple example shows that we cannot consider operators with property (1) from above and $T(qx) = qTx$ for any x in X and q an arbitrary quaternion.

DEFINITION 13.2.5. Let X be a left linear space over q. A seminorm on X is a function $p: X \to R$ such that:

(1) $p(x + y) \leqslant p(x) + p(y)$,

(2) $p(qx) = |q| p(x)$

for all x, y in X and q in Q. We say that p is a norm (and then we use the notation $(\| , \|)$ if $p(x) = 0$ iff $x = 0$.

DEFINITION 13.2.6. The metric induced by the norm p is defined by the formula

$$d(x, y) = p(x - y)$$

and the topology induced by the metric is called the norm topology (or the strong topology).

DEFINITION 13.2.7. A left linear space over Q with a norm which is a complete metric space with respect to the metric induced by the norm is called a left quaternionic Banach space.

We now introduce a special class of space, namely spaces on which there exists an inner product with values in Q.

DEFINITION 13.2.8. An inner product on a left linear space X over Q is a mapping $\langle , \rangle$ on $X_{\times} X$ with values in Q and such that the conditions

(1) $\langle x, x \rangle \geqslant 0, \quad \langle x, x \rangle = 0$ iff $x = 0$,

(2) $\langle px + qy, z \rangle = p\langle x, z \rangle + q\langle y, z \rangle$,

(3) $\langle x, y \rangle = \overline{\langle y, x \rangle}$,

(4) $\langle x, py + qz \rangle = \langle x, y \rangle \bar{p} + \langle x, z \rangle \bar{q}$

are satisfied for all x, y, z in X and p, q in Q.

PROPOSITION 13.2.9 (the Cauchy–Buniakowsky inequality). *Let X be a left linear space over Q with an inner product $\langle , \rangle$. Then for any x, y in X the following inequality holds:*

$$|\langle x, y \rangle|^2 \leqslant \langle x, x \rangle \cdot \langle y, y \rangle.$$

Proof. Since the proof is not essentially different from the case of complex spaces it is omitted.

COROLLARY 13.2.10. *Let X be a left linear space over Q and on X an inner product*

$\langle\,,\rangle$. *Then*

$$x \to \|x\| = \langle x, x\rangle^{1/2}$$

is a norm on X.

Proof. Exactly as in the complex case. □

DEFINITION 13.2.11. A left quaternionic complete inner product space is a left quaternionic Banach space whose norm is generated by an inner product.

DEFINITION 13.2.12. Let X be a left quaternionic complete inner product space (q.c.i.p.s.) with the inner product $\langle\,,\rangle$. If x, y are elements in X then we say that these are orthogonal if $\langle x, y\rangle = 0$. If M is a subset in X then $M^\perp$ denotes the set

$$(x, \langle x, m\rangle = 0).$$

REMARK 13.2.13. It is clear from the definition that $M^\perp$ is a closed linear subspace of X.

DEFINITION 13.2.14. A family $(e_i)_{i\in I}$ of elements in X is said to be orthonormal if

$$\langle e_i, e_j\rangle = \delta_{ij} = \begin{cases} 1, & i = j \\ 0, & i \neq j \end{cases}.$$

Let X be a left quaternionic complete inner product space and let $\mathscr{F}$ be the set of all orthonormal families of elements of X. Then we can define an order relation (a partial order relation) on $\mathscr{F}$ by inclusion. Then it is not difficult to see that we can apply Zorn's lemma and we get maximal families.

DEFINITION 13.2.15. A base in X is any orthonormal family which is a maximal element in $\mathscr{F}$.

PROPOSITION 13.2.16. *Every left quaternionic complete inner product space has a basis.*

We leave to the interested reader to translate to this setting properties of complex complete inner product spaces. We mention only two results which we need.

PROPOSITION 13.2.17. *For any closed left linear subspace* M *of* X *there exists the (unique) decomposition:*

$$X = M \oplus M^\perp$$

such that every element x *in* X *is of the form*

$$x = x_m + x_m^\perp$$

with $x_m \in M$, $x_m^\perp \in M^\perp$.

PROPOSITION 13.2.18. *If* F *is a continuous linear functional on* X *then there exists a unique element* x_0 *in* X *with the following properties:*

(1) $F(x) = \langle x, x_0\rangle$,

(2) $\|F\| = \|x_0\|$.

Thus the Fréchet–Riesz representation theorem holds in left quaternionic complete inner product spaces.

REMARK 13.2.19. Concerning the extension problem of linear functionals (which preserves the norm), we mention that the problem was solved, positively, by G. A. Sukhomlinov in 1938.

Suppose now that X, Y are two left linear quaternionic spaces and consider a mapping $T: X \to Y$.

DEFINITION 13.2.20. The mapping T is said to be linear if the properties

(1) $T(x + y) = Tx + Ty$ (we use the notation $T(x) = Tx$ whenever no confusion may be possible),
(2) $T(qx) = qTx$

hold for all x, y in X and q in $Q \cap R$.

We say that T is bounded if

$$\sup(\|Tx\|, \|x\| \leqslant 1) < \infty$$

and in this case the number is called the norm of T. We denote the set of all bounded linear operators on X with values in Y by $L(X, Y)$ and $L(X, X) = L(X)$.

Now we remark that $L(X, Y)$ is a real linear space, and $L(X)$ is a real Banach *-algebra.

DEFINITION 13.2.21. For any T in $L(X)$ an element T^* which has the property that for all x, y in X,

$$\langle Tx, y\rangle = \langle x, T^*y\rangle,$$

is called the adjoint of T.

We leave the interested reader to prove the analogue of properties from complex space to this setting.

13.3. The Symplectic Image of a Left Quaternionic Complete Inner Product Space

Let X be a left complete quaternionic inner product space with an inner product $\langle\ ,\ \rangle$. Then we can consider X as a complex linear space which, in this case, is denoted by X^c. This X^c coincides as a set with X but the scalar multiplication is restricted to complex numbers as well as to values of the inner product.

DEFINITION 13.3.1. The symplectic image of X is the complex complete inner product space X^c.

We remark also that the norms of x, as an element of X and as an element of X^c, coincide.

We now consider the map $K: X^c \to X^c$ defined by the formula

$$Kx = kx,$$

where k is the unit quaternion k. Then clearly the following assertions hold:

(1) $\langle Kx, Ky \rangle = \langle y, x \rangle_c$,

(2) $K^2 = -I$.

(We used the notation $\langle\ ,\ \rangle_c$ to denote the inner product on X^c.) We note that every operator T on X may be considered as an operator on X^c and in this case we use the notation T^c for T.

REMARK 13.3.2. The spectrum of T^c is symmetric with respect to the real axis.

We now consider an important class of operators on X.

DEFINITION 13.3.3 (Viswanath, 1971). The operator J on X is said to be imaginary if there exists an orthonormal basis $(e_i)_{i \in I}$ of X such that

$$Je_s = ie_s$$

for each s in I (in the form of J, i is the quaternion i). Then we have the following result.

THEOREM 13.3.4 (Viswanath, 1971). *For an element T in $L(x)$ the following assertions are equivalent:*

(1) *T is imaginary.*

(2) *$T = -T$ and $I - TT$ is the projection on the null space of T.*

(3) *T is normal (i.e. $TT = T\ T$) and $-T^2$ is a projection.*

(4) *T is normal and $T^2 + T^4 = 0$.*

(5) *$T^c = iP = KPK^{-1}$ for a unique projection O in X^c.*

Proof. The implications (1) → (2) → (3) → (4) are obvious. Suppose now that (4) holds. Thus J^c has a spectral measure concentrated at $(\pm i)$ and thus $Tx = ix$ is valid iff $T(ix) = -i(kx)$, which gives that $T^c = -iP - iKPK^{-1}$.

Suppose now that (5) holds. If $(e_i)_{i \in I}$ is an orthonormal basis for the range of P in X^c then this is a base for the range of $P + KPK^{-1}$ in X. Suppose that $(f_j)_{j \in J}$ is a basis for ker T in X. Then (e_i, f_j) is a basis for X and

$$Te_s = ie_s, \quad s \in I$$

$$Tf_j = 0, \quad j \in J.$$

This proves the assertion and the theorem. □

13.4. Classes of Operators on Left Quaternionic Inner Product Spaces

We now present several classes which are, to some extent, the analogue of the corresponding class from the case of complex complete inner product spaces.

Let X be a left quaternionic complete inner product space and let $L(X)$ denote the set of all bounded linear operators on X. For any T in $L(X)$, T denotes the adjoint. We recall that in the case of complex spaces a complex number z is said to be an eigenvalue for T if for some nonzero x in the space $Tx = zx$ holds. Then clearly for any complex number z_0 the element $y = z_0x$ corresponds to the same eigenvalue.

In the case of quaternionic spaces the situation is not the same and what is invariant is in fact the conjugation class of the eigenvalue. (We recall that two quaternions p and q are said to be conjugate if $p = sqs^{-1}$ with s a quaternion with $|s| = 1$; this is an equivlance relation and each equivalence relation is a conjugation class.)

DEFINITION 13.4.1. The element T in $L(X)$ is said to be:

(1) hermitian if $T = T^*$,
(2) normal if $TT^* = T^*T$,
(3) hyponormal if $T^*T - TT^* \geqslant 0$,
(4) of class N if $\|Tx\|^2 \leqslant \|T^2x\| \, \|x\|$.

It is not difficult to see that all the above classes are distinct and the following inclusions hold:

(hermitian) $\subset$ (normal) $\subset$ (hyponormal) $\subset$ (of class N).

DEFINITION 13.4.2. For any T in $L(X)$ the spectral radius is the number defined by

$$r_T = \lim \|T^n\|^{1/n}.$$

DEFINITION 13.4.3. The element T in $L(X)$ is said to be normaloid if $r_T = |T|$.

Exactly as in the case of complex spaces (see the author's book (1981) for details) we can prove the following result.

PROPOSITION 13.4.4. *Every operator of class N is normaloid.*

PROPOSITION 13.4.5. *If for some m,* $T^m x = 0$, *then* $Tx = 0$ *when T is of class N.*

PROPOSITION 13.4.6. *If T is hyponormal then* $T + rI$, *r in R is a hyponormal operator.*

DEFINITION 13.4.7. Let T be an element of $L(X)$. The numerical range of T is the set

$$W(T) = (\langle Tx, x \rangle, \|x\| = 1).$$

For results concerning the numerical ranges for operators on left quaternionic inner product spaces we refer to Powers (1976) and Torgašev (1976, 1978) where the reader may find positive and negative results as well as some interesting problems concerning $W(\, , \,)$ and related classes of operators.

13.5. Spectral Theory on Left Quaternionic Complete Inner Product Spaces

Let $(S, \mathscr{B}, \mu)$ be a measure space with $\mu \geqslant 0$. As usual, we use the term 'function' when we really mean classes of equivalence (two functions being in the same class whenever they differ on a set of measure zero).

We consider on S functions with values in Q. Further we consider E be a countable additive set function on $\mathscr{B}$ with valued projections on X with the property that $E(S) = I$. Then for each essentially bounded function f which is real we may define the mapping

$$(x, y) \to \int_S f(s)\,\mathrm{d}\langle E(s)x, y\rangle$$

and as in the complex case we find a unique hermitian operator A_f with the property that

$$\langle A_f\, x, y\rangle = \int_S f(s)\,\mathrm{d}\langle E(s)x, y\rangle.$$

For the operator A_f we use the symbol

$$A_f = \int_X f\mathrm{d}E.$$

Then the mapping $f \to A_f$ defines a functional calculus with properties similar to the complex case.

Now in the case of nonhermitianity we cannot use the above approach since the integral $\int_S f\,\mathrm{d}E$ is meaningless for f not essentially real due to the noncommutativity of Q. For this purpose we introduce the notion of a 'spectral system'.

DEFINITION 13.5.1 (Viswanath, 1971). Let J be an imaginary operator on the left quaternionic complete inner product space X, and let E be a spectral measure on $\mathscr{B}$ with values projections on X. We say that J is admissible with respect to E if the following conditions are satisfied:

(1) J commutes with each $E(M)$, $M \in \mathscr{B}$,

(2) for some M_0 in $\mathscr{B}$, $I - JJ^* = E(M_0)$.

DEFINITION 13.5.2 (Viswanath, 1971). The pair (E, J) is then said to be a spectral system.

We now consider the real *-Banach algebra $\mathscr{A}$ of all E-essentially bounded real-valued functions on $S = C^+ = (q, b \geqslant 0,\ q = a + bi + cj + dk)$.

DEFINITION 13.5.3 (Viswanath, 1971). For each f in $\mathscr{A}$, $f = f_1 + if_2$ the integral $\int_S f\,\mathrm{d}E$ is the operator given by the formula:

$$\int_S f_1\,\mathrm{d}E + J\left(\int_S f_2\,\mathrm{d}E\right).$$

Then, as in the complex case, we can prove the following result.

THEOREM 13.5.4 (Viswanath, 1971). *The mapping $f \to \int_S f\,\mathrm{d}E$ defines a functional calculus.*

Then the spectral theorem for normal operators is as follows.

THEOREM 13.5.5 (Viswanath, 1971). *Let N be a bounded normal operator on X. Then there exists a unique spectral system (E, J), E being a spectral measure, J an* *imaginary admissible operator,* $I - JJ = E(R)$, E *is on* C^+ *and*

$$N = \int_S z \, \mathrm{d}E(z).$$

COROLLARY 13.5.6 (Viswanath, 1971). *If* N *is a normal operator in* $L(X)$ *then* N *has the unique decomposition*

$$N = A + JB,$$

where J *is a imaginary operator,* A, B *are hermitian with* $B \geqslant 0$ *and* (A, J, B) *are mutually commuting.*

Using this result we can prove that Fuglede's theorem holds for the space $L(X)$. This positively answers the question about the validity of Fuglede's theorem for quaternion spaces posed by Berberian (1959).

We mention that a variant of Fuglede's theorem obtained from the complex case by complexification is mentioned by Viswanath (1978).

THEOREM 13.5.7 (Istrăţescu, 1984). *Let* N *be a normal element in* $L(X)$ *and suppose that for some* T *in* $L(X)$ *we have* $TN = NT$. *Then* $TN = NT$ *holds.*

Proof. For the operator N we have the decomposition $N = A + JB$ and also the relations $TN^m = N^m T$ hold, and consider now the function of a complex variable z in C defined by the formula:

$$\mathrm{e}^{zN} T \mathrm{e}^{-zN} = f(z) = \mathrm{e}^{zN} \mathrm{e}^{-\bar{z}N} \, T \, \mathrm{e}^{\bar{z}N} \mathrm{e}^{-zN}$$
$$= \int_S \mathrm{e}^{2i(ys - xt)} \, \mathrm{d}E(s, t) T \int_S \mathrm{e}^{2i(xt - ys)} \, \mathrm{d}E(s, t)$$

which implies that the function f has a constant norm. But in this case we get that $f = \text{const}$. This clearly implies our assertion. □

13.6. Functional Calculus for Operators of Left Quaternionic Complete Inner Product Spaces

We remarked that K. Viswanath has proposed a useful functional calculus for normal operators. Contributions in extending the functional calculus to certain classes of operators were made by Powers (1976), Suljagič (1979, 1983) and Torgašev (1978, 1979, 1980). In this section we present a functional calculus following Torgašev (1978).

We consider X to be a left quaternionic complete inner product space and we defined for T in $L(X)$ the notion of a spectrum.

DEFINITION 13.6.1. The spectrum $\sigma(T)$ of T is the spectrum of the operator T^c in $L(X^c)$.

Now if M is a set in the complex plane then we set M^* to be the set $(\bar{m}, m \in M)$. (Where $\bar{m}$ denotes the complex conjugate of the number m). We use the notation $K_0 = jI$ and $K(q) = qI$ for q in Q. Here j is the unit quaternion j.

For T in $L(X)$ we consider the family $\mathscr{F}_T^c$ of all complex-valued functions defined on an open subset G_f of the complex plane, and such that the following properties hold:

(1) $\sigma(T) \subset G_f$,

(2) f is locally analaytic in G_f.

We call $\mathscr{F}_T^{c,r}$ the subfamily of all functions in $\mathscr{F}_T^c$ with the property that

$$f(z) = \overline{f(\bar{z})}$$

for all z in $G_f \cap G_f^*$.

DEFINITION 13.6.2 (Torgašev, 1978). Let T be in $L(X)$ and let f be in $\mathscr{F}_T^c$, with D a bounded domain in the complex plane with

$$\sigma(T) \subseteq D \subset \bar{D} \subset G_f.$$

Then we set

$$f(T) = -1/(2\pi i) \int_{\partial D^+} K(f(z)) R(T, z) \, \mathrm{d}K(z).$$

(D^+ is the part of D in the upper complex plane and D^- is the part of D in the lower plane).

Now since $\sigma(T)$ is always an r-symmetric set (i.e. $M = M^*$) then we may suppose without loss of generality that our domain is also r-symmetric.

DEFINITION 13.6.3 (Torgašev, 1978). We say that f in $\mathscr{F}_T^c$ is equivalent to g in $\mathscr{F}_T^{c,r}$ if f is equal to g on an open r-symmetric set containing $\sigma(T)$.

Then we have

THEOREM 13.6.4 (Torgašev, 1978). *For any f in $\mathscr{F}_T^c$ the operator $f(T)$ is in $L(X)$ iff there exists g in $\mathscr{F}_T^{r,c}$ which is equivalent to f.*

Proof. The sufficiency part. To prove that $f(T)$ is in $L(X)$ it suffices to prove that $K_0(g(T)\bar{K}_0) = g(T)$. □

Now

$$K_0 g(T)\bar{K}_0 = (2\pi i)^{-1} \int_{D^+} K_0 K(g(z)) R(T, z) \, \mathrm{d}(z\bar{K}_0)$$

$$= (2\pi i)^{-1} \int_{D^+} K(\bar{g}(z) R(T, \bar{z}) \, \mathrm{d}K(\bar{z})$$

$$= (2\pi i)^{-1} \int_{D^+} K(g(\bar{z})) R(T, \bar{z}) \, \mathrm{d}K(\bar{z})$$

and since D is r-symmetric z is in D^+ iff the conjugate is in D^-, which implies further that

$$K_0 g(T)\bar{K}_0 = (2\pi i)^{-1} \int_{D^-} K(g(u)) R(T, u) \, \mathrm{d}K(u)$$

$$= -(2\pi i)^{-1} \int_{D^+} K(g(z)) R(T, z) \, \mathrm{d}K(z) = g(t).$$

Conversely, if $f(T)$ is in $L(x)$, i.e. if $K_0 f(T)\bar{K}_0 = f(t)$, then

$$f(T) = -(2\pi i)^{-1} \int_{D^+} K(\bar{f}(\bar{z}))R(T, z)\, \mathrm{d}K(z) = g(T)$$

with g given by the formula

$$g(z) = \tfrac{1}{2}(f(z) + \bar{f}(\bar{z})).$$

The notion of a spectral set in the quaternionic case was introduced by A. Torgašev in the following manner.

DEFINITION 13.6.5 (Torgašev, 1978). Let T be in $L(X)$. A set F in the extended complex plane is said to be a q-spectral set for T if the following conditions are satisfied:

(1) F is closed and contains (T).
(2) For every r-symmetric rational function f having no poles in F,

$$\| f(T)\| \leqslant \| f \|_F = \sup(|f(z)|, z \in F).$$

It is clear that if F is a spectral set for T^c then F is a spectral set for T. Then we have the following natural question posed by A. Torgašev.

QUESTION 13.6.6. Are the spectral sets for T and T^c the same?

In his (1978) paper A. Torgašev has remarked that the von Neumann characterization of inner product structures remains valid for the case of contraction spaces.

THEOREM 13.6.7. *Let X be a left quaternionic complete inner product space with the property that $(z, |z| \leqslant 1)$ is a spectral set for every contraction in $L(X)$. Then the norm of X is generated by an inner product.*

The proof of this theorem is modeled on the case of complex spaces. For other results concerning spectral sets for elements in $L(X)$, as well as for other related results, we refer to Torgašev (1978) and for an elegant study of quaternionic Banach algebras we refer to his paper (1978). The reader may find related interesting results in the paper of Viswanath (1978). We also mention that results about dilation theory, the Beurling theorem, etc., can be found in S. Suljagič's papers of 1980 and 1983. For unbounded operators on quaternionic spaces we refer to Yuang-Yung Tseng (1935) and Stone (1951). There is also an interesting paper of Goldstine and Horwitz (1964, 1966) in which a theory of inner product spaces over the algebra of Cayley numbers is developed.

The interest in quaternionic spaces is also due in part to their possible applications in physics. It is worth mentioning that Birkhoff and von Neumann (1935) suggested the possibility of a quaternionic quantum mechanics. In this sense the reader is referred to the paper by Finkelstein *et al.* (1962) where he may find the foundations of quaternionic quantum mechanics as well as a very interesting physical motivation for the use of quaternions.

At this point it may be of interest to know if the famous Wigner theorem holds for left quaternionic complete inner product spaces. For Wigner's theorem see Wigner (1929, 1959), Bargman (1964), Bracci *et al.* (1975) as well as the literature quoted there and at the end of our book.

CHAPTER 14

Inner Product Algebras

14.0. Introduction

The class of algebras over the complex field C, on which there is an inner product which has some properties related to the algebraic operations as well as to the involution on the algebras, plays an important role in several chapters of analysis, especially in the study of von Neumann algebras and in the extensions of the famous Plancherel formulas.

An interesting and important class of inner product algebras was introduced by W. Ambrose in 1945 and 1949 under the name H*-algebras. Several related classes of inner product algebras were introduced by Keown (1953, 1967), Yood (1969, 1974). For the applications to von Neumann algebras we refer to the book on von Neumann algebras of Dixmier (1957, 1969). For applications to Plancherel formulas we refer to Godement (1954), Lance (1975) and Southerland (1973). In this chapter we present an introduction to the study of such algebras.

14.1. Inner Product Algebras

We now consider the class of inner product algebras and we prove some of their properties.

DEFINITION 14.1.1. An algebra A over the complex field is said to be an inner product algebra if on A there exists an involution and an inner product $\langle\,,\rangle$ such that the following properties hold:

(1) $\langle x, y\rangle = \langle x^*, y^*\rangle$, $x, y \in A$,
(2) $\langle xy, z\rangle = \langle y, x^* z\rangle$, $x, y, z \in A$,
(3) For each $a \in A$, the operator $L_a\colon A \to A$, defined by the formula $L_a(x) = ax$, is continuous.
(4) The set $(xy, x, y \in A)$ is dense in A.

DEFINITION 14.1.2. We say that an inner product algebra is a pre-H*-algebra if the inner product $\langle\,,\rangle$ satisfies the inequality:

$$\langle xy, xy\rangle \leqslant \langle x, x\rangle \cdot \langle y, y\rangle$$

for all x, y in A; if A is complete then we say that it is an H*-algebra.

We now give some examples of inner product algebras.

EXAMPLE 14.1.3. Let A be the algebra of all continuous complex-valued functions on $[0,1]$ with the usual algebraic operations and the involution '*' defined by $(f^*)(t) = \bar{f}(t)$ (where the bar denotes the complex conjugate).

If for any f, g in A we set

$$\langle f, g\rangle = \int_0^1 f(t)\bar{g}(t)\,\mathrm{d}t$$

then A is an inner product algebra.

EXAMPLE 14.1.4. Let A be a C*-algebra and let φ be a special state of A (i.e. if $\varphi(x^*x) = 0$ then $x = 0$). If we set

$$\langle x, y\rangle = \varphi(y^*x)$$

then A is an inner product algebra.

EXAMPLE 14.1.5. Let AP be the set of all almost periodic continuous functions on R which is an algebra with the usual operations. If we set

$$\langle f, g\rangle = M(f\bar{g}),$$

where M is the mean value of $f\bar{g}$, then AP is an inner product algebra.

14.1.6. Let L^∞ be the Banach algebra of all essentially bounded (measurable) functions on $[0, 1]$. If μ is a measure on the ideal space of L^∞ such that

$$F(f) = \int_\Omega \hat{f}(s)\,\mathrm{d}\mu(s)$$

($\hat{f}$ is the Gelfand transform of f) is a special functional, then L^∞ is an inner product algebra if we set for f, g in L^∞,

$$\langle f, g\rangle = F(f\bar{g}).$$

If A is an inner product algebra (with the inner product $\langle\,,\rangle$ then R_a denotes the operator defined on A which is the right multiplication by a. If H_A is the completion of A with respect to the norm generated by the inner product $\langle\,,\rangle$ then the operators L_a and R_a extend to bounded operators on H_A which are denoted by $\bar{L}_a$ and $\bar{R}_a$ respectively.

We remark that the following relations hold:

(1) $\|L_a\| = \|\bar{L}_a\| = \|\bar{L}_{a^*}\| = \|L_{a^*}\|$.

(2) $\|R_a\| = \|\bar{R}_a\| = \|R_{a^*}\| = \|R_{a^*}\|$.

(3) $\|\bar{R}_a\| \leqslant \|\bar{L}_a\|$.

Indeed, the first relations in (1) and (2) are clear. To see that (3) holds we remark that

$$\|\bar{R}_a\| = \sup_{\|x\|\leqslant 1} \|(xa)^*\| = \sup_{\|x\|\leqslant 1} \|L_{a^*}(x^*)\| \leqslant \|L_{a^*}\|$$

and thus

$$\|\bar{R}_a\| \leqslant \|L_a\| = \|\bar{L}_a\|.$$

Similarly we prove that

$$\|\bar{L}_{a^*}\| \leqslant \|\bar{R}_a\|$$

and thus (3) follows.

DEFINITION 14.1.7. An element a of the inner product algebra A is called right (left) bounded if the mapping $R_a(L_a)$ defined by

$$R_a(x) = \bar{L}_x(a)$$

$(L_a(x) = \bar{R}_x(a))$ is bounded.

It is not difficult to see that an element a is right (left) bounded if and only if it is left (right) bounded, and in this case we call it bounded. The set of all bounded elements of A is denoted by A_b.

Also this is again an inner product algebra with the operations and the involution induced by A. This remark was first noted by Godement (1954). The algebra is said to be full if $A = A_b$. Sometimes A_b is said to be the fulfilment of A.

DEFINITION 14.1.8. An inner product algebra is said to be orthocomplemented if whenever I is a closed ideal (right or left) in A then

$$A = I \oplus I^{\perp}.$$

If A is an inner product algebra and S is any subset in A then we set:

$$S^{\perp} = (x, x \in A, \langle x, s \rangle = 0, s \in S)$$
$$L(S) = (x, x \in A, xs = 0, s \in S)$$
$$R(S) = (x, x \in A, sx = 0, s \in S).$$

DEFINITION 14.1.9. An inner product algebra is said to be a right-modular complemented algebra if the following conditions are satisfied:

(1) $x \to xy$, $y \to xy$ are continuous for each fixed y and each fixed x respectively.
(2) If I is a right or left ideal in A with $I^{\perp} = 0$ then I is dense in A.
(3) The intersection of the closed modular maximal right ideals is 0, and for each closed modular maximal ideal m, $m^{\perp}$ is a right ideal.

Similarly, we can define the notion of left-modular complemented algebra.

For the proof of the following result we refer to B. Yood.

THEOREM 14.1.10 (Yood, 1974). *For an inner product algebra A the following assertions hold:*

(1) *If A is orthocomplemented then it is a dual algebra.*
(2) *A is an anihilator algebra if and only if A is a dual algebra.*

There exist inner product algebras which are dual algebras and not orthocomplemented.

To see this we consider the inner product algebra of all continuous complex-valued functions on the torus (one-dimensional). Then I, the set of all functions which admit an analytic continuation in $(z, |z| < 1)$, is an ideal in A and $I \oplus I^{\perp}$ is dense in A but A is not equal to $I \oplus I^{\perp}$.

Another example can be given by using the algebra H^{∞}. For details of such a construction we refer to Hofmann (1962).

THEOREM 14.1.11 (Yood, 1974). *If A is an inner product algebra which is full then it is orthocomplemented.*

Proof. Let J be a closed right ideal in A and let J^c be its completion in H_A (=the completion of A). Then we can consider the orthogonal projection of H_A onto J^c, say P. We remark that J^c is a right invariant subspace of H_A. This means that if $b \in J^c$ and $a \in A$ then $R_a(c) \in J^c$. Indeed, since b is in J^c there exists a sequence (b_n) in I with the property that $b = \lim b_n$ (in fact the Cauchy sequence). Then we have

$$\lim \bar{R}_a(b_n) = \bar{R}_a(b) = \lim b_n b \in J^c.$$

Now we have

$$P(A) = A(J^c) = J$$

and since $H_A = J^c \oplus (J^c)^-$ for each x in A, we have the decomposition $x = x_J + x_J^{\perp}$ with $x_J \in J^c$, $x_J^{\perp} \in (J^c)^-$. But $x_J = Px \in A$ and thus $x_J^{\perp}$ is in A. The theorem is proved. □

REMARK 14.1.12. There are examples of orthocomplemented algebras which are not full. For example we can consider the algebras C_p with p in $(1, 2)$ with the inner product induced by C_2.

DEFINITION 14.1.13. Let A be an inner product algebra. Then the associated $C^*(A)$ of A is the norm closure of $(\bar{L}_a, a \in A)$ in $L(H_A)$.

DEFINITION 14.1.14. The Rieffel norm of an element x in an inner product algebra A is defined as follows:

$$\|x\|_R = \|x\| + \|\bar{L}_x\|.$$

The following assertion is obvious.

PROPOSITION 14.1.15. *$(A, \|\,,\|_R)$ is a normed algebra.*

DEFINITION 14.1.16. The inner product algebra A is said to be replete if the completion of A with respect to $\|\,,\|_R$ is equal to A.

PROPOSITION 14.1.17. *Suppose that the inner product algebra A is replete. Then A is a *-ideal in $C^*(A)$ and the involution is symmetric.*

Proof. First we remark that

$$\|xy\|_R \leqslant \|\bar{L}_x\| \, \|y\|_R.$$

We now consider y in A and w in $C^*(A)$. From the definition of $C^*(A)$ there exists (z_n) in A such that

$$w = \lim \bar{L}_{z_n}.$$

Then we have,

$$\|z_n y - z_m y\|_R \leqslant \|z_n - z_m\| \, \|y\|_R$$

which implies that $(z_n y)$ is a Cauchy sequence, with respect to the Rieffel norm. Since A is supposed to be replete, we find u in A with the property that $\lim(u - z_n y) = 0$ and thus

$$\lim(\bar{L}_u - \bar{L}_{z_n}) = 0.$$

But

$$w\bar{L}_y - \bar{L}_{z_n}\bar{L}_y = (w - \bar{L}_{z_n})\bar{L}_y$$

and thus

$$w\bar{L}_y = \lim \bar{L}_{z_n y} \quad \text{or} \quad w\bar{L}_y = \bar{L}_u.$$

Thus A is a left ideal in $C^*(A)$ and since A is a *-algebra in $C^*(A)$ it is a two-sided ideal of $C^*(A)$.

Since the symmetry of the involution implies the symmetry on its two-sided 8-ideals we have the assertion of the proposition. For other results we refer to Rieffel (1969) and Yood (1969, 1974).

14.2. Complete Inner Product Algebras with Identity

Suppose now that A is an inner product algebra which has the following properties:

(1) A is a complete inner product space.
(2) $\|xy\| \leqslant \|x\| \, \|y\|$.
(3) A has an identity, (which we denote by e).

It was conjectured by I. Kaplansky that any such algebra is the field of complex numbers. In fact, the conjecture of Kaplansky was more general; the only division algebra which is an inner product space and is complete satisfying (2) and (3) over the real is in fact the set of real numbers, the set of complex numbers or the set of quaternions.

The conjecture was proved in the affirmative by Ingelstam (1963) using some results from an earlier paper (1962) and some results from the geometry of the Banach unit sphere combined with complex function theory.

Our proof, which we present only for the complex case, seems to be simpler and has the advantage, in an appropriately modified form, of yielding an interesting result of B. Yood concerning the possibility of a B*-algebra being an inner product algebra (the result states that such a possibility does not exists).

14.2.1. *If A is a complete complex algebra which has properties* (2) *and* (3) *then A coincides with C (the set of complex numbers).*

Proof (Istrăţescu, 1983). Since A is a complete inner product space we consider the closed subspace $Ce = C$. Then we must show that the orthogonal complement of C is 0. Suppose that this is not the case. Then we find a nonzero element in $m = C^{\perp}$. The element x^*x is obviously positive and we can consider the element

$$T = \|x\|^2 e - x^*x$$

which is positive as an operator on A. Then for any y in A we have, by the generalized Cauchy–Buniakowsky inequality, that

$$|\langle(\|x\|^2 e - x^*x)e, y\rangle| \leqslant \langle(\|x\|^2 e - x^*x), e\rangle|\cdot\langle(\|x\|^2 - x^*x)y, y\rangle| = 0.$$

This implies that

$$\|x\|^2 e = x^*x$$

for all x in m. Now we suppose that $\|x\|^2 = 1$ and thus $x^*x = e$.

Now we remark that if x is in m then x^* is also in m. Thus we may suppose that x is a hermitian element. Thus we have the following interesting result: every hermitian element in m has the property that its square is a multiple of the identity of A.

Since $x^2 = e$ we have $x = e - 2\mathrm{p}$, where p is a hermitian projection. Since x is in m we must have $\langle h, e\rangle = 0$ and this implies that

$$1 = 2\langle p, e\rangle \quad \text{or} \quad \langle p, e\rangle = \tfrac{1}{2} = p, p\rangle = \|p\|^2.$$

Since p is a projection, this implies that $\|p\| = 1$. Thus we have a contradiction, and therefore every hermitian element in m is equal to 0. Since m is a selfadjoint set it is $\{0\}$. The Ingelstam theorem for the case of complex algebras is proved. □

REMARK 14.2.2. The above proof also works in the case of real algebras. We conjecture that an appropriate modification also works in the case of quaternions.

Using the fact that the norm of hermitian or normal elements in a C*-algebra is equal to the spectral radius, the above proof can be applied to obtain the following result of B. Yood.

THEOREM 14.2.3 (Yood, 1969). *If A is an infinite dimensional C^*-algebra, then there is not an inner product on A such that A is a pre-H^*-algebra with respect to the same involution.*

A pre-H^*-algebra is an inner product algebra which satisfies condition (2) *above, i.e. for all x, y in A,*

$$\|xy\| \leqslant \|x\|\,\|y\|.$$

14.3. H*-Algebras

An important and interesting class of algebras is the class in which the norm $x \to \langle x, x\rangle^{1/2} = |x|^2$ is a norm algebra, i.e.

$$\langle xy, xy\rangle \leqslant \langle x, x\rangle\langle y, y\rangle$$

holds for all x, y in the algebra A. In this case the algebra is called a pre-H*-algebra; if $(A, |\,,|)$ is complete, A is called a H*-algebra. The class of H*-algebras was first considered by Ambrose (1945).

In this section we present a structure theorem for algebras in this class, which says that such an algebra is completely determined by a family of (mutually) orthogonal minimal closed two-sided ideals, and this is applied to the algebra $L^2(G, \mathrm{d}x)$, where G is a compact group and $\mathrm{d}x$ is the normalized Haar measure.

To prove the structure theorem for H*-algebras we need some results which are given as lemmas.

LEMMA 14.3.1. *Let A be a H*-algebra and let x be in A with the properties:*

(1) $x^* = x$ *(i.e. x is hermitian).*
(2) $\|x\| = \sup_{y \in A,\, y \neq 0} \|xy\|/\|y\| = 1.$

Then the sequence (x^{2n}) converges to a Hermitian idempotent.

Proof. We recall that an element p of an algebra is called idempotent if $p^2 = p$. Since x is a hermitian element of A and satisfies property (2), all powers x^k have the same property. In this case $|x| \geqslant 1$ and for $m > n$, $m = 2p$, $n = 2q$ we have,

$$\langle x^m, x^n \rangle \leqslant \|x^{m-n}\| \langle x^n, x^n \rangle = \langle x^n, x^n \rangle$$

and

$$\langle x^m, x^m \rangle \leqslant \|x^{1/2(m-n)}\| \langle x^{n+1/2(m-n)}, x^{n+1/2(m-n)} \rangle = \langle x^n, x^n \rangle.$$

Then we have

$$1 \leqslant \langle x^m, x^m \rangle \leqslant \langle x^m, x^n \rangle \leqslant \langle x^n, x^n \rangle \leqslant \cdots \leqslant \langle x^2, x^2 \rangle = |x^2|^2$$

and since (x^n, x^n) is a bounded and monotone sequence, it converges, and this implies that

$$\lim_{n,m \to \infty} \langle x^m, x^n \rangle = \lim_{p \to \infty} \langle x^p, x^p \rangle.$$

But the following identity holds:

$$|x^m - x^n|^2 = \langle x^m - x^n, x^m - x^n \rangle = \langle x^m, x^m \rangle - \langle x^m, x^n \rangle - \langle x^n, x^m \rangle + \langle x^n, x^n \rangle$$

and the above results imply that

$$\lim_{n,m \to \infty} |x^m - x^n| = 0.$$

Thus (x^{2n}) is a Cauchy sequence and obviously $e = \lim x^{2n}$ is a hermitian idempotent. The lemma is proved. □

LEMMA 14.3.2. *If I_1 is a left ideal in A such that $I_1 \neq 0$, then there exists in I_1 a nonzero hermitian idempotent.*

Proof. We can consider an element x in I_1, $x \neq 0$ and since x^*x is a hermitian element, multiplying eventually with a scalar, we may suppose without loss of generality that $\|x^*x\| = 1$. Then, by Lemma 14.3.1 the sequence $((x^*x)^{2n})$ converges to a nonzero hermitian idempotent e which is clearly in I_1. □

For the structure theorem we need the class of idempotents considered in the following definition.

DEFINITION 14.3.3. An idempotent e in the H*-algebra A is called reducible if there exist idempotents e_1, e_2 such that:

(1) $e = e_1 + e_2$,
(2) $e_1 e_2 = e_2 e_1 = 0$

(and if $e^* = e$ then we require that $e_1^* = e_1$, $e_2^* = e_2$). Any idempotent which is not reducible is called irreducible.

Let e be a hermitian idempotent and suppose that e is reducible. Thus we have e_1, e_2, two hermitian idempotents such that:

$$e = e_1 + e_2, \qquad e_1 e_2 = e_2 e_1.$$

In this case we see that

$$\langle e_1, e_2 \rangle = \langle e_2 e_1, e_2 \rangle = 0.$$

Conversely, if e is a hermitian idempotent and e_1, e_2 are two hermitian idempotents such that

$$e = e_1 + e_2, \qquad \langle e_1, e_2 \rangle = 0$$

then

$$\langle e_1 e_2 \rangle = 0.$$

Indeed, we have

$$|e|^2 = |e, e| = |e_1|^2 + |e_2|^2$$

and

$$|e_1 e_2|^2 = \langle e_1 e_2, e_1 e_2 \rangle = \langle (e_2 e_1^2)^* e_2, e_2 \rangle = \langle e_1^2, e_2^2 \rangle = \langle e_1, e_2 \rangle$$

which implies that $|e_1 e_2| = 0$ and, similarly, we show that $e_2 e_1 = 0$.

Since the norm of any hermitian idempotent is at least 1 we see that any hermitian idempotent is of the form: for some integer n,

$$e = e_1 + \cdots + e_n,$$

where e_i are hermitian idempotents and are pairwise orthogonal.

LEMMA 14.3.4. *If I_1 is a minimal left ideal of A then it is of the form Ae, where e is an ireducible hermitian idempotent.*

Proof. From the result given above we find in I_1 a nonzero hermitian idempotent. We denote this idempotent by e. Clearly Ae is an ideal in I_1. If e is not irreducible then we find two hermitian idempotents e_1 and e_2 in I_1 such that $e = e_1 + e_2$ and obviously Ae_1 and Ae_2 are left ideals in A. Since I_1 is minimal, this is a contradiction. Thus we must have $Ae = I_1$ with e a hermitian irreducible idempotent.

Conversely, if e is an irreducible hermitian idempotent then we show that Ae is minimal. Indeed, in the contrary case we find another proper left ideal in $I_1 = Ae$. In this proper ideal I_1^1 contained in I_1 we find, by the above result, a hermitian idempotent, say e_1. Then

$$ee_1 = ee_1 e = e_2$$

is a hermitian idempotent and $e - e_2$ is a hermitian idempotent. This implies that e is not irreducible and this contradiction shows that Ae is minimal.

LEMMA 14.3.5. *The algebra A is generated by minimal left ideals.*

Proof. First we remark that any left ideal which is minimal is closed (the assertion follows from the fact that any left ideal which is minimal is of the form Ae with e a hermitian and irreducible idempotent). We now consider A_1 to be the closed subspace generated by all the minimal left ideals of A. We must show that $A_1 = A$. This is equivalent, using the inner product structure of A, to proving that $A_1^{\perp} = 0$. Suppose, on the other hand, that this is not true, i.e. $A_1^{\perp} \neq 0$. Since it is easy to see that $A_1^{\perp}$ is a left ideal (and closed) we find, again using the above result, that in A_1 there exist hermitian idempotents which clearly generate minimal left ideals. This is a contradiction with the definition of A_1 and thus we must have $A_1^{\perp} = 0$. □

In proving the structure theorem for H*-algebras the following result is of basic importance.

LEMMA 14.3.6. *If I_1 is a minimal left ideal then it generates a minimal two-sided ideal in A. If I_1, I_2 are two minimal two-sided ideals then they are mutually orthogonal and A is the direct sum of the closure of I_1 and I_2.*

Proof. Let I_1 be a minimal left ideal of A. Then $I_1 = Ae$, with e a hermitian irreducible idempotent. In this case the two-sided ideal generated by $AeA = N$ is selfadjoint, i.e.

$$N^* = (x^*, x \in N) = N$$

as is easy to see. Now we show that N is a minimal two-sided ideal of A. Indeed, in the contrary case we find a proper two-sided ideal N_1 contained in N. But we have $N_1 I_1 \subset N_1 \cap I_1$, and as I_1 is assumed to be minimal, we have either $I_1 \subset N_1$ (in which case $N_1 = N$) or $N_1 I_1 = 0$. If the second possibility is true, then

$$N_1 I_1 A = N_1 N = 0$$

and thus

$$N N_1^* \subset N_1 N = 0.$$

This implies that $N_1 = 0$ and thus N is minimal.

Now we prove that $N = N^2$. Indeed, $N^2 \subset N$ and from $N^2 = NN^*$ the assertion follows.

Now we prove that if N_1, N_2 are two minimal two-sided ideals then they are orthogonal. Indeed, if x, y are in N_1 and z is in N_2 then we have

$$\langle xy, z \rangle = \langle y, x^* z \rangle = 0$$

which implies that $N_1 \perp N_2$. But we know that A is generated by left minimal ideals, which implies that A is a direct sum of the closures of N_1 and N_2 since, as we proved, every minimal left ideal generates a minimal two-sided ideal. □

From the above considerations we have also the following result.

LEMMA 14.3.7. *If I is a closed two-sided ideal in A then it is generated by minimal two-sided ideals.*

LEMMA 14.3.8. *Let I_1 be a left minimal ideal, $I_1 = Ae$, where e is a hermitian idempotent. Then eAe is isomorphic to the field of complex numbers.*

Proof. It will suffice to show that every element in eAe has a left inverse (if it is not zero, of course). Since I_1 is assumed minimal for any x in eAe, $I_1 = Ax$ (Ax is

clearly a left ideal in I_1) and thus we find y in A with the property that $yx = e$.
In this case we have,

$$(eye)(exe) = eyexe = eyx = e$$

and thus (since $x = exe$), every nonzero element of eAe has a left inverse. This implies that eAe is isomorphic to C (the set of complex numbers).

LEMMA 14.3.9. *Let $(e_i)_{i\in I}$ be a family of elements of A and let N be a closed minimal two-sided ideal. Then the following assertions are equivalent:*

(1) *$(e_i)_{i\in I}$ is a maximal family of mutually orthogonal irreducible hermitian idempotents.*
(2) *$(Ae_i)_{i\in I}$ is a family of mutually orthogonal minimal left ideals spanning N.*
(3) *$(e_i A)_{i\in I}$ is a family of mutually orthogonal minimal right ideals spanning N.*

Proof. Since the proof is simple it is omitted. □

THEOREM 14.3.10. (The structure theorem of two-sided ideals of H*algebras). *Let N be a minimal two-sided ideal of A. Then it is isomorphic to the algebra of matrices $(c_{ij})_{(i,j)\in I\times I}$ with the property that $\Sigma_{i,j}|c_{ij}|^2 < \infty$.*

Proof. We consider the family $(e_i)_{i\in I}$ as given in Lemma 14.3.8 and consider e_1 to be fixed. Further, we consider the ideal Ae_iA which clearly generates N (N is assumed to be minimal). Thus we have $e_1Ae_iAe_1 \neq 0$. Thus we get that $e_1Ae_i \neq 0$. But we know that e_1Ae_1 is isomorphic to the set of complex numbers, and if m_{1i} is and element of e_1Ae_i we have that $m_{1i}m_{1i}^*$ is in e_1Ae_1, and (multiplying eventually with a scalar) we may suppose that $m_{1i}m_{1i}^* = e_1$ and thus m_{1i}^* is in $(e_1Ae_i)^* = e_iAe_1$. Similarly, $m_{1i}m_{1i}^*$ is in e_iAe_i and in fact equal to e_i. Then we define the following family of elements: $e_{i1} = e_{1i}^*$ and $e_{ij} = e_{i1}e_{1j}$ for all i, j in I. Then we have the relations:

(1) $e_{ij}e_{k1} = e_{i1}$ if $j = k$
$\quad = 0$ if $j \neq k$
(2) $e_{ij} = e_{ji}^*$, $e_{ii} = e_i$,
(3) $\langle e_{ij}, e_{kl}\rangle = 0 \quad i \neq k, \quad j \neq 1$
$\quad = \langle e_1, e_1\rangle, \quad i \leqslant k, \quad j \leqslant 1$

Since e_iAe_j is one-dimensional it is isomorphic to the space C. Thus if x is in N $x = \Sigma xe_i = \Sigma e_ixe_j = \Sigma c_{ij}e_{ij}$ which gives that (e_{ij}) is an orthonormal basis for N. Thus we can consider the mapping $x \to (c_{ij})$ of N into the set of matrices (c_{ij}) with

$$\sum_{(i,j)} |c_{ij}|^2 < \infty, \quad i,j\in I.$$

and this clearly is an isomorphism because (e_{ij}) is a basis. The theorem is proved. □

The following result is the structure theorem for H*-algebras and the proof is implicit in the above proofs.

THEOREM 14.3.1 (The structure theorem for H*-algebras). *If A is an H*-algebra then it is the direct sum of the closures of its minimal two-sided ideals; every minimal* *two-sided ideal of A is isomorphic to the algebra of matrices* $(c_{ij})_{(i,j)\in I\times I}$. *I is an index set, such that*

$$\sum_{(i,j)\in I\times I} |c_{ij}|^2 < \infty.$$

For an application to a group algebra we need the following result.

PROPOSITION 14.3.12. *For any two-sided ideal of A the following assertions are equivalent:*

(1) *N is finite-dimensional.*
(2) *N contains an identity (as a subalgebra of A).*
(3) *N contains a nonzero central element.*

Proof. (1) → (2). Since N is supposed to be finite dimensional we have

$$e = e_1 + \cdots + e_n$$

and for each x in N,

$$x = \sum_i xe_i = \sum e_i x = ex = xe$$

which implies that e is an identity of N.

(2) → (3). Let x in A and then we have

$$x = e(xe) = (ex)e = ex$$

and as e is nonzero the assertion is proved.

(3) → (1) Let x be a nonzero central element in N and thus

$$xe_i = (xe_i)e_i = e_i(xe_i) = e_i xe_i$$

which implies that the projection of x onto Ae_i is $x_i e_i$, with x_i a complex number. Clearly, $x = \Sigma\, x_i e_i$, and since

$$x_j e_{jk} = xe_{jk} = x_{jk}x = x_k e_{jk},$$

this gives that $x_i = x_j$. Thus all the coefficients x_i are equal. Since $x = x_i e_i$, the sum must be finite and thus N is finite dimensional. □

14.3.13. If an H*-algebra A is commutative then all the elements of A are central, and thus the two-sided ideals are finite dimensional (in fact, one-dimensional).

We now apply the structure results obtained above to an algebra associated with compact groups.

Let G be a compact group and let dx be its Haar measure normalized (i.e. $dx(G) = 1$). We consider the following spaces: $L_\infty(G)$, $L^1(G, dx) = L^1(G)$, $L^2(G, dx) = L^2(G)$. If f is in L_∞ then we have clearly the following inequality

$$\|f\|_1 \leqslant \|f\|_2 = \|f\|_\infty$$

and if g is another function in L_∞, the convolution is well defined:

$$(f \otimes g)(x) = \int_G f(xs)g(s^{-1})\,ds$$

with

$$\|(f \otimes g)\|_\infty \leqslant \|f\|_2\|g\|_2. \tag{*}$$

We have the following result.

PROPOSITION 14.3.14. *For any compact group G, $L^2(G)$ is an H^*-algebra.*

Proof. If we define the involution on $L^2(G)$ by $*: L^2 \to L^2$ with

$$(h^*)(s) = \bar{h}(s) = h^*(s),$$

then using Fubini's theorem we get that

$$\langle f \otimes g, h \rangle = \langle g, f^* \otimes h \rangle$$

and, combined with the above inequality (*), we get that L^2 is in fact an H*-algebra.

We now give a characterization of a class of central elements of L^2 which are useful in obtaining the structure theorem of L^2.

LEMMA 14.3.15. *The function $f \in (G)$ (the space of all continuous complex-valued functions on G with the sup norm) is central in L^2 if and only if $f(xy) = f(yx)$ holds for all x, y in G.*

Proof. The assertion follows by remarking that f is central in L^2 if and only if for all g in L^2 the following formula holds:

$$\int_G (f(xy))g(y^{-1}) - g(y)(f(y^{-1}x))\,\mathrm{d}y = 0.$$

In this formula, the second term of the integrand y can be replaced by y^{-1} (which is possible since integration is with respect to the Haar measure) and as g is assumed to be arbitrary the above formula implies our assertion. □

From the lemma we obtain the following important result.

RESULT 14.3.16. *Every closed ideal of L^2 contains a nonzero central element.*

Proof. Let I be an ideal and we assume that $I \neq 0$. Thus we find a nonzero element f in I and set $h = f \otimes f^*$ and since $h(x) = \langle f_x, f \rangle$ (where f_x is the translate of f) we see that f is continuous. In this case the function $x \to \int_G h(u^{-1}xu)\,\mathrm{d}u$ is continuous for each u in G and thus

$$x \to \int_G h(u^{-1}xu)\,\mathrm{d}u = h_1(x)$$

is continuous and as e is the unit of G,

$$h_1(e) = \|f\|_2^2$$

which implies that h_1 is nonzero. If we replace u by $u^{-1}y$ in the integral form for $h_1(xy)$ we get the relation $h_1(xy) = h_1(yx)$. Thus h_1 is a central element of L^2. Also we remark, from the integral form of h_1, that it is in I.

Using the structure theorems obtained for H*-algebras we have the following result.

THEOREM 14.3.17. *For any compact group G, $L^2 = L^2(G, \mathrm{d}x)$, where $\mathrm{d}x$ is the Haar* *measure, is the direct sum of finite dimensional (mutually) orthogonal minimal two-sided ideals.*

For other information concerning group algebras or the H*-algebra of a group, we refer to the excellent books of Dixmier (1969) and Loomis (1953).

14.4. Inner Product Algebras and H*-Algebras

Let A be an inner product algebra and let $E(A) = E$ be the set of all nonzero projections in A_b. There are examples of inner product algebras without nonzero projections. We consider the algebra of all continuous complex-valued functions on $[0, 1]$ with the property that for every element f in the algebra, $f(0) = 0$. Then with the inner product defined by the formula

$$\langle f, g\rangle = \int_0^1 f(s)\bar{g}(s)\,\mathrm{d}s$$

for any f, g, this is clearly an inner product algebra. Then this algebra does not have nonzero projections but in A_b they abund.

DEFINITION 14.4.1. The inner product algebra A is said to be projection bounded from below (above) if there exists $c > 0$ such that $\|p\| \geqslant c$ ($\|p\| \leqslant c$) for all $p \in E$.

The following result of Yood (1974) gives a characterization of full algebras which are renormable as H*-algebras.

THEOREM 14.4.2. *A full inner product algebra is renormable as an H*-algebra if and only if A is projection bounded from below.*

For a proof we refer to Yood (1974) which also contains other interesting characterizations of inner product algebras with $A_b = A$.

CHAPTER 15

Non-Archimedean, Nonstandard, Intuitionistic and Constructive Inner Product Spaces

15.0. Introduction

It is quite natural to try to extend the results obtained in the theory of linear spaces over the field R of C (the real numbers and the complex numbers, respectively) to linear spaces over the fields in certain classes of fields. In this respect some classes of fields have received increasing attention. Among these we mention the non-archimedean fields, the intuitionistic fields and nonstandard fields obtained from R and C.

Suppose that K is a field. Then a valuation v on K is a mapping $v: K \to R$ satisfying the following conditions:

(1) $v(0) = 0, \quad v(ab) = v(a)v(b)$,

(2) $v(a + b) \leqslant v(a) + v(b)$.

A field on which a valuation is defined is called a valued field. There are fields satisfying a stronger condition than condition (2) above, namely

(2′) $v(a + b) \leqslant \max(v(a), v(b))$

A field whose valuation satisfies conditions (1) and (2′) is said to be a non-archimedean field. It is clear that any field admits the following non-archimedean valuation:

$$v(x) = \begin{cases} 0, & x = 0 \\ 1, & x \neq 0 \end{cases}$$

This valuation is called the trivial valuation; a nontrivial valuation of a field K is such that for at least one element u, $v(u) \neq 0, 1$.

For the reader interested in such topics the books by Bachman (1964), Monna (1970), Rooij (1978) and the paper by Rooij and Schikhof (1971) may be useful.

In this chapter we present some results about non-archimedean linear spaces on which we can define something which we call an inner product. Several properties are derived concerning inner products and quadratic forms and some elements of operator theory.

An interesting and important point of view on the foundation of mathematics was introduced and developed by the great mathematician L. E. J. Brouwer and which is now called 'Intuitionism'. The construction of an intuitionistic analysis leads naturally to an intuitionistic functional analysis, particularly to the intuitionistic notion of inner product space which was considered and studied by A. Heyting. We present results about inner product spaces (intuitionistic inner product spaces) obtained essentially by A. Heyting and Ashvinikumar.

Another field of great interest is that constructed by A. Robinson – the so-called nonstandard extension of R (and of C); this field plays a fundamental role in the construction of the nonstandard analysis of Robinson (1961) (see also Robinson, 1966). The basic tool in these constructions is the so-called ultraproduct which essentially was given by J. Łos in 1955; it is worth mentioning that the idea behind the construction of ultraproducts goes back to the construction by T. Skolem in 1934 of a nonstandard model of arithmetic. The application of ultraproducts to inner product spaces in connection with problems in quantum mechanics was given by Takeuti (1962); this idea was pursued further by many authors (e.g. Farrukh, 1975; Kelemen and Robinson,1974; Popowicz, 1975). In this chapter we present the basic constructions of the nonstandard extensions of R(and C) and nonstandard linear spaces as well as ultraproducts. Two applications are mentioned: the first, by J. Lindenstrauss and L. Tzafriri, for the characterization of Banach spaces isomorphic to inner product spaces; and the second, by G. Takeuti, for quantum mechanics.

Another direction of research which has attracted great interest is the constructive mathematics, in the sense of E. Bishop, which has its roots in intuitionism. We shall also present some facts concerning constructive inner product spaces.

15.1. Non-Archimedean Normed Linear Spaces. Non-Archimedean Inner Product Spaces

Let K be a field with identity denoted by 1 (we also use the notation 1 for the identity of R).

DEFINITION 15.1.1. A valuation v on K is a function $v: K \to R^+$ with the properties

(1) $v(x) \geqslant 0$,

(2) $v(xy) \leqslant v(x)\, v(y)$,

(3) $v(x + y) \leqslant v(x) + v(y)$

for all x, y in K. If v is a valuation on K then

$$d(x, y) = \sigma(x - y)$$

is a metric on K. In general, (K, d) is not a complete metric space.

DEFINITION 15.1.2. If K is a field and v is a valuation on K then we say that v is an Archimedean valuation if the set $\{{}^{v}(n \cdot 1)\}$ is unbounded. In the contrary case we say that the valuation is non-archimedean.

DEFINITION 15.1.3. If K is a field and v is a valuation on K then we say that the valuation is dense if the set

$$|K| = (v(k), k \in K, k \neq 0)$$

is dense in R^+. In the contrary case we say that the valuation is a discrete valuation.

EXAMPLE 15.1.4. Let K be an arbitrary field and set $v: K \to R^+$ by the relation

$$v(x) = \begin{cases} 0, & x = 0 \\ 1, & x \neq 0. \end{cases}$$

Then this is a valuation on K (called the trivial valuation).

EXAMPLE 15.1.5. Let $K = R$, R being the set of all real numbers. If we set $v: R \to R^+$ by the relation

$$v(x) = \begin{cases} x, & x \geqslant 0 \\ -x, & x \leqslant 0 \end{cases}$$

then this is a valuation on R.

EXAMPLE 15.1.6. Let $K = Q$, the set of all rational numbers. Then this is a subfield of R and the valuation defined in 15.1.5 clearly induces a valuation on Q.

We remark that the valuation in 15.1.4 is non-Archimedean, the valuations in 15.1.5 and 15.1.6 are Archimedean valuations.

We now present an interesting and important non-Archimedean valuations on Q.

EXAMPLE 15.1.7. *The p-adic valuations* v_p. Let p be a prime number and let c be a strictly positive number. Then a valuation on Q, v_p, may be defined by the formula:

$$v_p(n) = 1, \quad n \text{ is not divisible by } p,$$
$$v_p(p) = c^a p^{-a}$$

for some fixed a.

It is not difficult to see that this is a valuation on Q which has the following property:

$$v_p(x + y) = \max(v_p(x), v_p(y)).$$

This clearly gives that v_p is a non-Archimedean valuation on Q.

We remark that the nontriviality of a valuation can be formulated as a requirement of at least one element k with $v(k) \neq 0$ and 1. Then it is not difficult to see that v_p are nontrivial valuations. The importance of v_p-adic valuations follows, to some extent, from the following famous result of A. Ostrowski (see Bachman, 1964, for a proof).

THEOREM 15.1.8 (Ostrowski, 1918). *Suppose that* v *is a nontrivial valuation on the field* Q. *Then the following assertions hold: either*

(1) v *is a non-archimedean valuation and then is of the form* $v(x) = v_p^t(x)$ *for some prime number* p *and some* $t > 0$, *or*
(2) v *is an Archimedean valuation and then is of the form* $v(x)^s$ *for some* $s > 0$ *and where* v *is the valuation in Example* 15.1.6.

We are now ready to define the notion of non-Archimedean space. This is given in the following definition.

DEFINITION 15.1.9. A linear space X over the non-archimedean field K is said to be a non-Archimedean space if it is a linear space over K and on X there exists a function $x \to \|x\|$ which satisfies the conditions:

(1) $\|x\| \geqslant 0, \quad \|x\| = 0$ iff $x = 0$.

(2) $\|kx\| = v(k)\|x\|$,

(3) $\|x + y\| \leqslant \max(\|x\|, \|y\|)$.

This is called a non-Archimedean norm. We say that $(X, \| \, , \|)$ is a non-Archimedean space.

Now if we set

$$d(x, y) = \|x - y\|$$

then it is easy to see that d satisfies the conditions of the so-called ultrametrics.

DEFINITION 15.1.10. We say that the non-Archimedean normed space X is a non-Archimedean Banach space if (X, d) is a complete metric space.

REMARK 15.1.11. Concerning the ultrametrics it is worth mentioning that Everett and Ulam (1966) have conjectured that distances between particles may behave in a non-Archimedean manner.

We now present some examples of non-Archimedean normed or Banach spaces.

EXAMPLE 15.1.12. Let K be a non-Archimedean field (i.e. a field with a non-Archimedean valuation). We consider the space m of all sequences $x = (x_i)$, $x_i \in K$, such that

$$\|x\| = \sup(v(x_i))$$

is bounded. An interesting and important subspace is the space c_0 of all sequences $x = (x_i)$, $x_i \in K$, $\lim v(x_i) = 0$.

It is not difficult to see that if K is complete with respect to the metric induced by the valuation, then m and c_0 are non-Archimedean Banach spaces.

EXAMPLE 15.1.13. Let S be a compact Hausdorff space and let K be a non-Archimedean field. We consider the space $C(S, K)$ of all continuous K-valued functions on S with the norm

$$\|f\| = \sup(v(f(s)), \quad s \in S)$$

which gives immediately that $C(S, K)$ is a non-Archimedean normed space; if K is a complete field then it is not difficult to see that $C(S, K)$ is a non-Archimedean Banach space.

Suppose now that X, Y are two non-Archimedean spaces over the same field K.

DEFINITION 15.1.14. A linear operator defined on X with values in Y is a mapping $T: X \to Y$ with the following property: for all x, y in X and k in K the following relation holds:

$$T(kx + y) = kTx + Ty.$$

If T is a continuous mapping (with respect to the topology induced by the metric defined using the norms of X and Y, then we say that T is a continuous linear operator.)

If $X = Y$ then the set of all continuous linear operators defined on X is denoted by $L(X)$; in the case $X = Y$ we use the notation $L(X, Y)$ for the set of all continuous linear operators on X with values in Y.

If $Y = K$ then any linear operator on X with values in Y is called a linear functional, and if this is continuous then we say that we have a continuous linear functional. The set of all continuous linear functionals over the non-archimedean space X is denoted by X^* and is called the dual of X. This can be organized as a Banach space (non-Archimedean) and, in general, as in the case of Banach spaces over R or C, the nth dual of X is defined as the dual on $X^{*(n-1)}$, $n \geqslant 2$.

As in the case of spaces over R or C, the basic problem concerning the duals is their nontriviality (i.e. the existence of elements not identically zero). Thus the problem is to determine to what extent a Hahn–Banach–Bohnenblust–Sobczyc–Sukhomlinov result holds for non-Archimedean normed spaces. In order to present such a result we need the notion of spherical completeness.

DEFINITION 15.1.15. A non-archimedean Banach space X is said to be spherically complete if for every sequence of balls

$$B_{r_1}(a_1) \supseteq B_{r_2}(a_2) \supseteq B_{r_3}(a_3) \supseteq \cdots \supseteq B_{r_n}(a_n) \supseteq B_{r_{n+1}}(a_{n+1}) \supseteq \cdots$$

with

$$r_1 > r_3 > \cdots r_n > r_{n+1} > \cdots$$

has a nonempty intersection.

The following result of A. W. Ingleton gives a characterization of those non-Archimedean Banach spaces with a nontrivial dual.

THEOREM 15.1.16 (Ingleton, 1952). *For a non-Archimedean Banach space over the field K the following assertions are equivalent*:

(1) *The field K is spherically complete.*
(2) *The extension property holds for every subspace of X, i.e. if X_1 is a (closed) linear subspace of X then any continuous functional on X_1 can be extended to a continuous functional on X with the same norm.*

For a proof of this result as well as for related results we refer to Rooij (1978).

We now present some results about quadratic forms over non-Archimedean linear spaces and inner products on such spaces.

DEFINITION 15.1.17. Let X be a linear space over the commutative field K. A quadratic form f on X is a function $f: X_\times X \to K$ with the following properties:

(1) $f(kx, kx) = k^2 f(x, x)$,
(2) $f(x + y, x + y) = f(x, x) + f(y, y) + g(x, y)$

where g is a bilinear form on X, i.e. it is a mapping defined on $X_\times X$ with values in

K and satisfying the conditions

(1) $g(ax + y, z) = ag(x, z) + g(y, z)$,

(2) $g(x, au + v) = ag(x, u) + g(x, v)$

for all x, y, u, v in X and a in K.

If f is a quadratic form on X then the dimension of f is, by definition, the dimension of X.

DEFINITION 15.1.18. The quadratic form f on X is said to be degenerate if for all x in X and some a in K, $a \neq 0$, the following relation holds:

$$f(a + x, a + x) = f(x, x).$$

If the characteristic of K is not 2 then the degeneracy of f is equivalent to the existence of some $x \neq 0$ with the property that for all y in X the following relations hold:

(1) $f(x, x) = 0$,

(2) $g(x, y) = 0$,

where g is as in Definition 15.1.17.

DEFINITION 15.1.19. We say that the quadratic form f is definite if $f(x, x)$ is not 0 for $x \neq 0$.

DEFINITION 14.1.20. If g is a bilinear form on X then for any subset m of X we set

$$m^{\perp} = (y, g(x, y) = 0, x \in m)$$

and we call $m^{\perp}$ the orthogonal of m.

DEFINITION 15.1.21. A subspace X_1 of X is called splitting (with respect to a bilinear form g) if $X = X_1 + X_1^{\perp}$.

DEFINITION 15.1.22. Let $T: D_T \to Y$, X is a non-Archimedean Banach space, X_1 is a dense subspace of X, Y is a non-Archimedean Banach space. The graph of T is the subset of $X_{\times} X$ defined as $\{(x, Tx), x \in D_T\}$.

The operator T is called closed if the following property holds: if (x_n) is a subset in D_T and if

$$\lim x_n = x \quad \text{and} \quad \lim Tx_n = y$$

then x is in D_T and $y = Tx$. Later we shall give an example of a class of operators which are closed (the adjoints).

Suppose now that K is a field of characteristic not equal to 2 and that K has a valuation which is non-Archimedean and K is complete. Then we have the following result.

THEOREM 15.1.23 (Springer, 1955). *Let f be a definite quadratic form over the non-*

Archimedean space X and set

$$\|x\| = v(f(x, x))^{1/2}.$$

Then $x \to \|x\|$ is a norm on X which is non-Archimedean.

Proof. All the properties of the norm are obvious except the fact that it is non-Archimedean. Let x, y be two arbitrary linearly independent elements of X and consider the polynomial

$$p(z) = f(zx + y, zx + y) = az^2 + bz + c$$

which is irreducible in $K(z)$. Then by Hensel's lemma we must have

$$v(b) = \max(v(a), v(b))$$

and then

$$\|x + y\| = v(x + y) = v(a + b + c) \leqslant \max(v(a), v(c))$$

which gives the assertion of the theorem. □

REMARK 15.1.24. In his 1962 paper, T. A. Springer has remarked that the irreducibility of $p(z)$ in $K(z)$ implies the following fact, which may be considered as an analogue of the Cauchy–Buniakowsky inequality:

$$v(b)^2 \leqslant v(a)v(c).$$

For other results about quadratic forms and different properties of the valuation of the field K we refer to Springer's papers; we mention the interesting result which generalizes a fact known in the case of Q_p (the completion of Q with respect to the metric induced by the p-adic valuation $v_p(\ ,\ )$) that any quadratic form in more than four variables represents zero nontrivially.

Concerning the inequalities

$$\|y + y\| \leqslant \max(\|x\|, \|y\|)$$

and

$$v(f(x, y))^2 \leqslant v(f(x, x)\, v(f(y, y))$$

it is of interest to have such inequalities for more general spaces. A class of spaces for which such inequalities are valid will now be defined.

DEFINITION 15.1.25. A non-Archimedean space over K is said to be a Springer space if on X there exists a definite symmetric quadratic form.

Then we have the following result.

THEOREM 15.1.26 (Bayod Bayod, 1976). *If X is a Springer space then the mapping $x \to v(f(x, x))^{1/2}$ is a non-Archimedean norm on X where f is the quadratic form on X.*

THEOREM 15.1.27 (Bayod Bayod, 1976). *If X is a Springer space and* char $K \neq 2$ *then the following Cauchy–Buniakowsky inequality holds*:

$$v(f(x, y)) \leqslant v(2)^{-1}(v(f(x, x))v(f(x, y))^{1/2}.$$

In order to prove these results we need some results about the roots of equations of the form $az^2 + bz + c = 0$. A classical result asserts that the equation

$$z^2 + bz + c = 0, \quad b, c \text{ in } K$$

has a root in K if $v(b^2) > 4v(c)$.
The following improvement of this result was obtained by J. M. Bayod Bayod in 1976.

PROPOSITION 15.1.28. *Let K be a non-Archimedean field with a valuation v. Then the equation*

$$z^2 - bz + c = 0,$$

b, c in K, has a root in K if $v(b^2) > v(c)$.

Proof. We define a sequence of elements in K as follows:

$$a_1 = a, \ldots, a_{n+1} = b - (c/a_n), n \geqslant 2. \qquad \square$$

We remark that (a_n) is well defined and

$$v(a_n) = v(b).$$

Indeed, for $n = 1$ the assertion is clear. Suppose that the relation is valid for all $i = n$. Now

$$v(a_{n+1}) = v(b - (c/a_n)) = \max(v(b), v(c/a_n)) = \max(v(a), v(c)/v(b))$$

and since $v(c)/v(b) < v(b)$ we get that

$$v(a_{n+1}) = v(b).$$

Thus the assertion follows for all n.
Set $q = v(c) \cdot v(b^2)^{-1}$, which is in $(0, 1)$ and then we remark that

$$\begin{aligned} v(a_{n+2} - a_{n+1}) &= v((b - (c/a_{n+1}) - b + (c/a_n))) \\ &= v(c) \cdot v(a_{n+1} - a_n)(v(a_n a_{n+1}))^{-2} \\ &= v(c) \cdot v(b^2)^{-1} \cdot v(a_{n+1} - a_n) \\ &= q \cdot v(a_{n+1} - a_n). \end{aligned}$$

This obviously implies that $\Sigma\, a_n$ is a convergent series in K. Thus (a_n) is a Cauchy sequence and as K is complete we find u in K with the property that $u = \lim a_n$. Then obviously u is a root of the equation

$$z^2 - bz + c = 0.$$

PROPOSITION 15.1.29 (Bayod Bayod, 1976). *Suppose that K is a non-Archimedean field and the equation*

$$az^2 + bz + c = 0, \quad a \neq 0$$

is without roots in K. Then the following inequalities hold:

(1) $v(b^2) \leqslant v(a)v(c)$,

(2) $v(b) = \max(v(a), v(c))$.

Proof. Suppose that the first inequality is not true. Then the above equation is equivalent to

$$z^2 - (-b/a)z + c/a = 0$$

which has the property that

$$v((b/a)^2) > v(c/a).$$

Thus by 15.1.28 in K we have roots for this equation, and this gives that

$$az^2 + bz + c = 0$$

has roots. This contradiction proves the inequality.

We note that inequality (2) follows from (1). Indeed,

$$v(b^2) = v(b)^2 \leqslant v(a)v(c) \leqslant \max(v(a)^2, v(c)^2) = (\max(v(a), v(c))^2$$

and this is equivalent to the required inequality.

Now the proof of Theorems 15.1.26 and 15.1.27 follows immediately from these inequalities. Indeed, if x, y are arbitrary elements in X then we consider the polynomial

$$p(z) = f(zx + y, zx + y) = f(x, x)z^2 + 2f(x, y)z + f(y, y) = az^2 + bz + c.$$

Applying the above inequalities to this polynomial we obtain the inequalities asserted in Theorems 15.1.26 and 15.1.27. It is worth mentioning that the above proof shows that if X is a non-Archimedean space with f a symmetric form which is definite, then $x \to v(f(x, x))^{1/2}$ defines a norm on X for arbitrary fields (i.e. of arbitrary characteristic).

COROLLARY 15.1.30. *Let X be a Springer space and suppose that the valuation v has the property that $v(2) = 1$. Then the mapping $h: X \to X^v$, defined by the relation*

$$h_x(x) = \langle x, y \rangle = f(x, y),$$

is an isometry.

We now present an example of a Springer space.

EXAMPLE 15.1.31. Let us consider K be a commutative field and let Q be the set of rationals. We consider the field F of all power series

$$s = \sum_{r \in Q} a_r t^r,$$

where $A_s = (r, a_r \neq 0)$ is well-ordered in Q, the mapping $r \leftarrow a_r$ is increasing on A_s. The set of power series can be organized as a field in the natural way. A valuation on F is defined by the formula:

$$\gamma(s) = \begin{cases} 2^{-a}, & a \text{ is the smallest element of } A_s, \\ 0, & \text{if } s = 0. \end{cases}$$

Then this is a spherically complete field. We now consider in F the subfield F_Z formed of all s with a_r integers. If we now consider the space c_0 associated with the fields F or F_Z, i.e. the set of all sequences $x = (x_n)$, x_n in the above fields, then we get a complete non-Archimedean space which is spherically complete.

Now if y is another element in c_0 then we set

$$\langle x, y\rangle = f(x, y) = \Sigma\, x_n y_n.$$

Then we have the following result.

THEOREM 15.1.32. *If the field is F or F_1 then c_0 is a Springer space.*

Another class of non-Archimedean spaces on which we have an inner product was considered by G. Kalisch. This class of spaces, which we call 'V spaces', is defined as follows.

DEFINITION 15.1.33 (Kalisch, 1947). Let X be a non-Archimedean linear space over the non-Archimedean field K satisfying the properties:

(1) $v(2) = 1$.
(2) Every unit of K (i.e. element u with $v(u) = 1$) has a square root in K.
(3) K is complete under the topology induced by the valuation.
(4) K is separable (i.e. a separable topological space).

Then we say that X is a V space if on X there exists a bilinear form $\langle\,,\rangle$ with values in K and which is symmetric, and on X there exists a norm $x \to \|x\|$ satisfying the properties:

(1) $\|x\| \geqslant 0$, $\|x\| = 0$ iff $x = 0$.

(2) $\|kx\| = v(k)\,\|x\|$.

(3) $\|x + y\| \leqslant \max(\|x\|, \|y\|)$.

(4) $v(\langle x, y\rangle) \leqslant \|x\|\,\|y\|$.

(5) For any x in X there exists k in K with the property that $\|x\| = v(k)$.

(6) For any x in X there exists $\bar{x}$ in X such that $\bar{x} \neq 0$ and

$$v(\langle x, \bar{x}\rangle) = \|x\|\,\|\bar{x}\|.$$

(7) $(X, \|\,,\|)$ is complete and separable.

In this paper G. Kalisch proved that in such a space we can consider orthogonal families, and proved their existence. For more details about these spaces we refer the reader to Kalisch (1947).

We now present some facts about operator theory on non-Archimedean spaces on which we have an inner product. Let X, Y be two non-Archimedean spaces over the same field K.

DEFINITION 15.1.34. A mapping $T\colon \mathrm{D}_T \to Y$, D_T being a dense linear subspace of X, is said to be a linear operator if for all x, y in D_T and k in K,

$$T(kx + y) = kTx + Ty.$$

The linear subspace D_T is called the domain of T. Of course, we can consider arbitrary domains (i.e. not necessarily dense) but these are too special.

An interesting class of operator, which is the analogue of the class of closed operators, is defined as follows.

DEFINITION 15.1.35. A linear operator $T: \mathrm{D}_T \to Y$ is said to be closed if whenever (x_n) is a sequence with $\lim x_n = x$ and $\lim Tx_n = y$ then x is in D_T and $Tx = y$.

It is not difficult to see that continuous operators are closed. We shall now present an interesting class of operators formed of closed operators. This is the class of adjoint operators.

Suppose that X, Y are non-Archimedean spaces over the same field and T is a linear operator. Then the adjoint of T is defined, as usual, using linear continuous functionals

$$y^*(Tx) = y_1(x).$$

We leave the reader to prove that this defines a linear operator $T^*: Y^* \to X^*$, more precisely defined on D_{T^*} in Y^*.

Now if normed non-Archimedean spaces have an inner product then we define, as in the case of complete inner product spaces, the complete inner product space adjoint. Then we have (we consider in what follows the inner product adjoint).

THEOREM 15.1.36. *The adjoint T^* is a closed operator.*

We now define some classes of operators.

DEFINITION 15.1.37. The operator T is called hermitian if for all x, y in X,

$$\langle Tx, y \rangle = \langle x, T^*y \rangle$$

(here $T: X \to X$).

DEFINITION 15.1.38. The operator T is called normal if $TT^* = TT^*$.

DEFINITION 15.1.39. For any operator $T: X \to X$, the numerical range is the set

$$W(T) = (\langle Tx, x \rangle, \|x\| = 1)$$

and the numerical radius is the number

$$w_T = \sup(v(a), a \in W(T)).$$

DEFINITION 15.1.40 (Kalisch, 1947). The operator $T: X \to X$ is said to be totally continuous if for any sequence (x_n) in X with

$$\lim \langle x_n, y \rangle = \langle x, y \rangle \ \lambda y \text{ in } X$$

we have that the sequence (Tx_n) admits a convergent subsequence.

DEFINITION 15.1.41. The operator $T: X \to X$ is said to be compact if whenever (x_n) is a bounded sequence in X the sequence (Tx_n) admits a convergent subsequence.

Then we have the following result.

THEOREM 15.1.42. *If X is a Kalisch space then for any hermitian operator T we* *have that*

$$\|T\| = w_T$$

i.e. T is normaloid.

REMARK 15.1.43. J. M. Bayod Bayod proved a more general result; see his Ph.D. thesis (1976).

PROBLEM 15.1.44. If T is a normal operator, does $\|T\| = w_T$?

Concerning orthocomplemented subspaces and projections we refer to Piziak (1972) and Bayod (1976).

We close with another problem.

PROBLEM 15.1.45. Let X be a non-Archimedean space with an inner product (assume, for example, a Springer or V space). Is it possible to construct a spectral theory for classes of operators on X. For example, is it possible to define a spectral resolution of the identity such that for normal or normal-like operators an integral decomposition can be obtained?

15.2. Nonstandard Inner Product Spaces

Nonstandard analysis is a relatively new mathematical field created by Abraham Robinson in 1961 and represents a beautiful example of an application of mathematical logic to the foundations of mathematical analysis.

As is well known, infinitesimals, i.e. numbers infinitely small and infinitely large, were used in the development of the mathematical analysis; these numbers were a subject of constant interest and controversy and Leibniz was one of those suggesting the use of infinitesimals. Gradually the use of infinitesimals fell into disrepute and the ε–δ method was introduced. The main reason why the infinitesimals were replaced by the rigorous ε–δ method was that not sufficiently precise rules were given for the system of infinitesimals. In this context we mention that were many attempts to solve the problem of constructing a system which includes infinitesimals and for the history and details about contributions we refer to Robinson's papers. It is worth considering an early contribution of Tulio Levi-Civita (1892/93) mentioned by A. Robinson.

Robinson has remarked that the notions of model theory can be used successfully to give a solid base of the use of infinitesimals. He constructs a proper extension of the reals R, $^*R$ which has, to some extent, the same properties as R. It is a well-known fact that $^*R$ must be a non-Archimedean field and this fact is used by Robinson to define in R the infinitesimals. Robinson's idea is to consider a specified collection of properties of R which can be formulated in a specified formal language in an appropriate interpretation in R and $^*R$ and the classical isomorphism theorem (for the determination of R up to an isomorphism) cannot be formulated. There are two approaches to construct $^*R$. The first is mentioned above and rests on the formalization of the properties of R in a first-order laguage; the second is by the use of ultraproducts. We begin with some fact concerning ultraproducts.

The notion of ultraproducts as well as the fundamental theorem was given by J. Łos in 1955. However, the idea behind ultraproducts goes back to a famous paper of Th. Skolem in 1934 where a nonstandard model for arithmetic is constructed. In 1948, in the case of fields, E. Hewitt studied a notion which turned out to be exactly the ultraproducts. For results about ultraproducts the reader is referred to Comfort and Negrepontis (1974); and for applications to Banach space theory reference is made to the excellent article by C. Ward Henson and L. C. Moore, Jr. (to appear).

Let I be a nonempty set and let $\mathscr{F}$ be an ultrafilter over I, i.e. a family of subsets of I with the following properties

(1) $\emptyset \notin \mathscr{F}$.
(2) If M, N are in $\mathscr{F}$ then $M \cap N \in \mathscr{F}$.
(3) If J is any subset of I then J or $I - J$ is in $\mathscr{F}$.

Also it is clear that the notion of ultrafilter may be defined using the order relation in the family of filters on I. We leave the interested reader to formulate the equivalence of the two definitions.

For each i in I we suppose that we have a set A_i and we form the Cartesian product $\pi_{i \in I} A_i$.

We define an equivalence relation $\overset{\mathscr{F}}{\sim}$ as follows:

$$a = (a_i)_{i \in I}, \quad b = (b_i)_{i \in I}$$

are said to be $a \overset{\mathscr{F}}{\sim} b$ if $(i, a_i = b_i)$ is in $\mathscr{F}$.

It is not difficult to see that this is indeed an equivalence relation on $\pi_i A_i$. The set $\pi_i A_i / \overset{\mathscr{F}}{\sim}$ is called the ultraproduct of the sets A_i. If $A_i = A$ for all i in I then the ultraproduct is called the ultrapower; is also called the nonstandard extension of A and is denoted by $*A$.

We consider now the set $A = R$ the reals and $I = N$ the set of integers. We consider an ultrafilter $\mathscr{F}$ on N with the following property that it does not contain any finite subset of N. We consider the nonstandard extension of R, $*R$, defined by the above filter. We remark that many properties of R are inherited by $*R$. We give some examples.

(1) $*R$ *is a field.* The operations '+' and '·' are clear. The additive identity is 0 and the multiplicative identity is 1. The inverse is defined as follows. Suppose that x is in R and is not zero. Then

$$(i, x(i) = 0) \notin \mathscr{F}$$

and

$$(i, x(i) \neq 0) \in \mathscr{F}.$$

Then we set

$$\bar{x}(i) = \begin{cases} 0, & x(i) = 0 \\ x(i)^{-1}, & x(i) \neq 0. \end{cases}$$

Clearly $\bar{x}$ is the inverse of x.

(2) $*R$ *is an ordered field.* If x, y are in $*R$ then we say that $x = y$ if

$$(i, x(i) = y(i)) \in \mathscr{F}.$$

(3) *R is properly embedded in $^*R$*. Indeed, if we set for x in R $x(i) = x$ then this defines clear the required embeddings.

(4) *$^*R$ is not complete*. This follows from the well-known fact that R is the only complete ordered field.

We remark that with respect to the order relation introduced above we have in $^*R$ infinitely large and infinitely small elements. For example, if we consider the element defined by $x(i) = i$ for all i in N, then clearly x is infinitely large; on the other hand if x is infinitely large then x^{-1} is infinitely small as the reader may verify.

DEFINITION 15.2.1 (Robinson, 1961). An element r of $R \subset {}^*R$ is said to be standard. Further, we say that a in $^*R$ is finite if $|a| \leqslant m$ for some m in R^+; a in $^*R$ is said to be infinite if m in R^+ we have $m \leqslant |a|$ and a is said to be infinitesimal if $|a| \leqslant m$ for all m in R^+, $x \approx y$ iff $x - y$ is infinitesimal.

We now consider the nonstandard complex number as elements of the form $x + iy$, where x, y are in $^*R$ and with the usual rules.

We consider an internal vector space over $^*K$ where K is $^*R$ or $^*C$ and we denote by $^*K_f$ the set of all finite elements of $^*K$. A seminorm p, the internal vector space X over K, is a map $p: X \to {}^*[0, \infty)$ with the properties

(1) $p(ax) = |a|\,p(x)$ for a in K_f and all x in X,

(2) $p(x + y) \leqslant p(x) + p(y)$

(where the value $+\infty$ is permitted). We say that on X we have a convex infinitesimal relation if we have a family of seminorms $(p_i)_{i \in I}$, card $I <$ card X and a relation '$\sim$' is defined as follows:

$$x \sim y \text{ iff } p_i(x - y) \text{ is infinitesimal for each } i \text{ in } I.$$

The infinitesimal vectors are denoted by

$$\inf_F(X) = (x, x \sim 0)$$

$\mathscr{B}$ is a family of internal absolutely convex sets of X with the property that for each B in $\mathscr{B}$, sup $p_i(x)$, $x \in B$) is finite, then we set

$$\mathscr{B}_d = UB, \quad B \in \mathscr{B}.$$

We say that $x \overset{\mathscr{B}}{\sim} y$ if $x - y$ is in $\mathscr{B}_d$ and

$$\inf_p(X) = (x, x \sim 0).$$

The set fin(X) is the set of all elements of X which are nearly annihilated by infinitesimal scalars, i.e. $x \in X$ is finite iff for each $\varepsilon \approx 0$, $\varepsilon x \sim 0$. The set of all finite elements of X is denoted by $\hat{X} = \text{fin}(X)$.

Then we have the hull completeness theorem: if B is an internal set such that sup$(p(x), x \in B)$ is finite for each seminorm p in the family, then the locally convex hull $\hat{B}$ of B is complete in $\hat{X}$ and if we have only a finite set of seminorms then $\hat{X}$ is complete.

For a proof of this results as well as for other interesting results we refer to Stroyan and Luxemburg (1976).

Suppose now that X is a complete inner product space with an inner product

$\langle , \rangle$ and an induced norm $\| , \|$. Then we can construct $^*X$ and the above functions $\langle , \rangle$ and $\| , \|$ are also extended as functions on $^*X$, and the properties are (to some extent) the same. Then $^*X$ is a complete space which contains a copy of X.

In $^*X$ we have the following classes of infinitesimals:

(1) norm infinitesimals,

$$x \overset{n}{\approx} y \text{ iff } \|x - y\| \approx 0,$$

(2) weak infinitesimals,

$$x \overset{w}{\approx} y \text{ iff for each } z \text{ in } X \text{(i.e. in the copy of } X \text{ in } {}^*X\text{)},$$

$$\langle x, z \rangle \approx \langle y, z \rangle,$$

(3) bounded weak infinitesimals,

$$x \overset{B}{\approx} y \text{ iff } \|x - y\| \text{ is finite and } x \overset{w}{\approx} y.$$

As is well known, one of the most spectacular success of nonstandard analysis was the proof of the existence of invariant subspaces for polynomially compact operators. For an exposition of this result we refer to Robinson (1966). For results about operators on nonstandard inner product spaces we refer to the papers by L. C. Moor, Jr quoted in the references.

We now present, following Takeuki (1962), an application to quantum mechanics. We consider the space $L^2(R, \mathrm{d}s)$,R is the set of real numbers and $\mathrm{d}s$ is the Lebesgue measure on R. On the space $L^2 = L^2(R, \mathrm{d}s)$ we have two unitary operators defined as follows:

$$(Uf)(t) = \lim_{n - \infty} (2\pi)^{-1/2} \int_{-n}^{n} \mathrm{e}^{-ist} f(s)\,\mathrm{d}s$$

$$(Vf)(s) = \lim_{n \to \infty} (2\pi)^{-1/2} \int_{-n}^{n} \mathrm{e}^{ist} f(t)\,\mathrm{d}t.$$

We consider an ultrafilter $\mathscr{F}$ as above and then the space $\chi = L^2(R, \mathrm{d}s)/\mathscr{F}$. Now if we denoted by U and V the operator induced on this space by U and V then we have the relation

$$i^{-1}\,\mathrm{d}/\mathrm{d}t = Vt \cdot U.$$

Indeed, we consider

$$f_n^a(t) = V(f_n(t - a)), \qquad g_n^a(t) = V\,\delta(t - a),$$

where δ is the element in χ corresponding to the sequence

$$n^{(t)} = \begin{cases} n^2 t + n, & -1/n < t \leqslant 0 \\ 0, & \text{otherwise} \\ -n^2 t + n, & -1/n > t > 0. \end{cases}$$

Since

$$i^{-1}(\mathrm{d}/\mathrm{d}t)(g^a(t)) = a g^a(t) \tag{+}$$

the assertion follows.

If $a \neq b$ then

$$*\langle (t-a),(t-b)\rangle = 0 \tag{++}$$

and then

$$\int_{-\infty}^{\infty} (Uf)(s)g^s(t)\,\mathrm{d}s = \int_{-\infty}^{\infty} (Uf)(s)V\,\delta(t-s)\,\mathrm{d}s = \langle (Uf)(s), U\,\delta(t-s)\rangle$$
$$= \langle f(s), \delta(t-s)\rangle \simeq f$$

and thus

$$f \approx (Uf)(s)g^s(t)\,\mathrm{d}s. \tag{+++}$$

The relations (+), (++), (+++) assert that the operators $i^{-1}(\mathrm{d}/\mathrm{d}t)$, t can be decomposed in a diagonal form in the space X, and this is the basic fact in quantum mechanics as presented in Dirac (1958). For other representations of the δ-symbol of Dirac we refer to Popowicz (1975).

15.3. Intuitionistic Complete Inner Product Spaces

We now consider an interesting class of spaces: the class of intuitionistic complete inner product spaces. The intuitionism is the creation of the famous mathematician L. E. J. Brouwer. In this section we do not enter into detail concerning the intuitionism, but we very briefly present the definitions. The reader interested in this area may wish to consult the excellent exposition of one of the main contributors, A. Heyting (1956, 1971). Also the reader may find in the references other very good accounts of intuitionism. First we present some definitions concerning the arithmetic from the intuitionistic point of view, following Heyting (1971).

DEFINITION 15.3.1. A sequence $a = \{a_n\}$ of rational numbers is said to be a Cauchy sequence if for every integer k we can find *effectively* an integer $n = n(k)$ such that $|a_{n+p} - a_n| < 1/k$ holds for all integers p.

A Cauchy sequence of rational numbers is said to be a 'real number generator' or, for simplicity, a 'number generator'.

DEFINITION 15.3.2. Let $a = \{a_n\}$, $b = \{b_n\}$ be two number generators. Then we say that:

(1) they are identical if $a_n = b_n$ for every n (and we write this as $a \equiv b$);
(2) they coincide if for every integer k we can find *effectively* an integer $n = n(k)$ such that for every integer p the following inequality holds:

$$|a_{n+p} - b_{n+p}| < 1/k$$

(and we write this as $a = b$).

It is proved that the 'coincidence', as defined above, is an equivalence relation.

If $a = b$ is contradictory (i.e. if the supposition that $a = b$ leads to a contradiction) then we write $a \neq b$.

DEFINITION 15.3.3. Let $a = \{a_n\}$ and $b = \{b_n\}$ be two real number generators. We say that a lies apart from b (and we write this is $a \neq b$) if we find *effectively* two integers n and k such that for every integer p the following inequality holds:

$$|a_{n+p} - b_{n+p}| > 1/k.$$

We are now ready to define the fundamental operations of the arithmetic.

DEFINITION 15.3.4. Let $a = \{a_n\}$, $b = \{b_n\}$ be two real number generators. Then

(1) $a + b$ is the sequence $\{a_n + b_n\}$,

(2) ab is the sequence $\{a_n b_n\}$,

(3) $-a$ is the sequence $\{-a_n\}$,

(4) if $a \neq 0$ then a^{-1} is the sequence $\{c_n\}$, where $c_n = a_n^{-1}$ if $a_n \neq 0$ and $c_n = 0$ if $a_n = 0$.

It is proved that $a + b$, ab, $-a$ and a^{-1} (when it exists) are real number generators.

Now if we set, for any rational number r, the sequence (a_n) with $a_n = r$ for all n, then (a_n) is clearly a real number generator and we identify r with this. Thus the system of real number generators can be considered as an extension of rationals.

We now define an important class of functions.

DEFINTION 14.3.5. A rational function f is defined by a finite number of applications of the fundamental operations considered in Definition 15.3.4. For example,

$$f(a, b, c, d) = (a + b)^{-1} cd$$

is defined whenever $a + b \neq 0$.

We now consider the order in the system of real number generators.

DEFINITION 14.3.6. Let $a = (a_n)$, $b = (b_n)$ be two real number generators. Then $a < b$ if we can find *effectively* the integers n and k such that $|b_{n+p} - a_{n+p}| > 1/k$ holds for every integer p. In this case we also use the notation $b > a$. We write $a \not> b$ if $a > b$ is impossible; also $a \not< b$ if $a < b$ is impossible.

We now define new real number generators connected with two real number generators.

DEFINITION 14.3.7. Let $a = \{a_n\}$, $b = \{b_n\}$ be two real number generators. Then

$\max(a, b)$ is the sequence $\{\max(a_n, b_n)\}$,

$\min(a, b)$ is the sequence $\{\min(a_n, b_n)\}$,

$a = \max\ (a, -a)$.

It is proved that max and min are again real number generators.

Now we define the very important notion of convergence for real number generators.

DEFINITION 14.3.8. We say that the sequence $\{x_n\}$ of real number generators is (positively) convergent to the limit a if whenever an integer k is given we find *effectively* an integer n such that for every integer p the inequality $|a - a_{n+p}| < 2^{-k}$ holds. Other notions can be similarly defined.

We have considered above a system which is an extension of the rationals and which consists of so-called 'real number generators'. A real number generator may be considered to be in a constant state of growth in the sense that at each moment we have only a finite part of it, i.e. for each integer n we have all prescriptions for the computation of the nth number of the sequence. In order to consider the real number generators as not being isolated entities, L. E. J. Brouwer has indicated how a satisfactory theory may be obtained without the use of the notion of arbitrary law. For this purpose he has considered the so-called 'infinitely proceeding sequence' (in short, 'i.p.s.') which means a sequence that can be continued *ad infinitum.*

The generation of an i.p.s. is not quite free and the restrictions to be made are such that at every stage we are able to decide which components we consider for the next stage and which not. In order to consider the intuitionistic concept of the continuum as a possibility of a gradual determination of points, L. E. J. Brouwer was led to consider the very important notion of a 'spread'.

The notion of spread is defined by two laws. In order to formulate them we need the following convention. Suppose that we have the finite sequence $a_1, a_2, \ldots, a_n, a_{n+1}$. Then we say that a_{n+1} is an immediate descendant of $a_1, \ldots, a_n$ and that $a_1, \ldots, a_n$ is the immediate ascendant of the sequence $a_1, a_2, \ldots, a_n, a_{n+1}$.

Now the introduction of the notion of spread is as follows.

DEFINITION 15.3.9. A spread law is a rule Λ such that:

(1) it divides the finite sequence of natural numbers into admissible and inadmissible sequences;
(2) using Λ we can decide for every k whether it is admissible or not;
(3) for every admissible sequence $a_1, a_2, \ldots, a_n, a_{n+1}$ the sequence $a_1, \ldots a_n$ is an admissible sequence;
(4) if $a_1, a_2, \ldots, a_n$ is an admissible sequence then using Λ we can decide whether $a_1, \ldots, a_n, k$ is an admissible sequence or not;
(5) for every admissible sequence $a_1, a_2, \ldots, a_n$ we can find at least one natural number k such that $a_1, \ldots, a_n, k$ is an admissible sequence.

DEFINITION 15.3.10. The complementary law Γ assigns to each finite sequence which is admissible (according to the spread law Λ) a definite mathematical entity.

REMARK 15.3.11. The entities are, in general, previously introduced mathematical entities.

DEFINITION 15.3.12. Let $a = \{a_n\}$ be an i.p.s. Suppose that for every integer n, $a_1, \ldots, a_n$ is an admissible sequence according to the spread law. Then we call a an admissible i.p.s.

The i.p.s. of the entities assigned by the complementary law to the sequence $\{A_n\}$, $A_n = a_1, \ldots, a_n$, is an element of the spread. If $a = \{a_n\}$ is an i.p.s. of the spread then a_n is said to be the nth component of a. Two i.p.s. of a spread are said to be equal if their components are equal for each n.

DEFINITION 15.3.13. Two spreads are equal if for every element in one we find an equal element in the other.

We are now ready to define the intuitionistic notion of a real number. From the intuitionistic point of view a set is defined if:

(1) we have a common mode to generate its elements;
(2) we have a characteristic property of its elements.

DEFINITION 15.3.14. Sets which are defined by (2) are called species. A species is a property which mathematical entities can be assumed to possess.

DEFINITION 15.3.15. Let S be a species. Then any mathematical entity which

(1) is defined before S or might have been defined before S, or
(2) satisfies the condition S

is a member of S.

Using this we can defined the notion of a real number.

DEFINITION 15.3.16. We consider the property S to be the coincidence with a given real number generator. Then this is a species and is called a real number. If x is a real number and if ξ is one of its real number generators then we say that ξ represents x. Also we say that ξ coincides with x.

DEFINITION 15.3.17. *The intuitionistic continuum.* The species of all real numbers which is not defined as a spread species is the (species) continuum.
The i.p.s. which are equal to the elements of a spread form a species which is called the spread species (of the spread).

The components of an i.p.s. u of natural numbers form a species (which, as a convention, is identified with u).

Concerning species, we consider also the following notions.

DEFINITION 15.3.18. Two species are said to be equivalent if they are in one-to-one correspondence.

DEFINITION 15.3.19. If S is a species then we say that it is a finite species if it is equivalent to the species consisting in building up successively the numbers from 1 to n for some n(in notation $1 \to n$). A species which is equivalent to the species N of all natural numbers is said to be denumerably infinite. Now it is clear that we can construct the arithmetic of real numbers and the arithmetic of real number generators may be used.

We now consider an important notion, the notion of a field.

DEFINITION 15.3.20. We say that a mathematical species K is a field if the following assertions hold:

(1) In K an apartness relation $\neq$ is defined.
(2) In K a commutative and associative 'addition' is defined. Also in K there exists an element called zero and the negative for any k in K.

(3) If z is in K and x, y are in K with $x \neq y$ then $x + z \neq y + z$.
(4) In K a commutative and associative 'multiplication' is defined with the following properties:
   (a) it is distributive with respect to addition;
   (b) there exists in K an element 1 which is the unit element and $1 \neq 0$.
(5) For any x in K with $x \neq 0$, x^{-1} exists and $x^{-1} \neq 0$.
(6) If z is in K, $z \neq 0$ then for any x, y in K, $x \neq y$ and we have $xz \neq yz$.

If K is a field then we now consider the intuitionistic analogue of the spaces R^n and C^n.

DEFINITION 15.3.21. Let K be a field. Then K^n is the species of sequences $(k_1, \ldots, k_n)$ for k_i in K, $i = 1, 2, 3, \ldots, n$. The vector space operations are defined as for R^n (or C^n). Linear independence has two meanings in the intuitionistic setting, namely the relation is given by

$$k_1 a_1 + \cdots + k_m a_m = 0,$$

where k_i are in the field K and a_i are in K^n, then we can require that $k_i \# 0$ or $k_i \neq 0$.

In the first case we say that the elements $a_1, \ldots, a_m$ are strong linearly dependent and in the second case they are weak linearly dependent. In what follows we use the strong dependence and we say simply that the elements are linearly dependent. If a system $a_1, \ldots, a_m$ of elements of K^n is not linearly dependent then we say that it is linearly independent.

Now we consider the notion of a complete inner product space (in the intuitionistic setting). The definition is as follows.

DEFINITION 15.3.22. A set L is called an inner product space on the field R (of real numbers) if the following conditions are satisfied:

(1) For each ordered pair (r, x), $r \in R$, $x \in L$, and element of L corresponds which is called the multiplication of x with r and is denoted by rx.
(2) For each ordered pair (x, y) of elements of L corresponds an element of L called the sum of x and y and is denoted by $x + y$.
(3) The axioms of linear spaces hold for the above operations.
(4) For each ordered pair (x, y) there exists a real number, called the inner product of x with y, and the usual axioms hold, the inner product is denoted by $\langle x, y \rangle$.
(5) For each x in L, $\langle x, x \rangle \not< 0$.
(6) If for x in L, $\langle x, x \rangle = 0$ then $x = 0$.
(7) For every integer n we find in L n mutually free elements.
(8) In L there exists a sequence (e_n) of vectors such that
   (a) every finite subsequence if free,
   (b) the linear span (i.e. the species of linear combinations of elements in (e_n) (only finite combinations are considered) is dense in L.
(9) If (x_n) is a sequence in L with the property that for any natural number k a natural number N can be found such that for every natural numbers p, q

$$\|x_{N+p} - x_{N+p}\| \leqslant 2^{-k},$$

then there exists in L an element x with the property that for every given natural number k a natural number N can be found such that for every natural number p,

$$\|x_{N+p} - x\| \leqslant 2^{-k}.$$

This axiom is the so-called completeness axiom.

We call a space with the above properties a complete inner product space in the sense of Heyting, or simple a Heyting space.

DEFINITION 15.3.23. If L is a Heyting space then we say that L has a denumerable infinite orthonormal basis if there exists an infinite system (e_i) of elements of L such that:

(1) $\langle e_i, e_j \rangle = \delta_{ij} = \begin{cases} 1, & i = j, \\ 0, & i \neq j, \end{cases}$

(2) for each x in L we can indicate, for each natural number k, a linear combination

$$r_1 e_{i_1} + \cdots r_m e_{i_m} = a_m$$

with the property that $\|a_m - x\| \leqslant 2^{-k}$.

DEFINITION 15.3.24. A sequence (x_n) in a Heyting space L is said to be complete if for each x in L and every natural number k we find an integer $n = n(k)$ such that for all $m \geqslant n$, $\|x_m - x\| \leqslant 2^{-k}$.

In his lecture notes (1949–1950) A. Heyting has proved the following important result.

THEOREM 15.3.25. *If L is a Heyting space having a denumerable orthonormal basis then L is isomorphic to the space of all sequences $x = (x_1, x_2, \ldots)$ such that*

$$|x_1|^2 + |x_2|^2 + \cdots + |x_n|^2 + \cdots$$

converges and where the operations '+' and '·' are defined in the known usual way, as well as the inner product.

One of the basic and interesting results in the theory of complete inner product spaces is the Cauchy–Buniakowsky inequality. The intuitionistic version was obtained by A. Heyting. The proof below is given by Ashvinikumar (1966).

PROPOSITION 15.3.26 (The Cauchy–Buniakowsky inequality). *Let L be a Heyting space. Then for any x, y in L the following assertion holds*:

$$\|x\|\|y\| \not< |\langle x, y \rangle|.$$

Proof (Ashvinikumar, 1966). If $y = 0$ then the assertion is clear. Suppose now that $y \neq 0$ and thus

$$0 \not> \big\| \|y\|^2 x - \langle x, y \rangle y \big\|^2 = \|y\|^4 \|x\|^2 - \|y\|^2 \overline{\langle x, y \rangle} \langle x, y \rangle - \|y\|^2 \langle x, y \rangle \overline{\langle x, y \rangle} + \\ + |\langle x, y \rangle|^2 \|y\|^2 = \|y\|^2 (\|x\|^2 - |\langle x, y \rangle|^2).$$

Now suppose that

$$\|x\|\|y\| < |\langle x, y\rangle|.$$

If y is apart from 0 then the above result gives a contradiction. Thus y must be 0. But then by the above remark and inequality is trivial. Thus the Cauchy–Buniakowsky inequality holds.

As we know, one of the basic results in the theory of complete inner product spaces is the representation theorem of Fréchet–Riesz of bounded linear functionals. In what follows we present an intuitionistic version of this result.

DEFINITION 15.3.27 (Ashvinikumar, 1966). A linear functional F on a Heyting space is a function on L with the following properties:

(1) For each x in L we can indicate a number $F(x)$ such that the following relation holds,

$$F(ax + y) = aF(x) + F(y),$$

(2) If $F(x)$ is apart from 0 then x is apart from 0.

DEFINITION 15.3.28. The linear functional F on X is called bounded if we can indicate a number M such that for every x in L

$$|F(x)| < M\|x\|.$$

DEFINITION 15.3.29 (Ashvinikumar, 1966). We say that the linear functional F on the Heyting space L has a supremum if we can indicate a real number $M > 0$ such that:

(1) for each unit element in L(i.e. $\|x\| = 1$), $|F(x)| \not> M$;
(2) for each natural number n we can indicate a unit element x_n such that

$$|F(x_n)| > M - 2^{-n}.$$

In this case we call M the supremum of F and it is denoted by $\|F\|$.

DEFINITION 15.3.30 (Ashvinikumar, 1966). The functional F is said to be apart form the null functional (denoted by 0) if we can indicate an element x of L such that $F(x)$ is apart from 0; we say that F assumes its supremum if we find an element u in X with the property that $\|F\| = |F(u)|$.

PROPOSITION 15.3.31 (Ashvinikumar, 1966). *Let F be a linear functional on the Heyting space L such that the following conditions are satisfied*:

(1) *F is apart from* 0.
(2) *F has a supremum.*

Then F assumes its supremum.

Proof. We consider the number $\|F\|$ and we suppose that it is apart from 0. Then we can indicate for any natural number k with the property that for $n \geqslant k$, we can indicate an element x_n in L with the property that

(1) $\|x_n\| < 1 + 2^{-n}$.

(2) $|F(x_n)| = \|F\|$. □

(We may assume without loss of generality that $F(x_n) = \|F\|$ since in the contrary case we set $\overline{F(x_n)}/\|F(x_n)|\cdot x_n$ as x_n.) But then we have,

$$\|x_n + x_m\|^2 = \|x_n\|^2 + \|x_m\|^2 + \langle x_n, x_m\rangle$$
$$+ \langle x_m, x_n\rangle \leqslant 1 + 2^{-n} + 1 + 2^{-m} + 2d,$$

where $2d = \langle x_n, x_m\rangle + \langle x_m, x_n\rangle$ and thus

$$F((x_n + x_m))^2/(\|x_n + x_m\|^2) > 4\|F\|^2/2(1 + 2^{-n-1} + 2^{-m-1+d}).$$

Now if d were less than $1 - 2^{-m-1} - 2^{-n-1}$ this gives a contradiction since this would exceed $\|F|^2$. Thus d is not less than the above number.

But

$$\|x_n - x_m\|^2 = \|x_n\|^2 + \|x_m\|^2 - \langle x_n, x_m\rangle - \langle x_m, x_n\rangle$$
$$\leqslant 1 + 2^{-m} + 1 + 2^{-n} - 2\mathrm{d} \not> 2^{-m+1} + 2^{-n+1}$$

which implies that the sequence (x_n) converges to an element of L, say u. Obviously

$$F(u) = \lim F(x_n) = \|F\|, \qquad \|u\| = \lim \|x_n\| = 1.$$

Thus F assumes its supremum and the proposition is proved.

The following presents a version of the Fréchet–Riesz representation theorem for Heyting spaces.

THEOREM 15.3.32 (Ashvinikumar, 1966). *Let L be a Heyting space and let L be a linear functional satisfying the following properties*:

(1) *F is apart from null.*
(2) *F assumes its supremum.*

Then there exists a unique element u in L such that $F(x) = \langle x, u\rangle$ for each x in L.

Proof. From Proposition 15.3.31 we find an element u in L with the property that $F(u) = \|F\|$. We have to consider several cases.

CASE 1. Suppose that v is an element of L orthogonal to u. Then $F(v) = 0$. Indeed, if the contrary is true then $|F(v)|$ is apart from zero. We may assume without loss of generality that $F(v)$ is a real number greater than zero. Clearly v is apart from zero and also we may assume that $\|v\| = 1$. In this case

$$(\|F(\|F\|u + F(v)v\|)) \left(\left(\big\| \|F\|u + F(v)v\big\|\right)\right)^2 = ((\|F\|F(u) + F(v)F(v))/(\|F\|^2 + F(v)^2)$$
$$= \|F\|^2 + F(v)^2$$

and this clearly contradicts the fact that F is the supremum for F. Thus we have that it is contradictory that $F(v)$ is apart from zero. Thus we must have $F(v) = 0$.

CASE 2. We consider the species of those elements of L with the property that $F(x) = 0$. This is denoted by L_1 which is a closed linear subspace of L. We show that any s in L_1 has the property that $\langle s, u\rangle = 0$.

Indeed, we consider the decomposition

$$s = \langle s, u\rangle u - u_1$$

and then clearly u_1 is orthogonal to u. From the result obtained above we have that $F(u_1) = 0$ and thus

$$F(s) = \langle s, u\rangle F(u) = \langle s, u\rangle \|F\|$$

which is apart from zero. But this contradicts the fact that s is in L_1 and thus we must have $\langle s, u\rangle = 0$. Now if x is an element of L we consider the element $F(x)u - F(u)x$ which is in L_1, and since u is orthogonal to each element of L_1 we obtain

$$F(x) = \langle x, \overline{F(u)}\, u\rangle.$$

Thus the element $w = Fu$ has the property that for each x in L,

$$F(x) = \langle x, w\rangle$$

Since the uniqueness follows easily, the proof is complete.

In his Ph.D. thesis (under A. Heyting) Ashvinikumar presents results about linear operators on Heyting spaces. The interested reader may wish to consult the lecture notes of Heyting (1949–1950), the Ph.D thesis of Ashvinikumar (1966) and the monograph of Heyting on 'Intuitionism' for further details and information.

In his book on intuitionism (Heyting, 1971 (1956)) develops a measure theory and correspondingly the integral associated. Using the integral he consider the metafunctions (a metafunction is a species of function which is equivalent to a given function) and further the space L of metafunctions f which are a.e. definite (in the intuitionistic sense) and such that f^2 is summable. He proves that this is a Heyting space. We refer to chapter 6 of this book for details.

REMARK 15.3.33. It may be of interest to consider classes of intuitionistic Banach spaces and to characterize the Heyting spaces, like the parallelogram law, orthogonality, etc.

15.4. Constructive Inner Product Spaces

In this section we present very briefly some facts about constructive real numbers and the construction of normed constructive spaces and constructive inner product spaces. We mention that constructivism, in the sense considered here, is that developed by E. Bishop in his well-known book. The reader interested in such problems should consult Bridges (1979) and Kušner (1973, in russian) as well as Bishop (1967). First we present the construction of constructive real numbers.

DEFINITION 15.4.1. A sequence is a rule which associates, to each positive integer n, an object, denoted by a_n, which is called the nth term of the sequence. We use the notation (a_n) for a sequence.

It is worth mentioning that the rule we considered in Definition 15.4.1 means that it is given explicitly or the terms of the sequence are written down until the rule of formation is quite clear. We mention that we may use for the sequence the notation $n \to a_n$ or $(a_1, a_2, \ldots)$.

If we have sets $S_1, \ldots, S_m$ then their Cartesian product $S_1 \times S_2 \times \cdots \times S_m$ is by definition the set of all ordered n-tuples $(s_1, \ldots, s_m)$ with s_i in S_i, $i = 1, 2, \ldots, m$, with the natural notion of equality of n-tuples. If Z is the set of integers, then we remark that $Z \times Z$ can be organized as a sequence. Indeed, we order the elements (i, j) in $Z \times Z$ first according to $|i| + |j|$, then regarding the value of i, and next the value of j. Thus we have the sequence:

$$((0,0),\ (-1,0),\ (0,-1),\ (1,0),\ (0,1),\ (-2,0),\ (-1,-1), \ldots)$$

in which each element of $Z \times Z$ appears exactly once.

Now from this sequence we delete the terms $(i, 0)$ and each term (p, q), $q \neq 0$ is replaced by p/q. Thus we have the sequence

$$(0/-1, 0/1, -1/-1, \ldots).$$

Further, we consider the sequence

$$(0/1,\ -1/1,\ 1/1, \ldots).$$

where we consider only $0/1$ and (p/q), $q \neq 0$ such that p, q are relatively prime integers and this clearly is the required sequence.

A set is a totality of objects constituted according to certain rules.

An equivalence relation is a rule having the following properties: if x, y, z are elements of the set M then

(1) $x = x$,

(2) if $y = x$ then $x = y$,

(3) if $x = y$ and $y = z$ then $x = z$.

As usual the symbol $\in$ denotes membership of an object to the set or that a construction is indicated which defines an object of the set M.

An operation from set A to set B (written $f: A \to B$) such that for each a in A associates an element denoted $f(a)$ of B and the rule of association must be an explicit, finite, mechanical reduction of a procedure to construct $f(a)$ from a.

The set A is the domain of the function f.

It is also clear that a sequence is a function on Z^+.

Now if A, B, C are sets and $f: A \to B$, $g: B \to C$ then we can define $h: A \to C$ by the composition $h(a) = g(f(a))$.

The function $f_1: B \to A$ is said to be the left inverse of $f: A \to B$ if for all a in A, $f_1(f(a)) = a$; similarly we define the right inverse.

For two sets A and B for which we have $f: A \to B$ which has a left and right inverse then we say that A and B are equivalent. A set A equivalent to Z^+ is said to be countably infinite and a set A is said to be finite if it is equivalent with $(1, 2, 3, \ldots, n)$ for some n.

We now define some of the basic notions of constructive real analysis.

DEFINITION 15.4.2. A sequence (x_n) of rational numbers is said to be regular if for all n, m in Z^+,

$$|x_n - x_m| \leqslant m^{-1} + n^{-1}.$$

Using the regular sequence we define the basic notion of 'constructive real number'.
DEFINITION 15.4.3. A real number is a sequence of rational numbers (x_n) which is a regular sequence.

DEFINITION 15.4.4. If (x_n), (y_n) are two real numbers then we say that they are equal if

$$|x_n - y_n| < 2/n.$$

Then we have the following result.

PROPOSITION 15.4.5. *The equality defined by Definition* 15.4.4 *on the set of constructive real numbers is an equivalence relation.*

Proof. The first two properties of an equivalence relation are clear. For (3) we need the following lemma. □

LEMMA 15.4.6. *The real number* (x_n) *is equal to the real number* (y_n) *iff for each positive integer* j *there exists an integer* N_j *such that*

$$|x_n - y_n| < j^{-1}, \quad n \geqslant N_j.$$

Proof. The necessity follows immediately since $j - 2j$ gives the assertion. Suppose that for each j we have the above property. Then if m is a positive integer with $m = \max(j, N_j)$,

$$|x_n - y_n| = |x_n - x_m + x_m - y_m + y_m - y_n|$$
$$= < n^{-1} + m^{-1} + j_+^{-1}(n^{-1} + m^{-1})\langle 2/n + 3/j$$

and since this holds for all j in Z^+ we obtain that (x_n) is equal to (y_n). From this lemma it is clear that (3) holds and the proposition is proved.

For each real number $x = (x_n)$ we associate an integer K_x such that

$$|x_n| \leqslant K_x, \quad n \in Z^+;$$

for example, if we take K_x to be the least integer greater than $x_1 + 2$, K_x is called the canonical bound of x.

PROPOSITION 15.4.7. If $(x_n) = x$, $(y_n) = y$ are real numbers and a is a rational number then:

$$x + y = (x_n + y_n),\ xy = (x_{2kn} y_{2kn}),\ -x = (-x_n),\ a^-$$
$$= (a, a, \ldots),\ \max(x, y) = (\max(x_n, y_n)).$$

For a proof we refer to Bishop (1967).

Then clearly the following holds: if z is another real number and b is another rational number:

(1) $x + y = y + x, \quad xy = yx.$

(2) $(x + y) + z = (x + (y + z)), \quad x(yz) = (xy)z.$

(3) $x(y + z) = xy + xz$.

(4) $x - x = 0^{-}$.

(5) $0^{-} + x = x, \quad 1^{-} \cdot x = x$.

(6) If $|x| = \max(x, x)$ then $|xy| = |x|\,|y|$.

(7) $(x + y)^{-} = x^{-} + y^{-}, \quad (xy)^{-} = x^{-}y^{-}$.

We now define an order relation on the set of real numbers.

DEFINITION 15.4.8. The real number $x = (x_n)$ is said to be positive if for some n in Z^{+}, $x_n > n^{-1}$; the real number x is said to be nonnegative (or to be in R^{0+}) if $x_n \geqslant -n^{-1}$.

Then we have that a real number x is positive iff there exists an integer N with the property that

$$x_m \geqslant N^{-1}, \quad m \geqslant N.$$

DEFINITION 15.4.9. If $x = (x_n)$, $y = (y_n)$ are two real numbers then we say that $x < y$ (or $y \geqslant x$) if $x - y$ is in R^{+} and we say that $x \leqslant y$ if $x - y$ is in R^{0+}.

Then we have the following basic result.

THEOREM 15.4.10. *If x, y are two real numbers then, at most, one of the relations $x < y$ or $y < x$ is valid.*

Again we refer to Bishop (1967) for a proof, as well as for other arithmetic properties of real numbers, especially for the definition of the inverse. The reader can find other interesting and important notions of constructive analysis in both Bishop (1967) Bridges (1979).

We now consider the usual metric spaces, but where the values of the metric are in real numbers (constructive real numbers). Then there we can define the notion of convergence, Cauchy sequences and completion. It is not difficult to see that completion can be defined using the regular sequences that are defined for the case of real numbers.

If X is a linear space over the real or complex numbers (the constructive real numbers or the complex constructive numbers) then the notion of 'inner product' and 'norm generated by an inner product' are defined after the usual model.

Then we have the following result.

THEOREM 15.4.11. *Let X be a complete normed space whose norm is generated by an inner product. Then the parallelogram law holds, i.e. for all x, y in X,*

$$\|x + y\|^2 + \|x - y\|^2 = 2(\|x\|^2 + \|y\|^2).$$

PROBLEM 15.4.12. If X is a Banach space whose norm satisfies the parallelogram law, is the norm of X generated by an inner product?

Also it may be of interest to discover if other metric characterizations can be extended to constructive setting; for example, those using the 'constructive modulus of convexity' and the 'constructive modulus of smoothness' seem to be quite difficult problems (see also Bishop's problem concerning the reflexivity of constructive uniform spaces).

At this point it may be of interest to mention the possible construction of constructive inner product modules.

CHAPTER 16

Indefinite Inner Product Structures

16.0. Introduction

The notion of indefinite inner product space was first considered by the famous physicist P. A. M. Dirac in 1942 and independently by the famous Soviet-Russian mathematician L. S. Pontrjagin 1944, who was motivated by some questions of S. L. Sobolev related to a problem in mechanics (published, much later, in 1960). The subject has attracted the interest of many mathematicians. For a detailed account of the subject we refer to the excellent survey articles by Krein and Iohvidov (1956, 1959) and to the book by Bognár (1974); for applications the readers may wish to consult the book by Daleckiĭ and Krein (1970) and the excellent papers by Synge (1954) and Bracci (1975).

In this chapter we present a very short introduction to the subject. After introducing some general notions about spaces with an indefinite inner product we consider some classes of linear operators on such spaces. Then we present some spaces which have special properties (such as Pontrjagin spaces, etc.). Further, we try to extend this setting to more general structures by considering indefinite inner product modules. This class is suggested by the inner product modules.

16.1. Indefinite Inner Product Linear Spaces

Let X be a linear space over K (R or C).

DEFINITION 16.1.1. An inner product on X is a mapping $\langle\,,\rangle$ defined on $X_\times X$ with values in K satisfying the following conditions:

(1) $\langle ax + y, z\rangle = a\langle x, z\rangle + \langle y, z\rangle$.
(2) $\langle x, y\rangle = \overline{\langle y, x\rangle}$.

(The conjugation appears, of course, only when $K = C$.)

REMARK 16.1.2. In the definition of an inner product given above we do not assume that $\langle x, x\rangle > 0$ for all x in X, $x \neq 0$.

We now consider several classes of elements of X related to possible values of an inner product.

DEFINITION 16.1.3. The element x of X is said to be:

(1) positive if $\langle x, x\rangle > 0$,
(2) negative if $\langle x, x\rangle < 0$,
(3) neutral if $\langle x, x\rangle = 0$.

The set $\{x, \langle x, x\rangle = 0\}$ is called the neutral set of the space L with an inner product $\langle\,,\rangle$. The set $\{x, \langle x, x\rangle > 0\}$ is called the positive set of L and $\{x, \langle x, x\rangle < 0\}$ is called the negative set of L.

DEFINITION 16.1.4. If X is a linear space with an inner product $\langle\,,\rangle$ then

$$\begin{aligned} X^{\infty} &= \{x, x \in L, x \neq 0, \langle x, x\rangle = 0\}, \\ X^{++} &= \{x, x \in L, \langle x, x\rangle > 0 \text{ or } x = 0\}, \\ X^{+} &= \{x, x \in L, \langle x, x\rangle \geqslant 0\}. \end{aligned}$$

Using these sets we now define the class of objects of study in this chapter.

DEFINITION 16.1.5. We say that a linear space over K with an inner product $\langle\,,\rangle$ is a space with an indefinite inner product if in X there are positive and negative elements.

We present several examples of linear spaces which are spaces with an indefinite inner product.

EXAMPLE 16.1.6. Let $u_i = \pm 1$, $i = 1, 2, 3, \ldots$, and consider the space l^2. A new inner product in l^2 may be defined as follows: if $x = (x_i)$, $y = (y_i)$ are in l^2 then we set

$$\langle x, y\rangle = \sum_{i=1}^{\infty} u_i x_i \bar{y}_i.$$

Then clearly $(l^2, \langle,\rangle)$ is an indefinite inner product space if we choose u_i not all of the same sign.

EXAMPLE 16.1.7. Let X, Y be two complete inner product spaces with the inner products $\langle\,,\rangle_X$, $\langle\,,\rangle_Y$. If we consider the Cartesian product $X_\times Y$, then this can be considered as a linear space with an indefinite inner product space if we set, for any two elements (x_1, y_1), (x_2, y_2) in $X_\times Y$,

$$\langle (x_1, y_1), (x_2, y_2)\rangle = \langle x_1, x_2\rangle_X - \langle y_1, y_2\rangle_Y.$$

EXAMPLE 16.1.8. Let X_n be a complete inner product space, for $n = 1, 2, 3, \ldots$. Then we consider the linear space of all sequences $x = (x_n)$, $x_n \in X_n$ and such that $\Sigma \|x_n\|^2 < \infty$. Then this can be organized as a linear space with an indefinite inner product if, when $y = (y_n)$ is another element in the space, we set

$$\langle x, y\rangle = \sum_{n=1}^{\infty} \langle x_n, y_n\rangle u_n,$$

where the u_n are as in Example 16.1.6.

Obviously, if each X_n is equal to K then we have Example 16.1.6.

Let X be a linear space with an inner product, then if the inner product is not indefinite we say that X is a linear space with a semidefinite inner product or that X is a semidefinite inner product space.

16.2. Orthogonality and Orthogonal Decomposition

If X is a space with an indefinite inner product $\langle\,,\rangle$ then we define the orthogonality exactly as in the case considered in Chapter 4.

DEFINITION 16.2.1. Let X be a linear space over K with an indefinite inner product. If M is a subset of X and N is a subset in X then we say that M is orthogonal to N if $\langle m, n \rangle = 0$ for all $m \in M$ and $n \in N$. For simplicity we write this as $M \perp N$.

If M and N reduce to a single point set, say $\{m\}$ and $\{n\}$, then we say that m is orthogonal to n and we write this as $m \perp n$.

DEFINITION 16.2.2. A family of subsets $\{M_i\}_{i \in I}$, I is an index set, is a family of pairwise orthogonal subsets if for any i, j in I, $i \neq j$, $M_i \perp M_j$ holds.

If X is of the form

$$X = M_1 + M_2 + \cdots + M_n, \quad M_i \text{ subspaces of } X$$

with

$$M_i \perp M_j, \quad i \neq j$$

then we say that X is the orthogonal direct sum of the subspaces M_i and we write this as

$$X = \sum_{i=1}^{n} + M_i.$$

If M is any subset of X then we set

$$M^{\perp} = (x, x \in X, \{x\} \perp M)$$

which is called the orthogonal set of M. We define $M^{\underbrace{\perp\perp\perp\perp\perp\cdots\perp}_{n \text{ times}}}$ iteratively, e.g. $M^{\perp\perp} = (M^{\perp})^{\perp}$.

Then we have the following result.

PROPOSITION 16.2.3. *If M is any subset in X then*

$$M^{\perp\perp} \supseteq M, \qquad M^{\perp\perp\perp} = M^{\perp}.$$

We now define an interesting subset of a linear space with an indefinite inner product.

DEFINITION 16.2.4. Let X be a linear space with an indefinite inner product $\langle\, , \rangle$ and let M be a linear subspace of X. The isotropic subspace of M is defined as the set $M \cap M^{\perp} = M^0$. If $M^0 \neq \{0\}$ then we say that M is degenerate (or that the inner product is degenerate on M).

REMARK 16.2.5. There exist simple examples which show the existence of degenerate subspaces.

The following proposition characterizes the isotropic subspace of a linear space with an indefinite inner product. For this we remark that the following result holds.

THEOREM 16.2.6 (The Cauchy–Buniakowsky inequality). *For any x, y in a linear space X with an inner product which is semidefinite, the following inequality holds:*

$$|\langle x, y \rangle|^2 \leqslant \langle x, x \rangle \langle y, y \rangle.$$

Since the proof does not differ from the case of inner product spaces it is omitted.

PROPOSITION 16.2.7. *The isotropic subspace of the linear space X with an indefinite inner product space $\langle\, ,\rangle$ is the set of all neutral elements of X.*

Proof. This follows immediately from the definition and the Cauchy–Buniakowsky inequality. □

As we know, for any linear space X and any linear subspace M_1 of X there exists another linear subspace M_2 such that for every x in X there exist x_i, $i = 1, 2$, $x_i \in M_i$, with the property that $x = x_1 + x_2$ and $M_1 \cap M_2 = \{0\}$. Then we have clearly the following result.

PROPOSITION 16.2.8. *If X is a linear space with an indefinite inner product and X^0 is the isotropic subspace then*

$$X = X^0 + X_1,$$

where X_1 is the complementary subspace of the neutral space of X.

Suppose now that X is a linear space with an indefinite inner product $\langle\, ,\rangle$ and we consider now some interesting classes of subspaces like maximal semidefinite, positive, negative, neutral, etc., in the sense that the respective subspace is not contained in another proper subspace of X of the same type.

PROPOSITION 16.2.9 (Bognár, 1966, 1974). *Let X be a linear space which has an indefinite inner product, and suppose that M_1, M_2 are two maximal subspaces which are either positive of negative. Then the neutral space of the subspace generated by M_1 and M_2 is $(M_1 \cap M_2)^0$.*

Proof. Let x be an element of $(M_1 \cap M_2)^0$ and thus

$$x \in M_1 \cap M_2, \quad \langle x, x\rangle = 0.$$

Then clearly x is in M_1^0 and x is in M_2^0 and we get easily that x is in $M_1 \cap M_2^{\perp}$ and in $M_1^{\perp} \cap M_2$. This gives, further, that

$$x \in (M_1 \cap M_2) \cap (M_1 \cap M_2)^{\perp} = (M_1 \cap M_2)^0.$$

Conversely, if x is in $(M_1 \cap M_2)^0$ then x is in M_i^0, $i = 1, 2$.

Suppose now that M_1 is positive (in the contrary case the argument is similar). Then (x, M_1) is a subspace which is positive and contains M_1. Since M_1 is maximal we get that x is in M_1. Similarly we prove that x is in M_2 and Proposition 16.2.9 is proved.

We leave the reader to prove the following assertion.

PROPOSITION 16.2.10. *Let X be a linear space with an indefinite inner product space and suppose that*

$$M_1 + M_2 = X$$

with M_1 a positive linear subspace and M_2 a negative subspace. Then M_i, $i = 1, 2$ are maximal subspaces.

PROPOSITION 16.2.11. *Let X be a linear space with an indefinite inner product and suppose that M is a positive linear subspace of X. Then M^- is a negative subspace.*

Let X be a linear space with an indefinite inner product $\langle , \rangle$ and suppose that X may be represented as an orthogonal direct sum of subspaces which are neutral, positive and negative, i.e.

$$X = M^0 + M^+ + M^-$$

where M^0 is a neutral subspace of X, and M^+ and M^- are positive and negative subspaces, respectively.

DEFINITION 16.2.12. If X has the above decomposition then we call such a decomposition the 'fundamental decomposition' of X.

REMARK 16.2.13. Not every linear space with an indefinite inner product $\langle , \rangle$ admits such decompositions and, in general, such a decomposition is not unique.

16.3. Linear Operators on Spaces with an Indefinite Inner Product

As usual we define the notion of a linear operator on the space X which is supposed to be a linear space (over K) and with an indefinite inner product, as a mapping $T: X \to X$ (or on a part D_T of X with values in X) such that the properties

(1) $T(x + y) = tx + Ty$,
(2) $T(kx) = kTx$

hold for all x, y in X and k in K.

The set $N(T) = (x, x \in X, Tx = 0)$ is called the null space of T (sometimes noted as Ker T); the subspace $R(T) = (Tx, x \in X)$ is called the range of T. It is clear that a similar definition can be given for operators defined on a space X and with values in another space Y (both supposed to be over the same field K and having an indefinite inner product).

The identity operator on X, denoted by I, is defined by the relation $Ix = x$, $x \in X$. If $T: X \to X$ is a linear operator then a number k is said to be an eigenvalue for T is for some $x_0 \neq 0$, $Tx_0 = kx_0$ holds; we say that x of X is a principal vector for the eigenvalue k if for some integer n, $(T - k)^n x = 0$ holds.

REMARK 16.3.1. Some restrictions are necessary if $D_T \neq X$ and we leave the reader to formulate them.

DEFINITION 16.3.2. The subspace generated by the principal vectors of an eigenvalue k of T is denoted by X_k; the dimension of X_k is called the multiplicity of the eigenvalue k.

We now consider several classes of linear operators on spaces with an indefinite inner product.

DEFINITION 16.3.3. Let X be a linear space with an indefinite inner product. The linear operator $T: X \to X$ is said to be an isometric operator if for all x, y in X the following relation holds:

$$\langle Tx, Ty \rangle = \langle x, y \rangle.$$

It is obvious that if T^{-1} exists then it is an isometric operator when T is an isometric operator. If X is a complete inner product space and T is an isometric operator then it is easy to prove that the eigenvalues of T are in $(z, |z| = 1)$. In the case of isometric operators on spaces with an indefinite inner product this property in not, in general, true.

EXAMPLE 16.3.4 (Bognár, 1974). Let X be a two-dimensional space with the basis (e_1, e_2) and the inner product on X is defined by the relations:

(1) $\langle e_j, e_j \rangle = 0$, $j = 1,2$.
(2) $\langle e_1, e_2 \rangle = i (i^2 = -1)$.

Then clearly $(X, \langle\ ,\ \rangle)$ is an indefinite inner product space over C. The operator $T: X \to X$, defined by the relations $Te_1 = e_1$, $Te_2 = e_1 + e_2$, is an isometric operator and

$$(T - I)^2 e_2 = 0.$$

If we set $Se_1 = \frac{1}{2}e_1$, $Se_2 = 2e_2$ then S is an isometric operator. The eigenvalues are $\frac{1}{2}$ and 2 and clearly are not in $(z, |z| = 1)$.

But we have the following analogue of a complete inner product space operators.

PROPOSITIONS 16.3.5. *Let $T: X \to X$ be an isometric operator on X which is an indefinite inner product. Suppose that T has two eigenvalues a, b, with $a\bar{b} = 1$. Then the corresponding eigenspaces are orthogonal.*

Proof. Since the proof is simple we omit it. □

Another important class of operators is considered in the following definition.

DEFINITION 16.3.6. Let $T: X \to X$, X is an indefinite inner product space over C. T is called symmetric if for all x, y in X,

$$\langle Tx, y \rangle = \langle x, Ty \rangle.$$

REMARK 16.3.7. Symmetric operators may have nonreal points in the spectrum.

EXAMPLE 16.3.8. We consider X to be a two-dimensional space over C with the basis (e_1, e_2) and an inner product defined by the relations:

(1) $\langle e_j, e_j \rangle = 0$, $j = 1, 2, e$
(2) $\langle e_1, e_2 \rangle = 1$.

We define $T: X \to X$ by the relations:

(1) $Te_1 = ie_1$,
(2) $Te_2 = -ie_2$.

Then T is symmetric and i, $-i$ are clearly in the spectrum of T.

PROPOSITION 16.3.9. *If T is symmetric and a, b are two eigenvalues of T with the property that $a \neq \bar{b}$, then the corresponding eigenspaces are orthogonal.*

The class of projections is defined as follows.

DEFINITION 16.3.10. A linear operator $P: X \to X$ is said to be a projection if

(1) P is symmetric,
(2) $P^2 = P$.

For further details concerning operators on indefinite inner product spaces we refer to the above-mentioned papers. We now define two important classes of spaces in the family of space with an indefinite inner product.

16.4. Some Classes of Spaces with an Indefinite Metric

One of the most interesting class of spaces with an indefinite metric was that considered by L. S. Pontrjagin. In this section we present two classes of spaces named after two of the main contributors to the theory of spaces with an indefinite metric: the class of Pontrjagin spaces and the class of Krein spaces.

DEFINITION 16.4.1. A Pontrjagin space Π_k is a set with the following properties:

(1) Π_k is a linear space over C.
(2) There exists a bilinear form on X having all the properties of an inner product except that $\langle x, x\rangle \geqslant 0$ for all x in Π_k.
(3) The inner product does not degenerate.
(4) In Π_k there exists at least one k-dimensional subspace whose elements are positive.
(5) For any $k+1$ elements in Π_k, $x_1, \ldots, x_{k+1}$ and for all complex numbers $c_1, \ldots, c_{k+1}$ we have that $\Sigma_{i,j}\langle x_i, x_j\rangle c_i\bar{c}_j$ contains no more than k positive squares.
(6) Among the decompositions of Π_k into the direct sum $\Pi_k^+ \oplus \Pi_k^-$, Π_k^+ is a positive k-dimensional subspace of Π_k, there exists one such that Π_k^- is a Banach space for the norm induced by the inner product (taken with $-$).

The class of Krein spaces is defined as follows.

DEFINITION 16.4.2. A Krein space is a space with an inner product which admits the decomposition:

$$\mathcal{M}^+ \oplus \mathcal{M}^-, \qquad \mathcal{M}^+ \subset \mathcal{M}^{++}, \qquad \mathcal{M}^- \subset \mathcal{M}^{--}$$

and $\mathcal{M}^+$ and $\mathcal{M}^-$ are complete with respect to the norms induced by the inner product (with the corresponding sgn). Clearly a Pontrjagin space is a Krein space.

EXAMPLE 16.4.3. Let X be the space of all sequences $x = (x_n)_{-\infty}^{\infty}$ of complex numbers with $\Sigma|x_n|^2 < \infty$. If x, y are in X then we set

$$\langle x, y\rangle = \sum_i a_i x_i \bar{y}_{-i-1},$$

where

$$0 < \inf a_i \leqslant \sup a_i < \infty.$$

Then this is a Krein space.

If P^+ and P^- denote the projections in the fundamental decomposition then we set $J = P^+ - P^-$ and

$$\langle x, y \rangle_J = \langle Jx, y \rangle.$$

An important class of subspaces in a Krein space is defined as follows.

DEFINITION 16.4.4. A subspace X_1 of the Krein space X is said to be uniformly positive if for all x in X_1 the inequality

$$\langle x, x \rangle \geqslant k\|x\|_J^2 = k\langle Jx, x \rangle.$$
$$(\langle x, x \rangle \leqslant k\langle Jx, x \rangle)$$

holds for some positive constant k. Singularly, we define uniformly negative subspaces. Then we have.

PROPOSITION 16.4.5. *A subspace X_1 of a Krein space is uniformly positive (uniformly negative) iff it has the fundamental decomposition*

$$X = X^+ \oplus X^-, \qquad X^+ \subset X^{++}, \qquad X^- \subset X^{--}$$

with $X_1 \subset X^+$ $(X_1 \subset X^{--})$.

Concerning operators on Krein spaces it is to be noted that many results known in the case of complete inner product spaces admit a version for these spaces. For example, we have the following result.

PROPOSITION 16.4.6. *If T is a linear operator on the Krein space X and*

(1) *T is closed,*
(2) *D_T is dense in X*

then if $R(T)$ is closed, $R(T^*)$ is closed.

For other very interesting results concerning operators on Pontrjagin and Krein spaces we refer to the literature mentioned in Section 16.0.

16.5. Modules with an Indefinite Inner Product

In this section we present the analogue in the case of modules over a B*-algebra with identity of the Pontrjagin and Krein types of spaces.

DEFINITION 16.5.1. A left Pontrjagin module over a B*-algebra A with identity is a set $\mathscr{M}_k$ with the following properties:

(1) $\mathscr{M}_k$ is a left module over A.
(2) In $\mathscr{M}_k$ we have a bilinear form $[\,,]$ with values in A such that the following relations hold:
    (a) $[x, y] = [y, x]^*$,
    (b) $[x + y, z] = [x, z] + [y, z]$,
    (c) $[ax, y] = [ax, y]$,
    (d) $\mathscr{M}_k$ contains no element other than zero with the property that $[u, x] = 0$ for all $x \in \mathscr{M}_k$,

(e) $\mathcal{M}_k$ contains at least one left module of dimension k, say $\mathcal{N}_k$, with the property that $[x, x] \geqslant 0$ for all x in $\mathcal{N}_k$ (i.e. these are positive elements),
(f) for any $k + 1$ elements $x_1, \ldots, x_{k+1}$ in φ and any state of A

$$\sum_{i,j}^{k+1} \varphi([x_i, x_j]) z_i \bar{z}_j$$

has no more than $k + 1$ positive squares,
(g) among all decompositions of $\mathcal{M}_k$ in a direct orthogonal sum, $\mathcal{U}_k^+ \oplus \mathcal{U}_k^-$, where $\mathcal{U}_k^+$ is a positive k-dimensional submodule and $\mathcal{U}_k^-$ is the orthogonal complement, there exists at least one with the property that $(\mathcal{U}_k^-, [\,,\,])$ is a complete inner product module.

Exactly as in the case of a Pontjagin space we define the class of positive, negative and null elements of the module.

Suppose now that $\mathcal{M}$ is a module over A and on $\mathcal{M}$ we have a bilinear form with values in A. Then if we set

$$\mathcal{M}^+ = (x, x \in \mathcal{M}, [x, x] \geqslant 0), \qquad \mathcal{M}^{++} = (x, x \in \mathcal{M}, [x, x] > 0, \text{ or } x = 0)$$
$$\mathcal{M}^- = (x, x \in \mathcal{M}, [x, x] \leqslant 0), \qquad \mathcal{M}^{--} = (x, x \in \mathcal{M}, [x, x] \leqslant 0, \text{ or } x = 0)$$

then we define the class of Krein module as follows.

DEFINITION 16.5.2. A Krein module is a set with the following properties:

(1) $\mathcal{M}$ is a left module over A.
(2) On $\mathcal{M}$ there is a bilinear form having the properties of an inner A-valued product, the values of the bilinear form are in A.
(3) $\mathcal{M}_k$ admits a decomposition

$$\mathcal{M}_k^+ \oplus \mathcal{M}_k^-, \mathcal{M}_k^+ \subset \mathcal{M}_k^{++}, \mathcal{M}_k^- \subset \mathcal{M}_k^{--},$$

(4) $\mathcal{M}_k^+$ and $\mathcal{M}_k^-$ are complete inner product modules with respect to the norm induced by the inner product $[\,,\,]$.

REMARK 16.5.3. It is clear that every Pontrjagin module is a Krein module.

The notion of adjoint and other classes for the Krein modules are defined exactly as in the case of inner product modules of Kaplansky. Much of the theory of Krein spaces holds in the case of Krein modules. However, the author was unable to prove the fundamental theorem for hermitian operators in the case of Pontrjagin spaces. See the author's paper (forthcoming) for further information.

CHAPTER 17

Some Applications of Inner Product Structures

17.0. Introduction

In this chapter we intend to present some applications of the inner product structures in various areas. Since there are many applications, the choice is, of course, highly arbitrary but we hope that the reader may find some of them interesting. Since the presentation of such applications requires many preliminaries that are quite different, we have decided to present some of the details very briefly and the literature we give may help the reader to glean the necessary information as well as other interesting facts.

Our aim is, first, to present applications which show how powerful and elegant are the methods based on inner product structures and, second, to show how many of the problems in different areas may be handled successfully using inner product structures. Both aspects are intended for people working in functional analysis (especially operator theory, geometry of complete inner product spaces and of Banach spaces) and for people who apply results from the theory of inner product structures. Of course, we leave the reader to judge the measure in which this has been realized.

17.1. Certain Applications of the Cauchy–Buniakowsky Inequality to Some Extremal Problems

There are numerous places where the use of the Cauchy–Buniakowsky inequality is a very important step in the proof of a mathematical fact. In this section we shall present, following Eaton and Olkin (1972), some applications of the Cauchy–Buniakowsky inequality to some extremal problems.

Consider the Banach space of all real-valued continuous functions on the interval $[0, 1]$, denoted in what follows by $C_{[0,1]}$, with the usual sup norm. We consider for any two elements f, g in $C_{[0,1]}$ the following number

$$\langle f, g \rangle = \int_0^1 f(s)g(s)\,\mathrm{d}s$$

($\mathrm{d}s$ is the Lebesgue measure). It is clear that this defines an inner product on $C_{[0,1]}$.

We consider the following problem:

$$\min\{\langle f, f \rangle + \langle f, g \rangle^2; \langle f, h \rangle = 1\}$$

with g, h fixed elements of $C_{[0,1]}$. In order to solve this problem we consider the complete inner product space $L^2_{([0,1],\,ds)}$ and if we set

$$f + g\langle f, g\rangle = (I + F_g)(f)$$

then our problem can be formulated as follows:

$$\min\{\langle f, (I + F_g)(f)\rangle; \langle f, h\rangle = 1\}.$$

It is not difficult to see that

$$\|I + F_g\| = 1 + \langle g, g\rangle$$

(the norm is computed as an operator on $L^2_{([0,1],\,ds)}$.

We remark that $I + F_g$ (as an operator on $L^2 = L^2_{([0,1],\,ds)}$ is positive and invertible. Define

$$(I + F_g)^{-1} = I - T_g$$

and then this last operator is again positive. Then it has a (unique) square root which is a positive operator. Set

$$B^2 = I - T_g$$

and then for all k and l in L^2, by the Cauchy–Buniakowsky inequality,

$$|\langle k, l\rangle|^2 \leqslant \langle Bl, Bl\rangle\langle B^{-1}k, B^{-1}k\rangle.$$

Thus

$$\langle B^{-2}k, k\rangle\langle B^2 l, l\rangle \geqslant |\langle k, l\rangle|^2$$

and since in the Cauchy–Buniakowsky inequality the equality holds iff $k = c(I - T_g)l$ for some constant c, if we set $k = h$, $l = f$ then the above inequality gives that

$$\langle (I + F_g)(f), f\rangle = (\langle (I - T_g)(h), h\rangle)^{-1}$$

with equality iff

$$f = c(I - T_g)h.$$

From the fact that $\langle f, h\rangle = 1$ we get that the value of the constant c is given by the formula:

$$c = (\langle (I - T_g)(h), h\rangle)^{-1}.$$

Thus the solution of the problem is (i.e. the minimizing function)

$$f_+(s) = (\langle (I - T_g)(h), h\rangle)^{-1}(I - T_g)(h)$$

and the minimum is

$$(1 + \langle g, g\rangle)/(\langle h, h\rangle + \langle g, g\rangle\langle h, h\rangle - \langle h, g\rangle^2).$$

An extension of the above problem can be formulated as follows. Let X be a complete inner product space and let H be an invertible positive element of $L(X)$. Consider, further, that $z_1, \ldots, z_n$ is a finite number of linearly independent elements in X. Our extremal problem is as follows:

$$\min\{\langle x, Hx\rangle, \langle x, z_i\rangle = 1, i = 1, 2, 3, \ldots, n\}.$$

The solution of this problem may be obtained in a similar manner as the solution of the above problem. Consider for this purpose the inverse H^{-1} of H and then, by the Cauchy–Buniakowsky inequality, for all x, y in X,

$$|\langle x, y\rangle|^2 \leqslant \langle x, Hx\rangle\langle y, H^{-1}y\rangle.$$

Define

$$W = \{u, \langle u, z_i\rangle = 1, i = 1, 2, 3, \ldots, n\}$$

and

$$V = \{v, v = k_1 z_1 + \cdots + k_n z_n, k_1 + \cdots + k_n = 1).$$

Then

$$\langle x, Hx\rangle \geqslant (\langle y, H^{-1}y\rangle)^{-1}$$

and the equality appears in the inequality iff

$$x = \pm H^{-1}y/(\langle y, H^{-1}y\rangle).$$

In order to obtain an explicit solution of the above problem we consider the matrix (a_{ij}), where

$$a_{ij} = \langle H^{-1}z_i, z_j\rangle.$$

Then the matrix $A = (a_{ij})$ is positive and invertible. If we set

$$m = A^{-1}e/(\langle e, A^{-1}e\rangle) = (m_1, \ldots, m_n)$$

where

$$e = (1, 1, \ldots, 1)$$

then

$$x_0 = H^{-1}y_0/(\langle y_0, H^{-1}y_0\rangle),$$

where

$$y_0 = m_1 z_1 + \cdots + m_n z_n$$

is the minimizing element and the minimum is $\langle x_0, Hx_0\rangle$.

We now present an interesting application of the above result to a problem considered by Hodges and Lehman (1956) on the Wilcoxon test. The problem is reduced to one of solving the following extremal problem:

$$\min\left\{\int f^2(s)\,\mathrm{d}s\right\}$$

with $f(s) \geqslant 0$ for all s and

$$\int f(s)\,\mathrm{d}s = \int (s+1)f(s)\,\mathrm{d}s = \int s^2 f(s)\,\mathrm{d}s = 1.$$

We consider a positive real number a and the space $L^2 = L^2_{([-a,a],\,\mathrm{d}s)}$ and

$$z_1(s) = 1, \quad z_2(s) = s + 1, \quad z_3(s) = s^2.$$

Further, we consider in $L(L^2)$ the operator $F = I$ (with the above notation). Then the matrix A is

$$2a\begin{pmatrix} 1 & 1 & a^2/3 \\ 1 & 1 + a^2/3 & a^2/3 \\ a^2/3 & a^2/3 & a^4/5 \end{pmatrix}$$

and m is

$$t^{-1}(3a^4 - 5a^2, 0, 15 - 5a^2), \qquad t = 3a^4 - 10a^2 + 15.$$

Then the minimizing function is

$$\begin{cases} c_0((3a^4 - 5a^2) + (15 - 5a^2)s^2, & s^2 \leqslant a^2 \\ 0, & s^2 > a^2. \end{cases}$$

The value of the constant c_0 is then $3/(8a^5)$ and further the value of a is determined from the other constraints on f. The fact that $f(s) \geqslant 0$ gives immediately that a^2 is in the interval $[\frac{5}{3}, 5]$. But then

$$\int f^2(s)\,ds = 3(3a^4 - 10a^2 + 15)/(8a^5) = g(a).$$

The function g has a minimum at $a = 5^{1/2}$ and then

$$f_+(s) = \begin{cases} c_1 - c_2 s^2, & s^2 \leqslant 5 \\ 0, & s^2 > 5 \end{cases}$$

with

$$c_1 = 3.5^{1/2}/20, \quad c_2 = 3/(20.5^{1/2})$$

as the minimizing function. Indeed, if f_1 is any function satisfying

$$\int f_1(s)\,ds = \int (s + 1) f_1(s)\,ds = \int s^2 f_1(s)\,ds = 1$$

we get that

$$\int (f - f_1)^2\,ds \leqslant \int f^2(s)\,ds - \int f_1^2(s)\,ds$$

and the assertion is proved.

Another interesting application is concerned with a problem of Chernoff and Savage (1958). The problem is as follows:

$$\min\{I(G)\}, \quad I(G) = \int_{-\infty}^{\infty} f(F^{-1}(G(s)))^{-1} g^2(s)\,ds$$

where F^{-1} is the inverse function of

$$F(s) = \int_{-\infty}^{s} f(t)\,dt,$$

where f is a strictly positive probability density (in the case considered by Chernoff and Savage this is for the normal distribution) and the above minimum is taken over all distribution functions G having as density the function g, mean 0 and variance 1.

The problem can be slightly generalized in the sense that we can consider f to be a twice continuously differentiable functions with the property that

$$f(s)f''(s) - (f'(s))^2 \leqslant 0.$$

Further, we suppose that

$$\int_{-\infty}^{\infty} f(s)|\log f(s)\, ds| < \infty.$$

If $\mathcal{G}$ is the family of all distribution functions having a density g and satisfying the inequality

$$\int_{-\infty}^{\infty} |\log(f(s))| g(s) < \infty$$

and

$$\int_{-\infty}^{\infty} (\log f(s)) g(s)\, ds = \int_{-\infty}^{\infty} (\log f(s)) f(s)\, ds.$$

The result of M. Eaton and L. Olkin is that for any G in $\mathcal{G}$,

$$I(G) \geqslant I(F) = 1$$

and when $\log f$ is strictly concave the inequality is strict. For details we refer to Eaton and Olkin (1972).

For the reader interested in the development around the Cauchy–Buniakowsky inequality, Wang (1979) contains a list of references which may be useful.

17.2. Invariant Subspaces for the Shift

One of the most interesting operator on a separable complete inner product space is the shift operator. For the reader's convenience we repeat the definition.

Suppose that $(e_i)_0^\infty$ is an orthonormal basis for H. Then the shift associated with this basis is the operator $S: H \to H$, defined by the relations

$$Se_i = e_{i+1}, \quad i = 0, 1, 2, \ldots .$$

It is clear that the adjoint S^* of S is defined by the relations

$$S^* e_i = \begin{cases} 0, & i = 0 \\ e_{i-1}, & i = 1, 2, 3, \ldots . \end{cases}$$

Let us consider for each n the subspace M_n of all x in H such that

$$\langle x, e_i \rangle = 0, \quad i = 0, 1, 2, \ldots, n.$$

Then clearly M_n is an invariant subspace for S. A. Beurling in his seminal paper (1949) has determined all invariant subspaces for S using some interesting classes of analytic functions.

Let us consider the open unit disk D,

$$D = (z, |z| < 1),$$

and $\mathscr{H}(D)$ to be the set of all analytic functions in D. $H^{\infty}(D)$ denotes the set of all function in $\mathscr{H}(D)$ which are bounded on D and have the norm

$$f \to \|f\|_{\infty} = \sup(|f(z)|, z \in D).$$

This is a Banach space.

We denote by T the unit circle in the complex plane and we identity the functions on T with the functions which are 2π-periodic: if f is a function on T we set

$$f_1(t) = f(e^{it})$$

and clearly f_1 is periodic with the period equal to 2π; conversely if f_1 is a 2π-periodic function then we can find f defined on T such that $f_1(t) = f(e^{it})$.

We consider the spaces $L^p(T)$ as the classes of functions such that

$$f \to \|f\|_p = \left(1/(2\pi) \int_{-\pi}^{\pi} |f(e^{it})^{\pi p} dt\right)^{1/p}, \quad 1 \leqslant p < \infty$$

and for simplicity we write this as

$$f \to \|f\|_p = \left(1/(2\pi) \int_{-\pi}^{\pi} |f(t)^{\pi p} dt\right)^{1/p}.$$

DEFINITION 17.2.1. The Hardy space H^p $(1 \leqslant p \leqslant \infty)$ consist of all functions f in $\mathscr{H}(D)$ such that

$$\|f\|_p = \lim_{r \to 1} M_p(f, r)$$

is finite and where

$$M_p(f, r) = \exp\left(1/(2\pi) \int_{-\pi}^{\pi} |f(r\,e^{it})^{\pi p} dt\right)^{1/p}, \quad 1 \leqslant p < \infty$$

$$M_{\infty} = \sup_t |f(r\,e^{it})|.$$

REMARK 17.2.2. Since the function $r \to M_p(f, r)$ is monotonically increasing on $[0, 1)$ (for fixed f) the limit in the above definition exists. Now from Minkowski's inequality we obtain that $f \to \|f\|_p$ is a norm.

The following lemma shows an important property of H^p spaces.

LEMMA 17.2.3. *H^p is a Banach space.*

Proof. We consider only the case $1 \leqslant p < \infty$. Let (f_n) be a Cauchy sequence in H^p and since for $|z| \leqslant r < R < 1$ the following formula holds

$$f_n(z) - f_m(z) = 1/(2\pi i) \int_{|z|=R} (f_n - f_m)(R\,e^{it})/(R\,e^{it} - z)\, d(R\,e^{it}) =$$

$$= R/2\pi \int_{-\pi}^{\pi} (f_n(R\,e^{it}) - f_m(R\,e^{it}))\, dt/(R\,e^{it} - z)$$

we get

$$|f_n(z) - f_m(z)| \leqslant M_1((f_n - f_m), R)(R - r)^{-1} \leqslant M_p((f_n - f_m), R)(R - r)^{-1}. \quad \square$$

This implies that (f_n) converges uniformly on compact subsets of U to, say, f which is clearly in $\mathscr{H}(U)$. We now show that in fact f is in H^p. Let $\varepsilon > 0$ and choose N such that for all $n \geqslant N$, $m \geqslant N$,

$$\|f_n - f_m\|^p \leqslant \varepsilon.$$

From the above inequalities we get, for $r \in (0, 1)$,

$$M_p((f - f_n), r) = \lim_{m \to \infty} M_p((f_m - f_n), r)$$

and this gives that f is in H^p.

REMARK 17.2.4. The case $p = \infty$ is easier and is left to the reader.

Now we recall that for a function f in $L^2(T)$ we define the numbers

$$a_n(f) = 1/(2\pi) \int_{-\pi}^{\pi} f(\mathrm{e}^{it})\, \mathrm{e}^{-int}\, \mathrm{d}t$$

which are called the 'Fourier coefficients' of f. This agrees with out terminology from bases in $L^2(T)$ since the family of functions $(e^{int})_{-\infty}^{\infty}$ is a basis for the space $L^2(T)$.

For our representation theorem of invariant subspaces for the shift operator we need more information about the elements of H^2: these are given in the following theorem. For a proof we refer to Rudin (1966) and Hoffman (1962).

THEOREM 17.2.5. *The following assertions are true:*

(1) *$f \in \mathscr{H}(D)$ of the form $f(z) = \Sigma_0^\infty a_n z^n$, $z \in D$ is in H^2 if and only if $\Sigma_0^\infty |a_n|^2 < \infty$ and then*

$$\|f\|_2^2 = \sum_0^\infty |a_n|^2.$$

(2) *For any $f \in H^2$ the radial limits $f^*(\mathrm{e}^{it})$ exist at almost all points of T and*

(α) *$f^* \in L^2(T)$,*

(β) *$a_n(f^*) = a_n$ for $n \geqslant 0$, $a_n(f^*) = 0$ for $n < 0$,*

(γ) $$\lim_{r \to 1} 1/(2\pi) \int_{-\pi}^{\pi} |f(r\,\mathrm{e}^{it}) - f^*(\mathrm{e}^{it})\, \mathrm{d}t = 0,$$

(δ) $$f(z) = 1/(2\pi) \int_{-\pi}^{\pi} P_r(s - t) f^*(\mathrm{e}^{it})\, \mathrm{d}t, \quad z = r\,\mathrm{e}^{is},$$

(γ) $f(z) = 1/(2\pi i) \int_\Gamma (f^*)(\xi)/(\xi - z)\, \mathrm{d}\xi$, Γ is the unit circle nontrivially oriented.

(3) *$f \to f^*$ is an isometry of H^2 onto a subspace of $L^2(T)$ which consists of those functions f such that $a_n(f) = 0$ for $n < 0$.*

For the representation theorem we now introduce, following Beurling, the notion of 'inner function'.

DEFINITION 17.2.6. An inner function f is an element of $\mathscr{H}^\infty(D)$ such that

$$|f^*| = 1 \qquad \text{a.e. on } T$$

(where f^* denotes, as above, the radial limits of f).

We mention that the class of inner functions has an interesting structure and the interested reader may consult the above-mentioned books of Hoffman and Rudin.

Suppose now that H is a separable complete inner product space and that $(e_i)_0^\infty$ is an orthonormal basis of H. Then for each x in H we have the decomposition

$$x = \sum_0^\infty a_i e_i, \qquad \|x\|^2 = \sum_0^\infty |a_i|^2.$$

We define a function on D by the formula

$$f_x(z) = \sum_0^\infty a_n z^n$$

and, according to the above-mentioned result, the mapping $x \to f_x$ is an isometric isomorphism of H onto H^2. Now it is easy to see that the shift operator associated with the orthonormal basis (e_i) corresponds to the multiplication by z. We again denote this operator (i.e. the multiplication by z) by S,

$$(Sf)(z) = zf(z).$$

Let us consider an inner function φ. Then the set

$$M_\varphi = (\varphi f, f \in H^2)$$

is a closed linear subspace which is invariant for S. The important result of A. Beurling asserts that every closed invariant subspace for the shift S is in fact an M for some inner function.

THEOREM 17.2.7 (Beurling, 1949). *Every closed linear subspace which is invariant for S is of the form M for some inner function.*

Proof. The proof which follows is that of Helson and Lowdenslager. Let M be a closed linear invariant subspace for S (and we suppose without loss of generality that it reduces not to (0)). Then there is a smallest integer k such that

$$f(z) = \sum_{n=k}^{0} a_n z^n, \quad a_k = 1$$

and $f \in M$. In this case $zf(z)$ is not in $Mz = (zf(z), f \in M)$. It is easy to see that Mz is a closed linear subspace of H^2 and thus is a closed subspace of M. Then we find φ in M such that $\varphi \perp Mz$. This implies that for all n, $\varphi \perp z^n\varphi$. □

Now, in H^2 we have the inner product defined via

$$\langle f, g \rangle = 1/(2\pi) \int_{-\pi}^{\pi} f^*(e^{it}) g^*(e^{it})\, dt$$

(where for any function f in $\mathscr{H}(D)$, f^* denotes the radial limits). Then we get that the following relations hold:

$$1/(2\pi)\int_{-\pi}^{\pi} |\varphi^*(e^{it})|^2 e^{-int}\,dt = 0, \quad n = 1, 2, 3, \ldots.$$

Taking the complex conjugate in these relations we get that

$$1/(2\pi)\int_{-\pi}^{\pi} |\varphi^*(e^{it})|\, e^{int}\,dt = 0, \quad n = 1, 2, 3, \ldots$$

and thus for $|\varphi|^*$ all the Fourier coefficients are zero except for $n = 0$. This implies that $|\varphi|^*$ is constant a.e. on T. But φ is in H^2 and φ^* has the property that $|\varphi^*| = 1$ a.e. on T. From the Poisson formula (17.2.5(2δ)) we have that $|\varphi(z)| \leqslant 1$. Thus φ is an inner function. We show now that $M = M_\varphi$. From the invariance of M under S we get that for any polynomial $p(z)$, $p(z)\varphi \in M$. But from the form of H^2 it is clear that the polynomials are dense in H^2 and the closedness of M implies that $\varphi H^2 \subset M$.

To prove the Beurling theorem it now remains to show that in the last inclusion we have not a proper inclusion. Indeed, as φH^2 is closed in H^2 and is different from M, there exists φ_1 in M which is orthogonal to φH^2, i.e.

$$1/(2\pi)\int_{-\pi}^{\pi} \varphi_1^*(e^{it})\bar{\varphi}^*(e^{it})\, e^{-int}\,dt = 0, \quad n = 0, 1, 2, \ldots$$

(and where the bar denotes the complex conjugate of φ^*). Since φ_1 is in M, from the invariance under S, we obtain that $z^n\varphi_1 \in M$. This implies that $z^n\varphi_1 \perp \varphi$, i.e.

$$1/(2\pi)\int_{-\pi}^{\pi} \varphi_1^*(e^{it})\bar{\varphi}^*(e^{it})\, e^{-int}\,dt = 0, \quad n = -1, -2, -3.$$

These relations show that all the Fourier coefficients of $\varphi_1^*\bar{\varphi}^*$ are zero. This implies that $\varphi_1^*\bar{\varphi}^*$ is zero a.e. on T. Since $|\varphi^*| = 1$ a.e. on T we get that $\varphi_1^* = 0$ a.e. on T. This implies, again using the Poisson formula, that $\varphi_1 = 0$. Thus the proof of Beurling's theorem is complete. □

For extensions of this result we refer to Lax (1959) and Helson (1964), and for a more general result see Section 11.9.

17.3. Fourier Transforms and the Plancheral Theorem

Let us consider on R^1 (the set of all real numbers) the Lebesgue measure dx and the space of functions (classes of functions) f such that

$$\|f\|_p^p = \int_{-\infty}^{\infty} |f(x)|^p\,dx < \infty.$$

We denote this space by $L^p(R^1, dx) = L^p$ and following a standard convention we use the measure $dm = 1/(2\pi)\,dx$ which has the advantage that it makes the Plancherel transformation symmetric (the precise sense of this will be seen in Theorem 17.3.12). Now, if f, g are in L^1 then we can define a new function $h \in L^1$ by the formula

$$h(x) = \int_{-\infty}^{\infty} f(x - s)g(s)\,\mathrm{d}m(s)$$

and for each f in L^1 we associate the function denoted by $\hat{f}$ and defined by

$$\hat{f}(+) = \int_{-\infty}^{\infty} f(s)e^{its}\,\mathrm{d}m(s)$$

which, briefly, we write as $h = f \otimes g$.

We remark that the fact that h is in L^1 follows from Fubini's theorem and we have clearly the following inequality:

$$|\hat{f}| \leqslant \|f\|_1.$$

The following proposition gives some of the most important properties of the mapping $f \to \hat{f}$.

PROPOSITION 17.3.1. *If f, g are in L^1 then the following assertions hold:*

(1) $\hat{f}_a(x) = \hat{f}(x - a) = \hat{f}(t)e^{-ita}$,
(2) $(e^{ita}f(x)) = \hat{f}(x - a)$,
(3) $(f \otimes g) = \hat{f} \cdot \hat{g}$.

Proof. Assertions (1) and (2) follow directly from the definition of the mapping $f \to f$. We now prove (3). We have, by Fubini's theorem,

$$\hat{h}(t) = \int_{-\infty}^{\infty} e^{-itx}\left(\int_{-\infty}^{\infty} f(x - s)g(s)\,\mathrm{d}m(s)\right)\mathrm{d}m(x)$$

$$= \int_{-\infty}^{\infty} g(s)\,e^{-its}\,\mathrm{d}m(s)\int_{-\infty}^{\infty} f(x - s)\,e^{-it(x-s)}\,\mathrm{d}m(x)$$

$$= \int_{-\infty}^{\infty} g(s)\,e^{-its}\,\mathrm{d}m(s)\int_{-\infty}^{\infty} f(x)\,e^{itx}\,\mathrm{d}m(x) = \hat{f}(t)\hat{g}(t). \qquad \square$$

We note that in the proof of these assertions two remarkable facts were used:

(1) if $\varphi(t) = e^{its}$ then $\varphi(t + t_1) = \varphi(t)\varphi(t_1)$,
(2) $\mathrm{d}m$(i.e. $1/(2\pi)\,\mathrm{d}x$) is translation invariant, i.e. for each measurable set E of finite measure and for each $s \in R^1$,

$$\mathrm{d}m(E + S) = \mathrm{d}m(E), \qquad E + s = (e + s, e \in E).$$

The following proposition adds some new properties of the mapping $f \to \hat{f}$.

PROPOSITION 17.3.2. *For the mapping $f \to \hat{f}$ the following assertions hold:*

(1) $(\hat{f}(x)) = \hat{f}(t)$,
(2) $(\hat{f}(x/a)) = a\hat{f}(t)$, $a > 0$,
(3) if $f(x) = -ixq(x)$, $g \in L^1$ then $\hat{f}$ is differentiable and $\hat{f}' = \hat{g}$.

Proof. Assertions (1) and (2) follow directly from the definition of $\hat{f}$. For (3) we remark that for each $s \neq t$ we have

$$(\hat{f}(s) - \hat{f}(t))/(s - t) = \int_{-\infty}^{\infty} f(x)(e^{-ix(t-s)} - 1)/(t - s)\, dm(x).$$

Since

$$|e^{-ix(t-s)} - 1|/|(t - s)| \leqslant |x|, \quad (t \neq s)$$

and

$$\lim_{t \to s} (e^{-ix(t-s)} - 1)/(t - s) = -ix$$

we get that

$$\hat{f}'(t) = -i \int_{-\infty}^{\infty} xf(x)\, e^{-ixt}\, dm(x).$$

This is exactly our assertion. □

REMARK 17.3.3. The reader familiar with the theory of Banach algebras may remark that L^1 is a Banach algebra where the multiplication is defined via '$\otimes$'. Since for any f, g in L^1, $f \otimes g = g \otimes f$ we get that L^1 is a commutative Banach algebra since it is immediate that the following inequality holds:

$$\|f \otimes g\|_1 \leqslant \|g\|_1 \|f\|_1.$$

The mapping $f \to \hat{f}$ is the so-called Gelfand map from the theory of commutative Banach algebras.

To obtain further results about the connection between the functions f and $\hat{f}$ (i.e. other properties of the Gelfand map) we need some auxiliary results which have also their own interest.

LEMMA 17.3.4. *Let $1 \leqslant p < \infty$ and for each $f \in L^p$ we consider the function*

$$f_y(x) = f(x - y)$$

(which is called the translate of f). Then the function defined on R by $y \to f_y$ is a uniformly continuous mapping of R^1 into L^p.

Proof. Let $\varepsilon > 0$ and f in L^p. Then we find a continuous function φ with the support in $[-k, k]$ and such that

$$\|f - \varphi\|_p \leqslant \varepsilon.$$

Since the support of φ is in $[-k, k]$ and since this is a compact set for ε (considered above) there exists $\delta(\varepsilon) > 0$ such that for all t, s, $|t - s| \leqslant \delta(\varepsilon)$

$$|\varphi(t) - \varphi(s)| \leqslant \varepsilon \cdot (3k)^{-1/p}.$$

Now suppose that $|t - s| \leqslant \delta(\varepsilon)$. Then we have

$$\int_{-\infty}^{\infty} |\varphi(x - s) - \varphi(x - t)|^p\, dx \leqslant (3k)^{-1}(2k + \delta)\varepsilon^p < \varepsilon^p$$

which implies that

$$\|\varphi_s - \varphi_t\|_p \leqslant \varepsilon.$$

From the translation invariance of dm we get that for all y in R and f in L^p

$$\|f_y\|_p = \|f\|_p.$$

Then, if $|t - s| \leqslant \delta(\varepsilon)$

$$\begin{aligned}\|f_s - f_t\|_p &= \|f_s - \varphi_s + \varphi_s - \varphi_t + \varphi_t - f_t\|_p \\ &= \|f_s - \varphi_s\|_p + \|\varphi_s - \varphi_t\|_p + \|\varphi_t - f_t\|_p \leqslant 3\varepsilon\end{aligned}$$

and thus the uniform continuity is proved. □

PROPOSITION 17.3.5. *If $f \in L^1$ then $\hat{f}$ is continuous and zero at ∞.*

Proof. The first assertion follows from the fact that we have for $s \neq t$,

$$|\hat{f}(t) - \hat{f}(s)| \leqslant \left| \int_{-\infty}^{\infty} f(x)(e^{itx} - e^{isx})\, dm(x) \right| \leqslant \int_{-\infty}^{\infty} |f(x)|\, |e^{-itx} - e^{-its}| dm(x)$$

and the dominated convergence theorem gives the assertion. □

The second assertion is proved as follows. First we remark that we have the relation

$$\hat{f}(t) = -\int_{-\infty}^{\infty} f(x)\, e^{it(x + \pi/t)}\, dm(x)$$

since $e^{i\pi} = -1$. Further, we have

$$\hat{f}(t) = -\int_{-\infty}^{\infty} f(x - \pi/t)\, e^{-itx}\, dm(x)$$

which implies that

$$2\hat{f}(t) = \int_{-\infty}^{\infty} (f(x) - f(x - \pi/t)\, e^{itx}\, dm(x)$$

and thus

$$|\hat{f}(t)| \leqslant \tfrac{1}{2}\|f - f\pi/t\|_1.$$

From the above theorem we obtain that for $|t| \to \infty, \hat{f}$ tends to zero, which gives our assertion. □

Following Rudin we set

$$H(t) = e^{-|t|} \quad \text{and} \quad h_\lambda(t) = \int_{-\infty}^{\infty} H(\lambda t)\, e^{itx}\, dm(x)$$

and we get the following relations:

(1) $h_\lambda(t) = (2/\pi)^{1/2} \lambda/(\lambda^2 + x)$.

(2) $\int_{-\infty}^{\infty} h_\lambda(x)\, dm(x) = 1$.

(3) $0 < H(t) \leqslant 1$.

(4) $\lim_{\lambda \to 0} H(\lambda t) = 1$.

PROPOSITION 17.3.6. *For any $f \in L^1$ the following relation holds:*

$$(f \otimes h_\lambda)(x) = \int_{-\infty}^{\infty} H(\lambda t) f(t)\, \mathrm{e}^{ixt}\, \mathrm{d}m(x).$$

Proof. From the definition of the convolution '$\otimes$' we get

$$(f \otimes h_\lambda)(x) = \int_{-\infty}^{\infty} f(x-s)\left(\int_{-\infty}^{\infty} H(\lambda t)\mathrm{e}^{ist}\, \mathrm{d}m(t)\right)\mathrm{d}m(s)$$

$$= \int_{-\infty}^{\infty} H(\lambda t)\, \mathrm{d}m(t) \int_{-\infty}^{\infty} f(x-s)\, \mathrm{e}^{its}\, \mathrm{d}m(s)$$

$$= \int_{-\infty}^{\infty} H(\lambda t)\, \mathrm{d}m(t) \int_{-\infty}^{\infty} f(x)\, \mathrm{e}^{it(x-s)}\, \mathrm{d}m(x)$$

$$= \int_{-\infty}^{\infty} H(\lambda t) \hat{f}(t)\, \mathrm{e}^{itx}\, \mathrm{d}m(t)$$

and the proposition is proved. □

PROPOSITION 17.3.7. *For any $f \in L^\infty$ continuous at x the following relation holds:*

$$\lim_{\lambda \to 0} (f \otimes h_\lambda)(x) = f(x).$$

Proof. Since

$$\int_{-\infty}^{\infty} h_\lambda(x)\, \mathrm{d}m(x) = 1$$

for all λ we get

$$(f \otimes h_\lambda)(x) - f(x) = \int_{-\infty}^{\infty} (f(x-s) - f(x))\, h_\lambda(s)\, \mathrm{d}m(s)$$

$$= \int_{-\infty}^{\infty} (f(x-s) - f(x))\lambda^{-1} h_1(s/\lambda)\, \mathrm{d}m(s)$$

$$= \int_{-\infty}^{\infty} (f(x-\lambda s) - f(x)) h_1(s)\, \mathrm{d}m(s)$$

and thus

$$|(f \otimes h_\lambda) - f| \leqslant 2 \int_{-\infty}^{\infty} \|f\|_\infty\, h_1(s)\, \mathrm{d}m(s).$$ □

From this, using the dominated convergence theorem, the assertion follows.

PROPOSITION 17.3.8. *Let p be in $[1, \infty)$ and let $f \in L^p$. Then the following relation holds*:

$$\lim_{\lambda \to 0} \|f \otimes h_\lambda - f\|_p = 0.$$

Proof. Let q be defined by the relation $1/p + 1/q = 1$, and in this case we have that $h \in L^2$. As the function f is in L^p the function $f \otimes \mathrm{h}$ is well defined, and using the relation

$$\int_{-\infty}^{\infty} h_\lambda(x)\, \mathrm{d}m(x) = 1$$

we obtain the relation

$$(f \otimes h)(x) - f(x) = \int_{-\infty}^{\infty} (f(x-s) - f(x))h_\lambda(s)\, \mathrm{d}m(s)$$

which implies that

$$\|f \otimes h - f\|_p^p \leqslant \int_{-\infty}^{\infty} \|f_s - f\|^p h_\lambda(s)\, \mathrm{d}m(s). \qquad \square$$

Since $s \to f_s$ is continuous and bounded (having the value f at $s = 0$) we obtain that the right-hand side of the above inequality tends to zero for $\lambda \to 0$ and this, of course, gives the assertion of the proposition.

PROPOSITION 17.3.9. *Suppose that for $f \in L^1$, $\hat{f}$ is in L^1 and we set*

$$f_1(x) = \int_{-\infty}^{\infty} \hat{f}(t)\, \mathrm{e}^{itx}\, \mathrm{d}m(t).$$

Then $f = f_1$ a.e. (dm).

Proof. We consider the function

$$(f \otimes \mathrm{h}_\lambda)(x) = \int_{-\infty}^{\infty} H(\lambda t)\, \hat{f}(t)\, \mathrm{e}^{ixt}\, \mathrm{d}m(t)$$

and since $H(\lambda t) \to 1$ for $\lambda \to 0$ we get, using the dominated convergence theorem, that $f \otimes \mathrm{h}_\lambda$ converges to f_1. But

$$\lim (f \otimes h_{\lambda_{n_n}})(x) = f(x) \quad \text{(a.e.(d}m\text{))}$$

for any sequence (λ_n) in R^1, $\lim \lambda_n = 0$. Then clearly $f = f_1$ a.e. (dm) and the proposition is proved. $\square$

From this we derive the following result.

COROLLARY 17.3.10. *Let $f \in L^1$ and suppose that $f = 0$ (a.e.(dm)). Then $\hat{f} = 0$ a.e.* (dm). We are now ready to consider the transformation introduced by the Italian mathematician Salvatore Pincherle.

We consider the space L^2 and we remark that $L^1 \cap L^2$ is a dense set in L^2. For each f in $L^1 \cap L^2$ we can define $\hat{f}$ and we have the following result.

PROPOSITION 17.3.11. *For each $f \in L^1 \cap L^2$,*

$$\|f\|_2 = \|\hat{f}\|_2.$$

Proof. We define for each f in $L^1 \cap L^2$ the functions

$$\hat{f}(x) = \bar{f}(-x), \qquad f_1 = f \otimes f. \qquad \square$$

Then we get

$$f_1(x) = \int_{-\infty}^{\infty} f(x-s)f(s)\,\mathrm{d}m(s) = \int_{-\infty}^{\infty} f(x+s)f(s)\,\mathrm{d}m(s) = \langle f_{-x}, f\rangle,$$

where $\langle\,,\rangle$ denotes the inner product on L^2.

Further, we have the following inequality

$$|f_1(x)| \leqslant \|f_{-x}\|_2\|f\|_2 = \|f\|_2^2$$

and we remark that as $x \to f_{-x}$ is continuous, the function f_1 is continuous. But f_1 is in L^1 and thus

$$(f_1 \otimes h_\lambda)(0) = f_1(0) = \|f\|_2^2. \tag{*}$$

Then $f_1 = |\hat{f}|_2^2$ and the monotone convergence theorem implies that

$$\lim_{\lambda \to 0} \int_{-\infty}^{\infty} H(\lambda t) f_1(t)\,\mathrm{d}m(t) = \int_{-\infty}^{\infty} |\hat{f}(t)|^2\,\mathrm{d}m(t)$$

and thus

$$\|f\|_2^2 = \|\hat{f}\|_2^2$$

and the proposition is proved.

THEOREM 17.3.12. *There exists a mapping $f \to \hat{f}$ defined on L^2 with values in L^2 which satisfies the following properties:*

(1) *For each $f \in L^1 \cap L^2$, $\hat{f}$ is the Fourier transform of f*

$$\hat{f}(t) = \int_{-\infty}^{\infty} f(x)\,\mathrm{e}^{itx}\,\mathrm{d}m(x).$$

(2) *$f \to \hat{f}$ is a unitary operator.*

Proof. We define the mapping $f \to \hat{f}$ as follows. If f is in $L^1 \cap L^2$ then $\hat{f}$ is the Fourier transform. □

For each $n \geqslant 0$ we define (for f in L^2) the function f_n as follows:

$$f_n(t) = \begin{cases} f(t), & -n \leqslant t \leqslant n \\ 0, & |t| > n \end{cases}$$

and we remark that $\lim \|f - f_n\|_2 = 0$. But f_n is in $L^1 \cap L^2$ and thus by Proposition 17.3.11 $\|f_n\|_2 = \|\hat{f}_n\|_2$, and thus for all n, m we have the relation $\|f_n - f_m\|_2 = \|\hat{f}_n - \hat{f}_m\|_2$ which implies that $(\hat{f}_n)$ is a Cauchy sequence in L^2. We denote by $\hat{f}$ the function in L^2 with $\hat{f} = \lim \hat{f}_n$. Then clearly we have $\|\hat{f}\|_2 = \|f\|_2$.

It is obvious that the mapping $f \to \hat{f}$ is a linear and isometric operator on L^2. If f is in L^2 then we denote by Tf the limit in L^2 of the Fourier transforms of f_n. If g is in L^2 then we can consider the sequence of functions defined by $\int_{-n}^{n} g(t)\,\mathrm{e}^{ixt}$ and we set Sg the limit in L^2 of this sequence. As we noted, this mapping $g \to Sg$ is again an isometry on L^2 and since $Sg(x) = Tg(-x)$ from the above results we get that for all f in $L^1 \cap L^2$, $\hat{f} \in L^1 \cap L^2$, $STf = f$. Now we remark that since for any λ the function $f \otimes h_\lambda$ is in $L^1 \cap L^2$ whenever f is in $L^1 \cap L^2$ we get $ST(f \otimes h_\lambda) = f \otimes h_\lambda$ since

$f \otimes h_\lambda$ satisfies the above property. But for $\lambda \to 0$, $\|f \otimes h_\lambda - f\|_2 \to 0$ and ST being an isometry the relation we have obtained asserts that ST is an isometry on $L^1 \cap L^2$. We know, however, that $L^1 \cap L^2$ is dense in L^2 and this clearly implies that ST is an isometry on L^2. Thus for all f in L^2 we have the relation

$$STf = f$$

and this implies that T and S are surjective operators. This means that $f \to \hat{f}$ is a unitary operator and the theorem is thus proved. □

The relation

$$\langle f, g \rangle = \langle \hat{f}, \hat{g} \rangle$$

is called the 'Parseval formula' and the mapping $f \to \hat{f}$ is called the Plancherel transform.

For further details about extensions to locally compact abelian groups or to locally compact groups the books by Rudin (1962), Loomis (1954) and Naimark (1968), and the literature quoted there, may be useful.

17.4. The Sturm–Liouville Problems and Inner Product Spaces

Suppose that we have a chord (nonhomogeneous) whose equation is

$$(pu_x)_x = \rho u_{tt} \tag{*}$$

(for any function of several variable H, H_s denotes the partial derivative of H with respect to the variable s) where p is the modulus of elasticity and $\rho(x)$ is the mass distribution. The problem is to find solutions of this equation satisfying certain boundary conditions. One of the most interesting methods of constructing solutions for equations originated in the work of Fourier and consists of looking at solutions of the form

$$u(x, t) = f(x)g(t)$$

which implies that the following relation holds:

$$(pf')'/f = g''/g.$$

This is the case if and only if for some constant $-\lambda$

$$g'' + \lambda g = 0$$

and

$$(pf')' + \lambda f = 0. \tag{**}$$

Since from physical reasons λ is strictly positive we define $\lambda = \nu^2$ and thus a possible solution of the above equation is as follows:

$$u(x, t) = f(x)(A \cos \nu t + B \sin \nu t),$$

where A and B are some constants. Thus the problem is now reduced to determining v such that we obtain a solution of equation (*) satisfying the boundary conditions.

The problem of finding solutions of equation (**) is the so-called Sturm–Liouville problem. A more general form of this problem is to find the solutions of the equation

$$(pf')' - qf + \lambda f = 0, \qquad (***)$$

where q is a continuous function. There are several important transformations which are of practical interest. For these we refer to books on differential equations (see especially Sansone, 1948 and Birkhoff and Rota, 1960–1961). The following seems to be the most important: let

$$f_1 = (p\rho)^{1/4}, \qquad t = \int_0^x (\rho/p)(s)\,ds, \qquad l = \int_0^\pi (\rho/p)(s)\,ds$$

and then equation (***) is transformed in equation

$$f_1'' - rf_1 + \lambda f_1 = 0. \qquad (****)$$

r being a continuous function. Thus we are led to consider equations of the form

$$u'' - qu + \lambda u = 0$$

and the boundary value conditions are

$$mf(0) + m_1 f'(0) = nf(\pi) + n_1 f'(\pi) = 0$$

where m, n, m_1, n_1 are real numbers and either mm_1 or nn_1 is not 0.

From the theory of differential equations it is known that the equation given above has a system of solutions (ψ_k) and a theorem of J. L. Liouville (see G. Sansone for a proof) asserts that the following asymptotic estimates hold:

$$\psi_k(x) = (2/\pi)^{1/2} \cos kx + O(k^{-1})$$

if m_1 and n_1 are both nonzero. If $m_1 = n_1 = 0$ then

$$\psi_k(x) = (2/\pi)^{1/2} \sin kx + O(k^{-1}).$$

This implies that if (e_k) is the orthonormal basis of $L^2_{[0,\pi]}$ with

$$e_k = (2/\pi)^{1/2} \cos kx$$

or

$$e_k = s(2/\pi)^{1/2} \sin kx$$

the following relation holds:

$$\sum_{n=0}^{\infty} \|e_k - \psi_k\|^2 < \infty.$$

Then the system (ψ_k) is an orthonormal basis of $L^2_{[0,\pi]}$ by the theorem 3.6.24.

For a discussion of the Sturm–Liouville's problem with some historical information we refer to the paper by Birkhoff and Rota (1960). For a detailed account on the theory of differential operators we refer to the excellent books of Naimark (1954) and Levitan and Sargasian (1975; translated from the Russian).

17.5. Measures of Dependence of Random Variables and Inner Product Spaces

In the application of mathematical statitics we are often faced with the problem of characterizing numerically the dependence (or independence) of random variables. In this respect various numbers associated with random variables were defined, among which we mention the correlation coefficient, the correlation ratios, the mean square contingency of K. Pearson, etc. In this section we consider a measure of dependence of random variables proposed by A. Renýi in 1959 and studied extensively by P. Csáki and J. Fisher 1960.

It is worth mentioning the interesting fact that Linfoot (1957) has introduced a measure of dependence of random variables based on information theory. In order to present Renýi's measure of dependence for random variables we need some preliminary notions from probability theory.

We consider a probability space (Ω, B, P), where Ω is an arbitrary set, B is a σ-algebra of subsets of Ω and P is a positive measure on B with $P(\Omega) = 1$. If ξ is a random variable, i.e. a measurable function defined on Ω, the expected value (if it exists) is defined by

$$E(\xi) = E_{\xi} = \int_{\Omega} \xi(w)\, \mathrm{d}P(w).$$

Associated with a random variable ξ we consider the σ-algebra B_{ξ} which is defined as the smallest σ-algebra of subsets in B such that ξ is measurable. It is clear that

$$B_{\xi} = (a, a = \xi^{-1}(E),\ E \text{ is a measurable subset of } R).$$

We recall also the following result related to the σ-algebra B_{ξ}, namely, that if η is a random variable then it is measurable with respect to B_{ξ} if and only if for some measurable (Borel) function $h, \eta = h(\xi)$. For a proof we refer to J. L. Doob's book on stochastic processes.

Also, we need the notion of the conditional expected value of a random variable defined by A. N. Kolmogorov. We assume that we have a σ-algebra in B, say B_1. Then the conditional expected value of the random variable ξ with respect to B_1 is a random variable ξ_1 with respect to B_1 such that for all b_1 in B_1 the following relation holds:

$$\int_{b_1} \xi_1(w)\, \mathrm{d}P(w) = \int_{b_1} \xi(w)\, \mathrm{d}P(w).$$

In what follows we are interested in conditional expected values of random variables with respect to σ-algebras of the form B_{ξ}, where ξ is a random variable. In this case if η is another random variable the conditional expected value of η with respect to B_{ξ} is denoted by $M(\eta/\xi)$. It is also clear that the conditional expected value of a random variable is uniquely determined with probability 1 by the Radon–Nikodym theorem.

We consider the space of all functions (classes of functions) f with the property that $\int_\Omega |f|^2 \, \mathrm{d}P < \infty$, and this can be organized as an inner product space with

$$\langle f, g\rangle = \int_\Omega f\bar{g} \, \mathrm{d}P.$$

Then this is a complete inner product space, which is denoted by $L^2(\Omega, B, P) = L^2$. For any ξ in L^2 the standard deviation D_ξ is the number defined by

$$\mathrm{D}_\xi = \|\xi - E\xi\|.$$

The random variable ξ^* (any element of L^2 will be called a random variable), defined by $(\xi - E_\xi)/\mathrm{D}_\xi$ when $\mathrm{D}_\xi \neq 0$ and where ξ is a random variable, is said to be the standardized random variable associated with ξ. The complete inner product space $L^2(\Omega, B_\xi, P)$, where ξ is a random variable, is denoted by L^2_ξ.

More generally, if we have a family of random variables $\mathscr{F} = (\xi_i)_{i \in I}$ then we can consider the smallest σ-algebra with the property that all $\xi_i, i \in I$, are measurable. We denote this σ-algebra by $B_{\mathscr{F}}$. Then we consider the complete inner product space $L^2(\Omega, B_{\mathscr{F}}, P) = L^2_{\mathscr{F}}$. It is also clear that L^2_ξ is a subspace in L^2 and, as we see immediately, L^2_ξ is closed in L^2, the space L^2 may be decomposed in the orthogonal sum:

$$L^2 = L^2_\xi \oplus (L^2_\xi)^\perp.$$

The orthogonal projection of L^2 onto L^2_ξ is denoted by A_ξ and from the definition it is a hermitian operator.

Then we have the following result of Bahadur (1955).

PROPOSITION 17.5.1. *For any random η variable in L^2 and any random variable ξ the following relation holds:*

$$M(\eta/\xi) = A_\xi \eta.$$

Proof. Since for any random variable ξ. □

$$\langle A_\xi \eta, f\rangle = \langle \eta, f\rangle$$

for all f in L^2_ξ, we can choose

$$f = \chi_a$$

for a in B, then we get

$$\int_a A_\xi \eta \, \mathrm{d}P = \int_a \eta \, \mathrm{d}P$$

and thus, since a is arbitrary in $B, A_\xi \eta = M(\eta/\xi)$ with probability 1, which is the required assertion.

We are now in a position to state the problem about the measure of dependence for random variables. Since such measures must serve for the comparison of random variables, their ranges are arbitrary and we choose to be the interval [0, 1]. We agree to attribute the value 1 to strict dependence and the value o to independence.

We now present, following A. Renýi, a list of axioms which must be satisfied by measures of dependence of random variables.

17.5.2. *Axioms for measures of dependence for random variables* (Renýi, 1959). The axioms which may be satisfied by a measure of dependence m are as follows:

(1) *m is defined for any pair of random variables ξ and η, neither of which is constant with probability 1. The value is denoted by $m(,)$.*
(2) $m(\xi,\eta) = m(\eta,\xi)$.
(3) $0 \leqslant m(\xi,\eta) \leqslant 1$.
(4) *$m(\xi,\eta) = 0$ if and only if ξ,η are independent random variables.*
(5) *$m(\xi,\eta) = 1$ if there is a strict dependence between ξ and η, i.e. either $\xi = h(\eta)$ or $\eta = k(\xi)$ for some measurable (Borel) function h and k.*
(6) *If F is one-to-one on R, and g is one-to-one on R (both measurable) then*

$$m(f(\xi),g(\eta)) = m(\xi,\eta).$$

(7) *If the joint distribution of ξ and η is normal then $m(\xi,\eta) = |R(\xi,\eta)|$, where $R(\xi,\eta)$ denotes the correlation coefficient of the random variables ξ and η.*

We now list some of the numerical functions defined on pairs of random variables and which were, and are, used to measure, in some sense, the dependence of random variables.

17.5.3. *The correlation coefficient.* If ξ and η are two random variables in L^2 then the correlation coefficient $R(\xi,\eta)$ is defined by the formula:

$$R(\xi,\eta) = (E(\xi,\eta) - E(\xi)E(\eta)/(\mathrm{D}\xi\ \mathrm{D}\eta).$$

Since the range is in $[-1,1]$ the absolute value is seen to satisfy the axioms given above, except possibly (4). Indeed, there are examples of random variables ξ and η which are not independent variables, but where $R(\xi,\eta) = 0$.

17.5.4. *The correlation ratios.* If ξ and η are two random variables then the correlation ratios $\theta_\xi(\eta)$ of η on ξ is defined by the relation

$$\theta_\xi(\eta) = \mathrm{D}(M(\eta/\xi))/\mathrm{D}\eta.$$

(Of course, we suppose that D exists and is strictly positive.) This is not generally symmetric (i.e. axiom (2) is not generally satisfied). Then, of course, we may utilize

$$\theta(\xi,\eta) = \sup(\theta_\xi(\eta),\theta_\eta(\xi)).$$

It is known that this does not satisfy axioms (1), (4) and (5).

17.5.5. *The mean square contingency.* Let us consider two random variables ξ and η and, following Renýi, we say that the dependence between ξ and η is regular if the joint distribution $F_{\xi,\eta}$ (defined by $F_{\xi,\eta}(e) = P(w; (\xi(w),\eta(w)) \in e$, e is a Borel subset in the plane) is absolutely continuous with respect to the direct product $F_{\xi*\eta}$ of the distributions of ξ and η (i.e. if e_1, e_2 are Borel subset of R then $F_{\xi*\eta}(e_1 * e_2) = P((w) \in e_1)) \cdot P(w,(w) \in e_2))$ and extended to the Borel subsets of the plane). If $k(s,t)$ is the Radon–Nikodym derivative and F_ξ and F_η are the distribution functions of ξ and η, respectively, then the mean square contingency of the random variables is defined by the formula:

$$C(\xi,\eta) = \left(\int_{-\infty}^{\infty}\int_{-\infty}^{\infty} (k(s,t) - 1)^2\, \mathrm{d}F_\xi(s)\, \mathrm{d}F_\eta(t)\right)^{1/2}.$$

It is clear that the range is in $(0, \infty)$ and it is known that C does not satisfy axioms (1), (3) and (5).

An interesting and important measure of dependence is the so-called maximal correlation coefficient.

17.5.6. *The maximal correlation coefficient.* If ξ and η are two random variables in L^2 then the maximal correlation coefficient $S(\xi,\eta)$ is defined by the relation

$$S(\xi,\eta) = \sup R(f(\xi), g(\eta)),$$

f,g measurable (Borel functions such that $R(f(\xi), g(\eta))$ has sense). It is not difficult to see that the maximal correlation coefficient $S(\xi,\eta)$ satisfies all the above axioms (1)–(7) of a measure of dependence. For the case of some classes of random variables, see Hirschfeld (1935), Renýi (1959) and Sarmanov (1958).

Again we mention the interesting measure of dependence introduced by Linfoot (1957) related to information theory and this satisfies all axioms (1)–(7) of a measure of dependence listed above.

Renýi (1959) mentions, and gives examples of, pairs of random variables such that for some measurable (Borel) functions f,q

$$S(\xi,\eta) = R(f(\xi), g(\eta)).$$

The first example was given by J. Czipszer and another is constructed in Renýi (1959).

In what follows we shall consider, for the space L_ξ^2, the following orthogonal decomposition:

$$L_\xi^2 = L_{\xi,0}^2 \oplus (L_{\xi,0}^2)^\perp,$$

where $L_{\xi,0}^2$ is the subspace of all elements in L_ξ^2 which are orthogonal to random variables which are constant; thus $(L_{\xi,0}^2$ is the subspace of L_ξ^2 generated by the random variables which are constant). We shall also consider the operators A_ξ and A_η defined only on $L_{\xi,0}^2$ and on $L_{\eta,0}^2$ respectively.

Now we consider the correlation ratio of the random variable η in L^2 on ξ which is

$$\theta_\xi(\eta) = \mathrm{D}(M(\eta/\xi)/\mathrm{D}\eta = E(\eta - M(\eta/\xi))/\|\eta - E\eta\|) = \|A_\xi\eta^*\|.$$

This gives further that

$$\theta_\xi^2(\eta) = \langle A_\xi\eta^*, A_\xi\eta^*\rangle = \langle \eta^*, A_\xi\eta^*\rangle$$

and thus

$$\theta_\xi(\eta) = \langle \eta^*, A_\xi\eta^*\rangle/\|A_\xi\eta^*\| = R(\eta, A_\xi\eta).$$

In this case, if η is in L^2 and f in L_ξ^2 we have

$$\|A_\xi\eta^* - \langle f^*,\eta^*\rangle f^*\|^2 = \|A_\xi\eta^* - \langle f^*,\eta^*\rangle f^*\|^2$$

and

$$\theta_\xi^2(\eta) = \langle f^*,\eta^*\rangle^2 + \|A_\xi\eta^* - \langle f^*,\eta^*\rangle f^*\|^2.$$

Since the conditional expected value satisfies the properties

$$E(M(\eta/\xi)) = E\eta, \qquad M(g(\xi)\eta/\xi) = g(\xi)M(\eta/\xi)$$

for any measurable (Borel) function g, it easily follows that the operators A_ξ and A_η are hermitian and positive.

The following result is an easy consequence of the definitions.

PROPOSITION 17.5.7. *For any random variables* ξ *and* η,

$$S(\xi, \eta) = \|A_\xi\| = \|A_\eta\|.$$

We have considered above the case when the operators A_ξ and A_η are, in fact, of a very general nature and thus it is of some interest to consider the case when they are in some of the classes that have some useful properties. Such classes are the classes of operators in C_2. Suppose that (e_i^ξ) and (e_j^η) are two orthonormal bases of the spaces $L^2_{\xi,0}$ and $L^2_{\eta,0}$ respectively. We denote the norm of A_ξ and A_η as operators in the class C_2 by $|||A_\xi|||$ and $|||A_\eta|||$ respectively. As we know, these are given by

$$|||A_\xi||| = \left(\sum_i \|A_\xi e_i^\xi\|^2\right)^{1/2}$$

$$|||A_\eta||| = \left(\sum_i \|A_\eta e_i^\eta\|^2\right)^{1/2}.$$

Now we are ready to define the contingency measure of dependence for some classes of random variables as follows.

DEFINITION 17.5.8. Suppose that ξ, η are two random variable with the property that the operators A_ξ and A_η, restricted to $L^2_{\xi,0}$ and $L^2_{\eta,0}$ respectively, are in class C_2. Then the Pearson contingency of the random variables ξ, η is the number defined by

$$C(\xi, \eta) = |||A_\xi||| = |||A_\eta|||.$$

The above definition of contingency was proposed by P. Csáki and J. Fisher.

DEFINITION 17.5.9. (Csáki and Fischer, 1960). The number λ is called an eigenvalue if for some $f \in L^2_{\xi,0}$, $g \in L^2_{\eta,0}$ we have the relations

$$A_\xi g = \lambda f, \qquad A_\eta f = \lambda g.$$

The elements f, g are eigenvectors belonging to λ.

THEOREM 17.5.10. (Csáki and Fischer, 1960). *For the standard random variables* $f \in L^2_{\xi,0}$, $g \in L^2_{\eta,0}$ *the following assertions are equivalent*:

(1) *f, g are eigenvectors.*
(2) $\theta_\xi(g) = \theta_\eta(f) = f, g$.
(3) *f, g are linearly correlated and* $\theta_f(g) = \theta_\xi(g)$; $\theta_g(f) = \theta_\eta(f)$.

Proof. Obviously (1) $\rightarrow$ (2). If (2) holds, from $A_f g = A_f A_\xi g$ we have

$$\|A_f g\| \leqslant \|A_\xi g\|$$

which gives that

$$\langle f,g\rangle \leqslant \theta_f(g) \leqslant \theta_\xi(g)$$

and then from our assumptions we have here only equalities and similarly

$$\langle f,g\rangle \leqslant \theta_g(f) \leqslant \theta_\eta(f),$$

where we have also only equalities. These equalities implies that f and g are linearly correlated and the other formulas in (3), hold. □

Suppose now that (3) holds. From the linear correlation we have that $A_f g = \lambda_f$ and $\theta_f(g) = \theta_\xi(g)$ implies that $A_f g = A_\xi g$ and then f,g are eigenvectors.

For further interesting results related to the operators A_ξ and A_η as well as to other operators which can be associated with the pair of random variables we refer to Csáki and Fischer (1960).

We now make remarks on some generalizations. As we noted above, we can consider the complete inner product spaces $L^2_{\mathscr{F}}$, where $\mathscr{F}$ is a family of random variables $(\xi_i)_{i\in I}$, and we consider the operator $A_{\mathscr{F}}$ which is the orthogonal projection of L^2 on the subspace $L^2_{\mathscr{F}}$. In the space $L^2_{\mathscr{F}}$ we can consider the subspace generated by constant random variables and then

$$L^2_{\mathscr{F}} = L^2_{\mathscr{F},0} \oplus (L^2_{\mathscr{F},0})^\perp.$$

Now we remark that we can formulate the axioms which may be satisfied for a measure of dependence for families of random variables (in L^2) as follows:

(1) *m is defined for any family of random variables which does not contain random variables equal to constants.*
(2) *If $\mathscr{F}$ is a finite family of random variables $(\xi_i)_1^n$ then for any permutation π,*

$$m(\xi_1,\ldots,\xi_n) = m(\xi_{\pi(1)},\ldots,\xi_{\pi(n)}).$$

(3) *The values of m are in* $[0,1]$.
(4) *$m = 0$ if and only if the family of random variables consists of random variables which are independent in the sense of Kolmogorov.*
(5) *$m = 1$ if and only if there exists two random variables in the family which are strictly dependent.*
(6) *If $(f_i)_{i\in I}$ is a family of measurable (Borel) functions then $m(\{f_i(\xi_i)\}) = m(\{\xi_i\})$.*
(7) *For any finite subfamily of random variables $\mathscr{F}_1$ of for which the joint distribution is normal, then m is the correlation coefficient of the random variables in the subfamily.*

We close by mentioning the very profound papers by Kagan (1975), Gerlein and Kagan (1979) and Gerlein (1979), where the inner product structures are used to derive, in a very elegant form, the basic of mathematical statistics as a linear theory.

17.6. Ergodic Theory and Complete Inner Product Spaces

In this section we present some facts about the connection between complete inner product spaces and ergodic theory. More precisely, we present the ergodic theorem of John von Neumann which, in contrast to the Individual Ergodic Theorem, considers the convergence in L^2, and next we present some facts about the characterizations of some classes of measure-preserving transformations, like strongly mixing and weakly mixing measure-preserving transformations. For the ergodic theory we refer to Cornfeld *et al.* (1982), Sinai (1970), and Friedman (1970).

We consider a measure space (Ω, B, μ) where Ω is a nonempty set, B is a σ-algebra of subsets of Ω, μ is a countable additive nonnegative extended real-valued function on B. If $\mu(\Omega) = 1$ then we say that (Ω, B, μ) is a probability space.

A measure space is a pair (F, B_F), where F is a nonempty set and B_F is a σ-algebra of subsets of F.

Suppose that (Ω_1, B_1, μ_1) and (Ω_2, B_2, μ_2) are two measure spaces. One of the most important class of mappings of the ergodic theory is defined as follows.

DEFINITION 17.6.1. A mapping $T: \Omega_1 \to \Omega_2$ is called a 'measure-preserving transformation' (shortly m.p.t.) where T is onto and one-to-one, T and T^{-1} are measurable, and $\mu_1(M) = \mu_2(TM)$, for all M in B_1.

The following theorem was proved by John von Neumann in 1932.

THEOREM 17.6.2 (The Mean Ergodic Theorem of J. von Neumann, 1932). *Let* (Ω, B, μ) *be a measure space with* $\mu(\Omega) < \infty$, *let* $T: \Omega \to \Omega$ *be a measure-preserving transformation and let* f *be an element of* $L^2(\Omega, B, \mu) = L^2$. *Then there is a function* $\bar{f}$ *in* L^2 *such that*

$$\lim_{n\to\infty} \left(\int_\Omega \left| 1/n \sum_{i=0}^{n-1} f_0 T^i - \bar{f} \right|^2 \mathrm{d}\mu \right)^{1/2} = 0.$$

REMARK 17.6.3. The condition $\mu(\Omega) < \infty$ may be relaxed to the property of Ω to be σ-finite. This is also not necessary for the validity of the theorem since then there exists a measurable subset of Ω' which can be decomposed into a countable family of subsets of finite measure, such that each $f_0 T^i$ vanishes outside of Ω'.

We do not prove the von Neumann theorem since we can give a more general result from which the assertion follows immediately. For this we remark that T determines a map $U: L^2 \to L^2$ which is defined by the formula:

$$(Uf)(w) = f(Tw).$$

We remark first that U is an isometry on L^2 since

$$\|Uf\| = \left(\int_\Omega |Uf(w)|^2 \, \mathrm{d}\mu(w) \right)^{1/2} = \left(\int_\Omega |f(Tw)|^2 \, \mathrm{d}\mu(w) \right)^{1/2} = \|f\|.$$

Since T is invertible we have that U is an invertible isometry. Then U is a unitary operator.

THEOREM 17.6.4 (The Mean Ergodic Theorem in complete inner product spaces). *Let* T *be an isometric operator defined on the complete inner product space* H. *Then if* $H_1 = (x, Tx = x)$ *then for each* x *in* H,

$$\lim_{n\to\infty} \left\| 1/n \left(\sum_{i=0}^{n-1} T^i x - Px \right) \right\| = 0,$$

where P *is the orthogonal projection of* H *onto* H_1.

Proof. First we remark that H_1 has the property that $H_1 = (x, T^*x = x)$. Indeed, if x is in H_1 then $Tx = x$ and thus, T being an isometry, we have that $T^*T = I$ which implies the relation

$$T^*Tx = T^*x = x.$$

Thus H_1 is a subspace of $(x, T^*x = x)$. Conversely, if x has the property that $T^*x = x$ then we have that

$$\|Tx - x\|^2 = \|Tx\|^2 + \|x\|^2 \langle Tx, x\rangle - \langle x, Tx\rangle \leqslant 0,$$

which implies our assertion. □

(*Note*: The relation between these subspaces is in fact valid for all contraction operators on H as is clear from the proof since we use only the fact that $Tx = x$.)

Now the proof of the theorem is as follows. First we remark that the theorem is obviously for elements of H_1. We now consider an element of the form $u - Tu$ with u arbitrary but fixed in H. Then, if $x = Tu - u$ then

$$1/n \sum_{i=0}^{n-1} T^i x = 1/n(T^n u - u)$$

and this obviously converges to the zero element of H. Thus the theorem is valid for such elements. Now we consider the closed subspace generated by the elements of the above form, i.e. we consider the closed subspace generated by the elements $u - Tu$ with u in H, which is denoted by H_2. Now we show that the theorem is true for elements in H_2. Indeed, we have to consider only elements of the form

$$x = \lim (u_n - Tu_n), \quad u_n \in H.$$

Let $\varepsilon > 0$ and then there exists N_ε such that

$$\|x - u_n + Tu_n\| \leqslant \varepsilon, \quad n \geqslant N_\varepsilon.$$

Then

$$\left\| 1/n \sum_{i=0}^{n-1} T^i x \right\| = \left\| 1/n \sum_{i=0}^{n-1} T^i(x - u_n + Tu_n) + 1/n \sum_{i=0}^{n-1} T^i(u_n - Tu_n) \right\| \leqslant \varepsilon$$

if $n \geqslant N_\varepsilon$

This implies that the theorem is true for elements in H_2. Since we have the decomposition

$$H = H_1 \oplus H_2$$

the assertion of the theorem follows from the results just obtained.

REMARK 17.6.5. There exists a great number of extensions of this theorem to certain classes of operators defined on Banach spaces or on certain classes of Banach spaces.

As an application of the Mean Ergodic Theorem we give the following beautiful result of A. I. Khinchin (=A. I. Hincin) which is a recurrence type theorem. First we need the following notion.

DEFINITION 17.6.6. A subset in $Z = (0, \pm 1, \pm 2, \dots)$ is said to be relatively dense if there exists a positive integer k_0 such that

$$M \cap (i, i+1, \ldots, i+k_0 - 1) \neq \emptyset$$

for each integer i in Z.

THEOREM 17.6.7 (The recurrence theorem of Khinchin, 1934). *Let $T:\Omega \to \Omega$ be a measure-preserving transformation on a probability space (Ω, B, μ). Then for any E in B and any $\varepsilon > 0$ the set*

$$M_\varepsilon = (i, i \in Z, \mu(T^i B \cap B) \geqslant (B)^2 - \varepsilon)$$

is relatively dense.

Proof. We consider the unitary operator induced by the transformation T which was defined above. Let E be an arbitrary set in B and let f_E be the characteristic function of the set E. Then we have the relation

$$f_{T^i B} = U^{-i} f_B$$

because $f_{T^i B}(w) = 1$ if and only if w is in $T^i B$. But this is equivalent to the fact that $T^{-i} w$ is in B, which is equivalent to the fact that $(U^{-i} f_B)(w) = 1$. Then we have, further, that

$$\mu(T^i B \cap B) = \int_\Omega f_B f_{T^i B}\, d\mu(w) = \int_\Omega f_B(w)(U^{-i} f_B)(w) = \langle f_B, U^{-i} f_B \rangle.$$

Now we apply the Mean Ergodic Theorem to the element f_B and thus for $\varepsilon > 0$ there exists an integer N such that for all $n \geqslant N$,

$$\left\| 1/n \sum_{i=0}^{n-1} U^i f_B - P f_B \right\| \leqslant \varepsilon / (\|f_B\| + 1).$$

This implies that for each i in Z,

$$\left\| 1/n \sum_{k=0}^{n-1} U^{k+i} f_B - P f_B \right\| \leqslant \varepsilon / (\|f_B\| + 1)$$

or

$$\left\| 1/n \sum_{k=i}^{n+i-1} U^k f_B - P f_B \right\| \leqslant \varepsilon / (\|f_B\| + 1).$$

Since we consider a probability space, the function equal to 1 (a.e.) is in L^2 (and we denote this function by 1). Then, according to the Mean Ergodic Theorem the space L^2 admits an orthogonal decomposition $H_1 \oplus H_2$. Obviously 1 is in H_1 and for f_B we have clearly the relations

$$\langle f_B, 1 \rangle = \langle f_B, P1 \rangle = \langle P f_B, 1 \rangle.$$

Then, by the Cauchy–Buniakowsky inequality,

$$\langle f_B, 1 \rangle^2 = \langle P f_B, 1 \rangle^2 \leqslant \|P f_B\|^2 \|1\|^2 = \langle P f_B, P f_B \rangle = \langle P f_B, 1 \rangle$$

which implies that

$$\left\| \left\langle 1/n \sum_{k=i}^{n+i-1} U^k f_B - Pf_B, f_B \right\rangle \right\| \leqslant \left\| 1/n \sum_{k=i}^{n+i-1} U^k f_B - Pf_B \right\| \|f_B\| \leqslant \varepsilon$$

for all i.

Thus

$$1/n \sum_{k=i}^{n+i-1} \mu(T^k B \cap B) = 1/n \sum_{k=i}^{n+i-1} \langle U^k f_B, f_B \rangle \geqslant \langle Pf_B, f_B \rangle - \varepsilon$$
$$= \langle f_B, 1 \rangle^2 - \varepsilon = \mu(B)^2 - \varepsilon.$$

for all i in Z. This proves Khinchin's recurrence theorem. □

We now present a characterization of a strongly mixing measure-preserving transformation on a probability space. First we define the notion of a strongly mixing m.p.t. T.

DEFINITION 17.6.8. The measure-preserving transformation T defined on the probability space (Ω, B, μ) is said to be strongly mixing if for each E, F in B

$$\lim_{n \to \infty} \mu(T^{-n} E \cap F) = \mu(E)\mu(F).$$

We remark that the following property holds.

PROPOSITION 17.6.9. *The measure-preserving transformation T is strongly mixing if and only if for any f, g in L^2 the following relation holds*:

$$\lim_{n \to \infty} \langle U^n f, g \rangle = \langle f, 1 \rangle \cdot \overline{\langle g, 1 \rangle}.$$

(Here U is the unitary operator induced by T defined above.)

Proof. If E, F are two subsets in B and f_E, f_F are the characteristic functions of E and F, respectively, then

$$\langle U^n f_E, f_F \rangle = \int_\Omega f_{T^{-n}E} f_B \, \mathrm{d}\mu(w) = \mu(T^{-n} E \cap F)$$

which converges to

$$\mu(E)\mu(F) = \langle f_E, 1 \rangle \langle f_F, 1 \rangle.$$

Thus the assertion about convergence is valid for characteristic functions and since they are dense in L^2 the assertion follows.

We now prove the following characterization of strongly mixing measure-preserving transformations following the A. Renýi's interesting paper of 1958.

THEOREM 17.6.10 (Renýi, 1958). *The measure-preserving transformation T on the probability space (Ω, B, μ) is strongly mixing if and only if for each E in B we have the following relation*:

$$\lim \mu(T^{-n} E \cap E) = \mu(E)^2.$$

Proof. It is obvious that the condition is necessary. To prove the sufficiency part we consider a fixed set E in B and the closed subspace in L^2 which is generated by the function 1 and the functions $U^k f_E$, k in Z. Since

$$\lim\langle U^n f_E, 1\rangle = \lim \mu(T^{-n}E \cap E) = \mu(E) = \langle f_E, 1\rangle \cdot \langle 1, 1\rangle$$

and

$$\lim \langle U^n f_E, U^k f_E\rangle = \lim\langle U^{-n+k} f_E, f_E\rangle = \lim \mu(T^{-n+k} E \cap E)$$
$$= \mu(E)^2 = \langle f_E, 1\rangle\langle U^k f_E, 1\rangle$$

we obtain that

$$\lim_{n\to\infty} \langle U^n f_E, g\rangle = \langle f_E, 1\rangle\overline{\langle g, 1\rangle}$$

for all g in H_1 (the subspace of the orthogonal decomposition in the proof of the Mean Ergodic Theorem).

If g is an arbitrary element of L^2 then $g = g_1 + g_2$ with g_i in H_i. Then we have that,

$$\lim\langle U^n f_E, g\rangle = \lim\langle U^n f_E, g_1\rangle + \lim\langle U^n f_E, g_2\rangle = \langle f_E, 1\rangle\overline{\langle g_1, 1\rangle} = \langle f_E, 1\rangle\overline{\langle g, 1\rangle}$$

and thus if we take $g = f_F$,

$$\lim \mu(T^{-n}E \cap F) = \lim\langle U^n f_E, f_F\rangle = \langle f_E, 1\rangle\cdot\langle f_F, 1\rangle = \mu(E)\mu(F).$$

Thus T is strongly mixing.

Another class of m.p.t. mappings is defined as follows.

DEFINITION 17.6.11. The measure-preserving transformation T defined on the probability space (Ω, B, μ) is said to be weakly mixing if for each E, F in B the following relation holds:

$$\lim_{n\to\infty} \left| 1/n\left(\sum_{i=0}^{n-1} \mu(T^{-k}E \cap F) - \mu(E)\mu(F)\right)\right| = 0.$$

It is easy to see that the following assertion holds:

PROPOSITION 17.6.12. *The measure-preserving transformation T is weakly mixing if and only if for each f, g in L^2*

$$\lim_{n\to\infty} \left| 1/\mathrm{n}\left(\sum_{i=0}^{n-1} \langle U^i f, g\rangle - \langle f, 1\rangle\overline{\langle g, 1\rangle}\right)\right| = 0.$$

(Here U is the unitary operator on L^2 induced by the m.p.t. T.)

Then we have the following characterization of m.p.t. mappings which are weakly mixing. For the proof we refer to the above-mentioned books on ergodic theory.
17.6.13. For the m.p.t. mapping T defined on the probability space (Ω, B, μ) the following assertions are equivalent:

(1) T is weakly mixing.
(2) The m.p.t. mapping $\tilde{T}$ defined on the probability space $(\Omega_\times\Omega,\ \mathrm{B}_\times\mathrm{B},\ \mu_\times\mu)$ by

the formula

$$\tilde{T}(x, y) = (Tx, Ty)$$

is weakly mixing,

(3) U has no eigenvectors other than the functions of the form $f(w) = k$, where k is a constant.

17.7. Classes of Stochastic Processes and Inner Product Structures

The contacts between the probability theory and complete inner product structures are encountered in probability theory when we are dealing with random variables having a finite second moment. The basic papers of Khinchin (1933, 1934) and Kolmogorov (1940, 1941) on stationary sequences and on helical curves were of fundamental importance. Further interesting and important results were obtained by Krein (1944), Cramer (1940), Wiener (1949), Wold (1938, 1954) and Zasuhin (1941). Further important advances are in the papers of Prohorov (1957), Yaglom (1955, 1960, 1962), Rozanov (1958, 1960, 1962), Masani (1962, 1968, 1969, 1972, 1973), Wiener and Masani (1957, 1958, 1959) and Rao (1967, 1978, 1979, 1982, 1983). See also the references in our bibliography for other papers related to the above subject.

An interesting connection between probability theory and complete inner product spaces was discovered by P. Masani which has presented a probability free proof of the famous theorem of Lévy–Khinchin on infinitely decomposable probability distributions.

In what follows we present some results which show the usefulness of complete inner product structures in the approach to some problems in probability theory. We are first interested in the theory of stationary sequences of random variables and their spectral representations. Some applications to prediction theory are indicated where we present the famous Wold–Zasuhin decomposition. Further, some elements of the theory of second-order stationary processes are presented. For a complete treatment the reader is referred to the books by Doob (1953), Rozanov (1967), and Cramer and Leadbetter (1967). We begin by presenting the problem of stationarity of stochastic processes.

Suppose that $\{\Omega, B, P\}$ is a probability space, i.e. $\{\Omega, B\}$ is a measure space and P is a measure on B with $P\{\Omega\} = 1$. If Ω is a topological space then B is the family of Borel sets of Ω, i.e. the smallest σ-algebra containing the closed (open) sets of Ω.

DEFINITION 17.7.1. A stochastic process over a set T is a mapping on T with values in the family of all measurable functions on $\{\Omega, B\}$.

The values of measurable functions which we consider are in R or C (R denotes the set of real numbers, C denotes the set of complex numbers) but sometimes we allow values in mores abstract sets such as Banach spaces (or complete inner product spaces), etc. For some of the most important cases the set T is $Z = \{0, \pm 1, \pm 2, \ldots\}$ or parts of it, or R or R^n. First we consider the case when the parameter is discrete, i.e. T is in this case Z or part of it.

DEFINITION 17.7.2. Let X be a complete inner product space and let $\{x_n\}_{n \in Z}$ be a subset of X. We say that this subset is a bisequence.

DEFINITION 17.7.3. A bisequence $\{x_n\}_{n\in Z}$ of elements of a complete inner product space X is said to be stationary if the relations

$$\langle x_{n+k}, x_{m+k}\rangle = \langle x_n, x_m\rangle$$

hold for all n, m and k in Z. The sequence $\{r_n\}_{n\in Z}$, $r_n = \langle x_n, x_0\rangle$, is called the covariance bisequence.

DEFINITION 17.7.4. A sequence $\{x_n\}_0^\infty$ of elements of a inner product space X is said to be stationary if the following relations hold for all n, m, $k = 0, 1, 2, 3, \ldots,$

$$\langle x_{n+k,} x_{m+k}\rangle = \langle x_n, x_m\rangle.$$

The sequence

$$\{r_n\}_0^\infty, \qquad r_n = \langle x_n, x_0\rangle$$

is called the covariance sequence.

The following is one of the basic results in the theory of bisequence of elements of a complete inner product space and which are stationary.

PROPOSITION 17.7.5. *Let* $\{x_n\}_{n\in Z}$ *be a stationary bisequence of elements of a complete inner product space* X. *Then the bisequence is positive definite, i.e. for all finite sets* $\{c_i\}_1^m$ *of complex numbers and any finite set of indices* $\{n_i\}_1^m$, *the following inequality holds*:

$$\sum_{i,j}^{m} \langle x_{n_i}, x_{n_j}\rangle c_i\bar{c}_j \geqslant 0.$$

(The bar denotes the complex conjugate.)

Proof. Since

$$\sum_{i,j}^{m} \langle x_{n_i}, x_{n_j}\rangle c_i\bar{c}_j = \left\langle \sum_{i}^{m} c_i x_{n_i}, \sum_{j}^{m} c_j x_{n_j}\right\rangle$$

the assertion follows.

We now present an example of a stationary bisequence.

EXAMPLE 17.7.6. Let X be a complete inner product space and let U be an unitary operator on X. We consider an arbitrary element x_0 in X and we set

$$x_n = U^n x_0, \quad n = \pm 1, \pm 2, \ldots.$$

Now

$$\langle x_{n+k,} x_{m+k}\rangle = \langle U^{n+k}x_0, U^{m+k}x_0\rangle = \langle U^n x_0, U^m x_0\rangle = \langle x_n, x_m\rangle.$$

This clearly gives that our bisequence is stationary.

Now we show that there exists a deep connection between covariance bisequences and stationary bisequences. First we give a definition.

DEFINITION 17.7.7. A bisequence of real numbers $\{r_n\}_{-\infty}^{\infty}$ is said to be positive definite if for all finite sets of complex numbers $\{c_i\}_1^m$ and all finite sets of indices

$\{n_i\}_1^m$ the following inequality holds:

$$\sum_{i,j}^{m} r_{i-j} c_i \bar{c}_j \geqslant 0.$$

We have the following basic result of A. N. Kolmogorov.

THEOREM 17.7.8 (Kolmogorov, 1941). *Let $\{r_n\}_{-\infty}^{\infty}$ be a bisequence of real numbers which is positive definite. Then there exists a stationary bisequence $\{x_n\}_{n\in Z}$ such that*

$$r_n = \langle x_n, x_0 \rangle.$$

Proof. Let us consider the $(2n+1) \times (2n+1)$ Toeplitz matrix $\{a_{ij}\}$ where $a_{ij} = r_{i-j}, i, j$ in $[-n, n]$. Then it is clear that this is a hermitian positive semidefinite matrix and thus it is of the form $H_n \times H_n$, where H_n is a lower triangular matrix and this is such that the upper $(2m+1) \times (2m+1)$ submatrix of H_n is H_m for all $1 \leqslant m < n$.

We consider now an orthogonal bisequence $\{u_n\}_{-\infty}^{\infty}$ of elements of a complete inner product space X. We consider further the sequence of elements in X defined as follows: x_n is given by the formula

$$x_n = \sum_{i=-n}^{\infty} h_{n\ i}^{(n)} u_i, \qquad H_n = \{h_{ij}^{(n)}\}.$$

Then for every n, p

$$r_n = \langle x_{n+p}, x_p \rangle$$

which proves the theorem. □

REMARK 17.7.9. In some problems it is useful to consider the following generalization of stationary bisequences. We consider a complete inner product space X and let q be an arbitrary but fixed positive integer. We construct in the usual way the complete inner product space

$$X^q = \underbrace{X \times X \times \cdots \times X}_{q-\text{times}}.$$

For $u = (u_1, \ldots, u_q)$, $v = (v_1, \ldots, v_q)$ elements in X^q we consider the matrix

$$\{\langle u_i, v_j \rangle\}$$

which we denote by $\langle u, v \rangle$.
We consider on X^q an inner product defined by the relation:

$$\langle\langle u, v \rangle\rangle = \operatorname{trace}\langle u, v \rangle.$$

Then it is not difficult to see that X^q with this inner product is a complete inner product space.

We can define then in the usual way the orthogonality in X^q as well as other notions.

DEFINITION 17.7.10. Let $\{u_n)_{-\infty}^{\infty}$ be a bisequence in X^q. We say that this is stationary if for all m, n, k in Z,

$$\langle\langle u_{n+k}, u_{m+k}\rangle\rangle = \langle\langle u_n, u_m\rangle\rangle.$$

The bi-sequence $\{r_n\}$ where

$$r_n = \langle\langle u_n, u_0\rangle\rangle$$

is called the covariance bi-sequence of the stationary sequence.

Then a slight extension of the Kolmogorov's argument shows that the following result is true.

PROPOSITION 17.7.11. Let $\{r_n\}_{-\infty}^{\infty}$ be a bisequence of matrices which is positive definite. Then there exists a stationary bisequence $\{u_n\}_{-\infty}^{\infty}$ of elements of X^q such that for all n,

$$r_n = \langle u_n, u_0\rangle.$$

Associated with the stationary bisequence $\{x_n\}_{-\infty}^{\infty}$ (or $\{u_n\}_{-\infty}^{\infty}$) is the closed subspace of X (or of X^q), denoted

$$X_{\{x_n\}} \text{ (or } X^q_{\{u_n\}})$$

generated by the sequence. We can define on the subspace $X_{\{x_n\}}$ (or $X^q{}_{\{u_n\}}$) by the formula:

$$Ux_n = x_{n+1}(Uu_n = u_{n+1}).$$

From just the definition of the subspace $X_{\{x_n\}}$ ($X^q{}_{\{u_n\}}$) we have the following result.

THEOREM 17.7.12 (A. N. Kolmogorov, 1941). The operator U is unitary.

Now using this fact we can consider some spectral representations associated with our sequence. Since U is a unitary operator we find a spectral measure on the unit circle $\Gamma = \{z, |z| = 1\}$, say $\{E_t\}$ such that

$$U = \int_\Gamma z\,\mathrm{d}E(z) = \int_\Gamma z\,\mathrm{d}E_z$$

or

$$U = \int_0^{2\pi} \mathrm{e}^{it}\,\mathrm{d}F_t,$$

where $\{F_t\}$ is a spectral measure on $[0, 2\pi]$. Further, we have the relations:

$$r_n = \langle x_n, x_0\rangle = \langle U^n x_0, x_0\rangle = \int_\Gamma z^n\,\mathrm{d}u,$$

where μ is a measure on Γ defined by the formula: if S is any Borel set in Γ then

$$\mu(S) = \langle E(S)x_0, x_0\rangle.$$

We call μ the spectral measure associated with the bisequence $\{x_n\}_{-\infty}^{\infty}$. Obviously

$$E_t = \langle E([1, \mathrm{e}^{it}])$$

has the property that

$$\langle x_n, x_0 \rangle = r_n = \int_0^{2\pi} e^{it} \, dE_t.$$

The derivative of E_t with respect to the Lebesgue measure dt on $[0, 2\pi]$ is called the (spectral) density of the bisequence $\{x_n\}_{-\infty}^{\infty}$. An interesting problem concerning bisequences is: Is E_t absolutely continuous with respect to the Lebesgue measure? We remark that similar statements can be made for stationary bisequences of elements in X^q.

Let us consider a stationary bisequence $\{x_n\}_{-\infty}^{\infty}$ in the complete inner product space X. We associate with this bisequence several closed subspaces of X, defined as follows:

$$M_n = \{x_i, i \leqslant n\}, \qquad M_{\infty} = \{x_i\}, \qquad M_{-\infty} = \bigcap_n M_n.$$

Using these classes of subspaces of X associated with the bisequence $\{x_n\}_{-\infty}^{\infty}$ (which is supposed to be stationary). We consider several subclasses of stationary bisequences.

DEFINITION 17.7.13. Let $\{x_n\}_{-\infty}^{\infty}$ be a stationary bisequence. We say that it is deterministic, nondeterministic, purely nondeterministic iff

$$M_{-\infty} = M_{\infty}, \qquad M_{-\infty} < M_{\infty}, \qquad M_{-\infty} = \{0\}$$

respectively.

We are now in a position to state the main problem concerning the class of stationary bisequences, the so-called 'prediction problem'.

PROBLEM 17.7.14 (The prediction problem for stationary bisequences). The prediction problem for the stationary bisequence $\{x_n\}_{-\infty}^{\infty}$ of elements of a complete inner product space X with the covariance bisequence $\{r_n\}$ is to find for each p in $(1, \infty)$ a sequence of numbers $\{a_n\}$ such that:

$$x'_p = \lim_{n \to \infty} \sum_0^n a_i x_{-i}$$

and compute

$$e_p = \langle x_p - x'_p, x_p - x'_p \rangle.$$

The number e_p is called the error for the lag p.

Of course, we can work with the space X^q and then we can define similarly the prediction problem. Then e_p is, in that case, a matrix which is called the prediction error matrix for lag p.

Following Zasuhin (1941) we call the rank of e_p, $p \neq 1$, the rank of the stationary bisequence. As was remarked by Masani (1966), the case of interest is when the rank of the stationary bisequence of elements of X^q is equal to q. We remark that the stationary bisequence is deterministic if the rank is 0.

Suppose that $\{x_n\}_{-\infty}^{\infty}$ is a stationary bisequence. Then we can define a function on the Borel sets of the unit circle and with values in X by the formula:

$$m_x(S) = E_t(S)x_0, \quad S \text{ is a Borel set,}$$

where E_t is the spectral measure considered above for the unitary operator U associated with the stationary bisequence. We remark that if S_1 and S_2 are two disjoint Borel sets of the unit circle then

$$\langle m_x(S_1), m_x(S_2)\rangle = 0$$

and also that if $\{S_i\}_1^\infty$ is a sequence of pairwise disjoint Borel sets on the unit circle then

$$m_x\left(\bigcup_1^\infty S_i\right) = \sum_{i=1}^\infty m_x(S_i).$$

This is a measure on the unit circle with values in X. It is called an X-valued countably additive, orthogonally scattered measure (on Γ). For simplicity we say that we have a c.a.o.s. measure on the unit circle. For an excellent exposition of the theory of c.a.o.s. measures the reader is referred to Masani (1968).

It is also clear that if we have a c.a.o.s. measure on the unit circle then the sequence $\{x_n\}_{-\infty}^\infty$ defined by the formulas

$$x_n = \int_\Gamma z^n \, d\mu(z),$$

μ being the c.a.o.s. measure, is a stationary bisequence. We note that similar considerations hold for X^q space. We refer for details to Masani (1966).

We now present the famous Wold–Zasuhin decomposition of stationary bisequences. In order to obtain this decomposition we give the so-called Wold–Zasuhin decomposition of an isometry defined on a complete inner product space. We remark that there exist some papers dealing with the extension to the Banach space case of this result.

If X is a complete inner product space then an operator T is $L(X)$, the set of all bounded linear operators on X, is called an isometry iff for all x, y in X,

$$\langle x, y\rangle = \langle Tx, Ty\rangle.$$

This is obviously equivalent to the relation

$$T^*T = I$$

I being the identity operator ($Ix = x$ for all $\mathbf{x}$ in X). Suppose now that X is separable and choose an orthonormal basis in X, say $\{e_i\}_0^\infty$. Then, associated with this basis, we can define the so-called 'shift operator', S, by the relations:

$$Se_i = e_{i+1}, \quad i = 0, 1, 2, \ldots.$$

Then it is obvious that

$$S^*e_0 = 0, \qquad S^*e_i = e_{i-1}, \qquad i \geqslant 1.$$

For the theory of this operator as well as for related classes of operators see the author's book *Topics in Operator Theory. The Volterra and the Shift Operators*

DEFINITION 17.7.15. An isometry T on X is said to be pure if

$$\bigcap_{n=0}^{\infty} T^n X = \{0\}.$$

It is easy to see that the shift S is a pure isometry.

The following result shows the importance of pure isometries.

THEOREM 17.7.16. *Let T be an isometry on X. Then $T = T_1 \oplus T_2$, where $T/m_1 = T_1$ is a unitary operator, $T/m_2 = T_2$ is a pure isometry, and m_2 is the orthogonal complement of m_1, m_1 is a reducing subspace for T.*

Proof. The space m_1 is defined as

$$m_1 = \cap T^n X$$

and obviously is invariant under T. We show that in fact m_1 is reducing.

Indeed, if x is in m_2 then Tx is orthogonal to Tm_1 and this gives that m_1 reduces T. Since T/m_1 is an isometry and onto we get that T_1 is unitary. From the definition of m_1 we get that

$$\bigcap_{0}^{\infty} T^n m_2 = \{0\}.$$

Thus T_2 is a pure isometry and the theorem is proved.

NOTE 17.7.17. Pure isometries are also called unilateral shifts.

Now we can apply the result obtained in Theorem 17.7.16 to the case of stationary bisequences of elements of a complete inner product space.

We apply the above results for the isometry defined by the operator U associated with the bisequence and $X = M_0$, $T = U^*$. It is obvious that T transforms isometrically M_0 onto M_{-1}. Then we have the following result.

COROLLARY 17.7.18. *In the stationary bisequence $\{x_n\}_{-\infty}^{\infty}$ the following decompositions hold*:

$$M_0 = M_{-\infty} + \sum_{i=0}^{\infty} U^{-i}(M_{-1}^{\perp} \cap M_0)$$

$$M_n = M_{-\infty} + \sum_{i=0}^{\infty} U^{n-i}(x_0 - P_{M_{-1}} x_0), \qquad M_{-\infty} \perp \sum_{i=-\infty}^{\infty} U^i(x_0 - P_{M_{-1}} x_0).$$

A similar decomposition holds for the case of stationary bisequences of elements in X^q.

REMARK 17.7.19. The above decomposition was proved for X by Wold (1938). For q greater than 1 it was conjectured by Zasuhin (1941). Full proofs were given by Doob (1953), Wiener and Masani (1957) and Masani (1962). Also the results about the structure of isometries was known to J. von Neumann but never published.

The element $U^n(x_0 - P_{M_{-1}} x_0)$ is called the nth innovation vector and

$$\{U^n(x_0 - P_{M_{-1}} x_0\}_{-\infty}^{\infty}$$

is called the innovation stationary bisequence. If we normalize the elements of this bisequence thenw e get the so-called normalized innovation bisequence.

We now consider a stationary bisequence $\{x_n\}_{-\infty}^{\infty}$ of elements of the complete inner product space X and we are interested in the computation of prediction elements y_n, $n \geqslant 1$, i.e. the elements y_n satisfying the property that for all y in M_0 the following inequality holds:

$$\|x_n - y_n\| \leqslant \|x_n - y\|.$$

The number

$$\delta_{1,n} = \|x_n - y_n\|$$

can be computed using the density of the bisequence. The basic result here is the famous formula obtained by G. Szegö, A. N. Kolmogorov and M. G. Krein:

$$\delta_{1,n}^2 = 2\pi \exp((2\pi)^{-1} \int_0^{2\pi} \log f(s)\, ds)$$

f being the spectral density of the stationary bisequence. From the definition of the n imber $\delta_{1,n}$ we have that

$$\begin{aligned}
\delta_{1,n}^2 &= \inf_y \{\|x_0 - y\|^2, y \in M_{-1}\} = \\
&= \inf\{\|x_0 - a_1 x_{-1} - \cdots - a_n x_n\|^2, a_i \text{ complex numbers}, n \geqslant 1\} \\
&= \inf_{a_i} \left\{ \int_\Gamma (1 - a_1 z^{-1} - a_2 z^{-2} - \cdots - a_n z^{-n})\, dE(z)^{\,2} \right\} \\
&= \inf \left\{ \int_\Gamma |(1 - a_1 z^{-1} - a_2 z^{-2} - \cdots - a_n z^{-n}|^2\, d\mu \right\},
\end{aligned}$$

where μ is the measure associted to the stationary bisequence which we considered above. Now the Szegö–Kolmogorov–Krein result asserts that the number in the last equality is equal to

$$\exp\, (2\pi)^{-1} \int_0^{2\pi} \log f(s)\, ds.$$

We refer for proofs of this important result to Hoffman (1962, Ch. 4) or other books on function algebras. We mention that in Hoffman's book the result is proved in such a way that the extension to more general settings is possible (for example, to the case of log-modular algebras, Dirichlet algebras, etc.).

Associated with the stationary bisequence $\{x_n\}_{-\infty}^{\infty}$ is the notion of 'filter' which we consider now.

DEFINITION 17.7.20. A filter is any linear and bounded operator T on the closed subspace generated by the stationary bisequence and which commutes with the operator U defined by the formulas:

$$Ux_n = x_{n+1}.$$

The filter T is said to be realizable (or causal) if for each n the subspaces M_n (defined for the stationary bisequence) are invariant subspaces for T. We say that $M_n \in \operatorname{Lat} T$, where Lat T denotes the family of all invariant subspaces of the operator T. It is known and easy to verify that this is a lattice.

For information about the filters in an abstract context we refer to Saeks (1973). For the reader interested in applications we refer to Bucy (1970), Kailath (1974), Parzen (1975), Koopmans (1974), Robinson (1967) and Yaglom (1962).

A beautiful exposition of the relationship of the Hardy space H^2 and the filter theory is given in the book of Dym and McKean (1972, Ch. 3) to which the reader is referred for many very interesting topics.

We now present some results for the case when the parameter T is R (the set of all real numbers).

DEFINITION 17.7.21 (Khinchin, 1934). The stochastic process $\{x_t\}_{t\in R}$ is said to be a stationary stochastic process of order 2 if the following conditions are satisfied:

(1) $E(|x_t|^2)$ is finite for each t in R.

(2) For all u, v, w in R,

$$\langle x_{u+w}, x_{v+w}\rangle = \langle x_u, x_v\rangle,$$

(3) For any t in R, $E(x_t) = 0$.

DEFINITION 17.7.22. A stationary second-order process $\{x_t\}_{t\in R}$ is said to be continuous if

$t \to r(t) = E(x_{t+s}\bar{x}_s)$ is a continuous function on R. (In the above formula the bar denotes a complex conjugate.)

We have the following simple result.

PROPOSITION 17.7.23. *A stationary second-order process is continuous iff it is continuous at* $t = 0$.

Proof. Since we have the relations

$$E(|x_{t+h} - x_t|^2 = 2r(0) - r(h) - r(-h)$$

$$|r(t) - r(s)|^2 = |E(x_t - x_s))x_0|^2 \leqslant E(|x_t - x_s|^2 \cdot E|x_0|^2$$

the assertion follows. □

DEFINITION 17.7.24 (Khinchin, 1934). Let $\{x_t\}_{t\in R}$ be a stationary second-order stochastic process. The function on R^2, defined by the formula

$$r(s, t) = E(x_t\bar{x}_s),$$

is called the covariance function of the stationary stochastic process $\{x_t\}_{t\in R}$.

REMARK 17.7.25. In the definition of the stochastic process given above we imposed the condition $E(x_t) = 0$. This is only a technical condition and in fact all the results hold for stochastic processes which do not satisfy this condition.

Associated with the stochastic process $\{x_t\}_{t\in R}$ of second order and stationary is a closed subspace in the space L^2 associated with the measure space $\{\Omega, B, P)$, namely the closed subspace generated by the functions x_t. We denote this subspace by L_x^2. Then we can consider the function $t \to x_t$ defined on R and with values in L_x^2, and this function can be considered as a curve in this space. For fundamental results concerning this point of view we refer to the papers of Kolmogorov (1940), Cramer (1940), Itô (1944), Krein (1944), Masani (1972) and Wiener (1923) as well as to the papers quoted in our references and the references of the above papers.

In the space L_x^2 we consider a family of operators defined as follows: for s fixed in R we set

$$T_s x_t = x_{t+s}, \quad t \text{ in } R.$$

It is easy to see that this is a bounded linear operator and in fact is a unitary operator on L_x^2. From the definition of the operator T_s we see that the relation

$$T_s(T_t) = T_{s+t}$$

holds for all t, s in R. This means that $t \to T_t$ is a group of unitary operators on L_x^2. For any fixed a in R we consider the closed subspace in L_x^2 which is generated by x_t with $t \leqslant a$. We denoted this space by $L_{x,a}^2$. We denote by P_a the projection onto the subspace $L_{x,a}^2$.

We now consider some important subclasses of stationary stochastic processes of second order.

DEFINITION 17.7.26. Let $\{x_t\}_{t \in R}$ be a stationary stochastic process of second order. We say that our stochastic process is deterministic if for all a in R,

$$L_x^2 = L_{x,a}^2.$$

The process is called completely nondeterministic if

$$M_{-\infty} = \bigcap L_{x,a} = \{0\}.$$

Using the projections P_a we can characterize the above classes of stochastic process as follows.

PROPOSITION 17.7.27. *Let $\{x_t\}_{t \in R}$ be a stochastic process which is stationary and of second order. Let us consider the elements*

$$y_t(a) = P_a x_t.$$

Then the stochastic process is deterministic iff

$$\|P_0 x_t\| = \|P_0 x_s\| \tag{*}$$

for all t and s in R.

Proof. If $L_x^2 = L_{x,a}^2$ then for all a in R,

$$P_a x_t = x_t$$

and thus the condition is necessary. Conversely, suppose that the above relation (*) holds. Then for all t in R,

$$\|P_0 x_t\| = \|x_0\|$$

and thus $y_t(0) = x_0$ for all t in R. This implies that x_0 is in $L^2_{x,t}$ for all t. Then

$$x_s = T_s x_0 \in T_s(L^2_{x,t-s}) = L^2_{x,t}.$$

Since t and s are arbitrary we get the relation $L^2_x = L^2_{x,t}$ for all t in R. The proposition is proved.

The proof of the following characterization is left to the reader

PROPOSITION 17.7.28. *Let $\{x_t\}_{t\in R}$ be a stationary stochastic process of second order. Then it is completely nondeterministic iff*

$$\lim_{t\to-\infty} \|P_0 x_t\| = 0.$$

We now present a decomposition result which gives us the decomposition of an arbitrary stochastic stationary process of second order in two processes of the type introduced in Definiton 17.7.26.

THEOREM 17.7.29 (Hanner, 1949). *Let $\{x_t\}_{t\in R}$ be a stationary stochastic process of second order. Then there exists two stationary stochastic processes of second order, say $\{x_t\}_{t\in R}$ and $\{v_t\}_{t\in R}$, such that:*

(1) $x_t = u_t + v_t, \quad t \in R,$

(2) $\{u_t\}_{t\in R}$ *is deterministic,*

(3) $\{v_t\}_{t\in R}$ *is completely nondeterministic,*

(4) *for each t in R,* $\langle u_t, v_t\rangle = 0$.

Proof. We consider the subspace $M_{-\infty} = \cap L^2_{x,a}$ and the subspace N_2 which is the orthogonal complement of the above subspace in L_x. Then for each s in R,

$$T_s = M_{-\infty} = M_{-\infty}$$

and thus $T_{1,s}$, the restriction of T_s to the above subspace, is unitary.

We get easily that T_s/N_2 is again a unitary operator. For each t in R we define

$$u_t = P_{M_{-\infty}} x_t, \qquad v_t = (I - P_{M_{-\infty}})x_t.$$

Properties (1) and (4) are clearly satisfied. Since

$$u_{t+h} + v_{t+h} = x_{t+h} = T_h(x_t)$$

we get that

$$u_{t+h} = T_h u_t, \qquad v_{t+h} = T_h v_t$$

and thus $\{u_t\}$, $\{v_t\}$ are stationary stochastic processes of second order. We can show that the first is deterministic and the second is completely nondeterministic.

We consider subspaces L^2_u *and* L^2_v which are closed subspaces in L^2_x generated by $\{u_t\}_{t\in R}$, $\{v_t\}_{t\in R}$ respectively. From the definition of $M_{-\infty}$ we get that

$$M_{-\infty} = L^2_u \quad \text{and} \quad M^{\perp}_{-\infty} = L^2_v.$$

We can prove that $L^2_{u,t} = L^2_u$ for all t in R and that $\cap_a L^2_{v,a} = \{0\}$. Since the proof is simple it is omitted.

DEFINITION 17.7.30. Let $\{x_t\}_{t\in R}$ be a stationary stochastic process of second order. The process is said to be continuous at $t = t_0$ if

$$\lim_{t\to t_0} E(|x_t - x_{t_0}|^2) = 0.$$

The process is said to be continuous if it is continuous at each t or R.

Then we have the following simple result whose proof is left to the reader.

PROPOSITION 17.7.31. *A stochastic stationary process of second order is continuous iff it is continuous at* $t = 0$.

DEFINITION 17.7.32. The stationary stochastic process $\{x_t\}_{t\in R}$ is said to have a derivative at t if

$$\lim_{h\to 0} h^{-1}(x_{t+h} - x_t)$$

exists in L_x^2. If it exsits it is denoted by dx_t/dt.

Then we have

PROPOSITION 17.7.33. *The stationary stochastic process of second order* $\{x_t\}_{t\in R}$ *has a derivative at t iff the second generalized derivative of the covariance function r exists at* (t, t), *i.e.*

$$\lim_{h,k\to 0} (hk)^{-1} E((x_{t+h} - x_t)(\bar{x}_{t+k} - \bar{x}_t))$$

exists and is finite.

Proof. Since the proof is not difficult we leave it to the reader. □

We now present an integral representation associated with the stationary stochastic processes of second order. First we recall that R^2 is a locally compact abelian group. Second, as is very easy to see, the function $r(s, t)$ is positive definite on R^2 and thus, by the Weil–Bochner theorem on the representation of positive definite functions on locally compact abelian groups, we get that there exists a positive measure μ on R such that

$$r(s, t) = \iint_{R\times R} e^{i(tu+sv)}\, d\mu(u, v).$$

Now as the process is stationary we have that

$$r(s, t) = \bar{r}(t - s),$$

where $\bar{r}$ is a positive definite function on R. Then by the same theorem of Weil–Bochner there exists a positive measure ν on R such that

$$\bar{r}(s) = \int_R e^{its}\, d\nu(s)$$

and thus

$$\bar{r}(t - s) = \int_R e^{i(t-s)u}\, d\nu(u)$$

which is called the Bochner–Khinchin representation.

We now present some decompositions of stationary stochastic processes of second order using certain operators which are intimately connected with stationary stochastic processes. For this purpose we consider the space $L^2(\mathrm{d}x\,\mathrm{d}y)$ for the space $I_\times I, I = [a, b]$. For each g in $L^2(I, \mathrm{d}x)$ we consider the function Tg defined by

$$(Tg)(t) = \int_I r(t,s)g(s)\,\mathrm{d}s$$

which is defined a.e. (dx). By the Cauchy–Buniakowsky inequality (and using Fubini's theorem) we get

$$|(Tg)(t)| \leqslant \left(\left(\int_I |r(t,s)|^2\,\mathrm{d}s\right)\left(\int_I |g(s)|^2\,\mathrm{d}s\right)\right)^{1/2}$$

which implies that T is a well-defined operator on $L^2(I, \mathrm{d}x)$. Obviously T is linear. Using the Lebesgue dominated theorem and the Ascoli–Arzelà theorem as well as the above inequality we get the following result.

PROPOSITION 17.7.34. *The operator T is a compact operator when the covariance function r is continuous in both variables.*

REMARK 17.7.35. The above assertion is, of course, valid for arbitrary continuous functions, and not only for covariance functions.

Concerning the operator T we have the following simple but important property.

PROPOSITION 17.7.35. *The operator T is hermitian when r is a real-valued function.*

This can be verified if we compute

$$\langle Tg, h\rangle, \qquad \langle g, Th\rangle$$

and we conclude that they are equal.

COROLLARY 17.7.37. *The operator T is positive if for all f in $L^2(I, \mathrm{d}x)$ the following inequality holds*:

$$\iint_{I_\times I} r(s,t)\,f(s)\,\bar{f}(t)\,\mathrm{d}t\,\mathrm{d}s \geqslant 0.$$

COROLLARY 17.7.38. *If $r(s,t) = r(t,s)$ is continuous and the condition in the above corollary is satisfied, then all the eigenvalues of T are nonnegative.*

In this case from the theory of normal and compact operators we conclude that for each f in $L^2(I, \mathrm{d}x)$ we have the decomposition

$$f = \sum_{i=0}^{\infty} a_i e_i + h,$$

where $\{e_i\}$ is the set of all eigenvectors of T and h is orthogonal to each e_i.

We mention that the following famous result holds used earlier in this book.

THEOREM 17.7.39 (Mercer's theorem). *Suppose that the operator T is defined for an arbitrary continuous function r (we use the same notation as for the covariance function) which is symmetric and the condition of Corollary 17.7.37 holds. Further, if*

$\{e_i\}$ is the set of eigenvectors of T corresponding to nonzero eigenvalues $\{\lambda_i\}$, then r has the following representation:

$$r(t,s)=\sum_{i=0}^{\infty}\lambda_i e_i(t)e_i(s)$$

where the series converges uniformly in both variables. If I is infinite then the convergence is uniform on each finite interval $I_\times I$. Also the following relation holds

$$\lim_{n\to\infty}\iint_{I\times I}\left|r(t,s)-\sum_{i=0}^{n}\lambda_i e_i(t)e_i(s)\right|^2 dt\,ds=0.$$

For a proof of this result we refer to books on integral equations.

We consider again a stationary stochastic process of second order $\{x_t\}_{t\in I}$, I being an interval in R. We note that the definitions can be formulated without difficulty in this case too. We consider the covariance function associated with the process

$$r(t,s)=E(x_t\bar{x}_s)$$

and we remark that it is positive definite, i.e. for any finite set of complex numbers (c_i) and any finite set of points in I, (t_i) we have the inequality

$$\sum c_i\bar{c}_j r(t_i,t_j)\geqslant 0.$$

Associated with the stationary stochastic process considered above we define an integral as follows. Let $\Delta=(t_i)_0^n$, where

$$a=t_0<t_1<\cdots<t_n=b,\qquad I=[a,b]$$

and set

$$s_\Delta=\sum f(t_i)x_{t_i}(t_{i+1}-t_i),$$

where f is a given real-valued funtion on I.

We set

$$v(\Delta)=\sup(t_i-t_{i-1}).$$

We say that f is integrable with respect to the stochastic process $\{x_t\}_{t\in I}$ if $\lim_{v(\Delta)\to 0}s_\Delta$ exists in the space $L^2(\Omega,B,P)$. The limit, if it exists, is denoted by $\int_a^b f(t)x_t\,dt$.

The following result presents a sufficient condition for integrability with respect to $\{x_t\}_{t\in I}$

PROPOSITION 17.7.40. *Suppose that the stochastic process $\{x_t\}_{t\in I}$ has the property that $r(s,t)$ is continuous on $I_\times I$. Then each continuous function on I (real-valued) is integrable, i.e. $\int_a^b f(t)x_t\,dt$ exists.*

Proof. Since the proof of this assertion can be modeled after the proof for the Cauchy integral we omit it. □

LEMMA 17.7.41. *Suppose that h,k are continuous real-valued functions on I and the* *stationary stochastic process $\{x_t\}_{t \in R}$ has $r(t,s)$ continuous. Then*

$$E\left(\int_a^b h(t)x_t\,\mathrm{d}t, \int_a^b k(t)x_t\,\mathrm{d}t\right) = \iint_{I \times I} h(t)k(s)r(t,s)\,\mathrm{d}t\,\mathrm{d}s.$$

Proof. Let

$$\Delta_1 = \{a = t_0 < t_1 < \cdots < t_n = b\}$$
$$\Delta_2 = \{a = s_0 < s_1 < \cdots < s_m = b\}$$

and

$$s_{\Delta_1} = \sum h(t_i)x_{t_i}(t_i - t_{i-1})$$
$$s_{\Delta_2} = \sum k(s_j)x_{s_j}(s_j - s_{j-1})$$

with

$$\lim s_{\Delta_1} = \int_a^b f(t)x_t\,\mathrm{d}t$$

$$\lim s_{\Delta_2} = \int_a^b k(t)x_t\,\mathrm{d}t \qquad \square$$

in $L^2(\Omega, B, P)$. Using the fact that the scalar product is continuous in both variables we get the following relation:

$$\lim_{\nu(\Delta_1),\,\nu(\Delta_2)\to 0} E(s_{\Delta_1}, s_{\Delta_2}) = E\left(\int_a^b h(t)x_t\,\mathrm{d}t, \int_a^b k(t)x_t\,\mathrm{d}t\right)$$

and this clearly gives the assertion of the lemma.

We are now in a position to obtain the decomposition of a stationary stochastic process of second order.

THEOREM 17.7.42 (Loeve, 1946). *Let $\{x_t\}_{t \in I}$ be a stationary process of second order with the property that $r(t,s)$ is a continuous function. Then the process has the decomposition*

$$x_t = \sum_0^\infty u_i e_i(t)$$

for t in I, where $\{e_i\}_0^\infty$ is the family of eigenvectors of the operator T defined by the function $r(t,s)$ as above, and also

$$u_i = \int_a^b e_i(t)x_t\,\mathrm{d}t$$

are random variables satisfying the relations

$$\langle u_i, u_j\rangle = \delta_{ij} = \begin{cases} 1, & i = j \\ 0, & i \neq j \end{cases}.$$

The series in the decomposition of x_t converges uniformly when I is a finite interval; if I is infinite then the convergence is uniform on any compact subset of I.

Proof. For each i we set

$$u_i = \int_a^b e_i(t)x_t\,\mathrm{d}t$$

which are clearly well defined, and also if we set

$$s_n(t) = u_0e_0(t) + u_1e_1(t) + \cdots + u_ne_n(t)$$

then

$$E(|s_n(t) - x_t|^2) = r(t, t) - \sum_0^n \lambda_i e_i(t)^2$$

which converges to zero uniformly on each compact subset of I. □

REMARK 17.7.43. If $I = [0, 1]$ and $r(s, t)$ is given by the formula

$$r(t, s) = \min(t, s)$$

then the stochastic process $\{x_t\}_{t\in R}$ is the so-called Brownian motion. In this case the eigenvalues λ_i are given by the formulas:

$$\lambda_i = 4\cdot(2i - 1)^{-2}\pi^{-2}.$$

The corresponding eigenvectors are given by the formulas:

$$e_i(t) = 2^{1/2}\sin(i - \tfrac{1}{2})(\pi t).$$

For extensions to the case when we consider the parameter T to be R we refer to the papers by Kelly *et al.* (1960) and Root and Pitcher (1955). For applications we refer to the paper by Fano (1961).

For some interesting extensions of the notion of stationary stochastic processes we refer to the papers by Loynes (1965) and Saworotnow (1973, 1979) as well as to the literature quoted there.

17.8. Inner Products and Differentials. Harmonic and Analytic Differentials

In this section we consider certain classes of differentials on Riemann surfaces and show how some of them can be organized as complete inner product spaces. Then we prove a decomposition which may be used to show the existence of harmonic and analytic differentials with prescribed properties. For the reader interested in such problems we refer for further results to W. V. D. Hodge's famous book (1941, second edition 1952) and the 1966 work of C. B. Morrey and the literature quoted there. For the reader's convenience we recall the notion of Riemann surfaces. For a detailed discussion we refer to Weyl (1923, 1951), Springer (1957) and Stoilow (1938).

DEFINITION 17.8.1. A manifold M is called an abstract Riemann surface if the following conditions are satisfied:

(1) There exists $(U_i)_{i\in I}$ an open covering of M, and if $(\varphi_i)_{i\in I}$, φ_i is a homeomorphism of U_i onto an open set of the complex plane C.

(2) If $U_i \cap U_j \neq \emptyset$ then $\varphi_i \varphi_j^{-1}$ is a conformal sense-preserving mapping of $\varphi_i(U_i \cap U_j)$ onto $\varphi_j(U_i \cap U_j)$.

We note that condition (2) means that $z \to \varphi_j(\varphi_i^{-1}(z))$ is an analytic function on $\varphi_i(U_i \cap U_j)$.

We now recall the definition and some properties of differentials. For more details we refer to Springer ($n = 2$) and to Rudin (1966).

If O is an open set in R^n then a k-surface in O is a C^1-mapping φ defined on a compact set K in R^k into O. The set K is called the parameter domain of φ.

DEFINITION 17.8.2. A differential form of order k ($\geqslant 1$) on the set O (O is an open set in R^n) is any function w denoted as

$$w = \sum a_{i_1 \cdots i_k}(x)\, \mathrm{d}x_{i_1} \wedge \cdots \wedge \mathrm{d}x_{i_k}$$

where the indices $i_1, \ldots, i_k$ range independently from 1 to n, on all k-surfaces φ on O whose value at the surface is

$$w = \int_K a_{i_1 \cdots i_k}(\varphi(u_1, \ldots, u_k))(\partial(x_{i_1}, \ldots, x_{i_k})/\partial(u_1, \ldots, u_k))\, \mathrm{d}u_1 \ldots \mathrm{d}u_k$$

and where

$$\partial(x_{i_1}, \ldots, x_{i_k})/\partial(u_1, \ldots, u_k)$$

is the Jacobian of the transformation

$$(u_1, \ldots, u_k) \to (\varphi_{i_1}(u_1, \ldots, u_k), \ldots, \varphi_{i_k}(u_1, \ldots, u_k)).$$

If w is a k-form then we say that it is of class C^k if all the functions $a_{i_1} \ldots i_k$ are of class C^k. The 0-form is a continuous function.

If w is a k-form of the form

$$w = a(x)\, \mathrm{d}x_{i_1} \wedge \cdots \wedge \mathrm{d}x_{i_k}$$

then we can consider a new form by interchanging some pair of subscripts; this new form, denoted by $\bar{w}$, has the property that $w = -\bar{w}$. Thus we have the anticommutative relations: for all i, j,

$$\mathrm{d}x_i \wedge \mathrm{d}x_j = -\mathrm{d}x_j \wedge \mathrm{d}x_i.$$

This clearly implies that $\mathrm{d}x_i \wedge \mathrm{d}x_i = 0$.

A basic k-form is a form of the following type:

$$\mathrm{d}x_I = \mathrm{d}x_{i_1} \wedge \cdots \wedge \mathrm{d}x_{i_k},$$

where $i_1 < i_2 < \cdots < i_{k-1} < i_k$.

From the remarks above we see that every differential form of order k (in short, k-form) can be written in the canonical form

$$w = \sum_I a_I\, \mathrm{d}x_I$$

where $\mathrm{d}x_I$ are basic k – forms.

Now if

$$I = [i_1, \ldots, i_k], \qquad J = [j_1, \ldots, j_j]$$

with

$$i_1 < i_2 < \cdots < i_k; \qquad j_1 < j_2 < \cdots < j_1$$

are two index sets and dx_I, dx_J are basic forms then $dx_I \wedge dx_J$ is the $(k+1)$ – form defined by

$$dx_I \, dx_J = (-1)^m \, dx_{[I,J]},$$

where m is the number of differences $j_p - i_q$ that are positive and $[I, J]$ is the set obtained by arranging the members of $I \cup J$ in increasing order. If w_1 and w_2 are p-form and q-form, respectively,

$$w_1 = \sum_I a_I \, dx_I, \qquad w_2 = \sum_J b_J \, dx_J$$

(the canonical form of w_1 and w_2) then $w_1 \wedge w_2$ is the $(p+q)$ form defined by:

$$w_1 \wedge w_2 = \sum_{I,J} a_I b_J \, dx_I \wedge dx_J.$$

Now if w is a 0-form of class C^1 then (since it is in fact a function of class C^1) we can define the following form

$$dw = \sum_i^n (D_i f)(x) \, dx_i$$

and then if w is a form of class C^1 we set

$$dw = \sum_I (da_I) \wedge dx_I.$$

Similarly, we define for any form of class m, $d^i w$, $i \leqslant m$.

17.8.3. *If w is a k-form on the open set O in R^n then w is exact if for some $(k-1)$-form on O we have $w = $ d; if w is on class C^1 and $dw = 0$ then it is called closed.*

Now if we set

$$w = w_1 + i w_2,$$

where w_i are differential forms, then we call w a complex differential, and the rules are clear for such objects.

We now recall the definition of the conjugation defined only for 1-forms. We consider only the case $n = 2$. Let w be the form

$$w = p \, dx + q \, dy$$

where p, q are complex-valued functions. The conjugate form corresponding to the form w is defined as the form $*w$ by the relation

$$*w = -q \, dx + p \, dy.$$

Then from the definition it follows that the following relations hold:

(1) $*(*w) = w$

(2) $*(w_1 + w_2) = *w_1 + *w_2$

(3) $w_1 \wedge (*w_2) = (p_1 p_2 + q_1 q_2)\,\mathrm{d}x \wedge \mathrm{d}y$

(where $w_i = p_i\,\mathrm{d}x + q_i\,\mathrm{d}y$, $i = 1, 2$),

(4) $w \wedge (*w) = (p^2 + q^2)\,\mathrm{d}x \wedge \mathrm{d}y$

(5) $*\mathrm{d}(w) = -\mathrm{d}(*w)$

(6) if f is a function of class C^2 and $\mathrm{d}f = \Sigma_{i=1}^{2} (D_i f)\,\mathrm{d}x_i$ then

$$\mathrm{d} \wedge *\mathrm{d}f = (\partial^2 f/\partial x^2 + \partial^2 f/\partial y^2)\,\mathrm{d}x\,\mathrm{d}y.$$

If w is a 1-form and c is a closed curve then $\int_c w$ is called the period of the closed form w.

We now consider differentials on Riemann surfaces. These are defined as forms w on regions of the Riemann surface in which the integration is defined by the above rule.

DEFINITION 17.8.4. A 1-differential form w of class C^1 is said to be a harmonic differential if it is of the form $w = \mathrm{d}f$, where f is harmonic (i.e. locally f satisfies the Laplace equation).

DEFINITION 17.8.5. A 1-differential form w is said to be regular analytic (or a holomorphic differential) if (locally, i.e. in each parametric disk) w is of the form

$$w = \mathrm{d}f,$$

where f is a holomorphic function in that disk.

We follow the usual convention and we use also the notation $\mathrm{d}x + i\,\mathrm{d}y = \mathrm{d}z$ and then if f is a holomorphic function then (locally),

$$\mathrm{d}f = f'(z)\,\mathrm{d}x = \partial f/\partial x\,\mathrm{d}x + \partial f/\partial y\,\mathrm{d}y$$

where f' is the derivative of f at $z = x + iy$.

If w is a harmonic differential then $\mathrm{d}w = 0$ and $\mathrm{d}(*w) = 0$, which gives that $\mathrm{d}*(*w) = 0$. This gives the important fact that $*w$ is also a harmonic differential, which implies that the form

$$w = w + i(*w)$$

is a holomorphic differential. The harmonic differential form $*w$ is called the conjugate harmonic differential to w. If w is a (complex) differential form (locally)

$$w = p\,\mathrm{d}x + q\,\mathrm{d}y$$

then we define the complex conjugate differential $\bar{w}$ of w by the formula (locally):

$$\bar{w} = \bar{p}\,\mathrm{d}x + \bar{g}\,\mathrm{d}y.$$

We now consider the set of all 1-differential forms of the form (locally)

$$w = p\,\mathrm{d}x + q\,\mathrm{d}y$$

where p, q are measurable functions; such a form will be called measurable and we say that the form w is (Lebesgue) square integrable if p^2 and q^2 are Lebesgue integrable. If w_1 and w_2 are two square integrable forms then we define the inner product by the formula:

$$\langle w_1, w_2 \rangle = \iint_S w_1 * \bar{w}_2.$$

In the set of all square integrable differentiable forms we define an equivalence relation as follows: two differential forms are equivalent if they are identical except on sets whose (Lebesgue) measure is zero. The symbol w will be used in what follows to denote the equivalence class of square integrable differential forms. In this case if $\langle w, w \rangle = 0$ then $w = 0$ (i.e. the axioms of the inner product are satisfied).

If w is a square integrable differential form then we define the number

$$w \to \left(\iint_S w * \bar{w} \right)^{1/2} = \|w\|$$

to be the norm of the differential form. The set of all square integrable differential forms on a Riemann surface S (in fact the set of all equivalence classes) is denoted by $L^2(S)$.

The basic information about this set is contained in the following result.

THEOREM 17.8.6. *$L^2(S)$ is a complete inner product space.*

Proof. The fact that $L^2(S)$ can be organized as an inner product space is clear and it remains to show that $L^2(S)$ is complete.

The proof is analogous to the proof of the fact that L^p-spaces are Banach spaces. We refer for details to books on integration and to G. Springer's book (1957). □

Since S is a Riemann surface it follows that there are some operators which are intrinsic to S and which act on $L^2(S)$. In what follows we consider such operators, known as smoothing operators.

First we consider the case of $L^2(D)$, $D = (z, |z| < 1)$.

Let $D_r = (z, |z| < 1 - r)$. We define on D the following function

$$s_r(x, y) = \begin{cases} k(r^2 - x^2 - y^2), & x^2 + y^2 < r^2 \\ 0 & x^2 + y^2 \geqslant r^2 \end{cases}$$

with the constant k given by the formula $k = 3/(r^6)$ (and in fact k is chosen such that the integral of s_r over D_{1-r} is 1 and this gives for k the above value). From the definition of the function s_r we see that it has continuous partial derivatives with respect to x and y. We now define the smoothing operator M_r on $L^2(D)$ by the following formula: if $f \in L^2$ then

$$(M_r f)(x, y) = \int_0^{2\pi} \int_0^r f(x + r\cos t, y + r\sin t) s_r(r\cos t, r\sin t) r \, dr \, dt. \qquad (*)$$

or

$$(M_r f)(x, y) = \iint\limits_{D_{1-r}} f(x + u, y + v)s_r(u, v)\, du\, dv$$

and

$$(M_r f)(x, y) = \iint\limits_{D} f(u, v)s_r(u - x, v - y)\, du\, dv.$$

Now if w is a differential form, $w = p(x, y) + q(x, y)\, dy$ in $L^2(D)$ then the operator M_r is defined by the formula:

$$w_r = (M_r p)(x, y)\, dx + (M_r q)(x, y)\, dy_2 = M_r p\, dx + M_r q\, dy.$$

We see that M_r is a linear operator from $L^2(D)$ in the space (linear) of differential form on D_r. The principal properties of this operator are listed below.

PROPOSITION 17.8.7. *For the operator M_r the following assertions hold*:

(1) *If w is in $L^2(D)$ then $M_r w$ is of class C^1 in D_r.*

(2) *If w is a harmonic differential form in D then $M_r w$ is also harmonic.*

(3) *For each w in $L^2(D)$, $\lim_{r\to 0} \|w - M_r w\| = 0$.*

(4) *If the support of γ is D_r then the support of $M_r \gamma$ is in D and*

$\langle M_r w, \gamma\rangle_{D_r} = \langle w, M_r^\gamma\rangle_D$.

Proof. We prove that $M_r f$ has continuous derivatives with respect to x and the same argument applies for the case of argument y. First we remark that the function s_r has the property that $\partial s_p/\partial x$ is continuous in $\bar{D}$ and bounded, i.e.

$|\partial s_r(x, y)/\partial x| \leqslant c$ (c is a constant).

Then we have that, for h sufficiently small,

$$|(s_r(x + h, y) - s_r(x, y))/h| = \left| h^{-1} \int_x^{x+h} (\partial s_r(x, y)/\partial x)\, dx \right| \leqslant c.$$

Then applying the Lebesgue dominated convergence theorem we obtain that

$$(M_r p)(x + h, y) - (M_r p)(x, y))/h = \iint\limits_{D} h^{-1} p(u, v)(s_r(u - x - h, v - y) - s_r(u - x, v - y))\, du\, dv$$

implies the following relation

$$\partial M_r/\partial x(x, y) = \iint\limits_{D} (p(u, v)\partial s_r(u - x, v - y)/\partial x\, du\, dv.$$

Since

$$\partial s_r(x, y)/\partial x$$

is uniformly continuous we find for any $\varepsilon > 0$ that there exists $\delta(\varepsilon) > 0$ such that whenever $d((x, y), (x', y')) < \delta(\varepsilon)$ then

$$|\partial s_r/\partial x(x, y) - \partial s_r/\partial x(x', y')| \leqslant \varepsilon.$$

In this case we obtain that

$$\partial(M_r)/\partial x(x, y) - \partial(M_r p)/\partial x(x', y')| \leqslant \left| \iint_D p(u, v)(\partial s_r(u - x, v - y)/\partial x) - \right.$$
$$\left. (\partial s_r(u - x', v - y')/\partial x)\, du\, dv \right| \leqslant \varepsilon \iint_D |p(u, v)|\, du\, dv.$$

This obviously implies continuity since in the last inequality we can apply the Cauchy–Buniakowky inequality and then the factor which multiplies ε is bounded.

Now we prove assertion (2). Since w is harmonic we find a harmonic function f such that $w = df$. The assertion follows if we prove the following facts:

(1) For any f of class C^1 (in D), $M_r\, df = dM_r f$.

(2) If f is harmonic then $M_r f = f$.

Let, us first prove fact (1). Let f be of class C^1 and then we have that

$$M_r(\partial f/\partial x) = (\partial/\partial x) M_r f.$$

Indeed, applying the rule of differentiating under an integral sign we have that

$$(\partial/\partial x) M_r f(x, y) = (\partial/\partial x) \iint_{D_{k-r}} f(x + u, y + v) s_r(u, v)\, du\, dv$$
$$= \iint_{D_{k-r}} (\partial/\partial x) f(x + u, y + v) s_r(u, v)\, du\, dv = M_r(\partial/\partial x) f(x, y).$$

The same holds for the argument y and these obviously imply the assertion of fact (1). For the assertion of the fact (2) we remark that if $r < r_1$ then

$$s_{r_1}(r \cos t, r \sin t) = k(r_1^2 - r^2)^2$$

and thus the form of $M_r f(x, y)$ is

$$M_r f(x, y) = \int_0^r tk(r^2 - t^2)^2 \int_0^{2\pi} f(x + t \cos u, y + t \cos u)\, du\, dt$$

and the mean values theorem for harmonic functions asserts that

$$2 f(x, y) = \int_0^{2\pi} f(x + t \cos u, y + t \sin u)\, du$$

which implies that

$$M_r f(x, y) \leqslant 2 \int_0^r k(r^2 - t^2)^2 \, \mathrm{d}t \, f(x, y) = f(x, y).$$

Thus (2) is proved.

For assertion (3) we use the fact that the set of all continuous functions is dense in $L^2(D)$, a fact which can be found in many texts on calculus. This implies immediately that if suffices to prove the assertion for the case of continuous functions. Indeed, let $\varepsilon > 0$ and then by the density of continuous functions mentioned above for p (and similarly for q) we find a continuous function, say h, with the property that $\|h - p\|_{L^2(D)} \leqslant \varepsilon$.
Then

$$D_r \|p - M_r p\| = \|p - h + h - M_r h + M_r h - M_r p\| \leqslant \varepsilon + \varepsilon + \|h - M_r h\|.$$

Now since we suppose without loss of generality that h is continuous on $\bar{D}$ it is in fact uniformly continuous and thus, for the given ε, from above we find $\delta(\varepsilon) > 0$ such that if $\mathrm{d}((x, y), (x', y')) \leqslant \delta$ then

$$|h(x, y) - h(x', y')| \leqslant \varepsilon.$$

Suppose that $r < \delta(\varepsilon)$ and thus

$$|M_r h(x, y) - h(x, y)| = \left| \iint_{D_{1-r}} (h(x + u, y + v) - h(x, y)) \, s_r(u, v) \, \mathrm{d}u \, \mathrm{d}v \right|$$

$$= \varepsilon \iint_{D_{1-r}} s_r(u, v) \, \mathrm{d}u \, \mathrm{d}v = \varepsilon.$$

Thus $\|h - M_r h\| \leqslant \varepsilon$ if $r < \delta$. This obviously implies that (3) holds.

The assertion in (4) follows by applying Fubini's theorem, and thus we omit it. The fact that S is a Riemann surface implies that in the space $L^2(S)$ there are some subspaces whose elements have important properties.

DEFINITION 17.8.8. The set of all continuous functions of class C^k with compact support is denoted by $C_0^k(S)$. We say that a differential form is in $C_0^k(S)$ if the coefficients a_i are in $C_0^k(S)$.

THEOREM 17.8.9. *The space $L^2(S)$ admits the following orthogonal decomposition*:

$$L^2(S) = D_0^e \oplus D_0^{co} \oplus M,$$

where D_0^e = the closure in $L^2(S)$ of the subspace of differentials of the form df *with f in $C_0^2(S)$; D_0^{co} = the closure in $L^2(S)$ of the subspace differential forms of the form* $*\mathrm{d}f$ *with f in $C_0^2(S)$*; and $M = (D_0)^7 4 \, (D_e^{co})^{\perp}$.

The differential forms on the form $*\mathrm{d}f$ are called coexact. From the form of M given above we see that M consists of differential forms which are orthogonal to D_0^e and to D_0^{co}.

Proof. First we show that D_0^e and D_e^{co} are orthogonal subspaces. From the definition it is clear that it suffices to prove the assertion for elements which generate the subspaces. Let w be an exact differential form of the form $\mathrm{d}f$ with f in $C_0^2(S)$ and let w_1 be a differential form of the form $*\mathrm{d}g$, where g is in $C_0^2(S)$. Then we have

$$\langle w, w_1 \rangle = \iint_S -\mathrm{d}f \wedge \mathrm{d}\bar{g} = -\iint_S f\,\mathrm{d} \wedge \mathrm{d}\bar{g} \qquad (*)$$

and since $\mathrm{d} \wedge \mathrm{d} = 0$ the assertion follows. □

To prove the assertion of the theorem it now suffices to show $M = (D_0^e \oplus D_0^{co})^\perp$. For this we consider a differential w in M and let w_1, w_2 be elements of subspaces which generate D_0^e and D_0^{co} respectively. Then we have that

$$\langle w, w_1 + w_2 \rangle = \langle w, w_1 \rangle + \langle w, w_2 \rangle$$

and, as in (*), we easily obtain that this number is zero. Thus we have the inclusion $M \subset (D_0^e \oplus D_0^{co})^\perp$.

For the converse inclusion we consider a differential form in $(D_0^e \oplus D_0^{co})^\perp$, say w, and thus for w_1 and w_2 in D_0^e and D_e^{co} respectively, we have

$$\langle w, w_1 \rangle = 0, \langle w, w_2 \rangle = 0$$

which imply the stated inclusion. Thus $M = (D_0^e \oplus D_e^{co})^\perp$.

In order to obtain more information about the properties of the elements of M we need the following result stated as

LEMMA 17.8.10. *Suppose that w is a differential form of class C^1. Then the following assertions hold*:

(1) $\mathrm{d}w = 0$ *if and only if w is in* (D_0^{co}).

(2) $\mathrm{d}(*w) = 0$ *if and only if w is in* $(D_0^e)^\perp$.

Proof. Suppose first that $\mathrm{d}w = 0$ and thus for any f in C_0^2 we have the relation

$$\langle w, *\mathrm{d}f \rangle = -\iint_S w\overline{(\mathrm{d}f)} = -\iint_S \bar{f}\,\mathrm{d}w = 0$$

If w_1 is in D_0^{co} and of the form $\mathrm{d}f$ with f in $C_0^2(S)$ then the above formula shows that $\langle w, w_1 \rangle = 0$. This implies obviously the fact that $w \in (D_0^{co})^\perp$. □

Suppose now that w is a differential form in (D_0^{co}). Then for all f in $C_0^2(S)$ we have that

$$\langle w, *\mathrm{d}f \rangle = 0 = \iint_S w\,\mathrm{d}\bar{f} = \iint_S \bar{f}\,\mathrm{d}w.$$

If $\mathrm{d}w$ is not zero then we find a disk sufficiently small such that on that disk $\mathrm{d}w = h(x, y)\,\mathrm{d}x\,\mathrm{d}y$. Then since f is arbitrary in the above formula we easily obtain a contradiction. Thus $\mathrm{d}w = 0$ and assertion (1) is proved. Since the proof of assertion (2) is similar it is omitted.

From this lemma we obtain an interesting property of the elements of M which are of class C^1.

COROLLARY 17.8.11. *The differential form w is in M if and only if w is a harmonic differential form.*

Proof. If w is harmonic then w is of class C^1 and $dw = d(^*w) = 0$. Then the above lemma implies that w is in M. Conversely, if w is in M and it is of class C^1 then $dw = 0$ and $d(^*w) = 0$, which gives that w is harmonic. □

The basic structure theorem for M is given in the following theorem.

THEOREM 17.8.12. *The differential form w is in M if and only if w is harmonic.*

For the proof we need the famous lemma of Weyl (1940). For interesting extensions of Weyl's Lemma we refer to Višik (1949).

Since the property to be of class C^1 is local, it is sufficient to consider only differential forms on a disk D in the complex plane, and in what follows we limit ourselves to this case. We consider the corresponding space $L^2(D)$ where we regard D as a Riemann surface. Then H. Weyl's lemma is as follows.

LEMMA 17.8.13 (Weyl, 1940). *Let w be a differential form in $L^2(D)$ and suppose that for all f in $C_0^2(D)$, $\langle w, df\rangle_D = 0$ and $\langle w, {}^*df\rangle_D = 0$. Then w is harmonic.*

Proof. Let w be a differential form with the above properties. In this case for any f in $C_0^2(D_r)$, $M_r w$ has the following properties:

(1) $\langle M_r w, df\rangle_{D_r} = \langle w, M_r\, df\rangle_D = \langle w, dM_r f\rangle_D = 0$

(2) $\langle M_r w, {}^*df\rangle_{D_r} = \langle w, M_r({}^*df)\rangle_D = \langle w, {}^*dM_r f\rangle_D = 0.$ □

Now $M_r w$ is in $C^1(D_r)$ and thus it is harmonic in D_r. Now we show that in fact $M_r w$ is independent of r, i.e. if r, r_1 are in $(0, \frac{1}{2})$ then $M_r w = M_{r_1} w$. If we consider $M_{r_1}(M_r w)$ (which is on D_{r+r_1}) to be of class C^1 it is thus harmonic there. This implies that $M_{r_1}(M_r w) = M_r w$ and the assertion follows if we show that the operators M_r commute (i.e. $M_r M_{r_1} = M_{r_1} M_r$). But this is an easy consequence of Fubini's theorem. Now we know that $\lim_{v\to 0} M_r w - w = 0$ since for any r, $D_{r+p} \supseteq D_{r+q}$ whenever $q < p$ we see that in D_{r+q}, $M_r w = w$. This implies that $M_r w$ is of class C^1 in D_r for any $r > 0$. This gives in fact that w is of class C^1 in D. Thus Weyl's lemma is proved, and at the same time Theorem 17.8.12.

From the above orthogonal decomposition of $L^2(S)$ we have that every differential form has a decomposition of the form:

$$w = w_e + w_{co} + w_h$$

with w_e in D_0^e (and called the exact component of w), w_{ce} in D_0^{co} (and called the coexact component of w) and w_n in M(and called the harmonic component of w).

Of course, it is of great interest to know more about the exact and coexact components of a differential form w. There are some sufficient conditions which assure that in fact these components are actually exact and coexact, for example, when these components are of class C^1. For a proof we refer to G. Springer's book (1957, Ch. 7, §3).

The results about the decomposition of $L^2(S)$ given above plays a fundamental role in establishing some existence theorems in the theory of Riemann surfaces, such as the existence of harmonic or analytic differential with prescribed poles of order 1 as well as prescribed residues, the fact that a Riemann surface has a countable basis, etc. For these and other results we refer to Springer (1957).

17.9. Differential Geometry in Complete Inner Product Spaces

In this section we intend to present an introduction to the differential geometry of curves and surfaces in complete inner product spaces. We note that the differential goemetry of curves and surfaces in inner product spaces is an incompletely developed chapter of mathematics although some interesting results have been obtained. Among the earliest contributors in extending the classical theory of curves to certain classes of infinite dimensional spaces were Giuseppe Vitali and the members of his school (whose results are in the remarkable book published by Vitali in 1929).

The most recent papers (to the best knowledge of the author) are those of Ravaska (1977) and Ragozin (1980) which are essentially dedicated to curve theory, and the beautiful paper of Ju. V. Titarenko on the Bonnet theorem and Codazzi's formulas for surfaces in complete inner product spaces.

We begin with some theory about curves in Banach spaces and than we specialize to the case of complete inner product spaces.

Let I be an open interval on R and let X be a Banach space. A function $f: I \to X$ is said to be continuous at $t = t_0$ of I if for any $\varepsilon > 0$ there exists $\delta(\varepsilon) > 0$ such that if $|t - t_0| \leqslant \delta(\varepsilon)$ then $\|f(t) - f(t_0)\| \leqslant \varepsilon$. We say that f is continuous on I if, at each point t of I, f is continuous.

The notion of derivative may be introduced as follows. We say that the function f is differentiable at $t = t_0$ if $\lim_{h \to 0} h^{-1}(f(t_0 + h) - f(t_0))$ exists. If this exists, then it is denoted by $f'(t_0)$. Similarly, to the classical case we define the derivative of higher order. As in the classical case we say that f is differentiable on I if f is differentiable at each point t of I.

The notion of integral may be defined following the model of real-valued functions, and we leave the reader to do this. Another way is as follows (which is sufficient for our purposes). If $f: I \to X$ is a continuous function then there exists F: $I \to X$ such that $F'(t) = f(t)$. Then we define the integral of f on I with respect to the Lebesgue measure $\mathrm{d}t$ by the formula

$$F(b) - F(a) = \int_a^b f(t)\,\mathrm{d}t.$$

Here we extended f to $\bar{I}$ by continuity.

The above integral (the Cauchy integral) has almost all the properties of the Cauchy integral for continuous functions with values in R (or C). For more details we refer to Dieudonné (1960).

DEFINITION 17.9.1. A parametrized differentiable curve in a Banach space X is a function $\alpha: I \to X$, where α is assumed to be of class C^m (i.e. α has derivative of all orders $i \leqslant m$ and these are continuous functions).

If α is a curve then $\alpha'(t)$ at the point t is called the tangent vector (or the velocity vector) of the curve α at the point t; the set $(\alpha(t), t \in I)$ is called the trace of the curve α.

EXAMPLE 17.9.2. Two parametrized differentiable curves have the same trace. Let us consider the complete inner product space $X = l^2$ of all sequence of real numbers $x = (x_n)$ such that

$$||x||^2 = \sum_{i=1}^{\infty} |x_i|^2 < \infty.$$

We consider the following curves:

$$i: (-\varepsilon, 2\pi + \varepsilon) \to l^2$$

with

$$\alpha_1(t) = (\cos t, \sin t, 2^{-3}, 2^{-4}, \dots, 2^{-n}, \dots)$$
$$\alpha_2(t) = (\cos 2t, \sin 2t, 2^{-3}, 2^{-4}, \dots, 2^{-n}, \dots)..$$

It is clear that these are parametrized differentiable curves with the same trace and clearly they are different curves.

DEFINITION 17.9.3. If $\alpha: I \to X$ is a parametrized differentiable curve then we say that the point $t = t_0$ of I is an ordinary point for the curve if $\alpha'(t_0) \neq 0$; in the contrary case, t_0 is called a singular point.

Suppose now that $\alpha: I \to X$ is a parametrized differentiable curve and t_0 is a point of I; then we can consider the line $\alpha(t_0) + \alpha'(t_0)(t - t_0)$, which is called the tangent line to the curve α at the point $t = t_0$.

We now consider another very important notion for curves, namely the notion of arc length. Suppose that $\alpha: I \to X$ is a parametrized differentiable curve in a Banach space X and suppose that we introduce a new parameter using a differentiable (real-valued) function $t = t(\tau)$. Then clearly we have the curve

$$\beta: J \to X$$

where

$$t: J \to I$$
$$\beta: J \to X, \quad \beta(\tau) = \alpha(t(\tau)).$$

Clearly

$$\beta'(\tau) = \alpha'(t(\tau))\, dt/d\tau$$

which means that the tangent vector for β has the same direction as the old tangent vector of α, and clearly they differ only from a numerical constant.

Suppose now that the parametrized differentiable curve is given and we are looking at the length of the arc of curve corresponding to two points of the interval I. We may consider an initial point and then we compute the length of the arc joining this initial point with the current point corresponding to the parameter t. Then clearly the length is

$$s(t) = \int_a^t ||\alpha'(u)||\, du.$$

The function $t \to s(t)$ defined on I is called the arc length of the curve α. It is obvious that

$$ds(t)/dt = \|\alpha'(t)\|$$

since the function $t \to \alpha'(t)$ is continuous (and this obviously implies the continuity of the function $t \to \|\alpha'(t)\|$).

Suppose now that the curve has only ordinary points. Then clearly the function $s \to s(t)$ is a continuously differentiable increasing function which has an inverse being continuously differentiable and increasing, $t = t(s)$. Then clearly we have the relation

$$dt/ds = (ds/dt)^{-1}.$$

Thus in this case we can choose an arc length s as the original parameter. If this is the case then s is called the natural parameter of the curve.

We remark that if X is a complete inner product space with an orthonormal basis $(e_i)_{i\in I}$ then $\alpha: I \to X$ has the form

$$\alpha(t) = \sum_{i\in I} \alpha_i(t) e_i$$

and the formula for the arc length from the points corresponding to $t = t_0$ as initial point to the current point is

$$s(t) = \left(\int_{t_0}^{t} \sum_{i\in I} |\alpha_i'(u)|^2 \, du \right)^{1/2},$$

where $\alpha_i'(\,)$ denotes the derivative of the function $\alpha_i(t)$.

Suppose now that the parametrized curve α is parametrized with the natural parameter s, the arc length. Associated with the curve $\alpha: I \to X$ is the curve

$$\beta: -I \to X, \quad I = (a, b) \quad \text{then} \quad -I = (-b, -a)$$

defined by the formula

$$\beta(-s) = \alpha(s).$$

Of course, this curve has the same trace as curve α, but is clearly described in the opposite direction. We say that the curves α and β differ by a change of orientation.

We now present the notion of osculating subspaces associated to a curve defined in a Banach space X. We assume that the parametrized differentiable curve $\alpha: I \to X$ is of class C^{m+1}. We consider an arbitrary point t_0 of I and then, by Taylor's formula, we have the relation

$$\alpha(t) = \alpha(t_0) + \alpha'(t_0)/1!\,(t - t_0) + \alpha''(t_0)/2!\cdot(t - t_0)^2 + \cdots +$$
$$\alpha^{(m)}(t_0)/m!\cdot(t - t_0) + \varepsilon_m(t)$$

with

$$\|\varepsilon_m(t)\| \leqslant |(t - t_0)|/(m + 1)!\|\alpha^{(m+1)}(t)\|.$$

Thus $\varepsilon_m(t) \to 0$ when $t \to t_0$.

From the above formula we see that the curve α lies in the subspace determined by $\alpha(t_0), \ldots, \alpha^{(m)}(t_0)$ with an accuracy up to terms of order m.

DEFINITION 17.9.4. The subspace generated by the elements $\alpha'(t_0), \ldots, \alpha^m(t_0)$ is called the osculating subspace O_m to the curve α at t_0.

We assumed in the above definition that the elements $\alpha'(t_0), \ldots, \alpha^m(t_0)$ are linearly independent. Suppose now that we introduce a new parameter $t = t(\tau)$. Then clearly we have

$$d\alpha/d\tau = d\alpha/dt \cdot dt/d\tau$$

and thus

$$d^2\alpha/d\tau^2 = d\alpha^2/dt^2 \cdot (dt/d\tau)^2 + d\alpha/dt \cdot (d^2 t d\tau^2).$$

This gives that the element $d^2\alpha/d\tau^2$ lies in the plane determined by the elements $d\alpha/dt$ and $d^2\alpha/dt^2$, which is thus independent of the choice of the parameter. This fact can also be proved for the higher derivative.

PROPOSITION 17.9.5. *For any integer i the element* $d^i\alpha/d\tau^i$ *lies in the subspace generated by* $d\alpha/dt$, $d^2\alpha/dt, \ldots, d^i\alpha/dt^i$.

Proof. For $i = 2$ this fact was noted above. Suppose now that it holds for all $j \leqslant i$ and we prove that it holds for $j + 1$.

Indeed, if

$$d^j\alpha/d\tau^j = \sum_{k=1}^{j} d^k\alpha/dt^k \varphi_k(\tau)$$

then, differentiating this relation, we get

$$d^{j+1}\alpha/d\tau^{j+1} = \sum_{k=1}^{j} d^{j+1}\alpha/dt^{j+1} \cdot dt/d\tau\, \varphi_k(\tau) + \sum_{k=1}^{j} d^j\alpha/dt^j \cdot d\varphi_k(\tau)/d\tau$$

and the assertion follows.

It follows from the above facts that the osculating subspace is independent of the parameter and thus justifies, in part, the name of osculating subspace for the subspace generated by the elements $\alpha'(t_0), \ldots, \alpha^{(m)}(t_0)$.

DEFINITION 17.9.6. The subspace generated by $\alpha'(t_0)$, $\alpha''(t_0)$ and $\alpha'''(t_0)$ is said to be a three-dimensional osculating subspace of the parametrized curve α at the point $t = t_0$ (we assume, of course, that the above elements are linearly independent).

Suppose now that $t = t_0$ is a singular point for the parametrized curve α, i.e. $\alpha'(t_0) = 0$. Suppose further that this singular point has the property that $\alpha''(t_0) \neq 0$. Then by Taylor's formula, in a neighborhood of t_0 we have the relation

$$\alpha(t) = \alpha(t_0) + \alpha''(t_0)/2!(t - t_0)^2 + \alpha'''(t_0)/3!(t - t_0)^3 + \varepsilon_3(t)$$

with $\lim_{t \to t_0} \varepsilon_3(t) = 0$. From this formula we see that our parametrized curve, with an accuracy of order 3, lies in the plane determined by $\alpha''(t_0)$ and $\alpha'''(t_0)$ and that there it has the equation $y = x^{3/2}$ corresponding to a cusp at the point $\alpha(t_0)$.

As is well known, one of the most interesting quantities associated with a curve in the space R^3(R is the set of real numbers) is the so-called curvature which is defined as follows.

DEFINITION 17.9.7. Suppose that $\alpha: I \to R^3$ is a parametrized differentiable curve parametrized by the arc length. Then the curvature of α at the point s is the number

$$k(s) = \|\alpha''(s)\|.$$

In what follows we define the analogue notion for parametrized curves in complete inner product spaces. We assume now that X is a complete inner product space and that $\alpha: I \to X$ is a parametrized differentiable curve. Then we have the following simple result.

PROPOSITION 17.9.8. *If $t \to \|\alpha(t)\|$ is a constant function then the elements $\alpha(t)$ and $\alpha'(t)$ are orthogonal.*

Proof. This follows from the following fact: if $u(t)$, $v(t)$ are two differentiable functions defined on I with values in X then $t \to \langle u(t), v(t)\rangle$ is a differentiable function and the following formula holds:

$$\langle u(t), v(t)\rangle' = \langle u'(t), v(t)\rangle + \langle u(t), v'(t)\rangle.$$

We leave the reader to prove this fact. □

Suppose now that the parametrized differentiable curve $\alpha: I \to X$ is parametrized with the arc length measured from i fixed point. Then

$$e_1(s) = \alpha'(s)$$

has a length equal to 1 (and assuming that the elements $\alpha'(s)$, $\alpha''(s)$ are linearly independent) we get that

$$e_1''(s) = \alpha''(s)$$

is in the osculating plane and is orthogonal to $e_1(s)$. Thus we have

$$e''(s) = k(s)e_2(s),$$

where $e_2(s)$ is a unit element orthogonal to $e_1(s)$. The number $k(s)$ in the above relation is called the curvature of the curve α at the point s.

We give now a formula for the curvature $k(t)$ in the case of an arbitrary parametrization of the curve.

Since we have the relation

$$ds/dt = \|d\alpha/dt\| = \langle d\alpha/dt, d\alpha/dt\rangle$$

and

$$k(s) = \lim_{h\to 0} h^{-1}\|e_1(s+h) - e_1(s)\|$$

we get

$$d^2s/dt^2 = \langle d^2\alpha/dt^2, d\alpha/dt\rangle/\|d\alpha/dt\|.$$

But since

$$d\alpha/ds = d\alpha/dt \cdot dt/ds$$

we have that

$$\begin{aligned}
\mathrm{d}^2\alpha/\mathrm{d}s^2 &= \mathrm{d}^2\alpha/\mathrm{d}t^2(\mathrm{d}t/\mathrm{d}s)^2 + \mathrm{d}\alpha/\mathrm{d}t\cdot \mathrm{d}^2t/\mathrm{d}s^2 \\
&= \mathrm{d}^2\alpha/\mathrm{d}t^2/(\mathrm{d}s/\mathrm{d}t)^2 + \mathrm{d}\alpha/\mathrm{d}t\cdot \mathrm{d}/\mathrm{d}t((\mathrm{d}s/\mathrm{d}t)^{-1})\cdot(\mathrm{d}s/\mathrm{d}t)^{-1} \\
&= \mathrm{d}^2\alpha/\mathrm{d}t^2/(\|\mathrm{d}\alpha/\mathrm{d}t\|)^2 - \mathrm{d}\alpha/\mathrm{d}t\cdot(\mathrm{d}s/\mathrm{d}t)^{-2}\mathrm{d}^2s/\mathrm{d}t^2\cdot(\mathrm{d}s/\mathrm{d}t)^{-1} \\
&= \mathrm{d}^2\alpha/\mathrm{d}t^2\cdot\|\mathrm{d}\alpha/\mathrm{d}t\|^{-2} - \mathrm{d}\alpha/\mathrm{d}t\cdot\langle \mathrm{d}\alpha/\mathrm{d}t, \mathrm{d}^2\alpha/\mathrm{d}t^2\rangle\cdot\|\mathrm{d}\alpha/\mathrm{d}t\|^{-4}
\end{aligned}$$

or

$$k(s) = \|\mathrm{d}^2\alpha/\mathrm{d}s^2\| = \|\mathrm{d}^2\alpha/\mathrm{d}t^2\cdot\|\mathrm{d}\alpha/\mathrm{d}t\|^{-2} - \mathrm{d}\alpha/\mathrm{d}t\cdot\langle \mathrm{d}\alpha/\mathrm{d}t, \mathrm{d}^2\alpha/\mathrm{d}t^2\rangle\|\mathrm{d}\alpha/\mathrm{d}t\|^{-4}\|$$

Suppose now that $\alpha: I \to X$ is a parametrized curve parametrized by the arc length and has at the point s an osculating plane. Suppose, further, that C is a circle of radius r lying in the osculating plane and tangent to the curve at the point s. In this case we show that we can choose a circle such that the accuracy with which the curve is in the osculating plane is of order 3.

Indeed, if $\alpha_1(s)$ represents the equation of the circle then

$$\alpha(s) = \alpha(s_0) + e_1(s_0)(s - s_0) + k_1(s_0)/2!e_2(s)(s - s_0)^2 + \varepsilon_2(s)$$
$$\alpha_1(s) = \alpha(s_0) + e_1(s_0)(s - s_0) + (2r)^{-1}e_2(s)(s - s_0)^2 + \tilde{\varepsilon}_2(s)$$

with

$$\lim_{s\to s_0} \varepsilon_2(\mathrm{s}) = \lim_{s\to s_0} \tilde{\varepsilon}_2(\mathrm{s}) = 0.$$

Now if we choose $(2r)^{-1} = k(s_0)/2$, the accuracy with which the curve lies in the osculating plane is clearly of order 3.

The center of the above circle is called the center of curvature of the curve α at the point s_0 and the radius of the above circle is called the radius of curvature of the curve at the point s_0.

REMARK 17.9.9. It is clear that if the curve is a circle then the above curvature coincides with the inverse of the radius.

One of the very useful formulas in the theory of curves (in R^3) is the set of formulas known as the formulas of Frénet–Serret. In what follows we present an analogue of these formulas.

Let $\alpha: I \to X$ be a parametrized differentiable (of class C^∞) curve. We suppose that for each s (the curve is parametrized by the arc length parameter) the set of elements in X, $(\alpha^m(s))$ is linearly independent. Then we consider for each n the subspace $E_n(s)$ which is generated by the elements $(\alpha^{(m)}(s), m \leqslant n)$. From the linear independence it follows that for each n we find a set of orthonormal elements $(e(s), \ldots, e(s))$ which generates $E_n(s)$. Clearly we have

$$e(s) = f_1(s)^{(1)}(s) + \cdots + f_n(s)^{(n)}(s), \quad j = 1, 2, \ldots, n,$$

where the functions f_i are differentiable.

From the form of e^j we get immediately that $de^j(s)/ds$ is in $E_{n+1}(s)$. Obviously

$$de^n(s)/ds = \sum_{i=1}^{n+1} a_{ni}e(s),$$

where $a_{i\,i+1}(s) > 0$ for all i and $a_{ii}(s) = 0$.

Since $\langle e_i(s), e_j(s)\rangle = \delta_{ij}$ we get the formula

$$\langle e_i'(s), e_j(s)\rangle + \langle e_i(s), e_j'(s)\rangle = 0$$

which implies that

$$a_{ij} = \langle e_i'(s), e_j(s)\rangle = -\langle e_i(s), e_j'(s)\rangle$$

because the element $e_i'(s)$ is in $E_{i+1}(s)$. This implies, further, that

$$\langle e_j'(s), e_i(s)\rangle = 0$$

for $j < i - 1$ and then

$$e_i'(s) = a_{i\,i-1}(s)e_{i-1}(s) + a_{i\,i+1}(s)e_{i+1}(s).$$

But we have also

$$a_{i\,i-1}(s) = \langle e_i'(s), e_{i-1}(s)\rangle = -\langle e_{i-1}'(s), e_i(s)\rangle = -a_{i-1\,i}.$$

Associated with the curve $\alpha: I \to X$ (parametrized with the arc length) we define the following functions:

$$(k_i(s))_{i=1}^{\infty}$$

where

$$\begin{aligned} k_1(s) &= k_1 = \langle e_1'(s), e_2(s)\rangle, \\ k_2(s) &= k_2 = \langle e_2'(s), e_3(s)\rangle, \\ &\vdots \\ k_n(s) &= k_n = \langle e_n'(s), e_{n+1}(s)\rangle, \\ &\vdots \end{aligned}$$

From the formulas for $e_i'(s)$ we have that

$$\begin{aligned} e_1'(s) &= k_1 e_2(s), \\ e_2'(s) &= -k_1 e_1(s) + k_2 e_3(s), \\ &\vdots \\ e_i'(s) &= -k_{i-1} e_{i-1}(s) + k_i e_{i+1}(s) \\ &\vdots \end{aligned}$$

These formulas are called the Frénet–Serret formulas for a Frénet curve, i.e. a curve such that $(\alpha^{(m)}(s))$ are linearly independent for each s.

The set of elements $(\alpha^m(s))$ is called the Frénet frame of the curve and the functions $(k_i(s))$ are called the curvatures. We also remark that the functions $k_i(s)$ are positive.

Using curvatures we can consider several classes of curves. One of the most important and interesting classes of curves defined using curvatures is the class of helices.

DEFINITION 17.9.10. A helix is a parametrized differential curve with the property that all the curvatures are constant. For example, in R^3 a helix is the curve defined by

$$\alpha:(-\pi,\pi)\to R^3$$
$$\alpha(t)=(a\cos t, a\sin t, bt).$$

Another class of curves is defined as follows.

DEFINITION 17.9.11. (G. E. Shilov). A curve is said to be self-congruent if it can be transformed into itself by a rigid motion which preserves the direction of increase of the parameter.

We now present an important example of such a curve. We consider for this purpose the space $X = L^2_{[0,\infty)}$ with respect to the Lebesgue measure. The curve is defined as follows:

$$t\to Z(t):(0,\infty)\to X,$$

where

$$Z(t)(s)=\begin{cases}1, & 0<s\leqslant t\\ 0, & \text{otherwise.}\end{cases}$$

DEFINITION 17.9.12. The curve $t\to Z(t)$ is called the Wiener spiral. If we set

$$T_s x(t) = x(s-t) + Z(s)(t)$$

then we get that the Wiener spiral is a self-congruent curve.

For some results concerning the determination of a curve if the curvatures $(k_i(s))$ are given, we refer to the papers mentioned in the introduction.

Also for a connection between certain curves and the geometry of the Banach space, we refer to Chapter 4.21 where we mention the interesting result of J. J. Schaeffer (1965) about the characterization of inner product spaces using the connection between a curve and its radial projection on $(x, x\in X, \|x\|=1)$ of the space.

We present now some facts concerning surfaces. For the reader's convenience we recall the notion of surface in R^3 as well as some related notions.

DEFINITION 17.9.13. A subset S in R^3 is said to be a regular surface if for each p in S we find a neighborhood V in R^3 and a mapping $x: U\to V\cap S$ for an open set $U\subset R^2$ such that the following properties are satisfied:

(1) x is a differentiable mapping.
(2) x is a homeomorphism (i.e. $x^{-1}: V\cap S\to U$ exists and is continuous).
(3) For each u in U, dx_u is one-to-one.

For examples and motivation we refer to books on differential geometry.

As the reader knows, for the local properties it is convenient to define the notion of a parametrized surface. This is given below.

DEFINITION 17.9.14. A parametrized surface is a mapping $x: U \to R^3$, U is an open set in R^2, which is differentiable. The set $(x(u), u \in U)$ is called the trace of the surface.

We say that the surface is regular if, for each u in U, the differential of x at u is an injective mapping; a point u for which dx_u is not injective is said to be a singular point of the surface.

Suppose that $x: U \to R^3$, U is an open set in R^2, is a parametrized surface. Then for each u in U we consider the set $dx_u(R^2)$ which is a plane in R^3 which passes through $x(u)$. This is called the tangent plane to S at $s(u)$, and is denoted by $T_w(S)$, $w = x(u)$. It is not difficult to see that the plane $T_w(S)$ does not depend on the parametrization of the surface.

We now consider the usual inner product in R^3 and then for p, q in $T_w(S)$ we can compute the inner product (as elements of R^3), which we denote by $\langle p, q\rangle_w$, and the quadratic form

$$I_w(p) = \langle p, p\rangle_w = \|p\|_w^2 \geqslant 0$$

defined on $T_w(S)$ is called the first fundamental form of the regular surface S.

We now introduce, following Ju. V. Titarenko, the notion of regular surfaces in real complete inner product spaces. Suppose that X, Y are two complete inner product spaces over the reals.

DEFINITION 17.9.15 (Titarenko, 1971). A mapping $x: U \to Y$, U is an open set in X, is said to be a regular surface if:

(1) x is of class $C^3(U, Y)$,
(2) for each u in U, dx_u (the Fréchet derivative of x at u) is invertible.

The subspace $dx_u(X)$ is called the tangent subspace to x at $u \in U$. We say that the regular surface x is an hypersurface if $(dx_u)^\perp$ is one-dimensional for each u in U. A unit vector (i.e. of norm 1) is said to be the normal vector to the hypersurface S (at $x(u)$). Now the first fundamental form of the hypersurface S at the point $u \in U$ is $\langle dr, dr\rangle$. Associated with this is the operator

$$G = (dx_u)^* \, dx_u$$

which is Hermitian (and defined on X).

The second fundamental form of the hypersurface S is defined by $\langle dn(u), dr\rangle^*$, where $n(u)$ is the normal vector to the surface at $x(u)$. Associated with this form is the operator

$$B(u): X \to X, \; B(u) = (dx_u)^* n'(u) = (dx_u)^* A \, dx_u.$$

The operator A in this formula is called the Rodrigues operator (after Olinde Rodrigues). We note that the operator G is also Hermitian.

Let us consider a fixed point u_0 in U and then

$$Y = dx_u(X) \oplus Rn(u).$$

Using this decomposition, Titarenko obtains some interesting formulas which are the analogue of the classical Codazzi formula. It is proved that, conversely, Codazzi's formulas (in the formulation of Titarenko) are sufficient, given the operators B and G, to assure the existence of a hypersurface having exactly, for the first and second fundamental forms, the objects given in the above formulas. For details we refer to the paper of Titarenko (1971).

17.10. Univalent Functions and Complete Inner Product Spaces

Let $f: D \to C$ be a function. We say that f is univalent in D if whenever $u \neq v$ are in D then $f(u) \neq f(v)$. An intensively studied class of univalent function is the so-called class S which consists of all univalent and analytic functions in the unit disk $(z, |z| < 1)$ normalized by the conditions:

(1) $f(0) = 0$,
(2) $f'(0) = 1$.

Thus every f in S has a Taylor development of the form

$$f(z) = z + a_2 z^2 + a_3 z^3 + \cdots$$

valid for $|z| < 1$.

Concerning this class of function one of its most interesting problems is the validity of the inequalities

$$|a_n| \leqslant n, \quad n = 2, 3, 4, 5, \ldots.$$

For partial results we refer to the 1983 book by P. L. Duren. In Notices A. M. S. 32 (1985), 2–6, it is asserted that the conjecture was settled in the positive.

Closely related to the class S is the class denoted by Σ and which consists of all functions of the form

$$h(z) = z + b_0 + b_1 z^{-1} + b_2 z^{-2} + \cdots$$

which are analytic and univalent in $(z, |z| < 1)$, except for a simple pole at infinity with residue 1.

Concerning the class Σ, there are some inequalities due to Gunsky and Goluzin (1947) and Lebedev (1951). We shall present shortly the derivation of these inequalities following the method given by Jenkins *et al.* (1974) using the inner product structures. For more details the reader is referred to the original paper. Also for some operator theoretic remarks in the case of class Σ we refer to Duren (1983).

In the class Σ we consider the subclass Σ_0 which consists of all g in the class Σ for which the image domain is a slit D domain with a slit s, s is a point set, the impression of ∂D which is the set of prime ends. For the theory of prime ends we refer to the 1983 papers of D. B. Epstein.

Further, let us consider $v: s \to C$ to be holomorphic on s, $\infty \neq z \in D$. Take t in s and we consider the branch of $\log(a - z)$ which is holomorphic on s and

$$\int_{\partial D} \log(t - z)\, dv(t) = 0.$$

We denote by C_1 the unit circle described negatively and set $u = v \circ g$, which is a function on C_1. The property of u which is important in what follows is that the set of all pairs on C_1 identified by u contains the pairs identified by g. The converse assertion is also true, in the sense that for any $u: C_1 \to C$ which, with finitely many exceptions, identifies pairs of points of C_1 there exists a decomposition

$$u = u_1 + u_2, \qquad (+)$$

where u_1 may be extended in the unit disk to an analytic function and u_2 to an analytic function in the set $(z, |z| > 1)$.

We consider the set of all functions

$$B = (u, u = u_1 + u_2),$$

where u_i are as in the above decomposition. We may suppose without loss of generality that

$$u_1(0) = u_2(\infty) = 0.$$

If

$$K_S(z, \xi) = \log((g(z) - g(\xi))/(z - \xi))$$

is the Schiffer kernel then

$$u_2(\xi) = (2\pi i)^{-1} \int_{C_1} K_S(z, \xi)\, du_1(z).$$

Now for φ, ψ arbitrary elements in B we set

$$\langle \varphi, \psi \rangle = (2\pi)^{-1} \int_{C_1} \varphi(z)\bar{\psi}(z)\, dz$$

which is an inner product on B. We denote by $\bar{B}$ the completion of B with respect to the norm induced by the above inner product. From the definition of the functions u_1 and u_2 we have

$$\langle u_1, u_2 \rangle = 0.$$

Now if φ is in B then we define the following functions:

$$(T\varphi)(z) = (2\pi i)^{-1} \int_{C_1} K_S(z, \xi)\, du(\xi),$$

$$(T_1\varphi)(z) = (2\pi i)^{-1} \int_{C_1} \bar{K}_S(z, \xi)\, du(\xi).$$

The basic properties of the mappings T and T_1 are:

(1) for any φ, ψ in B, $\langle T\varphi, T\psi \rangle = \langle \varphi, \psi \rangle$,
(2) $TB = B$,
(3) for any u in B we have $TT_1 u = T_1 T u = u$.

Now applying the Cauchy–Buniakowsky inequality we get the following inequality:

$$|\langle Tu_1, u_2\rangle|^2 \leqslant \langle u_1, u_1\rangle \cdot \langle u_2, u_2\rangle. \qquad (++)$$

Now if we choose u_2 such that for $z = 1$,

$$u_2(z) = \bar{u}_1(z)$$

then $(+)$ gives that

$$|\langle Tu_1, \bar{u}_1\rangle| \leqslant \langle u_1, u_1\rangle. \qquad (+++)$$

The evaluation of the power series coefficients gives the Gunsky inequalities.

If in $(++)$ we set

$$u_1(z) = \sum_{n=1}^{\infty} a_n \log(1 - t_n^{-1} z), \quad a_n \in C, \quad |t_n| > 1$$

then we get the Golusin inequalities.

REMARK 17.10.1. Since there are various generalizations of the Cauchy–Buniakowsky inequality it is of interest to know if we can obtain some inequalities for the functions in class Σ.

17.11. Complete Inner Product Spaces and Roots of Polynomials (and Analytic Functions)

The localization of roots of polynomials is an old and important problem. It has attracted the attention of many famous mathematicians, among whom we mention A. L. Cauchy, E. Laguerre, P. Montel and J. L. Walsh. For a detailed bibliography, as well as for further information, we refer to the paper of van Vleck (1929) and the books of Parodi (1959) and Marden (1966) and the author's note (forthcoming).

In this section, following an interesting paper by Ifantis and Kouris (1973), we present an approach to some results in the theory of localization of roots of polynomials and analytic functions based on certain operators on a complete inner product space.

The basic idea in the paper of Ifantis and Kouris is to use certain operators in the Hardy space H^2 for the unit disk (with respect to the Lebesgue measure). There are many ways, as we also noted in another chapter, to look at the space H^2. Here we consider H^2 to be the space of all complex-valued functions in the unit disk $(z, |z| < 1)$ with the property that if

$$f(z) = \sum_{0}^{\infty} a_i z^i$$

then

$$\|f\|^2 = |a_0|^2 + |a_1|^2 + \cdots + |a_n|^2 + \cdots$$

is finite.

We now consider a separable complete inner product space X with the orthonormal basis $(e_i)_1^{\infty}$. Then we can consider the shift operator associated with this basis, defined by the relations:

$$Se_i = e_{i+1}, \quad i = 1, 2, 3, \ldots .$$

Then clearly S is an isometric operator and satisfies the inequality

$$\|S^*x\| \leqslant \|Sx\|$$

for all x in X. This means that S is a hyponormal operator. Thus we can conclude that, for S, Weyl's theorem holds, i.e. the essential spectrum of S is equal to the spectrum of S from which we delete isolated eigenvalues of finite multiplicity. In what follows we need such a result for a compact perturbation of S and thus we know that the above assertion concerning the essential spectrum holds for such a perturbation.

From the definition of S it follows that the adjoint S^* is given by the formulas:

$$S^*e_1 = 0, \quad \text{and} \quad S^*e_i = e_{i-1}, \quad i = 2, 3, \ldots .$$

Suppose now that z is a complex number in the unit disk and consider the element in X of the form

$$x_z = \sum_0^\infty z^i e_{i+1}$$

and we remark that x_z is an eigenelement of S^*.

We now consider the following function defined on the unit disk:

$$z \to \langle x_z, x \rangle$$

for x in X. Then it is clear that

$$x \to \langle x_z, x \rangle = f(z)$$

is an isometric isomorphism of X onto H^2 corresponding to the unit disk.

Suppose now that h is an element in X of the form

$$h = \sum_1^\infty h_i e_i$$

with (h_i) in l^2. We consider h a fixed element in X. We consider, associated with h, the following operator on X,

$$Cx = \langle x, h \rangle e_1$$

which is obviously a bounded compact operator. Then we set $T = S + C$ which is a compact perturbation of the shift operator S. We remark that for each z in the unit disk we have the relation

$$zf(z) = \langle x_z, Sx \rangle.$$

Also, we have that

$$f(z) - f(0) = z\langle x_z, S^*x \rangle$$

and

$$\langle x_z, C^*x \rangle = \langle Cx_z, x \rangle = f(0) \sum_{n=1}^{\infty} \bar{h}_n z^{n-1}$$

which implies that the problem $Tx = \bar{s}x$ is equivalent with the problem of solving the equation

$$f(z) - f(0) + f(0) \sum_{n=1}^{\infty} \bar{h}_n z^n = szf(z)$$

in H^2. This formula may be written as

$$f(z) = f(0)\left(1 - \sum_{n=1}^{\infty} \bar{h}_n z^n\right) = -f(0)P(z)$$

when $s = 0$ and

$$f(z) = f(0)\left(1 - \sum_{n=1}^{\infty} \bar{h}_n z^n\right)\Big/(1 - sz) = f(0)P(z)(s(z - s^{-1}))^{-1}$$

when $s \neq 0$, with

$$P(z) = \sum_{1}^{\infty} \bar{h}_n z^n - 1.$$

It is obvious that every element of H^2 can be normalized to such an element like $P(z)$ for suitable h_n.

Using the above formulas we can prove the following result.

THEOREM 17.11.1. *The function $P(z)$ has roots in the unit disk* $(z, |z| < 1)$ *if and only if the operator T^* (the adjoint of T) has an eigenvalue $\bar{s}$ with* $|s| > 1$.

Proof. Suppose that s is a root of $P(z)$ and $|s| > 1$. In this case $\bar{s}^{-1} = u$ is an eigenvalue for T which follows from the formula for $f(z)$. Conversely, suppose that such a u exists with $|u| \leqslant 1$ and then $f(z)$, defined by the above formula, is in H^2 and clearly $p(s)$ (s is defined by the above equality) is zero. This proves the theorem. □

Another result useful for our purpose is:

THEOREM 17.11.2. *The following relation holds:*

$$(z, |z| < 1) \subseteq \sigma_p(T^*).$$

Proof. We recall that $\sigma_p(\ )$ means the point spectrum of ( ). Now from the definition of S we see that $S^*S = I$ and from the form of T we get that

$$S^*T = S^*S + S^*C = S^*S = I.$$

From this we obtain immediately that every point z is the (open) unit disk satisfies the relation

$$(T^* - z)S = z(z^{-1} - S)$$

(we assumed that $z \neq 0$, if $z = 0$ the assertion of the theorem is clear). Since z^{-1} is in the resolvent of S, from the above formula the assertion of the theorem follows. □

The following is a related result.

THEOREM 17.11.3. *If s is in $\sigma(T^*)$ and $|s| < 1$ then s is in $\sigma_p(T^*)$.*

Proof. We remark that the operator S is hyponormal and thus Weyl's theorem holds for S (as well as for any compact perturbation of it). This clearly implies the assertion of the theorem.

COROLLARY 17.11.4. *For the operator T the following inequalities hold:*

$$r_{T^*} = r_T \geqslant 1.$$

Proof. Since the essential spectrum of T is $(z, |z| \leqslant 1)$ the assertion follows. □

COROLLARY 17.11.5. *A lower bound ρ for the roots of the function $P(z)$ is*

$$\rho \geqslant r_T^{-1} = r_T^{-1}.$$

Proof. Suppose that s is a root of $P(z)$ with $|s| < 1$. Then $\bar{s}^{-1}$ is an eigenvalue for T and thus

$$1/|s| \geqslant r_T = r_T.$$

Conversely, if s is a root and $|s| < 1$, then $1/|s| \geqslant 1$ and

$$1/|s| \geqslant r_T = r_T$$

which completes the proof of the corollary.

An estimate of the lower bound for the roots of $P(z)$ can be obtained using the norm of T (or T^*). Since

$$\|T^*\| = \|S^* + C^*\| = \sup_{\|x\| \leqslant 1} (\|S^*x + C^*x\|) = \sup_{\|x\| \leqslant 1} (\|S^*x + \overline{\langle e_1, x\rangle} h\|)$$

$$= \sup(\big|\|S^*x\|^2 + |\langle x, e_1\rangle|^2\|h\|^2 + \langle S^*x, \overline{\langle e_1, x\rangle} h\rangle +$$

$$+ \langle \overline{\langle e_1, x\rangle} h, S^*x\rangle\big|)^{1/2} \leqslant (1 + \|h\|^2)^{1/2}$$

and because, for $x = e_1$, we have an equality, this gives that the norm of T^* is equal to $(1 + \|h\|^2)^{1/2}$ which, further, implies that a lower bound for the roots of $P(z)$ is $1/(1 + \|h\|^2)^{1/2}$.

COROLLARY 17.11.6. *The function $P(z)$ has at least one root in an (open) unit disk if and only if $r_T > 1$.*

Proof. The complex number $\bar{s}$ with $|s| > 1$ is an eigenvalue for T^* if and only if s^{-1} is a root of $P(z)$ and thus the assertion of the corollary follows. □

COROLLARY 17.11.7. *The function $P(z)$ has no roots in an (open) unit disk if and only if $r_T = 1$.*

From this we have another useful result, given as

COROLLARY 17.11.8. *If T and S are similar operators then $P(z)$ has no roots in an (open) unit disk.*

We are now in a position to give some applications to the problem of location of roots for polynomials.

Let us consider the polynomial $p(z)$ defined by

$$p(z) = a_0 + a_1 z/+ \cdots + a_n z^n$$

with $a_0 + 0 + a_n$.

Associated with this polynomial we consider the polynomial $r(z)$, defined by the formula

$$r(z) = a_n + a_{n-1} z + \cdots + a_1 z^{n-1} + a_0 z^n$$

which has the property that the roots of r are the inverse of the roots of p.

We now consider, on the complete separable inner product space X, two compact operators C_i, $i = 1, 2$, defined as follows:

$$C_1 x = \langle x, h_1 \rangle e_1, \qquad C_2 x = \langle x, h_2 \rangle e_1$$

where h_i, $i = 1, 2$, are given by the formulas:

$$h_1 = \bar{a}_0^{-1}(\bar{a}_1 e_1 + \bar{a}_2 e_2 + \cdots + \bar{a}_n e_n)$$
$$h_2 = \bar{a}_n^{-1}(\bar{a}_{n-1} e_1 + \bar{a}_{n-2} e_2 + \cdots + \bar{a}_0 e_n).$$

From the above results we have the following theorem:

THEOREM 17.11.9. *The polynomial p has all its roots in an (open) unit disk if and only if*

$$r_{S-C_1} = r_{S-C_2} = 1.$$

Suppose now that p is a symmetric polynomial, i.e.

$$p(z) = 1 + a_1 z + a_2 z^2 + \cdots + a_1 z^{n-1} + z^n = s(z)$$

(i.e. $a_i = a_{n-1}$). Then clearly $h_1 = h_2$ and thus $C_1 = C_2$. Then we have:

COROLLARY 17.11.10. *A symmetric polynomial $s(z)$ has all its roots in an (open) unit disk if and only if $r_{S-C} = 1$.*

Let us consider the polynomial

$$p(z) = -1 + \bar{c}_1 z + \cdots + \bar{c}_m z^m, \quad c_m \neq 0$$

and the polynomial

$$q(z) = \bar{c}_m + \bar{c}_{m-1} z + \cdots + \bar{c}_1 z^{m-1} - z^m$$

which has the same roots as the polynomial

$$r(z) = -1 - (\bar{c}_{m-1}/\bar{c}_m)z - \cdots - (\bar{c}_1/\bar{c}_m)z^{m-1} + 1/\bar{c}_m \cdot z^m.$$

From Corollary 17.11.4, a lower bound for the roots of the polynomial r is

$$(1 + |c_m|^{2-1} + |c_1|^2 |c_m^2|^{-1} + \cdots + |c_{m-1}^2| |c_m^2|^{-1})^{-1/2}$$

and thus an upper bound for the roots of p is

$$((1+|c_1^2|+\cdots+|c_{m-1}^2|+|c_m^2|)/|c_m|)^{-1}$$

which is essentially the Carmichael–Mason inequality (1914).

For the derivation of other inequalities we need an auxiliary result about the iterates of $S+C$. First we remark that

$$C^*x = \langle x, e_1 \rangle h.$$

LEMMA 17.11.11. *Suppose that h is an element of the form $h=\Sigma_{i=1}^m c_i e_i$ and x is an element of X, $\|x\|=1$ with $x=\Sigma_1^\infty x_i e_i$. Then we have the relation:*

$$(S^*+C^*)^n x = \sum_{i=1}^{m} a_n^i e_i + \sum_{k=1}^{\infty} x_{k+m+n} e_{m+k},$$

where

$$a_n^1 = \sum_{i=1}^{m} c_i a_{n-i}^1 + x_{n+1} \tag{i}$$

$$a_n^2 = \sum_{i=1}^{m-1} c_{i+1} a_{n-i}^1 + x_{n+2} \tag{ii}$$

$$\vdots$$

$$a_n^m = c_m a_{n-1}^1 + x_{n+m} \tag{iii}$$

$$a_0^1 = x_1, \quad a_1^1 = x_1 c_1 + x_2, \quad a_2^1 = (x_1 c_1 + x_2) c_1 + x_1 c_2 + x_3, \quad a_3^1 = \cdots \tag{iv}$$

Proof. The formula for the iterate of $S+C$ follows from the form of S and C. The other formulas follow by identification. □

COROLLARY 17.11.12. *The coefficients a_n^i satisfy the following inequalities:*

$$|a_1^1| \leqslant 1 + |c_1|,$$

$$|a_2^1| \leqslant 1 + |c_1| + |c_2| + |c_1^2|,$$

$$|a_3^1| \leqslant 1 + |c_1| + |c_2| + |c_3| + 2|c_1 c_2| + |c_1^2| + |c_1^3|,$$

$$\vdots$$

$$|a_n^1| \leqslant \sum_{i=1}^{m-1} |c_{i+1}|\,|a_{n-i}^1| + 1$$

$$\vdots$$

$$|a_n^m| \leqslant |c_m|\,|a_{n-1}^1| + 1.$$

Now we use these inequalities to obtain some classical inequalities. Suppose that h is of the form

$$h = \sum_{i=1}^{m} c_i e_i$$

and we define

$$R = |c_1| + \cdots + |c_m|.$$

Then from the above inequalities we have that

$$|a_1^1| \leqslant 1 + R,$$
$$|a_2^1| \leqslant 1 + R + R^2,$$
$$\vdots$$
$$|a_n^1| \leqslant 1 + R + R^2 + \cdots + R^n$$

which implies that

$$|a_n^1| \leqslant (R^{n+1} - 1)/(R - 1),$$
$$|a_n^2| \leqslant |c_2|((R^{n-} - 1)/(R - 1)) + |c_3|((R^{n-1} - 1)/(R - 1)) + \cdots + |c_m| \times$$
$$\times ((R^{n-m+2} - 1)/(R - 1))$$
$$\vdots$$
$$|a_n^m| \leqslant |c_m|(R^n - 1)/(R - 1) + 1.$$

Suppose now that $R > 1$. Since for $x = \Sigma_1^\infty x_i e_i$ the inequality

$$\|(S^* + c^*)^n x\| \leqslant \left(\sum_{i=1}^{m} |a_n^i|^2 + \sum_{i=n+m+1}^{\infty} |x_i|^2 \right)^{1/2}$$

holds, this implies that $r_T \leqslant R$.

If $R > 1$ we obtain, similarly, that $r_T = 1$.

If $R = 1$ we obtain that $r_T = 1$ and thus from 17.11.4 the lower bound p of the roots satisfies the inequality

$$1/\rho \leqslant r_{T^*} = \max(1, R).$$

Thus an upper bound for the roots of the polynomial q is $\max(1, R)$, which is the classical inequality of Montel (1934/1935).

If we set $|c| = \max(|c_i|)$ then clearly we have the inequalities:

$$a_1^1 = 1 + |c|,$$
$$a_2^1 = (1 + |c|)^2,$$
$$\vdots$$
$$a_n^1 = (1 + |c|)^n.$$

Thus an upper bound for the roots of the polynomial q is $\max_i(1 + |c_i|)$ which is the famous inequality of A. L. Cauchy.

17.12. Bohr's Basic Theorem of Almost Periodic Functions

As we know, each continuous functions f on $[-\pi, \pi]$, $f(-\pi) = f(\pi)$ is a limit of a sequence of trigonometric polynomials with respect to the metric induced by the norm

$$f \to \|f\| = \sup_{t \in R} |f(t)|$$

(Féjer's theorem). Of course, we can consider f defined on R with the property that for all t,

$$f(t+2\pi)=f(t).$$

On R there exists an important class of functions approximable uniformly by trigonometric polynomials. We recall that a trigonometric polynomial on R is a function of the form

$$x\to\sum_{k=1}^{n}c_k\,e^{i\lambda_k x},$$

where c_k are complex numbers.

For the reader's convenience we recall that an almost periodic function on R is a function having the following property: for each $\varepsilon>0$ there exists $l=l(\varepsilon)>0$ such that in each interval $[a,b]$ of R, $b-a=l$ there exists τ such that for all $x\in R$,

$$|f(x+\tau)-f(x)|<\varepsilon.$$

The number τ is called the almost period of f (with respect to ε). Some of the important properties of almost periodic functions are given below, and as the proof is simple it is omitted.

(1) *Each almost periodic function is bounded and uniformly continuous on R.*
(2) *If f, g are almost periodic functions and a, b are complex numbers, then the function $af+bg$ is almost periodic.*
(3) *If f is almost periodic and there exists $K>0$ such that $|f(x)|\geqslant K$ for all $x\in R$, then the function f^{-1} is almost periodic.*

For the proof of the following important property of almost periodic functions we refer to literature on the subject, such as Levitan (1953): for each almost periodic function f there exists a number denoted by $M(f)$ and defined as

$$M(f)=\lim_{T\to\infty}\frac{1}{2T}f(x)\,dx=\lim_{T\to\infty}\frac{1}{2T}\int_{-T+a}^{T+a}f(x)\,dx$$

(uniformly in a) and $M(|f|^2)=0$ if and only if $f\equiv 0$.

If we consider the space of all almost periodic functions, denoted by AP, then we can define an inner product on AP by the following formula: if f, g are in AP then we set

$$\langle f,g\rangle=M(f\bar{g}).$$

It is clear that AP is not complete under this inner product and we denote by B^2 the completion of AP with respect to the norm induced by the above inner product.

For each $\lambda\in R$ we consider the function

$$x\to e^{i\lambda x}$$

and we note the following relations:

$$\langle e^{i\lambda x},e^{i\mu x}\rangle=\delta_{\lambda\mu}=\begin{cases}1, & \lambda=\mu\\ 0 & \lambda\neq\mu\end{cases}.$$

This implies clearly that B^2 is nonseparable.

Now if f is almost periodic then we can consider the following numbers associated with f, called the 'mean Fourier coefficients' of f,

$$a_\lambda(f) = \lim_{T\to\infty} \frac{1}{2T}\int_{-T}^{T} f(x)\,\mathrm{e}^{-i\lambda x}\,\mathrm{d}x = M(f\,\mathrm{e}^{i\lambda x})$$

where $\lambda \in R$. Since the family of functions $(\mathrm{e}^{i\lambda x})_{\lambda\in R}$ is orthonormal in B^2 we obtain that for each f, the family of complex number $a_\lambda(f) \neq 0$ is countable, and thus is $(a_{\lambda_k}(f))_{k=1}^{\infty}$. Thus for each f in AP we can consider the f Fourier series

$$\sum_{k=1}^{\infty} a_{\lambda_k}(f)\,\mathrm{e}^{-i\lambda_k x}$$

and since f is in B^2 (the class of functions equivalent to f) we have the inequality

$$\sum_{k=1}^{\infty} |a_{\lambda_k}(f)|^2 \leqslant M(|f|)$$

The basic result in the theory of almost periodic functions is the following theorem of H. Bohr:

THEOREM 17.12.1. *For each almost periodic function f the following relation holds:*

$$\sum_{k=1}^{\infty} |a_{\lambda_k}(f)|^2 = M(|f|^2).$$

The proof we give below was found by H. Weyl and uses the theory of normal operators in complete inner product spaces.

Proof. For the proof we consider first the operator $N\colon B^2 \to B^2$ defined as follows: for each g in AP we set

$$(Ng)(y) = \lim_{T\to\infty} \frac{1}{2T}\int_{-T}^{-T} f(y-x)g(x) = M(f(y-x)g(x))$$

and this induces an operator on B^2 which we denote again by N. Obviously, N is a linear and bounded operator. The important properties for use of the operator N are given in the following proposition. □

PROPOSITION 17.12.2. *The operator N is normal and compact on B^2.*

Proof. We consider the operator $T\colon B^2 \to B^2$ defined as follows: we set for g in AP

$$(Tg)(y) = \lim_{T\to\infty} \frac{1}{2T}\int_{-T}^{T} \bar{f}(x-y)g(x)\,\mathrm{d}x$$

and the operator induced on B^2 is denoted again by T. We note that the following properties of T hold:

(1) $\langle Ng, h\rangle = \langle g, Th\rangle$,
(2) $NT = TN$.

To prove property (1) we note that

$$\langle Ng, h\rangle = M(M(f(y-x)g(x))\bar{h}(y)) = M(g(x)M(f(y-x)\bar{h}(y)))$$
$$= M(g(x)\bar{M}(\bar{f}(y-x)h(y))) = \langle g, Th\rangle$$

and this implies our assertion.

For the second property we remark that for g in AP,

$$(NTg)(x) = M(f(x-t)\bar{M}(f(s-t)g(t))) = M(g(t)M(f(x-t)\bar{f}(s-t)))$$

and if we set

$$h(x,s) = M(f(x-t)\bar{f}(s-t)) = \lim_{T\to\infty}\frac{1}{2T}\int_{-T}^{T} f(x-t)\bar{f}(s-t)\,\mathrm{d}t$$

using the change of variable $s - t = \sigma - s$ we have

$$h(x,s) = \lim_{T\to\infty}\frac{1}{2T}\int_{-T+\tau+s}^{T+\tau+s} f(\sigma-s)\bar{f}(\sigma-x)\,\mathrm{d}\sigma = M(f(\sigma-s)\bar{f}(\sigma-x))$$

which implies that for g in AP,

$$(NTg)(x) = M(g(s)M(f(\sigma-s)\bar{f}(\sigma-x))) = M(\bar{f}(\sigma-x)M(f(\sigma-s)g(s))) = (TNg)(x)$$

and this clearly implies relation (2). From this it is clear that N is a normal operator.

To prove the compactness property of N it is sufficient to prove that the image through N of the set $(h, \|h\| \leqslant 1)$ of B^2 is precompact. Now if h is an element of this set then

$$(Nh)(x) = \lim\frac{1}{2T}\int_{-T}^{T} f(x-t)h(t)\,\mathrm{d}t$$

and applying the Cauchy inequality 0 in the right side we obtain

$$|(Nh)(x)| \leqslant \lim_{T\to\infty}\left(\frac{1}{2T}\int_{-T}^{T}|f(x-t)|^2\,\mathrm{d}t\right)^{1/2}\left(\frac{1}{2T}\int_{-T}^{T}|h(t)|^2\,\mathrm{d}t\right)^{1/2}$$
$$\leqslant M(|f|^2).$$

This implies obviously that the set $(Nh, \|h\| \leqslant 1)$ is uniformly bounded.

We now show that this set is equicontinuous. Indeed, for h, $\|h\| \leqslant 1$,

$$(Nh)(x) - (Nh)(x_1) = \lim_{T\to\infty}\frac{1}{2T}\int_{-T}^{T}(f(x-t) - f(x_1-t))h(t)\,\mathrm{d}t$$

and applying the Cauchy inequality in the right side we get

$$|(Nh)(x) - (Hh)(x_1)| \leqslant (M(|f(x-t) - f(x_1-t)|)^2\,\mathrm{d}t)^{1/2}. \tag{*}$$

Now f is in AP and thus it is uniformly continuous. Thus for each $\varepsilon > 0$ there exists $\delta(\varepsilon) > 0$ such that for $|u - v| < \delta(\varepsilon)$, $|f(u) - f(v)| < \varepsilon$. This obviously implies that $(Nh, \|h\| \leqslant 1)$ is equicontinuous. Then by the Arzelà–Ascoli theorem, $(Nh, \|h\| \leqslant 1)$ is precompact in AP (AP with the sup norm). This easily implies that $(Nh, \|h\| \leqslant 1)$ is precompact in B^2, as follows from (*). Thus assertion (2) is proved. □

We know that the spectrum of any bounded compact operator is countable and has 0 as a (possible) limit point. To prove Bohr's theorem we note that if λ is an eigenvalue for N and

$$E_\lambda = (g, Ng = \lambda g)$$

is the corresponding eigenspace, then this is finite dimensional. Thus we can find the elements $g_{\lambda,1}, g_{\lambda,2}, \ldots, g_{\lambda,n} (n < \infty)$ such that this is an orthonormal basis for E_λ. Since each $g_{\lambda,k}$ is an element satisfying the relation

$$Ng_{\lambda,k} = \lambda g_{\lambda,k}$$

we obtain that the element $g_{\lambda,k,\sigma}$ defined by

$$g_{\lambda,k,\sigma}(t) = g_{\lambda,k}(t + \sigma) \tag{*}$$

is again in E_λ. This implies that

$$g_{\lambda,k,\sigma} = \sum_{r=1}^{n} c_{kr}(\sigma) g_{\lambda,r}, \quad k = 1, 2, 3, \ldots, n.$$

From the relation (*) we obtain that the functions $c_{kr}(\sigma)$ are differentiable and thus the following system of ordinary equations is satisfied by these functions:

$$\mathrm{d}g_r(t)/\mathrm{d}t = \sum_{k=1}^{n} (\mathrm{d}c_{rk}(\sigma)/\mathrm{d}\sigma)_{\sigma=0}\, g_k(t), \quad r = 1, 2, 3, \ldots, n.$$

The solutions of this system are functions of the form $p(t)\,\mathrm{e}^{at}$, where p is a polynomial in t and a is a scalar. These functions, being almost periodic, are uniformly bounded on R and this implies that the polynomial p reduces to a constant term and that a is of the form ib with b in R. Thus, the general form of the functions $g_{\lambda,k}$ is

$$g_{\lambda,k}(t) = \sum_{l=1}^{n} c_k^l\, \mathrm{e}^{i\lambda_l t}$$

with c_k^l as constants.

Since $(g_{\lambda,k})$ is an orthonormal basis for E_λ we can assume without loss of generality that there exists a new basis formed from functions of the form

$$\mathrm{e}^{i\lambda_1 t}, \mathrm{e}^{i\lambda_2 t}, \ldots, \mathrm{e}^{i\lambda_n t}.$$

As these functions are in E, we obtain that the mean Fourier coefficient of $\bar{\mathrm{e}}^{i\lambda_j t} f(t)$ with respect to the function $\mathrm{e}^{i\lambda_j t}$ is exactly a_{λ_j}. This means that the mean Fourier coefficients $(\mathrm{e}^{i\lambda_j t})$ are eigenvectors with their corresponding eigenvalues $(a_{\lambda_j})_1^n$.

Now we consider the set of all eigenelements $(\mathrm{e}^{i\lambda_n t})_1^\infty$ with eigenvalues $(a_{\lambda_n})_1^\infty$.

Associated with the function f we now consider the series

$$\sum_{k=1}^{\infty} a_{\lambda_k} \mathrm{e}^{i\lambda_k t}$$

and we prove that it converges in B^2. For this we remark that the following inequality holds: if h is any almost periodic function then

$$\left|\sum_{j=m}^{\infty} a_{\lambda_j}(h)\right|^2 \leqslant \left(\sum_{j=m}^{\infty} |a_{\lambda_j}|^2\right)^{1/2}\left(\sum_{j=m}^{\infty} |\langle h, e^{i\lambda_j t}\rangle|^2\right)^{1/2}$$

$$\leqslant \left(\sum_{j=m}^{\infty} |a_\lambda|^2\right)(M|h|^2)^{1/2}.$$

This implies that the above series converges uniformly (in t).

We now consider the function $M(f(t-s)\bar{f}(-s))$ which is of the form Nh for some h.

Since the operator N is normal and compact, each element h in the range of N is of the form $\Sigma\langle h, e_i\rangle e_i$, where (e_i) is a system of eigenvectors of N corresponding to the eigenvalues of N. Thus the function $t \to M(f(t-s)\bar{f}(-s))$ is of the stated form, i.e. it is of the form

$$\sum_{j=1}^{\infty} a_{\lambda_j} M(\bar{f}(-s)\,e^{-i\lambda_j s})\,e^{i\lambda_j t}$$

and applying the above inequality for $h(t) = \bar{f}(-t)$ we get, using the uniform convergence, that the following relation holds:

$$M(f(t-s)\bar{f}(-s)) = \sum_{j=1}^{\infty} a_{\lambda_j} M(\bar{f}(-s)\,e^{-i\lambda_j s})e^{i\lambda_j t}.$$

This relation for $t = 0$ implies that

$$M(|f(-s)|^2) = \sum_{j=1}^{\infty} a_{\lambda_j} M(\bar{f}(-s)\,e^{-i\lambda_j s})$$

which is in fact

$$M(|f|^2) = \sum_{j=1}^{\infty} |a_{\lambda_j}|^2.$$

Thus the theorem is proved. □

17.13. Inner Product Structures of Lie Algebras and Jordan Algebras

In this section we are interested in the structure theory of certain classes of algebras such as Lie algebras, L*-algebras, and Jordan algebras which have an inner product. For the reader's convenience we recall some definitions. For a detailed account of Lie algebras we refer to Jacobson (1962).

DEFINITION 17.13.1. A (nonassociative) algebra over the field K is said to be a Lie algebra if its multiplication '·' satisfies the following identities:

(1) $x \cdot x = x^2 = 0$,
(2) $(x \cdot y) \cdot z + (y \cdot z) \cdot x + (z \cdot x) \cdot y = 0$ (Jacobi's identity).

Instead of $x \cdot y$ we simply use xy. Then we have

COROLLARY 17.13.2. *If A is a Lie algebra then for any x, y in A we have the relation $xy = -yx$.*

Proof. Since $(x + y)^2 = 0$ the assertion follows. □

EXAMPLE 17.13.3. If X is a Banach space and $A = L(X)$ with the 'multiplication'

$$x, y = xy - yx$$

for any x, y in $L(X)$, then $L(X)$ is a Lie algebra.

Let A be a Lie algebra.

DEFINITION 17.13.4. A derivation D on the Lie algebra A is any application with the following properties:

(1) D is a linear operator on A.
(2) $D(xy) = xDy + (Dx)y$.

Then the following result holds:

PROPOSITION 17.13.5. *With the usual multiplication the set of derivations on A is a Lie algebra if we set*

$$[D_1, D_2] = D_1 D_2 - D_2 D_1.$$

Sometimes this is called the derivation algebra of A.

We recall that for derivations the Leibniz rule holds:

$$D^n(xy) = \sum_{k=0}^{n} \binom{n}{k} D^k(x) D^{n-k}(y).$$

DEFINITION 17.13.6. A derivation on an algebra is said to be inner if

$$D(x) = xy - yx$$

for some y in the algebra.

If A is an algebra then the set $(\Sigma\, a_i b_i)$ is an algebra and is denoted by A^2; if A is a Lie algebra the algebra A^2 is denoted by A'.

We now define the class of algebras which are related to inner product structures.

DEFINITION 17.13.7 (Schue, 1960). An L*-algebra is a Lie algebra A over complex numbers with the following properties:

(1) On A an inner product is define such that A is a complete inner product space.
(2) There exists a map $k: A \to A$ with the property that

$$\langle [x, y], z \rangle = \langle y, [x^*, z] \rangle$$

holds for all x, y, z in A.

DEFINITION 17.13.8. If A is a L*-algebra and M, N are subsets in A then we set

$$[M, N] = ([m, n]; m \in M, n \in N)$$

= the closed linear subspace of A generated by the elements $[m, n]$ with m in M and n in N.

We also use the notation M^* to denote the set $(m^*, m \in M)$.

If A is a Lie algebra then we define the following mapping on A:

$$D_x(y) = [x, y].$$

If A is an L*-algebra then we assume that $\|D_x\| \leqslant \|x\|$.

DEFINITION 17.13.9. If A is an L*-algebra then we say that it is semisimple if $[A, A] = A$.

We say that the L-algebra A is simple if no closed nontrivial semisimple ideal exists.

PROPOSITION 17.13.10. *Let A be an L*-algebra. Then the following assertions hold:*

(1) *$x \to x^*$ is an involution on A.*
(2) *For any x in A, x^* is the adjoint of x with respect to the inner product $\langle\,,\rangle$.*
(3) *For any x, y in the following identities hold:*
   (a) $[x, y] = [y^*, x^*]^*$,
   (b) $\langle x, y\rangle = \langle y^*, x^*\rangle$.

Proof. Since the proof is simple we omit it. □

We suppose now that A is an L*-algebra which is semisimple.

DEFINITION 17.13.11. A Cartan subalgebra of the L*-algebra A is a maximal selfadjoint abelian subalgebra of A.

THEOREM 17.13.12 (Schue, 1960). *If H is a Cartan subalgebra of the L*-algebra A then H is maximal abelian and $H^\perp = [H, A]$.*

Proof. Let x be in A such that $[H, x] = 0$. Then

$$[H, x^*] = [H^*, x]^* = [H, x]^* = 0$$

and thus

$$[H, x + x^*] = [H, x - x^*] = 0.$$

But H is maximal selfadjoint and thus

$$(x + x^*) \in H, \qquad (x - x^*) \in H.$$

Then x and x^* are in H. Let h_1, h_2 be in H and x in A. Since

$$\langle [h_2^*, h_1], x\rangle = \langle h_1, [h_2, x]\rangle$$

we get that $[h_2, x] \in H^\perp$. Thus $[H, A]$ is in $H^\perp$. Conversely, if x is in $[H, A]^\perp$ then

$$\langle x, [h^*, y]\rangle = 0 = \langle [h, x], y\rangle$$

which gives that $[H, x] = 0$ and thus x is in H.

The following theorem is a structure theorem for L*-algebras.

THEOREM 7.13.13 (Schue, 1960). *Let A be a semisimple L*-algebra. Then there exist simple closed ideals $(I_i)_{i \in I}$ such that $A = \Sigma_{i \in I} \oplus I_i$ and every closed ideal is of the form $\Sigma_{i \in J} \oplus I_i$, where J is a subset of I.*

Proof. Let H be a Cartan subalgebra of A and define

$$B = (D_h, h \in H)$$

as the C*-algebra generated by the derivations D_h, $h \in H$.

Since H is a Cartan algebra, B is a commutative C*-algebra (without identity). By the Gelfand–Naimark theorem it is isomorphic-isometric to the algebra of all continuous complex-valued functions on a locally compact space, and each function vanishes at infinity. Then for each homomorphism of B into the complex numbers we can define a linear functional on H by the relation:

$$\alpha(Dh) = f(h). \tag{*}$$

By the Fréchet–Riesz representation theorem we find h_α such that

$$\alpha(D_h) = \langle h, h_\alpha \rangle$$

since we assumed that $\|D_h\| \leqslant \|h\|$. Of course,

$$\alpha(D_h^*) = \overline{(D_h)}$$

and this gives that $h_\alpha^* = h_\alpha$, i.e. the representing elements are Hermitian. Let α, β be two distinct homomorphisms. We say that $\alpha \perp \beta$ if $h_\alpha \perp h_\beta$.

DEFINITION 17.13.14. A family of homomorphisms is called indecomposable if it is not a union of other families (nonempty) which are orthogonal.

Using Zorn's lemma we get that the family of all homomorphisms of B is a union of indecomposable families, say $\{M_i, i \in I\}$. Let

$$H_i = (h_\alpha, h_\alpha \in M_i).$$

Then we set

$$A_i = H_i + [H_i, A]$$

and we remark that:

(1) A_i is a simple closed ideal in A (i.e. A_i has no nontrivial closed ideals),
(2) $A_i \perp A_j$ if $i \neq j$.

Then $K = \Sigma_{i \in I} A_i$ is a closed ideal which contains H and $[K^\perp, A] = 0$, which gives that $K^\perp = 0$, i.e. $K = A$.

Now the second assertion of the theorem follows from the definition of the A_i's.

The above decomposition may be written as follows. Let (h) be a linear continuous functional on H and set

$$V_\alpha = (v_\alpha, [h, v_\alpha] = \alpha(h)v_\alpha \quad \text{for all } h \text{ in } H).$$

If $V_\alpha \neq (0)$ then $\alpha(\cdot)$ is called a nonzero root of A with respect to the Cartan subalgebra H. Let I be the set of all roots of A with respect to H. Then clearly

$$K = \sum \oplus V_\alpha,$$

where the sum is taken ever all distinct roots of α relative to H and clearly $H \subseteq K \subseteq A$.

DEFINITION 17.13.15 (Schue, 1960). We say that A has a Cartan decomposition relative to H if $K = A$.

For the existence of a Cartan decomposition for every L*-algebra we refer to J. R. Schue's paper (1961).

DEFINITION 17.13.16. An element h of a semisimple algebra A is said to be regular if

$$N(h) = (x, x \in A, xh = 0)$$

is a Cartan subalgebra of A.

Then we have the following result.

THEOREM 17.13.17 (Balachandran, 1966). *If A is a semisimple L*-algebra and H is a Cartan subalgebra of A then h in H is regular iff $\alpha(h) \neq 0$ for every root α of A with respect to H.*

Proof. For any h in H, $N(h) \supseteq H$. If h is regular then $N(h)$ is a Cartan subalgebra of A, and thus $N(h) = H$. Then

$$0 = [h, v_\alpha] = \alpha(h)v_\alpha.$$

Conversely, if h is in H and $\alpha(h) \neq 0$ for every α then for the regularity of h it suffices to show that $N(h) = H$. Suppose that this is not the case and thus H is a proper subset of $N(h)$. Then we find x in $N(h)$ with

$$x = h_1 + b$$

where h_1 is in H and b in $H^\perp$, $b \neq 0$. But $H^\perp$ is the orthogonal sum of all subspaces V_α and each V_α is on-dimensional. Thus

$$b = \sum v_\alpha, \quad v_\alpha \in V_\alpha.$$

Since $b \neq 0$ at least one v_α is not zero. Then

$$[h, x] = [h, h_1 + b] = [h, b] = \sum \alpha(h)v_\alpha \neq 0$$

which is a contradiction with the fact that x is in $N(h)$. Thus $N(h) = H$ and h is a regular element. □

COROLLARY 17.13.18 (Balachandran, 1966). *Every Cartan subalgebra of a separable L*-algebra which is semisimple contains regular elements which form a dense second category set in H.*

Proof. For each root α of A (with respect to H) we set

$$N_\alpha = (h, h \text{ in } H, \alpha(h) = 0).$$

Then N is a closed linear subspace of A and is not H. From the separability condition we get that (α) is a countable set, and from Baire's category theorem the assertion follows. □

REMARK 17.13.19. In his (1966) paper, V. K. Balachandran gives an example to show that the separability condition is essential for the existence of regular elements.

For results on Cartan subalgebras of L*-algebras we refer to Schue (1961),

Balachandran and Parthasarathy (1966), Unsain (1971) and for roots we refer to Balachandran (1968, 1982).

We now consider the problem of the uniqueness of the inner product topology on semisimple L*-algebras. The basic result concerning this question was obtained by Balachandran (1968). This result will be obtained as a consequence of results given as lemmas.

LEMMA 17.13.20. *If A is an L*-algebra then for some $k > 0$ and all x, y in X we have the inequality*

$$\|[x, y]\| \leqslant k\|x\|\,\|y\|.$$

Proof. This follows from the principle of uniform boundedness. □

DEFINITION 17.13.21. If $\langle\,,\rangle$ is the inner product on A then we set

$$r = \sup_{x \neq 0 \neq y} \|[x, y]\|/\|x\|\|y\|.$$

DEFINITION 17.13.22. We say that an inner product is standard if $r \leqslant 1$.

Clearly we have the following result.

PROPOSITION 17.13.23. *If an inner product $\langle\,,\rangle$ is not standard then*

$$\langle\langle x, y\rangle\rangle = r^2\langle x, y\rangle$$

is a standard inner product.

Suppose now that A is a semisimple L*-algebra with respect to the inner products $\langle\,,\rangle_i$, $i = 1, 2$. We denote A as a L*-algebra with an inner product $\langle\,,\rangle_i$ by A_i, $i = 1$, 2. Then we have

PROPOSITION 17.13.24. *If $A_1 = H_1\Sigma_\alpha \oplus V_\alpha^1$ is a Cartan decomposition of A_1 with respect to Cartan subalgebra H_1 with the components V^1, then*

$$A_2 = H_1 \sum_\alpha \oplus V_\alpha^1.$$

Proof. Since

$$A_1 = H_1 \sum_\alpha \oplus V_\alpha^1$$

from the definition of H_1 and V_α^1 the assertion follows. □

DEFINITION 17.13.25. An operator $TL(X)$, X is a complete inner product space, is said to be diagonal if with respect to an orthonormal basis $(e_i)_{i \in I}$ of X we have

$$Te_i = \lambda_i e_i$$

for each i in I.

Obviously we have the following result.

PROPOSITION 17.13.26. *If T is diagonal then*

$$\|T\| = \sup|\lambda_i|.$$

We now consider several classes of elements in a L*-algebra.

DEFINITION 17.13.27. An element x in a L*-algebra A is said to be

(1) Hermitian if $x = x^*$,
(2) normal if $x^*x = xx^*$,
(3) hyponormal if $x^*x - xx^* \geqslant 0$.

If u is an element in a L*-algebra A then we set

$$(\operatorname{ad} u)(x) = [x, u]$$

which is clearly a bounded linear operator on A.

PROPOSITION 17.13.28. *Let A_i be as above. If x is a normal element then*

$$\|\operatorname{ad} x\|_1 = \|\operatorname{ad} x\|_2,$$

where ad x *is the right multiplication with* x.

Proof. Since x is normal by Zorn's lemma, we find a Cartan subalgebra which contains x, say H_x. Then we have the Cartan decomposition with respect to H. Then we get that ad x is a diagonal operator on A_1 and on A_2. Proposition 17.13.27 implies the assertion. □

We now consider the Hermitian subspace of A, $H(A)$, which consists of all Hermitian elements of A. Then we have the following result.

PROPOSITION 17.13.29. *$H(A)$ is a real complete inner product space with respect to both the inner products $\langle\, , \rangle_i$.*

Proof. Clear. □

PROPOSITION 17.13.30. *Let A, A_i be as above and suppose that for some norm $\|\, , \|$ we have*

$$\|x\| \leqslant \|x\|_1, \qquad \|x\| \leqslant \|x\|_2.$$

Then $\|\, , \|_1$ and $\|\, , \|_2$ are equivalent norms.

Proof. We consider on A the following norm

$$\|x\|_3 = \|x\|_1 + \|x\|_2$$

and (x_n) is a sequence which is a Cauchy sequence with respect to $\|\, , \|_3$. Then (x_n) is a Cauchy sequence with respect to both norms $\|\, , \|_i$ as well as with respect to the norm $\|\, , \|$. By the closed graph theorem we then get that $\|\, , \|$ and $\|\, , \|_3$ are equivalent norms as well as $\|\, , \|_2$ and $\|\, , \|_1$. This implies the assertion of the proposition. □

THEOREM 17.13.31. *Suppose that A is a semisimple L-algebra with respect to two inner products $\|\, , \|_i$, $i = 1, 2$. Then the norms they generate are equivalent norms.*

Proof. Since $A = H(A) + iH(A)$

we have

$$\|\operatorname{ad} x\|_1 = \|\operatorname{ad} x\|_2$$

and this gives a norm on A since $\operatorname{ad} x$ is one-to-one. But

$$\|\operatorname{ad} x\|_1 \leqslant \|x\|_1, \qquad \|\operatorname{ad} x\|_2 \leqslant \|x\|_2$$

and thus

$$\|x\| \leqslant \|x\|_1, \qquad \|x\| \leqslant \|x\|_2$$

and by the above result we have that, for some $k > 0$,

$$\|x_1\| \leqslant k\|x\|_2$$

for all x in $H(A)$. But each x in A has the unique decomposition

$$x = x_1 + ix_2,$$

where x_i are in $H(A)$, and thus

$$\|x\|_1 = \|x_1\|_1^2 + \|x_2\|_1^2.$$

Similarly for the norm induced by $\|\,,\|_2$. Then the equivalence of the norms follows from the above result. □

We now give a result that we need for what follows.

PROPOSITION 17.13.32 (Schue, 1960). *Suppose that A is a semisimple L*-algebra and that $T: A \to \bar{A}$ is a bounded linear operator where $\bar{A}$ is a complete inner product space with the property that for all x, y in A*

$$\langle T([x, y]), Tz \rangle = \langle Ty, T([x^*, z]) \rangle$$

for all x, y, z in A. Then for some $\varepsilon > 0$ we have

$$\langle Tx, Ty \rangle = \varepsilon \langle x, y \rangle.$$

Proof. Since T is as above, we find $s \geqslant 0$ such that

$$\langle Tx, Ty \rangle = \langle Sx, y \rangle$$

and from the above property of T we conclude that S commutes with D_x and thus every projection in the spectral decomposition of S commutes with D_x. But the range of such projections of S is a closed ideal and thus is (0) or A. Then $S = I \cdot \varepsilon$ for some $\varepsilon > 0$ and the assertion follows. □

THEOREM 17.13.33 (Balachandran, 1968). *If A is a simple L*-algebra which is an L*-algebra with respect to an inner product $\langle\,,\rangle_1$ and with respect to an inner product $\langle\,,\rangle_2$ then*

$$\langle\,,\rangle_1 = \varepsilon \cdot \langle \cdot \rangle_2$$

for some $\varepsilon > 0$.

Proof. We apply the above result with $T = I$. □

Now we consider a class of Lie algebras suggested by the class of inner product algebras and L*-algebras.

DEFINITION 17.13.34 (Istrăţescu, 1984). An inner product Lie algebra is any Lie algebra A on which there exists:

(1) an inner product $\langle , \rangle$ $(A, \|,\|)$, $\|,\|$ is the norm induced by the inner product, is not necessarily complete,
(2) an involution $*: A \to A$,
(3) $\langle x, y \rangle = \langle y^*, x^* \rangle$,
(4) $\langle [x, y], z \rangle = \langle y, [x, z] \rangle$, $[\,,\,]$ is the Lie product of A,
(5) for each a in A, $L_a: A \to A$, $L_a(x) = [a, x]$ is continuous,
(6) the set of all elements of the form $[x, y]$ with x, y in A, is dense in A.

If $\mathscr{a}$ denotes the completion of A then the operators L_a and R_a, where $R_a(x) = [x, a]$, can be extended to bounded linear operators on the completion and the norms are preserved. We denote the corresponding operators on the completion by $\mathscr{L}_a$ and $\mathscr{R}_a$, respectively.

We remark that on $\mathscr{a}$ we can define a structure of a Lie algebra and the product on $\mathscr{a}$ we denote by $[\,,\,]$ (as in A).

DEFINITION 17.13.35 (Istrăţescu, 1984). An element a in $\mathscr{a}$ is said to be bounded if $x \to [a, x]$ is a bounded linear operator on $\mathscr{a}$.

DEFINITION 17.13.36 (Istrăţescu, 1984). We say that A is a full Lie algebra if

$$A = A_b = (b,\ b \text{ is a bounded element of } \mathscr{a}).$$

Sometimes we say that A_b is the fulfilment of A.

DEFINITION 17.13.37 (Istrăţescu, 1984). If A is an inner product Lie algebra then we say that it is orthocomplemented if for each closed right ideal (or losed left ideal) J,

$$A = J \oplus J^{-}.$$

Then we have

THEOREM 17.13.38 (Istrăţescu, 1984). *A full Lie Algebra with an inner product is orthocomplemented.*

Proof. Let J be an arbitrary closed right ideal (the proof is similar in the case of left ideals) and let $\bar{J}$ be the closed ideal corresponding to J in $\mathscr{a}$. We remark that $\bar{R}_a j \in \bar{J}$ for all a in A and j in $\bar{J}$. But j in $\bar{J}$ is of the form

$$j = \lim j_n,\ j_n \in J.$$

Then we have

$$\bar{R}_a j = \lim \bar{R}_a j_n = \lim j_n a \in \bar{J}$$

which proves our assertion.

Now if $P_{\bar{J}}$ denotes the orthogonal projection of $\mathscr{a}$ onto $\bar{J}$ then we have that

$$P(A) = A \cap \bar{J} = J.$$

But $\mathscr{a}$ is complete with a corresponding inner product and thus

$$\mathscr{a} = \bar{J} \oplus \bar{J}^{\perp}$$

and for each a in A we can write

$$a = u + v$$

with u in $\bar{J}$ and v in $\bar{J}^{\perp}$. Since $P_{\bar{J}}a = u$ we get that $v = a - u$ is in $J^{\perp}$ and the theorem is proved. □

REMARK 17.13.39. If we consider a complete inner product space X and the spaces C_p, $1 \leqslant p \leqslant 2$, of all bounded linear operators T on X with

$$\|T\|_p = \left(\sum_{i=1}^{\infty} \lambda_i p\right)^{1/p} < \infty,$$

where (λ_i) is the set of all eigenvalues of T^*T counted according to their multiplicity, then this is a Lie algebra with the Lie product

$$[T, S] = TS - ST$$

which is a inner product Lie algebra that is not full but is orthocomplemented.

Let A be an arbitrary algebra and let M be a subset in A. Then we set

$$A_l(M) = (x, x \in A, xM = 0), \qquad A_r(M) = (x, Mx = 0)$$

and we call $A_l(M)$ the left annihilator of M and $A_r(M)$ the right annihilator of M. The class of algebras related to annihilators was introduced by F. Bonsall and A. W. Goldie. This suggest that we should define the analogous class for Lie algebras as follows.

DEFINITION 17.13.40 (Istrăţescu, 1984). An inner product Lie algebra A is said to be an annihilator algebra if for any closed left ideal I_l and any closed right ideal I_r of A the following conditions are satisfied:

(1) $A_l(I_r) = (x, x \in A, [I_r, x] = 0) = 0$ iff $I_r = A$,
(2) $A_r(I_l) = (x, x \in A, [x, I_l] = 0) = 0$ iff $I_l = A$.

DEFINITION 17.13.41. If every closed right ideal in A is an annihilator ideal then A is said to be a dual Lie algebra.

We have the following result.

THEOREM 17.13.42 (Istrăţescu, 1984). *Let A be an inner product Lie algebra, then if A is a full Lie algebra, A is a dual algebra.*

Proof. Suppose that I_r is a closed right ideal in A and set I_r^c as the closure of I_r in the completion of A. Then $A_r(A_l)(I_r^c)$ is $I_r^{\perp\perp}$. We must prove that $I_r^{\perp\perp}$ is in I_r. Let x be in $I_r^{\perp\perp}$ and thus

$$x = \lim (u_n + v_n)$$

with $u_n \in I_r$, $v_n \in I_r^-$. But $u_n \in I_r^c$, $v_n \in (I_r^c)^-$ and we find u, v in I_r^c, $(I_r^c)^-$ respectively, such that

$$u = \lim u_n, \qquad v = \lim v_n.$$

But x is in $I_r^{\perp\perp}$ and thus $0 = \langle x, v_n\rangle - \langle x, v\rangle$, which implies that

$$x = uI_r^c \cap A = I_r.$$

This clearly proves the assertion as well as the theorem.

We close with some remarks about some problems connected with the above results.

QUESTION 17.13.43. Is it possible to construct a theory of inner product Lie algebras which recaptures many basic results from L*-algebras?

QUESTION 17.13.44. Does there exist an analogue of the notion of system of roots for an inner product Lie algebra?

QUESTION 17.13.45. Is there an analogue of Weyl's group?

For more details about L*-algebras and complete inner product Lie algebras we refer to the work of Balachandran (1964, 1966, 1968, 1969, 1970, 1972), Balachandran and Rema (1969), Schue (1960, 1961), Viola Devapakkian (1974), Viola Devapakkian and Rema (1976) as well as to the literature quoted there.

17.14. Potential Theory and Inner Product Structures

The use of inner product structure methods in potential theory is based on the properties of energy and has the origin in the work of Ch. De la Vallée Poussin, M. Riesz, O. Frostman and H. Cartan. A new and important step was made by A. Beurling who introduced the so-called Dirichlet spaces and developed the theory of such spaces with J. Deny. In this section we follow the excellent exposition of J. Deny in his CIME lectures (1969, 1970).

Let S be a locally compact space countable at infinity and ξ be a Radon measure on S. We suppose that ξ is positive.

DEFINITION 17.14.1. A functional complete inner product space H of base ξ is a complete inner product space of functions on S such that for each f in such a space is locally integrable and for each compact subset K in S there exists a constant $c(K) > 0$ such that for any f in H the following inequality holds:

$$\int |f(s)|\,\mathrm{d}\xi(s) \leqslant c(K)\|f\|.$$

We now give some examples of such spaces.

EXAMPLE 17.14.2. Let ξ be a Radon measure and let p be a function which is positive and locally integrable. We consider the space $L^2(p\,\mathrm{d}\xi)$ which consists of all functions g with the property that

$$\int |g(s)|^2 p(s)\,\mathrm{d}\xi(s) < \infty.$$

This is a space of base $p\,\mathrm{d}\xi$. Indeed if K is an arbitrary compact subset in K then we have

$$\int_K |f(s)|\,\mathrm{d}\xi = \int f(s)p(s)^{1/2}\cdot p(s)^{-1/2}\,\mathrm{d}\xi(s) \leqslant \|f\|\left(\int_K \mathrm{d}\xi(s)/p\right)^{1/2}$$

and the assertion is proved.

EXAMPLE 17.14.3. Let V be an open subset in R^m ($m \geqslant 1$) and D(V) denotes the set of all C^∞ functions with compact support (in V). We set

$$\|u\|^{\perp} = \int_V |\text{grad}\, u|^2 \, \mathrm{d}s$$

ds being the Lebesgue measure in R^m. H is the completion of $D(V)$ with respect to the above norm and this is a space of base ds.

Since f is in $D(V)$ we have

$$f(s_1, \ldots, s_n) = \int_{-a}^{s_1} (\partial f/\partial t)(t, s_2, \ldots, s_n)\, \mathrm{d}t$$

for a sufficiently large a such that V is contained in the set

$$(s, s = (s_1, \ldots, s_n), |s_1| \leqslant a).$$

Applying the Cauchy–Buniakowsky inequality we get Poincaré's inequality

$$\int |f(s)|^2 \, \mathrm{d}\xi(s) \leqslant 4a^2 \int |\text{grad}\, f|^2 \, \mathrm{d}s.$$

This implies that there exists a mapping $T: H \to D(V)$ which has the property that $Tf = f$ for f in $D(V)$. We now show that T is injective. Let f be an element in H with the property that $Tf = 0$. Then we find (f_n) in $D(V)$ with the property that $f_n \to f$ and for each g in $D(V)$ we have

$$\langle f, g \rangle = -\langle Tf, \Delta g \rangle$$

or

$$\langle f, g \rangle = 0.$$

But $D(V)$ is dense in H and thus f must be the zero element of H. Thus T is injective.

EXAMPLE 17.14.4. Let S be a locally compact abelian group with the Haar measure ξ. If v is another measure on S which is of positive type, i.e. for f continuous with compact support the following inequality holds:

$$\int_S (f \oplus \bar{f})\, \mathrm{d}\bar{v}(s) = 0$$

then

$$u_f = f \oplus v$$

is a.e. defined and locally integrable. The function u_f is called the v-potential generated by f. From the inequality

$$\int_S |f \oplus g|^2 \, \mathrm{d}\bar{v} \leqslant \int_S |f \oplus \tilde{f}|\, \mathrm{d}\bar{v} \int_S g \oplus \tilde{g} \, \mathrm{d}\bar{v}$$

we conclude that $u_f = 0$ (a.e.) iff

$$\int_S f \oplus \tilde{f} \, \mathrm{d}v = 0$$

and thus $u_f = u_g$ iff

$$\int_S f \oplus \tilde{f}\, d\bar{v} = \int_S g \oplus \tilde{g}\, d\bar{v}.$$

Then if we define

$$\|f\| = [(f \oplus f \oplus v)(0)]^{1/2}$$

then this defines a seminorm which is generated by an inner product on the space H_0 of all f in L^1_{loc} which are potentials. If u_f, u_g are two such functions then we set

$$\langle u_f, u_g \rangle = \int f \oplus \tilde{g}\, dv$$

and the completion of H_0 is denoted by H. This is a space H as in the above definition. Indeed, if K is a compact subset in S then for given u_f we set

$$g(s) = \begin{cases} K^{(s)u_f(s)/u_f(s)}, & u_f(s) \neq 0 \\ 0, & \text{otherwise.} \end{cases}$$

By the Cauchy–Buniakowsky inequality we get that the following inequality holds:

$$\int_K |u_f|\, d\xi \leqslant \|u_f\|\, \|u_g\|$$

and thus

$$\int_K |u_f(s)|\, d\xi \leqslant c(K)\|u_f\|$$

and the assertion is proved.

We now define the notion of Dirichlet space. First we consider the notion of normal contraction.

DEFINITION 17.14.5. Let u, v be two complex-valued functions on S. We say that v is a normal contraction of u if

(1) $|v(s) - v(t)| \leqslant |u(s) - u(t)|$,
(2) $|v(s)| \leqslant |u(s)|$
holds for all s, t in S.

EXAMPLES 17.14.6. If S in C, the set of complex numbers, then

(1) $z \to |z|$ is a normal contraction,
(2) if C is a closed convex set in C then P_C is a normal contraction.

DEFINITION 17.14.7. Let S be a locally compact space and let μ be a positive Radon measure on S. A Dirichlet space is a functional complete inner product space H with base ξ such that the following property holds: if v is a normal contraction of u in H then v is in H and $\|v\| \leqslant \|u\|$.

We note that some of the spaces considered above are clearly Dirichlet spaces. We now give another interesting example of a Dirichlet space. We take as space H the family of all functions on $(0, 1)$ with the property that f' is in $L^2_{[0,1]}$ and $f(0) = f(1) = 0$. Since

$$|f(t)| = (1 - t)\|f'\|_{L^2}$$

this space with the norm $f \to \|f'\|_{L^2}$ is a Dirichlet space.

REMARK 17.14.8. A similar construction can be made for $[0, \infty)$ and $[0, a]$ with a finite.

We now list some properties of Dirichlet spaces.

PROPOSITION 17.14.9. *Let H be a Dirichlet space. Then the following assertions hold:*

(1) *If u is in H then* $\operatorname{Re} u$, $\operatorname{Im} u$ *are in H.*
(2) *If u is a real function in H then u^+ and u^- are in H.*
(3) *If $C_r = (z, z = r)$ and $P_{C_r} = T_r$ then for each u in H*

$$\lim_{r \to 0} T_r u = 0.$$

Proof. Properties (1) and (2) are clear. For (3) we remark that $z \to T_r z$ is a normal contraction and thus

$$\|u - T_r u\| \leqslant \|u\|$$

which implies that

$$\|T_r u\|^2 \leqslant 2 \operatorname{Re}\langle u, T_r u\rangle$$

and this implies our assertion. □

PROPOSITION 17.14.10. *Let (u_n) be a sequence in H which is weakly convergent to u and let T be a normal contraction in the complex plane. Then (Tu_n) converges weakly to Tu; the convergence is strong if $Tu = u$.*

Proof. Since $Tu_n = u_n$, the sequence (Tu_n) is bounded. Now for each f bounded, measurable and with compact support we have

$$|\langle Tu_n - Tu, u_f\rangle = \left|\int (Tu_n - Tu)\bar{f}\, d\xi\right| \leqslant \int |u_n - u|\,|f|\, d\xi \leqslant \|u_n - u\|\,\|u_f\|$$

which proves the first assertion.

Suppose now that $Tu = u$. We prove that

$$\lim \|Tu_n\| = \|u\|$$

and the result follows since the weak convergence was proved above.

Since

$$\overline{\lim}\, \|Tu_n\| \leqslant \lim \|u_n\| = \|u\| = \|Tu\|$$

the convergence follows. □

REMARK 17.14.11. The above convergence result holds in more general spaces,.for example in locally uniformly convex spaces (see the author's book on strict convexity and complex strict convexity, 1984).

PROPOSITION 17.14.12. *If u is in H and u is bounded then for each integer m, u^m is in H.*

Proof. If K is a constant with the property that $u = K$ then

$$|u^m(s)| \leqslant K^{m-1}|u(s)|$$

and the assertion follows. □

PROPOSITION 17.14.13. *If u, v are in H and real-valued then*

(1) $\inf(u, v)$, $\sup(u, v)$ *are in* H,
(2) *if u, v are positive and in H then $\langle u, v\rangle \leqslant 0$ when $u(s)v(s) = 0$.*

Proof. The first assertion follows from the formulas.

$$\inf(u, v) = \tfrac{1}{2}(|u + v|) - \tfrac{1}{2}(|u - v|),$$
$$\sup(u, v) = \tfrac{1}{2}(u + v) + \tfrac{1}{2}|u - v|$$

and the second assertion follows from the formula

$$u + v = |u - v|.$$ □

PROPOSITION 17.14.14. *If u, v are in H and positive with*

(1) $0 \leqslant v \leqslant 1$,
(2) $v(s) = 1$ a.e. (ξ)

then

$$\langle u, v\rangle \geqslant 0.$$

Proof. Since for all $t > 0$

$$v = T(v + tu),$$

where T is the normal contraction $z \to \operatorname{Re} z$ on $(0, 1)$ taking the square of the norm and letting $t \to 0$ the inequality follows. Let H be of the base ξ and let f be a bounded, measurable with compact support. Then the function u_f is a unique element with the property

$$\langle u, u_f\rangle = \int_S u(s)\bar{f}(s)\,\mathrm{d}\xi(s).$$

DEFINITION 17.14.15. The element u_f is called the potential generated by f.

DEFINITION 17.14.16. A pure potential in H is an element of the adherence of the set of potentials (u_f), $f \geqslant 0$.

DEFINITION 17.14.17. The space H is said to have a real kernel if all pure potentials are positive.

DEFINITION 17.14.18. Let H be a space with a real kernel. We say that the principle of maximum is verified in H if for all functions f, g which are bounded, measurable and with compact support and positive then

$$u_f(s) \leqslant u_g(s) + 1$$

holds if it holds on the set $(s, f(s) \geqslant 0)$.

Now if $g = 0$ then the above definition reduces to the classical maximum principle and if (1) is deleted in the above inequality then we get the classical principle of domination. Then we have the following basic result.

THEOREM 17.14.19. *In any Dirichlet space the principle of maximum is verified.*

For a proof we refer to Deny's lectures mentioned above. For another expository paper on Dirichlet space we refer to Deny (1961/62).

17.15. Gravity Theory, Statistical Physics and Dynamics and Inner Product Structures

In this section we intend to present very briefly some applications of inner product structures to physics. First we begin with an interesting approach to gravitational fields as was proposed by Moshe Carmeli in 1969. Further, we present an application to correlation functions and finally some theory of dynamics in inner product spaces as was proposed by Synge and Yodzis (1972). Another very interesting application is in the theory of geometric quantization. We refer to the excellent paper by Vaisman (1979) and the literature quoted there for details.

We now present an application to gravitation theory. In 1968 R. Penrose developed the twistor algebra (for an explanation of this term we refer to Penrose (1969)) in order to obtain a description of zero rest-mass fields in curved space–time as well as flat. Useful tools in these very interesting investigations are certain operators defined on some complete inner product spaces of functions in the complex variables $(z_i)_0^3$ which are supposed to be homogeneous of degree $-2s-2$ and analytic in suitable domains and such that a certain integral is finite. (This integral is taken as a norm of the corresponding function which is generated by an inner product.) This approach has inspired Carmeli (1969) to introduce a complete inner product space of functions on SU_2 which is used to describe the gravitational field (the Weyl tensor) in most natural way. Results concerning this direction of investigation of the gravitational fields are in the papers by Carmeli (1969, 1980, 1983), Carmeli and Malin (1982) and Malin (1982).

Moshe Carmeli's approach consists in taking as natural variables the five complex functions $(\psi_m)_{-2}^2$ which describe the Weyl tensor. These are functions on SU_2 which satisfy the relations

$$\psi_m(cu) = e^{im\lambda}\, \psi_m(u)$$

with λ in R and

$$c = \begin{pmatrix} e^{-im/2} & 0 \\ 0 & e^{-im/2} \end{pmatrix}$$

and ψ_m are also assumed to depend of the radial distance r and time t, $\psi_m = \psi_m(r, t, u)$ $m = -2, -1, 0, 1, 2$.

Let us consider the usual complete inner product space $L^2(\mathrm{SU}_2)$ with respect to the Haar measure $\mathrm{d}u$. The inner product on this space is defined by the formula:

$$\langle f, g\rangle = \int f(u)\bar{g}(u)\,\mathrm{d}u.$$

We consider the subspace $L_2^{2s}(\mathrm{SU}_2)$ which consists of all f_s in the above space and satisfying the following relations:

$$f_s(cu) = \mathrm{e}^{i\lambda s} f_s(u)$$

for each $s = 0, \pm\frac{1}{2}, \pm 1, \ldots$. Then this is a closed subspace and thus is a complete inner product space with respect to the norm induced from the inner product on $L^2(\mathrm{SU}_2)$. Since if f_s, f_r are in the corresponding space we have

$$\langle f_s, f_r\rangle = \int f_s(u)\bar{f}_r(u)\,\mathrm{d}u = \int f_s(cu)\bar{f}_r(cu)\,\mathrm{d}u = \mathrm{e}^{i(s-r)}\langle f_s, f_r\rangle$$

the above subspaces are pairwise orthogonal. We consider now the orthogonal direct sum

$$\sum_{-2}^{2} \oplus L_{2\,2m}^{2}(\mathrm{SU}_2) = \mathscr{H}$$

with the usual inner product and then we get a complete inner product space which is a closed subspace of $L^2(\mathrm{SU}_2)$.

If $(T_{mn}^{j}(u))$ are the matrix elements of the irreducible representations of SU_2 in their canonical form, then they form a complete orthogonal system in $\mathscr{H}$, i.e. a basis.

M. Carmeli (1980) applies the space $\mathscr{H}$ for writing the gravitational field functions as state vectors in $\mathscr{H}$. For example, the Schwarzschild field has the representation:

$$\psi_{-2} = \psi_{-1} = 0, \qquad \psi_0 = -m/r^3, \qquad \psi_1 = \psi_2 = 0$$

and the Schwarzschild state vector is (here we use the notation in use in Physics)

$$\psi_{\mathrm{Schw}} = \left| \begin{matrix} 0 \\ 0 \\ -m/r^3 \\ 0 \\ 0 \end{matrix} \right\rangle$$

and the Kerr and Kerr–Newman fields are given respectively as follows:

$$\psi_{-2} = \psi_{-1} = 0, \qquad \psi_0 = m^3, \qquad \psi_1 = \psi_2 = 0$$

$$\psi_{-2} = \psi_{-1} = 0, \qquad \psi_0 = (m + e^2\bar{\rho})\rho^3, \qquad \psi_1 = \psi_2 = 0,$$

where e is the electric charge, $\rho = -1/(r\text{-}ia\cos\theta)$, a being the angular momentum per mass unit. For details we refer to M. Carmeli's book of 1977 and to the literature quoted in Carmeli (1980).

If ψ denotes the vector state (Schwarzschild, Kerr or Kerr–Newman) then

$$\psi(u) = \sum a_{mn}^{i} T_{mn}^{i}(u).$$

Another interesting fact remarked by M. Carmeli and S. Malin is that the Petrov classification of the gravitational field (based on a classification of the Weyl tensor) is in fact a classification of states in a complete inner product space. We refer to the details in Carmeli and Malin (1981).

We shall now present some facts about the application of inner product structures in statistical mechanics. The problem of central interest is to give conditions under which the so-called correlation functions characterize the system (i.e. the state of the system). A space is a set whose points are infinite particle configurations and we assume further that a measure is defined there. A state of the system means a probability measure over the space and we consider the quantities

$$\rho_n(H) = \int_{H^n} \rho_n \, \mathrm{d}s_1 \cdots \mathrm{d}s_n$$

H^n being a measurable subset in X^n, ρ_n, the correlation functions.

We consider the space K of all finite sequence of points of X. If H is in K and suppose that we enumerate the points of the configuration ξ, $(x_1, \ldots, x_p)$ and by N_H we denote the number of sequences $(i_1, \ldots, i_k)$ with the property that k is arbitrary and all i_k distinct and such that $(x_{i_1}, \ldots, x_{i_k})$ is in H. This number clearly depends on the configuration. Then the mapping on configurations $\xi \to N_H(\xi)$ is a measurable mapping.

If we set

$$H \to \int_X N_H(\xi) \, \mathrm{d}\mu$$

then this is a measure which is called the correlation measure associated with the measure μ.

Using in an ingenious way Carleman's theorem on quasianalytic functions, A. Lenard has obtained the relation

$$\rho(H) = \sum_n \int_{H^n} \rho_n \, \mathrm{d}s_1 \cdots \mathrm{d}s_n.$$

As is known, Carleman's theorem was used to obtain some interesting results about quasianalytic vectors of an operator on a complete inner product space. (See the author's book, 1981, concerning this problem. A configuration is a finite or countable system of points of X considered without regard to order and repetitions are permitted.) Later S. I. Levy considered modified correlation measures and a complete inner product space associated intrinsically with X from which he obtained the uniqueness of μ using semianalytic vectors. For details see the papers by Lenard (1973) and Levy (1974).

As mentioned above, another application in which we are interested is the development of some dynamics in complete inner product spaces. This was done in an elegant paper by Synge and Yodzis (1972) to which the reader is referred for details as well as for some very interesting historical notes.

Suppose that H is a complete inner product space with the orthonormal basis $(e_i(t))$, where t indicates that the basis is rotating. We use the notation '$\cdot$' for $\mathrm{d}/\mathrm{d}t$

(the dot notation) and thus

$$\dot{e}_i = \sum_j w_{ij} e_j.$$

We suppose also that orthogonality is preserved, and thus

$$0 = \langle \dot{e}_i, e_j \rangle + \langle e_i, \dot{e}_j \rangle$$

which gives that

$$w_{ij} + \bar{w}_{ji} = 0.$$

The operator on H, which has the matrix $w = (w_{ij})$ with respect to the above basis, is called the 'angular velocity' for the basis (e_i). If $(\bar{e}_i)$ is another orthonormal basis of H then

$$\dot{\bar{e}}_i = \sum_j a_{ij} e_j \qquad (+)$$

and the matrix (a_{ij}) defines a unitary operator on H, say U. This is independent of t. From the form of $(+)$ we have

$$\dot{\bar{e}}_i = \sum_j a_{ij} e_j$$

and thus taking into account the form of $\dot{e}_i$ we get the following relations:

$$\bar{e}_i = \sum_j a_{ij} \sum_k w_{jk} \dot{e}_j. \qquad (++)$$

If $(\bar{w}_{ij})$ is the angular velocity for the basis $(\bar{e}_i)$ then $(++)$ implies that

$$\bar{w} = (\bar{w}_{ij}) = UwU^*.$$

We now present an analogue of a basic result in classical dynamics (see Whittaker, 1927), the so-called Hayward theorem. In order to present it we need some preliminary facts.

Suppose that (f_i) is an orthonormal basis for H and we consider this fixed. Connected with this we consider a skew-hermitian operator, $f = (f_{ij})$. This is called by J. L. Synge and P. Yodzis the primary bivector base. Obviously, if (e_i) is another basis of H then $e_i = \Sigma_j b_{ij} f_j$ and (b_{ij}) is the matrix of a unitary operator U on H. Associated to (e_i) is the matrix given by $e = VfV^*$.

If x is a skew symmetric operator then it transforms with respect to the basis by the formula $x \to V_{\times} V^*$.

If we set $\hat{x} = \mathrm{tr}(xe)$ then this is a geometrical object because its value is independent of the orthonormal basis used. It is also clear that $\hat{x}$ is real. If we consider the angular velocity w then we have $\hat{w} = \mathrm{tr}(we)$, which is called the angular velocity of H. If x is an arbitrary bivector then

$$\mathrm{d}\hat{x}/\mathrm{d}t = \mathrm{d}/\mathrm{d}t(\mathrm{tr}(xe)) = \mathrm{tr}(\dot{x}e) + \mathrm{tr}(x\dot{e})$$

and if we compute $\dot{e}$ we get that

$$\dot{e} = (\dot{e}_{ij}) = we - ew$$

which implies that

$$\mathrm{tr}(x\dot{e}) = \mathrm{tr}(x(we - ew)) = \mathrm{tr}(w(ex - xe)) + \mathrm{tr}(e(xw - wx))$$

or

$$\mathrm{d}\hat{x}/\mathrm{d}t = \mathrm{tr}(e(x - wx + xw)) \qquad (+++)$$

which is

THEOREM 17.15.1 (Hayward's theorem; Synge and Yodzis, 1972). *If H rotates with angular velocity w then the rate of change of bivector x is given by formula* $(+++)$.

In their paper Synge and Yodzis (1972) present generalizations of the Euler equations of motions of a rigid body with fixed point, and they also present other paper that may be of interest to the reader.

17.16. Quantum Mechanics and Operators on Complete Inner Product Spaces

We now present a short introduction to axiomatic theory of quantum mechanics using the theory of unbounded linear operators on complete inner product spaces. For the reader interested in quantum mechanics using the theory of operators of quaternionic spaces, we refer to Finkelstein *et al.* (1962, 1963); we also refer to the fundamental book of von Neumann (1929) and to the works of Mackey (1963), Wigner (1959), van der Vaerden (1931), Weyl (1931), Dirac (1958, 1974), Blochintsev (1961), Schiff (1955), Feynman *et al.* (1965) and de Broglie (1929).

We consider a mechanical system and in what follows it is not important to know exactly what this is since we are interested in objects related to the system, the so-called states of the system. Our first axioms make it possible for us to make the states of a system accessible to mathematical tools of operator theory on inner product spaces.

AXIOM 17.16.1. *The set of all states of a quantum mechanical system is in one-to-one correspondence with the family of all subspaces of a complex complete inner product space X, which are of dimension* 1.

AXIOM 17.16.2. *The set of all observables of a quantum mechanical system is in one-to-one correspondence with the family of all selfadjoint operators on a complex complete inner product space X.*

The following axiom relates the contents of the two axioms above.

AXIOM 17.16.3. *Suppose that in a quantum mechanical system we have a state and an observable. From Axiom* 17.16.1 (*the state*) *we have an element of the complex complete inner product space X, say f, and according to Axiom* 17.16.2 (*the observable*) *we have a selfadjoint operator on X, say A. If the spectral decomposition of A is*

$$A = \int_{-\infty}^{\infty} t \, \mathrm{d}E_t$$

then the probability that the measurement of the observable is in

(1) (a, b) *is* $\|(E_b - E_{a+0})f\|^2$,
(2) $(a, b]$ *is* $\|(E_{b+0} - E_{a+0})f\|^2$,
(3) $[a, b)$ *is* $\|(E_b - E_a)f\|^2$,
(4) $[a, b]$ *is* $\|(E_{b+0} - E_a)f\|^2$.

REMARK 17.16.4. It is not difficult to see that if we define, for any interval (a, b) on R,

$$\mu(a, b) = \|(E_b - E_{a+0})f\|^2$$

then this can be extended to a probability measure on (R, B^1), B^1 being the family of Borel subsets of R.

The above group of axioms describe the static of a quantum mechanical system. Now we give a group of axioms describing the dynamic of the system, i.e. the evolution of the system with respect to the variable time (denoted in what follows by t).

AXIOM 17.16.5. *For any quantum mechanical system there exists a unique selfadjoint operator B on X (X is a complex complete inner product space which is the same for all axioms) such that if the system at $t = 0$ is in the state corresponding to f_0 in X $\|f_0\| = 1$ then at each time $t > 0$ the system is in the state corresponding to f_t with*

$$f_t = e^{-t/h \cdot B} f_0$$

where h is Planck's constant.

DEFINITION 17.16.6. The operator B is called the hamiltonian of the quantum mechanical system.

DEFINITION 17.16.7. The state of a quantum mechanical system is said to be stationary if it is invariant with respect to t (which means that $f_t = f_0$ with the notations of Axiom 17.16.5).

EXAMPLE 17.16.8. We now consider the classical example of a mechanical system of quantum mechanics, the so-called harmonic oscillator.

We consider the movement of a particle on R which, at time t, is at the point $x(t) \in R$. The energy of the particle is given by the formula

$$\tfrac{1}{2}\mathrm{m} \cdot \mathrm{d}^2x/\mathrm{d}t^2 = (2m)^{-1}p^2,$$

where p is the impulse of the particle with mass m. The potential energy is

$$V(x) = \frac{k}{2}x^2$$

with k a positive constant. The hamiltonian of the system is

$$h(x, p) = (2m)^{-1}p^2 + \frac{k}{2}x^2.$$

The operators A and B of Axioms 17.16.2 and 17.16.5 are defined as follows:

$$A: D_A \to L^2(R, \mathrm{d}x), \ (Af)(x) = xf(x) \text{ a.e. } (\mathrm{d}x)$$

$$D_A = (f, f \in L^2, xf(x) \in L^2).$$

Since the functions of class C^∞ with compact support are obviously in D_A we see that D_A is dense in L^2. Also it is clear from the examples of selfadjoint operators that A is a selfadjoint operator.

Let us consider the operator on L^2 defined by the formula

$$Tf = -h^2/2m \, \mathrm{d}^2 f/\mathrm{d}x^2 + \frac{k}{2} x^2 f(x)$$

$$D_T = (f, f \in L^2, f \text{ has a compact support}).$$

It is clear that D_T is a dense subset (subspace) in L^2. The operator B is defined as the closure of the operator T, and from the examples of essentially selfadjoint operators we conclude that B is essentially selfadjoint.

Suppose that we have a mechanical system and that we are interested in the stationary states of the system. The following result gives a characterization of these states using the hamiltonian of the system. From here on we identify completely the objects that are in correspondence as, in the axioms stated above.

THEOREM 17.16.9. *If a quantum mechanical system has a hamiltonian B then the state f is stationary if and only if f is an eigenvector of B.*

Proof. Suppose first that f is an eigenvector of B, i.e. for some real number z, $Bf = zf$, $\|f\| = 1$. Then clearly we have that

$$(E_{r+0} - E_r)f = f,$$

where (E_λ) is the decomposition of unity associated with B. Further,

$$\|\mathrm{e}^{-it/h \cdot B} f - \mathrm{e}^{itr/h} f\| = \int_{-\infty}^{\infty} |\mathrm{e}^{-i/h \cdot \lambda t} - \mathrm{e}^{-irt/h}|^2 \, \mathrm{d}\|E_\lambda f\|^2 =$$

since

$$E_\lambda f = 0, \quad \lambda \leqslant r$$

$$E_\lambda f = f, \quad \lambda > r.$$

Thus,

$$f_t = \mathrm{e}^{-irt/h} f$$

which shows that f is a stationary state of the system (here and in what follows we identify all elements of the form zf, $|z| = 1$).

Conversely, if f is a stationary state of the system whose hamiltonian is B then

$$f(t) = g(t) f$$

where $g: R \to C$ satisfies the following properties:

(1) $|g(t)| = 1$, $\quad g$ continuous,
(2) $g(t + s) = g(t)g(s)$.

But is a well-known result that a continuous function g satisfying the above properties is necessarily of the form $g(t) = e^{imt}$, with m a constant. Thus we can write

$$f_t = e^{ir/h \cdot t} f$$

and we show that f is an eigenvector of B corresponding to the eigenvalue r. To prove this it suffices to show that

$$E_\lambda f = 0, \quad \lambda \leqslant r$$
$$E_\lambda f = f, \quad \lambda > r.$$

Suppose that this is not true and thus there exists $\mu \neq r$ such that for some $\varepsilon > 0$

$$(E_{\mu+\varepsilon} - E_{\mu-\varepsilon})f \neq 0.$$

Then we have

$$0 = \| f_t - f_t \|^2 = \| e^{-it/h \cdot B} f - e^{-it/h \cdot r} f \|^2$$
$$= \int_{-\infty}^{\infty} |e^{-i\lambda t/h} - e^{-irt/h}|^2 \, d\| E_\lambda f \|^2$$
$$= \int_E |e^{-i\lambda t/h} - e^{-irt/h}|^2 \, d\| E_\lambda f \|^2, \quad E = [\mu - \varepsilon, \mu + \varepsilon].$$

We may suppose that λ is sufficiently small such that

$$|e^{-\lambda it/h} - e^{-irt/h}| \geqslant d > 0 \quad \text{for all } \lambda \text{ in } [\mu - \varepsilon, \mu + \varepsilon].$$

Then the above inequalities imply that

$$0 \geqslant d(\| E_{\mu+\varepsilon} f \|^2 - \| E_{\mu-\varepsilon} f \|^2) = d_1(\| E_{\mu+\varepsilon} - E_{\mu-\varepsilon}) f \|^2.$$

But this is clearly a contradiction and thus the assertion is proved. □

According to our axioms above, a fundamental role in the study of a system is played by the hamiltonian operator, and thus it is of fundamental importance to descirbe methods or procedures to obtain the hamiltonian of a system. We shall describe later the so-called 'quantification' of Louis de Broglie and E. Schrödinger to obtain the hamiltonian of a system in quantum mechanics.

We now indicate some 'interpretation rules' for the above notions connected with a system in quantum mechanics. These are given as 'rules', and are as follows.

RULE 17.16.10. *If we have a system which is invariant in time, then it is in a stationary state. This stationary state is an eigenvector of the hamiltonian and the eigenvalue is the energy of the system. The normal state of a system is the stationary state which has the smallest energy.*

RULE 17.16.11. *If the system moves from a stationary state with energy E_1 to another stationary state with energy E_2 then an electromagnetic radiation is generated. The frequency is given by the formula*

$$\nu = h^{-1}(E_2 - E_1)$$

where h is Planck's constant.

The following example may be useful in clarifying the meaning of these rules.

EXAMPLE 17.16.12. We consider a harmonic oscillator. Then the eigenvalues of the hamiltonian are

$$E_n = h(k/m)^{1/2}(n + \tfrac{1}{2}), \quad n = 0, 1, 2, 3, \ldots$$

and thus the frequencies are

$$\gamma_{n_1,n_2} = h^{-1}(E_{n_1} - E_{n_2}) = (2\pi)^{-1}(k/m)(n_1 - n_2).$$

If we set $\gamma = \gamma_{1,0} = (2\pi)^{-1}(k/m)^{1/2}$, then

$$E_n = h \cdot \nu(n + \tfrac{1}{2}), \quad n = 0, 1, 2, 3, \ldots$$

which are the Planck's hypotheses for a harmonic oscillator.

We now define the notion of mean value, which is very important in connection with the observables associated with a system.

DEFINITION 17.16.13. Let us consider a quantum mechanical system from a quantum mechanical point of view, and let A be a selfadjoint operator which represents an observable of the system. Suppose that the domain of A is D_A and consider a state f of the system in which the system exists. Then we say that $\langle Af, f \rangle$ is the mean value of the measurement of the observable in state f. The number $\delta(f, A) = \| Af - \langle Af, f \rangle f \|$ is called the variance of the observable (at state f).

DEFINITION 17.16.14. A measurement of the observable is called sharp if the following relation holds:

$$\delta(A, f) = 0.$$

From the definition it is clear that the following assertion holds.

PROPOSITION 17.16.15. *A measuring is sharp if and only if*

$$Af = \langle Af, f \rangle f.$$

We mention that the problem of physical interpretation in quantum mechanics is more difficult than in the case of classical mechanics and the observables which are objects of experimental procedures of measuring are those of position, momentum and spin. We mention that there are two types of measurements. In the first, the so-called determinative measurement, one can compute (within the bounds of accuracy related to the apparatus used) what the values of the respective observables would be if we assume that no disturbances arise due to the interaction of the apparatus and the system.

The second type of measurement is the so-called preparatory measurement. By this method we obtain not an exact value but a 'range' of values for the corresponding observable in which we are interested. Suppose now that we have an apparatus which is used to obtain a range for the position and also for the momentum of an atomic object, then classical mechanics tells us that by constructing sufficiently good devices we can make the linear dimension of the above ranges for the position and momentum as small as we wish. But due to diffraction effects it is a well-known fact that this is not the case. This has led to the formulation of the so-called 'uncertainty principle'. This principle says that there does not exist an apparatus which prepares a particle to have a position in an interval of length Δx and a component p_x of momentum in an interval of length Δp_x without satisfying the inequality

$$\Delta x \Delta p_x \geqslant h/2\pi.$$

For interesting discussions on this relation in quantum mechanics we refer the reader to books on the subject.

In what follows we give an abstract formulation of the uncertainty principle in the formalism considered above.

THEOREM 17.16.16 (The 'Uncertainty Principle'). *Suppose that we have a quantum mechanical system and A, B are two observables (and according to our convention these are identified with two selfadjoint operators which are denoted by the same letters). Let us consider further f be a state of the system with the following properties:*

(1) *f is in D_A and in D_B,*
(2) $Af \in D_B$, $Bf \in D_A$.

Then the following inequality holds:

$$\delta(A, f) \cdot \delta(B, f) \geqslant \tfrac{1}{2} |\langle (AB - BA)f, f \rangle|.$$

Proof. We consider two real numbers a and b and then we have the following identity:

$$\begin{aligned}\langle (BA - AB)f, f \rangle &= \langle (B - bI)(A - aI)f - (A - aI)(B - bI))f, f \rangle \\ &= \langle (B - bI)(A - aI)f, f \rangle - \langle f, (B - bI)(A - aI)f \rangle \\ &= 2i \operatorname{Im}\langle (B - bI)(A - aI)f, f \rangle.\end{aligned}$$

Now, the operators A and B are selfadjoint, a, b are real numbers, and then Af, f and Bf, f are also real numbers; the numbers a and b are arbitrary and thus we can set $a = \langle Af, f \rangle$, $b = \langle Bf, f \rangle$. Then from the above identity we obtain the relation

$$\langle (BA - AB)f, f \rangle = 2i \ \operatorname{Im}\langle (A - \langle Af, f \rangle)f, (B - \langle Bf, f \rangle)f \rangle$$

and in the right side we can apply the Cauchy–Buniakowsky inequality to obtain that

$$|\langle (AB - BA)f, f \rangle| \leqslant 2\delta(A, f) \cdot \delta(B, f)$$

and the uncertainty principle is proved. □

From this result we easily obtain the following.

COROLLARY 17.16.17. *Suppose that the operators A and B in Theorem 17.16.16 satisfy the relation $(AB - BA)f = -rf$ for some complex number z. Then*

$$\tfrac{1}{2} r \leqslant \delta(A, f) \cdot \delta(B, f).$$

EXAMPLE 17.16.18. We now consider the case of the harmonic oscillator. The position of a particle is described by the operator A where

$$Af = xf$$

$$D_A = (f, f \in L^2(R), xf(x) \in L^2).$$

The Hamiltonian is the operator B defined by the formula

$$Bf = h/i \cdot d/dx\, f, \quad h \text{ is Planck's constant.}$$

with the domain in the Sobolev space $D_B = W_2^1(R)$. Then B is a selfadjoint operator and for f we have

$$(BA - AB)f = h/i \cdot f.$$

Thus, in this case $r = -h/i$ and thus

$$r/2 \leqslant \delta(A, f) \cdot \delta(B, f).$$

This relation was originally obtained by W. Heisenberg for the position and momentum operators.

17.16.19. The above relation between the operators A and B is of course related to the famous canonical commutation relations: suppose that n is an integer and we have the selfadjoint operators $P_1, \ldots, P_n$ and $Q_1, \ldots, Q_n$, then we say that they satisfy the canonical commutation relations if for all k, $k = 1, 2, \ldots, n$, the following relations hold:

$$(P_k Q_k - Q_k P_k)f = -ihf, \quad f \in D_{(P_k Q_k - Q_k - P_k)}, \quad h \text{ is Planck's constant,}$$

and any other two operators in $(P_1, \ldots, P_n; Q_1, \ldots, Q_n)$ commute.

We note also that there are a great number of papers dealing with the representation of cannonical commutation relations.

As we noted above in the study of a quantum mechanical system, it is of fundamental importance to know the Hamiltonian of the system. The following quantification rule gives a procedure to obtain the Hamiltonian of a quantum mechanical system if we know the Hamiltonian of the system considered from the point of view of classical mechanics.

RULE 17.16.20 (Quantization). *Suppose that we have a quantum mechanical system of n particles described by the coordinates $x_1, \ldots, x_{3n}$ and with the momentum coordinates $p_1, \ldots, p_{3n}$ and the Hamiltonian of the system considered from the point of view of classical mechanics is H. We know that $H = H(x_1, \ldots, x_{3n}; p_1, \ldots, p_{3n})$.*

In the expression of H, instead of x_i we write a multiplication with x_i and instead of p_i we consider $h/i \cdot \partial/\partial x_i$ (where h is Planck's constant), and thus we obtain an expression

$$H(x_1, \ldots, x_{3n}; h/i\, \partial/\partial x_1, \ldots, h/i\, \partial/\partial x_{3n})$$

which is clearly a differential operator, and this will be the Hamiltonian of the system (and, of course, this operator acts on a space of functions in R^{3n}).

We now give an example to show how this procedure operates.

EXAMPLE 17.16.21. We consider a harmonic oscillator. The Hamiltonian of the system, from a classical mechanical point of view, is

$$h(x, p) = (2m)^{-1} p^2 + k/2 \cdot x^2$$

where k is a constant.

According to Rule 17.16.20, after quantization we obtain the following differential expression:

$$(2m)^{-1}(h/i\cdot\partial/x)^2 + k/2\cdot x^2 = -h^2/2m\cdot d^2/dx^2 + k/2\cdot x^2,$$

and thus the Hamiltonian of our quantum mechanical system is

$$Bf = -h^2/2m\cdot d^2f/dx^2 + k/2\cdot x^2 f(x)$$

and the domain of B is

$$D_B = (f, f'' \text{ exists, } x^2 f(x)\in L^2(R)).$$

Another example is now given by considering that the particle i is not in R but in R^3.

EXAMPLE 17.16.22. Suppose that we have a particle in R^3 and the Hamiltonian of the system from a classical mechanical point of view is

$$H(x, y, z, p_1, p_2, p_3) = (2m)^{-1}(p_1^2 + p_2^2 + p_3^2) + V(x, y, z).$$

Then the hamiltonian of the system from a quantum mechanical point of view is

$$-h^2/2m(\partial^2/\partial x^2 + \partial^2/\partial y^2 + \partial^2/\partial z^2)f + V(x, y, z)f = Bf$$

with the domain

$$D_B = (f, f\in L^2(R^3), f \text{ of class } C^\infty \text{ and with compact support}).$$

If the system has many particles which are independent, they we assume that the Hamiltonian of the entire system is equal to thē sum of the Hamiltonians of the particles of the system.

17.17. Number Theory and Complete Inner Product Spaces

It was remarked that certain classes of functions have very interesting connections with problems in number theory. We mention in this context the following example of such functions: suppose that we have the interval $[0, \frac{1}{2}]$ and we consider the sequence of functions

$$1, (t), (2t), \ldots, (nt), \ldots,$$

where (h) means the least nonnegative residue of h modulo 1. Then the above sequence is a basis for $L^2[[0, \frac{1}{2}], dt]$, dt being the Lebesgue measure. Results of this type, with very interesting connections with questions in number theory, are contained in the papers by Wintner (1944) and Hartman (1947). Results on some orthonormal systems related to number theory are also given in papers by Romanov (1946, 1951). In this section we present some of Romanov's result. The reader may consult these papers for other interesting results.

In order to present some of Romanov's theorems we need some notations and results from number theory. For further details we refer to A. Blanchard's book of 1969 and to the famous books by G. H. Hardy, E. M. Wright and I. M. Vinogradov of 1964 and 1954 (Russian).

The Möbius function μ is defined on a set of integers by

$$\mu(x) = \begin{cases} 1, & x = 1, \\ (-1)^r, & \text{if } x \text{ is the product of } r \text{ distinct prime numbers} \\ 0, & \text{otherwise.} \end{cases}$$

One of the basic properties of the Möbius function is that it is multiplicative (in the sense of number theory, i.e. if x, y are numbers with $(x, y) = 1$ then $\mu(xy) = \mu(x)\mu(y)$). The Möbius function appears in many places; we mention only the fact that Riemann's ξ function for $\mathrm{Re}\, s \geqslant 1$ has an inverse given by the Dirichlet series

$$\xi^{-1}(s) = \sum_{n=1}^{\infty} \mu(n) n^{-s}.$$

Some useful properties of the Möbius function are given in the following proposition.

PROPOSITION 17.17.1. *The Möbius function* μ *has the following properties:*

(1) $\sum_{k/n} \mu(k) = \begin{cases} 1, & n = 1 \\ 0, & n > 1, \end{cases}$

(2) *If* f *and* g *are two functions defined for* $t \geqslant 1$ *and*

$$g(t) = \sum_{n \leqslant t} f(t/n)$$

then

$$f(t) = \sum_{n \leqslant t} (n) g(t/n).$$

Proof. The Assertion in (1) is obvious if $n = 1$. Suppose that $n > 1$. We set $P(n) = \Pi_{p/n}\, p$ and then

$$\sum_{k/n} \mu(k) = \sum_{k/P(n)} \mu(k).$$

Now if $P(n)$ has r distinct prime factors, we have

$$\sum_{k/P(n)} \mu(k) = 1 + C_r^1 \cdot (-1) + C_r^2 \cdot (-1)^2 + \cdots + C_r^r \cdot (-1)^r = 0$$

and thus (1) is proved.

Now for the assertion in (2) we have for $\mathrm{Re}\, s > 1$ that

$$1 = \xi(s) \cdot \xi^{-1}(s) = \left(\sum_{n=1} n^{-s} \right) \left(\sum_{n=1} \mu(n) n^{-s} \right) = \sum_{n=1} \mathrm{d}(n) n^{-s}$$

with $\mathrm{d}(n) = \Sigma_{k/n}\, \mu(k)$. Since

$$\begin{aligned} f(t) &= \sum_{n \leqslant t} \mathrm{d}(n) f(t/n) = \sum_{n \leqslant t} f(t/n) \sum_{k/n} \mu(k) \\ &= \sum_{lk \leqslant t} \mu(k) f(t/(kl)) = \sum_{k \leqslant t} \mu(k) \sum_{l \leqslant t/k} f(t/(lk)) \\ &= \sum_{k \leqslant t} \mu(k) g(t/k) \end{aligned}$$

and thus (2) is proved.

DEFINITION 17.17.2. Two functions h, h_1 defined for $t \geqslant 1$ are said to be a Möbius couple (h, h_1) if

$$h_1(t) = \sum_{n \leqslant t} h(t/n).$$

COROLLARY 17.17.3. *If* (h, F) *is the Möbius couple defined by* $F(n) = \Sigma_{m/n} f(m)$ *(and* $f(t) = 0$ *if* t *is not equal to an integer) then*

$$f(n) = \sum_{m/n} \mu(m)F(n/m).$$

The above formula is known as the Möbius inversion formula.

COROLLARY 17.17.4. *Let* (b_n) *be a sequence and set*

$$a_n = \sum_{k/n} b_k, \qquad b_n = \sum_{k/n} (n/k)a_k.$$

For the result, which we give later, the following assertion is very important.

THEOREM 17.17.5 (Romanov, 1946). *Let* f *be a function defined on the integers and* $k < n$. *Then*

$$\sum_{d/n} (n/d)f((d, k)) = 0.$$

Proof. For any function f defined on the integers we set

$$Sf(n) = \sum_{d/n} f(d), \; \bar{S}f(n) = \sum_{d/n} (n/d)f(d),$$

$$Af(n) = f((n, N)), \qquad Bf(n) = \varepsilon(N/n)f(n), \qquad Cf(n) = f(n),$$

where

$$\varepsilon(n) = \begin{cases} 1, & n = 2k \\ 0, & n = 2k + 1 \end{cases}$$

and N is a given integer. Then clearly we have the following relations:

(1) $SBf(n) = S(N/n)f(n) = \sum_{d/n} (N/n)f(\mathrm{d}),$

$$= \sum_{d/(n,N)} f(d) = ASf(n),$$

(2) $SB = AS,\ B = S^{-1}AS,\ BS^{-1} = S^{-1}A,$

(3) $S^{-1}Af(n) = S^{-1}f((n, N)) = \sum_{d/N} (n/d)f((d, N))$

$$= BS^{-1}f(n) = (N/n)S^1 f(n).$$

If we set $N = k < n$ then the assertion of the theorem follows. □

In order to construct certain families of elements in a complete inner product space X we introduce the following notion.

DEFINITION 17.17.6 (Romanov, 1946). Let g be a function defined on the integers and let (x_n) be a family of elements in X. We say that (x_n) has property D_g if

$$\langle x_n, x_m \rangle = g((n, m)).$$

Using the property D_g we have the following method to generate orthogonal families of elements.

PROPOSITION 17.17.7 (Romanov, 1946). *Let (x_n) be a sequence in a complete inner product space X and suppose that (x_n) has property D_g. Then the sequence (y_n) with*

$$y_n = \sum_{d/n} (n/d) x_d$$

is an orthogonal sequence.

Proof. Suppose that $k < $ n. Then we have

$$\langle y_k, x_n \rangle = \sum_{d/n} \mu(n/n) \langle x_d, x_n \rangle = \sum_{d/n} (n/d) g((d, k))$$

which implies that for $m < n$, $\langle y_m, y_n \rangle = 0$. Now

$$\langle y_n, y_n \rangle = \left\langle y_n, \sum_{d/n} (n/d) x_d \right\rangle = \langle y_n, x_n \rangle = \sum_{d/n} \langle (n/d) x_d, x_n \rangle$$

$$= \sum_{d/n} (n/d) g(d).$$

We now assume that g is a function on the integers such that

$$\sum_{d/n} \mu(n/d) g(d) \neq 0.$$

Then we have

COROLLARY 17.17.8. *If we set*

$$G(n) = \sum_{d/n} \mu(n/d) g(d)$$

then if (x_n) is a sequence with property D_g then the sequence (y_n), with

$$y_n = \sum_{d/n} G(d)^{1/2} x_d / (G(n) \cdot {}^{1/2}),$$

is an orthonormal sequence.

Using the properties of the Möbius function given above we get that

$$x_n = \sum_{d/n} (G(d)^{1/2} y_d).$$

PROPOSITION 17.17.9 (Romanov, 1946). *Suppose that w is a complex-valued function on the integers and*

(1) $w(nm) = w(n) w(m)$,

(2) $\sum_{n=1}^{\infty} |w(n)|^2 < \infty$.

If (x_n) *is an orthonormal sequence of elements in a complex complete inner product space, then the sequence* (y_n), *where*

$$y_n = \bar{w}(n)^{-1} \sum_{k=1}^{\infty} w(k)x_{kn},$$

has property D_g, *where* g *is defined by the relation*

$$g(n) = r/|w(n)|^2 \cdot r = \sum_{k=1}^{\infty} |w(k)|^2.$$

Proof. For arbitrary n and m we have

$$\begin{aligned}\langle y_n, y_m \rangle &= \left\langle \bar{w}(n)^{-1} \sum_{k=1} w(k)x_{kn}, \bar{w}(n)^{-1} \sum_{s=1}^{\infty} w(s)x_{sn} \right\rangle \\ &= \left\langle \bar{w}(n)^{-1} \sum_{k=0(\mathrm{mod}\, n)} w(k/n)x_k, \bar{w}(m)^{-1} \sum_{l=0(\mathrm{mod}\, m)} w(l/m)x_1 \right\rangle \\ &= \bar{w}(n)^{-1}\bar{w}(m)^{-1} \sum_{k=0(\mathrm{mod}\,(n,m))} \bar{w}(k/n)\bar{w}(k/m) \\ &= |(\bar{w}(n)\bar{w}(m))|^{-2} \sum_{k=0(\mathrm{mod}\,(n,m))} |w(k)|^2 \\ &= \sigma(|w(n)||w(m)|)^{-2}|w((n,m))|^2\end{aligned}$$

and the assertion follows. □

COROLLARY 17.17.10 (Romanov, 1946). *Suppose that* w *is a complex-valued function on the integers and*

(1) $w(n) \neq 0$,
(2) $w(nm) = w(n)w(m)$,

(3) $r = \sum_{k=1}^{\infty} |w(k)|^2 < \infty.$

Then for any orthonormal sequence (x_n) *of a complete complex inner product space* X *the sequence* (y_n) *is an orthonormal sequence and where*

$$y_n = (r \cdot O(n))^{-1/2} \sum_{d/n} (n/d)|\bar{w}(d)|^{-1} \sum_{k=1}^{\infty} w(k)x_{kd}$$

and

$$O(n) = \sum_{d/n} (n/d)|w(d)|^{-2} = w(n)^{-2} \prod_{p/n} (1 - |w(p)|^2).$$

Concerning the orthonormal systems which generate an inner product space X, we have the following result.

THEOREM 17.17.11 (Romanov, 1946). *Suppose that* (x_n) *is an orthonormal system in the subspace* X_1 *which it generates. Then the system* (y_n) *from Corollary* 17.17.10 *is a basis for the same subspaces* X_1.

In his 1946 paper N. P. Romanov introduced the notion of a (infinite) matrix with property D_g: the matrix (a_{ij}) is said to have the property D_g if

$$a_{ij} = g((i,j)),$$

where g is a function on the integers; the matrix (a_{ij}) is said to have property D if it is of the form AA^* for some matrix A.

Suppose that g is a function on the integers and set

$$G(n) = \sum_{d/n} (n/d)g(d)$$

and

$$M = (m_{ij}), \qquad m_{ij} = G(i)^{-1/2}\mu(i/j)$$

$$A = (a_{ij}), \qquad a_{ij} = e_{ij}G(j)^{1/2}, \qquad e_{ij} = \begin{cases} 1, & \text{if } i = jp \text{ for some } p \\ 0 & \text{otherwise.} \end{cases}$$

Then using the properties of the Möbius function we obtain the relation

$$AM = I.$$

If $B = (b_{ij})$, $b_{ij} = g((i,j))$ then we also have the relations

$$AA^* = B, \qquad MBM^* = I.$$

Using some special functions g, some matrices are constructed and interesting identities are then derived. We refer to Romanov (1946) for details.

If for x real we set

$$f(x) = x - [x] - \tfrac{1}{2}$$

then in Landau's famous book (1927, 1955), pp. 170ff, the following interesting relations are proved:

$$\int_0^1 f(nt)f(mt)\,\mathrm{d}t = (n,m)^2(12nm)^{-1}.$$

In his 1951 paper, N. P. Romanov extends this result very much as follows: suppose that g satisfies the functional equation

$$g(mt) = m^{s-1}\sum_{0}^{m-1} g(t + k/m)$$

then, if we set

$$h_n(t) = c^{-1/2}n^s g(nt - [nt])$$

with

$$\int_0^1 g(nt)g(mt)\,\mathrm{d}t = c(mn)^{-s},$$

we obtain

$$\int_0^1 h_n(t)h_m(t)\,\mathrm{d}t = (n, m)^{2s}.$$

Other similar constructions can be made using the Bernoulli polynomials. For other constructions and applications we refer the reader to Romanov (1946, 1951, 1948) and Kotelianskiĭ (1953).

CHAPTER 18

A Collection of Problems

18.0. Introduction

In this very short chapter we present some problems connected with the inner product structures. This collection of problems is by no means complete and our purpose is to direct the attention of interested readers to problems that are mentioned by different authors. We give references to where the problem is first stated as well as references to where the latest results (to the best knowledge of the present author) can be found.

The author would be greatly indebted for information concerning the problems, as well as related problems, that may be used to improve the present collection.

18.1. Problems on Inner Product Structures

PROBLEM 18.1.1 (Mazur, 1938). Let X be a Banach space and suppose that each pair of elements x, y in X, $\|x\| = \|y\|$ has the property that for some rotation U of X, $Ux = y$. Then, is the norm of X generated by an inner product? [S. Mazur, *C. R. Acad. Sci. Paris* 207 (1938), 761–764.]

PROBLEM 18.1.2 (Belluce and Kirk, 1967). Let X be a Banach space. We say that X has property (R) if any bounded closed, convex subset C of X has a fixed property with respect to nonexpansive mappings. Then, is the norm of X generated by an inner product iff X has property (R) and every C as above is the convex closure of the fixed point set of some nonexpansive mapping on X? [L. P. Belluce and W. A. Kirk, *Illinois J. Math.* **11** (1967), 474–479; R. E. Burck, Jr, *Proc. A. M. S.* 43(1974), 173–175.]

Remark. If X is reflexive, the answer is 'yes'.

PROBLEM 18.1.3 (DeMarr, 1967). Let X be a partially ordered linear space which is order isomorphic to its order dual X'. Then is it possible to define an inner product $\langle\, , \rangle$ on X such that the following properties hold:

(1) If x is in X then $\langle x, \rangle$ is a positive linear functional in X'.
(2) If x' is in X' and $x' \geqslant 0$ then there exists x in X, $x \geqslant 0$ such that $\langle x, \rangle = x'$?

[Ralph DeMarr, *Illinois J. Math.* **11**(1967), 71–77.]

PROBLEM 18.1.4 (Pełczýnski, 1972). How do we define a Hilbert space 'intrinsically' in the language of normed spaces? [A. Pełczýnski, *Wiadom. Mat.* (2) **15** (1972), 3–11.]

PROBLEM 18.1.5 (Folklore). Let X be a complex Banach space and suppose that for each bounded linear operator on X, T, the numerical range, in Lumer's sense, is a convex set. Then, is the norm of X generated by an inner product?

PROBLEM 18.1.6 (Cohen, 1973). Let X, Y be two Banach spaces and let U be in $X \otimes Y$. Then do we have $\|u\|_\varepsilon = w_2(u)$ when either X or Y is a Banach space with a norm generated by an inner product? [J. S. Cohen, *Math. Ann.* 201(1973), 177–200.]

PROBLEM 18.1.7 (Lindenstrauss and Tzafriri, 1977). Let X be an infinite dimensional Banach space with the property that for each couple of quotient maps T_1 and T_2 on the same space Y, there exists an automorphism t of X such that $T_1 = T_2 t$. Does it follow that X is isomorphic to either l^1 or l^2? [J. Lindenstrauss and L. Tzafriri, *Lecture Notes in Math.* No. 338.]

PROBLEM 18.1.8 (Lindenstrauss and Tzafriri, 1977). Suppose that every infinite dimensional subspace (or quotient space) of X is isomorphic to X. X isomorphic to l^2? [J. Lindenstrauss and L. Tzafriri, *Lecture Notes in Math.* No. 338.]

PROBLEM 18.1.9 (Lindenstrauss and Tzafriri, 1977). Suppose that the Banach space X, up to equivalence, has a unique symmetric basis and this basis is minimal symmetirc. Is then X isomorphic to c_0 or $l^p(p$ in $[1, \infty))$? [J. Lindenstrauss and L. Tzafriri, *Lecture Notes in Math.* No. 338.]

PROBLEM 18.1.10 (Folklore). Suppose that a Banach space X has the property that it is isometric to a quotient space of L and to a subspace of an L_1-space. Then, is the norm of X generated by an inner product? [S. V. Kisliakov, in *Geometry of Linear Operator Space and Operator Theory* (Russian), ed. B. Mitiagin, 1977, Yaroslavl; 1977.]
Remark. This is the isometric version of a problem related to the famous Grothendieck result.

PROBLEM 18.1.11 (Blumenthal, 1975). Suppose that a Banach space X has a right triangle feeble four-point property. Then, is the norm of X generated by an inner product? [L. M. Blumenthal, *Lecture Notes in Math.* No. 490 (1975), 1–13.]

PROBLEM 18.1.12 (Blumenthal, 1975). Suppose that a Banach space X has an equilateral feeble four-point property. Then, is the norm of X generated by an inner product? [L. M. Blumenthal, *Lecture Notes in Math.* No. 490 (1975), 1–13.]

PROBLEM 18.1.13 (Blumenthal, 1975). Suppose that a Banach space X has the property that whenever x, y are elements in X with $\|x\| = \|y\| = \|x - y\| = 1$, then $\|x + y\| = 3^{1/2}$. Then, is the norm of X generated by an inner product? [L. M. Blumenthal, *Lecture Notes in Math.* No. 490 (1975), 1–13; L. M. Kelly, *Lecture Notes in Math.* No. 490 (1975), 14–16.]
Remark. This is a reformulation of Problem 18.1.12.

PROBLEM 18.1.14 (Martin and Valentine, 1976). Suppose that X is a complete convex, externally convex, metric space which admits angles, and their supplements are in Wilson's sense. Is then the cosine law homogeneous? [C. Martin and J. E. Valentine, *Colloq. Math.* **24** (1976), 209–217.]

PROBLEM 18.1.15 (C. Franchetti and coworkers). Let X be a Banach space. Then are the following assertions equivalent?

(1) The norm of X is generated by an inner product.
(2) $\pi_A(d) \subset D$ for any admissible set A and every diametral space D of A.
(3) $\|\pi_A^\circ(x) - \pi_D^\circ(\pi_A^\circ(x))\| \leqslant \|x - \pi_D^\circ(x)\|, \quad \forall x \in X$
for every admissible A and every diametral space D of A. [C. Franchetti and coworkers, *Rev. Roum. Math. Pure Appl.* **17** (1972), 1045–1048.]

PROBLEM 18.1.16 (Enflo, 1970). Let G be a connected subgroup of a metric linear space which is Lipschitz embeddable in a complete inner product space. Then it is not necessarily isomorphic to a complete inner product space. [P. Enflo, Israel *J. Math.* 8 (1970), 230–252; 253–272.]

PROBLEM 18.1.17 (Enflo, 1970). If G is a commutative local Banach space group where a Banach space B is uniformly convex or has roundness 1, then G is a local Banach space and if B is a complete inner product space then the group is a local complete inner product space. [P. Enflo, *Israel J. Math.* 8 (1970), 230–252; 253–272.]

PROBLEM 18.1.18 (Banach, 1932). Let X be a Banach space with the property that all its infinite dimensional separable subspaces are isomorphic to one another. Is then X isomorphic to a complete inner product space? [A. Pełczýnski, *Wiadom. Math.* 2 (15) (1972), 3–11.]

PROBLEM 18.1.19 (Folklore). Let X be a Banach space and suppose that X has property (K). Is X then either a complete inner product space or does it have a binary intersection property? [D. G. de Figueiredo, *Funcional Nao Linear*, Vol. 5, Universidade de Brasilia.]

PROBLEM 18.1.20 (Kwapień, 1972). Let X be a Banach space and suppose that the following property holds: if (f_n) is a complete orthonormal system in L^2 then for some universal constant $C > 0$ and any $x_1, \ldots, x_n$ in X,

$$\int_0^1 \left\| \sum_1^n f_i(t) x_i \right\|^2 \mathrm{d}t \leqslant C \sum_{i=1}^n \|x_i\|^2.$$

[S. Kwapień, *Studia Math.* 44 (1972), 383–395.]
Remarks. (1) The problem is reduced to the case where the system in L^2 is a Haar system (A. Pełczynski).
(2) The problem has a positive answer if the system is the Walsh system.

PROBLEM 18.1.21 (Freese, 1968). If X is a space which has an isosceles feeble Euclidean four-point property, has X then an external isosceles feeble Euclidean four-point property?
[R. W. Freese, *Proc. A. M. S.* 19 (1968), 953–958.]

PROBLEM 18.1.22 (Oman, 1969). A Banach space X has its norm generated by an inner product iff there exists $\varepsilon > 0$ such that for $\|x\| = \|y\| = 1$ and $\|x - y\| < \varepsilon$

there exists a, b in $(0, 1)$, $a \neq b$ such that

$$b(1-b)\|ax+(1-a)y\|^2 - a(1-a)\|bx+(1-b)y\|^2 \; R(a+b-1)(a-b)$$

where R is one of the relations $\geqslant$, $=$, $\leqslant$. [J. Oman, Ph.D. Thesis, Michigan State Univ., 1969.]

PROBLEM 18.1.23 (Oman, 1969). A Banach space X has its norm generated by an inner product iff there exists a nonempty set C and a relation R such that for x in C and y in X there are a, b in $(0, 1)$ such that $(x, y) \in A(a, b, R)$. [J. Oman, Ph.D. Thesis, Michigan State Univ., 1969.]

PROBLEM 18.1.24 (Oman, 1969). Let X be a real Banach space and let C be a closed, bounded, convex with symmetric orthogonality and C is a sphere in X iff there exists a closed subset C_1 of X which contains C in its interior and whose boundary is an E-set of C. [J. Oman, Ph.D. Thesis, Michigan State Univ., 1969.]

PROBLEM 18.1.25 (Istrăţescu, 1983). Let X be a Banach space. Then the norm of X is generated by an inner product iff for any infinitesimal generator of a contraction semigroup of class C_0 the following inequality holds:

$$\|Ax\|^2 \leqslant 2\|A^2x\|, \qquad \|x\| = 1, \quad x \in D(A^2).$$

(i.e. iff the Hardy–Littlewood–Pólya inequality holds with the constant 2).

PROBLEM 18.1.26 (Istrăţescu). Let X be a Banach space and $C(X)$ be the number defined by A. Lifschitz,

$$C(X) = \tfrac{1}{2} \sup_{\|x\| \leqslant 1, \|y\| \leqslant 1, \|x-y\| \geqslant 1} \|x+y\|.$$

X has the norm generated by an inner product iff

$$C(X) = \sqrt{3/2}.$$

PROBLEM 18.1.27 (G. W. Mackey). Let X be a normed linear space over K. Suppose that in $L_c(X)$, $(A, B)M$ and $(A, B)M^*$ are equivalent. Is then X a Banach space? [F. Maeda and S. Maeda, *Theory of Symmetric Lattices*, Springer Verlag, New York, 1970.]

PROBLEM 18.1.28 (S. Banach). Let X be a Banach space and $k \geqslant 2$. Suppose that all k-dimensional subspaces (closed) are isometrically isomorphic to one another and that dim $X > k$. Is it true that X is isomorphic and isometric to a complete inner product space? [H. Auerbach, S. Mazur et S. Ulam, *Monatsheft. Math. Phys.* 42 (1935), 45–48; A. Dvoretzky, *Proc. Coll. Convexity*, 61–66. København, 1967; M. L. Gromov, *Izv. Akad. Nauk SSSR* **31** (1967), 1105–1114.]

APPENDIX

Measures on Locally Compact, Spaces

Let X be a locally compact space and let $K(X)$ be the set of all continuous (real or complex valued) functions on X with compact support, i.e. f is in $K(X)$ iff

$$(s, f(s) \neq 0)$$

has a compact closure which is denoted by supp f.

DEFINITION A.1. A measure on X is any continuous linear functional on $K(X)$ (on $K(X)$ we consider the sup norm). If the values are real then we say that we have a real measure, and if the values are complex then we say that we have a complex measure.

DEFINITION A.2. A measure is said to be positive if, for positive functions, the values are in R^+.

If μ is a measure on X then for f in $K(X)$ we use the following notations to denote the value of μ at f,

$$\int f \,\mathrm{d}\mu = \langle f, \mu \rangle = \int f(s) \,\mathrm{d}\mu(s).$$

It is well known that the measure μ on X is continuous iff for any compact set M in X there exists a constant $K_M > 0$ such that for every f in $K(X)$, supp $f \subseteq M$ we have

$$\int f \,\mathrm{d}\mu \leqslant K_M \| f \|.$$

If μ is a measure, then we say that μ is bounded if the number

$$\| \mu \| = \sup_{f \in K(X), \| f \| \leqslant 1} | \mu(f) | < \infty.$$

We denote the set of all bounded measures on X by $\mathscr{M}^1(X)$. If we set $\mu \to \| \mu \|$ then $\mathscr{M}^1(X)$ is a Banach space.

For details concerning this subject we refer to the Bourbaki books (Fasc. 13, 21, 25), Hermann, Paris, 1967.

References

Abreu, J. L. (1970), A note on harmonizable and stationary sequences, *Bol. Soc. Mat. Mexicana* **15**, 48–51. MR 50#5913.

Abreu, J. L. (1976), H-valued generalized functions and orthogonally scattered measures, *Adv. Math.* **19**, 382–412. MR 53#19636.

Abreu, J. L. (1978), Transformation valued measures. *Adv. Math.* **27**, 1–11. MR 57#10426.

Abreu, J. L. (1983), Stationary prediction of non-stationary sequences in Hilbert spaces, in *Prediction Theory and Harmonic Analysis*, North-Holland, Amsterdam, pp. 1–12. MR 84i:60056.

Abdik, A. (1981), Espaces pre-hilbertiens, *Hacetepe Bull. Natur. Sci. Engr.* **10**, 1–5, R. Ž. 1982, 3B 715.

Accardi, L. (1982), Foundations of quantum probability, *Rend. Mat. Sem. Univ. Politechn. Torino*, Special Issue, pp. 249–270. MR 85i:81004.

Achiezer, N. I. and Glazman, I. M. (1950), *Theory of Linear Operators in Hilbert Space*, (Russian). Gostehizdat, Moscow, MR 13,358. English translation: Frederik Ungar, New York, Vol. 1961. MR 41#9015a. Vol. 2 (1963). MR#9015b.

Adler, R. J. (1981), *The Geometry of Random Fields*, J. Wiley, Chichester, 1981. MR 82h:60103.

Aharonov, Y., Albert, D. Z. and Au, C. K. (1981), New interpretation of the scalar product in Hilbert space, *Phys. Rev. Lett.* **47**, No. 15, 1029–1031. Errata. MR 82m:81004a,b.

Ahlberg, J., Nillson, E. and Walsh, J. L. (1967), *Theory of Splines and Their Applications*, Academic Press, New York. MR 39#684. Russian transl. MR 50#5280.

Ahuja, G. C., Narang, T. D. and Trehan, S. (1974), Best approximation on convex sets in metric spaces, *J. Approx. Theory*, **12**, 194–97. MR 50#10642.

Aitchison, P. W. (1970), A characterization of the ellipsoid, *J. Australian Math. Soc.* **11**, 385–394. MR 43#1042.

Aitchison, P. W. (1974), The determination of convex bodies by some local conditions, *Duke Math. J.* **41**, 193–209. MR 48#12293.

Aitchison, P. W., Petty, C. M. and Rogers, C. A. (1971), A convex body with a false center is an ellipsoid, *Mathematika* **18**, 50–59. MR 45#2572.

Aizenberg, L. A. and Yuzhakov, A. P. (1983), Integral representations and residues in multi-dimensional complex analysis, *Transl. Math. Monographs* Vol. 58. Amer. Math. Soc. Providence, R. I. 1983. MR 58#28593 (Russian edition).

Aizenberg, L. A. and Dautov, Sh. A. (1983), Differential forms orthogonal to holomorphic functions or forms, and their properties *AMS. Monographs.* No. 56 MR 84e:32005.

Akilov, G. P. and Kantorovich, L. V. (1967), *Functional Analysis in Normed Linear Spaces* (Russian). Izdat. Nauka, Moscow. MR 58#23465. English transl., Macmillan Co. New York, 1964. MR 35#4699.

Aldox, D. J. (1976), A characterization of Hilbert spaces using the central limit theorem, *J. London Math. Soc.* **14** (2), 376–380. MR 56#1390.

Alexander, R. (1979), Metric averaging in Euclidean and Hilbert spaces, *Pacif. J. Math.* **85**, 1–9. MR 82a:52006.

Aliev, B. I. (1971), Certain properties of Fourier series in Hilbert spaces (Russian. Azerbaidzani summary). *Izv. Akad. Nauk Azerbaidz. SSR Ser. Fiz. Techn. Nauk*, 1971, No. 2, 3–8. MR 45#4128.

Allen, G. D. and Narcowich, F. J. (1975), On the representation and approximation of a class of operator-valued analytic functions, *Bull. Amer. Math. Soc.* **81**, 410–412. MR 51#1461.

Allen, G. D. and Narcowich, F. J. (1976), *R*-operators. I. Representation theory and applications, *Indiana Univ. Math. J.* **25**, 945–963. MR 55#3861.

Alonzo, J. (1982), Una nota sobre ortogonalidad en espacios normados, *Actas de las IX Jornados Matematicas Hispano-Lusas*, **1**, 211–214. Universidad de Salamanca.

Alonzo, J. (1984), *Ortogonalidad en Espacios Normados*, Universidad de Estramadura, Faculdad de Ciencias.
Ambarzumian, V. Linear geometry in Hilbert space, *Phys. Z.* **30**, 651–652, F.M.55, 243.
Ambarzumian, V. On linear geometry of functional spaces, *Z. für Physik* **55**, 801–805. F.M. 55, 243.
Ambrose, W. (1945), Structure theorems for a special class of Banach algebras, *Trans. Amer. Math Soc.* **57**, 364–386. MR 7,126.
Ambrose, W. (1949), The L^2-system of a unimodular group, *Trans. Amer. Math. Soc.* **65**, 27–48. MR 10,429.
Amir, Dan (1978), Chebyshev centers and uniform convexity. *Pacif. J. Math.* **77**, 1–6. MR 80h:46017.
Amir, Dan and Deutsch, F. (1978), Another approximation theoretic characterization of inner product spaces. *Proc. Amer. Math. Soc.* **71**, 99–102. MR 80d:46039.
Amir, Dan, Deutch, F. and Ziegler, Z. (1980), Relative Chebyshev centers in normed linear spaces. I, *J. Approx. Theory* **29**, 235–252. MR 82c:41030.
Amir, Dan, Deutch, F. and Frachetti, C. (1983), A note on characterizations of Hilbert spaces, *Boll. Un. Mat. Ital. A* **2** (6), No. 3, 305–309. MR 85h:46035.
Amir-Moéz, Ali and Davis, C. (1960), Generalized Frobenius inner products, *Mathematische Annalit. Annalen* **141**, 107–112. MR 22#9508.
Ancona, A. (1972), Contraction module et principle de réduite dans les espaces ordonnées à forme coercive, *C. R. Acad. Sci. Paris Ser. A–B* **275**, 701–704. MR 49#5386.
Andalafte, E. Z. and Blumenthal, L. M. (1964), Metric characterization of Banach and Euclidean spaces, *Fund. Math.* **55**, 23–55. MR 29#2625.
Andalafte, E. Z. and Valentine, J. E. (1973), Criteria for unique metric lines in Banach spaces, *Proc. Amer. Math. Soc.* **39**, 367–370. MR 47#2499.
Andalafte, E. Z. and Valentine, J. E. (1973), Intrinsic four point properties which characterize hyperbolic and Euclidean spaces, *Bull. Acad. Polon. Sci. Ser. Sci. Math. Phys. Astron.* **21**, 1103–1106. MR 50#1133.
Andalafte, E. Z., Valentine, J. E. and Wayment, S. G. (1979), Triangle median properties which characterize Banach spaces. *Houston J. Math.* **5**, 307–312. MR 82c:46016.
Andalafte, E. Z., Freese, R. W. and Diminnie, C. R. (1982), New four point properties which characterize inner product spaces, *Math. Japonica* **27**, 253–261.
Anderson, R. D., Curtis, D. W. and van Mill, J. (1982), A fake topological Hilbert space, *Trans. Amer. Math. Soc.* **272**, 311–321. MR 83j:57009.
Antoine, J. P. (1976, 1976, 1980, 1980), Partial inner product spaces, *J. Funct. Anal.* **23**, No. 4, 369–378. MR 55#10991a. *J. Funct. Anal.* **23**, No. 4, 379–391. MR 55#10991b. *J. Math. Phys.* **21**, 268–279. *J. Math. Phys.* **21**, 267–279.
Antoine, J. P. and Karwowski, W. (1979), Countable Hilbert spaces and partial inner product spaces (Russian summary). *Bull. Acad. Polon. Sci. Ser. Sci. Math. Phys. Astron.* **27**, 223–232 (1980). MR 84f:46003.
Antoine, J. P. and Karwowski, W. (1982), Interpolation theory and refinement of nested Hilbert spaces, *J. Math. Phys.* **22**, 2489–2496. MR 84h:46099a.
Antoine, J. P. and Karwowski, W. (1982), Erratum. *Ibid.* **23**, No. 9, 1737. MR 84h:46099b.
Arens, R. (1947), Representation of *-algebras, *Duke Math. J.* **14**, 269–272. MR 9, 44.
Aronszajn, N. (1935), Characterisation metriques de l'éspace de Hilbert, des éspaces vectoriels et des certains groupes métriques. *C. R. Acad. Sci. Paris* **201**, 811–813; 873–875.
Aronszajn, N. (1943), La théorie des noyaux reproduisants et ses applications. *Proc. Camb. Phil. Soc.* **39**, 133–153. MR 5, 38.
Aronszajn, N. (1948), Reproducing kernels and pseudo-reproducing kernels and their applications to partial differential equations of physics. *Harvard University Graduate School of Engineering.* MR 11, 187.
Aronszajn, N. (1950), Theory of reproducing kernels, *Trans. Amer. Math. Soc.* **68**, 337–404. MR 14, 479.
Aronszajn, N. and Panitchpadki, P. (1956), Extension of uniformly continuous transformations and hyperconvex spaces. *Pacif. J. Math.* **6**, 405–439. MR 18, 917.
Aronszajn, N., Panitchpadki, P. and Smith, K. T. (1957), Characterization of positive reproducing kernels. Applications to Green's functions. *Amer. J. Math.* **79**, 611–622. MR 19, 566.
Ascoli, G. (1932), Sugli spazi lineari metrici e le loro varieta lineari. *Annali di Mat. Pura Appl.* **10** (4), 33–81.

Ash, R. (1965), *Information Theory*. Interscience Publishers, New York. MR 37# 5049.

Asplund, E. and Pták, V. (1971), A minimax inequality for operators and related numerical range. *Acta Math.* **126**, 53–62. MR 42# 2288.

Ashvinikumar, (1966), Hilbert spaces in intuitionism. Written in English and Esperanto, with a summary in Dutch and a concise version in Hindi. Amsterdam.

Atteia, M. (1965), Généralisation de la definition et des propriétés des «spline» fonctions. *C. R. Acad. Sci. Paris.* **260**, 3550–3553. MR 35# 3340.

Atteia, M. (1965), Spline fonctions généralisées. *C. R. Acad. Sci. Paris* **261**, 2149–2153. MR 35# 3441.

Atteia, M. (1966), Sur les fonctions spline généralisées. *Actes du Cinquieme Congres de l'AFIRO* (Lille) 113–116. Assoc. Française Informat. Recherche Operationelle (AFIRO). MR 45# 5640.

Atteia, M. (1968), Fonctions spline definies sur un ensemble convexe. *Numer. Math.* **12**, 192–210. MR 41# 2265.

Atteia, M. (1970), Fonctions spline et noyaux reproduisantes d'Aronszajn et Bergman, *Rev. Française Informat. Recherche Operationelle* **4**, R 3, 31–43. MR 45#9109.

Auerbach, H. (1932), Sur les groupes bornées de substitutions lineaires. *C. R. Acad. Sci. Paris* **195**, 1367–1369.

Auerbach, H. (1940), Sur une propriété characteristique de l'ellipsoïde, *Studia Math.* **9**, 17–22. MR 3, 89.

Auerbach, H., Mazur, S. and Ulam, S. (1935), Sur un propriété charactéristique de l'ellipsoïde, *Monatsh. Math. Phys.* **42**, 45–48. Zbl. 11 (1935), 222.

Azizov, T. and Iohvidov, I. S. (1979), Linear operators in spaces with an indefinite metric and their applications (Russian). *Mathematical Analysis*, **17**, pp. 113–205. Akad. Nauk SSSR, Vsesoiusz. Inst. Naučn. i Techn. Inform. Moscow. MR 81m:47052. Entire coll. MR 81d:00002.

Azoff, A. (1978), Kaplansky–Hilbert modules and the self-adjointness of operator algebras, *Amer. J. Math.* **100**, 957–972. MR 80b:46073.

Babenko, K. I. (1947), On bases in hilbert spaces (Russian), *DAN SSSR* **57**, 427–430. MR 9, 142.

Baidak, G. I., Braverman, M. S. and Petunin, Yu. I. (1983), Additivity of dispersion-*a* characteristic property of the Hilbert space L_2 (Ω, U,μ) (Russian). *Functsional. Anal. i Prilozen*, **17**, No. 3, 66–68. MR 84 h:46043. English transl.:*Functional Anal. Appl.* **17** (1983), 218–220. R. Ž. 1983, 12B 916. MR 84j:46043.

Bajasgalan, C. (1980), The neutral cone and its maximal subspaces. *Functions, Series, Operators*, Vol. 1, 2. Budapest, pp. 183–189. *Colloq. Math. Soc. J. Bolyai*, No. 35, North-Holland, Amsterdam, New York, 1983. MR 85i:46032.

Balabaev, V. E. (1976), The representation of homogeneous additive functionals on a quaternionic Hilbert space (Russian). *Jaroslavl. Gosud. Univ.* Qualitative and approx. methods for investigation of operator equations (Russian). MR 58# 30089.

Balachandran, V. K. (1964), The Weyl group of an L^*-algebra. *Math. Ann.* **154**, 157–165. MR 28# 4374.

Balachandran, V. K. (1966), Regular elements of L^*-algebras. *Math. Z.* **93**, 161–163. MR 33# 2689.

Balachandran, V. K. (1968), On the uniqueness of the inner product topology in a semisimple L^*-algebra. *Topology* **7**, 305–309. MR 37# 4137.

Balachandran, V. K. (1968), Simple systems of roots in L^*-algebras. *Trans. Amer. Math. Soc.* **130**, 513–524. MR 38# 199.

Balachandran, V. K. (1970), Fixed point free automorphisms in L^*-algebras. *Arch. d. Math.* **21**, 386–389. MR 42# 7721.

Balachandran, V. K. (1970), Automorphisms of L^*-algebras. *Tohoku Math. J.* **22**, 163–170. MR 42# 7722.

Balachandran, V. K. (1972), Real L^*-algebras. *Indian J. Pure Appl. Math.* **3** (6), 1224–1246. MR 50# 420.

Balachandran, V. K. and Parthasarathy, P. R. (1966), Cartan subalgebras of an L^*-algebra. *Math. Ann.* **166**, 300–301. MR 34# 3348.

Balachandran, V. K., Parthasarathy, P. R. and Rema, P. S. (1969), Uniqueness of the norm topology in certain Banach–Jordan algebras. *Publ. of the Ramanujan Inst.* No. 1, 283–289. MR 42# 4046.

Balachandran, V. K. (1983), Preprint.

Balakrishnan, A. V. (1957), A note on the sampling principle for continuous signals. *IRE Trans. Inform. Theory* IT–3, 143–146.

Banach, S. (1932), *Théorie des Opérations Linéaires*. Warszawa.

Baranger, J. (1970), Un théoreme de charactérisation de certain sous-espaces hilbertiennes de l^1. *Rev. Française Informat. Recherche Operationelle* **4**, Ser. R–3, 131–134. MR 45:7442.

Barbieri, F. and Facchinetti, G. (1973, 1974), Osservazioni sopra alcune definizioni de pseudoprodotti interni. *Atti Sem. Fis. Mat. Univ. Modena* **22**, 48–59. MR 52:6388. (See also a paper reviewed under MR 51:13658).

Barbu, V. (1982), Invariant manifolds for Hamiltonian systems in Hilbert spaces. *Evolution Systems and Their Applications*, pp. 1–15. *Res. Notes in Math.* No. 68. Pitman, Boston, Mass. and London. MR 84k:58070.

Bargman, V. (1961), On a hilbert space of analytic functions and an associated integral transform, *Comm. Pure Appl. Math.* **14**, 187–214. MR 28:486.

Bargman, V. (1962), Remarks on hilbert spaces of analytic functions. *Proc. Nat. Acad. Sci. USA* **48**, 199–204. MR 24:A2845.

Bargman, V. (1964), Note on Wigner's theorem on symmetry operators. *J. Math. Phys.* **5**, 862–868. MR 29:1917.

Bari, N. K. (= N. K. Bary) (1943), On stability properties of some orthogonal systems (French). *Mat. Sbornik* **12**, 3–27. MR 5, 5.

Bari, N. K. (1944), On the complete systems of orthogonal functions (French). *Mat. Sbornik* **14**, 51–108. MR 6, 267.

Bari, N. K. (1951), Biorthogonal systems and bases in hilbert spaces (Russian). *Ucen. Zap. Moscow. Gosud. Univ.* **148**, *Matematika* **4**, 69–107. MR 14, 289. MR 23# A3411.

Bari, N. K. (1961), *Trigonometrical Series* (Russian). Moscow, 1961. *A Treatise on Trigonometric Series* Vols. I, II. Macmillan Co. N. Y. (1964) MR 30# 1347. English translation.

Baronti, M. (1981), Su alcuni parametri degli spazi normati. *Boll. Un. Mat. Ital. B* **18** (3), 1065–1075. R. Ž. 1982, 6B 752.

Bayod, J. M. (1976), Productos internos en espacios normados no arquimedianos. Ph. D. Thesis, Univ. of Bilbao, Spain.

Bayod, J. M. (1980), Espacios autoduales. *Rev. Mat. Hispano-Amer.* **40**, 75–85.

Beauzamy, B. and Marrcey, B. (1976), Points minimaux et ensembles optimaux dans les éspaces de Banach. *See* Ch. VI of the Ph. D. Thesis of Beauzamy, Paris.

Bessaga, C. (1969), Topics from infinite dimensional topology; *K*-skeletons and Anderson's *Z*-sets. *Aarhus Univ. Lecture Notes Series* **8**. MR 46# 6388.

Bessaga, C. (1974), O kostce hilberta. *Roczniki Płoskiego Towa. Matematycznego Ser. II Wiadomosci Matematyczne* **18**, 5–18. MR 57# 13955.

Bessaga, C. and Pełczyński, A. (1975), *Selected topics in infinite dimensional topology*. Monograph. Math. No. 58. PWN. MR 57# 17657.

Beesack, P. R. (1975), Generalized Cauchy–Schwarz inequalities in linear spaces. *Univ. Beograd Publ. Elektrotehn. Fak. Ser. Mat. Fiz. No.* 461–497, 77–85, MR 51# 5869.

Bellman, R. (1944), Almost orthogonal series. *Bull. Amer Math. Soc.* **50**, 517–519.

Belobrov, P. K. (1968), Some criteria for the hilbertness of a space (Russian). *Izv. Vysš. Učebn. Zaved. Matematika*, No. 9(76), 13–15, MR 38# 6339.

Beltrami, E. (1865), Risoluzione del problema: riportare i punti di una superficie sopra us piano in modo che le linee geotetiche veggano rappresentate su linie rette. *Annali di Mat. Pura Appl.* **7**, 185–204. *See also* Opere matematiche I. Milano. 1902, pp. 262–280.

Bendat, J. and Sherman, S. (1955), Monotone and convex operator functions, *Trans. Amer. Math. Soc.* **79**, 56–71. MR 18, 588.

Benitez, C. (1975), A property of Birkhoff orthogonality and a characterization of pre-hilbert spaces (Spanish). *Collect. Math.* **26**, No. 3, 211–218. MR 56# 16343.

Berberian, S. K. (1959), Note on a theorem of Fuglede and Putnam. *Proc. Amer. Math. Soc.* **10**, 175–182.

Berezanskiĭ, Yu. M. (1980), Bilinear forms and hilbert riggings (Russian). *Spectral Analysis of Differential Operators*, pp. 83–106. Akad. Nauk Ukrain. SSR, Inst. Math. Kiev. MR 83h:47014.

Berg, G. (1973), Metric characterizations of euclidean spaces, *Pacif. J. Math.* **48**, 11–28. MR 48:12486.

Berger, C. A. (1965), A strange dilation theorem. *Not. Amer. Math. Soc.*, 50.

Bergman, S. (1941), *Sur les Fonctions Orthonormales des Plusieurs Variables avec les Applications a la Théorie des Fonctions Analytiques*, Interscience Publ. New York. MR 2, 359.

Bergman, S. (1947), Sur les fonctions orthogonales, *Mem. Sci. Math.* **106**, 1–63. MR 11, 344.

Bergman, S. (1950), The kernel functions and conformal mappings. *Math. Surveys* No. 5. Amer. Math. Soc. Providence, R. I. MR 12, 402.

Bergman, S. and Schiffer, M. (1953), *Kernel Functions and Elliptic Equations in Mathematical Physics*. Academic Press. MR 14:876.

Bergman, S., Schiffer, S. and Chalmers, B. (1967), A procedure for conformal mappings of triply-connected domains, *Math. of Comp.* **21**, No. 100, 527–542. MR 37:4243a, b.

Bennet, G., Goodman, V. G. and Newman, C. M. (1975), Norms of random matrices, *Pacif. J. Math.* **59**, 359–365. MR 52:13896.

Berkeson, E. (1965), Some types of Banach spaces, hermitian operators and Bade functionals, *Trans. Amer. Math. Soc.* **116**, 376–385. MR 32:4554.

Berkeson, E. (1970), A characterization of complex hilbert spaces, *Bull. London Math. Soc.* **2**, 313–315. MR 43:881.

Berkeson, E. (1972), Hermitan projections and orthogonality in Banach spaces, *Proc. London Math. Soc.* **24** (1), 101–118. MR 45:4191.

Betheel, A. (1980), Orthogonal matrix functions related to multivariable Nevanlinnapick problem, *Bull. Soc. Math. Belg. Ser. B* **32**, 149–170. MR 84e:93047.

Beurling, A. (1949), On two problems concerning linear transformation in Hilbert spaces, *Acta Math.* **81**, 239–255. MR 10, 381.

Beurling, A. and Deny, J. (1959), Dirichlet spaces, *Proc. Nat. Acad. Sci. USA* **45**, 208–215. MR 21:5098.

Beyer, W. A. (1969), Inner product magnitude preserving transformations in complex Hilbert spaces, *Compositio Math.* **21**, 137–142. MR 40:779.

Birkhoff, G. (1935), Orthogonality in linear metric spaces, *Duke Math. J.* **1**, 169–172.

Birkhoff, G. (1948), Lattice theory, *Amer. Math. Soc. Colloq. Publ.* No. 25. Second Ed. MR 23:A 5. First Ed. MR 1:325.

Birkhoff, G. and von Neumann, J. (1936/7), The Logic of Quantum Mechanics. *Ann. Math.* **37** (2), 823–847.

Birkhoff, G., von Neumann, J. and Rota, G. C. (1962), *Ordinary Differential Equations*. Ginn and Company, Boston, Mass. New York, Toronto. MR 25:2253.

Birkhoff, G. von Neumann, J., and Rota, G. G. (1970), On the completeness of Sturm–Liouville expansions, *Amer. Math. Monthly* **67**, 363–370. MR 23:A2734.

Bigelow, C. and Day, D. (1983), Digital topography, *Scientific American* **249** (2), 94–105.

Birman, M. S. (1956), On the method of Friederichs of extending positive definite operators to self-adjoint operators (Russian), *Leningrad. Gosud. Ped. Inst. Učen. Zapiski* **33**, 132–136.

Bishop, E. (1967), *Foundations of Constructive Analysis*. McGraw Hill Book Co. MR 36:4930.

Bishop, E. (1979), Schizophrenia in contemporary mathematics. *Ams, Misoula, University of Montana.*

Bishop, E. (1975), The crisis in contemporary mathematics, *Historia Math.* **2**, 507–517.

Bishop, E. and Bridge, D. S. (1985), *Constructive Analysis*. New York.

Blanusa, D. (1953), An isometric and without singularities embedding of the n-dimensional hyperbolic spaces in Hilbert spaces (German). *Monatsh. Math.* **57**, 102–108. MR 15, 61.

Blanuša, D. (1955), On the embedding of hyperbolic spaces into Euclidean spaces (German). *Monatsh. Math.* **59**, 217–227. MR 17, 188.

Blum, J. R. and Hansen, D. L. (1962), On a problem in Hilbert space with applications. *J. Math. Mech.* **11**, 497–501. MR 26:4416.

Blumenthal, L. M. (1933), A note on the four point property. *Bull. Amer. Math. Soc.* **39**, 423–426.

Blumenthal, L. M. (1936), Remarks concerning the Euclidean four point property. *Ergebnisse eines Math. Colloq. Vienna*, **7**, 8–10.

Blumenthal, L. M. (1943), Remarks on a weak four point property. *Rev. Cienc. Lima* **45**, 183–193. MR 6, 12.

Blumenthal, L. M. (1953, 1970), *Theory and Applications of Distance Geometry*. Oxford, Clarendon Press. MR 14, 1009. Second Ed. Chelsea Publ. Co. New York. MR 42:3678.

Blumenthal, L. M. (1950), Generalized Euclidean spaces in terms of a quasi-inner product. *Amer. J. Math.* **72**. MR 12, 436.

Blumenthal, L. M. (1955), An extension of a theorem of Jordan and von Neumann. *Pacif. J. Math.* **5**, 161–167. MR 16, 1139.

Blumenthal, L. M. (1967), An extension of a theorem concerning equilateral triples. *Amer. Math. Monthly* **74**, 295–296. MR 35:845.

Blumenthal, L. M. (1968), Note on normed linear spaces (Spanish). *Rev. Ci. Madrid* **62**, 307–310. MR 38:1502.

Blumenthal, L. M. (1942), Note on an arc without tangents. *Rep. Math. Colloq.* **4** (2), 3pp. MR 4, 112.

Blumenthal, L. M. (1975), Four point properties and norm postulates. The geometry of metric and linear spaces. *Lecture Notes in Math.* No. 490, pp. 1–13. MR 53:6432.

Blumenthal, L. M., and Ellis, D. (1949), Notes on lattices. *Duke Math. J.* **16**, 585–590. MR 11, 369.

Blumenthal, L. M., Ellis, D. and Menger, K. (1970), *Studies in Geometry*. W. H. Freeman and Co. San Francisco. MR 42:8370.

Blumenson, L. B. (1963), A generalized polarization identity. *Amer. Math. Monthly* **70**, 277–281. MR 36:6634.

Boas, R. P. (1940), General expansion theorems. *Proc. Nat. Acad. Sci. USA* **26**, 140–143. MR 1, 229.

Boas, R. P. (1941), A general moment problem. *Amer. J. Math.* **63**, 361–370. MR 2, 281.

Boas, R. P. (1954), *Entire Functions*. Academic Press, New York, MR 16, 914.

Boas, R. P. (1972), Summation formulas and band-limited signals. *Tohoku Math. J.* **24**, 121–125. MR 48:9252.

Bochner, S. (1942), Hilbert distances and positive definite functions. *Ann. of Math.* **42** (2), 647–656. MR 3, 206.

Bochner, S. (1956), Stationarity, boundedness, almost periodicity of random valued functions. *Proc. 3rd. Berkeley Sympos. Math. Statist. and Probability Theory*, **2**, 7–27. University of California Press. MR 18, 940.

Bohnenblust, H. F. (1940), An axiomatic characterization of L^p-spaces. *Duke Math. J.* **6**, 627–640. MR 8, 102.

Bohnenblust, H. F. (1942), A characterization of complex Hilbert spaces. *Portugal. Math.* **3** (2), 103–109. MR 4, 247.

Bohnenblust, H. F. and Karlin, S. (1965), Geometric properties of the unit sphere in Banach algebras. *Ann. of Math.* **62** (2), 217–269.

Bombieri, S. (1971), A note of the large sieve, *Acta Arith.* **18**, 401–404. MR 44:3982.

Bombieri, E. (1974), Le grand crible dans la théorie analytique des nombres. *Astérisque* **18**, Soc. Math. France, Paris. MR 51:8057.

Bond, F. E. and Cahn, C. R. (1958), On sampling the zeros of bandwidth limited signals. *IRE Transf. Inform. Theory IT*-4, 110–113.

Bognar, J. (1972), Involution as operator conjugation. *Colloq. Math. Soc. J. Bolyai*, **5**, pp. 53–64. Amsterdam. MR 50:10848.

Bognar, J. (1974), Indefinite inner product spaces. Bd. 78. New York. MR 57:7125.

Bognar, J. (1977), Star-algebras induced by non-degenerate inner products. *Acta Sci. Sveged* **39**, 15–26. MR 55:11012.

Bonic, R. and Frampton, J. (1960), Smooth functions on Banach manifolds. *J. Mech. Math.* **15**, 877–898. MR 33:6647.

Bonic, R. and Reiss, F. (1966), A characterization of Hilbert spaces. *An. Acad. Cienc. Brasil* **38**, 239–241. MR 36:1961.

Bonsall, F. F. and Goldie, A. W. (1954), Annihilator algebras. *Proc. London Math. Soc.* **4**, 154–167. MR 15, 881.

Bonsall, F. F., Goldie, A. W. and Duncan, J. (1971, 1973), *Numerical Ranges of Operators on Normed Spaces and of Elements of Normed Algebras*. Cambridge Univ. Press, London Math. Soc. Lecture Notes Series No. 2, No. 10. MR 44:5779. MR 56:1063.

Borel, E. (1897), Sur l'interpolation. *C. R. Acad. Sci. Paris* **124**, 673–676.

Borel, E. (1898), Sur la recherche des singularites d'un fonction définie par un developpement de Taylor. *C. R. Acad. Sci. Paris* **127**, 1001–1003.

Borel, E. (1899), Mémoire sur les séries divergentes. *Ann. Ecole Norm. Sup.* **16** (3), 9–131.

Bollé, D. and D'Hondt, J. (1981), On the Hilbert space approach to classical time delay. *J. Phys. A* **14** (7), 1663–1674. MR 82j:70012.

de Boor, C. and Lynch, R. F. (1966), On splines and their minimum property. *J. Math. Mech.* **15**, 953–966. MR 34:3159.

Borodjanskiĭ, B. M. (1974), Nonstandard analysis and enlargements of Hilbert spaces (Russian). *Moskow. Gosud. Zaočn. Ped. Inst. Sbornik Naucn. Trudy Vyp.* **31**, 265–269. R. Ž. 1975, 5B 614.

Bosznay, A. P. (1978), A remark on the linearity of metric projections. *Ann. Univ. Sci. Budapest Eötvos Sect. Math.* **21**, 37–39. MR 80d:46040.

Bosznay, A. P. (1978), Two characterizations of Hilbert spaces by approximation theoretical properties. *Ann. Univ. Sci. Budapest, Eötvos Sect. Math.* **21**, 27–36. MR 82e:46033.

Bourbaki, N. *Elements de Mathématiques*, Livres 1, 3, 5. Paris. MR 34:2425, MR 34:3277 MR 34:5044a, 5044б.

Bourbaki, N. (1967), *Integration*. Fasc. 13, 21, 25. Hermann, Paris.

Bourlet, C. (1897). *Ann. Ecole Norm. Sup.* **14** (3), 133–190.

Bracci, L., Morchio, G. and Strocchi, F. (1975), Wigner's theorem on symmetries in indefinite metric spaces. *Comm. Math. Phys.* **41**, 289–299. MR 52:4889.

Bracks, F. (1979), The inductive limit of quaternionic Hilbert spaces with reproducing kernel. *Bull. Soc. Roy. Sci. Liège* **43**, 568–574. MR 51:3665.

de Branges, L. (1968), *Hilbert Spaces of Entire Functions*. Prentice Hall, Englewood Cliffs, New Jersey. MR 37:4590. French transl. 1972. Masson et Cié. Paris. MR 52:11562.

Bretagnolle, J. *et al.* (1960), Lois stables et espaces L^p. *Ann Inst. H. Poincaré, Sect. B* **II** (3), 231–266. MR 34:3605.

Brodskiĭ, M. S. and Milman, D. (1948), On the center of a convex set (Russian). *DAN SSSR* (*N. S.*) **59**, 837–840. MR 9, 448.

Brodskiĭ, M. S., Milman, D. and Livšič, M. S. (1958), Spectral analysis of non-selfadjoint operators and intermediate systems (Russian). *Uspekhi Mat. Nauk* **11**, 1 (79), 3–85. English transl. *Amer. Math. Soc. Transl.* **13** (2), (1960), 265–346. MR 20:7221. MR 22:3982.

de Broglie, L. (1954, 1955), L'interprétation statistique de la mécanique quantique. *Conference Nobel, J. Phys. Radium* **8** (16), 737–743. MR 17, 328.

de Broglie, L. (1939), *La mécanique Ondulatoire des Systemes de Corpuscules*. Gauthier Villars, Paris. Second Ed. Paris, 1950. MR 11, 631.

Brown, J. L. Jr. (1967). On the error in reconstructing a non-bandlimited function by means of the bandpass sampling theorem. *J. Math. Anal. appl.* **18**, 75–84. Erratum ibid. **21**, (1968), 699. MR 34:4787.

Brown, J. L., Jr. (1981), Multi-channel sampling of low pass signals. *IEEE Trans. Circuit and Systems CAS* **28**, 101–106. MR 82b:94015.

Brown, T. A. (1915, 1916), Fourier's integral. *Proc. Edinburgh Math. Soc.* **34**, 3–10.

Braverman, M. Sh. and Semenov, E. M. (1982), Finite dimensional subspaces of symmetric spaces (Russian). *Sibirskiĭ Mat. Ž.* **23**, 12–24. MR 83e:46021.

Brucks, R. E. (1975), A characterization of Hilbert spaces. *Proc. Amer. Math. Soc.* **43**, 173–175. MR 49:5788.

Brunner, B. (1984), Hilbert spaces with amorphous bases (German). *Compsoitio Math.* **52**, 381–387. MR 85i:03113.

Bucy, R. C. (1970), Linear and nonlinear filtering. *IEEE* **58**, 854–864. MR 46 46:8738.

Budjanskiĭ, Ju. A. (1974), Nonstandard analysis enlargements of Hilbert spaces (Russian). *Harmonic Analysis on Groups. Moskow. Gosud. Zaocn. Ped. Inst. Sbornik Trudov Vyp.* **39**, 205–264. MR 58:10427. R. Ž. 1975, 5B 614.

Buniakowsky, V. (1859), Sur quelques inegalités concernant les integrales ordinaires et les intégrales aux différences finies. *Mém. Acad. St. Petersburg* (*Leningrad*), **1** (7), No. 9.

Burkholder, D. L. (1980), A geometrical characterization of Banach spaces in which martingale difference sequences are unconditional. *The Annal Prob.* **9**, 997–1011. MR 83f:60070.

Burkholder, D. L. (1981), Martingale transforms and the geometry of Banach space. *Lecture Notes in Math.* No. 860, 35–50. MR 83k:46021.

Burstin, C. (1931), A contribution to the problem of embedding of Riemannian spaces in Euclidean spaces. *Rec. Math.* (*Mat. Sbornik*) **38**, 74–85.

Burton, G. R. (1977), Some characterizations of ellipsoids. *Israel J. Math.* **28**, 339–349. R. Ž. 1978, 7A 841.

Busemann, H. (1941), On the Leibniz definition of planes. *Amer. J. Math.* **63**, 101–111. MR 2, 258.

Busemann, H. (1947), Note on a theorem on convex sets. *Mat. Tidsskrift* **13**, 32–34.

Busemann, H. (1955), *The geometry of Geodesics*. Academic Press, New York. MR 17, 779. Russian transl. Innostrannaja Lit. Moscow, 1962.

Byhovskiĭ, E. B. and Smirnov, N. V. (1960), Orthogonal decomposition of vector square summable functions on a given domain and the operators of vector analysis (Russian). *Trudy Mat. Inst. Steklov* **59**, 5–36. MR 22:12375.

Bynum, W. L. and Drew, J. H. (1972), A weak parallelogram law for 1^p. *Amer. Math. Monthly* **79**, 1012–1015. MR 47:5556.

Byrnes, J. S. (1972), Complete multipliers *Trans. Amer. Math. Soc.* **172**, 399–403. MR 46:7790.

Byrnes, J. S. and Newman, D. J. (1969), Completeness preserving multipliers. *Proc. Amer. Math. Soc.* **21**, 445–450. MR 39:1893.

Buzano, Maria Luiza (1971/3), Generalizzazione della diseguaglianza de Cauchy–Schwarz. (English summary). *Rend. Sem. Mat. Univ. e Politech. Torino* **31** pp. 405–409. MR 49:9596.

Calabi, E. (1953), Intrinsic embeddings of complex manifolds. *Ann. of Math.* **58** (2), 1–23, MR 15, 160.

Calderon, A. and Devinatz, A. (1955), Sur certain courbes dans l'espace de Hilbert. *C. R. Acad. Sci. Paris* **241**, No. 6, 539–541. MR 17, 178.

Calderon, A. and Devinatz, A. (1955), Sur certain courbes a courbure constant dans l'espace de Hilbert. *C. R. Acad. Sci. Paris Ser.* **241**, 586–587. MR 17, 178.

Cain, B. S., Sanders, S. B. and Schneider, H. (1976/77), On the geometry of dual pairs. *Stud. in Appl. Math.* **56**, 71–79. MR 58:17800.

Calkin, J. W. (1941), Two-sided ideals and congruences in the ring of bounded operators in Hilbert spaces. *Ann. of Math.* **42** (2), 829–873. MR 3, 208.

Callebaut, D. K. (1965), Generalizations of the Cauchy–Schwarz inequality. *J. Math. Anal. Appl.* **12**, 491–494. MR 33:237.

Rodriguez Cano, Jose Juan (1968, 1971), A necessary and sufficient condition for a Banach space to be a Hilbert space (Spanish). *Proc. Ninth Annual Spanish Mathematicians' Congres, Granada*, 251–260. *Inst. Jorge Juan, Madrid*, MR 51:6374.

Caraman, P. (1974), Quasi-conformal mappings in real normed spaces. *Rev. Roum. Math. Pures Appl.* **24**, 33–78.

Carleson, L. (1953), An interpolation problem for bounded analytic functions. *Amer. J. Math.* **14**, 921–930.

Carlson, J. W. and Hicks, T. L. (1978/9), A characterization of inner product spaces. *Math. Japonica* **23**, 371–373. MR 80b:46034.

Carlsson, S. O. (1962), Orthogonality in normed linear spaces. *Arkiv for Math.* **4**, 297–318. MR 25:5364.

Carlsson, S. O. (1964), A characteristic property of Euclidean spaces. *Arkiv für Math.* **5**, 327–330. MR 36:1962.

Carmeli, Moshe (1969), Hilbert space description for the gravitational field. *Physics Letter* **28A** No. 10, pp. 683–684.

Carmeli, Moshe, (1969), Hilbert space description for the gravitational field. General Physics Research Lab. Aeorspace Res. Lab. Wright-Patterson, AFB, Ohio.

Carmeli, Moshe (1969), Group-theoretic approach to the new conserved quantities in general relativity. *J. Math. Phys.* **10**, 569–574.

Carmeli, Moshe, Quantization of the gravitational field: the presentation of gravity as the Hilbert space of the quantized system. *Il Nuovo Cimento Ser. 11*, **55B**, 89–96.

Carmeli, Moshe (1982), *Classical Fields. General Relativity and Gauge Theory*, Wiley Interscience. MR 84e:83002.

Carmeli, Moshe (1983), Quantization of the gravitational field: the presentation of gravity as the Hilbert space of the quantized system. Preprint.

Carmichael, R. D. (1917/18), Elementary inequalities for the roots of an algebraic equation. *Bull. Amer. Math. Soc.* **24**, 286–296.

Carmichael, R. D. and Mason, T. E. (1914), Note on the roots of algebraic equations. *Bull. Amer. Math. Soc.* **21**, 14–22.

Cartan, E. (1927), Sur la possibilité de plonger un espace riemannien donné dans un espace euclidien. *Ann. Soc. Polon. Math.* **6**, 1–17.

Cartan, H. (1957), Théorie des filters. *C. R. Acad. Sci. Paris* **205**, 595–598.

Cassels, J. W. S. (1975), Measures of the non-convexity of sets and the Shapley–Folkman–Starr theorem. *Proc. Cambridge Phil. Soc.* **78**, 433–436. MR 52:6570.

Cauchy, A. L. (1921), Analyse algébrique. *Course d'analyse à l'École Polytechnique, Paris, le Partie.* Oeuvres complètes, II, III.

Čauš, N. (1970), The construction of Banach spaces with the aid of a generalized inner product (Russian). *Mat. Zametki* **8**, 67–76. R. Ž. 1971, 1B 667. *Math. Notes* **8**, 508–517.

Cartier, P. (1960/61), Classes de formes bilinéaires sur les espaces de Banach. *Séminaire Bourbaki 13e année*, no. 211.

Cernes, A. (1972), A characterization of hilbert spaces (French). Ph. D. Thesis, Paris. 1972.

Cernes, A. (1971), Ensembles maximaux accrétifs et *m*-accrétifs. *Israel J. Math.* **19**, 335–348. MR 52:4042.

Chalmers, B. L. (1969), On boundary behavior of the Bergman kernel function and related domain functionals. *Pacif. J. Math.* **29**, 243–250. MR 40:402.

Chalmers, B. L. (1969), Subspace kernels and minimum problems in Hilbert space with kernel functions. *Pacif. J. Math.* **31**, 619–628. MR 49:9603.

Chalmers, B. L. (1970), Interpolation series and sequences. *Archiv der Math.* (Basel) **21**, 609–616. MR 44:1824.

Chalmers, B. L. (1971), Some interpolation problems in Hilbert spaces. *Michigan Math. J.* **18**, 41–49. MR 43:2477.

Chalmers, B. L. (1975), Images of bounded linear operators in Hilbert space with kernel functions. *Archiv der Math.* (Basel) **26**, 313–316. MR 58:30105.

Chen, S. S. (1972), The Schwarz lemma and the Hadamard three-spheres theorem in Hilbert spaces. *Monatsh. Math.* **76**, 218–221. MR 46:4179.

Chern, S. S. and Kuiper, N. (1952), Some theorems on the isometric imbeddings of compact Riemannian manifolds in Euclidean spaces. *Ann. of Math.* **56** (2), 422–430. MR 14, 408.

Chernoff, H. C. and Savage, I. R. (1958), Asymptotic normality and efficiency of certain nonparametric tests. *Ann. of Math. Statist.* **29**, 972–994.

Chevalley, C. (1946), *Theory of Lie Groups.* I. Princeton Univ. Press, Princeton, MR 7, 412. MR 18, 583.

Chevalley, C. (1955), Théorie des groupes de Lie III. *Actualit. Sci. et Indust.* No. 1226. Paris, MR 16, 901.

Chevet, S., Chobanian, S. A., Tarieladze, V. I. *et al.* (1977), Characterization de certaines classes des espaces de Banach par les measures gaussiennes. *C. R. Acad. Sci. Paris Ser. A–B* **285**. A793–A796. MR 57:1587.

Chiţescu, I. (1974), Kothe spaces that are Hilbert spaces. *Bull. Math. Soc. Sci. Math. R. S. Roumanie* **18** (66), 25–28. R. Ž. 1976, 8B 764.

Chobanian, S. (1959), On a class of functions of a Banach space-valued stationary stochastic process (Russian). *Sakharth SSR, Mech. Mat. Acad. Moambe* **55**, 21–24.

Chobanian, S. and Tarieladze, V. I. (1977), Gaussian characterization of certain Banach spaces. *J. multivariate Anal.* **7**, 183–203. MR 55:6545.

Chojnacki, W. (1976), On Banach algebras which are Hilbert spaces. *Annales Soc. Mat. Polon. Ser. I. Prace Mat.* **20**, 279–281.

Cimino, G. (1954), Spazi hilbertiani di funzioni harmoniche connesse. *Convegno Intern. Equazioni lineari alle derivate parziale.* Ed. Cremonese, Roma, 1955, pp. 76–85. R. Ž. 1957, 7067. MR 17, 604.

Cjan Shi Sun, (1957), A generalization of Buniakowsky's inequality with applications to the theory of integral equations and Hilbert spaces. *Acta Math. Sinica* **7**, 200–228. English summary. R. Ž. 1961, 4B 427.

Clarkson, J. A. (1936), Uniformly convex spaces. *Trans. Amer. Math. Soc.* **40**, 396–414.

Cohen, J. S. (1970), A characterization of inner product spaces using absolutely 2-summing operators. *Studia Math.* **38**, 271–276.. MR 43:6707.

Cohen, J. S. (1973), Absolutely *p*-summing operators and their conjugates. *Math. Ann.* **201**, 177–200. MR 51:6472.

Cohen, S. (1973), On inner products for Banach algebras. *Boll. Un. Mat. Ital.* **7** (4), 35–41. MR 47:3997.

Comfort, W. W. and Gordon, H. (1961), Inner product spaces and the tri-spherical intersection property. *Proc. Amer. Math. Soc.* **12**, 327–329. MR 22:1263.

Constantin, Gheorghe (1973), Cesaro *p*-summing operators. *Atti Accad. Naz. dei Lincei, Rend.* **54** (8), 42–45. MR 50:8138.

Cook, J. M. (1953), The mathematics of second quantization. *Trans. Amer. Math. Soc.* **74**, 222–245. MR 14, 825.

Cook, J. M. (1966), Complex hilbertian structures on stable linear dynamical systems. *J. Math. Mech.* **16**, 339–349. MR 34:625.

Cook, J. M. (1967), Banach algebras and asymptotic mechanics. *Inst. d'Etudes Sci. de Cargese, École d'Eté.* Gordon and Breach Sci. Publ. New York, 1967.

Cooper, J. L. (1947), One parameter semigroups of isometries in hilbert spaces. *Ann. of Math.* **48** (2), 827–842. MR 10, 257.

Courant, R. (1950), *The Dirichlet Principle, Conformal Mappings and Minimal Surfaces.* Appendix by M. Scheffer. Interscience, New York. MR 12, 90.

Cornfeld, I. P., Fomin, S. V. and Sinai, Ya. G. (1982), *Ergodic Theory* (Russian). Moscow, English trans. Springer Verlag, New York.

Courtain, R. (1979), On the stochastic integral in Hilbert spaces. *Warwick Control Centre Reports* No. 20.

Courtain, R. and Keppler, Ray A survey of stochastic integration in infinite dimensions. *Warwick Control Center Reports* No. 34.

Courtain, R., Keppler, Ray, and Pritchard, A. J. (1976), A semigroup approach to infinite dimensional system theory. *Warwick Control Center Reports* No. 47.

Cramer, H. (1940), On the theory of stationary random processes. *Ann. of Math.* **41** (2), 215–240. MR 1, 150.

Cramer, H. (1951), A contribution to the theory of stochastic processes. *Proc. Second Berkeley Sympos. Math. Statist. Prob.* pp. 329–339. Univ. of California Press, 1951. MR 13, 475.

Cramer, H. and Leadbetter, M. (1967), *Stationary and Related Stochastic Processes.* J. Wiley & Sons, New York.

Csaki, P. and Fischer, J. (1960), On bivariate stochastic connection (Russian summary). *Magyar Tud. Mat. Kutato Inst. Kozl.* **5**, 311–323. MR 23:A4244.

Csaki, P. and Fischer, J. (1960), Contribution to the problem of maximal correlation (Russian summary). *Magyar Tud. Mat. Kutato Inst. Közl.* **5**, 325–337. MR 23:A4245.

Crabb, M. J. and Davie, A. M. (1975), Von Neumann inequality for Hilbert space operators. *Bull. London Math. Soc.* **7**, 49–50. MR 51:1432.

Cudia, D. F. (1963), Rotundity. *Proc. Sympos. Pure Math. No. 7, Convexity*, pp. 73–79. Amer. Math. Soc. Providence, R. I. MR 27:5106.

Cudia, D. F. (1964), The geometry of Banach spaces. Smoothness. *Trans. Amer. Math. Soc.* **110**, 284–314. MR 29:446.

Dacunha-Castelle, D. and Krivine, J. L. (1972), Applications des ultraproduits a l'étude des espaces et des algebres de Banach. *Studia Math.* **41**, 315–334. MR 46:4165.

Danelič, I. A. (Danielich) (1976), Normed spaces which satisfy Appolonius' theorem (Russian). *Mat. Zametki* **20** (2), 247–252. MR 55:8766. English transl. *Math. Notes* **20**, 696–699. MR 55:8766.

Daletskiĭ, Yu. L. (1982), Stochastic differential geometry. *Akad. Nauk Ukrain SSR, Inst. Math. Preprint No.* 46, 48*pp. MR* 84*j*:58132.

David, Mendel (1969), Commutators of two operators one of which is unbounded and seminormal. *Annali Mat. Pura Appl.* **83** (4), 185–194. MR 41:2438.

Davie, A. M. (1978), Quotient algebras of uniform algebras. *J. London Math. Soc.* **7** (2), 31–40. MR 48:2779.

Davies E. B. (1980), Nonunitary scattering and capture. I. Hilbert space theory. *Comm. Math. Phys.* **71** (3), 277–288. MR 84k:81125.

Davis, H. F. (1958), On isosceles orthogonality. *Math. Mag.* **32**, 129–131. MR 21:3754.

Day, M. M. (1947), Some characterizations of inner product spaces. *Trans. Amer. Math. Soc.* **62**, 320–337. MR 9, 192.

Day, M. M. (1959), On criteria of Kasahara and Blumenthal for inner product spaces. *Proc. Amer. Math. Soc.* **10**, 92–100. MR 21:5140.

Day, M. M. (1962), *Normed Linear Spaces.* Springer Verlag, New York. MR 26:2847.

Day, M. M. (1981), Comments on notes of Stanojević *et al.* Proc. *Amer. Math. Soc.* **82**, 105–106. MR 83h:46035.

Daykin, D. E. and Eliezer, C. J. (1968), Generalizations of Hölder's and Minkowski's inequalities. *Proc. Cambridge Phil. Soc.* **64**, 1023–1027. Mr 39:386.

Daykin, D. E. and Eliezer, C. J. (1969), Elementary proofs of basic inequalities. *Amer. Math. Monthly* **76**, 543–546.

Deavours, C. A. (1973), The quaternionic calculus. *Amer. Math. Monthly* **80**, 995–1008. MR 48:4332.

Degani, F. C. and Fiocchi, C. (1974), Problemi degli autovalori in spazi con prodotto interno. *Atti Sem. Mat. Fis. Univ. Modena* **23**, 51–64. MR 51:13657.

Delbosco, D. (1975/6), Caratterizzazzioni degli spazi normati con prodotto interno (English summary). *Rend. Sem. Mat. Univ. e Politech. Torino* **34**, 347–352. MR 56:1036.

Delbosco, D. (1977/8), Normed spaces with an inner product (Italian. English summary). *Rend. Sem. Nat. Univ. Politec. Torino* **36**, 421–425. MR 80g:46025.

de Marr, Ralph (1967), On order isomorphism of a partially ordered linear space and its order dual. *Illinois J. Math.* **11**, 71–77. MR 34:4881.

Deny, J. (1959/60), Formes et espaces de dirichlet. *Séminaire Bourbaki 12e année*, No. 187.

Deny, J. (1965), Principle complet du maximum et contractions. *Ann. Inst. Fourier (Grenoble)*, **15**, Fasc. 1, 259–272. MR 32:5913.

Deny, J. (1969), Methods hilbertiens en théorie du potentiel. *Potential Theory, CIME, Ciclo Stressa*, pp. 121–201. Edizione Cremonese, Roma, Italia (Italy). MR 44:1833.

Delanghe, R. and Bracks, F. (1978), Hypercomplex function theory and Hilbert modules with reproducing kernel. *Proc. London Math. Soc.* **37**, 545–576. MR 81j:46032.

Deprit, A. (1954), Algebre symmetrique et second quantification d'un systeme de bosons. *Ann. Soc. Sci. Bruxelles Sect. I* **68**, 23–33. MR 16, 49.

Diananda, P. H. (1962, 1963, 1963), On rearrangement of series: I. *Proc. Cambridge Phil. Soc.* **58**, 158–159. MR 25:356; II. *Colloq. Math.* **10**, 277–279. MR 27:5059; III. *Colloq. Math.* **10**, 288–289. MR 27:5060.

Diaz, J. B. and Metcalf, F. T. (1969), A complementary triangle inequality in Hilbert and Banach spaces. *Proc. Amer. Math. Soc.* **17**, 88–97. MR 32:6184.

Dijkstra, J. J. (1984), Fake topological Hilbert spaces and characterizations of dimensions in terms of negligibility. *CWI Tract. 2, Amsterdam, Math. Centrum, Wiskunde.* MR 85h:57017.

Deutsch, F. (1973), *Metric Projections.* Notes (Unpublished), Spring.

Dieudonné, J. (1960), *Foundations of Modern Analysis.* Academic Press. MR 22:11074. French transl. Gauthier Villars, 1963. MR 28:5149.

Diminnie, C. R. A new orthogonality relation for normed linear spaces. *Math. Nachr.* Preprint.

Diminnie, C. R. and White, A. G. (1975), 2-inner product spaces and Gateaux partial derivatives. *Comment. Math. Univ. Carolin.* **16**, 115–119. MR 51:6375.

Diminnie, C. R. and White, A. G. (1982), A characterization of 2-inner product spaces. *Math. Japonica* **27** (3), 275–277. MR 84h:46031.

Diminnie, C. R., Freese, R. W. and Andalafte, E. (1983), An extension of Pythagorean and isosceles orthogonality and a characterization of inner product spaces. *J. Approx. Theory* **39**, 295–298.

Dixmier, J. (1969), Les algèbres d'opérateurs dans l'espace hilbertien (*Algebres de von Neumann*). Gauthier Villars, Paris. MR, 20, 1234. Second Edition, Paris, 1969. MR 50:5482.

Dixmier, J. (1964), *Les C*-algebres et Leurs Representations.* Gauthier Villars, Paris. MR 30:1404. Second Ed. Paris, 1969. MR 39:7442.

Dixmier, J. (1952), Algebres quasi-unitaires. *Comment. Math. Helvet.* **26**, 275–322. MR 14, 660.

Dobys, L. N. and Nikolskiĭ, N. K. (1976), Two ways avoiding hereditary completeness (Russian, English summary). Investigation on linear operators and theory of functions. VII. *Zap. Naučn, Semin. Leningrad Otdel. Mat. Inst. Steklov (LOMI)* **65**, 183–188, 208–208. MR 58:17801.

Doicinov, P. B. (1960, 1961), Embedding of uniform spaces into Hilbert spaces and Euclidean spaces (Russian). *DAN SSSR* **135**, 1314–1317. *Soviet Math. Doklady* **1**, 1376–1379. MR 23:A1349.

Dolezal, V. (1979), Monotone operators and applications in control and network theory. *Studies in Automation and Control*, No. 2, Elsevier, Amsterdam. MR 82e:93050.

Donoghue, W. F. (1957), Continuous functions spaces isometric to Hilbert spaces. *Proc. Amer. Math. Soc.* **8**, 1–2. MR 18, 661.

Donoghue, W. F. (1974), *Monotone Matrix Functions and Analytic Continuation.* Bd. 207. Springer Verlag, New York. MR 58:6279.

Donoghue, W. F. (1980), Reproducing kernel spaces and analytic continuation. *Rocky Mountain J. Math.* **10**, 85–97. MR 81j:32027.

Doob, J. L. (1953), *Stochastic Processes.* J. Wiley, New York. MR 15, 445.

Doob, J. L. (1956), Appreciation of Khinchin. *Berkeley Sympos. on Math. Statist. and Prob.* University of California Press, pp. 17–23.

Dor, L. E. (1980), A geometric characterization of inner product spaces. *Israel J. Math.* **37**, 84–91. MR 83d:46029.

Dorfman, I. Ya (1970), Means and the Laplacian of functions on Hilbert spaces (Russian). *Mat. Sbornik* **81** (123), 192–208. MR 54:3403.

Dorfman, I. Ya. (1976), The divergence of vector fields on Hilbert space. *Sibirskii Mat. Z.* **17**, 1023–1031. *Siberian J. Math.* **17**, 757–763. MR 54:13970.

Drobot, V. (1969), Rearrangement series of functions. *Trans. Amer. Math. Soc.* **142**, 239–248. MR 39: 7328.

Driscoll, M. F. (1973), The reproducing kernel Hilbert space structure of the sample paths of a gaussian process. *Z. Wahrsch. verw. Geb.* **26**, 309–316. MR 51:6949.

Dubinsky, E. *et al.*, (1972), On Banach spaces for which $\pi_2(L_\infty, X) = B(L_\infty, X)$. *Studia Math.* **44**, 617–648. MR 51:1350.

Duffin, R. J. and Eachus, J. J. (1942), Some notes on an expansion theorem of Paley and Wiener. *Bull. Amer. Math. Soc.* **48**, 850–855. MR 4, 97.

Duffin, R. J. and Schaeffer, A. C. (1952), A class of nonharmonic Fourier series. *Trans. Amer. Math. Soc.* **72**. MR 13, 839.

Dumitrescu, C. (1981), Un produit scalaire probabiliste. *Rev. Roum. Math. Pures Appl.* **26**, 399–404. MR 82m:46021.

Dunford, N. and Schwartz, J. (1958), *Linear Operators. I. General Theory.* Interscience Publ. New York. MR 22:8302. Russian transl. Izdat. Inostr. Lit. Moscow, 1962.

Dunford, N. and Swartz, J. (1963), *Linear Operators. II. Selfadjoint Operators in Hilbert Spaces.* Interscience Publ. New York. MR 32:6181.

Dunn, J. C. (1973), On recursive averaging processes and Hilbert spaces. Extensions of the contraction principle. *J. Franklin Inst.* **295**, 117–133. MR 51:1499.

Dunn, J. C. (1976), Convexity, monotonicity and gradient processes in Hilbert spaces. *J. Math. Anal. Appl.* **53**, 145–158. MR 52:9013.

Duren, P. L. (1983), *Univalent Functions.* Springer Verlag, New York.

Dupré, M. J. *et al.* (1980), Triviality theorems for Hilbert modules. *Topics in Operator Theory.* pp. 71–79. MR 84h:46096.

Dvoretzky, A. (1959), A theorem on convex bodies and applications to Banach spaces. *Proc. Nat. Acad. Sci. U.S.A.* **45**, 223–226. MR 21:4391. *Erratum, Ibid.* p. 1554.

Dvoretzky, A. (1961), Some results on convex bodies and Banach spaces. *Proc. Internat. Sympos. Linear Spaces, Jerusalem*, pp. 123–160. Jerusalem, Academic Press; Pergamon Press, Oxford, 1961. MR 25:2518.

Dvoretzky, (1967), A characterization of Banach spaces isomorphic to inner product spaces. *Proc. Colloq. Pure Math. No. 7, Convexity*, Copenhagen, 1965, pp. 61–66. Københavns, Math. Inst. Copenhagen, MR 36:662.

Dym, Harry (1970), An introduction to de Branges spaces of entire functions with applications to differential equations of the Sturm–Liouville type. *Advances in Math.* **5**, 395–471. MR 43:2481.

Dym, Harry and McKean, H. P. Jr. (1970), Application of de Branges spaces of integral functions to the prediction of stationary gaussian processes. *Illinois J. Math.* **14**, 299–343. MR 41:4642.

Dym, Harry and McKean, H. P., Jr. (1972) *Fourier Series and Integrals.* Academic Press. MR 56:6829.

Dym, H. and McKean, H. P. Jr. (1976), *Gaussian Processes, Function Theory and the Inverse Spectral Problem.* Academic Press. N. Y. MR 56:6829.

Dzbrašjan, M. M. (1961, 1964), On quasi-isometric mapping of the function spaces $L^2_{\sigma_1}(a_1, b_1)$, $L^2_{\sigma_2}(a_2, b_2)$. *Izv. Akad. Nauk SSSR Ser. Mat.* **25**, 87–112. *Amer. Math. Soc. Transl.* **40**, 127–156.

Eaton, M. L. and Olkin, I. (1972), Applications of the Cauchy–Schwarz inequality to some extremal problems. *Inequalities III.* Academic Press, New York, London, pp. 83–101. MR 49:10042.

Efimova, E. I. and Uglanov, A. V. (1982), Green's formula on a Hilbert space (Russian). *Mat. Sbornik (N.S.)* **119** (161), No. 2, 225–232. MR 84k:46038.

Edstein, M. and Thompson, A. C. (1967), Contractions, isometries and some properties of inner product spaces. *Indag. Math.* **28**, 326–331. MR 36:5672.

Edwards, D. A. and Tonescu, C. (1959), Some remarks on commutative algebras of operators on Banach spaces. *Trans. Amer. Math. Soc.* **93**, 541–551.

Efimova, E. I. and Uglanov, A. V. (1982), Green's formula on a Hilbert space (Russian). *Mat. Sbornik (N.S.)* **119** (161), No. 2, 225–232. MR 84k:46038.

Eidelheit, M. (1959), On isomorphism of ring of linear operators. *Studia Math.* **9**, 97–104.

Einhorn, S. J. (1969), Functions positive definite in $C_{(0,1)}$. *Proc. Amer. Math. Soc.* **22**, 702–703. MR 39:7408.

Eliezer, C. J. and Daykin, D. E. (1967), Generalizations and applications of Cauchy–Schwarz inequality. *Quart. J. Math. Oxford Ser.* (2), **40**, 247–250. MR 37:1541.

Eliezer, C. J. and Mond, B. (1972), Generalizations of the Cauchy–Schwarz and Holder inequalities. *Inequalities III.* Ed. O. Sinha, pp. 97–101. Academic Press. MR 48:1422.

Ellis, D. (1963), A modification of the parallelogram law characterization of Hilbert spaces. *Math. Z.* **59**, 94–96. MR 15, 135.

Embry, E. (1975), Orthogonality and the numerical range. *J. Math. Soc. Japan* **27**, 405–411. MR 32:1352.

Enflo, P. (1970), Uniform structures and square roots in topological groups. I. & II. *Israel J. Math.* **8**, 230–252. MR 41:8568: **8**, 253–272. MR 41:8568.

Enflo, P. (1975), A counterexample to the approximation problem in Banach spaces. *Acta Math.* **130**, 309–317. MR 53:6388. Russian transl. *Matematika* **18** (1974), 146–155.

Enflo, P., Lindenstrauss, J. and Pisier, G. (1975), On the three space problem. *Math. Scandi.* **36**, 199–210. MR 52:3928.

Engelking, R. *Outline of general topology* (1968), Transl. from the Polish. North Holland Publ. Co. P.W.N. MR 58:18316b.

Epstein, B. (1958), The kernel function and conformal invariants. *J. Math. Anal. Appl.* **7**, 925–936. MR 21:1386.

Epstein, B. (1965), *Orthogonal Families of Analytic Functions*. Macmillan, New York, London. MR 31:4916.

Epstein, B. and Minker, J. (1960), Extremal interpolatory problems in the unit disc. *Proc. Amer. Math. Soc.* **11**, 777–784. MR 22:9622.

Erdös, P. (1940), The dimension of rational points in Hilbert space. *Ann. of Math.* **41**, 734–736. MR 1, 20.

Esser, M. (1958), A characterization of L^p-spaces. *Portugal. Math.* **17**, 19–39. MR 20:6033.

Everett, C. J. and Ulam, S. (1966), On some possibilities of generalizing the Lorentz group in the special relativity theory. *J. Comb. Theory* **1**, 248–270. MR 33:7075.

Evgrafov, M. A. (1956), The method of near systems in the space of analytic functions and its applications to interpolation (Russian). *Trudy Moskov. Mat. Obsč.* **5**, 89–201. *Amer. Math. Soc. Transl.* (2) **16** (1960), 195–314. MR 18, 472.

Eymard, P. (1964), The Fourier algebra of a locally compact group (French). *Bull. Soc. Math. France* **92**, 181–236. MR 37:4208.

Fabian, M., Whitfield, J. H. M. and Zizler, V. (1983), Norms with locally lipschitzian derivatives. *Israel J. Math.* **44** (3), 262–276. MR 84i:46028.

Fakhoury, H. Deux characterisations des espaces de Banach hilbertisables. *Exposé au Séminaire «Initiation a l'analyse.»* Unpublished manuscript.

Fano, R. M. (1961), *Transmission of Information*. MIT Press, Cambridge, Mass.

Falb, P. L. (1967), Infinite dimensional filtering: the Kalman–Bucy filter in Hilbert space. *Information and Control* **11**, 102–137. MR 37:2544.

Falb, P. L. (1968), *Direct Methods in Optimal Control Theory*. McGraw Hill, New York.

Faulkner, G. and Shonkwiller, R. (1978), Kernel dilation in reproducing kernel Hilbert space and its applications to moment problems. *Pacif. J. Math.* **77**, 103–115. MR 81a:46030.

Faulkner, G. and Shonkwiller, R. (1978), Cosine representations of abelian *-semigroups and generalized cosine operator functions. *Canad. J. Math.* **30**, 474–482. MR 58:17952.

Faulkner, G. and Huneycutt, J. E. Jr. (1978), Orthogonal decomposition of isometries in a Banach space. *Proc. Amer. Math. Soc.* **69**, 125–128. *R. Ž.* 1979, 2B 542. MR 57:3892.

Favini, Angelo (1970), Sulla teoria delle interpolazione negli spazi vettoriali topologici. *Rendiconti di Mat.* **3** (2), Serie VI, 1–30. MR 42:840.

Fedorov, V. S. (1945), Sur la monogénité. *C. R. Acad. (Doklady) Sci. URSS (N.S.)* **48**, 389–390. MR 8, 25.

Fedorov, V. S. (1946), Sur la monogénité. *Rec. de Math. (Mat. Sbornik) (N.S.)* **18** (60), (Russian). pp. 353–378. MR 8, 25.

Fedorov, V. S. (1953), On monogeneity for hypercomplex functions. *Math. Sbornik (N.S.)* **32** (74), 249–254. MR 15, 214.

Fedorov, V. S. (1958), Basic properties of generalized monogenic functions (Russian). *Izv. Vysš. Učebn, Zaved. Matematika*, **6** (7), 257–265. MR 24:A247.

Fedotov, A. M. (1981), An information inequality for operator equations in Hilbert space. *Theory of Probability and its Applications* **26**, 370–378. MR 83i:65046.

Feldman, G. M. (1972), On spectral subspaces of quasi-analytic operators (Russian). *Sb. Mat. Fizica and Functional. Analiz. Vyp.* 3, Harkov, pp. 81–87. R. Ž. 1973, 4B 822.

Feynman, R. R. and Hibbs, A. R. (1965). *Quantum mechanics and path integrals*. McGraw Hill, New York.

Feynman, R. R., Leighton, R. B. and Sands, M. (1965), *The Feynman Lectures on Physics. Quantum Mechanics.* Addison Wesley, Reading Mass.

Fearnley-Sander, D. and Saymons, J. S. V. (1974), Appolonius and inner products. *Amer. Math. Monthly* **81**, 990–993. MR 50:4637. R. Ž. 1976, 7B 611.

Feintuch, A. and Saeks, R. (1982), System theory. *Pure and applied Math. Series.* No. 102. Academic Press. MR 84e:93003.

Fichera, G. (1951), Geometria analitica degli spazi funcionali ed equazioni funzionali differenziali lineari. *Matematiche (Catania)*, **6**, 67–84. MR 13, 660.

Fichera, G. (1982) Sulla "Kernel function". *Boll. Un. Mat. Ital.* **7** (3), MR 14, 156.

Fichera, G. (1954), *Lezioni sulle transformazioni lineari-Vol. 1.* Introduzione all'analisi lineari. Istituto Matematica, Universita di Trieste. MR 16, 415.

Fichera, G. (1961), Teoria assiomatica delle forme armoniche. *Rend. Mat. Pura Appl.* **20** (5), 147–171. MR 25:3547.

Ficken, F. A. (1946), Note on the existence of scalar products in normed linear spaces. *Ann. of Math.* **45** (2), 362–366. MR 5, 270. (*See also Bull. Amer. Math. Soc.* **49** (1943), 363.)

Fiedler, M. (1977), Isodynamic systems in euclidean spaces and an n-dimensional analogue of a theorem of Pompeiu. *Casopis pest. Matematiky roc.* **102**, 370–381. MR 57:1253.

de Figueiredo, D. G. Extension of contractions and mappings of accretive type. *Atas da Semana de Analise Funcional nao linear*, Vol. 5, pp. 109–160.

de Figueiredo, D. G. (1967), On radial projection in normed linear spaces. *Bull. Amer. Math. Soc.* **73**, 364–368. MR 35:280.

de Figueireo, D. G. (1967), On the projection onto convex sets and the extension of contractions. *Proc. Conf. on Projections and Related Topics.* Clemson Univ. Clemson.

de Figueiredo, D. G. (1970), The extension of contraction in normed spaces. *Proc. Sympos. Pure. Math.* No. 18, Part. 1, 95–104. MR 43:877.

Finkelstein, D., Jauch, J. M., Siminovich, S. and Speiser, D. (1962), Foundations of quaternion quantum mechanics. *J. Math. Phys.* **3**, 207–220. MR 25:952.

Finkelstein, D., Jauch, J. M. and Speiser, D. (1959), Notes on quaternion quantum mechanics. II. III. CERN, 59–9, 59–17.

Finkelstein, D., Jauch, J. M. and Speiser, D. (1963), Quaternionic representation of compact groups. *J. Math. Physics* **4**, 136–140. MR 26:259.

Fitzpatrick, S. and Reznik, Bruces (1983), Skewness in Banach spaces. *Trans. Amer. Math. Soc.* **275**, 587–597. MR 84a:46029.

Fišman, K. M. (1956), On a class of Hilbert spaces of analytic functions (Russian). *DAN SSSR (N.S.)* **107**, 24–27. MR 17, 1070.

Fišman, K. M. (1965), On the completeness of some systems of analytic functions (Russian). *DAN SSSR (N.S.)* **107**, 205–208. MR 17, 1070.

Figueiredo, J. P. de (1983), A generalized Fock spaces framework for nonlinear systems and signal analysis. *IEEE Trans. Circuits and Systems* **30** (9), 637–647. MR 85f:93035a.

de Figueiredo, J. P. (1984), Correction. *Ibid.* **31** (2), 236. MR 85f: 93035b.

Fleming, R. J. and Jamison, J. E. (1974), Hermitian and adjoint abelian operators on certain Banach spaces. *Pacific J. Math.* **52**, 67–84. MR 50:10880.

Fock, V. (1932), Konfigurations Raum und zweite Quantelung, *Z. Phys.* **75**, 622–647.

Fogel, L. J. (1955), A note on the sampling theorem. *IRE Trans. Inform. Theory* **1**, 47–48.

Fusias, C. (1957), On some theorems of J. von Neumann related to spectral sets. *Ac. Szeged Sci.* **18**, 15–20. MR 19, 757.

Fowler, III. Northrup (1979), Elementary counterexamples in infinite dimensional inner product spaces. *Math. Mag.* **52** (2), 96–97. MR 80b:46035.

Fortet, R. (1941), Remarques sur les espaces uniformement convexes. *Bull. Soc. Math. France* **69**, 23–46. MR 7, 124.

Franchetti, C. (1975), Chebyshev centers and hypercircles. *Boll. Un. Mat. Ital. 11* (4), Suppl. Fasc. **3**, 565–573. MR 52:11440.

Franchetti, C. (1977), Relationship between the Jung constant and a projection constant in Banach spaces. *Ann. Univ. Ferrara* **23**, 39–44. MR 58:12292.

Franchett, C. (1980), On the radial projection in Banach spaces. In *Approximation Theory III.* (Ed. W. Cheney), pp. 425–428. Academic Press, New York. MR 826:46017.

Franchetti, C. and Fiuri, M. (1972), Some characterization properties of real Hilbert spaces. *Rev. Roum. Math. Pures Appl.* **17**, 1045–1048. MR 47:7392.

Franchetti, C. and Papini, P. L. (1977), Some properties of hyperlanes in real normed spaces. *Math. Chronicle* **6**, 82–89. MR 58:17793.

Franchetti, C. and Papini, P. L., (1981), Approximation properties of sets with bounded complements. *Proc. Royal Soc. Edinburgh* **89A**, 75–86. MR 82m:46016.

Franchetti, C. and Light, W. (1984), The alternating algorithm in uniformly convex spaces. *J. London Math. Soc.* **29** (2), 545–555. MR 85H:41064.

Franchetti, B. (1974), Spazi di Banach e forme quadratiche quasi-non singolari (English summary). *Boll. Un. Mat. Ital.* **10** (4), 444–450. MR 51:6373.

Fréchet, M. (1906), Sur quelques points du calcul fonctionel. *Rend. Circ. Mat. Palermo* **22**, 1–74.

Fréchet, M. (1907), Sur les ensembles des fonctions et des opérations linéaires. *C. R. Acad. Sci. Paris* **144**, 1414–1416.

Fréchet, M. (1928), *Espaces Abstraits*. Gauthier Villars, Paris.

Fréchet, M. (1934), Determination de la classe la plus generale des espaces abstraits vectoriels distanciés applicables vectorielement sur l'espace concret de Hilbert. *Rev. Mat. Hispano Amer. II* **9**, 193–201.

Fréchet, M. (1935), Sur la definition axiomatique d'une classe d'espaces vectoriels distancié applicable sur l'espace de Hilbert. *Ann. of Math.* **36** (2), 705–718.

Fréchet, M. (1958), Definition de la some et du produit par scalaires en terms de distance. *Ann. Ecole Norm. Sup.* **75** (3), 223–255. MR 21:4394.

Fredholm, I. (1903), On a class of functional equations (French). *Acta Math.* **27**, 365–390.

Freeman, J. M. (1965), Perturbation of the shift operator. *Trans. Amer. Math. Soc.* **114**, 251–260. MR 30:2342.

Freese, R. W. (1961), Ptolemaic metric spaces. *Ph. D. Thesis*, University of Missouri–Columbia.

Freese, R. W. (1968), Criteria for inner product spaces. *Proc. Amer. Math. Soc.* **19**, 953–958. MR 37:3452.

Freese, R. W. (1973), A 2-metric characterization of the euclidean plane. *Math. Anal.* **206**, 285–294. MR 48:8533.

Freese, R. W. and Andalafte, E. Z. (1982), A new class of four-point properties which characterize euclidean spaces. *J. of Geometry* **18**, 43–53. MR 83k:51027.

Freese, R. W. and Murphy, G. (1980), The CMP and Banach spaces with unique metric lines. *J. of Geometry* **14**, 50–59. MR 81g:51008.

Friedman, N. (1970), *Introduction to Ergodic Theory*. Van Nostrand Reinhold Co. New York. MR 55:8310.

Fubini, G. (1905), Sulle metriche definite da una forma hermitiana. *Atti Istituto Veneto* **6**, 501–513.

Fuglede, B. (1950), A commutativity theorem for normal operators. *Proc. Nat. Acad. Sci. USA* **36**, 35–40. MR 11, 371.

Fuhrman, P. (1981), *Linear Systems and Operators in Hilbert Spaces*. McGraw Hill, New York. MR 83d:47001.

Fuks, B. A. (1960), Kernel functions of regions (Russian). *Issledovannia po sovremmenom problemam teorii funtsiĭ komplexnogo peremenogo*, pp. 287–294. *Izdat. Gosud. Izdat. Fiz. Mat. Lit Moscow*, 1960 (Russian). MR 22:6898.

Furstenberg, H. (1960), Stationary processes and prediction theory. *Ann. of Math. Studies No.* **44**, Princeton Univ. Press, New Jersey. MR 25:3573.

Furstenberg, H., Katznelson, Y. and Ornstein, D. (1982), The ergodic theoretical proof of Szemeredi's theorem. *Bull. Amer. Math. Soc.* **7**, 527–553. MR 846:28016.

Gafney, M. (1955), Hilbert space methods in the theory of harmonic integrals. *Trans. Amer. Math. Soc.* **78**, 426–444. MR 16,957.

Gaissinger, L. (1975), Pythagoras and the Cauchy–Schwarz inequality. *Amer. Math. Monthly* **83**, 40–41. MR 52:6552.

Gajda, Z. (1983), On *H*-smooth and *H*-convex sets in linear spaces (Serbo–Croatian summary). *Glasnik Mat. Ser. III* **18** (38), 119–131. MR 84i:46016.

Gao, J. (1983) *Nanjing Daxue Xuebao* No. 1, 5–12. (Not available to the writer).

Garabedian, P. (1948), Schwarz's lemma and the Szegö kernel function. *Ph. D. Thesis*, Harvard Univ.

Garabedian, P. (1949), Schwarz' lemma and the Szegö kernel function. *Trans. Amer. Math. Soc.* **67**, 1–35. MR 11,340.

Garabedian, P. A new proof of the Riemann mapping theorem. *Proc. Sympos. Appl. Math. Ser. No.* **18**, 207–213. Bureau Nat. Standards. MR 14,860.

Garabedian, P. and Schiffer, I. M. (1949), Identities in the theory of conformal mappings. *Trans. Amer. Math. Soc.* **65**, 187–238. MR 10,522.

Garabedian, P. and Schiffer, L. M. (1950), On the existence theorems of potential theory and conformal mappings. *Ann. of Math.* **52** (2). MR 12,89.

Gangolly, R. (1967), Positive definite kernels on homogeneous spaces and stochastic processes related to Levy's Brownian motion of several parameters. *Ann. Inst. H. Poincaré Sect. B* **3**, 121–225. MR 35:6172.

Garding, L. (1953), The Dirichlet problem for linear elliptic partial differential equations. *Math. Scand.* **1**, 55–72. MR 16,366.

Garding, L. (1954), Applications of the theory of direct integrals of Hilbert spaces to some integral and differential operators. *Lecture series* No. 11. *Inst. Fluid Dynamics*, Univ. of Maryland, College Park, Md. MR 17,159.

Garkavi, A. L. (1964), On the Chebyshev center and the convex hull of a set (Russian). *Uspehi Mat. Nauk* **19**, No. 6 (120), 139–145. MR 30:5221.

Garkavi, A. L. (1962), The best possible net and the best possible cross section of a set in a normed linear space (Russian). *Izv. Akad. Nauk SSSR Ser. Mat.* **26**, 87–106. MR 25,429.

Garkavi, A. L. (1970), The theory of best approximation in normed linear spaces. *Progress in Math.* No. 8, Plenum Press, New York, pp. 83–152.

Garling, D. J. H. (1967), On ideals of operators in Hilbert spaces. *Proc. London Math. Soc.* **17** (3), 115–138. MR 34:8208.

Garling, D. J. H. (1970), Absolutely summing operators in Hilbert space. *Studia Math.* **38**, 319–331. MR 43:6769.

Gastinel, N. and Joly, J. L. (1970), Condition numbers and general projection method. *Linear Algebra and its Applications* **3**, 185–224. MR 42:823.

Gel'fand, I. M. (1939), On the theory of normed rings. *C. R. Acad. (Doklady) Sci. USSR* **25**, 570–572. MR 1,330.

Gel'fand, I. M. (1939), On the theory of normed rings. On the theory of almost periodic functions. *C. R. Acad. (Doklady) Sci. USSR*, **25**, 573–574. MR 1,331.

Gel'fand, I. M. (1941), Normed rings. *Rec. Math. (Mat. Sbornik) (N.S.)* **9** (51), 3–24. MR 3, 51.

Gel'fand, I. M. (1956). Remark on the work of N. K. Bari 'Biorthogonal systems and bases in Hilbert spaces' (Russian), *Učen. Zapiski Moskov. Gosud. Univ. 148, Mathematika* **4**, 224–225. MR 14,289.

Gel'fand, I. M. and Naimark, M. A. (1943), On the imbedding of normed rings into the ring of operators in Hilbert space. *Réc. Math. (Mat. Sbornik) (N.S.)* **12** (54), 197–203. MR 5,147.

Gel'fand, I. M. and Naimark, M. A. (1948), Normed rings with involution and their applications. *Izv. Akad. Nauk SSSR, Ser. Mat.* **12**, 445–490. MR 10,199.

Gel'fand, I. M. and Yaglom, A. Y. (1957), Computation of the amount of information about a stochastic function contained in another (Russian). *Uspehi Mat. Nauk (N.S.)* **12**, No. 1 (73), 3–52. MR 18,980.

Gel'fand, I. M., Raikov, D. A. and Šilov, G. E. (1960), Commutative normed rings (Russian). *Sovremmenye Problemy Mat. Gosud. Lit. Fiz. Mat. Moscow.* MR 23:A1242.

Gel'fand, I. M. and Vilenkin, N. Ja. (1961), *Generalized Functions* (Russian). Vol. 4. Some applications of harmonic analysis. Equipped Hilbert spaces. Fizmatgiz, Moscow. MR 26:4173.

Gebelein, H. (1941), Das stat. probl. der Korr. als Variations-und...., *ZAMP* **21**, pp. 364–379.

Gerish, W. (W. Geriš) (1971), A remark on the work of N. K. Bari on biorthogonal systems and bases in Hilbert spaces (Russian). *DAN SSSR*. English transl. *Sov. Math. Doklady* **12**, 1517–1519. MR 45:869.

George, M. D. (1965), The spectrum of an operator in Banach spaces. *Proc. Amer. Math. Soc.* **16**, 980–982. MR 32:4542.

Getoor, R. (1957), On characteristic functions on Banach spaces-random variables. *Pacif. J. Math.* **7**, 885–887. MR 19, 584.

Gilbert, R. P. and Hille, G. N. (1977), Reproducing kernels for generalized Hilbert function spaces. *Appl. Anal.* **6**, 159–161. *R. Ž.* 1977, 12B 752. MR 55:8776.

Gilbert, R. P. and Hille, G. N. (1977), Hilbert function modules with reproducing kernels. *Nonlinear Anal.* **1**, pp. 135–150. MR 58:30106.

Ghosh, N. N. (1970), Infinitesimal geometry in Hilbert spaces. I. II. III. *Indian J. Pure Appl. Math.* **1**, No. 1, 87–93; *ibid.* **1**, No. 1, 94–94; *ibid.* **1**, No. 2, 269–275. MR 45:6037.

Giellis, G. R. (1972), A characterization of Hilbert modules. *Proc. Amer. Math. Soc.* **36**, 440–442. MR 47:9297.

Gil de Lamadrid, J. (1962), On finite dimensional approximations of mappings in Banach spaces. *Proc. Amer. Math. Soc.* **13**, 163–169. MR 25:2412.

Giles, J. R. Classes of semi-inner products spaces. *Trans. Amer. Math. Soc.* **129**, 436–447. MR 36:663.

Gilis, R. (= R. Gylyš) (1980), Linear operators on Hilbert spaces with reproducing kernel (Russian. Lithuanian and English summaries). *Litovsk Mat. Sb.* **20**, No. 2, 15–25. MR 83a:47001.

Gleason, A. M. (1957), Measures on the closed subspaces of a Hilbert space. *J. Math. Mech.* **6**, 885–893. MR 20:2609.

Gleason, A. M. (1966), The definition of a quadratic form. *Amer. Math. Monthly* **73**, 1049–1050. MR 34:7543.

Godement, R. G. (1954), Theorie des caracters. I. Algebres unitaires. *Ann. of Math.* **59** (2), 47–62. MR 15,441.

Godement, R. G. (1954), Theorie des caracters. II. Definition et properties generales des caracters. *Ann, of Math.* **59** (2), 63–85. MR 15,441.

Glushkin, E. D. (1980), A generalization of Khintschine's inequality and its applications in the theory of operator ideals. *Studia Math.* **67**, 149–155. MR 81m:47033.

Goffman, C. (1962), Preliminaries to functional analysis. *MAA Studies in Math.* Vol. 1, pp. 138–180 Prentice Hall, Inc., Englewood Cliffs, New Jersey.

Gohberg, I. C. and Marcus, A. S. (1962), On the stability of bases in Banach spaces and Hilbert spaces (Russian). *Izv. Akad. Nauk SSR Moldovenească Ser. Mat.* **5**, 17–35. MR 37:1955.

Gohberg, I. C., Marcus, A. S. and Krein, M. G. (1965), Introduction to the theory of linear non-selfadjoin operators (Russian). *Izdat. Nauka.* English transl. *Math. Monographs* No. 18, Amer, Math. Soc. Providence, R. I. 1969. MR 36:3137. MR 39:7447.

Gołab, S. (1973), Ein Beitrag zur axiomatizierung des Begriffes des Skalarproducts. *Demonstratio Math.* **5**, 273–278. MR 51:4018.

Gołab, S. (1974), On a generalization of Schwarz inequality. *Publ. Math. Debrecen* **21**, 155–158. MR 50:1493.

Gołab, S. (1975), On the Carnot scaler product in weakly normed spaces. *Aequationes Math.* **13**, No. 1/2, 9–13. MR 53:3664.

Gołab, S. and Swiatak, H. (1976), Note on inner product spaces in vector spaces. *Aequationes Math.* **20**, No. 1/2, 696–699. MR 55:8766.

Goldman, Malcolm (1958), Structures of AW*-algebras. I. *Quart. Math. J.* **23**, pp. 23–34. MR 17,512.

Goldstein, A. A. (1969), Convex programming in Hilbert spaces. *Bull. Amer. Math. Soc.* **70**, 709–710.

Goldstine, H. H. and Horovitz, L. P. (1964), Hilbert spaces with non-associative scalars. I. *Math. Ann.* **154**, 1–27. MR 28:4331. II. *Math. Ann.* 164, 291–316. MR 33:6341.

Golomb, M. (1972), Complete orthonormal systems in pre-Hilbert spaces. *Amer. Math. Monthly* **72**, 263–267. MR 45:9111.

Golovin, V. D. (1963), Stability of a basis of exponential functions (Russian. Armjan summary.) *Akad. Nauk Armjan SSR Doklady* **36**, 65–70. MR 27:2769.

Goodey, P. R. (1971), On Hausdorff measures in Hilbert spaces. *Quart. Math. J. Oxford* Ser. (2) **22**, 271–276. MR 46:319.

Goodner, D. B. (1950), Projections in normed linear spaces. *Trans. Amer. Math. Soc.* **69**, 89–108. MR 12,266.

Goluzin, G. M. (1963), A method of variation in conformal mappings. II (Russian). *Mat. Sbornik* **21**, 83–117.

Gorjačev, A. P. (1970), Undisorted linear normed spaces (Russian). *Vestnik Moskov. Gosud. Univ. Ser. Mat. Mech.* **25**, No. 1, 21–28. MR 43:2471.

Górňiak, J. (1979), Remarks on positive definite operator-valued functions in linear spaces. *Prob. theory on vector spaces.* Lecture Notes in Math. No. 656, pp. 37–44. MR 80d:46086.

Górňiak, J. and Weron, A. (1981), Aronszajn–Kolmogorov type theorems for positive definite kernels in locally convex spaces. *Studia Math.* **69**, 235–246. MR 83f:46002.

Górňiak, J. and Weron, A. (1976), Analogue of Sz. Niga's dilation theorem. *Bull. Acad. Polon. Sci. Math. Astronom. Phys.* **24**, 867–872. MR 54:13598.

Gould, S. H. (1957), Variational methods for eigenvalue problems. An introduction to the methods of Rayleigh, Ritz, Weinstein and Aronszajn. *Expositions* No. 20, Univ. of Toronto Press, Toronto. MR 19,287.

Greenfeld, S. J. and Walach, N. R. (1971), The Hilbert ball and bi-ball are holomorphically inequivalent. *Bull. Amer. Math. Soc.* **77**, 261–263.

Greenfeld, S. J. and Walach, N. R. (1972), Automorphism groups of bounded domains in Banach spaces. *Trans. Amer. Math. Soc.* **166**, 45–57. MR 45:5420.

Grenander, U. and Szegö, G. (1958), *Toeplitz Forms and Their Applications.* Univ. of California Press, Calif. MR 20:1349. Russian transl. MR 24:A349.

Gould, G. G. (1968), The spectral representation theorem of normal operators on a rigged Hilbert space. *J. London Math. Soc.* **43**, 745–754. MR 37:5713.

Gould, G. G. (1970), On Hirschfeld's spectrum-preserving embedding. *J. London Math. Soc.* **2** (2), 171–178. MR 40:4790.

Gribanov, Ju. I. (1968), A majorizing test for convergence in norm in Hilbert space (Russian). *Kazan Gosud. Univ. Učen. Zapiski* **128**, kniga 5, 49–51. MR 46:647.

Grinblat, L. S. (1979), Circles in the unit sphere of a Hilbert space (Russian). *J. Funct. Anal.* **31**, 272–274. *R. Ž.* 1979, 10B 756.

Grinblyum, M. M. (1950), On the representation of spaces of type B in the form of a direct sum of subspaces. *DAN SSSR (N.S.)* **70**, 749–752. MR 11,525.

Grinblyum, M. M. and Gurevich, L. A. (1941), On a property of bases in Hilbert spaces (French). *C. R. Acad. (Doklady) URSS (N.S.)* **30**, 289–291. MR 2,313.

Gromov, M. (1967), On a geometrical conjecture of Banach (Russian). *Izv. Akad. Nauk SSSR, Ser. Mat.* **21**, 1105–1114. MR 36:655.

Gerlein, O. V. and Kagan, A. M. (1979), Hilbert space methods in classical problems of mathematical statistics. *J. of Sov. Math.* **12**, 184–214. MR 53:6913.

Gross, L. (1963), Harmonic analysis on Hilbert spaces. *Memoirs Amer. Math. Soc.* No. 46, Amer. Math. Soc. 1963. MR 28:4304.

Gross, L. (1967), Potential theory on Hilbert spaces. *J. Functional Anal.* **1**, 123–183. MR 37:3331.

Gross, L. (1965/6), Abstract Wiener spaces. *Fifth Berkeley Sympos. Math. Statist. Prob.* Berkeley Calif. Vol. 2, Part. 1, pp. 31–42. Univ. of California Press, Calif. 1967. MR 35:3027.

Grossman, Alexander (1964), Nested Hilbert spaces in quantum mechanics I. *J. Math. Phys.* **5**, 1025–1037.

Grothendieck, A. (1953), Sur certaines classes des suites dans les espaces de Banach et le théoreme de Dvoretzky–Rogers. *Bol. Soc. Mat. Sao Paulo* **8**, 81–110. MR 20:1195.

Grothendick, A. (1953), Resumé de la théorie métrique des produits tensoriels topologiques. *Bol. Soc. Mat. Sao Paulo* **8**, 1–79. MR 20:1194.

Grothendieck, A. (1955), Produits tensoriels topologiques et espaces nucleaires. *Memoirs Amer. Math. Soc.* No. 16. MR 17,763.

Grothendieck, R. (1954), Espaces vectoriels topologiques. *Inst. Mat. Appl. Univ. Sao Paulo.* MR 17,1110. Republished by Gordon and Breach Publishers, 1969.

Grove, L. G. (1965), Tensor products over H^*-algebras. *Pacif. J. Math.* **15**, 857–863. MR 32:2925.

Grünbaum, F. A. (1974), The orthogonal invariants of a curve in Hilbert space. *Proc. Amer. Math. Soc.* **42**, 268–271. MR 48:6901.

Guccione, S. J. and Stanojević, S. V. (1976), Probabilistic foundations of quantum theories and Rubin–Stone spaces. *Bull. Amer. Math. Soc.* **82** (3), 481–483. MR 53:14096.

Gudder, S. P. (1974), Inner product spaces. *Amer. Math. Monthly* **81**, 29–36. MR 48:9362. Correction, *Ibid.* **82** (1975), 251–252. Mr 54:8248. Second correction with S. S. Holland, Jr. *Ibid.* **82** (1975), 818. MR 52:1274.

Gudder, S. P. and Strawther, D. (1974), Orthogonality and nonlinear functionals. *Bull. Amer. Math. Soc.* **80**, 946–950. MR 50:5434.

Gudder, S. P. and Strawther, D. (1975), Orthogonally additive and orthogonally increasing functions on vector spaces. *Pacif. J. Math.* **58**, 427–436. MR 52:11542.

Gudder, S. P. *Stochastic Methods in Quartum Mechanics.* North Holland, N. Y., Oxford. MR 84j:81003.

Guichardet, A. (1972), Symmetric Hilbert spaces and related topics. *Lecture Notes in Math.* No. 261. MR 58:12416.

Guichardet, A. (1971/2), Espaces hilbertiens symmetriques et produits tensoriels continu. *Sem. Goulauic-*

Schwartz. Exposé No. 15–16. 12 pp. Ecole Politechn. Centre de Math. Paris, 1972. MR 52:14947.

Gulevich, N. M. (1982), Estimation of the deviation of fixed points from a nonconvex set (Russian. English summary). Studies in topology. IV. *Zap. Naučn. Sem. Leningrad. Otdel. Mat. Inst. Steklov, (LOMI)*, **122**, 13–16. MR 84d:47075.

Gunning, R. C. and Rossi, H. *Analytic Functions of Several Complex Variables*. Prentice Hall N.J. 1965 MR 31:4927.

Gurariĭ, V. I. (1962), On the inclination of subspaces and conditional bases in Banach spaces. (Russian). *DAN SSSR* **45**, 504–506.

Gurariĭ, V. I. (1965), On bases for sets in Banach spaces. *Rev. Roum. Math. Pures Appl.* **10**, 1235–1240.

Gurariĭ, V. I. and Meletidi, M. A. (1970), Stability of completeness of sequences in Banach spaces (Russian). *Bull. Acad. Polon. Sci. Ser. Sci. Math. Astronom. Phys.* **18**, 535–536. MR 43:3771.

Gurariĭ, N. I. and Sazonov, Ju. I. (1970), Normed spaces that do not have distortion of the unit sphere (Russian). *Mat. Zamtki* **7**, 307–310. MR 41:8971. Math. Notes 7 (1970), 187–189.

Gürsey, F, (1980), Quaternion methods in field theory. *Proc. Johns Hopkins Workshop in Current Problems of Particle Theory*. **4**, 255–288. John Hopkins Univ. Md. MR 83c:81059. Entire coll. MR 83b.81010b.

Gusev, V. A. (1963), On quaternionic functions monogenic in the sense of Fedorov (Russian. Romanian and English summaries). *An. Stiint. Univ. A. I. Cuza Iaşi, Sect. Ia* **9**, 33–40 MR 32:2600.

Gutmann, S. and Sepp, L. A. (1978), Orthogonal bases of two subspaces with all mutual angles acute. *Indiana Univ. Math. J.* **27** (1), 79–90. MR 56:16342.

Haantjes, J. (1947), Distance geometry, curvature in abstract metric spaces. *Indag. Math.* **9**, 302–314. MR 9,53.

Haantjes, J. (1951), A characteristic property of geodesic in certain metric spaces. *Indag. Math.* **13**, 66–73. MR 12,732.

Hahn, H. (1922), On sequences of linear operators (German). *Monatsh. Math.* **32**, 3–88.

Hahn, H. (1932), *Real Functions* (German). Leipzig. Chelsea Publ. Co. New York, reprint.

Hahn, K. T. (1975), Trigonometry on the unit ball of a complex Hilbert space. *Bull. Amer. Math. Soc.* **81**, 183–186. MR 51:4305.

Halmos, P. (1967), *A Hilbert Space Problem Book*. Van Nostrand. MR 34:8178.

de la Harpe, P. (1972), Sur le groupe spinoriel associé a un espace de Hilbert. *C. R. Acad. Sci. Paris Ser. A–B* **274**, *A* 1484–*A*1487. MR 45:6040.

de la Harpe, P. (1972), Classical Banach–Lie algebras and Banach–Lie Groups of Operators in Hilbert space *Lecture Notes in Math.* No. 285. MR 57:16372.

Hausdorff, F. (1962), *Set Theory*. 2nd edition, Chelsea Publ. Co. New York. MR 25:4999.

Hannan, E. J. (1947), The concept of a filter. *Proc. Camb. Philos. Soc.* **63**, 221–227. MR 36:2207.

Hanner, Olof (1947), Deterministic and non-deterministic stationary random processes. *Arkiv for Math. Bd.* **1**, 161–177. MR 11,606.

Hartman, P. (1947), Multiplicative sequences and toplerian (L^2)-bases. *Duke Math. J.* **14**, 755–767. MR 9,426.

Hartman, P. and Wintner, A. (1948), Toplerian (L^2)-bases. *Trans. Amer. Math. Soc.* **63**, 207–225. MR 9,982.

Hain, R. (1979), A characterization of smooth functions on a Banach space. *Proc. Amer. Math. Soc.* **77**, 63–67. R. Z. 1980, 5B 965.

Hardy, G. H. (1940), Notes on special systems of orthogonal functions. III. *Proc. Cambridge Math. Soc.* **36**, 1–8. MR 1,141.

Hardy, G. H. (1941), Notes on special systems of orthogonal functions. IV. The orthogonal functions of Whittaker's cardinal series. *Proc. Cambridge Phil. Soc.* **37**, 331–348. MR 3,108.

Havrda, J. (1973), On the generalization of the concept of the projection (Czech. English summary). *Cašopis Pešt. Mat.* **98**, 265–268. MR 51:13656.

Hayman, W. (1958), Interpolation by bounded functions. *Ann. Inst. Fourier (Grenoble)* **13**, 277–290. MR 22:8128.

Heider, L. J. (1955), A note on a theorem of K. G. Wolfson. *Proc. Amer. Math. Soc.* **6**, 305–309. MR 16,935.

Helms, L. J. (1969), *Introduction to Potential Theory*. Wiley Interscience, N. Y. MR 41:5638.

Helson, H. (1964), *Lectures on Invariant Subspaces*. Academic Press, New York. MR 30:1409.

Helson, H. and Lowdenslager, D. (1958), Prediction theory and Fourier series in several variables. I. *Acta Math.* **99**, 165–202. MR 20:4155. II. *Acta Math.* **106** (1961), 171–213. MR 31:562.

Helson, H. and Lowdenslager, D. (1961), Invariant subspaces. *Proc. Internat. Sympos. Linear Spaces, Jerusalem, 1960*. pp. 251–262. Academic Press, Jerusalem, Pergamon Press, Oxford.

Helton, J. W. (1976), Systems with infinite dimensional state space; the Hilbert space approach. *Proc. IEEE* **64**, 145–160. MR 54:4764.

Helton, J. W. (1982), Non-euclidean functional analysis and electronics. *Bull. Amer. Math. Soc.* **7**, 1–64. MR 84c:94001.

Henkin, G. M. (1963), On the stability of unconditional bases in uniformly Banach spaces (Russian). *Uspekhi Mat. Nauk* **18**, 219–223. R. Ž. 1964, 7B 520. MR 29:451.

Henkin, G. M. (1967), The non-isomorphy of certain spaces of different numbers of variables (Russian). *Funkcional. Analiz. i Priložen.* **1**, 57–68. MR 36:4331.

Henkin, G. M. (1968), The Banach space of analytic functions in a ball and in a bicylinder are not isomorphic (Russian). *Funkcional. Analiz, i Priložen.* **2**, 82–91. MR 54:3378.

Herz, C. S. (1971), Theory of *p*-spaces with an application to convolution operators. *Trans. Amer. Math. Soc.* **154**, 69–82. MR 42:7833.

Herz, C. S. (1974), A generalization of the notion of Fourier transform (French. English summary). *Ann. Inst. Fourier (Grenoble)* **24**, No. 3, 145–157. MR 54:13466.

Helly, E. (1921), On systems of linear equations with an infinite set of unknowns (German). *Monatsh. Math.* **31**, 60–69.

Henson, C. Ward (1972), The nonstandard hull of a uniform space. *Pacif. J. Math.* **43**, 115–137. MR 47:2559.

Hestenes, R. M. (1956), Hilbert space methods in variational theory and numerical analysis. *Proc. Internat. Congress of Math. Amsterdam, 1954, Vol. III*, pp. 229–236. North-Holland. MR 19,43.

Heuvres, K. J. (1973), On types of functions which can serve as scalar products on linear spaces. *Linear Algebra and Appl.* **6**, 83–96. MR 47:788.

Heyting, A. (1949–50), Intuitionisitische Theorie der Hilbert-Ruimten. Unpublished lecture notes (Dutch.)

Heyting, A. (1950), Espaces de hilbert et intuitionisme. *Les methodes formelles en axiomatique*. Centre Internationaux du Centre National de la Recherche Sci. No. 26, Paris, pp. 59–63. Centre Nat. Recherche Sci. Paris, 1953. MR 15,91.

Heyting, A. (1962), *After Thirty Years*: Logic and methodology and philosophy of science, pp. 194–197. Stanford Univ., Calif. 1962. MR 27:1366.

Hida, T. (1960/61), Canonical representation of Gaussian processes and their applications. *Mem. College Sci. Univ. Kyoto Ser. A. Math.* **33**, 109–155. MR 22:10012.

Hida, T. and Ikeda, N. (1965/6), Analysis on Hilbert spaces with reproducing kernel arising from multiple Wiener integral. *Proc. Fifth Berkeley Sympos. Math. Statist. Probability Theory*, Part. I, pp. 117–143. Univ. of Calif. Press, Berkeley, 1967, MR 36:2214.

Higgins, J. R. (1972), An interpolation series associated with Bessel–Hankel transform. *J. London Math. Soc.* **5**, 707–714.

Higgins, J. R. (1976), A sampling theorem for irregularly spaces sample points. *IEEE Trans. Inform. Theory IT* **22**, 621–622.

Higgins, J. R. (1977), Completeness and basis properties of sets of special functions. *Cambridge Tracts in Math.* No. 72. Cambridge Univ. Press. MR 58:17240.

Higgins, J. R. (1985), Five short stories about the cardinal series. *Bull Amer. Math. Soc.* **12**, 45–89.

Hilding, S. (1946), On the closure of disturbed complete orthonormal sets in Hilbert space. *Arkiv för Math. Astronom. Fys.* **32** B, No. 7, 3 pp. MR 8,151.

Hilding, S. (1948), Linear methods in the theory of complete sets in Hilbert spaces. *Arkiv för Math. Astronom. Fys* **35**, A No. 38. MR 11,174.

Hilding, S. (1948), Note on the completeness theorems of Paley–Wiener type. *Ann. of Math.* **49** (2), 953–955. MR 10,198.

Hille, E. (1972), Introduction to general theory of reproducing kernels. *Rocky Mountain J. Math.* **2** (3), 321–368. MR 47:2611.

Hille, E. *et al.* (1957), Functional analysis and semigroups. *Amer. Math. Soc. Colloq. Publ.* MR 19,664.

Hirschfeld, A. O. (1935), A connection between correlation and contingency. *Proc. Camb. Philos. Soc.* **31**, 520–524.

Hirschfeld, R. A. (1958), Sur la théorie génerale des meilleures approximations. *C. R. Acad. Sci. Paris* **246**, 1485–1488. MR 20:2602.

Hirschfeld, R. A. (1958), On best approximation in normed vector spaces. II. *Nieuw Archief voor Wisk.* **VI** (3), 99–107. MR 21:5141.

Hirschfeld, R. A. (1973), On hilbertizable Banach algebras. *Bull. Soc. Math. Belg.* **25**, No. 4, 331–334. MR 56:16377.

Hinčin, A. Ya. (Khinckine) (1943), *Mathematical Principles of Statistical Mechanics* (Russian). OGIZ, Moscow–Leningrad. MR 8,187.

Hoare, A. H. M. (1984), On groups with real valued length function. *Math. Proc. Cambridge Phil. Soc.* **95**, 223–227.

Hodge, J. L. and Lehmann, H. E. (1956), The efficiency of some non-parametric competitors of the t-test. *Ann. of Math. Statist.* **27**, 324–335. MR 18,78.

Hodge, W. V. (1941), *Theory and Applications of Harmonic Integrals.* Univ. of Cambridge Press. MR 2, 296. Second Edition Cambridge Univ. Press, 1952. MR 14,500.

Hoffman, K. (1962), *Banach Spaces of Analytic Functions.* Prentice-Hall, Englewood Cliffs, New Jersey. MR 24:A2844.

Holevo, A. S. (= A. S. Kholevo) (1980), *Probabilistic and Statistical Aspects of Quantum Theory.* (Russian). Nauka Moscow. MR 85i:81038b. English transl. by the author, North Holland, 1982. MR 85i:81038a. *See also* the review in *Bull. Amer. Math. Soc.* **13** (1985), 80–85.

Holladay, J. (1957), A smoothest curve approximation. *Math. Tables Aid. Comput.* **11**, 233–243. MR 20:414.

Holmes, R. D. (1964), *Ph. D. Thesis*, MIT, Cambridge, Mass.

Homes, R. D. (1979), Mathematical foundations of signal processing. *SIAM Review* **21**, 361–388. MR 81:94001.

Holmes, R. D. (1980), Applied approximation theory. *Approximation Theory III.* Academic Press, pp. 523–539. MR 82b:41001.

Holmes, R. D. and Lambert, J. (1973), A geometrical approach to property. (*SAIN*) *J. Approx. Theory* **7**, 132–142. MR 49:9508.

Holland, S. S. (1969), Partial solution of Mackey's problem about modular pairs and completeness. *Canad. J. Math.* **21**, 1518–1525. MR 40:6240.

Holub, J. R. (1974), Tensor product mappings. I. *Math. Ann.* **188**, 1–12. MR 44:2052. II. Proc. A. M. S. 42 (1974), 437–441. MR 48:9438.

Holub. J. R. (1973), On the metric geometry of ideals of operators on Hilbert space. *Math. Ann.* **201**, 157–163. MR 48:4757.

Holub, J. R. (1973), Bases and norm ideals of compact operators. Proc. *Amer. Math. Soc.* **38**, 343–346. MR 47:5560.

Holub, J. R. (1974), Schauder bases and norm ideals of compact operators. II. *Indiana Univ. Math. J.* **24** (1), 555–564. MR 50:2933.

Holub. J. R. (1975), A characterization of the norm ideal of compact operators in Hilbert spaces. *J. Math. Anal. Appl.* **50** (3), 596–606. MR 51:14191.

Holub, J. R. (1975), Rotundity, orthogonality and characterizations of inner product spaces. Bull. *Amer. Math. Soc.* **81**, 1087–1089. MR 51:1263.

Holub, J. R. (1981), On bases and the shift operator. *Studia Math.* **71**, 191–202.

Holub, J. R. (1983), Shift basic sequences in the Wiener disc algebra. *Proc. Amer. Math. Soc.* Preprint.

Holub, J. R. (1983), Shift operators on $C(K)$. (Preprint).

Honda, H. (1961, 1963), A characteristic property of L^p-spaces. ($p > 1$). II. III. *Proc. Japan Acad. Sci.* **36**, 123–127; **39**, 348–351. MR 22:12381; MR 28:3321.

Honda, H. and Yamamuro, S. (1959), A characteristic property of L^p-spaces. I. *Proc. Japan Acad. Sci.* **35**, 446–448. MR 22:2894.

Hongo, E. (1957), On quasi-unitary algebras with semi-finite left rings. *Bull. Kyushu Inst. Techn. Mat. Nat. Sci.* **3**, 1–10, MR 19,872.

Hongo, E. (1958), On some properties of quasi-unitary algebras. *Ibid.* **4**, 1–6. MR 21:7455.

Hongo, E. (1959), On left-multiplicative operators in a quasi-unitary algebra. *Ibid.* **5**, 19–22. MR21: 7456.

Hongo, E. and Orihara, M. (1954), A remark on a quasi-unitary algebra. *Yokohama Math. J.* **2**, 68–72. MR 17,66.
Hörmander, L. (1965), L^2-estimates and the existence theorems for the ∂-operator. *Acta Math.* **113**, 89–152. MR 31:3691.
Huhnaišvili, G. E. (1956), Continuous images of Hilbert spaces (Russian). *Izv. Akad. Nauk SSSR Ser. Mat.* **20**, 275–288. R. Ž. 1957, 3969. MR 19,565.
Hurewicz, W. (1928), On infinite dimensional point sets. *Proc. Nederl. Akad. Wetensch.* **31**, 916–922.
Hutton, C. V. (1974), On the approximation numbers of an operator and its adjoint. *Math. Ann.* **210**, 277–280. MR 49:11296.
Hutton, C. V. (1975), p-factorable operators. *Trans. Amer. Math. Soc.* **205**, 167–180. MR 53:1321.
Hutton, C. V., Morrell, J. S. and Retherford, J. R. (1974), Approximation numbers and Kolmogorov diameters. *Bull. Amer. Math. Soc.* **80**, 462–466. MR 49:1166.
Hutton, C. V., Morrell, J. S. and Retherford, J. R. (1976), Diagonal operators, approximation numbers and Kolmogorov diameters. *J. Approx. Theory* **16**, 48–80. MR 56:12952.
Hunt, R. A. (1966), On $L(p, q)$ spaces. *Enseign. Math.* **12**, 249–276.
Hruščev, S. V., Nikolskiĭ, N. K. and Pavlov, B. (1981), Uniconditional bases of exponentials and of reproducing kernel. *Lecture Notes in Math.* No. 864, 214–335. R. Ž. 1982, 2B 135.
Hsu, L. C. (1948), On Romanoff's device of orthogonalization. *Sci. Rep. Nat. Tsing Hua Univ.* **5**, 1–12. MR 10,198.
Hsu, L. C. (1948), A generalization of Romanoff's method for the construction of orthogonal systems. *Acad. Sinica Science Record* **2**, 178–182. MR 10,198.
Huff, M. (1976), Conformal maps on Hilbert space. *Bull. Amer. Math. Soc.* **82**, 147–149. R. Ž. 1977, 1B 942.

Ifantis, E. K. and Kouris, C. B. (1973), A Hilbert space approach to the localization problem of the roots of analytic functions. *Indiana Univ. Math. J.* **23**, 11–22. MR 47:3644.
Illusie, L. (1964/5), Contractibilité du groupe linéaire des espaces de Hilbert de dimension infinie. *Séminaire Bourbaki.*
Ingelstam, L. (1962), A vertex property for Banach algebras with identity. *Math. Scandi.* **11**, 22–32. MR 26:4199.
Inglestam, L. (1963), Hilbert algebras with identity. *Bull. Amer. Math. Soc.* **69**, 794–796. MR 27:4096.
Inglestam, L. (1964), Real Banach algebras. *Akriv f. Math.* **5**, 239–270. MR 30:2358.
Inglestam, L. (1966), Real algebras with a Hilbert space structure. *Ark. f. Math.* **6**, 459–465. MR 33:7875.
Ingleton, A. W. (1952), The Hahn–Banach theorem for non-archimedean valued fields. *Proc. Cambridge Phil. Soc.* **48**, 41–45. MR 13,659.
Istrăţescu, V. I. (1981), *Linear Operator Theory. An Introduction.* Marcel Dekker, Inc. New York.
Istrăţescu, V. I. (1984), *Strict Convexity and Complex Strict Convexity. Theory and Applications.* Marcel Dekker, Inc. New York.
Istrăţescu, V. I. (1981), *Fixed Point Theory. An Introduction.* D. Reidel Publ. Co. Dordrecht-Boston, Mass.
Istrăţescu, V. I. (1983), Another proof that a complex Hilbert algebra is isomorphic to the complex numbers, *Indian J. Pure Appl. Math.* **14** (2), 179–180.
Istrăţescu, V. I. On Banach spaces with a generalized inner product. I. II. (To appear.)
Istrăţescu, V. I., An algebraic characterization of complete inner product spaces. (To appear.)
Istrăţescu, V. I. (1982), A simple proof of Kakutani–Bohnenblust characterization of inner product spaces. (1982).
Istrăţescu, V. I. (1981–2), On c_p-classes of a semi-simple Banach algebra.
Istrăţescu, V. I. (1981–2), A simple proof of one of Blumenthal characterizations of complete inner product spaces.
Istrăţescu, V. I. (1986), Notes on operator theory in quaternionic spaces. (Unpublished notes.)
Istrătescu, V. I. (1987), Von Neumann modules and Kaplansky module AW^*-algebras. (To appear.)
Istrăţescu, V. I. (1987), Remarks on dilation theory of Banach modules. (To appear.)
Istrăţescu, V. I. (1987), Rigged Kaplansky inner product modules. (To appear.)
Istrăţescu, V. I. (1987), On ideal sets of operators in Banach spaces. (To appear.)
Istrăţescu, V. I. (1980), On a generalization of absolutely summing operators.

Istrăţescu, V. I. (1983), Generalizing normality for operators on Banach spaces. Hyponormality. I. *Collectanea Mathematica* **XXXIV**, 247–259. II.

Istrăţescu, V. I. (1987), on Holemy's inequality and chordal metrics. (To appear.)

Istrăţescu, V. I. (1987), On some characterizations of complete inner product spaces properties of semi-orthogonalities. (To appear.)

Istrăţescu, V. I. (1987), Some parameter associated with a normed space and inner product structures. (To appear.)

Istrăţescu, V. I. (1987), On linear operators for which T*T and T + T* commute and related classes of operators. (To appear.)

Istrăţescu, V. I. (1978), Some characterizations of hermitian operators and related classes of operators. Unpublished manuscript. II.

Istrăţescu, V. I. (1987), On some classes of elements on Banach algebras. (To appear.)

Istrăţescu, (1987), A remark on a paper by Kyle. (To appear.)

Istrăţescu, V. I. (1987), Some characterizations of complete inner product spaces by properties of semi-orthogonalities. (To appear.)

Istrăţescu, V. I. (1986), Gravity centers and inner product structures. (To appear.)

Istrăţescu, V. I. (1986), Topics in the geometry of zeros of polynomials. (Unpublished Notes.)

Istrăţescu, V. I. (1986), Irrational and transcendental numbers. Special chapters. (Unpublished manuscript.)

Ingham, A. E. (1936), Some trigonometrical series with applications to the theory of series. *Math. Z.* **41**, 367–379.

Iseki, K. (1953), Sur les anneaux normés de Hilbert. *I. C. R. Acad. Sci. Paris* **236**, 1123–1125. II. *Ibid.* 237 (1953), 545–546.

Ito, K. (1944), A screw line in Hilbert space and its applications to probability theory. *Proc. Imperial Acad. Tokyo* **20**, 203–209. MR 7,252.

Ito, K. (1953), *Introduction to Stochastic Processes*, (Japanese). Iwanami Shôten, Tokyo. MR 15,448. (Russian transl. Moscow, 2 vols. 1960).

Iynaga, M. (1979/80), A differential geometry on the Hilbert manifolds. *Math. Report Kyushu Univ.* **12**, 29–45. MR 82d:58010.

Ito, S. (1980), A note on dilations in modules over C^*-algebras. *J. London Math. Soc.* **22**, 117–126.

Ionescu Tulcea, C. (1965), Scaler dilations and scalar extensions of operators on Banach spaces. *J. Math. Mech.* **14**, 841–856.

Ionescu Tulcea, C. (1956), *Hilbert Spaces* (In Roumanian). Editura Acad. R. P. R. Bucharest.

Iohvidov, I. S. and Krein, M. G. (1956), Spectral theory of operators in spaces with an indefinite metric. I. II. *Trudy Moskov. Obscestva* **5**, 367–432: **8** (1959), 413–496. Amer. Math. Soc. **13** (2); **34** (2).

Jajte, R. and Lebl R. J. (1981), Gleason measures and some inner products on the algebra of bounded operators on a Hilbert space. *Studia Math.* **70**, 139–145.

Jakstaš, Z. I. (1967), O perekose sfery v prostranstve. *Učen. Zap. Kalinin. Gosud. Ped. Inst.* **52**, 208–213. R. Ž. 1968, 8B 631.

Jakstaš, Z. I. (1968), Ob adnom criteriĭ gilbertogo prostranstva. *Učen. Zap. Kalinin. Gosud. Ped. Inst.* **61**, 190–192.

James, R. C. (1943), Orthogonality in normed linear spaces. *Bull. Amer. Math. Soc.* **49**, 373.

James, R. C. (1945), Orthogonality in normed linear spaces. *Duke Math. J.* **12**, 291–302. MR 6,273.

James, R. C. (1946), Inner products in normed linear spaces. *Bull. Amer. Math. Soc.* **52**, 1008.

James, R. C. (1947), Inner products in normed linear spaces. *Bull. Amer. Math. Soc.* **53**, 559–566. MR 9,42.

James, R. C. (1947), Orthogonality and linear functionals in normed linear spaces. *Trans. Amer. Math. Soc.* **61**, 265–292. MR 9,42.

Janet, M. (1926), Sur la possibilité de plonger un espace riemannien donné dans un espace Euclidien. *Ann. Soc. Math. Polon.* **5**, 38–43.

Janussis, A. D. (1982), New interpretation of the scalar product in Hilbert spaces. *Lett. Nuovo Cimento* **35** (2), No. 5, 145–145. MR 84m:81059. R. Ž. 1983, 2B 799.

Jarvinen, R. D. (1981), *Finite and Infinite Dimensional Linear Spaces*. Marcel Dekker. MR 83k:46001.

Jenkins, J. *et al.* (1974), A Hilbert space approach method in the theory of Schlicht functions. *Proc. Sympos. Complex Analysis, 1973*. London Math. Soc. Lecture Notes Series No. 12, 75–79.

Jenssen, B. (1940), Two theorems on convex sets. *Tidsskrift B*, 66–70. MR 2,261.

Johnson, G. G. (1973), Inner product characterizations by difference equations. *Proc. Amer. Math. Soc.* **37**, 535–536. MR 47:785.

Johnson, W. B. *et al.* (1979), Eigenvalues of p-summing and l^p-type operators in Banach spaces. *J. Funct. Anal. i Priložen.* **32** (3), 353–380 (Russian). MR 80i:47033. (There is a transl. (perhaps) in *J. Functional Anal.* **32** (1979).)

Joichi, J. T. (1966), Normed linear spaces equivalent to inner product spaces. *Proc. Amer. Math. Soc.* **17**, 423–426. MR 33:6346.

Joichi, J. T. (1968), More characterizations of inner product spaces. *Proc. Amer. Math. Soc.* **19**, 1185–1186. MR 38:4974.

Jolly, J. L. (1969), La constante rectangle d'un espace vectoriel normé. *C. R. Acad. Sci. Paris Ser. A–B* **268**, 36–38. MR 39:6056.

Jolly, J. L. (1969), Characterization d'espaces hilbertiens aux moyen de la constante rectangle. *J. Approx. Theory* **2**, 301–311. MR 42:5019.

Jordan, P., von Neumann, J. and Wigner, E. (1934), On algebraic generalizations of the quantum mechanical formalism. *Ann. of Math.* **34** (2), 29–64.

Jordan, P. and von Neumann, J. (1935), On inner products in linear metric spaces. *Am. of Math.* **36** (2), 719–723.

Jurgelas, V. V. (1982), On some geometrical characterizations of Banach spaces and accretive operators (Russian). *Izv. Vysš. Učebn. Zaved. Matematika* **2** (240), 63–69.

Kabaila, V. (1958), On interpolation by bounded functions of class H_δ (Russian). *Uspekhi Mat. Nauk* **13** (1), 181–188. MR 20:117.

Kaczmarz, S. and Steinhaus, H. (1929), Le systeme orthonormal de M. Rademacher. *Studia Math.* **1**, 231–247.

Kadets, M. I. (= M. I. Kadec) (1964), The exact value of the Paley–Wiener constant (Russian). *DAN SSSR* **155**, 1253–1254. MR 28:5289. *Soviet Math. Doklady* **5**, 559–561.

Kadets, M. I. and Mityagin, B. S. (1973), Complemented subspaces in Banach spaces (Russian). *Uspekhi Mat. Nauk* **28** (6), 77–94. MR 53:3649.

Kagan, M. M. (1975), On a measure of divergency between two scalar products and its statistical application. *Sankya Ser. A* **37** (4), 492–501. MR 58:4672.

Kagan, M. M. and Gerlein, O. V. (1975), Hilbert space methods in classical problems of mathematical statistics (Russian. English summary). *Zap. Naučn. Semin. Leningrad Otdel. Mat. Inst. Steklov (LOMI)* **53**, 64–100. 142–143. MR 53:6913.

Kagan, M. M. and Gerlein, O. V. (1975), Hilbert space methods in classical problems of mathematical statistics. *Proc. Third Japan-URSS Sympos. Prob. Theory, Tašhkent 1975*, 134–159. *Lecture Notes in Math. No.* 550 (1976). MR 58:3216.

Kahane, J. P. (1968), *Some Random Series of Functions.* D. C. Heath and Co., Lexington, Mass. MR 40:8095. Russian transl. *Izdat. Mir*, Moscow, 1973, MR 52:14797.

Kahane, J. P. (1981), Hélices et quasi-hélices. *Mathematical analysis. Part. B. Advances in Math. Suppl. Stud. Vol. 7B*, pp. 417–433. Academic Press, 1981.

Kakihara, Yuichiro, (1983), Semivariation and operator semivariation of Hilbert space valued measures. *Proc. Japan Acad.* **59**, Ser. A No. 10, 456–458.

Kakihara, Yuichiro (1983), On a Hilbert module over an operator algebra and its applications to harmonic analysis. *Kodai Math. J.* **6**, 289–300. MR 85c:46044.

Kakihara, Yuichiro (1980), Hilbert $B(H)$-modules with applications. II. *Res. Rep. Inst. Inf. Sci. Techn.* (Tokyo Denki Univ.) **6**, 33–45. MR 83b:46065.

Kakihara, Yuichiro(1982), Hilbert $B(H)$-modules with applications. III. *Ibid.* **7**, 11–22. MR 846:46055.

Kakihara, Yuichiro (1983), Hilbert $B(H)$-modules with applications. IV. *Ibid.* **5**, 27–30.

Kakihara, Yuichiro (1981), Stationary processes and information channels. *Ibid.* **3**, 13–20. MR 824:60076.

Kakihara, Yuichiro and Terasaki, T. (1979), Hilbert $B(H)$-modules with applications. I. *Res. Rep. Inst. Inf. Sci. Techn.* (Tokyo Denki Univ.) **5** (1979), 23–32. MR 83b:46069.

Kailath, T. (1974), A view of three decades of linear filtering theory. *IEEE Trans. Inf. Theory* **20**, 145–181.

Kakutani, S. (1939), Some characterizations of real euclidean spaces. *Japan J. Math.* **16**, 93–97. MR 1,146. MR 57:5337.

Kakutani, S. (1941), Concrete representations of abstract (L)-spaces and the mean ergodic theorems. *Ann. of Math.* **42** (2), 523–537. MR 3,319.

Kakutani, S. and Mackey, G. W. (1944), Two characterizations of real Hilbert spaces. *Ann. of Math.* **45** (2), 50–58. MR 5,146.

Kakutani, S. and Mackey, G. W. (1946), Ring and lattice characterizations of complex Hilbert spaces. *Bull. Amer. Math. Soc.* **52**, 727–733. MR 8,31.

Kalašnikov, A. L. and Slugin, S. N. (1970), S. L. Sobolev's Hilbert space as a variational structure (Russian). *DAN SSSR* **194**, 261–264. MR 42:5008.

Kalish, G. K. (1947), on 1-adic Hilbert spaces. Ann. of Math. **48** (2), 180–192 MR 8,388.

Kallianpur, G. (1970), Zero-one-laws for gaussian processes. *Trans. Amer. Math. Soc.* **149**, 199–211. MR 42:1200.

Kallianpur, G. (1971), Abstract Wiener processes and their reproducing kernel Hilbert space. *Z. Wahrschein. Theorie* **17**, 113–123.

Kallianpur, G. (1970), The role of reproducing kernel Hilbert spaces in the study of gaussian processes. *Advances in Probability*. Vol. 2. Marcel Dekker, Inc. New York. pp. 49–84.

Kalman, R. E. and Bucy, R. S. (1961), New results in linear prediction theory and filtering theory. *J. Basic Engr. (Trans. ASME Ser. D)* **83D**, 95–100.

Kalman, R. E., Falb, P. L. and Arbib, M. (1969), *Topics in Mathematical System Theory*. MacGraw-Hill Books Co. MR 40:8465.

Kalton, N. J. and Ruckle, W. H. (1973), A series characterization of subspaces of $L_p(\mu)$ spaces. *Bull. Amer. Math. Soc.* **79**, 1019–1022. MR 47:9247.

Kalton, N. J. and Wood, G. V. (1976), Orthonormal systems in Banach spaces and their applications. *Math. Proc. Cambridge Phil. Soc.* **79**, 493–510. MR 53:6291.

Kandelaki, N. P., Kartzivadze, J. N. and Cantladze, T. L. (1976), An orthogonal multiplication in Hilbert space (Russian). *Trudy Tbilisi Univ.* **179**, 45–57. (Gruzian and English summary). R. Z. 1977, 6B 565.

Kantor, M. A representation theorem for l^p spaces. *Proc. Amer. Math. Soc.* **31**, 472–474.

Kantorovich, L. V. (1935), Sur les prolongements des fonctionelles linéaires. *C. R. Acad. Sci. URSSS* **1**, 207–210.

Kaplansky, I. (1948), Dual rings. *Ann. of Math.* **49** (2), 689–701. MR 10, 7.

Kaplansky, I. (1949), Normed algebras. *Duke Math. J.* **16**, 399–418. MR 11,115.

Kaplansky, I. (1951), A theorem on ring of operators. *Pacif. J. Math.* **1**, 227–232. MR. 14,291.

Kaplansky, I. (1953), Modules over operator algebras. *Amer. J. Math.* **75**, 839–858. MR 15, 327.

Kaplansky, I. (1951), Projections in Banach algebras. *Ann. of Math.* **53** (2), 235–249.

Kapoor, P. and Prasad, J. (1978), Orthogonality and characterizations of inner product spaces. *Bull. Austral. Math. Soc.* **19**, 403–416. MR 80h:46028.

Kapoor, P. and Mathur, S. B. (1981), Some geometric characterizations of inner product spaces. *Bull. Austral. Math. Soc.* **24**, 239–246.

Kapoor, P. and Mathur, S. B. (1983), Metric projection bound and the Lipschitz constant of the radial retraction. *Journ. approx. Thy.* **38**, 66–70.

Karhunen, K. (1950), On the structure of stationary random functions (German). *Arkiv.f. Math.* **1**, 141–150. MR 11,607.

Kasimov, V. A. (1983), On the Kuiper theorem for Hilbert modules $l^2(H)$ (Russian). *DAN Akad. Nauk Azerbaidzan SSR* **38**, No. 6, 8–10. R. Ž. 1983, 5B 754.

Kasahara, S. (1954), A characterization of Hilbert spaces. *Proc. Japan Acad.* **30**, 846–848. MR 16,1032.

Kasparov, G. G. (1980), The operator K functor and extensions of c^*-algebras (Russian). *Izv. Akad, Nauk SSSR Ser. Math.* **44**, No. 3, 571–636. MR 81m:58075.

Kawada, Y. (1943), On the ring of operators of a Banach space (German). *Proc. Imperial Acad. Tokyo* **19**, 616–621.

Kay, D. C. (1978), A remark on Freese' characterization of inner product spaces. Unpublished manuscript.

Kay, D. C. (1967), A parallelogram law for certain L^p-spaces. *Amer. Math. Monthly* **74**, 140–147. MR 34:8148.

Kaz, I. (1950), On Hilbert spaces generated by positive hermitan matrix functions (Russian). *Zapiski Naučn. Issled. Inst. Mat. Mech. i Harkov Mat. Obsč.* **22**.

Kelly, E. J., Reed, I. S. and Root, W. L. (1960), The detection of Radar Echoes in Noise. *J. Soc. Ind. Appl. Math.* **8**, 309–341.

Kelly, J. B. (1975), Hypermetric spaces. The geometry of metric and linear spaces, pp. 17–31. *Lecture Notes in Math.* No. 490. MR 53:9161.

Kelley, J. L. (1950), Convergence in topology. *Duke Math. J.* **17**, 277–283. MR 12,194.

Kelley, J. L. (1952), Banach spaces with the extension property. *Trans. Amer. Math. Soc.* **72**, 323–326. MR 13,659.

Kelley, J. L. (1955), *General Topology*. Van Nostrand. MR 16,1136.

Kelly, L. M. (1966), The Massera–Schaeffer inequality. *Amer. Math. Monthly* **73**, 1102–1103. MR 34:4873.

Kelly, L. M. (1975), On the equilateral feeble four point property. The geometry of metric and linear spaces. *Lecture Notes in Math.* No. 490. MR 53:6433.

Kerwin, M. R. (1973), The trace class of a full Hilbert algebra. *Trans Amer. Math. Soc.* **178**, 259–270. MR 47:7446.

Kellog, C. N. (1968), Centralizers and H^*-algebras. *Pacif. J. Math.* **17**, 121–129. MR 33:1749.

Khavin, V. P. (1970), Spaces of analytic functions (Translated from the Russian). *Progress in Math.* No. 1, 76–169. Plenum Press, New York, 1970.

Khinchin, A. I. (=A. I. Hinčin) (1923), On dyadic fractions. *Math. Z.* **18**, 109–113.

Khinchin, A. I. (1934), Correlatinstheorie der stationaire stochastische Prozesse. *Math. Ann.* **109**.

Khinchin, A. I. (1938), Correlation theory of stationary stochastic processes (Russian). *Uspekhi Mat. Nauk* **4**, 42–51.

Khinchin, A. I. (1951), *The Foundations of Quantum Mechanics* (Russian). Izdat. Techn. Teor. Lit. Moscow–Leningrad, 1951. English transl. Dover Publ. Co. 1957. MR 13,894.

Kimeldorf, G. and Whaba, G. (1971), Some results on Tschebicheffian spline functions. *J. Math. Anal. Appl.* **33**, 82–95. MR 44:7198.

Kirčev, K. P. and Trojansky, S. L. (1975), A certain characterization of spaces with scalar products (Russian). *C. R. Acad. Sci. Bulgare* **28**, 445–447. MR 58:7046.

Kirk, W. A. and Royalty, W. D. (1971), Fixed point theorems for certain nonlinear non-expansive mappings. *Illinois J. Math.* **15**, 656–663. MR 45:1007.

Kisliakov, S. V. (1977), p-absolutely summing operators. *Geometry of Linear Spaces and Theory of Linear Operators* (Russian). Ed. B. Mitiagin. Jaroslavl. pp. 114–175. MR 58:17804.

Kisliakov, S. V. (1979), Reflexive supspaces as a C_A^*, (Russian). *Functional Anal. Appl.* **13**, 16–23. MR 80h:46033.

Kizner, Z. I. (1971), Families of planes in Hilbert spaces and developable surfaces (Russian). *Uspehi Mat. Nauk.* **26** (3) (159), 201–202. MR 46:648.

Klamkin, M. S. (1975), A vector norm inequality. *Amer. Math. Monthly* **82** (8), 829–830. MR 52:684.

Klee, V. (1949), A characterization of convex sets. *Amer. Math. Monthly* **56**, 247–249. MR 10, 621.

Klee, V. (1952), *Math. Reviews* **13**, 661.

Klee, V. (1960). Circumspheres and inner products. *Maths. Scand.* **8**, 363–370. MR 23:A2734.

Klee, V. Infinite dimensional intersection theorem. *Amer. Math. Soc. Proc. Sympos. Pure Math. Vol. 7, Convexity*, pp. 349–360.

Klee, W. (1982), Inspheres and inner products. *Abstracts* **3**, 275.

Koehler, D. O. (1970), G_{2n} spaces. *Trans. Amer. Math. Soc.* 150, 507–518. MR 41:7411.

Koehler, D. O. (1975), The analytic properties of G_{2n} spaces. *Proc. Amer. Math. Soc.* **35**, 201–206. MR 46:644.

Kolmogorov, A. N. (1932), Sulla una forma generale di un processo stochastico omogeno. *Atti Accad. Nazionale dei Lincei, Rend.* **15**, 805–808; 866–869.

Kolmogorov, A. N. (1935), La trasformation de Laplace dans les espaces linéaires. *C. R. Acad. Sci. Paris* **200**, 1717–1718.

Kolmogorov, A. N. (1940), Curves in Hilbert spaces which are invariant for a group of mouvements (German). *C. R. Acad. (Doklady) URSS (N.S.)* **26**, 6–9. MR 2,200.

Kolmogorov, A. N. (1940). Wiener spirals and other interesting curves in Hilbert spaces. *C. R. Acad. (Doklady) URSS* **26**, 115–118. MR 2,220.

Kolmogorov, A. N. (1941), Stationary sequences in Hilbert spaces. *Bull. Moscov. Gosud. Univ.* **2**, No. 6. MR 5,101.

Kolmogorov, A. N. (1941), Interpolation and extrapolation of stationary random sequences (Russian). *Izv. Akad. Nauk SSSR Ser. Mat.* **5**, 3–14. MR 3,4.

Kolmogorov, A. N. (1936) *Foundations of Probability* (Russian). Moscow. English transl. Chelsea Publ. Co. New York, 1950. MR 11,374. *Izdat. Nauva Koscow*, 1974. MR 50:5878.

Koopman, B. O. (1931), Hamiltonian systems and transformations in Hilbert spaces. *Proc. Nat. Acad. Sci. USA* **17**, No. 5,315–318.

Koopman, B. O. and von Neumann, John (1952) Dynamical systems of continuous spectra. *Proc. Nat. Acad. Sci. USA* **18**, 255–263.

Koopmans, L. H. (1974), *The Spectral Analysis of Time Series*. Academic Press, New York. MR 50:1447.

Kopperman, Ralph (1967), Application of infinitary languages to metric spaces. *Pacif. J. Math.* **23**, 299–310. MR 36:1307.

Kopperman, Ralph (1967), The $L_{w_1 w_1}$ theory of Hilbert spaces. *J. Symbolic Logic* **32**, 295–304. MR 35:6534.

Kopperman, Ralph (1967), On the axiomatizability of uniform spaces. *J. Symbolic Logic* **32**, 289–294. MR 35:6535.

Korányi, A. (1956), On a theorem of Lowner and its connections with resolvents of selfadjoint transformations. *Acta Sci. Math. (Szeged)*, **17**, 63–70. MR 18,588.

Körezlioglu, H. (1968), Reproducing kernels in separable Hilbert spaces. *Pacif. J. Math.* **25**, 305–314. MR 37:5683.

Kostyučenko, A. G. and Skorohod, A. V. (1953), On a theorem of N. K. Bari (Russian). *Uspekhi Mat. Nauk (N.S.)* **8** (5) (57), 165–166. MR 15,632.

Kostyučenko, A. K. and Mitiagin, B. S. (1960), Positive definite functions on nuclear spaces (Russian). *DAN SSSR* **13**, 13–16, *Soviet Math. Dokl.* **1**, 177–180. MR 22:12391.

Kotelianskiĭ, D. M. (1953), On an arithmetic method of orthogonalization in nonseparable Hilbert spaces (Russian). *Mat. Sbornik* **33** (75), 181–192. MR 15,236.

Kotelianskiĭ, D. M. (1953), On N. P. Romanov's method of obtaining identities for arithmetic functions (Russian). *Ukrain. Mat. Ž.* **5**, 453–458. MR 15,779.

Krein, M. G. (1950), Hermitian positive definite kernels in homogeneous spaces (Russian). I. II. *Ukrain. Mat. Ž.* **1**. II. **2**, 10–59. MR 12,719.

Krein, M. G. (1947), The theory of selfadjoint extension of semibounded hermitian transformations and its applications. I. II (Russian). *Rec. Math. (Matematiceskiĭ Sbornik) (N.S.)* I **20** (62), 431–495. MR 9,515. II. **21** (63), 365–404. MR 9,515.

Krein, M. G. and Krasnoselskiĭ, M. A. (1947), Fundamental theorems on the extension of hermitian operators and certain of their applications to the theory of orthonormal polynomials and the problem of moments (Russian). *Uspehi Mat. Nauk (N.S.)* **2** (19), 60–106. MR 10,198.

Krein, M. G. (1957), On Bari bases of a Hilbert space (Russian). *Uspehi Mat. Nauk* **12**, vyp. **3** (75), 333–341. MR 19,1184.

Krein, M. G. On the basic approximation problem in the theory of interpolation and filtering of stationary random processes (Russian). *DAN SSSR*, **94**, pp. 13–16.

Krein, M. G. and Rechtmann, P. (1938), Sur une question de Nevanlinna–Pick. *Travaux de l'Université d'Odessa*, **2**, 63–68.

Krein, M. G., Milman, D. P. and Rutman, M. A. (1940), On a property of a base in Banach spaces (Russian). *Zap. Mat. Obsčestva (5) Kharkov*, **16**, 106–110.

Kramar, E. (1981), Locally convex topological vector spaces with hilbertian seminorms. *Rev. Roum. Math. Pures Appl.* **26**, 55–62. MR 82k:46 008a.

Kramar, E. (1981), Linear operators in *H*-locally convex spaces. *Rev. Roum. Math. Pures Appl.* **26**, 62–77. MR 82k:46008b.

Krivonosov, L. N. (1970), A certain method of prescribing a pre-Hilbert space norm (Russian). *Izv. Vysš. Učebn. Zaved. Mat.* No. 1 (116), 52–54. MR 45:7452.

Kubota, T. (1939), A new construction of the Euclidean geometry in the affine space. *Monatsh. Math. Phys.* **18**, 96–102.

Kučenko, E. N. and Koševnikova, I. G. (1970), On the definition of Hilbert spaces over semifields (Russian). *Taskent. Politechn. Inst. Naučn. Trudy (N.S.)* **56**, 40–44. R. Ž. 1971, 4 B 634.

Kuelbs, J., Larkin, F. M. and Williamson, J. A. (1972), Weak probability distributions on reproducing kernel Hilbert spaces. *Rocky Mountain J. Math.* **2**, 369–378. MR 45:8796.

Kuiper, N. (1965), The homotopy type of unitary group of Hilbert spaces. *Topology* **3**, 19–30. MR 31:4034.

Kuiper, N. (1969), Hilbert manifolds. Geometrical aspects. *Seminaire Mathematiques Superieures*. Montreal, Canada, été.

Kuratowski, K. (1969), *Introduction to Set Theory and Topology*. Warszawa, Oxford.

Kuratowski, K. (1960), *Topologie*. PWN, Warszawa.

Kurepa, S. G. (1964), The Cauchy functional equation and scalar product in vector spaces. *Glasnik Mat. Fiz. i Astr.* **19**, 23–36.

Kurepa, S. G. (1965), Quadratic and sesquilinear functionals. *Glasnik Mat. Fiz. i Astr.* **20**, 79–92.

Kruepa, S. G. (1966), On the Cauchy–Schwarz inequality (Serbo-Croatian summary). *Glasnik Mat. Ser. III* **1** (2), 147–158. MR 36:1963.

Kušner, B. A. (1973), Lectures on constructive mathematical analysis. (Russian). *Izdat. Nauka, Moscow*. MR 52:53.

Kutateladze, S. S. (1982), On a criterion of hilbertness of a space. (Russian). *Optimizacija*, Novosibirsk **28**/**29**, 146–147. R. Ž. 1982, 98 620.

Kuzmin, R. O. (1928), On a problem of gauss. *DAN SSSR Ser. A* **1928**, 375–380.

Kuzmin, V. V. (1966), A metric space with middle elements. *Leningrad Inz. Ekonom. Inst. Trudy Vyo.* **63**, 18–30. MR 35:3417.

Kwapien, S. (1970), A linear topological characterization of inner product spaces. *Studia Math.* **38**, 277–278. MR 43:2478.

Kwapien, S. (1970), On a theorem of L. Schwartz and its applications to absolutely summing operators. *Studia Math.* **38**, 193–201. MR 43:3822.

Kwapien, S. (1972), isomorphic characterization of inner product spaces by orthogonal series with vector valued coefficients. *Studia Math.* **44**, 583–585. MR 49:5789.

Kwapien, S. (1972), *Exposé No. 8, Seminaire L Schwartz*, Ecole Politechniques de Paris, Paris. MR 53:14095.

Kyle, J. (1977), Norms of derivations. *J. London Math. Soc.* **16** (2), 297–312. MR 58:7113.

Lacey, E. H. and Bernau, S. J. (1974), Characterization and classifications of some classical Banach spaces. *Advances in Math.* **12**, 367–401. MR 49:5794.

Lagenhop, C. E. (1973), Differentiability of the distance to a set in Hilbert space. *J. Math. Anal. Appl.* **44**, 620–624. MR 48:12008.

Lalescu, Traian (1912), *Introduction à la Theorie des Equations Integrales*. Paris.

Lance, E. C. (1975), Direct integrals of left Hilbert algebras. *Math. Ann.* **216**, 11–28. MR 51:8833.

LaDuke, Jeanne (1970), On a certain generalization of l^p spaces. *Pacif. J. Math.* **35**, 155–168.

Lambert, H. B. (1968), Topological characterization of inner product spaces. *Ph.D. Thesis*, Texas Univ.

Landau, E. D. and Lifschitz, E. M. (1958), *Quantum Mechanics*. Transl. from the Russian. Addison-Wesley. Reading, Mass.

Landau, H. J. and Polack, H. O. (1961), Prolate spheroidal wave functions. Fourier analysis and uncertainty. II. *Bell. System J.* **40**, 68–84. MR 25:4147.

Landau, H. J. and Polack, H. O. (1962), Prolate spheroidal wave functions. Fourier analysis and uncertainty. III. *Ibid.* **41**, 1295–1336. MR 26:5200.

Landau, H. J. and Shepp, L. A. (1970), On the supremum of a gaussian process. *Sankya. The Indian Journal of Statistics Ser. A* **32**, 369–378. MR 44:3381.

Land, S. (1962), *Introduction to Differentiable Manifolds*. Interscience.

Larkin, F. M. (1972), Gaussian measures in Hilbert spaces and applications in numerical analysis. *Rocky Mountain J. Math.* **2**, 379–421. MR 46:2331.

Lau, L. S. (1958), On a characterization of Hilbert spaces. *Advancement in Math.* **4**, 462–464. (Chinese. English summary). MR 20:4178.

Lax, P. D. Translation invariant subspaces. *Acta Math.* **101** (159), 163–178. MR 21:4359.

Leach, E. B. and Whitfields, J. H. U. (1977), Differentiable functions and rough norms on Banach spaces. *Proc. Amer. Math. Soc.* **33**, 120–126. MR 45:2471.

Lebedev, N. A. (1951), Some estimates and extremal problems in the theory of conformal mappings. *Diss.* Leningrad State Univ. (Russian).

Lebow, A. *et al.* (1971), Semigroups of operators and measures of noncompactness. *J. Functional Anal.* **7**, 1–26. MR 42:8301.

Le Cam, L. (1957), Convergence in distribution of stochastic processes. *Univ. of California Publ. Statist.* Vol. 2 no. 11, 207–236. MR 19,128.

Leduc, M. (1971), Derivabilité dans les espaces normés. *Ph.D. Thesis*, Univ. Paris Sud, Paris, **52**, pp. R. Ž. 1975, 9B 848.

Leduc, M. (1969), Charactérisation des espaces euclidiens. *C.R. Acad. Sci. Paris Ser. A-B* **268**, 943–946. MR 41:2367.

Leduc, M. (1967/8), Characterisation des espaces euclidiens. *Seminaire Choquet, Paris*, Exp. No. 9. Paris, 1969. MR 43:6708.

Lee, D. T. and Morf, M. (1980), Recursive square root ladder estimation algorithms. *Proc. Internat. Conf. on automatic systems and sequential processing.* Denver, Colo.

Legiša, P. (1976/7), Well embedded Hilbert subspaces in C^*-algebras. *Studia Math.* **59**, 37–40. MR 55:6197a.

Legiša, P. (1976), Well embedded Hilbert subspaces in C^*-algebras. II. *Glasnik Mat. Ser. III* **11** (31), No. 1, 63–57. MR 55:6197b.

Legiša, P. (1976), Hilbert subspaces in the C^*-algebra $B(H)$. (Serbo-Croatian summary). *Glasnik Mat. Ser.* III **11** (31) (1), 59–62. MR 55:6198.

Lehtö, O. (1949), Applications of certain orthogonal systems to some function-theoretic extremal problems. *Ann. Fenn. Acad. Ser. A I*, **1** (59), MR 11,170.

Lehtö, O. (1950), On the existence of analytic continuation. *Ann. Fenn. Acad. ser. A I.* **1** (70). MR 12,89.

Lehtö, O. (1949), On the existence of analytic functions with a finite Dirichlet integral. *Ann. Fenn. Acad. Ser. A I* **1** (67). MR 11,3338.

Lehtö, O. (1950), On Hilbert space with kernel function. *Ann. Fenn. Acad. Ser. A I* **1**, (74). MR 12,337.

Lehtö, O. (1950), Some remarks on kernel functions in Hilbert spaces. *Ann. Fenn. Acad. Ser. A I* **1** (109). MR 14,185.

Lelong-Ferrand, J. (1957), Sur l'application des methodes de Hilbert a l'etude des transformations infinitesimales d'une variete differentiable. *Bull. Soc. Math. Belg.* **9**, 59–73. MR 12,802.

Lelong-Ferrand, J. (1957), Sur les transformations infinitestimales d'une varité, con derées comme des opérateurs hilbertiens. *C.R. Acad. Sci. Paris* **245** 1585–1588. MR 20:2023.

Lelong-Ferrand, J. (1958), Applications des methodes de Hilbert a l'etude des transformations infinitesimales d'une varieté. *Bull. Soc. Math. France* **86**, 1–28. MR 21:2274.

Lelong-Ferrand, J. (1957), Applications of Hilbert space methods to Lie groups acting on a differentiable manifold. *Proc. Nat. Acad. Sci. USA* **43**, 249–252. MR 19,444.

Levi, Beppo (1906), Sul principio di Dirichlet. *Rend. Circ. mat. Palermo* **22**, 293–360.

Levi Civita, T. Sugli infiniti ed infinitesimi attuali quali dementi aualitici. (1892–1893). *Opere matematiche*, Bologna, 1954, pp. 1–39.

Levin, A. Yu. and Petunin, Yu. I. (1967), Some questions connected with the concept of orthogonality in Banach spaces (Russian). *Uspehi Mat. Nauk* **18** (3), 167–170. MR 27:2833.

Levinson, N. (1936), On non-harmonic Fourier series. *Ann. of Math.* **37** (2), 919–936.

Levinson, N. (1940), Gap and density theorems. *Amer. Math. Soc. Colloq. Publ. New York.* MR 2,180.

Levitan, B. M. and Sargasjan, I. S. (1975), Introduction to spectral theory: selfadjoint ordinary differential operators. Transl. from the Russian. AMS Transl. No. 39. AMS Providence, R. I. MR 51:6026.

Levý, P. (1956), Sur une classe de courbes de l'espace de Hilbert et sur une equation integrale nonlineaire. *Ann. Ecole. Norm. Sup.* **73**, 121–156. MR 20:2787.

Levý, P. (1965), *Processes Stochastiques et Mouvement Brownien.* Gautheir Villars, 1948. Second Ed.

Lewittes, J. (1968), An extension of hyperbolic geometry and Julia's theorem. *Duke Math. J.* **35**, 777–782. MR 39:5792.

Lin, Bor-Luh and Singer, I. (1971), On conditional bases of l^2. *Comment. Math Prace Math.* **15**, 135–139. MR 47:1079.

Lindenstrauss, J. (1963), On the modulus of smoothness and divergent series in Banach spaces. *Michigan Nath. J.* **10**, 241–252. MR 29:6316.

Lindenstrauss, J. (1971), Decomposition of Banach spaces. *Indian Math. Journal* **20**, 917–919. MR 53:8870.

Lindenstrauss, J. (1975/6), Some open problems in Banch space theory. *Semiaire Choquet, 15e année*, No. 18, 9p.

Lindenstrauss, J. and Pelczynski, A. (1968), Absolutely summing operators in L_p spaces and their applications. *Studia Math.* **29**, 275–326. MR 37:6473.

Lindenstrauss, J. and Rosenthal, H. (1969), The L^p-spaces. *Israel J. Math.* **7**, 325–349. MR 42:5012.

Lindenstrauss, J. and Tzafriri, L. (1971), On the complemented subspace problem. *Israel J. Math.* **9**, 263–269. MR 43:2474.

Lindenstrauss, J. and Zippin, M. (1969), Banach spaces with a unique unconditional basis. *J. Functional Anal.* **3**, 115–125.

Lindenstrauss, J. and Perles, M. A. (1969), On extreme operators in finite dimensional spaces. *Duke Math. J.* **36**, 301–314. MR 39:1938.

Lindenstrauss, J. and Tzafriri, L. (1973), Classical Banach spaces. *Lecture Notes in Math.* No. 338. MR 54:3344.

Lindenstrauss, J. and Tzafriri, L. (1977), Classical Banach spaces. *Ergebn. der Math. No.* 92. New York, 1977. Mr 58:17766.

Lindhal, R. and Maserik, P. H. (1971), Positive definite functions on involution semigroups. *Duke Math. J.* **38**, 771–782. MR 45:916.

Linfoot, E. H. (1954), An informational measure of correlation. *Information and Control* **1**, 85–89. MR 19,1148.

Linnik, Ju. V. (1949), Quaternions and Cayley numbers. Some applications of quaternionic arithmetic. *Uspekhi Mat. Nauk* **4**, 49–98.

Liusternik, L. A. and Sobolev, V. I. (L. A. Lusternik and V. I. Sobolev) (1959), *Elements of Functional Analysis* (Russian). Fizmatgiz, Moscow. MR 31:1536. English transl. Ungar, New York, 1962.

Laomont, L. S. and Mandelson, P. (1963), The Wigner unitary-antiunitary theorem. *Ann. of Math.* **73** (2), 549–559. Mr 27:5130.

Loomis, L. H. (1953), *An Introduction to Abstract Harmonic Analysis.* van Nostrand Co. Princeton, N.J. MR 14,883.

Lorch, Edgar R. (1945), The Caucy–Schwarz inequality and selfadjoint spaces. *Ann. of Math.* **46** (2), 468–473. MR 7,125.

Lorch, Edgar R. (1948), On certain implications which characterize Hilbert spaces. *Annals of Math.* **49** (2), 523–532. MR 10,129.

Lorch, Edgar R. (1953), A curvature study of convex bodies in Banach spaces. *Annali di Math. Pura Appl.* **34**, 105–112. MR 14,657.

Loeve, M. (1945), Fonctions aléatoires de second ordre, *C.R. Acad. Sci. Paris* **220**, MR 7,458.

Loeve, M. (1946), Fonctions aléatoires de second ordre. II. *C.R. Acad. Sci. Paris* **222**, MR 7,458.

Loeve, M. (1945), Fonctions aléatoires de second ordre. *Rev. Sci.* **83**, 297–303. MR 8,38.

Loeve, M. (1946), Fonctions aléatoires de second ordre. *Rev. Sci.* **84**.

Loeve, M. (1955), *Probability Theory. Foundations. Random Processes.* van Nostrand Co. Inc. New York. MR 16,598–599.

Loveland, L. D. and Valentine, J. E. (1978), Metric criteria for Banach spaces and Euclidean spaces. *Fund. Math.* **100**, 75–81. MR 58:2587.

Löwig, H. (1934), Complex Euclidean spaces of arbitrary finite dimension or transfinite dimension (German). *Acta Lit. Sci. Szeged* **7**, 1–33.

Löwner, K. (1934), Über monotone Matrix functionen. *Math. Z.* **38**, 177–216.

Löwner, K. (1939), Grundzuge einer Inhaltslehre im hilbertschen Raume. *Ann. of Math.* **40** (2), 816–833.

Loynes, R. M. (1965), VH-spaces. *Trans. Amer. Math. Soc.* **116**, 167–180. MR 33:584.

Loynes, R. M. (1965), On generalized positive definite functions. *Proc. London Math. Soc.* **15** (3), 373–384. MR 30:4140.

Loynes, R. M. (1965), On generalization of second order stationarity. *Proc. London Math. Soc.* **15** (3), 385–398. MR 31:803.

Lubell, D. (1973), Homogeneous and multilinear forms related to a theorem of Jordan and von Neumann. *Advances in Math.* **10**, 291–296. MR 47:5563.

Luchins, E. H. (1959), On strictly semi-simple Banach algebras. *Pacif. J. Math.* **9**, 551–554. MR 21:7449.

Lumer, G. (1961), Semi-inner product spaces. *Trans. Amer. Math. Soc.* **100**, 29–43. MR 24:A2860.

Lumer, G. (1964), Spectral operators, hermitian operators and bounded groups. *Acta Math. Sci.* (*Szeged*), **25**, 75–85. MR 29:6329.

Lykova, O. B. (1966), Generalization of a theorem of Bogoljubov to the case of Hilbert space (Russian). *Ukrain. Mat. Ž.* **18**, No. 5, 53–59. MR 34:3050. *Amer. Math. Soc. Transl.* **77** (2). MR 39:2.

Lenard, A. (1973), Correlation functions and the uniqueness of the state in classical statistical mechanics. *Comm. Math. Phys.* **30**, 35–44.

Levy, S. I. (1974), Hilbert space theoreitc methods in the study of correlation measures. *Comm. Math. Phys.* **37**, 41–62. MR 49:12018.

Macbeath, A. M. (1951), A compactness theorem for affine equivalence classes of convex regions. *Canad. J. Math.* **3**, 54–61. MR 13,577.

Mackey, G. W. (1957), Quantum mechanics and Hilbert spaces. *Amer. Math. Monthly* **6**, 885–893. MR 20:2608.

Mackey, G. W. (1963), *The Mathematical Foundations of Quantum Mechanics.* W. Benjamin New York. MR 27:5501.

Maczynski, M. J. (1973), On a functional representation of the lattice of projections on a Hilbert space. *Studia Math.* **47**, 253–259. MR 52:3935.

Maczynski, M. J. (1974), On a lattice characterization of Hilbert spaces. *Colloq. Math.* **31**, 243–248. MR 51:1353.

Maczynski, M. J. (1983), A functional characterization of inner product spaces. *Demonstratio Math.* **16** (3), 797–803. MR 85f:46048.

Maczynski, M. J. (1976), Orthomodularity and lattice characterization of Hilbert spaces (Russian summary). *Bull. Acad. Polon. Sci. Ser. Sci. Math. Astronom. Phys.* **24**, 481–487. MR 53:14097.

Maczynski, M. J. (1977), A remark on Mackey's problem about modular pairs and completeness. *Bull. Acad. Polon. Sci. Ser. Sci. Math. Astronom. Phys.* **25**, 27–31.

Madeiros, Adanto de Justa, L. (1961), Existence of scalar products in normed linear spaces (Portuguese). *Gaz. Mat. (Lisboa)*, **22**, (85/86) 1–6. MR 24:A2829.

Maeda, F. and Maeda, S. (1970), Theory of symmetric lattices. *Die Grund.* No. 173 (1970). Springer Verlag, New York.

Majumder, N. Bose (1977), Cantor set in the l^2 space and its properties. *Bull. Calcutta Math. Soc.* **69**, 21–25. MR 58:27496a. *Errata, Ibid.* **69**, No. 2. loose page. MR 58:2746b.

Makagon, A. and Weron, A. (1976), Wold–Cramer concordance theorems for interpolation of q-variate processes over locally compact abelian groups. *J. Multivariate Analysis* **6**, 123–137. MR 54:14072.

Makarov, B. M. (1979), A characterization of Hilbert spaces. *Mat. Zametki* **20**, 739–746 (Russian). MR 80m:46021.

Malengreau, Julien, (1945), *Etude Critique du Téoreme de Pythagora.* 2nd. Edition, Lausanne. MR 8,168.

Malin, S. (1982), Carmeli's Hilbert space for quantum gravity: a group theoretical approach. *Physica* **114A**, 154–156. North Holland Publ. Co.

Malin, S. and Carmeli, M. The Hilbert space of quantum gravity. *Il Nuovo Cimento* **66B** (2), 164–172. (Italian summary).

Mandrekar, V. (1978), Characterization of Banach spaces through validity of Bochner theorem. *Lecture Notes in Math. No.* 644, 314–326. MR 80k:60011.

Marden, M. (1949), Geometry of polynomials. *Math. Surveys* No. 3, Amer. Math. Soc. 1949. Second edition 1966. MR 37:1562.

Markin, S. P. (1975), Hermitian sesquilinear forms on a Banach space (Russian). *Voronez. Gosud. Univ. Trudy Naučn. Issled. Inst. Mat. VGU Vyp.* **17**. *Teorja Operatorov. Operator Uravn.* pp. 57–62. MR 58:23522. R. Ž. 1976, 6B 638.

Martin, C. and Valentine, J. E. (1975/6), Angles in metric and normed linear spaces. *Colloq. Math.* **34** (2), 209–217. MR 54:5981.

Martinelli, E. (1981), Quaternions. Classical algebra, modern geometry. *Proc. of the Conf. held in honor of the 80th Anniversary of the Birth of R. Calapso, Messina, Taormina.* pp. 249–264. Veschi, Roma, 1981. MR 84j:57018. Entire coll. MR 84f:00012.

Martucci, G. (1982), *Spazi di Hilbert con Elementi di Meccanica Quantica.* Bologna, Pitagora, R. Ž. 1983, 3B 878K.

Maruyama, G. (1959), The harmonic analysis of stationary stochastic processes. *Mem. Fac. Sci. Kyushu Univ.* **4**.

Marchaud, André (1959), Un theoreme sur les corps convexes. *Ann. Sci. Ecole Norm. Sup.* **76** (3), 283–304. MR 23:A2126.

Marques, Manuel D. P. Monteiro (1982), Produit de comesures sur un localement compact et definition d'un espace hilbertien de mesures bornees. *Seminaire d'analyse convexe Montpellier.* Expose No. 20.

Marvall Casenoves, D. (1968), Random spaces and generalized Hilbert spaces and Banach spaces (Spanish). *Rev. Acad. Ci. Madrid* **62**, 11–68. MR 38:6332.

Masani, P. (1960), the prediction theory of multivariate stochastic processes. III. *Acta Math.* **104**, 141–162. MR 22:12679.

Masani, P. (1972), Isometric flows on Hilbert spaces. *Bull. Amer. Math. Soc.* **68**, 624–632. MR 26:2887.

Masani, P. (1966), Recent trends in multivariate prediction theory. *Multivariate Analysis*, pp. 351–382. Academic Press, New York. MR 35:5079.

Masani, P. (1968), Orthogonally scattered measures. *Advances in Math.* **2**, 61–117. MR37:4231.

Masani, P. (1969), Explicit form of the Fourier–Plancherel transform over locally compact abelian groups. In *Abstract Spaces and Approximation Theory*, pp. 162–182. Birkhauser Press, Basal. MR 41:2303.

Masani, P. (1970), Quasi-isometric measures and their applications. *Bull. Amer. Math. Soc.* **76**, 427–428. MR 41:3704.

Masani, P. (1972), On helixes in Hilbert space. *Theory of Probability and its Applications* **17** (1), 1–19. MR 45:7520.

Masani, P. (1976), On helixes in Banach spaces. *Sankya Ser. A* **38**, 1–27. MR 57:10792.

Masani, P. (1978), Dilations as propagators of hilbertian varieties. *SIAM J. Math. Anal.* **9**, 414–456. MR 58:17897.

Masani, P. (1978), Propagators and dilations. In *Probability Theory on Vector Spaces*, Lectures Notes in Math. No. 656, 95–117. MR 80g:47004.

Masani, P. (1984), *Banach and Hilbert spaces of Vector-Valued Functions.* Pitman, London.

Masani, P. and Robertson, J. (1962), The time domain analysis of continuous parameter weakly stationary stochastic processes. *Pacif. J. Math.* **12**, 1361–1378. MR 26:7047.

Maschler, M. (1956), Minimal domains and their Bergman kernel. *Pacif. J. Math.* **6**, 501–516. MR 18,473.

Maurica, Javier José Martinez (1977), Diagramas des estados de operadores entre espacios normados no arqimedianos. *Ph.D. Thesis*, Universidad de Bilbao.

Maury, S. (1969), Sous espaces hilbertiens d'un espace vectoriels topologiques. *Seminaire d'Analyse Unilaterale*, Montpellier, Vol. 2, Exposé No. 7.

Mazun, S. (1938), Quelques proprietes characteristiques des espaces euclidiens. *C. R. Acad. Sci. Paris* **207**, 761–764.

Maurin, K. (1967), *Methods of Hilbert Spaces.* PWN, Warsaw. MR 36:6957.

McCarthy, C. A. (1967), c_p. *Israel J. Math.* **5**, 249–271. MR 37:735.

McCarthy, C. A. (1971), Some remarks on operator convex functions. *Indiana Univ. Math. J.* **20**, 921–923. MR 54:3456.

McIntosh, Alan (1978), The Toeplitz–Hausdorff theorem and ellipticity conditions. *Amer. Math. Monthly* **85**, 475–477. MR 58:22127.

McLaughlin, H. W. and Zaretzki, P. M. (1971), Simultaneous approximation and interpolation with norm preserving. *J. Approx. Theory* **4**, 54–58.

McClaughtin, H. W. and Metcalf, F. T. (1967), A remark on a recent generalization of the Cauchy–Schwarz inequality. *J. Math. Anal. Appl.* **18**, 522–523.

McShane, E. J. (1962), A theory of limits, *MAA Studies in Math. Vol.* 1, pp. 7–29.

McShane, E. J. (1955), A theory of convergence. *Canad. Math. J.* **22**, 323–332, MR 15,641.

Meir, A. and Klamkin, M. S. (1982), Ptolemy's inequality, chordal metric, multiplicity metric. *Pacif. J. Math.* **101**, No. 2,389–392. MR 83k:51028.

Menger, K. (1928), The metric of Hilbert spaces (German). *Anzeiger Akad. Wiss. Vienna*, **65**, 159–160.

Menger, K. (1931), New foundations of Euclidean Geometry. *Amer. J. Math.* **53**, 721–745.

Meshkov, V. Z. (1974), On smooth functions in the space of James (Russian). *Vestnik MGU Ser. Fiz-Mat.* **4**, 9–13.

Meshkov, V. Z. (1975), Hilbertness criteria of Banach spaces (Russian). *Trudy 8th Zimn. Skoly po Math. Program. i smezn.-voprosy, Drogobici*, Moscow, 1976, pp. 151–153. R. Z. 1978, 9B 556.

Meshkov, V. Z. (1978), Smoothness properties in Banach spaces. *Studies Math.* **63**, 111–123.

Miamee, A. G. and Salehi, H. (1978), Harmonizability, *V*-boundedness and stationary dilation of stochastic processes. *Indiana Univ. Math. J.* **27**, 37–50.

Mickle, E. (1949), On the extension of a transformation. *Bull. Amer. Math. Soc.* **55**, 160–164. MR 10,691.

Milićić, P. M. (1971), Sur la transversalité dans des espaces normées. *Mathematika Balkanica* **1**, 171–176. MR 45:4129.

Milićić, P. M. (1973), Sur l'existence du produit scalaire. *Mat. Vesnik* **10** (25), 3–7. MR 52:1275.

Milićić, P. M. (1973), Sur le semi-inner produit escalaire generalisé. *Mat. Vesnik* **10** (25), 325–329. MR 50:958.

Milićić, P. M. (1977), A representation of Wigner type operators on *H*-spaces over the system of quaternions (Serbo-Croatian. English summary). *Mat. Vesnik* **4** (19), 370–375. MR 37:4581.

Milićić, P. M. (1977), Les systemes ortonormaux dans les espaces vectoriels normées. *Mat. Vesnik* **1** (14) (29), No. 3, 243–249. MR 58:23517.

Milićić, P. M. (1977), *C*-complete systems in normed spaces. *Publ. Inst. Math.* **21** (35), 145–150.

Milićić, P. M. (1976), Sur le complement orthogonal du sous-espace fermé d'un ėspace vectoriel normé. *Mathematika Balkanica* **621**, 119–124.

Milićić, P. M. (1980), Systemes orthonormaux dans des espaces avec semi-produit scalaire et dans leurs completes metriques. *Matemaicki Vesnik* **4** (17) (32), 175–180.

Milićić, P. M. (1981), Sur les suites orthonormaux dans des espaces normés. *Matematicki Vesnik* **5** (18) (33), 95–102.

Milićić, P. M. (1981), Les points extremes et les points unis de la boule unitée etudies au moyen du semi-produit scalaire. *Matematicki Vesnik* **5** (18) (33), 383–388.

Milman, V. D. (1970), Geometric theory of Banach spaces. I. The theory of bases and minimal systems (Russian). *Uspekhi Mat. Nauk* **25**, 113–173. MR 43:6704. Russian Surveys **25** (3), 111–170.

Milman, V. D. (1971), Geometric theory of Banach spaces. II. Geometry of the unit sphere (Russian). *Uspekhi Mat. Nauk* **26**, 73–149. MR 54:8240. Russian Surveys **26**, 79–163.

Minaksundaran, S. (1960), Hilbert algebras. *Proc. Internat. Congress of Math.* 1958, pp. 407–411. Cambridge, Harvard Univ. Press, Mass. MR 22:8357.

Mingo, A. and Phillips, W. J. (1984), Equivariant triviality theorem for Hilbert *C**-modules. *Proc. Amer. Math. Soc.* **91** (2), 225–230. MR 85f:46111.

Minlos, R. (1959), Generalized stochastic processes and their extensions with respect to measure (Russian). *Trudy Moskovskogo Mat. Obscestva* **8**, 497–518.

Mitiagin, B. S. (1964), Normed ideals of intermediate type (Russian). *Izv. Akad. Nauk SSSR Ser. Mat.* **28**, 819–832. *Amer. Math. Soc. Transl.* **63** (2), 180–194. MR 30:4142.

Mitiagin, B. S. (1977), Editor. *Geometry of Linear Spaces and Theory of Operators* (Russian). Jaroslavl.

Mitiagin, B. S. and Pelczynski, A. (1966), Nuclear operators and approximate dimension. *Proc. Internat. Congress of Math. Moscow*, pp. 366–372. Izdat. Mir, Moscow, 1968. MR 39:6046.

Mlak, W. and Weron, A. (1980), Dilations of Banach space operator valued functions. *Ann. Polon. Math.* **38**, 295–303. MR 82b:47003.

Mjicielski, J. (1977), A conjecture of Ulam on the invariance of measure in Hilbert's cube. *Studia Math.* **60**, 1–10. MR 55:3222.

Amir-Moez, Ali R. (1968), Metric properties of unitary spaces. *Math. Series No.* 3, Texas Technological College, Lubock, Texas. MR 41:2518.

Moisil, Grigore C. (1931), Sur les quaternions monogenes. *Bull. Sci. Math.* **II**, 168–174.

Moisil, Grigore C. (1949), On a generalization of the idea of monogeneity given by V. S. Fedorov (Romanian). *Acad. Rep. Pop. Romane Bull. Sti. A* **1**, 959–964. MR 13,455.

Moisil, Grigore C. (1950), On monogenic functions in the sense of V. S. Fedorov. *Acad. Rep. Pop. romane Bull. Sti. Ser. Mat. Fiz. Chim.* **2**, 545–556. MR 13,455.

van Mill, Jan (1982), Representing countable groups by homeomorphism groups in Hilbert spaces. *Math. Ann.* **259**, 321–329.

Monauri, Luigi (1980), Hilbert space relaxation systems. *Warwick Control Center Rep.* No. 96.

Montel, P. (1932), Sur la limite superieur du module des racines d'une equation algebrique. *C. R. Soc. Sci. Varsovie* **24**, 317–326.

Montel,P. (1934/5), Sur quelques limites pour les modules de zéros des polynomes. *Comm. Math. Helvet.* **7**, 178–200. *See also C. R. Acad. Sci. Paris* **199** (1934), 651–653; 760–762.

Monna, A. F. (1971), *Analyse non-archimedienne.* Springer Verlag, New york. MR 45:4101.

Moore, E. H. (1935), General analysis. With the cooperation of R. W. Barnard. Part. I. *Mem. of the American Phil. Soc. Vol.* 1, Philadelphia.

Moore, M. H. (1973), An inner product inequality. *SIAM Math. Anal.* **4**, 514–518. MR 48:6903.

Morgan, C. L. (1974), Embedding metric spaces in Euclidean spaces. *J. Geometry* **5**, 101–107.

Morita, K. (1956), On the kernel function for symmetric domains. *Sci. Rep. Tokyo Kyoitu Univ. Daigaku Sect. I* **5**, 190–212. MR 19,541.

Morrey, C. B. (1966), Multiple integrals in the calculus of variations. *Grund. Math. Wiss.* **130**. MR 34:2380.

Moskovitz, D. and Dines, L. L. (1939), Convexity in a linear space with an inner product. *Duke Math. J.* **5**, 520–534.

Motzkin, Th. (1935), Sur quelques charactéristiques des ensembles convexes. *Atti Accad. Naz. dei Lincei, Rend.* **6** (21), 562–567.

Motzkin, Th. (1935), Sur quelques propriétés charactéristiques des ensembles borné convexes. *Atti Accad. Naz. dei Lincei, Rend.* **6** (21), 773–779.

Mourier, E. (1951), Elements aléatoires dans un éspace de Banach. *Ann. Inst. H. Poincaré* **12**, 215–230.

Moussatat, W. (1976), On the asymptotic theory of statistical experiments and some of its applications. *Ph.D. Thesis*, Univ. of California, Berkeley.

Mukhamadieva, V. S. (1983), Bounded endomorphisms of Hilbert modules over semifields (Russian). *DAN UzSSR* (1), 11–12. MR 85k:47092.

Munroe, M. E. (1946). Absolute and conditional convergence in Banach spaces. *Duke Math. J.* **12**, 351–356. MR 8,387.

Murphy, T. and West, T. T. (1970), Inner products on Banach spaces and supports of Radon measures. *Proc. Irish Acad. Sci. Ser. A* **69**, 55–61. MR 41:4213.

Murray, F. D. (1973), Certain generalizations in the theory of probability measures in linear spaces *Theory of Probability and Applications* **18**, 64–75.

Muštari, D. (= D. Mousthari) (1976), On a probabilistic characterization of Hilbert spaces (Russian). *Teoriya Veroiatnostei i Primen.* **21**, 421–422. MR 54:6220. English trans. *Theory of Prob. Appl.* MR 47:7793.

Muštari, D. (1973), Certain general questions of the theory of probability measures in linear spaces. *Theory of Prob. Appl.* **18**, 64–75.

Muštari, D. (1976), La topologie du type Sazonov pour des espaces de Banach et les supports hilbertienne. *Ecole d'été de Calcul des Probabilites de Saint-Flour Sain-Flour.* Ann. Sci. Univ. Clermont-Ferrand No. 61. Matr. No. 14 (1976), 77–87. MR 57:17749.

Muštari, D. (1980), Spaces of cotype p $(0 \leqq p \leqq 2)$ (Russian. French summary). *Teoriya Veroiatnostei i Primen.* **25**, 105–118. MR 82d:46029.

Naftalevich, A. G. (1950), On interpolation by functions of bounded characteristic (Russian). *Učen. Zap. Vilnius Gosud. Univ.* **5**, 5–27.

Nagumo, M. (1936), Characerisierung der Allgemeinen Euclidischen Raume durch eine Postulate fur Schwerpunkte. *Japan J. Math.* **12**, 123–128.

Nagumo, M. (1941), On spectral theory (German). *Proc. Math. Soc. Japan* **23** (3), 485–511. MR 3,210.

Nagumo, M. (1941), On a characterization of general spaces (German). I. II. *Proc. Imperial Acad. Tokyo* **17**, 301–307. MR 7,249: II. *Ibid.* **18** (1942), 280–286. MR 7,249.

Naimark, M. A. (= M.A. Neumark) (1943), Positive definite operator functions on a commutative group. *Izv. Akad. Nauk. SSSR, Ser. Math.* **7**, 237–244. MR 5,272.

Naimark, M. A. (1943), On a representation of additive operator functions. *Comptes Rendus (Doklady) Acad. URSS* **41**, 359–361. MR 6,71.

Naimark, M. A. (1964), *Normed Rings* (Russian). GITTL, Moscow, 1956. MR 19,870. English transl. P. Noordhof, Groningen. MR 34:4928. Second Russian Ed. Moscow, 1968. MR 50:8076. English transl. 1970. MR 50:8075.

Naimark, M. A. (1954) *Differential Linear Operators* (Russian). GITTL, Nauka, Moscow. MR 16702. MR 50:5547.

Nakano, H. (1950), Hilbert algebras. *Tohoku Math. J.* **2** (2), 4–23. MR 12,837.

Nanta, B. (1971), Topologies on semi-inner product spaces. *Compositio Math.* **23**, 309–316. MR 45:9112.

Narici, L. and Beckenstein, E. (1981), Strange terrain-nonarchimedean spaces. *Amer. Math. Monthly* **88**, 667–676.

Nehari, Z. (1949), The kernel function and canonical map. *Duke Math. J.* **16**, 165–178. MR 10,440.

Nehari, Z. (1951), Extremal problems in the theory of bounded analytic functions. *Amer. J. Math.* **73**, 78–106.

Nehari, Z. (1952), On weighted kernels. *J. D'Analise Mathematique* **2**, 126–149. MR 14,742.

Nehari, Z. (1952), *Conformal Mappings*. McGraw-Hill Book Co. Inc. MR 13,640.

Nehari, Z. The kernel function and the canonical Schlicht-mapping. *National Bureau of Standards. Proc. Sympos. Appl. Math.* No. 19, 215–224. MR 14,861.

von Neumann John (1929), Allgemeine Eigenwerttheorie Hermitischer Funktionaloperatoren. *Math. Ann.* **102**, 49–131.

von Neumann, John (1929), Zur Algebra der Funktionaloperatoren. *Math. Ann.* **102**, 370–427.

von Neumann, John (1932), Zur Operatorenmethode in der klassischen Mechanik. *Ann. of Math.* **33** (2), 587–642; 789–791.

von Neumann, John (1942), Approximate properties of matrices of high finite order. *Portugaliae Math.* **3**, 1–62.

von Neumann, John (1950), Functional operators. *I. Ann. of Math. Stud.* No. 21. Princeton Univ. Press, Princeton, N. J. MR 11,240.

von Neumann, John (1950), Functional operators. II. *Ann. of Math. Stud.* No. 22 Princeton, Univ. Press, Princeton, N.J. MR 11,599.

von Neumann, John (1951), A spectral theory for general operators of a unitary space. *Math. Nachr.* **4**, 258–281. MR 13,254.

von Neumann, John (1955), *The Mathematical Foundation of Quantum Mechanics.* Princeton Univ. Press. Princeton, N.J. MR 16,654.

MacNerney, J. S. (1955), Weakly reproducing kernels. *Bull. Amer. Math. Soc.* **61**, Abstract 61-2-266.

MacNerney, J. S. (1960), Hellinger integrals in inner product spaces. *J. Elisha Mitchell Sci. Soc.* **76**, 252–273.

MacNerney, J. S. (1970), Continuous embeddings of Hilbert spaces. *Rend. Circ. Mat. Palermo* **19** (2), 109–112. MR 46:2405.

MacNerney, J. S. (1970), Dense embeddings of Hilbert spaces. *Proc. Amer. Math. Soc.* **24**, 92–94. MR 40:7781.

Nevanlinna, R. (1919), Uber beschränkte funktionen die in gegebenen punkte vergeschriebene werte annehmen. *Ann. Acad. Fenn. Ser. B* **13**.

Newman, D. J. (1959), Interpolation in H_∞. *Trans. Amer. Math. Soc.* **92**, 501–507. MR 22:8130.

Newman, D. J. (1961), The nonexistence of projections from L^1 to H^1. *Proc. Amer. Math. Soc.* **12**, 98–99.

Newman, D. J. and Shapiro, H. S. (1964), A Hilbert space of entire functions related to operational calculus. Univ. of Mich. Ann Arbor. Mim. Notes. 92 pp.

Newman, D. J. and Shapiro, H. S. (1966), Certain Hilbert spaces of entire functions. *Bull. Amer. Math. Soc.* pp. 971–977.

Newman, D. J. and Wohlgelernter, D. K. (1972), Two Hilbert spaces in which polynomials are not dense. *Trans. Amer. Math. Soc.* **168**, 67–72. MR 45:3723.

Nguyen Van Thu (1980), Prediction problems. *Dissert. Math.* (*Rozprawy Mat.*) **163**, 69 pp. MR 81f:60059.

Nguen, Thanh Van (1972), Bases de Schauder dans certain espaces de fonctions holomorphes. *Ann. Inst. Fourier* (*Grenoble*) **22**, 169–253. MR 52:8895.

Nelson, E. (1959), Analytic vectors. *Ann. of Math.* **70** (2), 572–615. MR 21:5901.

Newburg, J. D. 1951), The variation of spectra. *Duke Math. J.* **18**, 165–176. MR 14,481.

Niemi, H. (1975), On stationary dilations and the linear prediction of certain stochastic processes. *Comm. Phys. Math.* **45**, 111–130.

Niemi, H. (1975), Stochastic processes as Fourier transforms of stochastic measures. *Ann. Acad. Sci. Fenn. A I* **591**, 1–47.

Niemi, H. (1975), On the support of a bimeasure and orthogonally scattered vector measures. *Ann. Acad. Fenn. A I* **591**, 249–275.

Niemi, H. (1983), Orthogonally scattered dilations of finitely additive vector measures with values in a Hilbert space. *Prediction theory and Harmonic Analysis*, Eds. V. Mandrekar and H. Salehi. pp. 233–251. North Holland.

Nguyen, Van Lam (1980), On Möbius transformation in a Hilbert space. *Rev. Roum. Math. Pures Appl.* **25**, 765–769. MR 82f:46024.

Nishioka, M. (1981), Geometric quantization of curved phase spaces. *Hadronic J.* **5**, No. 1, 207–213. MR 83g:58025.

Neveu, J. (1965), Le probleme d'échantillonage et de l'interpolation d'un signal. *C. R. acad. Sci. Paris* **260**, 49–51. MR 30:2951.

Nikaido, G. (1980), Remarks on the lower bounds of a linear operator. *Proc. Japan Acad. Sci.* **56** *Ser. A*, 321–323.

Nikolskiĭ, S. M. (1954), On some problems of approximation of functions through polynomials. *Proc. Internat. Congress of Math.* Vol. 1, 371–382. Amsterdam, Holland.

Nikolskiĭ, N. K. (1977), Lectures on the shift operator. *J. of Soviet Math.* **8**. 7 Lectures.

Nussbaum, A. E. (1965), Quasi-analytic vectors. *Arkiv for Math.* **6**, 179–191. MR 33:3105.

Nussbaum, A. E. (1969), A note on quasi-analytic vectors. *Studia Math.* **33**, 306–309. MR 40:4781.

Obi, G. M. (1978), An algebraic closed graph theorem. *Pacif. J. Math.* **74**, 197–207. MR 57:17181.

Obi, G. M. (1980), Closed graph theorem for qudratic spaces. *J. London Math. Soc.* **22** (2), 245–250. MR 82c:15030.

Odinets, V. P. (1976), On the norm of linear and nonlinear metric projections (Russian).

Odinets, V. P. (1980), An algebraic extension of the Lax-Milgram theorem. *Pacif. J. Math.* **86**, 1543–522. MR 82d:15016, *Izv. Vyss. Učebn. Zaved. Matematika*, No. 7(170), 56–64.

Odinets, V. P. (1979), On elementary rotations in Banach spaces. (Russian). *Izv. VUZ. Uinniti* No. 2755 79, R.Ž. 1979, 12B 656.

Ohira, K. (1951), On a certain complete separable and metric space. *Mem. Fac. Sci. Kyushu Univ. Ser. A* **6**, 9–15. MR 13,850.

Oheira, K. (1952), On some characterizations of abstract Euclidean spaces by properties of orthogonality. *Kumamoto J. Sci. Ser. A. 1, No. 1* 23–26. MR 14,881.

O'Brien, R. (1977), A note on contraction extension property in dual Banach spaces. *J. London Math. Soc.* **16**, 490–492. R. Ž. 1978, 8B 581.

Ogasawara, T. (1955), A theorem on operator algebras. *J. Sci. Hiroshima Univ.* **18**, 207–309, MR 17:1227.

Ogasawara, T. (1955), A structure theorem for quasi-complete unitary algebras. *Ibid.* **19**, 79–85.

Okikiolu, G. O. (1982), Inner products and related norms in L^p spaces and estimates for bounded operators. *Bull. Math.*, No. 7, 1–13. MR 84j:46048a.

Okikiolu, G. O. (1982), Minimum value estimates for bounded linear operators in inner product spaces and L^p spaces. *Bull. Math.*, No. 7, 14–40. MR 84j:46048b.

O'Connel, R. F. and Rajagopal, A. K. (1982), New interpretation of the scalar product in Hilbert space. *Phys. Rev. Lett.* **48**, 525–526. MR 83c:81006.

Olagunju, P. A. (1967), On Banach spaces that cannot be made into BIP space. *Proc. Cambridge Phil. Soc.* **63**, 949–950. MR 35:5906.

Oman, J. (1969), Characterization of inner product spaces. *Ph.D. Thesis*, Michigan State Univ.

Omar, J. (1975), Norm identities which characterize inner product spaces. In *The Geometry of Metric and Linear Spaces:* Proc. Conf. Michigan State Univ. East Lansing, Michigan, 1974, pp. 116–133. *Lecture Notes in Math. No. 490.* MR 53:1236.

Omar, J. (1981), Localization of Ficken's characterization of inner product spaces. Preprint.

Ovčinnikov, V. I. (1980), Interpolation of operators of class s_p in Hilbert spaces (Russian). *Mat. Zametki* **27**, 273–282. MR 82d:46098. *Math. Notes* **27**, 135–140.

Ozawa, M. (1956), Some estimates of the Szego kernel function. *Kodai Math. Sem. Report* **8**, 71–78 Mr 18,200.

Ozawa, M. (1980), Hilbert $B(H)$ modules and stationary processes. *Kodai Math. J.* **3**, 26–39. MR 82b:46086.

Ozawa, M. (1983), Boolean valued analysis and Type I AW^*-algebras. *Proc. Japan Acad.* **59**, *Ser. A No. 8*, 368–371. MR 85m:46068.

Ozawa, M. (1983), Boolean valued interpretation of Hilbert space theory. *J. Math. Soc. Japan* **35**, 609–627. MR 85d:46094.

Ozawa, M. (1983). Classification of Type I AW^*-algebras and Boolean valued analysis. Preprint.

Packel, E. (1974), *Functional Analysis. A Short Course.* Intext, New York and London.

Packel, E. (1974), Hilbert space operators and quantum mechanics. *Amer. Math. Monthly* **81**, 863–873, R.Ž. 1975, bc 1150.

Padke, S. V. and Thakare, N. K. (1974), When is an s.i.p. space a Hilbert space?. *Math. Student* **42**, 193–194. MR 52:14946.

Palais, R. S. (1963), Morse theory on Hilbert manifolds. *Topology* **2**, 299–340. MR 28:1633.

Paley, R. E. A. C. and Wiener, N. (1933), *Fourier Transforms and the Complex Domain*. Cambridge Univ. Press.

Panda, B. B. and Kahoon, O. P. (1974), On equidistant sets in normal linear spaces. *Bull. Austral. Math. Soc.* **11**, 443–454. MR 50:10763.

Papini, P. L. (1974), Some questions related to the concept of orthogonality in Banach spaces (Italian summary). *Boll. Un. Mat. Ital.* **9** (4), 386–401. MR 50:8014.

Papini, P. L. (1975), Proximity maps. *Bases. Boll. Un. Mat. Ital.* **11** (4), 44–63. MR 52:1266.

Papini, P. L. (1983), Inner products and norm derivativs. *J. Math. Anal. Appl.* **91**, 592–598. MR 84f:46029.

Papini, P. L. (1982), Approximation and norm derivatives in real normed spaces. *Resultate vonder Math.* **5**, 81–94. MR 84a:41044.

Papini, P. L. (1983), Minimal and closest points, nonexpansive and quasi-nonexpansive retractions in real Banach spaces. *Convexity and its Applications*, pp. 248–263. Berkhauser, Boston, MR 85g:46022. Entire coll. MR 84m:52001.

Parodi, M. (1959), *La Localisation des Valeurs Characteristiques des Matrices et ses Applications*. Paris, 1959.

Parthasarathy, K. R. (1966), Probability theory on the closed subspaces of a Hilbert space (French summary). *Colloq. Internat. CNRS No. 186. Clermont Ferrand*, pp. 265–292. Edition Centre National Res. Sci. Paris, 1970. MR 54:9494.

Parthasarathy, K. R. (1977), *Introduction to Probability and Measure*. Macmillan Co. India. Macmillan. London, 1977. MR 58:313221.

Parzen, E. (1956), A simple proof and some extensions of sampling theorems. *Techn. Report 7*. Stanford Univ. Stanford.

Parzen, E. (1970), Statistical inference on time series by RKHS methods. *Proc. Twelfth Biennial Semin. Canad. Math. Congress*. pp. 1–37.

Parzen, E. (1959), Statistical inference on time series by Hilbert space methods. *Techn. Report No. 23*. Stanford Univ. Stanford.

Parzen, E. (1961), An approach to time series analysis. *Ann. of Math. Statist.* **32**, 951–989, MR 26:874.

Parzen, E. (1962), Probability density functions and reproducing kernel Hilbert spaces. *Proc. Sympos. Time Series Analysis*, John Wiley, N.Y. 1963. pp. 155–169. MR 26:7119.

Parzen, E. (1967), *Time Series Analysis Papers*. Holden-Day, San Francisco, California. MR 36:6091.

Parzen, E. (1975), Some solutions of time series modeling and prediction problem. The *Search of Oil*. Ed. D. Owen. Marcel Dekker, Inc. New York.

Paschke, W. L. (1972), Inner product modules over B^*-algebras. *Trans. Amer. Math. Soc.* **182**, 443–468. MR 50:8087.

Paschke, N. L. (1976), Inner product modules arising from compact automorphism groups of von Neumann algebras. *Trans. Amer. Math. Soc.* **224**, 87–102. MR 54:8308.

Paschke, W. L. (1977), Integrable actions on von Neumann algebras. *Math. Scandi.* **40**, 234–248. MR 57:1145.

Pasqua, D. Del (1955), Una nozione di varieta lineare disjunte di uno spazio di Banach. *Rend. Di Mat. Pura Appl.* **13** (5), 406–422. MR 17,986.

Pavlovic, M. (1979), Two three dimensional characterizations of a scalar product (Russian). *Mat. Vesnik* **3** (16) (31), 493–495. MR 82j:46028.

Pedersen, G. K. (1972), Some operator monotone functions. *Proc. Amer. Math. Soc.* 309–310. MR 46:6078.

Pedersen, R. N. (1968), On unitary properties of Grunsky's matrix. *Archiv. Rational Mech. Anal.* **29**, 370–377. MR 37:2972.

Pedrick, G. B. (1957), Theory of reproducing kernels in Hilbert spaces of vector-valued functions. *Univ. of Kansas, Techn. Report No. 19*.

Peetre, J. (1969), Exact interpolation theorems for Lipschitz continuous functions. *Ricerche Mat.* **18**, 239–259. R.Ž. 1971, 1B 683.

Pełczynski, A. (1960), Projections in certain Banach spaces. *Studia Math.* **19**, 209–228.

Pełczynski, A. (1966/7), A characterization of Hilbert–Schmidt operators. *Studia Math.* **28**, 355–350. MR 35:7163.

Pełczynski, A. (1972), Certain problems of Banach (Polish). *Wiadom. Mat.* **15** (2), 3–11. MR 58:12298.

Pełczynski, A. (1976), Banach spaces of analytic functions and absolutely summing operators. *Lecture Notes*, Columbus, Ohio.

Pełczynski, A. (1975). On unconditional bases and Rademacher averages. *Lecture Notes in Math.* No. 472, 119–129.

Pełezynski, A. and Tovunczyk, H. (1978), Certain non-locally convex linear metric spaces homeomorphic to Hilbert spaces. *Comment. Math.* MR 267–285. R.Ž. 1980, 3B 626.

Peller, V. V. and Khruschev, S. V. (1982), Hankel operators, operators and stationary processes. *Uspekhi Mat. Nauk* **37**, 53–124. *Russian Math. Surveys* **37**, 61–144.

Penico, J. and Stanojevič, S. V. (1980), An integral analogue to parallelogram law. *Proc. Amer. Math. Soc.* **79**, 427–430. MR 81j:46027.

Penny, R. C. (1976), Self-dual cones in Hilbert spaces. *J. Functional Anal.* **21**, 305–315. MR 53:6295.

Penrose, R. (1968), Twistor quantisation and curved spaces-time. *International Journal of Theoretical Physics* **1**, 61–99.

Perfect, H. (1958), Pythagorean orthogonality in normed linear spaces. *Proc. Edinburgh Math. Soc.* **9** (2), 168–169. MR 20:7212.

Petunin, Ju. I. and Pličko, A. N. (1973), Interpolation of Holder scales. *Trudy 6th Zimn. Skoly po mat. progr. i zmezn. voprosy, Drogobici.*

Pevac, L. (1980), Characterizations of Hilbert spaces by means of best N-nets (Serbo-Croatian. English summary). *Mat. Vesnik* **4** (17) (32), 73–76. MR 82g:51018.

Phelps, R. (1957), Convex sets and nearest points. *Proc. Amer. Math. Soc.* **8**, 790–797. MR 19,432.

Phelps, R. (1958), Convex sets and nearest points, *Proc. Amer. Math. Soc.* **9**, 867–873. MR 21:2900.

Philips, R. S. (1940), A characterization of Euclidean spaces. *Bull. Amer. Math. Soc.* **46** (2), 930–933.

Phillips, J. (1974), Complementation for right ideals in generalized Hilbert algebras. *Trans. Amer. Math. Soc.* **197**, 409–417.

Pincherle, Salvator con Amaldi, U. (1901), Le operazioni distributive e le loro applicazioni all'analisi. Bologna, Zanichelli.

Piziak, R. (1969), An algebraic characterization of Hilbert space geometry. *Ph.D. Thesis*, Univ. of Mass.

Piziak, R. (1972), Sesquiliear forms in infinite dimensions. *Pacif. J. Math.* **43**, MR 47:7396.

Plymen, R. J. (1976), Spinors in Hilbert space. *Math. Proc. Cambridge Phil. Soc.* **80**, 337–347. MR 54:7707.

Pollard, H. (1944), Completeness theorems of Paley–Wiener type. *Ann. of Math.* **45** (2), 738–739. MR 6,127.

Pontrjagin, L. (1944), Hermitian operators in spaces with indefinite metric (Russian. English summary). *Bulletin Acad. Sci. URSSS Ser. Math.* (*Izvestija Akad. Natuk SSSR Ser. Math.*) **8**, 243–280. MR 6,273

Potepun, L. J. (1974), Hilbertian symmetric spaces (Russian). *Izv. Vyss. Ucebn. Zaved. Mat.* **140**, 90–95. MR 50#14196. *Sov. Math. VUZ* **18**, No. 1, 73–77.

Powers, N. C. (1973), Real linear operators on quaternionic Hilbert spaces. *Proc. Amer. Math. Soc.* **40**, 1–8. MR 47:7511.

Powers, N. C. (1976), A functional calculus for sesquihermitian operators on quaternionic Hilbert spaces. *J. Australian Math. Soc. Ser. A* **21**, 96–107. MR 53:3758.

Precupanu, T. (1968), Sur les produits scalaires dans les espaces vectoriels topologiques *Rev. Roum. Math-Pures Appl.* **13**, 85–90. MR 38:3707.

Precupanu, T. (1970), Bases orthogonales dans des espaces lineaires a semi-normes hilbertiennes. *Rev. Roum. Math. Pures Appl.* **15**, 1035–1038.

Precupanu, T. (1969), Espaces lineaires a semi-normes hilbertiennes. *An. Stiint. Univ. Al. I. Cuza, Iasi* **15**, 83–93.

Precupanu, T. (1972), Orthogonality in linear spaces (Romanian. French summary), *Studi si Cercet. Mat.* **24**, 881–889. MR 50:10775.

Precupanu, T. (1976), On a generalization of Caucy–Buniakowsky–Schwartz inequality. *An. Stiint. Univ. Al. I. Cuza Sect. I a Mat.* (*N.S.*) **22** (2), 173–175. MR 56:1038.

Precupanu, T. (1978), Characterizations of hilbertian norms (Italian summary). *Boll. Un. Mat. Ital. B* **15** (5) (1), 161–169. MR 80b:46036.

Pregovecki, E. (1969), Topologies on generalized inner product spaces. *Canad. J. Math.* **21**, 158–169. MR 39:3294.

Prijately, Nico (1964), Algebra characterization of Hilbert spaces. *Publ. of the Dept. of Math. Univ. of Ljubljana, Inst. of Math. Phys. and Mechanics* **1**, 43–57. MR 31:2598.

Pritchard, A. J. and Crouch, P. E. Sensitivity analysis for the linear quadratic control problems in Hilbert space. *Warwick Control Centre Rep.* No. 27.

Prigorski, V. A. (1965), On some classes of bases in Hilbert spaces (Russian). *Uspekhi Mat. Nauk* **20** (5), 231–256. R. Ž. 1966, 5B 454.

Privalov, I. I. (1950), Boundary properties of analytic functions (Russian), *Izdat. Tehn. Teor. Lit. Moscow*, Leningrad. MR 13,927.

Prhorov, Ju. V. (1957), A characterization of normal distributions in Hilbert spaces. *Teoria Prot. Appl.* **2**, 468–470.

Prhorov, Ju, V. (1961), The method of characteristic functionals. *Proc. 4th Sumpos. Berkeley, Univ. of Calif.* Berkeley, pp. 403–419. MR 24:A3670.

Prhorov, Ju. V. and Sazonov, V. (1961), Some results associated with Bochner's theorem. *Theory of Prob. Appl.* **6**, 82–87.

Prugovecki, E. (1982), *Quantum Mechanics in Hilbert Space.* Second Ed. Academic Press, New York. MR 84k:81005. First Ed. MR 58:14457.

Pulatova, M. J. (1975/6), On a class of orthonormal systems of arithmetical character. *Math. J. Okayama Univ.* **18**, 81–86. MR 53:8873.

Purice, R. A new proof of Wigner's theorem. *Cent. Inst. Phys. (Report)* No. FT-222, 14 pp. R.Ž. 1983, 11B 1078.

Poisson, S. D. (1820), Mémoire sur la maniere d'exprimer les fonctions, par des séries de quantités périodiques, et sur l'usage de cette transformation dans la résolution de différens problemes. *J. Ecole Roy. Polytechnique* **11**, 417–489.

Puritz, C. W. (1970), Skies and monads in nonstandard analysis. *Ph.D. Thesis*, Glasgow Univ.

Puritz, C. W. (1971), Ultrafilters and standard functions in nonstand and arithmetic. *Proc. London Math. Soc.* **22** (3), 705–733.

Puritz, C. W. (1972), Almost perpendicular vectors. Studies in logic and the foundations of mathematics, Vol. 69, pp. 115–121. North Holland. MR 58:1321.

Puritz, C. W. (1972), Skies, constellations and monads. *Studies on Logic and the Foundations of Mathematics.* Vol. 69, pp. 215–243. North Holland. MR 58:31038.

Quadling, D. (1979), Q for quaternions. *Math. Gazette* **63**, 98–110.

Radu, V. (1975), On linear operators and orthogonality in nonarchimedean spaces. *Math. Balkanica* **5**, 247–252. MR 58:23651.

Ragozin, D. (1980), Curvatures and curves in Hilbert spaces. *Math. Ž.* **173**, 51–67.

Rakov, S. A. (1976), Lorentz sequence spaces. *Mat. Zametki* **20**, 501–510. MR 55:11000.

Rakov, S. A. (1977), Ultraproducts and the three space problem (Russian). *Funktsionalnyi Analiz. i Priložen.* **11** (3), 88–89. MR 57:1090.

Rao, D. K. (1976), A triangle inequality for angle in a Hilbert space. *Rev. Columbiana Mat.* **10** (3), 95–97. MR 56:1037.

Rao, G. S. and Jothibai, A. (1984), SAIN in normed linear spaces. *Indian J. Pure Appl. Math.* **15** (4), 355–360. MR 85m: 41048.

Rao, M. M. (1967), Characterizing Hilbert spaces by smoothness. *Indag. Math.* **70**, 132–135. MR 35:2131.

Rao, M. M. (1972), Notes on characterizing Hilbert spaces by smoothness and smoothness in Orlicz spaces. *J. Math. Anal. Appl.* **76**, 228–234. MR 52:1286.

Rao, M. M. (1981), Representation of weakly harmonizable processes. *Proc. Nat. Acad. Sci. USA.* **78**, 5288–5289. MR 82i:60059.

Rao, M. M. (1978), Covariance analysis of nonstationary time series. *Developments in Statistics* **1**, 171–225. MR 80a:62149.

Rao, M. M. (1982), Harmonizable processes: structure theory. *L'Enseignment Math.* **28**, 295–351. MR 85c:60052.

Rao, M. M. and Chang, D. K. (1983), Bimeasures and sampling theorems for weakly harmonizable processes. *Stochastic Analysis and Applications.* **1** (1), 21–55.

Rassias, T. M. (1979), On the theory of inner product spaces. *Eleutheria* **2**, 481–484. MR 82f:46023.

Rassias, T. M. (1984), New characterizations of inner product spaces. *Bull. Sci. Math.* **108**, 95–99. MR 85h:46037.

Rätz, J. (1978), Additive mappings on inner product spaces (Serbo-Croatian summary). *Glasnik Mat. Ser. III* **13** (33), 63–68. MR 58:12307.

Rätz, J. (1981), On the characterization of the scalar product (Ger.) *Elem. Math.* **36**, 94–97. 82i:15026.

Redhefer, R. M. and Folsmen, P. (1983), On an inequality of Schur (Ger). *Montsber. Math.* **95** (2), 137–148. MR 84m:26028.

Redhefer, R. M. and Folsmen, P. (1983), Schur generalization of Hilbert's inequality. *Proc. Amer. Math. Soc.* **87** (3), 567–568. MR 84d:26017.

Redhefer, R. M. and Folsmen, P. Bilinear functions with Hilbert space operators as variables. *Mat. Ann.* MR 84i:47036.

Rejtö, L. (1971), On a generalization of Farkas theorem. *Studia Sci. Math. Hungarica* **6**, 361–365. MR 45:9113.

Renýi, A. (1958), On mixing sequences of sets. *Acta Math. Hungar.* **9**, 215–228. MR 20:4623.

Renýi, A. (1969), On measure of dependence. *Acta Math. Hungar.* **10**, 441–451. (English summary). MR 22:6005.

Renýi, A. (1959), A new version of the probabilistic generalization of the large sieve. *Acta Math. Hungar.* **10**, 217–226. MR 22:1938.

Renýi, A. (1960), *Probability Theory.* Budapest.

Reid, W. T. (1951), Symmetrizable completely continuous linear transformations in Hilbert space. *Duke Math. J.* **18**, 41–56.

Restrepo, G. (1964), Differentiable norms in Banach spaces. *Bull. Amer. Math. Soc.* **70**, 413–414. MR 28:4338.

Retherford, J. R. (1964), Basic sequences and the Paley–Wiener criterion. *Pacif. J. Math.* **14**, 1019–1027.

Retherford, J. R. (1975), Applications of Banach ideals of operators. *Bull. Amer. Math. Soc.* **81**, 961–972. MR 54:955.

Retherford, J. R. and Holub, J. R. (1971), The stability of bases in Banach and Hilbert spaces. *J. Reine und Angew. Math.* **246**, 136–146. MR 45:867.

Reznick B. (1978), Banach spaces which satisfy linear identities. *Pacif. J. Math.* **74**, 221–233.

Reznick, B. (1979), Banach spaces with polynomial norms. *Pacif. J. Math.* **82**, 223–235.

de Rham, G. (1955), Varietes differentiable. Formes, courants, formes harmoniques. *Actualites Sci. Ind. No.* 1222. Publ. Inst. Math. Nancago, III. Herman et Cie, Paris. MR 16,957.

Riefel, M. (1969), Square integrable representations of Hilbert algebras. *J. Functional Anal.* **3**, 265–300. MR 39:6094.

Riefel, M. (1974), Induced representations of C^*-algebras. *Advances in Math.* **13**, 176–257.

Riefel, M. and van Daele, A. (1977), A bounded operator approach to Tomita–Takesaki theory. *Pacif. J. Math.* **69**, 187–221. MR 55:11066.

Riesz, F. (1913), *Leçons sur les Equations avec une Infinité d'Inconnues*, Gauthier Villars, Paris.

Riesz, F. (1934), On the theory of Hilbert spaces (German). *Acta Math. Sci. Szeged* **7**, 34–38.

Riesz, F. (1918), On linear functional equations (German). *Acta Math.* **41**, 71–98.

Riesz, F. (1953), *Leçons d'Analyse Fonctionneles.* 3 ème Edition, Gauthier-Villars, Paris. MR 15,132. English transl. F. Ungar, New York, 1962.

Riesz, M. (1958), Clifford numbers and Spinors. *Inst. for fluid Dynamics and Applied Math. Lecture Series* No. 38. Univ. of Maryland, College Park, Maryland.

Rankin, R. A. (1955), On packing of spheres in Hilbert spaces. *Proc. Glasgow. Math. Assoc.* **2**, 145–146. MR 17,523.

del Rio, M. (1977), The rectangular constant for two-dimensional spaces. *J. Approx. Theory* **19**, 15–21.

Rijkov, S. S. (1957), Combinatorial topology of Hilbert spaces. *Cand. diss.* Moscow State Univ. *See also DAN SSSR* **114** (1957), 494–497. R.Ž. 1961, 2A 203.

Ritt, R. K. (1955), A generalization of inner products. *Michigan Math. J.* **3**, 23–26. MR 17,63.

Roach, F. A. (1975), Continued fractions over an inner product space. *Proc. Amer. Math. Soc.* **24**, 576–582. MR 54:776.

Roberts, D. B. (1934), On the geometry of abstract vector spaces. *Tohoku Math. J.* **39**, 42–59.

Roberts, J. H. (1956). The rational points in Hilbert space. *Duke Math. J.* **23**, 489–491. MR 19,565.

Rodrigues Cano, J. J. (1968), A necessary and sufficient condition for a Banach space to be a Hilbert space (Spanish). *Proc. Ninth Annual Conf. Spanish Mathematicians Granada, Spain.* pp. 251–260. Inst. Jorge Juan, Mat. Madric, 1971. MR 51:6374.

Robert, A. (1982), Meilleure approximation lineaire et espaces Euclidiens. *L'Enseignement Math.* **28**, 281–293. MR 84c:46020.

Robinson, E. (1962), Structural properties of stationary stochastic processes with applications. *Time Series*. M. Rosenblatt, Editor. J. Wiley, New York.

Robinson, E. (1967), *Multichannel Time Series Analysis*. Holden Day, San Francisco.

Rohlin, V. (1948), *Unitary Rings* (Russian). *DAN SSSR* (*N.S.*) **59**, 643–646. MR 9,446.

Rolewicz, S. (1972), *Metric Linear Spaces* PWN, Warsaw. MR 55:10933.

Romanov, N. P. (1946), The applications of functional analysis to the questions for the distributions of prime numbers (Russian). *Bull.* (*Izvestija*) *Math. Inst. Univ. Tomsk* **3**, 145–173. MR 8,445.

Romanov, N. P. (1946), On a special family of infinite unitary matrices. *C. R. Doklady Acad. Sci. URSS* (*N.S.*) **52**, 295–297. MR 8,137.

Romanov, N. P. (1946), Hilbert space and number theory (Russian). *Izv. Akad. Nauk SSSR* **10**, 2–34. MR 8, 9.

Romanov, N. P. (1945), On a special orthonormal system and its connection with the theory of primes. *Rec. Math.* (*Math. Sbornik*) (*N.S.*) **16** (58), 353–364. MR 7,365.

Romanov, N. P. (1948), Concerning the distribution of prime numbers. *Mat. Sbornik* (*N.S.*) **23** (65), 259–278. MR 10,285.

Romanov, N. P. (1951), Hilbert space and number theory (Russian). *Izv. Akad. Nauk SSSR Ser. Mat.* **15**, 131–152. MR 13,208.

Root, W. L. and Pitcher, T. S. (1955), Some remarks on signal detection. *IRE Trans. Inform. Theory*, **IT-1**, 33–38.

Rozanov, Ju. (1963), Stationary stochastic process (Russian). *Izdat. Lit. Mat. Moscow*. MR 28:2580.

Rozanov, Ju. (1967), *Stationary Random Processes*. Fizmatgiz, Moscow. English transl. Holden-Day San Francisco. MR 38#4985.

Rozanov, Ju. (1959), Spectral analysis of abstract functions. *Theory of Prob. Appl.* **4**, 271–287. MR 23:A 685; **4** (1959), 291–310. MR 22#8565.

Rozewski, Z. (1979), A contribution to the theme of orthogonality on sets (German, Czech and Russian summaries). *Cašopis Pešt. Mat.* **104** (3), 245–256. MR 80i:04003.

Rousseau, R. (1978), Construction of a left Hilbert algebra with respect to a Minkowski form. *Report on Math. Phys.* **14**, 321–323.

Rousseau, R. (1977), The left Hilbert algebra associated to a semi-direct product. *Math. Proc. Cambridge Phil. Soc.* **82**, 411–418. MR 56:16386.

Routledge, N. A. (1952), A result in Hilbert space. *Quart. J. Math. Oxford* **3** (2), 12–18. MR 13,661.

Ravaska, T. (1977), Theory of curves in a complex Hilbert space. *Anal. Acad. Fenn. Sci. Ser. A Math. Dissert.* **14**, 1–42. MR 57:4215.

Ravaska, T. (1979), Frenet theory of surfaces. *Ann. Acad. Sci. Fenn. Ser. A I Nat.* **4** (1), 151–168 MR 80g:53003.

Ravaska, T. (1980), Differential geometry of C^∞ helices in Hilbert space. *J. Math. Anal. Appl.* **75**, 385–401.

Ravaska, T. (1980), On Helical and Δ-helical surface in Hilbert space. *J. Math: Anal.* **76**, 18–36.

Rubin, H. and Stone, M. H. (1953), Postulates for generalizations of Hilbert spaces. *Proc. Amer. Math. Soc.* **4**, 611–616. MR 15,39.

Ruckle, W. (1983), Absolutely divergent series and classes of Banach spaces. *Math. Proc. Cambridge Phil. Soc.* **94**, 281–289.

Rudin, W. (1962), Projections on invariant subspaces. *Proc. Amer. Math. Soc.* **13**, 429–432.

Rudin, W. (1962), *Fourier Analysis on Groups*. Interscience Publ. Co. New York. MR 27:2808.

Rudin, W. (1974), *Principles of Mathematical Analysis*. McGraw-Hill Book Co. MR 49#8783.

Rudin, W. and Smith, K. T. (1961), Linearity of best approximations: a characterization of ellipsoids. *Proc. Nederl. Akad. Wetensch. Ser. A* **64**, 97–103. MR 23:A 2028.

Rzewuski, J. (1972), On Hilbert structure of certain linear spaces. *Bull. Acad. Polon. Sci. Ser. Sci. Math. Astronom. Phys.* **20**, 85–94. MR 46:2406.

van Rossum, H. (1968), Hilbert space methods for generalized continued fractions. *Nieuw. Arch. Wisk.* **16** (3), 152–164. MR 39:135.

Sadovnichi̇, V. A. and Lyabshkin, V. A. (1983), *Geometry of Hilbert Spaces and Three Principles of Functional Analysis* (Russian). Znanie, Moscow. MR 84i:46003.

Saeks, R. (1970), Causality in Hilbert spaces. *SIAM Rev.* **12**, 357–383. MR 54:1014.

Saeks, R. (1973), Resolution spaces, operators and systems. *Lecture Notes in Economics and System Theory* No. 82. MR 57:5211.

Saeks, R. (1973), State in Hilbert space. *SIAM Rev.* **15**, 283–308. MR 57:7249.

Saeks, R. (1973), Fourier Analysis on Hilbert space. *SIAM Rev.* **15**, 604–638. MR 57:7250.

Saeks, R. (1976), Reproducing kernel resolution spaces and applications. *J. Franklin Inst.* **302** (4), 331–355. MR 57:19035. R.Z. 1977, 10B 630.

Saénz, R. (1974), On the K-completion of a Hilbert module (Spanish). *Rev. Columbiana Mat.* **8**, 253–265. MR 51:3905.

Saitoh, S. (1983), Hilbert spaces induced by Hilbert space-valued functions. *Proc. Amer. Math. Soc.* **89**, 74–78. MR 84h:44001.

Saitoh, S. (1982), Integral transforms in Hilbert space. *Proc. Japan Acad.* **58**, *Ser. A*, 361–364.

Saitoh, S. (1983), The Weierstrass Transform and Isometry in the Heat Equation. *Appl. Anal.* **16**, 1–6.

Sakai, S. (1971), C^*-algebras and W^*-algebras. *Erge. der Math.* **60**, MR 56:1082.

Sanz de Bremand, Antonio Plans (1969), *Hilbert Space* (Spanish). *Rev. Acad. Ci. Zaragoza* **24** (2), 5–25. MR 45:20a.

Sansone, G. (1948, 1949), *Equazioni Differenziali nel Campo Reale.* Vol. 1 Second Ed. Nicola Zanichelli, Bologna Vol. 2 Second Ed-Nicola Zanichelli, Bologna. MR 11,32; MR 10,193.

Sard, A. (1963), *Linear Approximation Theory.* Amer. Math. Soc. Providence, R. I. MR 28:1429.

Saphar, P. (1972), Applications p-decomposantes et p-absolument sommantes. *Israel J. Math.* **11**, 164–179. MR 45:5779.

Sargasjan, I. S. and Levitan, B. M. (1975), Introduction to spectral theory: selfadjoint ordinary differential operators. Transl. from the Russian. *Amer. Math. Soc. Math. Monograph. Transl.* No. 39. MR 45#8911.

Sarmanov, O. V. (1946), Sur les solutions monotones des equations integrales de correlation. *C. R. (Doklady) Acad. Sci. URSS (N.S.)* **53**, 773–776. MR 8,467.

Sarmanov, O. V. (1958), The maximal correlation coefficient. (The symmetric case.) (Russian). *DAN SSSR* **120**, 715–718. MR 21:2334.

Sarmanov, O. V. (1958), The maximal correlation coefficient. (The non-symmetric case.) (Russian). *DAN SSSR* 121, 52–55, MR 20:5539.

Savage, L. J. (1946), The application of vectorial methods to metric geometry. *Duke math. J.* **13**, 521–528.

Saworotnow, P. (1968), On generalized Hilbert spaces. *Duke Math. J.* **35**, 191–197. MR 37:3333.

Saworotnow, P. (1970), Representation of a topological group on a Hilbert module. *Duke Math. J.* **37**, 145–150. MR 41:7442.

Saworotnow, P. (1972), Integral as a certain type of positive definite function. *Proc. Amer. Math. Soc.* **35**, 93–95. MR 45:7982.

Saworotnow, P. (1973), Abstract staionary processes. Proc. *Amer. Math. Soc.* **40**, 585–589. MR 48:3111.

Saworotnow, P. (1976), A characterization of Hilbert modules. *Bull. Calcutta Math. Soc.* **68**, 39–41. MR 56:16387.

Saworotnow, P. (1976), H^*-algebra valued measure on a locally compact space. *Bull. Calcutta Math. Soc.* **68**, 155–166.

Saworotnow, P. (1979), Stationary processes on a semi-group. *Bull. Calcutta Math. Soc.* **71**, 143–156. MR 82a:60050.

Saworotnow, P. (1981), Irreducible representations of a Banach algebra on a Hilbert module. *Houston J. Math.* **7**, 275–281.

Saworotnow, P. (1980), Linear spaces with an H^*-algebra valued inner product. *Proc. Amer. Math. Soc.* **262**, 543–549. MR 81m:46080.

Sazonov, V. V., (1958), A remark on characteristic functionals. *Theory of Prob. Appl.* **3**, 188–192.

Salehi, H. (1972), Stone's theorem for a group of unitary operators over a Hilbert space. *Proc. Amer. Math. Soc.* **31**, 480–484.

Salinas, N. (1972), Ideal sets and ideals of compact operators on Hilbert spaces. *Indiana Univ. Math. J.* **22**, 505–521.

Samn, S. (1972), Existence of a seminormal basis in $C_{(0,1)}$. *Proc. Amer. Math. Soc.* **35**, 183–186.

Schaffer, J. J. (1965), Another characterization of Hilbert spaces. *Studia Math.* **25**, 271–276. MR 31:3834.

Schatten, R. (1957), The space of completely continuous operators on a Hilbert space. *Math. Ann.* **134**, 47–49.

Schatten, R. (1960), Norm ideals of completely continuous operators. *Erge. der Math. Wiss.* **27**. MR 22:9878.

Schatten, R. and von Neumann, John (1948), The cross-space of linear transformations. II. *Ann. of Math.* **49** (2), 557–582. MR 10,256.

Schoenberg, I. J. (1935), Remarks to Maurice Fréchet's article 'Sur la definition axiomatique d'une classe d'espaces distancies vectoriellement applicable sur l'espace de Hilbert'. *Ann. of Math.* **35** (2), 724–732.

Schoenberg, I. J. (1938), Metric spaces and completely monotonic functions. *Trans. Amer. Math. Soc.* **44**, 522–536.

Schoenberg, I. J. (1952), A remark on M. M. Day's characterization of inner product spaces and a conjecture of L. M. Blumenthal. *Proc. Amer. Math. Soc.* **3**, 961–964. MR 14,564.

Schoenberg, I. J. (1953), On a theorem of Kirszbraun and Valentine. *Amer. Math. Monthly* **60**, 620–622. MR 15,341.

Schoenberg, I. J. (1964), Spline functions and least quadrature formulae. *Bull. Amer. Math. Soc.* **70**, 143–148. MR 28:394.

Schoenberg, I. J. and von Neumann, John, Fourier integrals and metric geometry. *Ann. of Math.* **39** (2), 811–841.

Schönbeck, S. O. (1966), Extensions of nonlinear contractions. *Bull. Amer. Math. Soc.* **72**, 98–101. MR 37:1960.

Schönbeck, S. O. (1966), On the extension of Lipschitz *m–ps*, *Arkif fo ör Math.* **15**, 201–208. MR 39:3297.

Schreiber, M. (1972), Quelques remarques sur les caractérisations des espaces L^p, $p \in [0, 1)$. *Ann. Inst. Henri Poincaré*, **8**, Sect. B, 83–92.

Schue, J. R. (1960), Hilbert space methods in the theory of Lie algebras. *Trans. Amer. Math. Soc.* **95**, 69–80. MR 22:8352.

Schue, J. R. (1961), Cartan decomposition for L^*-algebras. *Trans. Amer. Math. Soc.* **98**, 334–349. MR 24:A3242.

Schumacher, L. (1981), *Spline Functions.* J. Wiley, N.Y.

Schwartz, J. (1967), *W^*-Algebras.* Gordon and Breach, New York. MR 38:547.

Schwartz, Laurent (1961/2), Sous espaces hilbertiens et antinoyaux associés. *Semin. Bourbaki 14e année*, no. 238.

Schwartz, L. (1974), Les espaces de type et cotype 2,... *Ann. Inst. Fourier (Grenoble)* **24**, 179–188.

Schwartz, L. (1981), Geometry and probability in Banach spaces. *Bull. Amer. Math. Soc.* **4**, 135–141.

Schwartz, L. (1957), Generalisations des espaces L^p. *Publ. Inst. Statist. Univ. Paris* **6**, 241–250. MR 20:6034.

Schwartz, L. (1964), Sous-espaces hilberienne et noyaux associées; applications aux representations des groupes de Lie. *Deuxieme Colloq. Anal. Fonc.* pp. 153–163. Centre Belgique Recherches Math. Libraire Univ. MR 32:2890.

Schwartz, L. (1964), Sous-espaces hilbertienne d'espace vectoriels topologiques et noyaux reproduisants. *J. Analyse Math.* **13**, 115–256. MR 31:3835.

Segal, I. (1949) Two sided ideals in operator algebras. *Ann. of Math.* **50** (2), 856–865. MR 11,187.

Segal, I. (1956), Tensor algebras over Hilbert spaces. I. *Trans. Amer. Math. Soc.* **81**, 106–134. MR 17,880.

Segal, I. (1956), Tensor algebras over Hilbert spaces. II. *Ann. of Math.* **63** (2), 160–175. MR 17,1114.

Segal, I. (1958), Distributions on Hilbert space and canonical systems of operators. *Trans. Amer. Math. Soc.* **88**, 12–41. MR 21:1545.

Seidman, T. (1968), On an inequality for operators on Hilbert spaces. *Report* 68-Carnegie Mellon Univ.

Seidman, T. (1969), On an identity for normal-like operators. *Israel J. Math.* **7**, 249–253. MR 40:7845.

Senechalle, D. A. (1968), A characterization of inner product spaces. *Proc. Amer. Math. Soc.* **19**, 1306–1312. MR 39:761.

Senechalle, D. A. (1973), Equations which characterize inner product space. *Proc. Amer. Math. Soc.* **40**, 209–214, R.Z. 1974, 4B:602.

Senechalle, D. A. (1971), Euclidean and noneuclidean norms in a plane. *Illinois J. Math.* **15**, 281–289. MR 43:6709.

Senechalle, D. A. (1978), Isometries between two dimensional subspaces implies an inner product. (Unpublished manuscript).

Sengupta, H. M. (1950), On rearrangement series. *Proc. Amer. Math. Soc.* **1**, 71–75. MR 11,346.

Sengupta, H. M. (1956), On rearrangement series. *Proc. Amer. Math. Soc.* **7**, 347–350. MR 17,1199.

Serre, J. P. (1962), Homomorphisms cempletement continus des espaces de Banach *p*-adiques. *Inst. Hautes Etudes Sci. Publ. Math.* **12**, 69–85. MR 26:1733.

Šerstnev, A. N. (1974), The concept of charge in a noncommutative framework of measurement theory (Russian). *Probability Methods and Cybernetics* no. X–XI, pp. 68–72. Izdat. Kazan Univ. Kazan. MR 57:17306a.

Šerstnev, A. N. (1974), States in a product of spaces (Russian). *Ibid.* No. X–XII, pp. 142–146. Izdat. Kazan Univ. Kazan. MR 57:17306b.

Shapiro, H. S. (1964), Reproducing kernel and Beurling's theorem. *Trans. Amer. Math. Soc.* **110**, 448–458. MR 28:2225.

Shapiro, H. S. and Shields, A. L. (1963), Interpolation in Hilbert spaces of analytic functions. *Studia Math.* **1**, 109–110. MR 26:4168.

Shenton, L. R. (1953 etc.), A determinantal expansion for a class of definite integral. I. (1953), *Proc. Edinburgh Math. Soc.* **9** (2), 44–52. MR 15,781. II. (1954), *Ibid.* **10** (2), 78–91. MR 16,575. III. (1954) *Ibid.* **10** (2), 134–140. MR 17,844. IV. (1957) *Ibid.* **10** (2), 153–166. MR 19,410. V. (1957); *Ibid.* **10** (2), 167–188. MR 19,410.

Shepherd, G. C. (1966), A prehilbert space consisting of closed convex sets. *Israel J. Math.* **4**, 1–10. MR 34,3435.

Sherman, T. (1965), Representation of Lie algebras by normal operators. *Proc. Amer. Math. Soc.* **16**, 1125–1129. MR 32:1293.

Shonkwiller, R. (1980), Dilations of reproducing kernels. *Proc. Internat. Conf. Blazejeko, 1979*, pp. 259–268. *Lecture Notes in Math. No.* 828. MR 82k:47013.

Sierpinski, W. (1965), *Cardinal and Ordinal Numbers.* Second Edition, Warszawa.

Singer, I. (1960), On the set of the best approximations of an element in a normed linear space. *Rev. Roum. Math. Pures Appl.* **5**, 383–402.

Singer, I. (1967), Best approximation in normed linear spaces by elements of linear subspaces (Romanian). Ed. Acad. RSR, Bucharest. MR 38:3677, Engl. Transl. MR 42:4937.

Singer, I. (1970, 1979), Bases in Banach spaces. I. II. New York. MR 45:7451.

Sinai, Ya. G. (1971), Dynamical systems. I. *Ergodic Theory.* Warsaw 1970. Aarhus Univ.

Singh, S, and Prasad, R. L. (1974), An aspect of the geometry of Hilbert spaces. *Progress Math.* **8**, 31–40. MR 51:3871.

Shul'man, V. S. (1983), Modules over operator algebras (Russian). *Functsional Analiz i Priložen.* No. 2, 94–95. MR 84k:46035. *Functional Anal. Appl.* **17** (2), 161–163.

Sibiriakov, G. V. (1974), Normed spaces with maximal bias (Russian). *Trudy Tomsk Gosud. Univ.* **253**, 23–28. MR 55:3755.

Skorohod, A. V. (1970), A remark on Gaussian measures on Banach spaces. *Teoria Veroiat. i Primen.* **15**, 519–520.

Skorohod, A. V. (1970), Surface integrals and the Green formula in Hilbert space (Russian). *Teoria Veroiat. i Primen. Vyp.* **2**, 172–175. MR 46:1997.

Skorohod, A. V. (1970), Admissible shifts of measures in Hilbert spaces (Russian). *Teoria Veroiat. i. Primen.* **15**, 577–598. MR 45:499. *Theory of Prob. Appl.* **15**, 557–580.

Skorohod, A. V. (1970), Certain questions on the theory of measure and integral in Hilbert space (Russian). *Proc. of the Eighth Math. Summer School*, Kaciveli pp. 102–127.

Skorohod, A. V. (1971), *Izdanie Int. Math. Acad. Nauk Ukrain SSR, Kiev*, MR 52:12025.

Skorohod, A. V. (1974), *Integration in Hilbert Space.* Springer Verlag N.Y. MR 57:6360.

Simpson, J. E. (1972), Locally convex algebras. I. *Rev. Roum. Math. Pures Appl.* **17**, 935–941.

Simpson, J. E. (1977), Locally Convex algebras. II. *Spectra. Rev. Roum. Math. Pures Appl.* **22**, 1155–1165.

Simpson, J. E. (1982), Dilations on locally convex spaces. *Pacif. J. Math.* **101**, 185–192. MR 84j:47007.

Slugin, S. N. (1962), Complex semiordered spaces and modules on them (Russian). *DAN SSSR* **147**, 306–309.

Slugin, S. N. (1965), *Complex Hilbert Structures* (Russian). Izvestija Akad. Nauk SSSR *Ser. Math.* **29**, 215–226. R. Ž. 1965, 8B 443.

Slutsky, E. (1939), Sur les fonctions aleatoires presque periodiques et sur la decomposition des fonctions aléatoires stationaires en composantes. *Actualités Sci. Indust.* No. 738 (1939), Paris.

Smiley, M. F. (1953), Right H^*-algebras. *Proc. Amer. Math. Soc.* **4**, 1–4. MR 14,600.

Smiley, M. F. (1965), Real Hilbert algebras with identity. *Proc. Amer. Math. Soc.* **16**, 440–441. MR 31:591.

Smiley, M. F. and Kirk, W. A. (1964), Another characterization of inner product spaces. *Amer. Math. Monthly* **71**, 890–891.

Smatkov, V. A. (1971), On one-dimensional approximation and interpolation preserving the norm. *Trudy Tsentral Zonal Ob. Mat. Kafedr. Kalinin Gosud. Ped. Inst.* Vyp. 2, pp. 182–194, R. Ž. 1971, 12B 1027.

Smith, J. F. (1972), The p-classes of an H^*-algebra. *Pacif. J. Math.* **42**, 777–793. MR 48:879.

Smith, J. F. (1972), The p-classes of a Hilbert module. *Proc. Amer. Math. Soc.* **36**, 428–434. MR 46:9753.

Smith, J. F. (1974), The structure of Hilbert modules. *J. London Math. Soc.* **8** (2), 741–749. MR 50:10826.

Smith, M. A. (1981), On the norm of metric projections. *J. Approx. Theory* **31**, 224–229. MR 83b:46024.

Smolianov, O. G. and Uglanov, A. V. (1973), Every Hilbert subspace of a Wiener space has measure zero (Russian). *Mat. Zametki*, **14**, 369–374. MR 49:7421. *Math. Notes* **14**, 772–774 (1974). MR 49:7421.

Smolyanov, O. G. and Fomin, S. V. (1976), Measures on linear topological spaces (Russian). *Uspekhi Mat. Nauk* **31** (4), 3–56. *Russian Math. Surveys* **31** (4), 1–53. MR 54:8776.

Solomjak, M. Z. (1957), On orthogonal bases in Banach spaces (Russian). *Vestnik Leningrad. Univ.* No. 1, 27–36.

Sova, M. (1966), Cosine operator functions. *Diss. Mat.* (*Roszprawy Mat*) **49**.

Springer, G. (1967), *Introduction to Riemann Surfaces.* Addison–Wesley, Reading Mass. MR 19,1169.

Springer, T. A. (1955), Quadratic forms over fields with discrete valuations. I. *Proc. Koninkl. Akad. Nederl. Wetensch.* **58**, 352–362. MR 17,17.

Springer, T. A. (1956), Quadratic forms over fields with discrete valuations. II. *Ibid.* (= *Indag. Math.*) **28**, 238–246. MR 17,945.

Springer, T. A. *Quaternions.* (Dutch). Math. Centrum, Amsterdam, ZW–001, 8pp. MR 33:1375.

Srinivasan, S., Sastry, M. A. and Wiswanath, K. (1979), Orthogonality in Banach spaces and approximation theory. *J. Madurai Kamaraj Univ.* **8**, 93–97. MR 82a:41035.

Stamney, W. L. (1957), On generalized euclidean and noneuclidean spaces. *Pacif. J. Math.* **7**, 1505–1511. MR 19,1192.

Stampfli, J. (1979), The norm of a derivation. *Pacif. J. Math.* **33**, 737–746.

Stanojević, S. V. (1970), Mielnik's probability manifolds and inner product spaces. *Bull. Acad. Polon. Sci. Ser. Sci. Math. Astronom. Phys.* **18**, 531–532. MR 42:8209.

Stanojević, S. V. (1973), Mielnik's probability spaces and characterizations of inner spaces. *Trans. Amer. Math. Soc.* **183**, 441–448. MR 48:6904.

Stanojević, S. V. and Suchanek, A. M. (1981), Integral identities of norms and characterizations of inner product spaces. *Proc. Amer. Math. Soc.* **81**, 101–103. MR 82i:46036.

Slepian, D. and Pollack, H. O. (1961), Prolate spheroidal wave functions. Fourier analysis and uncertainty. I. *Bell. System Techn. J.* **40**, 43–68. MR 25:4146.

Stesin, M. I. (1973), On the Aleksandrov-radii of finite-dimensional octahedrons (Russian). *Vestnik Moscov. Univ. No.* 3, 30–35. MR 48:7219.

Stesin, M. I. (1974), The Aleksandrov diameters of spheres (Russian). *DAN SSSR* **217**, 31–33. MR 51:1354.

Stechkin, S. B. (1967), Problem. In the *2nd All-Union Sympos. Integral Geometry* (Russian), Akad. Nauk SSSR, Petrozavodsk.

Stilles, W. J. (1965), A solution of Hirschfeld problem. *Nieuw Archiv voor Wisk.* **13** (3), 116–119. MR 32:6195.

Stilles, W. J. (1965), Closest point maps and their products. *Nieuw Archiv voor Wisk.* **13** (3), 19–29. MR 31:2601.

Stilles, W. J. (1965), Closest point maps and their products. *Nieuw Archiv voor Wisk.* **13** (3), 212–215. MR 32:8119.

Stilles, W. J. (1969), On inversions in normed linear spaces. *Proc. Amer. Math. Soc.* **20**, 505–508. MR 38:3718.

Stinespring, W. F. (1955), Positive functions on C^*-algebras. *Proc. Amer. Math. Soc.* **5**, 211–216.

Stone, M. H. (1932), On one parameter groups in Hilbert spaces. *Ann. of Math.* **33** (2), 643–648.

Stone, M. H. (1932), Linear operators on Hilbert spaces. *Amer. Math. Soc. Colloq. Publ.*

Stone, M. H. (1951), Unbounded operators in Hilbert space. *J. Indian Math. Soc.* **15**, 155–162. MR 14, 565.

Stone, M. H. (1961), Hilbert space methods in conformal mappings. *Proc. Sympos. Linear spaces Jerusalem, 1960*. pp. 409–425. Academic Press, Pergamon Press. MR 25:178.

Straus, V. A. (1973), *G*-orthogonal systems and bases in Hilbert spaces (Russian). *Izv. Vysš. Učebn. Zaved. Mat.* **136** (9), 108–117. MR 50:960.

Stroescu, E. (1972), Spectral dilations of operators on Banach spaces (Romanian. English summary). *Studii şi Cercet. Mat.* **24**, 919–972. MR 51:3926.

Stroescu, E. (1972), *U*-spectral dilations and *U*-spectral extensions of operators on Banach spaces (Romanian. English summary). *Studii şi Cercet. Mat.* **24**, 817–824. MR 51:1474.

Stroescu, E. (1972), *U*-spectral dilations for operators on Banach spaces. *J. Math. Anal. Appl.* **39**, 279–295. MR 47:5624.

Stroescu, E. (1972), A dilation theorem for operators on Banach spaces. *Actes du Colloq. d'Analyse Fonctionelle, Univ. Bordeaux, 1971*, pp. 365–373. Bull. Soc. Math. France, Mem. No. 31–32. Soc. Math. France, Paris. MR 51:13755.

Stroescu, E. (1973), Isometric dilations of contractions on Banach spaces. *Pacif. J. Math.* **47**, 257–262. MR 48:4776.

Stroescu, E. (1974), Functorial theory of the dilation on Banach spaces. *Rev. Roum. Math. Pures Appl.* **19**, 97–104. MR 50:14285.

Stroescu, E. (1983), Functorial behavior of the spectral dilation for operators on Banach spaces. *Rev. Roum. Math. Pures Appl.* **28**, 321–332.

Suciu, I, (1973), Dilation and extension theorems for operator-valued mappings (Romanian. English summary). *The theory of operators and operator algebras* (Romanian). MR 93–130. Ed. Acad. R.S.R. Bucharest, MR 51:6453.

Suciu, I. (1973), *The Theory of Operator Algebras* (Romanian), pp. 93–130. Ed. Acad. R.S.R. MR 51:6453. English transl. Noordhoff, Leyden, 1975. MR 51:6428.

Sukhomlinov, G. A. (1937), Analytic functionals. *Bull. Moskov. Gosud. Univ.* A_1, **2**.

Sukhomlinov, G. A. (1938), Functionals on linear complex spaces and quaternionic spaces. *Mat. Sbornik N.S.* **3** (45), 353–358. (Russian).

Steiger, W. L. (1969), On a generalization of the Cauchy–Schwarz inequality. *Amer. Math. Monthly* **76**, 815–816.

Stern, J. (1979), Le groupe des isometries d'un espace de Banach. *Studia Math.* **64**, 139–149.

Stern, J. (1978), Ultrapowers and local properties of Banach spaces. *Trans. Amer. Math. Soc.* **249**, 231–252.

Sudbery, A. (1979), Quaternionic analysis. *Math. Proc. Cambridge Phil. Soc.* **85**, 199–225. MR 80g:30031.

Suljagić, Salin (1980), On contractions in a quaternionic Hilbert space (Serbo-Croatian summary). *Glasnik Math. III* **15** (35), 87–96. MR 82c:47011.

Suljagić, S. (1980), Quaternionic Beurling's theorem. *Glasnik Math.* **15** (35), 327–339. MR 82i:47007.

Suljagić, S. (1980), Shift and the class C_0 (1) in a quaternionic Hilbert space. *Colloq. Math. Janoš Bolyai* **35**, pp. 1133–1137.

Suljagić, S. (1983), Operators in quaternionic Hilbert spaces. *Berch. Forschung Graz. Research Center Austria*, Graz, No. 198, pp. 20.

Sullivan, F. E. (1970), Structure of real L^p-spaces. *J. Math. Anal. Appl.* **32**, 621–629. MR 42:5031.

Sullivan, F. E. (1968), Norm characterization of real L^p spaces. Bull. *Amer. Math. Soc.* **74**, 153–154.

Sullivan, T. F. (1973), The geometry of solids in Hilbert spaces. *Notre Dame Journal of Formal Logic* **14**, 575–580. MR 49:1081.

Sundaresan, K. (1958), Orthogonality in normed linear spaces. *J. Indian Math. Soc.* **12**, 99–107.

Sundaresan, K. (1961), Characterization of inner product spaces. *Math. Student* **29**, 41–45.

Sundaresan, K. (1967), Smooth Banach spaces. *Math. Ann.* **173**, 191–199.

Sundaresan, K. (1966), Smooth Banach spaces. *Bull. Amer. Math. Soc.* **72**, 520–521.

Sundaresan, K. (1972), Orthogonality and nonlinear functionals on Banach spaces. *Proc. Amer. Math. Soc.* **34**, 187–190.

Swart, J. (1972), An axiomatic characterization of Hilbert spaces. *Atti Accad. Naz. dei Lincei, Rend.* **52** (8), 166–174. MR 48:4985.

Synge, J. L. (1958), La geometrie elementaire de l'espace fonctionnel avec des applications a la physique classique. *Colloq. Henri Poincaré*, Inst. Henri Poincaré, Paris, pp. 1–22.

Synge, J. L. Lecture No. II, Function-space with indefinite metric. *Ibid.* pp. 9–16.

Synge, J. L. and Yodzis, P. (1972), Kinematics, angular momentum and Eulerian dynamics in Hilbert space. *Proc. Irish Roy. Acad. Sci.* **72**, 121–147.

Szarek, S. J. (1976), On the best constants in the Khinchin inequality. *Studia Math.* **58**, 197–298. MR 55:3672.

Szegö, G. (1967), *Orthogonal Polynomials.* Third Ed. A.M.S. Math. Colloq. Publ. No. 23, MR 46:9631.

Szankowski, A. (1974), On Dvoretzky's theorem on almost spherical sections of convex bodies. *Israel J. Math.* **17**, 325–338. MR 50:2881.

Szepticki, P. and Van Vleck, F. S. (1982), Centers and nearest points of sets. *Proc. Amer. Math. Soc.* **85**, 27–31. MR 83c:46021.

Takahashi, A. (1971), Fields of Hilbert modules. *Ph.D. Thesis*, Tulane Univ.

Takahashi, A. (1979), Hilbert modules and their representations. *Rev. Columbiana Mat.* **13**, 1–39. MR 81k:46056a.

Takahashi, A. (1979), A duality between Hilbert modules and fields of Hilbert spaces. *Rev. Columbiana Mat.* **13**, 93–120. MR 81k:46056b.

Takemoto, H. (1973), A characterization of AW^*-modules and representation of Gel'fand type of noncommutative operator algebras. *Michigan Math. J.* **20**, 115–127. MR 47:7457.

Takemoto, H. (1975), Decomposable operators in continuous fields of Hilbert spaces. *Tohoku Math. J.* **27**, 413–435. MR 56:12922.

Takemoto, H. (1976), On the weakly continuous constant field of Hilbert spaces and its applications to the reduction theory of von Neumann algebras. *Tohoku Math. J.* **28**, 479–496. MR 54:5851.

Takenouki, O. (1951), On the maximal Hilbert algebras. *Tohoku Math. J.* **3** (2), 123–131.

Takenouki, O. (1952), On the structure of maximal Hilbert algebras. *Math. J. Okayama Univ.* **1**, 1–32.

Takeuti, G. (1962), Dirac spaces. *Proc. Japan Acad. Sci.* **38**, 414–418. MR 26:2651.

Takeuti, G. (1979), Quantum set theory. *Current Issues in Quantum Logic.* Erice, pp. 303–332. Ettore Majorana International Sci. Ser. Phys. Sci. Plenum Press, N.Y. London, 1981. MR 84j:03136. Entire coll. MR 84i:03005.

Takesaki, M. (1970), Tomita's theory of modular Hilbert algebras and its applications. *Lecture Notes in Math.* No. 128. MR 42:5061.

Talalian, F. A. (1972), Rearrangement of series in Hilbert spaces (Russian). *Mat. Zametki* **12**, 275–280. MR 47:7398. *Math. Notes* **12**, 599–602.

Taldykin, A. T. (1974), Tests for a basis in Hilbert space (Russian). *DAN SSSR* **216**, 505–508. MR 50:2885.

Tam, K. W. (1969), Isometries of certain function spaces. *Pacif. J. Math.* **31**, 233–246.

Tan, K. K. (1984), A generalization of a theorem of Routledge. *J. Math. Anal. Appl.* **102** (1), 58–61. MR 85k:46019.

Tapia, R. S. (1973), A characterization of inner product spaces. *Bull. Amer. Math. Soc.* **79**, 530–531. MR 47:3964.

Tapia, R. S. (1973), A characterization of inner product spaces. *Proc. Amer. Math. Math. Soc.* **41**, 569–574. MR 49:5791.

Taylor, A. E. (1935), The Hilbert space postulates, a further reduction. Bull. *Amer. Math. Soc.* **41**, 847–848.

Taylor, A. E. (1947), A geometric theorem and its applications to biorthogonal systems. *Bull. Amer. Math. Soc.* **53**, 614–616.

Taylor, P. D. (1973), A Banach space characterization of the space of affine continuous functions on a compact convex set. *Amer. Math. Monthly* **80**, 911–915.

Textorius, (1974), A remark on Schwarz's inequality (Swedish. English summary). *Norsk Mat. Tidsskr.* **22**, 63–65. MR 50:5438.

Thang Dand Hung and Tien Nguyen Duy (1981), On convergence of martingales and geometric properties of Banach spaces. *Theory of Prob. Appl.* **26**, 379–385.

Thele, R. L. (1974), Some results on the radial projection in Banach spaces. *Proc. Amer. Math. Soc.* **42**, 483–486. MR 6892.

Thomas, E. T. (1964/5), Une axiomatique des espaces de Dirichlet. *Seminaire Theorie du Potentiel*, Paris, 90. annee No. 9.

Thomspon, A. C. (1977), Isometric embeddings in Hilbert spaces. *Nieuw Arch. Wisk.* **25** (3), 161–168. MR 58:7047.

Timan, M. F. and Tukhliev, K. (1983), Properties of some orthonormal systems (Russian). *Izv. Vysš. Učebn. Zaved. Matematika* (9), 65–73. MR 85f:42045.

Tineo, A. (1981), Isometric characterization of Hilbert spaces (Spanish). *Notes Mat. Univ. Andas* (5), 11. R. Ž. 1983, 3B 900.

Tits, J. (1955), Sur certaines classes d'espaces homogenes de groupes de Lie. *Acad. Roy. Belg. Cl. Sci.* **8** (29), 268 pp. MR 17,874.

Toita, M. (1967), Quasi-standard von Neumann algebras. Mimeographed Notes.

Toita, M. (1967), Standard forms of von Neumann algebras. *Fifth Functional Anal. Sympos.* Math. Soc. of Japan.

Tonelli, L. (1921, 3), *Fondamenti di Calcolo Delle Variazioni.* Vol. I. II. Nicola Zanichelli, Bologna.

Toranzos, F. A. (1945), On projectivity in Hilbert spaces (Spanish). *Publ. Inst. Math. Univ. Nacional Litoral* **7**, 187–197. MR 7,452.

Toranzos, F. A. (1966/7), Immersion of convex metric spaces in E^n (Spanish). *Math. Notae* **21**, 29–53. MR 39:870.

Toranzos, F. A. (1967), Metric characterization of the *n*-cell. *Univ. Nac. Tucuman. Rev. Ser. A* **17**, 149–154. MR 38:2734.

Titarenko, Ju. V. (1971), The Bonnet theorem for surfaces in Hilbert spaces (Russian). *Vestnik Moscov. Univ.* **5**, 45–49. MR 45:1090.

Tomaszewki, B. (1982), Two remarks on the Khintchine–Kahane inequality. *Colloq. Math.* **46**, 283–288.

Torgašev, A. (1974), A generalization of spaces with inner products and semi-inner products (Serbo-Croatian summary). *Mat. Vesnik* **11** (26), 301–313. MR 51:6377.

Torgašev, A, (1978), The symmetric quaternionic Banach algebras. I. Gel'fand theory. *Publ. Inst. Math. Beograd N.S.* **24** (33), 173–188. MR 80b:46067.

Torgašev, A. (1979), Characteristic metric functions of Wachs spaces. I. *Publ. Inst. Math. Beograd* **35** (39), 225–239. MR 80i:47016.

Torgašev, A. (1980), Characteristic operator functions on Wachs spaces. *Ibid.* **27** (49), 287–296. MR 82k:47027.

Torgašev, A. (1976), The numerical range of an operator in a Wachs space. *Math. Vesnik* **13** (28), 101–117. MR 58:17888.

Torgašev, A. (1976), Extensions of two classic kinds of Wachs space operators. *Mat. Vesnik* **13**, 216–226. MR 57:1187.

Torgašev, A. (1978), Concepts of spectral sets in Wachs spaces. I. *Publ. Inst. Math. Beograd N.S.* **23** (37), 213–222. MR 80b:47008.

Torunczyk, H. (1981), Characterizing Hilbert space topology. *Fund. Math.* **111**, 247–262.

Torunczyk, H. and West, J. (1983), A Hilbert space limit for the iterated hyperspace functor. *Proc. Amer. Math. Soc.* **89**, 329–335. MR 84m:54010.

Tseng, Y. Y. (1935), Spatial representation of selfadjoint functional transformations. in a nonhilbertian space. *Sci. Rep. Nat. Tsing Hua Univ. Peiping A* **3**, 113–125.

Tseng, Y. Y. (1933), *Ph.D. Thesis* (under M. H. Stone), Univ. of Chicago, Chicago (Quaternionic spaces and linear operators).

Tudor, C. (1965), On a theorem of Aronszajn (Romanian). *Studii si Cercet. Mat.* **17**, 295–299. R. Ž. 1966, 2B 363.

Tzafriri, L. (1971), An isomorphic characterization of L^p and c_0 spaces. *Michigan Math. J.* **18**, 21–31. MR 58:2169.

Umegaki, H. (1955), Positive definite functions and direct product of Hilbert spaces. *Tohoku Math. J.* **7**, 206–211. MR 17,1112.

Unni, K. R. and Puttmadaiah, C. (1981), On orthogonality in sip spaces. *Tsukuba J. Math.* **5**, 15–19. MR 82k:46036.

Unsain, I. (1970), Classification of the simple separable real L^*-algebras. *Ph.D. Thesis*, Univ. of California, Berkeley, Calif.

Unsain, I. (1971), Classification of the simple separable real L^*-algebras. *Bull. Amer. Math. Soc.* **77**, 462–466. MR 43:304.

Unsain, I. (1971), Cartan–Ambrose theorem for infinite dimensional Riemannian manifolds. *Univ. of Maryland Preprint.*

Unsain, I. (1972), Classification of the simple separable real L^*-algebras. *Journal of Diff. Geometry* **7**, 423–451. MR 48:4068.

Urysohn, P. (1951), Hilbert spaces as images of metric spaces (Russian). *Papers on Topology and Related Domains*, Tom. 1, Moscow-Leningrad. pp. 147–150. (*See also Math. Ann.* **92**, 302–304.)

Vakhania, N. N. (1968), The covariance operator of a probability distribution in a Banach space (Russian). *Soobsč. Akad. Naux. Grat. SSR* **53**, 17–20.

Vakhania, N. N. (1971), *Probability Distributions on Linear Spaces* (Russian). Mecnieraba, Tbilisi, 1971.

Vakhania, N. N. (1973), On some questions of the theory of probability measures on Banach spaces. *Lecture Notes* Nagoya University.

Vaisman, Izu, (1979), Basic ideas of geometric quantization. *Rend. Sem. Mat. Univers. Politecn. Torino* **37**, 31–41. MR 85g:58040.

Valentine, F. A. (1945), A Lipschitz condition preserving extension for vector functions. *Amer. J. Math.* **67**, 83–93. MR 6,203.

Valentine, F. A. (1946), *Convex Sets*. McGraw-Hill. *Reprinted*, R. F. Krieger Publ. Co. 1976. MR 53:14306.

Valentine, J. E. (1971), On criteria of Blumenthal for inner product spaces. *Fund. Math.* **72**, 265–269. MR 45:2579.

Valentine, J. E. (1973), Intrinsic four point properties which characterize hyperbolic and Euclidean spaces (Russian summary). *Bull. Acad. Polon. Sci. Ser. Sci. Math. Astronom. Phys.* **21**, 1103–1106. MR 50:1133.

Valentine, J. E. (1975), Angles in metric spaces. In *The Geometry of Metric and Linear Spaces*, pp. 66–73. *Lecture Notes in Math.* No 490. MR 53:6435.

Valentine, J. E. (1978), The Menelaus property characterizes real rotund spaces. *Fund. Math.* **100**, 41–44 MR 58:18167.

Valentine, J. E. and Wayment S. G. (1971), Wilson angles in linear normed spaces. *Pacif. J. Math.* **31**, 239–243. MR 43:3775.

Valentine, J. E. and Wayment, S. G. (1972), Metric characterization of hyperbolic and Euclidean spaces. *Colloq. Math.* **25**, 259–265. MR 48:12474.

Valentine, J. E. and Wayment, S. G. (1972), The Euclidean isosceles and queasy four point property. *Proc. Amer. Math. Soc.* **34**, 233–238. MR 45:5882.

Valentine, J. E. and Wayment, S. G. (1975), Local Euclidean four point property which characterizes inner product spaces. *Proc. Amer. Math. Soc.* **50**, 337–343. MR 51:6602.

Valentine, J. E. (1980), Banach–Euclidean four point properties. *Fund. Math.* **106**, 227–230. MR 81i:52002.

Vasudeva, H. (1973), On monotone matrix functions of two variables. *Trans. Amer. Math. Soc.* **176**, 305–318. MR 47:2408.

Veic, B. E. (= B. E. Veits) (1965), Bessel and Hilbert systems in Banach spaces and questions of stability (Russian). *Izv. Vysš. Učebn. Zaved. Matematika*, **45** (2), 7–23. English transl. *Amer. Math. Soc. Transl.* **93** (2), (1970), 83101. MR 33:549.

Veic, B. E. (1966), Some characteristic properties of unconditional bases (Russian). *DAN SSSR* **155**, 509–512. *Soviet Math. Dokl.* **5** (1964), 426–439. MR 29:3865.

Veic, B. E. (1971), Criteria for hilbertness of Banach spaces in terms of smoothness (Russian). *Mat. Zametki* **9**, 385–390. *Math. Notes* **9**, 221–224. MR 44:782.

Venkov, B. A. (1929), On the arithmetic of quaternions. *Izv. Akad. Nauk 1922*, 205–220; 221–246; 489–504, 532–562; 607–622.

Vershik, A. M. (1964), The general theory of gaussian measures in linear spaces (Russian). *Usnehi Mat. Wark* **19**, 210–212.

Vidav, I. (1956), Eine metrische kennzeichnung der selbstadjungierte Operatoren. *Math. Z.* **66**, 121–128.

Vidav, I, (1959), On the axiomatic of Hilbert spaces. *Math. Z.* **71**, 458–460. MR 21:7424.

Vinala, W. (1974), Splines in Hilbert spaces. *Lecture Notes in Math.* No. 319, 538–560.

Vinogradov, I. M. (1944), *Foundations of the Theory of Numbers* (Russian). OGIZ, Moscow, Leningrad. MR 7,413. English translation.

Viola Devapakkiam, C. (1974), Hilbert space methods in the theory of Jordan algebras. I. *Math. Proc. Camb. Phil. Soc.* **78**, 293–300. MR 52:1330.

Viola, C. Devapakkiam, and Rema, P. S. (1976), Hilbert space methods in the theory of Jordan algebras. II. *Math. Proc. Camb. Phil. Soc.* **79**, 307–319. MR 53:8162.

Viswanath, K. (1971), Normal operators on quaternionic Hilbert spaces. *Trans. Amer. Math. Soc.* **162**, 337–350.
Viswanath, K. (1977), Quaternionic representations of topological groups. *Journal Indian Math. Soc.* **41**, 149–162.
Viswanath, R. (1978), Operators on real Hilbert spaces. *Journal Indian Math. Soc.* **42**, 1–13.
Višik, M. I. (1949), The method of orthogonal projection and direct decomposition in the theory of differential elliptic operators (Russian). *Mat. Sbornik* (*N.S.*) **25** (67), 184–194. MR 11,250.
Vitali, G. (1929), *Geometria nello spazio hilbertiano*. Bologna.
Vitali, G. (1931), Alcuni elementi di mecanica negli spazi curvi. *Annali di Mat. Pura Appl.* **9** (4), 75–89.
Vitali, G. and Sansone, G. (1935), *Moderna teoria delle funzioni di variabile reali*. Part. 1, 2. Bologna.
Van Vleck, E. B. (1929), On the location of roots of polynomials and entire functions. *Bull. Amer. Math. Soc.* **35**, 643–683.
Volodin, E. A. and Smulian, Ju. L. (1973), Certain questions of the geometry of rigged Hilbert spaces (Russian). *Mat. Zametki* **13**, 396–402. MR 47:5567.
Volterra, Vito, (1887), Sopra le funzioni che dipendono da altre funzioni. *Atti Accad. Naz. Dei Lincei Rend.* **3** (4), 97–105; 141–146; 153–158.
Volterra, V. (1913), Leçons sur les équations integrales et les equations integro-differentiales. *Collection Borel*, Paris.
Volterra, V. (1929), *Theory of Functionals and of Integral and Integro-Differential Equations*. Edited by L. Fantappié. Transl. by M. Long, London. Dover, 1959.
Volterra, V. (with J. Pérès) (1926), *Théorie Générale des Fonctionneles*. Tome 1, *Collection Borel*, Paris.
Volterra, V. (with B. Hostinsky), (1938), Operations infinitesimales lineaires, applications aux equations differentielles et fonctionneles. *Collection Borel*, Paris.
Vrbová, Pavla, (1973), Quadratic functionals and bilinear forms. *Casopis pro pest. mat.* **98**, 159–161. MR 49:11082.
de Vries, J. (1970), A note on topological linearization of locally compact transformation groups in Hilbert spaces. *Math. Centrum, Amsterdam.* MR 44:5412.
Vukman, J. (1980), Characterization of Hilbert spaces. *Glasnik Mat. III* **15** (35) 301–304. MR 82k:46035.
Vukman, J. (1981), Real symmetric *-algebras. *Glasnik Mat.* **16** (36), 91–103. MR 83a:46060.
Vukman, J. (1982), Involutions on $L(X)$. *Glasnik Mat.* **17** (37), 65–72. MR 847:46028.
Vukman, J. (1982), A note on normability of locally convex spaces. *Comment. Math. Univ. Carolin.* **24**, 17–22. MR 85b:46002.
Vukman, J. (1983), A characterization of real and complex Hilbert spaces. *Glasnik Mat.* **18** (38), 103–106. MR 85d:46029.
Vukman, J. (1983), A characterization of real and complex Hilbert spaces among all normed spaces. *Bull. Australian Math. Soc.* **27**, 339–345. MR 84i:46033.
Vukman, J. (1984), On irreducible operator *-algebras on Banach spaces. *Acta Math. Hung.* **43** (1–2), 43–46. MR 85g:46070.
Vukman, J. (1983), Bilinear and quadratic functionals (Slovenian. English summary). *Obzornik Mat. Fiz.* **30** (3), 80–85. MR 85f:15022.
Vukman, J. (1984), Complementation in Banach spaces (Slovenian. English summary). *Obzornik Mat. Fiz.* **31** (1), 1–3. MR 85j:46027.

Walsh, J. L. (1924), An inequality for the roots of an algebraic equation. *Ann. of Math.* **25**, 285–286.
Walter, Vimala (1974), Splines in Hilbert spaces. *Lecture Notes in Math. No.* 399, 538–560.
Wang, Chung-Lie, (1979), On development of inverses of the Cauchy and Holder inequalities. *SIAM REV.* **21**, 550–557.
Warren, P. and Howell, J., A strong law of large numbers for orthogonal Banach space random variables. *Lecture Notes in Math.* No. 526, pp. 253–262.
Weil, G. G. (1972), Vector space decomposition and the abstract imitation problem. *Pacif. J. Math.* **41**, 263–272. MR 46:4181.
Weron, A. (1973), On positive definite operator valued functions in Banach spaces (Russian). *Bull. Acad. Sci. Georgian SSR* **71**, 297–300.
Weron, A. (1974), On characterization of interpolable and minimal stationary processes. *Studia Math.* **49**, 165–183.

Weron, A. (1976), Remarks on positive definite operator valued functions in Banach spaces. *Bulletin Acad. Polon. Sci. Ser. Sci. Math. Astronom. Phys.* **24**, 873–876. MR 55:6211.

Weron, A. (1975), Prediction theory in Banach spaces. *Lecture Notes in Math. No.* 472, 207–228. MR 52:15648.

Weston, J. D. (1949), A note on the theory of communication. *Phil. Mag. Ser.* 7 **11**, 449–451.

Weston, J. D. (1949), The cardinal series in Hilbert spaces. *Proc. Cambidge Math. Soc.* **45**, 335–341.

Weyl, H. (1940), The method of orthogonal projection in potential theory. *Duke Math. J.* **7**, 411–444. MR 2,202.

Weyl, H. (1949), Inequalities between the two kinds of eigenvalues of a linear transformation. *Proc. Nat. Acad. Sci. USA* **35**, 408–411. MR 11,37.

Weyl, H. (1950), Ramifications, old and new, of the eigenvalue problem. *Bull. Amer. Math. Soc.* **56**, 115–139. MR 11,666.

Whittaker, E. (1915), On the functions which are represented by the expansions of interpolation theory. *Proc. Edinbourgh Roy. Soc.* **35**, 181–194.

Whittaker, E. T. (1927), *Analytical Dynamics*. Cambridge Univ. Press.

Wigner, E. (1949), Nuclear reactions and level widths. *Amer. J. Physics* **17**, 99–109.

Wigner, E. (1951), On a class of analytic functions from the quantum theory of collisions. *Ann. of Math.* **53** (2), 36–67.

Wigner, E. (1964), Causality, *R*-matrix and collision matrix. In *Proc. Internat. School of Physics. Enrico Fermi*, Course 29. Academic Press, New York.

Wigner, E. (1959), *Group Theory and its Applications to the Quantum Theory of Atomic Spectra*. Academic Press, New York.

Wigner, E. and von Neumann John, (1954), Significance of Loewner's theorem in the quantum theory of collisions. *Ann. of Math.* **59** (2), 418–433. MR 16,25.

Wigner, E. and Eisenbud, L. (1947), Higher angular momenta and long range interaction in resonance reactions. *Phys. Rev.* **72**, 29–41.

Wiener, N. (1923), Note on a paper by M. Banach. *Fund. Math.* 4, 136–143.

Wiener, N. (1949), *Extrapolation, Interpolation and Smoothing of Stationary Time Series with Engineering Applications*. J. Wiley New York. MR 11,118.

Wiener, N. and Masani, P. (1957, 1958), The prediction theory for multivariate stochastic processes. I. II. *Acta Math.* **98**, 111–150. MR 20:4323. II. *Ibid.* **99**, 93–157. MR 20:4323.

Wiener, N. and Masani, P. On bivariate stationary stochastic processes and the factorization of matrix-valued functions. *Theory of Prob. and Appl.* (*URSS*). English Transl. **4**, pp. 300–308.

Wilbur, W. J. (1977), On characterizing the standard quantum logics. *Trans. Amer. Math. Soc.* **233**, 265–282. MR 57:8531.

Wells, A. E. (1917), *Amer. J. Math.* **39**, 163–184. FM 46,653.

Williams, J. P. (1968), Schwarz norms for operators. *Pacif. J. Math.* **24**, 181–189. MR 37:761.

Wils, W. W. (1970), Direct integrals of Hilbert spaces. I. *Math. Scand.* **26**, 73–78. MR 41:9010.

Wilson, W. A. (1932), On angles in certain metric spaces. *Bull. Amer. Math. Soc.* **38**, 580–588.

Wilson, W. A. (1932), A relation between metric and Euclidean spaces. *Amer. J. Math.* **54**, 505–517.

Wilson, W. A. (1932), On the imbedding of metric sets in Euclidean spaces. *Amer. J. Math.* **54**, 322–326.

Wilson, W. H. (1918), On a certain general class of functional equations. *Amer. J. Math.* **40**, 263–282.

Wintner, A. (1933), Dynamical systems and unitary matrices. *Math. Z.* **36**, 630–637.

Wintner, A. (1944), Diophantine approximations and Hilbert space. *Amer. J. Math.* **66**, 564–578.

Wold, H. (1954), *A Study in the Analysis of Stationary Time Series*. 2nd Edition, Almquist and Wiksell, Stockholm. MR 15,811.

Wolfson, K. G. (1953), The algebra of bounded operators on Hilbert space. *Duke Math. J.* **20**, 533–538.

Wong, E. and Thomas, J. B. (1961), On the multidimensional prediction and filtering problem and the factorization of spectral matrices. *J. Franklin Inst.* **272**, 87–99. MR 25:1977.

Wong, Pak-ken (1974), The *p*-class in a dual B^*-algebra. *Trans. Amer. Math. Soc.* **200**, 355–368. MR 50:10837.

Wong, Pak-ken (1976), A minimax formula for dual B^*-algebras. *Trans. Amer. Math. Soc.* **224**, 281–298.

Wooters, W. K. (1981), Statistical distance and Hilbert space. *Phys. Rev.* D 3, **23**, 357–362. MR 82a:81007.

Woyczyński, Wojbor, A. (1978), Geometry and Martingales in Banach spaces. In *Probability on Banach spaces. Advances in prob. and related topics.* Marcel Dekker, Inc. New York, pp. 269–517.

Wright, J. D. Maitland (1969), A spectral theorem for normal operators on a Kaplansky–Hilbert module. *Proc. London Math. Soc.* **19** (3), 258–268.

Yadrenko, M. I. (1983), *Spectral Theory of Random Fields.* Optimization Software Inc. New York. Review in *Bull. Amer. Math. Soc.* **13**, 56–62.

Yaglom, A. (1953), Correlation theory of processes with random stationary *n*th increments (Russian). *Mat. Sbornik* (*N.S.*) **37**, 141–196.

Yaglom, A. (1961), Second-order homogeneous random fields. *Proc. Fourth Berkeley Sympos. Math. Statist. Prob.* Vol. 2, Univ. of California, Berkeley, Calif. pp. 593–622. MR26:4399.

Yaglom, A. (1962), *An Introduction to the Theory of Stationary Random Functions.* Transl. from the Russian. Prentice-Hall, Englewood Cliffs N.J.

Yang, L. and Zhang, J. Z. (1984), Metric spaces which cannot be isometrically embedded in Hilbert space. *Bull. Austral. Math. Soc.* **30** (2), 161–167. MR 85j:51020.

Yang, Con-Je (= Yang Ts'ung-Jen) (1963), Characterizations of inner product spaces (Chinese). *Shue Jinzan* **6**, 197–200. MR 32:4527.

Yeronimus, G. (1962), *Orthogonal Polynomials on the Unit Circle* (Russian). Moscow.

Young, R. M. (1974), Inequalities for a perturbation theorem of Paley and Wiener. *Proc. Amer. Math. Soc.* **43**, 320–321.

Young, R. M. (1974), Interpolation in a classical Hilbert space of entire functions. *Trans. Amer. Math. Soc.* **192**, 97–114. MR 50:10290.

Young, R. M. G. (1976), On the best possible constant in the Khinchine inequality. *J. London Math. Soc.* **14** (2), 496–504. R. Ž. 1977, 10B 625.

Young, R. M. G. (1980), *An Introduction to Nonharmonic Series.* Academic Press, New York.

Yin, Hong Sheng, (1981), Three geometric parameters of normed linear spaces (Chinese. English summary). *J. Math. Exposition* **1** (2), 39–48. MR 84j:46027.

Young, W. H. (1911), On the analytic basis of non-Euclidean geometry. *Amer. J. Math.* **33**, 249–286.

Yood, B. (1969), On algebras which are prehilbert algebras. *Duke Math. J.* **36**, 261–272. MR 41:4242.

Yood, B. (1974), Hilbert algebras as topological algebras. *Arkiv f. Math.* **12**, 131–151. MR 52:1329.

Yurgelas, K. V. (= Jurgelas) (1982), Some geometric characteristics of Banach spaces and accretive operators (Russian). *Izv. Vysš. Učebn. Zaved. Matematika* No. 5, 63–71. MR 84j:46028.

Zabreiko, P. P. (1969), A theorem on subadditive functionals. *Funcktional Analiz. i Priložen.* **3**, 86–88. MR 39:3283.

Zaranonello, E. H. (1976), The meaning of the Cauchy–Schwarz–Buniakowsky inequality. *Proc. Amer. Math. Soc.* **59**, 133–137. MR 57:13589.

Zahariuta, V. P. (1967), Continuable bases in spaces of analytic functions of one and several variables (Russian). *Sibirsk. Mat. Ž.* **8**, 277–292. MR 35:5915.

Zaremba, S. (1907), L'equation biharmonique et une classe remarquable de fonctions fundamentales harmoniques. *Bull. Intermat. de l'Acad. Gu. Cracovic.* pp. 147–196.

Zaremba, S. (1908), Sur le calcul des fonctions demandées dans le problème de Dividilet et le problème hydrodynamique. *Ibid.* pp. 125–195.

Zasuhin, V. (1941), On the theory of multidimensional stationary random processes (Russian). *DAN SSSR* **33**, 435–437.

Zelinskiĭ, Ju. B. (1980), Various generalizations of the concept of convexity (Russian). *Metric Questions of the Theory of Functions* (Russian), pp. 38–42. Naukova Dumka, Kiev. MR 82c:52002.

Zeng, W. L. (1982), An extension of the parallelogram law. Characterization of inner product spaces. *J. Math. Res. Expositions* **2** (3), 23–25. MR 83m:46030.

Žmud, E. M. (= Jmud) (1973), On a theorem of Wigner (Russian). *Teoryia Functsii i Funcktionalnyi Analiž. i ih primenen.* **17**, 71–82. MR 50:959.

Zygmund, A. (1972), *Trigonometric Series.* Vols. 1. 2. Cambridge Univ. Press. MR 58:2971.

Zyla, L. V. and de Figueiredo, R. J. P. (1983), Nonlinear systems identification based on a Fock space framework. *SIAM J. Control Optim.* **21** (6), 931–939. MR 84m:93031.

List of Symbols and Abbreviations

Author Index

Index

GPSR Compliance

The European Union's (EU) General Product Safety Regulation (GPSR) is a set of rules that requires consumer products to be safe and our obligations to ensure this.

If you have any concerns about our products, you can contact us on ProductSafety@springernature.com

In case Publisher is established outside the EU, the EU authorized representative is:

Springer Nature Customer Service Center GmbH
Europaplatz 3
69115 Heidelberg, Germany

Batch number: 10832080

Printed by Printforce, the Netherlands